McGraw-Hill's HAZARDOUS CHEMICAL SAFETY GUIDE FOR THE PLASTICS INDUSTRY

McGraw-Hill's HAZARDOUS CHEMICAL SAFETY GUIDE FOR THE PLASTICS INDUSTRY

Richard P. Pohanish

Stanley A. Greene

McGRAW-HILL

New York San Francisco Washington, D.C. Auckland Bogota
Caracas Lisbon London Madrid Mexico City Milan
Montreal New Delhi San Juan Singapore
Sidney Tokyo Toronto

Library of Congress Cataloging-in-Publication Data

Pohanish, Richard P.
McGraw-Hill's hazardous chemical safety guide for the plastics industry / Richard P. Pohanish, Stanley A. Greene.
p. cm.
Includes indexes.
ISBN 0-07-135607-X
1. Hazardous substances—Handbooks, manuals, etc. 2. Plastics industry and trade—Safety measures—Handbooks, manuals, etc. I. Greene, Stanley A. II. Title.
TP1130.P64 2000
668.4'199—dc21 00-024656
CIP

McGraw-Hill
A Division of The McGraw-Hill Companies

1 2 3 4 5 6 7 8 9 0 DOC/DOC 9 0 9 8 7 6 5 4 3 2 1 0 9

ISBN 0-07-135607-X

The sponsoring editor of this book was Bob Esposito, the editing supervisor was Sally Glover, and the production supervisor was Pamela Pelton. It was set in Times New Roman by the author.

Printed and bound by R. R. Donnelley & Sons Company.

CONTENTS

NOTICE

This reference is intended solely for guidance, and to provide data about hazardous chemicals that are used in the plastic products industry. It cannot be assumed that all necessary information is contained in this work and that other, or additional, information or assessments may not be required. Extreme care has been taken in the preparation of this work and, to the best knowledge of the publisher and authors, the information presented is accurate. Information may not be available for some chemicals; consequently, absence of data does not necessarily mean that a substance is hazardous or unregulated under a specific law or statue.

Neither the publisher nor the authors assume any liability or responsibility for completeness or accuracy of the information presented or for any alleged damages resulting in connection with, or arising from, the use of any information appearing in this book. The publisher and authors strongly encourage all readers to use this book as a guide only, as it is not a formal interpretation or clarification of the statutes and regulations. If data are required for legal purposes, original source documents and the appropriate agencies, which are referenced, should be consulted.

Dedication

To Dina Pohanish and Katherine Greene

Acknowledgments

The authors wish to acknowledge their indebtedness to all those who leant assistance in the preparation of this work, including Dawn Shellenberger, Hazardous Waste Librarian, EPA, Region 3, Dr. Les Smith, Chief of Toxicology Information at the Agency for Toxic Substances and Disease Registry (ASTDR) in Atlanta, Georgia, Seth Heminway at the EPA Office of Compliance, Office of Enforcement and Compliance Assurance, Brad Pabody, Fire Management, Federal Emergency Management Agency, and the many government scientists and contract employees who developed the various documents and public databases that provided information appearing in this work. The authors appreciate the efforts of Richard W. Johnson, Ph.D. of Rohm & Haas Corporation and others who reviewed and constructively commented on this work during its development.

We are grateful to Heidi Crane and Jennifer Pohanish for their technical and editorial support and to Mr. Robert Esposito, Sally Glover, and others of the staff of McGraw-Hill for their help and advice in the preparation of this work.

Some of the work upon which this handbook was based initially appeared in sections of U.S. Environmental Protection Agency Publications 305-B-97-002, 310-R-95-016, 310-R-95-012, 530-SW-89-051A, 530-R-95-042, 530-F-94-009, and 310-R-97-006.

The authors also acknowledge and thank the American Conference of Governmental Industrial Hygienists, Inc. (ACGIH) for granting permission to use data from *1999 Threshold Limit Values (TLVs®) for Chemical Substances and Physical Agents and Biological Exposure Indices (BEIs®)* and the National Fire Protection Association (NFPA) for granting permission to use material from NFPA 704-1990, *Identification of the Fire Hazards of Materials.*

INTRODUCTION

Plastics today are one of the most used materials in U.S. industrial and commercial life. The first plastics were invented in the 1800s when people experimented to produce everyday objects out of alternative materials. The first plastic was developed in 1851 when hard rubber, or ebonite, was synthesized. This was the first material that involved a distinct chemical modification of a natural material. The following table lists selected plastic resins and synthetic fibers by year of development and their principal uses.

Introduction of Selected Plastic Resins and Manmade Fibers

Year	Material	Example
1868	Cellulose Nitrate	Eyeglass frames
1900	Viscose Rayon	Lining in clothing, curtains
1909	Phenol-Formaldehyde	Telephone handset
1927	Cellulose Acetate	Toothbrushes, lacquers
1927	Polyvinyl Chloride	Wall covering, pipe, siding
1936	Acrylic	Brush backs, display signs
1938	Polystyrene	Housewares, toys
1939	Nylon	Fibers, films, gears
1942	Low-Density Polyethylene	Packaging, squeeze bottles
1942	Unsaturated Polyester	Boat hulls
1952	Polyethylene terephthalate	Clothing, fiberfill
1957	Polypropylene	Safety helmets
1964	Polyimide	Bearings
1970	Thermoplastic polyester	Electrical/electronic parts
1978	Linear Low-Density Polyethylene	Extruded film
1985	Liquid Crystal Polymers	Electrical/electronic Parts

Source: This table has been adapted from *Facts and Figures of the U.S. Plastics Industry*, prepared annuallyby The Society of the Plastics Industry, Inc., Washington, DC. Please refer to that document for a more complete listing of plastic resin development.

The first plastics in the U.S. were developed while John Wesley Hyatt was experimenting to produce a billiard ball from materials other than ivory. In 1870, John and his brother Isaiah took out a patent for a process producing a hornlike material using cellulose nitrate and camphor.

Another important precursor to modern plastics was the development of formaldehyde resins. Early experiments to produce white chalkboards in Germany around the turn of the twentieth century led to the development of formaldehyde resins. These resins were first produced by reacting casein (milk protein) with formaldehyde.

During the 1930s, the initial commercial development of today's major thermoplastics took place. These included polyvinyl chloride, low-density polyethylene, polystyrene, and polymethyl methacrylate. Demand for plastics escalated during World War II, when substitutes for scarce natural materials, like rubber, were in high demand. Large-scale production for synthetic rubbers triggered extensive research into polymer chemistry and new plastic materials.

In the 1940s, polypropylene and high-density polyethylene were developed, and in 1978, linear low-density polyethylene was developed. Large-scale production of these materials reduced their cost substantially, which allowed these new plastics materials to compete with traditional materials like wood and metal. The introduction of alloys and blends of various polymers has made it possible to tailor properties to fit certain performance requirements that a single resin could not provide. Demand for plastics has steadily increased, and now plastics are accepted as basic materials along with the more traditional materials in designs and engineering plans (SPI, 1995).

In 1664, Robert Hooke first suggested that manmade yarn could be produced. He speculated, in *Micrographia,* that synthetic fibers could be patterned after the excretion of silk by silkworms.

During the nineteenth century, scientists were busy making precursor solutions of the first manmade fibers, cellulosic fibers. In 1840, F. Gottlob Keller of Germany devised a technique for making pulp for paper by squeezing powdered wood taken from a grindstone. This enabled the future production of rayon and other cellulosic items. During that same year, Louis Schwabe, an English silk manufacturer, developed the first spinnerette through which a spinning solution could be extruded (Linton, 1966).

The first manmade fibers commercially manufactured in the U.S. were the cellulosics, led by rayon in 1910 and acetate in 1924. Cellulosic fibers are manufactured by first treating cellulose with chemicals, dissolving, and then regenerating the fibers. Cellulose is an abundant naturally occurring organic compound which makes up a large portion of the world's vegetable matter. Often referred to as artificial silk, rayon retained many of the same physical properties as cotton, such as high moisture absorption and subsequent swelling of the fibers. While cellulose acetate was first developed as a plastic in 1865, it was not successfully spun into a fiber until the 1920s.The first U.S. acetate production took place at the Cumberland, Maryland, plant of British Celanese (now Hoechst Celanese).

In 1926, Du Pont Laboratories began a chemical research program that led to the advent of the synthetic, or noncellulosic, fiber industry. Unlike cellulosic fibers, synthetic fibers are wholly compounded from chemicals. The first synthetic fiber that Du Pont developed was Fiber 66. Now known as nylon-6,6, the fiber began widespread production for markets, such as nylon hosiery, in 1939. During World War II, nylon was used in producing parachutes, uniforms, and a host of other military equipment. Started primarily as a hosiery yarn, the use of nylon spread after the war into other applications like carpeting and woven fabrics.

Wrinkle-resistant and strong, the first polyester fiber, Terylene, was developed by a British scientist group called the Calico Printers Association. In 1946, Du Pont secured exclusive rights to produce this polyester fiber in the U.S. In December 1950, Du Pont announced plans to build its first plant at

Kinston, North Carolina, at a capacity of 36 million pounds a year and a cost of $40 million. Du Pont first unveiled the new fiber, named Dacron, at a famous press conference where it was displayed in a swimsuit that had been worn 67 days continuously without ironing. After polyester fibers were first produced commercially in the U.S. in 1953, the fibers were rapidly used to make men's suits, women's blouses, and men's shirts.

Since then, most technological advances in manmade fibers have occurred in synthetics, which now make up almost all of the U.S. production of manmade fibers. Synthetic fibers have many advantages to cellulosic fibers, such as controlled shrinkage, crease retention, and wrinkle resistance. Synthetic fibers have developed to seem more natural, softer, easier to care for, more lustrous, and more comfortable.

Plastics Materials, Synthetic Resins, and Nonvulcanizable Elastomers (as listed under SIC 2821)

- acetal resins
- acetate, cellulose (plastics)
- acrylic resins
- acrylonitrile-butadiene-styreneresins
- alcohol resins, polyvinyl
- alkyd resins
- allyl resins
- butadiene copolymers, containing less than 50% butadiene
- carbohydrate plastics
- casein plastics
- cellulose nitrate resins
- cellulose propionate (plastics)
- coal tar resins
- condensation plastics
- coumarone-indene resins
- cresol resins
- cresol-furfural resins
- dicyandiamine resins
- diisocyanate resins
- elastomers, nonvulcanizable (plastics)
- epichlorohydrin bisphenol
- epichlorohydrin diphenol
- epoxy resins
- ester gum
- ethyl cellulose plastics
- ethylene-vinyl acetate resins
- fluorohydrocarbon resins
- ion exchange resins
- ionomer resins
- isobutylene polymers
- lignin plastics
- melamine resins
- methyl acrylate resins
- methyl cellulose plastics
- methyl methacrylate resins molding compounds, plastics
- nitrocellulose plastics (pyroxylin)
- nylon resins
- petroleum polymer resins
- phenol-furfural resins
- phenolic resins
- phenoxy resins
- phthalic alkyd resins
- phthalic anhydride resins
- polyacrylonitrile resins
- polyamide resins
- polycarbonate resins
- polyesters
- polyethylene resins
- polyhexamethylenediamineadipamide resins
- polyisbutylenes
- polymerization plastics, except fibers
- polypropylene resins
- polystyrene resins
- polyurethane resins
- polyvinyl chloride resins
- polyvinyl halide resins
- polyvinyl resins
- protein plastics
- pyroxylin
- resins, synthetic
- rosin modified resins
- silicone fluid solution (fluid for sonar transducers)
- silicone resins
- soybean plastics
- styrene resins
- styrene-acrylonitrile resins
- tar acid resins
- urea resins
- vinyl resins

Plastic Resins

Plastic resins are typically broken down into two categories: thermoplastics and thermosets. Thermoplastic resins are malleable resins that can be heated and molded into shapes repeatedly. Once these resins have been recovered from products, they can be melted and reprocessed into resin feedstocks. These feedstocks are then used to manufacture new products. Commercially important examples of thermoplastic resins include all forms of polyethylene (PE), polyvinyl chloride (PVC) and polyethylene, polypropylene, and polystyrene and are shown in the table above.

Thermoset resins are resins that tend to be rigid and insoluble, and can be heated and molded only once. Since thermosets cannot be converted back to resin feedstock, recycling opportunities are limited. Thermosets can be shredded and compressed into other products. Thermoplastic resins dominate plastic resin sales and production. The leading thermosets in sales were phenolic resins, urea resins, and unsaturated polyester resins. Specialty plastic resins, which often include thermosets, are produced on a customized basis in small production runs and typically involve significant research and development costs.

Resin Characteristics, Markets, and Products

Resin	Characteristics	Primary Product Markets	Product Examples
THERMOPLASTICS			
Low-density polyethylene (LDPE)	Moisture -proof; inert. Largest volume resin used for packaging.	Packaging	High-clarity extruded film, wire and cable coatings, garbage bags, coated papers
Polyvinyl chloride (PVC)	Strength and clarity; brittle unless modified with plasticizers	Building and construction; packaging	Construction pipe, meat wrap, blister packs, cooking oil bottles, phono records, wall covering, flooring
High-density polyethylene (HDPE)	Tough, flexible, translucent	Packaging	Milk and detergent bottles, heavy-duty films (e.g., boil-in-bag pouches), liners, wire and cable insulation
Polypropylene (PP)	Stiff, heat and chemical resistant	Furniture and furnishings, packaging, other	Syrup bottles, yogurt and margarine tubs, fish nets, drinking straws, auto battery cases, carpet backing, office machines and furniture, auto fenders

THERMOPLASTICS			
Other styrenics	Strong, stretchable	Adhesives, coatings and inks	Assembly and construction adhesives, pressure-sensitive labels and tapes, footwear soles, roof coatings
Polyethylene terephthalate (PET)	Tough, shatter resistant	Packaging, consumer products	Soft drink bottles and other beverage containers, food and medicine containers, synthetic textiles, X-ray and photographic film, magnetic tape
Acrylonitrile/butadiene/ styrene (ABS)	Tough, abrasion-resistant	Transportation, electrical and electronic products	Pipe, refrigerator door linings, telephones, sporting goods, automotive brake parts
THERMOSETS			
Phenolic	Heat resistant, strength, shatter resistant	Building and construction	Handles, knobs, electrical connectors, appliances, automotive parts
Polyurethane	Malleable for rigid or flexible foams	Furniture and furnishings, building and construction, transportation	Cushioning, auto bumpers and door panels, varnishes
Urea and melamine	Rigid, chemically resistant	Building and construction, consumer products	Plywood binding, knobs, handles, dinnerware, toilet seats
Polyester, unsaturated	Malleable for fabrication of large parts	Building and construction, transportation	Electrical components, automobile parts, coatings, cast shower/bath units

Manmade Fibers

Manmade fibers are produced primarily for use as raw materials for the textile industry, for the carpets and rugs market, the industrial and miscellaneous consumer products market, and the apparel market. The increasing use of manmade fibers in a variety of markets has enabled manmade fibers to account for nearly 60% of all fibers, natural and manmade, consumed in the U.S.

There are two main types of manmade fibers: noncellulosic (SIC 2824) and cellulosic (SIC 2823). Noncellulosic, or synthetic, fibers consist of fibers that are formed by the polymerization and subsequent fiber formation of synthetic organic chemicals and refined petroleum products.

Plastics production and use has grown because of the many advantages plastics offer over other more traditional materials. A few of the desirable intrinsic properties of plastics include: (1). design flexibility-plastics can be modified for a wide variety of end uses, (2). High resistance to corrosion, (3). Low weight, and (4). Shatter resistance.

Plastics are resins, or polymers, that have been synthesized from petroleum or natural gas derivatives. The production of plastics products, both solid and foam, is a relatively diverse industry. Simpler processes consist of : (1). Using chemical compounds to impart the appropriate characteristics to the plastic resin with chemical additives; (2). Converting plastic materials in the form of pellets, granules, powders, sheets, fluids, or preforms into either immediate or final-formed plastic shapes or parts via molding operations; and (3). Finishing products. There are also several methods of reacting plastic resin and catalyst materials to form a thermoset plastic material into its final shape.

Chemical additives such as colorants, stabilizers, and plasticizers are often mixed with resins to modify their properties. Different combinations of resins and additives have allowed the creation of products meeting a wide variety of specifications, and give the final product certain characteristics (some of these additives can also be applied to the shaped product during the finishing process). Plastic additives may contain toxic constituents such as lead and cadmium. Plastics contribute 28 percent of cadmium in municipal solid waste and approximately two percent of all lead, but data are too limited to determine whether these and other plastic additives contribute significantly to leachate produced in municipal solid waste landfills. Plastics that contain heavy metal-based additives may also contribute to the metal content of incineration ash.

Plastic Additives

Plastic additives and their functions, in terms of their effect on the final product, are listed below. Also listed are some of the typical chemicals used as additives in the plastics products manufacturing process.

Additive Lubricants assist in easing the flow of the plastic in the molding and extruding proces by lubricating the metal surfaces that come in contact with the plastic. Additives include stearic acid, waxes, fatty acid esters, fatty acid amines. Among compounds used are didecyl ether and glyceryl tristerate.

Antioxidants inhibit the oxidation of plastic materials that are exposed to oxygen or air at normal or high temperatures. Additives include alkylated phenols, amines, organic phosphites and phosphates, and esters. Among compounds used are 1,3,5-trimethyl-2,4,6-tris(3,5-di-tert-butyl-4-hydroxybenzyl) benzene, phenyldidecyl phosphite, distearyl thiodipropionate, 2,6-di-tert-butyl-p-cresol, dibutyltin sulfide, and tris(nonylphenyl)phosphite.

Antistats (antistatic agents) reduce or eliminate the tendency of thermoplastic polymers to accumulate static charges which result in adherent particles of foreign matter such as dust. Additives include quarternary ammonium compounds, anionics, phosphate ester, and fatty amines. Among compounds used are polyethylene glycols, tetracyanoquinodimethane.

Blowing Agents (foaming agents) produce a cellular structure within the plastic mass and can include compressed gases that expand upon pressure release, soluble solids that leach out and leave pores, or liquids that change to gases and in the process develop cells. Additives include azodicarbonamide, modified azos, and OBSH. Among compounds used are trifluoromethane, oxybis(benzenesulfonyl hydrazide), and potassium borohydride

Colorants impart color to the plastic resin. Among compounds used are titanium dioxide, iron oxides, anthraquinones, carbon black, and strontium chromate

Flame Retardants reduce the tendency of the plastic product to burn. Among compounds used are antimony trioxide, chlorinated paraffins, bromophenols, bis(tribromophenyl)ethane, 2,5-dibromoterephthalic acid, and tetrabromothalic anhydride.

Heat Stabilizers assist in maintaining the chemical and physical properties of the plastic by protecting it from the effects of heat such as color changes, undesirable surface changes, and decreases in electrical and mechanical properties. Among compounds used are lead, barium-cadmium, tin, calcium-zinc, barium stearate, ethylamine, leadphthalate, stannic chloride, and zinc stearate.

Impact Modifiers prevent brittleness and increase the resistance of the plastic to cracking. Among compounds used are cellulose acetate, methylene diphenyl isocyanate, chlorotrifluoroethylene, and polypropylene.

Organic Peroxides initiate or control the rate of polymerization in thermosets and many thermoplastics. Among compounds used are MEK peroxide, benzoyl peroxide, alkyl peroxide, and peresters. Peroxides are one of the most hazardous class of compounds. Many organic peroxides can explode violently due to one or a combination of factors including heat, chemical contact, mechanical shock, or friction. Some peroxides are great fire hazards, and firefighting can be very difficult.

Plasticizers increase the plastic's flexibility, toughness and workability. Additives include adipates, azelates, trimellitates, and dioctyl phthalate/diisooctyl phthalate/diisodecyl phthalate (DOP/DIOP/DIDP). Among compounds used are those containing esters of phthalate, adipate, and sebacate, tricresyl phosphate, castor oil, and ethylene glycols, its derivitives, and other polyols.

Ultraviolet Stabilizers (UV light absorbers) absorb or screen out ultraviolet radiation, thereby preventing the premature degradation of the plastic product. Among compounds used are benzophenones, benzotriazoles, salicyclates, phenol-nickel complexes, and substituted acrylonitriles.

Environmental Concerns

The plastic resin and manmade fiber industries are affected by nearly all federal environmental statutes. In addition, the industries are subject to numerous laws and regulations from state and local governments.

Raw material inputs to plastic resin and manmade fibers industries primarily consist of synthetic organic chemicals, such as ethylene glycol and acrylonitrile, and refined petroleum products, such as ethylene. The majority of these chemicals are used either as monomers or as monomer precursors. Other uses are as solvents, catalysts, and additives. Because chemical processes rarely convert 100 percent of raw materials to desired products, byproducts and unreacted monomer may constitute a large part of facilities' waste streams. Pollutant outputs generally include VOCs, off-spec or contaminated polymer, and wastewater from equipment cleaning. Typical waste streams associated with processes involved in plastic resin and manmade fiber manufacture are listed in the following

Table , "Potential Releases During Manufacturing," Waste streams vary depending on what polymer is being synthesized, what fiber spinning method is used, and whether a batch or continuous process is used. According to a report from the New Jersey Hazardous Waste Facilities Siting Commission, small-scale batch facilities that make polymers to order often have complex and variable wastestreams.

Potential Releases during Manufacturing

Media	Potential Sources of Emissions
Air	Point source emissions: stack, vent, distillation unit, reactor, storage tank vent), material loading/unloading operations (including rail cars, tank trucks, and marine vessels). Fugitive emissions: pumps, valves, flanges, sample collection, mechanical seals, relief devices, tanks. Secondary emissions: waste and wastewater treatment units, cooling tower, process sewer, sump, spill/leak areas.
Liquid wastes (organic or aqueous)	Equipment wash solvent/water, lab samples, surplus chemicals, product washes/purifications, seal flushes, scrubber blowdown, cooling water, steam jets, vacuum pumps, leaks, spills, spent/used solvents, housekeeping (pad washdown), waste oils/lubricants from maintenance.
Solid wastes	Spent catalysts, spent filters, sludges, wastewater treatment biological sludge, contaminated soil, old equipment/insulation, packaging material, reaction by-products, spent carbon/resins, drying aids.
Ground Water Contamination	Unlined ditches, process trenches, sumps, pumps/valves/fittings, wastewater treatment ponds, product storage areas, tanks and tank farms, aboveground and underground piping, loading/unloading areas/racks, manufacturing maintenance facilities.

Air Emissions

Over 70 percent of TRI releases for plastic resin and manmade fiber plants are in the form of air emissions., Commonly released chemicals include carbon disulfide, methanol, and other volatile solvents and monomers. Typical chemicals released are listed in the following section on TRI releases and transfers. Air emissions from plastic resin and manmade fiber plants arise from point sources and fugitive emission sources, such as valves, pumps, tanks, compressors, etc. Point sources of air emissions may include monomer storage and feed dissolver tanks and reactors.

While individual leaks are typically small, the sum of all fugitive leaks at a plant can be one of its largest emission sources. Fugitive emissions can be emitted continuously or intermittently. Continuous air emissions may be emitted from monomer recovery systems, dryer stacks and miscellaneous solid handling vents, centrifuge vents, and blending operations. Fugitive emissions can also result from volatilization of monomers, solvents, and other volatile organic compounds during polymerization; sublimation of solids during resin production; wastewater treatment; and volatilization of solvents during storage and handling of resins. These emissions are largely controlled by solvent and monomer recovery systems.

Summary of Potential Releases Emitted During Plastic Resin and Manmade Fiber Manufacturing

Process	Air Emissions	Process Wastewater	Residual Wastes
Preparing Reactants	volatilized monomer, solvents	little or no wastewater produced	raw material drum residuals
Polymerization	volatilized monomer, solvents, reaction byproducts	little or no wastewater produced	off-specification or contaminated polymer, reaction byproducts, spent equipment oil, spent solvent, catalyst manufacture waste, gas purification catalyst waste
Polymer Recovery	volatilized solvents and unreacted monomer	little or no wastewater produced	little or no residual waste produced
Polymer Extrusion	volatilized solvents and unreacted monomer	extruder quench water	off-specification or contaminated polymer
Equipment Cleaning	volatilized solvents and unreacted monomer	reactor and floor wash water contaminated with organics, acids, and salts; equipment rinse water	little or no residual waste produced
Unloading and Storage of Reactants	volatilized monomer and solvents	rinse water from cleaning out transport vehicles containing solvents, monomers, and other reactants	little or no residual waste produced
Conveyance and Pellet Storage	volatilized residual monomer or solvents from plastic pellets	little or no wastewater produced	plastic pellets from leaks or spills
Spinning	volatilized residual monomer solvents, additives, other organics, volatilized finishes	water contaminated with residual monomer solvents, additives, other organics, finishes	off-spec polymer, off-spec fiber, and residual finishes
Fiber Processing	volatilized residual monomer solvents, additives, other organics, volatilized finishes	water contaminated with residual monomer, solvents, additives, other organics	residual monomer and solvents; off-spec fibers
Pollution Control Systems	volatilized solvents and unreacted monomer	water contaminated with residual solvents and unreacted monomer; air stripper water	little or no residual waste produced

Source: U.S. EPA, *AP-42*, Office of Air and Radiation, 1993; U.S. EPA, ***Best Management Practices for Pollution Prevention in the Textile Industry***, Office of Research and Development, 1995; *SOCMA Pollution Prevention Study*, Prepared for SOCMA, Washington, DC, 1993; Randall, P.M., "Pollution Prevention Strategies for Minimizing of Industrial Wastes in the Vinyl Chloride Monomer - Polyvinyl Chloride Industry," in *Environmental Progress*, volume 13, no.4, November 1994; AFMA, 1997; Wellman, 1997.

In addition to pollutants emitted during polymerization, fiber finishes are sources of volatile and hazardous air pollutants emitted from manmade fiber processing operations. Because melt spinning does not require the use of solvents, melt spinning emits significantly less VOCs than dry or wet spinning processes. Dry spinning typically emits the largest amounts of VOC per pound of fiber produced of the three main spinning methods. Dry spinning can emit from 5 to 150 kg total non-methane organic carbons (TNMOC) per Mg of product, while melt spinning can emit less than 5 kg TNMOC per Mg product. Wet spun fibers typically emit 5 to 20 kg TNMOC per Mg of product. Air pollutant emissions include volatilized residual monomer, fiber lubricants, organic solvents, additives, and other organic compounds used in fiber processing (EPA, 1993).

Unrecovered solvent accounts for some of the VOC emissions from fiber spinning processes, particularly for acetate production. Typically, 94 to 98 percent of the solvents used in fiber spinning processes is recovered. The largest amounts of unrecovered solvent are emitted from the fiber spinning and drying steps. Other emission sources include dope preparation (dissolving the polymer, blending the spinning solution, and filtering the dope), fiber processing (drawing, washing, crimping), and solvent recovery. Other pollutants emitted during manufacturing include air pollutants emitted during combustion. Criteria air pollutants, such as SO_x, NO_x, CO, and CO_2, are emitted from combustion equipment used to heat reactors, dryers, and other process equipment.

Wastewater

Contaminated waste water is another concern in the plastic product industry. Water used in the plastic molding and forming processes falls into three main categories: (1) water to cool or heat the plastics products; (2) water to clean the surface of both the plastics products and the equipment used in production; (3) and water to finish the plastics products.

Cooling and heating water usually comes into contact with raw materials or plastics products during molding and forming operations for the purpose of heat transfer. The only pollutant found in some waste water discharged by contact cooling and heating in a treatable concentration is bis(2-ethylhexyl) phthalate (BEHP) and it is toxic. Many facilities do not process materials containing BEHP, making this pollutant output irrelevant for many manufactures.

Cleaning water includes water that is used to clean the surface of the plastic product or the molding equipment that is or has been in contact with the formed plastic product. The types of pollutants found in cleaning water in treatable concentrations are biochemical oxygen demand (BOD), oil and grease, total suspended solids (TSS), chemical oxygen demand (COD), total organic carbon (TOC), total phenols, phenol, and zinc.

Finishing water consists of water used to carry away waste plastic material or to lubricate the product during the finishing operation. TSS, bis(2-ethylhexyl) phthalate, di-n-butyl phthalate, and dimethyl phthalate are the pollutants identified in finishing water in treatable concentrations.

Of the pollutants found in all three types of process water, BOD5, oil and grease, TSS, and pH are considered conventional pollutants, TOC and COD are considered nonconventional pollutants, and bis (2-ethylhexyl) phthalate, di-n-butyl phthalate, dimethyl phthalate, phenol, and zinc are considered priority toxic pollutants.

Plastic resin and manmade fiber facilities generate relatively large amounts of wastewater from processes, cooling operations, utilities and maintenance, and air pollution control systems. Unless solvents are used in polymerization processes, wastewater contaminants are usually restricted to off-spec polymer, polymer, and raw materials. Wastewater streams from polymerization operations typically contain dilute concentrations of organics, acids, and salts. Process wastewater may be generated from water that comes into direct contact with raw materials, intermediate products, finished products, byproducts, or waste product. Process wastewater may also be generated from indirect contact process water discharged from vacuum jets and steam ejectors. Cooling water makes up a large portion of water used in the industries and can either be generated from water that contains contaminants or from water used in noncontact processes, such as water treatment wastes and boiler blowdown (EPA, 1987).

Effluent containing contaminants may also be discharged from batch operations during equipment cleaning. Wastes generated from cleaning operations include vessel wash waters, floor wash waters, equipment draining, sump draining, and air stripper water effluent. According to SOCMA, these discharges can be minimized by initiating water conservation programs and by cleaning reactors using high-pressure water or process solvents which can be recycled into the reactor.

Wastewater is also generated during monomer and polymer recovery processes, such as centrifuging, monomer stripping, and slurry tanks. Process sources generate liquid wastes with relatively high concentrations of contaminants, including equipment oil, spent solvent, and raw material drum residuals. Leaks and spills also generate waste and often occur at pumps, flanges, valves, and agitator seals. Loading/unloading operations and bag filling operations also are common sources of leaks and spills (Randall, 1994).

In addition to pollutants emitted during polymerization, fiber finishes are sources of volatile and hazardous pollutants found in manmade fiber plant wastewater. Spin finishes may increase biological oxygen demand (BOD) and chemical oxygen demand (COD) and some may be toxic to aquatic life (EPA, 1995).

Residual Wastes

Residual wastes make up a significant portion of wastes from plastic resin and manmade fiber facilities. Unless solvents are used in polymerization processes, residual wastes are usually restricted to off-spec polymer, polymer, and raw material chemicals (EPA, 1987).Typical contaminants include contaminated polymer, catalyst manufacture waste, gas purification catalyst waste, reaction by-products, waste oil, and general plant wastes (Clements and Thompson, 1993). Although properly run and maintained plants with new technology may be capable of obtaining 95 percent or higher polymer yields, off-spec and contaminated polymer is still generated and makes up a sizeable portion of the wastestream. Unreacted or improperly reacted polymer synthesis or regeneration residues may include monomers, oligomers, metals, degradation products, solvents, and coagulants (EPA, 1995). Other sources of residual waste include cleanup absorbents, spent activated carbon, laboratory wastes, and air pollution control residues (SOCMA, 1993). Process-related residual waste can be reduced by implementing better inventory control practices, personnel training, and enhanced process control systems. Process changes and raw material substitutions can also be used to reduce residual waste pollution.

There are other general types of pollution and resource material outputs that can occur at one or more stages of the plastics product manufacturing process. In addition, there are some plastics products disposal concerns. Manufacturing outputs include spills, leaks, and fugitive emissions of chemicals during the application of additives prior to molding or during finishing; waste water discharges during cooling and heating, cleaning, and finishing operations; plastic pellet releases to the environment prior to molding, and fugitive emissions from molding and extruding machines. Each of these is discussed below.

One concern during the plastic product manufacturing process is the potential release of the additive chemicals prior to molding and during the finishing process. Releases could be in the form of spills during weighing, mixing, and general handling of the chemicals, in the form of leaks from chemical containers and molding machines, or in the form of fugitive dust emissions from open chemical containers. It should be noted that not all plastic product manufacturers use additives because many purchased pellets already contain the necessary additives, making this pollution output irrelevant for many facilities. The chemicals used in the plastic product manufacturing process are usually added in such small amounts that most manufacturers do not consider them to be a problem; however, some of the additives could be toxic and therefore even small amounts could present significant problems.

Pellet Release

Plastic pellets and granules used to mold intermediate and final plastics products are often lost to floor sweepings during transport or while being loaded into molding machines, and may end up in waste water. Although they are inert, plastic pellets are an environmental concern because of the harm they can cause if runoff carries them to wetlands, estuaries, or oceans where they may be ingested by seabirds and other marine species. EPA storm water regulations classify plastic pellets as "significant materials," and therefore the discovery of a single pellet in storm water runoff is subject to federal regulatory action. Fugitive emissions from the molding processes may be an environmental concern because of the many additives, including cadmium and lead, which can be released during the application of high heat and pressure.

Plastics products also pose disposal concerns. Discarded plastics products and packaging make up a growing portion of municipal and solid waste. Because only a small percentage of plastic is recycled (less than one percent), virtually all disposed plastics products are put into landfills or incinerated. By the year 2000, the amount of disposed plastic will increase by 50 percent from present levels. Current estimates show that plastic constitutes 14 to 21 percent of the waste stream by volume and seven percent of the waste stream by weight. Because of its resistance to degradation, improper plastic disposal can have a particularly serious effect in the marine environment in terms of ecological risks and aesthetics.

In terms of landfill disposal, the slow degradation of plastic is not a significant factor in landfill capacity; research has shown that other constituents (e.g., paper, wood, food wastes) also degrade very slowly. However, the additives contained in plastic, such as colorants, stabilizers, and plasticizers, may include toxic constituents such as lead and cadmium which can leach out into the environment as the plastic degrades. Plastic contributes nearly 30 percent of all cadmium found in municipal solid waste and approximately two percent of all lead. Data are too limited to determine whether these and other plastic additives contribute significantly to the leachate produced in municipal solid waste landfills. Plastic that contains heavy metal-based additives may also contribute to the metal content of incinerator ash.

INTRODUCTION to the CHEMICAL SECTION

Each record in the chemical section is presented in a uniform format containing up to 50 data elements. Data items include record number, record name, trade names and synonyms, EEC numbers, CAS number, ERG (1996 North American Emergency Response Guide) four-digit ID, three digit ERG Guide, ERG Guide Hazard Category, molecular formula, RTECS number, physical properties, health and safety information including IDLH, odor threshold, OSHA air contaminant lists, workplace regulations including OSHA Air Contaminant lists (Tables Z-1, Z-2, Z-3), OSHA Specifically Regulated Substances and Process Safety Management of Highly Hazardous Chemicals including Threshold Quantities, exposure limits (ACGIH, NIOSH, and OSHA), minimal recommendations for respirators, including maximum use concentrations (MUCs), emergency use and escape, chemical reactivity, fire information, including NFPA 704 Hazard Ratings, flash point, flammable limits in air, autoignition temperature, first aid, and environmental information.

The environmental information section includes data relevant to each compound cited in the Clean Air Act (CAA); Clean Water Act (CWA); Resource Conservation and Recovery Act (RCRA); Safe Drinking Water Act (SDWA); Comprehensive Environmental Response, Compensation, and Liability Act (CERCLA), the law commonly known as Superfund; Emergency Planning and Community Right-to-Know Act (EPCRA, also known as SARA Title III); Marine Pollutants from DOT Hazardous Materials Transportation Act; and California's Proposition 65 carcinogens and reproductive toxins. Many law or regulatory references in this book have been abbreviated. For example, Title 40 of the Code of Federal Regulations, Part 261, subpart 32 has been abbreviated as 40 CFR 261.32.

Following is a explanation of the individual fields and special terms found in each chemical record. Where information was not found in the available literature or government documents, the expression "N/A" is used, or the fields are not included in the chemical record to save space.

NAME AND RECORD NUMBER SECTION

Record Number: The first line of each hazardous substance record contains a unique alphanumeric record number in the format A:xxx, B:xxx, C:xxx, etc. The letter indicates the alphabetical section of the record. The numbering system is not sequential and is designed to allow for the addition of new records in future editions of this work. The CAS and Synonym/Trade Name indexes found in the back of this book are keyed to these record numbers.

Record Name: Each record is arranged alphabetically by a commonly used chemical name.

Synonyms: Lists scientific names, synonym names, trade names, foreign names, generic names, and "trivial" names frequently encountered in industry. For many records the synonym section may contain codes or identification numbers: European Economic Community (EEC), European Inventory of Existing Commercial Substances (EINECS), National Cancer Institute (NCI), etc. These codes and identifying numbers can be useful to readers who wish to locate additional information about a hazardous substance.

IDENTIFICATION SECTION

CAS Registry Number: A unique identifier is assigned to each chemical registered with the Chemical Abstracts Service (CAS), a division of the American Chemical Society (ACS). Mixtures usually do not have CAS numbers. This number is used to identify chemicals on the basis of their molecular structure. CAS numbers, in the format xxx-xx-x, are usually found on Material Safety Data Sheets (MSDS) and may be found on containers. CAS numbers can be used in conjunction with chemical names for positive identification and to allow quick reference to other databases. To avoid confusion with like-sounding names, the CAS number should always be used in conjunction with substance names.

ERG ID numbers: Also called the DOT ID number, this is a four–digit hazard identification number in the format UNxxxx or NAxxxx. This ID number must appear on shipping documents, the exterior of packages, and on specified containers, including cargo tanks, vehicles, and rail cars. Identification numbers containing a UN prefix are also known as United Nations numbers and are authorized for use with international shipments of hazardous materials. The "NA" prefix is used for shipments between Canada, Mexico, and/or the United States and may not be used for international shipments. The data in this section are taken from North American Emergency Response Guide 1996 (NAERG96). The letter "P" following the guide number identifies those materials that present a dangerous polymerization hazard under certain conditions.

ERG Guide Number: This is a three-digit number used in the new North American Emergency Response Guide 1996 developed jointly by the U.S. Department of Transportation (DOT), Transport Canada (TC), and the Secretariat of Communications and Transportation of Mexico (SCT). This number is a guide to initial first aid actions to be taken by professional first responders to stabilize an emergency situation. ERG Guide numbers are not unique for specific substances. Where a material may be shipped in more than a single physical state, both designations are given.

ERG Guide Category: This is a brief description of the hazard(s) associated with specific materials. Where a material may be shipped in more than a single physical state, both designations are given.

Molecular Formula: Following the Hill system, formulas are ordered by number of carbons (if present), number of hydrogens (if present), and then alphabetically by element.

RTECS Number: The NIOSH Registry of Toxic Effects of Chemical Substances identification number, in the format ABxxxxxxx, can be useful for obtaining additional toxicologic information on specific substances.

Properties: This field provides a general summary of properties and possible dangers of a material including physical state (solid, liquid or gas), color, combustibility or flammability, odor description, solubility or miscibility in water, and other special properties. Color descriptions are listed for pure materials. Occasionally the color of a chemical changes when it dissolves in water or becomes a gas. Odor descriptions are listed for pure materials. The expression "characteristic" may be used only when no other reasonable description is found. Specific properties such as autoignition temperature, odor thresholds, IDLH, and flash points can be found in other sections of the chemical record.

HEALTH AND SAFETY INFORMATION SECTION

IARC Carcinogen Risk: This presents the evaluation by International Agency for Research on Cancer (IARC), a scientific group that classifies and evaluates chemicals according to their cancer-causing potential.

IDLH Value: This value stands for "Immediately Dangerous to Life or Health." An atmospheric concentration of any toxic, corrosive, or asphyxiant substance that poses an immediate threat to life or would cause irreversible or delayed adverse health effects or would interfere with an individual's ability to escape from a dangerous environmental condition within 30 minutes. However, the 30-minute period is meant to represent a *Margin of Safety* and is *not* meant to imply that any person should stay in the work environments any longer than necessary. In fact, *EVERY EFFORT SHOULD BE MADE TO EXIT IMMEDIATELY* from an IDLH condition. The concentrations are reported in either parts per million (ppm) or milligrams per cubic meter (mg/m^3), or both.

Odor Threshold: This is the lowest concentration in air that most humans can detect by smell. Some value ranges are reported. The published values cannot be relied on to prevent overexposure because human sensitivity to odors varies over wide limits, some chemicals cannot be smelled at toxic concentrations, odors can be masked by other odors, and some compounds rapidly deaden the sense of smell.

OSHA Table Z Air Contaminant (29 CFR, Section 1910.1000): If a chemical substance appears on one of the three OSHA air contaminant tables, it is noted in this section.

OSHA Specifically Regulated Substances (29CFR 1910.1001-1048): If a chemical substance appears in OSHA, it is noted in this section.

OSHA Process Safety Management of Highly Hazardous Chemicals (29 CFR Part 1910.119, Appendix A): If a chemical substance appears on this list of highly hazardous chemicals, toxics, and reactives, it is noted in this section, along with threshold quantity (TQ) in pounds.

Exposure Limits: Where available, this field contains legally enforceable airborne Permissible Exposure Limits (PELs) from OSHA, recommended airborne exposure limits from NIOSH or ACGIH, and special warnings when a chemical substance is a Special Health Hazard Substance Each are described below.

TLVs have *not* been developed as legal standards, and the ACGIH does not advocate their use as such. The TLV is defined as the time weighted average (TWA) concentration for a normal 8-hour workday and a 40-hour workweek, to which nearly all workers may be repeatedly exposed, day after day, without adverse effects. A ceiling value (TLV-C) is the concentration that should not be exceeded during any part of the working exposure. If instantaneous monitoring is not feasible, then the TLV-C can be assessed by sampling over a 15-minute periods except for those substances that may cause immediate irritation when exposures are short. As some people become ill after exposure to concentrations lower than the exposure limits, this value cannot be used to define exactly what is a "safe" or "dangerous" concentration.

ACGIH threshold limit values (TLVs) are reprinted with permission of the American Conference of Governmental Industrial Hygienists, Inc., from the booklet entitled, *1999 Threshold Limit Values for Chemical Substances and Physical Agents and Biological Exposure Indices.* This booklet is revised on an annual basis. No entry appears when the chemical is a mixture; it is possible to calculate the TLV for a mixture only when the TLV for each component of the mixture is known and the composition of the mixture by weight is also known. According to ACGIH, *Documentation of the Threshold Limit Values and Biological Exposure Indices, 6th Edition* is necessary to fully interpret and implement the TLVs.

OSHA permissible exposure limits (PELs) are found in Tables Z-1, Z-2, and Z-3 of OSHA General Industry Air Contaminants Standard (29 CFR 1910.1000) that were effective on July 1, 1993, and that are currently enforced by OSHA. Unless otherwise noted, PELs are the time weighted average (TWA) concentrations that must not be exceeded during any 8-hour shift of a 40-hour workweek. An OSHA ceiling concentration must not be exceeded during any part of the workday; if instantaneous monitoring is not feasible, the ceiling must be assessed as a 15-minute TWA exposure. In addition there are a number of substances from Table Z-2 that have PEL ceiling values that must not be exceeded except for a maximum peak over a specified period (e.g., a 5-minute maximum peak in any 2 hours).

The NIOSH Recommended Exposure Limit (REL) is a time weighted average (TWA) concentration for up to a 10-hour workday during a 40-hour workweek. A ceiling REL should not be exceeded at any time. Exposure limits are usually expressed in units of parts per million (ppm), i.e., the parts of vapor (gas) per million parts of contaminated air by volume at 25°C (77°F) and one atmosphere pressure. For a chemical that forms a fine mist or dust, the concentration is given in milligrams per cubic meter (mg/m^3).

Short-term exposure limits (15-minute TWA): This field contains short-term exposure limits (STELs) from ACGIH, NIOSH, and OSHA. The parts of vapor (gas per million parts pf contaminated air by volume at 25°C (77°F) and one atmospheric pressure is given. The limits are given in milligrams per cubic meter (mg/m^3) for chemicals that can form a fine mist or dust. Unless otherwise specified, the STEL is a 15-minute TWA exposure that should not be exceeded at any time during the workday.

Recommendations for Respirator Selection: This section provides instant access to NIOSH and/or OSHA allowable respirators along with maximum use concentration (MUC), expressed in ppm or mg/m^3, or condition (i.e., emergency or planned entry into unknown concentrations or IDLH conditions, or escape from suddenly occurring respiratory hazards). For each MUC, the section lists only those respirators with the minimum required protection factor and other use restrictions. More protective respirators can be worn. All recommended respirators of a given class can be used at any concentration equal to or less than the MUC listed for that class.

A complete respiratory protection program must be implemented by the employer and must fulfill all requirements of 29 CFR 1910.134. At a minimum, a respiratory protection program should include regular training, fit testing, periodic environmental monitoring, maintenance, inspection, cleaning, and storage. Selection of a specific respirator within a given class of recommended respirators depends on the specific situation; this choice should be made by a knowledgeable person. SCBAs must be inspected monthly.

Remember: Air-purifying respirators will not protect users against oxygen-deficient atmospheres, and they are not to be used in IDLH conditions. The only respirators recommended for firefighting are self-contained breathing apparatuses that have a full facepieces and are operated in a pressure-demand or other positive-pressure modes.

Where OSHA or NIOSH have made no specific respirator recommendations, this field contains NIOSH minimal recommendations for any concentrations above the NIOSH REL, or where there is no REL, at any detectable concentration: SCBAF:PD,PP (any self-contained breathing apparatus that has a full facepiece and is operated in a pressure-demand or other positive-pressure mode); or SAF:PD,PP:ASCBA (any supplied-air respirator that has a full facepiece and is operated in a pressure-demand or other positive-pressure mode in combination with an auxiliary, self-contained breathing apparatus operated in a pressure-demand or other positive-pressure mode).

ESCAPE: GMFOV (any air-purifying, full-facepiece respirator (gas mask) with a chin-style, front- or back-mounted organic vapor canister; or SCBAE (any appropriate escape-type, self-contained breathing apparatus).

Note: (1) Air purifying respirators will not protect users against oxygen-deficient atmospheres, and they are not to be used in IDLH conditions. (2) Respirators containing charcoal sorbents must not be used in conditions involving oxidizing agents.

CHEMICAL REACTIVITY SECTION

This field describes important hazardous chemical incompatibilities or reactions of each substance with other materials, including structural materials such as metal, wood, plastics, cement, and glass. The nature of the hazard, such as severe corrosion or formation of a flammable gas, is described. This list is by no means complete or all inclusive. In some cases a very small quantity of material can act as a catalyst and produce violent reactions such as polymerization, disassociation, and condensation. Catalysts, when known, are listed.

The accidental mixing of one chemical group with another can in some cases be expected to result in a vigorous and hazardous chemical reaction. The generation of toxic gases; the heating, overflow, and rupture of containers; and fire explosion are possible consequences of such reactions.

FIRE INFORMATION SECTION

NFPA Hazard Classifications: The NFPA 704 Hazard Ratings (Classifications) are reprinted with permission from NFPA 704-1990 *Identification of the Fire Hazards of Materials*, Copyright (c) 1990, National Fire Protection Association, Quincy, MA 02269. This reprinted material is not the complete and official position of the National Fire Protection Association on the referenced subject, which is represented only by the standard in its entirety. The classifications are defined in Table 1. Some materials do not have ANSI/NFPA 49 or 325. In these cases values were assigned by the Federal Emergency Management Agency, United States Fire Administration using ANSI/NFPA definitions and data on the material, or on structurally similar materials. These non-NFPA 704 designations are labeled FEMA Hazard Classifications.

Table 1. EXPLANATION OF NFPA HAZARD CLASSIFICATIONS

Classification	Definition
	Health Hazard (blue)
4	Materials that on very short exposure could cause death or major residual injury (even though prompt medical treatment was given), including those that are too dangerous to be approached without specialized protective equipment.
3	Materials that on short exposure could cause serious temporary or residual injury (even though prompt medical treatment was given), including those requiring protection from all bodily contact.
2	Materials that on intense or continued, but not chronic, exposure could cause temporary incapacitation or possible residual injury, including those requiring the use of protective equipment that has an independent air supply.
1	Materials that on exposure would cause irritation but only minor residual injury, including those requiring the use of an approved air-purifying respirator.

0 Materials that on exposure under fire conditions offer no hazard beyond that of ordinary combustible material.

Flammability (red)

4 This degree includes flammable gases, pyrophoric liquids, and Class IA flammable liquids. Materials that will rapidly or completely vaporize at atmospheric pressure and normal ambient temperature or that are readily dispersed in air and that will burn readily.

3 Includes Class IB and IC flammable liquids and materials that can be easily ignited under almost all normal temperature conditions.

2 Materials that must be moderately heated before ignition will occur and includes Class II and Class IIIA combustible liquids and solids and semisolids that readily give off ignitible vapors.

1 Materials that must be preheated before ignition will occur, such as Class IIIB combustible liquids, and solids and semisolids whose flash point exceeds 200°F (93.4°C), as well as most ordinary combustible materials.

0 Materials that will not burn.

Reactivity (yellow)

4 Materials that, in themselves, are readily capable of detonation, explosive decomposition, or explosive reaction at normal temperatures and pressures.

3 Materials that, in themselves, are capable of detonation or explosive reaction but require a strong initiating source or heating under confinement; includes materials that are sensitive to thermal and mechanical shock at elevated temperatures and pressures and materials that react explosively with water.

2 Materials that are normally unstable and readily undergo violent chemical change but are not capable of detonation; includes materials that can undergo chemical change with rapid release of energy at normal temperatures and pressures. This also includes materials that may react violently with water or that may form potentially explosive mixtures in water.

1 Materials that are normally stable but that may become unstable at elevated temperatures and pressures and materials that will react with water with some release of energy, but not violently.

0 Materials that are normally stable, even under fire exposure conditions, and that do not react with water.

Other (white)

WATER Materials that react so violently with water that a possible hazard results when they come in contact with water, as in a fire situation; similar to Reactivity classification 2.

OXY Oxidizing material; any solid or liquid that readily yields oxygen or other oxidizing gas, or that readily reacts to oxidize combustible materials.

Flash Point: This is defined as the lowest temperature at which vapors above a volatile, combustible substance will ignite in air when exposed to a flame. Depending on the test method used, the values given are either Tag Closed Cup (c.c.) (ASTM D56) or Cleveland Open Cup (o.c.) (ASTM D93). The values, along with those in "Flammable Limits in Air" and "Autoignition Temperature" below, give an indication of the relative flammability of the chemical. In general, the open-cup value is about 10°F to 15°F (−12°C to −9°C) higher than the closed-cup value.

Flammable Limits in Air: The percentage concentration in air (by volume) is given for the LEL (lower explosive-flammable-limit in air, % by volume) and UEL (upper explosive-flammable-limit in air, % by volume), at room temperature, unless other specified. The values, along with those in Flash Point and Autoignition Temperature, give an indication of the relative flammability of the chemical.

Lower Explosive Limit (LEL): The lowest concentration of gas or vapor (% by volume in air) that will burn or explode when a source of ignition is present. Mixtures below the LEL are too "lean" to burn.

Upper Explosive Limit (UEL): The highest concentration of gas or vapor (% by volume in air) that will burn or explode when a source of ignition is present. Mixtures above the UEL are too "rich" to burn.

Autoignition Temperature: This is the minimum temperature at which the material will ignite without the presence of spark or flame. Values given are only approximate and may change substantially with changes in geometry, gas, or vapor concentrations, presence of catalysts, or other factors.

FIRST AID: First aid for exposure to the vapor (gas), liquid, or solid and includes inhalation, ingestion (swallowing), and contact with eyes or skin. The instruction "Do not induce vomiting" is given if an unusual hazard is associated with the chemical being sucked into the lungs (aspiration) while the patient is vomiting. "Seek medical attention" or "Call a doctor" is recommended in those cases in which only competent medical personnel can treat the injury properly. In all cases of human exposure, seek medical assistance as soon as possible.

ENVIRONMENTAL REGULATION SECTION

The descriptions within this section are intended solely for general information. Depending on the nature or scope of the activities at a particular facility, these summaries may or may not necessarily describe all applicable environmental requirements. For example, these potentially applicable regulations do not specify state requirements that may be applicable (and more stringent) to an individual facility, nor do they cite permitting regulations under which EPA or states issue facility-specific permits. Moreover, they do not constitute formal interpretations of the statutes and regulations. For further information, readers should consult the Code of Federal Regulations (CFR) and other state or local regulatory agencies. EPA Hotline numbers are provided for each major statute.

CLEAN AIR ACT (CAA)

The Clean Air Act (CAA) and its amendments, including the Clean Air Act Amendments (CAAA) of 1990, are designed to “protect and enhance the nation's air resources so as to promote the public health and welfare and the productive capacity of the population."The CAA consists of six sections, known as Titles, which direct EPA to establish national standards for ambient air quality and for EPA and the states to implement, maintain, and enforce these standards through a variety of mechanisms. Under the CAAA, many facilities will be required to obtain permits for the first time. state and local governments oversee, manage, and enforce many of the requirements of the CAAA. CAA regulations appear at 40 CFR Parts 50-99.

Pursuant to Title I of the CAA, EPA has established national ambient air quality standards (NAAQSs) to limit levels of "criteria pollutants," including carbon monoxide, lead, nitrogen dioxide, particulate matter, ozone, and sulfur dioxide. Geographic areas that meet NAAQSs for a given pollutant are classified as attainment areas; those that do not meet NAAQSs are classified as non-attainment areas. Under §110 of the CAA, each state must develop a State Implementation Plan (SIP) to identify sources of air pollution and to determine what reductions are required to meet federal air quality standards.

Title I also authorizes EPA to establish New Source Performance Standards (NSPSs), which are nationally uniform emission standards for new stationary sources falling within particular industrial categories. NSPSs are based on the pollution control technology available to that category of industrial source but allow the affected industries the flexibility to devise a cost-effective means of reducing emissions.

Under Title I, EPA establishes and enforces National Emission Standards for Hazardous Air Pollutants (NESHAPs), nationally uniform standards oriented towards controlling particular hazardous air pollutants (HAPs). Title III of the CAAA further directed EPA to develop a list of sources that emit any of 189 HAPs, and to develop regulations for these categories of sources. To date EPA has listed 174 categories and developed a schedule for the establishment of emission standards. The emission standards will be developed for both new and existing sources based on "maximum achievable control technology" (MACT). The MACT is defined as the control technology achieving the maximum degree of reduction in the emission of the HAPs, taking into account cost and other factors.

Title II of the CAA pertains to mobile sources, such as cars, trucks, buses, and planes. Reformulated gasoline, automobile pollution control devices, and vapor recovery nozzles on gas pumps are a few of the mechanisms EPA uses to regulate mobile air emission sources.

Title IV establishes a sulfur dioxide emissions program designed to reduce the formation of acid rain. Reduction of sulfur dioxide releases will be obtained by granting to certain sources limited emissions allowances, which, beginning in 1995, will be set below previous levels of sulfur dioxide releases.

Title V of the CAAA of 1990 created a permit program for all "major sources" (and certain other sources) regulated under the CAA. One purpose of the operating permit is to include in a single document all air emissions requirements that apply to a given facility. States are developing the permit programs in accordance with guidance and regulations from EPA. Once a state program is approved by EPA, permits will be issued and monitored by that state.

Title VI is intended to protect stratospheric ozone by phasing out the manufacture of ozone-depleting chemicals and restrict their use and distribution. Production of Class I substances, including 15 kinds of chlorofluorocarbons (CFCs), are to be phased out entirely by the year 2000, while certain hydrochlorofluorocarbons (HCFCs) will be phased out by 2030.

The original CAA authorized EPA to set limits on plastic resin and manmade fiber plant emissions. In its new source performance standards (NSPS) for polymer manufacturing facilities (40 CFR Part 60 Subpart DDD), EPA set minimum standards for the lowest achievable emissions rates (LAER) and best available control technologies (BACT). The NSPS for Polymers requires air emission controls on new and existing facilities that manufacture polypropylene, polyethylene, polystyrene and poly(ethylene terephthalate). Included are standards on controlling intermittent and continuous sources of emissions from processes. EPA also published an NSPS for synthetic fiber production facilities (40 CFR Part 60 Subpart HHH). The NSPS for Synthetic Fibers regulates VOC emissions from facilities that use solvents in manufacturing fibers. There are additional NSPS that apply to plastic resin and synthetic fiber manufacturers including those for flares (40 CFR Part 60 Subpart A), storage vessels (40 CFR Part 60 Subpart K), equipment leaks (40 CFR Part 60 Subpart VV), air oxidation processes (40 CFR Part 60 Subpart III), distillation operations (40 CFR Part 60 Subpart NNN), and reactor processes (40 CFR Part 60 Subpart RRR).

Industry Specific Requirements

The Clean Air Act Amendments of 1990 set National Emission Standards for Hazardous Air Pollutants (NESHAP) from industrial sources for 41 pollutants to be met by 1995 and for 148 other pollutants to be reached by 2003. Several provisions affect the plastic resin and manmade fiber industries.

In April 1994, the EPA published Hazardous Organic National Emissions Standards for Hazardous Air Pollutants, also known as HON, in a rule aimed at reducing air toxics emissions from chemical and allied product plants. This rule, which consists of four subparts, affects hundreds of plastic resin and manmade fiber plants and thousands of chemical process units since potential organic hazardous air pollutants are widely used as reactants. Processes covered include heat exchange systems and maintenance operations (40 CFR Part 63 Subpart F); process vents, storage vessels, transfer operations, and wastewater (40 CFR Part 63 Subpart G); equipment leaks (40 CFR Part 63 Subpart H); and equipment leaks for polycarbonate plants (40 CFR Part 63 Subpart I). Another NESHAP that may affect plastic resin and manmade fiber manufacturers is that for treatment, storage, and disposal facilities (40 Part CFR 63 Subpart AA). The HON also includes innovative provisions, such as emissions trading, that offer industry flexibility in complying with the rule's emissions goals.

Subsets of the plastic resin and manmade fiber industries are regulated under other NESHAPs. EPA published a final rule for epoxy resins and nonnylon polyamide resins in March 1995. The rule was expected to reduce epichlorohydrin emissions from process vents and storage tank emissions. In September 1996, EPA published a final rule for Group I Polymers and Resins (61 FR 46906) under 40 CFR part 63, Subpart U. This rule focused on reducing emissions from facilities that make certain elastomers used in the manufacture of synthetic rubber products. The rule was expected to reduce emissions of styrene, hexane, toluene, and other toxics. Provisions on pollution prevention, as well as a market-based provision on emissions averaging, were also included in the rule.

In September 1996, EPA also published a final rule for Designated Group IV Polymers and Resins (61 FR 48208) under 40 CFR part 63, Subpart JJJ. This rule was expected to reduce emissions of air toxics from poly (ethylene terephlate), nitrile, and styrene-based resins facilities. The rule was expected to reduce styrene, butadiene, and methanol emissions from storage vessels, process vents, equipment leaks, and wastewater operations. A direct final notice (62 FR 1869) was published on January 14, 1997, which extended the heat exchange system compliance date for the Group I rule and the equipment leak compliance dates for both the Group I and Group IV rules. Other NESHAPs that apply to the industry cover vinyl chloride manufacturers (40 CFR Part 61 Subpart F), benzene equipment leaks (40 CFR Part 61 Subpart J), fugitive emissions (40 CFR Part 61 Subpart V), benzene emissions from benzene storage vessels (40 CFR Part 61 Subpart Y), benzene emissions from benzene transfer operations (40 CFR Part 61 Subpart BB), and benzene waste operations (40 CFR Part 61 Subpart FF).

In addition to the applicable general requirements of the CAA, rubber and plastics products manufacturing facilities are subject to numerous industry-specific air regulations. Several source performance standards (NSPS) affect facilities in the rubber and plastics manufacturing industries. One NSPS is 40 CFR Part 60, Subpart BBB, "Standards of Performance for the Rubber Tire Manufacturing Industry. This standard applies to new, modified, or reconstructed facilities which mass produce passenger car and light-duty truck tires and similar tires with a bead diameter less than or equal to 0.5 meters (19.7 inches) and a cross-section dimension less than 0.325 meters (12.98 inches). The emission limits in the standard are for volatile organic compound (VOC) emissions from undertread cementing, sidewall cementing, tread end cementing, bead cementing, and green tire spraying operations.

Another NSPS that affects certain facilities in the rubber and plastics manufacturing industries is 40 CFR Part 60, Subpart TTT, "Standards of Performance for Industrial Surface Coating: Surface Costing of Plastic Parts for Business Machines. This standard applies to new, modified, or reconstructed facilities that apply coatings to plastic parts that will be used in the manufacture of business machines. The emission limits in the standards are for VOC emissions from prime coats, color coats, texture coats, and touch-up coats. Other NSPS standards such as NSPS DDD-For VOC for the Polymer Manufacturing Industry, and NSPS VVV-For Polymeric Coating of Supporting Substrates Facilities, may also apply to some facilities under SIC 30.

Under Title III of the Clean Air Act Amendments (CAAA), the EPA is developing many national emission standards for hazardous air pollutants (NESHAPs). Several of these will affect certain facilities in the rubber and plastics products manufacturing industries. These include standards for reinforced plastics and composites, rubber tire manufacturing, and plastic parts coating. Development of the standard for reinforced plastics and composites started in 1995. The development of the other two standards has not yet started.

Under Title I of the CAAA and under previous legislation, the EPA has provided guidance and other information to state and local agencies on reducing VOC emissions from existing sources in ozone nonattainment areas. These documents are referred to as Control Techniques Guidelines (CTGs) and Alternative Control Techniques (ACTs). An ACT for coating of plastic parts was issued in 1994 (Alternative Control Techniques Document: Surface Coating of Automotive/Transportation and Business Machine Plastic Parts, EPA-453/R-94/017).

EPA's Control Technology Center, at (919) 541-0800, provides general assistance and information on CAA standards. The Stratospheric Ozone Information Hotline, at (800) 296-1996, provides general information about regulations promulgated under Title VI of the CAA, and EPA's EPCRA Hotline, at (800) 535-0202, answers questions about accidental release prevention under CAA §112(r). In addition, the Technology Transfer Network Bulletin Board System (modem access (919) 541-5742)) includes recent CAA rules, EPA guidance documents, and updates of EPA activities.

CLEAN WATER ACT (CWA)

The primary objective of the Federal Water Pollution Control Act, commonly referred to as the Clean Water Act (CWA), is to restore and maintain the chemical, physical, and biological integrity of the nation's surface waters. Pollutants regulated under the CWA include "priority" pollutants, including various toxic pollutants; "conventional" pollutants, such as biochemical oxygen demand (BOD), total suspended solids (TSS), fecal coliform, oil and grease, and pH; and "nonconventional" pollutants, including any pollutant not identified as either conventional or priority.

The CWA regulates both direct and indirect discharges. The National Pollutant Discharge Elimination System (NPDES) program (CWA §402) controls direct discharges into navigable waters. Direct discharges or "point source" discharges are from sources such as pipes and sewers. NPDES permits, issued by either EPA or an authorized state (EPA has presently authorized forty states to administer the NPDES program), contain industry-specific, technology-based and/or water-quality-based limits, and establish pollutant monitoring and reporting requirements. A facility that intends to discharge into the nation's waters must obtain a permit prior to initiating its discharge. A permit applicant must provide quantitative analytical data identifying the types of pollutants present in the facility's effluent. The permit will then set forth the conditions and effluent limitations under which a facility may make a discharge.

A NPDES permit may also include discharge limits based on federal or state water quality criteria or standards that were designed to protect designated uses of surface waters, such as supporting aquatic life or recreation. These standards, unlike the technological standards, generally do not take into account technological feasibility or costs. Water quality criteria and standards vary from state to state, and site to site, depending on the use classification of the receiving body of water. Most states follow EPA guidelines, which propose aquatic life and human health criteria for many of the 126 priority pollutants.

Storm Water Discharges

In 1987 the CWA was amended to require EPA to establish a program to address storm water discharges. In response, EPA promulgated the NPDES storm water permit application regulations. Storm water discharge associated with industrial activity means the discharge from any conveyance which is used for collecting and conveying storm water and which is directly related to manufacturing, processing, or raw materials storage areas at an industrial plant (40 CFR 122.26(b)(14)). These regulations require that facilities with the following storm water discharges apply for a NPDES permit: (1) a discharge associated with industrial activity; (2) a discharge from a large or medium municipal storm sewer system; or (3) a discharge which EPA or the state determines to contribute to a violation of a water quality standard or is a significant contributor of pollutants to waters of the United States.

The term "storm water discharge associated with industrial activity" means a storm water discharge from one of 11 categories of industrial activity defined at 40 CFR 122.26. Six of the categories are defined by SIC codes, while the other five are identified through narrative descriptions of the regulated industrial activity. If the primary SIC code of the facility is one of those identified in the regulations, the facility is subject to the storm water permit application requirements. If any activity at a facility is covered by one of the five narrative categories, storm water discharges from those areas where the activities occur are subject to storm water discharge permit application requirements.

Those facilities/activities that are subject to storm water discharge permit application requirements are identified below. To determine whether a particular facility falls within one of these categories, the regulation should be consulted.

Category i: Facilities subject to storm water effluent guidelines, new source performance standards, or toxic pollutant effluent standards.

Category ii: Facilities classified as SIC 24-lumber and wood products (except wood kitchen cabinets); SIC 26-paper and allied products (except paperboard containers and products); SIC 28-chemicals and allied products (except drugs and paints); SIC 29-petroleum refining; and SIC 311-leather tanning and finishing.

Category iii: Facilities classified as SIC 10-metal mining; SIC 12-coal mining; SIC 13-oil and gas extraction; and SIC 14-nonmetallic mineral mining.

Category iv: Hazardous waste treatment, storage, or disposal facilities.

Category v: Landfills, land application sites, and open dumps that receive or have received industrial wastes.

Category vi: Facilities classified as SIC 5015-used motor vehicle parts; and SIC 5093-automotive scrap and waste material recycling facilities.

Category vii: Steam electric-power-generating facilities.

Category viii: Facilities classified as SIC 40-railroad transportation; SIC 41-local passenger transportation; SIC 42-trucking and warehousing (except public warehousing and storage); SIC 43-U.S. Postal Service; SIC 44-water transportation; SIC 45-transportation by air; and SIC 5171-petroleum bulk storage stations and terminals.

Category ix: Sewage treatment works.

Category x: Construction activities except operations that result in the disturbance of less than five acres of total land area.

Category xi: Facilities classified as SIC 20-food and kindred products; SIC 21-tobacco products; SIC 22-textile mill products; SIC 23-apparel related products; SIC 2434-wood kitchen cabinets manufacturing; SIC 25-furniture and fixtures; SIC 265-paperboard containers and boxes; SIC 267-converted paper and paperboard products; SIC 27-printing, publishing, and allied industries; SIC 283-drugs; SIC 285-paints, varnishes, lacquer, enamels, and allied products; SIC 30-rubber and plastics; SIC 31-leather and leather products (except leather and tanning and finishing); SIC 323-glass products; SIC 34-fabricated metal products (except fabricated structural metal); SIC 35-industrial and commercial machinery and computer equipment; SIC 36-electronic and other electrical equipment and components; SIC 37-transportation equipment (except ship and boat building and repairing); SIC 38-measuring, analyzing, and controlling instruments; SIC 39-miscellaneous manufacturing industries; and SIC 4221-4225-public warehousing and storage.

Pretreatment Program

Another type of discharge that is regulated by the CWA is one that goes to a publicly owned treatment works (POTWs). The national pretreatment program (CWA §307(b)) controls the indirect discharge of pollutants to POTWs by "industrial users. Facilities regulated under §307(b) must meet certain pretreatment standards. The goal of the pretreatment program is to protect municipal wastewater treatment plants from damage that may occur when hazardous, toxic, or other wastes are discharged into a sewer system and to protect the quality of sludge generated by these plants. Discharges to a POTW are regulated primarily by the POTW itself, rather than the state or EPA.

EPA has developed technology-based standards for industrial users of POTWs. Different standards apply to existing and new sources within each category."Categorical" pretreatment standards applicable to an industry on a nationwide basis are developed by EPA. In addition, another kind of pretreatment standard, "local limits," are developed by the POTW in order to assist the POTW in achieving the effluent limitations in its NPDES permit.

Regardless of whether a state is authorized to implement either the NPDES or the pretreatment program, if it develops its own program, it may enforce requirements more stringent than federal standards.

Industry-Specific Requirements

The Clean Water Act, first passed in 1972 and amended in 1977 and 1987, gives EPA the authority to regulate effluents from sewage treatment works, chemical plants, and other industrial sources into waters. The act sets "best available" technology standards for treatment of wastes for both direct and indirect (discharged to a Publicly Owned Treatment Work (POTW)) discharges. EPA originally promulgated effluent limitations guidelines and standards for the plastic resin and manmade fiber industries in two phases. Phase I, covering 13 products and processes, was promulgated on April 5, 1974 (39 FR 12502), and Phase II, covering 8 additional products and processes, was promulgated on January 23, 1975 (40 FR 3716). In 1976, these regulations were challenged and eventually remanded by the federal circuit court in FMC Corp. versus Traine, 539F.2d 973 (4th Cir. 1976). As a result, EPA withdrew both the Phase I and II plastic resin and manmade fiber regulations on August 4, 1976 (41 FR 32587) (EPA, 1987).

On November 5, 1987, EPA proposed final effluent guidelines (52FR42522) for the organic chemical, plastics, and synthetic fiber industries (OCPSF) (40 CFR Part 414).The effluent guidelines include limits for biological oxygen demand (BOD), total suspended solids (TSS), and acidity (pH). In this rule, limits are specified for facilities that manufacture rayon fibers, other synthetic fibers, thermoplastic resins, and thermoset resins.

The majority of this rule was upheld by the federal courts in 1989 when the Chemical Manufacturers Association sued the EPA. The court left the rule in effect pending further rulemaking but remanded three aspects of the OCPSF guidelines. The court remanded the New Source Performance Standards (NSPS) and the Pretreatment Standards for New Sources (PSNS) for consideration of whether zero discharge limits were appropriate for the industries; the subcategorization of the industries into two subcategories imposing differing limitations based on Best Available Technology Economically Achievable (BAT); and limitations for BAT Subpart J pollutants that were based upon in-plant biological treatment technology.

The EPA decided not to revise the NSPS and PSNS standards or the BAT subcategorization scheme and promulgated two sets of amendments to the rule in 1992 and 1993. On September 11, 1992, EPA promulgated a first set of amendments (57 FR 41836) to the OCPSF rule. These amendments allowed regulatory authorities to establish alternative cyanide limitations and standards for cyanide resulting from complexing of cyanide at the process source and to establish alternative metals limitations and standards to accommodate low background levels of metals in non-"metal-bearing waste streams." These amendments also allowed regulatory authorities to specify the method for determining five-day biochemical oxygen demand and total suspended solids effluent limitations for direct discharge plants (FR, September 11, 1992).

On July 9, 1993, EPA promulgated the remaining portions of the OCPSF rule in a second set of amendments (58 FR 36872) which added Subpart J limitations based on BAT and NSPS for 19 additional pollutants. These amendments also established Pretreatment Standards for Existing Sources (PSES) and PSNS for 11 of these 19 pollutants. EPA also corrected the criteria for designating "metal-" and "cyanide-bearing" waste streams. In this rulemaking, phenol and 2,4-dimethylphenol pretreatment standards were not promulgated, since EPA concluded that they did not pass through POTWs. The implementation of the guidelines is left to the states who issue NPDES permits for each facility. The compliance date for PSES was no later than July 23, 1996 (FR, July 9, 1993).

The Storm Water Rule (40 CFR §122.26(b)(14) Subparts (i, ii)) requires the capture and treatment of stormwater at facilities producing chemicals and allied products, including plastic resin and synthetic fiber manufacture. Required treatment will remove from stormwater flows a large fraction of both conventional pollutants, such as suspended solids and biological oxygen demand (BOD), as well as toxic pollutants, such as certain metals and organic compounds.

Plastics products manufacturers are subject to applicable general CWA requirements and to the specific requirements contained in 40 CFR Part 463, "Plastic Molding and Forming Point Source Category Effluent Limitations Guidelines; Pretreatment Standards and New Source Performance Standards. This regulation establishes effluent limitations guidelines and standards that limit the discharge of pollutants into navigable waters by existing and new sources engaged in plastic molding and forming. The regulated pollutants include BOD5, TSS, oil and grease, and pH.

EPA's Office of Water, at (202) 260-5700, will direct callers with questions about the CWA to the appropriate EPA office. EPA also maintains a bibliographic database of Office of Water publications which can be accessed through the Ground Water and Drinking Water resource center, at (202) 260-7786.

SAFE DRINKING WATER ACT (SDWA)

The Safe Drinking Water Act (SDWA) mandates that EPA establish regulations to protect human health from contaminants in drinking water. The law authorizes EPA to develop national drinking water standards and to create a joint federal-state system to ensure compliance with these standards. The SDWA also directs EPA to protect underground sources of drinking water through the control of underground injection of liquid wastes.

EPA has developed primary and secondary drinking water standards under its SDWA authority. EPA and authorized states enforce the primary drinking water standards, which are contaminant-specific concentration limits that apply to certain public drinking water supplies. Primary drinking water standards consist of maximum contaminant level goals (MCLGs), which are nonenforceable health-based goals, and maximum contaminant levels (MCLs), which are enforceable limits set as close to MCLGs as possible, considering cost and feasibility of attainment.

The SDWA Underground Injection Control (UIC) program (40 CFR Parts 144-148) is a permit program which protects underground sources of drinking water by regulating five classes of injection wells. UIC permits include design, operating, inspection, and monitoring requirements. Wells used to inject hazardous wastes must also comply with RCRA corrective action standards in order to be granted a RCRA permit, and must meet applicable RCRA land disposal restrictions standards. The UIC permit program is primarily state-enforced, since EPA has authorized all but a few states to administer the program.

The SDWA also provides for a federally implemented Sole Source Aquifer program, which prohibits federal funds from being expended on projects that may contaminate the sole or principal source of drinking water for a given area, and for a state-implemented Wellhead Protection program, designed to protect drinking water wells and drinking water recharge areas.

EPA promulgated regulations contained in the Federal Register, Vol. 55 No. 222, "National Pollutants Discharge Elimination System Permit Application Regulations for Storm Water Discharge; Final Rule" on November 16, 1990. These regulations require permit applications for storm water discharges from selected municipal and industrial point sources.

EPA's Safe Drinking Water Hotline, at (800) 426-4791, answers questions and distributes guidance pertaining to SDWA standards. The hotline operates from 9:00 a.m. through 5:30 p.m., EST, excluding Federal holidays.

RESOURCE CONSERVATION AND RECOVERY ACT (RCRA)

The Resource Conservation and Recovery Act (RCRA) of 1976, which amended the Solid Waste Disposal Act, addresses solid (Subtitle D) and hazardous (Subtitle C) waste management activities. The Hazardous and Solid Waste Amendments (HSWA) of 1984 strengthened RCRA's waste management provisions and added Subtitle I, which governs underground storage tanks (USTs).

Regulations promulgated pursuant to Subtitle C of RCRA (40 CFR Parts 260-299) establish a "cradle-to-grave" system governing hazardous waste from the point of generation to disposal. RCRA hazardous wastes include the specific materials listed in the regulations ("F," "K," "U,"or "P"), or materials which exhibit a hazardous waste characteristic as shown in the following tables.

Characteristic Hazardous Wastes	
Ignitability	Flashpoint below 140°F §261.21
Corrosivity	Liquids with a pH equal to or below 2 or equal to or above 12.5 or which corrode steel at a specified rate §261.22
Reactivity	Reacts violently with water or other substances to create toxic gases §261.23

I Introduction

Characteristic Hazardous Wastes	
Toxicity	A waste that leaches specified amounts of metals, pesticides, or organic chemicals using Toxicity Characteristic Leaching Procedure (TCLP) §261.24
Listed Hazardous Wastes	
"F" wastes	Hazardous wastes from nonspecific sources §261.31
"K" Wastes	Hazardous wastes from specific sources §261.32
"U" Wastes	Hazardous wastes from discarded commercial products, off-specification species, container residues§261.34
"P" Wastes	Acutely hazardous wastes from discarded commercial products, off-specification species, container residues§261.33

Note: If a waste is not found on any of these lists, it may be found on a state hazardous waste list.

Regulated entities that generate hazardous waste are subject to waste accumulation, manifesting, and recordkeeping standards. Facilities that treat, store, or dispose of hazardous waste must obtain a permit, either from EPA or from a state agency which EPA has authorized to implement the permitting program. Subtitle C permits contain general facility standards such as contingency plans, emergency procedures, recordkeeping and reporting requirements, financial assurance mechanisms, and unit-specific standards. RCRA also contains provisions (40 CFR Part 264 Subpart S and §264.10) for conducting corrective actions which govern the cleanup of releases of hazardous waste or constituents from solid waste management units at RCRA-regulated facilities.

Although RCRA is a federal statute, many states implement the RCRA program. Currently, EPA has delegated its authority to implement various provisions of RCRA to 46 of the 50 states.

Most RCRA requirements are not industry specific but apply to any company that transports, treats, stores, or disposes of hazardous waste. Here are some important RCRA regulatory requirements:

- Identification of Solid and Hazardous Wastes (40 CFR Part 261) lays out the procedure every generator should follow to determine whether the material created is considered a hazardous waste, solid waste, or is exempted from regulation.

- Standards for Generators of Hazardous Waste (40 CFR Part 262) establishes the responsibilities of hazardous waste generators including obtaining an ID number, preparing a manifest, ensuring proper packaging and labeling, meeting standards for waste accumulation units, and recordkeeping and reporting requirements. Generators can accumulate hazardous waste for up to 90 days (or 180 days, depending on the amount of waste generated) without obtaining a permit.

- Land Disposal Restrictions (LDRs) are regulations prohibiting the disposal of hazardous waste on land without prior treatment. Under the LDRs (40 CFR 268), materials must meet land disposal restriction (LDR) treatment standards prior to placement in a RCRA land disposal unit (landfill, land treatment unit, waste pile, or surface impoundment). Wastes subject to the LDRs include solvents, electroplating wastes, heavy metals, and acids. Generators of waste subject to the LDRs must provide notification of such to the designated TSD facility to ensure proper treatment prior to disposal.

- Used Oil Management Standards (40 CFR Part 279) impose management requirements affecting the storage, transportation, burning, processing, and re-refining of the used oil. For parties that merely generate used oil, regulations establish storage standards. For a party considered a used oil marketer (one who generates and sells off-specification used oil directly to a used oil burner), additional tracking and paperwork requirements must be satisfied.

- Tanks and Containers used to store hazardous waste with a high volatile organic concentration must meet emission standards under RCRA. Regulations (40 CFR Part 264-265, Subpart CC) require generators to test the waste to determine the concentration of the waste, to satisfy tank and container emissions standards, and to inspect and monitor regulated units. These regulations apply to all facilities who store such waste, including generators operating under the 90-day accumulation rule.

- Underground Storage Tanks (USTs) containing petroleum and hazardous substance are regulated under Subtitle I of RCRA. Subtitle I regulations (40 CFR Part 280) contain tank design and release detection requirements, as well as financial responsibility and corrective action standards for USTs. The UST program also establishes increasingly stringent standards, including upgrade requirements for existing tanks, that must be met by 1998.

- Boilers and Industrial Furnaces (BIFs) that use or burn fuel containing hazardous waste must comply with strict design and operating standards. BIF regulations (40 CFR Part 266, Subpart H) address unit design, provide performance standards, require emissions monitoring, and restrict the type of waste that may be burned.

Industry-Specific Requirements

Products, intermediates, and off-specification products generated at plastic resin and synthetic fiber facilities that are considered hazardous wastes are listed in 40 CFR Part 261.33(f). Some of the handling and treatment requirements for RCRA hazardous waste generators are covered under 40 CFR Part 262 and include the following: determining what constitutes a RCRA hazardous waste (Subpart A); manifesting (Subpart B); packaging, labeling, and accumulation time limits (Subpart C); and recordkeeping and reporting (Subpart D).

Many plastic resin and synthetic fiber facilities store some hazardous wastes at the facility for more than 90 days, and therefore are a storage facility under RCRA. Storage facilities are required to have a RCRA treatment, storage, and disposal facility (TSDF) permit (40 CFR Part 262.34). Some plastic resin and synthetic fiber facilities are considered TSDF facilities and are subject to the following regulations covered under 40 CFR Part 264: contingency plans and emergency procedures (40 CFR Part 264 Subpart D); manifesting, recordkeeping, and reporting (40 CFR Part 264 Subpart E); use and management of containers (40 CFR Part 264 Subpart I); tank systems (40 CFR Part 264 Subpart J); surface impoundments (40 CFR Part 264 Subpart K); land treatment (40 CFR Part 264 Subpart M); corrective action of hazardous waste releases (40 CFR Part 264 Subpart S); air emissions standards for process vents of processes that process or generate hazardous wastes (40 CFR Part 264 Subpart AA); emissions standards for leaks in hazardous-waste-handling equipment (40 CFR Part 264 Subpart BB); and emissions standards for containers, tanks, and surface impoundments that contain hazardous wastes (40 CFR Part 264 Subpart CC).

A number of RCRA wastes have been prohibited from land disposal unless treated to meet specific standards under the RCRA Land Disposal Restriction (LDR) program. The wastes covered by the RCRA LDRs are listed in 40 CFR Part 268 Subpart C and include a number of wastes commonly generated at plastic resin and synthetic fiber facilities. Standards for the treatment and storage of restricted wastes are described in Subparts D and E, respectively.

Many plastic resin and synthetic fiber facilities are also subject to the underground storage tank (UST) program (40 CFR Part 280). The UST regulations apply to facilities that store either petroleum products or hazardous substances (except hazardous waste) identified under the Comprehensive Environmental Response, Compensation, and Liability Act. UST regulations address design standards, leak detection, operating practices, response to releases, financial responsibility for releases, and closure standards.

Facilities engaged in rubber product use RCRA-regulated unused commercial chemical products which, if spilled or sent for disposal, are considered hazardous waste. These include ethylene thiourea, phenol, guanidines, and some lead, selenium, and cadmium compounds. Because these are all compounding agents which are added to the rubber mixture in their original form, spills are a reasonable possibility and RCRA requirements are likely to apply. Also, according to Toxic Release Inventory (TRI) data, rubber and miscellaneous plastics products manufacturing facilities use many solvents that are regulated by RCRA. These solvents include toluene, methyl ethyl keytone, 1,1,1-trichloroethane, acetone, methanol, xylene, methyl isobutyl keytone, trichlorofluoromethane, freon 113, trichloroethylene, and n-butyl alcohol.

EPA's RCRA/Superfund/UST Hotline, at (800) 424-9346, responds to questions and distributes guidance regarding all RCRA regulations. The RCRA Hotline operates weekdays from 8:30 a.m. to 7:30 p.m., EST, excluding federal holidays.

COMPREHENSIVE ENVIRONMENTAL RESPONSE, COMPENSATION, AND LIABILITY ACT (CERCLA)

The Comprehensive Environmental Response, Compensation, and Liability Act (CERCLA), a 1980 law commonly known as Superfund, authorizes EPA to respond to releases, or threatened releases, of hazardous substances that may endanger public health, welfare, or the environment. CERCLA also enables EPA to force parties responsible for environmental contamination to clean it up or to reimburse the Superfund for response costs incurred by EPA. The Superfund Amendments and Reauthorization Act (SARA) of 1986 revised various sections of CERCLA, extended the taxing authority for the Superfund, and created a free-standing law, SARA Title III, also known as the Emergency Planning and Community Right-to-Know Act (EPCRA).

The CERCLA hazardous substance release reporting regulations (40 CFR Part 302) direct the person in charge of a facility to report to the National Response Center (NRC) any environmental release of a hazardous substance which exceeds a reportable quantity. Reportable quantities are defined and listed in 40 CFR § 302.4. A release report may trigger a response by EPA, or by one or more federal or state emergency response authorities.

EPA implements hazardous substance responses according to procedures outlined in the National Oil and Hazardous Substances Pollution Contingency Plan (NCP) (40 CFR Part 300). The NCP includes provisions for permanent cleanups, known as remedial actions, and other cleanups referred to as "removals." EPA generally takes remedial actions only at sites on the National Priorities List (NPL), which currently includes approximately 1300 sites. Both EPA and states can act at other sites; however, EPA provides responsible parties the opportunity to conduct removal and remedial actions and encourages community involvement throughout the Superfund response process.

EPA's RCRA/Superfund/UST Hotline, at (800) 424-9346, answers questions and references guidance pertaining to the Superfund program. The CERCLA Hotline operates weekdays from 8:30 a.m. to 7:30 p.m., EST, excluding federal holidays.

Emergency Planning And Community Right-To-Know Act

The Superfund Amendments and Reauthorization Act (SARA) of 1986 created the Emergency Planning and Community Right-to-Know Act (EPCRA, also known as SARA Title III), a statute designed to improve community access to information about chemical hazards and to facilitate the development of chemical emergency response plans by state and local governments. EPCRA required the establishment of state emergency response commissions (SERCs), responsible for coordinating certain emergency response activities and for appointing local emergency planning committees (LEPCs).

EPCRA and its regulations (40 CFR Parts 350-372) establish four types of reporting obligations for facilities which store or manage specified chemicals:

- EPCRA §302 requires facilities to notify the SERC and LEPC of the presence of any "extremely hazardous substance" (the list of such substances is in 40 CFR Part 355, Appendices A and B) if it has such substance in excess of the substance's threshold planning quantity, and directs the facility to appoint an emergency response coordinator.

- EPCRA §304 requires the facility to notify the SERC and the LEPC in the event of a release exceeding the reportable quantity of a CERCLA hazardous substance or an EPCRA extremely hazardous substance.

- EPCRA §§311 and 312 require a facility at which a hazardous chemical, as defined by the Occupationai Safety and Health Act, is present in an amount exceeding a specified threshold to submit to the SERC, LEPC, and local fire department material safety data sheets (MSDSs) or lists of MSDSs and hazardous chemical inventory forms (also known as Tier I and II forms). This information helps the local government respond in the event of a spill or release of the chemical.

- EPCRA §313 requires manufacturing facilities included in SIC codes 20 through 39, which have ten or more employees, and which manufacture, process, or use specified chemicals in amounts greater than threshold quantities, to submit an annual toxic chemical release report. This report, commonly known as the Form R, covers releases and transfers of toxic chemicals to various facilities and environmental media, and allows EPA to compile the national Toxic Release Inventory (TRI) database.

All information submitted pursuant to EPCRA regulations is publicly accessible, unless protected by a trade secret claim.

EPA's EPCRA Hotline, at (800) 535-0202, answers questions and distributes guidance regarding the emergency planning and community right-to-know regulations. The EPCRA Hotline operates weekdays from 8:30 a.m. to 7:30 p.m., EST, excluding federal holidays.

SAFE DRINKING WATER ACT (SDWA)

The Safe Drinking Water Act (SDWA) mandates that EPA establish regulations to protect human health from contaminants in drinking water. The law authorizes EPA to develop national drinking water standards and to create a joint federal-state system to ensure compliance with these standards. The SDWA also directs EPA to protect underground sources of drinking water through the control of underground injection of liquid wastes.

EPA has developed primary and secondary drinking water standards under its SDWA authority. EPA and authorized states enforce the primary drinking water standards, which are contaminant-specific concentration limits that apply to certain public drinking water supplies. Primary drinking water standards consist of maximum contaminant level goals (MCLGs), which are nonenforceable health-based goals, and maximum contaminant levels (MCLs), which are enforceable limits set as close to MCLGs as possible, considering cost and feasibility of attainment.

The SDWA Underground Injection Control (UIC) program (40 CFR Parts 144-148) is a permit program, which protects underground sources of drinking water by regulating five classes of injection wells. UIC permits include design, operating, inspection, and monitoring requirements. Wells used to inject hazardous wastes must also comply with RCRA corrective action standards in order to be granted a RCRA permit, and must meet applicable RCRA land disposal restrictions standards. The UIC permit program is primarily state-enforced, since EPA has authorized all but a few states to administer the program.

The SDWA also provides for a federally-implemented Sole Source Aquifer program, which prohibits federal funds from being expended on projects that may contaminate the sole or principal source of drinking water for a given area, and for a state-implemented Wellhead Protection program, designed to protect drinking water wells and drinking water recharge areas.

EPA's Safe Drinking Water Hotline, at (800) 426-4791, answers questions and distributes guidance pertaining to SDWA standards. The Hotline operates from 9:00 a.m. through 5:30 p.m., EST, excluding federal holidays.

MARINE POLLUTANTS FROM DOT HAZARDOUS MATERIALS TRANSPORTATION ACT

The Environmental Protection Agency (EPA) in conjunction with the Department of Transportation (DOT) has promulgated regulations governing the transportation of hazardous wastes and hazardous substances. A marine polutant is a material listed in Appendix B to 49CFR Section 172.101 and when a mixture or solution of one or more marine pollutants is packaged in a concentration which equals or exceeds (1) 10% by weight of the solution or mixturel or (2) 1% by weight of the solution or mixture for those materials identified as severe marine pollutants in Appendix B. Severe marine pollutants contain the letters "PP" next to the chemical name.

CALIFORNIA'S PROPOSITION 65 CARCINOGENS AND REPRODUCTIVE TOXINS

Proposition 65, also known as the Safe Drinking Water and Toxic Enforcement Act of 1986, convenes meetings of a scientific advisory panel responsible for identifying chemicals to be listed, reviews risk assessments on listed chemicals, and compiles the list of chemicals known to the state to cause cancer or reproductive toxicity.

TOXIC SUBSTANCES CONTROL ACT

The Toxic Substances Control Act (TSCA), passed in 1976, gives the Environmental Protection Agency comprehensive authority to regulate any chemical substance whose manufacture, processing, distribution in commerce, use, or disposal may present an unreasonable risk of injury to human health or the environment. Four sections are of primary importance to the plastic resin and manmade fiber industries. TSCA §5 (new chemicals) mandates that plastic resin and manmade fiber companies submit premanufacture notices that provide information on health and environmental effects for each new product and test existing products for these effects (40 CFR Part 720). TSCA §4 (existing chemicals) authorizes the EPA to require testing of certain substances (40 CFR Part 790). TSCA §6 gives the EPA authority to prohibit, limit, or ban the manufacture, process, and use of chemicals (40 CFR Part 750). For certain chemicals, TSCA §8 also imposes record-keeping and reporting requirements, including substantial risk notification; record-keeping for data relative to adverse reactions; and periodic updates to the TSCA Chemical Inventory.

Under §5(h)(4), which grants EPA authority to promulgate rules granting exemptions to some or all of the premanufacture requirements for new chemicals, EPA published an exemption rule in 1984 and an amendment to the rule in 1995. The amendment, entitled Premanufacture Notification Exemptions (PMN) rule, contains a section on polymers (40 CFR Part 723.250) that allows polymers that meet certain restrictions to be exempt from some of the reporting requirements for new chemicals. Two exemptions {40 CFR Part 723.250(e)(1) and (e)(2)} exempt polymers based on molecular weight and oligomer content. The third exemption (40 CFR Part 723.250(e)(3)) exempts certain polyester polymers, which use particular monomers and reactants.

In addition to meeting the specific criteria of one of the three exemption types, the new polymer must also not fall into one of the prohibited categories. This section (40 CFR Part 723.250(d)) excludes certain polymers from reduced reporting requirements, namely: certain cationic polymers; polymers that do not meet elemental restrictions; polymers that are reasonably predicted to decompose, degrade, or depolymerize; and polymers which are produced from monomers and/or other reactants which are not on the TSCA inventory or otherwise exempted from reporting under a §5 exemption.

CHECKLIST OF POTENTIALLY APPLICABLE ENVIRONMENTAL REGULATIONS FOR THE PLASTIC AND ORGANIC CHEMICAL MANUFACTURING INDUSTRY

Common Production Unit Processes

Potentially significant releases/emissions are identified for each of the common production unit processes found in the organic chemicals, plastics, and synthetic fibers manufacturing industry. Potentially applicable environmental regulations are indicated by 40 CFR Part, and Subpart, for each corresponding release/emission type. The universe of medial-specific regulations potentially applicable to organic chemical manufacturing facilities are identified in the following tables for the Clean Air Act (CAA), the Clean Water Act (CWA), and the Resource Conservation and Recovery Act (RCRA). Additional environmental statutes such as the Safe DrinkingWater Act (SDWA), Federal Insecticide, Fungicide, and Rodenticide Act (FIFRA), Emergency Planning and Community Right-to-Know Act (EPCRA), and Toxic Substance Control Act (TSCA) that may apply to manufacturing facilities are not summarized in this section, since regulations developed under these statutes are not unit-process or waste-treatment-type specific. The reader should note that this checklist is intended solely as guidance. Because applicable regulations are specific to each individual facility, the reader is advised to use the Federal Register or the Code of Federal Regulations to determine applicable requirements.

Potentially Applicable CAA Regulations

Part 60 - Standards of Performance for New Sources	
Subpart D	**Steam Generation** *Applicability D_b* : Steam generating units with a heat input capacity from fuels combusted in steam generating unit>29 MW(100 millio.ı BTU/hr. *Standards for D_b*: particulate matter 60.43b; sulfur dioxide 60.42b; nitrogen dioxides 60.44b; *Test Methods:* 60.45b; *Monitoring:* 60.47b, 60.48b; *Reporting/ Record keeping:* 60.49b *Applicability D_c* : Steam generating units with maximum design heat input capacity of ≤29 MW (100 mmBtu/hr) but ≥2.9 MW (10mmBtu/hr). *Standards for D_c*: particulate matter 60.43c; sulfur dioxide 60.42c;*Test Methods:* 60.44c; 60.45c; *Monitoring:* 60.46c, 60.47c; *Reporting/Record keeping:* 60.48c
Subpart G	**Nitric Acid Plants** *Applicability:* Nitric acid production plants. The discharge of gases is prohibited which: *contain NO_2* in excess of 1.5 kg per metric tpm acid produced; Exibit 10% opacity or greater. *Test Methods:* 60.74; *Monitoring:* 60.73
Subpart H	**Sulfuric Acid Plants** *Applicability: Sulfuric acid production plants. The discharge of gases is prohibited which contains sulfur dioaxide in excess of 2 kg per metric ton of acid produced; contain H_2SO_4 in excess of 0.075 kg per metric ton of acid produced; Exhibits 10% opacity or greater.* *Test Methods:* 60.85; *Monitoring:* 60.84
Subpart K	**Petroleum Storage Vessels (K_b-VOC Storage Vessels)** *Applicability:* Storage vessels with design capacity ≥151 m³. Containing a VOL with TVP ≥5.2 lPa but less than 76.6 kPa; Storage vessels with design capacity ≥75 m³, *but ≤ 151* m³, *containing VOL with TVP ≥27.6 kPa but less than 76.6 kPa; Storage vessels with design capacity ≥75* m³ *and TVP ≥76.6 kPa.* *K_b: Test Methods:* 60.113b; *Monitoring:* 60.116b; *Reporting/Record keeping:* 60.115b
Subpart GG	**Stationary Gas Turbines** *Applicability:* Stationery gas turbines with heat input at peak load≥10.7 gigajoules/hr., based on lower heating value of the fuel fired. *Standards for:* sulfur dioxide 60.333; nitrogen oxides 60.332;*Test Methods:* 60.335; *Monitoring:* 60.334; *Reporting/Record keeping:* 60.334

Part 60 - Standards of Performance for New Sources	
Subpart VV	**VOC Equipment Leaks (SOCMI)** *Applicability:* Desing capacity of facility to produce as final product or intermediate 1,000 Mg/yr., or more of any chemicals identified. *Standards for:* 60.482, 60.483; *Monitoring:* 60.485; *Reporting:* 60.486; *Record keeping:* 60.486
Subpart III	**Air Oxidation Processes VOC Emissions (SOCMI)** *Applicability:* Each affected facility that produces any of the chemicals listed in 60.616 as a product, co-product, by-product, or intermediate. *Test Methods:* 60.614; *Monitoring:* 60.613; *Reporting/Record keeping:* 60.615
Subpart DDD	**Standards of Performance for VOC Emissions from the Polymer Manufacturing Industry** *Test Methods:* 60.664; *Monitoring:* 60.663; *Reporting/Record keeping:* 60.665
Subpart HHH	**Standards of Performance for Synthetic Fiber Production Facilities**
Subpart NNN	**Distillation Processes VOC Emissions (SOCMI)** *Applicability:* Each affected facility that produces any of the chemicals listed in 60.667 as a product, co-product, or intermediate.
Subpart RRR	**Reactor Processes VOC Emissions (SOCMI)** *Applicability:* Each affected facility that produces any of the chemicals listed in 60.707 as a product, co-product, or intermediate. *Test Methods:* 60.704; *Monitoring:* 60.703; *Reporting/Record keeping:* 60.705

Part 61 - Hazardous Air Pollutant (HAP) Emission Standards	
Subpart F	**Vinyl Chloride** *Applicability:* Plants producing: 1). Ethylene dichloride by reaction of oxygen and hydrogen chloride with ethylene, 2) Vinyl; chloride by any process, and/or 3) One or more polymers containing any fraction of polymerized vinyl chloride. *Monitoring:* 61.67, 61.68; *Reporting:* 61.69, 61.70 Standards for VOC Storage Vessels; *Record keeping:* 61.71
Subpart J	**Equipment Leaks (Fugitive Emission Sources of Benzene)** *Applicability:* Sources intended to operate in benzene service, including pumps, compressors, pressure relief devices, sampling connections, system open-ended valves or lines, valve flanges and other connectors, product accumulator vessels, and control devices; Required to comply with Part 61, Subpart V
Subpart M	**Asbestos** *Applicability:* 61.145 is applicable to owners or operators of a demolition or renovation activity; 61.146 applicable to owners or operators of an operation in which asbestos-containing materials are spray applied. Standard for Demolition and Renovation 61.145; Standard for Spraying 61.146; Standard for Insulating Materials 61.148; Standard for Waste Disposal for Manufacturing, fabricating, Demolition, Renovation, and Spraying Operations

Part 61 - Hazardous Air Pollutant (HAP) Emission Standards	
Subpart V	**Equipment Leaks (Fugitive Emission Sources)** *Applicability:* sources intended to operate in VHAP service, including pumps, compressors, pressure relief devices, sampling connection systems, open-ended valves or lines, valve flanges or other connectors, product accumulation vessels, and control devices. *Monitoring:* 61.242, 61.245; *Reporting:* 61.247; *Record keeping:* 61.246
Subpart Y	**Benzene Emissions from Benzene Storage Vessels** *Applicability:* All benzene storage vessels with a design capacity ≥ 38 m^3 (10,000 gals) *Monitoring:* 61.272; *Reporting:* 61.274, 61.275; *Record keeping:* 61.276
Subpart BB	**Benzene Emissions from Benzene Transfers** *Applicability:* Facilities loading benzene onto tank trucks, rail cars, or marine vessels at each benzene production facility and bulk termal. *Monitoring:* 61.303; *Reporting:* 61.305; *Record keeping:* 61.304
Subpart FF	**Benzene Waste Operations** *Applicability:* All existing sources and facilities handling benzene wastes, including: chemical manufacturing plant, coke by-product recovery plants, petroleum refineries, and hazardous waste treatment storage and disposal facilities. *Monitoring:* 61.354; *Reporting:* 61.357; *Record keeping:* 61.356

Part 63 - Source Category HAP Emission Standards	
Subpart F	**Organic HAPs from Synthetic Organic Chemical Manufacturing Industry (SOCMI)** *Applicability:* SOCMI facility that meets all of the following: 1) Manufactures as a primary product one or more of the chemicals listed in Table I, 2) Uses a reactant or manufacture as a product, by-product, or coproduct, one or more of the organic HAPs listed in Table 2, and 3) is located at a plant site that emits more than 10 TPY of any individual HAP or more than 25 TPY of any combination of HAP in any 12-month period. *Monitoring:* 63.103; *Reporting/Record keeping:* 63.103
Subpart G	**Organic HAPs from Process Vents, Storage Vessels, Transfer Operations, and Wastewater (SOCMI)** *Applicability:* Affects all Group I and Group 2 process vents (defined in 63.111) at affected facilities as determined under Subpart F. *Process Vent Monitoring:* 63.114; *Reporting:* 63.118; *Record keeping:* 63.117,63.118 *Applicability:* Affects all Group I and Group 2 storage vessels (defined in 63.111) at affected facilities as determined under Subpart F. *Storage Vessels Monitoring:* 63.119; *Reporting:* 63.112; *Record keeping:* 63.123 *Applicability:* Affects all Group I and Group 2 transfer operations (defined in 63.111) at affected facilities as determined under Subpart F. *Transfer Operations Monitoring:* 63.127; *Reporting/Record keeping:* 63.129 *Applicability:* Affects all Group I and Group 2 process wastewater [defined in 63.111 and 63.132(d)] at affected facilities as determined under Subpart F. *Wastewater Monitoring:* 63.143; *Reporting:* 63.146; *Record keeping:* 63.147

Part 63 - Source Category HAP Emission Standards	
Subpart H	**Organic HAPs for Equipment Leaks (SOCMI)** *Applicability:* Applies to pumps, compressors, agitators, pressure relief devices, sampling connection systems, open-ended valves or lines, valves, connectors, surge control vessels, bottoms receivers, instrumentation systems, and control devices or systems used to operate an organic HAP for 300 hours or more during a calendar year. *Monitoring:* 63.180; *Reporting:* 63.182; *Record keeping:* 63.181
Subpart I	**Organic HAPs for Equipment Leaks for Certain Processes (Negotiated Regulation)** *Applicability:* Major sources (including pumps, compressors, agitators, pressure relief devices, sampling connection systems, open-ended valves or lines, valves, connectors, surge control vessels, bottoms receivers, and instrumentation systems that are intended to operate in organic hazardous air pollutant services for 300 hours or more during a calendar year).
Subpart Q	**HAPs for Industrial Process Cooling Towers** *Applicability:* All new and existing industrial process cooling towers (IPCTs) which use chromium-based water treatment chemicals and are either a major source or are an integral part of a facility which is a major major source (defined in 64.401).
Subpart W	**National Emission Standards for Hazardous Air Pollutants for Epoxy Resins Production and Nonnylon Polyamides Production**

Additional Regulations that may Apply (CAA)	
Subpart 68	**Chemical Accident Prevention Provisions** *Applicability:* Owners or operators of stationery sources that have more than a threshold quantity (TQ) of a regulated substance in a process, as determined under §68.115.
Subpart 82	**Protection of Stratospheric Ozone** *Applicability:* any individual, corporate, or government entity that produces, transforms, imports, or exports these controlled substances. *Subpart A:* Production and Consumption Controls *Subpart E:* The Labelling of Products Using Ozone-Depleting Substances *Subpart F:* Recycling and Emissions Reduction

Note: Facilities should refer to their state environmental agency to identify any differences between state and federal program requirements.

Potentially Applicable CWA Regulations

Effluent Limitations Guidelines and Standards	
Part 414	**Organic Chemicals Plastics and Synthetic Fibers** *Applicability:* Discharges from manufacturers of products under SIC codes 2821 (Plastic Materials, Sythetic Resins, and Nonvulcanizable Elastomers), 2823 (Cellulosic Man-Made Fibers), 2824 (Synthetic Organic Fibers, except Cellulosic), 2865 (Cyclic Crudes and Intermediates, Dyes and Organic Pigments), and 2869 (Industrial Organic Chemicals, not elsewhere classified).

Effluent Limitations Guidelines and Standards	
Part 110	**Discharge of Oil** *Applicability:* Prohibited discharges include certain discharges to U.S. navigable water, adjoining shorelines. Or to waters of the contiguous zone, occuring in connection with the activities under the Outer Continental Shelf Lands Act of the Deepwater Port Act, or those that may affect U.S. natural resources. Manufacturing facilities using oil and that are either located by a municipal storm sewer that discharges to waters or near streams or bodies of water.
Part 112	**Oil Pollution Prevention** *Applicability:* Nontransportation related onshore and off-shore facilities engaged in drilling, producing, gathering, storing, processing, refining, transferring, distributing, or consuming oil and oil products that could resonably discharge oil in harmful quantities, as defined in Part 110.
Part 116	**Designation of Hazardous Substances**
Part 117	**Determination of Reportable Quantities (Rqs) for Hazardous Substances** *Applicability:* Does not apply to facilities that discharge the substance under and NPDES permit to a POTW, as long as any applicable effluent limit or pretreatment standard is met.
Additional Regulations that may Apply (CWA)	
Part 415	**Inorganic Chemicals**
Part 428	**Rubber Manufacturing**
Part 443	**Paving and Roofing Materials**
Part 446	**Paint Formulating**
Part 457	**Explosives Manufacturing**
Part 463	**Plastics, Molding, and Forming**

Potentially Applicable RCRA Regulations

Part 261	**Hazardous Waste Identification**
Part 262	**Hazardous Waste Generation**
Part 263	**Hazardous Waste Transporters**
Part 264 & 265	**Hazardous Waste Treatment, Storage, and Disposal**
Part 266 Subpart H	**Hazardous Waste Burned in Boilers and Industrial Furnaces**
Part 268	**Land Disposal Restrictions**

Part279	**Standards for the Management of Used Oil**
Part 280	**Underground Storage Tanks**

Federal and state permitting regulations and requirements are described in the following Section of 40 CFR

CAA	**Parts 70-72**
CWA	**Part 122-125 and 403**
RCRA	**Part 270-272**

MATERIALS HANDLING

➤Pipes, Valves, and Connections

Joint and Valve Head Leaks ***Sampling valves may yield additional waste material before and after sampling as a means of assuring that material is representative or as a result of pressure changes; the impact of such losses can usually be minimized by collecting the material for reuse or redesigning the sampling systems to minimize or eliminate the need for such waste.***	
Part 60, Subpart VV	for equipment leaks of VOCs
Part 61, Subpart J	for equipment leaks of benzene
Part 61, Subpart V	for equipment leaks of volitile HAPs
Part 61, Subpart BB	for benzene emissions from benzene transfer operations
Part 63, Subpart F	for hazardous organic NESHAPs
Part 63, Subpart H	for hazardous organic NESHAPs from equipment leaks
Equipment Cleanout ***The use of air or water to purge or decontaminate equipment can produce airborne streams and wastewater that escape with no control or contribute to fluctuations in wastewater characteristics.***	
Part 262	for listed and characteristic hazardous waste generation
Part 405-471	effluent guidelines for process wastewaters (e.g., Part 414 for organic chemicals)

Maintenance Operations ***Paint stripping, sandblasting, painting, welding and cutting, and lubrication may result in the disposal of equipment (e.g., valve gaskets, lengths of pipe, etc.); decontamination is desirable but is often not done since the piece is being discarded. This results in the item being hazardous or contributing to the characterization of other wastes as hazardous.***	
Part 63, Subpart F	for hazardous organic NESHAPs
Part 63, Subpart H	for hazardous organic NESHAPs for equipment leaks
Part 262	for listed and characteristic hazardous waste generators
Part 405-471	effluent guidelines for process wastewater (e.g., Part 414 for organic chemicals)

Failures/Malfunctions ***Air emissions may be immediately hazardous to the health of workers, forcing evacuation until properly equipped personnel can shut equipment and control the problem. Any delay which can increase the amount of material discharged into the environment can be minimized if automatic shutoff controls or suppressive equipment are in place.***	
Part 262	for listed and characteristic hazardous waste generators
Part 405-471	effluent guidelines for process wastewaters (e.g., Part 414 for organic chemicals)

➤Pumps, Compressors, and Steam Jet Ejectors

Vapor Losses ***Emissions of the volatile components in the stream being processed may escape around seals in pumps and compressors. Overpressurization may also cause relief valves to release. The agitation caused by rapid movement of liquids may aerate liquids and "aparge" volatiles into the headspace of the pump/compressor or the process vessel. These vapors may be emitted through relief valves. Certain high-speed pumps may actually cavitate, causing volatile components to vaporize and escape into the gas stream. If heat generated by the pump/compressor is not dissipated by some form of heat exchange fluid, the temperature of the fluid may increase, at least locally, causing vaporization of volatiles.***	
Part 60, Subpart VV	for equipment leaks of VOCs
Part 61, Subpart J	for equipment leaks of benzene
Part 61, Subpart V	for equipment leaks of volatile HAPs
Part 63, Subpart F	for hazardous organic NESHAPs
Part 63, Subpart H	for hazardous organic NESHAPs from equipment leaks

Liquid Leaks ***May occur at seals and ports if not properly tightened. Lubricant, liquid seals, and heat exchange fluid may become contaminated with process liquid but remain within the pump/compressor until maintenance work is done. An emulsion stream of lubricant oil and aqueous process liquid may be removed continuously from compressor for separation in an oil/water separator; other process chemicals will also be present, at least in the aqueous phase.***	
Part 262	for listed characteristic hazardous waste generation
Part 405-471	effluent guidelines for process wastewaters (e.g., Part 414 for organic chemicals)
Condensate ***Steam jet ejectors produce condensate contaminated with volatile components of the process fluid. This aqueous wastewater is removed at steam traps and can be collected and treated in a central facility. Water heated to make steam may be pretreated, or a blowdown may be removed to minimize scaling, etc. Steam jet ejectors will produce condensate with low concentrations of process chemicals during startup.***	
Part 405-471	effluent guidelines for process wastewaters (e.g., Part 414 for organic chemicals)
Startup/Shutdown ***Leaks of volatiles during startup and shutdowm will usually be small compared to those generated during normal operation; such leaks may be found during startup. Initial or final purging of a pump/compressor and associated processing system may produce vapor emissions from vents and relief valves, particularly when conditions deviate from operating conditions.***	
Part 60, Subpart VV	for equipment leaks of VOCs
Part 61, Subpart J	for equipment leaks of benzene
Part 61, Subpart V	for equipment leaks of volatile HAPs
Part 63, Subpart F	for hazardous organic NESHAPs
Part 63, Subpart H	for hazardous organic NESHAPs from equipment leaks
Maintenance Operations ***Purging fluids (inert gas, water, etc.) will become contaminated and require treatment and disposal. Lubricant fluids drained from pumps and compressors will usually be contaminated with process fluid and require treatment and/or disposal. Any gaskets, mechanical seals, etc. removed may require decontamination before they can be disposed..***	
Part 262	for listed characteristic hazardous wasted operations.
Part 405-471	effluent guidelines for process wastewaters (e.g., Part 414 for organic chemicals).

Failures/Malfunctions ***Failure of pump or compressor seals can cause the rapid emission of vapors, usually through relief vents. Loss of vacuum could allow air to mix with process liquid and vapor. If the material is oxygen sensitive, emissions of decomposition products could occur through relief valves. Failures of diaphragms and mechanical seals in pumps can cause process liquid and lubrican to mix, necessitating disposal of one or both fluids. Mechanical or electrical failure of pumps or compressors could subject process fluid to excess heat, resulting in decomposition.***	
Part 262	for listed and characteristic hazardous waste generators
Part 405-471	effluent guidelines for process wastewaters (e.g., Part 414 for organic chemicals)

➤Storage tanks, Containers, and Vessels

Vapor Losses ***The primary environmental impacts under normal operating conditions include volatile organics (VOCs) and inorganic emissions.***	
Part 61, Subpart Y	for benzene emissions from benzene storage vessels
Part 60, Subpart K, K_a and K_b	for VOC storage vessels
Part 63, Subpart G	for hazardous organic NESHAPs from storage vessels
Inert Gases ***Occasionally, inert gas blankets are used to prevent contamination of tanks and to minimize fire/explosion risks.***	
Part 60, Subpart K, K_a and K_b	for VOC storage vessels
Part 61, Subpart Y	for benzene emissions from benzene storage vessels
Part 63, Subpart G	for hazardous organic NESHAPs from storage vessels
Leaks/Spills ***Liquid wastes are largely limited to tank contents that may leak or spill into secondary containment, particularly during filling or draining of tanks. Contaminated rainwater collected in secondary containment may require treatment.***	
Part 262	for listed and characteristic hazardous waste generators
Part 405-471	effluent guidelines for process wastewaters (e.g., Part 414 for organic chemicals)

Solid Waste ***Solid and/or hazardous wastes may include contaminated soil from secondary containment after a spill or ongoing leaks, vent condensate, absorber solids, or scrubber solutions that are not recycled.***	
Part 262	for listed and characteristic hazardous waste generators
Tank Filling ***As the uncontaminated gas in the tank space is replaced by the liquid and vapor, gaseous losses (both VOCs and inorganics) through vents and valves can be significant, particularly as the tank approaches equilibrium.. Leaks at valves, flanges, etc. are most likely found during these periods.***	
Subpart Y	Benzene Emissions from Benzene Storage Vessels
Part 60, Subpart K, K_a and K_b	for VOC storage vessels
Part 63, Subpart G	for hazardous organic NESHAPs from storage vessels
Tank Draining ***Contaminants may escape form valves/flanges. The floating roof may have a residual ,thin film of liquid that might escape if not properly collected and/or treated. Vent control systems will also be in flux during filling and draining and would be subject to upset during these periods.***	
Part 61,Subpart Y	Benzene Emissions from Benzene Storage Vessels
Part 60, Subpart K, K_a and K_b	for VOC storage vessels
Part 63, Subpart G	for hazardous organic NESHAPs from storage vessels
Part 405-471	effluent guidelines for process wastewaters (e.g., Part 414 for organic chemicals)
Venting ***Tanks will require venting with air or inert gas before entry can be made safely; these emissions will contain vapors of the contents and will require collection/treatment. Residual contents in disconnected valves and piping may vaporize and require collection/treatment.***	
Part 61,Subpart Y	Benzene Emissions from Benzene Storage Vessels
Part 60, Subpart K, K_a and K_b	for VOC storage vessels
Part 63, Subpart G	for hazardous organic NESHAPs from storage vessels

<table>
<tr><td colspan="2">Maintenance Operations
Wastes: Sand blasting media, old paint, and VOCs. The use of certain paints will generate different and unique VOCs. The use of welding and cutting torches cause formation of metal oxide fumes, and residuals from the tank may decompose from elevated temperatures.</td></tr>
<tr><td>Part 262</td><td>for listed characteristic hazardous waste operations</td></tr>
<tr><td>Part 405-471</td><td>effluent guidelines for process wastewaters (e.g., Part 414 for organic chemicals)</td></tr>
<tr><td colspan="2">Tank Bottoms
The use of steam, water, and special detergents/solvents may mix with residual contents of the tank and require treatment/disposal.</td></tr>
<tr><td>Part 262</td><td>for listed and characteristic hazardous waste generators</td></tr>
<tr><td>Part 405-471</td><td>effluent guidelines for process wastewaters (e.g., Part 414 for organic chemicals)</td></tr>
<tr><td colspan="2">Failures/Malfunctions
Overfilling, leaking, and catastrophic ruptures can cause major emergency situations. Fire and suppressants are another source of potential environmental contamination.</td></tr>
<tr><td>Part 262</td><td>for listed and characteristic hazardous waste generators</td></tr>
<tr><td>Part 405-471</td><td>effluent guidelines for process wastewaters (e.g., Part 414 for organic chemicals)</td></tr>
</table>

➤**Blending and Milling**

<table>
<tr><td colspan="2">Particulate
The major environmental problem is the emissions of particulates at all stages, particularly when material is transferred from storage piles, bags, or bins and the solids are "dropped" into the reactor or other form of mixing chamber. With large operation, the escape of emissions can be minimized by the use of automated delivery conveyances and particulate control/collection systems. Such control systems are not often practical for small or nonroutine operations, and solids may be transferred from bags and boxes with little or no control. Similarly, the particulate captured in various environmental control systems, if not recycled, may have to be discarded as solid or hazardous wastes.</td></tr>
<tr><td>Part 262</td><td>for listed and characteristic hazardous waste generators</td></tr>
<tr><td colspan="2">Empty Containers
Can create secondary particulate emissions when they are removed to a disposal area and the containers may have to disposed of as hazardous waste.</td></tr>
<tr><td>Part 262</td><td>for listed and characteristic hazardous waste generators</td></tr>
</table>

Scrubber Water ***Milling wet material will increase water pollution concerns.***	
Part 262	for listed and characteristic hazardous waste generators
Part 405-471	effluent guidelines for process wastewaters (e.g., Part 414 for organic chemicals)

Mists/Vapors ***If the system is open to the atmosphere, incomplete absorption of the gas can result in gaseous emissions as well as mist and vapor of the liquid material from vents, safety releases, etc.; in rarer cases, fine particulate matter may also be produced and carried out of the chamber. Milling and mixing accompanied by heat may increase pollution concerns and add the possibility of decomposition products from material adhering to heated surfaces. Fine, dry solids can become charged with static electricity. Static discharges (sparks) can ignite flammable vapors, and certain dust clouds can detonate explosively.***	
Part 60, Subpart VV	for equipment leaks of VOCs
Part 61, Subpart J	for equipment leaks of benzene
Part 61, Subpart V	for equipment leaks of volatile HAPs
Part 63, Subpart F	for hazarsous organic NESHAPs
Part 63, Subpart H	for hazardous organic NESHAPs from equipment leaks

Spills ***The mixing of solid and liquid ingredients on exposed mixing rollers can result in material running off the sides of the rollers ; this matter can become waste, and if heated, either intentionally or due to friction, can cause material decomposition, escape of volatiles and even fires. Where runaway heating is a problem, fire suppression system are usually employed; this may produce additional wastewater and thermally damaged products that may have to be discarded.***	
Part 262	for listed and characteristic hazardous waste generators
Part 405-471	effluent guidelines for process wastewaters (e.g., Part 414 for organic chemicals)

Maintenance Operations ***Abrasive material and the resulting dust in the air can require frequent maintenance with the replacement of seals, gaskets, and lubricants and will contribute significant solid waste contaminated with the materials being processed. When mixing liquids or gases, the problems should be less serious and normal maintenance requirements probably can be anticipated.***	
Part 60, Subpart VV	for equipment leaks of VOCs
Part 61, Subpart J	for equipment leaks of benzene

Maintenance Operations ***Abrasive material and the resulting dust in the air can require frequent maintenance with the replacement of seals, gaskets, and lubricants and will contribute significant solid waste contaminated with the materials being processed. When mixing liquids or gases, the problems should be less serious and normal maintenance requirements probably can be anticipated.***	
Part 61, Subpart V	for equipment leaks of volatile HAPs
Part 63, Subpart F	for hazarsous organic HAPs
Part 63, Subpart G	for organic HAPs from process vents
Part 262	for listed and characteristic hazardous waste generators
Part 405-471	effluent guidelines for process wastewaters (e.g., Part 414 for organic chemicals)

Failures ***The major failure mode that must be considered is potential dust explosions, with resulting fire, equipment destruction , and widespread contamination of equipment with decomposed material. Failure of mixing equipment containing liquids would normally result in spills or leaks in the process area that should be anticipated with containment and/or cleanup equipment. Captured material could then be recovered or safely evaluated and properly discarded.***	
Part 60, Subpart VV	for equipment leaks of VOCs
Part 61, Subpart J	for equipment leaks of benzene
Part 61, Subpart V	for equipment leaks of volatile HAPs
Part 63, Subpart F	for hazardous organic HAPs
Part 63, Subpart G	for organic HAPs from process vents
Part 262	for listed and characteristic hazardous waste generators
Part 405-471	effluent guidelines for process wastewaters (e.g., Part 414 for organic chemicals)

REACTIONS

➤Batch, Continuous, and Fluidized Bed Reactors

<table>
<tr><td colspan="2">Vapor Losses
Fugitive leaks of solvent, reagent, product and byproduct vapors, as both liquid and gas, can occur around head and port seals, stirrer glands, pump and valve packings, and pipe flanges.</td></tr>
<tr><td>Part 60, Subpart III</td><td>for VOC emissions from SOCMI Air Oxidation Unit Processes</td></tr>
<tr><td>Part 60, Subpart RRR</td><td>for VOC emissions from SOCMI Reactor Processes</td></tr>
<tr><td>Part 63, Subpart F</td><td>for hazardous organic NESHAPs</td></tr>
<tr><td>Part 63, Subpart G</td><td>for hazardous organic NESHAPs from process vents</td></tr>
<tr><td>Part 262</td><td>for listed and characteristic hazardous waste generators</td></tr>
<tr><td colspan="2">Leaks
If contaminated, then steam, water, oils and silicone fluids, and brine solutions require treatment before heat transfer liquid or gas can be used or disposed of.</td></tr>
<tr><td>Part 60, Subpart III</td><td>for VOC emissions from SOCMI Air Oxidation Unit Processes</td></tr>
<tr><td>Part 60, Subpart VV</td><td>for equipment leaks of VOCs</td></tr>
<tr><td>Part 60, Subpart RRR</td><td>for VOC emissions from SOCMI Reactor Processes</td></tr>
<tr><td>Part 61, Subpart J</td><td>for equipment leaks of benzene</td></tr>
<tr><td>Part 61, Subpart V</td><td>for equipment leaks of volatile HAPs</td></tr>
<tr><td>Part 262</td><td>for listed and characteristic hazardous waste generators</td></tr>
<tr><td>Part 405-471</td><td>effluent guidelines for process wastewaters (e.g., Part 414 for organic chemicals)</td></tr>
<tr><td colspan="2">Sampling and Analysis
Liquid waste may be generated by leaking sampling valves, lines may need to be drained before and after sampling to remove holdup, and excess sample may be collected. Such material may require disposal if it cannot be recharged to the reactor. Sampling may also disturb the reactor equilibrium and cause "breathing" losses as re-equilibration occurs.</td></tr>
<tr><td>Part 60, Subpart III</td><td>for VOC emissions from SOCMI Air Oxidation Unit Processes</td></tr>
<tr><td>Part 60, Subpart RRR</td><td>for VOC emissions from SOCMI Reactor Processes</td></tr>
<tr><td>Part 63, Subpart F</td><td>for hazardous organic NESHAPs</td></tr>
</table>

Sampling and Analysis ***Liquid waste may be generated by leaking sampling valves, lines may need to be drained before and after sampling to remove holdup, and excess sample may be collected. Such material may require disposal if it cannot be recharged to the reactor. Sampling may also disturb the reactor equilibrium and cause "breathing" losses as re-equilibration occurs.***	
Part 63, Subpart G	for hazardous organic NESHAPs from process vents
Part 262	for listed and characteristic hazardous waste generators
Part 405-471	effluent guidelines for process wastewaters (e.g., Part 414 for organic chemicals)
Fluidizing Gas ***The fluidizing gas, which may be a process chemical, air or, an inert gas, may become an emission when discharged.***	
Part 60, Subpart III	for VOC emissions from SOCMI Air Oxidation Unit Processes
Part 60, Subpart RRR	for VOC emissions from SOCMI Reactor Processes
Part 63, Subpart F	for hazardous organic NESHAPs
Part 63, Subpart G	for hazardous organic NESHAPs from process vents
Startup ***Contamination would usually be limited to emission of volatiles and particulates as materials are charged and the reaction mass is heated. Gas streams contaminated with solvent or reagent vapors may escape while the reactor is being charged and as temperature increases. Leaks of liquids and gasses at valves, vents, joints, etc. are most apt to be discovered during startup; thermal expansion may create additional leaks.***	
Part 60, Subpart VV	for equipment leaks of VOCs
Part 60, Subpart III	for VOC emissions from SOCMI Air Oxidation Unit Processes
Part 60, Subpart RRR	for VOC emissions from SOCMI Reactor Processes
Part 61, Subpart J	for equipment leaks of benzene
Part 61, Subpart V	for equipment leaks of volatile HAPs
Part 63, Subpart F	for hazardous organic HAPs

<table>
<tr><td colspan="2">Off-Spec Product
With continuous and fluidized bed reactors, the materials initially charged may not meet product specifications due to incorrect flow ratios, low temperature or pressure, incorrect pH, etc., and must be recirculated (if possible) or discarded as startup wastes. Product generated during shutdown of a continuous reactor will often be off-spec and require reprocessing or disposal.</td></tr>
<tr><td>Part 262</td><td>for listed and characteristic hazardous waste generators</td></tr>
<tr><td colspan="2">Spills
During charging, liquid and solid spill may occur; transfer of fine solids from bags and tote-bins may cause dust emissions.</td></tr>
<tr><td>Part 262</td><td>for listed and characteristic hazardous waste generators</td></tr>
<tr><td>Part 405-471</td><td>effluent guidelines for process wastewaters (e.g., Part 414 for organic chemicals)</td></tr>
<tr><td colspan="2">Shutdown
When a batch reactor is shutdown, additional vapor losses occur as the reactor is opened or vented. The physical transfer of liquid product for a batch reactor may result in spills and solvent vapor losses, such as at valves, particularly if transfer is made while material is still warm. It is difficult to capture the vapors in the head space of a reactor. Residual nonvolatile material, possibly containing byproduct (e.g., salts) will often remain in the reactor as "stickage" after the initial draining and require tedious removal; this material may be added to the transferred material, discarded as wast, or incorporated in succeeding batches if compatible.</td></tr>
<tr><td>Part 60, Subpart VV</td><td>for equipment leaks of VOCs</td></tr>
<tr><td>Part 60, Subpart III</td><td>for VOC emissions from SOCMI Air Oxidation Unit Processes</td></tr>
<tr><td>Part 60, Subpart RRR</td><td>for VOC emissions from SOCMI Reactor Processes</td></tr>
<tr><td>Part 61, Subpart J</td><td>for equipment leaks of benzene</td></tr>
<tr><td>Part 61, Subpart V</td><td>for equipment leaks of volatile HAPs</td></tr>
<tr><td>Part 63, Subpart F</td><td>for hazardous organic HAPs</td></tr>
<tr><td>Part 63, Subpart G</td><td>for hazardous organic NESHAPs from process vents</td></tr>
<tr><td colspan="2">Separation
Will leave reactant/product-contaminated solids that require further processing (e.g., solvent extraction), reactivation, or disposal as solid/hazardous wastes.</td></tr>
<tr><td>Part 262</td><td>for listed and characteristic hazardous waste generators</td></tr>
</table>

Reactivation ***The material in a fluidized bed becomes contaminated and must be cleaned or replaced. Thermal reactivation may release new gaseous pollutants; chemical reactivation/cleaning may produce a liquid waste stream of undefined character. Discarded spent media will be solid waste and may exhibit hazardous waste characteristics.***	
Part 262	for listed and characteristic hazardous waste generators
Part 405-471	effluent guidelines for process wastewaters (e.g., Part 414 for organic chemicals)
Equipment flushing ***Purging gas, steam, or aqueous streams may be necessary to facilitate safe handling or as part of decontamination. These streams, contaminated with the contents of the system, will require disposal, usually as hazardous wastes.***	
Part 262	for listed and characteristic hazardous waste generators
Part 405-471	effluent guidelines for process wastewaters (e.g., Part 414 for organic chemicals)
Gaskets, Packing, Piping, Filters, etc. May become contaminated and need to be disposed of as waste that may be hazardous.	
Part 262	for listed and characteristic hazardous waste generators
Maintenance Operations ***Paint stripping, painting, welding, lubrication, etc., will generate additional airborne emissions, liquid contamination, or solid wastes that must be removed. As reactor and associated equipment are dismantled, vapor and liquid holdup must be captured and disposed of as wastes.***	
Part 262	for listed and characteristic hazardous waste generators
Part 405-471	effluent guidelines for process wastewaters (e.g., Part 414 for organic chemicals)

Failures/Malfunctions ***Emergency situations may make it necessary to discard the reaction mixture at almost any point in the process. This mixture may have to be disposed of as waste. In the event of "runaway" conditions or a major failure in a reactor system, it may be necessary to "dump" an entire batch of material, although certain ingredients (e.g., catalysts, solvents) may be recoverable. Various materials used to control runaway reactions will usually make the reaction mixture unsuitable for reuse, thereby generating a larger volume of a liquid waste of indefinite character. In failure situations there can be sudden and LARGE releases of steam, volatile organic compounds (e.g., solvents), inorganic gases, or particulates and fluidizing media, and unidentifiable secondary products (e.g., tars, polymers) from thermal and/or oxidative degradation. Backup into heat exchange systems (e.g., checkvalve failure, coil failure) could occur and generate large volumes of contaminated heat exchange fluid. Large-scale contamination of the area around equipment is also likely to occur during an emergency; with little or no containment in most cases this is likely to require the use of large quantities of absorbent materials and subsequent disposal. In the case of major failures, the process equipment and the facility may be damaged and require extensive maintenance or disposal.***	
Part 60, Subpart III	for VOC emissions from SOCMI Air Oxidation Unit Processes
Part 60, Subpart RRR	for VOC emissions from SOCMI Reactor Processes
Part 63, Subpart F	for hazardous organic NESHAPs
Part 63, Subpart G	for hazardous organic NESHAPs from process vents
Part 262	for listed and characteristic hazardous waste generators
Part 405-471	effluent guidelines for process wastewaters (e.g., Part 414 for organic chemicals)

HEAT TRANSFER

➤Heat Exchangers

Blowdown ***Heat exchange fluid to minimize corrosion and/or buildup of solids on the barrier surface may be to only discharge under normal operating conditions. Where heat exchange liquid is pretreated to reduce corrosion, foulding, scaling, etc., blowdown of heat exchange fluid could contain such additives.***	
Part 63, Subpart F	for hazardous organic NESHAPs
Part 63, Subpart G	for hazardous organic NESHAPs from process vents
Venting ***Heat exchange gases may also occur to control pressure or contaminant buildup.***	
Part 63, Subpart F	for hazardous organic NESHAPs

<table>
<tr><td colspan="2">Venting
Heat exchange gases may also occur to control pressure or contaminant buildup.</td></tr>
<tr><td>Part 63, Subpart G</td><td>for hazardous organic NESHAPs from process vents</td></tr>
<tr><td colspan="2">Leaks
Depending on the pressure on the two sides of the barrier, leaks from corrosion can cause either the process fluid or the heat exchange fluid to flow across the barrier and contaminate the other fluid. This could result in gaseous emissions, contaminated waste or water/organic streams where heat transfer fluids are used. Leaks can also occur at flanges, valves, and, particularly at compressors used to move process or heat exchange fluid. Even today, organic heat transfer fluids may still contain hazardous components such as PCBs.</td></tr>
<tr><td>Part 60, Subpart VV</td><td>for equipment leaks of VOCs</td></tr>
<tr><td>Part 61, Subpart J</td><td>for equipment leaks of benzene</td></tr>
<tr><td>Part 61, Subpart V</td><td>for equipment leaks of volatile HAPs</td></tr>
<tr><td>Part 63, Subpart F</td><td>for hazardous organic HAPs</td></tr>
<tr><td>Part 63, Subpart G</td><td>for hazardous organic NESHAPs from process vents</td></tr>
<tr><td colspan="2">Cooling Towers
A major additional wastestream is the vapor and mist discharged to the atmosphere; it may contain any volatile or dispersible components of the process fluid, ranging from ammonia and metallic corrosion inhibitors to bacteria. Another unique impact of cooling towers is the production of local fog that can affect visibility and safety due to icing. Cooling towers are not usually suitable for process fluids where hazardous contaminants would be emitted to the atmosphere.</td></tr>
<tr><td>Part 405-471</td><td>effluent guidelines for process wastewaters (e.g., Part 414 for organic chemicals)</td></tr>
<tr><td colspan="2">Startup/Shutdown
Potential or minor leaks may become exaggerated as a result of thermal changes in piping and vessels. Compressors may leak excessively until seals and gaskets seat into place.</td></tr>
<tr><td>Part 63, Subpart F</td><td>for hazardous organic HAPs</td></tr>
<tr><td>Part 63, Subpart G</td><td>for hazardous organic NESHAPs from process vents</td></tr>
</table>

<table>
<tr><td colspan="2"> Equipment Cleanout Scale and other sediments would be removed from the heat exchanger, generating either a solid waste or waste sludge containing metals and other contaminants, such as carbonized process chemicals. If high-pressure water or steam is used to accomplish this, a wastewater would be generated. Cleaning solvents may be used to remove organic heat exchange fluids, thus generating either nonaqueous, emulsion, or aqueous waste streams that would require treatment. Organic (e.g., chelating agents) used in place of or in addition to mechanical means to remove solids from heat exchangers will also be present in the wastewater and require treatment and disposal. These may also change the nature of the material removed from the exchanger.</td></tr>
<tr><td>Part 262</td><td>for listed and characteristic hazardous waste generators</td></tr>
<tr><td>Part 405-471</td><td>effluent guidelines for process wastewaters (e.g., Part 414 for organic chemicals)</td></tr>
<tr><td colspan="2"> Maintenance Operations Replacement of damaged sections of a heat exchanger may generate a waste of flush water containing the process fluid and/or any chemicals used to treat the exchange fluid. In addition, plant maintenance will generate the normal sand blasting, paint stripping, painting, welding ,and cutting emissions (particulates, VOCs, metallic fumes, etc.) when a unit is refurbished.</td></tr>
<tr><td>Part 63, Subpart F</td><td>for hazardous organic NESHAPs</td></tr>
<tr><td>Part 63, Subpart H</td><td>for hazardous organic NESHAPs for equipment leaks</td></tr>
<tr><td>Part 262</td><td>for listed and characteristic hazardous waste generators</td></tr>
<tr><td>Part 405-471</td><td>effluent guidelines for process wastewater (e.g., Part 414 for organic chemicals)</td></tr>
<tr><td colspan="2"> Failures/Malfunctions Heat exchanger failures can result in the gross contamination of one fluid with the other, as opposed to minor contamination that might occur from small leaks. If the leak is into a heat exchange gas, uncontrolled venting to the atmosphere would probably result since control or collection systems are not common. With water as the heat exchange fluid, the facility may be able to contain the contaminated fluid and gradually feed it into a central treatment system. Because of their cost, organic fluid systems are often equipped with separation/recovery systems. Leaks of heat exchange fluid into the process fluid may only dilute the process fluid or may contaminate it or thermally degrade it so that it is off-grade and must be discarded as waste. Obviously ,PCB contamination would require disposal in accordance with stringent regulations. Failure of any part of a heat exchange system (e.g., relief valve) could discharge hot gas or liquid into the local environment. The impact of venting hot air or steam may be minor unless significant contamination was present. Similarly, the impact of a aqueous discharge to the ground surface would be dependent on the hazards of the process chemical present.</td></tr>
<tr><td>Part 63, Subpart F</td><td>for hazardous organic NESHAPs</td></tr>
<tr><td>Part 63, Subpart H</td><td>for hazardous organic NESHAPs for equipment leaks</td></tr>
</table>

<table>
<tr><td colspan="2"> Failures/Malfunctions Heat exchanger failures can result in the gross contamination of one fluid with the other, as opposed to minor contamination that might occur from small leaks. If the leak is into a heat exchange gas, uncontrolled venting to the atmosphere would probably result since control or collection systems are not common. With water as the heat exchange fluid, the facility may be able to contain the contaminated fluid and gradually feed it into a central treatment system. Because of their cost, organic fluid systems are often equipped with separation/recovery systems. Leaks of heat exchange fluid into the process fluid may only dilute the process fluid or may contaminate it or thermally degrade it so that it is off-grade and must be discarded as waste. Obviously ,PCB contamination would require disposal in accordance with stringent regulations. Failure of any part of a heat exchange system (e.g., relief valve) could discharge hot gas or liquid into the local environment. The impact of venting hot air or steam may be minor unless significant contamination was present. Similarly, the impact of a aqueous discharge to the ground surface would be dependent on the hazards of the process chemical present.</td></tr>
<tr><td>Part 262</td><td>for listed and characteristic hazardous waste generators</td></tr>
<tr><td>Part 405-471</td><td>effluent guidelines for process wastewater (e.g., Part 414 for organic chemicals)</td></tr>
</table>

➤Condensers

<table>
<tr><td colspan="2"> Heat Exchange Fluids "Direct" implies the heat exchange fluid and the process fluid are in intimate contact; consequently, any process liquid will mix with the heat exchange fluid. The water, steam , or other heat exchange liquid leaving the system will be contaminated with volatiles from the process fluid. The cooled heat exchange fluid will condense and may become part of the bulk process liquid. Where it cannot be incorporated because of product quality, conditions can be controlled so that the heat exchange fluid evaporates and is either exhausted to the atmosphere (e.g., as steam) or is condensed in a secondary heat exchanger. The steam will be contaminated with process volatiles and/or process liquid mist.</td></tr>
<tr><td>Part 63, Subpart F</td><td>for hazardous organic NESHAPs</td></tr>
<tr><td>Part 63, Subpart G</td><td>for hazardous organic NESHAPs from processing vents</td></tr>
<tr><td>Part 262</td><td>for listed and characteristic hazardous waste generators</td></tr>
<tr><td>Part 405-471</td><td>effluent guidelines for process wastewater (e.g., Part 414 for organic chemicals)</td></tr>
<tr><td colspan="2"> Blowdown Any recirculation of the heat exchange fluid, such as by the use of secondary heat exchanger used to recover heat, may require a blowdown to control contaminant [and temperature] buildup. The blowdown will be a liquid waste that may or may not be suitable for recycle.</td></tr>
<tr><td>Part 405-471</td><td>effluent guidelines for process wastewater (e.g., Part 414 for organic chemicals)</td></tr>
</table>

Leaks *"Indirect" contact means that a barrier separates the two fluids. With an indirect or "no-contact" condenser, the most common environmental problem occurs when vapors from the process fluid are drawn into leaks in the barrier and are removed with the heat exchange fluid. Because gaseous heat exchange is relatively inefficient, large volumes of air would normally be used, diluting any process components and making capture and treatment of any contaminants less attractive. A monitoring device is often used to alert operators to the unexpected presence of contaminants in the gas stream. To an extent, the same problem exists during normal operation when using liquid heat exchange fluid such as water. The direction in which a leak will travel is dependent on which side of the barrier is at high pressure. If the process fluid cannot tolerate contamination with the heat exchange fluid, then any leaks would be into the heat exchanger/condenser. Vapors of the process liquid leaking into the heat exchange fluid can be discharged with the fluid. Monitoring is often used to detect such leaks. If heat exchange fluid leaks into the process fluid, it will, at least, dilute the liquid and retard removal of volatiles; at the extreme, unacceptable contamination of the process liquid could result, requiring disposal of the process material. In a related sense, where the water is recirculated (e.g., holding basin or cooling tower), the blowdown that would eventually be necessary will be at least as badly contaminated with any process components, plus any impurities that were in the incoming water. Any pumps or compressors used to circulate the heat exchange fluid will experience similar contamination.*	
Part 60, Subpart VV	for equipment leaks of VOCs
Part 61, Subpart J	for equipment leaks of benzene
Part 61, Subpart V	for equipment leaks of volatile HAPs
Part 63, Subpart F	for hazardous organic HAPs
Part 63, Subpart G	for organic HAPs from process vents
Part 262	for listed and characteristic hazardous waste generators
Part 405-471	effluent guidelines for process wastewaters (e.g., Part 414 for organic chemicals)
Startup/Shutdown *During startup, high flow of the heat exchange fluid will dilute any contamination with process fluid; however, the low concentrations may make recovery or recycle unattractive; discharge is often more cost-effective and simpler. Leaks into the condenser also may not be easily detectable during startup because of low concentrations. Similar conditions will exist at shutdown as the process fluid temperature is decreased if the flow of the heat exchange fluid is maintained.*	
Part 60, Subpart VV	for equipment leaks of VOCs
Part 61, Subpart J	for equipment leaks of benzene
Part 61, Subpart V	for equipment leaks of volatile HAPs

<table>
<tr><td colspan="2">Startup/Shutdown
During startup, high flow of the heat exchange fluid will dilute any contamination with process fluid; however, the low concentrations may make recovery or recycle unattractive; discharge is often more cost-effective and simpler. Leaks into the condenser also may not be easily detectable during startup because of low concentrations. Similar conditions will exist at shutdown as the process fluid temperature is decreased if the flow of the heat exchange fluid is maintained.</td></tr>
<tr><td>Part 262</td><td>for listed and characteristic hazardous waste generators</td></tr>
<tr><td>Part 405-471</td><td>effluent guidelines for process wastewaters (e.g., Part 414 for organic chemicals)</td></tr>
<tr><td colspan="2">Equipment Flushing
Gas purging or washing on the process side of the condenser with solvent or water will usually be carried out to assure safe handling during maintenance. The heat exchange fluid side will usually require decontamination when fluids other than steam/water (e.g., when using ammonia, brine, Freons or other organic liquids) and disposal of contaminated heat exchange fluids may then be necessary. Any pumps or compressors handling liquids would also need to be decontaminated, generating additional emissions or liquid wastes.</td></tr>
<tr><td>Part 262</td><td>for listed and characteristic hazardous waste generators</td></tr>
<tr><td>Part 405-471</td><td>effluent guidelines for process wastewaters (e.g., Part 414 for organic chemicals)</td></tr>
<tr><td colspan="2">Piping, Tubes, etc
Any piping, heat exchanger tubes, or compressor/pump parts that are being replaced may need to be disposed of as hazardous waste, particularly if the process liquid is hazardous.</td></tr>
<tr><td>Part 262</td><td>for listed and characteristic hazardous waste generators</td></tr>
<tr><td colspan="2">Failures/Malfunctions
Sudden failures during operation of pumps, compressors or heat exchange fluid, piping, valves, etc. can allow relatively concentrated vapors to escape from the condenser unless there are appropriate scrubbers or other secondary collection systems. Less frequently, a major leak of heat exchange fluid into the process liquid will occur. This could make the process liquid unsuitable for further use, requiring that it be disposed of as concentrated liquid waste. Catastrophic failure of the process system can, rarely, occur if solids crystallize (e.g., sublimation, decomposition, freezing) and plug narrow passages. Without proper controls, the pressure within the system can continue to increase until hot, concentrated vapor is rapidly discharged through either a pressure relief valve or through a rupture, if it does not first melt the plugging material. It may simply exceed the cooling capacity of the condenser and escape through vents, but it also may damage the heat exchanger and create leaks.</td></tr>
<tr><td>Part 63, Subpart F</td><td>for hazardous organic NESHAPs</td></tr>
<tr><td>Part 63, Subpart G</td><td>for hazardous organic NESHAPs from process vents</td></tr>
</table>

Failures/Malfunctions ***Sudden failures during operation of pumps, compressors or heat exchange fluid, piping, valves, etc. can allow relatively concentrated vapors to escape from the condenser unless there are appropriate scrubbers or other secondary collection systems. Less frequently, a major leak of heat exchange fluid into the process liquid will occur. This could make the process liquid unsuitable for further use, requiring that it be disposed of as concentrated liquid waste. Catastrophic failure of the process system can, rarely, occur if solids crystallize (e.g., sublimation, decomposition, freezing) and plug narrow passages. Without proper controls, the pressure within the system can continue to increase until hot, concentrated vapor is rapidly discharged through either a pressure relief valve or through a rupture, if it does not first melt the plugging material. It may simply exceed the cooling capacity of the condenser and escape through vents, but it also may damage the heat exchanger and create leaks.***	
Part 262	for listed and characteristic hazardous waste generators
Part 405-471	effluent guidelines for process wastewaters (e.g., Part 414 for organic chemicals)

➤**Evaporators**

Exit Gases ***During normal operations, an evaporator is usually under reduced pressure and the major loss is VOC (and inorganic) emissions (and mist) drawn off in the exit gas. If steam or water jets are used to generate that vacuum, they will become contaminated with those contaminants and may require treatment; if mechanical pumps are used, any lubricant or compressor oil also may become contaminated and require disposal.***	
Part 63, Subpart F	for hazardous organic HAPs
Part 63, Subpart G	for hazardous organic NESHAPs from process vents
Vapor Losses ***Ambient pressure evaporation of solvents of gas (e.g., air, inert gas) contaminated with low concentrations of VOCs from the dilute stream tends to be more difficult and more costly.***	
Part 60, Subpart VV	for equipment leaks of VOCs
Part 61, Subpart J	for equipment leaks of benzene
Part 61, Subpart V	for equipment leaks of volatile HAPs
Part 63, Subpart F	for hazardous organic HAPs
Part 63, Subpart G	for organic HAPs from process vents

Off-spec Product	
External heating sources (steam, hot oil, etc.) would normally leak into an evacuated evaporator and not create a new contaminated stream; however, the product may become contaminated if such a leak occurs and require disposal or reprocessing.	
Part 262	for listed and characteristic hazardous waste generators
Startup	
During startup, volatile components will be lost, usually at low rate, into the existing gas stream or into ejector steam or water until operating conditions are achieved. Transfer pipes, valves, etc. may leak as material is introduced into the evaporator if these portions of the system are not under reduced pressure.	
Part 60, Subpart VV	for equipment leaks of VOCs
Part 61, Subpart J	for equipment leaks of benzene
Part 61, Subpart V	for equipment leaks of volatile HAPs
Part 63, Subpart F	for hazardous organic HAPs
Part 63, Subpart G	for organic HAPs from process vents
Part 405-471	effluent guidelines for process wastewaters (e.g., Part 414 for organic chemicals)
Inert Gases	
When evaporation is complete, inert gas may be introduced to avoid thermal decontamination, but this should not create emission sources.	
Part 63, Subpart F	for hazardous organic HAPs
Part 63, Subpart G	for hazardous organic NESHAPs from process vents
Equipment Flushing	
When maintenance is being carried out on an evaporation system, gas and water will often be used to purge the system. These streams will be contaminated with the process fluids and, depending on the nature and concentrations, may require treatment for disposal. With proper planning, it may be possible to recycle such wastewaters into future batches.	
Part 63, Subpart F	for hazardous organic NESHAPs
Part 63, Subpart H	for hazardous organic NESHAPs for equipment leaks
Part 262	for listed and characteristic hazardous waste generators
Part 405-471	effluent guidelines for process wastewater (e.g., Part 414 for organic chemicals)

Maintenance Operations ***Normal maintenance wastes could include sandblast solids, paint strippers, paints, lubricants, welding and cutting fumes, etc., in addition to gaskets, pipes, etc., that are replaced.***	
Part 262	for listed and characteristic hazardous waste generators
Part 405-471	effluent guidelines for process wastewaters (e.g., Part 414 for organic chemicals)

Failures/Malfunctions ***Emergency situations could arise if an evaporator were overheated or became involved in a fire. In either case, thermally damaged process fluid would probably be discarded as waste. Particularly with temperature-, or oxygen-sensitive materials, failure of vacuum seals during operation may require premature shutdown and the need to discard or even dilute exposed process fluid to minimize further reaction.***	
Part 262	for listed and characteristic hazardous waste generators
Part 405-471	effluent guidelines for process wastewaters (e.g., Part 414 for organic chemicals)

SEPARATION

➤Distillation

Cooling/Condensation Contamination ***The primary environmental impacts under normal operating conditions relate to cooling and condensation, including overhead vapor (organic and inorganic) contamination of contact or noncontact cooling water or steam jet condensate, water jet discharge, or vacuum pump discharge where vacuum is used. Water used for cooling or steam heat may also require pretreatment and may produce concentrate wastes requiring treatment; antifoulants, corrosion protection agents, etc. may all be present in these waters.***	
Part 60, Subpart NNN	for VOC emissions from distillation operations
Part 405-471	effluent guidelines for process wastewaters (e.g., Part 414 for organic chemicals)

Leaks ***All valves, flanges, sampling points, vents, and circulating pump seals are also potential sources of airborne emission of the liquid and/or vapor components of the material being distilled. Leaks of liquids and gases at valves, vents, joints, etc. are most apt to be discovered during startup; thermal expansion may create additional, ananticipated leaks.***	
Part 60, Subpart VV	for equipment leaks of VOCs
Part 61, Subpart J	for equipment leaks of benzene

<table>
<tr><td colspan="2">Leaks
All valves, flanges, sampling points, vents, and circulating pump seals are also potential sources of airborne emission of the liquid and/or vapor components of the material being distilled. Leaks of liquids and gases at valves, vents, joints, etc. are most apt to be discovered during startup; thermal expansion may create additional, ananticipated leaks.</td></tr>
<tr><td>Part 61, Subpart V</td><td>for equipment leaks of volatile HAPs</td></tr>
<tr><td>Part 63, Subpart H</td><td>for hazardous organic NESHAPs from equipment leaks</td></tr>
<tr><td colspan="2">Inert Gas
Inert gas blankets used for some distillations may become contaminated with volatile materials, and vents bearing these materials may require treatment/control.</td></tr>
<tr><td>Part 262</td><td>for listed and characteristic hazardous waste generators</td></tr>
<tr><td colspan="2">Cooling/Heating During Startup
Contact cooling water may initially become contaminated with "light ends" of the chemicals being processed. Excess steam may condense until process temperature is attained and may be contaminated with the chemicals being processed.</td></tr>
<tr><td>Part 60, Subpart NNN</td><td>for VOC emissions from distillation operations</td></tr>
<tr><td>Part 63, Subpart F</td><td>for hazarsous organic HAPs</td></tr>
<tr><td>Part 63, Subpart G</td><td>for organic HAPs from process vents</td></tr>
<tr><td>Part 262</td><td>for listed and characteristic hazardous waste generators</td></tr>
<tr><td>Part 405-471</td><td>effluent guidelines for process wastewaters (e.g., Part 414 for organic chemicals)</td></tr>
<tr><td colspan="2">Off-spec Product
Off-spec material contaminated with light ends, cleaning agents, etc., may have to be recirculated or discarded as wastes.</td></tr>
<tr><td>Part 262</td><td>for listed and characteristic hazardous waste generators</td></tr>
<tr><td colspan="2">Heavy Ends/Still Bottoms
During shutdown, heavy ends and still bottoms will be present, may require manual discharge, and will frequently require disposal. These wastes may be listed or characteristic hazardous wastes, e.g., still bottoms from distillation of chlorinated organics will be listed hazardous wastes.</td></tr>
<tr><td>Part 262</td><td>for listed and characteristic hazardous waste generators</td></tr>
<tr><td>Part 405-471</td><td>effluent guidelines for process wastewaters (e.g., Part 414 for organic chemicals)</td></tr>
</table>

Equipment Cleanout ***Any equipment being dismantled, (e.g., pumps, vents, piping) containing gaseous or liquid material can produce wastes requiring special handling for environmental safety reasons.***	
Part 262	for listed and characteristic hazardous waste generators
Part 405-471	effluent guidelines for process wastewaters (e.g., Part 414 for organic chemicals)
Equipment Flushing ***Equipment will also require flushing with gaseous (purge gas) or aqueous systems (detergents, neutralizing agents, etc.) to facilitate handling or as part of decontamination. These streams, contaminated with the contents of the system, will require further management.***	
Part 262	for listed and characteristic hazardous waste generators
Part 405-471	effluent guidelines for process wastewaters (e.g., Part 414 for organic chemicals)
Gaskets, Packing, Piping, etc. ***Gaskets, packing, piping, etc. may be contaminated and require disposal as wastes that may be hazardous.***	
Part 262	for listed and characteristic hazardous waste generators
Maintenance Operations ***Paint stripping, painting welding, lubrication, etc., will also generate additional airborne emissions or liquid contamination that must be removed as part of a final cleaning or during the restartup of the distillation operation.***	
Part 262	for listed and characteristic hazardous waste generators
Part 405-471	effluent guidelines for process wastewaters (e.g., Part 414 for organic chemicals)
Failures/Malfunctions ***In failure or similar situations , there can be sudden and LARGE releases of steam, volatile organic and/or inorganic materials being distilled/processed, and even secondary products from the thermal and/or oxidative degradation of heated materials in the unit. Without proper check valves, backup into cooling water systems and processing equipment could occur and require more extensive control and cleanup outside the distillation unit. Similarly, contamination of the area around equipment is likely to occur during an emergency, but may also occur to a smaller extent during maintenance and even normal operations.***	
Part 61, Subpart V	for equipment leaks of volatile HAPs
Part 63, Subpart F	for hazarsous organic HAPs

<table>
<tr><td colspan="2"> Failures/Malfunctions In failure or similar situations , there can be sudden and LARGE releases of steam, volatile organic and/or inorganic materials being distilled/processed, and even secondary products from the thermal and/or oxidative degradation of heated materials in the unit. Without proper check valves, backup into cooling water systems and processing equipment could occur and require more extensive control and cleanup outside the distillation unit. Similarly, contamination of the area around equipment is likely to occur during an emergency, but may also occur to a smaller extent during maintenance and even normal operations.</td></tr>
<tr><td>Part 262</td><td>for listed and characteristic hazardous waste generators</td></tr>
<tr><td>Part 405-471</td><td>effluent guidelines for process wastewaters (e.g., Part 414 for organic chemicals)</td></tr>
</table>

➤Ion Exchange

<table>
<tr><td colspan="2"> Vapor Losses During routine operation, emissions of VOCs and organic gases would be limited to those escaping from a tank or transfer system.</td></tr>
<tr><td>Part 63, Subpart F</td><td>for hazardous organic HAPs</td></tr>
<tr><td>Part 63, Subpart G</td><td>for hazardous organic NESHAPs from process vents</td></tr>
<tr><td colspan="2"> Liquid Process Wastes The major discharges would be aqueous streams, including process liquid, refrigerant solution, rinse water, and backwash water, i.e., those streams passing through the system during operation or during the regeneration cycle. Small amounts of resin fines may also be swept into any of the aqueous streams. Other, smaller sources of wastewater would include leaks at pipe flanges, pump seals, and valves, that would be present in any containment/transfer operation.</td></tr>
<tr><td>Part 262</td><td>for listed and characteristic hazardous waste generators</td></tr>
<tr><td>Part 405-471</td><td>effluent guidelines for process wastewaters (e.g., Part 414 for organic chemicals)</td></tr>
<tr><td colspan="2"> Off-spec Products Makeup of regenerant solution, often from stored concentrate (e.g., acid, alkalai), could also produce off-spec or excess regenerant that would require disposal.</td></tr>
<tr><td>Part 262</td><td>for listed and characteristic hazardous waste generators</td></tr>
<tr><td colspan="2"> Leaks The transfer of regenerant from storage would be a source of leaks from pipes, flanges, valves.</td></tr>
<tr><td>Part 60, Subpart VV</td><td>for equipment leaks of VOCs</td></tr>
<tr><td>Part 61, Subpart J</td><td>for equipment leaks of benzene</td></tr>
</table>

<table>
<tr><td colspan="2">Leaks
The transfer of regenerant from storage would be a source of leaks from pipes, flanges, valves.</td></tr>
<tr><td>Part 61, Subpart V</td><td>for equipment leaks of volatile HAPs</td></tr>
<tr><td>Part 63, Subpart H</td><td>for hazardous organic NESHAPs from equipment leaks</td></tr>
<tr><td colspan="2">Regeneration
Depending on the nature of the process solution being exchanged, the wastewater from regeneration might require secondary treatment, such as neutralization/precipitation, that would produce sludge or solid waste requiring disposal. Changes in the character and concentration of the process solution being ion exchanged would usually lead to more frequent cycling (i.e., more rapid consumption of exchange sites) rather than a need to discontinue operations and discard the resin; however, increased cycling will lead to more frequent resin replacement. Powdered resins are something of the exception in that single use, followed by separation from the treated process liquid and disposal, is the standard approach. The increase in solids disposal requirements would be compensated by the elimination of all regenerant wastewater streams. Wastewater streams generated during ion exchange at a chemical manufacturing facility would be transported to a central facility for treatment along with other plant-generated wastewaters or would be pretreated in the process area before discharge to the central facility.</td></tr>
<tr><td>Part 262</td><td>for listed and characteristic hazardous waste generators</td></tr>
<tr><td>Part 405-471</td><td>effluent guidelines for process wastewaters (e.g., Part 414 for organic chemicals)</td></tr>
<tr><td colspan="2">Startup
Leaks in pipes and flanges, valves and pumps are most likely to be discovered during startup or shutdown due to changes in pressure, vibration, and thermal changes. Change in headspace volume and pressure as a exchanger vessel is filled or emptied may produce vapor emissions for a short duration. Although resins are usually shipped and used in a water-wetted state, minor temperature changes may occur as a result of resin solvation with solutes in the process solution, activation, and regeneration. This could increase volatilization of any VOCs present in the solution. Resin fines are removed during the initial charging and startup of an exchanger, and these would become solid waste if they cannot be recycled. The fines are slow to settle.</td></tr>
<tr><td>Part 60, Subpart VV</td><td>for equipment leaks of VOCs</td></tr>
<tr><td>Part 61, Subpart J</td><td>for equipment leaks of benzene</td></tr>
<tr><td>Part 61, Subpart V</td><td>for equipment leaks of volatile HAPs</td></tr>
<tr><td>Part 63, Subpart F</td><td>for hazardous organic HAPs</td></tr>
<tr><td>Part 63, Subpart G</td><td>for organic HAPs from process vents</td></tr>
<tr><td>Part 63, Subpart H</td><td>for hazardous organic NESHAPs from equipment leaks</td></tr>
</table>

<table>
<tr><td colspan="2">Startup
Leaks in pipes and flanges, valves and pumps are most likely to be discovered during startup or shutdown due to changes in pressure, vibration, and thermal changes. Change in headspace volume and pressure as a exchanger vessel is filled or emptied may produce vapor emissions for a short duration. Although resins are usually shipped and used in a water-wetted state, minor temperature changes may occur as a result of resin solvation with solutes in the process solution, activation, and regeneration. This could increase volatilization of any VOCs present in the solution. Resin fines are removed during the initial charging and startup of an exchanger, and these would become solid waste if they cannot be recycled. The fines are slow to settle.</td></tr>
<tr><td>Part 262</td><td>for listed and characteristic hazardous waste generation</td></tr>
<tr><td>Part 405-471</td><td>effluent guidelines for process wastewaters (e.g., Part 414 for organic chemicals)</td></tr>
<tr><td colspan="2">Equipment Flushing
Process and regenerant solutions and wash water trapped in valves and pipes when dismantling equipment for maintenance. These would usually be diluted with water used to flush the system. All streams would also be expected to contain some amount of resin and resin fines. It is during maintenance operations that resin may be replaced. If this large volume of solid material, contaminated with process chemicals, cannot be returned to the vendor for processing it must be discarded; depending on the characteristics of the resin at that point, disposal may range from landfilling to incineration. The latter approach may be used to recover valuable metals or simply to recovery the energy value of the resins.</td></tr>
<tr><td>Part 262</td><td>for listed and characteristic hazardous waste generators</td></tr>
<tr><td>Part 405-471</td><td>effluent guidelines for process wastewaters (e.g., Part 414 for organic chemicals)</td></tr>
<tr><td colspan="2">Maintenance Operations
Can generate all of the normal gaseous, waterborne, and solid emissions expected from area and equipment washdown, paint stripping, sandblasting, painting, welding and cutting operations, and lubrication.</td></tr>
<tr><td>Part 262</td><td>for listed and characteristic hazardous waste generators</td></tr>
<tr><td>Part 405-471</td><td>effluent guidelines for process wastewaters (e.g., Part 414 for organic chemicals)</td></tr>
</table>

<table>
<tr><td colspan="2">Failures/Malfunctions
Catastrophic failure of ion exchange vessels and pipes due to flange/joint failure and corrosive or erosive failure can produce rapid discharge of large volumes of the process liquid together with the resin. Secondary containment is not usually present, resulting in the contamination of the immediate process area. It is common practice to wash down such spills, as more minor leaks, into central wastewater collection system unless precluded by specific hazardous aspects. Overfilling and overflow from the ion exchange vessel, due to operator inattention or to instrument failure, would be a source of relatively large volumes of wastewater, although this would be expected to occur with low frequency.</td></tr>
<tr><td>Part 262</td><td>for listed and characteristic hazardous waste generators</td></tr>
<tr><td>Part 405-471</td><td>effluent guidelines for process wastewaters (e.g., Part 414 for organic chemicals)</td></tr>
</table>

➤Filtration

<table>
<tr><td colspan="2">Vapor Losses
Many filtrations are carried out in essentially open systems, consequently vapor can be lost to the work environment. With vacuum filtration, loss of volatiles into the vacuum stream may be greater.</td></tr>
<tr><td>Part 63, Subpart F</td><td>for hazardous organic NESHAPs</td></tr>
<tr><td>Part 63, Subpart G</td><td>for hazardous organic NESHAPs from processing vents</td></tr>
</table>

<table>
<tr><td colspan="2">Liquid Leaks
Many filters, such as filter presses, will leak around seals, producing a liquid waste. While this waste may be the same as the feedstream, it often will be disposed as waste. Compressor seal fluids and steam jet ejector condensate will also become contaminated.</td></tr>
<tr><td>Part 60, Subpart VV</td><td>for equipment leaks of VOCs</td></tr>
<tr><td>Part 61, Subpart J</td><td>for equipment leaks of benzene</td></tr>
<tr><td>Part 61, Subpart V</td><td>for equipment leaks of volatile HAPs</td></tr>
<tr><td>Part 63, Subpart H</td><td>for hazarsous organic NESHAPs from equipment leaks</td></tr>
<tr><td>Part 262</td><td>for listed and characteristic hazardous waste generators</td></tr>
<tr><td>Part 405-471</td><td>effluent guidelines for process wastewaters (e.g., Part 414 for organic chemicals)</td></tr>
</table>

Backflushing ***During backflushing of filters to prevent blinding, the process liquid will become contaminated with solids; this slurry may be discarded as waste.***	
Part 262	for listed and characteristic hazardous waste generators
Part 405-471	effluent guidelines for process wastewaters (e.g., Part 414 for organic chemicals)
Liquid Process Wastes ***If the solid is the desired product, the liquid filtrate, containing essentially all of the process chemicals, will be a waste requiring disposal. Additional contaminated liquid will usually be produced while washing the filter cake.***	
Part 262	for listed and characteristic hazardous waste generators
Part 405-471	effluent guidelines for process wastewaters (e.g., Part 414 for organic chemicals)
Solid Process Wastes ***If the filter cake is not the desired product, it will need to be treated as waste since it will contain process chemicals and may also contain filter-aids. The decision as to whether to wash the filter cake will depend on the value of the liquid material and the impact of the liquid on the characterization of the cake.***	
Part 262	for listed and characteristic hazardous waste generators
Startup ***Volatiles may be lost to the vacuum system. Liquid passing through the filter before a proper blanket is developed may be wasted, although it can often be recirculated.***	
Part 60, Subpart VV	for equipment leaks of VOCs
Part 61, Subpart J	for equipment leaks of benzene
Part 61, Subpart V	for equipment leaks of volatile HAPs
Part 63, Subpart F	for hazardous organic HAPs
Part 63, Subpart G	for organic HAPs from process vents
Part 63, Subpart H	for hazardous organic NESHAPs from equipment leaks
Part 262	for listed and characteristic hazardous waste generators
Part 405-471	effluent guidelines for process wastewaters (e.g., Part 414 for organic chemicals)

Equipment Flushing ***Purging of a filter with air or inert gas or washing with water or solvent to remove hazardous (toxic, corrosive, flammable) vapors or liquids is good practice to protect workers involved in maintenance. Pumps, compressors, valves, and pipes would be similarly decontaminated before they are handled.***	
Part 262	for listed and characteristic hazardous waste generators
Part 405-471	effluent guidelines for process wastewaters (e.g., Part 414 for organic chemicals)

Maintenance Operations ***In addition to any residual solids in or on the filter supports, fabrics, blankets, or membranes may be disposed during maintenance, along with lubricants and pump or compressor fluids.***	
Part 262	for listed and characteristic hazardous waste generators
Part 405-471	effluent guidelines for process wastewaters (e.g., Part 414 for organic chemicals)

Failures/Malfunctions ***Fires can occur on filters when air passes through a filter containing organic liquids. This would usually result in the need to dispose of all material as waste and may produce excess, uncontrolled volatile emissions. In other failures, such as a torn filter, liquid and solid may enter the vacuum system, contaminating pump fluids and steam jet ejector condensate. If the slurry is captured, it usually can be recycled.***	
Part 63, Subpart F	for hazarsous organic HAPs
Part 63, Subpart G	for organic HAPs from process vents
Part 262	for listed and characteristic hazardous waste generators
Part 405-471	effluent guidelines for process wastewaters (e.g., Part 414 for organic chemicals)

➤Drying

Vapor Losses ***Products and byproducts are often dried as intermediate or final stages in processing. Indirect heating (steam or electric) may be used to remove organic solvents and/or water during these operations. Vapors exiting the dryer can be captured by a condenser, absorber, or scrubber. Ambient temperature drying may be used for temperature-sensitive material; volatiles lost in these operations may be more difficult to capture because of the low temperature and high air flow. Freeze drying and vacuum drying are two versions of low-temperature drying that minimize the volume of offgasses. Drying with direct heating may also produce dilute VOC/air emissions.***	
Part 63, Subpart F	for hazardous organic HAPs

Vapor Losses *Products and byproducts are often dried as intermediate or final stages in processing. Indirect heating (steam or electric) may be used to remove organic solvents and/or water during these operations. Vapors exiting the dryer can be captured by a condenser, absorber, or scrubber. Ambient temperature drying may be used for temperature-sensitive material; volatiles lost in these operations may be more difficult to capture because of the low temperature and high air flow. Freeze drying and vacuum drying are two versions of low-temperature drying that minimize the volume of offgasses. Drying with direct heating may also produce dilute VOC/air emissions.*	
Part 63, Subpart G	for hazardous organic NESHAPs from process vents
Liquid Process Wastes *The displaced solvent (usually water) would be discharged as part of the vacuum system and may contain any volatile contaminants present in the wet solid. Where water is the solvent removed, or steam is used for direct heating, even non-volatile organics may be distilled or carried over as mist into the existing gas stream. These emissions can be captured by condensation.*	
Part 262	for listed and characteristic hazardous waste generators
Part 405-471	effluent guidelines for process wastewaters (e.g., Part 414 for organic chemicals)
Solid Process Wastes *Solids adhering to the walls of the dryer may be discarded as waste at the completion of batch drying operations if they cannot be used in subsequent batches.*	
Part 262	for listed and characteristic hazardous waste generators
Leaks *Where steam or hot fluid is used to provide heat indirectly, contamination with soluble process chemicals could occur due to leaks, although it is more likely that leaks would contaminate the process material being dried.*	
Part 60, Subpart VV	for equipment leaks of VOCs
Part 61, Subpart J	for equipment leaks of benzene
Part 61, Subpart V	for equipment leaks of volatile HAPs
Part 63, Subpart H	for hazardous organic NESHAPs from equipment leaks
Part 262	for listed and characteristic hazardous waste generators
Part 405-471	effluent guidelines for process wastewaters (e.g., Part 414 for organic chemicals)

Particulates ***Some solids and certain drying procedures produce fine particulates that can become airborne; filters, scrubbers, venturis, and precipitators may be used to collect the particulate. It is usually discarded, along with contaminated filters, bags, etc.; recycling is a waste minimization possibility.***	
Part 262	for listed and characteristic hazardous waste generators
Off-spec Product ***Dried solid product also may be discarded as waste if it becomes contaminated, such as by falling on the floor during transfer operations. The nature of the solid will determine whether it is hazardous or not.***	
Part 262	for listed and characteristic hazardous waste generators
Startup ***Leaks in pipe flanges, valves, pumps, and dust collection systems are most likely to be discovered during startup of a dryer due to vibration and thermal changes. Elevated temperature will increase volatilization of VOCs in the solids.***	
Part 63, Subpart F	for hazarsous organic HAPs
Part 63, Subpart G	for organic HAPs from process vents
Part 405-471	effluent guidelines for process wastewaters (e.g., Part 414 for organic chemicals)
Equipment Cleanout ***Cleaning of a dryer usually generates air or waterborne streams containing the solids being dried. Depending on the nature of the solid, these may be blown out into air or sluiced to the plant wastewater treatment facility as solids-contaminated wastewater. Solids trapped at joints and bends may be manually removed and discarded as solid waste; their presumed long residence time in the dryer may have altered their properties so that they are no longer recyclable as product.***	
Part 262	for listed and characteristic hazardous waste generators
Part 405-471	effluent guidelines for process wastewaters (e.g., Part 414 for organic chemicals)
Maintenance Operations ***Maintenance operations can generate all of the normal gaseous, waterborne, and solid emissions expected from paint stripping, sandblasting, painting, welding and cutting operations, and lubrication.***	
Part 262	for listed and characteristic hazardous waste generators
Part 405-471	effluent guidelines for process wastewaters (e.g., Part 414 for organic chemicals)

Failures/Malfunctions

Fires are one of the more common causes for catastrophic failure of dryers. Product remaining is usually unsuitable for recovery and is discarded as solid or hazardous waste, depending on the characteristics. Decomposition products may escape into the air during the fire. Water used to extinguish fire would be contaminated with the solid and its degradation products. Ideally, such water would be collected and treated on-site. Equipment damage by fire or overheating of drying equipment could need to be disposed of as waste if it cannot be reconditioned.

Part 262	for listed and characteristic hazardous waste generators

➤Crystallization

Vapor Losses

Volatile components of the process liquid will volatilize into crystallizer head space as material is introduced and while the temperature of the unit is charging, particularly when crystallization is brought about by heating to evaporate excess solvent. Where crystallization is brought about by cooling the process liquid, volatilization will be less significant.

Part 63, Subpart F	for hazardous organic HAPs
Part 63, Subpart G	for hazardous organic NESHAPs from process vents

Solid Process Wastes

When the solid is the desired product, the major liquid waste generated from a crystallizer would be the concentrated mother liquor remaining after the desired solid material has crystallized and been separated. This liquor may be discharged as a strong waste, but it also may be used in other process sequences or recycled as saturated solvent for succeding batches of product.

Part 262	for listed and characteristic hazardous waste generators

Liquid Process Wastes

When the solid material is not the desired product, decolorizing agents (e.g., carbon) and filter-aids may be present in the liquid and be removed at the same time, producing a solid waste contaminated with the mother liquor. Solids that become contaminated (e.g., by falling outside the crystallizer) will be solid waste unless they can be recycled.

Part 262	for listed and characteristic hazardous waste generators
Part 405-471	effluent guidelines for process wastewaters (e.g., Part 414 for organic chemicals)

<table>
<tr><td colspan="2">Leaks
Less common would be leaks into or out of heat exchanger fluid. Cooling may use cooling fluids such as brine, ammonia, or Freons; the first two of these would be corrosive to heat exchangers and could result in frequent leaks. When heating or cooling using an heat exchanger, cross-contamination of process and heat exchanger fluids may occur, depending on the relative pressure in the heat exchanger.</td></tr>
<tr><td>Part 60, Subpart VV</td><td>for equipment leaks of VOCs</td></tr>
<tr><td>Part 61, Subpart J</td><td>for equipment leaks of benzene</td></tr>
<tr><td>Part 61, Subpart V</td><td>for equipment leaks of volatile HAPs</td></tr>
<tr><td>Part 63, Subpart H</td><td>for hazardous organic NESHAPs from equipment leaks</td></tr>
<tr><td>Part 262</td><td>for listed and characteristic hazardous waste generators</td></tr>
<tr><td>Part 405-471</td><td>effluent guidelines for process wastewaters (e.g., Part 414 for organic chemicals)</td></tr>
<tr><td colspan="2">Startup
During the startup of a crystallizer, introduction of liquid into a vessel, evaporation of excess solvent or cooling of the process fluid from elevated temperatures will allow vapors to escape into head space of the vessel; they will be vented with any exhaust air moving through the system for worker or product safety. Heat exchanger leaks also may be detected during the startup. Depending on the heat exchanger fluid, such leaks could require entire batches of product liquor to be dumped as contaminated.</td></tr>
<tr><td>Part 60, Subpart VV</td><td>for equipment leaks of VOCs</td></tr>
<tr><td>Part 61, Subpart J</td><td>for equipment leaks of benzene</td></tr>
<tr><td>Part 61, Subpart V</td><td>for equipment leaks of volatile HAPs</td></tr>
<tr><td>Part 63, Subpart F</td><td>for hazardous organic NESHAPs</td></tr>
<tr><td>Part 63, Subpart G</td><td>for organic HAPs from process vents</td></tr>
<tr><td>Part 262</td><td>for listed and characteristic hazardous waste generators</td></tr>
<tr><td>Part 405-471</td><td>effluent guidelines for process wastewaters (e.g., Part 414 for organic chemicals)</td></tr>
</table>

<table>
<tr><td colspan="2"> Separation On completion of the crystallization stage, separation of the solid from the liquid would be accomplished by settling, centrifugation, or filtration. Inefficient separation need not generate a waste, since it should be practical to recycle the material to a succeeding batch or recirculate it through the separator.</td></tr>
<tr><td>Part 405-471</td><td>effluent guidelines for process wastewaters (e.g., Part 414 for organic chemicals)</td></tr>
<tr><td colspan="2"> Equipment Flushing During maintenance the major waste would be wash water (or solvent) used to clean the crystallizer; however, with good planning such waste liquids often can be reused in subsequent runs. Pumps and compressors used to transfer process fluid or heat exchange fluid also may need to be decontaminated, resulting in additional liquid waste which may not be suitable for reuse. Residual solids adhering to the walls often are scraped off and, depending on their value, either recycled or disposed of as waste. As with all heat exchanger applications, maintenance often includes the removal of scale and the replacement of damaged or worn tubes, which then becomes solid or hazardous waste.</td></tr>
<tr><td>Part 63, Subpart F</td><td>for hazardous organic HAPs</td></tr>
<tr><td>Part 63, Subpart G</td><td>for organic HAPs from process vents</td></tr>
<tr><td>Part 262</td><td>for listed and characteristic hazardous waste generators</td></tr>
<tr><td>Part 405-471</td><td>effluent guidelines for process wastewaters (e.g., Part 414 for organic chemicals)</td></tr>
<tr><td colspan="2"> Failures/Malfunctions Sudden failures of crystallizer walls, piping, valves, pumps, etc. certainly can occur and would allow process liquid to spill outside the crystallizer, making cleanup/washdown of the area necessary. Except for cross-contamination of the process liquid, such as with incompatible heat exchange fluid, it often is practical to recycle any process liquid that does not undergo the desired solid/liquid separation. Plugging of ports, valves, or piping with solids can cause shutdown of a crystallizer until solids can be melted or forced out; however, except for upstream damage to the system by pressure buildup, or the contamination of solids by fluids used to dislodge such plugs (e.g., steam), it would again, normally be possible to recover and recycle the process liquid.</td></tr>
<tr><td>Part 262</td><td>for listed and characteristic hazardous waste generators</td></tr>
<tr><td>Part 405-471</td><td>effluent guidelines for process wastewaters (e.g., Part 414 for organic chemicals)</td></tr>
</table>

➤**Centrifugation**

Vapor Losses ***High turbulence of liquid streams entering a centrifuge, coupled with relatively large and fluctuation headspace, can increase the loss of volatile materials; these losses can be partially controlled by appropriate capture hoods on or over the centrifuge.***	
Part 63, Subpart F	for hazardous organic HAPs
Part 63, Subpart G	for hazardous organic NESHAPs from process vents

Leaks ***Leaks of liquids and gases may also occur around valves, pumps , and piping flanges due to vibration.***	
Part 60, Subpart VV	for equipment leaks of VOCs
Part 61, Subpart J	for equipment leaks of benzene
Part 61, Subpart V	for equipment leaks of volatile HAPs
Part 63, Subpart H	for hazardous organic NESHAPs from equipment leaks
Part 405-471	effluent guidelines for process wastewaters (e.g., Part 414 for organic chemicals)

Liquid Process Wastes ***If the solid is the product, the liquid discharged will contain relatively high levels of suspended solids and all dissolved species in the liquid. Disposal or recycle will depend on constituents present. A wash cycle, commonly used to remove residual solvent and soluble species from the solid, will produce secondary, more dilute waste liquid.***	
Part 262	for listed and characteristic hazardous waste generators
Part 405-471	effluent guidelines for process wastewaters (e.g., Part 414 for organic chemicals)

Solid Process Wastes ***If the liquid is the desired product, then the solid "cake" is a waste and will usually retain significant quantities of the liquid phase.***	
Part 262	for listed and characteristic hazardous waste generators

Cooling Water/Fluids ***Any cooling water or steam will usually be in the centrifuge jacket and is not likely to be contaminated, although monitoring of the discharge may be desirable.***	
Part 405-471	effluent guidelines for process wastewaters (e.g., Part 414 for organic chemicals)

Cooling Fluids ***Fluids may be used to cool bearings, and these may be contaminated with the process liquid and will usually be considered waste because of undefined contamination.***	
Part 262	for listed and characteristic hazardous waste generators
Part 405-471	effluent guidelines for process wastewaters (e.g., Part 414 for organic chemicals)
Startup/Shutdown ***During the startup and shutdown, liquid streams may be discharged or recycled until proper speed or solids blanket are achieved. The liquid will contain the soluble ingredients and relatively high levels of solids. Ideally, this liquid is recirculated rather than discarded. Washing and removal of the solid cake will often generate an additional liquid waste containing lower concentrations of solids and solute present in the liquid phase. Vibration while coming to speed may produce leaks of liquids and gases at valves, vents, flanges, etc.***	
Part 63, Subpart F	for hazardous organic HAPs
Part 63, Subpart G	for organic HAPs from process vents
Part 262	for listed and characteristic hazardous waste generators
Part 405-471	effluent guidelines for process wastewaters (e.g., Part 414 for organic chemicals)
Equipment Dismantling ***Minor sources of gaseous or liquid wastes may be created when equipment is dismantled, (e.g., housing, pumps, vents, piping).***	
Part 262	for listed and characteristic hazardous waste generators
Part 405-471	effluent guidelines for process wastewaters (e.g., Part 414 for organic chemicals)
Equipment Flushing ***Flushing with purge gas or aqueous streams (detergents, neutralizing agents, etc.) to facilitate handling or as part of decontamination will produce relatively minor streams contaminated with the contents of the system; collection and/or treatment may be necessary.***	
Part 63, Subpart F	for hazardous organic HAPs
Part 63, Subpart G	for organic HAPs from process vents
Part 262	for listed and characteristic hazardous waste generators
Part 405-471	effluent guidelines for process wastewaters (e.g., Part 414 for organic chemicals)

Gaskets, Packing, Piping, etc. ***Contaminated gaskets, packing, piping, etc. will require disposal.***	
Part 262	for listed and characteristic hazardous waste generators
Maintenance Operations ***Paint stripping, painting, welding., lubrication, etc. will also generate additional airborne or liquid emissions.***	
Part 262	for listed and characteristic hazardous waste generators
Part 405-471	effluent guidelines for process wastewaters (e.g., Part 414 for organic chemicals)
Failures/Malfunctions ***In failure situations there can be sudden and LARGE releases of the material being centrifuged if it becomes necessary to stop the centrifuge rapidly. Usually, this material can be reprocessed, although there often be some discharge until the situation is brought under control. Damaged parts can usually be decontaminated by washing and then discarded as scrap.***	
Part 262	for listed and characteristic hazardous waste generators
Part 405-471	effluent guidelines for process wastewaters (e.g., Part 414 for organic chemicals)

➤Extraction

Vapor Losses ***Volatiles, primary organic vapors, may be lost during normal extraction operations from the headspace over an extraction. If heat is applied, the loss of vapors will increase, even if a condenser is used. Vapors can be directed through a vent and captured by a scrubber or adsorbent. Vapors removed during ambient temperature extraction with air or inert gas, such as to remove dissolved volatile organics, may be more difficult to control because the vapors are present at low vapor pressures. When, at some point in the extraction cycle the pressure is released, some of the extractant (solvent) may be lost as a gas (e.g., propane, carbon dioxide) and some of the extract may be lost as mist, leaving a liquid or solid that may be product or waste.***	
Part 63, Subpart F	for hazardous organic HAPs
Part 63, Subpart G	for hazardous organic NESHAPs from process vents
Liquid Process Wastes ***With either water or organic fluids as the extractant, the extract will become increasingly rich during processing. If the extract (aqueous or organic) is not the desired product or coproduct, then it will be waste that must be treated as part of disposal.***	
Part 262	for listed and characteristic hazardous waste generators

Liquid Process Wastes With either water or organic fluids as the extractant, the extract will become increasingly rich during processing. If the extract (aqueous or organic) is not the desired product or coproduct, then it will be waste that must be treated as part of disposal.	
Part 405-471	effluent guidelines for process wastewaters (e.g., Part 414 for organic chemicals)
Solid Process Wastes If the solution of the extracted material is the product, then the residual material, containing residual extracting solvent reagents ,will be a waste liquid or solid requiring disposal. On completion of an extraction cycle, the nonproduct phase, extract, or residue (liquid or solid) will contain process chemicals and require disposal. Recovery and recycle of solvent extractants is commonly practiced.	
Part 262	for listed and characteristic hazardous waste generators
Leaks When using a heat exchanger to provide or remove heat, leaks into the heat exchange fluid could make monitoring, treatment, and disposal necessary.	
Part 262	for listed and characteristic hazardous waste generators
Part 405-471	effluent guidelines for process wastewaters (e.g., Part 414 for organic chemicals)
Startup The initial filling of a vessel with the process material and extractant can allow volatiles to escape from the headspace through a vent, open reactor, or a pressure relief valve. Valves, pump seals, and pipe flanges also may leak vapors or liquid.	
Part 60, Subpart VV	for equipment leaks of VOCs
Part 61, Subpart J	for equipment leaks of benzene
Part 61, Subpart V	for equipment leaks of volatile HAPs
Part 63, Subpart H	for hazardous organic NESHAPs from equipment leaks
Part 405-471	effluent guidelines for process wastewaters (e.g., Part 414 for organic chemicals)
Equipment Flushing A process vessel and associated equipment (e.g., pipes, pumps, valves) would normally be purged to remove volatile materials and washed with water and/or solvent before maintenance is undertaken. These streams would be contaminated with both the extractant and the process chemical.	
Part 262	for listed and characteristic hazardous waste generators

<table>
<tr><td colspan="2" align="center">Equipment Flushing
A process vessel and associated equipment (e.g., pipes, pumps, valves) would normally be purged to remove volatile materials and washed with water and/or solvent before maintenance is undertaken. These streams would be contaminated with both the extractant and the process chemical.</td></tr>
<tr><td>Part 405-471</td><td>effluent guidelines for process wastewaters (e.g., Part 414 for organic chemicals)</td></tr>
<tr><td colspan="2" align="center">Packing, Seals, Gaskets
Valve packing, pump seals, and any gasket material being replaced would probably contain some contamination, and proper disposal would require some consideration.</td></tr>
<tr><td>Part 262</td><td>for listed and characteristic hazardous waste generators</td></tr>
<tr><td colspan="2" align="center">Failures/Malfunctions
Overpressurization or excessive temperatures might cause pressure relief valves to activate, discharging vapors or even liquid. Oxygen or heat-sensitive material may be damaged under such failures and require disposal as off-spec material. Heat exchanger leaks could result in contamination of the process fluid that would make it unusable or contamination of the heat exchange fluid. System parts damaged in a failure need to be decontaminated (e.g., with steam, water, or solvent) before they are discarded.</td></tr>
<tr><td>Part 262</td><td>for listed and characteristic hazardous waste generators</td></tr>
<tr><td>Part 405-471</td><td>effluent guidelines for process wastewaters (e.g., Part 414 for organic chemicals)</td></tr>
</table>

KEY TO ACRONYMS AND ABBREVIATIONS

ABS	Acrylonitrile-Butadiene-Styrene
ACGIH	American Conference of Governmental Industrial Hygenists
ACS	American Chemical Society
ACTs	Alternative Control Techniques
AFMA	American Fiber Manufacturers Association
ANSI	American National Standards Institute
APF	Assigned Protection Factor
ASTM	American Society for Testing and Materials
ANSI	American National Standards Institute
BEHP	Bis(2-Ethylhexyl)Phthalate
BIF	Boilers and Industry Furnaces (RCRA)
BOPP	Biaxially Oriented Polypropylene
BOD	Biochemical Oxygen Demand
C	Centigrade
CAA	Clean Air Act
CAAA	Clean Air Act Amendments of 1990

carc.	carcinogen
CAS	Chemical Abstract Service
cc	Cubic Centimeter
c.c.	Closed Cup
CERCLA	Comprehensive Environmental Response, Compensation and Liability Act
CERCLIS	CERCLA Information System
CFCs	Chlorofluorocarbons
CFR	*Code of Federal Regulations*
CHEMTREC	Chemical Manufacturers Association (CMA) Transportation Emergency Center
cis-	Refers to a particular arrangement of elements within a chemical molecule
CO	Carbon Monoxide
COD	Chemical Oxygen Demand
CR	Polychloroprene Rubber
CTG	Control Technology Guidance
CWA	Clean Water Act
DOT	U.S. Department of Transportation
DOT ID	U.S. Department of Transportation Identification Numbers.
E	Modulus of Elasticity
EEA	Ethylene Acrylic Acid
EEC	European Economic Community
EHS	Extremely Hazardous Substances
EINECS	European Inventory of Existing Commercial Substances
ELP	Enviromental Leadership Program
EMAA	Ethylene Methyl Acrylic Acid (copolymer)
EnBA	Ethylene-n-Butyl Acrylate (copolymer)
EPA	United States Environmental Protection Agency
EPCRA	Emergency Planning and Community Right-to-Know Act
ERG96	1996 North American Emergency Response Guide
EVA	Ethylene Vinyl Acetate (copolymer)
EVOH	Ethylene Vinyl Alcohol (copolymer)
F	Fahrenheit
FEMA	Federal Emergency Management Agency
FINDS	Facility Indexing System
FPA	Flexible Packaging Association
FR	*Federal Register*
FR	Flame Retardant
FRP	Fiber-Reinforced Plastics
HALS	Hindered Amine Light Stabilizer
HAPs	Hazardous Air Pollutants (CAA)
HCFC	Hydrochlorofluorocarbons
HCS	Hazard Communication Standard
HDPE	High-Density Polyethylene
HMW	High Molecular Weight
HP	Hypersensitivity Pneumonitis
HSDB	Hazardous Substances Data Bank
IARC	International Agency for Research on Cancer
IBM	Injection Blowing Machine

IDEA	Integrated Data for Enforcement Analysis
IDLH	Immediately Dangerous to Life or Health (concentrations)
kg or kgs	Kilogram or Kilograms
lb or lbs	Pound or Pounds
LCP	Liquid Crystal Polymer Plastic
LDPE	Low-Density Polyethylene
LDR	Land Disposal Restrictions (RCRA)
ILMA	Independent Lubricant Manufacturers Association
LEL	Lower explosive (flammable) limit in air, % by volume at room temperature or other temperture as noted
LEPCs	Local Emergency Planning Committees
m^3	Cubic Meter
MACT	Maximum Achievable Control Technology (CAA)
MCLGs	Maximum Contaminant Level Goals
MCLs	Maximum Contaminant Levels
MDPE	Medium-Density Polethylene
mg	Milligrams
MI	Melt Index
min	Minute
MSDS	Material Safety Data Sheets
MVTR	Moisture Vapor Transmission Rate
MW	Molecular Weight
n–	Abbreviation for "normal," referring to the arrangement of carbon atoms in a chemical molecule
N-	Symbol used in some chemical names, indicating that the next section of the name refers to a chemical group attached to a nitrogen atom
NAAQS	National Ambient Air Quality Standards (CAA)
NAFTA	North American Free Trade Agreement
NAICS	North American Industrial Classification System
NCDB	National Compliance Database (for TSCA, FIFRA, EPCRA)
NCI	National Cancer Institute
NCP	National Oil and Hazardous Substances Pollution Contingency Plan
NEIC	National Enforcement Investigation Center
NESHAP	National Emission Standards for Hazardous Air Pollutants
NFPA	National Fire Protection Association
NIOSH	National Institute of Occupational Safety and Health
NO_2	Nitrogen Dioxide
NO_X	Nitrogen Oxide
NOV	Notice of Violation
NPDES	National Pollution Discharge Elimination System (CWA)
NPL	National Priorities List
NR	Natural Rubber
NRC	National Response Center
NSPS	New Source Performance Standards (CAA)
NTP	National Toxicology Program
OAQPS	Office of Air Quality Planning and Standards
OAR	Office of Air and Radiation

o.c.	Open Cup
OECA	Office of Enforcement and Compliance Assurance
OPA	Oil Pollution Act
OPP	Oriented Polypropylene
OPPTS	Office of Prevention, Pesticides, and Toxic Substances
OSHA	Occupational Safety and Health Administration
OSW	Office of Solid Waste
OSWER	Office of Solid Waste and Emergency Response
Oxy	Oxidizer or oxidizing agent
p-	Abbreviation for "para" referring to a particular arrangement of elements within a chemical molecule.
P	Polymerization hazard (following DOT Guide number)
P2	Pollution Prevention
PA	Polyamide
PAN	Polycrylonitrile Plastic
PAR	Polyarylite Plastic
PBT	Polybutylene Terephthalate
PCB	Polychlorinated Biphenyl
PCS	Permit Compliance System (CWA Database)
PE	Polyethylene
PEL	Permissible Exposure Limit (OSHA)
PET	Polyethylene Terephthalate
PI	Polyimide Plastic
pot. carc	Potential Carcinogen
POTW	Publicly Owned Treatments Works
PP	Polypropylene
PP	Severe Marine Pollutant
PPE	Personal Protective Equipment
ppm	Parts Per Million
PS	Polystyrene Plastic
PUR	Polyurethane Plastic
PVP	Polyvinyl Plastic
PVDC	Polyvinylidene Chloride
PVDF	Polyvinylidene Fluoride Plastic
RCRA	Resource Conservation and Recovery Act
RCRIS	RCRA Information System
REL	Recommended Exposure Limits (NIOSH)
RIM	Reaction Injection Molding
RP	Reinforced Plastic
RQ	Reportable Quantity
RTECS	Registry of Toxic Effects of Chemical Substances
RTK	Right-to-Know
RTM	Resin Transfer Molding
SAN`	Styrene-Acrylonitrile (copolymer)
SARA	Superfund Amendments and Reauthorization Act
SBM	Stretch Blow Molding
s. carc	Suspected Carcinogen

SCBA	Self-Contained Breathing Apparatus
SDWA	Safe Drinking Water Act
SEP	Supplementary Environmental Project
SERC	State Emergency Response Commission
SIC	Standard Industrial Classification
SIP	State Implementation Plan
SPE	Society of Plastics Engineers
SPI	Society of the Plastics Industry, The
SO_2	Sulfur Dioxide
STCC	Standard Transportation Commodity Code
STEL	Short-Term Exposure Limit
sus. carc.	Suspected Carcinogen
sym-	Abbreviation for "symmetrical," referring to a particular arrangement of elements within a chemical molecule
TC	Toxic Characteristic
TCRIS	Toxic Chemical Release Inventory System
temp.	Temperature
tert-	Abbreviation for "tertiary," referring to a particular arrangement of elements within a chemical molecule
TLV	Threshold Limit Value (ACGIH)
TLV-C	Threshold Limit Value-Ceiling
Tm	Melt Temperature
TM	Transfer Molding
TNMOC	Total Non-Methane Organic Carbons
TOC	Total Organic Carbon
TP	Thermoplastic
TPO	Thermoplastic Olefin
TPQ	Threshold Planning Quantity
TRI	Toxic Release Inventory
TRIS	Toxic Release Inventory System
TQ	Threshold Quantity
TSCA	Toxic Substances Control Act
TSS	Total Suspended Solids
TWA	Time-Weighted Average
UEL	Upper explosive (flammable) limit in air, % by volume at room temperature or other temperature as noted
UHMW	Ultra-High Molecular Weight HDPE
UF	Urea Formaldehyde Plastic
UIC	Underground Injection Control (SDWA)
UST	Underground Storage Tanks (RCRA)
VAM	Vinyl Acetate Monomer
VOCs	Volatile Organic Compounds
>:	The symbol for "greater than"
<:	The symbol for "less than"

ACENAPHTHENE

A:003

SYNONYMS: ACENAFENO (Spanish); ACENAPHTHYLENE, 1,2-DIHYDRO; 1,8-DIHYDROACENAPHTHALENE; 1,8-DIHYDROACENAPHTHYLENE; ETHYLENENAPHTHALENE; 1,8-ETHYLENENAPHTHALENE; NAPHTHYLENEETHYLENE; PERIETHYLENENAPHTHALENE

IDENTIFICATION

CAS:	83-32-9
ERG ID:	UN 2811
ERG Guide:	TOXIC SOLID, ORGANIC, N.O.S.
ERG Guide Category:	154
Molecular Formula:	$C_{12}H_{10}$
RTECS No.:	AB1000000
Properties:	Combustible needlelike crystalline solid. White. Sinks in water; insoluble.

HEALTH AND SAFETY INFORMATION

A polynuclear aromatic hydrocarbon (PAH). Irritates skin and respiratory tract. Mutagenic. Attacks kidney, liver, skin.

Odor Threshold: 0.08 ppm (chemically pure)

RECOMMENDATIONS FOR RESPIRATOR SELECTION

At any concentrations above the NIOSH REL, or where there is no REL, at any detectable concentration: SCBAF:PD,PP (any self-contained breathing apparatus that has a full facepiece and is operated in a pressure-demand or other positive-pressure mode); or SAF:PD,PP:ASCBA (any supplied-air respirator that has a full facepiece and is operated in a pressure-demand or other positive-pressure mode in combination with an auxiliary self-contained breathing apparatus operated in a pressure-demand or other positive-pressure mode).

Escape: GMFOV [any air-purifying, full-facepiece respirator (gas mask) with a chin-style, front-or back-mounted organic vapor canister]; or SCBAE (any appropriate escape-type, self-contained breathing apparatus).

CHEMICAL REACTIVITY

Incompatible with oxidizers, ozone.

FIRE INFORMATION

Decomposition products upon heating include irritating vapors and acrid smoke. If material or contaminated runoff enters waterways, notify downstream users of potentially contaminated waters. Notify local health and fire officials and pollution control agencies. Use dry chemicals, foam, carbon dioxide.

FIRST AID

Move victim to fresh air; call emergency medical care. Apply artificial respiration if victim is not breathing. Do not use mouth-to-mouth method if victim ingested or inhaled the substance; induce artificial respiration with the aid of a pocket mask equipped with a one-way valve or other proper respiratory medical device. Administer oxygen if breathing is difficult. Remove and isolate contaminated clothing and shoes. In case of contact with material, immediately flush skin or eyes with running water for at least 20 minutes. For minor skin contact, avoid spreading material on unaffected skin. Keep victim warm and quiet. Effects of exposure (inhalation, ingestion, or skin contact) to substance may be delayed. Ensure that medical personnel are aware of the material(s) involved and take precautions to protect themselves.

ENVIRONMENTAL INFORMATION

CLEAN WATER ACT: Section 307 Priority Pollutants (subject to effluent limitations); Section 307 Toxic Pollutants.
RCRA Land Ban Waste.
RCRA Universal Treatment Standards: Wastewater (mg/L), 0.059; Nonwastewater (mg/kg), 3.4.
RCRA Ground Water Monitoring List: Suggested methods (PQL ug/L): 8100(200); 8270(10).
EPCRA Section 304 Reportable Quantity (RQ): CERCLA, 100 lb (45.4 kg).

ACETAL A:005

SYNONYMS: ACETAAL (Dutch); ACETAL (Spanish); ACETAL DIETHYLIQUE (French); ACETALE (Italian); ACETALDEHYDE DIETHYLACETAL; ACETEHYDE; 1,1-DIAETHOXY-AETHAN (German); DIAETHYLACETAL (German); 1,1-DIETHOXYACETAL; 1,1-DIETHOXY-ETHAAN (Dutch); 1,1-DIETHOXYETHANE; DIETHYL ACETAL; 1,1-DIETOSSIETANO (Italian); EEC No. 605-015-00-1; EINECS No. 203-310-6; ETHYLIDENE DIETHYL ETHER; USAF DO-45

IDENTIFICATION

CAS:	105-57-7
ERG ID:	UN 1088
ERG Guide:	127
ERG Guide Category:	FLAMMABLE LIQUIDS (POLAR/WATER MISCIBLE)
Molecular Formula:	$C_6H_{14}O_2$
RTECS No.:	AB2800000
Properties:	Highly flammable liquid. Colorless. Pungent, agreeable, woody odor; nutty aftertaste. Floats on the surface of water; slightly soluble in water. Vapors are heavier than air; will collect in low areas.

HEALTH AND SAFETY INFORMATION

Irritating to skin, eyes, nose, and respiratory tract. A narcotic; hypnotic.

RECOMMENDATIONS FOR RESPIRATOR SELECTION

At any concentrations above the NIOSH REL, or where there is no REL, at any detectable concentration: SCBAF:PD,PP (any self-contained breathing apparatus that has a full facepiece and is operated in a pressure-demand or other positive-pressure mode); or SAF:PD,PP:ASCBA (any supplied-air respirator that has a full facepiece and is operated in a pressure-demand or other positive-pressure mode in combination with an auxiliary self-contained breathing apparatus operated in a pressure-demand or other positive-pressure mode).

Escape: GMFOV [any air-purifying, full-facepiece respirator (gas mask) with a chin-style, front-or back-mounted organic vapor canister]; or SCBAE (any appropriate escape-type, self-contained breathing apparatus).

CHEMICAL REACTIVITY

Forms explosive mixture with air. Violent reaction with oxidizers. Forms unstable and explosive peroxides with heat and light. May accumulate static electrical charges, and may cause ignition of its vapors.

FIRE INFORMATION

Vapors in confined areas and containers may explode in fire. Vapors are heavier than air and will collect in low areas. Vapors in confined areas may explode when exposed to fire. Vapors may travel long distances to ignition sources and flashback. Storage containers and parts of containers may rocket great distances, in many directions. If material or contaminated runoff enters waterways, notify downstream users of potentially contaminated waters. Notify local health and fire officials and pollution control agencies. Use dry chemicals, foam, carbon dioxide. From a secure explosion-proof location, use water spray to cool exposed containers. If cooling streams are ineffective (venting sound increases in volume and pitch, tank discolors or shows any signs of deforming), withdraw immediately to a secure position.

NFPA Hazard Classifications:

Health Hazards (Blue) 2
Flammability (Red) 3
Reactivity (Yellow) 0

Flash Point: –5°F (–21°C) c.c.
Flammable Limits in Air: LEL 1.65%; UEL 10.4%
Autoignition Temperature: 446°F (230°C)

FIRST AID

Move victim to fresh air. Call emergency medical care. Apply artificial respiration if victim is not

breathing. Administer oxygen if breathing is difficult. Remove and isolate contaminated clothing and shoes. In case of contact with substance, immediately flush skin or eyes with running water for at least 20 minutes. Wash skin with soap and water. Keep victim warm and quiet. Ensure that medical personnel are aware of the material(s) involved and take precautions to protect themselves.

ENVIRONMENTAL INFORMATION

Marine Pollutant (49CFR, Subchapter 172.101, Appendix B).
TSCA: .40CFR716.120(*d*)1 as aldehydes

ACETALDEHYDE A:010

SYNONYMS: ACETALDEHIDO (Spanish); ACETIC ALDEHYDE; ACETEHYD (German); ACETEHYDE; ACETIC EHYDE; EEC No. 605-003-00-6; EINECS No. 200-836-8; EHYDE ACETIQUE (French); EIDE ACETICA (Italian); ETHANAL; ETHYL ALDEHYDE; ETHYL EHYDE; FEMA No.2003; NCI-C56326; OCTOWY EHYD (Polish)

IDENTIFICATION

CAS:	75-07-0
ERG ID:	UN 1089
ERG Guide:	129
ERG Guide Category:	FLAMMABLE LIQUIDS (POLAR/WATER-MISCIBLE/NOXIOUS)
Molecular Formula:	C_2H_4O
RTECS No.:	AB 1925000
Properties:	Extremely flammable, colorless liquid. Gas above 69◦F (21°C). Penetrating fruity, strong, sharp pungent odor. Soluble in water.

HEALTH AND SAFETY INFORMATION

Very irritating to skin and eyes; prolonged contact can cause burns.

IARC Evaluation: Possible carcinogen, rating 2B; sufficient animal evidence
NTP: Anticipated carcinogen
IDLH:. 2,000 ppm; potential carcinogen
Odor Threshold: 0.186 ppm
OSHA Table Z-1-A Air Contaminant
OSHA Process Safety Management of Highly Hazardous Chemicals: (29 CFR Part 1910.119, Appendix A) TQ = 2,500 lb

Exposure Limits:	**TWA**	**CEILING**
ACGIH:		25 ppm; 45 mg/m^3

NIOSH: [(1)]
OSHA: 200 ppm; 360 mg/m^3

[(1)] Potential occupational carcinogen. Reduce exposure to lowest feasible level.

RECOMMENDATIONS FOR RESPIRATOR SELECTION: NIOSH

At any concentrations above the NIOSH REL, or where there is no REL, at any detectable concentration: SCBAF:PD,PP (Any self-contained breathing apparatus that has a full facepiece and is operated in a pressure-demand or other positive-pressure mode) or SAF:PD,PP:ASCBA (Any supplied-air respirator that has a full facepiece and is operated in a pressure-demand or other positive-pressure mode in combination with an auxiliary self-contained breathing apparatus operated in a pressure-demand or other positive pressure mode).

Escape: GMFOV [Any air-purifying, full-facepiece respirator (gas mask) with a chin-style, front-or back-mounted organic vapor canister]; or SCBAE (Any appropriate escape-type, self-contained breathing apparatus).

CHEMICAL REACTIVITY

Reacts with strong oxidizers, strong acids, strong bases, alcohols, ammonia, and amines, halogens, phenols, ketones, hydrogen cyanide, and hydrogen sulfide. May dissolve rubber. Slightly corrosive to mild steel. May explode without warning when exposed to heat, dust, corrosive, or oxidizers.

FIRE INFORMATION

Firefighting gear (including SCBA) does not provide adequate protection. If exposure occurs, remove and isolate gear immediately and thoroughly decontaminate personnel. Vapors are heavier than air and will collect in low areas. Vapors in confined areas may explode when exposed to fire. Vapors may travel long distances to ignition sources and flashback. Storage containers and parts of containers may rocket great distances, in many directions. If material or contaminated runoff enters waterways, notify downstream users of potentially contaminated waters. Notify local health and fire officials and pollution control agencies. Use dry chemicals, foam, carbon dioxide. From a secure explosion-proof location, use water spray to cool exposed containers. If cooling streams are ineffective (venting sound increases in volume and pitch, tank discolors or shows any signs of deforming), withdraw immediately to a secure position.

NFPA Hazard Classifications:

Health Hazards (Blue) 3
Flammability (Red) 4
Reactivity (Yellow) 2

Flash Point: –36°F (–38°C); c.c.; –58°F (–50°C) o.c.
Flammable Limits in Air: LEL 4%; UEL 60%
Autoignition Temperature: 347°F (175°C)

FIRST AID

Move victim to fresh air. Call emergency medical care. Apply artificial respiration if victim is not breathing. Administer oxygen if breathing is difficult. Remove and isolate contaminated clothing and shoes. In case of contact with substance, immediately flush skin or eyes with running water for at least 20 minutes. Wash skin with soap and water. Keep victim warm and quiet. Effects of exposure (inhalation, ingestion, or skin contact) to substance may be delayed. Ensure that medical personnel are aware of the material(s) involved and take precautions to protect themselves.

ENVIRONMENTAL INFORMATION

Clean Air Act: Hazardous Air Pollutants (Title I, Part A, Section 112); Accidental Release Prevention/Flammable Substances (Section 112[r], Table 3), TQ = 10,000 lb (4540 kg)
Clean Water Act: Section 311 Hazardous Substances/RQ (same as CERCLA) Section 313 Priority Chemicals.
EPA Hazardous Waste Number (RCRA No.): U001; D001
RCRA Section 261 Hazardous Constituents
RCRA Section 266, Appendix 5: Air concentrations
EPCRA Section 304: Reportable Quantity (RQ); CERCLA, 1,000 lb (454 kg)
EPCRA Section 313: Form R *de minimis* Concentration Reporting Level: 0.1%.
California's Proposition 65: Carcinogen.
TSCA: 40CFR716.120(*d*)1 as aldehydes

ACETAMIDE A:015

SYNONYMS: ACETAMIDO (Spanish); ACETIC ACID AMIDE; ACETIMIDIC ACID ACETIC ACID AMIDE; ETHANAMIDE; METHANECARBOXAMIDE; NCI-C02108

IDENTIFICATION

CAS:	60-35-5
ERG ID:	UN 2811
ERG Guide:	TOXIC SOLID, ORGANIC, N.O.S.
ERG Guide Category:	154
Molecular Formula:	C_2H_5NO
RTECS No.:	AB4025000
Properties:	Noncombustible, deliquescent crystalline solid. Colorless. Odorless (pure); mousy odor (impure). Mixes with water; decomposes slowly in cold water releasing ammonia.

HEALTH AND SAFETY INFORMATION

Mild skin, eye, and mucous membrane irritant. Possible human carcinogen.

IARC Evaluation: Group 2B, animal evidence sufficient.
Odor Threshold: 140-160 mg/m^3

RECOMMENDATIONS FOR RESPIRATOR SELECTION: NIOSH

At any concentrations above the NIOSH REL, or where there is no REL, at any detectable concentration: SCBAF:PD,PP (any self-contained breathing apparatus that has a full facepiece and is operated in a pressure-demand or other positive-pressure mode); or SAF:PD,PP:ASCBA (any supplied-air respirator that has a full facepiece and is operated in a pressure-demand or other positive-pressure mode in combination with an auxiliary self-contained breathing apparatus operated in a pressure-demand or other positive-pressure mode).

Escape: GMFOV [any air-purifying, full-facepiece respirator (gas mask) with a chin-style, front-or back-mounted organic vapor canister]; or SCBAE (any appropriate escape-type, self-contained breathing apparatus).

CHEMICAL REACTIVITY

Reacts with nonoxidizing mineral acids, sulfuric acid, nitric acid, ammonia, isocyanates, phenols, cresols.

FIRE INFORMATION

Combustion products include toxic nitrogen oxides and ammonia. If material or contaminated runoff enters waterways, notify downstream users of potentially contaminated waters. Notify local health and fire officials and pollution control agencies. Use extinguishing agent suitable for surrounding fire.

FIRST AID

Move victim to fresh air; call emergency medical care. Apply artificial respiration if victim is not breathing. Do not use mouth-to-mouth method if victim ingested or inhaled the substance; induce artificial respiration with the aid of a pocket mask equipped with a one-way valve or other proper respiratory medical device. Administer oxygen if breathing is difficult. Remove and isolate contaminated clothing and shoes. In case of contact with material, immediately flush skin or eyes with running water for at least 20 minutes. For minor skin contact, avoid spreading material on unaffected skin. Keep victim warm and quiet. Effects of exposure (inhalation, ingestion, or skin contact) to substance may be delayed. Ensure that medical personnel are aware of the material(s) involved and take precautions to protect themselves.

ENVIRONMENTAL INFORMATION

Clean Air Act: Hazardous Air Pollutants (Title I, Part A, Section 112)
EPCRA Section 304: Reportable Quantity (RQ): CERCLA RQ, 1 lb (0.454 kg).
EPCRA Section 313: Form R *de minimis* concentration reporting level: 0.1%.
California's Proposition 65: Carcinogen.
TSCA: 40CFR 704.225(*a*); CAIR list reporting required.

ACETIC ACID A:020

SYNONYMS: ACETIC ACID (AQUEOUS SOLUTION); ACETIC ACID, GLACIAL; ACIDE ACETIQUE (French); ACIDO ACETICO (Italian, Spanish); ALUMINUM ETCH 16-1-1-2; ALUMINUM ETCH 82-3-15-0; AS-1; AS-1CE; AS-5CE; AS-CZ5E; AS-18CZ6E; AS-1400; AS-18CZ10A; AZIJNZUUR (Dutch); CEA-100 MICRO-CHROME ETCHANT; COPPER, BRASS BRITE DIP 1127; COPPER, BRASS BRITE DIP 127; DAZZLENS CLEANER; EEC No. (85% in water) 607-002-00-6; EEC No. (100%) 607-002-01-3; EINECS No. 200-580-7; EPF B20 FIXER; ESSIGSAEURE (German); 777 ETCH; ETHANOIC ACID; ETHYLIC ACID; FEMA No. 2006; FRECKLE ETCH; GLACIAL ACETIC ACID; GLASS ETCH; KODAK 33 STOP BATH; KOVAR BRIGHT DIP (412X); KTI ALUMINUM ETCH I/II; LENS CLEANER M6015; MAE ETCHANTS; METAL ETCH; METHANE CARBOXYLIC ACID; MIXED ACID ETCH (5-2-2); MIXED ACID ETCH (6-1-1); OCTOWY KWAS (Polish); PAD ETCH; PFC; POLY ETCH 95%; PROCESSOR FIXER CONCENTRATE; RAPID FILM FIX; RDH LIME SOLVENT; SILICON ETCH SOLUTION; STRESS RELIEF ETCH; VINEGAR ACID; VINEGAR (4-6% SOLUTION IN WATER); WET K-ETCH; WRIGHT ETCH

IDENTIFICATION

CAS:	64-19-7
ERG ID:	UN 2789 (solution greater than 80% acid, by mass) UN 2790 (solution greater than 10% but not more than 80%, by mass)
ERG Guide:	132 (solution greater than 80% acid, by mass) 153 (solution greater than 10% but not more than 80%, by mass)
ERG Guide Category:	Solution greater than 80% acid, by mass: FLAMMABLE LIQUIDS -CORROSIVE Solution greater than 10% but not more than 80%, by mass: TOXIC AND/OR CORROSIVE (COMBUSTIBLE)
Molecular Formula:	$C_2H_4O_2$
RTECS No.:	AF1225000
Properties:	Very flammable, colorless, watery liquid or crystals. Combustible and corrosive. Pungent, sour, sharp, vinegarlike odor. Sinks and mixes with water; produces heat when mixed with water. Irritating vapor is produced. Freezing point is 62°F (17°C) (acid may freeze and burst its container). Vapors are heavier than air; collects in low areas.

HEALTH AND SAFETY INFORMATION

Very irritating to skin, eyes, nose, and lungs; prolonged skin contact can cause burns. Eye contact may result in permanent corneal opacification. Ingestion of concentrated solution may result in esophageal perforation, hemoglobinuria, hematuria, and uremia. Repeated or prolonged exposure to acetic acid may cause a darkening, thickening, or irritation of the skin or hands, erosion of the enamel of exposed front teeth; and chronic inflammation of the nose, throat, and bronchial tubes.

IDLH: 50 ppm
Odor Threshold: 0.016 ppm
OSHA Table Z-1-A Air Contaminant

Exposure Limits:	TWA	STEL
ACGIH:	10 ppm; 25 mg/m^3	15 ppm; 37 mg/m^3
NIOSH:	10 ppm; 25 mg/m^3	15 ppm; 37 mg/m^3
OSHA:	10 ppm; 25 mg/m^3	

RECOMMENDATIONS FOR RESPIRATOR SELECTION: NIOSH/OSHA

50 ppm: SA:CF (any supplied-air respirator operated in a continuous-flow mode); or PAPROV [any powered, air-purifying respirator with organic vapor cartridge(s)]; CCRFOV [any air-purifying, full-facepiece respirator (gas mask) with a chin-style, front-or back-mounted acid gas canister]; or GMFOV [any air-purifying, full-facepiece respirator (gas mask) with a chin-style, front-or back-mounted organic vapor canister]; or SCBAF (any self-contained breathing apparatus with a full facepiece); or SAF (any supplied-air respirator with a full facepiece).

Emergency or Planned Entry into Unknown Concentrations or IDLH Conditions: SCBAF:PD,PP (any self-contained breathing apparatus that has a full facepiece and is operated in a pressure-demand or other positive-pressure mode); or SAF:PD,PP:ASCBA (any supplied-air respirator that has a full facepiece and is operated in a pressure-demand or other positive-pressure mode in combination with an auxiliary self-contained breathing apparatus operated in a pressure-demand or other positive-pressure mode).

Escape: GMFOV [any air-purifying, full-facepiece respirator (gas mask) with a chin-style, front-or back-mounted organic vapor canister] or SCBAE (any appropriate escape-type, self-contained breathing apparatus).

Note: Substance causes eye irritation or damage; eye protection needed.

CHEMICAL REACTIVITY

Vapor forms explosive mixture with air. Violent reaction with oxidizers, organic amines, and bases such as hydroxides and carbonates. Incompatible with strong acids, aliphatic amines, alkanolamines, isocyanates, alkylene oxides, epichlorohydrin, acetaldehyde, 2-aminoethanol, ammonia, ammonium nitrate, chlorosulfonic acid, chromic acid, ethylene diamine, ethyleneimine, halides, peroxides, perchlorates, perchloric acid, permanganates, phosphorus isocyanate, phosphorus trichloride, potassium tert-butoxide, and xylene. Attacks cast iron, stainless steel, and other metals forming flammable hydrogen gas. Will attack many forms of rubber or plastic.

FIRE INFORMATION

Firefighting gear (including SCBA) does not provide adequate protection. If exposure occurs, remove and isolate gear immediately and thoroughly decontaminate personnel. Vapors are heavier than air and will collect in low areas. Vapors in confined areas may explode when exposed to fire. Vapors may travel long distances to ignition sources and flashback. Storage containers and parts of containers may rocket great distances, in many directions. If material or contaminated runoff enters waterways, notify downstream users of potentially contaminated waters. Notify local health and fire officials and pollution control agencies. Use dry chemicals, foam, carbon dioxide. From a secure explosion-proof location, use water spray to cool exposed containers. If cooling streams are ineffective (venting sound increases in volume and pitch, tank discolors, or shows any signs of deforming), withdraw immediately to a secure position.

NFPA Hazard Classifications:

Health Hazards (Blue) 3
Flammability (Red) 2
Reactivity (Yellow) 0

Flash Point: 104°F (40°C), c.c.
Flammable Limits in Air: LEL 5.4%; UEL 16.0%
Autoignition Temperature: 867°F (463°C)

FIRST AID

FOR MORE THAN 10% BUT NOT MORE THAN 80% ACID SOLUTION: Move victim to fresh air; call emergency medical care. Apply artificial respiration if victim is not breathing. Do not use mouth-to-mouth method if victim ingested or inhaled the substance; induce artificial respiration with the aid of a pocket mask equipped with a one-way valve or other proper respiratory medical device. Administer oxygen if breathing is difficult. Remove and isolate contaminated clothing and shoes. In case of contact with substance, immediately flush skin or eyes with running water for at least 20 minutes. For minor skin contact, avoid spreading material on unaffected skin. Keep victim warm and quiet. Effects of exposure (inhalation, ingestion, or skin contact) to substance may be delayed. Ensure that medical personnel are aware of the material(s) involved and take precautions to protect themselves.

FOR MORE THAN 80% ACID SOLUTION: Move victim to fresh air and call emergency medical care. Apply artificial respiration if victim is not breathing. Do not use mouth-to-mouth method if victim swallowed or inhaled the substance; induce artificial respiration with the aid of a pocket mask equipped with a one-way valve or other proper respiratory medical device. If breathing is difficult, give oxygen. In case of contact with substance, immediately flush skin or eyes with running water for at least 20 minutes. Remove and isolate contaminated clothing and shoes. Keep victim warm and quiet. Effects of exposure (inhalation, ingestion or skin contact) to substances may be delayed. Ensure that medical personnel are aware of the material(s) involved, and take precautions to protect themselves.

ENVIRONMENTAL INFORMATION

Clean Water Act: Section 311 Hazardous Substances/RQ (same as CERCLA)
EPCRA Section 304: Reportable Quantity (RQ): CERCLA, 5,000 lb (2,270 kg)

ACETIC ANHYDRIDE A:030

SYNONYMS: ACETIC ACID, ANHYDRIDE; ACETIC OXIDE; ACETYL ANHYDRIDE; ACETYL ETHER; ACETYL OXIDE; ANHIDRIDO ACETICO (Spanish); ANHYDRIDE ACETIQUE (French); ANIDRIDE ACETICA (Italian); AZIJNZUURANHYDRIDE (Dutch); EEC No. 607-008-00-9; EINECS No. 203-564-8; ESSIGSAEUREANHYDRID (German); ETHANOIC ANHYDRATE; ETHANOIC ANHYDRIDE; OCTOWY BEZWODNIK (Polish)

IDENTIFICATION

CAS:	108-24-7
ERG ID:	UN 1715
ERG Guide:	137
ERG Guide Category:	WATER REACTIVE -CORROSIVE
Molecular Formula:	$C_4H_6O_3$
RTECS No.:	AK1925000
Properties:	Watery, combustible liquid. Colorless. Strong pungent vinegar odor. Sinks and reacts slowly with water; forms acetic acid and heat. Irritating vapor is produced. Vapors are heavier than air and will collect in low areas.

HEALTH AND SAFETY INFORMATION

Vapors and liquid are extremely irritating to skin, eyes, nose, and lungs; skin and eye contact causes severe burns and possible blindness.

IDLH: 200 ppm
Odor Threshold: 0.029 ppm
OSHA Table Z-1-A Air Contaminant

Exposure Limits:	**TWA**	**CEILING**
ACGIH:	5 ppm; 21 mg/m^3	
NIOSH:		5 ppm; 20 mg/m^3
OSHA:	5 ppm; 20 mg/m^3	

RECOMMENDATIONS FOR RESPIRATOR SELECTION: NIOSH /OSHA

125 ppm: SA:CF (any supplied-air respirator operated in a continuous-flow mode); or PAPROV [any powered, air-purifying respirator with organic vapor cartridge(s)]
200 ppm: CCRFOV [any air-purifying, full-facepiece respirator (gas mask) with a chin-style, front-or back-mounted acid gas canister]; or GMFOV [any air-purifying, full-facepiece respirator (gas mask) with a chin-style, front-or back-,mounted organic vapor canister]; or PAPRTOV [any powered, air-purifying respirator with a tight-fitting facepiece and organic vapor cartridge(s)]; or SCBAF (any self-contained breathing apparatus with a full facepiece); or SAF (any supplied-air respirator with a full facepiece)

Emergency or Planned Entry into Unknown Concentrations or Idlh Conditions:
SCBAF:PD,PP (any self-contained breathing apparatus that has a full facepiece and is operated in a pressure-demand or other positive-pressure mode); or SAF:PD,PP:ASCBA (any supplied-air respirator that has a full facepiece and is operated in a pressure-demand or other positive-pressure mode in combination with an auxiliary self-contained breathing apparatus operated in a pressure-demand or other positive pressure mode).

Escape: GMFOV[any air-purifying, full-facepiece respirator (gas mask) with a chin-style, front-or back-, mounted organic vapor canister]; or SCBAE (any appropriate escape-type, self-contained breathing apparatus).

Note: Substance causes eye irritation or damage; eye protection needed.

CHEMICAL REACTIVITY

Reacts slowly with cold water to form acetic acid; considerable heat liberated when water spray is used. Corrosive in the presence of water or moisture. Reacts explosively with a large number of chemicals. Reacts with alcohols, strong oxidizers (especially chromic acid), amines, alcohols, strong caustics. Attacks iron, steel, and other metals.

FIRE INFORMATION

Firefighting gear (including SCBA) does not provide adequate protection. If exposure occurs, remove and isolate gear immediately and thoroughly decontaminate personnel. Vapors are heavier than air and will collect in low areas. Vapors in confined areas may explode when exposed to fire. Vapors may travel long distances to ignition sources and flashback. Storage containers and parts of containers may rocket great distances, in many directions. If material or contaminated runoff enters waterways, notify downstream users of potentially contaminated waters. Notify local health and fire officials and pollution control agencies. Use dry chemicals, foam, carbon dioxide. From a secure explosion-proof location, use water spray to cool exposed containers. If cooling streams are ineffective (venting sound increases in volume and pitch, tank discolors or shows any signs of deforming), withdraw immediately to a secure position.

NFPA Hazard Classifications:

Health Hazards (Blue) 3
Flammability (Red) 2
Reactivity (Yellow) 1

Flash Point: 136° F (58° C), o.c.; 120° F (49° C) c.c.
Flammable Limits in Air: LEL 2.7%; UEL 10.3%
Autoignition Temperature: 626°F (330°C)

FIRST AID

Move victim to fresh air. Call emergency medical care. Apply artificial respiration if victim is not breathing. Do not use mouth-to-mouth method if victim ingested or inhaled the substance; induce artificial respiration with the aid of a pocket mask equipped with a one-way valve or other proper respiratory medical device. Administer oxygen if breathing is difficult. Remove and isolate contaminated clothing and shoes. In case of contact with substance, immediately flush skin or eyes with running water for at least 20 minutes. For minor skin contact, avoid spreading material on unaffected skin. Removal of solidified molten material from skin requires medical assistance. Keep victim warm and quiet. Effects of exposure (inhalation, ingestion, or skin contact) to substance may be delayed. Ensure that medical personnel are aware of the material(s) involved and take precautions to protect themselves.

ENVIRONMENTAL INFORMATION

Clean Water Act: Section 311 Hazardous Substances/RQ (same as CERCLA)
EPCRA Section 304 Reportable Quantity (RQ): CERCLA, 5,000 lb (2,270 kg)

ACETONE A:040

SYNONYMS: ACETON (Dutch, German, Polish); ACETONA (Spanish); ARC CLEANER; β-KETOPROPANE; DAZZLENS CLEANER (LENS CLEANER M6015); DIMETHYL KETONE; DIMETHYLFORMALDEHYDE; DIMETHYLFORMEHYDE; DIMETHYLKETAL; EEC No. 606-001-00-8; EINECS No. 200-662-2; FEMA No. 3326; GENESOLV DFX; GENESOLV 404 AZEOTROPE; KETONE; KETONE, DIMETHYL; KETONE PROPANE; KTI PBR I CLEANING COMPOUND; 3M SCOTCHKOTE ELECTRICAL COATING; METHYL KETONE; MS-114 CONFORMAL COATING STRIPPER; 2-PROPANONE; PROPANONE; PYROACETIC ACID; PYROACETIC ETHER; REDUCER DTR602; SASETONE; WAXIVATION COMPOUND

IDENTIFICATION

CAS:	67-64-1
ERG ID:	UN 1090
ERG Guide:	127
ERG Guide Category:	FLAMMABLE LIQUIDS (POLAR/WATER MISCIBLE)
Molecular Formula:	C_3H_6O
RTECS No.:	AL3150000
Properties:	Highly flammable, watery liquid. Colorless. Sweet odor resembling fingernail polish remover; pungent; sharp, penetrating residual; ketonic. Floats and mixes with water. Flammable, irritating vapor is produced. Forms cyanide in the body. Vapors are heavier than air; will collect in low areas.

HEALTH AND SAFETY INFORMATION

Acetone is irritating to the skin, eyes, nose, and lungs. Symptoms of exposure to large quantities of acetone may include headache, unsteadiness, confusion, lassitude, drowsiness, vomiting, and respiratory depression. Repeated skin exposure can cause chronic dryness and irritation. Long-term exposure may result in liver, kidney, brain, and/or nerve damage. Levels above 500 ppm can produce eye irritation after three hours. Daily exposure at this level has resulted in irritation of throat and lungs, as well as dizziness, and inflammation of stomach and intestines.

IDLH: 2,500 ppm [LEL]
Odor Threshold: 4.58 ppm; AIHA geometric mean air odor threshold, 62 ppm (detectable); 130 ppm (recognizable).
OSHA Table Z-1-A Air Contaminant
NIOSH Criteria Document: NIOSH 78-173 KETONES

Exposure Limits:	TWA	STEL
ACGIH:	500 ppm; 1,188 mg/m^3	750 ppm; 1,782 mg/m^3
NIOSH:	250 ppm; 590 mg/m^3	
OSHA:	1,000 ppm; 2400 mg/m^3	

RECOMMENDATIONS FOR RESPIRATOR SELECTION: NIOSH

2,500 ppm: CCRFOV [any chemical cartridge respirator with a full facepiece and organic vapor cartridge(s)]; or PAPROV [any powered, air-purifying respirator with organic vapor cartridge(s)]; or GMFOV [any air-purifying, full-facepiece respirator (gas mask) with a chin-style, front-or back-mounted acid gas canister]; or SA (any supplied-air respirator); or SCBAF (any self-contained breathing apparatus with a full facepiece).

Emergency or Planned Entry into Unknown Concentrations or IDLH Conditions: SCBAF:PD,PP (any self-contained breathing apparatus that has a full facepiece and is operated in a pressure-demand or other positive-pressure mode); or SAF:PD,PP:ASCBA (any supplied-air respirator that has a full facepiece and is operated in a pressure-demand or other positive-pressure mode in combination with an auxiliary self-contained breathing apparatus operated in a pressure-demand or other positive-pressure mode).

Escape: GMFOV [any air-purifying, full-facepiece respirator (gas mask) with a chin-style, front-or back-mounted organic vapor canister]; or SCBAE (any appropriate escape-type, self-contained breathing apparatus).

Note: Substance reported to cause eye irritation or damage.

CHEMICAL REACTIVITY

Forms explosive mixture with air. Violent reaction with chloroform. Incompatible with strong acids, aliphatic amines, chloroform, chromic anhydride, chromyl chloride, hexachloromelamine, hydrogen peroxide, nitrosyl chloride, nitrosyl perchlorate, nitryl perchlorate, permonosulfuric acid, and potassium tert-butoxide. Unstable and explosive peroxides formed with strong oxidizers. May accumulate static electrical charges and may cause ignition of its vapors. Dissolves most rubber, resins, and plastics.

FIRE INFORMATION

Vapors are heavier than air and will collect in low areas. Vapors in confined areas may explode when exposed to fire. Vapors may travel long distances to ignition sources and flashback. Storage containers and parts of containers may rocket great distances, in many directions. If material or contaminated runoff enters waterways, notify downstream users of potentially contaminated waters. Notify local health and fire officials and pollution control agencies. Use dry chemicals, foam, carbon dioxide. From a secure explosion-proof location, use water spray to cool exposed containers. If cooling streams are ineffective (venting sound increases in volume and pitch, tank discolors or shows any signs of deforming), withdraw immediately to a secure position.

NFPA Hazard Classifications:

Health Hazards (Blue) 1
Flammability (Red) 3
Reactivity (Yellow) 0

Flash Point: –4°F (–20°C)
Flammable Limits in Air: LEL 2.5%; UEL 12.8%
Autoignition Temperature: 869°F (465°C)

FIRST AID

Move victim to fresh air. Call emergency medical care. Apply artificial respiration if victim is not breathing. If breathing is difficult, give oxygen. Remove and isolate contaminated clothing and shoes. In case of contact with substance, immediately flush skin or eyes with running water for at least 20 minutes. Wash skin with soap and water. Keep victim warm and quiet. Ensure that medical personnel are aware of the material(s) involved and take precautions to protect themselves.

ENVIRONMENTAL INFORMATION

Acetone is highly volatile, and once it reaches the troposphere (lower atmosphere), it will react with other gases, contributing to the formation of ground-level ozone, a major component of urban smog and other pollutants. If released into water, acetone will be degraded by microorganisms or will evaporate into the atmosphere. Degradation by microorganisms will be the primary removal mechanism. Reactions of acetone in the lower atmosphere contribute to the formation of ground-level ozone. Ozone (a major component of urban smog) can affect the respiratory system, especially in sensitive individuals such as asthmatics or allergy sufferers. Capturing, purifying, and recycling solvents can be an effective method of reducing pollution. Facilities can reduce chemical releases and save money by recycling solvents used in polymerization, fiber manufacture, and supporting operations. One major firm installed carbon adsorption solvent recovery units to recover and recycle acetone back to the acetate fiber spinning process. Using carbon adsorption, overall plant acetone recovery efficiency reaches nearly 99 percent. This firm plans to achieve additional reductions by revamping air handling and ventilation systems to improve acetone capture.

EPA Hazardous Waste Number (RCRA No.): U002
RCRA Section 261 Hazardous Constituents.
RCRA Land Ban Waste.
RCRA Universal Treatment Standards: Wastewater (mg/L), 0.28; Nonwastewater (mg/kg), 160.
RCRA Ground Water Monitoring List: Suggested methods (PQL mg/L): 8240(100).
EPCRA Section 304: Reportable Quantity (RQ): CERCLA, 5,000 lb (2,270 kg).
EPCRA Section 313: Acetone deleted from this list, Federal Register, Vol. 60, No. 116, 6/16/95.
TSCA: 40CFR799.5000
TSCA: 40CFR716.120(*d*)1 as aldehydes

ACETONITRILE A:060

SYNONYMS: ACETONITRIL (German, Dutch); ACETONITRILO (Spanish); CYANOMETHANE; CYANURE de METHYL (French); EEC No. 608-001-00-3; EINECS No. 200-835-2; ETHANENITRILE; ETHYL NITRIL; ETHYL NITRILE; METHANECARBONITRIL; METHANECARBONITRILE; METHANE, CYANO-; METHYL CYANIDE; NCI-C60822; USAF EK-488

IDENTIFICATION

CAS:	75-05-8
ERG ID:	UN 1648
ERG Guide:	131
ERG Guide Category:	FLAMMABLE LIQUIDS -TOXIC
Molecular Formula:	C_2H_3N
RTECS No.:	AL7700000
Properties:	Flammable, watery liquid. Colorless. Sweet, etherlike odor. Floats and mixes with water. Flammable, irritating vapor is produced. Vapors are heavier than air; can collect and stay in low areas.

HEALTH AND SAFETY INFORMATION

Acetonitrile can cause fatal cyanide poisoning. Effects are delayed and may be due to slow formation of cyanide ion. Also, long-term exposure may damage the central nervous system, liver, kidneys, lungs; and may cause thyroid gland enlargement.

IDLH: 500 ppm
Odor Threshold: 97.7 ppm
NIOSH Criteria Document: NIOSH 78-212 NITRILES

Exposure Limits:	**TWA**	**STEL**
ACGIH:*	40 ppm; 67 mg/m^3	60 ppm; 101 mg/m^3
NIOSH:	20 ppm; 34 mg/m^3	
OSHA:	40 PPM; 70 mg/m^3	

*Skin contact contributes significantly to overall exposure.

RECOMMENDATIONS FOR RESPIRATOR SELECTION: NIOSH

200 ppm-CCRFOV [Any air-purifying, full-facepiece respirator (gas mask) with a chin-style, front-or back-mounted acid gas canister]; SA (any supplied-air respirator); or SCBA (any self-contained breathing apparatus)
500 ppm-SA:CF (any supplied-air respirator operated in a continuous-flow mode); or PAPROV [any powered, air-purifying respirator with organic vapor cartridge(s)]; or CCRFOV [any chemical cartridge respirator with a full facepiece and organic vapor cartridge(s)]; or GMFOV [Any air-purifying, full-facepiece respirator (gas mask) with a chin-style, front-or back-mounted acid gas canister]; or SCBAF

(any self-contained breathing apparatus with a full facepiece); or SAF (any supplied-air respirator with a full facepiece)

Emergency or Planned Entry into Unknown Concentrations or Idlh Conditions: SCBAF:PD,PP (any self-contained breathing apparatus that has a full facepiece and is operated in a pressure-demand or other positive-pressure mode); or SAF:PD,PP:ASCBA (Any supplied-air respirator that has a full facepiece and is operated in a pressure-demand or other positive-pressure mode in combination with an auxiliary self-contained breathing apparatus operated in a pressure-demand or other positive pressure mode.

Escape: GMFOV [Any air-purifying, full-facepiece respirator (gas mask) with a chin-style, front-or back-,mounted organic vapor canister]; or SCBAE (Any appropriate escape-type, self-contained breathing apparatus)

CHEMICAL REACTIVITY

Forms explosive mixture with air (flash point 42°F/6°C). Reacts with water (especially if acid or alkaline), acids, caustics, nitrating agents, indium, nitrogen tetroxide, n-fluoro compounds, sulfur trioxide, iron (III) salts of perchlorate, indium, perfluorourea, nitrogen-fluorine compounds. Violent reaction with strong oxidizers. May accumulate static electrical charges, and may cause ignition of its vapors.

FIRE INFORMATION

Vapors are heavier than air and will collect in low areas. Vapors in confined areas may explode when exposed to fire. Vapors may travel long distances to ignition sources and flashback. Storage containers and parts of containers may rocket great distances, in many directions. If material or contaminated runoff enters waterways, notify downstream users of potentially contaminated waters. Notify local health and fire officials and pollution control agencies. Use dry chemicals, foam, carbon dioxide. From a secure explosion-proof location, use water spray to cool exposed containers. If cooling streams are ineffective (venting sound increases in volume and pitch, tank discolors or shows any signs of deforming), withdraw immediately to a secure position.

NFPA Fire Classifications:

Health Hazards (Blue) 2
Flammability (Red) 3
Reactivity (Yellow) 0

Flash Point: 42°F (6°C) o.c.
Flammable Limits in Air: LEL 3.0%; UEL 16.0%
Autoignition Temperature: 975°F (524°C)

FIRST AID

Move victim to fresh air. Call emergency medical care. Apply artificial respiration if victim is not breathing. Do not use mouth-to-mouth method if victim ingested or inhaled the substance; induce artificial respiration with the aid of a pocket mask equipped with a one-way valve or other proper respiratory

medical devise. Administer oxygen if breathing is difficult. Remove and isolate contaminated clothing and shoes. In case of contact with substance, immediately flush skin or eyes with running water for at least 20 minutes. Wash skin with soap and water. Keep victim warm and quiet. Effects of exposure (inhalation, ingestion, or skin contact) to substance may be delayed. Ensure that medical personnel are aware of the material(s) involved and take precautions to protect themselves.

ENVIRONMENTAL INFORMATION

CLEAN AIR ACT: Hazardous Air Pollutants (Title I, Part A, Section 112)
EPA HAZARDOUS WASTE NUMBER (RCRA No.): U003
RCRA Section 261 Hazardous Constituents.
RCRA Land Ban Waste.
RCRA Universal Treatment Standards: Wastewater (mg/L), 5.6; Nonwastewater (mg/kg), 38
RCRA Ground Water Monitoring List, Suggested methods (PQL ug/L): 8015(100)
EPCRA Section 304 Reportable Quantity (RQ): CERCLA, 5,000 lb (2270 kg).
EPCRA Section 313: Form R *de minimus* concentration reporting level: 1.0%.
TSCA: 716.120a

ACETOPHENONE A:065

SYNONYMS: ACETOFENONA (Spanish); ACETYL BENZENE; BENZOYL METHIDE HYPNONE; DYMEX; EINECS No. 202-708-7; ETHANONE,1-PHENYL-; FEMA No. 2009; HYPNONE; KETONE METHYL PHENYL; METHYL PHENYL KETONE; 1-PHENYLETHANONE; PHENYL METHYL KETONE; USAF EK-496

IDENTIFICATION

CAS:	98-86-2
ERG ID:	NA 1993
ERG Guide:	128
ERG Guide Category:	FLAMMABLE LIQUIDS (NONPOLAR/WATER-IMMISCIBLE)
Molecular Formula:	C_8H_8O
RTECS No.:	AM5250000
Properties:	Flammable liquid. Colorless. Flowery, sweet odor. Sinks slowly in water; slightly soluble.

HEALTH AND SAFETY INFORMATION

Irritating to the eyes, skin, and respiratory tract. Mutagenic.

Odor Threshold: 0.363 ppm

Exposure Limits: **TWA**

ACGIH: 10 ppm; 49 mg/m^3

CHEMICAL REACTIVITY

May form an explosive mixture with air at temperatures above 170°F (77°C)

FIRE INFORMATION

Vapors are heavier than air and will collect in low areas. Vapors in confined areas may explode when exposed to fire. Vapors may travel long distances to ignition sources and flashback. Storage containers and parts of containers may rocket great distances, in many directions. If material or contaminated runoff enters waterways, notify downstream users of potentially contaminated waters. Notify local health and fire officials and pollution control agencies. Use dry chemicals, foam, carbon dioxide. From a secure explosion-proof location, use water spray to cool exposed containers. If cooling streams are ineffective (venting sound increases in volume and pitch, tank discolors or shows any signs of deforming), withdraw immediately to a secure position.

NFPA Hazard Classifications:

Health Hazards (Blue) 1
Flammability (Red) 2
Reactivity (Yellow) 0

Flash Point: 180°F (82°C) o.c.; 170°F (77°C) c.c.
Autoignition Temperature: 1058°F (570°C)

FIRST AID

Move victim to fresh air. Call emergency medical care. Apply artificial respiration if victim is not breathing. If breathing is difficult, give oxygen. Remove and isolate contaminated clothing and shoes. In case of contact with substance, immediately flush skin or eyes with running water for at least 20 minutes. Wash skin with soap and water. Keep victim warm and quiet. Ensure that medical personnel are aware of the material(s) involved and take precautions to protect themselves.

ENVIRONMENTAL INFORMATION

Clean Air Act: Hazardous Air Pollutants (Title I, Part A, Section 112)
Clean Water Act: Section 307 Priority Pollutants; Section 311 Hazardous Substances/RQ (same as CERCLA); Section 313 Priority Chemical
EPA Hazardous Waste Number (RCRA No.): U004
RCRA Section 261 Hazardous Constituents
RCRA Land Ban Waste
RCRA Universal Treatment Standards: Wastewater (mg/L), 0.010; Nonwastewater (mg/kg), 9.7 (Note: Listed as CAS 96-86-2).

RCRA Ground Water Monitoring List: Suggested test methods(s) (PQL μg/L): 8270 (10)
EPCRA Section 304 Reportable Quantity (RQ): CERCLA, 5,000 lb (2,270 kg)
EPCRA Section 313: Form R *de minimis* concentration reporting level: 1.0%.

ACETYL BROMIDE A:073

SYNONYMS: ACETIC ACID BROMIDE; ACETIC BROMIDE; ACETILO de BROMURO (Spanish); ETHANOYL BROMIDE; STCC No. 4931705

IDENTIFICATION

CAS:	506-96-7
ERG ID:	UN 1716
ERG Guide:	156
ERG Guide Category:	SUBSTANCES-TOXIC and/or CORROSIVE (COMBSTIBLE/WATER-SENSITIVE)
Molecular Formula:	C_2H_3BrO
RTECS No.:	AO5955000
Properties:	Highly flammable liquid. Colorless; turns yellow on contact with air. Sharp, acrid, unpleasant odor. Sinks in water; violent reaction, forming toxic hydrogen bromide. Vapors are heavier than air and will collect in low areas.

HEALTH AND SAFETY INFORMATION

Irritating to the eyes, nose, and respiratory tract. Corrosive. Liquid will burn skin and eyes. Poison. Breathing the vapors can cause death. Skin and eye contact can cause severe burns and blindness.

RECOMMENDATIONS FOR RESPIRATOR SELECTION

At any concentrations above the NIOSH REL, or where there is no REL, at any detectable concentration: SCBAF:PD,PP (any self-contained breathing apparatus that has a full facepiece and is operated in a pressure-demand or other positive-pressure mode); or SAF:PD,PP:ASCBA (any supplied-air respirator that has a full facepiece and is operated in a pressure-demand or other positive-pressure mode in combination with an auxiliary, self-contained breathing apparatus operated in a pressure-demand or other positive-pressure mode).

Escape: GMFOV [any air-purifying, full-facepiece respirator (gas mask) with a chin-style, front-or back-mounted organic vapor canister]; or SCBAE (any appropriate escape-type, self-contained breathing apparatus).

CHEMICAL REACTIVITY

Reacts violently with water. Corrodes or attacks most metals and wood in the presence of moisture.

FIRE INFORMATION

Combustion products include toxic hydrogen bromide, carbonyl bromide and bromine fumes. Firefighting gear (including SCBA) provide NO protection. If exposure occurs, remove and isolate gear immediately and thoroughly decontaminate personnel. Vapors are heavier than air and will collect in low areas. Vapors in confined areas may explode when exposed to fire. Vapors may travel long distances to ignition sources and flash back. If material or contaminated runoff enters waterways, notify downstream users of potentially contaminated waters. Notify local health and fire officials and pollution control agencies. Storage containers and parts of containers may rocket great distances, in many directions. Extinguish with dry chemicals or carbon dioxide.

NFPA Hazard Classifications:

Health Hazards (Blue) 3
Flammability (Red) 3
Reactivity (Yellow) 2
Special Situations . . WATER REACTIVE

FIRST AID

Move victim to fresh air. Call emergency medical care. Apply artificial respiration if victim is not breathing. Do not use mouth-to-mouth method if victim ingested or inhaled the substance; induce artificial respiration with the aid of a pocket mask equipped with a one-way valve or other proper respiratory medical device. Administer oxygen if breathing is difficult. Remove and isolate contaminated clothing and shoes. In case of contact with substance, immediately flush skin or eyes with running water for at least 20 minutes. For minor skin contact, avoid spreading material on unaffected skin. Keep victim warm and quiet. Effects of exposure (inhalation, ingestion, or skin contact) may be delayed. Ensure that medical personnel are aware of the material(s) involved and take precautions to protect themselves.

ENVIRONMENTAL INFORMATION

Clean Water Act: Section 311 Hazardous Substances/RQ (same as CERCLA).
EPCRA Section 304: CERCLA, 5,000 lb (2270 kg).
TSCA: 40CFR 716.120 (*a*); Health and Safety Reporting List, 6/1/87; CHIP available

ACETYL BUTYL CITRATE A:077

SYNONYMS: ACETYL TRI-n-BUTYL CITRATE; ACETYLCITRIC ACID, TRIBUTYL ESTER; ACETY TRIBUTYL CITRATE; BLO-TROL; CITRATO de ACETIL TRIBUTILO (Spanish); CITROFLEX A; CITROFLEX A-4; 1,2,3-PROPANETRICARBOXYLIC ACID, 2-(ACETYLOXY)-,

22 Acetyl Butyl Citrate

TRIBUTYL ESTER; TRIBUTYL ACETYLCITRATE; TRIBUTYL O-ACETYLCITRATE; TRIBUTYL 2-(ACETYLOXY)-1,2,3-PROPANETRICARBOXYLATE; TRIBUTYL CITRATE ACETATE

IDENTIFICATION

CAS:	77-90-7
ERG ID:	UN 2810
ERG Guide:	TOXIC LIQUID, ORGANIC, N.O.S.
ERG Guide Category:	154
Molecular Formula:	$C_{20}H_{34}O_8$
Properties:	Combustible liquid. Odorless. Mixes with water.

HEALTH AND SAFETY INFORMATION

Low toxicity by skin contact and if swallowed.

At any concentrations above the NIOSH REL, or where there is no REL, at any detectable concentration: SCBAF:PD,PP (any self-contained breathing apparatus that has a full facepiece and is operated in a pressure-demand or other positive-pressure mode); or SAF:PD,PP:ASCBA (any supplied-air respirator that has a full facepiece and is operated in a pressure-demand or other positive-pressure mode in combination with an auxiliary, self-contained breathing apparatus operated in a pressure-demand or other positive-pressure mode).

Escape: GMFOV [any air-purifying, full-facepiece respirator (gas mask) with a chin-style, front-or back-mounted organic vapor canister]; or SCBAE (any appropriate escape-type, self-contained breathing apparatus).

CHEMICAL REACTIVITY

Contact with strong oxidizers may cause fire and explosions.

FIRE INFORMATION

Decomposition products upon heating include irritating vapors and acrid smoke. Vapors are heavier than air and will collect in low areas. Vapors in confined areas may explode when exposed to fire. If material or contaminated runoff enters waterways, notify downstream users of potentially contaminated waters. Notify local health and fire officials and pollution control agencies. Use dry chemicals, alcohol-resistant foam, carbon dioxide. From a secure explosion-proof location, use water spray to cool exposed containers. If cooling streams are ineffective (venting sound increases in volume and pitch, tank discolors or shows any signs of deforming), withdraw immediately to a secure position.

Flash Point: 400°F (204°C)

FIRST AID

Move victim to fresh air; call emergency medical care. Apply artificial respiration if victim is not breathing. Do not use mouth-to-mouth method if victim ingested or inhaled the substance; induce artificial respiration with the aid of a pocket mask equipped with a one-way valve or other proper respiratory medical device. Administer oxygen if breathing is difficult. Remove and isolate contaminated clothing and shoes. In case of contact with material, immediately flush skin or eyes with running water for at least 20 minutes. For minor skin contact, avoid spreading material on unaffected skin. Keep victim warm and quiet. Effects of exposure (inhalation, ingestion, or skin contact) to substance may be delayed. Ensure that medical personnel are aware of the material(s) involved and take precautions to protect themselves.

ACETYLENE A:080

SYNONYMS: ACETILENO (Spanish); ACETYLEN; ACETYLENE, DISSOLVED; EEC No. 601-015-00-0; ETHINE; ETHYNE; NARCYLEN; WELDING GAS

IDENTIFICATION

CAS:	74-86-2
ERG ID:	UN 1001
ERG Guide:	116
ERG Guide Category:	GASES -FLAMMABLE (UNSTABLE)
Molecular Formula:	C_2H_2
RTECS No.:	AO9600000
Properties:	Flammable, compressed gas. Colorless. Etherlike when pure; mild, garliclike odor due to impurities. This flammable gas is slightly lighter than air and will disperse slowly unless confined. Shipped in pressurized cylinders, under pressure, partially filled with acetone. Soluble in water. Not shipped by rail tank cars.

HEALTH AND SAFETY INFORMATION

Contaminated acetylene, produced by mixing water and calcium carbide, may contain arsine, phosphine, or hydrogen sulfide. Exposure may result in long-term effects from contaminants.

NIOSH Criteria Document: NIOSH 76-195

Exposure Limits:	**TWA**	**CEILING**
ACGIH:*		
NIOSH:		2,500 ppm; 2662 mg/m^3
OSHA:	None	

*A simple asphyxiant

CHEMICAL REACTIVITY

A strong reducing agent that reacts violently with oxidizers. Forms explosive mixture with air. Forms shock-sensitive mixture with copper and copper salts, mercury and mercury salts, and silver and silver salts. Reacts with brass, bromine, cesium hydride, chlorine, cobalt, cuprous acetylise, fluorine, iodine, mercuric nitrate, nitric acid, potassium, rubidium hydride, trifluoromethyl hypofluorite, and sodium hydride.

FIRE INFORMATION

Gases are heavier than air and will collect in low areas. Gases in confined areas may explode when exposed to fire. Gases may travel long distances to ignition sources and flash back. Storage containers and parts of containers may rocket great distances, in many directions. If material or contaminated runoff enters waterways, notify downstream users of potentially contaminated waters. Notify local health and fire officials and pollution control agencies. Do not extinguish the fire unless the flow of the gas can be stopped and any remaining gas is out of the line. Specially trained personnel may use fog lines to cool exposures and let the fire burn itself out. Use dry chemicals, carbon dioxide. From a secure explosion-proof location, use water spray to cool exposed containers. If cooling streams are ineffective (venting sound increases in volume and pitch, tank discolors or shows any signs of deforming), withdraw immediately to a secure position.

NFPA Hazard Classifications:

Health Hazards (Blue) 0
Flammability (Red) 4
Reactivity (Yellow) 3

Flash Point: 0°F (–18°C) (Flammable gas)
Flammable Limits in Air: LEL 2.5%; UEL 82%
Autoignition Temperature: 581°F (305°C)

May be generated by mixing calcium carbide and water. Acetylene produced by the calcium carbide process may contain other hazardous materials: arsine, phosphine, or hydrogen sulfide. Levels of 100,000 ppm may produce slight intoxication; levels of 200,000 ppm may produce a staggering walk; at 300,000 ppm, general incoordination may occur; 350,000 ppm for 5 minutes can cause unconsciousness; 800,000 ppm results in complete anesthesia with increased blood pressure and rapid breathing.

FIRST AID

Move victim to fresh air. Call emergency medical care. Apply artificial respiration if victim is not breathing. If breathing is difficult, give oxygen. Remove and isolate contaminated clothing and shoes. In case of contact with liquefied gas, thaw frosted parts with lukewarm water. Keep victim warm and quiet. Ensure that medical personnel are aware of the material(s) involved and take precautions to protect themselves.

ENVIRONMENTAL INFORMATION

Clean Air Act: Section 112(r), Accidental Release Prevention/Flammable substances (Section 68.130) TQ = 10,000 lb (4,540 kg)
EPA Hazardous Waste Number (RCRA No.): D001

ACETYL PEROXIDE SOLUTION A:081

SYNONYMS: ACETYL PEROXIDE; ACETYL PEROXIDE SOLUTION, NOT OVER 25% PEROXIDE; DIACETYL PEROXIDE SOLUTION; EINECS No. 203-748-8; ETHANOYL PEROXIDE; PEROXIDO de ACETILO (Spanish)

IDENTIFICATION

CAS:	110-22-5
ERG ID:	UN 2084 (less than 25% solution); Forbidden (solid and solutions of more than 25%)
ERG Guide:	148
ERG Guide Category:	ORGANIC PEROXIDES (HEAT AND CONTAMINATION SENSITIVE/TEMPERATURE CONTROLLED)
Molecular Formula:	$C_4H_6O_4$
Properties:	Combustible liquid or crystalline solid. Colorless. Sharp, pungent odor. Sinks and mixes with water; slightly soluble. Stored and handled in a 25% solution in dimethyl phthalate.

HEALTH AND SAFETY INFORMATION

Severe eye hazard in a concentrated solution. Skin irritant.

OSHA Process Safety Management of Highly Hazardous Chemicals: (29 CFR Part 1910.119, Appendix A) TQ = 5, 000 lb (concentration > 70%).

CHEMICAL REACTIVITY

A powerful oxidizer; may ignite organic or other combustible materials on contact. Violent reaction with reducing agents. Explosion hazard, shock and heat sensitive. Keep away from all other chemicals. Storage temperatures are critical and should not go below 23°F (-5°C), or above 81°F (27°C). Self-activating decomposition takes place at 120°F (49°C). Produces heat upon contact with water, vapor, or steam. Contact with acid forms toxic and corrosive fumes.

FIRE INFORMATION

Remove all ignition sources. Vapors are heavier than air and will collect in low areas. Vapors in confined areas may explode when exposed to fire. Vapors may travel long distances to ignition sources and flashback. Storage containers and parts of containers may rocket great distances, in many directions. If material or contaminated runoff enters waterways, notify downstream users of potentially contaminated waters. Notify local health and fire officials and pollution control agencies. Use dry chemicals, foam, carbon dioxide. From a secure explosion-proof location, use water spray to cool exposed containers. If cooling streams are ineffective (venting sound increases in volume and pitch, tank discolors or shows any signs of deforming), withdraw immediately to a secure position.

NFPA Hazard Classifications: (25% solution in dimethyl phthalate)

Health Hazards (Blue) 1
Flammability (Red) 2
Reactivity (Yellow) 4

Flash Point: 113°F (45°C) o.c.
Autoignition Temperature: Explodes when heated (see Reactivity section).

FIRST AID

Move victim to fresh air. Call emergency medical care. Apply artificial respiration if victim is not breathing. Administer oxygen if breathing is difficult. Remove and isolate contaminated clothing and shoes. Remove material from skin immediately. In case of contact with substance, immediately flush skin or eyes with running water for at least 20 minutes. Keep victim warm and quiet. Ensure that medical personnel are aware of the material(s) involved and take precautions to protect themselves.

4-ACETYLRESORCINOL A:083

SYNONYMS: 4-ACETILRESORCINOL (Spanish); ACETYL RESORCINOL(4-); 2',4'-DIHYDROXYACETOPHENONE; 1-(2,4-DIHYDROXYPHENYL)ETHANONE; ETHANONE, 1-(2,4-DIHYDROXYPHENYL)-; RESACETOPHENONE; β-RESACETOPHENONE; RESOACETOPHENONE

IDENTIFICATION

CAS:	89-84-9
ERG ID:	UN 2811
ERG Guide:	TOXIC SOLID, ORGANIC, N.O.S.
ERG Guide Category:	154
Molecular Formula:	$C_8H_8O_3$
RTECS No.:	AM7525000
Properties:	Light tan crystals, leaflets, or needles. Slightly soluble in water.

HEALTH AND SAFETY INFORMATION

Causes severe irritation of the eyes. Moderately toxic if swallowed.

At any concentrations above the NIOSH REL, or where there is no REL, at any detectable concentration: SCBAF:PD,PP (any self-contained breathing apparatus that has a full facepiece and is operated in a pressure-demand or other positive-pressure mode); or SAF:PD,PP:ASCBA (any supplied-air respirator that has a full facepiece and is operated in a pressure-demand or other positive-pressure mode in combination with an auxiliary self-contained breathing apparatus operated in a pressure-demand or other positive-pressure mode).

Escape: GMFOV [any air-purifying, full-facepiece respirator (gas mask) with a chin-style, front-or back-mounted organic vapor canister]; or SCBAE (any appropriate escape-type, self-contained breathing apparatus).

FIRE INFORMATION

Decomposition products upon heating include toxic gases. If material or contaminated runoff enters waterways, notify downstream users of potentially contaminated waters. Notify local health and fire officials and pollution control agencies. Use dry chemicals, foam, carbon dioxide. From a secure explosion-proof location, use water spray to cool exposed containers. If cooling streams are ineffective (venting sound increases in volume and pitch, tank discolors, or shows any signs of deforming), withdraw immediately to a secure position.

FIRST AID

Move victim to fresh air; call emergency medical care. Apply artificial respiration if victim is not breathing. Do not use mouth-to-mouth method if victim ingested or inhaled the substance; induce artificial respiration with the aid of a pocket mask equipped with a one-way valve or other proper respiratory medical device. Administer oxygen if breathing is difficult. Remove and isolate contaminated clothing and shoes. In case of contact with material, immediately flush skin or eyes with running water for at least 20 minutes. For minor skin contact, avoid spreading material on unaffected skin. Keep victim warm and quiet. Effects of exposure (inhalation, ingestion, or skin contact) to substance may be delayed. Ensure that medical personnel are aware of the material(s) involved and take precautions to protect themselves.

ACETYL TRI-2-ETHYLHEXYL CITRATE A:085

SYNONYMS: CITRATO de ACETIL TRI-2-ETILHEXILO (Spanish)

IDENTIFICATION

CAS:	N/A
ERG ID:	UN 2810
ERG Guide:	TOXIC LIQUID, ORGANIC, N.O.S.
ERG Guide Category:	154
Molecular Formula:	$C_{32}H_{58}O_8$
Properties:	Combustible liquid. Insoluble in water.

HEALTH AND SAFETY INFORMATION

At any concentrations above the NIOSH REL, or where there is no REL, at any detectable concentration: SCBAF:PD,PP (any self-contained breathing apparatus that has a full facepiece and is operated in a pressure-demand or other positive-pressure mode); or SAF:PD,PP:ASCBA (any supplied-air respirator that has a full facepiece and is operated in a pressure-demand or other positive-pressure mode in combination with an auxiliary, self-contained breathing apparatus operated in a pressure-demand or other positive-pressure mode).

Escape: GMFOV [any air-purifying, full-facepiece respirator (gas mask) with a chin-style, front-or back-mounted organic vapor canister]; or SCBAE (any appropriate escape-type, self-contained breathing apparatus).

CHEMICAL REACTIVITY

Contact with strong oxidizers may cause fire and explosions.

FIRE INFORMATION

Decomposition products upon heating may include irritating vapors and acrid smoke. If material or contaminated runoff enters waterways, notify downstream users of potentially contaminated waters. Notify local health and fire officials and pollution control agencies. Use dry chemicals, foam, carbon dioxide. From a secure explosion-proof location, use water spray to cool exposed containers. If cooling streams are ineffective (venting sound increases in volume and pitch, tank discolors, or shows any signs of deforming), withdraw immediately to a secure position.

FIRST AID

Move victim to fresh air. Call emergency medical care. Apply artificial respiration if victim is not breathing. Administer oxygen if breathing is difficult. Remove and isolate contaminated clothing and shoes. In case of contact with substance, immediately flush skin or eyes with running water for at least 20 minutes. Wash skin with soap and water. Keep victim warm and quiet. Ensure that medical personnel are aware of the material(s) involved and take precautions to protect themselves.

ACROLEIN (INHIBITED) A:086

SYNONYMS: ACREHYDE; ACROLEINA (Italian, Spanish); ACROLEINE (Dutch, French); ACRYLEHYD (German); ACRYLEHYDE; ACRYLIC ALDEHYDE; ACRYLALDEHYDE; ACRYLIC EHYDE; AKROLEIN (Czech); AKROLEINA (Polish); ALLYLALDEHYDE; ALLYLEHYDE; ALLYL EHYDE; AQUALIN; AQUALINE; BIOCIDE; EEC No. 607-061-00-8; EINECS No. 203-453-4; EHYDE ACRYLIQUE (French); EIDE ACRILICA (Italian);ETHYLENE ALDEHYDE; ETHYLENE EHYDE; NSC 8819; PROPENAL; PROPENAL (Czech); 2-PROPENAL; PROP-2-EN-1-AL; 2-PROPEN-1-ONE; SLIMICIDE

IDENTIFICATION

CAS:	107-02-8
ERG ID:	UN 1092
ERG Guide:	131 P
ERG Guide Category:	FLAMMABLE LIQUIDS -TOXIC
Molecular Formula:	C_3H_4O
RTECS No.:	AS1050000
Properties:	Highly flammable, watery liquid. Colorless to light yellow.

Sharp, piercing, irritating odor. Causes tearing. Floats on and mixes with water.

HEALTH AND SAFETY INFORMATION

Severe health hazard. A sensitizer. Breathing poisonous vapors can cause death. May cause illness if absorbed through the skin. Corrosive and irritating to the eyes, nose and, respiratory system. Prolonged contact can cause skin burns and chronic bronchitis. May cause pulmonary edema, which can be delayed for several hours.

IARC Evaluation: Group 3 rating.
IDLH: 2 ppm
Odor Threshold: 0.174 ppm
OSHA Table Z-1-A Air Contaminant
OSHA Process Safety Management of Highly Hazardous Chemicals: (29 CFR Part 1910.119, Appendix A) TQ = 150 lb

Exposure Limits:	**TWA**	**STEL**
ACGIH:*	0.1 ppm; 0.23 mg/m^3	0.3 ppm; 0.69 mg/m^3
NIOSH:	0.1 ppm; 0.25 mg/m^3	0.3 ppm; 0.8 mg/m^3
OSHA:	0.1 ppm; 0.25mg/m^3	

*Skin contact contributes significantly to overall exposure.

RECOMMENDATIONS FOR RESPIRATOR SELECTION: NIOSH/OSHA

2 ppm: SA:CF (any supplied-air respirator operated in a continuous-flow mode); or PAPROV [any powered, air-purifying respirator with organic vapor cartridge(s)]; or CCRFOV [any air-purifying, full-facepiece respirator (gas mask) with a chin-style, front-or back-mounted acid gas canister]; or GMFOV [any air-purifying, full-facepiece respirator (gas mask) with a chin-style, front-or back-mounted organic vapor canister]; or SCBAF (any self-contained breathing apparatus with a full facepiece); or SAF (any supplied-air respirator with a full facepiece)

Emergency or Planned Entry into Unknown Concentrations or Idlh Conditions:
SCBAF:PD,PP (any self-contained breathing apparatus that has a full facepiece and is operated in a pressure-demand or other positive-pressure mode); or SAF:PD,PP:ASCBA (any supplied-air respirator that has a full facepiece and is operated in a pressure-demand or other positive-pressure mode in combination with an auxiliary, self-contained breathing apparatus operated in a pressure-demand or other positive pressure mode.

Escape: GMFOV [any air-purifying, full-facepiece respirator (gas mask) with a chin-style, front-or back-mounted organic vapor canister]; or SCBAE (any appropriate escape-type, self-contained breathing apparatus).

Note: Substance reported to cause eye irritation or damage; may require eye protection.

CHEMICAL REACTIVITY

Forms explosive mixture with air. Elevated temperatures or sunlight may cause explosive polymerization. A strong reducing agent; reacts violently with oxidizers. Reacts with acids, alkalis, ammonia, amines, peroxides. Shock-sensitive peroxides or acids may be formed over time. Attacks zinc and cadmium.

FIRE INFORMATION

Firefighting gear (including SCBA) does not provide adequate protection. If exposure occurs, remove and isolate gear immediately and thoroughly decontaminate personnel. Vapors are heavier than air and will collect in low areas. Vapors in confined areas may explode when exposed to fire. Vapors may travel long distances to ignition sources and flash back. Storage containers and parts of containers may rocket great distances in many directions. May react with itself, blocking relief valves and leading to tank explosion. If material or contaminated runoff enters waterways, notify downstream users of potentially contaminated waters. Notify local health and fire officials and pollution control agencies. Use dry chemicals, alcohol-resistant foam, or carbon dioxide. From a secure explosion-proof location, use water spray to cool exposed containers. If cooling streams are ineffective (venting sound increases in volume and pitch, tank discolors, or shows any signs of deforming), withdraw immediately to a secure position.

NFPA Hazard Classifications:

Health Hazards (Blue) 4
Flammability (Red) 3
Reactivity (Yellow) 3

Flash Point: Less than 0°F (–18°C) o.c.; –15°F (–26°C) c.c.
Flammable Limits in Air: LEL 2.8%; UEL 31%
Autoignition Temperature: 455°F (235°C), unstable.

FIRST AID

Move victim to fresh air. Call emergency medical care. Apply artificial respiration if victim is not breathing. Do not use mouth-to-mouth method if victim ingested or inhaled the substance; induce artificial respiration with the aid of a pocket mask equipped with a one-way valve or other proper respiratory medical device. Administer oxygen if breathing is difficult. Remove and isolate contaminated clothing and shoes. In case of contact with substance, immediately flush skin or eyes with running water for at least 20 minutes. Wash skin with soap and water. Keep victim warm and quiet. Effects of exposure (inhalation, ingestion, or skin contact) to substance may be delayed. Ensure that medical personnel are aware of the material(s) involved and take precautions to protect themselves.

ENVIRONMENTAL INFORMATION

Clean Air Act: Hazardous Air Pollutants (Title I, Part A, Section 112); Section 112[r], Accidental Release Prevention/Flammable substances (Section 68.130); TQ = 5,000 lb (1,275 kg)
Clean Water Act: Section 311 Hazardous Substances/RQ (same as CERCLA); Section 307 Priority Pollutants; Section 313 Priority Chemicals; Toxic Pollutant (Section 401.15)

EPA Hazardous Waste Number (RCRA No.): P003
RCRA Section 261 Hazardous Constituents
RCRA Land Ban Waste
RCRA Universal Treatment Standards: Wastewater (mg/L), 0.29; Nonwastewater, N/A
RCRA Ground Water Monitoring List Suggested Methods (PQL μg/L: 8030 (5); 8240 (5)
EPCRA Section 302: Extremely Hazardous Substances: TPQ = 500 lb (228 kg)
EPCRA Section 304: Reportable Quantity (RQ): EHS/CERCLA, 1 lb (0.454 kg)
EPCRA Section 313: Form R *de minimis* concentration reporting level: 1.0%.
Marine Pollutant (49 CFR, Subchapter 172.101, Appendix B)
TSCA: 40CFR716.120(*d*)1 as aldehydes

ACRYLAMIDE A:087

SYNONYMS: ACRILAMIDA (Spanish); ACRYLAMIDE MONOMER; ACRYLIC ACID AMIDE (50%); ACRYLIC AMIDE; ACRYLIC AMIDE 50%; AKRYLAMID (Czech); EEC No. 616-003-00-0; EINECS No. 201-173-7; ETHYLENECARBOXAMIDE; ETHYLENE MONOCLINIC TABLETS CARBOXAMIDE; PROPENAMIDE; 2-PROPENAMIDE; PROPENAMIDE; VINYL AMIDE

IDENTIFICATION

CAS:	79-06-1
ERG ID:	UN 2074
ERG Guide:	153 P
ERG Guide Category:	SUBSTANCES-TOXIC AND/OR CORROSIVE (COMBUSTIBLE)
Molecular Formula:	C_3H_5NO
RTECS No.:	AS3325000
Properties:	Combustible white crystals or powder. Odorless. Initially sinks in water; soluble.

HEALTH AND SAFETY INFORMATION

Irritating to eyes, skin, and respiratory system. May be absorbed through the skin. Affects nervous system and may cause liver damage.

IARC Evaluation: Possible carcinogen; Rating 2B
IDLH: 60 mg/m^3; potential human carcinogen
OSHA Table Z-1-A Air Contaminant

Exposure Limits:	**TWA**
ACGIH: (1)	0.03 mg/m^3
NIOSH: (2)	0.03 mg/m^3

OSHA: [3] 0.3 mg/m^3

[1] Animal carcinogen. Skin contact contributes significantly in overall exposure.
[2] Potential human carcinogen; reduce exposure to lowest feasible level; skin contact contributes significantly in overall exposure.
[3] Skin contact contributes significantly in overall exposure.

RECOMMENDATIONS FOR RESPIRATOR SELECTION: NIOSH

At any concentrations above the NIOSH REL, or where there is no REL, at any detectable concentration: SCBAF:PD,PP (any self-contained breathing apparatus that has a full facepiece and is operated in a pressure-demand or other positive-pressure mode); or SAF:PD,PP:ASCBA (any supplied-air respirator that has a full facepiece and is operated in a pressure-demand or other positive-pressure mode in combination with an auxiliary self-contained breathing apparatus operated in a pressure-demand or other positive-pressure mode)

Escape: GMFOV [any air-purifying, full-facepiece respirator (gas mask) with a chin-style, front-or back-,mounted organic vapor canister]; or SCBAE (any appropriate escape-type, self-contained breathing apparatus).

CHEMICAL REACTIVITY

May decompose with heat and polymerize at temperatures above 184°F (84°C), or exposure to light, releasing ammonia gas. Reacts violently with strong oxidizers. Reacts with reducing agents, peroxides, acids, bases, and vinyl polymerization initiators. Fine particles of dust form explosive mixture with air.

FIRE INFORMATION

Combustion products include toxic nitrogen oxides. Vapors are heavier than air and will collect in low areas. If material or contaminated runoff enters waterways, notify downstream users of potentially contaminated waters. Notify local health and fire officials and pollution control agencies. Use dry chemicals, alcohol-resistant foam, carbon dioxide. From a secure explosion-proof location, use water spray to cool exposed containers. If cooling streams are ineffective (venting sound increases in volume and pitch, tank discolors or shows any signs of deforming), withdraw immediately to a secure position.

Health Hazards (Blue) 2
Flammability (Red) 2
Reactivity (Yellow) 2

Flash Point: 280°F (138°C) c.c.
Autoignition Temperature: 464°F (240°C)

FIRST AID

Move victim to fresh air and call emergency medical care. Apply artificial respiration if victim is not breathing. Do not use mouth-to-mouth method if victim swallowed or inhaled the substance; induce artificial respiration with the aid of a pocket mask equipped with a one-way valve or other proper respiratory medical device. If breathing is difficult, give oxygen. In case of contact with substance, immediately flush skin or eyes with running water for at least 20 minutes. Remove and isolate contaminated clothing and shoes. Keep victim warm and quiet. Effects of exposure (inhalation, ingestion or skin contact) to substances may be delayed. Ensure that medical personnel are aware of the material(s) involved and take precautions to protect themselves.

ENVIRONMENTAL INFORMATION

Clean Air Act: Hazardous Air Pollutants (Title I, Part A, Section 112)
EPA Hazardous Waste Number (RCRA No.): U007
RCRA Section 261 Hazardous Constituents
RCRA Land Ban Waste
RCRA Universal Treatment Standards: Wastewater (mg/L), 19; Nonwastewater (mg/kg), 23.
Safe Drinking Water Act: MCL, treatment technique; MCGL, zero; Regulated Chemical (47 FR 9352)
EPCRA Section 302, Extremely Hazardous Substances: TPQ = 1,000/10,000 lb (454/4,540 kg)
EPCRA Section 304 Reportable Quantity (RQ): EHS/CERCLA, 5,000 lb (2,270 kg)
EPCRA Section 313: Form R *de minimis* concentration reporting level: 0.1%.
TSCA: 716.120a listed chemical
California's Proposition 65: Carcinogen.

ACRYLIC ACID A:089

SYNONYMS: ACIDO ACRILICO (Spanish); ACROLEIC ACID; ACRYLIC ACID, INHIBITED; ACRYLIC ACID, GLACIAL; AQUEOUS ACRYLIC ACID (technical grade is 94%); EEC No. 607-061-00-8; EINECS No. 201-177-9; ETHYLENECARBOXYLIC ACID; GLACIAL ACRYLIC ACID; KYSELINA AKRYLOVA; PROPENE ACID; PROPENOIC ACID; 2-PROPENOIC ACID; VINYLFORMIC ACID

IDENTIFICATION

CAS:	79-10-7
ERG ID:	UN 2218
ERG Guide:	131P (inhibited)
ERG Guide Category:	FLAMMABLE LIQUIDS -TOXIC
Molecular Formula:	$C_3H_4O_2$
RTECS No.:	AS4375000
Properties:	Very flammable, watery liquid. Colorless. Irritating, rancid, acrid odor. Corrosive. Sinks and mixes with water. Irritating vapor is produced. Freezing point is 55°F (13°C). Poisonous gas may be produced in fire.

HEALTH AND SAFETY INFORMATION

Corrosive to eyes, skin, and repsiratory system.

IARC Evaluation: Unclassifiable Substance, Group 3; no human data; no animal data.
Odor Threshold: 0.4 ppm
OSHA Table Z-1-A Air Contaminant

Exposure Limits:	**TWA**
ACGIH:*	2 ppm; 5.9 mg/m^3
NIOSH:*	2 ppm; 6 mg/m^3
OSHA:	None

*Skin contact contributes significantly to overall exposure.

RECOMMENDATIONS FOR RESPIRATOR SELECTION: NIOSH

At any concentrations above the NIOSH REL, or where there is no REL, at any detectable concentration: SCBAF:PD,PP (any self-contained breathing apparatus that has a full facepiece and is operated in a pressure-demand or other positive-pressure mode); or SAF:PD,PP:ASCBA (any supplied-air respirator that has a full facepiece and is operated in a pressure-demand or other positive-pressure mode in combination with an auxiliary self-contained breathing apparatus operated in a pressure-demand or other positive-pressure mode).

Escape: GMFOV [any air-purifying, full-facepiece respirator (gas mask) with a chin-style, front-or back-mounted organic vapor canister] or SCBAE (any appropriate escape-type, self-contained breathing apparatus).

CHEMICAL REACTIVITY

Forms explosive mixture with air. Forms explosive peroxides; light, heat, and peroxides can cause polymerization. Violent reaction with strong oxidizers. Incompatible with sulfuric acid, caustics, ammonia, amines, isocyanates, alkylene oxides, epichlorohydrin, toluene diamine, oleum, pyridine, methyl pyridine, n-methyl pyrrolidone, 2-methyl-6-ethyl aniline, aniline, ethylene diamine, ethyleneimine, and 2-aminoethanol. Severely corrodes carbon steel and iron; attacks other metals. May accumulate static electrical charges and may cause ignition of its vapors.

FIRE INFORMATION

Firefighting gear (including SCBA) does not provide adequate protection. If exposure occurs, remove and isolate gear immediately and thoroughly decontaminate personnel. Vapors are heavier than air and will collect in low areas. Vapors in confined areas may explode when exposed to fire. Vapors may travel long distances to ignition sources and flashback. Storage containers and parts of containers may rocket great distances, in many directions. If material or contaminated runoff enters waterways, notify downstream users of potentially contaminated waters. Notify local health and fire officials and pollution control agencies. Use dry chemicals, foam, carbon dioxide. From a secure explosion-proof location, use water spray to cool exposed containers. If cooling streams are ineffective (venting sound increases in volume and pitch, tank discolors or shows any signs of deforming), withdraw immediately to a secure position.

NFPA Hazard Classifications:

Health Hazards (Blue) 3
Flammability (Red) 2
Reactivity (Yellow) 2

Flash Point: 122°F (50°C) o.c.(glacial); 130°F (54°C) o.c.
Flammable Limits in Air: LEL 2.4%; UEL 8.0% (glacial); LEL 5.3%; UEL 26.0%
Autoignition Temperature: 820°F (438°C)

FIRST AID

Move victim to fresh air. Call emergency medical care. Apply artificial respiration if victim is not breathing. Do not use mouth-to-mouth method if victim ingested or inhaled the substance; induce artificial respiration with the aid of a pocket mask equipped with a one-way valve or other proper respiratory medical device. Administer oxygen if breathing is difficult. Remove and isolate contaminated clothing and shoes. In case of contact with substance, immediately flush skin or eyes with running water for at least 20 minutes. Wash skin with soap and water. Keep victim warm and quiet. Effects of exposure (inhalation, ingestion, or skin contact) to substance may be delayed. Ensure that medical personnel are aware of the material(s) involved and take precautions to protect themselves.

ENVIRONMENTAL INFORMATION

Clean Air Act: Hazardous Air Pollutants (Title I, Part A, Section 112)
EPA Hazardous Waste Number (RCRA No.): U008, D002
RCRA Section 261 Hazardous Constituents
EPCRA Section 304: Reportable Quantity (RQ): CERCLA, 5,000 lb (2,270 kg)
EPCRA Section 313: Form R *de minimis* concentration reporting level: 1.0%
TSCA: 799.5000

ACRYLONITRILE A:090

SYNONYMS: ACRILONITRILO (Spanish); ACRYLNITRIL (German); ACRYLNITRIL (Dutch); ACRYLON; ACRYLONITRILE MONOMER; AKRYLONITRYL (Polish); AN; CARBACRYL; CIANURO di VINILE (Italian); CYANOETHYLENE; CYANURE de VINYLE (French); EEC No. 608-00300-4; EINECS No. 203-466-5; ENT 54; FUMIGRAIN; MILLER'S FUMIGRAIN; NITRILE ACRILICO (Italian); NITRILE ACRYLIQUE (French); PROPENENITRILE; 2-PROPENENITRILE; TL 314; VCN; VENTOX; VINYL CYANIDE; VINYL CYANIDE, PROPENENITRILE

IDENTIFICATION

CAS:	107-13-1
ERG ID:	UN 1093
ERG Guide:	131P

36 Acrylonitrile

ERG Guide Category:	FLAMMABLE LIQUIDS -TOXIC
Molecular Formula:	C_3H_3N
RTECS No.:	AT5250000
Properties:	Highly flammable, watery liquid. Colorless to light yellow. Irritating odor resembling onions or garlic. Floats on water; moderately soluble. Vapors are heavier than air and will collect in low areas. Combustion products include toxic cyanide gas and nitrogen oxide.

HEALTH AND SAFETY INFORMATION

Poison. Produces cyanide in the body. Breathing the vapors, skin contact, or swallowing the material can kill you. Irritates the eyes and respiratory tract. Prolonged contact with the skin will cause burns, which can be delayed for several hours. Affects the nervous system.

IARC Evaluation: Probable carcinogen.
NTP: Anticipated carcinogen.
IDLH: 85 ppm; potential human carcinogen.
Odor Threshold: 1.66 ppm (sense of smell fatigues rapidly). Odor can only be detected above the PEL
OSHA Table Z-1-A Air Contaminant
OSHA Specifically Regulated Substances (29 CFR 1910.1001-1048)

Exposure Limits:	**TWA**	**CEILING**
ACGIH: [1]	1 ppm	10 ppm/15 min.
OSHA: [2]	2 ppm	10pp m
NIOSH: [2]	1ppm/15 min	

[1] Suspected human carcinogen. Skin contact contributes significantly in overall exposure.
[2] Potential human carcinogen. Reduce exposure to lowest feasible level. Skin contact contributes significantly in overall exposure.

RECOMMENDATIONS FOR RESPIRATOR SELECTION: NIOSH

At any concentrations above the NIOSH REL, or where there is no REL, at any detectable concentration: SCBAF:PD,PP (any self-contained breathing apparatus that has a full facepiece and is operated in a pressure-demand or other positive-pressure mode); or SAF:PD,PP:ASCBA (any supplied-air respirator that has a full facepiece and is operated in a pressure-demand or other positive-pressure mode in combination with an auxiliary, self-contained breathing apparatus operated in a pressure-demand or other positive-pressure mode)

Escape: GMFOV [any air-purifying, full-facepiece respirator (gas mask) with a chin-style, front-or back-mounted organic vapor canister]; or SCBAE (any appropriate escape-type, self-contained breathing apparatus).

CHEMICAL REACTIVITY

Forms explosive mixture with air. Heat or light can cause polymerization. Reacts violently with strong bases and peroxides, possibly causing polymerization. Reacts with oxidizers, acids, bromine, amines. Attacks copper and copper alloys. Attacks aluminum in high concentrations. Penetrates leather, so contaminated leather shoes and gloves should be destroyed.

FIRE INFORMATION

Firefighting gear (including SCBA) does not provide adequate protection. If exposure occurs, remove and isolate gear immediately and thoroughly decontaminate personnel. May react with itself without warning, blocking relief valves, and leading to container explosion. Vapors are heavier than air and will collect in low areas. Vapors in confined areas may explode when exposed to fire. Vapors may travel long distances to ignition sources and flash back. Storage containers and parts of containers may rocket great distances, in many directions. If material or contaminated runoff enters waterways, notify downstream users of potentially contaminated waters. Notify local health and fire officials and pollution control agencies. Use dry chemicals, foam, carbon dioxide. From a secure explosion-proof location, use water spray to cool exposed containers. If cooling streams are ineffective (venting sound increases in volume and pitch, tank discolors or shows any signs of deforming), withdraw immediately to a secure position.

NFPA Hazard Classifications:

Health Hazards (Blue) 4
Flammability (Red) 3
Reactivity (Yellow) 2

Flash Point: 31°F (1°C) c.c.; 32°F (0°C) c.c.
Flammable Limits in Air: LEL 3.05%; UEL 17.0%
Autoignition Temperature: 896°F (480°C)

FIRST AID

Move victim to fresh air. Call emergency medical care. Apply artificial respiration if victim is not breathing. Do not use mouth-to-mouth method if victim ingested or inhaled the substance; induce artificial respiration with the aid of a pocket mask equipped with a one-way valve or other proper respiratory medical device. Administer oxygen if breathing is difficult. Remove and isolate contaminated clothing and shoes. In case of contact with substance, immediately flush skin or eyes with running water for at least 20 minutes. Wash skin with soap and water. Keep victim warm and quiet. Effects of exposure (inhalation, ingestion, or skin contact) to substance may be delayed. Ensure that medical personnel are aware of the material(s) involved and take precautions to protect themselves.

ENVIRONMENTAL INFORMATION

Clean Air Act: Hazardous Air Pollutants (Title I, Part A, Section 112); Section 112[r], Accidental Release Prevention/Flammable Substances (Section 68.130), TQ = 20,000 lb (9,150 kg)

Clean Water Act: Section 307 Priority Pollutants; Section 311 Hazardous Substances/RQ (same as CERCLA); Section 313 Priority Chemicals
EPA Hazardous Waste Number (RCRA No.): U009
RCRA Section 261 Hazardous Constituents
RCRA Land Ban Waste
RCRA Universal Treatment Standards: Wastewater (mg/L), 0.24; Nonwastewater (mg/kg), 84.
RCRA Ground Water Monitoring List, Suggested Methods (PQL μg/L): 8030 (5); 8240 (5).
Safe Drinking Water Act: Priority List (55 FR 1470)
EPCRA Section 302, Extremely Hazardous Substances: TPQ = 10,000 lb (4,540 kg)
EPCRA Section 304 Reportable Quantity (RQ): EHS/CERCLA, 100 lb (45.5 kg)
EPCRA Section 313: Form R *de minimis* concentration reporting level: 0.1%.
California's Proposition 65: Carcinogen.

ADIPIC ACID — A:091

SYNONYMS: ACIDO ADIPICO (Spanish); ACIFLOCTIN; ACINETTEN; ADILAC-TETTEN; ADIPINIC ACID; 1,4-BUTANEDICARBOXYLIC ACID; DICARBOXYLIC ACID C_6; EEC No. 607-144-00-9; EINECS No. 204-673-3; FEMA No. 2011; HEXANEDIOIC ACID; 1,6-HEXANEDIOIC ACID; KYSELINA ADIPOVA (Czech), MOLTEN ADIPIC ACID

IDENTIFICATION

CAS:	124-04-9
ERG ID:	UN 9077
ERG Guide:	153
ERG Guide Category:	SUBSTANCES-TOXIC and/or CORROSIVE (COMBUSTIBLE)
Molecular Formula:	$C_6H_{10}O_4$
RTECS No.:	AU8400000
Properties:	Combustible, crystalline powder. White. Odorless. Sinks and mixes slowly with water.

HEALTH AND SAFETY INFORMATION

Irritates eyes and respiratory system. Skin contact can cause redness and pain.

Exposure Limits:	**TWA**
ACGIH:	5 mg/m^3

At any concentrations above the NIOSH REL, or where there is no REL, at any detectable concentration: SCBAF:PD,PP (any self-contained breathing apparatus that has a full facepiece and is operated in a pressure-demand or other positive-pressure mode); or SAF:PD,PP:ASCBA (any supplied-air respirator that has a full facepiece and is operated in a pressure-demand or other positive-pressure mode in

combination with an auxiliary, self-contained breathing apparatus operated in a pressure-demand or other positive-pressure mode).

Escape: GMFOV [any air-purifying, full-facepiece respirator (gas mask) with a chin-style, front-or back-mounted organic vapor canister]; or SCBAE (any appropriate escape-type, self-contained breathing apparatus).

CHEMICAL REACTIVITY

Dust forms an explosive mixture with air. Friction from stirring, pouring, or pneumatic transfer can form electrostatic charge on dry material.

FIRE INFORMATION

If material or contaminated runoff enters waterways, notify downstream users of potentially contaminated waters. Notify local health and fire officials and pollution control agencies. Use dry chemicals, alcohol-resistant foam, carbon dioxide. From a secure explosion-proof location, use water spray to cool exposed containers. If cooling streams are ineffective (venting sound increases in volume and pitch, tank discolors, or shows any signs of deforming), withdraw immediately to a secure position.

NFPA Hazard Classifications:

Health Hazards (Blue) . . . . -
Flammability (Red) 1
Reactivity (Yellow) 0

Flash Point: Combustible solid. 300°F (149°C) o.c.; 385°F (196°C) c.c.
Autoignition Temperature: 788°F (420°C)

FIRST AID

Move victim to fresh air and call emergency medical care. Apply artificial respiration if victim is not breathing. Do not use mouth-to-mouth method if victim swallowed or inhaled the substance; induce artificial respiration with the aid of a pocket mask equipped with a one-way valve or other proper respiratory medical device. If breathing is difficult, give oxygen. In case of contact with substance, immediately flush skin or eyes with running water for at least 20 minutes. Remove and isolate contaminated clothing and shoes. Keep victim warm and quiet. Effects of exposure (inhalation, ingestion or skin contact) to substances may be delayed. Ensure that medical personnel are aware of the material(s) involved and take precautions to protect themselves.

ENVIRONMENTAL INFORMATION

Clean Water Act: Section 311 Hazardous Substances/RQ (same as CERCLA)
EPCRA Section 304 Reportable Quantity (RQ): CERCLA, 5,000 lb (2,270 kg)

ADIPONITRILE A:092

SYNONYMS: ADIPIC ACID DINITRILE; ADIPIC ACID NITRILE; ADIPODINITRILE; ADIPONITRILO (Spanish); 1,4-DICYANOBUTANE; HEXANEDINITRILE; HEXANEDIOIC ACID, DINITRILE; NITRILE ADIPICO (Italian); TETRAMETHYLENE CYANIDE; RTECS No. AV2625000; STCC 4915185

IDENTIFICATION

CAS:	111-69-3
ERG ID:	UN 2205
ERG Guide:	153
ERG Guide Category:	SUBSTANCES-TOXIC AND/OR CORROSIVE (COMBUSTIBLE)
Molecular Formula:	$C_6H_8N_2$
RTECS No.:	AV2625000
Properties:	Combustible water-white, oily liquid. A solid below 34°F (1.1°C). Practically odorless. Floats on water and is slightly soluble. Vapors are heavier than air and will collect in low areas.

HEALTH AND SAFETY INFORMATION

Poison; produces cyanide in the body. Breathing the material or skin contact can cause death.

IDLH: Not Determined
Odor Threshold: Odor is not a reliable indicator of the presence of toxic amounts of vapor.
NIOSH Criteria Document: NIOSH 111-69-3; NIOSH 78-212 NITRILES

Exposure Limits:	**TWA**
ACGIH:+	2 ppm
NIOSH:	4 ppm; 18 mg/m^3
OSHA:	None

+Skin contact contributes significantly to overall exposure.

RECOMMENDATIONS FOR RESPIRATOR SELECTION: NIOSH

40 ppm: SA (any supplied-air respirator)
100 ppm: SA:CF (any supplied-air respirator operated in a continuous-flow mode)1
200 ppm: SCBAF (any self-contained breathing apparatus with a full facepiece); or SAF (any supplied-air respirator with a full facepiece)
250 ppm: SAF:PD,PP (Any supplied-air respirator that has a full facepiece and is operated in a pressure-demand or other positive-pressure mode)

EMERGENCY OR PLANNED ENTRY INTO UNKNOWN CONCENTRATIONS OR IDLH CONDITIONS: SCBAF:PD,PP (any self-contained breathing apparatus that has a full facepiece and is operated in a pressure-demand or other positive-pressure mode); or SAF:PD,PP:ASCBA (Any supplied-air respirator that has a full facepiece and is operated in a pressure-demand or other positive-pressure mode in combination with an auxiliary self-contained breathing apparatus operated in a pressure-demand or other positive pressure mode.

ESCAPE: GMFOV [Any air-purifying, full-facepiece respirator (gas mask) with a chin-style, front-or back-mounted organic vapor canister]; or SCBAE (Any appropriate escape-type, self-contained breathing apparatus)

CHEMICAL REACTIVITY

Contact with oxidizers (i.e., perchlorates, nitrates) can cause fire and explosions. Incompatible with strong acids. Decomposes above 194°F (90°C), forming toxic hydrogen cyanide.

FIRE INFORMATION

Combustion or decomposition products upon heating include toxic hydrogen cyanide and nitrogen oxides. Firefighting gear (including SCBA) does not provide adequate protection. If exposure occurs, remove and isolate gear immediately and thoroughly decontaminate personnel. Vapors are heavier than air and will collect in low areas. Vapors in confined areas may explode when exposed to fire. Vapors may travel long distances to ignition sources and flash back. Storage containers and parts of containers may rocket great distances, in many directions. If material or contaminated runoff enters waterways, notify downstream users of potentially contaminated waters. Notify local health and fire officials and pollution control agencies. Use dry chemicals, foam, or carbon dioxide. From a secure explosion-proof location, use water spray to cool exposed containers. If cooling streams are ineffective (venting sound increases in volume and pitch, tank discolors or shows any signs of deforming), withdraw immediately to a secure position.

NFPA Hazard Classifications:

Health Hazards (Blue) 2
Flammability (Red) 2
Reactivity (Yellow) 1

Flash Point: 199 °F (92.7°C) o.c. (Technical grade); 261°F (163°C) (pure) o.c.
Flammable Limits in Air: LEL 1.7%; UEL 5.0%
Autoignition Temperature: 1022°F (550°C)

FIRST AID

Move victim to fresh air; call emergency medical care. Apply artificial respiration if victim is not breathing. Do not use mouth-to-mouth method if victim ingested or inhaled the substance; induce artificial respiration with the aid of a pocket mask equipped with a one-way valve or other proper respiratory medical device. Administer oxygen if breathing is difficult. Remove and isolate contaminated clothing and shoes. In case of contact with substance, immediately flush skin or eyes with running water for at least 20 minutes. For minor skin contact, avoid spreading material on unaffected skin. Keep victim warm and quiet.

Effects of exposure (inhalation, ingestion, or skin contact) to substance may be delayed. Ensure that medical personnel are aware of the material(s) involved and take precautions to protect themselves.

ENVIRONMENTAL INFORMATION

EPCRA (Section 313): As cyanide compounds. Form R de minimus concentration reporting level: 1.0%.
EPCRA Section 302: Extremely Hazardous Substances: TPQ = 1,000 lb (454 kg).
EPCRA Section 304: Reportable Quantity (RQ): EHS, 1 lb (0.454 kg).
MARINE POLLUTANT: (49CFR, Subchapter 172.101, Appendix B) as cyanide mixtures or solutions.

ALLYL ALCOHOL A:093

SYNONYMS: AA; ALCOOL ALLILCO (Italian); ALCOOL ALLYLIQUE (French); ALILICO ALCOHOL (Spanish); ALLILOWY ALKOHOL (Polish); ALLYL AL; ALLYL ALCOHOL; ALLYLALKOHOL (German); ALLYLIC ALCOHOL; EEC No. 603-015-00-6; 3-HYDROXYPROPENE; ORVINYLCARBINOL; PROPENOL; 2-PROPENOL; 2-PROPEN-1-OL; PROPEN-1-OL-3; 1-PROPEN-3-OL; PROPENYL ALCOHOL; 2-PROPENYL ALCOHOL; SHELL UNKRAUTTED A; VINYL CARBINOL; VINYL CARBINOL,2-PROPENOL; WEED DRENCH

IDENTIFICATION

CAS:	107-18-6
ERG ID:	UN 1098
ERG Guide:	131
ERG Guide Category:	FLAMMABLE LIQUIDS-TOXIC
Molecular Formula:	C_3H_6O
RTECS No.:	BA5075000
Properties:	Flammable liquid. Clear, colorless. Pungent, mustardlike odor. Soluble in water. Vapors are heavier than air and will collect in low areas.

HEALTH AND SAFETY INFORMATION

Corrosive to eyes, skin, and respiratory tract. Inhalation may cause pulmonary edema which can be delayed for several hours; there is a risk of death in serious cases. Affects the central nervous system, kidney, and liver. Poison. Breathing the vapors, ingestion, or skin contact can cause severe irritation, and health problems.

IDLH: 20 ppm
Odor Threshold: 0.47 ppm; 1.7 ppm [AIHA, (detectable)]
OSHA Table Z-1-A Air Contaminant

Exposure Limits:	TWA	STEL
ACGIH:[1]	2 ppm	4 ppm
NIOSH:[1]	2 ppm; 5 mg/m^3	4 ppm; 10 mg/m^3
OSHA:	2 ppm; 5 mg/m^3	

[3] Skin contact contributes significantly in overall exposure

RECOMMENDATIONS FOR RESPIRATOR SELECTION: NIOSH

20 ppm: SA:CF* (any supplied-air respirator operated in a continuous-flow mode); or PAPROV *[any powered, air-purifying respirator with organic vapor cartridge(s)]; or CCRFOV [Any air-purifying, full-facepiece respirator (gas mask) with a chin-style, front-or back-mounted acid gas canister]; or GMFOV [Any air-purifying, full-facepiece respirator (gas mask) with a chin-style, front-or back-,mounted organic vapor canister]; or SCBAF (any self-contained breathing apparatus with a full facepiece); or SAF (any supplied-air respirator with a full facepiece); or SAF (any supplied-air respirator with a full facepiece).

EMERGENCY OR PLANNED ENTRY INTO UNKNOWN CONCENTRATIONS OR IDLH CONDITIONS: SCBAF:PD,PP (any self-contained breathing apparatus that has a full facepiece and is operated in a pressure-demand or other positive-pressure mode); or SAF:PD,PP:ASCBA (Any supplied-air respirator that has a full facepiece and is operated in a pressure-demand or other positive-pressure mode in combination with an auxiliary self-contained breathing apparatus operated in a pressure-demand or other positive-pressure mode.

ESCAPE: GMFOV [Any air-purifying, full-facepiece respirator (gas mask) with a chin-style, front-or back-,mounted organic vapor canister]; or SCBAE (Any appropriate escape-type, self-contained breathing apparatus).

* *Note:* Substance reported to cause eye irritation or damage; may require eye protection.

CHEMICAL REACTIVITY

Forms explosive mixture with air. Incompatible with strong oxidizers, acids, carbon tetrachloride. Prolonged exposure to air may cause the formation of peroxides; elevated temperatures, oxidizers, or peroxides may cause explosive polymerization.

FIRE INFORMATION

Firefighting gear (including SCBA) does not provide adequate protection. If exposure occurs, remove and isolate gear immediately and thoroughly decontaminate personnel. Vapors are heavier than air and will collect in low areas. Vapors in confined areas may explode when exposed to fire. Vapors may travel long distances to ignition sources and flash back. Storage containers and parts of containers may rocket great distances, in many directions. If material or contaminated runoff enters waterways, notify downstream users of potentially contaminated waters. Notify local health and fire officials and pollution control agencies.

NFPA Hazard Classifications:

Health Hazards (Blue) 4
Flammability (Red) 3
Reactivity (Yellow) 1

Flash Point: 70°F (21°C)
Flammable Limits in Air: LEL 2.5%; UEL 18.0%
Autoignition Temperature: 713°F (378°C)

FIRST AID

Move victim to fresh air. Call emergency medical care. Apply artificial respiration if victim is not breathing. Do not use mouth-to-mouth method if victim ingested or inhaled the substance; induce artificial respiration with the aid of a pocket mask equipped with a one-way valve or other proper respiratory medical device. Administer oxygen if breathing is difficult. Remove and isolate contaminated clothing and shoes. In case of contact with substance, immediately flush skin or eyes with running water for at least 20 minutes. Wash skin with soap and water. Keep victim warm and quiet. Effects of exposure (inhalation, ingestion or skin contact) to substance may be delayed. Ensure that medical personnel are aware of the material(s) involved and take precautions to protect themselves.

ENVIRONMENTAL INFORMATION

EPA priority pollutant: groundwater and solid waste matrices.
CLEAN AIR ACT: Section 112(r), Accidental Release Prevention/Flammable substances (Section 68.130), TQ = 15,000 lb (5825 kg)
CLEAN WATER ACT: Section 311 Hazardous Substances/RQ (same as CERCLA).
EPA HAZARDOUS WASTE NUMBER (RCRA No.): P005
RCRA Section 261 Hazardous Constituents
RCRA Land Ban Waste
EPCRA Section 302, Extremely Hazardous Substances: TPQ = 1,000 lb (454 kg).
EPCRA Section 304 Reportable Quantity (RQ): CERCLA, 100 lb (45.4 kg).
EPCRA Section 313 Form R: *de minimus* concentration reporting level: 1.0%.
TSCA: Health and Safety Reporting List

ALLYL BROMIDE A:094

SYNONYMS: BROMALLYLENE; 3-BROMO-1-PROPENE; 3-BROMOPROPENE; 3-BROMOPROPYLENE; BROMURO de ALILO (Spanish)

IDENTIFICATION

CAS: 106-95-6
ERG ID: UN 1099

ERG Guide: 131
ERG Guide Category: FLAMMABLE LIQUIDS-TOXIC
Molecular Formula: C_3H_5Br
RTECS No.: UC7090000
Properties: Highly flammable liquid. Colorless to light yellow. Pungent, unpleasant odor. Sinks in water and is slightly soluble. Vapors are heavier than air and will collect in low areas.

HEALTH AND SAFETY INFORMATION

Corrosive. Severely irritating to skin, eyes, nose, and lungs; prolonged contact with skin causes burns. Poison. Breathing the vapors can cause death. May cause pulmonary edema, which can be delayed for several hours following exposure.

RECOMMENDATIONS FOR RESPIRATOR SELECTION: NIOSH

At any concentrations above the NIOSH REL, or where there is no REL, at any detectable concentration: SCBAF:PD,PP (any self-contained breathing apparatus that has a full facepiece and is operated in a pressure-demand or other positive-pressure mode); or SAF:PD,PP:ASCBA (any supplied-air respirator that has a full facepiece and is operated in a pressure-demand or other positive-pressure mode in combination with an auxiliary self-contained breathing apparatus operated in a pressure-demand or other positive-pressure mode).

Escape: GMFOV [any air-purifying, full-facepiece respirator (gas mask) with a chin-style, front-or back-mounted organic vapor canister]; or SCBAE (any appropriate escape-type, self-contained breathing apparatus).

CHEMICAL REACTIVITY

Forms explosive mixture with air. Contact with oxidizers may cause fire and explosions. Heat or light exposure can cause decomposition and corrosive vapors.

FIRE INFORMATION

Firefighting gear (including SCBA) does not provide adequate protection. If exposure occurs, remove and isolate gear immediately and thoroughly decontaminate personnel. Vapors are heavier than air and will collect in low areas. Vapors in confined areas may explode when exposed to fire. Vapors may travel long distances to ignition sources and flash back. Storage containers and parts of containers may rocket great distances, in many directions. If material or contaminated runoff enters waterways, notify downstream users of potentially contaminated waters. Notify local health and fire officials and pollution control agencies. Use dry chemicals, foam, carbon dioxide. From a secure explosion-proof location, use water spray to cool exposed containers. If cooling streams are ineffective (venting sound increases in volume and pitch, tank discolors, or shows any signs of deforming), withdraw immediately to a secure position.

NFPA Hazard Classifications:

Health Hazards (Blue) 3

Flammability (Red) 3
Reactivity (Yellow) 1

Flash Point: 30°F (–1°C)
Flammable Limits in Air: LEL 4.4%; UEL 7.3%
Autoignition Temperature: 563°F (295°C)

FIRST AID

Move victim to fresh air. Call emergency medical care. Apply artificial respiration if victim is not breathing. Do not use mouth-to-mouth method if victim ingested or inhaled the substance; induce artificial respiration with the aid of a pocket mask equipped with a one-way valve or other proper respiratory medical devise. Administer oxygen if breathing is difficult. Remove and isolate contaminated clothing and shoes. In case of contact with substance, immediately flush skin or eyes with running water for at least 20 minutes. Wash skin with soap and water. Keep victim warm and quiet. Effects of exposure (inhalation, ingestion, or skin contact) to substance may be delayed. Ensure that medical personnel are aware of the material(s) involved and take precautions to protect themselves.

ENVIRONMENTAL INFORMATION

Marine Pollutant (49 CFR, Subchapter 172.101, Appendix B)

ALLYL CHLORIDE A:095

SYNONYMS: ALLILE (CLORURO DI) (Italian); ALLYLCHLORID (German); ALLYL CHLORIDE; ALLYLE (CHLORURE D') (French); CHLORALLYLENE; 3-CHLOROPRENE; 1-CHLORO PROPENE-2; 1-CHLORO-2-PROPENE; 3-CHLOROPROPENE-1; 3-CHLOROPROPENE; 3-CHLORO-1-PROPENE; 3-CHLOROPROPYLENE; 3-CHLORPROPEN (German); CLORURO de ALILO (Spanish); EEC No. 602-029-00-x; NCI-C04615; 1-CHLORO-2-PROPENE

IDENTIFICATION

CAS:	107-05-1
ERG ID:	UN 1100
ERG Guide:	131
ERG Guide Category:	FLAMMABLE LIQUIDS -TOXIC
Molecular Formula:	C_3H_5Cl
RTECS No.:	UC7350000
Properties:	Highly flammable liquid. Colorless to yellowish brown or red. Sharp, pungent, irritating odor. Floats on water; insoluble; decomposes and forms hydrochloric acid; flammable, irritating vapor is produced. Vapors are heavier than air and will collect in low areas.

HEALTH AND SAFETY INFORMATION

Severely irritating to eyes, skin, and respiratory tract; prolonged contact with skin can cause burns. May cause pulmonary edema, which can be delayed for several hours. Can cause liver and kidney damage.

IARC Evaluation: Rating 3, inadequate animal evidence.
IDLH: 250 ppm
Odor Threshold: 0.489 ppm
OSHA Table Z-1-A Air Contaminant
OSHA Process Safety Management of Highly Hazardous Chemicals: (29 CFR Part 1910.119, Appendix A) TQ = 1,000 lb

Exposure Limits:	**TWA**	**STEL**
ACGIH:	1 ppm; 3 mg/m^3	2 ppm; 6 mg/m^3
NIOSH:	1 ppm; 3 mg/m^3	2 ppm; 6mg/m^3
OSHA:	1 ppm; 3 mg/m^3	

RECOMMENDATIONS FOR RESPIRATOR SELECTION: NIOSH/OSHA

25 ppm: SA:CF (any supplied-air respirator operated in a continuous-flow mode).
50 ppm: SCBAF (any self-contained breathing apparatus with a full facepiece); or SAF (any supplied-air respirator with a full facepiece).
250 ppm: SAF:PD,PP (any supplied-air respirator that has a full facepiece and is operated in a pressure-demand or other positive-pressure mode).

Emergency or Planned Entry into Unknown Concentrations or Idlh Conditions: SCBAF:PD,PP (any self-contained breathing apparatus that has a full facepiece and is operated in a pressure-demand or other positive-pressure mode); or SAF:PD,PP:ASCBA (any supplied-air respirator that has a full facepiece and is operated in a pressure-demand or other positive-pressure mode in combination with an auxiliary, self-contained breathing apparatus operated in a pressure-demand or other positive-pressure mode.

Escape: GMFOV [any air-purifying, full-facepiece respirator (gas mask) with a chin-style, front-or back-mounted organic vapor canister]; or SCBAE (any appropriate escape-type, self-contained breathing apparatus).

Note: Substance reported to cause eye irritation or damage; may require eye protection.

CHEMICAL REACTIVITY

When heated or exposed to light, may react with itself without warning with explosive violence. Slow decomposition forming hydrochloric acid. Violent reaction with strong oxidizers. Reacts with acids, amines, iron and aluminum chlorides, magnesium, zinc. Attacks steel and aluminum in presence of moisture.

FIRE INFORMATION

Firefighting gear (including SCBA) does not provide adequate protection. If exposure occurs, remove and isolate gear immediately and thoroughly decontaminate personnel. Vapors are heavier than air and will collect in low areas. Vapors in confined areas may explode when exposed to fire. Vapors may travel long distances to ignition sources and flash back. Storage containers and parts of containers may rocket great distances, in many directions. If material or contaminated runoff enters waterways, notify downstream users of potentially contaminated waters. Notify local health and fire officials and pollution control agencies. Use dry chemicals, foam, carbon dioxide. From a secure explosion-proof location, use water spray to cool exposed containers. If cooling streams are ineffective (venting sound increases in volume and pitch, tank discolors or shows any signs of deforming), withdraw immediately to a secure position.

NFPA Hazard Classifications:

Health Hazards (Blue) 3
Flammability (Red) 3
Reactivity (Yellow) 1

Flash Point: –25°F (–32°C) c.c.
Flammable Limits in Air: LEL 29%; UEL 11.1%
Autoignition Temperature: 737°F (485°C)

FIRST AID

Move victim to fresh air. Call emergency medical care. Apply artificial respiration if victim is not breathing. Do not use mouth-to-mouth method if victim ingested or inhaled the substance; induce artificial respiration with the aid of a pocket mask equipped with a one-way valve or other proper respiratory medical device. Administer oxygen if breathing is difficult. Remove and isolate contaminated clothing and shoes. In case of contact with substance, immediately flush skin or eyes with running water for at least 20 minutes. Wash skin with soap and water. Keep victim warm and quiet. Effects of exposure (inhalation, ingestion, or skin contact) to substance may be delayed. Ensure that medical personnel are aware of the material(s) involved and take precautions to protect themselves.

ENVIRONMENTAL INFORMATION

Clean Air Act: Hazardous Air Pollutants (Title I, Part A, Section 112)
Clean Water Act: Section 311 Hazardous Substances/RQ (same as CERCLA); Section 313 Priority Chemicals
Hazardous Constituents: (40 CFR/262, Appendix VII, waste number not listed.
RCRA Land Ban Waste
RCRA Universal Treatment Standards: Wastewater (mg/L), 0.036; Nonwastewater (mg/kg), 30.
RCRA Ground Water Monitoring List, Suggested Methods (PQL μg/L): 8010 (5); 8240 (5).
EPCRA Section 304 Reportable Quantity (RQ): CERCLA, 1,000 lb (454 kg)
EPCRA Section 313: Form R *de minimis* concentration reporting level: 1.0%.
California's Proposition 65: Carcinogen.
Marine Pollutant (49 CFR, Subchapter 172.101, Appendix B)

ALLYL CHLOROFORMATE A:096

SYNONYMS: ALLYL CHLOROCARBONATE; CHLOROFORMIC ACID, ALLYL ESTER; FORMIC ACID, CHLORO-, ALLYL ESTER; 2-PROPENYL CHLOROFORMATE

IDENTIFICATION

CAS:	2937-50-0
ERG ID:	UN 1722
ERG Guide:	155
ERG Guide Category:	SUBSTANCES-TOXIC and/or CORROSIVE (FLAMMABLE/WATER-SENSITIVE)
Molecular Formula:	$C_4H_5Cl\ O_2$
RTECS No.:	LQ5775000
Properties:	Highly flammable liquid. Colorless, watery. Extremely irritating odor. Sinks in water; not soluble; decomposes to form allyl alcohol and chloroformic acid. Vapors are heavier than air and will collect in low areas.

HEALTH AND SAFETY INFORMATION

Serious health hazard. Severely irritating to eyes, skin, and respiratory tract. May be absorbed through the skin.

RECOMMENDATIONS FOR RESPIRATOR SELECTION:

At any concentrations above the NIOSH REL, or where there is no REL, at any detectable concentration: SCBAF:PD,PP (any self-contained breathing apparatus that has a full facepiece and is operated in a pressure-demand or other positive-pressure mode); or SAF:PD,PP:ASCBA (any supplied-air respirator that has a full facepiece and is operated in a pressure-demand or other positive-pressure mode in combination with an auxiliary, self-contained breathing apparatus operated in a pressure-demand or other positive-pressure mode).

Escape: GMFOV [any air-purifying, full-facepiece respirator (gas mask) with a chin-style, front-or back-mounted organic vapor canister]; or SCBAE (any appropriate escape-type, self-contained breathing apparatus).

CHEMICAL REACTIVITY

Reacts with water forming toxic allyl alcohol and chloroformic acid. Contact with oxidizers can cause fire and explosions. Elevated temperatures, peroxides, or oxidizers may cause explosive polymerization. Reacts with acids, alkalies, alcohols, and amines.

FIRE INFORMATION

Firefighting gear (including SCBA) does not provide adequate protection. If exposure occurs, remove and isolate gear immediately and thoroughly decontaminate personnel. Vapors are heavier than air and will collect in low areas. Vapors in confined areas may explode when exposed to fire. Vapors may travel long distances to ignition sources and flash back. Storage containers and parts of containers may rocket great distances, in many directions. If material or contaminated runoff enters waterways, notify downstream users of potentially contaminated waters. Notify local health and fire officials and pollution control agencies. Use dry chemicals, foam, carbon dioxide. From a secure explosion-proof location, use water spray to cool exposed containers. If cooling streams are ineffective (venting sound increases in volume and pitch, tank discolors or shows any signs of deforming), withdraw immediately to a secure position.

NFPA Hazard Classifications:

Health Hazards (Blue) 3
Flammability (Red) 3
Reactivity (Yellow) 1

Flash Point: 88°F (31°C)

FIRST AID

Move victim to fresh air. Call emergency medical care. Apply artificial respiration if victim is not breathing. If breathing is difficult, give oxygen. Remove and isolate contaminated clothing and shoes. In case of contact with substance, immediately flush skin or eyes with running water for at least 20 minutes. Wash skin with soap and water. Keep victim warm and quiet. Ensure that medical personnel are aware of the material(s) involved and take precautions to protect themselves.

ALLYL GLYCIDYL ETHER — A:097

SYNONYMS: AGE; ALIL GLICIDIKICO ETER (Spanish); ALLIL-GLICIDIL ETERE (Italian); ALLYL-2,3-EPOXYPROPYL ETHER; ALLYLGLYCIDAETHER (German); 1-ALLYLOXY-2,3-EPOXY-PROPAAN (Dutch); 1-ALLYLOXY-2,3-EPOXY-PROPAN (German); 1-(ALLYLOXY)-2,3-EPOXYPROPANE; EEC No. 603-038-00-1; NCI-C56666; [(2-PROPENYLOXY)METHYL]OXIRANE

IDENTIFICATION

CAS:	106-92-3
ERG ID:	UN 2219
ERG Guide:	129
ERG Guide Category:	FLAMMABLE LIQUIDS (POLAR/WATER MISCIBLE/NOXIOUS)
Molecular Formula:	$C_6H_{10}O_2$
RTECS No.:	RR0875000
Properties:	Flammable liquid. Colorless. Pleasant odor. Vapors are heavier than air and will collect in low areas.

HEALTH AND SAFETY INFORMATION

Irritates eyes, skin, and respiratory system. Inhalation may cause pulmonary edema which can be delayed for several hours; there is a risk of death in serious cases. High or repeated exposures can cause skin disorders.

IDLH: 50 ppm
OSHA Table Z-1-A Air Contaminant

Exposure Limits:	TWA	CEILING	STEL
ACGIH:	1 ppm		
NIOSH:[(1)]	5 ppm; 22 mg/m^3		10 ppm; 44 mg/m^3
OSHA:		10ppm; 45 mg/m^3	

[(1)] Skin contact contributes significantly in overall exposure.

RECOMMENDATIONS FOR RESPIRATOR SELECTION: NIOSH

50 ppm: CCROV [any chemicall cartridge respirator with organic vapor cartridge(s)]; or PAPROV [any powered, air-purifying respirator with organic vapor cartridge(s)]; or GMFOV [any air-purifying, full-facepiece respirator (gas mask) with a chin-style, front-or back-mounted acid gas canister]; or SA (any supplied-air respirator); or SCBAF (any self-contained breathing apparatus with a full facepiece)

At any concentrations above the NIOSH REL, or where there is no REL, at any detectable concentration: SCBAF:PD,PP (any self-contained breathing apparatus that has a full facepiece and is operated in a pressure-demand or other positive-pressure mode); or SAF:PD,PP:ASCBA (any supplied-air respirator that has a full facepiece and is operated in a pressure-demand or other positive-pressure mode in combination with an auxiliary self-contained breathing apparatus operated in a pressure-demand or other positive-pressure mode).

Escape: GMFOV [any air-purifying, full-facepiece respirator (gas mask) with a chin-style, front-or back-mounted organic vapor canister]; or SCBAE (any appropriate escape-type, self-contained breathing apparatus).

CHEMICAL REACTIVITY

Forms explosive mixture with air. Contact with acids or bases may cause explosive polymerization. Contact with oxidizers or amines may cause fire and explosions.

FIRE INFORMATION

Vapors are heavier than air and will collect in low areas. Vapors in confined areas may explode when exposed to fire. Vapors may travel long distances to ignition sources and flash back. Storage containers and parts of containers may rocket great distances, in many directions. If material or contaminated runoff enters waterways, notify downstream users of potentially contaminated waters. Notify local health and fire officials and pollution control agencies. Use dry chemicals, foam, carbon dioxide. From a secure explosion-proof location, use water spray to cool exposed containers. If cooling streams are ineffective (venting sound increases in volume and pitch, tank discolors, or shows any signs of deforming), withdraw immediately to a secure position.

Flash Point: 135°F (57°C)

FIRST AID

Move victim to fresh air. Call emergency medical care. Apply artificial respiration if victim is not breathing. Administer oxygen if breathing is difficult. Remove and isolate contaminated clothing and shoes. In case of contact with substance, immediately flush skin or eyes with running water for at least 20 minutes. Wash skin with soap and water. Keep victim warm and quiet. Effects of exposure (inhalation, ingestion, or skin contact) to substance may be delayed. Ensure that medical personnel are aware of the material(s) involved and take precautions to protect themselves.

ENVIRONMENTAL INFORMATION

Marine Pollutant (49 CFR, Subchapter 172.101, Appendix B)

ALLYL ISOCYANATE A:098

SYNONYMS: ISOCIANATO de ALILO (Spanish); ISOCYANIC ACID, ALLYL ESTER

IDENTIFICATION

CAS:	1476-23-9
ERG ID:	UN 2478
ERG Guide:	155
ERG Guide Category:	ISOCYANATES, FLAMMABLE, TOXIC, n.o.s.
Molecular Formula:	C_4H_5NO
RTECS No.:	NQ8175000
Properties:	Combustible liquid. Vapors are heavier than air and will collect in low areas.

HEALTH AND SAFETY INFORMATION

Causes eyes to water (lacramator). Irritates skin, eyes, respiratory tract.

Odor Threshold: 0.035

RECOMMENDATIONS FOR RESPIRATOR SELECTION:

At any concentrations above the NIOSH REL, or where there is no REL, at any detectable concentration: SCBAF:PD,PP (any self-contained breathing apparatus that has a full facepiece and is

operated in a pressure-demand or other positive-pressure mode); or SAF:PD,PP:ASCBA (any supplied-air respirator that has a full facepiece and is operated in a pressure-demand or other positive-pressure mode in combination with an auxiliary self-contained breathing apparatus operated in a pressure-demand or other positive-pressure mode).

Escape: GMFOV [any air-purifying, full-facepiece respirator (gas mask) with a chin-style, front-or back-mounted organic vapor canister]; or SCBAE (any appropriate escape-type, self-contained breathing apparatus).

CHEMICAL REACTIVITY

Very reactive; may undergo many addition-type reactions. Contact with oxidizers may cause fire and explosions. Reacts with nonoxidizing mineral acids, sulfuric acid, nitric acid, organic acids, caustics, ammonia, aliphatic amines, alkanolamines, aromatic amines, amides, alcohols/glycols, caprolactum solution.

FIRE INFORMATION

Combustion products include toxic nitrogen oxides. Vapors are heavier than air and will collect in low areas. Vapors in confined areas may explode when exposed to fire. Vapors may travel long distances to ignition sources and flash back. Storage containers and parts of containers may rocket great distances, in many directions. If material or contaminated runoff enters waterways, notify downstream users of potentially contaminated waters. Notify local health and fire officials and pollution control agencies. Use dry chemicals, foam, carbon dioxide. From a secure explosion-proof location, use water spray to cool exposed containers. If cooling streams are ineffective (venting sound increases in volume and pitch, tank discolors, or shows any signs of deforming), withdraw immediately to a secure position.

Flash Point: 110°F (43°C)

FIRST AID

Move victim to fresh air. Call emergency medical care. Apply artificial respiration if victim is not breathing. If breathing is difficult, give oxygen. Remove and isolate contaminated clothing and shoes. In case of contact with substance, immediately flush skin or eyes with running water for at least 20 minutes. Wash skin with soap and water. Keep victim warm and quiet. Ensure that medical personnel are aware of the material(s) involved and take precautions to protect themselves.

ENVIRONMENTAL INFORMATION as isocyanates

TSCA: 40CFR712.30 (e8); 40CFR716.120 (d8)

ALLYL METHACRYLATE A:099

SYNONYMS: AGEFLEX AMA; ALLYLESTER KYSELINY METHAKRYLOVE (Czech); EINECS

No. 202-473-0; METHACRYLIC ACID, ALLYL ESTER; METRACRILO de ALILO (Spanish); PROPENOIC ACID(2-), 2-METHYL-, 2-PROPENYL ESTER (TSCA)

IDENTIFICATION

CAS: 96-05-9
Molecular Formula: $C_7H_{10}O_2$
RTECS No.: UD3483000

HEALTH AND SAFETY INFORMATION

Moderately toxic by skin, eye contact, or if inhaled or swallowed.

RECOMMENDATIONS FOR RESPIRATOR SELECTION:

At any concentrations above the NIOSH REL, or where there is no REL, at any detectable concentration: SCBAF:PD,PP (any self-contained breathing apparatus that has a full facepiece and is operated in a pressure-demand or other positive-pressure mode); or SAF:PD,PP:ASCBA (any supplied-air respirator that has a full facepiece and is operated in a pressure-demand or other positive-pressure mode in combination with an auxiliary self-contained breathing apparatus operated in a pressure-demand or other positive-pressure mode).

Escape: GMFOV [any air-purifying, full-facepiece respirator (gas mask) with a chin-style, front-or back-mounted organic vapor canister]; or SCBAE (any appropriate escape-type, self-contained breathing apparatus).

CHEMICAL REACTIVITY

Contact with oxidizers may cause fire and explosions. May form dangerous peroxides when exposed to air. Esters are incompatible with nitrates, nitric acid, sulfuric acid.

FIRE INFORMATION

Combustion products include toxic vapors. If material or contaminated runoff enters waterways, notify downstream users of potentially contaminated waters. Notify local health and fire officials and pollution control agencies. Use dry chemicals, alcohol-resistant foam, carbon dioxide. From a secure explosion-proof location, use water spray to cool exposed containers. If cooling streams are ineffective (venting sound increases in volume and pitch, tank discolors or shows any signs of deforming), withdraw immediately to a secure position.

FIRST AID

Move victim to fresh air. Call emergency medical care. Apply artificial respiration if victim is not breathing. Administer oxygen if breathing is difficult. Remove and isolate contaminated clothing and shoes. In case of contact with substance, immediately flush skin or eyes with running water for at least 20 minutes. Wash skin with soap and water. Keep victim warm and quiet. Effects of exposure (inhalation,

ingestion, or skin contact) to substance may be delayed. Ensure that medical personnel are aware of the material(s) involved and take precautions to protect themselves.

ENVIRONMENTAL INFORMATION

TSCA: 40CFR712.30(d). Chemicals listed by reporting date.

ALLYLTRICHLOROSILANE A:101

SYNONYMS: ALILTRICLOROSILANO (Spanish); ALLYL TRICHOROSILANE, STABILIZED; ALLYLSILICONE TRICHLORIDE; ALLYLTRICHLOROSILANE; SILANE, ALLYLTRICHLORO-; SILANE, TRICHLOROALLYL-; SILANE,TRICHLORO-2-PROPENYL-; ALLYLSILICONE TRICHLORIDE; TRICHLOROALLYLSILANE

IDENTIFICATION

CAS:	107-37-9
ERG ID:	UN 1724
ERG Guide:	155
ERG Guide Category:	SUBSTANCES-TOXIC and/or CORROSIVE (FLAMMABLE/WATER-SENSITIVE)
Molecular Formula:	$C_3H_5Cl_3Si$
RTECS No.:	VV1530000
Properties:	Highly flammable liquid. Colorless. Pungent, irritating odor. Reacts vigorously with water, forming toxic and corrosive hydrochloric acid.

HEALTH AND SAFETY INFORMATION

Skin and eye contact can cause severe burns and blindness. Poison. Breathing the vapors can cause death.

RECOMMENDATIONS FOR RESPIRATOR SELECTION:

At any concentrations above the NIOSH REL, or where there is no REL, at any detectable concentration: SCBAF:PD,PP (any self-contained breathing apparatus that has a full facepiece and is operated in a pressure-demand or other positive-pressure mode); or SAF:PD,PP:ASCBA (any supplied-air respirator that has a full facepiece and is operated in a pressure-demand or other positive-pressure mode in combination with an auxiliary, self-contained breathing apparatus operated in a pressure-demand or other positive-pressure mode).

Escape: GMFOV [any air-purifying, full-facepiece respirator (gas mask) with a chin-style, front-or back-mounted organic vapor canister]; or SCBAE (any appropriate escape-type, self-contained breathing apparatus).

CHEMICAL REACTIVITY

Contact with oxidizers may cause fire and explosions. Reacts vigorously with water, forming hydrochloric acid. Attacks metals in the presence of moisture.

FIRE INFORMATION

Combustion products include toxic and corrosive hydrogen chloride. Firefighting gear (including SCBA) does not provide adequate protection. If exposure occurs, remove and isolate gear immediately and thoroughly decontaminate personnel. Vapors are heavier than air and will collect in low areas. Vapors in confined areas may explode when exposed to fire. Vapors may travel long distances to ignition sources and flash back. If material or contaminated runoff enters waterways, notify downstream users of potentially contaminated waters. Notify local health and fire officials and pollution control agencies. Storage containers and parts of containers may rocket great distances, in many directions. Material reacts with water but can be extinguished with low- or medium-expansion AFFF foam or dry chemicals if available in sufficient amounts.

NFPA Hazard Classifications:

Health Hazards (Blue) 3
Flammability (Red) 3
Reactivity (Yellow) 2
Special situations .... Water

Flash Point: 95°F (35°C) o.c.

FIRST AID

Move victim to fresh air. Call emergency medical care. Apply artificial respiration if victim is not breathing. If breathing is difficult, give oxygen. Remove and isolate contaminated clothing and shoes. In case of contact with substance, immediately flush skin or eyes with running water for at least 20 minutes. Wash skin with soap and water. Keep victim warm and quiet. Ensure that medical personnel are aware of the material(s) involved and take precautions to protect themselves.

ALUMINUM CHLORIDE A:120

SYNONYMS: ALLUMINIO (CLORURO DI); ALUMINUMCHLORID (German); ALUMINUM CHLORIDE (1:3); ALUMINUM CHLORIDE, ANHYDROUS; ALUMINUM CHLORIDE SOLUTION; ALUMINUM TRICHLORIDE; ANHYDROL FORTE; ANHYDROUS ALUMINUM CHLORIDE; CHLORURE d'ALUMINUM (French); CLORATO ALUMINICO (Spanish); DRICLOR; EEC No. 013-003-00-7; EINECS No. 231-208-1; PAC; PAC (VAN); PEARSALL; PRAESTOL K2001; TRICHLORO ALUMINUM

IDENTIFICATION

CAS:	7446-70-0
ERG ID:	UN 1726 (anhydrous); UN 2581 (solution)
ERG Guide:	137 (anhydrous); 154 (solution)
ERG Guide Category:	Anhydrous: WATER REACTIVE -CORROSIVE Solution: TOXIC AND/OR CORROSIVE (NONCOMBUSTIBLE)
Molecular Formula:	$AlCl_3$
RTECS No.:	BDO525000
Properties:	Solid crystals or powder. White, gray, yellow, or green. Strong, acidic, irritating odor like hydrochloric acid. Sinks in water and reacts, forming toxic and corrosive hydrogen chloride gas.

HEALTH AND SAFETY INFORMATION

Long-term exposure may affect the developing fetus. Estimated lethal dose is about 8 oz for an average 150-lb adult.

IDLH: 50 ppm as hydrogen chloride
Odor Threshold: 6.31ppm as hydrogen chloride

Exposure Limits:	**TWA**
ACGIH:	2 mg/m^3 as Al
NIOSH:	2 mg/m^3
OSHA:	None

RECOMMENDATIONS FOR RESPIRATOR SELECTION: NIOSH/OSHA as hydrogen chloride

50 ppm: CCRS (any chemical cartridge respirator with cartridge(s) providing protection against the compound of concern); or GMFS [any air-purifying, full-facepiece respirator (gas mask) with a chin-style, front-or back-mounted canister providing protection against the compound of concern]; or PAPRS [any powered, air-purifying respirator with cartridge(s) providing protection against the compound of concern]; or SA (any supplied-air respirator); or SCBAF (any self-contained breathing apparatus with a full facepiece).

Emergency or Planned Entry into Unknown Concentrations or IDLH Conditions: SCBAF:PD,PP (any self-contained breathing apparatus that has a full facepiece and is operated in a pressure-demand or other positive-pressure mode); or SAF:PD,PP:ASCBA (any supplied-air respirator that has a full facepiece and is operated in a pressure-demand or other positive-pressure mode in combination with an auxiliary, self-contained breathing apparatus operated in a pressure-demand or other positive-pressure mode).

Escape: GMFAG [any air-purifying, full-facepiece respirator (gas mask) with a chin-style, front-or

back-mounted organic vapor canister]; or SCBAE (any appropriate escape-type, self-contained breathing apparatus).

Note: Substance reported to cause eye irritation or damage; may require eye protection.

CHEMICAL REACTIVITY

Contact with air or water forms corrosive hydrochloric acid and toxic hydrogen chloride gas. Reaction with water may be violent. Water, alcohol, and alkenes cause polymerization. Incompatible with nitrobenzene, organic material, and bases. Attacks metal in presence of moisture, forming flammable hydrogen gas.

FIRE INFORMATION

May burn but does not readily ignite. If material or contaminated runoff enters waterways, notify downstream users of potentially contaminated waters. Notify local health and fire officials and pollution control agencies. Use dry chemicals, sand, carbon dioxide. Do not use water (forms poisonous and corrosive fumes of hydrochloric acid.

FEMA Hazard Classifications:

Health Hazards (Blue) 3
Flammability (Red) 0
Reactivity (Yellow) 2
Special situations . . . . Water

FIRST AID

ANHYDROUS: Move victim to fresh air. Call emergency medical care. Apply artificial respiration if victim is not breathing. Do not use mouth-to-mouth method if victim ingested or inhaled the substance; induce artificial respiration with the aid of a pocket mask equipped with a one-way valve or other proper respiratory medical device. Administer oxygen if breathing is difficult. Remove and isolate contaminated clothing and shoes. In case of contact with substance, immediately flush skin or eyes with running water for at least 20 minutes. For minor skin contact, avoid spreading material on unaffected skin. Removal of solidified molten material from skin requires medical assistance. Keep victim warm and quiet. Effects of exposure (inhalation, ingestion, or skin contact) to substance may be delayed.

SOLUTION: Move victim to fresh air; call emergency medical care. Apply artificial respiration if victim is not breathing. Do not use mouth-to-mouth method if victim ingested or inhaled the substance; induce artificial respiration with the aid of a pocket mask equipped with a one-way valve or other proper respiratory medical device. Administer oxygen if breathing is difficult. Remove and isolate contaminated clothing and shoes. In case of contact with material, immediately flush skin or eyes with running water for at least 20 minutes. For minor skin contact, avoid spreading material on unaffected skin. Keep victim warm and quiet. Effects of exposure (inhalation, ingestion, or skin contact) to substance may be delayed. Ensure that medical personnel are aware of the material(s) involved and take precautions to protect themselves.

ALUMINUM HYDRIDE A:140

SYNONYMS: ALANE; ALUMINUM TRIHYDRIDE; α-ALUMINUM TRIHYDRIDE; HIDURO ALUMINICO (Spanish)

IDENTIFICATION

CAS:	7784-21-6
ERG ID:	UN 2463
ERG Guide:	138
ERG Guide Category:	SUBSTANCES-WATER REACTIVE (EMITTING FLAMMABLE GAS)
Molecular Formula:	AlH_3
RTECS No.:	BD0930000
Properties:	Flammable powder or microcrystalline solid. White. Sinks in water; reacts violently, forming flammable and explosive hydrogen gas. Spontaneously flammable in air.

HEALTH AND SAFETY INFORMATION

Although no toxic effects have been reported, metal hydrides of some metals are extremely toxic.

Exposure Limits: **TWA** (as soluble salts)

ACGIH:	2 mg(Al)/m^3
NIOSH:	2 mg(Al)/m^3
OSHA:	None

RECOMMENDATIONS FOR RESPIRATOR SELECTION:

At any concentrations above the NIOSH REL, or where there is no REL, at any detectable concentration: SCBAF:PD,PP (any self-contained breathing apparatus that has a full facepiece and is operated in a pressure-demand or other positive-pressure mode); or SAF:PD,PP:ASCBA (any supplied-air respirator that has a full facepiece and is operated in a pressure-demand or other positive-pressure mode in combination with an auxiliary self-contained breathing apparatus operated in a pressure-demand or other positive-pressure mode).

Escape: GMFOV [any air-purifying, full-facepiece respirator (gas mask) with a chin-style, front-or back-mounted organic vapor canister]; or SCBAE (any appropriate escape-type, self-contained breathing apparatus).

CHEMICAL REACTIVITY

Dangerously unstable material. Spontaneously flammable in air or oxygen. Contact with moisture, acid or acid fumes forms flammable and explosive hydrogen gas. Contact with oxidizers or tetrazole compounds (e.g., aminotetrazole) causes fire and explosions.

FIRE INFORMATION

Vapors are heavier than air and will collect in low areas. Vapors in confined areas may explode when exposed to fire. If material or contaminated runoff enters waterways, notify downstream users of potentially contaminated waters. Notify local health and fire officials and pollution control agencies. Use dry chemicals, alcohol-resistant foam, carbon dioxide. From a secure explosion-proof location, use water spray to cool exposed containers. If cooling streams are ineffective (venting sound increases in volume and pitch, tank discolors, or shows any signs of deforming), withdraw immediately to a secure position.

FIRST AID

Move victim to fresh air. Call emergency medical care. Apply artificial respiration if victim is not breathing. Administer oxygen if breathing is difficult. Remove and isolate contaminated clothing and shoes. In case of contact with substance, immediately flush skin or eyes with running water for at least 20 minutes. Keep victim warm and quiet. Ensure that medical personnel are aware of the material(s) involved and take precautions to protect themselves.

ALUMINUM METAL A:160

SYNONYMS: A 999; A 00; A 995; A 99; A 95; AA 1199; AA 1099; AD 1; AD1M; ADO; AE; ALAUN (German); ALLBRI ALUMINUM PASTE AND POWDER; ALUMINA FIBRE; ALUMINIO (Spanish); ALUMINIUM FLAKE; ALUMINIUM; ALUMINIUM BRONZE; ALUMINUM DEHYDRATED; ALUMINUM, METALLIC, POWDER; ALUMINUM POWDER; ALUMINUM POWDER, UNCOATED, NON-PYROPHORIC (DOT); ALUMINUM 27; AO A1; AR2; AV00; AV000; C.I. 77000; EMANAY ATOMIZED ALUMINUM POWDER; JISC 3108; JISC 3110; L16; METANA; METANA ALUMINUM PASTE; NORAL ALUMINUM; NORAL EXTRA FINE LINING GRADE; NORAL NON-LEAFING GRADE; NORAL INK GRADE ALUMINUM; PAP-1

IDENTIFICATION

CAS:	7429-90-5
ERG ID:	UN 1309 (aluminum powder, coated); UN 1396 (aluminum powder, uncoated); UN 9260 (molten)
ERG Guide:	170 (aluminum powder, coated); 138 (aluminum powder, uncoated); 169 (molten)
ERG Guide Category:	Aluminum powder, coated: METALS (POWDERS, DUST, SHAVINGS, BORINGS, TURNINGS, CUTTINGS, ETC.) Aluminum powder, uncoated: WATER-REACTIVE (EMITTING FLAMMABLE GASES) Aluminum (molten): ALUMINUM (MOLTEN)
Molecular Formula:	Al
RTECS No.:	BD0330000
Properties:	Silvery-white, malleable, ductile, odorless metal. Powder and dust have metallic odor. Flammable solid if finely divided; powder easily ignited.

HEALTH AND SAFETY INFORMATION

Powder can irritate respiratory system. Aluminum has been linked to Alzheimer's disease. It is unknown whether aluminum causes the disease or whether the buildup of aluminum occurs in people who already have the disease. May cause scarring of the lungs (pulmonary fibrosis) cough, shortness of breath, bone disease, or brain damage.

OSHA Table Z-1-A Air Contaminant

Exposure Limits:	**TWA**
ACGIH:	10 mg/m^3 as metal dust
NIOSH:	10 mg/m^3 (total); 5 mg/m^3 (respiratory fraction)
OSHA:	15 mg/m^3 (total); 5 mg/m^3 (respiratory fraction)

RECOMMENDATIONS FOR RESPIRATOR SELECTION: NIOSH

At any concentrations above the NIOSH REL, or where there is no REL, at any detectable concentration: SCBAF:PD,PP (any self-contained breathing apparatus that has a full facepiece and is operated in a pressure-demand or other positive-pressure mode); or SAF:PD,PP:ASCBA (any supplied-air respirator that has a full facepiece and is operated in a pressure-demand or other positive-pressure mode in combination with an auxiliary, self-contained breathing apparatus operated in a pressure-demand or other positive-pressure mode).

Escape: GMFOV [any air-purifying, full-facepiece respirator (gas mask) with a chin-style, front-or back-mounted organic vapor canister]; or SCBAE (any appropriate escape-type, self-contained breathing apparatus).

CHEMICAL REACTIVITY

Aluminum powder is a strong reducing agent; violent reaction with many compounds, including oxidizers, acids, caustics, halogens, halogenated hydrocarbons, nitrates, sulfates, mercury and mercury compounds, methyl chloride, metal oxides, trichloroethylene, and oxygen. May accumulate static electrical charges from friction or stirring, and self-ignite.

FIRE INFORMATION: (Fume or dust)

If material or contaminated runoff enters waterways, notify downstream users of potentially contaminated waters. Notify local health and fire officials and pollution control agencies. Use dry chemicals, alcohol-resistant foam, or carbon dioxide. From a secure explosion-proof location, use water spray to cool exposed containers.

FEMA Hazard Classifications:

Health Hazards (Blue) 0

Flammability (Red) 3
Reactivity (Yellow) 1

Autoignition Temperature: 1,400°F (760°C)

FIRST AID

Move victim to fresh air. Call emergency medical care. Apply artificial respiration if victim is not breathing. Administer oxygen if breathing is difficult. Remove and isolate contaminated clothing and shoes. In case of contact with substance, immediately flush skin or eyes with running water for at least 20 minutes. Keep victim warm and quiet. Ensure that medical personnel are aware of the material(s) involved and take precautions to protect themselves.

ALUMINUM (MOLTEN): Move victims to fresh air. Call emergency medical care. Apply artificial respiration if victim is not breathing. Administer oxygen if breathing is difficult. For severe burns, immediate medical attention is required. Removal of solidified molten material from skin requires medical assistance. Remove and isolate contaminated clothing and shoes. In case of contact with substance, immediately flush skin or eyes with running water for at least 20 minutes. In case of contact with liquefied gas, thaw frosted parts with lukewarm water. Keep victim warm and quiet. Keep victim under observation. Ensure that medical personnel are aware of the material(s) involved and take precautions to protect themselves.

ENVIRONMENTAL INFORMATION: (Fume or dust)

Clean Water Act: Section 313 Priority Chemicals.
EPA Hazardous Waste Number (RCRA No.): D003.
Safe Drinking Water Act: SMLC, 0.05 to 0.2 mg/L; Priority List (55 FR1470).
EPCRA Section 313: (Reportable only in fume or dust form): Form R de minimis concentration reporting level: 1.0%.

ALUMINUM OXIDE A:180

SYNONYMS: A-1 (SORBENT); A-2; ACTIVATED ALUMINUM OXIDE; ALCAN AA-100; ALCAN C-70; ALCAN C-71; ALCAN C-72; ALCAN C-73; ALCOA F1; ALEXITE; ALMITE; ALON; ALON C; ALOXITE; ALUFRIT; ALUMINA; α-ALUMINA; β-ALUMINA; γ-ALUMINA; ALUMINITE 37; ALUMINUM OXIDE (2:3); α-ALUMINUM OXIDE; β-ALUMINUM OXIDE; γ-ALUMINUM OXIDE; ALUMINUM OXIDE C; ALUMINUM SESQUIOXIDE; ALUMINUM TRIOXIDE; ALUMITE; ALUNDUM; ALUNDUM 600; BACKLAP SLURRY; BAUXITE; BAYERITE; BOEHMITE; BRASIVOL; BROCKMANN, ALUMINUM OXIDE; C-1; CAB-O-GRIP; CATAPAL S; COMPALOX; CONOPAL; CORUNDUM; D 201; DIALUMINUM TRIOXIDE; DIASPORE DIRUBIN; DISPAL; DISPAL ALUMINA; DISPAL M; DOTMENT 324; DOTMENT 358; DURAL; DYCRON; EINECS No. 215-691-6; EMERY; EXOLON XW 60; F 360 (ALUMINA); FASERTON; FASERTONERDE; FAST CURE 45 EPOXY; FLAME GUARD; G2 (OXIDE); GIBBSITE; GK (OXIDE); GO (OXIDE); HYPALOX II; ITACLOR; JUBENON R; KA 101; KETJEN B; KHP 2; LA 6; LUCALOX; LUDOX CL;

MAFTECMARTIPOL; MARTISORB; MARTOXIN; MICROGRIT WCA; MICROPOLISH ALUMINA; OXIDO ALUMINICO (Spanish); PS-1; PS-1 (ALUMINA); PURDOX; Q-LOID A 30; RC 172DBM; REALOX

IDENTIFICATION

CAS:	1344-28-1
ERG Guide Category:	IRRITANT
Molecular Formula:	Al_2O_3
RTECS No.:	BD1200000
Properties:	Crystalline powder, or solid. White. Odorless. Insoluble in water.

HEALTH AND SAFETY INFORMATION

Irritates eyes and respiratory system. Exposure is usually accompanied by exposure to silicates and iron compounds that are often linked to respiratory diseases, including emphysema. Repeated exposure to high levels may cause pulmonary fibrosis with shortness of breath; this condition can be fatal.

OSHA Table Z-1-A Air Contaminant (fibrous form)

Exposure Limits:	**TWA**
ACGIH:	10 mg/m^{3*}
NIOSH:	10 mg/m^3 (Al) as total dust; 5 mg/m^3 respirable fraction.
OSHA:	15 mg/m^3 (Al) as total dust; 5 mg/m^3 respirable fraction.

* The value for inhalable (total) particulate matter containing no asbestos and less than 1% crystalline silica.

RECOMMENDATIONS FOR RESPIRATOR SELECTION: NIOSH

At any concentrations above the NIOSH REL, or where there is no REL, at any detectable concentration: SCBAF:PD,PP (any self-contained breathing apparatus that has a full facepiece and is operated in a pressure-demand or other positive-pressure mode); or SAF:PD,PP:ASCBA (any supplied-air respirator that has a full facepiece and is operated in a pressure-demand or other positive-pressure mode in combination with an auxiliary self-contained breathing apparatus operated in a pressure-demand or other positive-pressure mode).

Escape: GMFOV [any air-purifying, full-facepiece respirator (gas mask) with a chin-style, front-or back-mounted organic vapor canister]; or SCBAE (any appropriate escape-type, self-contained breathing apparatus).

CHEMICAL REACTIVITY

Reacts violently with chlorine trifluoride and may cause violent polymerization of ethylene oxide. Reacts slowly with water to form aluminum.

FIRE INFORMATION

If material or contaminated runoff enters waterways, notify downstream users of potentially contaminated water.

FEMA Hazard Classifications:

Health Hazards (Blue) 0
Flammability (Red) 0
Reactivity (Yellow) 0

FIRST AID

EYE: If this chemical contacts the eyes, immediately wash the eyes with large amounts of water, occasionally lifting the lower and upper lids. Get medical attention immediately. Contact lenses should not be worn when working with this chemical.
SKIN: If this chemical contacts the skin, immediately wash the contaminated skin with soap and water. If this chemical penetrates the clothing, immediately remove clothing, wash the skin with soap and water, and get medical attention promptly.
BREATH: If a person breathes large amounts of this chemical move exposed person to fresh air at once. If breathing has stopped, perform mouth-to-mouth resuscitation. Keep the affected person warm and at rest. Get medical attention as soon as possible.
SWALLOW: If this chemical has been swallowed, get medical attention immediately.

ENVIRONMENTAL INFORMATION

EPCRA Section 313: (Reportable in fibrous forms only): Form R *de minimis* concentration reporting level: 1.0%

AMINOETHYLETHANOLAMINE A:186

SYNONYMS: 2-(2-AMINOETHOXY) ETHANOL; N-(2-AMINOETHYL) ETHANOLAMINE; 2-[(2-AMINOETHYL)AMINO]ETHANOL; N-AMINOETHYLETHANOL AMINE; AMINOETILETANOLAMINA (Spanish); EINECS No. 203-867-5; 1-(2-HYDROXYETHYLAMINO)-2-AMINOETHANE; 2-(2-HYDROXYETHYLAMINO)ETHYLAMINE; HYDROXYETHYLETHYLENEDIAMINE, N-B-; N-HYDROXETHYL-1,2-ETHANEDIAMINE; N-B-HYDROXYETHYLETHYLENEDIAMINE

IDENTIFICATION

CAS:	111-41-1
ERG ID:	UN 3055
ERG Guide:	153

ERG Guide Category: SUBSTANCES-TOXIC AND/OR CORROSIVE (COMBUSTIBLE)
Molecular Formula: $C_4H_{12}N_2O$
RTECS No.: KJ6300000
Properties: Combustible liquid. Colorless. Mild ammonia odor. Sinks and mixes with water.

HEALTH AND SAFETY INFORMATION

Corrosive. Irritates the skin, eyes, and respiratory tract. May cause pulmonary edema, which can be delayed for several hours. Prolonged or repeated exposure may cause skin problems.

At any concentrations above the NIOSH REL, or where there is no REL, at any detectable concentration: SCBAF:PD,PP (any self-contained breathing apparatus that has a full facepiece and is operated in a pressure-demand or other positive-pressure mode); or SAF:PD,PP:ASCBA (any supplied-air respirator that has a full facepiece and is operated in a pressure-demand or other positive-pressure mode in combination with an auxiliary, self-contained breathing apparatus operated in a pressure-demand or other positive-pressure mode).

Escape: GMFOV [any air-purifying, full-facepiece respirator (gas mask) with a chin-style, front-or back-mounted organic vapor canister]; or SCBAE (any appropriate escape-type, self-contained breathing apparatus).

CHEMICAL REACTIVITY

Reacts with oxidizers. Incompatible with copper, copper alloys, aluminum, and zinc.

FIRE INFORMATION

Heat causes decomposition, giving off toxic vapors. Vapors are heavier than air and will collect in low areas. Vapors in confined areas may explode when exposed to fire. If material or contaminated runoff enters waterways, notify downstream users of potentially contaminated waters. Notify local health and fire officials and pollution control agencies. Use dry chemicals, alcohol-resistant foam, carbon dioxide. From a secure explosion-proof location, use water spray to cool exposed containers. If cooling streams are ineffective (venting sound increases in volume and pitch, tank discolors, or shows any signs of deforming), withdraw immediately to a secure position.

NFPA Hazard Classifications:

Health Hazards (Blue) 2
Flammability (Red) 1
Reactivity (Yellow) 0

Flash Point: 275°F (135°C) o.c.
Flammable Limits in Air: LEL 15%; UEL 27%
Autoignition Temperature: 695°F (368°C)

FIRST AID

Move victim to fresh air and call emergency medical care. Apply artificial respiration if victim is not breathing. Do not use mouth-to-mouth method if victim swallowed or inhaled the substance; induce artificial respiration with the aid of a pocket mask equipped with a one-way valve or other proper respiratory medical device. If breathing is difficult, give oxygen. In case of contact with substance, immediately flush skin or eyes with running water for at least 20 minutes. Remove and isolate contaminated clothing and shoes. Keep victim warm and quiet. Effects of exposure (inhalation, ingestion, or skin contact) to substances may be delayed. Ensure that medical personnel are aware of the material(s) involved and take precautions to protect themselves.

AMINOETHYLPIPERAZINE A:192

SYNONYMS: N-AMINOETILPIPERAZINA (Spanish); N-AMINOETHYLPIPERAZINE; N-(β-AMINOETHYL)PIPERAZINE; N-(2-AMINOETHYL)PIPERAZINE; 1-(2-AMINOETHYL)PIPERAZINE; EINECS No. 205-411-0

IDENTIFICATION

CAS:	140-31-8
ERG ID:	UN 2815
ERG Guide:	153
ERG Guide Category:	SUBSTANCES-TOXIC AND/OR CORROSIVE (COMBUSTIBLE)
Molecular Formula:	$C_6H_{15}N_3$
RTECS No.:	TK8050000
Properties:	Combustible liquid. Light colored. Soluble in water.

HEALTH AND SAFETY INFORMATION

May cause irritation to the eyes and respiratory system. Mutation data reported.

RECOMMENDATIONS FOR RESPIRATOR SELECTION:

At any concentrations above the NIOSH REL, or where there is no REL, at any detectable concentration: SCBAF:PD,PP (any self-contained breathing apparatus that has a full facepiece and is operated in a pressure-demand or other positive-pressure mode); or SAF:PD,PP:ASCBA (any supplied-air respirator that has a full facepiece and is operated in a pressure-demand or other positive-pressure mode in combination with an auxiliary self-contained breathing apparatus operated in a pressure-demand or other positive-pressure mode).

Escape: GMFOV [any air-purifying, full-facepiece respirator (gas mask) with a chin-style, front-or back-mounted organic vapor canister]; or SCBAE (any appropriate escape-type, self-contained breathing apparatus).

CHEMICAL REACTIVITY

Contact with oxidizers may cause fire and explosions. Solution is a strong base. Forms explosive mixture with air. Incompatible with nonoxidizing mineral acids, strong acids, organic acids, organic anhydrides, isocyanates, vinyl acetate, acrylates, substituted allyls, alkylene oxides, epichlorohydrin, ketones, aldehydes, alcohols, glycols, phenols, cresols, caprolactum solution, strong oxidizers. Contact with copper alloys, zinc, or galvanized steel may cause violent reaction.

FIRE INFORMATION

Vapors are heavier than air and will collect in low areas. Vapors in confined areas may explode when exposed to fire.If material or contaminated runoff enters waterways, notify downstream users of potentially contaminated waters. Notify local health and fire officials and pollution control agencies. Use dry chemicals, alcohol-resistant foam, carbon dioxide. From a secure explosion-proof location, use water spray to cool exposed containers. If cooling streams are ineffective (venting sound increases in volume and pitch, tank discolors or shows any signs of deforming), withdraw immediately to a secure position.

NFPA Hazard Classifications:

Health Hazards (Blue) 2
Flammability (Red) 2
Reactivity (Yellow) 0

Flash Point: 200°F (93°C) o.c.

FIRST AID

Move victim to fresh air; call emergency medical care. Apply artificial respiration if victim is not breathing. Do not use mouth-to-mouth method if victim ingested or inhaled the substance; induce artificial respiration with the aid of a pocket mask equipped with a one-way valve or other proper respiratory medical device. Administer oxygen if breathing is difficult. Remove and isolate contaminated clothing and shoes. In case of contact with substance, immediately flush skin or eyes with running water for at least 20 minutes. For minor skin contact, avoid spreading material on unaffected skin. Keep victim warm and quiet. Effects of exposure (inhalation, ingestion, or skin contact) to substance may be delayed. Ensure that medical personnel are aware of the material(s) involved and take precautions to protect themselves.

AMMONIA A:200

SYNONYMS: AM-FOL; AMMONIA, ANHYDROUS; AM-FOL; AMMONIAC (French); AMMONIACA (Italian); AMMONIA GAS; AMMONIALE (German); AMONIACO (Spanish); AMONIAK (Polish); ANHYDROUS AMMONIA; AQUA AMMONIA; CLEARVIEW GLASS CLEANER; DAXAD-32S; EEC No. 007-001-00-5 (anhydrous); EINECS No. 231-635-3; HIGH SPEED ENDURA-ETCH STARTER; LIQUID AMMONIA; NF SOLDER STRIPPER 3114B; NITRO-SIL; PRO 330 CLEAR THIN SPREAD; R717; REFRIGERANT R717; SPIRIT OF HARTSHORN; SUPER SOLDER STRIP 1805

IDENTIFICATION

CAS:	7664-41-7 (anhydrous); 1336-21-6 (25% solution in water)
ERG ID:	UN 1005 (anhydrous; more than 50% ammonia solution) UN 2672 (10-35% ammonia solution) UN 2073 (more than 35% but not more than 50% ammonia solution)
ERG Guide:	125 (anhydrous; more than 50% ammonia solution); 154 (10-35% ammonia solution); 125 (more than 35% but not more than 50% ammonia solution)
ERG Guide Category:	Anhydrous; more than 50% ammonia: GASES-CORROSIVE 10-35% ammonia: TOXIC AND/OR CORROSIVE (NONCOMBUSTIBLE); More than 35% but not more than 50% ammonia: GASES CORROSIVE
Molecular Formula:	H_3N
RTECS No.:	BO0875000
Properties:	Anhydrous ammonia is a colorless, highly irritating gas at room temperature with a pungent, suffocating odor. It is lighter than air and flammable at high concentrations and temperatures. Easily compressed, it forms a clear, colorless liquid under pressure. Readily dissolves in water, forming ammonium hydroxide, an alkaline, corrosive solution. Floats and "boils" on water. Poisonous, visible vapor cloud is produced. Gas is lighter than air; hugs the ground when cool.

HEALTH AND SAFETY INFORMATION

Contact with liquid may cause frostbite. Corrosive to skin, eyes, nose, throat, and lungs. There is no antidote for ammonia poisoning. Chemical-protective clothing is recommended because ammonia can cause skin irritation and burns. Repeated exposure may cause chronic irritation of the respiratory tract. Chronic inflammation of the bronchi and airway hyperactivity have been reported. Chronic irritation of the eye membranes has also been reported.

IDLH: 300 ppm
Odor Threshold: 5.75 ppm
OSHA Table Z-1-A Air Contaminant
OSHA Process Safety Management of Highly Hazardous Chemicals: (29 CFRPart 1910.119, Appendix A) TQ = anhydrous 10,000 lb; solutions(>44% ammonia by weight) 15,000 lb
NIOSH Criteria Document: NIOSH 74-136

Exposure Limits:	**TWA**	**STEL**
ACGIH:	25 ppm; 17 mg/m^3	35 ppm; 24 mg/m^3
NIOSH:	25 ppm; 18 mg/m^3	35 ppm; 27 mg/m^3
OSHA:	50 ppm; 35 mg/m^3	

RECOMMENDATIONS FOR RESPIRATOR SELECTION: NIOSH

250 ppm: CCRS (any chemical cartridge respirator with cartridge(s) providing protection against the compound of concern); or SA (any supplied-air respirator).
300 ppm: SA:CF (any supplied-air respirator operated in a continuous-flow mode); or PAPRS (any powered, air-purifying respirator with cartridge(s) providing protection against the compound of concern); or CCRFS (any chemical cartridge respirator with a full facepiece and cartridge(s) providing protection against the compound of concern); or GMFS [any air-purifying, full-facepiece respirator (gas mask) with a chin-style, front-or back-mounted canister providing protection against the compound of concern]; or SCBAF (any self-contained breathing apparatus with a full facepiece); or SAF (any supplied-air respirator with a full facepiece).

Emergency or Planned Entry into Unknown Concentrations or IDLH Conditions: SCBAF:PD,PP (any self-contained breathing apparatus that has a full facepiece and is operated in a pressure-demand or other positive-pressure mode); or SAF:PD,PP:ASCBA (any supplied-air respirator that has a full facepiece and is operated in a pressure-demand or other positive-pressure mode in combination with an auxiliary, self-contained breathing apparatus operated in a pressure-demand or other positive-pressure mode).

Escape: GMFS [any air-purifying, full-facepiece respirator (gas mask) with a chin-style, front-or back-mounted canister providing protection against the compound of concern]; or SCBAE (any appropriate escape-type, self-contained breathing apparatus).

Note: Substance reported to cause eye irritation or damage; may require eye protection.

CHEMICAL REACTIVITY

Violent reaction with strong oxidizers and acids. Shock-sensitive compounds may be formed with gold, halogens, mercury, mercury oxide, and silver oxide. Fire and explosions may be caused by trimethylammonium amide, 1-chloro-2,4-dinitrobenzene, o-chloronitrobenzene, platinum, trioxygen difluoride, selenium difluoride dioxide, boron halides, mercury, chlorine, iodine, bromine, hypochlorites, chlorine bleach, amides, organic anhydrides, isocyanates, vinyl acetate, alkylene oxides, epichlorohydrin, and aldehydes. Attacks some coatings, plastics, and rubber. Attacks copper, brass, bronze, aluminum, steel, tin, zinc, and their alloys.

FIRE INFORMATION

Firefighting gear (including SCBA) does not provide adequate protection. If exposure occurs, remove and isolate gear immediately and thoroughly decontaminate personnel. Vapors are heavier than air and will collect in low areas. Vapors in confined areas may explode when exposed to fire. Vapors may travel long distances to ignition sources and flash back. Storage containers and parts of containers may rocket great distances, in many directions. If material or contaminated runoff enters waterways, notify downstream users of potentially contaminated waters. Notify local health and fire officials and pollution control agencies. DO NOT PUT WATER ON LIQUID AMMONIA; WILL INCREASE EVAPORATION. Use dry chemicals, carbon dioxide. From a secure explosion-proof location, use water spray to cool exposed containers. If cooling streams are ineffective (venting sound increases in volume and pitch, tank discolors or shows any signs of deforming), withdraw immediately to a secure position.

NFPA Hazard Classifications:

Health Hazards (Blue) 3
Flammability (Red) 1
Reactivity (Yellow) 0

Flash Point: Gas. Indefinite below 32°F (0°C); sometimes difficult to ignite
Flammable Limits in Air: LEL 15%; UEL 28%
Autoignition Temperature: 1204°F (651°C)

FIRST AID

AMMONIA; AMMONIA, ANHYDROUS; MORE THAN 50% AMMONIA: Move victim to fresh air. Call emergency medical care. Apply artificial respiration if victim is not breathing. Do not use mouth-to-mouth method if victim ingested or inhaled the substance; induce artificial respiration with the aid of a pocket mask equipped with a one-way valve or other proper respiratory medical device. Administer oxygen if breathing is difficult. Remove and isolate contaminated clothing and shoes. In case of contact with liquefied gas, thaw frosted parts with lukewarm water. Keep victim warm and quiet. Keep victim under observation. Effects of contact or inhalation may be delayed. Ensure that medical personnel are aware of the material(s) involved and take precautions to protect themselves.

AMMONIA SOLUTION, 10%-35%: Move victim to fresh air; call emergency medical care. Apply artificial respiration if victim is not breathing. Do not use mouth-to-mouth method if victim ingested or inhaled the substance; induce artificial respiration with the aid of a pocket mask equipped with a one-way valve or other proper respiratory medical device. Administer oxygen if breathing is difficult. Remove and isolate contaminated clothing and shoes. In case of contact with substance, immediately flush skin or eyes with running water for at least 20 minutes. For minor skin contact, avoid spreading material on unaffected skin. Keep victim warm and quiet. Effects of exposure (inhalation, ingestion, or skin contact) to substance may be delayed. Ensure that medical personnel are aware of the material(s) involved and take precautions to protect themselves.

AMMONIA SOLUTION, MORE THAN 35% BUT NOT MORE THAN 50% AMMONIA: Move victim to fresh air. Call emergency medical care. Apply artificial respiration if victim is not breathing. Do not use mouth-to-mouth method if victim ingested or inhaled the substance; induce artificial respiration with the aid of a pocket mask equipped with a one-way valve or other proper respiratory medical device. Administer oxygen if breathing is difficult. Remove and isolate contaminated clothing and shoes. In case of contact with liquefied gas, thaw frosted parts with lukewarm water. In case of contact with substance, immediately flush skin or eyes with running water for at least 20 minutes. Keep victim warm and quiet. Keep victim under observation. Effects of contact or inhalation may be delayed. Ensure that medical personnel are aware of the material(s) involved and take precautions to protect themselves.

ENVIRONMENTAL INFORMATION

Clean Air Act: Section 112(r), Accidental Release Prevention/Flammable substances (Section 68.130); (anhydrous) TQ = 10,000 lb (4,540 kg); (concentration equal to or greater than 20%) TQ =20,000 lb (9,150 kg).

Clean Water Act: Section 311 Hazardous Substances/RQ (same as CERCLA); Section 313 Priority Chemicals.
EPCRA Section 302, Extremely Hazardous Substances: TPQ = 500 lb (228 kg).
EPCRA Section 304: Reportable Quantity (RQ): CERCLA, 100 lb (45.4 kg).
EPCRA Section 313: Form R *de minimis* concentration reporting level: 1.0%; includes anhydrous ammonia and aqueous ammonia from water dissociable ammonium salts and other sources; 10% of total aqueous ammonia, and 100% of anhydrous forms of ammonia is reportable under this listing. If a facility manufactures, processes, or otherwise uses anhydrous ammonia or aqueous ammonia, they must report under the ammonia listing. Solutions containing aqueous ammonia at a concentration in excess of 1% of the 10% reportable under this listing should be factored into threshold and release determinations.

AMMONIUM ACETATE A:206

SYNONYMS: ACETATO AMONICO (Spanish); ACETIC ACID, AMMONIUM SALT; EINECS No. 211-468-9

IDENTIFICATION

CAS:	631-61-8
ERG ID:	UN 9079
ERG Guide:	171
ERG Guide Category:	SUBSTANCES (LOW TO MODERATE HAZARD)
Molecular Formula:	$C_2H_7NO_2$
Properties:	Solid. White. Weak ammonia odor. Sinks and mixes with water.

HEALTH AND SAFETY INFORMATION

At any concentrations above the NIOSH REL, or where there is no REL, at any detectable concentration: SCBAF:PD,PP (any self-contained breathing apparatus that has a full facepiece and is operated in a pressure-demand or other positive-pressure mode); or SAF:PD,PP:ASCBA (any supplied-air respirator that has a full facepiece and is operated in a pressure-demand or other positive-pressure mode in combination with an auxiliary self-contained breathing apparatus operated in a pressure-demand or other positive-pressure mode).

Escape: GMFOV [any air-purifying, full-facepiece respirator (gas mask) with a chin-style, front-or back-mounted organic vapor canister]; or SCBAE (any appropriate escape-type, self-contained breathing apparatus).

FIRE INFORMATION

Not flammable . Decomposes in heat, forming toxic oxides of nitrogen and ammonia. If material or contaminated runoff enters waterways, notify downstream users of potentially contaminated waters. Notify local health and fire officials and pollution control agencies. Use extinguishing agent suitable for surrounding fire.

FEMA Hazard Classifications:

Health Hazards (Blue) 1
Flammability (Red) 0
Reactivity (Yellow) 0

FIRST AID

Move victim to fresh air. Call emergency medical care. Apply artificial respiration if victim is not breathing. Administer oxygen if breathing is difficult. In case of contact with substance, immediately flush skin or eyes with running water for at least 20 minutes. Remove and isolate contaminated clothing and shoes. Ensure that medical personnel are aware of the material(s) involved and take precautions to protect themselves.

ENVIRONMENTAL INFORMATION

Clean Water Act: Section 311 Hazardous Substances/RQ (same as CERCLA)
EPCRA Section 304 Reportable Quantity (RQ): CERCLA, 5,000 lb (2,270 kg)
EPCRA Section 313: Source of aqueous ammonia. Molecular weight 77.08. NH_3 equivalent weight 22.09.

AMMONIUM BENZOATE — A:212

SYNONYMS: BENZATO AMONICO (Spanish); BENZOIC ACID, AMMONIUM SALT; EINECS No. 212-468-9; VULNOC AB

IDENTIFICATION

CAS:	1863-63-4
ERG ID:	UN 9080
ERG Guide:	171
ERG Guide Category:	SUBSTANCES (LOW to MODERATE HAZARD)
Molecular Formula:	$C_7H_9NO_2$
RTECS No.:	DG3378000
Properties:	Combustible solid. Colorless. Odorless. Sinks in water and is moderately soluble.

HEALTH AND SAFETY INFORMATION

Irritating to the eyes, skin and respiratory systems.

RECOMMENDATIONS FOR RESPIRATOR SELECTION: NIOSH

At any concentrations above the NIOSH REL, or where there is no REL, at any detectable

concentration: SCBAF:PD,PP (any self-contained breathing apparatus that has a full facepiece and is operated in a pressure-demand or other positive-pressure mode); or SAF:PD,PP:ASCBA (any supplied-air respirator that has a full facepiece and is operated in a pressure-demand or other positive-pressure mode in combination with an auxiliary, self-contained breathing apparatus operated in a pressure-demand or other positive-pressure mode).

Escape: GMFOV [any air-purifying, full-facepiece respirator (gas mask) with a chin-style, front-or back-mounted organic vapor canister]; or SCBAE (any appropriate escape-type, self-contained breathing apparatus).

CHEMICAL REACTIVITY

Contact with oxidizers may cause fire and explosions.

FIRE INFORMATION

Combustion or decomposition products upon heating include toxic nitrogen oxides and ammonia. Vapors are heavier than air and will collect in low areas. Vapors in confined areas may explode when exposed to fire. If material or contaminated runoff enters waterways, notify downstream users of potentially contaminated waters. Notify local health and fire officials and pollution control agencies. Use dry chemicals, alcohol-resistant foam, carbon dioxide. From a secure explosion-proof location, use water spray to cool exposed containers. If cooling streams are ineffective (venting sound increases in volume and pitch, tank discolors, or shows any signs of deforming), withdraw immediately to a secure position.

FEMA Hazard Classifications:

Health Hazards (Blue) 2
Flammability (Red) 1
Reactivity (Yellow) 0

FIRST AID

Move victim to fresh air. Call emergency medical care. Apply artificial respiration if victim is not breathing. Administer oxygen if breathing is difficult. In case of contact with substance, immediately flush skin or eyes with running water for at least 20 minutes. Remove and isolate contaminated clothing and shoes. Ensure that medical personnel are aware of the material(s) involved and take precautions to protect themselves.

ENVIRONMENTAL INFORMATION

Clean Water Act: Section 311 Hazardous Substances/RQ (same as CERCLA)
EPCRA Section 304 Reportable Quantity (RQ): CERCLA, 5,000 lb (2,270 kg)
EPCRA Section 313: Form R *de minimis* concentration reporting level:1.0%. Source of aqueous ammonia. Molecular weight 139.15. NH_3 equivalent weight 12.24.

AMMONIUM BIFLUORIDE A:220

SYNONYMS: ACID AMMONIUM FLUORIDE; AMMONIUM ACID FLUORIDE; AMMONIUM BIFLUORIDE, HYDROGEN; AMMONIUM DIFLUORIDE, HYDROGEN; AMMONIUM FLUORIDE, HYDROGEN; AMMONIUM HYDROGEN FLUORIDE; AMMONIUM HYDROGEN DIFLUORIDE; AMMONIUM HYDROFLUORIDE; BIFLUORURO AMONICO (Spanish); EINECS No. 215-676-4; SAFE STEP CLEANER DEGREASER

IDENTIFICATION

CAS:	1341-49-7
ERG ID:	UN 1727 (solid); UN 2817 (solution)
ERG Guide:	154 (solid and solution)
ERG Guide Category:	TOXIC AND/OR CORROSIVE (NONCOMBUSTIBLE)
Molecular Formula:	F_2H_5N
RTECS No.:	BQ9200000
Properties:	Corrosive solid or cloudy liquid. White. Odorless solid or pungent solution. Sinks and mixes with water. Poisonous gases may be produced in fire.

HEALTH AND SAFETY INFORMATION

Extremely corrosive. Repeated exposure to fluoride chemicals may cause stiffness in muscles or ligaments and even crippling.

IDLH: 250 mg/m^3 as F
NIOSH Criteria Document: NIOSH 76-103 INORGANIC FLUORIDE

Exposure Limits: TWA

ACGIH:	2.5 mg/m^3 as F
NIOSH:	2.5 mg/m^3 as F
OSHA:	2.5 mg/m^3 as F

RECOMMENDATIONS FOR RESPIRATOR SELECTION: NIOSH/OSHA as fluorine

1 ppm: SA* (any supplied-air respirator).
2.5 ppm: SA:CF* (any supplied-air respirator operated in a continuous-flow mode).
5 ppm: SCBAF (any self-contained breathing apparatus with a full facepiece); or SAF (any supplied-air respirator with a full facepiece).
25 ppm: SAF:PD,PP (any supplied-air respirator that has a full facepiece and is operated in a pressure-demand or other positive-pressure mode).

Emergency or Planned Entry into Unknown Concentrations or IDLH Conditions:
SCBAF:PD,PP (any self-contained breathing apparatus that has a full facepiece and is operated in a pressure-demand or other positive-pressure mode); or SAF:PD,PP:ASCBA (any supplied-air respirator

that has a full facepiece and is operated in a pressure-demand or other positive-pressure mode in combination with an auxiliary self-contained breathing apparatus operated in a pressure-demand or other positive-pressure mode).

Escape: GMFS [end of service life indicator (ESLI) required] (any air-purifying, full-facepiece respirator (gas mask) with a chin-style, front-or back-mounted canister providing protection against the compound of concern]; or SCBAE (any appropriate escape-type, self-contained breathing apparatus).

*Substance reported to cause eye irritation or damage; may require eye protection.

CHEMICAL REACTIVITY

Incompatible with alkalies and water (forms hydrochloric acid). Attacks glass, cement, and most metals. Flammable hydrogen gas may collect in enclosed spaces.

FIRE INFORMATION

If material or contaminated runoff enters waterways, notify downstream users of potentially contaminated waters. Notify local health and fire officials and pollution control agencies. Use any extinguishing agent suitable for surrounding fire. From a secure explosion-proof location, use water spray to cool exposed containers. If cooling streams are ineffective (venting sound increases in volume and pitch, tank discolors or shows any signs of deforming), withdraw immediately to a secure position.

FEMA Hazard Classifications:

Health Hazards (Blue) 3
Flammability (Red) 0
Reactivity (Yellow) 0

FIRST AID

Move victim to fresh air; call emergency medical care. Apply artificial respiration if victim is not breathing. Do not use mouth-to-mouth method if victim ingested or inhaled the substance; induce artificial respiration with the aid of a pocket mask equipped with a one-way valve or other proper respiratory medical device. Administer oxygen if breathing is difficult. Remove and isolate contaminated clothing and shoes. In case of contact with substance, immediately flush skin or eyes with running water for at least 20 minutes. For minor skin contact, avoid spreading material on unaffected skin. Keep victim warm and quiet. Effects of exposure (inhalation, ingestion, or skin contact) to substance may be delayed. Ensure that medical personnel are aware of the material(s) involved and take precautions to protect themselves.

ENVIRONMENTAL INFORMATION

Clean Water Act: Section 311 Hazardous Substances/RQ (same as CERCLA).
EPCRA Section 304: Reportable Quantity (RQ): CERCLA, 100 lb (45.4 kg).
EPCRA Section 313: Form R *de minimis* concentration reporting level:1.0%.Source of aqueous ammonia. Molecular weight 57.04. NH_3 equivalent weight 29.86.

AMMONIUM HYDROXIDE — A:320

SYNONYMS: AMMONIA WATER; AQUEOUS AMMONIA; BURMAR LAB CLEAN; ENPLATE NI-418B; ENSTRIP NP-1; HOUSEHOLD AMMONIA; HIDROXIDO AMONICO (Spanish); PPD 5932 DEVELOPER; POLY SILICON ETCH; PRE-METAL ETCH; RCA CLEAN (STEP 1); SCAN KLEEN; SCRUBBER-VAPOX

IDENTIFICATION

CAS:	1336-21-6
ERG ID:	UN 2672 (10-35% ammonia in water) UN 2073 (35-50% ammonia in water)
ERG Guide:	154 (10-35%); 125 (35-50%)
ERG Guide Category:	SUBSTANCES-TOXIC AND/OR CORROSIVE (NON-COMBUSTIBLE) (10-35%) GASSES-CORROSIVE (35-50%)
Molecular Formula:	H_5NO
RTECS No.:	BQ9625000
Properties:	Watery liquid. Colorless to milky-white. Ammonia odor. Floats and mixes with water. Irritating vapor is produced. Vapor is lighter than air; hugs the ground when cool.

HEALTH AND SAFETY INFORMATION

Severely irritating to eyes, skin, and respiratory system. Skin and eye contact can cause severe burns and blindness. Inhalation may cause pulmonary edema which can be delayed for several hours; there is a risk of death in serious cases. May cause chronic bronchitis. Repeated or prolonged skin contact may cause dryness, itching, and redness.

IDLH: 300 ppm as ammonia
Odor Threshold: 50 ppm

Exposure Limits:	**TWA**	**STEL**
ACGIH:	25 ppm; 17 mg/m^3	35 ppm; 24 mg/m^3
NIOSH:	25 ppm; 17 mg/m^3	35 ppm; 35 mg/m^3
OSHA:	50 ppm; 35 mg/m^3	

CHEMICAL REACTIVITY

Violent reaction with strong oxidizers, acids. Shock-sensitive compounds may be formed with halogens, mercury oxide, silver oxide. Fire and explosions may be caused by contact with trimethylammonium amide, 1-chloro-2,4-dinitrobenzene, o-chloronitrobenzene, platinum, trioxygen difluoride, selenium difluoride dioxide, boron halides, mercury, chlorine, iodine, bromine, hypochlorites, chlorine bleach, amides, organic anhydrides, isocyanates, vinyl acetate, alkylene oxides, epichlorohydrin, aldehydes. Attacks some coatings, plastics and rubber. Attacks copper, brass, aluminum, steel, and their alloys.

FIRE INFORMATION

FEMA Hazard Classifications:

Health Hazards (Blue) 2
Flammability (Red) 1
Reactivity (Yellow) 0

FIRST AID

Move victim to fresh air; call emergency medical care. Apply artificial respiration if victim is not breathing. Do not use mouth-to-mouth method if victim ingested or inhaled the substance; induce artificial respiration with the aid of a pocket mask equipped with a one-way valve or other proper respiratory medical devise. Administer oxygen if breathing is difficult. Remove and isolate contaminated clothing and shoes. In case of contact with substance, immediately flush skin or eyes with running water for at least 20 minutes. For minor skin contact, avoid spreading material on unaffected skin. Keep victim warm and quiet. Effects of exposure (inhalation, ingestion, or skin contact) to substance may be delayed. Ensure that medical personnel are aware of the material(s) involved and take precautions to protect themselves.

ENVIRONMENTAL INFORMATION

CLEAN WATER ACT: Section 311 Hazardous Substances/RQ (same as CERCLA).
EPCRA Section 304 Reportable Quantity (RQ): CERCLA, 1,000 lb (454 kg).
EPCRA Section 313: Form R *de minimus* concentration reporting level: 1.0%. Source of aqueous ammonia. Molecular weight 35.05. NH_3 equivalent weight 48.59.

AMMONIUM NITRATE A:340

SYNONYMS: AMMONIUM(I) NITRATE(1:1); AMMONIUM SALTPETER; ANSAX; EINECS No. 229-347-8; HERCO PRILLS; NITRAM; NITRATO AMONICO (Spanish); NITRIC ACID, AMMONIUM SALT; NORWAY SALTPETER; SALTPETER; VARIOFORM I

IDENTIFICATION

CAS:	6484-52-2
ERG ID:	UN 1942 with not more than 0.2% of combustible substance; or with organic coating; UN 2426 liquid (hot concentrated solution); UN 2072 (fertilizers)
ERG Guide:	140 (with not more than 0.2% combustible substance); (liquid, hot concentrated solution)
ERG Guide Category:	OXIDIZERS
Molecular Formula:	$H_4N_2O_3$
RTECS No.:	BR9050000

Properties: Solid: flakes, beads, or pellets. Clear, colorless (pure) to gray or brown (fertilizer grades). Odorless. Soluble in water. Toxic nitrogen oxide fumes produced in fire.

HEALTH AND SAFETY INFORMATION

Irritates eyes, skin and respiratory system. Long-term exposure may cause allergic reaction; blood effects.

RECOMMENDATIONS FOR RESPIRATOR SELECTION:

At any concentrations above the NIOSH REL, or where there is no REL, at any detectable concentration: SCBAF:PD,PP (any self-contained breathing apparatus that has a full facepiece and is operated in a pressure-demand or other positive-pressure mode); or SAF:PD,PP:ASCBA (any supplied-air respirator that has a full facepiece and is operated in a pressure-demand or other positive-pressure mode in combination with an auxiliary self-contained breathing apparatus operated in a pressure-demand or other positive-pressure mode).

Escape: GMFS* [any air-purifying, full-facepiece respirator (gas mask) with a chin-style, front-or back-mounted canister providing protection against the compound of concern]; or SCBAE (any appropriate escape-type, self-contained breathing apparatus).

*Only nonoxidiziable sorbents are allowed (not charcoal).

CHEMICAL REACTIVITY

Contact with water forms a weak acid. A strong oxidizer; violent reaction with reducing agents, ammonium chloride and heat, chlorides, combustible and organic materials, acids, powdered metals, phosphorus, sodium hypochlorite, sodium chlorate, and sulfur. Contact with strong alkalies releases toxic ammonia gas. When contaminated with oil, charcoal, or flammable liquids, can be considered an explosive that can be detonated by combustion or shock.

FIRE INFORMATION

Not flammable but will support combustion. If material or contaminated runoff enters waterways, notify downstream users of potentially contaminated waters. Notify local health and fire officials and pollution control agencies. Use dry chemicals, alcohol-resistant foam, carbon dioxide. From a secure explosion-proof location, use water spray to cool exposed containers. If cooling streams are ineffective (venting sound increases in volume and pitch, tank discolors, or shows any signs of deforming), withdraw immediately to a secure position.

NFPA Hazard Classifications:

Nonfire Situations

Health Hazards (Blue) 0
Flammability (Red) 0

Reactivity (Yellow) 3
Special situations . . . . OXY

Fire Situations

Health Hazards (Blue) 2
Flammability (Red) 0
Reactivity (Yellow) 3
Special situations . . . . OXY

FIRST AID

Move victim to fresh air. Call emergency medical care. Apply artificial respiration if victim is not breathing. Administer oxygen if breathing is difficult. Remove and isolate contaminated clothing and shoes. In case of contact with substance, immediately flush skin or eyes with running water for at least 20 minutes. Keep victim warm and quiet. Ensure that medical personnel are aware of the material(s) involved and take precautions to protect themselves.

ENVIRONMENTAL INFORMATION (solution)

EPCRA Section 313: Reportable as ammonia and nitrate compounds. *Ammonia compounds:* Form R *de minimis* concentration reporting level: 1.0%. (includes anhydrous ammonia and aqueous ammonia from water dissociable ammonium salts and other sources; 10% of total aqueous ammonia is reportable under this listing). *Nitrate compounds:* Form R *de minimis* concentration reporting level: 1.0%. Source of aqueous ammonia. Molecular weight 80.04. NH_3 equivalent weight 21.28. Toxic Chemical Category as nitrate compound: N511.

AMMONIUM PERSULFATE A:360

SYNONYMS: AMMONIUM PEROXYDISULFATE; AMMONIUM PEROXYSULFATE; AMMONIUM PERSULFATE 843; AMMONIUM PERSULFATE 6566; BURMAR SA-80; DIAMMONIUM PEROXYDISULFATE; DIAMMONIUM PERSULFATE; PERSULFATE d'AMMONIUM (French); PERSULFATO AMONICO (Spanish); PEROXYDISULFURIC ACID, DIAMMONIUM SALT

IDENTIFICATION

CAS: 7727-54-0
ERG ID: UN 1444
ERG Guide: 140
ERG Guide Category: OXIDIZERS
Molecular Formula: $H_8N_2O_8S_2$

RTECS No.: SE0350000

Properties: Solid. Light straw to colorless. Mild unpleasant odor. Sinks and mixes with water. Poisonous gases may be produced in fire.

HEALTH AND SAFETY INFORMATION

Corrosive to eyes, skin, and respiratory tract. Inhalation may cause pulmonary edema, which can be delayed for several hours; there is a risk of death in serious cases. Long-term exposure may cause allergy-like reaction; dermatitis, urticaria, rhinitis,"baker's" asthma.

IDLH: 300 ppm as ammonia

RECOMMENDATIONS FOR RESPIRATOR SELECTION:

At any concentrations above the NIOSH REL, or where there is no REL, at any detectable concentration: SCBAF:PD,PP (any self-contained breathing apparatus that has a full facepiece and is operated in a pressure-demand or other positive-pressure mode); or SAF:PD,PP:ASCBA (any supplied-air respirator that has a full facepiece and is operated in a pressure-demand or other positive-pressure mode in combination with an auxiliary self-contained breathing apparatus operated in a pressure-demand or other positive-pressure mode).

Escape: GMFS* [any air-purifying, full-facepiece respirator (gas mask) with a chin-style, front-or back-mounted canister providing protection against the compound of concern]; or SCBAE (any appropriate escape-type, self-contained breathing apparatus).

*Only nonoxidiziable sorbents are allowed (not charcoal).

CHEMICAL REACTIVITY

A strong oxidizer. Reacts with reducing agents and combustibles (e.g., grease and wood) and finely divided and powdered metals. Risk of fire and explosion.

FIRE INFORMATION

If material or contaminated runoff enters waterways, notify downstream users of potentially contaminated waters. Notify local health and fire officials and pollution control agencies. Use any extinguishing agent appropriate for surrounding fire. From a secure explosion-proof location, use water spray to cool exposed containers. If cooling streams are ineffective (venting sound increases in volume and pitch, tank discolors or shows any signs of deforming), withdraw immediately to a secure position.

FEMA Hazard Classifications:

Health Hazards (Blue) 1
Flammability (Red) 0
Reactivity (Yellow) 1
Special situations . . . . OXY

FIRST AID

Move victim to fresh air. Call emergency medical care. Apply artificial respiration if victim is not breathing. Administer oxygen if breathing is difficult. Remove and isolate contaminated clothing and shoes. In case of contact with substance, immediately flush skin or eyes with running water for at least 20 minutes. Keep victim warm and quiet. Ensure that medical personnel are aware of the material(s) involved and take precautions to protect themselves.

ENVIRONMENTAL INFORMATION

EPCRA Section 313: Form R *de minimis* concentration reporting level: 1.0%. Source of aqueous ammonia. Molecular weight 228.19. NH_3 equivalent weight 14.93.

AMMONIUM SULFAMATE A:400

SYNONYMS: AMCIDE; AMICIDE; AMMATE; AMMONIUM AMIDOSULFONATE; AMMONIUMSALZ der AMIDOSULFONSAEURE (German); AMMONIUMSALZ AMIDOSULFONSAEURE(German); AMMONIUM SULPHAMATE; AMS; EINECS No. 231-871-7; IKURIN; MONOAMMONIUM SULFAMATE; SULFAMATE; SULFAMATO AMONICO (Spanish); SULFAMIC ACID, MONOAMMONIUM SALT; SULFAMINSAEURE (German)

IDENTIFICATION

CAS:	7773-06-0
ERG ID:	UN 9089
ERG Guide:	171
ERG Guide Category:	SUBSTANCES (Low to Moderate Hazard)
Molecular Formula:	$H_6N_2O_3S$
RTECS No.:	WO6125000
Properties:	Non-flammable solid. Deliquescent, hygroscopic, colorless to white crystalline. Odorless. Mixes with water.

HEALTH AND SAFETY INFORMATION

Irritating to the eyes, nose, and respiratory tract.

IDLH: 1,500 mg/m^3
OSHA Table Z-1-A Air Contaminant

Exposure Limits: TWA

ACGIH:	10 mg/m^3
NIOSH:	10 mg/m^3 (total); 5mg/m^3 (resp)
OSHA:	15 mg/m^3 (total); 5mg/m^3 (resp)

RECOMMENDATIONS FOR RESPIRATOR SELECTION: NIOSH

50 mg/m^3: DM (any dust and mist respirator).
100 mg/m^3: DMXSQ (any dust and mist respirator except single-use and quarter mask respirators); or SA (any supplied-air respirator).
250 mg/m^3: SA:CF (any supplied-air respirator operated in a continuous-flow mode); or PAPRDM, if not present as a fume (any powered, air-purifying respirator with a dust and mist filter).
500 mg/m^3: SAT:CF (any supplied-air respirator that has a tight-fitting facepiece and is operated in a continuous-flow mode); or PAPRTHiE (any powered, air-purifying respirator with a tight-fitting facepiece and a high-efficiency particulate filter); or HiEF (any air-purifying, full-facepiece respirator with a high-efficiency particulate filter); or SCBAF (any self-contained breathing apparatus with a full facepiece); or SAF (any supplied-air respirator with a full facepiece).
1500 mg/m^3: SA:PD,PP (any supplied-air respirator operated in a pressure-demand or other positive-pressure mode).

EMERGENCY OR PLANNED ENTRY INTO UNKNOWN CONCENTRATIONS OR IDLH CONDITIONS: SCBAF:PD,PP (any self-contained breathing apparatus that has a full facepiece and is operated in a pressure-demand or other positive-pressure mode); or SAF:PD,PP:ASCBA (Any supplied-air respirator that has a full facepiece and is operated in a pressure-demand or other positive-pressure mode in combination with an auxiliary, self-contained breathing apparatus operated in a pressure-demand or other positive-pressure mode.

ESCAPE: HiEF (Any air-purifying, full-facepiece respirator with a high-efficiency particulate filter); or SCBAE (Any appropriate escape-type, self-contained breathing apparatus).

CHEMICAL REACTIVITY

A strong oxidizer; keep away from organic materials, flammable materials, reducing agents. Incompatible with acids, hot water. Elevated temperatures causes a highly exothermic reaction with water.

FIRE INFORMATION

Combustion products include toxic fumes of nitrogen oxides, ammonia, sulfur oxides, sulfamic acid, and sulfonates. If material or contaminated runoff enters waterways, notify downstream users of potentially contaminated waters. Notify local health and fire officials and pollution control agencies. Use any extinguishing agents appropriate for surrounding fire.

FEMA Hazard Classifications:

Health Hazards (Blue) 0
Flammability (Red) 0
Reactivity (Yellow) 0

FIRST AID

Move victim to fresh air. Call emergency medical care. Apply artificial respiration if victim is not breathing. Administer oxygen if breathing is difficult. In case of contact with substance, immediately flush skin or eyes with running water for at least 20 minutes. Remove and isolate contaminated clothing and shoes. Ensure that medical personnel are aware of the material(s) involved and take precautions to protect themselves.

ENVIRONMENTAL INFORMATION

Clean Water Act: Section 311 Hazardous Substances/RQ (same as CERCLA).
EPCRA Section 304: Reportable Quantity (RQ): CERCLA, 5,000 lb (2270 kg).
EPCRA Section 313: (ammonium sulfamate is a chemical source of aqueous ammonia): Form R *de minimis* concentration reporting level: 1.0%. (includes anhydrous ammonia and aqueous ammonia from water dissociable ammonium salts and other sources; 10 percent of total aqueous ammonia and 100 percent of anhydrous forms of ammonia is reportable under this listing. Molecular weight: 114.12; NH_3 Equivalent: 14.92.

AMMONIUM SULFATE A:420

SYNONYMS: AMMONIUM HYDROGEN SULFATE; AMMONIUM SULFATE (2:1); AMMONIUM SULFATE SOLUTION; AMMONIUM SULPHATE; DIAMMONIUM SULFATE; EINECS No. 231-984-1; NIPRO; NIPRO (II); SULFATO AMONICO (Spanish); SULFURIC ACID, DIAMMONIUM SALT

IDENTIFICATION

CAS:	7783-20-2
ERG ID:	UN 2506
ERG Guide:	154
ERG Guide Category:	SUBSTANCES -TOXIC AND/OR CORROSIVE (NONCOMBUSTIBLE)
Molecular Formula:	$H_8N_2O_4S$
RTECS No.:	BS4500000
Properties:	White crystalline solid. Odorless. Sinks and mixes with water. Poisonous gases may be produced in fire.

HEALTH AND SAFETY INFORMATION

Irritates eyes, skin and respiratory tract. Inhalation may cause pulmonary edema which can be delayed for several hours; there is a risk of death in serious cases.

RECOMMENDATIONS FOR RESPIRATOR SELECTION: NIOSH

At any concentrations above the NIOSH REL, or where there is no REL, at any detectable concentration: SCBAF:PD,PP (any self-contained breathing apparatus that has a full facepiece and is operated in a pressure-demand or other positive-pressure mode); or SAF:PD,PP:ASCBA (any supplied-air respirator that has a full facepiece and is operated in a pressure-demand or other positive-pressure mode in

combination with an auxiliary self-contained breathing apparatus operated in a pressure-demand or other positive-pressure mode).

Escape: GMFOV [any air-purifying, full-facepiece respirator (gas mask) with a chin-style, front-or back-mounted organic vapor canister]; or SCBAE (any appropriate escape-type, self-contained breathing apparatus).

CHEMICAL REACTIVITY

Contact with acids forms ammonia gas. Incompatible with chlorates (sodium hypochlorate forms explosive material), nitrates, and nitrites. Aqueous solution is acidic and attacks metals.

FIRE INFORMATION

Not flammable. If material or contaminated runoff enters waterways, notify downstream users of potentially contaminated waters. Notify local health and fire officials and pollution control agencies. Use any extinguishing agents appropriate for surrounding fire.

FIRST AID

Move victim to fresh air. Call emergency medical care. Apply artificial respiration if victim is not breathing. Do not use mouth-to-mouth method if victim ingested or inhaled the substance; induce artificial respiration with the aid of a pocket mask equipped with a one-way valve or other proper respiratory medical device. Administer oxygen if breathing is difficult. Remove and isolate contaminated clothing and shoes. In case of contact with substance, immediately flush skin or eyes with running water for at least 20 minutes. For minor skin contact, avoid spreading material on unaffected skin. Keep victim warm and quiet. Effects of exposure (inhalation, ingestion, or skin contact) may be delayed. Ensure that medical personnel are aware of the material(s) involved and take precautions to protect themselves.

ENVIRONMENTAL INFORMATION

EPCRA Section 313: Form R *de minimis* concentration reporting level: 1.0%. Source of aqueous ammonia. Molecular weight 132.13. NH_3 equivalent weight 25.78.

AMYL ACETATE A:425

SYNONYMS: ACETATE d'AMYLE (French); ACETATO DE AMILO (Spanish); ACETIC ACID n-AMYL ESTER; ACETIC ACID PENTYL ESTER; AMYL ACETIC ACID; n-AMYL ACETATE; AMYL ACETATE, MIXED ISOMERS; AMYL-ACETATE (n-); AMYL ACETIC ESTER; AMYL ACETIC ETHER; AMYAZETAT (German); BANANA OIL; BIRNENOEL; EEC No. 607-13-00-2; EINECS No. 211-047-3; OCTAN AMYLU (Polish); PEAR OIL; PENT ACETATE; PENTYL ACETATES; n-PENTYL ETHANOATE; PENTYL ESTER OF ACETIC ACID; 1-PENTANOL ACETATE; PRIMARY AMYL ACETATE

IDENTIFICATION

CAS:	628-63-7
ERG ID:	UN 1104
ERG Guide:	129
ERG Guide Category:	FLAMMABLE LIQUIDS (POLAR/WATER-MISCIBLE/NOXIOUS)
Molecular Formula:	$C_7H_{14}O_2$
RTECS No.:	AJ1925000
Properties:	Flammable, watery liquid. Colorless to yellow. Bananalike odor. Floats on water; negligible solubility; flammable, irritating vapor is produced.

HEALTH AND SAFETY INFORMATION

Irritates eyes, skin, and respiratory system. Affects central nervous system; high exposure can cause headache, dizziness and unconsciousness. Destroys skin's natural oils, causing redness, itching and inflammation.

IDLH: 1,000 ppm
Odor Threshold: 0.163 ppm
OSHA Table Z-1-A Air Contaminant

Exposure Limits:	**TWA**
ACGIH:	100 ppm, 532 mg/m^3
NIOSH:	100 ppm; 525 mg/m^3
OSHA:	100 ppm; 525 mg/m^3

RECOMMENDATIONS FOR RESPIRATOR SELECTION: NIOSH/OSHA

1,000 ppm: CCRFOV [any air-purifying, full-facepiece respirator (gas mask) with a chin-style, front-or-back-mounted acid gas canister]; or GMFOV [any air-purifying, full-facepiece respirator (gas mask) with a chin-style, front-or back-,mounted organic vapor canister]; or PAPROV [any powered, air-purifying respirator with organic vapor cartridge(s)]; or SA (any supplied-air respirator); or SCBAF (any self-contained breathing apparatus with a full facepiece); or SAF (any supplied-air respirator with a full facepiece).

Emergency or Planned Entry into Unknown Concentrations or IDLH Conditions: SCBAF:PD,PP (any self-contained breathing apparatus that has a full facepiece and is operated in a pressure-demand or other positive-pressure mode); or SAF:PD,PP:ASCBA (any supplied-air respirator that has a full facepiece and is operated in a pressure-demand or other positive-pressure mode in combination with an auxiliary self-contained breathing apparatus operated in a pressure-demand or other positive-pressure mode).

Escape: GMFOV [any air-purifying, full-facepiece respirator (gas mask) with a chin-style, front-or-back-mounted organic vapor canister]; or SCBAE (any appropriate escape-type, self-contained breathing apparatus).

Note: Substance reported to cause eye irritation or damage; may require eye protection.

CHEMICAL REACTIVITY

Incompatible with strong acids, nitrates, strong alkalies, strong oxidizers, and heat. May soften or dissolve plastics.

FIRE INFORMATION

Vapors are heavier than air and will collect in low areas. Vapors in confined areas may explode when exposed to fire. Vapors may travel long distances to ignition sources and flash back. Storage containers and parts of containers may rocket great distances, in many directions.If material or contaminated runoff enters waterways, notify downstream users of potentially contaminated waters. Notify local health and fire officials and pollution control agencies. Use dry chemicals, foam, carbon dioxide. From a secure explosion-proof location, use water spray to cool exposed containers. If cooling streams are ineffective (venting sound increases in volume and pitch, tank discolors or shows any signs of deforming), withdraw immediately to a secure position.

NFPA Hazard Classifications:

Health Hazards (Blue) 1
Flammability (Red) 3
Reactivity (Yellow) 0

Flash Point: (n-) 60°F (16°C) c.c.; (iso-) 89°F (32°C) c.c.
Flammable Limits in Air: (n-) LEL 1.1%; UEL 7.5%
Autoignition Temperature: (n-) 572°F (300°C)

FIRST AID

Move victim to fresh air. Call emergency medical care. Apply artificial respiration if victim is not breathing. Administer oxygen if breathing is difficult. Remove and isolate contaminated clothing and shoes. In case of contact with substance, immediately flush skin or eyes with running water for at least 20 minutes. Wash skin with soap and water. Keep victim warm and quiet. Effects of exposure (inhalation, ingestion, or skin contact) to substance may be delayed. Ensure that medical personnel are aware of the material(s) involved and take precautions to protect themselves.

ENVIRONMENTAL INFORMATION

Clean Water Act: Section 311 Hazardous Substances/RQ (same as CERCLA)
EPCRA Section 304: Reportable Quantity (RQ): CERCLA, 5,000 lb (2,270 kg)
TSCA: 712.30d10; 716.120d 10; 799.5000

AMYL TRICHLOROSILANE A:438

SYNONYMS: AMILTRICLOROSILANO (Spanish); AMYLTRICHLOROSILANE; EINECS No. 203-515-0; PENTYLTRICHLOROSILANE; TRICHLOROAMYLSILANE

IDENTIFICATION

CAS:	107-72-2
ERG ID:	UN 1728
ERG Guide:	155
ERG Guide Category:	SUBSTANCES-TOXIC and/or CORROSIVE (FLAMMABLE/WATER-SENSITIVE)
Molecular Formula:	$C_5H_{11}Cl_3Si$
RTECS No.:	VV4725000
Properties:	Flammable liquid. Colorless to yellow. Pungent odor. Sinks in water; reacts vigorously forming hydrochloric acid. Vapors are heavier than air and will collect in low areas.

HEALTH AND SAFETY INFORMATION

Corrosive. Skin and eye contact causes severe burns and possible blindness. Poison. Breathing the vapors can cause death.

RECOMMENDATIONS FOR RESPIRATOR SELECTION:

At any concentrations above the NIOSH REL, or where there is no REL, at any detectable concentration: SCBAF:PD,PP (any self-contained breathing apparatus that has a full facepiece and is operated in a pressure-demand or other positive-pressure mode); or SAF:PD,PP:ASCBA (any supplied-air respirator that has a full facepiece and is operated in a pressure-demand or other positive-pressure mode in combination with an auxiliary self-contained breathing apparatus operated in a pressure-demand or other positive-pressure mode).

Escape: GMFOV [any air-purifying, full-facepiece respirator (gas mask) with a chin-style, front-or back-mounted organic vapor canister]; or SCBAE (any appropriate escape-type, self-contained breathing apparatus).

CHEMICAL REACTIVITY

Forms explosive mixture with air. Contact with oxidizers may cause fire and explosions. Incompatible with strong acids, organic anhydrides, isocyanates, aldehydes. Attacks metals in the presence of moisture.

FIRE INFORMATION

Firefighting gear (including SCBA) does not provide adequate protection. If exposure occurs, remove and isolate gear immediately and thoroughly decontaminate personnel. Vapors are heavier than air and will collect in low areas. Vapors in confined areas may explode when exposed to fire. Vapors may travel long distances to ignition sources and flash back. Storage containers and parts of containers may rocket great distances, in many directions. If material or contaminated runoff enters waterways, notify downstream users of potentially contaminated waters. Notify local health and fire officials and pollution control agencies. Use dry chemicals, foam, carbon dioxide. From a secure explosion-proof location, use water spray to cool exposed containers. If cooling streams are ineffective (venting sound increases in volume and pitch, tank discolors or shows any signs of deforming), withdraw immediately to a secure position.

NFPA Hazard Classifications:

Health Hazards (Blue) 3
Flammability (Red) 2
Reactivity (Yellow) 2
Special situations .... Water

Flash Point: 145°F (63°C) o.c.

FIRST AID

Move victim to fresh air. Call emergency medical care. Apply artificial respiration if victim is not breathing. If breathing is difficult, give oxygen. Remove and isolate contaminated clothing and shoes. In case of contact with substance, immediately flush skin or eyes with running water for at least 20 minutes. Wash skin with soap and water. Keep victim warm and quiet. Ensure that medical personnel are aware of the material(s) involved and take precautions to protect themselves.

ANILINE A:440

SYNONYMS: AMINOBENZENE; AMINOPHEN; ANILIN (Czech and German); ANILINA (Italian, Polish, Spanish); ANILINE OIL; ARYLAMINE; BENZENEAMINE; BENZENE AMINO; BENZIDAM; BLUE OIL; C.I. 76000; C.I. OXIDATION BASE 1; CYANOL; EINECS No. 200-539-3; EEC No. 612-008-00-7; HUILE d'ANILINE (French); KRYSTALLIN; KYANOL; NCI-C03736; PHENYLAMINE

IDENTIFICATION

CAS:	62-53-3; 142-04-1
ERG ID:	UN 1547
ERG Guide:	153
ERG Guide Category:	TOXIC AND/OR CORROSIVE (COMBUSTIBLE)
Molecular Formula:	$C_6H_2N_6$
RTECS No.:	BW6650000
Properties:	Very flammable, oily liquid. Clear, colorless to yellowish; darkens to brown on exposure to air. Strong, aromatic, musty, fishy odor; not unpleasant. Sinks slowly in water; slightly soluble. Toxic nitrogen oxides produced when heated.

HEALTH AND SAFETY INFORMATION

Aniline, the simplest aromatic amine, is a prototypical inducer of methemoglobinemia and impaired oxygen transport to tissues. Chemical protective clothing is recommended because aniline vapor and liquid can be dermally absorbed and may contribute to systemic toxicity. Direct contact with liquid aniline can cause skin burns. Chronic effects may be due to acute damage to the brain, heart, and kidneys. Loss of appetite, dizziness, insomnia, tremors, malignant bladder growths, liver damage, and jaundice. Anemia has been reported. Has been linked to bladder cancer according to a NIOSH study. Signs and symptoms include (1) blood in the urine, (2) other changes in the appearance of the urine, (3) changes in urinary habits, (4) lumps in the groin and lower abdomen, and (5) pain in the lower abdomen or back. Aniline can cross the placental barrier. Because fetal hemoglobin is more easily oxidized to methemoglobin than is adult hemoglobin and is less easily reduced back to normal hemoglobin, methemoglobin theoretically may be at higher levels in fetuses than in exposed mothers.

IARC Evaluation: Group 3, unclassifiable substance; human evidence, inadequate; animal evidence, limited.
IDLH: 100 ppm; potential human carcinogen.
Odor Threshold: 0.676 ppm
OSHA Table Z-1-A Air Contaminant

Exposure Limits:	TWA
ACGIH:(1)(2)	2 ppm; 7.6 mg/m^3
NIOSH:(1)(3)	
OSHA:(1)	5 ppm; 19 mg/m^3

(1)Skin contact contributes significantly to overall exposure.
(2)Animal carcinogen
(3)Potential human carcinogen. Reduce exposure to lowest feasible level.

RECOMMENDATIONS FOR RESPIRATOR SELECTION: NIOSH

At any concentrations above the NIOSH REL, or where there is no REL, at any detectable concentration: SCBAF:PD,PP (any self-contained breathing apparatus that has a full facepiece and is operated in a pressure-demand or other positive-pressure mode); or SAF:PD,PP:ASCBA (any supplied-air respirator that has a full facepiece and is operated in a pressure-demand or other positive-pressure mode in combination with an auxiliary self-contained breathing apparatus operated in a pressure-demand or other positive-pressure mode).

Escape: GMFOV [any air-purifying, full-facepiece respirator (gas mask) with a chin-style, front-or back-mounted organic vapor canister]; or SCBAE (any appropriate escape-type, self-contained breathing apparatus).

CHEMICAL REACTIVITY

Forms explosive mixture with air. Contact with water forms a strong alkaline solution. Unless inhibited (usually methanol), readily able to polymerize. Fires and explosions may result from contact with halogens, strong acids, strong oxidizers (chlorine, permanganates), strong alkalies (sodium hydroxide), organic anhydrides, ozone, acetic anhydride, isocyanates, aldehydes, and sodium peroxide. Reacts with alkali metals and alkali earth metals. Attacks some plastics, rubber, and coatings, copper, and alloys.

FIRE INFORMATION

Firefighting gear (including SCBA) does not provide adequate protection. If exposure occurs, remove and isolate gear immediately and thoroughly decontaminate personnel.Vapors are heavier than air and will collect in low areas. Vapors in confined areas may explode when exposed to fire. Vapors may travel long distances to ignition sources and flash back. Storage containers and parts of containers may rocket great distances, in many directions.If material or contaminated runoff enters waterways, notify downstream users of potentially contaminated waters. Notify local health and fire officials and pollution control agencies. Use dry chemicals, foam, carbon dioxide. From a secure explosion-proof location, use water spray to cool exposed containers. If cooling streams are ineffective (venting sound increases in volume and pitch, tank discolors, or shows any signs of deforming), withdraw immediately to a secure position.

NFPA Hazard Classifications:

Health Hazards (Blue) 3
Flammability (Red) 2
Reactivity (Yellow) 0

Flash Point: 168°F (76°C) o.c.; 158°F (70°C) c.c.
Flammable Limits in Air: LEL 1.3%; UEL 11%
Autoignition Temperature: 1140°F (616°C)

FIRST AID

Move victim to fresh air. Call emergency medical care. Apply artificial respiration if victim is not breathing. Do not use mouth-to-mouth method if victim ingested or inhaled the substance; induce artificial respiration with the aid of a pocket mask equipped with a one-way valve or other proper respiratory medical device. Administer oxygen if breathing is difficult. Remove and isolate contaminated clothing and shoes. In case of contact with substance, immediately flush skin or eyes with running water for at least 20 minutes. For minor skin contact, avoid spreading material on unaffected skin. Keep victim warm and quiet. Effects of exposure (inhalation, ingestion, or skin contact) may be delayed. Ensure that medical personnel are aware of the material(s) involved and take precautions to protect themselves.

ENVIRONMENTAL INFORMATION

Clean Air Act: Hazardous Air Pollutants (Title I, Part A, Section 112)
Clean Water Act: Section 311 Hazardous Substances/RQ (same as CERCLA); Section 313 Priority Chemicals
EPA Hazardous Waste Number (RCRA No.): U012
RCRA Section 261 Hazardous Constituents
RCRA Land Ban Waste
RCRA Universal Treatment Standards: Wastewater (ug/L), 0.81; Nonwastewater (mg/kg), 14

RCRA Ground Water Monitoring List: Suggested methods (PQL mg/L): 8270(10)
EPCRA Section 302, Extremely Hazardous Substances: TPQ = 1,000 lb (454 kg)
EPCRA Section 304: Reportable Quantity (RQ): CERCLA, 5,000 lb (2,270 kg)
EPCRA Section 313: Form R *de minimis* concentration reporting level: 1.0%
TSCA: 799.18
California's Proposition 65: Carcinogen

ANTIMONY — A:460

SYNONYMS: ANTIMONIO (Spanish); ANTIMONY BLACK; ANTIMONY METAL; ANTIMONY REGULUS; ANTYMON (Polish); C.I. 77050; EINECS No. 231-146-5; SILVER GLO BP; SILVER GLO 3KBP; SILVER GLO 33BP; STIBIUM; THERMOGUARD CPA

IDENTIFICATION

CAS:	7440-36-0
ERG ID:	UN 1549 (inorganic, n.o.s.); UN 2871 (powder); UN 3141 (inorganic liquid compounds, n.o.s.)
ERG Guide:	157 (inorganic, n.o.s.; inorganic liquid compounds, n.o.s.) 170 (powder)
ERG Guide Category:	157: SUBSTANCES -TOXIC AND/OR CORROSIVE (NONCOMBUSTIBLE/WATER-SENSITIVE) 170: METALS (POWDERS, DUST, SHAVINGS, BORINGS, TURNINGS, CUTTINGS, Etc.)
Molecular Formula:	Sb
RTECS No.:	CC4025000
Properties:	Brittle solid; scalelike crystals; or powder. Silver-white lustrous solid; dark gray lustrous powder. Moderate explosion hazard in the form of dust when exposed to flame. Poisonous gases released when exposed to heat or fire.

HEALTH AND SAFETY INFORMATION

Poisonous. Repeated exposure may cause headache, sleeplessness, dizziness, metallic taste, ulcers, weight loss, nausea, vomiting, diarrhea, and impairment of sense of smell; pain and tightness in chest; heart and liver damage; and skin sores and ulcers.

IDLH: 50 mg/m^3
OSHA Table Z-1-A Air Contaminant
NIOSH Criteria Document: NIOSH 78-216

Exposure Limits:	**TWA**
ACGIH:	0.5 mg/m^3
NIOSH:	0.5 mg/m^3
OSHA:	0.5 mg/m^3

RECOMMENDATIONS FOR RESPIRATOR SELECTION: NIOSH/OSHA

5 mg/m^3: DMXSQ [if not present as a fume] (any dust and mist respirator except single-use and quarter mask respirators); or SA (any supplied-air respirator).
12.5 mg/m^3: SA:CF [any supplied-air respirator operated in a continuous-flow mode)]; PAPRDM [if not present as a fume] (any powered, air-purifying respirator with a dust and mist filter).
25 mg/m^3: HiEF (any air-purifying, full-facepiece respirator with a high-efficiency particulate filter); or SAT:CF (any supplied-air respirator that has a tight-fitting facepiece and is operated in a continuous-flow mode); or PAPRTHiE (any powered, air-purifying respirator with a tight-fitting facepiece and a high-efficiency particulate filter); or SCBAF (any self-contained breathing apparatus with a full facepiece); or SAF (any supplied-air respirator with a full facepiece).
50 mg/m^3: SA:PD,PP (any supplied-air respirator operated in a pressure-demand or other positive-pressure mode).

Emergency or Planned Entry into Unknown Concentrations or IDLH Conditions: SCBAF:PD,PP (any self-contained breathing apparatus that has a full faceplate and is operated in a pressure-demand or other positive-pressure mode); or SAF:PD,PP:ASCBA (any supplied-air respirator that has a full facepiece and is operated in a pressure-demand or other positive-pressure mode in combination with an auxiliary self-contained breathing apparatus operated in a pressure-demand or other positive-pressure mode).

Escape: HiEF (any air-purifying, full-facepiece respirator with a high-efficiency particulate filter); or SCBAE (any appropriate escape-type, self-contained breathing apparatus).

CHEMICAL REACTIVITY

Violent reaction with strong oxidizers, strong acids. Forms explosive mixtures with chloric and perchloric acid. Oxidizers, acids, especially halogenated acids, produce toxic stibine gas (antimony hydride). Mixtures with nitrates or halogenated compounds may cause combustion.

FIRE INFORMATION

Decomposition products upon heating include toxic gases of antimony. If material or contaminated runoff enters waterways, notify downstream users of potentially contaminated waters. Notify local health and fire officials and pollution control agencies. Use any extinguishing agent suitable for surrounding fire. From a secure explosion-proof location, use water spray to cool exposed containers. If cooling streams are ineffective (venting sound increases in volume and pitch, tank discolors or shows any signs of deforming), withdraw immediately to a secure position.

FEMA Hazard Classifications: (powder)

Health Hazards (Blue) 2
Flammability (Red) 2
Reactivity (Yellow) 1

FIRST AID

Inorganic, n.o.s.; inorganic liquid compounds, n.o.s.:Move victim to fresh air. Call emergency medical care. Apply artificial respiration if victim is not breathing. Do not use mouth-to-mouth method if victim ingested or inhaled the substance; induce artificial respiration with the aid of a pocket mask equipped with a one-way valve or other proper respiratory medical device. Administer oxygen if breathing is difficult. Remove and isolate contaminated clothing and shoes. In case of contact with substance, immediately flush skin or eyes with running water for at least 20 minutes. For minor skin contact, avoid spreading material on unaffected skin. Keep victim warm and quiet. Effects of exposure (inhalation, ingestion, or skin contact) may be delayed. Ensure that medical personnel are aware of the material(s) involved and take precautions to protect themselves.

Powder: Move victim to fresh air. Call emergency medical care. Apply artificial respiration if victim is not breathing. Administer oxygen if breathing is difficult. Remove and isolate contaminated clothing and shoes. In case of contact with substance, immediately flush skin or eyes with running water for at least 20 minutes. Keep victim warm and quiet. Ensure that medical personnel are aware of the material(s) involved and take precautions to protect themselves.

ENVIRONMENTAL INFORMATION

Clean Air Act: Section 307 Priority Pollutants.
Clean Water Act: Section 313 Priority Chemicals; Toxic Pollutant (Section 401.15).
RCRA Section 261 Hazardous Constituents: Waste number not listed.
RCRA Land Ban Waste
Safe Drinking Water Act: MCL, 0.006 mg/L; MCLG, 0.006 mg/L.
RCRA Universal Treatment Standards: Wastewater (mg/L), 1.9; Nonwastewater (mg/L), TCLP
RCRA Ground Water Monitoring List: Suggested methods (PQL ug/L): 6010(300); 7040(2,000); 7041(30)
EPCRA Section 304: Reportable Quantity (RQ): CERCLA, 5,000 lb (2,270 kg) only if the diameter of pieces of solid metal has a diameter equal to or greater than 0.004 in.
EPCRA Section 313: Form R *de minimis* concentration reporting level: 1.0%. Form R, Toxic Chemical Category Code: N010
TSCA: 716.20

ANTIMONY PENTACHLORIDE A:467

SYNONYMS: ANTIMONIC CHLORIDE; ANTIMONIO (PENTACLORURO di) (Italian); ANTIMONPENTACHLORID (German); ANTIMONY(V) CHLORIDE; ANTIMONY PERCHLORIDE;

ANTIMOONPENTACHLORIDE (Dutch); EEC No. 051-022-00-3; EINECS No. 231-601-8; BUTTER OF ANTIMONY; PENTACHLOROANTIMONY; PENTACLORURO de ANTIMONIO (Spanish); TENTACHLORURE D'ANTIMOINE (French); PERCHLORURE D'ANTIMOINE (French); STCC 4932310

IDENTIFICATION

CAS:	7647-18-9
ERG ID:	UN 1730 (liquid); UN 1731 (solution)
ERG Guide:	157
ERG Guide Category:	SUBSTANCES-TOXIC and/or CORROSIVE (NON-COMBUSTIBLE/WATER-SENSITIVE)
Molecular Formula:	Cl_5Sb
RTECS No.:	CC5075000
Properties:	Noncombustible oil or liquid. Colorless to yellowish-red. Offensive odor. Sinks and mixes with water; decomposes: Heat, hydrogen chloride and antimony oxides.

HEALTH AND SAFETY INFORMATION

Poisonous if inhaled or swallowed. Corrosive; extremely irritating to the eyes, nose, and respiratory tract. Can cause pulmonary edema, which can be delayed several hours following exposure. Can cause kidney, liver, and heart damage.

IDLH: 50 mg/m^3 as antimony and compounds (as Sb)
OSHA Table Z-1-A Air Contaminant, antimony and compounds (as Sb)
NIOSH Criteria Document: NIOSH 78-216

Exposure Limits:	**TWA**
ACGIH:	0.5 mg(Sb)/m^3
NIOSH:	0.5 mg(Sb)/m^3
OSHA:	0.5 mg(Sb)/m^3

RECOMMENDATIONS FOR RESPIRATOR SELECTION: NIOSH as Sb

5 mg/m^3: DMXSQ [if not present as a fume] (any dust and mist respirator except single-use and quarter mask respirators); or SA (any supplied-air respirator).
12.5 mg/m^3: SA:CF [any supplied-air respirator operated in a continuous-flow mode)]; PAPRDM [if not present as a fume] (any powered, air-purifying respirator with a dust and mist filter).
25 mg/m^3: HiEF (any air-purifying, full-facepiece respirator with a high-efficiency particulate filter); or SAT:CF (any supplied-air respirator that has a tight-fitting facepiece and is operated in a continuous-flow mode); or PAPRTHiE (any powered, air-purifying respirator with a tight-fitting facepiece and a high-efficiency particulate filter); or SCBAF (any self-contained breathing apparatus with a full facepiece); or SAF (any supplied-air respirator with a full facepiece).
50 mg/m^3: SA:PD,PP (any supplied-air respirator operated in a pressure-demand or other positive-pressure mode).

Emergency or Planned Entry into Unknown Concentrations or IDLH Conditions: SCBAF:PD,PP (any self-contained breathing apparatus that has a full faceplate and is operated in a pressure-demand or other positive-pressure mode); or SAF:PD,PP:ASCBA (any supplied-air respirator that has a full facepiece and is operated in a pressure-demand or other positive-pressure mode in combination with an auxiliary, self-contained breathing apparatus operated in a pressure-demand or other positive-pressure mode).

Escape: HiEF (any air-purifying, full-facepiece respirator with a high-efficiency particulate filter); or SCBAE (any appropriate escape-type, self-contained breathing apparatus).

CHEMICAL REACTIVITY

Reacts with water to form toxic and corrosive hydrogen chloride gas. Reacts violently with ammonia, caustics, and many other chemicals. Attacks metals, especially in the presence of moisture. Contact with moist air produces heavier-than-air corrosive vapor; will collect in low areas.

FIRE INFORMATION

Not flammable. Decomposition products include toxic gases of antimony and chlorine. If material or contaminated runoff enters waterways, notify downstream users of potentially contaminated waters. Notify local health and fire officials and pollution control agencies. Use any extinguishing agent suitable for surrounding fire.

NFPA Hazard Classifications:

Health Hazards (Blue) . 4
Flammability (Red) 0
Reactivity (Yellow) 1

FIRST AID

Move victim to fresh air. Call emergency medical care. Apply artificial respiration if victim is not breathing. Do not use mouth-to-mouth method if victim ingested or inhaled the substance; induce artificial respiration with the aid of a pocket mask equipped with a one-way valve or other proper respiratory medical device. Administer oxygen if breathing is difficult. Remove and isolate contaminated clothing and shoes. In case of contact with substance, immediately flush skin or eyes with running water for at least 20 minutes. For minor skin contact, avoid spreading material on unaffected skin. Keep victim warm and quiet. Effects of exposure (inhalation, ingestion, or skin contact) may be delayed. Ensure that medical personnel are aware of the material(s) involved and take precautions to protect themselves.

ENVIRONMENTAL INFORMATION

Clean Air Act: Hazardous Air Pollutants (Title I, Part A, Section 112), as antimony compounds.
Clean Water Act: Section 311 Hazardous Substances/RQ (same as CERCLA); Section 307 Priority Pollutants; Section 313 Priority Chemicals; Section 311 Hazardous Substances/RQ (same as CERCLA); Toxic Pollutant (Section 401.15) as antimony compounds.

RCRA Section 261 Hazardous Constituents: Waste number not listed, as antimony compounds, n.o.s.
EPCRA Section 304 Reportable Quantity (RQ): CERCLA, 1,000 lb (454 kg).
EPCRA Section 313: Form R *de minimis* concentration reporting level: 0.1%. Form R, Toxic Chemical Category Code: N010

ANTIMONY TRICHLORIDE A:473

SYNONYMS: STIBINE, TRICHLORO-; ANTIMONIUS CHLORIDE; ANTIMONY(III) CHLORIDE; TRICHLORO STIBINE; ANTIMOINE (TRICHLORURE D') (French); ANTIMONIO (TRICHLORURO di) (Italian); ANTIMONY BUTTER; ANTIMOONTRICHLRIDE (Dutch); BUTTER OF ANTIMONY; CHLORID ANTIMONITY; C.I. 77056; EINECS No. 233-047-2; STCC No. 4932316; STIBINE, TRICHLORO-; TRICLORURO de ANTIMONIO (Spanish); TRICHLOROSTIBINE; TRICHLORURE d' ANTIMOINE (French)

IDENTIFICATION

CAS:	10025-91-9
ERG ID:	UN 1733 (liquid, solid or solution)
ERG Guide:	157
ERG Guide Category:	SUBSTANCES-TOXIC and/or CORROSIVE (NON-COMBUSTIBLE/WATER-SENSITIVE)
Molecular Formula:	Cl_3Sb
RTECS No.:	CC4900000
Properties:	Noncombustible hygroscopic crystalline solid; fumes in air. Colorless to pale yellow. Sharp, pungent, unpleasant odor. Sinks in water and decomposes violently: Heat, hydrogen chloride, and antimony oxychloride is produced.

HEALTH AND SAFETY INFORMATION

Poisonous if inhaled or swallowed. Corrosive; extremely irritating to the eyes, nose, and respiratory tract. Can cause pulmonary edema, which can be delayed several hours following exposure. Can cause heart, kidney, and liver damage.

IDLH: 50 mg/m^3 as antimony and compounds (as Sb)
OSHA Table Z-1-A Air Contaminant, antimony and compounds (as Sb)
NIOSH Criteria Document: NIOSH 78-216

Exposure Limits:	**TWA**
ACGIH:	0.5 mg(Sb)/m^3
NIOSH:	0.5 mg(Sb)/m^3
OSHA:	0.5 mg(Sb)/m^3

RECOMMENDATIONS FOR RESPIRATOR SELECTION: NIOSH as Sb

5 mg/m³: DMXSQ *if not present as a fume* (any dust and mist respirator except single-use and quarter mask respirators); or SA (any supplied-air respirator).
12.5 mg/m³: SA:CF [any supplied-air respirator operated in a continuous-flow mode)]; PAPRDM, *if not present as a fume* (any powered, air-purifying respirator with a dust and mist filter).
25 mg/m³: HiEF (any air-purifying, full-facepiece respirator with a high-efficiency particulate filter); or SAT:CF (any supplied-air respirator that has a tight-fitting facepiece and is operated in a continuous-flow mode); or PAPRTHiE (any powered, air-purifying respirator with a tight-fitting facepiece and a high-efficiency particulate filter); or SCBAF (any self-contained breathing apparatus with a full facepiece); or SAF (any supplied-air respirator with a full facepiece).
50 mg/m³: SA:PD,PP (any supplied-air respirator operated in a pressure-demand or other positive-pressure mode).

Emergency or Planned Entry into Unknown Concentrations or IDLH Conditions: SCBAF:PD,PP (any self-contained breathing apparatus that has a full faceplate and is operated in a pressure-demand or other positive-pressure mode); or SAF:PD,PP:ASCBA (any supplied-air respirator that has a full facepiece and is operated in a pressure-demand or other positive-pressure mode in combination with an auxiliary, self-contained breathing apparatus operated in a pressure-demand or other positive-pressure mode).

Escape: HiEF (any air-purifying, full-facepiece respirator with a high-efficiency particulate filter); or SCBAE (any appropriate escape-type, self-contained breathing apparatus).

CHEMICAL REACTIVITY

Reacts with water to form toxic and corrosive hydrogen chloride gas. Reacts violently with aluminum, potassium , sodium. Attacks metals, especially in the presence of moisture, giving off flammable hydrogen gas. Contact with moist air produces heavier-than-air corrosive hydrogen chloride vapor.

FIRE INFORMATION

Not flammable. Decomposition products upon heating include toxic gases of antimony and chlorine. .If material or contaminated runoff enters waterways, notify downstream users of potentially contaminated waters. Notify local health and fire officials and pollution control agencies. Use any extinguishing agent suitable for surrounding fire.

FIRST AID

Move victim to fresh air. Call emergency medical care. Apply artificial respiration if victim is not breathing. Do not use mouth-to-mouth method if victim ingested or inhaled the substance; induce artificial respiration with the aid of a pocket mask equipped with a one-way valve or other proper respiratory medical device. Administer oxygen if breathing is difficult. Remove and isolate contaminated clothing and shoes. In case of contact with substance, immediately flush skin or eyes with running water for at least 20 minutes. For minor skin contact, avoid spreading material on unaffected skin. Keep victim warm and quiet. Effects of exposure (inhalation, ingestion, or skin contact) may be delayed. Ensure that medical personnel are aware of the material(s) involved and take precautions to protect themselves.

ENVIRONMENTAL INFORMATION

Clean Air Act: Hazardous Air Pollutants (Title I, Part A, Section 112), as antimony compounds.
Clean Water Act: Section 311 Hazardous Substances/RQ (same as CERCLA); Section 307 Priority Pollutants; Section 313 Priority Chemicals; Section 311 Hazardous Substances/RQ (same as CERCLA); Toxic Pollutant (Section 401.15) as antimony compounds.
RCRA Section 261 Hazardous Constituents: Waste number not listed, as antimony compounds, n.o.s.
EPCRA Section 304 Reportable Quantity (RQ): CERCLA, 1,000 lb (454 kg).
EPCRA Section 313: Form R *de minimis* concentration reporting level: 0.1%. Form R, Toxic Chemical Category Code: N010

ARGON A:520

SYNONYMS: ARGON, COMPRESSED; ARGON G-6; ARGON GAS; ELEMENTAL ARGON; LIQUID ARGON

IDENTIFICATION

CAS:	7440-37-1
ERG ID:	UN 1006 (compressed); UN 1951 (liquid, cooled)
ERG Guide:	121 (compressed); 120 (liquid, cooled)
ERG Guide Category:	GASES -INERT
Molecular Formula:	Ar
RTECS No.:	CF2300000
Properties:	Colorless, odorless, nonflammable, inert gas.

HEALTH AND SAFETY INFORMATION

A simple asphyxiant. Gas is heavier than air and can accumulate at ground level, causing oxygen deficiency and risk of suffocation. Liquid can cause frostbite.

RECOMMENDATIONS FOR RESPIRATOR SELECTION

At any concentrations above the NIOSH REL, or where there is no REL, at any detectable concentration: SCBAF:PD,PP (any self-contained breathing apparatus that has a full facepiece and is operated in a pressure-demand or other positive-pressure mode); or SAF:PD,PP:ASCBA (any supplied-air respirator that has a full facepiece and is operated in a pressure-demand or other positive-pressure mode in combination with an auxiliary self-contained breathing apparatus operated in a pressure-demand or other positive-pressure mode).

Escape: GMFOV [any air-purifying, full-facepiece respirator (gas mask) with a chin-style, front-or back-mounted organic vapor canister]; or SCBAE (any appropriate escape-type, self-contained breathing apparatus).

FIRE INFORMATION

Storage containers and parts of containers may rocket great distances, in many directions. Notify local health and fire officials and pollution control agencies. Use any extinguishing agent suitable for surrounding fires. From a secure explosion-proof location, use water spray to cool exposed containers. If cooling streams are ineffective (venting sound increases in volume and pitch, tank discolors or shows any signs of deforming), withdraw immediately to a secure position.

NFPA Hazard Classifications: Not listed

FIRST AID

BREATH: If a person breathes large amounts of this chemical, move exposed person to fresh air at once. If breathing has stopped, perform mouth-to-mouth resuscitation. Keep the affected person warm and at rest. Get medical attention as soon as possible.
SKIN: If frostbite has occurred, seek medical attention immediately; do *NOT* rub the affected areas or flush them with water. In order to prevent further tissue damage, do *NOT* attempt to remove frozen clothing from frostbitten areas. If frostbite has *NOT* occurred, immediately and thoroughly wash contaminated skin with soap and water.

ARSENIC A:540

SYNONYMS: ACCUSPIN ASX-10 SPIN-ON DOPANT; ARSEN (German, Polish); ARSENIC-75; ARSENIC BLACK; ARSENIC, METALLIC; ARSENICO (Spanish); ARSENIC, SOLID; AS-120; AS-217; BRIGHTENER E-3; COLLOIDAL ARSENIC; EINECS No. 231-148-6; GREY ARSENIC; METALLIC ARSENIC; REALGAR; RUBY ARSENIC; BUTTER OF ARSENIC

IDENTIFICATION

CAS:	7440-38-2
ERG ID:	UN 1558 (metal); UN 1562 (dust)
ERG Guide:	152 (metal and dust)
ERG Guide Category:	TOXIC (COMBUSTIBLE)
Molecular Formula:	As
RTECS No.:	CG0525000
Properties:	Brittle, solid crystals. Silver-gray or tin-white. Odorless. Sinks in water. Poisonous gases are produced in fire.

HEALTH AND SAFETY INFORMATION

May cause weakness, loss of appetite, irritation and inflammation of the eyes, nose and throat; breakdown of nasal tissue, excessive itching; development of thickened skin and warts on hands and feet, skin discoloration; paralysis of the hands and feet; and skin, lung and liver (tumors) cancer. Also changes in carbohydrate metabolism and peroxide levels, and increased systolic blood pressure may be detected in exposed workers. There has been an increase of chromosomal abnormalities in workers exposed to arsenic.

IARC Evaluation: Group 1, known carcinogen; human evidence, sufficient.
NTP: Known carcinogen.
IDLH: 5 mg/m^3 as arsenic; potential human carcinogen.
OSHA Specifically Regulated Substances (29 CFR 1910.1018) as inorganic arsenic.

Exposure Limits:	**TWA**	**CEILING**
ACGIH:*	0.01 mg/m^3	
NIOSH:**		0.002 mg/m^3/15 min
OSHA:	0.010 mg/m^3	

*Confirmed human carcinogen.
**Potential human carcinogen. Reduce exposure to lowest feasible level.

RECOMMENDATIONS FOR RESPIRATOR SELECTION: NIOSH

At any concentrations above the NIOSH REL, or where there is no REL, at any detectable concentration: SCBAF:PD,PP (any self-contained breathing apparatus that has a full faceplate and is operated in a pressure-demand or other positive-pressure mode); or SAF:PD,PP:ASCBA (any supplied-air respirator that has a full facepiece and is operated in a pressure-demand or other positive-pressure mode in combination with an auxiliary self-contained breathing apparatus operated in a pressure-demand or other positive-pressure mode).

Escape: GMFAGHiE [any air-purifying, full-facepiece respirator (gas mask) with a chin-style, front-or back-mounted acid gas canister having a high-efficiency particulate filter]; or SCBAE (any appropriate escape-type, self-contained breathing apparatus).

CHEMICAL REACTIVITY

Incompatible with strong acids, strong oxidizers, peroxides, bromine pentafluoride, bromine trifluoride, cesium acetylene carbide, chromium trioxide, nitrogen trichloride, and silver nitrate. Hydrogen gas can react with inorganic arsenic to form highly toxic arsine gas. Forms highly toxic fumes on contact with acids or active metals (aluminum, iron, and zinc).

FIRE INFORMATION

Noncombustible solid in bulk form, but a slight explosion hazard in the form of dust when exposed to flame. If material or contaminated runoff enters waterways, notify downstream users of potentially contaminated waters. Notify local health and fire officials and pollution control agencies. Use any extinguishing agent suitable for surrounding fire.

FEMA Hazard Classifications:

Health Hazards (Blue) 3
Flammability (Red) 1
Reactivity (Yellow) 0

FIRST AID

Move victim to fresh air. Call emergency medical care. Apply artificial respiration if victim is not breathing. Do not use mouth-to-mouth method if victim ingested or inhaled the substance; induce artificial respiration with the aid of a pocket mask equipped with a one-way valve or other proper respiratory medical device. Administer oxygen if breathing is difficult. Remove and isolate contaminated clothing and shoes. In case of contact with substance, immediately flush skin or eyes with running water for at least 20 minutes. For minor skin contact, avoid spreading material on unaffected skin. Keep victim warm and quiet. Effects of exposure (inhalation, ingestion, or skin contact) may be delayed. Ensure that medical personnel are aware of the material(s) involved and take precautions to protect themselves.

ENVIRONMENTAL INFORMATION

Clean Air Act: Hazardous Air Pollutants (Title I, Part A, Section 112).
Clean Water Act: Section 307 Priority Pollutants; Section 313 Priority Chemicals; Toxic Pollutant (Section 401.15).
RCRA Maximum Concentration of Contaminants: Regulatory level, 5.0 mg/L
RCRA Section 261 Hazardous Constituents: Waste number not listed.
RCRA Land Ban Waste
RCRA Universal Treatment Standards: Wastewater (mg/L), 1.4; Nonwastewater (mg/L), 5.0 TCLP.
RCRA Ground Water Monitoring List: Suggested methods (PQL mg/L): (total) 6010(500), 7060(10), 7061(20).
Safe Drinking Water Act: Regulated chemical (47 FR 9352); MCL, 0.05 mg/L (Section 141.11) applies only to community water systems.
EPCRA Section 304: Reportable Quantity (RQ): CERCLA, 1 lb (0.454 kg).
Note: No reporting required if diameter of metal is equal to or exceeds 0.004 in.
EPCRA Section 313: Form R *de minimis* concentration reporting level: 0.1%; Form R, Toxic Chemical Category Code: N020.
Marine Pollutant (49CFR, Subchapter 172.101, Appendix B).
California's Proposition 65: Carcinogen.

ARSENIC ACID A:543

SYNONYMS: ACIDO ARSENICO (Spanish); ARSENATE; ARSENIC PENTOXIDE; o-ARSENIC ACID; METAARSENIC ACID; ORTHOARSENIC ACID; ZOTOX

IDENTIFICATION

CAS:	7778-39-4; 1327-52-2
ERG ID:	UN 1553 (liquid); UN 1554 (solid)
ERG Guide:	154 (liquid and solid)
ERG Guide Category:	SUBSTANCES -TOXIC AND/OR CORROSIVE (NONCOMBUSTIBLE)
Molecular Formula:	AsH_3O_4

RTECS No.: CG070000

Properties: Solid crystals or solution. White or colorless. Odorless. Sinks and mixes with water. Freezing point is 95○F. Dust poisonous if inhaled. Solution or solid poisonous if swallowed.

HEALTH AND SAFETY INFORMATION

IDLH: 5 mg/m^3 as arsenic inorganic compounds; human carcinogen

OSHA Table Z-1-A Air Contaminant

Exposure Limits:	TWA	CEILING
ACGIH:[(1)]	0.01 mg/m^3 as As	
NIOSH:[(2)]		0.002 mg/m^3/15 min as As
OSHA:	0.010 mg/m^3 as As	

[(1)] Confirmed carcinogen.
[(2)] Potential carcinogen. Reduce exposure to lowest feasible level.

RECOMMENDATIONS FOR RESPIRATOR SELECTION: NIOSH as arsenic

At any concentrations above the NIOSH REL, or where there is no REL, at any detectable concentration: SCBAF:PD,PP (any self-contained breathing apparatus that has a full faceplate and is operated in a pressure-demand or other positive-pressure mode); or SAF:PD,PP:ASCBA (any supplied-air respirator that has a full facepiece and is operated in a pressure-demand or other positive-pressure mode in combination with an auxiliary, self-contained breathing apparatus operated in a pressure-demand or other positive-pressure mode).

Escape: GMFAGHiE [any air-purifying, full-facepiece respirator (gas mask) with a chin-style, front-or back-mounted acid gas canister having a high-efficiency particulate filter]; or SCBAE (any appropriate escape-type, self-contained breathing apparatus).

CHEMICAL REACTIVITY

Incompatible with sulfuric acid, caustics, ammonia, amines, isocyanates, alkylene oxides, oxidizers, epichlorohydrin, vinyl acetate, and amides. Corrodes metals.

FIRE INFORMATION

Not flammable. If material or contaminated runoff enters waterways, notify downstream users of potentially contaminated water. Notify local health and fire officials and pollution control agencies. Use extinguishing agent suitable for surrounding fire.

FIRST AID

Move victim to fresh air. Call emergency medical care. Apply artificial respiration if victim is not

breathing. Do not use mouth-to-mouth method if victim ingested or inhaled the substance; induce artificial respiration with the aid of a pocket mask equipped with a one-way valve or other proper respiratory medical device. Administer oxygen if breathing is difficult. Remove and isolate contaminated clothing and shoes. In case of contact with substance, immediately flush skin or eyes with running water for at least 20 minutes. For minor skin contact, avoid spreading material on unaffected skin. Keep victim warm and quiet. Effects of exposure (inhalation, ingestion, or skin contact) may be delayed. Ensure that medical personnel are aware of the material(s) involved and take precautions to protect themselves.

ENVIRONMENTAL INFORMATION

Clean Air Act: Hazardous Air Pollutants (Title I, Part A, Section 112) as arsenic compounds.
Clean Water Act: Toxic Pollutant (Section 401.15) as arsenic and compounds.
RCRA Section 261 Hazardous Constituents, as arsenic compounds, n.o.s.
EPCRA Section 304: Reportable Quantity (RQ): CERCLA, 1 lb (0.454 kg).
EPCRA Section 313: Form R *de minimis* concentration reporting level: 0.1%; Form R, Toxic Chemical Category Code: N020.
Marine Pollutant (49CFR, Subchapter 172.101, Appendix B).
California's Proposition 65: Carcinogen.

ARSENIC TRIOXIDE A:560

SYNONYMS: ACIDE ARSENIEUX (French); ANHYDRIDE ARSENIEUX (French); ARSENIC BLANC (French); ARSENIC (III) OXIDE; ARSENIC SESQUIOXIDE; ARSENICUM ALBUM; ARSENIGEN SAURE (German); ARSENIOUS ACID; ARSENIOUS OXIDE; ARSENIOUS TRIOXIDE; ARSENITE; ARSENOLITE; ARSENOUS ACID; ARSENOUS ACID ANHYDRIDE; ARSENOUS ANHYDRIDE; ARSENOUS OXIDE; ARSENOUS OXIDE ANHYDRIDE; ARSENIC SESQUIOXIDE; ARSENTRIOXIDE; ARSODENT; CLAUDELITE; CLAUDETITE; CRUDE ARSENIC; DIARSENIC TRIOXIDE; EEC No. 033-003-00-0; EINECS No. 215-481-4; SPINRITE ARSENIC; TRIOXIDO de ARSENICO (Spanish); WHITE ARSENIC

IDENTIFICATION

CAS:	1327-53-3
ERG ID:	UN 1561
ERG Guide:	151
ERG Guide Category:	TOXIC (NONCOMBUSTIBLE)
Molecular Formula:	As_2O_3
RTECS No.:	CG3325000
Properties:	Solid crystals or powder. White. Garliclike odor. Sinks and mixes slowly with water.

HEALTH AND SAFETY INFORMATION

Poisonous. Corrosive to eyes, skin and respiratory tract. Inhalation may cause pulmonary edema which can be delayed for several hours; there is a risk of death in serious cases. High or prolonged exposures may cause kidney, liver damage, skin damage or disorders, nasal septum damage. May cause cancer. Urinanalysis may be helpful in monitoring exposure.

IDLH: 5 mg/m^3 as arsenic; potential human carcinogen.
OSHA Specifically Regulated Substances (29 CFR 1910.1018)

Exposure Limits:	**TWA**	**CEILING**
ACGIH:*	0.01 mg/m^3 as As	
NIOSH:**		0.002 mg/m^3/15 min as As
OSHA:	0.010 mg/m^3 as As	

* Confirmed carcinogen.
** Potential carcinogen. Reduce exposure to lowest feasible level.

RECOMMENDATIONS FOR RESPIRATOR SELECTION: NIOSH as arsenic

At any concentrations above the NIOSH REL, or where there is no REL, at any detectable concentration: SCBAF:PD,PP (any self-contained breathing apparatus that has a full faceplate and is operated in a pressure-demand or other positive-pressure mode); or SAF:PD,PP:ASCBA (any supplied-air respirator that has a full facepiece and is operated in a pressure-demand or other positive-pressure mode in combination with an auxiliary self-contained breathing apparatus operated in a pressure-demand or other positive-pressure mode).

Escape: GMFAGHiE [any air-purifying, full-facepiece respirator (gas mask) with a chin-style, front-or back-mounted acid gas canister having a high-efficiency particulate filter]; or SCBAE (any appropriate escape-type, self-contained breathing apparatus).

CHEMICAL REACTIVITY

Incompatible with acids, fluorine, fluorides, sodium chlorate, aluminum, halogens, and chemically active metals.

FIRE INFORMATION

If material or contaminated runoff enters waterways, notify downstream users of potentially contaminated waters. Notify local health and fire officials and pollution control agencies. Use any extinguishing agent suitable for surrounding fire.

NFPA Hazard Classifications:

Health Hazards (Blue) 3
Flammability (Red) 0
Reactivity (Yellow) 0

FIRST AID

Move victim to fresh air. Call emergency medical care. Apply artificial respiration if victim is not breathing. Do not use mouth-to-mouth method if victim ingested or inhaled the substance; induce artificial respiration with the aid of a pocket mask equipped with a one-way valve or other proper respiratory medical device. Administer oxygen if breathing is difficult. Remove and isolate contaminated clothing and shoes. In case of contact with substance, immediately flush skin or eyes with running water for at least 20 minutes. For minor skin contact, avoid spreading material on unaffected skin. Keep victim warm and quiet. Effects of exposure (inhalation, ingestion, or skin contact) may be delayed. Ensure that medical personnel are aware of the material(s) involved and take precautions to protect themselves.

Urinary arsenic may be a useful indicator of exposure.

ENVIRONMENTAL INFORMATION

Clean Air Act: Hazardous Air Pollutants (Title I, Part A, Section 112) as arsenic compounds.
Clean Water Act: Section 311 Hazardous Substances/RQ (same as CERCLA); Section 313 Priority Chemicals; Toxic Pollutant (Section 401.15) as arsenic and compounds.
EPA Hazardous Waste Number (RCRA No.): PO12
RCRA Section 261 Hazardous Constituents
RCRA Section 302, Extremely Hazardous Substances: TPQ = 100/10,000 lb (45.4/4,540 kg) solid.
EPCRA Section 304: Reportable Quantity (RQ): EHS, CERCLA, 1 lb (0.454 kg).
EPCRA Section 313: Form R *de minimis* concentration reporting level: 0.1%; Form R, Toxic Chemical Category Code: N020.
California's Proposition 65: Carcinogen.

ASBESTOS A:600

SYNONYMS: ACTINOLITE; AMIANTHUS; AMOSITE; AMOSITE ASBESTOS; AMPHBOLE; ANTHOPHYLITE; ANTHOPHYLLITE; ASBEST (German); ASBESTO (Spanish); ASBESTOS FIBER; CHRYSOTILE; CROCIDOLITE (RIEBECKITE); FIBROUS GRUNERITE; KROKYDOLITH (German); NCI C08991; SERPENTINE; TREMOLITE

IDENTIFICATION

CAS:	1332-21-4 (amianthus, amosite, amphibole) 12001-28-4 (crocidolite) 12001-29-5 (white asbestos, chrysotile) 12172-73-5 (amosite) 77536-66-4 (actinolite) 77536-67-5 (antophylite) 77536-68-6 (tremolite)
ERG ID:	UN 2212 (asbestos, crocidolite, blue, brown, asbestos) UN 2590 (asbestos, chrysotile, white asbestos)
ERG Guide:	171

ERG Guide Category: LOW TO MODERATE HAZARD
RTECS No.: CI6475000
Properties: Fibrous solids. Odorless. White or greenish (chrysotile); blue (crocidolite), gray-green (amosite).

HEALTH AND SAFETY INFORMATION

IARC Evaluation: Group 1; sufficient human evidence.
NTP: Known human carcinogen.
IDLH: Potential human carcinogen.
OSHA Table Z-1-A Air Contaminant
OSHA Specifically Regulated Substances (29 CFR 1910.1001-1048) as amosite.

Exposure Limits:	**TWA**
ACGIH:(1)	0.1 fibers/cc
NIOSH:(1)	
OSHA:(2)	

(1)Potential carcinogen. Reduce exposure to lowest feasible level.
(2)As found in 29 CFR 1910.1001, the OSHA PEL for asbestos fibers (i.e., actinolite asbestos, amosite, anthophyllite asbestos, chrysotile, crocidolite, and tremolite asbestos) is an 8-hour TWA airborne concentration of 0.1 fibers (longer than 5 micrometers and having a length-to-diameter ratio of at least 3:1) per cubic centimeter of air (0.1 fiber/cm^3), as determined by the membrane filter method at approximately 400x magnification with phase contrast illumination. No worker should be exposed in excess of 1 fiber/cm^3 (excursion limit) as averaged over a sampling period of 30 minutes. As found in 29 CFR 1910.1101, the OSHA PEL for asbestos is an 8-hour airborne concentration of 2 fibers (longer than 5 micrometers and having a length-to-diameter ratio of at least 3:1) per cubic centimeter of air (2 fibers/cm^3, as determined by the membrane filter method at approximately 400–450x magnification (4-millimeter objective) with phase-contrast illumination. This standard applies in lieu of the revised standard governing occupational exposure to asbestos (as defined in 1910.1001) and the nonfibrous forms of tremolite, anthophyllite, and actinolite during the period and to the extent that the revised standards have been partially stayed.

RECOMMENDATIONS FOR RESPIRATOR SELECTION: NIOSH/OSHA

At any concentrations above the NIOSH REL, or where there is no REL, at any detectable concentration: SCBAF:PD,PP (any self-contained breathing apparatus that has a full facepiece and is operated in a pressure-demand or other positive-pressure mode); or SAF:PD,PP:ASCBA (any supplied-air respirator that has a full facepiece and is operated in a pressure-demand or other positive-pressure mode in combination with an auxiliary, self-contained breathing apparatus operated in a pressure-demand or other positive-pressure mode).

Escape: HiEF (any air-purifying, full-facepiece respirator with a high-efficiency particulate filter); or SCBAE (any appropriate escape-type, self-contained breathing apparatus).

FIRE INFORMATION

If material or contaminated runoff enters waterways, notify downstream users of potentially contaminated waters. Notify local health and fire officials and pollution control agencies. Use any extinguishing agent suitable for surrounding fire.

FIRST AID

Move victim to fresh air. Call emergency medical care. Apply artificial respiration if victim is not breathing. Administer oxygen if breathing is difficult. In case of contact with substance, immediately flush skin or eyes with running water for at least 20 minutes. Remove and isolate contaminated clothing and shoes. Ensure that medical personnel are aware of the material(s) involved and take precautions to protect themselves.

ENVIRONMENTAL INFORMATION: (friable form)

Clean Air Act: Hazardous Air Pollutants (Title I, Part A, Section 112).
Clean Water Act: Section 307 Priority Pollutants; Section 313 Priority Chemicals; Toxic Pollutant (Section 401.15).
Safe Drinking Water Act: MCL, 7 million fibers/L (longer than 10 microns); MCLG, 7 million fibers/L (longer than 10 microns); Regulated chemical (47 FR 9352).
EPCRA Section 304: Reportable Quantity (RQ): CERCLA, 1 lb (0.454 kg).
EPCRA Section 313: (Reportable only in friable form): Form R *de minimis* concentration reporting level: 0.1%.
TSCA: 716.120c6
California's Proposition 65: Carcinogen.

AZOBISISOBUTYLONITRILE A:700

SYNONYMS: ACETO AZIB; AIBN; AZOBISISOBUTIRONITRILO (Spanish); α,α'-AZOBISISOBUTYLONITRILE; AZOBISISOBUTYRONITRILE, (2,2'); 2,2'-AZOBIS(ISOBUTYRONITRILE); 2 2 AZOBIS ISOBUTYRONITRILE; 2,2'-AZOBIS(2-METHYLPROPIONITRILE); 2 2 AZOBIS-2-METHYLPROPIONITRILE; AZODIISOBUTYRONITRILE; α,α'-AZODIISOBUTYRONITRILE; AZODIISOBUTYRONITRILE; 2,2'-AZODIISOBUTYRONITRILE; AZODIISOBUTYRONITRILE; 2,2'-DICYANO-2,2'-AZOPROPANE; POLY ZOLE AZDN; POROFOR 57; VAZO 64

IDENTIFICATION

CAS:	78-67-1
ERG ID:	UN 2952
ERG Guide:	150
ERG Guide Category:	SUBSTANCES (SELF-REACTIVE/TEMPERATURE CONTROLLED)
Molecular Formula:	$C_8H_{12}N_4$

108 Azobisisobutylonitrile

RTECS No.: UG0800000
Properties: Flammable crystalline solid. White.

HEALTH AND SAFETY INFORMATION

Can pass through the unbroken skin. Inhalation or skin contact can cause weakness, difficult breathing, convulsions, coma and death. High or repeated exposures can cause liver damage; eye and respiratory system irritation, loss of appetite, weakness, dizziness. A highly reactive explosion hazard.

IDLH: 25 mg/m^3 as CN

Exposure Limits:	TWA	CEILING
ACGIH: [1]	5 mg/m^3 as cyanide	
NIOSH: [1]		4.7 ppm; 5 mg/m^3 per 10 minutes as cyanides
OSHA: [1]	5 mg/m^3 as cyanides	

[1] Skin contact contributes significantly in overall exposure

RECOMMENDATIONS FOR RESPIRATOR SELECTION: NIOSH as cyanides

50 mg/m^3: SA (any supplied-air respirator); or SCBAF (any self-contained breathing apparatus with full facepiece).

Emergency or Planned Entry into Unknown Concentrations or IDLH Conditions: SCBAF:PD,PP (any self-contained breathing apparatus that has a full facepiece and is operated in a pressure-demand or other positive-pressure mode); or SAF:PD,PP:ASCBA (any supplied-air respirator that has a full facepiece and is operated in a pressure-demand or other positive-pressure mode in combination with an auxiliary self-contained breathing apparatus operated in a pressure-demand or other positive-pressure mode).

Escape: GMFSHiE [any air-purifying, full-facepiece respirator (gas mask) with a chin-style, front-or back-mounted canister providing protection against the compound of concern and having a high efficiency particulate filter]; or SCBAE (any appropriate escape-type, self-contained breathing apparatus).

CHEMICAL REACTIVITY

Unstable. Contact with oxidizers or solvents may cause fire and explosions. Incompatible with acetone, aluminum hydride, lithium ; can cause explosion. Keep cool and away from heat; decomposition forms N_2 buildup of pressure and can cause closed containers to explode. Store under dry ice or nitrogen blanket.

FIRE INFORMATION

Decomposes at 225°F (107°C). Combustion products include toxic nitrogen oxides and cyanide. Firefighting gear (including SCBA) may not provide adequate protection. If exposure occurs, remove and isolate gear immediately and thoroughly decontaminate personnel. Vapors are heavier than air and will collect in low areas. Vapors in confined areas may explode when exposed to fire. Vapors may travel long distances to ignition sources and flash back. Storage containers and parts of containers may rocket great distances, in many directions. If material or contaminated runoff enters waterways, notify downstream users of potentially contaminated waters. Notify local health and fire officials and pollution control agencies. Use dry chemicals, foam, carbon dioxide. From a secure explosion-proof location, use water spray to cool exposed containers. If cooling streams are ineffective (venting sound increases in volume and pitch, tank discolors or shows any signs of deforming), withdraw immediately to a secure position.

NFPA Hazard Classifications:

Health Hazards (Blue) 3
Flammability (Red) .... -
Reactivity (Yellow) 2

Autoignition Temperature: 147°F (64°C)

FIRST AID

Move victim to fresh air. Call emergency medical care. Apply artificial respiration if victim is not breathing. Administer oxygen if breathing is difficult. Remove and isolate contaminated clothing and shoes. In case of contact with substance, immediately flush skin or eyes with running water for at least 20 minutes. Keep victim warm and quiet. Ensure that medical personnel are aware of the material(s) involved and take precautions to protect themselves.

ENVIRONMENTAL INFORMATION as cyanide compound

EPCRA Section 313: Form R *de minimus* concentration reporting level: 1.0%.
Marine Pollutant (49CFR, Subchapter 172.101, Appendix B) as cyanide mixtures, cyanide solutions.

AZODICARBAMIDE — A:800

SYNONYMS: 1,1'-AZOBISCARBAMIDE; AZOBISCARBONAMIDE; AZOBISCARBOXAMIDE; 1,1'-AZOBIS(FORMAMIDE); 1 1 AZOBIS FORMAMIDE; AZODICARBOAMIDE; AZODICARBOAMIDA (Spanish); AZODICARBONAMIDE (DOT); AZODICARBOXAMIDE; AZODICARBOXYLIC ACID DIAMIDE; D(1,1')-BIUREA; D 1 1 BIUREA; CELOSEN AZ; ChKhZ 21; C K Z 21; DIAZENEDICARBOXAMIDE; GENITRON AC; GENITRON AC 2; GENITRON AC 4; KEMPORE; KEMPORE 125; KEMPORE R 125; LUCEL ADA; NCI-C55981; NITROPORE; PINHOLE AK 2; POROFOR 505; POROFOR ADC R; POROFOR ChKhZ 21; POROFOR C K Z 21; UNIFOAM AZ; UNIFORM AZ; YUNIHOMU AZ

IDENTIFICATION

CAS:	123-77-3
ERG ID:	UN 3242
ERG Guide:	149

110 Azodicarbamide

ERG Guide Category: SUBSTANCES (SELF-REACTIVE)
Molecular Formula: $C_2H_4N_4O_2$
RTECS No.: LQ1040000
Properties: Flammable solid. Orange to orange-red powder or crystals. Insoluble in cold water; slightly soluble in hot water.

HEALTH AND SAFETY INFORMATION

May produce irritating, toxic, and/or corrosive gases. Inhalation, or contact with vapors, substance, or composition products may cause severe injury or death.

RECOMMENDATIONS FOR RESPIRATOR SELECTION:

At any concentrations above the NIOSH REL, or where there is no REL, at any detectable concentration: SCBAF:PD,PP (any self-contained breathing apparatus that has a full facepiece and is operated in a pressure-demand or other positive-pressure mode); or SAF:PD,PP:ASCBA (any supplied-air respirator that has a full facepiece and is operated in a pressure-demand or other positive-pressure mode in combination with an auxiliary, self-contained breathing apparatus operated in a pressure-demand or other positive-pressure mode).

Escape: GMFOV [any air-purifying, full-facepiece respirator (gas mask) with a chin-style, front-or back-mounted organic vapor canister]; or SCBAE (any appropriate escape-type, self-contained breathing apparatus).

CHEMICAL REACTIVITY

Dust may form explosive mixture with air. Contact with oxidizers may cause fire and explosions. May be ignited by heat, sparks, or flame. Heat, chemical reaction, friction, or impact may cause self decomposition or self-ignition.

FIRE INFORMATION

Decomposition products upon heating include toxic nitrogen oxides. Vapors are heavier than air and will collect in low areas. Vapors in confined areas may explode when exposed to fire. If material or contaminated runoff enters waterways, notify downstream users of potentially contaminated waters. Notify local health and fire officials and pollution control agencies. Use dry chemicals, alcohol-resistant foam, carbon dioxide. From a secure explosion-proof location, use water spray to cool exposed containers. If cooling streams are ineffective (venting sound increases in volume and pitch, tank discolors or shows any signs of deforming), withdraw immediately to a secure position.

FIRST AID

Move victim to fresh air. Call emergency medical care. Apply artificial respiration if victim is not breathing. Administer oxygen if breathing is difficult. Remove and isolate contaminated clothing and shoes. Remove material from skin immediately. In case of contact with substance, immediately flush skin or eyes with running water for at least 20 minutes. Keep victim warm and quiet. Ensure that medical personnel are aware of the material(s) involved and take precautions to protect themselves.

BARIUM

B:001

SYNONYMS: BARIO (Spanish); BARIUM, DENDRITIC PIECES; BARIUM, ELEMENTAL; BARIUM METAL; BARIUM, NONPYROPHORIC; BARIUM, PYROPHORIC; BARYUM (French); EINECS No. 231-149-1

IDENTIFICATION

CAS:	7440-39-3
ERG ID:	UN 1399, Barium alloy; UN 1400, Barium and barium metal; UN 1854, Barium alloy, pyrophoric
ERG Guide:	135 (barium alloy, pyrophoric); 138 (barium, barium alloy, barium metal)
ERG Guide Category:	135 SUBSTANCES-SPONTANEOUSLY COMBUSTIBLE 138 SUBSTANCES-WATER-REACTIVE (EMITTING FLAMMABLE GASES)
Molecular Formula:	Ba
RTECS No.:	CQ8370000
Properties:	Combustible solid. Lustrous, yellowish-white, malleable. Odorless. Extremely reactive; readily oxidized. Dangerous when wet.

HEALTH AND SAFETY INFORMATION

Poisonous. Dust irritates respiratory tract. High or repeated exposures may cause weakness, paralysis of limbs, respiratory system, heart; circulatory system effects.

IDLH: 50 mg/m^3 as barium.

Exposure Limits:	**TWA**
ACGIH:	0.5 mg/m^3
NIOSH:	0.5 mg/m^3 as Ba
OSHA:	0.5 mg/m^3 as Ba

RECOMMENDATIONS FOR RESPIRATOR SELECTION: NIOSH

At any concentrations above the NIOSH REL, or where there is no REL, at any detectable concentration: SCBAF:PD,PP (any self-contained breathing apparatus that has a full facepiece and is operated in a pressure-demand or other positive-pressure mode); or SAF:PD,PP:ASCBA (any supplied-air respirator that has a full facepiece and is operated in a pressure-demand or other positive-pressure mode in combination with an auxiliary, self-contained breathing apparatus operated in a pressure-demand or other positive-pressure mode).

Escape: GMFOV [any air-purifying, full-facepiece respirator (gas mask) with a chin-style, front- or back-mounted organic vapor canister]; or SCBAE (any appropriate escape-type, self-contained breathing apparatus).

CHEMICAL REACTIVITY

Powder or dust can self-ignite in air at room temperature. Usually stored under a blanket of inert gas, kerosene, petroleum, or other oxygen-free liquid. Rapid decomposition and possible ignition with water. Extremely reactive. Fire and explosions may be caused by ammonia, alcohol, halogenated solvents, carbon tetrachloride, trichlorotrifluoroethane, acids, oxygen, or oxidizers.

FIRE INFORMATION

Powder or dust is pyrophoric. If material or contaminated runoff enters waterways, notify downstream users of potentially contaminated waters. Notify local health and fire officials and pollution control agencies. Use any extinguishing agent suitable for surrounding fire.

FEMA Hazard Classifications:

Health Hazards (Blue) 1
Flammability (Red) 4
Reactivity (Yellow) 3
Special situations .... WATER

FIRST AID

SUBSTANCES-SPONTANEOUSLY COMBUSTIBLE (135): Move victim to fresh air. Call emergency medical care. Apply artificial respiration if victim is not breathing. Administer oxygen if breathing is difficult. Remove and isolate contaminated clothing and shoes. In case of contact with substance, immediately flush skin or eyes with running water for at least 20 minutes. Keep victim warm and quiet. Ensure that medical personnel are aware of the material(s) involved and take precautions to protect themselves.

SUBSTANCES-WATER-REACTIVE (EMITTING FLAMMABLE GASES) (138): Move victim to fresh air. Call emergency medical care. Apply artificial respiration if victim is not breathing. Administer oxygen if breathing is difficult. Remove and isolate contaminated clothing and shoes. In case of contact with substance, wipe from skin immediately; flush skin or eyes with running water for at least 20 minutes. Keep victim warm and quiet. Ensure that medical personnel are aware of the material(s) involved and take precautions to protect themselves.

ENVIRONMENTAL INFORMATION

EPA Hazardous Waste Number (RCRA No.): D005.
RCRA Toxicity Characteristic (Section 261.24): Maximum Concentration of Contaminants, regulatory level, 100.0 mg/L.
RCRA Section 261 Hazardous Constituents
EPA Hazardous Waste Number (RCRA No.): D005
RCRA Maximum Concentration Limit for Ground Water Protection: 1.0 mg/L.
RCRA Land Ban Waste
RCRA Universal Treatment Standards: Wastewater (mg/L), 1.2; Nonwastewater (mg/L), 7.6 TCLP.

RCRA Ground Water Monitoring List: Suggested methods (PQL ug/L): 6010(20); 7080(1,000).
Safe Drinking Water Act: MCL, 2 mg/L; MCLG, 2 mg/L; Regulated chemical (47 FR 9352).
EPCRA Section 313: Form R *de minimis* concentration reporting level: 1.0%; Form R Toxic Chemical Category Code: N040.
Marine Pollutant (49CFR, Subchapter 172.101, Appendix B).

BARIUM CARBONATE B:010

SYNONYMS: BARIUMCARBONAAT, NATUURLIQK (Dutch); BARIUM CARBONATE (1:1); CARBONATE de BARYUM (French); CARBONATO BARICO (Spanish); CARBONATO di BARIO (Italian); CARBONIC ACID, BARIUM SALT (1:1); C.I. 77099; C.I. PIGMENT WHITE 10; EEC No. 056-002-00-7; EINECS No. 208-167-3

IDENTIFICATION

CAS:	513-77-9
ERG ID:	UN 1564
ERG Guide:	154
ERG Guide Category:	SUBSTANCES-TOXIC AND/OR CORROSIVE (NONCOMBUSTIBLE)
Molecular Formula:	$CBaO_3$
RTECS No.:	CQ8600000
Properties:	Solid crystals or powder. White to cream color. Odorless. Sinks in water; slightly soluble.

HEALTH AND SAFETY INFORMATION

Highly poisonous; potentially lethal to humans. May cause skin problems. High or repeated exposures may cause weakness, paralysis of limbs, respiratory system, heart.

IDLH: 50 mg/m^3 as barium, soluble compounds
OSHA Table Z-1-A Air Contaminant as barium soluble compound

Exposure Limits:	**TWA**
ACGIH:	0.5 mg/m^3 as Ba
NIOSH:	0.5 mg/m^3 as Ba
OSHA:	0.5 mg/m^3 as Ba

RECOMMENDATIONS FOR RESPIRATOR SELECTION: NIOSH/OSHA as barium

5 mg/m^3: DMXSQ (any dust and mist respirator except single-use and quarter mask respirators); or SA (any supplied-air respirator).

12.5 mg/m^3: SA:CF (any supplied-air respirator operated in a continuous-flow mode); or PAPRDM (any powered, air-purifying respirator with a dust and mist filter).
25 mg/m^3: HiEF (any air-purifying, full-facepiece respirator with a high-efficiency particulate filter); or SAT:CF (any supplied-air respirator that has a tight-fitting facepiece and is operated in a continuous-flow mode); or PAPRTHiE (any powered, air-purifying respirator with a tight-fitting facepiece and a high-efficiency particulate filter); or SCBAF (any self-contained breathing apparatus with a full facepiece); or SAF (any supplied-air respirator with a full facepiece).
50 mg/m^3: SAF:PD,PP (any supplied-air respirator that has a full facepiece and is operated in a pressure-demand or other positive-pressure mode).

Emergency or Planned Entry into Unknown Concentrations or IDLH Conditions: SCBAF:PD,PP (any self-contained breathing apparatus that has a full facepiece and is operated in a pressure-demand or other positive-pressure mode); or SAF:PD,PP:ASCBA (any supplied-air respirator that has a full facepiece and is operated in a pressure-demand or other positive-pressure mode in combination with an auxiliary, self-contained breathing apparatus operated in a pressure-demand or other positive-pressure mode).

Escape: HiEF (any air-purifying, full-facepiece respirator with a high-efficiency particulate filter); or SCBAE (any appropriate escape-type, self-contained breathing apparatus).

CHEMICAL REACTIVITY

Incompatible with acids, aluminum, fluorine, magnesium, silicon, and oxidizers.

FIRE INFORMATION

Not flammable. If material or contaminated runoff enters waterways, notify downstream users of potentially contaminated waters. Notify local health and fire officials and pollution control agencies. Use any extinguishing agent suitable for surrounding fire.

FEMA Hazard Classifications:

Health Hazards (Blue) 2
Flammability (Red) 0
Reactivity (Yellow) 0

FIRST AID

Move victim to fresh air. Call emergency medical care. Apply artificial respiration if victim is not breathing. Do not use mouth-to-mouth method if victim ingested or inhaled the substance; induce artificial respiration with the aid of a pocket mask equipped with a one-way valve or other proper respiratory medical device. Administer oxygen if breathing is difficult. Remove and isolate contaminated clothing and shoes. In case of contact with substance, immediately flush skin or eyes with running water for at least 20 minutes. For minor skin contact, avoid spreading material on unaffected skin. Keep victim warm and quiet. Effects of exposure (inhalation, ingestion, or skin contact) may be delayed. Ensure that medical personnel are aware of the material(s) involved and take precautions to protect themselves.

ENVIRONMENTAL INFORMATION as barium compounds

RCRA Section 261 Hazardous Constituents: As barium compounds, n.o.s.
EPCRA Section 313: Includes any unique chemical substance that contains barium as part of that chemical's infrastructure.This category does not include barium sulfate (7727-43-7). Form R *de minimis* concentration reporting level: 0.1%. Form R Toxic Chemical Category Code: N040.
Marine Pollutant (49CFR, Subchapter 172.101, Appendix B).

BARIUM CHLORIDE B:020

SYNONYMS: BA-01083; BARIUM DICHLORIDE; CLORURO BARICO (Spanish); EEC No. 056-002-00-7; NCI-C61074

IDENTIFICATION

CAS:	10361-37-2
ERG ID:	UN 1564
ERG Guide:	154
ERG Guide Category:	TOXIC AND/OR CORROSIVE (NON-COMBUSTIBLE)
Molecular Formula:	$BaCl_2$
RTECS No.:	CQ8750000
Properties:	Colorless water-soluble, crystalline powder. Odorless. Soluble in water.

HEALTH AND SAFETY INFORMATION

May cause skin irritation. Toxic if swallowed or inhaled. Some barium compounds may be contaminated with silica. Silica can cause gradual lungs scarring. High or repeated exposures may cause weakness, paralysis of limbs, respiratory system, heart.

IDLH: 50 mg/m^3 as barium, soluble compounds.
OSHA Table Z-1-A Air Contaminant as barium soluble compound

Exposure Limits: TWA

ACGIH:	0.5 mg/m^3 as barium (soluble compounds)
NIOSH:	0.5 mg/m^3 as barium (soluble compounds)
OSHA:	0.5 mg/m^3 as barium (soluble compounds)

RECOMMENDATIONS FOR RESPIRATOR SELECTION: NIOSH/OSHA as barium (soluble compounds)

5 mg/m^3: DMXSQ (any dust and mist respirator except single-use and quarter mask respirators); or SA (any supplied-air respirator)

12.5 mg/m³: SA:CF (any supplied-air respirator operated in a continuous-flow mode); or PAPRDM (any powered, air-purifying respirator with a dust and mist filter).
25 mg/m³: HiEF (Any air-purifying, full-facepiece respirator with a high-efficiency particulate filter); or SAT:CF (Any supplied-air respirator that has a tight-fitting facepiece and is operated in a continuous-flow mode); or PAPRTHiE (any powered, air-purifying respirator with a tight-fitting facepiece and a high-efficiency particulate filter); or SCBAF (any self-contained breathing apparatus with a full facepiece); or SAF (any supplied-air respirator with a full facepiece).
50 mg/m³: SAF:PD,PP (Any supplied-air respirator that has a full facepiece and is operated in a pressure-demand or other positive-pressure mode).

Emergency or Planned Entry into Unknown Concentrations or IDLH Conditions:
SCBAF:PD,PP (any self-contained breathing apparatus that has a full facepiece and is operated in a pressure-demand or other positive-pressure mode); or SAF:PD,PP:ASCBA (Any supplied-air respirator that has a full facepiece and is operated in a pressure-demand or other positive-pressure mode in combination with an auxiliary, self-contained breathing apparatus operated in a pressure-demand or other positive-pressure mode.

Escape: HiEF (Any air-purifying, full-facepiece respirator with a high-efficiency particulate filter); or SCBAE (Any appropriate escape-type, self-contained breathing apparatus).

CHEMICAL REACTIVITY

Corrosive substance. May attack some metals in the presence of moisture.

FIRE INFORMATION

Combustion products include toxic chlorine fumes. If material or contaminated runoff enters waterways, notify downstream users of potentially contaminated waters. Notify local health and fire officials and pollution control agencies. Use any extinguishing agent suitable for surrounding fire.

FIRST AID

Move victim to fresh air; call emergency medical care. Apply artificial respiration if victim is not breathing. Do not use mouth-to-mouth method if victim ingested or inhaled the substance; induce artificial respiration with the aid of a pocket mask equipped with a one-way valve or other proper respiratory medical device. Administer oxygen if breathing is difficult. Remove and isolate contaminated clothing and shoes. In case of contact with substance, immediately flush skin or eyes with running water for at least 20 minutes. For minor skin contact, avoid spreading material on unaffected skin. Keep victim warm and quiet. Effects of exposure (inhalation, ingestion, or skin contact) to substance may be delayed. Ensure that medical personnel are aware of the material(s) involved and take precautions to protect themselves.

ENVIRONMENTAL INFORMATION

RCRA Section 261 Hazardous Constituents, as barium compounds, n.o.s.
EPCRA Section 313: Includes any unique chemical substance that contains barium as part of that chemical's infrastructure. This category does not include barium sulfate (7727-43-7). Form R *de minimus* concentration reporting level: 0.1%. Form R Toxic Chemical Category Code: N040.

MARINE POLLUTANT (49CFR, Subchapter 172.101, Appendix B).
TSCA: PAIR Reporting; cited 40CFR 712.30(*w*).

BARIUM CHROMATE B:030

SYNONYMS: BARIUMCHROMAAT (Dutch); BARIUM CHROMATE(VI); BARIUM CHROMATE(1:1); BARIUM CHROMATE, A0735; BARIUM CHROMATE OXIDE; BARYTA YELLOW; CHROMATE de BARYUM (French); CHROMATO BARICO (Spanish); CHROMATO de BARIO (Spanish); CHROMATO di BARIO (Italian); CHROMIC ACID, BARIUM SALT(1:1); C.I. 77103; C.I. PIGMENT YELLOW 31; EINECS No. 233-660-5; LEMON CHROME; LEMON YELLOW; PERMANENT YELLOW; STEINBUHL YELLOW; ULTRAMARINE YELLOW.

IDENTIFICATION

CAS:	10294-40-3
ERG ID:	UN 1564
ERG Guide:	154
ERG Guide Category:	SUBSTANCES-TOXIC AND/OR CORROSIVE (NONCOMBUSTIBLE)
Molecular Formula:	$Ba{\bullet}CrH_2O_4$
RTECS No.:	CQ876000
Properties:	Combustible crystalline powder. Yellow. Odorless. Insoluble in water.

HEALTH AND SAFETY INFORMATION

High or repeated exposures may cause weakness, paralysis of limbs, respiratory system, heart.

OSHA Table Z-1-A Air Contaminant as chromium metal and insoluble salts (as Cr)

Exposure Limits:	**TWA**
ACGIH:*	0.5 mg/m^3 as Ba
	0.05 mg/m^3 as Cr
NIOSH:	0.001 mg/m^3 as Cr
OSHA:	0.5 mg/m^3 as Ba
	0.1 mg/m^3 as CrO_3

*Confirmed human carcinogen. Chromates are occupational carcinogens; limit exposure to lowest feasible concentration.

RECOMMENDATIONS FOR RESPIRATOR SELECTION: NIOSH

At any concentrations above the NIOSH REL, or where there is no REL, at any detectable concentration: SCBAF:PD,PP (any self-contained breathing apparatus that has a full facepiece and is

operated in a pressure-demand or other positive-pressure mode); or SAF:PD,PP:ASCBA (any supplied-air respirator that has a full facepiece and is operated in a pressure-demand or other positive-pressure mode in combination with an auxiliary, self-contained breathing apparatus operated in a pressure-demand or other positive-pressure mode).

Escape: GMFOV [any air-purifying, full-facepiece respirator (gas mask) with a chin-style, front- or back-mounted organic vapor canister]; or SCBAE (any appropriate escape-type, self-contained breathing apparatus).

CHEMICAL REACTIVITY

Reacts vigorously with reducing agents, accelerators, combustible organic materials.

FIRE INFORMATION

Not flammable. Decomposition products upon heating include toxic gases. If material or contaminated runoff enters waterways, notify downstream users of potentially contaminated waters. Notify local health and fire officials and pollution control agencies.
Use any extinguishing agent suitable for surrounding fire.

FIRST AID

Move victim to fresh air. Call emergency medical care. Apply artificial respiration if victim is not breathing. Do not use mouth-to-mouth method if victim ingested or inhaled the substance; induce artificial respiration with the aid of a pocket mask equipped with a one-way valve or other proper respiratory medical device. Administer oxygen if breathing is difficult. Remove and isolate contaminated clothing and shoes. In case of contact with substance, immediately flush skin or eyes with running water for at least 20 minutes. For minor skin contact, avoid spreading material on unaffected skin. Keep victim warm and quiet. Effects of exposure (inhalation, ingestion, or skin contact) may be delayed. Ensure that medical personnel are aware of the material(s) involved and take precautions to protect themselves.

ENVIRONMENTAL INFORMATION as barium compounds.

RCRA Section 261 Hazardous Constituents: As barium compounds, n.o.s.
EPCRA Section 313: Includes any unique chemical substance that contains barium as part of that chemical's infrastructure. This category does not include barium sulfate (7727-43-7). Form R *de minimis* concentration reporting level: 0.1%. Form R Toxic Chemical Category Code: N040.
Marine Pollutant (49CFR, Subchapter 172.101, Appendix B).

ENVIRONMENTAL INFORMATION as chromium compounds.

Clean Air Act: Hazardous Air Pollutants (Title I, Part A, Section 112).
Clean Water Act: Toxic Pollutant (Section 401.15); Section 307 Toxic Pollutants as chromium and compounds.
RCRA Section 261 Hazardous Constituents., waste number not listed.

EPCRA (Section 313): Includes any unique chemical substances that contains chromium as part of that chemical's infrastructure. Form R *de minimis* concentration reporting level: Chromium(VI).compounds: 0.1%; Chromium(III) compounds: 1.0%. Form R Toxic Chemical Category Code: N090.
California's Proposition 65: Carcinogen as chromium (hexavalent compounds).

BARIUM CYANIDE B:035

SYNONYMS: BARIUMCYANID (German); BARIUM CYANIDE, SOLID; BARIUM DICYANIDE; CIANURO BARICO (Spanish); CIANURO di BARIO (Italian); CYANURE de BARYUM (French); EINECS No. 208-822-3

IDENTIFICATION

CAS:	542-62-1
ERG ID:	UN 1565
ERG Guide:	157
ERG Guide Category:	SUBSTANCES-TOXIC AND/OR CORROSIVE (NONCOMBUSTIBLE/WATER-SENSITIVE)
Molecular Formula:	C_2BaN_2
RTECS No.:	CQ8785000
Properties:	Noncombustible crystals or powder. White. Sinks in water; soluble.

HEALTH AND SAFETY INFORMATION

Extremely poisonous via any route. High or repeated exposures may cause weakness, paralysis of limbs, respiratory system, heart.

IDLH: 50 mg/m^3 as barium, soluble compounds.
OSHA Table Z-1-A Air Contaminant as barium soluble compound

Exposure Limits:	**TWA**
ACGIH:	0.5 mg/m^3 as Ba
NIOSH:	0.5 mg/m^3 as Ba
OSHA:	0.5 mg/m^3 as Ba

RECOMMENDATIONS FOR RESPIRATOR SELECTION: NIOSH/OSHA as barium

5 mg/m^3: DMXSQ (any dust and mist respirator except single-use and quarter mask respirators); or SA (any supplied-air respirator).
12.5 mg/m^3: SA:CF (any supplied-air respirator operated in a continuous-flow mode); or PAPRDM (any powered, air-purifying respirator with a dust and mist filter).

25 mg/m^3: HiEF (any air-purifying, full-facepiece respirator with a high-efficiency particulate filter); or SAT:CF (any supplied-air respirator that has a tight-fitting facepiece and is operated in a continuous-flow mode); or PAPRTHiE (any powered, air-purifying respirator with a tight-fitting facepiece and a high-efficiency particulate filter); or SCBAF (any self-contained breathing apparatus with a full facepiece); or SAF (any supplied-air respirator with a full facepiece).
50 mg/m^3: SAF:PD,PP (any supplied-air respirator that has a full facepiece and is operated in a pressure-demand or other positive-pressure mode).

Emergency or Planned Entry into Unknown Concentrations or IDLH Conditions: SCBAF:PD,PP (any self-contained breathing apparatus that has a full facepiece and is operated in a pressure-demand or other positive-pressure mode); or SAF:PD,PP:ASCBA (any supplied-air respirator that has a full facepiece and is operated in a pressure-demand or other positive-pressure mode in combination with an auxiliary, self-contained breathing apparatus operated in a pressure-demand or other positive-pressure mode).

Escape: HiEF (any air-purifying, full-facepiece respirator with a high-efficiency particulate filter); or SCBAE (any appropriate escape-type, self-contained breathing apparatus).

CHEMICAL REACTIVITY

Reacts with nitric acid, nitrates, nitrites, chlorates, magnesium, and strong oxidizers. Corrosive to metals in the presence of moisture. Barium will precipitate as sulfate or carbonate. Cyanide will slowly convert to the less toxic cyanate. The cyanide ion is in equilibrium with HCN, a very weak acid. Cyanides in water are corrosive to metals.

FIRE INFORMATION

Not flammable. Very poisonous; combustion products include toxic nitrogen oxide and barium cyanide fumes. If material or contaminated runoff enters waterways, notify downstream users of potentially contaminated waters. Notify local health and fire officials and pollution control agencies.

FEMA Hazard Classifications:

Health Hazards (Blue) 3
Flammability (Red) 0
Reactivity (Yellow) 0

FIRST AID

Move victim to fresh air. Call emergency medical care. Apply artificial respiration if victim is not breathing. Do not use mouth-to-mouth method if victim ingested or inhaled the substance; induce artificial respiration with the aid of a pocket mask equipped with a one-way valve or other proper respiratory medical device. Administer oxygen if breathing is difficult. Remove and isolate contaminated clothing and shoes. In case of contact with substance, immediately flush skin or eyes with running water for at least 20 minutes. For minor skin contact, avoid spreading material on unaffected skin. Keep victim warm and quiet. Effects of exposure (inhalation, ingestion, or skin contact) may be delayed. Ensure that medical personnel are aware of the material(s) involved and take precautions to protect themselves.

ENVIRONMENTAL INFORMATION as barium compounds

Clean Water Act: Section 311 Hazardous Substances/RQ (same as CERCLA); Section 313 Priority Chemicals.
EPA Hazardous Waste Number (RCRA No.): P013
RCRA Section 261 Hazardous Constituents
EPCRA Section 304: Reportable Quantity (RQ): CERCLA, 10 lb (4.54 kg).
EPCRA Section 313: (as barium compounds). Form R *de minimis* concentration reporting level: 0.1%. Form R Toxic Chemical Category Code: N040.
EPCRA Section 313: (As cyanide compounds): Form R *de minimis* concentration reporting level: 1.0%. Form R Toxic Chemical Category Code: N106.

ENVIRONMENTAL INFORMATION as cyanide compounds

Clean Air Act: Hazardous Air Pollutants (Title I, Part A, Section 112).
Clean Water Act: Section 307 Priority Pollutants as cyanide, total.
EPA Hazardous Waste Number (RCRA No.): P030 as cyanides soluble salts and complexes, n.o.s.
RCRA Section 261 Hazardous Constituents as cyanides, soluble salts, and complexes, n.o.s.
EPCRA (Section 313): X+CN-where X = H+ or any other group where a formal dissociation may occur. For example, KCN or Ca(CN)2. Form R *de minimis* concentration reporting level: 1.0%. Form R Toxic Chemical Category Code: N016.
Marine Pollutant (49CFR, Subchapter 172.101, Appendix B) as cyanide mixtures, cyanide solutions, or cyanides, inorganic, n.o.s.

BARIUM NITRATE B:060

SYNONYMS: BARIUM DINITRATE; BARIUMNITRAAT (Dutch); BARIUMNITRAT (German); BARIUM(1l) NITRATE(1:1); BARIUM SALT OF NITRIC ACID; DUSICNAN BARNATY (Czech); EEC No. 056-002-00-7 as barium; EINECS No. 233-020-5; NITRATE de BARYUM (French); NITRATO BARICO (Spanish); NITRIC ACID, BARIUM SALT

IDENTIFICATION

CAS:	10022-31-8
ERG ID:	UN 1446
ERG Guide:	141
ERG Guide Category:	OXIDIZERS-TOXIC (SOLID)
Molecular Formula:	BaN_2O_6
RTECS No.:	CQ9625000
Properties:	Colorless to clear crystalline solid. Odorless. Soluble in water.

HEALTH AND SAFETY INFORMATION

Irritates eyes, skin, and respiratory system. High or repeated exposures may cause weakness, paralysis of limbs, respiratory system, heart problems.

IDLH: 50 mg/m^3 as barium, soluble compounds.
OSHA Table Z-1-A Air Contaminant as barium soluble compound.

Exposure Limits:	TWA
ACGIH:	0.5 mg/m^3 as barium (soluble compounds)
NIOSH:	0.5 mg/m^3 as barium (soluble compounds)
OSHA:	0.5 mg/m^3 as barium (soluble compounds)

RECOMMENDATIONS FOR RESPIRATOR SELECTION: NIOSH/OSHA as barium (soluble compounds).

5 mg/m^3: DMXSQ (any dust and mist respirator except single-use and quarter mask respirators); or SA (any supplied-air respirator).
12.5 mg/m^3: SA:CF (any supplied-air respirator operated in a continuous-flow mode); or PAPRDM (any powered, air-purifying respirator with a dust and mist filter).
25 mg/m^3: HiEF (any air-purifying, full-facepiece respirator with a high-efficiency particulate filter); or SAT:CF (any supplied-air respirator that has a tight-fitting facepiece and is operated in a continuous-flow mode); or PAPRTHiE (any powered, air-purifying respirator with a tight-fitting facepiece and a high-efficiency particulate filter); or SCBAF (any self-contained breathing apparatus with a full facepiece); or SAF (any supplied-air respirator with a full facepiece).
50 mg/m^3: SAF:PD,PP (any supplied-air respirator that has a full facepiece and is operated in a pressure-demand or other positive-pressure mode).

Emergency or Planned Entry into Unknown Concentrations or IDLH Conditions: SCBAF:PD,PP (any self-contained breathing apparatus that has a full facepiece and is operated in a pressure-demand or other positive-pressure mode); or SAF:PD,PP:ASCBA (any supplied-air respirator that has a full facepiece and is operated in a pressure-demand or other positive-pressure mode in combination with an auxiliary, self-contained breathing apparatus operated in a pressure-demand or other positive-pressure mode).

Escape: HiEF (any air-purifying, full-facepiece respirator with a high-efficiency particulate filter); or SCBAE (any appropriate escape-type, self-contained breathing apparatus).

Note: Only nonoxidiziable sorbents should be used (not charcoal).

CHEMICAL REACTIVITY

A strong oxidizer. Incompatible with strong acids, combustibles, aluminum-magnesium alloys, and sulfur; finely divided metals may form friction and shock-sensitive compounds.

FIRE INFORMATION

If material or contaminated runoff enters waterways, notify downstream users of potentially contaminated waters. Notify local health and fire officials and pollution control agencies. Use any extinguishing agent suitable for surrounding fire.

FEMA Hazard Classifications:

Health Hazards (Blue) 2
Flammability (Red) 0
Reactivity (Yellow) 0

FIRST AID

Move victim to fresh air. Call emergency medical care. Apply artificial respiration if victim is not breathing. Administer oxygen if breathing is difficult. Remove and isolate contaminated clothing and shoes. In case of contact with substance, immediately flush skin or eyes with running water for at least 20 minutes. Keep victim warm and quiet. Ensure that medical personnel are aware of the material(s) involved and take precautions to protect themselves.

ENVIRONMENTAL INFORMATION as barium compounds

RCRA Section 261 Hazardous Constituents
EPCRA Section 313: Form R *de minimis* concentration reporting level: 0.1%. Form R Toxic Chemical Category Code: N040.
Marine Pollutant (49CFR, Subchapter 172.101, Appendix B).

ENVIRONMENTAL INFORMATION as nitrate water dissociable compounds

EPCRA Section 313: Reportable only when in aqueous solution. Form R *de minimis* concentration reporting level: 1.0%. Form R Toxic Chemical Category Code: N511.
Marine Pollutant (49CFR, Subchapter 172.101, Appendix B) as nitrates, inorganic, n.o.s.

BARIUM STEARATE B:075

SYNONYMS: BARIUM DISTEARATE; ESTEARATO BARICO (Spanish); OCTADECANOIC ACID, BARIUM SALT; STAVINOR 40; STEARIC ACID, BARIUM SALT

IDENTIFICATION

CAS:	6865-35-6
ERG ID:	UN 1564
ERG Guide:	154
ERG Guide Category:	SUBSTANCES-TOXIC AND/OR CORROSIVE (NONCOMBUSTIBLE)
Molecular Formula:	$C_{36}H_{70}BaO_4$
RTECS No.:	WI2840000
Properties:	Fine white, crystalline powder. Mild fatty odor. Insoluble in water.

HEALTH AND SAFETY INFORMATION

High or repeated exposures may cause weakness, paralysis of limbs, respiratory system, heart.

RECOMMENDATIONS FOR RESPIRATOR SELECTION

At any concentrations above the NIOSH REL, or where there is no REL, at any detectable concentration: SCBAF:PD,PP (any self-contained breathing apparatus that has a full facepiece and is operated in a pressure-demand or other positive-pressure mode); or SAF:PD,PP:ASCBA (any supplied-air respirator that has a full facepiece and is operated in a pressure-demand or other positive-pressure mode in combination with an auxiliary, self-contained breathing apparatus operated in a pressure-demand or other positive-pressure mode).

Escape: GMFOVHiE [any air-purifying, full-facepiece respirator (gas mask) with a chin-style, front- or back-mounted organic vapor canister having a high-efficiency particulate filter]; or SCBAE (any appropriate escape-type, self-contained breathing apparatus)

CHEMICAL REACTIVITY

Contact with oxidizers may cause fires and explosions.

FIRE INFORMATION

If material or contaminated runoff enters waterways, notify downstream users of potentially contaminated waters. Notify local health and fire officials and pollution control agencies. Use dry chemicals, alcohol-resistant foam, carbon dioxide. From a secure explosion-proof location, use water spray to cool exposed containers. If cooling streams are ineffective (venting sound increases in volume and pitch, tank discolors, or shows any signs of deforming), withdraw immediately to a secure position.

NFPA Hazard Classifications: not listed
Flash Point: 500°F (260°C) c.c.

FIRST AID

Move victim to fresh air. Call emergency medical care. Apply artificial respiration if victim is not breathing. Do not use mouth-to-mouth method if victim ingested or inhaled the substance; induce artificial respiration with the aid of a pocket mask equipped with a one-way valve or other proper respiratory medical device. Administer oxygen if breathing is difficult. Remove and isolate contaminated clothing and shoes. In case of contact with substance, immediately flush skin or eyes with running water for at least 20 minutes. For minor skin contact, avoid spreading material on unaffected skin. Keep victim warm and quiet. Effects of exposure (inhalation, ingestion, or skin contact) may be delayed. Ensure that medical personnel are aware of the material(s) involved and take precautions to protect themselves.

ENVIRONMENTAL INFORMATION as barium compounds

RCRA Section 261 Hazardous Constituents: As barium compounds, n.o.s.

EPCRA Section 313: Includes any unique chemical substance that contains barium as part of that chemical's infrastructure. This category does not include barium sulfate (7727-43-7). Form R *de minimis* concentration reporting level: 0.1%. Form R Toxic Chemical Category Code: N040.
Marine Pollutant (49CFR, Subchapter 172.101, Appendix B).

BARIUM SULFATE B:076

SYNONYMS: ACTYBARYTE; ARTIFICIAL BARITE; ARTIFICIAL HEAVY SPAR; BAKONTAL; BARIDOL; BARITE; BARITOP; BAROSPERSE; BAROTRAST; BARYTA WHITE; BARYTES; BASOFAR: BAYRITE; BIANCO FISSO (Italian); BLANC FIXE (French); C.I. 77120; C.I. PIGMENT WHITE 21; CITOBARYUM; COLONATRAST; EINECS No. 231-784-4; ENAMEL WHITE; ESOPHOTRAST; EWEISS (German); E-Z-PAQUE; FINEMEAL; LACTOBARYT; LIQUIBARINE (French); MACROPAQUE; NEOBAR; ORATRAST; PRECIPITATED BARIUM SULPHATE; RAYBAR; REDI-FLOW; SOLBAR; SULFATE de BARYM, AUTRE QUE NATUREL (French); SULFATO BARICO (Spanish); SULFURIC ACID, BARIUM SALT (1:1); SUPRAMIKE; TRAVAD; UNIBARYT; ZWAARSPAAT (Dutch)

IDENTIFICATION

CAS:	7727-43-7
ERG ID:	UN 1584
ERG Guide:	154
ERG Guide Category:	SUBSTANCES-TOXIC AND/OR CORROSIVE (NONCOMBUSTIBLE)
Molecular Formula:	BaO_4S
RTECS No.:	CR0600000
Properties:	Noncombustible powder or crystals. White or yellowish. Odorless. Practically insoluble in water.

HEALTH AND SAFETY INFORMATION

Irritates eyes, skin, and respiratory system. High or repeated exposures may cause weakness, paralysis of limbs, respiratory system, heart.

IDLH: 50 mg/m^3 as barium, soluble compounds.
OSHA Table Z-1-A Air Contaminant

Exposure Limits: TWA

ACGIH:	10 mg/m^3 of total dust when toxic impurities are not present, e.g., quartz < 1%.
NIOSH:	10 mg/m^3 total dust; 5 mg/m^3 respiratory fraction
OSHA:	15 mg/m^3 total dust; 5 mg/m^3 respiratory fraction

RECOMMENDATIONS FOR RESPIRATOR SELECTION: NIOSH

At any concentrations above the NIOSH REL, or where there is no REL, at any detectable concentration: SCBAF:PD,PP (any self-contained breathing apparatus that has a full facepiece and is operated in a pressure-demand or other positive-pressure mode); or SAF:PD,PP:ASCBA (any supplied-air respirator that has a full facepiece and is operated in a pressure-demand or other positive-pressure mode in combination with an auxiliary, self-contained breathing apparatus operated in a pressure-demand or other positive-pressure mode).

Escape: GMFOV [any air-purifying, full-facepiece respirator (gas mask) with a chin-style, front- or back-mounted organic vapor canister]; or SCBAE (any appropriate escape-type, self-contained breathing apparatus).

CHEMICAL REACTIVITY

Reacts with phosphorus and aluminum (aluminum, in the presence of heat, can cause an explosion).

FIRE INFORMATION

Noncombustible solid. Decomposition products upon heating include toxic gases of sulfur oxides. If material or contaminated runoff enters waterways, notify downstream users of potentially contaminated waters. Notify local health and fire officials and pollution control agencies. Use any extinguishing agent suitable for surrounding fire.

FIRST AID

Move victim to fresh air; call emergency medical care. Apply artificial respiration if victim is not breathing. Do not use mouth-to-mouth method if victim ingested or inhaled the substance; induce artificial respiration with the aid of a pocket mask equipped with a one-way valve or other proper respiratory medical device. Administer oxygen if breathing is difficult. Remove and isolate contaminated clothing and shoes. In case of contact with substance, immediately flush skin or eyes with running water for at least 20 minutes. For minor skin contact, avoid spreading material on unaffected skin. Keep victim warm and quiet. Effects of exposure (inhalation, ingestion, or skin contact) to substance may be delayed. Ensure that medical personnel are aware of the material(s) involved and take precautions to protect themselves.

ENVIRONMENTAL INFORMATION

MARINE POLLUTANT (49CFR, Subchapter 172.101, Appendix B) as barium compounds, soluble, n.o.s.

BENZENE B:078

SYNONYMS: (6)ANNULENE; BENCENO (Spanish); BENZEEN (Dutch); BENZELENE; BENZEN (Polish); BENZOL; BENZOLE; BENZOLENE; BICARBURET OF HYDROGEN; CARBON

NAPHTHA; CARBON OIL; COAL NAPHTHA; COAL NAPHTHA, PHENYL HYDRIDE; COAL TAR NAPHTHA; CYCLOHEXATRIENE; EEC No. 601-020-00-8; EINECS No. 200-753-7; FENZEN (Czech); MINERAL NAPHTHA; MOTOR BENZOL; PHENYL HYDRIDE; PHENE; PYROBENZOL; PYROBENZOLE

IDENTIFICATION

CAS:	71-43-2
ERG ID:	UN 1114
ERG Guide:	130
ERG Guide Category:	FLAMMABLE LIQUIDS (NONPOLAR/WATER-IMMISCIBLE/NOXIOUS)
Molecular Formula:	C_6H_6
RTECS No.:	CY1400000
Properties:	Highly flammable, watery liquid. Colorless. Gasolinelike odor. Floats on water; insoluble in water. Freezing point is 42°F (6°C).

HEALTH AND SAFETY INFORMATION

Irritates eyes, skin, and respiratory system. Destroys skin's natural oils; caused redness, dryness. Affects the central nervous system; red blood cell production. Known to cause cancer in humans following long-term exposure; contact should be avoided. High or repeated exposure can cause liver, kidney, blood disease; leukemia.

IARC Evaluation: Carcinogen; Group 1; sufficient human evidence; sufficient animal evidence.
IDLH: 500 ppm. Potential human carcinogen
Odor Threshold: 8.65 ppm
OSHA Table Z-1-A Air Contaminant
OSHA Table Z-2 Air Contaminant
OSHA Specifically Regulated Substances (29 CFR 1910.1001-1048) (liquefied or gaseous).

Exposure Limits:	**TWA**	**STEL**
ACGIH:[+ *]	10 ppm; 32 mg/m^3	
NIOSH:[**]	0.1 ppm	1 ppm
OSHA:	1 ppm	5 ppm

[+]Notice for intended change to a confirmed human carcinogen; skin designation; TWA of 0.5 ppm and 1.6 mg/m^3; STEL of 2.5 ppm and 8 mg/m^3.
[*]Suspected carcinogen.
[**]Potential carcinogen. Reduce exposure to lowest feasible level.

RECOMMENDATIONS FOR RESPIRATOR SELECTION: NIOSH

At any concentrations above the NIOSH REL, or where there is no REL, at any detectable

concentration: SCBAF:PD,PP (any self-contained breathing apparatus that has a full facepiece and is operated in a pressure-demand or other positive-pressure mode); or SAF:PD,PP:ASCBA (any supplied-air respirator that has a full facepiece and is operated in a pressure-demand or other positive-pressure mode in combination with an auxiliary, self-contained breathing apparatus operated in a pressure-demand or other positive-pressure mode).

Escape: GMFOV [any air-purifying, full-facepiece respirator (gas mask) with a chin-style, front- or back-mounted organic vapor canister]; or SCBAE (any appropriate escape-type, self-contained breathing apparatus).

CHEMICAL REACTIVITY

Forms explosive mixture with air. May accumulate static electrical charges and may cause ignition of its vapors. Incompatible with strong oxidizers, nitric acid, fluorides, oxygen, ozone, and perchlorates. Attacks most rubber products. Attacks many forms of plastics, rubber products, and coatings.

FIRE INFORMATION

Vapors are heavier than air and will collect in low areas. Vapors are heavier than air and will collect in low areas. Vapors in confined areas may explode when exposed to fire. Vapors may travel long distances to ignition sources and flash back. Storage containers and parts of containers may rocket great distances, in many directions. If material or contaminated runoff enters waterways, notify downstream users of potentially contaminated waters. Notify local health and fire officials and pollution control agencies. Use dry chemicals, foam, carbon dioxide. From a secure explosion-proof location, use water spray to cool exposed containers. If cooling streams are ineffective (venting sound increases in volume and pitch, tank discolors, or shows any signs of deforming), withdraw immediately to a secure position.

NFPA Hazard Classifications:

Health Hazards (Blue) 2
Flammability (Red) 3
Reactivity (Yellow) 0

Flash Point: 12°F (11°C) c.c. Benzene is solid at 12°F (⁻11°C)
Flammable Limits in Air: LEL 1.3%; UEL 7.8%
Autoignition Temperature: 928°F (498°C)

FIRST AID

Move victim to fresh air. Call emergency medical care. Apply artificial respiration if victim is not breathing. Administer oxygen if breathing is difficult. Remove and isolate contaminated clothing and shoes. In case of contact with substance, immediately flush skin or eyes with running water for at least 20 minutes. Wash skin with soap and water. Keep victim warm and quiet. Effects of exposure (inhalation, ingestion, or skin contact) to substance may be delayed. Ensure that medical personnel are aware of the material(s) involved and take precautions to protect themselves.

ENVIRONMENTAL INFORMATION

Clean Air Act: Hazardous Air Pollutants (Title I, Part A, Section 112)
Note: Including benzene from gasoline.
Clean Water Act: Section 311 Hazardous Substances/RQ (same as CERCLA); Section 307 Priority Pollutants; Section 313 Priority Chemicals; Toxic Pollutant (Section 401.15).
EPA Hazardous Waste Number (RCRA No.): D019.
RCRA Toxicity Characteristic (Section 261.24): Maximum Concentration of Contaminants, regulatory level, 0.5 mg/L
RCRA Section 261 Hazardous Constituents
RCRA Land Ban Waste
RCRA Universal Treatment Standards: Wastewater (mg/L), 0.14; Nonwastewater (mg/kg), 10
RCRA Ground Water Monitoring List: Suggested test method(s) (PQL ug/L): 8020(2); 8240(5)
Safe Drinking Water Act: MCL, 0.005 mg/L; MCLG, zero; Regulated chemical (47 FR 9352)
EPCRA Section 304: Reportable Quantity (RQ): CERCLA, 10 lb (4.54 kg).
EPCRA Section 313: Form R *de minimis* concentration reporting level: 0.1%
Marine Pollutant (49CFR, Subchapter 172.101, Appendix B)
California's Proposition 65: Carcinogen; Reproductive toxin; Male reproductive toxin

BENZOTHIOZOLE DISULFIDE B:079

SYNONYMS: ALTAX®; BENZOTHIAZOLYL DISULFIDE; 2-BENZOTHIAZOLYL DISULFIDE; BIS(BENZOTHIAZOLYL)DISULFIDE; BIS(2-BENZOTHIAZYL) DISULFIDE; DI-2-BENZOTHIAZOLYLDISULFIDE; DIBENZOTHIAZYL DISULFIDE; 2,2'-DIBENZOTHIAZYL DISULFIDE; DIBENZOYLTHIAZYL DISULFIDE; DIBENZTHIAZYL DISULFIDE; 2,2'-DITHIOBIS(BENZOTHIAZOLE); DWUSIARCZEK (Polish); EINECS No. 204-424-9; MBTS; MBTS RUBBER ACCELERATOR; 2-MERCAPTOBENZOTHIAZOLEDISULFIDE; 2-MERCAPTOBENZOTHIAZYLDISULFIDE; PERKACIT® MBTS; ROYAL MBTS; THIOFIDE; USAF B-33; USAF CY-5; USAF EK-5432; VULKACIT DM; VULKACIT® DMC

IDENTIFICATION

CAS:	120-78-5
ERG ID:	UN 2811 as toxic solid, organic, n.o.s.
ERG Guide:	154
ERG Guide Category:	SUBSTANCES-TOXIC AND/OR CORROSIVE (NONCOMBUSTIBLE)
Molecular Formula:	$C_{14}H_8N_2S_4$
RTECS No.:	DL4550000
Properties:	Powder. Pale yellow. Odorless. Insoluble in water.

HEALTH AND SAFETY INFORMATION

Sulfides are toxic and corrosive to eyes, skin, and respiratory tract.

RECOMMENDATIONS FOR RESPIRATOR SELECTION:

At any concentrations above the NIOSH REL, or where there is no REL, at any detectable concentration: SCBAF:PD,PP (any self-contained breathing apparatus that has a full facepiece and is operated in a pressure-demand or other positive-pressure mode); or SAF:PD,PP:ASCBA (any supplied-air respirator that has a full facepiece and is operated in a pressure-demand or other positive-pressure mode in combination with an auxiliary, self-contained breathing apparatus operated in a pressure-demand or other positive-pressure mode).

Escape: GMFOV [any air-purifying, full-facepiece respirator (gas mask) with a chin-style, front- or back-mounted organic vapor canister]; or SCBAE (any appropriate escape-type, self-contained breathing apparatus).

CHEMICAL REACTIVITY

Contact with oxidizers may cause spontaneous explosion. Contact with water, steam, or acids can produce toxic and flammable hydrogen sulfide vapors; the evolved hydrogen sulfide forms an explosive mixture with air.

FIRE INFORMATION

Combustion products include toxic sulfur oxides and nitrogen oxides. If material or contaminated runoff enters waterways, notify downstream users of potentially contaminated waters. Notify local health and fire officials and pollution control agencies. Use dry chemicals, alcohol-resistant foam, carbon dioxide. From a secure explosion-proof location, use water spray to cool exposed containers. If cooling streams are ineffective (venting sound increases in volume and pitch, tank discolors, or shows any signs of deforming), withdraw immediately to a secure position.

NFPA Hazard Classifications: Not listed

FIRST AID

Move victim to fresh air; call emergency medical care. Apply artificial respiration if victim is not breathing. Do not use mouth-to-mouth method if victim ingested or inhaled the substance; induce artificial respiration with the aid of a pocket mask equipped with a one-way valve or other proper respiratory medical device. Administer oxygen if breathing is difficult. Remove and isolate contaminated clothing and shoes. In case of contact with substance, immediately flush skin or eyes with running water for at least 20 minutes. For minor skin contact, avoid spreading material on unaffected skin. Keep victim warm and quiet. Effects of exposure (inhalation, ingestion, or skin contact) to substance may be delayed. Ensure that medical personnel are aware of the material(s) involved and take precautions to protect themselves.

BENZYL ALCOHOL B:080

SYNONYMS: ALCOHOL BENCILICO (Spanish); BENATOL; BENTALOL; BENZENECARBINOL;

EEC No. 603-057-00-5; EINECS No. 202-859-9; FEMA No. 2137; α-HYDROXYTOLUENE; NCI-C06111; PD ACTIVATOR; PHENYLCARBINOL; PHENYLMETHANOL; PHENYLMETHYL ALCOHOL; α-TOLUENOL

IDENTIFICATION

CAS:	100-51-6
ERG ID	UN 3082
ERG Guide:	171
ERG Guide Category:	LOW TO MODERATE HAZARD
Molecular Formula:	C_7H_8O
RTECS No.:	DN315000
Properties:	Combustible, colorless liquid. Mild, pleasant odor; sharp taste. Partially soluble in water.

HEALTH AND SAFETY INFORMATION

Irritates eyes, skin, and respiratory system. Removes skin's natural oils; may cause redness, dryness. Can affect central nervous system and cause headache, dizziness, unconsciousness.

Odor Threshold: 5.5 ppm

Exposure Limits: TWA

AIHA (WEEL) 10 ppm

RECOMMENDATIONS FOR RESPIRATOR SELECTION:

At any concentrations above the NIOSH REL, or where there is no REL, at any detectable concentration: SCBAF:PD,PP (any self-contained breathing apparatus that has a full facepiece and is operated in a pressure-demand or other positive-pressure mode); or SAF:PD,PP:ASCBA (any supplied-air respirator that has a full facepiece and is operated in a pressure-demand or other positive-pressure mode in combination with an auxiliary, self-contained breathing apparatus operated in a pressure-demand or other positive-pressure mode)

Escape: GMFOV [any air-purifying, full-facepiece respirator (gas mask) with a chin-style, front- or back-mounted organic vapor canister]; or SCBAE (any appropriate escape-type, self-contained breathing apparatus).

CHEMICAL REACTIVITY

Forms explosive mixture with air. Slowly oxidizes in air and oxygen. Incompatible with mineral acids, caustics, aliphatic amines, isocyanates. Violent reaction with strong oxidizers. Corrodes aluminum at high temperature. Attacks some nonfluorinated plastics; may not attack polypropylene.

FIRE INFORMATION

Vapors are heavier than air and will collect in low areas. Vapors in confined areas may explode when exposed to fire. Vapors may travel long distances to ignition sources and flash back. Storage containers and parts of containers may rocket great distances, in many directions. If material or contaminated runoff enters waterways, notify downstream users of potentially contaminated waters. Notify local health and fire officials and pollution control agencies. Use dry chemicals, foam, carbon dioxide. From a secure explosion-proof location, use water spray to cool exposed containers. If cooling streams are ineffective (venting sound increases in volume and pitch, tank discolors, or shows any signs of deforming), withdraw immediately to a secure position.

NFPA Fire Classifications:

Health Hazards (Blue) 2
Flammability (Red) 1
Reactivity (Yellow) 0

Flash Point: 220°F (105°C) o.c.
Flammable Limits in Air: LEL 1.3%; UEL 13%
Autoignition Temperature: 817°F (436°C)

FIRST AID

Move victim to fresh air. Call emergency medical care. Apply artificial respiration if victim is not breathing. Administer oxygen if breathing is difficult. In case of contact with substance, immediately flush skin or eyes with running water for at least 20 minutes. Remove and isolate contaminated clothing and shoes. Ensure that medical personnel are aware of the material(s) involved and take precautions to protect themselves.

BENZYLAMINE B:082

SYNONYMS: BENZENEMETHANAMINE; α-AMINOTOLUENE; BENCILAMINA (Spanish); EEC No. 612-047-00-X; EINECS No. 202-854-1; PHENYLMETHYL AMINE

IDENTIFICATION

CAS:	100-46-9
ERG ID:	UN 2734
ERG Guide:	132
ERG Guide Category:	FLAMMABLE LIQUIDS-CORROSIVE
Molecular Formula:	C_7H_9N
Properties:	Combustible liquid. Colorless to light yellow. Strong ammonia odor. Floats and mixes with water. Poisonous gases may be produced in fire.

HEALTH AND SAFETY INFORMATION

Highly irritating to eyes, skin, and respiratory tract.

IDLH: 300 ppm as ammonia
Odor Threshold: NA

Exposure Limits:	**TWA**	**STEL**
NIOSH:	25 ppm; 18 mg/m^3 as ammonia	35 ppm; 27 mg/m^3 as ammonia
OSHA:	50 ppm; 35 mg/m^3 as ammonia	

RECOMMENDATIONS FOR RESPIRATOR SELECTION: NIOSH as ammonia

250 ppm: CCRS [any chemical cartridge respirator with cartridge(s) providing protection against the compound of concern)]; or SA (any supplied-air respirator).
300 ppm: SA:CF (any supplied-air respirator operated in a continuous-flow mode); or PAPRS [any powered, air-purifying respirator with cartridge(s) providing protection against the compound of concern]; or CCRFS (any chemical cartridge respirator with a full facepiece and cartridge(s) providing protection against the compound of concern); or GMFS [any air-purifying, full-facepiece respirator (gas mask) with a chin-style, front- or back-mounted canister providing protection against the compound of concern]; or SCBAF (any self-contained breathing apparatus with a full facepiece); or SAF (any supplied-air respirator with a full facepiece).

Emergency or planned entry into unknown concentrations or idlh conditions:
SCBAF:PD,PP (any self-contained breathing apparatus that has a full facepiece and is operated in a pressure-demand or other positive-pressure mode); or SAF:PD,PP:ASCBA (any supplied-air respirator that has a full facepiece and is operated in a pressure-demand or other positive-pressure mode in combination with an auxiliary, self-contained breathing apparatus operated in a pressure-demand or other positive-pressure mode.

Escape: GMFS [any air-purifying, full-facepiece respirator (gas mask) with a chin-style, front- or back-mounted canister providing protection against the compound of concern]; or SCBAE (any appropriate escape-type, self-contained breathing apparatus).
Note: Substance reported to cause eye irritation or damage; may require eye protection.

CHEMICAL REACTIVITY

May ignite on contact with cellulose nitrate of high surface area. Incompatible with acrolein, acrylonitrile, tert-butyl nitroacetylene, ethylene oxide, isopropyl chlorocarbonate, maleic anhydride, triisobutylaluminum. In presence of moisture may weakly corrode some metals. Liquid will attack some plastics.

FIRE INFORMATION

Vapors are heavier than air and will collect in low areas. Vapors in confined areas may explode when

exposed to fire. If material or contaminated runoff enters waterways, notify downstream users of potentially contaminated waters. Notify local health and fire officials and pollution control agencies. Use dry chemicals, alcohol-resistant foam, carbon dioxide. From a secure explosion-proof location, use water spray to cool exposed containers. If cooling streams are ineffective (venting sound increases in volume and pitch, tank discolors, or shows any signs of deforming), withdraw immediately to a secure position.

NFPA Hazard Classifications:

Health Hazards (Blue) 2
Flammability (Red) 1
Reactivity (Yellow) 0

Flash Point: 168°F (76°C) o.c

FIRST AID

Move victim to fresh air; call emergency medical care. Apply artificial respiration if victim is not breathing. Do not use mouth-to-mouth method if victim ingested or inhaled the substance; induce artificial respiration with the aid of a pocket mask equipped with a one-way valve or other proper respiratory medical device. Administer oxygen if breathing is difficult. Remove and isolate contaminated clothing and shoes. In case of contact with substance, immediately flush skin or eyes with running water for at least 20 minutes. For minor skin contact, avoid spreading material on unaffected skin. Keep victim warm and quiet. Effects of exposure (inhalation, ingestion, or skin contact) to substance may be delayed. Ensure that medical personnel are aware of the material(s) involved and take precautions to protect themselves.

BENZYL BENZOATE B:084

SYNONYMS: ASCABIN; ASCABIOL; BENYLATE; BENZOATO de BENCILO (Spanish); BENZOIC ACID, BENZYL ESTER; BENZOIC ACID, PHENYLMETHYL ESTER; BENZYL ALCOHOL, BENZOIC ESTER; BENZYL BENZENECARBOXYLATE; BENZYL PHENYLFORMATE; COLEBENZ; EINECS No. 204-402-9; FEMA No. 2138; NOVOSCABIN; PERUSCABIN; SCABANCA; VANZOATE; VENZONATE

IDENTIFICATION

CAS:	120-51-5
ERG ID:	UN 2810 (liquid); UN 2811 (solid)
ERG Guide:	153 (liquid); 154 (solid)
ERG Guide Category:	TOXIC LIQUID, ORGANIC, N.O.S.; TOXIC SOLID, ORGANIC, N.O.S.
Molecular Formula:	$C_{14}H_{12}0_2$
RTECS No.:	DG4200000
Properties:	Oily thick liquid or white solid. Colorless. Slight aromatic, characteristic odor. Insoluble in water.

HEALTH AND SAFETY INFORMATION

Can irritate eyes, nose, and respiratory tract. Repeated or prolonged contact can cause skin problems.

RECOMMENDATIONS FOR RESPIRATOR SELECTION:

At any concentrations above the NIOSH REL, or where there is no REL, at any detectable concentration: SCBAF:PD,PP (any self-contained breathing apparatus that has a full facepiece and is operated in a pressure-demand or other positive-pressure mode); or SAF:PD,PP:ASCBA (any supplied-air respirator that has a full facepiece and is operated in a pressure-demand or other positive-pressure mode in combination with an auxiliary, self-contained breathing apparatus operated in a pressure-demand or other positive-pressure mode).

Escape: GMFOV [any air-purifying, full-facepiece respirator (gas mask) with a chin-style, front- or back-mounted organic vapor canister]; or SCBAE (any appropriate escape-type, self-contained breathing apparatus).

CHEMICAL REACTIVITY

Contact with oxidizers may cause fire and explosions.

FIRE INFORMATION

Decomposition products upon heating include acrid smoke and irritating fumes. Vapors are heavier than air and will collect in low areas. Vapors in confined areas may explode when exposed to fire. If material or contaminated runoff enters waterways, notify downstream users of potentially contaminated waters. Notify local health and fire officials and pollution control agencies. Use dry chemicals, alcohol-resistant foam, carbon dioxide. From a secure explosion-proof location, use water spray to cool exposed containers. If cooling streams are ineffective (venting sound increases in volume and pitch, tank discolors, or shows any signs of deforming), withdraw immediately to a secure position.

NFPA Hazard Classifications:

Health Hazards (Blue) 1
Flammability (Red) 1
Reactivity (Yellow) 0

Flash Point: 298°F (148°C) c.c.
Autoignition Temperature: 898°F (481°C)

FIRST AID

Move victim to fresh air and call emergency medical care. Apply artificial respiration if victim is not breathing. Do not use mouth-to-mouth method if victim swallowed or inhaled the substance; induce artificial respiration with the aid of a pocket mask equipped with a one-way valve or other proper respiratory medical device. If breathing is difficult, give oxygen. In case of contact with substance, immediately flush skin or eyes with running water for at least 20 minutes. Remove and isolate contaminated clothing and shoes. Keep victim warm and quiet. Effects of exposure (inhalation, ingestion or, skin contact) to substances may be delayed. Ensure that medical personnel are aware of the material(s) involved and take precautions to protect themselves.

BENZYL BUTYRATE B:088

SYNONYMS: BENZYL n-BUTANOATE; BENZYL n-BUTYRATE; BUTANOIC ACID, PHENYLMETHYL ESTER;
BUTIRATO de BENCILO (Spanish); EINECS No. 203-105-1; FEMA No. 2140

IDENTIFICATION

CAS:	103-37-7
ERG ID:	UN 3272
ERG Guide:	127
ERG Guide Category:	FLAMMABLE LIQUIDS (POLAR/WATER-MISCIBLE)
Molecular Formula:	$C_{11}H_{14}O_2$
RTECS No.:	ES7350000
Properties:	Combustible liquid. Colorless. Floral odor. Irritating fumes when heated.

HEALTH AND SAFETY INFORMATION

RECOMMENDATIONS FOR RESPIRATOR SELECTION:

At any concentrations above the NIOSH REL, or where there is no REL, at any detectable concentration: SCBAF:PD,PP (any self-contained breathing apparatus that has a full facepiece and is operated in a pressure-demand or other positive-pressure mode); or SAF:PD,PP:ASCBA (any supplied-air respirator that has a full facepiece and is operated in a pressure-demand or other positive-pressure mode in combination with an auxiliary, self-contained breathing apparatus operated in a pressure-demand or other positive-pressure mode).

Escape: GMFOV [any air-purifying, full-facepiece respirator (gas mask) with a chin-style, front- or back-mounted organic vapor canister]; or SCBAE (any appropriate escape-type, self-contained breathing apparatus).

CHEMICAL REACTIVITY

Reacts with nitrates and strong oxidizing agents (contact with oxidizers may cause fire and explosions).

FIRE INFORMATION

Decomposition products upon heating include irritating vapors and acrid smoke. If material or contaminated runoff enters waterways, notify downstream users of potentially contaminated waters. Notify local health and fire officials and pollution control agencies.

NFPA Hazard Classifications: Not listed

Flash Point: 212°F (100°C)

FIRST AID

Move victim to fresh air. Call emergency medical care. Apply artificial respiration if victim is not breathing. Administer oxygen if breathing is difficult. Remove and isolate contaminated clothing and shoes. In case of contact with substance, immediately flush skin or eyes with running water for at least 20 minutes. Wash skin with soap and water. Keep victim warm and quiet. Ensure that medical personnel are aware of the material(s) involved and take precautions to protect themselves.

BENZYL DIMETHYLAMINE B:090

SYNONYMS: ARALDITE ACCELERATOR 062; BDMA; n-BENCILDIMETILAMINA (Spanish); BENZYL-N,N-DIMETHYLAMINE; n-BENZYLDIMETHYLAMINE; n,n-DIMETHYLBENZENEMETHANAMINE; n,n-DIMETHYLBENZYLAMINE; EINECS No. 203-149-1; n-(PHENYLMETHYL)DIMETHYLAMINE; SUMINE 2015

IDENTIFICATION

CAS:	103-83-3
ERG ID:	UN 2619
ERG Guide:	132
ERG Guide Category:	FLAMMABLE LIQUIDS-CORROSIVE
Molecular Formula:	$C_9H_{13}N$
Properties:	Flammable liquid. Poisonous gases produced when heated.

HEALTH AND SAFETY INFORMATION

Corrosive to eyes, skin, and respiratory tract. Inhalation may cause pulmonary edema, which can be delayed for several hours; there is a risk of death in serious cases.

RECOMMENDATIONS FOR RESPIRATOR SELECTION:

At any concentrations above the NIOSH REL, or where there is no REL, at any detectable concentration: SCBAF:PD,PP (any self-contained breathing apparatus that has a full facepiece and is operated in a pressure-demand or other positive-pressure mode); or SAF:PD,PP:ASCBA (any supplied-air respirator that has a full facepiece and is operated in a pressure-demand or other positive-pressure mode in combination with an auxiliary, self-contained breathing apparatus operated in a pressure-demand or other positive-pressure mode).

Escape: GMFOV [any air-purifying, full-facepiece respirator (gas mask) with a chin-style, front- or back-mounted organic vapor canister]; or SCBAE (any appropriate escape-type, self-contained breathing apparatus).

CHEMICAL REACTIVITY

Forms explosive mixture with air. Reacts with acids, organic anhydrides, isocyanates, aldehydes, strong oxidizers, and calcium hypochlorite. Attacks some plastics, rubber, or coatings.

FIRE INFORMATION

Vapors are heavier than air and will collect in low areas. Vapors in confined areas may explode when exposed to fire. If material or contaminated runoff enters waterways, notify downstream users of potentially contaminated waters. Notify local health and fire officials and pollution control agencies. Use dry chemicals, alcohol-resistant foam, carbon dioxide. From a secure explosion-proof location, use water spray to cool exposed containers. If cooling streams are ineffective (venting sound increases in volume and pitch, tank discolors, or shows any signs of deforming), withdraw immediately to a secure position.

FEMA Hazard Classifications:

Health Hazards (Blue) 2
Flammability (Red) 2
Reactivity (Yellow) 0

Flash Point: 170°F (77°C)

FIRST AID

Move victim to fresh air. Call emergency medical care. Apply artificial respiration if victim is not breathing. Do not use mouth-to-mouth method if victim swallowed or inhaled the substance; induce artificial respiration with the aid of a pocket mask equipped with a one-way valve or other proper respiratory medical device. Administer oxygen if breathing is difficult. Remove and isolate contaminated clothing and shoes. In case of contact with substance, immediately flush skin or eyes with running water for at least 20 minutes. Keep victim warm and quiet. Effects of exposure (inhalation, ingestion, or skin contact) to substances may be delayed. Ensure that medical personnel are aware of the material(s) involved and take precautions to protect themselves.

BENZYL TRIMETHYL AMMONIUM CHLORIDE B:095

SYNONYMS: BTMAC; EINECS No. 200-300-3; TRIMETHYLBENZYLAMMONIUM CHLORIDE

IDENTIFICATION

CAS: 56-93-9

ERG ID: UN 3082
ERG Guide: 171
ERG Guide Category: SUBSTANCES (LOW TO MODERATE HAZARD)
Molecular Formula: $C_{10}H_{16}NCl—H_2O$
RTECS No.: BO8400000
Properties: Combustible liquid. Light yellow. Mild almond odor. May float or sink in water.

HEALTH AND SAFETY INFORMATION

Poisonous if swallowed. May cause eye irritation.

CHEMICAL REACTIVITY

Reacts with oxidizers and strong acids.

FIRE INFORMATION

Combustion products include toxic nitrogen oxides, ammonia, and chlorides. Vapors are heavier than air and will collect in low areas. Vapors in confined areas may explode when exposed to fire. If material or contaminated runoff enters waterways, notify downstream users of potentially contaminated waters. Notify local health and fire officials and pollution control agencies. Use dry chemicals, alcohol-resistant foam, carbon dioxide. From a secure explosion-proof location, use water spray to cool exposed containers. If cooling streams are ineffective (venting sound increases in volume and pitch, tank discolors, or shows any signs of deforming), withdraw immediately to a secure position.

NFPA Hazard Classifications:

Health Hazards (Blue) 1
Flammability (Red) 1
Reactivity (Yellow) 0

FIRST AID

Move victim to fresh air. Call emergency medical care. Apply artificial respiration if victim is not breathing. Administer oxygen if breathing is difficult. In case of contact with substance, immediately flush skin or eyes with running water for at least 20 minutes. Remove and isolate contaminated clothing and shoes. Ensure that medical personnel are aware of the material(s) involved and take precautions to protect themselves.

BERYLLIUM B:100

SYNONYMS: BERILIO (Spanish); BERYLLIUM DUST; BERYLLIUM METAL POWDER; BERYLLIUM-9; BERYLLIUM POWDER;EEC No. 004-001-00-7;EINECS No. 231-150-7; GLUCINIUM; GLUCUNUM

140 Beryllium

IDENTIFICATION

CAS:	7440-41-7
ERG ID:	UN 1567
ERG Guide:	134
ERG Guide Category:	FLAMMABLE SOLIDS-TOXIC AND/OR CORROSIVE
Molecular Formula:	Be
RTECS No.:	DS1750000
Properties:	Brittle grayish metal. Odorless. Sinks in water. Poisonous gases may be produced in fire.

HEALTH AND SAFETY INFORMATION

May cause cancer. Long-term exposure can cause bone damage, lung damage, cough, pain in the joints, general weakness, weight loss, "clubbing" of fingers, shortness of breath, cyanosis, and allergic contact dermatitis; possible skin and eye allergy; scars on the lungs and possibly other organs; development of kidney stones.

IARC Evaluation: Group 2A, probable human carcinogen; human evidence, sufficient; animal evidence, sufficient.
NTP: NTP anticipated human carcinogen.
IDLH: Potential human carcinogen; 4 mg/m^3 as Be
NIOSH criteria Document: NIOSH 72-10268

Exposure Limits:	**TWA**	**Ceiling**
ACGIH: (1)	0.002 mg/m^3 as Be	0.01 mg/m^3
NIOSH: (2)	not to exceed 0.0005 mg/m^3	
OSHA:	0.002 mg/m^3	0.005 mg/m^3; 0.025 mg/m^3 [30-min. maximum peak]

(1) Confirmed carcinogen
(2) Potential carcinogen; limit exposure to lowest feasible concentration.

RECOMMENDATIONS FOR RESPIRATOR SELECTION:

At any concentrations above the NIOSH REL, or where there is no REL, at any detectable concentration: SCBAF:PD,PP (any self-contained breathing apparatus that has a full facepiece and is operated in a pressure-demand or other positive-pressure mode); or SAF:PD,PP:ASCBA (any supplied-air respirator that has a full facepiece and is operated in a pressure-demand or other positive-pressure mode in combination with an auxiliary, self-contained breathing apparatus operated in a pressure-demand or other positive-pressure mode).

Escape: GMFOV [any air-purifying, full-facepiece respirator (gas mask) with a chin-style, front- or back-mounted organic vapor canister]; or SCBAE (any appropriate escape-type, self-contained breathing apparatus).

FIRE INFORMATION

Noncombustible solid in bulk form; combustible hazard in the form of powder or dust.

CHEMICAL REACTIVITY

Contact with acids evolves flammable hydrogen gas. Incompatible with alkalies, chlorinated hydrocarbons, oxidizable agents, carbon tetrachloride, trichloroethylene, lithium, phosphorous.

FIRE INFORMATION

Vapors are heavier than air and will collect in low areas. Vapors in confined areas may explode when exposed to fire. If material or contaminated runoff enters waterways, notify downstream users of potentially contaminated waters. Notify local health and fire officials and pollution control agencies. Use dry chemicals, alcohol-resistant foam, carbon dioxide. From a secure explosion-proof location, use water spray to cool exposed containers. If cooling streams are ineffective (venting sound increases in volume and pitch, tank discolors, or shows any signs of deforming), withdraw immediately to a secure position.

NFPA Fire Classifications (dust or powder):

Health Hazards (Blue) 3
Flammability (Red) 1
Reactivity (Yellow) 0

FIRST AID

Move victim to fresh air. Call emergency medical care. Apply artificial respiration if victim is not breathing. Do not use mouth-to-mouth method if victim swallowed or inhaled the substance; induce artificial respiration with the aid of a pocket mask equipped with a one-way valve or other proper respiratory medical devise. If breathing is difficult, give oxygen. Remove and isolate contaminated clothing and shoes. In case of contact with substance, immediately flush skin or eyes with running water for at least 20 minutes. For minor skin contact, avoid spreading material on unaffected skin. Keep victim warm and quiet. Effects of exposure (inhalation, ingestion, or skin contact) to substances may be delayed. Ensure that medical personnel are aware of the material(s) involved and take precautions to protect themselves..

ENVIRONMENTAL INFORMATION

CLEAN WATER ACT: Section 307 Priority Pollutants; Toxic Pollutant (Section 401.15)
EPA HAZARDOUS WASTE NUMBER (RCRA No.): P015
RCRA Section 261 Hazardous Constituents.
RCRA Land Ban Waste.
RCRA Universal Treatment Standards: Wastewater (mg/L), 0.82; Nonwastewater (mg/L), 0.014 TCLP
RCRA Ground Water Monitoring List. Suggested test method(s) (PQL ug/L): (total) 6010(3); 7090(50); 7091(2)

SAFE DRINKING WATER ACT: MCL, 0.004 mg/L; MCLG, 0.004 mg/L

EPCRA Section 304 Reportable Quantity (RQ): CERCLA, 10 lb (4.54 kg).
Note: No report required if the diameter of the pieces of solid metal is equal to or exceeds 0.004 inches.
EPCRA Section 313: Form R *de minimus* concentration reporting level: 0.1%. Form R Toxic Chemical Category Code: N050
CALIFORNIA'S PROPOSITION 65: Carcinogen.

BISPHENOL A B:135

SYNONYMS: BISFENOL A (Spanish); BISFENOLO A (Italian); BISFEROL A (German); 2,2-BIS-4'-HYDROXYFENYLPROPAN (Czech); BIS(4-HYDROXYPHENYL)DIMETHYLMETHANE; 2,2-BIS(4-HYDROXYPHENYL)PROPANE; p,p'-BISPHENOL A; β,β'-BIS(p-HYDROXYPHENYL)PROPANE; 2,2-BIS(p-HYDROXYPHENYL)PROPANE; BIS(p-HYDROXYPHENYL)PROPANE; DIAN; DIANO; p,p'-DIHYDROXYDIPHENYLDIMETHYLMETANE; 4,4'-DIHYDROXYDIPHENYLDIMETHYLMETANE; p,p'-DIHYDROXYDIPHENYLPROPANE; 2,2-(4,4'-DIHYDROXYDIPHENYL)PROPANE; 4,4'-DIHYDROXYDIPHENYLPROPANE; 4,4'-DIHYDROXYDIPHENYL-2,2-PROPANE; 4,4'-DIHYDROXY-2,2-DIPHENYLPROPANE; β-DI-p-HYDROXYPHENYLPROPANE; 2,2-DI(4-HYDROXYPHENYL)PROPANE; DIMETHYL BIS(p-HYDROXYPHENYL)METHANE; DIMETHYLMETHYLENE p,p'-DIPHENOL; 2,2-DI(4-PHENYLOL)PROPANE; DIPHENYLOLPROPANE; EINECS No. 201-245-8; IPOGNOX 88; p,p'-ISOPROPILIDENDIFENOL (Spanish); ISOPROPYLIDENEBIS(4-HYDROXYBENZENE); 4,4'-ISOPROPYLIDENEBIS[PHENOL]; p,p'-ISOPROPYLIDENEBISPHENOL; p,p'-ISOPROPYLIDENEDIPHENOL; 4,4'-ISOPROPYLIDENEDIPHENOL; 4,4'-(1-METHYLETHYLIDENE)BISPHENOL; NCI-C50635; PARABIS A; PHENOL, 4,4'-ISOPROPYLIDENEDI-; PHENOL, 4,4'-(1-METHYLETHYLIDENE)BIS-; PLURACOL 245; RIKABANOL; UCAR BISPHENOL HP

IDENTIFICATION

CAS:	80-05-7
ERG ID:	Not listed
Molecular Formula:	$C_{15}H_{16}O_2$
RTECS No.:	SL6300000; SL1970000
Properties:	Combustible solid flakes or powder. White to light brown. Mild medicine, phenol-like odor. Sinks in water; insoluble.

HEALTH AND SAFETY INFORMATION

Dust irritates eyes, skin, and respiratory tract. Skin contact can cause dermatitis. Long-term exposure may result in sensitization; allergic response. Epoxy resins containing this substance may cause allergic reactions.

RECOMMENDATIONS FOR RESPIRATOR SELECTION

At any concentrations above the NIOSH REL, or where there is no REL, at any detectable concentration: SCBAF:PD,PP (any self-contained breathing apparatus that has a full facepiece and is operated in a pressure-demand or other positive-pressure mode); or SAF:PD,PP:ASCBA (any supplied-air respirator that has a full facepiece and is operated in a pressure-demand or other positive-pressure mode in combination with an auxiliary, self-contained breathing apparatus operated in a pressure-demand or other positive-pressure mode)

Escape: GMFOV [any air-purifying, full-facepiece respirator (gas mask) with a chin-style, front- or back-mounted organic vapor canister]; or SCBAE (any appropriate escape-type, self-contained breathing apparatus)

CHEMICAL REACTIVITY

Contact with strong oxidizers may cause fire and explosion.

FIRE INFORMATION

Vapors are heavier than air and will collect in low areas. Vapors in confined areas may explode when exposed to fire. If material or contaminated runoff enters waterways, notify downstream users of potentially contaminated waters. Notify local health and fire officials and pollution control agencies. Use dry chemicals, alcohol-resistant foam, carbon dioxide. From a secure explosion-proof location, use water spray to cool exposed containers. If cooling streams are ineffective (venting sound increases in volume and pitch, tank discolors, or shows any signs of deforming), withdraw immediately to a secure position.

FEMA Hazard Classifications:

Health Hazards (Blue) 1
Flammability (Red) 1
Reactivity (Yellow) 0

Flash Point: 415°F (213°C) o.c.
Autoignition Temperature: 1112°F (600°C)

FIRST AID

Move victim to fresh air. Call emergency medical care. Apply artificial respiration if victim is not breathing. Do not use mouth-to-mouth method if victim swallowed or inhaled the substance; induce artificial respiration with the aid of a pocket mask equipped with a one-way valve or other proper respiratory medical devise. If breathing is difficult, give oxygen. Remove and isolate contaminated clothing and shoes. In case of contact with substance, immediately flush skin or eyes with running water for at least 20 minutes. For minor skin contact, avoid spreading material on unaffected skin. Keep victim warm and quiet. Effects of exposure (inhalation, ingestion, or skin contact) to substances may be delayed. Ensure that medical personnel are aware of the material(s) involved and take precautions to protect themselves.

ENVIRONMENTAL INFORMATION

EPCRA Section 313: Form R *de minimis* concentration reporting level: 1.0%
TSCA: 40CFR716.120(*a*)

BORON TRIFLUORIDE B:240

SYNONYMS: BORANE, TRIFLUORO-; BORON FLUORIDE; FLUORURE de BORE (French); LEECURE B; LEECURE B SERIES; TRIFLUOROBORANE; TRIFLUOROBORON; TRIFLOURURO de BORO (Spanish)

IDENTIFICATION

CAS:	7637-07-2
ERG ID:	UN 1008
ERG Guide:	125
ERG Guide Category:	GASES-CORROSIVE
Molecular Formula:	BF_3
RTECS No.:	ED2275000
Properties:	Compressed gas. Colorless. Pungent, irritating, suffocating odor; may be considered pleasant at low concentrations. Forms dense, white aerosol mist in moist air. Reacts with water or moist air, forming highly corrosive hydrofluoric acid. Poisonous gases may be produced in fire. Vapors are heavier than air; can collect in low areas.

HEALTH AND SAFETY INFORMATION

Poison. Breathing the gas can cause death. Skin and eye contact causes severe burns and blindness. Contact with liquid may cause frostbite. Long-term exposure may cause damage to the kidneys, lungs, teeth, and bones. Repeated exposure may cause damage to the nose lining with dryness and nose bleeding; bleeding gums, dry and scaly skin, joint pain, osteosclerosis, increased bone density and brittleness; possible crippling; mottled tooth enamel, weight loss, anemia, weakness, and decreased pulmonary function.

IDLH: 25 ppm
Odor Threshold: 1.5 ppm. Odor is not a reliable indicator of the presence of toxic amounts of gas.
NIOSH criteria Document: NIOSH 77-122

Exposure Limits:	Ceiling
ACGIH:	1 ppm; 2.8 mg/m^3
NIOSH:	1 ppm; 3 mg/m^3
OSHA:	1 ppm; 3 mg/m^3

RECOMMENDATIONS FOR RESPIRATOR SELECTION: NIOSH/OSHA

10 ppm: SA * (any supplied-air respirator)
25 ppm: SA:CF * (any supplied-air respirator operated in a continuous-flow mode); or SCBAF (any self-contained breathing apparatus with a full facepiece); or SAF (any supplied-air respirator with a full facepiece)

Emergency or Planned Entry into Unknown Concentrations or IDLH Conditions: SCBAF:PD,PP (any self-contained breathing apparatus that has a full facepiece and is operated in a pressure-demand or other positive-pressure mode); or SAF:PD,PP:ASCBA (any supplied-air respirator that has a full facepiece and is operated in a pressure-demand or other positive-pressure mode in combination with an auxiliary, self-contained breathing apparatus operated in a pressure-demand or other positive-pressure mode.

Escape: GMFS [any air-purifying, full-facepiece respirator (gas mask) with a chin-style, front- or back-mounted canister providing protection against the compound of concern]; or SCBAE (any appropriate escape-type, self-contained breathing apparatus)

* Substance reported to cause eye irritation or damage; may require eye protection.

CHEMICAL REACTIVITY

Decomposes in heat, water and moist air. Fire and explosions may result from contact with alkali metals, alkaline earth metals (except magnesium), alkyl nitrate, calcium oxide. Incandesces when heated with alkali metals or alkaline earth metals (except magnesium). Corrodes most metals in the presence of moisture. Will attack some plastics and rubber coatings. May be stored as a highly flammable ethyl ether complex.

FIRE INFORMATION

Combustion and decomposition products include toxic hydrogen fluorides and fluorine. Firefighting gear (including SCBA) does not provide adequate protection. If exposure occurs, remove and isolate gear immediately and thoroughly decontaminate personnel. Gases are heavier than air and will collect in low areas. Storage containers and parts of containers may rocket great distances, in many directions. Notify local health and fire officials and pollution control agencies. From a secure explosion-proof location, use water spray to cool exposed containers. If cooling streams are ineffective (venting sound increases in volume and pitch, tank discolors, or shows any signs of deforming), withdraw immediately to a secure position.

NFPA Fire Classifications:

Health Hazards (Blue) 4
Flammability (Red) 0
Reactivity (Yellow) 3
Special situations .... Water, OXY

FIRST AID

Move victim to fresh air. Call emergency medical care. Apply artificial respiration if victim is not breathing. Do not use mouth-to-mouth method if victim ingested or inhaled the substance; induce artificial respiration with the aid of a pocket mask equipped with a one-way valve or other proper respiratory medical devise. Administer oxygen if breathing is difficult. Remove and isolate contaminated clothing and shoes. In case of contact with liquefied gas, thaw frosted parts with lukewarm water. In case of contact with substance, immediately flush skin or eyes with running water for at least 20 minutes. Keep victim warm and quiet. Keep victim under observation. Effects of contact or inhalation may be delayed. Ensure that medical personnel are aware of the material(s) involved and take precautions to protect themselves.

ENVIRONMENTAL INFORMATION

CLEAN AIR ACT: Accidental Release Prevention/Flammable substances, (Section 112[r], Table 3), TQ = 5,000 lb (2270 kg).
SAFE DRINKING WATER ACT as boron: Priority List (55 FR 1470)
EPCRA Section 302 Extremely Hazardous Substances: TPQ = 500 lb (227 kg).
EPCRA Section 304 Reportable Quantity (RQ): EHS, 500 lb (227 kg).
EPCRA Section 313: Form R *de minimus* concentration reporting level: 1.0%.

BROMINE B:242

SYNONYMS: BROMINE; BROM (German); BROME (French); BROMO (Italian, Spanish); BROOM (Dutch); EEC No. 035-001-00-5; EINECS No. 231-778-1; MOLECULAR BROMINE

IDENTIFICATION

CAS:	7726-95-6
ERG ID:	UN 1744
ERG Guide:	154
ERG Guide Category:	SUBSTANCES-TOXIC AND/OR CORROSIVE (NONCOMBUSTIBLE)
Molecular Formula:	Br_2
RTECS No.:	EF9100000
Properties:	Watery, fuming liquid. Reddish-brown. Sharp, irritating odor.

HEALTH AND SAFETY INFORMATION

Poison. Breathing the vapor can cause death. Skin and eye contact causes severe burns and blindness.

IDLH: 3 ppm
Odor Threshold: 0.066 ppm
OSHA Table Z-1-A Air Contaminant

Exposure Limits:	TWA	STEL
ACGIH:	0.1 ppm; 0.66 mg/m^3	0.2 ppm; 1.3 mg/m^3
NIOSH:	0.1 ppm; 0.7 mg/m^3	0.3 ppm; 2 mg/m^3
OSHA:	0.1 ppm; 0.7 mg/m^3	

RECOMMENDATIONS FOR RESPIRATOR SELECTION: NIOSH/OSHA as bromine

2.5 ppm: SA:CF (any supplied-air respirator operated in a continuous-flow mode); or PAPRS (any powered, air-purifying respirator with cartridge(s) providing protection against the compound of concern.
3 ppm: CCRFS (any chemical cartridge respirator with a full facepiece and cartridge(s) providing protection against the compound of concern) organic vapor and acid gas cartridge(s); or GMFS [any air-purifying, full-facepiece respirator (gas mask) with a chin-style, front- or back-mounted canister providing protection against the compound of concern]; or PAPRTS (any powered, air-purifying respirator with a tight-fitting facepiece and cartridge(s) providing protection against the compound of concern); or SCBAF (any self-contained breathing apparatus with a full facepiece); or SAF (any supplied-air respirator with a full facepiece).

Emergency or Planned Entry into Unknown Concentrations or Idlh Conditions:
SCBAF:PD,PP (any self-contained breathing apparatus that has a full facepiece and is operated in a pressure-demand or other positive-pressure mode); or SAF:PD,PP:ASCBA (any supplied-air respirator that has a full facepiece and is operated in a pressure-demand or other positive-pressure mode in combination with an auxiliary, self-contained breathing apparatus operated in a pressure-demand or other positive-pressure mode.

Escape: GMFS [any air-purifying, full-facepiece respirator (gas mask) with a chin-style, front- or back-mounted canister providing protection against the compound of concern]; or SCBAE (Any appropriate escape-type, self-contained breathing apparatus).
Note: Substance causes eye irritation or damage; eye protection needed.
Note: Only nonoxidizable sorbents are allowed (not charcoal).

CHEMICAL REACTIVITY

Reacts with water to form acidic hydrobromic acid solution and toxic fumes. Reacts violently with aluminum. May cause fire in contact with wood, cotton, straw. Iron, steel, stainless steel, and copper are corroded by bromine and are
especially subject to attack by wet bromine. Of the plastics, only those which are highly fluorinated resist bromine attack.

FIRE INFORMATION

Firefighting gear (including SCBA) does not provide adequate protection. If exposure occurs, remove and isolate gear immediately and thoroughly decontaminate personnel. Strong oxidizer; will increase the intensity of fire. Vapors are heavier than air and will collect in low areas. Storage containers and parts of containers may rocket great distances, in many directions.If material or contaminated runoff enters waterways, notify downstream users of potentially contaminated waters. Notify local health and fire officials and pollution control agencies.

NFPA Hazard Classifications:

Health Hazards (Blue) 3
Flammability (Red) 0
Reactivity (Yellow) 0
Special Notice .... OXY

Flash Point: Not flammable

FIRST AID

Move victim to fresh air; call emergency medical care. Apply artificial respiration if victim is not breathing. Do not use mouth-to-mouth method if victim ingested or inhaled the substance; induce artificial respiration with the aid of a pocket mask equipped with a one-way valve or other proper respiratory medical device. Administer oxygen if breathing is difficult. Remove and isolate contaminated clothing and shoes. In case of contact with substance, immediately flush skin or eyes with running water for at least 20 minutes. For minor skin contact, avoid spreading material on unaffected skin. Keep victim warm and quiet. Effects of exposure (inhalation, ingestion, or skin contact) to substance may be delayed. Ensure that medical personnel are aware of the material(s) involved and take precautions to protect themselves.

ENVIRONMENTAL INFORMATION

Clean Air Act: Accidental Release Prevention/Flammable Substances (Section 112 [r], TQ = 10,000 lb (4,540 kg)
EPCRA Section 302, Extremely Hazardous Substances: TPQ = 500 lb (227 kg)
EPCRA Section 304 Reportable Quantity (RQ): EHS/CERCLA, 500 lb (227 kg)
EPCRA Section 313: Form R *de minimis* concentration reporting level: 1.0%.

BROMOACETIC ACID B:244

SYNONYMS: ACETIC ACID, BROMO-; ACIDE BROMOAC.ACTE.ETIQUE (French); ACIDO BROMOACETICO (Spansih); BROOMAZIJNZUUR (German); BROMESSIGSAEURE (German); a-BROMOETHANOIC ACID; BROMOETHANOIC ACID; EEC No. 607-065-00-X; EINECS No. 201-175-8

IDENTIFICATION

CAS:	79-08-3
ERG ID:	UN 1938
ERG Guide:	156
ERG Guide Category:	SUBSTANCES-TOXIC AND/OR CORROSIVE (COMBUSTIBLE/WATER-SENSITIVE)
Molecular Formula:	$C_2H_3BrO_2$

RTECS No.: AF5950000
Properties: Solid crystals. Soluble in water.

HEALTH AND SAFETY INFORMATION

Corrosive to eyes, skin, and respiratory tract. Inhalation may cause pulmonary edema, which can be delayed for several hours; there is a risk of death in serious cases.

RECOMMENDATIONS FOR RESPIRATOR SELECTION:

At any concentrations above the NIOSH REL, or where there is no REL, at any detectable concentration: SCBAF:PD,PP (any self-contained breathing apparatus that has a full facepiece and is operated in a pressure-demand or other positive-pressure mode); or SAF:PD,PP:ASCBA (any supplied-air respirator that has a full facepiece and is operated in a pressure-demand or other positive-pressure mode in combination with an auxiliary, self-contained breathing apparatus operated in a pressure-demand or other positive-pressure mode).

Escape: GMFOVHiE [any air-purifying, full-facepiece respirator (gas mask) with a chin-style, front- or back-mounted organic vapor canister having a high-efficiency particulate filter]; or SCBAE (any appropriate escape-type, self-contained breathing apparatus)

CHEMICAL REACTIVITY

Contact with oxidizers may cause fire and explosions.

FIRE INFORMATION

Vapors are heavier than air and will collect in low areas. Vapors in confined areas may explode when exposed to fire.If material or contaminated runoff enters waterways, notify downstream users of potentially contaminated waters. Notify local health and fire officials and pollution control agencies. Use dry chemicals, dry sand, carbon dioxide. From a secure explosion-proof location, use water spray to cool exposed (an closed) containers. If cooling streams are ineffective (venting sound increases in volume and pitch, tank discolors, or shows any signs of deforming), withdraw immediately to a secure position.

FEMA Hazard Classifications:

Health Hazards (Blue) 2
Flammability (Red) 1
Reactivity (Yellow) 0

Flash Point: 212°F (100°C)

FIRST AID

Move victim to fresh air. Call emergency medical care. Apply artificial respiration if victim is not breathing. Do not use mouth-to-mouth method if victim ingested or inhaled the substance; induce artificial respiration with the aid of a pocket mask equipped with a one-way valve or other proper respiratory medical device. Administer oxygen if breathing is difficult. Remove and isolate contaminated clothing and shoes. In case of contact with substance, immediately flush skin or eyes with running water for at least 20 minutes. For minor skin contact, avoid spreading material on unaffected skin. Keep victim warm and quiet. Effects of exposure (inhalation, ingestion, or skin contact) may be delayed. Ensure that medical personnel are aware of the material(s) involved and take precautions to protect themselves.

BROMOFORM B:246

SYNONYMS: BROMOFORME (French); BROMOFORMIO (Italian, Spanish); EEC No. 602-007-00-X; EINECS No. 200-854-6; METHENYL TRIBROMIDE; NCI-C55130; METHANE, TRIBROMO-; METHYL TRIBROMIDE; TRIBROMMETHAAN (Dutch); TRIBROMMETHAN (German); TRIBROMOMETAN (Italian); FORMYL TRIBROMIDE; METHENYL TRIBROMIDE; METHYLENE TRIBROMIDE; TRIBROMOMETHANE

IDENTIFICATION

CAS:	75-25-2
ERG ID:	UN 2515
ERG Guide:	159
ERG Guide Category:	SUBSTANCES (IRRITATING)
Molecular Formula:	$CHBr_3$
RTECS No.:	PB5600000
Properties:	Colorless to yellow liquid. Sweetish, chloroformlike odor. Sinks and very slowly mixes with water.

HEALTH AND SAFETY INFORMATION

Irritates eyes, skin, and respiratory tract. Inhalation or ingestion can affect central nervous system with dizziness, headache, tremors, difficult breathing, and unconsciousness. Mutation data reported.

IDLH: 850 ppm
Odor Threshold: 0.447 ppm
OSHA Table Z-1-A Air Contaminant

Exposure Limits:	**TWA**
ACGIH:	0.5 ppm; 5.2 mg/m^3
NIOSH:	0.5 ppm; 5 mg/m^3
OSHA:	0.5 ppm; 5 mg/m^3

Note: Skin contact contributes significantly in overall exposure.

RECOMMENDATIONS FOR RESPIRATOR SELECTION: NIOSH/OSHA

12.5 ppm: SA:CF (any supplied-air respirator operated in a continuous-flow mode); or PAPROV [any powered, air-purifying respirator with organic vapor cartridge(s)}
25 ppm: CCRFOV [any chemical cartridge respirator with a full facepiece and organic vapor cartridge(s)]; or GMFOV [Any air-purifying, full-facepiece respirator (gas mask) with a chin-style, front- or back-mounted acid gas canister]; or PAPRTOV [any powered, air-purifying respirator with a tight-fitting facepiece and organic vapor cartridge(s)]; or SCBAF (any self-contained breathing apparatus with a full facepiece); or SAF (any supplied-air respirator with a full facepiece)
850 ppm: SAF:PD,PP (any supplied-air respirator that has a full facepiece and is operated in a pressure-demand or other positive-pressure mode)

Emergency or Planned Entry into Unknown Concentrations or Idlh Conditions: SCBAF:PD,PP (any self-contained breathing apparatus that has a full facepiece and is operated in a pressure-demand or other positive-pressure mode); or SAF:PD,PP:ASCBA (any supplied-air respirator that has a full facepiece and is operated in a pressure-demand or other positive-pressure mode in combination with an auxiliary, self-contained breathing apparatus operated in a pressure-demand or other positive-pressure mode.

Escape: GMFOV [any air-purifying, full-facepiece respirator (gas mask) with a chin-style, front- or back-mounted organic vapor canister]; or SCBAE (any appropriate escape-type, self-contained breathing apparatus).
Note: Substance causes eye irritation or damage; eye protection needed.

CHEMICAL REACTIVITY

Reacts, possibly violently, with chemically active metals, lithium, sodium, potassium, calcium, powdered aluminum, zinc, magnesium, strong caustics, and acetone. Gradually decomposes, acquiring yellow color; air and light accelerate decomposition. Store in a cool place, away from heat and light. Reacts with some forms of plastics, rubber, and coatings.

FIRE INFORMATION

Not flammable. If material or contaminated runoff enters waterways, notify downstream users of potentially contaminated waters. Notify local health and fire officials and pollution control agencies.

FIRST AID

Move victim to fresh air. Call emergency medical care. Apply artificial respiration if victim is not breathing. Do not use mouth-to-mouth method if victim ingested or inhaled the substance; induce artificial respiration with the aid of a pocket mask equipped with a one-way valve or other proper respiratory medical device. Administer oxygen if breathing is difficult. Remove and isolate contaminated clothing and shoes. In case of contact with substance, immediately flush skin or eyes with running water for at least 20 minutes. For minor skin contact, avoid spreading material on unaffected skin. Keep victim warm and quiet. Effects should disappear after individual has been exposed to fresh air for approximately 10 minutes. Ensure that medical personnel are aware of the material(s) involved and take precautions to protect themselves.

ENVIRONMENTAL INFORMATION

Clean Air Act: Hazardous Air Pollutants (Title I, Part A, Section 112)
Clean Water Act: Section 307 Priority Pollutants; Section 313 Priority Chemicals
EPA Hazardous Waste Number (RCRA No.): U255
RCRA Section 261 Hazardous Constituents
RCRA Land Ban Waste
RCRA Universal Treatment Standards: Wastewater (mg/L), 0.63; Nonwastewater (mg/kg), 15
RCRA Ground Water Monitoring List. Suggested test method(s) PQL μg/L): 8010 (2); 8240 (5)
Safe Drinking Water Act: Priority List (55 FR 1470)
EPCRA Section 304 Reportable Quantity (RQ): CERCLA, 100 lb (45.4 kg)
EPCRA Section 313: Form R *de minimis* concentration reporting level: 1.0%.
California's Proposition 65: Carcinogen.
Marine Pollutant (49CFR, Chapter 172.101, Appendix B).

1–BROMOPROPANE B:248

SYNONYMS: n-PROPYLBROMIDE; PROPYL BROMIDE; EEC No. 602-019-00-5; EINECS No. 203-445-0

IDENTIFICATION

CAS:	106-94-5
ERG ID:	UN 2344
ERG Guide:	132
ERG Guide Category:	FLAMMABLE LIQUIDS-CORROSIVE
Molecular Formula:	C_3H_7Br
Properties:	Flammable liquid. Colorless. Sinks in water.

HEALTH AND SAFETY INFORMATION

May cause eye irritation. Moderately toxic if inhaled or swallowed.

RECOMMENDATIONS FOR RESPIRATOR SELECTION:

At any concentrations above the NIOSH REL, or where there is no REL, at any detectable concentration: SCBAF:PD,PP (any self-contained breathing apparatus that has a full facepiece and is operated in a pressure-demand or other positive-pressure mode); or SAF:PD,PP:ASCBA (any supplied-air respirator that has a full facepiece and is operated in a pressure-demand or other positive-pressure mode in combination with an auxiliary, self-contained breathing apparatus operated in a pressure-demand or other positive-pressure mode).

Escape: GMFOV [any air-purifying, full-facepiece respirator (gas mask) with a chin-style, front- or back-mounted organic vapor canister]; or SCBAE (any appropriate escape-type, self-contained breathing apparatus).

CHEMICAL REACTIVITY

Contact with oxidizers may cause fire and explosions.

FIRE INFORMATION

Combustion products include toxic fumes of bromine. Vapors are heavier than air and will collect in low areas. Vapors are heavier than air and will collect in low areas. Vapors in confined areas may explode when exposed to fire. Vapors may travel long distances to ignition sources and flash back. Storage containers and parts of containers may rocket great distances, in many directions. If material or contaminated runoff enters waterways, notify downstream users of potentially contaminated waters. Notify local health and fire officials and pollution control agencies. Use dry chemicals, foam, carbon dioxide. From a secure explosion-proof location, use water spray to cool exposed containers. If cooling streams are ineffective (venting sound increases in volume and pitch, tank discolors, or shows any signs of deforming), withdraw immediately to a secure position.

NFPA Hazard Classifications: (as n-propyl bromide)

Health Hazards (Blue) 2
Flammability (Red) 3
Reactivity (Yellow) 0

Flash Point: < 72°F(< 22 °C)
Flammable Limits in Air: LEL 4.6 %; UEL Unknown
Autoignition Temperature: 914°F (490°C)

FIRST AID

Move victim to fresh air; call emergency medical care. Apply artificial respiration if victim is not breathing. Do not use mouth-to-mouth method if victim ingested or inhaled the substance; induce artificial respiration with the aid of a pocket mask equipped with a one-way valve or other proper respiratory medical device. Administer oxygen if breathing is difficult. Remove and isolate contaminated clothing and shoes. In case of contact with substance, immediately flush skin or eyes with running water for at least 20 minutes. For minor skin contact, avoid spreading material on unaffected skin. Keep victim warm and quiet. Effects of exposure (inhalation, ingestion, or skin contact) to substance may be delayed. Ensure that medical personnel are aware of the material(s) involved and take precautions to protect themselves.

1,3–BUTADIENE B:250

SYNONYMS: BIETHYLENE; BIVINYL; BUTADIEEN (Dutch); BUTA-1,3-DIEEN (Dutch); BUTADIEN (Polish); BUTADIENE; BUTA-1,3-DIEN (German); BUTA-1,3-DIENE; α-γ-BUTADIENE; BUTADIENO (Spanish); DIVINYL; EEC No. 601-013-00-X; EINECS No. 203-450-8; ERYTHRENE; NCI-C50602; PYRROLYLENE; VINYLETHYLENE

154 1,3-Butatdiene

IDENTIFICATION

CAS:	106-99-0
ERG ID:	UN 1010
ERG Guide:	116P
ERG Guide Category:	GASES-FLAMMABLE (UNSTABLE)
Molecular Formula:	C_4H_6
RTECS No.:	EI9275000
Properties:	Liquefied, compressed gas. Colorless. Gasolinelike odor. Floats and boils on water. Flammable visible vapor cloud is produced.

HEALTH AND SAFETY INFORMATION

Irritates eyes, skin, and respiratory tract. High levels can cause unconsciousness and death. Contact with liquid can cause frostbite. Possible human carcinogen.

IARC Evaluation: IARC possible carcinogen, rating 2B;
NTP: NTP anticipated carcinogen
IDLH: 2,000 ppm [10% LEL]; potential human carcinogen.
Odor Threshold: 0.455 ppm
OSHA Table Z-1-A Air Contaminant

Exposure Limits:	**TWA**	**STEL**
ACGIH:	2 ppm; 4.4 mg/m^3	
NIOSH:*		
OSHA:	1 ppm; 2.21 mg/m^3	5 ppm; 11.05 mg/m^3

*Potential human carcinogen. Reduce exposure to lowest feasible level.

RECOMMENDATIONS FOR RESPIRATOR SELECTION: NIOSH

At any concentrations above the NIOSH REL, or where there is no REL, at any detectable concentration: SCBAF:PD,PP (any self-contained breathing apparatus that has a full facepiece and is operated in a pressure-demand or other positive-pressure mode); or SAF:PD,PP:ASCBA (any supplied-air respirator that has a full facepiece and is operated in a pressure-demand or other positive-pressure mode in combination with an auxiliary, self-contained breathing apparatus operated in a pressure-demand or other positive-pressure mode)

Escape: GMFS [any air-purifying, full-facepiece respirator (gas mask) with a chin-style, front- or back-mounted canister providing protection against the compound of concern]; or SCBAE (any appropriate escape-type, self-contained breathing apparatus).

CHEMICAL REACTIVITY

Reacts with oxidizers, phenol, chlorine dioxide, crotonaldehyde. Unsafe in contact with acetylide-forming materials such as monel, copper, and copper alloys.

FIRE INFORMATION

Gases are heavier than air and will collect in low areas. Gases in confined areas may explode when exposed to fire. Gases may travel long distances to ignition sources and flash back. Storage containers and parts of containers may rocket great distances, in many directions. If material or contaminated runoff enters waterways, notify downstream users of potentially contaminated waters. Notify local health and fire officials and pollution control agencies. Do not extinguish the fire unless the flow of the gas can be stopped and any remaining gas is out of the line. Specially trained personnel may use fog lines to cool exposures and let the fire burn itself out. Use dry chemicals, foam, carbon dioxide. From a secure explosion-proof location, use water spray to cool exposed containers. If cooling streams are ineffective (venting sound increases in volume and pitch, tank discolors, or shows any signs of deforming), withdraw immediately to a secure position.

NFPA Hazard Classifications:

Health Hazards (Blue) 2
Flammability (Red) 0
Reactivity (Yellow) 3

Flash Point: -105°F (-76°C) (liquid)
Flammable Limits in Air: LEL 2.0%; UEL 11.5%
Autoignition Temperature: 788°F (420°C)

FIRST AID

Move victim to fresh air. Call emergency medical care. Apply artificial respiration if victim is not breathing. If breathing is difficult, give oxygen. Remove and isolate contaminated clothing and shoes. In case of contact with liquefied gas, thaw frosted parts with lukewarm water. Keep victim warm and quiet. Ensure that medical personnel are aware of the material(s) involved and take precautions to protect themselves.

ENVIRONMENTAL INFORMATION

Clean Air Act: Hazardous Air Pollutants (Title I, Part A, Section 112); Accidental Release Prevention/Flammable Substances (Section 112[r], Table 3), TQ = 10,000 lb (4,540 kg)
EPCRA Section 304 Reportable Quantity (RQ): CERCLA, 10 lb (4.54 kg)
EPCRA Section 313: Form R *de minimis* concentration reporting level: 0.1%.

BUTANE B:260

SYNONYMS: A-17; BU-GAS; n-BUTANE; BUTANEN (Dutch); BUTANI (Italian); BUTANO (Spanish); BUTYL HYDRIDE; CLEARVIEW GLASS CLEANER; DIETHYL; EEC No. 601-004-00-0; EINECS No. 203-448-7; LIQUIFIED PETROLEUM GAS; METHYLETHYLMETHANE; METHYL ETHYL METHANE; TWINKLE STAINLESS STEEL CLEANER

156 Butane

IDENTIFICATION

CAS:	106-97-8
ERG ID:	UN 1011
ERG Guide:	115
ERG Guide Category:	GASES-FLAMMABLE (INCLUDING REFRIGERATED LIQUIDS)
Molecular Formula:	C_4H_{10}
RTECS No.:	EJ4200000
Properties:	Extremely flammable, liquefied, compressed gas. A liquid below 30°F (−1.1°C). Colorless. Natural gaslike odor. Floats and boils on water. Flammable, visible vapor cloud is formed. Vapors are heavier than air and will collect in low areas.

HEALTH AND SAFETY INFORMATION

Chronic exposure can result in visual hallucinations, irritability. Social withdrawal has been reported. Cardiac sensitization may also be possible.

Odor Threshold: 204 ppm

Exposure Limits: TWA

ACGIH:	800 ppm; 1900 mg/m^3
NIOSH:	800 ppm; 1900 mg/m^3
OSHA:	None

RECOMMENDATIONS FOR RESPIRATOR SELECTION:

At any concentrations above the NIOSH REL, or where there is no REL, at any detectable concentration: SCBAF:PD,PP (any self-contained breathing apparatus that has a full facepiece and is operated in a pressure-demand or other positive-pressure mode); or SAF:PD,PP:ASCBA (any supplied-air respirator that has a full facepiece and is operated in a pressure-demand or other positive-pressure mode in combination with an auxiliary, self-contained breathing apparatus operated in a pressure-demand or other positive-pressure mode).

Escape: GMFOV (any air-purifying, full-facepiece respirator (gas mask) with a chin-style, front- or back-mounted organic vapor canister); or SCBAE (any appropriate escape-type, self-contained breathing apparatus).

CHEMICAL REACTIVITY

Forms explosive mixture with air. Reacts with oxidizers, chlorine, fluorine; nickel carbonyl plus oxygen. May accumulate static electric charges that can result in ignition of its vapors.

FIRE INFORMATION

Gases are heavier than air and will collect in low areas. Gases in confined areas may explode when exposed to fire. Gases may travel long distances to ignition sources and flash back. Storage containers and parts of containers may rocket great distances, in many directions. Notify local health and fire officials and pollution control agencies. Do not extinguish the fire unless the flow of the gas can be stopped and any remaining gas is out of the line. Specially trained personnel may use fog lines to cool exposures and let the fire burn itself out. Use dry chemicals, foam, carbon dioxide. From a secure, explosion-proof location, use water spray to cool exposed containers. If cooling streams are ineffective (venting sound increases in volume and pitch, tank discolors, or shows any signs of deforming), withdraw immediately to a secure position.

NFPA Fire Classifications:

Health Hazards (Blue) 1
Flammability (Red) 4
Reactivity (Yellow) 0

Flash Point: -76°F (-60°C)
Flammable Limits in Air: LEL 1.6%; UEL 8.4%
Autoignition Temperature: 550°F (287°C)

FIRST AID

Move victim to fresh air. Call emergency medical care. Apply artificial respiration if victim is not breathing. Administer oxygen if breathing is difficult. Remove and isolate contaminated clothing and shoes. Clothing frozen to the skin should be thawed before being removed. In case of contact with liquefied gas, thaw frosted parts with lukewarm water. Keep victim warm and quiet. Keep victim under observation. Ensure that medical personnel are aware of the material(s) involved and take precautions to protect themselves.

ENVIRONMENTAL INFORMATION

CLEAN AIR ACT: Accidental Release Prevention/Flammable substances, (Section 112[r], Table 3), TQ = 10,000 lb (4,540 kg).

1,4–BUTANEDIOL B:268

SYNONYMS: BUTANE-1,4-DIOL; BUTANODIOL (Spanish); 1,4-BUTYLENEGLYCOL; 1,4-DIHYDROXYBUTANE; EINECS No. 203-786-5; TETRAMETHYLENE GLYCOL; 1,4-TETRAMETHYLENE GLYCOL

IDENTIFICATION

CAS: 110-63-4

ERG ID:	UN 1987
ERG Guide:	127
ERG Guide Category:	FLAMMABLE LIQUIDS (POLAR/WATER-MISCIBLE)
Molecular Formula:	$C_4H_{10}O_2$
RTECS No.:	EK0525000
Properties:	Combustible, thick liquid. Colorless. Odorless. Sinks and mixes with water. Freezing point is 68°F (20°C)

HEALTH AND SAFETY INFORMATION

Poison. Irritates eyes. High or repeated exposure may cause kidney damage.

RECOMMENDATIONS FOR RESPIRATOR SELECTION

At any concentrations above the NIOSH REL, or where there is no REL, at any detectable concentration: SCBAF:PD,PP (any self-contained breathing apparatus that has a full facepiece and is operated in a pressure-demand or other positive-pressure mode); or SAF:PD,PP:ASCBA (any supplied-air respirator that has a full facepiece and is operated in a pressure-demand or other positive-pressure mode in combination with an auxiliary, self-contained breathing apparatus operated in a pressure-demand or other positive-pressure mode).

Escape: GMFOV [any air-purifying, full-facepiece respirator (gas mask) with a chin-style, front- or back-mounted organic vapor canister]; or SCBAE (any appropriate escape-type, self-contained breathing apparatus)

CHEMICAL REACTIVITY

Forms explosive mixture with air. Contact with oxidizers may cause fire and explosions.

FIRE INFORMATION

Vapors are heavier than air and will collect in low areas. Vapors in confined areas may explode when exposed to fire. If material or contaminated runoff enters waterways, notify downstream users of potentially contaminated waters. Notify local health and fire officials and pollution control agencies. Use dry chemicals, alcohol-resistant foam, carbon dioxide. From a secure explosion-proof location, use water spray to cool exposed containers. If cooling streams are ineffective (venting sound increases in volume and pitch, tank discolors, or shows any signs of deforming), withdraw immediately to a secure position.

NFPA Hazard Classifications:

Health Hazards (Blue) 1
Flammability (Red) 1
Reactivity (Yellow) 0

Flash Point: 250°F (121°C) o.c.
Flammable Limits in Air: NA

Autoignition Temperature: 756°F (402°C)

FIRST AID

Move victim to fresh air. Call emergency medical care. Apply artificial respiration if victim is not breathing. Administer oxygen if breathing is difficult. Remove and isolate contaminated clothing and shoes. In case of contact with substance, immediately flush skin or eyes with running water for at least 20 minutes. Wash skin with soap and water. Keep victim warm and quiet. Ensure that medical personnel are aware of the material(s) involved and take precautions to protect themselves.

2,3–BUTANEDIONE B:274

SYNONYMS: BIACETYL; BUTADIONE; 2,3-BUTADIONE; BUTANEDIONE (DOT); BUTANODIONA (Spanish); DIACETYL (FCC); 2,3-DIKETOBUTANE; DIMETHYL DIKETONE; DIMETHYLGLYOXAL; EINECS No. 207-069-8; EINECS No. 207-069-8; GLYOXAL, DIMETHYL-; FEMA No. 2370

IDENTIFICATION

CAS:	431-03-8
ERG ID:	UN 2346
ERG Guide:	127
ERG Guide Category:	FLAMMABLE LIQUIDS (POLAR/WATER-MISCIBLE)
Molecular Formula:	$C_4H_6O_2$
RTECS No.:	EK2625000
Properties:	Extremely flammable liquid. Green-yellow color. Strong, characteristic chlorine-like, rancid-butter odor. Soluble in water.

HEALTH AND SAFETY INFORMATION

Irritates the eyes, skin, and respiratory tract. May have a narcotic effect on the central nervous system.

NIOSH Criteria Document: NIOSH 78-173 KETONES

RECOMMENDATIONS FOR RESPIRATOR SELECTION:

At any concentrations above the NIOSH REL, or where there is no REL, at any detectable concentration: SCBAF:PD,PP (any self-contained breathing apparatus that has a full facepiece and is operated in a pressure-demand or other positive-pressure mode); or SAF:PD,PP:ASCBA (any supplied-air respirator that has a full facepiece and is operated in a pressure-demand or other positive-pressure mode in combination with an auxiliary, self-contained breathing apparatus operated in a pressure-demand or other positive-pressure mode).

Escape: GMFOV [any air-purifying, full-facepiece respirator (gas mask) with a chin-style, front- or back-mounted organic vapor canister]; or SCBAE (any appropriate escape-type, self-contained breathing apparatus).

CHEMICAL REACTIVITY

Forms an explosive mixture with air. Contact with oxidizers may cause fire and explosions.

FIRE INFORMATION

Decomposition products upon heating may include toxic and irritating gases. Vapors are heavier than air and will collect in low areas. Vapors in confined areas may explode when exposed to fire. Vapors may travel long distances to ignition sources and flash back. Storage containers and parts of containers may rocket great distances, in many directions. Use dry chemicals, alcohol-resistant foam, carbon dioxide, large quantities of water.

NFPA Hazard Classifications:

Health Hazards (Blue) 1
Flammability (Red) 3
Reactivity (Yellow) 0

Flash Point: 80°F (27°C)

FIRST AID

Move victim to fresh air. Call emergency medical care. Apply artificial respiration if victim is not breathing. Administer oxygen if breathing is difficult. Remove and isolate contaminated clothing and shoes. In case of contact with substance, immediately flush skin or eyes with running water for at least 20 minutes. Wash skin with soap and water. Keep victim warm and quiet. Ensure that medical personnel are aware of the material(s) involved and take precautions to protect themselves.

2-BUTOXYETHANOL B:280

SYNONYMS: BUCS; BUTOKSYETYLOWY ALKOHOL (Polish); 2-BUTOXIETANOL (Spanish); 2-BUTOSSI-ETANOLO (Italian); 2-BUTOXY-AETHANOL (German); BUTOXYETHANOL(2-); BUTYL OXITOL; BUTYL CELLOSOLVE; o-BUTYL ETHYLENE GLYCOL; BUTYL ICINOL; BUTYLGLYCOL (French, German); CLEARVIEW GLASS CLEANER; DOWANOL EB; EEC No. 603-014-00-0; EGBE; EINECS No. 203-905-0; EKTASOLVE EB; ETHYLENE GLYCOL MONOBUTYL ETHER; FORMULA 409 CLEANER; GAFCOL EB; GLYCOL MONOBUTYL ETHER; GLYCOL BUTYL ETHER; HD DETERGENT; JEFFERSOL EB; LAMINAR DEVELOPER KB-1B/KB-113; LOCOTENGE 520; MICROPOSIT REMOVER 1112A; MS CLEANER; OAKITE 33; 3-OXA-1-HAPTANOL; PERMA DRY INK 555 RED; POLY-SOLV EB; PROBE RK; RISTON II D200 DEVELOPER; RS-4 STRIPPER, POSITIVE; SIMPLE GREEN (SUNSHINE MAKERS); TRUE BLUE GLASS CLEANER; WHIRLWIND GLASS CLEANER; ZEP ONCE OVER WALL CLEANER

IDENTIFICATION

CAS:	111-76-2
ERG ID:	UN 2369
ERG Guide:	152
ERG Guide Category:	TOXIC (COMBUSTIBLE)
Molecular Formula:	$C_6H_{14}O_2$
RTECS No.:	KJB575000
Properties:	Combustible liquid. Clear color. Pleasant, etherlike odor. Soluble in water

HEALTH AND SAFETY INFORMATION

May cause anemia, liver and kidney damage, lung effects, and dermatitis.

IDLH: 700 ppm
Odor Threshold: 0.001 ppm
OSHA Table Z-1-A Air Contaminant

Exposure Limits:	**TWA**
ACGIH:*	25 ppm; 121 mg/m^3
NIOSH:*	5 ppm; 24 mg/m^3
OSHA:*	50 ppm; 240 mg/m^3

*Skin contact contributes significantly to overall exposure.

RECOMMENDATIONS FOR RESPIRATOR SELECTION: NIOSH/OSHA

50 ppm: CCROV [any chemical cartridge respirator with organic vapor cartridge(s)]; or SA (any supplied-air respirator)
125 ppm: SA:CF (any supplied-air respirator operated in a continuous-flow mode); or PAPROV [any powered, air-purifying respirator with organic vapor cartridge(s)]
250 ppm: CCRFOV [any chemical cartridge respirator with a full facepiece and organic vapor cartridge(s)]; or GMFOV [any air-purifying, full-facepiece respirator (gas mask) with a chin-style, front- or back-mounted acid gas canister]; or PAPRTOV [any powered, air-purifying respirator with a tight-fitting facepiece and organic vapor cartridge(s)]; or SCBAF (any self-contained breathing apparatus with a full facepiece); or SAF (any supplied-air respirator with a full facepiece).

Emergency or Planned Entry into Unknown Concentrations or IDLH Conditions:

SCBAF:PD,PP (any self-contained breathing apparatus that has a full facepiece and is operated in a pressure-demand or other positive-pressure mode); or SAF:PD,PP:ASCBA (any supplied-air respirator that has a full facepiece and is operated in a pressure-demand or other positive-pressure mode in combination with an auxiliary, self-contained breathing apparatus operated in a pressure-demand or other positive-pressure mode).

Escape: GMFOV [any air-purifying, full-facepiece respirator (gas mask) with a chin-style, front- or back-mounted organic vapor canister]; or SCBAE (any appropriate escape-type, self-contained breathing apparatus).
Note: Substance reported to cause eye irritation or damage; may require eye protection.

CHEMICAL REACTIVITY

Forms explosive mixture with air. Violent reaction with strong caustics and strong oxidizers. Attacks some coatings, plastics, and rubber. Attacks metallic aluminum at high temperatures.

FIRE INFORMATION

Vapors are heavier than air and will collect in low areas. Vapors in confined areas may explode when exposed to fire. Vapors may travel long distances to ignition sources and flash back. Storage containers and parts of containers may rocket great distances, in many directions. If material or contaminated runoff enters waterways, notify downstream users of potentially contaminated waters. Notify local health and fire officials and pollution control agencies. Use dry chemicals, foam, carbon dioxide. From a secure explosion-proof location, use water spray to cool exposed containers. If cooling streams are ineffective (venting sound increases in volume and pitch, tank discolors, or shows any signs of deforming), withdraw immediately to a secure position.

NFPA Hazard Classifications:

Health Hazards (Blue) 2
Flammability (Red) 2
Reactivity (Yellow) 0

Flash Point: 143°F (62°C)
Flammable Limits in Air: LEL 1.1%; UEL 127%

FIRST AID

Move victim to fresh air. Call emergency medical care. Apply artificial respiration if victim is not breathing. Do not use mouth-to-mouth method if victim ingested or inhaled the substance; induce artificial respiration with the aid of a pocket mask equipped with a one-way valve or other proper respiratory medical device. Administer oxygen if breathing is difficult. Remove and isolate contaminated clothing and shoes. In case of contact with substance, immediately flush skin or eyes with running water for at least 20 minutes. For minor skin contact, avoid spreading material on unaffected skin. Keep victim warm and quiet. Effects of exposure (inhalation, ingestion, or skin contact) may be delayed. Ensure that medical personnel are aware of the material(s) involved and take precautions to protect themselves.

n-BUTYL ACETATE B:340

SYNONYMS: ACETATE de BUTYLE (French); ACETATO de n-BUTILO (Spanish); ACETIC ACID,

BUTYL ESTER; ACETIC ACID n-BUTYL ESTER; ARISTOLINE (+); AZ 1312-SFD (+); AZ 1350J (+); AZ 1370 (+); AZ THINNER; AZ 1310-SF (+); AZ 1370-SF (+); AZ 1375 (+); AZ 1470 (+); AZ 4140 (+); AZ 4210 (+); AZ 4330 (+); AZ 4620 (+); BUTILE (ACETATI di) (Italian); BUTYLACETAT (German); BUTYLACETATEN (Dutch); 1-BUTYL ACETATE; BUTYLE (ACETATE de) (French); BUTYL ACETATE; n-BUTYL ESTER OF ACETIC ACID; BUTYL ETHANOATE; EEC No. 607-025-00-1; EINECS No. 208-760-7; 6-6 EPOXY CHEM RESIN FINISH, CLEAR CURING AGENT; FEMA No. 2174; GOODRITE NR-R; KTI 1300 THINNER; KTI 1350J (+); KTI 1370/1375 (+); KTI 1470 (+); KTI II (+); MICROPOSIT 111S (+); MICROPOSIT 119S (+); MICROPOSIT 119 THINNER; MICROPOSIT 1375 (+); MICROPOSIT 1400-33 (+); MICROPOSIT 1470 (+); MICROPOSIT 6009 (+); MICROPOSIT 1400S (+); MICROPOSIT SAL 601-ER7 (+); MICROPOSIT XP-6012 (+); OCTAN n-BUTYLU (Polish); TSMR 8800 (+); TSMR 8800 BE; ULTRAMAC SOLVENT EPA; ULTRAMAC PR-1024 MB-628 RESIN; WAYCOAT 204 (+); WAYCOAT RX 507 (+); WAYCOAT HPR 205/207 (+); XANTHOCHROME (+); XIR-3000-T RESIN

IDENTIFICATION

CAS:	123-86-4
ERG ID:	UN 1123
ERG Guide:	129
ERG Guide Category:	FLAMMABLE LIQUIDS (POLAR/WATER-MISCIBLE/NOXIOUS)
Molecular Formula:	$C_6H_{12}O_2$
RTECS No.:	AF7350000
Properties:	Highly flammable, watery liquid. Colorless. Pleasant, fruity odor in low concentrations, disagreeable odor in higher concentrations. Floats on water; insoluble. Flammable, irritating vapor is produced in fire.

HEALTH AND SAFETY INFORMATION

IDLH: 1,700 ppm [LEL]
Odor Threshold: 0.007 ppm

Prolonged or repeated exposure to skin may cause drying and cracking; skin allergy may develop. May cause birth defects or miscarriage. Similar solvents and petroleum-based chemicals can cause nerve and brain damage, memory loss, personality changes, fatigue, sleep disturbances, reduced coordination, and a feeling of "pins and needles" in the arms and legs.

Exposure Limits:	**TWA**	**STEL**
ACGIH:	150 ppm; 713 mg/m^3	200 ppm; 950 mg/m^3
NIOSH:	150 ppm; 710 mg/m^3	200 ppm; 950 mg/m^3
OSHA:	150 ppm; 710 mg/m^3	

RECOMMENDATIONS FOR RESPIRATOR SELECTION: NIOSH/OSHA

1,500 ppm: CCROV [any chemical cartridge respirator with a full facepiece and organic vapor cartridge(s)]; or SA (any supplied-air respirator)
1,700 ppm: SA:CF (any supplied-air respirator operated in a continuous-flow mode); or PAPROV [any powered, air-purifying respirator with organic vapor cartridge(s)]; or CCRFOV [any air-purifying, full-facepiece respirator (gas mask) with a chin-style, front- or back-mounted acid gas canister]; or GMFOV [any air-purifying, full-facepiece respirator (gas mask) with a chin-style, front- or back-mounted acid gas canister]; or SCBAF (any self-contained breathing apparatus with a full facepiece); or SAF (any supplied-air respirator with a full facepiece).

Emergency or Planned Entry into Unknown Concentrations or IDLH Conditions: SCBAF:PD,PP (any self-contained breathing apparatus that has a full facepiece and is operated in a pressure-demand or other positive-pressure mode); or SAF:PD,PP:ASCBA (any supplied-air respirator that has a full facepiece and is operated in a pressure-demand or other positive-pressure mode in combination with an auxiliary, self-contained breathing apparatus operated in a pressure-demand or other positive-pressure mode).

Escape: GMFOV [any air-purifying, full-facepiece respirator (gas mask) with a chin-style, front- or back-mounted organic vapor canister]; or SCBAE (any appropriate escape-type, self-contained breathing apparatus).

Note: Substance reported to cause eye irritation or damage; may require eye protection.

CHEMICAL REACTIVITY

Forms explosive mixture with air. Reacts with water on standing to form acetic acid and n-butyl alcohol. Violent reaction with strong oxidizers and potassium-tert-butoxide. Incompatible with caustics, strong acids, and nitrates. Dissolves rubber, many plastics, resins, and some coatings. May accumulate static electrical charges, and may cause ignition of its vapors.

FIRE INFORMATION

Vapors are heavier than air and will collect in low areas. Vapors in confined areas may explode when exposed to fire. Vapors may travel long distances to ignition sources and flash back. Storage containers and parts of containers may rocket great distances, in many directions. If material or contaminated runoff enters waterways, notify downstream users of potentially contaminated waters. Notify local health and fire officials and pollution control agencies. Use dry chemicals, foam, carbon dioxide. From a secure explosion-proof location, use water spray to cool exposed containers. If cooling streams are ineffective (venting sound increases in volume and pitch, tank discolors, or shows any signs of deforming), withdraw immediately to a secure position.

NFPA Hazard Classifications:

Health Hazards (Blue) 1
Flammability (Red) 3
Reactivity (Yellow) 0

Flash Point: 72°F (22°C) c.c.
Flammable Limits in Air: LEL 1.7%; UEL 15%
Autoignition Temperature: 797°F (425°C)

FIRST AID

Move victim to fresh air. Call emergency medical care. Apply artificial respiration if victim is not breathing. If breathing is difficult, give oxygen. Remove and isolate contaminated clothing and shoes. In case of contact with substance, immediately flush skin or eyes with running water for at least 20 minutes. Wash skin with soap and water. Keep victim warm and quiet. Effects of exposure (inhalation, ingestion, or skin contact) to substance may be delayed. Ensure that medical personnel are aware of the material(s) involved and take precautions to protect themselves.

ENVIRONMENTAL INFORMATION

Clean Water Act: Section 311 Hazardous Substances/RQ (same as CERCLA)
EPCRA Section 304: Reportable Quantity (RQ): CERCLA, 5,000 lb (2,270 kg)

iso–BUTYL ACRYLATE B:348

SYNONYMS: ACRILATO de ISOBUTILO (Spanish); ACRYLIC ACID, ISO-BUTYL ESTER; EEC No. 607-115-00-0; EINECS No. 203-417-8; ISOBUTYL ACRYLATE, INHIBITED; ISOBUTYL PROPENOATE; ISOBUTYL-2-PROPENOATE; 2-METHYL-1-PROPYL ACRYLATE

IDENTIFICATION

CAS:	106-63-8
ERG ID:	UN 2527
ERG Guide:	130P
ERG Guide Category:	FLAMMABLE LIQUIDS (NONPOLAR/WATER-IMMISCIBLE/NOXIOUS)
Molecular Formula:	$C_7H_{12}O_2$
Properties:	Flammable, watery liquid. Colorless. Sharp, fragrant odor. Floats on water. Irritating vapor is produced.

HEALTH AND SAFETY INFORMATION

Corrosive to eyes, skin, and respiratory tract. Inhalation may cause pulmonary edema, which can be delayed for several hours; there is a risk of death in serious cases. High or repeated exposure may cause skin problems.

RECOMMENDATIONS FOR RESPIRATOR SELECTION

At any concentrations above the NIOSH REL, or where there is no REL, at any detectable concentration: SCBAF:PD,PP (any self-contained breathing apparatus that has a full facepiece and is

operated in a pressure-demand or other positive-pressure mode); or SAF:PD,PP:ASCBA (any supplied-air respirator that has a full facepiece and is operated in a pressure-demand or other positive-pressure mode in combination with an auxiliary, self-contained breathing apparatus operated in a pressure-demand or other positive-pressure mode).

Escape: GMFOV [any air-purifying, full-facepiece respirator (gas mask) with a chin-style, front- or back-mounted organic vapor canister]; or SCBAE (any appropriate escape-type, self-contained breathing apparatus)

CHEMICAL REACTIVITY

Reacts with strong acids, aliphatic amines, alkanolamines.

FIRE INFORMATION

Vapors are heavier than air and will collect in low areas. Vapors in confined areas may explode when exposed to fire. Vapors may travel long distances to ignition sources and flash back. Storage containers and parts of containers may rocket great distances, in many directions.If material or contaminated runoff enters waterways, notify downstream users of potentially contaminated waters. Notify local health and fire officials and pollution control agencies. Use dry chemicals, foam, carbon dioxide. From a secure explosion-proof location, use water spray to cool exposed containers. If cooling streams are ineffective (venting sound increases in volume and pitch, tank discolors, or shows any signs of deforming), withdraw immediately to a secure position.

NFPA Hazard Classifications:

Health Hazards (Blue) 2
Flammability (Red) 2
Reactivity (Yellow) 2

Flash Point: 94°F (34°C) o.c.
Flammable Limits in Air: LEL 1.9%; UEL 8.0%
Autoignition Temperature: 644°F (340°C)

FIRST AID

Move victim to fresh air. Call emergency medical care. Apply artificial respiration if victim is not breathing. Administer oxygen if breathing is difficult. Remove and isolate contaminated clothing and shoes. In case of contact with substance, immediately flush skin or eyes with running water for at least 20 minutes. Wash skin with soap and water. Keep victim warm and quiet. Effects of exposure (inhalation, ingestion, or skin contact) to substance may be delayed. Ensure that medical personnel are aware of the material(s) involved and take precautions to protect themselves.

n-BUTYL ACRYLATE B:350

SYNONYMS: ACRILATO de n-BUTILO (Spanish); ACRILATO di BUTILE (Italian); ACRYLATE de BUTYLE (French); ACRYLIC ACID, BUTYL ESTER; ACRYLIC ACID n-BUTYL ESTER; BUTYLACRYLAAT (Dutch); BUTYLACRYLATE, INHIBITED; BUTYL ACRYLATE; BUTYL ESTER OF ACRYLIC ACID; BUTYL 2-PROPENOATE; 2-PROPENOIC ACID, BUTYL ESTER; n-BUTYL 2-PROPENOATE; EEC No. 607-062-00-3; EINECS No. 205-480-7

IDENTIFICATION

CAS:	141-32-2
ERG ID:	UN 2348
ERG Guide:	129P
ERG Guide Category:	FLAMMABLE LIQUIDS (POLAR/WATER-MISCIBLE/NOXIOUS)
Molecular Formula:	$C_7H_{12}O_2$
RTECS No.:	UD3150000
Properties:	Very flammable liquid. White, waterlike. Sharp, fragrant odor. Floats on water; insoluble.

HEALTH AND SAFETY INFORMATION

Corrosive to eyes, skin, and respiratory tract. Inhalation may cause pulmonary edema, which can be delayed for several hours; there is a risk of death in serious cases. High or repeated exposure may cause skin problems.

IARC Evaluation: Unclassified substance; Group 3; no human data; inadequate animal evidence
Odor Threshold: 0.003 ppm

Exposure Limits:	**TWA**
ACGIH:	10 ppm; 52 mg/m^3
NIOSH:	10 ppm; 55 mg/m^3
OSHA:	None

RECOMMENDATIONS FOR RESPIRATOR SELECTION: NIOSH

At any concentrations above the NIOSH REL, or where there is no REL, at any detectable concentration: SCBAF:PD,PP (any self-contained breathing apparatus that has a full facepiece and is operated in a pressure-demand or other positive-pressure mode); or SAF:PD,PP:ASCBA (any supplied-air respirator that has a full facepiece and is operated in a pressure-demand or other positive-pressure mode in combination with an auxiliary, self-contained breathing apparatus operated in a pressure-demand or other positive-pressure mode).

Escape: GMFOV [any air-purifying, full-facepiece respirator (gas mask) with a chin-style, front- or

back-mounted organic vapor canister]; or SCBAE (any appropriate escape-type, self-contained breathing apparatus).

CHEMICAL REACTIVITY

Forms explosive mixture with air. Heat, sparks, open flame, light, or peroxides may cause explosive polymerization. Incompatible with strong acids, amines, halogens, hydrogen compounds, oxidizers, sunlight, or other catalysts.

FIRE INFORMATION

Vapors are heavier than air and will collect in low areas. Vapors in confined areas may explode when exposed to fire. Vapors may travel long distances to ignition sources and flash back. Storage containers and parts of containers may rocket great distances, in many directions. If material or contaminated runoff enters waterways, notify downstream users of potentially contaminated waters. Notify local health and fire officials and pollution control agencies. Use dry chemicals, foam, carbon dioxide. From a secure explosion-proof location, use water spray to cool exposed containers. If cooling streams are ineffective (venting sound increases in volume and pitch, tank discolors, or shows any signs of deforming), withdraw immediately to a secure position.

NFPA Hazard Classifications:

Health Hazards (Blue) 2
Flammability (Red) 2
Reactivity (Yellow) 2

Flash Point: 118°F (48°C) o.c.; 103°F (39°C) c.c.
Flammable Limits in Air: LEL 1.5%; UEL 9.9%
Autoignition Temperature: 559°F (292°C)

FIRST AID

Move victim to fresh air. Call emergency medical care. Apply artificial respiration if victim is not breathing. Administer oxygen if breathing is difficult. Remove and isolate contaminated clothing and shoes. In case of contact with substance, immediately flush skin or eyes with running water for at least 20 minutes. Wash skin with soap and water. Keep victim warm and quiet. Effects of exposure (inhalation, ingestion, or skin contact) to substance may be delayed. Ensure that medical personnel are aware of the material(s) involved and take precautions to protect themselves.

ENVIRONMENTAL INFORMATION

EPCRA Section 313: Form R *de minimis* concentration reporting level: 1.0%

n-BUTYL ALCOHOL B:360

SYNONYMS: ALCOHOL BUTILICO-n (Spanish); ALCOHOL C-4; ALCOOL BUTYLIQUE (French); BUTANOL; 1-BUTANOL; n-BUTANOL; BUTYL ALCOHOL; BUTAN-1-OL; BUTANOLEN (Dutch); BUTANOLO (Italian); BUTYL HYDROXIDE; BUTYLOWY ALKOHOL (Polish); CEM 420; CCS 203; DAG 154; EEC No. 603-004-00-6; EINECS No. 200-751-6; 6-6 EPOXY CHEM RESIN FINISH, CLEAR CURING AGENT; EPOXY SOLVENT CURE AGENT; FEMA No. 2178; 1-HYDROXYBUTANE; ISANOL; KESTER 5612 PROTECTO; METHYLOLPROPANE; NBA; PROPYLCARBINOL; PROPYL CARBINOL; n-PROPYL CARBINOL; PROPYLMETHANOL; PROTECTO 5612; TEBOL-88; TEBOL-99

IDENTIFICATION

CAS:	71-36-3
ERG ID:	UN 1120
ERG Guide:	129
ERG Guide Category:	FLAMMABLE LIQUIDS (POLAR/WATER-MISCIBLE/NOXIOUS)
Molecular Formula:	$C_4H_{10}O$
RTECS No.:	EO1400000
Properties:	Flammable, watery liquid. Colorless. Mild, alcohol odor. Floats and mixes slowly with water. Vapors are heavier than air; will collect in low areas.

HEALTH AND SAFETY INFORMATION

Prolonged or repeated contact may cause blurred vision and other eye problems, skin drying and cracking, and central nervous system changes. May cause liver and kidney damage; vertigo; hearing problems; and may effect victim's sense of balance. May cause birth defects. Similar solvents and petroleum-based chemicals can cause nerve and brain damage; memory loss, personality changes, fatigue, sleep disturbances, reduced coordination, and a feeling of "pins and needles" in the arms and legs.

IDLH: 1,400 ppm [LEL]
Odor Threshold: 0.03 ppm
OSHA Table Z-1-A Air Contaminant

Exposure Limits:	**CEILING**
ACGIH:* **	50 ppm; 152 mg/m^3
NIOSH:*	50 ppm; 150 mg/m^3
OSHA:	100 ppm; 300 mg/m^3

*Skin contact contributes significantly in overall exposure.
**ACGIH 1998 notice of intended change to CEILING of 25 ppm and 76 mg/m^3.

RECOMMENDATIONS FOR RESPIRATOR SELECTION: NIOSH/OSHA

1,250 ppm: SA:CF * (any supplied-air respirator operated in a continuous-flow mode); or PAPROV * [any powered, air-purifying respirator with organic vapor cartridge(s)]
1,400 ppm: CCRFOV [any chemical cartridge respirator with a full facepiece and organic vapor cartridge(s)]; or GMFOV [any air-purifying, full-facepiece respirator (gas mask) with a chin-style, front- or back-mounted acid gas canister]; or PAPRTOV * (any powered, air-purifying respirator with a tight-fitting facepiece and organic vapor cartridge(s)]; SCBAF (any self-contained breathing apparatus with a full facepiece); or SAF (any supplied-air respirator with a full facepiece)

Emergency or Planned Entry into Unknown Concentrations or IDLH Conditions: SCBAF:PD,PP (any self-contained breathing apparatus that has a full facepiece and is operated in a pressure-demand or other positive-pressure mode); or SAF:PD,PP:ASCBA (any supplied-air respirator that has a full facepiece and is operated in a pressure-demand or other positive-pressure mode in combination with an auxiliary, self-contained breathing apparatus operated in a pressure-demand or other positive-pressure mode)

Escape: GMFOV [any air-purifying, full-facepiece respirator (gas mask) with a chin-style, front- or back-mounted organic vapor canister]; or SCBAE (any appropriate escape-type, self-contained breathing apparatus).

* Substance causes eye irritation or damage; eye protection needed.

CHEMICAL REACTIVITY

Forms explosive mixture with air. May react with aluminum above 120°F (49°C). Attacks some plastics, rubber, and coatings. Incompatible with strong acids, halogens, caustics, alkali metals, aliphatic amines, and isocyanates.

FIRE INFORMATION

Vapors are heavier than air and will collect in low areas. Vapors in confined areas may explode when exposed to fire. Vapors may travel long distances to ignition sources and flash back. Storage containers and parts of containers may rocket great distances, in many directions. If material or contaminated runoff enters waterways, notify downstream users of potentially contaminated waters. Notify local health and fire officials and pollution control agencies. Use dry chemicals, foam, carbon dioxide. From a secure explosion-proof location, use water spray to cool exposed containers. If cooling streams are ineffective (venting sound increases in volume and pitch, tank discolors, or shows any signs of deforming), withdraw immediately to a secure position.

NFPA Hazard Classifications:

Health Hazards (Blue) 1
Flammability (Red) 3
Reactivity (Yellow) 0

Flash Point: 98°F (37°C) c.c

Flammable Limits in Air: LEL 1.4%; UEL 11.2%
Autoignition Temperature: 650°F (343°C)

FIRST AID

Move victim to fresh air. Call emergency medical care. Apply artificial respiration if victim is not breathing. If breathing is difficult, give oxygen. Remove and isolate contaminated clothing and shoes. In case of contact with substance, immediately flush skin or eyes with running water for at least 20 minutes. Wash skin with soap and water. Keep victim warm and quiet. Effects of exposure (inhalation, ingestion, or skin contact) to substance may be delayed. Ensure that medical personnel are aware of the material(s) involved and take precautions to protect themselves.

ENVIRONMENTAL INFORMATION

EPA HAZARDOUS WASTE No. (RCRA No.): U031.
RCRA Section 261 Hazardous Constituents
RCRA Land Ban Waste
RCRA Universal Treatment Standards: Wastewater (mg/L), 5.6; Nonwastewater (mg/kg), 2.6.
EPCRA Section 304: Reportable Quantity (RQ): CERCLA, 5,000 lb (2,270 kg).
EPCRA Section 313: Form R *de minimis* concentration reporting level: 1.0%.

tert-BUTYL ALCOHOL B:364

SYNONYMS: ALCOHOL terc-BUTILICO (Spanish); ALCOOL BUTYLIQUE TERTIAIRE (French); tert-BUTANOL; BUTANOL TERTIAIRE (French); tert-BUTYL HYDROXIDE; 1,1-DIMETHYLETHANOL; EEC No. 603-005-00-1; EINECS No. 200-889-7; METHANOL, TRIMETHYL-; 2-METHYLPROPAN-2-OL; 2-METHYL-2-PROPANOL; NCI-C55367; 2-PROPANOL,2-METHYL-; TBA; tert-BUTYLALKOHOL (German); TRIMETHYLCARBINOL; TRIMETHYL METHANOL

IDENTIFICATION

CAS:	75-65-0
ERG ID:	UN 1120
ERG Guide:	129
ERG Guide Category:	FLAMMABLE LIQUIDS (POLAR/WATER-MISCIBLE/NOXIOUS)
Molecular Formula:	$C_4H_{10}O$
RTECS No.:	EO1925000
Properties:	Highly flammable, oily liquid or combustible solid. Colorless. Camphorlike odor. Soluble in water. Freezes below 75°F (24°C). Vapors are heavier than air; will collect in low areas. Poisonous gases produced in fire and on contact with acids.

HEALTH AND SAFETY INFORMATION

Irritates eyes, skin, and respiratory tract.

IDLH: 1,600 ppm
Odor Threshold: 21.5 ppm
OSHA Table Z-1-A Air Contaminant

Exposure Limits:	TWA	STEL
ACGIH:	100 ppm; 303 mg/m^3	
NIOSH:	100 ppm; 300 mg/m^3	150 ppm; 450 mg/m^3
OSHA:	100 ppm; 300 mg/m^3	

RECOMMENDATIONS FOR RESPIRATOR SELECTION: NIOSH/OSHA

1,600 ppm: SA:CF * (any supplied-air respirator operated in a continuous-flow mode); or PAPROV * [any powered, air-purifying respirator with organic vapor cartridge(s)]; or CCRFOV [any chemical cartridge respirator with a full facepiece and organic vapor cartridge(s)]; or GMFOV [any air-purifying, full-facepiece respirator (gas mask) with a chin-style, front- or back-mounted acid gas canister]; or SCBAF (any self-contained breathing apparatus with a full facepiece); or SAF (any supplied-air respirator with a full facepiece)

Emergency or Planned Entry into Unknown Concentrations or IDLH Conditions: SCBAF:PD,PP (any self-contained breathing apparatus that has a full facepiece and is operated in a pressure-demand or other positive-pressure mode); or SAF:PD,PP:ASCBA (any supplied-air respirator that has a full facepiece and is operated in a pressure-demand or other positive-pressure mode in combination with an auxiliary, self-contained breathing apparatus operated in a pressure-demand or other positive-pressure mode)

Escape: GMFOV [any air-purifying, full-facepiece respirator (gas mask) with a chin-style, front- or back-mounted organic vapor canister]; or SCBAE (any appropriate escape-type, self-contained breathing apparatus).

* Substance causes eye irritation or damage; eye protection needed.

CHEMICAL REACTIVITY

Forms explosive mixture with air. May accumulate static electrical charges, and may cause ignition of its vapors. Incompatible with strong acids, including mineral acids, strong oxidizers or caustics, aliphatic amines, isocyanates, and alkali metals (e.g., lithium, sodium, potassium, rubidium, cesium, and francium). Attacks many plastics and some coatings.

FIRE INFORMATION

Forms explosive mixture with air. Vapors are heavier than air and will collect in low areas. Vapors in confined areas may explode when exposed to fire. If material or contaminated runoff enters waterways, notify downstream users of potentially contaminated waters. Notify local health and fire officials and pollution control agencies. Use dry chemicals, alcohol-resistant foam, carbon dioxide. From a secure explosion-proof location, use water spray to cool exposed containers. If cooling streams are ineffective (venting sound increases in volume and pitch, tank discolors, or shows any signs of deforming), withdraw immediately to a secure position.

NFPA Hazard Classifications:

Health Hazards (Blue) 1
Flammability (Red) 3
Reactivity (Yellow) 0

Flash Point: 52°F (11°C)
Flammable Limits in Air: LEL 2.4%; UEL 8.0%
Autoignition Temperature: 896°F (480°C)

FIRST AID

Move victim to fresh air. Call emergency medical care. Apply artificial respiration if victim is not breathing. Administer oxygen if breathing is difficult. Remove and isolate contaminated clothing and shoes. In case of contact with substance, immediately flush skin or eyes with running water for at least 20 minutes. Wash skin with soap and water. Keep victim warm and quiet. Effects of exposure (inhalation, ingestion, or skin contact) to substance may be delayed. Ensure that medical personnel are aware of the material(s) involved and take precautions to protect themselves.

ENVIRONMENTAL INFORMATION

EPCRA Section 313: Form R *de minimis* concentration reporting level: 1.0%

BUTYL BENZOATE B:373

SYNONYMS: ANTHRAPOLE AZ; BENZOIC ACID-n-BUTYL ESTER; BENZOIC ACID, BUTYL ESTER; n-BUTYL BENZOATE; EINECS No. 205-252-7

IDENTIFICATION

CAS:	136-60-7
ERG ID:	UN 3272
ERG Guide:	127
ERG Guide Category:	FLAMMABLE LIQUIDS (POLAR/WATER-MISCIBLE)
Molecular Formula:	$C_{11}H_{14}O_2$
RTECS No.:	DG4925000
Properties:	Flammable, viscous liquid. Colorless. Characteristic odor. Insoluble in water.

HEALTH AND SAFETY INFORMATION

Irritates eyes (severe) and skin.

RECOMMENDATIONS FOR RESPIRATOR SELECTION: NIOSH

At any concentrations above the NIOSH REL, or where there is no REL, at any detectable concentration: SCBAF:PD,PP (any self-contained breathing apparatus that has a full facepiece and is operated in a pressure-demand or other positive-pressure mode); or SAF:PD,PP:ASCBA (any supplied-air respirator that has a full facepiece and is operated in a pressure-demand or other positive-pressure mode in combination with an auxiliary, self-contained breathing apparatus operated in a pressure-demand or other positive-pressure mode).

Escape: GMFOV [any air-purifying, full-facepiece respirator (gas mask) with a chin-style, front- or back-mounted organic vapor canister]; or SCBAE (any appropriate escape-type, self-contained breathing apparatus).

CHEMICAL REACTIVITY

Contact with oxidizers can cause fire and explosions.

FIRE INFORMATION

Vapors are heavier than air and will collect in low areas. Storage containers and parts of containers may rocket great distances, in many directions. If material or contaminated runoff enters waterways, notify downstream users of potentially contaminated waters. Notify local health and fire officials and pollution control agencies. Use dry chemicals, foam, carbon dioxide. From a secure explosion-proof location, use water spray to cool exposed containers. If cooling streams are ineffective (venting sound increases in volume and pitch, tank discolors, or shows any signs of deforming), withdraw immediately to a secure position.

NFPA Hazard Classifications:

Health Hazards (Blue) 1
Flammability (Red) 1
Reactivity (Yellow) 0

Flash Point: 225°F (107°C) o.c.

FIRST AID

Move victim to fresh air. Call emergency medical care. Apply artificial respiration if victim is not breathing. Administer oxygen if breathing is difficult. Remove and isolate contaminated clothing and shoes. In case of contact with substance, immediately flush skin or eyes with running water for at least 20 minutes. Wash skin with soap and water. Keep victim warm and quiet. Ensure that medical personnel are aware of the material(s) involved and take precautions to protect themselves.

BUTYL BENZYL PHTHALATE B:375

SYNONYMS: BBP; BENZYLBUTYLFTALAAT (Dutch); BENZYLBUTYLPHTHALAT (German); BENZYL n-BUTYL PHTHALATE; n-BUTYL BENZYL PHTHALATE; EINECS No. 201-622-7; FTALATO de BUTILBENCILO (Spanish); NCI-C54375; PALATINOL BB; PHTHALIC ACID, BENZYL BUTYL ETHER; PHTALATE de BENZYLE et de BUTYLE (French); SATICIZER 160; SICOL 160; UNIMOLL BB

IDENTIFICATION

CAS:	85-68-7
ERG ID:	UN 3082
ERG Guide:	171
ERG Guide Category:	SUBSTANCES (LOW TO MODERATE HAZARD)
Molecular Formula:	$C_{19}H_{20}O_4$
RTECS No.:	TH9990000
Properties:	Combustible liquid. Colorless. Slight odor. Sinks in water.

HEALTH AND SAFETY INFORMATION

Causes mild skin irritation. Questionable carcinogen.

IARC Evaluation: Unclassified substance: Group 3; no human data; inadequate animal evidence

RECOMMENDATIONS FOR RESPIRATOR SELECTION

At any concentrations above the NIOSH REL, or where there is no REL, at any detectable concentration: SCBAF:PD,PP (any self-contained breathing apparatus that has a full facepiece and is operated in a pressure-demand or other positive-pressure mode); or SAF:PD,PP:ASCBA (any supplied-air respirator that has a full facepiece and is operated in a pressure-demand or other positive-pressure mode in combination with an auxiliary, self-contained breathing apparatus operated in a pressure-demand or other positive-pressure mode)

Escape: GMFOV [any air-purifying, full-facepiece respirator (gas mask) with a chin-style, front- or back-mounted organic vapor canister]; or SCBAE (any appropriate escape-type, self-contained breathing apparatus)

CHEMICAL REACTIVITY

Incompatible with strong acids, nitrates, and oxidizers. Destructive to rubber and paint. Store in a refrigerator or other cool, dry place.

FIRE INFORMATION

Vapors are heavier than air and will collect in low areas. Storage containers and parts of containers may rocket great distances, in many directions. If material or contaminated runoff enters waterways, notify downstream users of potentially contaminated waters. Notify local health and fire officials and pollution control agencies. Use dry chemicals, foam, carbon dioxide. From a secure explosion-proof location, use water spray to cool exposed containers. If cooling streams are ineffective (venting sound increases in volume and pitch, tank discolors, or shows any signs of deforming), withdraw immediately to a secure position.

NFPA Hazard Classifications:

Health Hazards (Blue) 1
Flammability (Red) 1
Reactivity (Yellow) 0

Flash Point: 390°F (199°C) o.c.

FIRST AID

Move victim to fresh air. Call emergency medical care. Apply artificial respiration if victim is not breathing. Administer oxygen if breathing is difficult. Remove and isolate contaminated clothing and shoes. In case of contact with substance, immediately flush skin or eyes with running water for at least 20 minutes. Ensure that medical personnel are aware of the material(s) involved and take precautions to protect themselves.

ENVIRONMENTAL INFORMATION

Clean Water Act: Section 307 Priority Pollutants; Section 313 Priority Chemicals.
RCRA Section 261 Hazardous Constituents: Waste number not listed.
RCRA Land Ban Waste
RCRA Universal Treatment Standards: Wastewater (mg/L), 0.017; Nonwastewater (mg/kg), 28.
RCRA Ground Water Monitoring List: Suggested test method(s) (PQL ug/L): 8060(5); 8270(10).
EPCRA Section 304: Reportable Quantity (RQ): CERCLA, 100 lb (45.4 kg).
Marine Pollutant (49CFR, Subchapter 172.101, Appendix B).

p–tert–BUTYLCATECHOL B:420

SYNONYMS: 1,2-BENZENEDIOL, 4-(1,1-DIMETHYLETHYL)-; 4-terc-BUTILPIROCATECOL (Spanish); 4-tert-BUTYLCATECHOL; p-tert-BUTYLPYROCATECHOL; 4-tert-BUTYLPYROKATECHIN (Czech); 4-(1,1-DIMETHYLETHYL)-1,2-BENZENEDIOL; EINECS No. 202-653-9; SYNOX TBC

IDENTIFICATION

CAS:	98-29-3
ERG ID:	UN 2811 toxic solid, organic, n.o.s.
ERG Guide:	154

ERG Guide Category: SUBSTANCES-TOXIC AND/OR CORROSIVE (NON-COMBUSTIBLE)

Molecular Formula: $C_{10}H_{14}O_2$

RTECS No.: UX1400000

Properties: Combustible crystalline solid. White. Slightly soluble in water.

HEALTH AND SAFETY INFORMATION

Irritates eyes, skin, and respiratory tract. Repeated or prolonged skin contact can cause blisters and other skin disorders.

Exposure Limits: TWA

AIHA (WEEL): [1] 5 mg/m^3

[1] Skin contact contributes significantly to overall exposure.

RECOMMENDATIONS FOR RESPIRATOR SELECTION: NIOSH

At any concentrations above the NIOSH REL, or where there is no REL, at any detectable concentration: SCBAF:PD,PP (any self-contained breathing apparatus that has a full facepiece and is operated in a pressure-demand or other positive-pressure mode); or SAF:PD,PP:ASCBA (any supplied-air respirator that has a full facepiece and is operated in a pressure-demand or other positive-pressure mode in combination with an auxiliary, self-contained breathing apparatus operated in a pressure-demand or other positive-pressure mode).

Escape: GMFOV [any air-purifying, full-facepiece respirator (gas mask) with a chin-style, front- or back-mounted organic vapor canister]; or SCBAE (any appropriate escape-type, self-contained breathing apparatus).

CHEMICAL REACTIVITY

Reacts violently with strong oxidizers (contact may cause fire and explosions).

FIRE INFORMATION

Decomposition products upon heating include toxic gases. Vapors are heavier than air and will collect in low areas. Vapors in confined areas may explode when exposed to fire. Vapors may travel long distances to ignition sources and flash back. Storage containers and parts of containers may rocket great distances, in many directions. If material or contaminated runoff enters waterways, notify downstream users of potentially contaminated waters. Notify local health and fire officials and pollution control agencies. Use dry chemicals, foam, carbon dioxide. From a secure explosion-proof location, use water spray to cool exposed containers. If cooling streams are ineffective (venting sound increases in volume and pitch, tank discolors, or shows any signs of deforming), withdraw immediately to a secure position.

NFPA Hazard Classifications:

Health Hazards (Blue) 2
Flammability (Red) 1
Reactivity (Yellow) 0

Flash Point: 266°F (130°C)

FIRST AID

Move victim to fresh air. Call emergency medical care. Apply artificial respiration if victim is not breathing. Do not use mouth-to-mouth method if victim ingested or inhaled the substance; induce artificial respiration with the aid of a pocket mask equipped with a one-way valve or other proper respiratory medical device. Administer oxygen if breathing is difficult. Remove and isolate contaminated clothing and shoes. In case of contact with substance, immediately flush skin or eyes with running water for at least 20 minutes. For minor skin contact, avoid spreading material on unaffected skin. Keep victim warm and quiet. Effects of exposure (inhalation, ingestion, or skin contact) may be delayed. Ensure that medical personnel are aware of the material(s) involved and take precautions to protect themselves.

6–tert–BUTYL–m–CRESOL B:440

SYNONYMS: 6-terc-BUTIL-m-CRESOL (Spanish); EINECS No. 201-842-3; PHENOL, 2-(1,1-DIMETHYLETHYL)-5-METHY; PHENOL, (1,1-DIMETHYLETHYL)-3-METHYL-

IDENTIFICATION

CAS:	88-60-8; 1333-13-7
ERG ID:	UN 2076
ERG Guide:	153
ERG Guide Category:	SUBSTANCES-TOXIC AND/OR CORROSIVE (COMBUSTIBLE)
Molecular Formula:	$C_{11}H_{16}O$
RTECS No.:	SK1578180
Properties:	Flammable liquid.

HEALTH AND SAFETY INFORMATION

Poisonous. May cause eye and skin irritation.

Odor Threshold: 250 ppm as m-cresol
OSHA Table Z-1-A Air Contaminant as cresol (all isomers)

Exposure Limits:	**TWA**
ACGIH:	5 ppm

NIOSH: 2.3 ppm; 10 mg/m^3
OSHA: 5 ppm; 22 mg/m^3

RECOMMENDATIONS FOR RESPIRATOR SELECTION: NIOSH as m-cresol

23 ppm: CCROVDM (any chemical cartridge respirator with organic vapor cartridge(s) in combination with a dust and mist filter); SA (any supplied-air respirator).
57.5 ppm: SA:CF (any supplied-air respirator operated in a continuous-flow mode; or PAPROVDM (any powered, air-purifying respirator with organic vapor cartridge(s) in combination with a dust and mist filter).
115 ppm: CCRFOVHiE (any chemical cartridge respirator with a full facepiece and organic vapor cartridge(s) in combination with a high-efficiency particulate filter; or GMFOVHiE [any air-purifying, full-facepiece respirator (gas mask) with a chin-style, front- or back-mounted organic vapor canister having a high-efficiency particulate filter]; or PAPRTHiE* (any powered, air-purifying respirator with a tight-fitting facepiece and a high-efficiency particulate filter; or SAT:CF* (Any supplied-air respirator that has a tight-fitting facepiece and is operated in a continuous-flow mode; or SCBAF (any self-contained breathing apparatus with a full facepiece; or SAF (any supplied-air respirator with a full facepiece).
250 ppm: SAF:PD,PP (Any supplied-air respirator that has a full facepiece and is operated in a pressure-demand or other positive-pressure mode).

EMERGENCY OR PLANNED ENTRY INTO UNKNOWN CONCENTRATIONS OR IDLH CONDITIONS: SCBAF:PD,PP (any self-contained breathing apparatus that has a full facepiece and is operated in a pressure-demand or other positive-pressure mode; or SAF:PD,PP:ASCBA (Any supplied-air respirator that has a full facepiece and is operated in a pressure-demand or other positive-pressure mode in combination with an auxiliary, self-contained breathing apparatus operated in a pressure-demand or other positive-pressure mode).

ESCAPE: GMFOVHiE [any air-purifying, full-facepiece respirator (gas mask) with a chin-style, front- or back-mounted organic vapor canister having a high-efficiency particulate filter];or SCBAE (Any appropriate escape-type, self-contained breathing apparatus).

* Substance reported to cause eye irritation or damage; may require eye protection.

CHEMICAL REACTIVITY

Forms explosive mixture with air. Contact with oxidizers can cause fire and explosions.

FIRE INFORMATION

Vapors are heavier than air and will collect in low areas. Vapors in confined areas may explode when exposed to fire. Vapors may travel long distances to ignition sources and flash back. Storage containers and parts of containers may rocket great distances, in many directions. If material or contaminated runoff enters waterways, notify downstream users of potentially contaminated waters. Notify local health and fire officials and pollution control agencies.

NFPA Hazard Classifications:

Health Hazards (Blue) 2
Flammability (Red) 2
Reactivity (Yellow) 0

Flash Point: 116°F (47°C)

FIRST AID

Move victim to fresh air and call emergency medical care. Apply artificial respiration if victim is not breathing. Do not use mouth-to-mouth method if victim swallowed or inhaled the substance; induce artificial respiration with the aid of a pocket mask equipped with a one-way valve or other proper respiratory medical device. If breathing is difficult, give oxygen. In case of contact with substance, immediately flush skin or eyes with running water for at least 20 minutes. Remove and isolate contaminated clothing and shoes. Keep victim warm and quiet. Effects of exposure (inhalation, ingestion, or skin contact) to substances may be delayed. Ensure that medical personnel are aware of the material(s) involved and take precautions to protect themselves.

ENVIRONMENTAL INFORMATION as cresols

CLEAN AIR ACT: Hazardous Air Pollutants (Title I, Part A, Section 112)
CLEAN WATER ACT: Section 311 Hazardous Substances/RQ 100 lb (4.54 kg); Section 313 Priority Chemicals.
EPA HAZARDOUS WASTE NUMBER (RCRA No.): U052; D026
RCRA Toxicity Characteristic (Section 261.24), Maximum Concentration of Contaminants, regulatory level, 200 mg/L. *Note*: if o-, m- and p-Cresol concentrations cannot be differentiated, the total cresol (D026) concentration is used. The regulatory level of total cresol is 200 mg/L.
RCRA Section 261 Hazardous Constituents
RCRA Land Ban Waste
EPCRA Section 304 Reportable Quantity (RQ): CERCLA, 1,000 lb (454 kg).
EPCRA Section 313: Form R *de minimus* concentration reporting level: 1.0%.
MARINE POLLUTANT (49CFR, Subchapter 172.101, Appendix B) as cresols (o-; m; p-)

1,2–BUTYLENE OXIDE B:460

SYNONYMS: α-BUTYLENE OXIDE; 1-BUTENE OXIDE; 1-BUTYLENE OXIDE; 1,2-BUTYLENE OXIDE, stabilized; EINECS No. 203-438-2; 1,2-EPOXYBUTANE; 2-ETHYLOXIRANE; NCI-C5527; OXIDO de 1,2-BUTILENO (Spanish); OXIRANE, ETHYL-; PROPYL OXIRANE

IDENTIFICATION

CAS: 106-88-7
ERG ID: UN 3022

ERG Guide: 127P
ERG Guide Category: FLAMMABLE LIQUIDS (POLAR/WATER-MISCIBLE)
Molecular Formula: C_4H_8O
Properties: Flammable liquid. Colorless. Sharp odor. Mixes with water.

HEALTH AND SAFETY INFORMATION

IARC Evaluation: IARC unclassified, rating 3; limited animal evidence.
Odor Threshold: 0.06 ppm

CHEMICAL REACTIVITY

Reacts with oxidizers, hydroxides, acids, and metal chlorides

FIRE INFORMATION

Vapors are heavier than air and will collect in low areas. Vapors in confined areas may explode when exposed to fire. Vapors may travel long distances to ignition sources and flash back. Storage containers and parts of containers may rocket great distances, in many directions. If material or contaminated runoff enters waterways, notify downstream users of potentially contaminated waters. Notify local health and fire officials and pollution control agencies. Use dry chemicals, foam, carbon dioxide. From a secure explosion-proof location, use water spray to cool exposed containers. If cooling streams are ineffective (venting sound increases in volume and pitch, tank discolors, or shows any signs of deforming), withdraw immediately to a secure position.

NFPA Hazard Classifications:

Health Hazards (Blue) 2
Flammability (Red) 3
Reactivity (Yellow) 2

Flash Point: less than -20°F (-29°C) o.c.; -7°F (-22°C) c.c.
Flammable Limits in Air: LEL 1.5%; UEL 18.3%
Autoignition Temperature: 822°F (439°C)

FIRST AID

Move victim to fresh air. Call emergency medical care. Apply artificial respiration if victim is not breathing. Administer oxygen if breathing is difficult. Remove and isolate contaminated clothing and shoes. In case of contact with substance, immediately flush skin or eyes with running water for at least 20 minutes. Wash skin with soap and water. Keep victim warm and quiet. Ensure that medical personnel are aware of the material(s) involved and take precautions to protect themselves.

ENVIRONMENTAL INFORMATION

Clean Air Act: Hazardous Air Pollutants (Title I, Part A, Section 112)

EPCRA Section 304 Reportable Quantity (RQ): CERCLA, 100 lb (45.4 kg)
EPCRA Section 313: Form R *de minimis* concentration reporting level: 1.0%.

n–BUTYL FORMATE B:480

SYNONYMS: BUTYL FORMATE; BUTYL METHANOATE; EEC No. 607-017-00-8; EINECS No. 209-772-5; FORMIATO de BUTILO (Spanish); FORMIC ACID, BUTYL ESTER

IDENTIFICATION

CAS:	592-84-7
ERG ID:	UN 1128
ERG Guide:	129
ERG Guide Category:	FLAMMABLE LIQUIDS (POLAR/WATER-MISCIBLE/NOXIOUS)
Molecular Formula:	$C_5H_{10}O_2$
RTECS No.:	LQ5500000
Properties:	Flammable liquid. Colorless. Floats on water.

HEALTH AND SAFETY INFORMATION

Irritates eyes, skin, and respiratory tract. Inhalation may affect central nervous system; narcotic effect.

RECOMMENDATIONS FOR RESPIRATOR SELECTION

At any concentrations above the NIOSH REL, or where there is no REL, at any detectable concentration: SCBAF:PD,PP (any self-contained breathing apparatus that has a full facepiece and is operated in a pressure-demand or other positive-pressure mode); or SAF:PD,PP:ASCBA (any supplied-air respirator that has a full facepiece and is operated in a pressure-demand or other positive-pressure mode in combination with an auxiliary, self-contained breathing apparatus operated in a pressure-demand or other positive-pressure mode).

Escape: GMFOV [any air-purifying, full-facepiece respirator (gas mask) with a chin-style, front- or back-mounted organic vapor canister]; or SCBAE (any appropriate escape-type, self-contained breathing apparatus).

CHEMICAL REACTIVITY

Forms explosive mixture with air. Contact with oxidizers can cause fire and explosions. Reacts with strong acids.

FIRE INFORMATION

Vapors are heavier than air and will collect in low areas. Vapors in confined areas may explode when

exposed to fire. Vapors may travel long distances to ignition sources and flash back. Storage containers and parts of containers may rocket great distances, in many directions. If material or contaminated runoff enters waterways, notify downstream users of potentially contaminated waters. Notify local health and fire officials and pollution control agencies. Use dry chemicals, foam, carbon dioxide. From a secure explosion-proof location, use water spray to cool exposed containers. If cooling streams are ineffective (venting sound increases in volume and pitch, tank discolors, or shows any signs of deforming), withdraw immediately to a secure position.

NFPA Hazard Classifications:

Health Hazards (Blue) 2
Flammability (Red) 3
Reactivity (Yellow) 0

Flash Point: 64°F (18°C) c.c.
Flammable Limits in Air: LEL 1.7%; UEL 8.2%
Autoignition Temperature: 612°F (322°C)

FIRST AID

Move victim to fresh air. Call emergency medical care. Apply artificial respiration if victim is not breathing. Administer oxygen if breathing is difficult. Remove and isolate contaminated clothing and shoes. In case of contact with substance, immediately flush skin or eyes with running water for at least 20 minutes. Wash skin with soap and water. Keep victim warm and quiet. Effects of exposure (inhalation, ingestion, or skin contact) to substance may be delayed. Ensure that medical personnel are aware of the material(s) involved and take precautions to protect themselves.

BUTYL LACTATE B:500

SYNONYMS: BUTYL a-HYDROXYPROPIONATE; n-BUTYL LACTATE; EINECS No. 205-316-4; LACTATE de BUTYLE (French); LACTATO di BUTILE (Italian); LACTIC ACID, BUTYL ESTER; LACTATO de n-BUTILO (Spanish); PROPANOIC ACID, 2-HYDROXY-, BUTYL ESTER

IDENTIFICATION

CAS:	138-22-7
ERG ID:	UN 1993
ERG Guide:	128
ERG Guide Category:	FLAMMABLE LIQUIDS (NONPOLAR/WATER-IMMISCIBLE)
Molecular Formula:	$C_7H_{14}O_3$
RTECS No.:	OD4025000
Properties:	Combustible liquid. Clear. Mild, characteristic odor.

HEALTH AND SAFETY INFORMATION

Irritates eyes and skin. Toxic if inhaled.

Odor Threshold: 7.06 ppm
OSHA Table Z-1-A Air Contaminant

Exposure Limits:	**TWA**
ACGIH:	5 ppm; 30 mg/m^3
NIOSH:	5 ppm; 25 mg/m^3
OSHA:	None

RECOMMENDATIONS FOR RESPIRATOR SELECTION

At any concentrations above the NIOSH REL, or where there is no REL, at any detectable concentration: SCBAF:PD,PP (any self-contained breathing apparatus that has a full facepiece and is operated in a pressure-demand or other positive-pressure mode); or SAF:PD,PP:ASCBA (any supplied-air respirator that has a full facepiece and is operated in a pressure-demand or other positive-pressure mode in combination with an auxiliary, self-contained breathing apparatus operated in a pressure-demand or other positive-pressure mode).

Escape: GMFOV [any air-purifying, full-facepiece respirator (gas mask) with a chin-style, front- or back-mounted organic vapor canister]; or SCBAE (any appropriate escape-type, self-contained breathing apparatus)

CHEMICAL REACTIVITY

Avoid contact with strong oxidizing agents and strong bases.

FIRE INFORMATION

Vapors are heavier than air and will collect in low areas. Vapors in confined areas may explode when exposed to fire. Vapors may travel long distances to ignition sources and flash back. Storage containers and parts of containers may rocket great distances, in many directions.If material or contaminated runoff enters waterways, notify downstream users of potentially contaminated waters. Notify local health and fire officials and pollution control agenciès. Use dry chemicals, foam, carbon dioxide. From a secure explosion-proof location, use water spray to cool exposed containers. If cooling streams are ineffective (venting sound increases in volume and pitch, tank discolors, or shows any signs of deforming), withdraw immediately to a secure position.

NFPA Hazard Classifications:

Health Hazards (Blue) 1
Flammability (Red) 2
Reactivity (Yellow) 0

Flash Point: 160°F (71°C)o.c.
Autoignition Temperature: 720°F (382°C)

FIRST AID

Move victim to fresh air. Call emergency medical care. Apply artificial respiration if victim is not breathing. If breathing is difficult, give oxygen. Remove and isolate contaminated clothing and shoes. In case of contact with substance, immediately flush skin or eyes with running water for at least 20 minutes.

Wash skin with soap and water. Keep victim warm and quiet. Ensure that medical personnel are aware of the material(s) involved and take precautions to protect themselves.

BUTYL METHACRYLATE B:525

SYNONYMS: n-BUTYL METHACRYLATE; BUTYL 2-METHACRYLATE; N-BUTYL α-METHYLACRYLATE; BUTYL 2-METHYL-2-PROPENOATE; N-BUTYL METHACRYLATE; 2-PROPENIC ACID, 2-METHYL-, BUTYL ESTER; 2-METHYL-BUTYLACRYLATE; BUTYL-2-METHYL-2-PROPENOATE; BUTILMETACRILATO (Italian); BUTYL METHACRYLAAT (Dutch); EEC No. 607-033-00-5; EINECS No. 202-615-1; METACRILATO de N-BUTILO (Spanish); METHACRYLATE de BUTYLE (French); METHACRYLIC ACID, BUTYL ESTER; METHACRYLSAEURE BUTYL ESTER (German)

IDENTIFICATION

CAS:	97-88-1
ERG ID:	UN 2227
ERG Guide:	129P
ERG Guide Category:	FLAMMABLE LIQUIDS (POLAR/WATER-MISCIBLE/NOXIOUS)
Molecular Formula:	$C_8H_{14}O$
RTECS No.:	OZ3675000
Properties:	Combustible liquid. colorless. Mild odor. Floats on water.

HEALTH AND SAFETY INFORMATION

Irritates eyes, skin. Toxic by skin contact.

RECOMMENDATIONS FOR RESPIRATOR SELECTION:

At any concentrations above the NIOSH REL, or where there is no REL, at any detectable concentration: SCBAF:PD,PP (any self-contained breathing apparatus that has a full facepiece and is operated in a pressure-demand or other positive-pressure mode); or SAF:PD,PP:ASCBA (any supplied-air respirator that has a full facepiece and is operated in a pressure-demand or other positive-pressure mode in combination with an auxiliary, self-contained breathing apparatus operated in a pressure-demand or other positive-pressure mode).

Escape: GMFOV [any air-purifying, full-facepiece respirator (gas mask) with a chin-style, front- or back-mounted organic vapor canister]; or SCBAE (any appropriate escape-type, self-contained breathing apparatus)

CHEMICAL REACTIVITY

Contact with oxidizers may cause polymerization, fire or explosions..

FIRE INFORMATION

Vapors are heavier than air and will collect in low areas. Vapors in confined areas may explode when exposed to fire. Vapors may travel long distances to ignition sources and flash back. Storage containers and parts of containers may rocket great distances, in many directions. If material or contaminated runoff enters waterways, notify downstream users of potentially contaminated waters. Notify local health and fire officials and pollution control agencies. Use dry chemicals, foam, carbon dioxide. From a secure explosion-proof location, use water spray to cool exposed containers. If cooling streams are ineffective (venting sound increases in volume and pitch, tank discolors, or shows any signs of deforming), withdraw immediately to a secure position.

NFPA Hazard Classifications:

Health Hazards (Blue) 2
Flammability (Red) 2
Reactivity (Yellow) 0

Flash Point: 126°F (52°C) o.c.
Flammable Limits in Air: LEL 2%: UEL 8%
Autoignition Temperature: 562°F (294°C)

FIRST AID

Move victim to fresh air. Call emergency medical care. Apply artificial respiration if victim is not breathing. Administer oxygen if breathing is difficult. Remove and isolate contaminated clothing and shoes. In case of contact with substance, immediately flush skin or eyes with running water for at least 20 minutes. Wash skin with soap and water. Keep victim warm and quiet. Effects of exposure (inhalation, ingestion, or skin contact) to substance may be delayed. Ensure that medical personnel are aware of the material(s) involved and take precautions to protect themselves.

iso–BUTYL METHACRYLATE B:550

SYNONYMS: EEC No. 607-113-00-X; EINECS No. 202-613-0; ISOBUTYL α-METHACRYLATE; ISOBUTYL 2-METHYL-2-PROPENOATE; METHACRYLIC ACID, ISOBUTY ESTER; 2-METHYLPROPYL METHACRYLATE

IDENTIFICATION

CAS: 97-86-9
ERG ID: UN 2283
ERG Guide: 130P
ERG Guide Category: FLAMMABLE LIQUIDS (NONPOLAR/WATER-IMMISCIBLE/NOXIOUS)
Molecular Formula: $C_8H_{14}O_2$
Properties: Combustible liquid. Colorless. Acrylic odor. Floats on water.

CHEMICAL REACTIVITY

Reacts with strong acids, amines.

FIRE INFORMATION

Vapors are heavier than air and will collect in low areas. Vapors in confined areas may explode when exposed to fire. Vapors may travel long distances to ignition sources and flash back. Storage containers and parts of containers may rocket great distances, in many directions.If material or contaminated runoff enters waterways, notify downstream users of potentially contaminated waters. Notify local health and fire officials and pollution control agencies. Use dry chemicals, foam, carbon dioxide. From a secure explosion-proof location, use water spray to cool exposed containers. If cooling streams are ineffective (venting sound increases in volume and pitch, tank discolors, or shows any signs of deforming), withdraw immediately to a secure position.

NFPA Hazard Classifications:

Health Hazards (Blue) 2
Flammability (Red) 2
Reactivity (Yellow) 0

Flash Point: 120°F (88°C) o.c.; 112°F (44°C) c.c.

FIRST AID

Move victim to fresh air. Call emergency medical care. Apply artificial respiration if victim is not breathing. Administer oxygen if breathing is difficult. Remove and isolate contaminated clothing and shoes. In case of contact with substance, immediately flush skin or eyes with running water for at least 20 minutes. Wash skin with soap and water. Keep victim warm and quiet. Effects of exposure (inhalation, ingestion, or skin contact) to substance may be delayed. Ensure that medical personnel are aware of the material(s) involved and take precautions to protect themselves.

BUTYL OLEATE B:575

SYNONYMS: EINECS No. 205-559-6; 9-OCTADECENOIC ACID BUTYL ESTER, (Z)-;

188 Butyl Oleate

(Z)-9-OCTADECENOIC ACID BUTYL ESTER; OLEATE de BUTYLE (French); OLEATO de BUTILO (Spanish); OLEIC ACID, BUTYL ESTER; PLASTHALL 503; UNIFLEX BYO

IDENTIFICATION

CAS:	142-77-8
ERG ID:	UN 3272
ERG Guide:	127
ERG Guide Category:	FLAMMABLE LIQUIDS (POLAR/WATER-MISCIBLE)
Molecular Formula:	$C_{22}H_{42}O_2$
RTECS No.:	RG3711000
Properties:	Combustible liquid. Insoluble in water.

HEALTH AND SAFETY INFORMATION

May cause skin irritation.

RECOMMENDATIONS FOR RESPIRATOR SELECTION:

At any concentrations above the NIOSH REL, or where there is no REL, at any detectable concentration: SCBAF:PD,PP (any self-contained breathing apparatus that has a full facepiece and is operated in a pressure-demand or other positive-pressure mode); or SAF:PD,PP:ASCBA (any supplied-air respirator that has a full facepiece and is operated in a pressure-demand or other positive-pressure mode in combination with an auxiliary, self-contained breathing apparatus operated in a pressure-demand or other positive-pressure mode).

Escape: GMFOV [any air-purifying, full-facepiece respirator (gas mask) with a chin-style, front- or back-mounted organic vapor canister]; or SCBAE (any appropriate escape-type, self-contained breathing apparatus).

CHEMICAL REACTIVITY

Contact with oxidizers may cause fire and explosions.

FIRE INFORMATION

If material or contaminated runoff enters waterways, notify downstream users of potentially contaminated waters.Notify local health and fire officials and pollution control agencies. Use dry chemicals, alcohol-resistant foam, carbon dioxide. From a secure explosion-proof location, use water spray to cool exposed containers. If cooling streams are ineffective (venting sound increases in volume and pitch, tank discolors, or shows any signs of deforming), withdraw immediately to a secure position.

NFPA Hazard Classifications:

Health Hazards (Blue) 0

Flammability (Red) 1
Reactivity (Yellow) 0

Flash Point: 356°F (180°C) o.c.

FIRST AID

Move victim to fresh air. Call emergency medical care. Apply artificial respiration if victim is not breathing. Administer oxygen if breathing is difficult. Remove and isolate contaminated clothing and shoes. In case of contact with substance, immediately flush skin or eyes with running water for at least 20 minutes. Wash skin with soap and water. Keep victim warm and quiet. Ensure that medical personnel are aware of the material(s) involved and take precautions to protect themselves.

tert–BUTYL PERBENZOATE B:600

SYNONYMS: BENZENECARBOPEROXOIC ACID, 1,1-DIMETHYLETHYL ESTER; terc.BUTYLPERBENZOAN (Czech); tert-BUTYL PEROXY BENZOATE; EINECS No. 210-382-2; ESPEROX 10; NOVOX; PERBENZOATE de BUTYLE TERTIAIRE (French); PERBENZOATO de TERCBUTILO (Spanish); PEROXYBENZOATE de tert-BUTYLE (French); TRIGONOX C

IDENTIFICATION

CAS:	614-45-9
ERG ID:	UN 2097 (technical pure or concentrated); UN 2098 (concentrated > 75% in solution); UN 2890 (<50% with inorganic solid)
ERG Guide:	146 (UN 2097); 145 (UN 2098; UN 2890)
ERG Guide Category:	145-ORGANIC PEROXIDES (HEAT AND CONTAMINATION SENSITIVE). 146-ORGANIC PEROXIDES (HEAT, CONTAMINATION AND FRICTION SENSITIVE)
Molecular Formula:	$C_{11}H_{14}O_3$
RTECS No.:	SD9460000
Properties:	Very flammable, highly explosive liquid. Slightly yellow. Mild, aromatic odor. Insoluble in water.

HEALTH AND SAFETY INFORMATION

Irritating to the eyes, skin, and lungs.

OSHA Process Safety Management of Highly Hazardous Chemicals: (29 CFR Part 1910.119, Appendix A) TQ = 7,500 lb

RECOMMENDATIONS FOR RESPIRATOR SELECTION: NIOSH

At any concentrations above the NIOSH REL, or where there is no REL, at any detectable concentration: SCBAF:PD,PP (any self-contained breathing apparatus that has a full facepiece and is operated in a pressure-demand or other positive-pressure mode); or SAF:PD,PP:ASCBA (any supplied-air respirator that has a full facepiece and is operated in a pressure-demand or other positive-pressure mode in combination with an auxiliary, self-contained breathing apparatus operated in a pressure-demand or other positive-pressure mode).

Escape: GMFOV [any air-purifying, full-facepiece respirator (gas mask) with a chin-style, front- or back-mounted organic vapor canister]; or SCBAE (any appropriate escape-type, self-contained breathing apparatus).

CHEMICAL REACTIVITY

Highly reactive, flammable, oxidizing compound. Reacts explosively with flammable, oxidizable, combustibles, organic materials and other materials. Heat sensitive; explodes above 140°F (60°C); may spontaneously explode if exposed to heat or fire.

FIRE INFORMATION

May spontaneously explode if exposed to heat or fire. Will increase the intensity of a fire. Vapors are heavier than air and will collect in low areas. Vapors in confined areas may explode when exposed to fire. Vapors may travel long distances to ignition sources and flash back. Storage containers and parts of containers may rocket great distances, in many directions. If material or contaminated runoff enters waterways, notify downstream users of potentially contaminated waters. Notify local health and fire officials and pollution control agencies. Use dry chemicals, foam, carbon dioxide. Use water spray to cool exposed containers. If cooling streams are ineffective (venting sound increases in volume and pitch, tank discolors, or shows any signs of deforming), withdraw immediately to a secure position.

NFPA Hazard Classifications:

Health Hazards (Blue) 1
Flammability (Red) 3
Reactivity (Yellow) 4
Special situations .... OXY

Flash Point: 190°F (88°C) o.c.

FIRST AID

Move victim to fresh air. Call emergency medical care. Apply artificial respiration if victim is not breathing. Administer oxygen if breathing is difficult. Remove and isolate contaminated clothing and shoes. Remove material from skin immediately. In case of contact with substance, immediately flush skin or eyes with running water for at least 20 minutes. Keep victim warm and quiet. Ensure that medical personnel are aware of the material(s) involved and take precautions to protect themselves.

tert–BUTYL PEROXYACETATE B:625

SYNONYMS: tert-BUTYL PERACETATE; EINECS No. 203-514-5; ETHANEPEROXOIC ACID, 1,1-DIMETHYLETHYL ESTER; LUPERSOL 70; PEROXIACETATO de terc-BUTILO (Spanish); PEROXYACETIC ACID, tert-BUTYL ESTER; TRIGONOX F-C50

IDENTIFICATION

CAS:	107-71-1
ERG ID:	UN 2095 (52-76% solutions); UN 2096 (<52% solutions); FORBIDDEN above 76%
ERG Guide:	146
ERG Guide Category:	ORGANIC PEROXIDES (HEAT, CONTAMINATION AND FRICTION SENSITIVE)
Molecular Formula:	$C_6H_{12}O_3$
RTECS No.:	SD8925000
Properties:	Highly flammable heat-, and shock-sensitive liquid. Colorless. Insoluble in water.

HEALTH AND SAFETY INFORMATION

Contact with eyes and skin will cause irritation.

RECOMMENDATIONS FOR RESPIRATOR SELECTION:

At any concentrations above the NIOSH REL, or where there is no REL, at any detectable concentration: SCBAF:PD,PP (any self-contained breathing apparatus that has a full facepiece and is operated in a pressure-demand or other positive-pressure mode); or SAF:PD,PP:ASCBA (any supplied-air respirator that has a full facepiece and is operated in a pressure-demand or other positive-pressure mode in combination with an auxiliary, self-contained breathing apparatus operated in a pressure-demand or other positive-pressure mode).

Escape: GMFOV [any air-purifying, full-facepiece respirator (gas mask) with a chin-style, front- or back-mounted organic vapor canister]; or SCBAE (any appropriate escape-type, self-contained breathing apparatus).

CHEMICAL REACTIVITY

Forms explosive mixture with air. Highly reactive oxidizer. Reacts violently when exposed to reducing agents, combustible and/or organic materials, heat, or flames [explodes on heating at or above 200°F (93°C)]. Sensitive to shock or friction.

FIRE INFORMATION

May spontaneously explode if exposed to heat or fire. Will increase the intensity of a fire. Cool containers with large quantities of water from an explosion-resistant location. Vapors in confined areas may explode when exposed to fire. Storage containers and parts of containers may rocket great distances, in many directions.

NFPA Hazard Classifications: (diluted with 25% of benzene)

Vapors are heavier than air and will collect in low areas. Vapors in confined areas may explode when exposed to fire. Vapors may travel long distances to ignition sources and flash back. Storage containers and parts of containers may rocket great distances, in many directions. If material or contaminated runoff enters waterways, notify downstream users of potentially contaminated waters. Notify local health and fire officials and pollution control agencies. Use dry chemicals, foam, carbon dioxide. From a secure explosion-proof location, use water spray to cool exposed containers. If cooling streams are ineffective (venting sound increases in volume and pitch, tank discolors, or shows any signs of deforming), withdraw immediately to a secure position.

Health Hazards (Blue) 2
Flammability (Red) 3
Reactivity (Yellow) 4
Special situations .... OXY

Flash Point: <80°F (<27°C) o.c.

FIRST AID

Move victim to fresh air. Call emergency medical care. Apply artificial respiration if victim is not breathing. Administer oxygen if breathing is difficult. Remove and isolate contaminated clothing and shoes. Remove material from skin immediately. In case of contact with substance, immediately flush skin or eyes with running water for at least 20 minutes. Keep victim warm and quiet. Ensure that medical personnel are aware of the material(s) involved and take precautions to protect themselves.

tert–BUTYL PEROXYPRIVALATE B:650

SYNONYMS: tert-BUTYL PERPIVALATE; tert-BUTYL TRIMETHYLPEROXYACETATE; EINECS No. 213-147-2; ESPEROX 31M; LUPERSOL 11; PEROXIPIVALATO de terc-BUTILO (Spanish); TRIGONOX 25/75; TRIGONOX 25-C75

IDENTIFICATION

CAS:	927-07-1
ERG ID:	UN 2110 (not > 77% in solution); pure compound FORBIDDEN
ERG Guide:	148
ERG Guide Category:	ORGANIC PEROXIDES (HEAT AND CONTAMINATION SENSITIVE/TEMPERATURE CONTROLLED)

Molecular Formula:	$C_9H_{18}O_3$
RTECS No.:	SE0950000
Properties:	Combustible oily liquid. Colorless. Insoluble in water.

HEALTH AND SAFETY INFORMATION

RECOMMENDATIONS FOR RESPIRATOR SELECTION:

At any concentrations above the NIOSH REL, or where there is no REL, at any detectable concentration: SCBAF:PD,PP (any self-contained breathing apparatus that has a full facepiece and is operated in a pressure-demand or other positive-pressure mode); or SAF:PD,PP:ASCBA (any supplied-air respirator that has a full facepiece and is operated in a pressure-demand or other positive-pressure mode in combination with an auxiliary, self-contained breathing apparatus operated in a pressure-demand or other positive-pressure mode).

Escape: GMFOV [any air-purifying, full-facepiece respirator (gas mask) with a chin-style, front- or back-mounted organic vapor canister]; or SCBAE (any appropriate escape-type, self-contained breathing apparatus).

CHEMICAL REACTIVITY

Forms explosive mixture with air. Highly reactive oxidizer. Reacts violently when exposed to reducing agents, acids, accelerators, combustible and/or organic materials, heat, or flames [rapid decomposition at 90°F (32°C)]. Sensitive to shock or friction.

FIRE INFORMATION

May spontaneously explode if exposed to heat or fire. Will increase the intensity of a fire. Cool containers with large quantities of water from an explosion-resistant location. Vapors in confined areas may explode when exposed to fire. Storage containers and parts of containers may rocket great distances, in many directions. Use dry chemical, carbon dioxide, or regular foam.

NFPA Hazard Classifications: (diluted with 25% of mineral spirits)

Health Hazards (Blue) 0
Flammability (Red) 3
Reactivity (Yellow) 4
Special situations .... OXY

Flash Point: 155°F (68°C) o.c.

FIRST AID

Move victim to fresh air. Call emergency medical care. Apply artificial respiration if victim is not breathing. Administer oxygen if breathing is difficult. Remove and isolate contaminated clothing and shoes. Remove material from skin immediately. In case of contact with substance, immediately flush skin or eyes with running water for at least 20 minutes. Keep victim warm and quiet. Ensure that medical personnel are aware of the material(s) involved and take precautions to protect themselves.

o–sec–BUTYLPHENOL B:675

SYNONYMS: o-sec-BUTILFENOL (Spanish); 2-sec.-BUTYLFENOL (Czech); 2-sec-BUTYLPHENOL; EINECS No. 201-933-8; 2-(1-METHYLPROPYL)PHENOL; PHENOL, 2-(1-METHYLPROPYL)-

IDENTIFICATION

CAS:	89-72-5
ERG ID:	UN 2228 (liquid); UN 2229 (solid)
ERG Guide:	153
ERG Guide Category:	SUBSTANCES-TOXIC AND/OR CORROSIVE (COMBUSTIBLE)
Molecular Formula:	$C_{10}H_{14}O$
RTECS No.:	SJ9100000
Properties:	Combustible liquid or solid [below 61°F (16°C)]. Colorless. Insoluble in water.

HEALTH AND SAFETY INFORMATION

May cause irritation of the eyes, skin, respiratory system. Prolonged skin contact will cause skin burns.

OSHA Table Z-1-A Air Contaminant

Exposure Limits:	**TWA**
ACGIH: (*)	5 ppm; 31 mg/m^3
NIOSH: (*)	5 ppm; 30 mg/m^3
OSHA:	None

(*) Skin contact contributes significantly to overall exposure.

RECOMMENDATIONS FOR RESPIRATOR SELECTION:

At any concentrations above the NIOSH REL, or where there is no REL, at any detectable concentration: SCBAF:PD,PP (any self-contained breathing apparatus that has a full facepiece and is operated in a pressure-demand or other positive-pressure mode); or SAF:PD,PP:ASCBA (any supplied-air respirator that has a full facepiece and is operated in a pressure-demand or other positive-pressure mode in combination with an auxiliary, self-contained breathing apparatus operated in a pressure-demand or other positive-pressure mode).

Escape: GMFOV [any air-purifying, full-facepiece respirator (gas mask) with a chin-style, front- or

back-mounted organic vapor canister]; or SCBAE (any appropriate escape-type, self-contained breathing apparatus).

REACTIVITY

Contact with oxidizers may cause fire and explosions.

FIRE INFORMATION

If material or contaminated runoff enters waterways, notify downstream users of potentially contaminated waters. Notify local health and fire officials and pollution control agencies.

Flash Point: 226°F (108°C)

FIRST AID

Move victim to fresh air and call emergency medical care. Apply artificial respiration if victim is not breathing. Do not use mouth-to-mouth method if victim swallowed or inhaled the substance; induce artificial respiration with the aid of a pocket mask equipped with a one-way valve or other proper respiratory medical device. If breathing is difficult, give oxygen. In case of contact with substance, immediately flush skin or eyes with running water for at least 20 minutes. Remove and isolate contaminated clothing and shoes. Keep victim warm and quiet. Effects of exposure (inhalation, ingestion or, skin contact) to substances may be delayed. Ensure that medical personnel are aware of the material(s) involved and take precautions to protect themselves.

ENVIRONMENTAL INFORMATION

TSCA: Health and Safety Reporting List

p–tert–BUTYLPHENOL B:700

SYNONYMS: p-tert-BUTILFENOL (Spanish); p-tert-BUTYLFENOL (Czech); 4-tert-BUTYLPHENOL; 4-T-BUTYLPHENOL; 4-(1,1-DEMETHYLETHYL)PHENOL; BUTYLPHEN; EINECS No. 202-679-0; 1-HYDROXY-4-tert-BUTYLBENZENE; UCAR BUTYLPHENOL 4-T

IDENTIFICATION

CAS:	98-54-4
ERG ID:	UN 2229
ERG Guide:	153
ERG Guide Category:	SUBSTANCES-TOXIC AND/OR CORROSIVE (COMBUSTIBLE)
Molecular Formula:	$C_{10}H_{14}O$
Properties:	Combustible, needlelike, or crystalline solid. White. Disinfectant-like odor. May float or sink in water; insoluble.

HEALTH AND SAFETY INFORMATION

Poisonous. Severe eye and skin irritant. Skin contact contributes significantly to overall exposure.

RECOMMENDATIONS FOR RESPIRATOR SELECTION

At any concentrations above the NIOSH REL, or where there is no REL, at any detectable concentration: SCBAF:PD,PP (any self-contained breathing apparatus that has a full facepiece and is operated in a pressure-demand or other positive-pressure mode); or SAF:PD,PP:ASCBA (any supplied-air respirator that has a full facepiece and is operated in a pressure-demand or other positive-pressure mode in combination with an auxiliary, self-contained breathing apparatus operated in a pressure-demand or other positive-pressure mode).

Escape: GMFOV [any air-purifying, full-facepiece respirator (gas mask) with a chin-style, front- or back-mounted organic vapor canister]; or SCBAE (any appropriate escape-type, self-contained breathing apparatus).

CHEMICAL REACTIVITY

Contact with oxidizers may cause fire and explosions. Reacts with acids, caustics, amines, amides.

FIRE INFORMATION

Vapors are heavier than air and will collect in low areas. Vapors in confined areas may explode when exposed to fire. If material or contaminated runoff enters waterways, notify downstream users of potentially contaminated waters. Notify local health and fire officials and pollution control agencies. Use dry chemicals, alcohol-resistant foam, carbon dioxide. From a secure explosion-proof location, use water spray to cool exposed containers.

FEMA Hazard Classifications: (solid)

Health Hazards (Blue) 1
Flammability (Red) 1
Reactivity (Yellow) 0

Flash Point: 235°F (113°C) c.c. (liquid)

FIRST AID

Move victim to fresh air and call emergency medical care. Apply artificial respiration if victim is not breathing. Do not use mouth-to-mouth method if victim swallowed or inhaled the substance; induce artificial respiration with the aid of a pocket mask equipped with a one-way valve or other proper respiratory medical device. If breathing is difficult, give oxygen. In case of contact with substance, immediately flush skin or eyes with running water for at least 20 minutes. Remove and isolate contaminated clothing and shoes. Keep victim warm and quiet. Effects of exposure (inhalation, ingestion or, skin contact) to substances may be delayed. Ensure that medical personnel are aware of the material(s) involved and take precautions to protect themselves.

BUTYL STEARATE B:740

SYNONYMS: APEX 4; BS; BUTYL OCTADECANOATE; n-BUTYL OCTADECANOATE; n-BUTYL STEARATE; EINECS No. 204-666-6; EMEREST 2325; ESTEARATO de BUTILO (Spanish); GROCO 5810; KESSCO BSC; KESSCOFLEX BS; POLYCIZER 332; OCTADECANOIC ACID, BUTYL ESTER (9CI); RC PLASTICIZER B-17; STEARATE de BUTYLE, (French); STARFOL BS-100; TEGESTER BUTYL STEARATE; UNIFLEX BYS; WICKENOL 122; WITCIZER 200; WITCIZER 201

IDENTIFICATION

CAS:	123-95-5
ERG ID:	UN 3272
ERG Guide:	127
ERG Guide Category:	FLAMMABLE LIQUIDS (POLAR/WATER-MISCIBLE)
Molecular Formula:	$C_{22}H_{44}O_2$
RTECS No.:	WI2900000
Properties:	Combustible oily or waxy liquid [above 68°F (20°C)]; crystals from alcohol, propanol, or ether. Colorless to pale yellow. Odorless or slight fruity, fatty odor. Floats on water; very slightly soluble.

HEALTH AND SAFETY INFORMATION

RECOMMENDATIONS FOR RESPIRATOR SELECTION:

At any concentrations above the NIOSH REL, or where there is no REL, at any detectable concentration: SCBAF:PD,PP (any self-contained breathing apparatus that has a full facepiece and is operated in a pressure-demand or other positive-pressure mode); or SAF:PD,PP:ASCBA (any supplied-air respirator that has a full facepiece and is operated in a pressure-demand or other positive-pressure mode in combination with an auxiliary, self-contained breathing apparatus operated in a pressure-demand or other positive-pressure mode).

Escape: GMFOV [any air-purifying, full-facepiece respirator (gas mask) with a chin-style, front- or back-mounted organic vapor canister]; or SCBAE (any appropriate escape-type, self-contained breathing apparatus).

CHEMICAL REACTIVITY

Contact with oxidizers may cause fire and explosions.

FIRE INFORMATION

Use dry chemical, carbon dioxide, water mist.

NFPA Hazard Classifications:

Health Hazards (Blue) 1
Flammability (Red) 1
Reactivity (Yellow) 0

Flash Point: 320°F (160°C)
Autoignition Temperature: 671°F (355°C)

FIRST AID

Move victim to fresh air. Call emergency medical care. Apply artificial respiration if victim is not breathing. Administer oxygen if breathing is difficult. Remove and isolate contaminated clothing and shoes. In case of contact with substance, immediately flush skin or eyes with running water for at least 20 minutes. Wash skin with soap and water. Keep victim warm and quiet. Ensure that medical personnel are aware of the material(s) involved and take precautions to protect themselves.

BUTYLTRICHLOROSILANE B:780

SYNONYMS: n-BUTYLTRICHLOROSILANE; EINECS No. 231-381-3; TRICHLOROBUTYLSILANE

IDENTIFICATION

CAS:	7521-80-4
ERG ID:	UN 1747
ERG Guide:	155
ERG Guide Category:	SUBSTANCES-TOXIC AND/OR CORROSIVE (FLAMMABLE/WATER-SENSITIVE)
Molecular Formula:	$C_4H_9Cl_3Si$
Properties:	Combustible liquid. Colorless. Sharp, irritating odor. Reacts violently with water. Irritating, visible vapor cloud is produced.

HEALTH AND SAFETY INFORMATION

IDLH: 50 ppm as hydrogen chloride
Odor Threshold: 0.255-10.6 as hydrogen chloride

Exposure Limits: CEILING

ACGIH: 5 ppm; 7.5 mg/m^3 as hydrogen chloride

NIOSH: 5 ppm; 7.0 mg/m^3 as hydrogen chloride
OSHA: 5 ppm; 7.0 mg/m^3 as hydrogen chloride

RECOMMENDATIONS FOR RESPIRATOR SELECTION: NIOSH/OSHA as hydrogen chloride

50 ppm: CCRS [any chemical cartridge respirator with cartridge(s) providing protection against the compound of concern)]; or GMFS [any air-purifying, full-facepiece respirator (gas mask) with a chin-style, front- or back-mounted canister providing protection against the compound of concern]; or PAPRS (any powered, air-purifying respirator with cartridge(s) providing protection against the compound of concern); or SA (any supplied-air respirator); SCBAF (any self-contained breathing apparatus with a full facepiece).

Emergency or planned entry into unknown concentrations or IDLH conditions: SCBAF:PD,PP (any self-contained breathing apparatus that has a full facepiece and is operated in a pressure-demand or other positive-pressure mode); or SAF:PD,PP:ASCBA (any supplied-air respirator that has a full facepiece and is operated in a pressure-demand or other positive-pressure mode in combination with an auxiliary, self-contained breathing apparatus operated in a pressure-demand or other positive-pressure mode.

Escape: GMFAG [any air-purifying, full-facepiece respirator (gas mask) with a chin-style, front-or back-,mounted organic vapor canister]; or SCBAE (any appropriate escape-type, self-contained breathing apparatus)
Note: Substance reported to cause eye irritation or damage; may require eye protection.

CHEMICAL REACTIVITY

Reacts vigorously with water to generate hydrogen chloride (hydrochloric acid). Will react with common metals to evolve hydrogen chloride and cause severe corrosion.

FIRE INFORMATION

Combustion products include corrosive hydrogen chloride. Gases are heavier than air and will collect in low areas. Gases in confined areas may explode when exposed to fire. Gases may travel long distances to ignition sources and flash back. Storage containers and parts of containers may rocket great distances, in many directions. If material or contaminated runoff enters waterways, notify downstream users of potentially contaminated waters. Notify local health and fire officials and pollution control agencies. Do not extinguish the fire unless the flow of the gas can be stopped and any remaining gas is out of the line. Specially trained personnel may use fog lines to cool exposures and let the fire burn itself out. Use dry chemicals pr medoum expansion AFFF foam. From a secure explosion-proof location, use water spray to cool exposed containers. If cooling streams are ineffective (venting sound increases in volume and pitch, tank discolors, or shows any signs of deforming), withdraw immediately to a secure position.

NFPA Hazard Classifications:

Health Hazards (Blue) 2
Flammability (Red) 2
Reactivity (Yellow) 0

Flash Point: 130°F (55°C) o.c.; 12°6F (52°C) c.c.

FIRST AID

Move victim to fresh air; call emergency medical care. Apply artificial respiration if victim is not breathing. Do not use mouth-to-mouth method if victim ingested or inhaled the substance; induce artificial respiration with the aid of a pocket mask equipped with a one-way valve or other proper respiratory medical device. Administer oxygen if breathing is difficult. Remove and isolate contaminated clothing and shoes. In case of contact with substance, immediately flush skin or eyes with running water for at least 20 minutes. For minor skin contact, avoid spreading material on unaffected skin. Keep victim warm and quiet. Effects of exposure (inhalation, ingestion, or skin contact) to substance may be delayed. Ensure that medical personnel are aware of the material(s) involved and take precautions to protect themselves.

ENVIRONMENTAL INFORMATION as hydrogen chloride

Clean Air Act: Hazardous Air Pollutants (Title I, Part A, Section 112); (anhydrous or gas) Accidental Release Prevention/Flammable Substances (Section 112[r], Table 3), TQ = 5,000 lb (2,270 kg)
Clean Water Act: Section 311 Hazardous Substances/RQ (same as CERCLA); Section 313 Priority Chemicals
EPCRA Section 304 Reportable Quantity (RQ): CERCLA, 5,000 lb (2,270 kg)
EPCRA Section 313: Form R *de minimis* concentration reporting level: 1.0%.
Note: Non-areosol forms of hydrochloric acid have been deleted from EPCRA/SARA 313 reporting, 7/29/96 (FR vol. 61, No. 146, pp. 39356-39357)

ENVIRONMENTAL INFORMATION as hydrogen chloride (concentration 30% or more)

Clean Air Act: Hazardous Air Pollutants (Title I, Part A, Section 112); (conc. 30% or greater) Accidental Release Prevention/Flammable Substances (Section 112[r], Table 3), TQ = 15,000 lb (6,810 kg)
Clean Water Act: Section 311 Hazardous Substances/RQ (same as CERCLA); Section 313 Priority Chemicals
EPCRA Section 304 Reportable Quantity (RQ): CERCLA, 5,000 lb (2,270 kg)
EPCRA Section 313: Form R *de minimis* concentration reporting level: 1.0%. Note: Non-areosol forms of hydrochloric acid have been deleted from EPCRA/SARA 313 reporting, 7/29/96 (FR vol. 61, No. 146, pp. 39356-39357)

BUTYRALDEHYDE B:820

SYNONYMS: ALDEHYDE BUTYRIQUE (French); ALDEIDE BUTIRRICA (Italian); BUTAL; n-BUTALDEHYDE; BUTANALDEHYDE; n-BUTYRALDEHYDE; BUTALDEHYDE; BUTALYDE; BUTANAL; n-BUTANAL (Czech); BUTIRALDEHIDO (Spanish); BUTYRAL; BUTYRAL BUTYRIC ALDEHYDE; BUTYRALDEHYD (German); BUTYL ALDEHYDE; n-BUTYL ALDEHYDE; BUTYRIC ACID; BUTYRIC ALDEHYDE; EEC No. 605-006-00-2; EINECS No. 204-646-6; FEMA No. 2219; NCI-C56291

IDENTIFICATION

CAS:	123-72-8
ERG ID:	UN 1129
ERG Guide:	129
ERG Guide Category:	FLAMMABLE LIQUIDS (POLAR/WATER-MISCIBLE/NOXIOUS)
Molecular Formula:	C_4H_8O
RTECS No.:	ES2275000
Properties:	Flammable, watery liquid, colorless. Pungent odor. Floats and mixes slowly with water. Flammable, irritating vapor is produced.

HEALTH AND SAFETY INFORMATION

Odor Threshold: 0.009 ppm

Exposure Limits: **TWA**

AIHA WEEL: 25 ppm

RECOMMENDATIONS FOR RESPIRATOR SELECTION

At any concentrations above the NIOSH REL, or where there is no REL, at any detectable concentration: SCBAF:PD,PP (any self-contained breathing apparatus that has a full facepiece and is operated in a pressure-demand or other positive-pressure mode); or SAF:PD,PP:ASCBA (any supplied-air respirator that has a full facepiece and is operated in a pressure-demand or other positive-pressure mode in combination with an auxiliary, self-contained breathing apparatus operated in a pressure-demand or other positive-pressure mode).

Escape: GMFOV [any air-purifying, full-facepiece respirator (gas mask) with a chin-style, front- or back-mounted organic vapor canister]; or SCBAE (any appropriate escape-type, self-contained breathing apparatus).

CHEMICAL REACTIVITY

Possible self-reaction in air; undergoes rapid oxidation to butyric acid in air. Violent reaction with strong oxidizers. May corrode steel due to corrosive action of butyric acid.

FIRE INFORMATION

Vapors are heavier than air and will collect in low areas. Vapors in confined areas may explode when exposed to fire. Vapors may travel long distances to ignition sources and flash back. Storage containers and parts of containers may rocket great distances, in many directions. If material or contaminated runoff enters waterways, notify downstream users of potentially contaminated waters. Notify local health and fire officials and pollution control agencies. Use dry chemicals, foam, carbon dioxide. From a secure explosion-proof location, use water spray to cool exposed containers. If cooling streams are ineffective (venting sound increases in volume and pitch, tank discolors, or shows any signs of deforming), withdraw immediately to a secure position.

NFPA Hazard Classifications:

Health Hazards (Blue) 3
Flammability (Red) 3
Reactivity (Yellow) 0

Flash Point: 10°F (12°C) c.c.
Flammable Limits in Air: LEL 2.5%; UEL 12.5%
Autoignition Temperature: 446°F (230°C)

FIRST AID

Move victim to fresh air. Call emergency medical care. Apply artificial respiration if victim is not breathing. Administer oxygen if breathing is difficult. Remove and isolate contaminated clothing and shoes. In case of contact with substance, immediately flush skin or eyes with running water for at least 20 minutes. Wash skin with soap and water. Keep victim warm and quiet. Effects of exposure (inhalation, ingestion, or skin contact) to substance may be delayed. Ensure that medical personnel are aware of the material(s) involved and take precautions to protect themselves.

ENVIRONMENTAL INFORMATION

EPCRA Section 313: Form R *de minimis* concentration reporting level: 1.0%.
TSCA: 40CFR716.120(*d*)1 as aldehydes

BUTYROLACTONE B:860

SYNONYMS: γ-6480; AGRISYNTH BLO; AGSOL EX BLO; γ-BL; BLO; BLON; BURMAR POSISTRIP LE; BUTIROLACTONA (Spanish); 4-BUTANOLIDE; BUTYRIC ACID LACTONE; α-BUTYROLACTONE; γ-BUTYROLACTONE (FCC); BUTYRYL LACTONE; 4-DEOXYTETRONIC ACID; DIHYDRO-2(3-H)-FURANONE; DYNASOLVE 699; EINECS No. 202-509-5; FEMA No. 3291; GBL; 4-HYDROXYBUTANOIC ACID; INSTAPAK PORT CLEANER; LACTONE; γ-HYDROXYBUTRIC ACID GLYCLIC ESTER; 4-HYDROXYBUTYRIC ACID γ-LACTONE; γ-HYDROXYBUTYROLACTONE; NCI-C55878; PR-21 RESIST; PR-55 RESIST; TETRAHYDRO-2-FURANONE

IDENTIFICATION

CAS:	96-48-0
ERG ID:	UN 2810, toxic liquid, n.o.s.
ERG Guide:	153

ERG Guide Category: SUBSTANCES-TOXIC AND/OR CORROSIVE
Molecular Formula: $C_4H_6O_2$
RTECS No.: LU3500000
Properties: Combustible liquid. Water-white. Pleasant caramel odor. Soluble in water.

HEALTH AND SAFETY INFORMATION

May irritate eyes, respiratory tract.

IARC Evaluation: Group 3, unclassifiable; no human data; animal evidence, inadequate.

RECOMMENDATIONS FOR RESPIRATOR SELECTION: NIOSH

At any concentrations above the NIOSH REL, or where there is no REL, at any detectable concentration: SCBAF:PD,PP (any self-contained breathing apparatus that has a full facepiece and is operated in a pressure-demand or other positive-pressure mode); or SAF:PD,PP:ASCBA (any supplied-air respirator that has a full facepiece and is operated in a pressure-demand or other positive-pressure mode in combination with an auxiliary, self-contained breathing apparatus operated in a pressure-demand or other positive-pressure mode).

ESCAPE: GMFOV [any air-purifying, full-facepiece respirator (gas mask) with a chin-style, front- or back-mounted organic vapor canister]; or SCBAE (any appropriate escape-type, self-contained breathing apparatus).

CHEMICAL REACTIVITY

Forms explosive mixture with air. Contact with oxidizers may cause fire and explosions.

FIRE INFORMATION

Decomposition products upon heating include acrid and irritating fumes. Vapors are heavier than air and will collect in low areas. Vapors in confined areas may explode when exposed to fire. Vapors may travel long distances to ignition sources and flash back. Storage containers and parts of containers may rocket great distances, in many directions. Fight fire from an explosion-proof position; use dry chemicals, alcohol-resistant foam, or carbon dioxide; cool containers with water spray.

NFPA Fire Classifications:

Health Hazards (Blue) 0
Flammability (Red) 1
Reactivity (Yellow) 0

Flash Point: 209°F (98°C) o.c.

FIRST AID

Move victim to fresh air. Call emergency medical care. Apply artificial respiration if victim is not breathing. Do not use mouth-to-mouth method if victim ingested or inhaled the substance; induce artificial respiration with the aid of a pocket mask equipped with a one-way valve or other proper respiratory medical device. Administer oxygen if breathing is difficult. Remove and isolate contaminated clothing and shoes. In case of contact with substance, immediately flush skin or eyes with running water for at least 20 minutes. For minor skin contact, avoid spreading material on unaffected skin. Keep victim warm and quiet. Effects of exposure (inhalation, ingestion, or skin contact) may be delayed. Ensure that medical personnel are aware of the material(s) involved and take precautions to protect themselves.

CADMIUM C:003

SYNONYMS: CADMIO (Spanish); CADMIUM METAL; C. I. 77180; COLLOIDAL CADMIUM; EINECS No. 231-152-8; KADMIUM (German)

IDENTIFICATION

CAS:	7440-43-9
ERG ID:	UN 2570
ERG Guide:	154
ERG Guide Category:	SUBSTANCES-TOXIC AND/OR CORROSIVE (NONCOMBUSTIBLE)
Molecular Formula:	Cd
RTECS No.:	EU9800000
Properties:	Silver-white crystalline solid, tinged with blue; or lustrous, malleable metal. Odorless. Powder form is combustible.

HEALTH AND SAFETY INFORMATION

Corrosive to eyes, skin, gastrointestinal tract, and respiratory tract. Inhalation may cause pulmonary edema, which can be delayed for several hours; there is a risk of death in serious cases. High or prolonged exposure can cause kidney, blood damage. Can affect sense of smell.

IARC Evaluation: Probable carcinogen, Group 2A; limited human evidence; sufficient animal evidence.
NTP: Anticipated carcinogen.
IDLH: 9 mg/m^3 as cadmium dust or fume; potential carcinogen.
OSHA Table Z-1 Air Contaminant (see cadmium standard 1910.1027)
OSHA Table Z-2 Air Contaminant (see cadmium standard 1910.1027)

Exposure Limits:	**TWA**
ACGIH:	0.01 mg/m^3 (dusts/salts); 0.002 mg/m^3 (respirable dust)

NIOSH: [1]
OSHA: 0.005 mg/m^3

[1] Potential carcinogen. Limit exposure to lowest feasible level.

RECOMMENDATIONS FOR RESPIRATOR SELECTION: NIOSH

At any concentrations above the NIOSH REL, or where there is no REL, at any detectable concentration: SCBAF:PD,PP (any self-contained breathing apparatus that has a full facepiece and is operated in a pressure-demand or other positive-pressure mode);or SAF:PD,PP:ASCBA (any supplied-air respirator that has a full facepiece and is operated in a pressure-demand or other positive-pressure mode in combination with an auxiliary, self-contained breathing apparatus operated in a pressure-demand or other positive-pressure mode).

Escape: HiEF (any air-purifying, full-facepiece respirator with a high-efficiency particulate filter);or SCBAE (any appropriate escape-type, self-contained breathing apparatus).

CHEMICAL REACTIVITY

Air exposure, especially with powdered form, may self-ignite. Violent reaction with strong oxidizers, hydrozoic acid, and tellurium. Acid contact forms hydrogen gas. May react with selenium, elemental sulfur, and zinc.

FIRE INFORMATION

Decomposition products upon heating include toxic cadmium fumes. If material or contaminated runoff enters waterways, notify downstream users of potentially contaminated waters. Notify local health and fire officials and pollution control agencies. Use special dry chemicals appropriate for metal fires, sand. Do not use water. From a secure explosion-proof location, use water spray to cool exposed containers.

FEMA Hazard Classifications: (powder)

Health Hazards (Blue) 2
Flammability (Red) 2
Reactivity (Yellow) 0

FIRST AID

Move victim to fresh air. Call emergency medical care. Apply artificial respiration if victim is not breathing. Do not use mouth-to-mouth method if victim ingested or inhaled the substance; induce artificial respiration with the aid of a pocket mask equipped with a one-way valve or other proper respiratory medical device. Administer oxygen if breathing is difficult. Remove and isolate contaminated clothing and shoes. In case of contact with substance, immediately flush skin or eyes with running water for at least 20 minutes. For minor skin contact, avoid spreading material on unaffected skin. Keep victim warm and quiet. Effects of exposure (inhalation, ingestion, or skin contact) may be delayed. Ensure that medical personnel are aware of the material(s) involved and take precautions to protect themselves.

ENVIRONMENTAL INFORMATION

Clean Air Act: Hazardous Air Pollutants (Title I, Part A, Section 112). Includes any unique chemical substance that contains cadmium as part of that chemical's infrastructure.
Clean Water Act: Section 307 Priority Pollutants; Section 313 Priority Chemicals; Toxic Pollutant (Section 401.15).
EPA Hazardous Waste Number (RCRA No.): D006
RCRA Section 261 Hazardous Constituents: Waste number not listed.
RCRA Toxicity Characteristic (Section 261.24): Maximum Concentration of Contaminants, regulatory level, 1.0 mg/L
RCRA Land Ban Waste
RCRA Universal Treatment Standards: Wastewater (mg/L), 0.69; Nonwastewater (mg/L), 0.19 TCLP.
RCRA Ground Water Monitoring List: Suggested test method(s) (PQL ug/L): 6010(40); 7130(50); 7131(1).
Safe Drinking Water Act: MCL, 0.005 mg/L; MCLG, 0.005 mg/L; Regulated chemical (47 FR 9352); Priority List (55 FR 1470).
EPCRA Section 304: Reportable Quantity (RQ): CERCLA, 10 lb (4.54 kg).
Note: No release report required if diameter of pieces is equal to or exceeds 0.004 in.
EPCRA Section 313: Form R *de minimis* concentration reporting level: 0.1%. Form R Toxic Chemical Category Code: N078
Marine Pollutant (49CFR, Subchapter 172.101, Appendix B).: Severe pollutant, as cadmium compounds.
California's Proposition 65: Carcinogen; Reproductive toxin; Female reproductive toxin.

CADMIUM ACETATE C:004

SYNONYMS: A13-01414; ACETIC ACID, CADMIUM SALT; ACETATO CADMIO (Spanish); BIS(ACETOXY)CADMIUM; CADMIUM (II) ACETATE; CADMIUM ACETATE DIHYDRATE; CADMIUMDI (ACETAAT) (Dutch); CADMIUM DIACETATE; C. I. 77185; EEC No. 048-001-00-5; EINECS No. 208-853-2

IDENTIFICATION

CAS:	543-90-8
ERG ID:	UN 2570
ERG Guide:	151
ERG Guide Category:	SUBSTANCES-TOXIC (NONCOMBUSTIBLE)
Molecular Formula:	$C_4H_6CdO_4 \cdot 3H_2O$
RTECS No.:	EU9810000
Properties:	Noncombustible crystalline solid. Colorless. Slight, acetic acid odor. Sinks and mixes with water.

HEALTH AND SAFETY INFORMATION

Corrosive to eyes, skin, gastrointestinal tract, and respiratory tract. Inhalation may cause pulmonary edema, which can be delayed for several hours; there is a risk of death in serious cases. High or prolonged exposure can cause kidney, blood damage. Can affect sense of smell.

IARC Evaluation: Probable carcinogen, Group 2A; limited human evidence; sufficient animal evidence.
NTP: Anticipated carcinogen.
IDLH: 9 mg/m^3 as cadmium dust or fume; potential carcinogen.
OSHA Table Z-1 Air Contaminant (see cadmium standard 1910.1027)
OSHA Table Z-2 Air Contaminant (see cadmium standard 1910.1027)

Exposure Limits:	**TWA**
ACGIH: [(1)]	0.01 mg/m^3 (dusts/salts); 0.002 mg/m^3 (respirable dust)
NIOSH: [(2)]	
OSHA:	0.005 mg/m^3

[(1)] Suspected carcinogen.
[(2)] Potential carcinogen. Limit exposure to lowest feasible level.

RECOMMENDATIONS FOR RESPIRATOR SELECTION: NIOSH

At any concentrations above the NIOSH REL, or where there is no REL, at any detectable concentration: SCBAF:PD,PP (any self-contained breathing apparatus that has a full facepiece and is operated in a pressure-demand or other positive-pressure mode);or SAF:PD,PP:ASCBA (any supplied-air respirator that has a full facepiece and is operated in a pressure-demand or other positive-pressure mode in combination with an auxiliary, self-contained breathing apparatus operated in a pressure-demand or other positive-pressure mode).

Escape: HiEF (any air-purifying, full-facepiece respirator with a high-efficiency particulate filter);or SCBAE (any appropriate escape-type, self-contained breathing apparatus).

CHEMICAL REACTIVITY

Incompatible with strong oxidizers, elemental sulfur, selenium, tellurium, strong acids, and nitrates

FIRE INFORMATION:

Not flammable. Decomposition products upon heating include toxic cadmium fumes. If material or contaminated runoff enters waterways, notify downstream users of potentially contaminated waters. Notify local health and fire officials and pollution control agencies. Use extinguishing agents appropriate for surrounding fire. From a secure explosion-proof location, use water spray to cool exposed containers.

FEMA Hazard Classifications:

Health Hazards (Blue) 2

Flammability (Red) 0
Reactivity (Yellow) 0

FIRST AID

Move victim to fresh air. Call emergency medical care. Apply artificial respiration if victim is not breathing. Do not use mouth-to-mouth method if victim ingested or inhaled the substance; induce artificial respiration with the aid of a pocket mask equipped with a one-way valve or other proper respiratory medical device. Administer oxygen if breathing is difficult. Remove and isolate contaminated clothing and shoes. In case of contact with substance, immediately flush skin or eyes with running water for at least 20 minutes. For minor skin contact, avoid spreading material on unaffected skin. Keep victim warm and quiet. Effects of exposure (inhalation, ingestion, or skin contact) may be delayed. Ensure that medical personnel are aware of the material(s) involved and take precautions to protect themselves.

ENVIRONMENTAL INFORMATION

Clean Air Act: Hazardous Air Pollutants (Title I, Part A, Section 112). Includes any unique chemical substance that contains cadmium as part of that chemical's infrastructure.
Clean Water Act: Section 311 Hazardous Substances/RQ (same as CERCLA); Section 313 Priority Chemicals.
EPCRA Section 304: Reportable Quantity (RQ): CERCLA, 10 lb (4.54 kg).
EPCRA Section 313: Form R *de minimis* concentration reporting level: 1.0%. Form R Toxic Chemical Category Code: N078.
Marine Pollutant (49CFR, Subchapter 172.101, Appendix B): Severe pollutant, as cadmium compounds.
California's Proposition 65: Carcinogen as cadmium compounds.

CADMIUM CHLORIDE C:006

SYNONYMS: CADDY; CADMIUM DICHLORIDE; CLORURO de CADMIO (Spanish); DICHLOROCADMIUM; EEC No. 048-008-00-3; EINECS No. 233-296-7; KADMIUM CHLORID (German); VI-CAD

IDENTIFICATION

CAS:	10108-64-2
ERG ID:	UN 2570
ERG Guide:	154
ERG Guide Category:	SUBSTANCES-TOXIC AND/OR CORROSIVE (NONCOMBUSTIBLE)
Molecular Formula:	$CdCl_2$
RTECS No.:	EV0178000
Properties:	Hygroscopic crystalline solid. White. Odorless. Sinks and mixes with water.

HEALTH AND SAFETY INFORMATION

Corrosive to eyes, skin, gastrointestinal tract, and respiratory tract. Inhalation may cause pulmonary edema, which can be delayed for several hours; there is a risk of death in serious cases. High or prolonged exposure can cause kidney, blood damage. Can affect sense of smell.

IARC Evaluation: Probable carcinogen. Group 2A; limited human evidence; sufficient animal evidence.
NTP: Anticipated carcinogen.
IDLH: 9 mg/m^3 as cadmium dust or fume; potential carcinogen.
OSHA Table Z-1 Air Contaminant (see cadmium standard 1910.1027)
OSHA Table Z-2 Air Contaminant (see cadmium standard 1910.1027)

Exposure Limits:	**TWA**
ACGIH: (1)	0.01 mg/m^3 (dusts/salts); 0.002 mg/m^3 (respirable dust)
NIOSH: (2)	
OSHA:	0.005 mg/m^3

(1) Suspected carcinogen.
(2) Potential carcinogen. Limit exposure to lowest feasible level.

RECOMMENDATIONS FOR RESPIRATOR SELECTION: NIOSH

At any concentrations above the NIOSH REL, or where there is no REL, at any detectable concentration: SCBAF:PD,PP (any self-contained breathing apparatus that has a full facepiece and is operated in a pressure-demand or other positive-pressure mode);or SAF:PD,PP:ASCBA (any supplied-air respirator that has a full facepiece and is operated in a pressure-demand or other positive-pressure mode in combination with an auxiliary, self-contained breathing apparatus operated in a pressure-demand or other positive-pressure mode).

Escape: HiEF (any air-purifying, full-facepiece respirator with a high-efficiency particulate filter);or SCBAE (any appropriate escape-type, self-contained breathing apparatus).

CHEMICAL REACTIVITY

Cadmium compounds may react with strong oxidizers; elemental sulfur, selenium, and tellurium. Forms impact-sensitive explosive with potassium.

FIRE INFORMATION:

Not flammable. Decomposition products upon heating include toxic cadmium fumes. If material or contaminated runoff enters waterways, notify downstream users of potentially contaminated waters. Notify local health and fire officials and pollution control agencies. Use extinguishing agents appropriate for surrounding fire. From a secure explosion-proof location, use water spray to cool exposed containers.

FEMA Hazard Classifications:

Health Hazards (Blue) 2
Flammability (Red) 0
Reactivity (Yellow) 0

FIRST AID

Move victim to fresh air. Call emergency medical care. Apply artificial respiration if victim is not breathing. Do not use mouth-to-mouth method if victim ingested or inhaled the substance; induce artificial respiration with the aid of a pocket mask equipped with a one-way valve or other proper respiratory medical device. Administer oxygen if breathing is difficult. Remove and isolate contaminated clothing and shoes. In case of contact with substance, immediately flush skin or eyes with running water for at least 20 minutes. For minor skin contact, avoid spreading material on unaffected skin. Keep victim warm and quiet. Effects of exposure (inhalation, ingestion, or skin contact) may be delayed. Ensure that medical personnel are aware of the material(s) involved and take precautions to protect themselves.

ENVIRONMENTAL INFORMATION

Clean Air Act: Hazardous Air Pollutants (Title I, Part A, Section 112). Includes any unique chemical substance that contains cadmium as part of that chemical's infrastructure.
Clean Water Act: Section 311 Hazardous Substances/RQ (same as CERCLA); Section 313 Priority Chemicals.
EPCRA Section 304: Reportable Quantity (RQ): CERCLA, 10 lb (4.54 kg).
EPCRA Section 313: Form R *de minimis* concentration reporting level: 0.1%. Form R Toxic Chemical Category Code: N078.
Marine Pollutant (49CFR, Subchapter 172.101, Appendix B).: Severe pollutant, as cadmium compounds.
California's Proposition 65: Carcinogen as cadmium compounds.

CADMIUM CYANIDE C:008

SYNONYMS: CIANURO de CADMIO (Spanish)

IDENTIFICATION

CAS:	N/A
ERG ID:	UN 2570
ERG Guide:	154
ERG Guide Category:	SUBSTANCES-TOXIC AND/OR CORROSIVE (NONCOMBUSTIBLE)
Molecular Formula:	C_2CdN_2
RTECS No.:	EV0260000 (cadmium compounds)
Properties:	Noncombustible white solid. Odorless.

HEALTH AND SAFETY INFORMATION

Corrosive to eyes, skin, gastrointestinal tract, and respiratory tract. Inhalation may cause pulmonary edema, which can be delayed for several hours; there is a risk of death in serious cases. High or prolonged exposure can cause kidney, blood damage. Can affect sense of smell.

IARC Evaluation: Probable carcinogen, Group 2A; limited human evidence; sufficient animal evidence.
NTP: Anticipated carcinogen.
IDLH: 9 mg/m^3 as cadmium dust or fume; potential carcinogen.
OSHA Table Z-1 Air Contaminant (see cadmium standard 1910.1027)
OSHA Table Z-1 Air Contaminant (see cadmium standard 1910.1027)

Exposure Limits:	**TWA**
ACGIH: (1)	0.01 mg/m^3 (dusts/salts); 0.002 mg/m^3 (respirable dust)
NIOSH: (2)	
OSHA:	0.005 mg/m^3

(1) Suspected carcinogen.
(2) Potential carcinogen. Limit exposure to lowest feasible level.

RECOMMENDATIONS FOR RESPIRATOR SELECTION: NIOSH

At any concentrations above the NIOSH REL, or where there is no REL, at any detectable concentration: SCBAF:PD,PP (any self-contained breathing apparatus that has a full facepiece and is operated in a pressure-demand or other positive-pressure mode);or SAF:PD,PP:ASCBA (any supplied-air respirator that has a full facepiece and is operated in a pressure-demand or other positive-pressure mode in combination with an auxiliary, self-contained breathing apparatus operated in a pressure-demand or other positive-pressure mode).

Escape: HiEF (any air-purifying, full-facepiece respirator with a high-efficiency particulate filter);or SCBAE (any appropriate escape-type, self-contained breathing apparatus).

CHEMICAL REACTIVITY

Cadmium compounds may react with strong oxidizers; elemental sulfur, selenium, and tellurium.

FIRE INFORMATION

Not flammable. Decomposition products upon heating include toxic cadmium fumes. If material or contaminated runoff enters waterways, notify downstream users of potentially contaminated waters. Notify local health and fire officials and pollution control agencies. Use extinguishing agents appropriate for surrounding fire. From a secure explosion-proof location, use water spray to cool exposed containers.

FIRST AID

Move victim to fresh air. Call emergency medical care. Apply artificial respiration if victim is not

breathing. Do not use mouth-to-mouth method if victim ingested or inhaled the substance; induce artificial respiration with the aid of a pocket mask equipped with a one-way valve or other proper respiratory medical device. Administer oxygen if breathing is difficult. Remove and isolate contaminated clothing and shoes. In case of contact with substance, immediately flush skin or eyes with running water for at least 20 minutes. For minor skin contact, avoid spreading material on unaffected skin. Keep victim warm and quiet. Effects of exposure (inhalation, ingestion, or skin contact) may be delayed. Ensure that medical personnel are aware of the material(s) involved and take precautions to protect themselves.

ENVIRONMENTAL INFORMATION as cadmium compounds

Clean Air Act: Hazardous Air Pollutants (Title I, Part A, Section 112). Includes any unique chemical substance that contains cadmium as part of that chemical's infrastructure.
Clean Water Act: Toxic Pollutant (Section 401.15).
RCRA Section 261 Hazardous Constituents., waste number not listed.
EPCRA (Section 313): Includes any unique chemical substance that contains cadmium as part of that chemical's infrastructure. Form R *de minimis* concentration reporting level: (inorganic compounds: 0.1%;organic compounds: 1.0%). Form R Toxic Chemical Category Code: N078.
Marine Pollutant (49CFR, Subchapter 172.101, Appendix B): Severe pollutant, as cadmium compounds.
California's Proposition 65: Carcinogen, as cadmium compounds.

CADMIUM FLUOROBORATE C:009

SYNONYMS: BORATE(1-), TETRAFLUORO-, CADMIUM; CADMIUM FLUOBORATE; CADMIUM FLUOBORATE SOLUTION; CADMIUM FLUOROBORATE SOLUTION; CADMIUM TETRAFLUOROBORATE; EINECS No. 238-490-5

IDENTIFICATION

CAS:	14486-19-2
ERG ID:	UN 2570
ERG Guide:	154
ERG Guide Category:	SUBSTANCES-TOXIC AND/OR CORROSIVE (NONCOMBUSTIBLE)
Molecular Formula:	B_2CdF_8
RTECS No.:	EV0525000
Properties:	Noncombustible, hygroscopic liquid. Colorless. Odorless. Sinks and mixes with water.

HEALTH AND SAFETY INFORMATION

Corrosive to eyes, skin, gastrointestinal tract, and respiratory tract. Skin contact may cause burns. Inhalation may cause pulmonary edema, which can be delayed for several hours; there is a risk of death in serious cases. High or prolonged exposure can cause kidney, blood damage. Can affect sense of smell.

IARC Evaluation: Probable carcinogen, Group 2A; limited human evidence; sufficient animal evidence.
NTP: Anticipated carcinogen.
IDLH: 9 mg/m^3 as cadmium dust or fume; potential carcinogen
OSHA Table Z-1 Air Contaminant (see cadmium standard 1910.1027)
OSHA Table Z-2 Air Contaminant (see cadmium standard 1910.1027)

Exposure Limits:	**TWA**
ACGIH: [(1)]	0.01 mg/m^3 (dusts/salts); 0.002 mg/m^3 (respirable dust)
NIOSH: [(2)]	
OSHA:	0.005 mg/m^3

[(1)] Suspected carcinogen.
[(2)] Potential carcinogen. Limit exposure to lowest feasible level.

RECOMMENDATIONS FOR RESPIRATOR SELECTION: NIOSH

At any concentrations above the NIOSH REL, or where there is no REL, at any detectable concentration: SCBAF:PD,PP (any self-contained breathing apparatus that has a full facepiece and is operated in a pressure-demand or other positive-pressure mode);or SAF:PD,PP:ASCBA (any supplied-air respirator that has a full facepiece and is operated in a pressure-demand or other positive-pressure mode in combination with an auxiliary, self-contained breathing apparatus operated in a pressure-demand or other positive-pressure mode).

Escape: HiEF (any air-purifying, full-facepiece respirator with a high-efficiency particulate filter);or SCBAE (any appropriate escape-type, self-contained breathing apparatus).

CHEMICAL REACTIVITY

Cadmium compounds may react with strong oxidizers; elemental sulfur, selenium, and tellurium.

FIRE INFORMATION

Not flammable. Decomposition products upon heating include toxic cadmium fumes. If material or contaminated runoff enters waterways, notify downstream users of potentially contaminated waters. Notify local health and fire officials and pollution control agencies. Use extinguishing agents appropriate for surrounding fire. From a secure explosion-proof location, use water spray to cool exposed containers.

FEMA Hazard Classifications:

Health Hazards (Blue) 2
Flammability (Red) 0
Reactivity (Yellow) 0

FIRST AID

Move victim to fresh air. Call emergency medical care. Apply artificial respiration if victim is not

breathing. Do not use mouth-to-mouth method if victim ingested or inhaled the substance; induce artificial respiration with the aid of a pocket mask equipped with a one-way valve or other proper respiratory medical device. Administer oxygen if breathing is difficult. Remove and isolate contaminated clothing and shoes. In case of contact with substance, immediately flush skin or eyes with running water for at least 20 minutes. For minor skin contact, avoid spreading material on unaffected skin. Keep victim warm and quiet. Effects of exposure (inhalation, ingestion, or skin contact) may be delayed. Ensure that medical personnel are aware of the material(s) involved and take precautions to protect themselves.

ENVIRONMENTAL INFORMATION as cadmium compounds

Clean Air Act: Hazardous Air Pollutants (Title I, Part A, Section 112). Includes any unique chemical substance that contains cadmium as part of that chemical's infrastructure.
Clean Water Act: Toxic Pollutant (Section 401.15).
RCRA Section 261 Hazardous Constituents., waste number not listed.
EPCRA (Section 313): Includes any unique chemical substance that contains cadmium as part of that chemical's infrastructure. Form R *de minimis* concentration reporting level: (inorganic compounds: 0.1%;organic compounds: 1.0%). Form R Toxic Chemical Category Code: N078.
Marine Pollutant (49CFR, Subchapter 172.101, Appendix B): Severe pollutant, as cadmium compounds.
California's Proposition 65: Carcinogen as cadmium compounds.

CADMIUM HYDROXIDE C:011

SYNONYMS: CADMIUM HYDRATE; HIDROXIDO de CADMIO (Spanish)

IDENTIFICATION

CAS:	N/A
ERG ID:	UN 2570
ERG Guide:	154
ERG Guide Category:	SUBSTANCES-TOXIC AND/OR CORROSIVE (NONCOMBUSTIBLE)
Molecular Formula:	CdH_2O_2
RTECS No.:	EV0260000 (cadmium compounds)
Properties:	White powder. Sinks in water; insoluble.

HEALTH AND SAFETY INFORMATION

Corrosive to eyes, skin, gastrointestinal tract, and respiratory tract. Inhalation may cause pulmonary edema, which can be delayed for several hours; there is a risk of death in serious cases. High or prolonged exposure can cause kidney, blood damage. Can affect sense of smell.

IARC Evaluation: Probable carcinogen, Group 2A; limited human evidence; sufficient animal evidence.
NTP: Anticipated carcinogen.

IDLH: 9 mg/m^3 as cadmium dust or fume; potential carcinogen.
OSHA Table Z-1 Air Contaminant (see cadmium standard 1910.1027)
OSHA Table Z-2 Air Contaminants (as cadmium fumes and dust)

Exposure Limits:	**TWA**
ACGIH: [(1)]	0.01 mg/m^3 (dusts/salts); 0.002 mg/m^3 (respirable dust)
NIOSH: [(2)]	
OSHA:	0.005 mg/m^3

[(1)] Suspected carcinogen.
[(2)] Potential carcinogen. Limit exposure to lowest feasible level.

RECOMMENDATIONS FOR RESPIRATOR SELECTION: NIOSH

At any concentrations above the NIOSH REL, or where there is no REL, at any detectable concentration: SCBAF:PD,PP (any self-contained breathing apparatus that has a full facepiece and is operated in a pressure-demand or other positive-pressure mode);or SAF:PD,PP:ASCBA (any supplied-air respirator that has a full facepiece and is operated in a pressure-demand or other positive-pressure mode in combination with an auxiliary, self-contained breathing apparatus operated in a pressure-demand or other positive-pressure mode).

Escape: HiEF (any air-purifying, full-facepiece respirator with a high-efficiency particulate filter);or SCBAE (any appropriate escape-type, self-contained breathing apparatus).

CHEMICAL REACTIVITY

Cadmium compounds may react with strong oxidizers; elemental sulfur, selenium, and tellurium.

FIRE INFORMATION

Not flammable. Decomposition products upon heating include toxic cadmium fumes. If material or contaminated runoff enters waterways, notify downstream users of potentially contaminated waters. Notify local health and fire officials and pollution control agencies. Use extinguishing agents appropriate for surrounding fire. From a secure explosion-proof location, use water spray to cool exposed containers.

FIRST AID

Move victim to fresh air. Call emergency medical care. Apply artificial respiration if victim is not breathing. Do not use mouth-to-mouth method if victim ingested or inhaled the substance; induce artificial respiration with the aid of a pocket mask equipped with a one-way valve or other proper respiratory medical device. Administer oxygen if breathing is difficult. Remove and isolate contaminated clothing and shoes. In case of contact with substance, immediately flush skin or eyes with running water for at least 20 minutes. For minor skin contact, avoid spreading material on unaffected skin. Keep victim warm and quiet. Effects of exposure (inhalation, ingestion, or skin contact) may be delayed. Ensure that medical personnel are aware of the material(s) involved and take precautions to protect themselves.

ENVIRONMENTAL INFORMATION as cadmium compounds

Clean Air Act: Hazardous Air Pollutants (Title I, Part A, Section 112). Includes any unique chemical substance that contains cadmium as part of that chemical's infrastructure.
Clean Water Act: Toxic Pollutant (Section 401.15)
RCRA Section 261 Hazardous Constituents, waste number not listed.
EPCRA (Section 313): Includes any unique chemical substance that contains cadmium as part of that chemical's infrastructure. Form R *de minimis* concentration reporting level: (inorganic compounds: 0.1%;organic compounds: 1.0%). Form R Toxic Chemical Category Code: N078
Marine Pollutant (49CFR, Subchapter 172.101, Appendix B): Severe pollutant, as cadmium compounds
California's Proposition 65: Carcinogen as cadmium compounds.

CADMIUM IODIDE C:012

SYNONYMS: CADMIUMJODIDE (Dutch); EINECS No. 232-223-6; IODURE de CADMIUM (French); YUDURO de CADMIO (Spanish)

IDENTIFICATION

CAS:	7990-80-9
ERG ID:	UN 2570
ERG Guide:	154
ERG Guide Category:	SUBSTANCES-TOXIC AND/OR CORROSIVE (NONCOMBUSTIBLE)
Molecular Formula:	CdI_2
RTECS No.:	EV1290000
Properties:	Powder. Pale yellow.

HEALTH AND SAFETY INFORMATION

Corrosive to eyes, skin, gastrointestinal tract, and respiratory tract. Inhalation may cause pulmonary edema, which can be delayed for several hours; there is a risk of death in serious cases. High or prolonged exposure can cause kidney, blood damage. Can affect sense of smell.

IARC Evaluation: Probable carcinogen, Group 2A; limited human evidence; sufficient animal evidence.
NTP: Anticipated carcinogen.
IDLH: 9 mg/m^3 as cadmium dust or fume; potential carcinogen.
OSHA Table Z-1 Air Contaminant (see cadmium standard 1910.1027)
OSHA Table Z-2 Air Contaminant (see cadmium standard 1910.1027)

Exposure Limits:	**TWA**
ACGIH: (1)	0.01 mg/m^3 (dusts/salts); 0.002 mg/m^3 (respirable dust)

NIOSH: [(2)]
OSHA: 0.005 mg/m^3

[(1)] Suspected carcinogen.
[(2)] Potential carcinogen. Limit exposure to lowest feasible level.

RECOMMENDATIONS FOR RESPIRATOR SELECTION: NIOSH

At any concentrations above the NIOSH REL, or where there is no REL, at any detectable concentration: SCBAF:PD,PP (any self-contained breathing apparatus that has a full facepiece and is operated in a pressure-demand or other positive-pressure mode);or SAF:PD,PP:ASCBA (any supplied-air respirator that has a full facepiece and is operated in a pressure-demand or other positive-pressure mode in combination with an auxiliary, self-contained breathing apparatus operated in a pressure-demand or other positive-pressure mode).

Escape: HiEF (any air-purifying, full-facepiece respirator with a high-efficiency particulate filter);or SCBAE (any appropriate escape-type, self-contained breathing apparatus).

CHEMICAL REACTIVITY

Cadmium compounds may react with strong oxidizers; elemental sulfur, selenium, and tellurium.

FIRE INFORMATION

Not flammable. Decomposition products upon heating include toxic cadmium fumes. If material or contaminated runoff enters waterways, notify downstream users of potentially contaminated waters. Notify local health and fire officials and pollution control agencies. Use extinguishing agents appropriate for surrounding fire. From a secure explosion-proof location, use water spray to cool exposed containers.

FIRST AID

Move victim to fresh air. Call emergency medical care. Apply artificial respiration if victim is not breathing. Do not use mouth-to-mouth method if victim ingested or inhaled the substance; induce artificial respiration with the aid of a pocket mask equipped with a one-way valve or other proper respiratory medical device. Administer oxygen if breathing is difficult. Remove and isolate contaminated clothing and shoes. In case of contact with substance, immediately flush skin or eyes with running water for at least 20 minutes. For minor skin contact, avoid spreading material on unaffected skin. Keep victim warm and quiet. Effects of exposure (inhalation, ingestion, or skin contact) may be delayed. Ensure that medical personnel are aware of the material(s) involved and take precautions to protect themselves.

ENVIRONMENTAL INFORMATION as cadmium compounds

Clean Air Act: Hazardous Air Pollutants (Title I, Part A, Section 112). Includes any unique chemical substance that contains cadmium as part of that chemical's infrastructure.
Clean Water Act: Toxic Pollutant (Section 401.15).
RCRA Section 261 Hazardous Constituents., waste number not listed.

EPCRA (Section 313): Includes any unique chemical substance that contains cadmium as part of that chemical's infrastructure. Form R *de minimis* concentration reporting level: (inorganic compounds: 0.1%;organic compounds: 1.0%). Form R Toxic Chemical Category Code: N078.
Marine Pollutant (49CFR, Subchapter 172.101, Appendix B): Severe pollutant, as cadmium compounds.
California's Proposition 65: Carcinogen as cadmium compounds.

CADMIUM OXIDE C:014

SYNONYMS: ASKA RID; CADMIUM FUME; CADMIUM MONOXIDE; CADMIUMOXID (German); CADMIUM OXIDE BROWN; CADMIUM OXIDE FUME; CASWELL No. 136AA; EEC No. 048-002-00-0; EINECS No. 215-146-2; KADMU TLENEK (Polish); NCI-C02551; OSSIDO di CADMIO (Italian); OXIDO de CADMIO (Spanish); OXYDE de CADMIUM (French)

IDENTIFICATION

CAS:	1306-19-0
ERG ID:	UN 2570
ERG Guide:	154
ERG Guide Category:	SUBSTANCES-TOXIC AND/OR CORROSIVE (NONCOMBUSTIBLE)
Molecular Formula:	CdO
RTECS No.:	EV1930000
Properties:	Brown-red to yellow-brown crystalline solid or amorphous powder. Odorless. Sinks in water; insoluble.

HEALTH AND SAFETY INFORMATION

Corrosive to eyes, skin, gastrointestinal tract, and respiratory tract. Inhalation may cause pulmonary edema, which can be delayed for several hours; there is a risk of death in serious cases. High or prolonged exposure can cause kidney, blood damage. Can affect sense of smell.

IARC Evaluation: Probable carcinogen, Group 2A; limited human evidence; sufficient animal evidence.
NTP: Anticipated carcinogen.
IDLH: 9 mg/m^3 as cadmium dust or fume; potential carcinogen
OSHA Table Z-1 Air Contaminant (see cadmium standard 1910.1027)
OSHA Table Z-2 Air Contaminant (see cadmium standard 1910.1027)

Exposure Limits:	**TWA**
ACGIH: (1)	0.01 mg/m^3 (dusts/salts); 0.002 mg/m^3 (respirable dust)
NIOSH: (2)	
OSHA:	0.005 mg/m^3

(1) Suspected carcinogen.
(2) Potential carcinogen. Limit exposure to lowest feasible level.

RECOMMENDATIONS FOR RESPIRATOR SELECTION: NIOSH

At any concentrations above the NIOSH REL, or where there is no REL, at any detectable concentration: SCBAF:PD,PP (any self-contained breathing apparatus that has a full facepiece and is operated in a pressure-demand or other positive-pressure mode);or SAF:PD,PP:ASCBA (any supplied-air respirator that has a full facepiece and is operated in a pressure-demand or other positive-pressure mode in combination with an auxiliary, self-contained breathing apparatus operated in a pressure-demand or other positive-pressure mode).

Escape: HiEF (any air-purifying, full-facepiece respirator with a high-efficiency particulate filter);or SCBAE (any appropriate escape-type, self-contained breathing apparatus).

CHEMICAL REACTIVITY

May form an explosive mixture with magnesium and heat. May react with sulfur, selenium, and zinc. Acid contact forms hydrogen gas.

FIRE INFORMATION

Not flammable. Decomposition products upon heating include toxic cadmium fumes. If material or contaminated runoff enters waterways, notify downstream users of potentially contaminated waters. Notify local health and fire officials and pollution control agencies. Use extinguishing agents appropriate for surrounding fire. From a secure explosion-proof location, use water spray to cool exposed containers.

FEMA Hazard Classifications:

Health Hazards (Blue) 2
Flammability (Red) 0
Reactivity (Yellow) 0

FIRST AID

Move victim to fresh air. Call emergency medical care. Apply artificial respiration if victim is not breathing. Do not use mouth-to-mouth method if victim ingested or inhaled the substance; induce artificial respiration with the aid of a pocket mask equipped with a one-way valve or other proper respiratory medical device. Administer oxygen if breathing is difficult. Remove and isolate contaminated clothing and shoes. In case of contact with substance, immediately flush skin or eyes with running water for at least 20 minutes. For minor skin contact, avoid spreading material on unaffected skin. Keep victim warm and quiet. Effects of exposure (inhalation, ingestion, or skin contact) may be delayed. Ensure that medical personnel are aware of the material(s) involved and take precautions to protect themselves.

ENVIRONMENTAL INFORMATION as cadmium compounds

Clean Air Act: Hazardous Air Pollutants (Title I, Part A, Section 112). Includes any unique chemical substance that contains cadmium as part of that chemical's infrastructure.
EPCRA Section 304: Reportable Quantity (RQ): EHS, 1 lb (0.454 kg).

EPCRA Section 313: Form R *de minimis* concentration reporting level: 0.1%. Form R Toxic Chemical Category Code: N078
EPCRA Section 302 Extremely Hazardous Substances: TPQ = 10/10,000 lb (4.54/4,540 kg)
Marine Pollutant (49CFR, Subchapter 172.101, Appendix B): Severe pollutant, as cadmium compounds
California's Proposition 65: Carcinogen as cadmium compounds

CADMIUM PERCHLORATE C:015

SYNONYMS: PERCLORATO de CADMIO (Spanish)

IDENTIFICATION

CAS:	10326-28-0
ERG ID:	UN 2570
ERG Guide:	154
ERG Guide Category:	SUBSTANCES-TOXIC AND/OR CORROSIVE (NONCOMBUSTIBLE)
Molecular Formula:	$Cd_2Cl_2O_8 \cdot {}_6H_2O$
RTECS No.:	CP1010417
Properties:	Damp crystals. White. Odorless. Soluble in water.

HEALTH AND SAFETY INFORMATION

Corrosive to eyes, skin, gastrointestinal tract, and respiratory tract. Inhalation may cause pulmonary edema, which can be delayed for several hours; there is a risk of death in serious cases. High or prolonged exposure can cause kidney, blood damage. Can affect sense of smell.

IARC Evaluation: Probable carcinogen, Group 2A; limited human evidence; sufficient animal evidence.
NTP: Anticipated carcinogen.
IDLH: 9 mg/m^3 as cadmium dust or fume; potential carcinogen.
OSHA Table Z-1 Air Contaminant (see cadmium standard 1910.1027).
OSHA Table Z-2 Air Contaminant (see cadmium standard 1910.1027)

Exposure Limits:	**TWA**
ACGIH: (1)	0.01 mg/m^3 (dusts/salts); 0.002 mg/m^3 (respirable dust)
NIOSH: (2)	
OSHA:	0.005 mg/m^3

(1) Suspected carcinogen.
(2)* Potential carcinogen. Limit exposure to lowest feasible level.

RECOMMENDATIONS FOR RESPIRATOR SELECTION: NIOSH

At any concentrations above the NIOSH REL, or where there is no REL, at any detectable

concentration: SCBAF:PD,PP (any self-contained breathing apparatus that has a full facepiece and is operated in a pressure-demand or other positive-pressure mode);or SAF:PD,PP:ASCBA (any supplied-air respirator that has a full facepiece and is operated in a pressure-demand or other positive-pressure mode in combination with an auxiliary, self-contained breathing apparatus operated in a pressure-demand or other positive-pressure mode).

Escape: HiEF (any air-purifying, full-facepiece respirator with a high-efficiency particulate filter);or SCBAE (any appropriate escape-type, self-contained breathing apparatus).

CHEMICAL REACTIVITY

Cadmium compounds may react with strong oxidizers, elemental sulfur, selenium, and tellurium. May cause fire in contact with organic matter, wood, paper, etc., especially when hot.

FIRE INFORMATION

Not flammable. Decomposition products upon heating include toxic cadmium fumes. If material or contaminated runoff enters waterways, notify downstream users of potentially contaminated waters. Notify local health and fire officials and pollution control agencies. Use extinguishing agents appropriate for surrounding fire. From a secure explosion-proof location, use water spray to cool exposed containers.

FIRST AID

Move victim to fresh air. Call emergency medical care. Apply artificial respiration if victim is not breathing. Do not use mouth-to-mouth method if victim ingested or inhaled the substance; induce artificial respiration with the aid of a pocket mask equipped with a one-way valve or other proper respiratory medical device. Administer oxygen if breathing is difficult. Remove and isolate contaminated clothing and shoes. In case of contact with substance, immediately flush skin or eyes with running water for at least 20 minutes. For minor skin contact, avoid spreading material on unaffected skin. Keep victim warm and quiet. Effects of exposure (inhalation, ingestion, or skin contact) may be delayed. Ensure that medical personnel are aware of the material(s) involved and take precautions to protect themselves.

ENVIRONMENTAL INFORMATION as cadmium compounds

Clean Air Act: Hazardous Air Pollutants (Title I, Part A, Section 112). Includes any unique chemical substance that contains cadmium as part of that chemical's infrastructure.
Clean Water Act: Toxic Pollutant (Section 401.15).
RCRA Section 261 Hazardous Constituents, waste number not listed.
EPCRA Section 313: Includes any unique chemical substance that contains cadmium as part of that chemical's infrastructure. Form R *de minimis* concentration reporting level: (inorganic compounds: 0.1%;organic compounds: 1.0%). Form R Toxic Chemical Category Code: N078.
Marine Pollutant (49CFR, Subchapter 172.101, Appendix B): Severe pollutant, as cadmium compounds.
California's Proposition 65: Carcinogen as cadmium compounds.

CADMIUM STEARATE C:016

SYNONYMS: ALAIXOL II; CADMIUM DISTEARATE; CADMIUM(II) STEARATE; CADMIUM OCTADECANOATE; EINECS No. 218-743-6; ESTEARATO de CADMIO (Spanish); KADMIUMSTEARAT (German); OCTADECANOIC ACID, CADMIUM SALT; STABILISATOR SCD; STABILIZER SCD; STEARIC ACID, CADMIUM SALT; SYNPRO® CADMIUM STEARATE

IDENTIFICATION

CAS:	2223-93-0
ERG ID:	UN 2570
ERG Guide:	154
ERG Guide Category:	SUBSTANCES-TOXIC AND/OR CORROSIVE (NONCOMBUSTIBLE)
Molecular Formula:	$C_{36}H_{72}O_4{\bullet}Cd$
RTECS No.:	RG1050000

HEALTH AND SAFETY INFORMATION

Corrosive to eyes, skin, gastrointestinal tract, and respiratory tract. Inhalation may cause pulmonary edema, which can be delayed for several hours; there is a risk of death in serious cases. High or prolonged exposure can cause kidney, blood damage. Can affect sense of smell.

IARC Evaluation: Probable carcinogen. Group 2A; limited human evidence; sufficient animal evidence.
NTP: Anticipated carcinogen.
IDLH: 9 mg/m^3 as cadmium dust or fume; potential carcinogen.
OSHA Table Z-1 Air Contaminant (see cadmium standard 1910.1027)
OSHA Table Z-2 Air Contaminant (see cadmium standard 1910.1027)

Exposure Limits:	**TWA**
ACGIH: (1)	0.01 mg/m^3 (dusts/salts); 0.002 mg/m^3 (respirable dust)
NIOSH: (2)	
OSHA:	0.005 mg/m^3

(1) Suspected carcinogen.
(2) Potential carcinogen. Limit exposure to lowest feasible level.

RECOMMENDATIONS FOR RESPIRATOR SELECTION: NIOSH

At any concentrations above the NIOSH REL, or where there is no REL, at any detectable concentration: SCBAF:PD,PP (any self-contained breathing apparatus that has a full facepiece and is operated in a pressure-demand or other positive-pressure mode);or SAF:PD,PP:ASCBA (any supplied-air respirator that has a full facepiece and is operated in a pressure-demand or other positive-pressure mode in combination with an auxiliary, self-contained breathing apparatus operated in a pressure-demand or other positive-pressure mode).

Escape: HiEF (any air-purifying, full-facepiece respirator with a high-efficiency particulate filter);or SCBAE (any appropriate escape-type, self-contained breathing apparatus).

CHEMICAL REACTIVITY

Cadmium compounds may react with strong oxidizers; elemental sulfur, selenium, and tellurium. Forms impact-sensitive explosive with potassium.

FIRE INFORMATION:

Not flammable. Decomposition products upon heating include toxic cadmium fumes. If material or contaminated runoff enters waterways, notify downstream users of potentially contaminated waters. Notify local health and fire officials and pollution control agencies. Use extinguishing agents appropriate for surrounding fire. From a secure explosion-proof location, use water spray to cool exposed containers.

FEMA Hazard Classifications:

Health Hazards (Blue) 2
Flammability (Red) 0
Reactivity (Yellow) 0

FIRST AID

Move victim to fresh air; call emergency medical care. Apply artificial respiration if victim is not breathing. Do not use mouth-to-mouth method if victim ingested or inhaled the substance; induce artificial respiration with the aid of a pocket mask equipped with a one-way valve or other proper respiratory medical device. Administer oxygen if breathing is difficult. Remove and isolate contaminated clothing and shoes. In case of contact with substance, immediately flush skin or eyes with running water for at least 20 minutes. For minor skin contact, avoid spreading material on unaffected skin. Keep victim warm and quiet. Effects of exposure (inhalation, ingestion, or skin contact) to substance may be delayed. Ensure that medical personnel are aware of the material(s) involved and take precautions to protect themselves.

ENVIRONMENTAL INFORMATION

Clean Air Act: Hazardous Air Pollutants (Title I, Part A, Section 112). Includes any unique chemical substance that contains cadmium as part of that chemical's infrastructure.
Clean Water Act: Section 311 Hazardous Substances/RQ (same as CERCLA); Section 313 Priority Chemicals.
EPCRA Section 302 Extremely Hazardous Substances: TPQ = 1,000/10,000 lb (454/4,540 kg). Reportable Quantity (RQ): CERCLA, 10 lb (4.54 kg).
EPCRA Sectio EPCRA Section 304: n 313: Form R *de minimis* concentration reporting level: 0.1%. Form R Toxic Chemical Category Code: N078.
Marine Pollutant (49CFR, Subchapter 172.101, Appendix B).: Severe pollutant, as cadmium compounds.
California's Proposition 65: Carcinogen as cadmium compounds.

CALCIUM CARBIDE C:018

SYNONYMS: ACETYLENOGEN; ACTYLENOGEN, CALCIUM ACETYLIDE; CARBURO CALCICO (Spanish); CALCIUM DICARBIDE; CARBIDE; CARBIDE, ACETYLENOGEN; EEC No. 006-004-00-9

IDENTIFICATION

CAS:	75-20-7
ERG ID:	UN 1402
ERG Guide:	138
ERG Guide Category:	WATER-REACTIVE (EMITTING FLAMMABLE GASES)
Molecular Formula:	C_2Ca
RTECS No.:	EV9400000
Properties:	Highly flammable, hygroscopic powder or granules. Grey to bluish black. Faint, garliclike odor. Sinks in water; reacts violently; bubbles appear on surface as flammable acetylene gas and toxic calcium hydroxide is produced.

HEALTH AND SAFETY INFORMATION

Severely irritating to skin, eyes, respiratory tract, and lungs. Skin and eye contact can cause severe burns and blindness. Inhalation may cause pulmonary edema, which can be delayed for several hours; there is a risk of death in serious cases.

RECOMMENDATIONS FOR RESPIRATOR SELECTION:

At any concentrations above the NIOSH REL, or where there is no REL, at any detectable concentration: SCBAF:PD,PP (any self-contained breathing apparatus that has a full facepiece and is operated in a pressure-demand or other positive-pressure mode);or SAF:PD,PP:ASCBA (any supplied-air respirator that has a full facepiece and is operated in a pressure-demand or other positive-pressure mode in combination with an auxiliary, self-contained breathing apparatus operated in a pressure-demand or other positive-pressure mode).

Escape: GMFOV [any air-purifying, full-facepiece respirator (gas mask) with a chin-style, front- or back-mounted organic vapor canister];or SCBAE (any appropriate escape-type, self-contained breathing apparatus).

CHEMICAL REACTIVITY

Reacts violently with water forming highly flammable acetylene gas (which may ignite spontaneously), and toxic calcium hydroxide. Acetylene gas may be contaminated with phosgene. Reacts with copper and brass to form explosive compound. Incompatible with hydrogen chloride (gas), selenium. magnesium, silver nitrate, sodium peroxide. Flammable gas formed in moist air.

FIRE INFORMATION

Firefighting gear (including SCBA) does not provide adequate protection. If exposure occurs, remove and isolate gear immediately and thoroughly decontaminate personnel. Vapors are heavier than air and will collect in low areas. Vapors in confined areas may explode when exposed to fire. Vapors may travel long distances to ignition sources and flash back. Storage containers and parts of containers may rocket great distances, in many directions. If material or contaminated runoff enters waterways, notify downstream users of potentially contaminated waters. Notify local health and fire officials and pollution control agencies. Do not use water or water-based extinguishers. Use dry sand, dry chemicals, lime, soda ash. From a secure explosion-proof location, use water spray to cool exposed containers.

NFPA Hazard Classifications:

Health Hazards (Blue) 3
Flammability (Red) 3
Reactivity (Yellow) 2
Special Situations . . . . WATER

FIRST AID

Move victim to fresh air. Call emergency medical care. Apply artificial respiration if victim is not breathing. Administer oxygen if breathing is difficult. Remove and isolate contaminated clothing and shoes. In case of contact with substance, immediately flush skin or eyes with running water for at least 20 minutes. Keep victim warm and quiet. Ensure that medical personnel are aware of the material(s) involved and take precautions to protect themselves.

ENVIRONMENTAL INFORMATION

Clean Water Act: Section 311 Hazardous Substances/RQ (same as CERCLA)
EPCRA Section 304 Reportable Quantity (RQ): CERCLA, 10 lb (4.54 kg)

CALCIUM CARBONATE C:020

SYNONYMS: AEROMATT; AGRICULTURE LIMESTONE; AGSTONE; AMICAL SC; AMICAL 101; ARAGONITE; ATOMIT; ATOMITE; BELL MINE PULVERIZED LIMESTONE; CALBUX; CALCITE; CALCIUM SALT OF CARBONIC ACID; CALCIUM SALT (1:1); CALCIUM(II) CARBONATE (1:1); CALCIUMCARBONAAT (Dutch); CALOFIL; CAMEL-FIL; CAMEL FINE; CARBITAL 35; CARBITAL 50; CARBITAL; CARBONATO CALCICO (Spanish); CARBONIC ACID, CALCIUM SALT (1:1); CARBONIC ACID; CC-103; CHALK; CP FILLER; DOLOMITE; DRIKALITE; DURAMITE; EINECS No. 207-439-9; FRANKLIN VATERITE; HALLCOTE; KOTAMITE; LIMESTONE; LITHOGRAPHIC STONE; MACROMITE; MARBLE; MARBLEMITE; MICROWHITE; NATURAL CALCIUM CARBONATE; NUMBER 1 WHITE; NUMBER 3 WHITE; OPACICOAT; PORTLAND STONE; PROTEXULATE; SC-53; SNOWFLAKE WHITE; SOCAL; SOHNHOFEN STONE; SUPER-P FLEX 100; SUPERCOAT; SUPERMITE; TRUCAL; TRUCARB; ULTRAMITE

IDENTIFICATION

CAS: 471-34-1; 1317-65-3 (Limestone)
ERG ID: Not listed
Molecular Formula: $CCaO_3$
RTECS No.: EV9580000
Properties: Noncombustible powder or crystalline solid. White, but may be colored. Odorless. Inorganic salt. Practically insoluble in water. Occurs in nature as limestone, chalk, marble, dolomite, aragonite, calcite, and oyster shells.

HEALTH AND SAFETY INFORMATION

Irritates the eyes and respiratory tract. Quartz contamination of limestone in excess of 2% may pose a silicosis risk.

OSHA Table Z-1-A Air Contaminant

Exposure Limits: **TWA**

ACGIH:[1] 10 mg/m^3 [1]
NIOSH:[1] 10 mg/m^3 total dust; 5 mg/m^3 respirable fraction
OSHA: 15 mg/m^3 total dust; 5 mg/m^3 respirable fraction

[1] The value is for inhalable (total) particulate matter containing no asbestos and <1% crystalline silica.

RECOMMENDATIONS FOR RESPIRATOR SELECTION: NIOSH

At any concentrations above the NIOSH REL, or where there is no REL, at any detectable concentration: SCBAF:PD,PP (any self-contained breathing apparatus that has a full facepiece and is operated in a pressure-demand or other positive-pressure mode);or SAF:PD,PP:ASCBA (any supplied-air respirator that has a full facepiece and is operated in a pressure-demand or other positive-pressure mode in combination with an auxiliary, self-contained breathing apparatus operated in a pressure-demand or other positive-pressure mode).

Escape: GMFOV [any air-purifying, full-facepiece respirator (gas mask) with a chin-style, front- or back-mounted organic vapor canister] or SCBAE (any appropriate escape-type, self-contained breathing apparatus).

CHEMICAL REACTIVITY

Incompatible with acids, ammonium salts, and fluorine.

FIRE INFORMATION

Noncombustible. If material or contaminated runoff enters waterways, notify downstream users of potentially contaminated waters. Notify local health and fire officials and pollution control agencies. Use any extinguishing agents suitable for surrounding fire. From a secure explosion-proof location, use water spray to cool exposed containers.

FIRST AID

Move victim to fresh air. Call emergency medical care. Apply artificial respiration if victim is not breathing. Administer oxygen if breathing is difficult. Remove and isolate contaminated clothing and shoes. In case of contact with substance, immediately flush skin or eyes with running water for at least 20 minutes. Wash skin with soap and water. Keep victim warm and quiet. Effects of exposure (inhalation, ingestion, or skin contact) to substance may be delayed. Ensure that medical personnel are aware of the material(s) involved and take precautions to protect themselves.

CALCIUM CHLORITE C:045

SYNONYMS: CHLOROUS ACID, CALCIUM SALT; CLORITO CALCICO (Spanish)

IDENTIFICATION

CAS:	14674-72-7
ERG ID:	UN 1453
ERG Guide:	140
ERG Guide Category:	OXIDIZERS
Molecular Formula:	$CaCl_2O_4$
RTECS No.:	EV9850000
Properties:	Noncombustible solid. White. Poisonous fumes emitted when heated.

HEALTH AND SAFETY INFORMATION

Irritates eyes, skin, and respiratory system.

RECOMMENDATIONS FOR RESPIRATOR SELECTION

At any concentrations above the NIOSH REL, or where there is no REL, at any detectable concentration: SCBAF:PD,PP (any self-contained breathing apparatus that has a full facepiece and is operated in a pressure-demand or other positive-pressure mode);or SAF:PD,PP:ASCBA (any supplied-air respirator that has a full facepiece and is operated in a pressure-demand or other positive-pressure mode in combination with an auxiliary, self-contained breathing apparatus operated in a pressure-demand or other positive-pressure mode).

Escape: GMFOV [any air-purifying, full-facepiece respirator (gas mask) with a chin-style, front- or back-mounted organic vapor canister];or SCBAE (any appropriate escape-type, self-contained breathing apparatus).

CHEMICAL REACTIVITY

A strong oxidizer; contact with flammable materials, reducing agents may cause fire and explosions. Contact with acids or chlorine forms explosive chlorine dioxide gas. Under moist conditions, contact with sulfur or phosphorus can cause fire and explosions. Ignites on contact with potassium thiocyanate.

FIRE INFORMATION

Noncombustible, but will increase the intensity of an existing fire. Decomposition products upon heating include toxic chlorine fumes. If material or contaminated runoff enters waterways, notify downstream users of potentially contaminated waters. Notify local health and fire officials and pollution control agencies. Use any extinguishing agent suitable for surrounding fire. From a secure explosion-proof location, use water spray to cool exposed containers.

NFPA Hazard Classifications: Not listed

FIRST AID

Move victim to fresh air. Call emergency medical care. Apply artificial respiration if victim is not breathing. Administer oxygen if breathing is difficult. Remove and isolate contaminated clothing and shoes. In case of contact with substance, immediately flush skin or eyes with running water for at least 20 minutes. Keep victim warm and quiet. Ensure that medical personnel are aware of the material(s) involved and take precautions to protect themselves.

CALCIUM HYPOCHLORITE C:060

SYNONYMS: B-K POWDER; BLEACHING POWDER; CALCIUM CHLOROHYDROCHLORITE; CAL HYPO; CALCIUM HYPOCHLORIDE; CALCIUM OXYCHLORIDE; CAPORIT; CCH; CHLORIDE OF LIME; CHLORINATED LIME; EEC No. 017-012-00-7; EINECS No. 231-908-7; HIPOCLORITO CALCICO (Spanish); HTH; HTH DRY CHLORINE; HY-CHLOR; HYPOCHLOROUS ACID, CALCIUM SALT; HYPORIT; INDUCLOR; IPOCLORITO di CALCIO (Italian); LIME CHLORIDE; LO-BAX; LOSANTIN; NEUTRAL ANHYDROUS CALCIUM HYPOCHLORITE; PERCHLORON; PITTCIDE; PITTCLOR; PITTCHLOR; PITTABS; PRESTOCHLOR; PULSAR; SENTRY; STELLOS; SWIM CLEAR

IDENTIFICATION

CAS:	7778-54-3
ERG ID:	UN 1748 (dry); UN 2208 (mixture, dry, with more than 10% but not more than 39% available chlorine); UN 2880 (hydrated)
ERG Guide:	140 (dry): 140 (mixture, dry with more than 10%, but not more than 39% available chlorine); 140 (hydrated)
ERG Guide Category:	OXIDIZERS
Molecular Formula:	$CaCl_2O_2$

RTECS No.: NH3485000

Properties: Powder, granules, pellets, or solid tablets or sticks; also available as a water solution. White solid, yellowish liquid. Pungent, household bleaching powder odor; chlorinelike. Sinks and mixes with water. Poisonous gases are produced when heated.

HEALTH AND SAFETY INFORMATION

Corrosive to the eyes, skin, and respiratory tract. Inhalation can cause pulmonary edema, which can be delayed for several hours; there is a risk of death in serious cases. Prolonged or repeated skin contact may cause blisters (vesicular eruptions), eczematoid dermatitis, bronchitis, chronic cough, and shortness of breath..

IDLH: 10 ppm as chlorine
IARC Evaluation: Group 3, unclassifiable substance; no human data; animal evidence, inadequate.
Exposure Limits: as chlorine gas (a breakdown product)

	TWA	CEILING	STEL
ACGIH:	0.5 ppm; 1.5 mg/m3		1 ppm; 2.9 mg/m^3
NIOSH:		0.5 ppm;1.45 mg/m^3 [15 min]	
OSHA:		1 ppm; 3 mg/m^3	

RECOMMENDATIONS FOR RESPIRATOR SELECTION: NIOSH/OSHA as chlorine

5 ppm: CCRS (any chemical cartridge respirator with cartridge(s) providing protection against the compound of concern);or SA (any supplied-air respirator).
12.5 ppm: SA:CF (any supplied-air respirator operated in a continuous-flow mode);or PAPRS (any powered, air-purifying respirator with cartridge(s) providing protection against the compound of concern);or CCRFS (any chemical cartridge respirator with a full facepiece and cartridge(s) providing protection against the compound of concern);or GMFS [any air-purifying, full-facepiece respirator (gas mask) with a chin-style, front- or back-mounted canister providing protection against the compound of concern];or SCBAF (any self-contained breathing apparatus with a full facepiece);or SAF (any supplied-air respirator with a full facepiece).

Emergency or Planned Entry into Unknown Concentrations or IDLH Conditions:
SCBAF:PD,PP (any self-contained breathing apparatus that has a full facepiece and is operated in a pressure-demand or other positive-pressure mode);or SAF:PD,PP:ASCBA (any supplied-air respirator that has a full facepiece and is operated in a pressure-demand or other positive-pressure mode in combination with an auxiliary, self-contained breathing apparatus operated in a pressure-demand or other positive-pressure mode).

Escape: GMFS [any air-purifying, full-facepiece respirator (gas mask) with a chin-style, front- or back-mounted canister providing protection against the compound of concern;or SCBAE (any appropriate escape-type, self-contained breathing apparatus).
Note: Substance reported to cause eye irritation or damage; may require eye protection.

CHEMICAL REACTIVITY

A strong oxidizer. Decomposes in heat or sunlight [decomposes at 212 °F (100 °C)]. Fire, explosions, or release of toxic gases may be caused by contact with acids, moisture, reducing agents, combustible materials, all other chemicals, especially acetylene, aniline and all other amines, anthracene, carbon tetrachloride, iron oxide, manganese oxide, mercaptans, diethylene glycol monomethyl ether, nitromethane, organic matter, organic sulfides, phenol, 1-propanethiol, propyl mercaptan, sulfur, turpentine, and organic sulfur compounds. Reacts with acids forming toxic and corrosive chlorine.

FIRE INFORMATION

Will increase the intensity of existing fires. If material or contaminated runoff enters waterways, notify downstream users of potentially contaminated waters. Notify local health and fire officials and pollution control agencies. Use large quantities of water.

NFPA Hazard Classifications:

Health Hazards (Blue) 3
Flammability (Red) 0
Reactivity (Yellow) 1
Special situations . . . . OXY

FIRST AID

Move victim to fresh air. Call emergency medical care. Apply artificial respiration if victim is not breathing. Administer oxygen if breathing is difficult. Remove and isolate contaminated clothing and shoes. In case of contact with substance, immediately flush skin or eyes with running water for at least 20 minutes. Keep victim warm and quiet. Ensure that medical personnel are aware of the material(s) involved and take precautions to protect themselves.

ENVIRONMENTAL INFORMATION

Clean Water Act: Section 311 Hazardous Substances/RQ (same as CERCLA).
EPCRA Section 304: Reportable Quantity (RQ): CERCLA, 1,000 lb (454 kg).

CALCIUM STEARATE C:095

SYNONYMS: AQUACAL; CALCIUM DISTEARATE; CALSTAR; DIESTEARATO de CALCIO (Spanish); EINECS No. 216-472-8; FLEXICHEM; FLEXICHEM CS; G 339 S; NOPCOTE C 104; OCTADECANOIC ACID, CALCIUM SALT; STAVINOR 30; STEARIC ACID, CALCIUM SALT; SYNPRO® 15F; WITCO G 339S

IDENTIFICATION

CAS: 1592-23-0

RTECS No.: WI3000000

Properties: Fine powdery solid. White. Slight characteristic odor. Floats in water; insoluble.

HEALTH AND SAFETY INFORMATION

Exposure Limits: **TWA** as inert dusts, particulates, not otherwise regulated (PNOR)

OSHA: 15mg/m^3 (total); 5mg/m^3 (respirable fraction)

RECOMMENDATIONS FOR RESPIRATOR SELECTION: NIOSH

At any concentrations above the NIOSH REL, or where there is no REL, at any detectable concentration: SCBAF:PD,PP (any self-contained breathing apparatus that has a full facepiece and is operated in a pressure-demand or other positive-pressure mode);or SAF:PD,PP:ASCBA (any supplied-air respirator that has a full facepiece and is operated in a pressure-demand or other positive-pressure mode in combination with an auxiliary, self-contained breathing apparatus operated in a pressure-demand or other positive-pressure mode).

Escape: GMFOVHiE [any air-purifying, full-facepiece respirator (gas mask) with a chin-style, front- or back-mounted organic vapor canister having a high-efficiency particulate filter];or SCBAE (any appropriate escape-type, self-contained breathing apparatus).

FIRE INFORMATION

Decomposition products upon heating include toxic gases. If material or contaminated runoff enters waterways, notify downstream users of potentially contaminated waters. Notify local health and fire officials and pollution control agencies. Use and extinguishing agent appropriate for surrounding fire.

NFPA Hazard Classifications: Not listed

FIRST AID

Move victim to fresh air. Call emergency medical care. Apply artificial respiration if victim is not breathing. Administer oxygen if breathing is difficult. Remove and isolate contaminated clothing and shoes. In case of contact with substance, immediately flush skin or eyes with running water for at least 20 minutes. Wash skin with soap and water. Keep victim warm and quiet. Effects of exposure (inhalation, ingestion, or skin contact) to substance may be delayed. Ensure that medical personnel are aware of the material(s) involved and take precautions to protect themselves.

CALCIUM THIOCYANATE C:105

SYNONYMS: CALCIUM DITHIOCYANATE; CALCIUM RHODANID (German); CALCIUM SULFOCYANATE; EINECS No. 218-244-3; THIOCYANIC ACID, CALCIUM SALT (2:1); TIOCIANATO CALCICO (Spanish)

232 Calcium Thiocyanate

IDENTIFICATION

CAS:	2092-16-2
ERG ID:	UN 2811 as poisonous solid, organic, n.o.s.
ERG Guide:	154
ERG Guide Category:	SUBSTANCES-TOXIC AND/OR CORROSIVE
Molecular Formula:	$C_2N_2S_2 \cdot Ca$
RTECS No.:	XK8540000
Properties:	Crystalline solid. White. Highly soluble in water.

HEALTH AND SAFETY INFORMATION

Poisonous if swallowed. May cause irritation of eyes and respiratory system.

RECOMMENDATIONS FOR RESPIRATOR SELECTION:

At any concentrations above the NIOSH REL, or where there is no REL, at any detectable concentration: SCBAF:PD,PP (any self-contained breathing apparatus that has a full facepiece and is operated in a pressure-demand or other positive-pressure mode);or SAF:PD,PP:ASCBA (any supplied-air respirator that has a full facepiece and is operated in a pressure-demand or other positive-pressure mode in combination with an auxiliary, self-contained breathing apparatus operated in a pressure-demand or other positive-pressure mode).

Escape: GMFOV [any air-purifying, full-facepiece respirator (gas mask) with a chin-style, front- or back-mounted organic vapor canister];or SCBAE (any appropriate escape-type, self-contained breathing apparatus).

CHEMICAL REACTIVITY

May react violently with nitrates, chlorates, strong acids, peroxides, and other oxidizers.

FIRE INFORMATION

Combustion products include toxic cyanide, sulfur, and nitrogen oxides. If material or contaminated runoff enters waterways, notify downstream users of potentially contaminated waters. Notify local health and fire officials and pollution control agencies. Use any extinguishing agent appropriate for surrounding fire. From a secure explosion-proof location, use water spray to cool exposed containers.

NFPA Hazard Classifications: Not listed

FIRST AID

Move victim to fresh air. Call emergency medical care. Apply artificial respiration if victim is not breathing. Administer oxygen if breathing is difficult. Remove and isolate contaminated clothing and shoes. In case of contact with substance, immediately flush skin or eyes with running water for at least 20 minutes. Wash skin with soap and water. Keep victim warm and quiet. Effects of exposure (inhalation,

ingestion, or skin contact) to substance may be delayed. Ensure that medical personnel are aware of the material(s) involved and take precautions to protect themselves.

CAPROLACTAM C:113

SYNONYMS: AMINOCAPROIC LACTAM; epsilon-CAPROLACTAM; CAPROLACTAMA (SPANISH); 1,6-HEXANOLACTAM; 2-KETOHEXAMETHYLENIMINE; 2-OXOHEXAMETHYLENIMINE; HEXAHYDRO-2H-AZEPINE-2-ONE; NCI-C50646

IDENTIFICATION

CAS:	105-60-2
ERG ID:	UN 3082
ERG Guide:	171
ERG Guide Category:	SUBSTANCES (LOW TO MODERATE HAZARD)
Molecular Formula:	$C_6H_{11}NO$
RTECS No.:	CM3675000
Properties:	Liquid or solid crystals in solution. Colorless or white. Mild odor. Sinks and mixes with water.

HEALTH AND SAFETY INFORMATION

Irritates eyes and respiratory system. Can cause eyes to become light-sensitive. Can affect central nervous system. Skin contact can cause burns.

IARC Evaluation: IARC Group 4, probably noncarcinogenic substance. Mutation data reported.
Odor Threshold: 0.064 ppm
OSHA Table Z-1-A Air Contaminant

Exposure Limits:	**TWA**	**STEL**
ACGIH:	(Dust) 1 mg/m^3	3 mg/m^3
	(Vapor) 5 ppm; 23 mg/m^3	10 ppm; 46 mg/m^3
NIOSH:.	(Dust) 1 mg/m^3	3 mg/m^3
	(Vapor) 0.22 ppm; 1 mg/m^3	0.66 ppm; 3mg/m^3
OSHA:	None	

CHEMICAL REACTIVITY

Reacts with strong oxidizers forming toxic vapors. Incompatible with sulfuric acid, caustics, aliphatic amines, isocyanates.

FIRE INFORMATION

Decomposition products upon heating include toxic gases of nitrous oxides and anhydrous ammonia.

Vapors are heavier than air and will collect in low areas. Vapors in confined areas may explode when exposed to fire. If material or contaminated runoff enters waterways, notify downstream users of potentially contaminated waters. Notify local health and fire officials and pollution control agencies. Use water spray or dry chemicals. From a secure explosion-proof location, use water spray to cool exposed containers.

FEMA Hazard Classifications:

Health Hazards (Blue) 1
Flammability (Red) 1
Reactivity (Yellow) 0

Flash Point: 25 °F (125 °C) o.c; 283 °F (139 °C) c.c.
Flammable Limits in Air: LEL 1.4%; UEL 8.0%
Autoignition Temperature: 707°F (375 °C)

FIRST AID

Move victim to fresh air. Call emergency medical care. Apply artificial respiration if victim is not breathing. Administer oxygen if breathing is difficult. In case of contact with substance, immediately flush skin or eyes with running water for at least 20 minutes. Remove and isolate contaminated clothing and shoes. Ensure that medical personnel are aware of the material(s) involved and take precautions to protect themselves.

ENVIRONMENTAL INFORMATION

Clean Air Act: Hazardous Air Pollutants (Title I, Part A, Section 112)
EPCRA Section 304 Reportable Quantity (RQ): CERCLA, 5,000 lb (2,270 kg)
TSCA: 40 CFR 716.120(*a*) list of substances; Health and Safety Reporting List 1/1/87.

CAPTAN C:117

SYNONYMS: AMERCIDE; CAPTAF; CAPTAF 85W; CAPTANCAPTENEET 26,538; CAPTANE; CAPTEX; EEC No. 613-044-00-6; EINECS No. 205-087-0; ENT 26,538; ESSOFUNGICIDE 406; FLIT 406; FUNGUS BAN TYPE II; GLYODEX 3722; LE CAPTANE (French); MERPAN; NCI-0077;orTHOCIDE;orTHOCIDE 7.5;orTHOCIDE 50;orTHOCIDE 406; SR406; STAUFFER CAPTAN; N-TRICHLOROMETHYLMERCAPTO-4-CYCLOHEXENE-1,2-DICARBOXIMIDE; N-(TRICHLOROMETHYLMERCAPTO)-DELTA(SUP 4)-TETRAHYDROPHTHALIMIDE; N-TRICHLOROMETHYLTHIOCYCLOHEX-4-ENE-1,2-DICARBOXIMIDE; N-TRICHLOROMETHYLTHIO-CIS-DELTA(SUP4)-CYCLOHEXENE-1,2-DICARBOXIMIDE; N-((TRICHLOROMETHYL)THIO)-4-CYCLOHEXENE-1,2-DICARBOXIMIDE; N-((TRICHLOROMETHYL)THIO)TETRAHYDROPHTHALIMIDE; N-TRICHLOROMETHYLTHIO-3A,4,7,7A-TETRAHYDROPHTHALIMIDE; VANCIDE; VANCIDE 89; VANGARD K; VONDCAPTAN

IDENTIFICATION

CAS:	133-06-2
ERG ID:	UN 9188
ERG Guide:	171
ERG Guide Category:	SUBSTANCES (LOW TO MODERATE HAZARD)
Molecular Formula:	$C_9H_8Cl_3NO_2S$
RTECS No.:	GW5075000
Properties:	Combustible crystalline solid or powder; may be dissolved in a flammable liquid. White (commercial product is a yellow powder). Slight pungent odor. Sinks in water; slightly soluble.

HEALTH AND SAFETY INFORMATION

Irritating to the eyes, skin, and respiratory tract; may cause bronchitis. Skin sensitizer; allergy may develop. Possible liver and kidney problems.

IARC Evaluation: Group 3 (not classifiable)
NTP: Carcinogen
OSHA Table Z-1-A Air Contaminant

Exposure Limits:	**TWA**
ACGIH:	5 mg/m^3
NIOSH:[1]	5 mg/m^3
OSHA:	None

[1] Potential occupational carcinogen.

RECOMMENDATIONS FOR RESPIRATOR SELECTION: NIOSH

At any concentrations above the NIOSH REL, or where there is no REL, at any detectable concentration: SCBAF:PD,PP (any self-contained breathing apparatus that has a full facepiece and is operated in a pressure-demand or other positive-pressure mode);or SAF:PD,PP:ASCBA (any supplied-air respirator that has a full facepiece and is operated in a pressure-demand or other positive-pressure mode in combination with an auxiliary, self-contained breathing apparatus operated in a pressure-demand or other positive pressure mode)

Escape: GMFOV [any air-purifying, full-facepiece respirator (gas mask) with a chin-style, front- or back-mounted organic vapor canister];or SCBAE (any appropriate escape-type, self-contained breathing apparatus).

CHEMICAL REACTIVITY

Contact with water or heat may give off poisonous gas. Reacts with strong alkaline materials such as hydrated lime; corrosive to metals.

FIRE INFORMATION

Decomposition products upon heating include toxic gases. If material or contaminated runoff enters waterways, notify downstream users of potentially contaminated waters. Notify local health and fire officials and pollution control agencies. Use any extinguishing agent suitable for surrounding fire. From a secure explosion-proof location, use water spray to cool exposed containers.

FEMA Hazard Classifications:

Health Hazards (Blue) 3
Flammability (Red) 0
Reactivity (Yellow) 0

FIRST AID

Move victim to fresh air. Call emergency medical care. Apply artificial respiration if victim is not breathing. Administer oxygen if breathing is difficult. In case of contact with substance, immediately flush skin or eyes with running water for at least 20 minutes. Remove and isolate contaminated clothing and shoes. Ensure that medical personnel are aware of the material(s) involved and take precautions to protect themselves.

ENVIRONMENTAL INFORMATION

Clean Air Act: Hazardous Air Pollutants (Title I, Part A, Section 112)
Clean Water Act: Section 311 Hazardous Substances/RQ (same as CERCLA); Section 313 Priority Chemicals
EPCRA Section 304 Reportable Quantity (RQ): CERCLA, 10 lb (4.54 kg)
EPCRA Section 313: Form R *de minimis* concentration reporting level: 1.0%.
California's Proposition 65: Carcinogen.

CARBON BLACK C:120

SYNONYMS: ACETYLENE BLACK; ADDIPAST; BLACK PEARLS 1100; CHANNEL BLACK; CONTINEX LH-10; CONTINEX N351; DERUSSOLE; DIABLACK-A; ELFTEX 675; EME-X8778; FLEX CARBON; FURNEX; FURNACE BLACK; FW 18; FW200 BEADS AND POWDER; KODAK EKTAPRINT K/L DEVELOPER; LAMP BLACK; MICROSPERSE; MODULEX; MOGUL L; NEGRO de CARBON (Spanish); NEOSPECTRA; NEOTEX; OCTOJET-104; PRINTEX; PRINTEX-P; RAVEN; SEVACARB; STATEX; STERLING; SUMIKON EME 6300 (EPOXY MOLDING COMPOUND); SUMIKON EME 6300H (EPOXY MOLDING COMPOUND); SUMIKON EME 6300HC (EPOXY MOLDING COMPOUND); SUMIKON EME 8150 (EPOXY MOLDING COMPOUND); SUMIKON EME 9100X (EPOXY CRESOL NOVOLAC RESIN SYSTEM); SUMIKON EME 9300H; SUPERBA; SUPERJET; TACK; THERMAL BLACK THERMAX STAINLESS; VELVETEX; VULCAN XC72R; XEROX 1040/1045/1048 DEVELOPER; XEROX 1075/1090/4050 DEVELOPER; XEROX 1020/1035/2830 TONER; XEROX 1040/1045/1048 TONER; XEROX 1075/1090/4050 TONER; XEROX CE-PROCESS/ECP-PROCESS BLACK PREMIX

IDENTIFICATION

CAS:	1333-86-4
ERG ID:	UN 1362
ERG Guide:	133
ERG Guide Category:	FLAMMABLE SOLIDS
Molecular Formula:	C
RTECS No.:	FF5800000
Properties:	Combustible solid powder, pellets, or paste. Dark brown to black. Odorless. May float or sink in water. May contain flammable hydrocarbons.

HEALTH AND SAFETY INFORMATION

May cause irritation to eyes, skin, and respiratory system. Long term exposure may cause skin, respiratory tract, heart damage. *Note:* Employer should monitor workplace for both carbon black levels and analyze samples for PAH (polycyclic aromatic hydrocarbon) levels. Carbon black containing PAH greater than 0.1% should be considered a suspect carcinogen. Chronic exposure can cause cough, phlegm, tiredness, chest pain, and headache; lung effects, bronchitis. Certain contaminants may cause skin rashes.

IARC: Evaluation: Group 3, unclassifiable; human evidence, inadequate; animal evidence, inadequate.
IDLH: 1750 mg/m^3
OSHA Table Z-1-A Air Contaminant as carbon black extracts
NIOSH Criteria Document: NIOSH 78-204

Exposure Limits: TWA

ACGIH:	3.5 mg/m^3
NIOSH: pot. carcinogen [(1)]	3.5 mg/m^3
OSHA:	3.5 mg/m^3

[(1)] 0.1 mg PAHs/m^3 [carbon black in presence of polycyclic aromatic hydrocarbons (PAHs)]

RECOMMENDATIONS FOR RESPIRATOR SELECTION: NIOSH/OSHA

17.5 mg/m^3: DM (any dust and mist respirator)
35 mg/m^3: DMXSQ (any dust and mist respirator except single-use and quarter mask respirators);or SA (any supplied-air respirator).
87.5 mg/m^3: SA:CF (any supplied-air respirator operated in a continuous-flow mode);or PAPRDM (any powered, air-purifying respirator with a dust and mist filter).
HiEF [any air-purifying, full-facepiece respirator with a high-efficiency particulate filter)];or PAPRTHiE (any powered, air-purifying respirator with a tight-fitting facepiece and a high-efficiency particulate filter);or SCBAF (any self-contained breathing apparatus with a full facepiece);or SAF (any supplied-air respirator with a full facepiece);or SA:PD,PP (any supplied-air respirator operated in a pressure-demand or other positive-pressure mode).

Emergency or planned entry into unknown concentrations or IDLH conditions: SCBAF:PD,PP (any self-contained breathing apparatus that has a full facepiece and is operated in a pressure-demand or other positive-pressure mode);or SAF:PD,PP:ASCBA (any supplied-air respirator that has a full facepiece and is operated in a pressure-demand or other positive-pressure mode in combination with an auxiliary, self-contained breathing apparatus operated in a pressure-demand or other positive pressure mode.

Escape: HiEF (any air-purifying, full-facepiece respirator with a high-efficiency particulate filter);or SCBAE (any appropriate escape-type, self-contained breathing apparatus).

CHEMICAL REACTIVITY

Dust can form an explosive mixture in air. Strong oxidizers such as chlorates, bromates, and nitrates may cause fire and explosions.

FIRE INFORMATION

If material or contaminated runoff enters waterways, notify downstream users of potentially contaminated waters. Notify local health and fire officials and pollution control agencies. Use any extinguishing agent suitable for surrounding fire. From a secure explosion-proof location, use water spray to cool exposed containers.

NFPA Hazard Classifications:

Health Hazards (Blue) 1
Flammability (Red) 1
Reactivity (Yellow) 0

FIRST AID

Move victim to fresh air. Call emergency medical care. Apply artificial respiration if victim is not breathing. Administer oxygen if breathing is difficult. Remove and isolate contaminated clothing and shoes. In case of contact with substance, immediately flush skin or eyes with running water for at least 20 minutes. Removal of solidified molten material from skin requires medical assistance. Keep victim warm and quiet. Ensure that medical personnel are aware of the material(s) involved and take precautions to protect themselves.

ENVIRONMENTAL INFORMATION (as polycyclic aromatic compounds, if found in carbon black)

Note: Aromatic hydrocarbons such as benzo(a)pyrene (50-32-8) are formed during the making of carbon black, and their adsorption on the carbon black could pose a cancer risk. Mixtures containing listed poycyclic aromatic compounds are reportable under EPCRA Section 313. Threshold calculations for the polycyclic aromatic compounds category should account only for the percentage of the polycyclic aromatic component(s) contained in the mixture.

EPCRA Section 313: Form R *de minimus* concentration reporting level: 0.1%. Form R Toxic Chemical Category Code for polycyclic aromatic compounds: N590.

CARBON DIOXIDE C:140

SYNONYMS: ACID GAS; ANHYDRIDE CARBONIQUE (French); BLUE SHIELD 6; BLUE SHIELD 7; BLUE SHIELD 8; CARBONIC ACID GAS; CARBONIC ANHYDRIDE; CARDICE; CHEMTRANIC FLUX OFF; DIOXIDO de CARBONO (Spanish); DRICOLD; DRY ICE (SOLID); EINECS No. 204-696-9; KOHLENDIOXYD (German); KOHLENSAURE (German); 882 LASER GAS; LASER GAS 920L; POLY I GAS; POLY II GAS

IDENTIFICATION

CAS:	124-38-9
ERG ID:	UN 1013 (gas or compressed gas); UN 1845 (solid, dry ice); UN 2187 (refrigerated liquid)
ERG Guide:	120
ERG Guide Category:	GASES-INERT (INCLUDING REFRIGERATED LIQUIDS)
Molecular Formula:	CO_2
RTECS No.:	FF6400000
Properties:	Nonflammable, liquefied compressed gas, cryogenic liquid, or white solid (dry ice; flakes or cubes) at low temperature. Colorless. Odorless; poor warning qualities. Solid sinks and "boils" in water; insoluble. Visible vapor cloud is produced. Gas is heavier than air; will collect in low areas.

HEALTH AND SAFETY INFORMATION

Irritates eyes, skin, and respiratory tract. Inhalation can cause severe breathing difficulties and unconsciousness. Can cause frostbite. Chronic exposures to increased levels of carbon dioxide in the atmosphere produced changes in blood and impaired function of the adrenal glands.

IDLH: 40,000 ppm (Atmospheric carbon dioxide above 12% can cause unconsciousness and death).
Odor Threshold: Less than 1–135,420 mg/m^3.
OSHA Table Z-1-A Air Contaminant
NIOSH Criteria Document: NIOSH 76-194

Exposure Limits:	**TWA**	**STEL**
ACGIH:	5,000 ppm; 9,000 mg/m^3	30,000 ppm; 54,000 mg/m^3
NIOSH:	5,000 ppm; 9,000 mg/m^3	30,000 ppm; 54,000 mg/m^3
OSHA:	None	

RECOMMENDATIONS FOR RESPIRATOR SELECTION: NIOSH/OSHA

40,000 ppm: SA (any supplied-air respirator);or SCBAF (any self-contained breathing apparatus with a full facepiece).

Emergency or Planned Entry into Unknown Concentrations or IDLH Conditions: SCBAF:PD,PP (any self-contained breathing apparatus that has a full facepiece and is operated in a pressure-demand or other positive-pressure mode);or SAF:PD,PP:ASCBA (any supplied-air respirator that has a full facepiece and is operated in a pressure-demand or other positive-pressure mode in combination with an auxiliary, self-contained breathing apparatus operated in a pressure-demand or other positive-pressure mode).

Escape: SCBAE (any appropriate escape-type, self-contained breathing apparatus)

CHEMICAL REACTIVITY

Contact with water forms carbonic acid. Incompatible with acrylaldehyde, amines, anhydrous ammonia, cesium monoxide, lithium, metal dusts, potassium, sodium, sodium carbide, sodium-potassium alloy, sodium peroxide, and titanium. Dusts of various metals such as magnesium, zirconium, titanium, aluminum, chromium, and manganese are ignitable and explosive when suspended in carbon dioxide. Dry ice can form shock-sensitive mixtures with sodium, potassium, or sodium-potassium alloys. Liquid or solid carbon dioxide will attack some forms of plastics, rubber, and coatings.

FIRE INFORMATION

Nonflammable. Gas is heavier than air and will collect in low areas. Storage containers and parts of containers may rocket great distances, in many directions. If material or contaminated runoff enters waterways, notify downstream users of potentially contaminated waters. Notify local health and fire officials and pollution control agencies. Use any extinguishing agent suitable for surrounding fire.

FIRST AID

Move victims to fresh air. Call emergency medical care. Apply artificial respiration if victim is not breathing. Administer oxygen if breathing is difficult. Clothing frozen to the skin should be thawed before being removed. In case of contact with liquefied gas, thaw frosted parts with lukewarm water. Keep victim warm and quiet. Ensure that medical personnel are aware of the material(s) involved and take precautions to protect themselves. *Note:* Acidosis may occur.

CARBON DISULFIDE C:160

SYNONYMS: CARBON BISULFIDE; CARBON BISULPHIDE; CARBON DISULPHIDE; CARBONE (SUFURE DE) (French); CARBONIO (SOLFURO di) (Italian); CARBON SULFIDE; DISULFURO de CARBONO (Spanish); DITHIOCARBONIC ANHYDRIDE; DWUSIARCZEK (Polish); EEC No. 006-003-00-3; EINECS No. 200-843-6; KOHLENDISULFID (SCHWEFELKOHLENSTOFF) (German);

KOOLSTOFDISULFIDE (ZWAVELKOOLSTOF) (Dutch); SCHWEFELKOHLENSTOFF (German); SULPHOCARBONIC ANHYDRIDE; SOLFURO di CARBONIO (Italian); WEEVILTOX; WEGLA

IDENTIFICATION

CAS:	75-15-0
ERG ID:	UN 1131
ERG Guide:	131
ERG Guide Category:	FLAMMABLE LIQUIDS-TOXIC
Molecular Formula:	CS_2
RTECS No.:	FF6650000
Properties:	Extremely flammable, watery liquid. Clear, colorless to faint yellow. Pure form has a sweet odor; reagent or commercial grade has an offensive odor of decaying cabbage or rotten eggs. Sinks in water; insoluble. Boiling point of 116°F (47 °C).

HEALTH AND SAFETY INFORMATION

Irritates eyes, skin, and respiratory system. Destroys skin's natural oils; may cause redness, inflammation. Can affect the central nervous system. Long-term effects may include nerve damage, unsteady gait, and speech impairment. Damage to the liver, kidney, and heart may result. Psychological and behavioral changes, including uncontrollable anger and suicidal tendencies, may also occur. Repeated pr prolonged exposure may cause fertility problems in both men and women; damage to the developing fetus; skin allergy with itching and rash; an increase in cholesterol and atherosclerosis with possible heart and other organ damage; nervous system and brain damage with various symptoms of coordination, muscle weakness, numbness, unsteady walking, difficulty in swallowing, nervousness, fatigue, irritability, and depression; mood and personality changes; insanity; and eye damage, blind spots, narrowing vision and decreased ability to see in the dark.

IDLH: 500 ppm
Odor Threshold: 0.096 ppm
OSHA Table Z-1-A Air Contaminant
OSHA Table Z-2 Air Contaminant
NIOSH Criteria Document: NIOSH 77-156

Exposure Limits:	**TWA**	**STEL**
ACGIH:	10 ppm; 31 mg/m^3	
NIOSH: [(1)]	1 ppm; 3 mg/m^3	10 ppm; 30 mg/m^3
OSHA:	None	

[(1)] Skin contact contributes significantly to overall exposure.

RECOMMENDATIONS FOR RESPIRATOR SELECTION: NIOSH

10 ppm: CCROV [any chemical cartridge respirator with organic vapor cartridge(s)];or SA (any supplied-air respirator).

25 ppm: SA:CF (any supplied-air respirator operated in a continuous-flow mode);or PAPROV [any powered, air-purifying respirator with organic vapor cartridge(s)].
50 ppm: CCRFOV [any chemical cartridge respirator with a full facepiece and organic vapor cartridge(s)];or GMFOV [any air-purifying, full-facepiece respirator (gas mask) with a chin-style, front- or back-mounted acid gas canister];or PAPRTOV [any powered, air-purifying respirator with a tight-fitting facepiece and organic vapor cartridge(s)];or SCBAF (any self-contained breathing apparatus with a full facepiece);or SAF (any supplied-air respirator with a full facepiece).
500 ppm: SA:PD,PP (any supplied-air respirator operated in a pressure-demand or other positive-pressure mode).

Emergency or Planned Entry into Unknown Concentrations or IDLH Conditions: SCBAF:PD,PP (any self-contained breathing apparatus that has a full facepiece and is operated in a pressure-demand or other positive-pressure mode);or SAF:PD,PP:ASCBA (any supplied-air respirator that has a full facepiece and is operated in a pressure-demand or other positive-pressure mode in combination with an auxiliary, self-contained breathing apparatus operated in a pressure-demand or other positive-pressure mode).

Escape: GMFOV [any air-purifying, full-facepiece respirator (gas mask) with a chin-style, front- or back-mounted acid gas canister];or SCBAE (any appropriate escape-type, self-contained breathing apparatus).

CHEMICAL REACTIVITY

Highly reactive. Contact with many substances can cause fire and explosions. Forms explosive mixture with air. Shock can cause explosive decomposition. Alkali metals, aliphatic amines, alkanolamines, aluminum, azides, chlorine monoxide, combustible substances, ethylene diamine, ethyleneimine, lead azide, lithium azide, nitric oxide, nitrogen dioxide, potassium, potassium azide, reducing agents, rubidium azide, sodium azide, and zinc. May be corrosive to metals of construction due to impurities. Liquid will attack some forms of plastics, rubber, and coatings.

FIRE INFORMATION

Firefighting gear (including SCBA) does not provide adequate protection. If exposure occurs, remove and isolate gear immediately and thoroughly decontaminate personnel. Vapors are heavier than air and will collect in low areas. Vapors in confined areas may explode when exposed to fire. Vapors may travel long distances to ignition sources and flash back. Storage containers and parts of containers may rocket great distances, in many directions. If material or contaminated runoff enters waterways, notify downstream users of potentially contaminated waters. Notify local health and fire officials and pollution control agencies. Use dry chemicals, foam, carbon dioxide. From a secure explosion-proof location, use water spray to cool exposed containers.

NFPA Hazard Classifications:

Health Hazards (Blue) 3
Flammability (Red) 4
Reactivity (Yellow) 0

Flash Point: -22°F (-30°C) c.c.
Flammable Limits in Air: LEL 1.3%; UEL 50%
Autoignition Temperature: 203°F (95°C)

FIRST AID

Move victim to fresh air. Call emergency medical care. Apply artificial respiration if victim is not breathing. Do not use mouth-to-mouth method if victim ingested or inhaled the substance; induce artificial respiration with the aid of a pocket mask equipped with a one-way valve or other proper respiratory medical device. Administer oxygen if breathing is difficult. Remove and isolate contaminated clothing and shoes. In case of contact with substance, immediately flush skin or eyes with running water for at least 20 minutes. Wash skin with soap and water. Keep victim warm and quiet. Effects of exposure (inhalation, ingestion, or skin contact) to substance may be delayed. Ensure that medical personnel are aware of the material(s) involved and take precautions to protect themselves.
Note: Carbon disulfide can be determined in expired air and in blood and urine.

ENVIRONMENTAL INFORMATION

Clean Air Act: Hazardous Air Pollutants (Title I, Part A, Section 112); Accidental Release Prevention/Flammable substances, (Section 112[r], Table 3), TQ = 20,000 lb (9,080 kg).
Clean Water Act: Section 311 Hazardous Substances/RQ (same as CERCLA); Section 313 Priority Chemicals.
EPA Hazardous Waste Number (RCRA No.): P022.
RCRA Section 261 Hazardous Constituents
RCRA Land Ban Waste
RCRA Universal Treatment Standards: Wastewater (mg/L), 3.8; Nonwastewater (mg/L), 4.8 TCLP.
RCRA Ground Water Monitoring List: Suggested test method(s) (PQL ug/L): 8240(5).
EPCRA Section 302 Extremely Hazardous Substances: TPQ = 10,000 lb (4,540 kg).
EPCRA Section 304: Reportable Quantity (RQ): CERCLA, 100 lb (45.4 kg).
EPCRA Section 313: Form R *de minimis* concentration reporting level: 1.0%.
Marine Pollutant (49CFR, Subchapter 172.101, Appendix B) as carbon bisulphide.
California's Proposition 65: Reproductive toxin (male, female).
TSCA: 40CFR 712.30(*e*)10; 40CFR 716.120 (*d)*10, chemicals in need of dermal absorption testing

CARBON MONOXIDE C:180

SYNONYMS: CARBONE (OXYDE de) (French); CARBONIC OXIDE; CARBONIO (OSSIDO di) (Italian); CARBON OXIDE; CO; EEC No. 006-001-00-2; EXHAUST GAS; FLUE GAS; KOHLENMONOXID (German); KOOLMONOXYDE (Dutch); 882 LASER GAS; LASER GAS 920L; MONOXIDE; MONOXIDO de CARBONO (Spanish); OXYDE de CARBONE (French); POLY II GAS; WEGLA TLENEK (Polish)

IDENTIFICATION

CAS: 630-08-0

ERG ID:	UN 1016; UN 9202 [refrigerated (cryogenic) liquid]
ERG Guide:	119 (gas); 168 [refrigerated (cryogenic) liquid]
ERG Guide Category:	119 GASES-TOXIC-FLAMMABLE 168 CARBON MONOXIDE (REFRIGERATED LIQUID)
Molecular Formula:	CO
RTECS No.:	FG3500000
Properties:	Flammable, colorless gas or cryogenic liquid. Odorless. Liquid floats and "boils" on water; slightly soluble; poisonous, flammable, visible vapor cloud is produced. Under certain conditions the gas is heavier than air; can collect in low areas.

HEALTH AND SAFETY INFORMATION

Affects the blood; combines with the red blood cells, making them unable to carry oxygen. This results in increased stress on the heart and circulatory system. Alcoholics, smokers, obese individuals and those with heart disease are more sensitive than a healthy person to the effects of carbon monoxide. Can cause brain damage. Skin contact with liquid may cause frostbite. Affects the central nervous system

IDLH: 1,200 ppm
Odor Threshold: 100,000 ppm
OSHA Table Z-1-A Air Contaminant
NIOSH Criteria Document: NIOSH 73-11000

Exposure Limits:	**TWA**	**CEILING**	**STEL**
ACGIH:	25 ppm; 29 mg/m^3		
NIOSH:	35 ppm; 40 mg/m^3	200 ppm; 229 mg/m^3	400 ppm, 15 min.
OSHA:	None		

RECOMMENDATIONS FOR RESPIRATOR SELECTION: NIOSH/OSHA

350 ppm: SA (any supplied-air respirator).
875 ppm: SA:CF (any supplied-air respirator operated in a continuous-flow mode).
1,500 ppm: GMFS *end of service life indicator (ESLI) required.* [any air-purifying, full-facepiece respirator (gas mask) with a chin-style, front- or back-mounted canister providing protection against the compound of concern];or SCBAF (any self-contained breathing apparatus with a full facepiece);or SAF (any supplied-air respirator with a full facepiece).

Emergency or Planned Entry into Unknown Concentrations or IDLH Conditions:
SCBAF:PD,PP (any self-contained breathing apparatus that has a full facepiece and is operated in a pressure-demand or other positive-pressure mode);or SAF:PD,PP:ASCBA (any supplied-air respirator that has a full facepiece and is operated in a pressure-demand or other positive-pressure mode in combination with an auxiliary, self-contained breathing apparatus operated in a pressure-demand or other positive-pressure mode).

Escape: GMFS *end of service life indicator (ESLI) required.* [any air-purifying, full-facepiece respirator (gas mask) with a chin-style, front- or back-mounted canister providing protection against the compound of concern];or SCBAE (any appropriate escape-type, self-contained breathing apparatus).

CHEMICAL REACTIVITY

Forms extremely explosive mixture with air. Violent reaction with strong oxidizers: bromine trifluoride, chlorine trifluoride, dichromates, lithium, halogenated compounds, and permanganates.

FIRE INFORMATION

Gases are heavier than air and will collect in low areas. Gases in confined areas may explode when exposed to fire. Gases may travel long distances to ignition sources and flash back. Storage containers and parts of containers may rocket great distances, in many directions. If material or contaminated runoff enters waterways, notify downstream users of potentially contaminated waters. Notify local health and fire officials and pollution control agencies. Do not extinguish the fire unless the flow of the gas can be stopped and any remaining gas is out of the line. Specially trained personnel may use fog lines to cool exposures and let the fire burn itself out. Use dry chemicals, foam, carbon dioxide. From a secure explosion-proof location, use water spray to cool exposed containers. If cooling streams are ineffective (venting sound increases in volume and pitch, tank discolors, or shows any signs of deforming), withdraw immediately to a secure position.

NFPA Hazard Classifications:

Health Hazards (Blue) 3
Flammability (Red) 4
Reactivity (Yellow) 0

Flash Point: Flammable gas
Flammable Limits in Air: LEL 12%; UEL 74%
Autoignition Temperature: 1128°F (609°C)

FIRST AID

GAS: Move victim to fresh air. Call emergency medical care. Apply artificial respiration if victim is not breathing. Do not use mouth-to-mouth method if victim ingested or inhaled the substance; induce artificial respiration with the aid of a pocket mask equipped with a one-way valve or other proper respiratory medical device. Administer oxygen if breathing is difficult. Remove and isolate contaminated clothing and shoes. In case of contact with substance, immediately flush skin or eyes with running water for at least 20 minutes. In case of contact with liquefied gas, thaw frosted parts with lukewarm water. Keep victim warm and quiet. Keep victim under observation. Effects of contact or inhalation may be delayed. Ensure that medical personnel are aware of the material(s) involved and take precautions to protect themselves.

REFRIGERATED LIQUID: Move victims to fresh air. Call emergency medical care. Apply artificial respiration if victim is not breathing. Administer oxygen if breathing is difficult. Remove and isolate contaminated clothing and shoes. In case of contact with substance, immediately flush skin or eyes with running water for at least 20 minutes. In case of contact with liquefied gas, thaw frosted parts with lukewarm water. Keep victim warm and quiet. Keep victim under observation. Effects of contact or inhalation may be delayed. Ensure that medical personnel are aware of the material(s) involved and take precautions to protect themselves.

CARBON TETRACHLORIDE C:200

SYNONYMS: BENZINOFORM; CARBONA; CARBON CHLORIDE; CARBON TET; CZTEROCHLOREK WEGLA (Polish); EEC No. 602-008-00-5; EINECS No. 200-262-8; ENT 4705; FASCIOLIN; FLUKOIDS; FREON 10; HALON 104; KATHARIN; METHANE TETRACHLORIDE; METHANE, TETRACHLORO-; NECATORINA; NECATORINE; PERCHLOROMETHANE; R 10; STCC; TETRACHLOORKOOLSTOF (Dutch); TETRACHLOORMETAN; TETRACHLORKOHLENSTOFF, TETRA (German); TETRACHLORMETHAN (German); TETRACHLOROCARBON; TETRACHLOROMETHANE; TETRACHLORURE de CARBONE (French); TETRACLOROMETANO (Italian); TETRACLORURO de CARBONO (Spanish); TETRACLORURO di CARBONIO (Italian); TETRAFINOL; TETRAFORM; TETRASOL; THAWPIT; UNIVERM; VERMOESTRICID

IDENTIFICATION

CAS: 56-23-5
ERG ID: UN 1846
ERG Guide: 151
ERG Guide Category: TOXIC (NONCOMBUSTIBLE)
Molecular Formula: CCl_4
RTECS No.: FG4900000
Properties: Clear, colorless, heavy liquid. Sweet, aromatic, etherlike odor. Sinks in water; insoluble. Toxic hydrogen chloride and phosgene vapors are produced when heated. Vapors are heavier than air; can collect in low areas.

HEALTH AND SAFETY INFORMATION

Irritating to eyes, skin, and respiratory tract. Preclude from exposure anyone with diseases of liver, kidneys, or alcoholism. Vapor is narcotic and can cause severe liver and kidney damage. May cause liver cancer. Has the potential for damaging the unborn child. In persons who regularly consume alcoholic beverage, liver damage may be increased. Repeated exposure may cause skin problems. Chronic exposure has also been associated with nausea, vomiting, dizziness, drowsiness, headache, visual disturbances, and poor eyesight.

IARC Evaluation: Group 2B, possible human carcinogen; human evidence, inadequate; animal evidence, sufficient.
NTP: Anticipated human carcinogen.
IDLH: 200 ppm; potential human carcinogen.
Odor Threshold: 40.7 ppm

OSHA Table Z-1-A Air Contaminant
OSHA Table Z-2 Air Contaminant
NIOSH Criteria Document: NIOSH 76-133

Exposure Limits:	**TWA**	**CEILING**	**STEL**
ACGIH: [(1)]	5 ppm; 31 mg/m^3		10 ppm; 63 mg/m^3
NIOSH: [(2)]			2 ppm; 12.6 mg/m^3 [60 min.]
OSHA:	10 ppm	25 ppm;200 ppm [(3)]	

[(1)] Suspected carcinogen.
[(2)] NIOSH REL potential human carcinogen; NIOSH STEL suspected human carcinogen.
[(3)] 5-minute maximum peak in any 4–hour period.

RECOMMENDATIONS FOR RESPIRATOR SELECTION: NIOSH

At any concentrations above the NIOSH REL, or where there is no REL, at any detectable concentration: SCBAF:PD,PP (any self-contained breathing apparatus that has a full facepiece and is operated in a pressure-demand or other positive-pressure mode);or SAF:PD,PP:ASCBA (any supplied-air respirator that has a full facepiece and is operated in a pressure-demand or other positive-pressure mode in combination with an auxiliary, self-contained breathing apparatus operated in a pressure-demand or other positive-pressure mode).

Escape: GMFOV [any air-purifying, full-facepiece respirator (gas mask) with a chin-style, front- or back-mounted organic vapor canister];or SCBAE (any appropriate escape-type, self-contained breathing apparatus).

CHEMICAL REACTIVITY

Becomes corrosive when in contact with water. Violent reaction with many compounds; severe reaction with oxidizers such as permanganates and dichromates, allyl alcohol, and silanes. Decomposes and may react explosively with chemically active metals such as sodium, potassium, and magnesium. Reacts with fluorine gas, alkali metals, and aluminum. Corrosive to some coatings, plastics, and rubber and to most metals, including iron, aluminum, and copper-based alloys. Forms highly poisonous and corrosive phosgene gas, hydrochloric acid, and chlorine when exposed to high temperatures, flames, or welding arcs.

FIRE INFORMATION

Combustion products include toxic hydrogen chloride and phosgene. Vapors are heavier than air and will collect in low areas. Vapors in confined areas may explode when exposed to fire. Storage containers and parts of containers may rocket great distances, in many directions. If material or contaminated runoff enters waterways, notify downstream users of potentially contaminated waters. Notify local health and fire officials and pollution control agencies. Use any extinguishing agents appropriate for surrounding fire. From a secure explosion-proof location, use water spray to cool exposed containers. If cooling streams are ineffective (venting sound increases in volume and pitch, tank discolors, or shows any signs of deforming), withdraw immediately to a secure position.

NFPA Hazard Classifications:

Health Hazards (Blue) 3
Flammability (Red) 0
Reactivity (Yellow) 0

FIRST AID

Move victim to fresh air. Call emergency medical care. Apply artificial respiration if victim is not breathing. Do not use mouth-to-mouth method if victim ingested or inhaled the substance; induce artificial respiration with the aid of a pocket mask equipped with a one-way valve or other proper respiratory medical device. Administer oxygen if breathing is difficult. Remove and isolate contaminated clothing and shoes. In case of contact with substance, immediately flush skin or eyes with running water for at least 20 minutes. For minor skin contact, avoid spreading material on unaffected skin. Keep victim warm and quiet. Effects of exposure (inhalation, ingestion, or skin contact) may be delayed. Ensure that medical personnel are aware of the material(s) involved and take precautions to protect themselves.

ENVIRONMENTAL INFORMATION

Clean Air Act: Hazardous Air Pollutants (Title I, Part A, Section 112); Stratospheric ozone protection (Title VI, Subpart A, Appendix A), Class I, Ozone Depletion Potential = 1.1.
Clean Water Act: Section 311 Hazardous Substances/RQ (same as CERCLA); Section 307 Priority Pollutants; Section 313 Priority Chemicals; Toxic Pollutant (Section 401.15).
EPA Hazardous Waste Number (RCRA No.): U211, D019.
RCRA Toxicity Characteristic (Section 261.24): Maximum Concentration of Contaminants, regulatory level, 0.5 mg/L
RCRA Section 261 Hazardous Constituents
RCRA Land Ban Waste
RCRA Universal Treatment Standards: Wastewater (mg/L), 0.057; Nonwastewater (mg/kg), 6.0.
RCRA Maximum Concentration Limit for Ground Water Protection (Section 264.94): 8010(1); 8240(5).
Safe Drinking Water Act: MCL, 0.005 mg/L; MCLG, zero; Regulated Chemical (47 FR 9352).
EPCRA Section 304: Reportable Quantity (RQ): CERCLA, 10 lb (4.54 kg).
EPCRA Section 313: Form R *de minimis* concentration reporting level: 0.1%.
California's Proposition 65: Carcinogen.
Marine Pollutant (49CFR, Subchapter 172.101, Appendix B).

CARBOXYBENZENE C:222

SYNONYMS: ACIDE BENZOIQUE (French); ACIDO BENZOICO (Spanish); BENZENECARBOXYLIC ACID; BENZOESAEURE (German); BENZOIC ACID; DRACYCLIC ACID; EINECS No. 200-618-2

IDENTIFICATION

CAS:	65-85-0
ERG ID:	UN 9094
ERG Guide:	153
ERG Guide Category:	SUBSTANCES-TOXIC AND/OR CORROSIVE (COMBUSTIBLE)
Molecular Formula:	$C_7H_6O_2$
RTECS No:	DG0875000
Properties:	Combustible crystals or powder. White. Faint, pleasant odor. Sinks in water.

HEALTH AND SAFETY INFORMATION

Irritates eyes (severely), skin, and respiratory system.

RECOMMENDATIONS FOR RESPIRATOR SELECTION

At any concentrations above the NIOSH REL, or where there is no REL, at any detectable concentration: SCBAF:PD,PP (any self-contained breathing apparatus that has a full facepiece and is operated in a pressure-demand or other positive-pressure mode);or SAF:PD,PP:ASCBA (any supplied-air respirator that has a full facepiece and is operated in a pressure-demand or other positive-pressure mode in combination with an auxiliary, self-contained breathing apparatus operated in a pressure-demand or other positive-pressure mode).

Escape: GMFOV [any air-purifying, full-facepiece respirator (gas mask) with a chin-style, front- or back-mounted organic vapor canister];or SCBAE (any appropriate escape-type, self-contained breathing apparatus).

CHEMICAL REACTIVITY

Contact with strong oxidizers may cause fire and explosions.

FIRE INFORMATION

If material or contaminated runoff enters waterways, notify downstream users of potentially contaminated waters. Notify local health and fire officials and pollution control agencies. Use dry chemicals, alcohol-resistant foam, carbon dioxide. From a secure explosion-proof location, use water spray to cool exposed containers.

NFPA Hazard Classifications:

Health Hazards (Blue) 2
Flammability (Red) 1
Reactivity (Yellow) . . . . --

Flash Point: 250 °F (121 °C) c.c.
Autoignition Temperature: 1058 °F (570 °C)

FIRST AID

Move victim to fresh air and call emergency medical care. Apply artificial respiration if victim is not breathing. Do not use mouth-to-mouth method if victim swallowed or inhaled the substance; induce artificial respiration with the aid of a pocket mask equipped with a one-way valve or other proper respiratory medical device. If breathing is difficult, give oxygen. In case of contact with substance, immediately flush skin or eyes with running water for at least 20 minutes. Remove and isolate contaminated clothing and shoes. Keep victim warm and quiet. Effects of exposure (inhalation, ingestion, or skin contact) to substances may be delayed. Ensure that medical personnel are aware of the material(s) involved and take precautions to protect themselves.

ENVIRONMENTAL INFORMATION

Clean Water Act: Section 311 Hazardous Substances/RQ (same as CERCLA)
EPCRA Section 304 Reportable Quantity (RQ): CERCLA, 5,000 lb (2,270 kg)

CASTOR OIL C:224

SYNONYMS: ACEITE de RICINO (Spanish); ACEITE de RICINO SULFONADO (Spanish); CASTER OIL; CASTOR OIL; CASTOR OIL, HYDROGENATED; NEOLID; EINECS No. 232-293-8; OIL of PALMA CHRISTI; RICINUS OIL; TANGANTANGAN OIL; TURKEY-RED OIL (SULFATED CASTOR OIL)

IDENTIFICATION

CAS:	8001-79-4; 1323-38-8
ERG ID:	UN 9277
ERG Guide:	171
ERG Guide Category:	SUBSTANCES (LOW TO MODERATE HAZARD)
RTECS No.:	YX1850000 (vegetable oil mist)
Properties:	Combustible oily liquid. Light yellow to green. Weak odor. Floats on water; insoluble.

HEALTH AND SAFETY INFORMATION

OSHA Table Z-1-A Air Contaminant as vegetable oil mists

Exposure Limits:	**TWA**
ACGIH:	10 mg/m^3 (vegetable oil mist)

NIOSH: 10 mg/m^3 (total)
5 mg/m^3 (respirable fraction as vegetable oil mist)

OSHA: 15 mg/m^3 (total)
5 mg/m^3 (respirable fraction as vegetable oil mist)

RECOMMENDATIONS FOR RESPIRATOR SELECTION

At any concentrations above the NIOSH REL, or where there is no REL, at any detectable concentration: SCBAF:PD,PP (any self-contained breathing apparatus that has a full facepiece and is operated in a pressure-demand or other positive-pressure mode);or SAF:PD,PP:ASCBA (any supplied-air respirator that has a full facepiece and is operated in a pressure-demand or other positive-pressure mode in combination with an auxiliary, self-contained breathing apparatus operated in a pressure-demand or other positive-pressure mode).

Escape: GMFOV [any air-purifying, full-facepiece respirator (gas mask) with a chin-style, front- or back-mounted organic vapor canister];or SCBAE (any appropriate escape-type, self-contained breathing apparatus)

CHEMICAL REACTIVITY

Reacts with strong oxidizers, strong acids.

FIRE INFORMATION

If material or contaminated runoff enters waterways, notify downstream users of potentially contaminated waters.Notify local health and fire officials and pollution control agencies. Use dry chemicals, alcohol-resistant foam, carbon dioxide.

NFPA Hazard Classifications:

Health Hazards (Blue) 0
Flammability (Red) 1
Reactivity (Yellow) 0

Flash Point: 445°F (229°C) c.c. (ricinus); 401°F (205°C) c.c. (hydrogenated)
Autoignition Temperature: 840°F (449°C) (ricinus)

FIRST AID

Move victim to fresh air. Call emergency medical care. Apply artificial respiration if victim is not breathing. Administer oxygen if breathing is difficult. In case of contact with substance, immediately flush skin or eyes with running water for at least 20 minutes. Remove and isolate contaminated clothing and shoes. Ensure that medical personnel are aware of the material(s) involved and take precautions to protect themselves.

CELLULOSE C:226

SYNONYMS: ABICEL; β-AMYLOSE; ARBOCEL; ARBOCEL BC 200; ARBOCELL B 600/30; AVICEL; CELLEX MX; α-CELLULOSE; CELLULOSE 248; CELLULOSE CRYSTALLINE; CELUFI; CELULOSA (Spanish); CEPO; CHROMEDIA CC 31; CUPRICELLULOSE; EINECS No. 232-674-9; ELCEMA F 150; ELCEMA G 250; FRESENIUS D 6; HEWETEN 10; HYDROXYCELLULOSE; KINGCOT; LA 01; MN-CELLULOSE; ONOZUKA P 500; PYROCELLULOSE; RAYOPHANE; RAYWEB Q; REXCEL; SIGMACELL; SOLKA-FIL; SPARTOSE OM-22; SULFITE CELLULOSE; TOMOFAN; TUNICIN; WHATMAN CC-31

IDENTIFICATION

CAS: 9004-34-6
Molecular Formula: $(C_6H_{10}O_5)_n$
RTECS No.: FJ5691460
Properties: Combustible fibrous particles. White. Odorless. Floats in water.

HEALTH AND SAFETY INFORMATION

A nuisance dust. Can cause irritation to eyes, skin, and respiratory tract.

Exposure Limits: TWA
OSHA Table Z-1-A Air Contaminant

ACGIH: 10 mg/m^3
NIOSH: 10 mg/m^3 (total); 5 mg/m^3 (respirable fraction)
OSHA: 15 mg/m^3 (total); 5 mg/m^3 (respiraable fraction

RECOMMENDATIONS FOR RESPIRATOR SELECTION: NIOSH

At any concentrations above the NIOSH REL, or where there is no REL, at any detectable concentration: SCBAF:PD,PP (any self-contained breathing apparatus that has a full facepiece and is operated in a pressure-demand or other positive-pressure mode);or SAF:PD,PP:ASCBA (any supplied-air respirator that has a full facepiece and is operated in a pressure-demand or other positive-pressure mode in combination with an auxiliary, self-contained breathing apparatus operated in a pressure-demand or other positive-pressure mode).

Escape: GMFOV [any air-purifying, full-facepiece respirator (gas mask) with a chin-style, front- or back-mounted organic vapor canister];or SCBAE (any appropriate escape-type, self-contained breathing apparatus).

CHEMICAL REACTIVITY

Contact with strong oxidizers can cause fire and explosions. Reacts with water, bromine pentafluoride, sodium nitrate, fluorine.

FIRE INFORMATION

Combustible solid. f material or contaminated runoff enters waterways, notify downstream users of potentially contaminated waters. Notify local health and fire officials and pollution control agencies. Use dry chemicals, alcohol-resistant foam, carbon dioxide.

NFPA Hazard Classifications: Not listed

CHLORACETONE C:235

SYNONYMS: ACETONE, CHLORO-; ACETONYL CHLORIDE; A-STOFF (German); CHLOROACETONE; CHLOROACETONE, stabilized (DOT); CHLOROMETHYL METHYL KETONE; 1-CHLORO-2-OXOPROPANE; CHLOROPROPANONE; 1-CHLORO-2-PROPANONE; CLOROACETONA (Spanish); EINECS No. 201-161-1; MONOCHLORACETONE; MONOCHLOROACETONE; MONOCHLOROACETONE, inhibited (DOT); MONOCHLOROACETONE, stabilized (DOT); MONOCHLOROACETONE, unstabilized (DOT); 2-PROPANONE, 1-CHLORO-; TONITE

IDENTIFICATION

CAS:	78-95-5
ERG ID:	UN 1695
ERG Guide:	131
ERG Guide Category:	FLAMMABLE LIQUIDS-TOXIC
Molecular Formula:	C_3H_5ClO
RTECS No.:	UC0700000; FM7525000
Properties:	Flammable liquid. Colorless, but can turn dark with age. Pungent, suffocating odor. Soluble in water

HEALTH AND SAFETY INFORMATION

Toxic; may be fatal if inhaled, ingested, or absorbed through the skin. Corrosive to the eyes, skin, and respiratory tract. May cause pulmonary edema, which can be delayed for several hours; there is a risk of death in serious cases.

Exposure Limits: CEILING

ACGIH:* 1 ppm; 3.8 mg/m^3

*Skin contact contributes significantly to overall exposure.

RECOMMENDATIONS FOR RESPIRATOR SELECTION: NIOSH

At any concentrations above the NIOSH REL, or where there is no REL, at any detectable concentration: SCBAF:PD,PP (any self-contained breathing apparatus that has a full facepiece and is operated in a pressure-demand or other positive-pressure mode);or SAF:PD,PP:ASCBA (any supplied-air

respirator that has a full facepiece and is operated in a pressure-demand or other positive-pressure mode in combination with an auxiliary, self-contained breathing apparatus operated in a pressure-demand or other positive-pressure mode).

Escape: GMFOV [any air-purifying, full-facepiece respirator (gas mask) with a chin-style, front- or back-mounted organic vapor canister];or SCBAE (any appropriate escape-type, self-contained breathing apparatus).

CHEMICAL REACTIVITY

Forms explosive mixture with air. Flammable when exposed to oxidizers, heat, or flame. Reacts with alkalis and sodium ethoxide. May become explosive with age.

FIRE INFORMATION

Decomposition products upon heating include toxic and corrosive vapors that include phosgene. Vapors are heavier than air and will collect in low areas. Vapors in confined areas may explode when exposed to fire. Vapors may travel long distances to ignition sources and flash back. Storage containers and parts of containers may rocket great distances, in many directions. If material or contaminated runoff enters waterways, notify downstream users of potentially contaminated waters. Notify local health and fire officials and pollution control agencies. Use dry chemicals, carbon dioxide, foam, AFFF.

FEMA Hazard Classifications:

Health Hazards (Blue) 2
Flammability (Red) 2
Reactivity (Yellow) 0

FLASH POINT: 45°F (7°C)

FIRST AID

Move victim to fresh air. Call emergency medical care. Apply artificial respiration if victim is not breathing. Do not use mouth-to-mouth method if victim ingested or inhaled the substance; induce artificial respiration with the aid of a pocket mask equipped with a one-way valve or other proper respiratory medical device. Administer oxygen if breathing is difficult. Remove and isolate contaminated clothing and shoes. In case of contact with substance, immediately flush skin or eyes with running water for at least 20 minutes. Wash skin with soap and water. Keep victim warm and quiet. Effects of exposure (inhalation, ingestion, or skin contact) to substance may be delayed. Ensure that medical personnel are aware of the material(s) involved and take precautions to protect themselves.

CHLORIC ACID C:250

SYNONYMS: ACIDE CHLORIQUE (French); ACIDO CLORICO (Spanish); CHLORSAEURE (German); EINECS No. 232-233-0; HYPURE C

IDENTIFICATION

CAS:	7790-93-4
ERG ID:	UN 2626
ERG Guide:	140
ERG Guide Category:	OXIDIZERS
Molecular Formula:	$ClHO_3$
RTECS No.:	FN9750000
Properties:	Liquid. Colorless. Odorless. Mixes with water.

HEALTH AND SAFETY INFORMATION

Poisonous. Corrosive to eyes, skin, and respiratory tract. Inhalation may cause pulmonary edema, which can be delayed for several hours; there is a risk of death in serious cases. Toxic and highly irritating if inhaled or swallowed.

RECOMMENDATIONS FOR RESPIRATOR SELECTION

At any concentrations above the NIOSH REL, or where there is no REL, at any detectable concentration: SCBAF:PD,PP (any self-contained breathing apparatus that has a full facepiece and is operated in a pressure-demand or other positive-pressure mode);or SAF:PD,PP:ASCBA (any supplied-air respirator that has a full facepiece and is operated in a pressure-demand or other positive-pressure mode in combination with an auxiliary, self-contained breathing apparatus operated in a pressure-demand or other positive-pressure mode).

Escape: GMFOV [any air-purifying, full-facepiece respirator (gas mask) with a chin-style, front- or back-mounted organic vapor canister];or SCBAE (any appropriate escape-type, self-contained breathing apparatus).

CHEMICAL REACTIVITY

A powerful oxidizer. Reacts violently with combustible materials, mineral acids, reducing agents, organic or other readily oxidizable substances, powdered metals, metal hydrides, and many other substances. Forms explosive mixture with some metals. Solutions of more than 40% are unstable; aqueous solutions can decomposes explosively during evaporation.

FIRE INFORMATION

Combustion products include toxic chlorine. Will increase the intensity of fire. Vapors are heavier than air and will collect in low areas. If material or contaminated runoff enters waterways, notify downstream users of potentially contaminated waters. Notify local health and fire officials and pollution control agencies. Use extinguishing agent suitable for surrounding fires.

NFPA Hazard Classifications: Not listed

FIRST AID

Move victim to fresh air. Call emergency medical care. Apply artificial respiration if victim is not breathing. Administer oxygen if breathing is difficult. Remove and isolate contaminated clothing and shoes. In case of contact with substance, immediately flush skin or eyes with running water for at least 20 minutes. Keep victim warm and quiet. Ensure that medical personnel are aware of the material(s) involved and take precautions to protect themselves.

CHLORINE C:260

SYNONYMS: BERTHOLITE; CHLOOR (Dutch); CASWELL No. 179; CHLOR (German); CHLORE (French); CHLORINE MOLECULAR; CLORO (Italian, Spanish); DICHLORINE; EEC No. 017-001-00-7; EPA PESTICIDE CHEMICAL CODE 020501; MOLECULAR CHLORINE; POLY I GAS

IDENTIFICATION

CAS:	7782-50-5
ERG ID:	UN 1017
ERG Guide:	124
ERG Guide Category:	GASES-TOXIC AND/OR CORROSIVE-OXIDIZING
Molecular Formula:	Cl_2
RTECS No.:	FO2100000
Properties:	Greenish-yellow gas. Amber liquid under high pressure or refrigeration. Strongly suffocating, irritating, bleachlike odor. Soluble in water.

HEALTH AND SAFETY INFORMATION

Severely irritating to eyes, skin, and lungs. Skin and eye contact can cause severe burns and blindness. Breathing the gas can cause death. Long-term effects include impairment of breathing functions, erosion of exposed front teeth, anxiety, and the formation of acnelike lesions on the skin. Contact with liquid may cause frostbite. May cause pulmonary edema, which can be delayed for several hours; there is a risk of death in serious cases.

IDLH: 10 ppm
Odor Threshold: 0.05 ppm
OSHA Table Z-1-A Air Contaminant
OSHA Process Safety Management of Highly Hazardous Chemicals (29CFR, Part 1910.119, Appendix A): Threshold Quantity: 1,500 pounds.
NIOSH Criteria Document: NIOSH 76-170

Exposure Limits:	**TWA**	**CEILING**	**STEL**
ACGIH:	0.5 ppm; 1.5 mg/m^3		1 ppm; 2.9 mg/m^3

NIOSH:	0.5 ppm; 1.45 mg/m^3 (15 min)
OSHA:	1 ppm; 0.3 mg/m^3

RECOMMENDATIONS FOR RESPIRATOR SELECTION: NIOSH/OSHA

5 ppm: CCRS (any chemical cartridge respirator with cartridge(s) providing protection against the compound of concern);or SA (any supplied-air respirator).
12.5 ppm: SA:CF (any supplied-air respirator operated in a continuous-flow mode);or PAPRS (any powered, air-purifying respirator with cartridge(s) providing protection against the compound of concern);or CCRFS (any chemical cartridge respirator with a full facepiece and cartridge(s) providing protection against the compound of concern);or GMFS [any air-purifying, full-facepiece respirator (gas mask) with a chin-style, front- or back-mounted canister providing protection against the compound of concern];or SCBAF (any self-contained breathing apparatus with a full facepiece);or SAF (any supplied-air respirator with a full facepiece).

Emergency or Planned Entry into Unknown Concentrations or IDLH Conditions:
SCBAF:PD,PP (any self-contained breathing apparatus that has a full facepiece and is operated in a pressure-demand or other positive-pressure mode);or SAF:PD,PP:ASCBA (any supplied-air respirator that has a full facepiece and is operated in a pressure-demand or other positive-pressure mode in combination with an auxiliary, self-contained breathing apparatus operated in a pressure-demand or other positive-pressure mode).

Escape: GMFS [any air-purifying, full-facepiece respirator (gas mask) with a chin-style, front- or back-mounted canister providing protection against the compound of concern];or SCBAE (any appropriate escape-type, self-contained breathing apparatus).

Note: Substance reported to cause eye irritation or damage; may require eye protection.

CHEMICAL REACTIVITY

A powerful oxidizer. Reacts with water, forming toxic hypochlorous acid. Reacts violently with combustible materials, reducing agents, and many other substances. Forms explosive mixtures with gasoline and petroleum products, ammonia, ether, turpentine, alcohols, acetylene, carbon disulfide, hydrogen, anhydrous ammonia, finely divided metals, hydrocarbons, organic compounds, and phosphorus. Fire, explosion, and the formation of toxic fumes may result from contact with many substances including alkylphosphines, aluminum, antimony, arsenic compounds, arsine, bismuth, boron, brass, calcium compounds, carbon, diethyl zinc, fluorine, germanium, hydrocarbons, and rubber. Attacks some plastics, synthetic rubber, and coatings. In the presence of moisture, extremely corrosive to iron, steel, copper, bronze, and zinc. Aqueous solutions of chlorine will react on long standing, especially if exposed to light or moisture, giving off oxygen and hydrochloric acid.

FIRE INFORMATION

Nonflammable but will increase the intensity of a fire. Firefighting gear (including SCBA) provides NO protection. If exposure occurs, remove and isolate gear immediately and thoroughly decontaminate personnel. Gases are heavier than air and will collect in low areas. Storage containers and parts of containers may rocket great distances, in many directions. If material or contaminated runoff enters waterways, notify downstream users of potentially contaminated waters. Notify local health and fire officials and pollution control agencies. Do not extinguish the fire unless the flow of the gas can be stopped and any remaining gas is out of the line. Specially trained personnel may use fog lines to cool exposures and let the fire burn itself out. Use extinguishing agent suitable for surrounding fires. From a secure, explosion-proof location, use water spray to cool exposed containers. If cooling streams are ineffective (venting sound increases in volume and pitch, tank discolors, or shows any signs of deforming), withdraw immediately to a secure position.

NFPA Hazard Classifications:

Health Hazards (Blue) 4
Flammability (Red) 0
Reactivity (Yellow) 0
Special situations .... OXY

FIRST AID

Move victim to fresh air. Call emergency medical care. Apply artificial respiration if victim is not breathing. Do not use mouth-to-mouth method if victim ingested or inhaled the substance; induce artificial respiration with the aid of a pocket mask equipped with a one-way valve or other proper respiratory medical device. Administer oxygen if breathing is difficult. Clothing frozen to the skin should be thawed before being removed. Remove and isolate contaminated clothing and shoes. In case of contact with substance, immediately flush skin or eyes with running water for at least 20 minutes. Keep victim warm and quiet. Keep victim under observation. Effects of contact or inhalation may be delayed. Ensure that medical personnel are aware of the material(s) involved and take precautions to protect themselves. *Note:* Chest radiographs should be followed and pulmonary function tests performed.

ENVIRONMENTAL INFORMATION

Clean Air Act: Hazardous Air Pollutants (Title I, Part A, Section 112); Accidental Release Prevention/Flammable substances, (Section 112[r], Table 3), TQ = 2,500 lb (1,135 kg).
Clean Water Act: Section 311 Hazardous Substances/RQ (same as CERCLA); Section 313 Priority Chemicals.
Safe Drinking Water Act: SMCL, 250 mg/L; Priority List (55 FR 1470).
EPCRA Section 302 Extremely Hazardous Substances: TPQ = 100 lb (45.4 kg).
EPCRA Section 304: Reportable Quantity (RQ): CERCLA, 10 lb (4.54 kg).
EPCRA Section 313: Form R *de minimis* concentration reporting level: 1.0%.
Marine Pollutant (49CFR, Subchapter 172.101, Appendix B).
TSCA: 40CFR 704.225

p–CHLOROANILINE C:290

SYNONYMS: 1-AMINO-4-CHLOROBENZENE; p-AMINOCHLOROBENZENE; BENZENEAMINE,

4-CHLORO-; 4-CHLORANILINE (Czech); p-CHLOROAMINOBENZENE; 4-CHLORO-1-AMINOBENZENE; 4-CHLORANILINE; 4-CHLOROANILINE; 4-CHLOROBENZENAMINE; 4-CHLOROBENZENEAMINE; 4-CHLOROPHENYLAMINE; p-CLOROANILINA (Spanish); EEC No. 612-010-00-8; NCI-C02039

IDENTIFICATION

CAS:	106-47-8; 27134-26-5
ERG ID:	UN 2018
ERG Guide:	152
ERG Guide Category:	SUBSTANCES-TOXIC (COMBUSTIBLE)
Molecular Formula:	C_6H_6ClN
RTECS No.:	BX0700000
Properties:	Combustible crystals or powder. Colorless, white or yellowish white. Mild sweet odor. Sinks and mixes very slowly with water.

HEALTH AND SAFETY INFORMATION

Irritates eyes, skin, and respiratory tract. Long- term effects includes liver, kidney, and blood damage. May be fatal if swallowed, inhaled, or absorbed through the skin.

DFG confirmed carcinogen.

Exposure Limits: The United Nations Environmental Program and the Centre of International Projects, Moscow set workplace air standards (MAC limits) for p-chloroaniline at 0.01 mg/m^3 (average daily basis) and 0.04 mg/m^3 (temporary basis). In addition, some U.S. states such as New York (6.0 μg/m^3) and Connecticut (0.06 μg/m^3) have issued guidelines or standards for this the para isomer of chloroaniline. (Sittig, 3rd edition).

RECOMMENDATIONS FOR RESPIRATOR SELECTION: NIOSH

At any concentrations above the NIOSH REL, or where there is no REL, at any detectable concentration: SCBAF:PD,PP (any self-contained breathing apparatus that has a full facepiece and is operated in a pressure-demand or other positive-pressure mode);or SAF:PD,PP:ASCBA (any supplied-air respirator that has a full facepiece and is operated in a pressure-demand or other positive-pressure mode in combination with an auxiliary, self-contained breathing apparatus operated in a pressure-demand or other positive-pressure mode).

Escape: GMFOV [any air-purifying, full-facepiece respirator (gas mask) with a chin-style, front- or back-mounted organic vapor canister];or SCBAE (any appropriate escape-type, self-contained breathing apparatus).

CHEMICAL REACTIVITY

Contact with oxidizers may cause fire and explosions. Incompatible with strong acids, acetic anhydride, chlorosulfonic acid, and ozone.

FIRE INFORMATION

Decomposes in fire, forming toxic nitrous oxide vapors and corrosive hydrochloric acid. Poison. If material or contaminated runoff enters waterways, notify downstream users of potentially contaminated waters. Notify local health and fire officials and pollution control agencies.

FEMA Hazard Classifications:

Health Hazards (Blue) 2
Flammability (Red) 1
Reactivity (Yellow) 0

Flash Point: Combustible solids. More than 220 °F (104 °C) o.c.

FIRST AID

Move victim to fresh air. Call emergency medical care. Apply artificial respiration if victim is not breathing. Do not use mouth-to-mouth method if victim ingested or inhaled the substance; induce artificial respiration with the aid of a pocket mask equipped with a one-way valve or other proper respiratory medical device. Administer oxygen if breathing is difficult. Remove and isolate contaminated clothing and shoes. In case of contact with substance, immediately flush skin or eyes with running water for at least 20 minutes. For minor skin contact, avoid spreading material on unaffected skin. Keep victim warm and quiet. Effects of exposure (inhalation, ingestion, or skin contact) may be delayed. Ensure that medical personnel are aware of the material(s) involved and take precautions to protect themselves.

ENVIRONMENTAL INFORMATION

EPA Hazardous Waste Number (RCRA No.): P024.
RCRA Section 261 Hazardous Constituents
RCRA Land Ban Waste
RCRA Universal Treatment Standards: Wastewater (mg/L), 0.46; Nonwastewater (mg/kg), 16.
RCRA Ground Water Monitoring List: Suggested test method(s) (PQL ug/L): 8270(20).
EPCRA Section 304: Reportable Quantity (RQ): CERCLA, 1,000 lb (454 kg).
EPCRA Section 313: Form R *de minimis* concentration reporting level: 1.0%.
California's Proposition 65: Carcinogen.
TSCA: 40CFR 799.18

CHLOROFORM C:320

SYNONYMS: CHLOROFORME (French); CLOROFORMIO (Italian); CHLOROFORMO (Spanish); EEC No. 602-006-00-4; EINECS No. 200-663-8; FORMYL TRICHLORIDE; FREON 20; METHANE TRICHLORIDE; METHANE, TRICHLORO-; METHENYL TRICHLORIDE; METHYL TRICHLORIDE; NCI-C02686; R 20 (REFRIGERANT); TCM; TRICHLOORMETHAAN (Dutch); TRICHLORMETHAN (Czech); TRICHLOROFORM; TRICHLOROMETHANE; TRICLOROMETANO (Italian)

IDENTIFICATION

CAS:	67-66-3
ERG ID:	UN 1888
ERG Guide:	151
ERG Guide Category:	TOXIC (NONCOMBUSTIBLE)
Molecular Formula:	$CHCl_3$
RTECS No.:	FS9100000
Properties:	Clear, colorless, watery liquid. Sweet ether-like odor. Sinks in water; slightly soluble. Toxic and irritating gases of phosgene and hydrogen chloride are produced when heated.

HEALTH AND SAFETY INFORMATION

Irritates eyes, skin, and respiratory tract. Destroys the skins natural oils, causing redness, dryness, and pain. Long-term exposure may include sluggishness, depression, hallucinations, loss of appetite, fatigue, and liver and kidney damage. Effects the nervous system and heart. Chloroform is a suspected carcinogenic agent because high levels cause kidney, liver, and thyroid cancer in rats and mice. Abnormal sperm found in laboratory tests.

IARC Evaluation: Group 2B, possible human carcinogen; human evidence, limited; animal evidence, sufficient.
NTP: Anticipated human carcinogen.
IDLH: 500 ppm; potential human carcinogen.
Odor Threshold: 11.7 ppm
OSHA Table Z-1-A Air Contaminant
NIOSH Criteria Document: NIOSH 75-114

Exposure Limits:	**TWA**	**STEL**
ACGIH: (1)	10 ppm; 49 mg/m^3	
NIOSH: (2)		2 ppm; 9.78 mg/m^3 [60 min]

(1) Animal carcinogen.
(2) Potential carcinogen. Reduce exposure to lowest feasible concentration.

RECOMMENDATIONS FOR RESPIRATOR SELECTION: NIOSH

At any concentrations above the NIOSH REL, or where there is no REL, at any detectable concentration: SCBAF:PD,PP (any self-contained breathing apparatus that has a full facepiece and is operated in a pressure-demand or other positive-pressure mode);or SAF:PD,PP:ASCBA (any supplied-air respirator that has a full facepiece and is operated in a pressure-demand or other positive-pressure mode in combination with an auxiliary, self-contained breathing apparatus operated in a pressure-demand or other positive-pressure mode).

Escape: GMFOV [any air-purifying, full-facepiece respirator (gas mask) with a chin-style, front- or back-mounted organic vapor canister];or SCBAE (any appropriate escape-type, self-contained breathing apparatus).

CHEMICAL REACTIVITY

Decomposes in the presence of excess water, sunlight, or high temperatures, forms highly toxic and corrosive phosgene and hydrogen chloride. Strong alkalies like lye and potassium hydroxide decompose chloroform to chloride salts and formates. Can become explosive in the presence of strong alkalies and water. In contact with water and high temperatures, becomes corrosive; attacks iron and other metals. Incompatible with acetone, aluminum, strong oxidizers, potassium, sodium, chemically active metals, and strong bases. Attacks plastics and rubber. Can accumulate static electrical charges.

FIRE INFORMATION

I If material or contaminated runoff enters waterways, notify downstream users of potentially contaminated waters. Notify local health and fire officials and pollution control agencies. Use any extinguishing agent suitable for surrounding fire. From a secure explosion-proof location, use water spray to cool exposed containers.

NFPA Hazard Classifications:

Health Hazards (Blue) 2
Flammability (Red) 0
Reactivity (Yellow) 0

FIRST AID

Move victim to fresh air. Call emergency medical care. Apply artificial respiration if victim is not breathing. Do not use mouth-to-mouth method if victim ingested or inhaled the substance; induce artificial respiration with the aid of a pocket mask equipped with a one-way valve or other proper respiratory medical device. Administer oxygen if breathing is difficult. Remove and isolate contaminated clothing and shoes. In case of contact with substance, immediately flush skin or eyes with running water for at least 20 minutes. For minor skin contact, avoid spreading material on unaffected skin. Keep victim warm and quiet. Effects of exposure (inhalation, ingestion, or skin contact) may be delayed. Ensure that medical personnel are aware of the material(s) involved and take precautions to protect themselves.

ENVIRONMENTAL INFORMATION

Clean Air Act: Hazardous Air Pollutants (Title I, Part A, Section 112); Accidental Release Prevention/Flammable substances, (Section 112[r], Table 3), TQ = 20,000 lb (9,080kg).
Clean Water Act: Section 311 Hazardous Substances/RQ (same as CERCLA); Section 307 Priority Pollutants; Section 313 Priority Chemicals; Toxic Pollutant (Section 401.15).
EPA Hazardous Waste Number (RCRA No.): U044; D022.
RCRA Toxicity Characteristic (Section 261.24): Maximum Concentration of Contaminants, regulatory level, 6.0 mg/L

RCRA Section 261 Hazardous Constituents
RCRA Land Ban Waste
RCRA Universal Treatment Standards: Wastewater (mg/L), 0.046; Nonwastewater (mg/kg), 6.0.
RCRA Ground Water Monitoring List: Suggested test method(s) (PQL ug/L): 8010(0.5); 8240(5).
Safe Drinking Water Act: Priority List (55 FR 1470).
EPCRA Section 302 Extremely Hazardous Substances: TPQ = 10,000 lb (4,540 kg).
EPCRA Section 304: Reportable Quantity (RQ): CERCLA, 10 lb (4.54 kg).
EPCRA Section 313: Form R *de minimis* concentration reporting level: 0.1%.
California's Proposition 65: Carcinogen.

CHLOROHYDRIN C:323

SYNONYMS: α-CHLORHYDRIN; CHLORODEOXYGLYCEROL; 1-CHLORO-2,3-DIHYDROXYPROPANE; 3-CHLORO-1,2-DIHYDROXYPROPANE; α-CHLOROHYDRIN; 1-CHLOROPROPANE-2,3-DIOL; 1-CHLORO-2,3-PROPANEDIOL; 3-CHLOROPROPANE-1,2-DIOL; 3-CHLORO-1,2-PROPANEDIOL; 3-CHLOROPROPYLENE GYLCOL; CLORHIDRINA (Spanish); ββ'-DIHYDROXYISOPROPYL CHLORIDE; 2,3-DIHYDROXYPROPYL CHLORIDE; EINECS No. 202-492-4; EPIBLOC; GLYCERIN-α-MONOCHLORHYDRIN; GLYCEROL CHLOROHYDRIN; GLYCEROL-α-CHLOROHYDRIN; GLYCEROL-α-MONOCHLOROHYDRIN (DOT); GLYCERYL CHLOROHYDRIN; MONOCHLORHYDRIN; MONOCHLOROHYDRIN; α-MONOCHLOROHYDRIN; U-5897

IDENTIFICATION

CAS:	96-24-2
ERG ID:	UN 2689
ERG Guide:	153
ERG Guide Category:	SUBSTANCES-TOXIC AND/OR CORROSIVE (COMBUSTIBLE)
Molecular Formula:	$C_3H_7ClO_2$
RTECS No.:	TY4025000
Properties:	Combustible, slightly viscous liquid. Colorless. Faint, pleasant odor. Soluble in water.

HEALTH AND SAFETY INFORMATION

Highly toxic. Eye and skin irritant. A possible carcinogen. A teratogen. May cause thyroid, kidney, and liver problems.

RECOMMENDATIONS FOR RESPIRATOR SELECTION

At any concentrations above the NIOSH REL, or where there is no REL, at any detectable concentration: SCBAF:PD,PP (any self-contained breathing apparatus that has a full facepiece and is

operated in a pressure-demand or other positive-pressure mode);or SAF:PD,PP:ASCBA (any supplied-air respirator that has a full facepiece and is operated in a pressure-demand or other positive-pressure mode in combination with an auxiliary, self-contained breathing apparatus operated in a pressure-demand or other positive-pressure mode).

Escape: GMFOV [any air-purifying, full-facepiece respirator (gas mask) with a chin-style, front- or back-mounted organic vapor canister];or SCBAE (any appropriate escape-type, self-contained breathing apparatus).

CHEMICAL REACTIVITY

Contact with oxidizers may cause fire and explosions.

FIRE INFORMATION

Decomposition products include toxic hydrogen chloride. If material or contaminated runoff enters waterways, notify downstream users of potentially contaminated waters. Notify local health and fire officials and pollution control agencies. Use dry chemicals, alcohol-resistant foam, carbon dioxide. From a secure explosion-proof location, use water spray to cool exposed containers.

Flash Point: >212°F (>100°C)

FIRST AID

Move victim to fresh air and call emergency medical care. Apply artificial respiration if victim is not breathing. Do not use mouth-to-mouth method if victim swallowed or inhaled the substance; induce artificial respiration with the aid of a pocket mask equipped with a one-way valve or other proper respiratory medical device. If breathing is difficult, give oxygen. In case of contact with substance, immediately flush skin or eyes with running water for at least 20 minutes. Remove and isolate contaminated clothing and shoes. Keep victim warm and quiet. Effects of exposure (inhalation, ingestion, or skin contact) to substances may be delayed. Ensure that medical personnel are aware of the material(s) involved and take precautions to protect themselves.

CHLOROPHENYLTRICHLOROSILANE C:326

SYNONYMS: CHLOROFENILTRICLOROSILANO (Spanish); CHLOROPHENYL TRICHLOROSILANE

IDENTIFICATION

CAS:	26571-79-9
ERG ID:	UN 1753
ERG Guide:	156

ERG Guide Category: SUBSTANCES-TOXIC AND/OR CORROSIVE (COMBUSTIBLE/WATER-SENSITIVE)

Molecular Formula: $C_6H_4Cl_4Si$

RTECS No.: VV2650000

Properties: Combustible liquid. Pale-yellow to colorless. Mixes with water, forming hydrochloric acid.

HEALTH AND SAFETY INFORMATION

Causes eye, skin, and respiratory tract irritation. May cause pulmonary edema, which can be delayed for several hours; there is a risk of death in serious cases.

RECOMMENDATIONS FOR RESPIRATOR SELECTION: as HCl

50 ppm: CCRS (any chemical cartridge respirator with cartridge(s) providing protection against the compound of concern);or GMFS (any air-purifying, full-facepiece respirator (gas mask) with a chin-style, front- or back-mounted canister providing protection against the compound of concern);or PAPRS (any powered, air-purifying respirator with cartridge(s) providing protection against the compound of concern);or SA (any supplied-air respirator);or SCBAF (any self-contained breathing apparatus with a full facepiece).

Emergency or Planned Entry into Unknown Concentrations or IDLH Conditions: SCBAF:PD,PP (any self-contained breathing apparatus that has a full facepiece and is operated in a pressure-demand or other positive-pressure mode);or SAF:PD,PP:ASCBA (any supplied-air respirator that has a full facepiece and is operated in a pressure-demand or other positive-pressure mode in combination with an auxiliary, self-contained breathing apparatus operated in a pressure-demand or other positive-pressure mode).

Escape: GMFAG (any air-purifying, full-facepiece respirator (gas mask) with a chin-style, front- or back-mounted acid gas canister);or SCBAE (any appropriate escape-type, self-contained breathing apparatus).

Note: Substance reported to cause eye irritation or damage; may require eye protection.

CHEMICAL REACTIVITY

Reacts with oxidizers. Contact with moisture forms heat and corrosive hydrogen chloride. Attacks metals in the presence of moisture.

FIRE INFORMATION

Combustion products include toxic chlorine. If material or contaminated runoff enters waterways, notify downstream users of potentially contaminated waters. Notify local health and fire officials and pollution control agencies. Use dry chemicals, alcohol-resistant foam, carbon dioxide. From a secure explosion-proof location, use water spray to cool exposed containers.

NFPA Hazard Classifications: Not listed
Flash Point: 255°F (124°C) o.c.

FIRST AID

Move victim to fresh air. Call emergency medical care. Apply artificial respiration if victim is not breathing. Do not use mouth-to-mouth method if victim ingested or inhaled the substance; induce artificial respiration with the aid of a pocket mask equipped with a one-way valve or other proper respiratory medical device. Administer oxygen if breathing is difficult. Remove and isolate contaminated clothing and shoes. In case of contact with substance, immediately flush skin or eyes with running water for at least 20 minutes. For minor skin contact, avoid spreading material on unaffected skin. Keep victim warm and quiet. Effects of exposure (inhalation, ingestion, or skin contact) may be delayed. Ensure that medical personnel are aware of the material(s) involved and take precautions to protect themselves.

ENVIRONMENTAL INFORMATION

Marine Pollutant (49CFR, Subchapter 172.101, Appendix B).

CHLOROPRENE C:329

SYNONYMS: CHLOROBUTADIENE; 2-CHLOROBUTADIENE; 2-CHLORO-1,3-BUTADIENE; 2-CHLORO-1,3-DIENE; β-CHLOROPRENE; CLOROPRENO (Spanish); EEC No. 602-036-00-8; EINECS No. 204-818-0; NEOPRENE

IDENTIFICATION

CAS:	126-99-8
ERG ID:	UN 1991
ERG Guide:	131P
ERG Guide Category:	FLAMMABLE LIQUIDS-TOXIC
Molecular Formula:	C_4H_5Cl
RTECS No.:	EI9625000
Properties:	Highly flammable liquid. Colorless. Pungent, ether-like odor. Floats and mixes slowly with water; slightly soluble. Flammable, irritating vapor is produced.

HEALTH AND SAFETY INFORMATION

Irritates eyes, skin, and respiratory tract. May cause pulmonary edema, which can be delayed for several hours; there is a risk of death in serious cases. Prolonged or repeated contact can cause liver and kidney damage and skin disorders. Do not wear contact lenses when working with this chemical.

IDLH: 300 ppm; potential human carcinogen.
Odor Threshold: 14.9 ppm
NIOSH Criteria Document: NIOSH 77-1210

Exposure Limits:	TWA	CEILING
ACGIH: [(1)]	10 ppm; 36 mg/m^3	
NIOSH: [(2)]		1 ppm; 3.6 mg/m^3 (15-minutes)
OSHA: [(1)]	25 ppm; 90 mg/m^3	

[(1)] Skin contact contributes significantly in overall exposure
[(2)] Potential human carcinogen

RECOMMENDATIONS FOR RESPIRATOR SELECTION: NIOSH

At any concentrations above the NIOSH REL, or where there is no REL, at any detectable concentration: SCBAF:PD,PP (any self-contained breathing apparatus that has a full facepiece and is operated in a pressure-demand or other positive-pressure mode);or SAF:PD,PP:ASCBA (any supplied-air respirator that has a full facepiece and is operated in a pressure-demand or other positive-pressure mode in combination with an auxiliary, self-contained breathing apparatus operated in a pressure-demand or other positive-pressure mode).

Escape: GMFOV [any air-purifying, full-facepiece respirator (gas mask) with a chin-style, front- or back-mounted organic vapor canister];or SCBAE (any appropriate escape-type, self-contained breathing apparatus).

CHEMICAL REACTIVITY

Reacts with peroxides and other oxidizers, alkaline earth and alkali metals, finely divided metals. Attacks some plastic materials. Readily forms peroxides and will polymerize at room temperature on standing.

FIRE INFORMATION

May react with itself without warning, blocking relief valves and leading to container explosions. Combustion products include toxic hydrogen chloride and phosgene. Vapors are heavier than air and will collect in low areas. Vapors in confined areas may explode when exposed to fire. Vapors may travel long distances to ignition sources and flash back. Storage containers and parts of containers may rocket great distances, in many directions. If material or contaminated runoff enters waterways, notify downstream users of potentially contaminated waters. Notify local health and fire officials and pollution control agencies. Use dry chemicals, alcohol-resistant foam, carbon dioxide. From a secure explosion-proof location, use water spray to cool exposed containers.

NFPA Hazard Classifications:

Health Hazards (Blue) 2
Flammability (Red) 3
Reactivity (Yellow) 0

Flash Point:-4°F (-20°C) c.c.
Flammable Limits in Air: LEL 4.0%; UEL 20.0%
Autoignition Temperature: 104°F (40°C)

FIRST AID

Move victim to fresh air. Call emergency medical care. Apply artificial respiration if victim is not breathing. Do not use mouth-to-mouth method if victim ingested or inhaled the substance; induce artificial respiration with the aid of a pocket mask equipped with a one-way valve or other proper respiratory medical device. Administer oxygen if breathing is difficult. Remove and isolate contaminated clothing and shoes. In case of contact with substance, immediately flush skin or eyes with running water for at least 20 minutes. Wash skin with soap and water. Keep victim warm and quiet.,Effects of exposure (inhalation, ingestion, or skin contact) to substance may be delayed. Ensure that medical personnel are aware of the material(s) involved and take precautions to protect themselves.

ENVIRONMENTAL INFORMATION

Clean Air Act: Hazardous Air Pollutants (Title I, Part A, Section 112)
RCRA Section 261 Hazardous Constituents, waste number not listed
RCRA Land Ban Waste
RCRA Universal Treatment Standards: Wastewater (mg/L), 0.057; Nonwastewater (mg/kg), 0.28.
RCRA Ground Water Monitoring List: Suggested test methods(s) (PQL μg/L); 8010 (50); 8240 (5)
EPCRA Section 304 Reportable Quantity (RQ): CERCLA, 100 lb (45.4 kg)
EPCRA Section 313: Form R *de minimis* concentration reporting level: 1.0%.
TSCA: 40CFR 716.120 (*a*); Health and Safety Reporting List; 40CFR 712.30(*e*)10 as chloroprene(beta-)

2–CHLOROPROPENE C:331

SYNONYMS: CHLORO PROPENE (2-); 2-CHLORO-1-PROPENE; 2-CHLOROPROPYLENE; β-CLOROPROPYLENE; β-CLOROPRENO (Spanish); 2-CHLORPRENE; EINECS No. 209-187-5; 1-PROPENE,2-CHLORO-

IDENTIFICATION

CAS:	557-98-2
ERG ID:	UN 2456
ERG Guide:	130P
ERG Guide Category:	FLAMMABLE LIQUIDS (NONPOLAR/WATER-IMMISCIBLE/NOXIOUS)
Molecular Formula:	C_3H_5Cl
RTECS No.:	UC7200000
Properties:	Flammable liquid or gas. Colorless. Very slight solubility in water.

HEALTH AND SAFETY INFORMATION

Irritates eyes, skin, and respiratory system. Long-term effects include impairment of breathing functions, erosion of exposed front teeth, anxiety, and the formation of acnelike lesions on the skin. Contact with liquid may cause frostbite.

RECOMMENDATIONS FOR RESPIRATOR SELECTION:

At any concentrations above the NIOSH REL, or where there is no REL, at any detectable concentration: SCBAF:PD,PP (any self-contained breathing apparatus that has a full facepiece and is operated in a pressure-demand or other positive-pressure mode);or SAF:PD,PP:ASCBA (any supplied-air respirator that has a full facepiece and is operated in a pressure-demand or other positive-pressure mode in combination with an auxiliary, self-contained breathing apparatus operated in a pressure-demand or other positive-pressure mode).

Escape: GMFOV [any air-purifying, full-facepiece respirator (gas mask) with a chin-style, front- or back-mounted organic vapor canister];or SCBAE (any appropriate escape-type, self-contained breathing apparatus)

CHEMICAL REACTIVITY

Contact with heat or oxidizers may cause fire and explosions.

FIRE INFORMATION

Decomposition products upon heating include toxic chlorine gas. Vapors are heavier than air and will collect in low areas. Vapors in confined areas may explode when exposed to fire. Vapors may travel long distances to ignition sources and flash back. Storage containers and parts of containers may rocket great distances, in many directions. If material or contaminated runoff enters waterways, notify downstream users of potentially contaminated waters. Notify local health and fire officials and pollution control agencies. Use dry chemicals, foam, carbon dioxide. From a secure explosion-proof location, use water spray to cool exposed containers. If cooling streams are ineffective (venting sound increases in volume and pitch, tank discolors, or shows any signs of deforming), withdraw immediately to a secure position.

NFPA Hazard Classifications:

Health Hazards (Blue) 2
Flammability (Red) 4
Reactivity (Yellow) 2

Flash Point: -4°F (-20°C)
Flammable Limits in Air: LEL 4.5%; UEL 16%

FIRST AID

Move victim to fresh air. Call emergency medical care. Apply artificial respiration if victim is not breathing. Administer oxygen if breathing is difficult. Remove and isolate contaminated clothing and shoes. In case of contact with substance, immediately flush skin or eyes with running water for at least 20 minutes. Wash skin with soap and water. Keep victim warm and quiet. Effects of exposure (inhalation, ingestion, or skin contact) to substance may be delayed. Ensure that medical personnel are aware of the material(s) involved and take precautions to protect themselves.

ENVIRONMENTAL INFORMATION

Clean Air Act: Accidental Release Prevention/Flammable Substances (Section 112[r], Table 3), TQ = 10,000 lb (4,540 kg)
Marine Pollutant (49CFR, Subchapter 172.101, Appendix B).

3-CHLOROPROPIONITRILE C:333

SYNONYMS: 3-CHLOROPROPANENITRILE; 3-CHLOROPROPANONITRILE; β-CHLOROPROPIONITRILE; CHLOROPROPIONITRILE (3-); 3-CLOROPROPIONITRILO (Spanish); EINECS No. 208-827-0; PROPANENITRILE,2-CHLORO-; USAF A-8798

IDENTIFICATION

CAS:	542-76-7
ERG ID:	UN 3272
ERG Guide:	131
ERG Guide Category:	FLAMMABLE LIQUIDS-TOXIC
Molecular Formula:	C_3H_4ClN
RTECS No.:	UG1400000
Properties:	Flammable liquid. Colorless. Soluble in water.

HEALTH AND SAFETY INFORMATION

IDLH: 25 mg/m^3 as cyanides
OSHA Table Z-1-A Air Contaminant as cyanides
NIOSH Criteria Document: NIOSH 78-212 NITRILES

Exposure Limits:	**TWA**	**CEILING**
ACGIH: (1)	5 mg/m^3 as cyanide	
NIOSH: (1)		4.7 ppm; 5 mg/m^3 per 10 minutes as cyanides
OSHA: (1)	5 mg/m^3 as cyanides	

(1) Skin contact contributes significantly in overall exposure

RECOMMENDATIONS FOR RESPIRATOR SELECTION: NIOSH as cyanides

50 mg/m^3: SA (any supplied-air respirator);or SCBAF (any self-contained breathing apparatus with full facepiece).

Emergency or Planned Entry into Unknown Concentrations or IDLH Conditions:
SCBAF:PD,PP (any self-contained breathing apparatus that has a full facepiece and is operated in a pressure-demand or other positive-pressure mode);or SAF:PD,PP:ASCBA (any supplied-air respirator that

has a full facepiece and is operated in a pressure-demand or other positive-pressure mode in combination with an auxiliary, self-contained breathing apparatus operated in a pressure-demand or other positive-pressure mode).

Escape: GMFSHiE [any air-purifying, full-facepiece respirator (gas mask) with a chin-style, front- or back-mounted canister providing protection against the compound of concern and having a high-efficiency particulate filter];or SCBAE (any appropriate escape-type, self-contained breathing apparatus).

CHEMICAL REACTIVITY

Contact with oxidizers can cause fire and explosions. Incompatible with sulfuric acid.

FIRE INFORMATION

Combustion products include toxic cyanide, chlorine, and nitrogen oxides. Vapors are heavier than air and will collect in low areas. Vapors in confined areas may explode when exposed to fire. Vapors may travel long distances to ignition sources and flash back. Storage containers and parts of containers may rocket great distances, in many directions. If material or contaminated runoff enters waterways, notify downstream users of potentially contaminated waters. Notify local health and fire officials and pollution control agencies.

NFPA Hazard Classifications:

Health Hazards (Blue) .... -
Flammability (Red) 2
Reactivity (Yellow) 1

Flash Point: 168°F (76°C) c.c.

FIRST AID

Move victim to fresh air. Call emergency medical care. Apply artificial respiration if victim is not breathing. Do not use mouth-to-mouth method if victim ingested or inhaled the substance; induce artificial respiration with the aid of a pocket mask equipped with a one-way valve or other proper respiratory medical device. Administer oxygen if breathing is difficult. Remove and isolate contaminated clothing and shoes. In case of contact with substance, immediately flush skin or eyes with running water for at least 20 minutes. Wash skin with soap and water. Keep victim warm and quiet. Effects of exposure (inhalation, ingestion, or skin contact) to substance may be delayed. Ensure that medical personnel are aware of the material(s) involved and take precautions to protect themselves.

ENVIRONMENTAL INFORMATION

EPA Hazardous Waste Number (RCRA No.): P027
RCRA Toxicity Characteristic (Section 261.24): RCRA Section 261 Hazardous Constituents
RCRA Land Ban Waste
EPCRA Section 302, Extremely Hazardous Substances: TPQ = 1,000 lb (454 kg)

EPCRA Section 304 Reportable Quantity (RQ): CERCLA, 1,000 lb (454 kg)
EPCRA Section 313: Form R *de minimis* concentration reporting level: 1.0%.

CHLOROSULFONIC ACID C:336

SYNONYMS: ACIDO CLOROSULFONICO (Spanish); CHLOROSULFURIC ACID; 4-CHLORO-o-TOLUIDINE, HYDROCHLORIDE; EEC No. 016-017-00-1; MONOCHLOROSULFURIC ACID; SULFONIC ACID, MONOCHLORIDE; SULFURIC CHLOROHYDRIN

IDENTIFICATION

CAS:	7790-94-5
ERG ID:	UN 1754
ERG Guide:	137
ERG Guide Category:	WATER REACTIVE-CORROSIVE
Molecular Formula:	$ClHO_3S$
RTECS No.:	FX5730000
Properties:	Nonflammable liquid. Colorless to light yellow. Sharp, pungent, choking odor. Reacts violently with water.

HEALTH AND SAFETY INFORMATION

Poison. Breathing the vapors or swallowing the liquid can cause death. Skin and eye contact causes severe burns and blindness.

IARC Evaluation: Group 2A unspecified as o-toluidine
Odor Threshold: 1-5 ppm
OSHA Table Z-1-A Air Contaminant as o-toluidine

Exposure Limits:	TWA	STEL
AIHA WEEL	0.3	5 ppm for 5 mins., 30 ppm for 10 mins., 10 ppm for 60 mins.

RECOMMENDATIONS FOR RESPIRATOR SELECTION:

At any concentrations above the NIOSH REL, or where there is no REL, at any detectable concentration: SCBAF:PD,PP (any self-contained breathing apparatus that has a full facepiece and is operated in a pressure-demand or other positive-pressure mode);or SAF:PD,PP:ASCBA (any supplied-air respirator that has a full facepiece and is operated in a pressure-demand or other positive-pressure mode in combination with an auxiliary, self-contained breathing apparatus operated in a pressure-demand or other positive pressure mode).

Escape: GMFOV [any air-purifying, full-facepiece respirator (gas mask) with a chin-style, front- or back-mounted organic vapor canister];or SCBAE (any appropriate escape-type, self-contained breathing apparatus).

CHEMICAL REACTIVITY

Reacts violently with water, forming hydrochloric acid (vapor) and sulfuric acid. A strong oxidizer; contact with reducing agents and combustible materials may cause fire and explosions. Reacts with air, forming corrosive fumes. Reacts with organic materials, acids, acrolein, alcohols, ammonia, anhydrides, esters, ketones, hydrogen peroxide, finely divided metals or powders, nitriles. Hydrogen, a highly flammable and explosive gas, is generated by the action of the acid on most metals. Corrosive to most metals.

FIRE INFORMATION

Firefighting gear (including SCBA) provides NO protection. If exposure occurs, remove and isolate gear immediately and thoroughly decontaminate personnel. Vapors are heavier than air and will collect in low areas. Vapors in confined areas may explode when exposed to fire. Storage containers and parts of containers may rocket great distances, in many directions. If material or contaminated runoff enters waterways, notify downstream users of potentially contaminated waters. Notify local health and fire officials and pollution control agencies.

NFPA Hazard Classifications:

Health Hazards (Blue) 4
Flammability (Red) 0
Reactivity (Yellow) 2
Special situations . Water Reactive, OXY

FIRST AID

Move victim to fresh air. Call emergency medical care. Apply artificial respiration if victim is not breathing. Do not use mouth-to-mouth method if victim ingested or inhaled the substance; induce artificial respiration with the aid of a pocket mask equipped with a one-way valve or other proper respiratory medical device. Administer oxygen if breathing is difficult. Remove and isolate contaminated clothing and shoes. In case of contact with substance, immediately flush skin or eyes with running water for at least 20 minutes. For minor skin contact, avoid spreading material on unaffected skin. Removal of solidified molten material from skin requires medical assistance. Keep victim warm and quiet. Effects of exposure (inhalation, ingestion, or skin contact) to substance may be delayed. Ensure that medical personnel are aware of the material(s) involved and take precautions to protect themselves.

ENVIRONMENTAL INFORMATION

Clean Water Act: Section 311 Hazardous Substances/RQ (same as CERCLA)
EPCRA Section 304 Reportable Quantity (RQ): CERCLA, 1,000 lb (454 kg)

m–CHLOROTOLUENE C:339

SYNONYMS: 3-CHLOROTOLUENE; 1-CHLORO-3-METHYLBENZENE; 3-CHLORO-1-METHYLBENZENE; m-CLOROTOLUENO (Spanish); m-TOLYL CHLORIDE

274 m-Chlorotoluene

IDENTIFICATION

CAS:	108-41-8
ERG ID:	UN 2238
ERG Guide:	130
ERG Guide Category:	FLAMMABLE LIQUIDS (NONPOLAR/WATER-IMMISCIBLE/NOXIOUS)
Molecular Formula:	C_7H_7Cl
Properties:	Combustible liquid. Colorless. Pungent odor. Sinks in water.

HEALTH AND SAFETY INFORMATION

Irritates eyes, skin, and respiratory tract.

Exposure Limits:	**TWA**	**STEL**
ACGIH:	50 ppm; 259 mg/m^3 (as o-chlorotoluene)	
NIOSH:	50 ppm; 250 mg/m^3	75 ppm; 375 mg/m^3 (as o-chlorotoluene)
OSHA:	None	

RECOMMENDATIONS FOR RESPIRATOR SELECTION:

At any concentrations above the NIOSH REL, or where there is no REL, at any detectable concentration: SCBAF:PD,PP (any self-contained breathing apparatus that has a full facepiece and is operated in a pressure-demand or other positive-pressure mode);or SAF:PD,PP:ASCBA (any supplied-air respirator that has a full facepiece and is operated in a pressure-demand or other positive-pressure mode in combination with an auxiliary, self-contained breathing apparatus operated in a pressure-demand or other positive-pressure mode).

Escape: GMFOV [any air-purifying, full-facepiece respirator (gas mask) with a chin-style, front- or back-mounted organic vapor canister];or SCBAE (any appropriate escape-type, self-contained breathing apparatus).

CHEMICAL REACTIVITY

A halogenated hydrocarbon. Contact with oxidizers can cause fire and explosions. Reacts with acids, alkalis, reducing materials, and may other materials.

FIRE INFORMATION

Vapors are heavier than air and will collect in low areas. Vapors in confined areas may explode when exposed to fire. Vapors may travel long distances to ignition sources and flash back. Storage containers and parts of containers may rocket great distances, in many directions. If material or contaminated runoff enters waterways, notify downstream users of potentially contaminated waters. Notify local health and fire officials and pollution control agencies.

NFPA Hazard Classifications:

Health Hazards (Blue) 2
Flammability (Red) 2
Reactivity (Yellow) 0

Flash Point: 123°F (51°C) c.c.

FIRST AID

Move victim to fresh air. Call emergency medical care. Apply artificial respiration if victim is not breathing. Administer oxygen if breathing is difficult. Remove and isolate contaminated clothing and shoes. In case of contact with substance, immediately flush skin or eyes with running water for at least 20 minutes. Wash skin with soap and water. Keep victim warm and quiet. Effects of exposure (inhalation, ingestion, or skin contact) to substance may be delayed. Ensure that medical personnel are aware of the material(s) involved and take precautions to protect themselves.

o–CHLOROTOLUENE C:340

SYNONYMS: BENZENE, 1-CHLORO-2-METHYL- (9CI); 1-CHLORO-2-METHYLBENZENE; 2-CHLORO-1-METHYLBENZENE; 2-CHLOROTOLUENE; o-CLOROTOLUENO (Spanish); 1-METHYL-2-CHLOROBENZENE; 2-METHYLCHLOROBENZENE; MICROPOSIT 119 THINNER; TOLUENE, o-CHLORO-; o-TOLYLCHLORIDE; BENZENE, o-TOLYL CHLORIDE

IDENTIFICATION

CAS:	95-49-8
ERG ID:	UN 2238
ERG Guide:	130
ERG Guide Category:	FLAMMABLE LIQUIDS (NON-POLAR/WATER IMMISCIBLE/NOXIOUS)
Molecular Formula:	C_7H_7Cl
RTECS No.:	XS9000000
Properties:	Highly flammable liquid. Colorless. Aromatic odor. Sinks in water; insoluble. Corrosive and toxic hydrogen chloride gases produced in fire.

HEALTH AND SAFETY INFORMATION

Irritating to the eyes, skin, and respiratory tract. Long-term or prolonged exposure may affect the liver.

Odor Threshold: 0.219 ppm

Exposure Limits:	**TWA**	**STEL**
ACGIH:	50 ppm; 259 mg/m^3	
NIOSH:	50 ppm; 250 mg/m^3	75 ppm; 375 mg/m^3
OSHA:	None	

RECOMMENDATIONS FOR RESPIRATOR SELECTION:

At any concentrations above the NIOSH REL, or where there is no REL, at any detectable concentration: SCBAF:PD,PP (any self-contained breathing apparatus that has a full facepiece and is operated in a pressure-demand or other positive-pressure mode);or SAF:PD,PP:ASCBA (any supplied-air respirator that has a full facepiece and is operated in a pressure-demand or other positive-pressure mode in combination with an auxiliary, self-contained breathing apparatus operated in a pressure-demand or other positive pressure mode)

Escape: GMFOV [any air-purifying, full-facepiece respirator (gas mask) with a chin-style, front- or back-mounted organic vapor canister];or SCBAE (any appropriate escape-type, self-contained breathing apparatus).

CHEMICAL REACTIVITY

Reactivity with water: Forms explosive mixture with air. Reacts with water. Contact with oxidizers may cause fire or explosions.

FIRE INFORMATION

Combustion products include toxic hydrogen chloride.Vapors are heavier than air and will collect in low areas. Vapors in confined areas may explode when exposed to fire. Vapors may travel long distances to ignition sources and flash back. Storage containers and parts of containers may rocket great distances, in many directions. If material or contaminated runoff enters waterways, notify downstream users of potentially contaminated waters. Notify local health and fire officials and pollution control agencies.

NFPA Hazard Classifications:

Health Hazards (Blue) 2
Flammability (Red) 2
Reactivity (Yellow) 0

Flash Point: 126°F (52°C) o.c.; 96°F (36°C) c,c.
Flammable Limits in Air: LEL 1.36% (calculated); UEL NA.

FIRST AID

Move victim to fresh air. Call emergency medical care. Apply artificial respiration if victim is not breathing. Administer oxygen if breathing is difficult. Remove and isolate contaminated clothing and shoes. In case of contact with substance, immediately flush skin or eyes with running water for at least 20

minutes. Wash skin with soap and water. Keep victim warm and quiet. Effects of exposure (inhalation, ingestion, or skin contact) to substance may be delayed. Ensure that medical personnel are aware of the material(s) involved and take precautions to protect themselves.

ENVIRONMENTAL INFORMATION

TSCA: 40CFR 712.30(*e*)10; 40CFR 716.120(*d*)10 dermal absorption testing; 40CFR 716.120(*a*)

p-CHLOROTOLUENE C:341

SYNONYMS: 4-CHLORO-1-METHYLBENZENE; 4-CHLOROTOLUENE;1-CHLORO-4-METHYLBENZENE; 4-CHLORO-1-METHYLBENZENE; 4-CHLOROTOLUENE; 4-CLOROTOLUENO (Spanish); p-CLOROTOLUENO (Spanish); p-TOLYL CHLORIDE;

IDENTIFICATION

CAS:	106-43-4
ERG ID:	UN 2238
ERG Guide:	130
ERG Guide Category:	FLAMMABLE LIQUIDS (NONPOLAR/WATER-IMMISCIBLE/NOXIOUS)
Molecular Formula:	C_7H_7Cl
Properties:	Very flammable liquid. Colorless. Aromatic odor. Sinks slowly in water; insoluble. Freezes at 46°F (8°C).

HEALTH AND SAFETY INFORMATION

Very irritating to the eyes, skin, and respiratory tract.

RECOMMENDATIONS FOR RESPIRATOR SELECTION:

At any concentrations above the NIOSH REL, or where there is no REL, at any detectable concentration: SCBAF:PD,PP (any self-contained breathing apparatus that has a full facepiece and is operated in a pressure-demand or other positive-pressure mode);or SAF:PD,PP:ASCBA (any supplied-air respirator that has a full facepiece and is operated in a pressure-demand or other positive-pressure mode in combination with an auxiliary, self-contained breathing apparatus operated in a pressure-demand or other positive-pressure mode).
Escape: GMFOV [any air-purifying, full-facepiece respirator (gas mask) with a chin-style, front- or back-mounted organic vapor canister];or SCBAE (any appropriate escape-type, self-contained breathing apparatus).

CHEMICAL REACTIVITY

A halogenated hydrocarbon. Contact with oxidizers can cause fire and explosions. Reacts with acids, alkalis, reducing materials, and many other materials.

FIRE INFORMATION

Combustion products include toxic hydrogen chloride. Vapors are heavier than air and will collect in low areas. Vapors in confined areas may explode when exposed to fire. Vapors may travel long distances to ignition sources and flash back. Storage containers and parts of containers may rocket great distances, in many directions. If material or contaminated runoff enters waterways, notify downstream users of potentially contaminated waters. Notify local health and fire officials and pollution control agencies.

NFPA Hazard Classifications:

Health Hazards (Blue) 2
Flammability (Red) 2
Reactivity (Yellow) 0

Flash Point: 120°F (49°C) o.c.

FIRST AID

Move victim to fresh air. Call emergency medical care. Apply artificial respiration if victim is not breathing. Administer oxygen if breathing is difficult. Remove and isolate contaminated clothing and shoes. In case of contact with substance, immediately flush skin or eyes with running water for at least 20 minutes. Wash skin with soap and water. Keep victim warm and quiet. Effects of exposure (inhalation, ingestion, or skin contact) to substance may be delayed. Ensure that medical personnel are aware of the material(s) involved and take precautions to protect themselves.

CHROMIC ANHYDRIDE C:400

SYNONYMS: ACIDO CROMICO (Spanish); ACIDE CHROMIQUE (French); ANHYDRIDE CHROMIQUE (French); ANIDRIDE CROMICA (Italian); CHROME(TRIOXYDE de) (French); CHROMIC ACID; CHROMIC(VI) ACID; CHROMIC(6+) ACID; CHROMIC OXIDE; CHROMIC TRIOXIDE; CHROMIUM(VI) OXIDE (1:3); CHROMIUM(6+) OXIDE (1:3); CHROMIHYDROXID (German); CHROMIUM HYDROXIDE; CHROMIUM TRIOXIDE; CROMO(TRIOSSIDO de) (Italian); CHROMSAEUREANHYDRID (German); CHROMTRIOXID (German); CROOMTRIOXYDE (Dutch); CROOMZUURANHYDRIDE (Dutch); DICHROMIUM TRIOXIDE HYDRATE; DICHROOMTRIOXIDEHYDRAAT (Dutch); EINECS No. 215-158-8; HIDROXIDO CROMICO (Spanish); HYDROXYDE CHROMIQUE (French); IDROSSIDO CROMICO (Italian); MONOCHROMIUM OXIDE; MONOCHROMIUM TRIOXIDE; PURATRONIC CHROMIUM TRIOXIDE; TRIOXYDE de DICHROMEHYDRATE (French)

IDENTIFICATION

CAS:	1333-82-0
ERG ID:	UN 1463
ERG Guide:	141
ERG Guide Category:	OXIDIZERS-TOXIC (SOLID)
Molecular Formula:	CrO_3
RTECS No.:	GB6650000
Properties:	Nonflammable dark red or brown solid flakes or powder. Odorless. Sinks and mixes slowly with water.

HEALTH AND SAFETY INFORMATION

IARC Evaluation: Unassigned substance, limited animal evidence.
NTP: Anticipated carcinogen.
IDLH: 15 mg/m^3 as chromium(VI); potential carcinogen.
OSHA Table Z-1-A Air Contaminant as chromic acid (CrO_3).

Exposure Limits:	**TWA**	**CEILING**
ACGIH:	0.05 mg/m^3 as Cr(VI)	
NIOSH:[(1)]	0.001 mg/m^3 as CR(VI)	
OSHA: [(1)]		0.1 mg/m^3 as CrO_3

[(1)] Potential human carcinogen. Reduce exposure to lowest feasible level.

RECOMMENDATIONS FOR RESPIRATOR SELECTION: NIOSH

At any concentrations above the NIOSH REL, or where there is no REL, at any detectable concentration: SCBAF:PD,PP (any self-contained breathing apparatus that has a full facepiece and is operated in a pressure-demand or other positive-pressure mode);or SAF:PD,PP:ASCBA (any supplied-air respirator that has a full facepiece and is operated in a pressure-demand or other positive-pressure mode in combination with an auxiliary, self-contained breathing apparatus operated in a pressure-demand or other positive-pressure mode).

Escape: HiEF (any air-purifying, full-facepiece respirator with a high-efficiency particulate filter);or SCBAE (any appropriate escape-type, self-contained breathing apparatus).

Note: Only nonoxidiziable sorbents should be used (not charcoal).

CHEMICAL REACTIVITY

A strong oxidizer. Reacts with acetic acid, acetic anhydride, acetone, anthracene, chromous sulfide, diethyl ether, dimethyl formamide, ethanol, hydrogen sulfide, methanol, naphthalene, camphor, glycerol, potassium ferricyanide, pyridine, turpentine, combustibles, and organic compounds. Aqueous solution is strongly acidic. Attacks metals in presence of moisture or water vapor. Reacts with bases. May react with organic materials rapidly enough to generate sufficient heat to cause ignition. Prolonged contact, particularly on wood floors, may produce a fire hazard.

FIRE INFORMATION:

Not flammable but will increase the intensity of an existing fire. Containers may explode in fire. If material or contaminated runoff enters waterways, notify downstream users of potentially contaminated waters. Notify local health and fire officials and pollution control agencies. Use any extinguishing agent suitable for surrounding fire. dry chemicals, alcohol-resistant foam, carbon dioxide. From a secure explosion-proof location, use water spray to cool exposed containers.

FEMA Hazard Classifications:

Health Hazards (Blue) 1
Flammability (Red) 0
Reactivity (Yellow) 2
Special situations . . . . .OXY

FIRST AID

Move victim to fresh air. Call emergency medical care. Apply artificial respiration if victim is not breathing. Administer oxygen if breathing is difficult. Remove and isolate contaminated clothing and shoes. In case of contact with substance, immediately flush skin or eyes with running water for at least 20 minutes. Keep victim warm and quiet. Ensure that medical personnel are aware of the material(s) involved and take precautions to protect themselves.

ENVIRONMENTAL INFORMATION as chromium compounds

Clean Air Act: Hazardous Air Pollutants (Title I, Part A, Section 112) as chromium compounds.
Clean Water Act: Toxic Pollutant (Section 401.15); Section 307 Toxic Pollutants as chromium and compounds.
RCRA Section 261 Hazardous Constituents: waste number not listed.
EPCRA Section 304: Reportable Quantity (RQ): CERCLA, 10 lb (4.54 kg).
EPCRA Section 313: Includes any unique chemical substances that contain chromium as part of that chemical's infrastructure. Form R *de minimis* concentration reporting level: Chromium(VI) compounds: 0.1%. Form R Toxic Chemical Category Code: N090.
California's Proposition 65: Carcinogen as chromium (hexavalent compounds).

CHROMIUM C:425

SYNONYMS: CHROM (German); CHROME (English, French); CHROMIUM, ELEMENT; CHROMIUM METAL; CROMO (Spanish); EINECS No. 231-157-5

IDENTIFICATION

CAS:	7440-47-3; 12105-81-6
Molecular Formula:	Cr
RTECS No.:	GB4200000
Properties:	Noncombustible blue-white to steel-gray lustrous, brittle, hard metal; silvery metal. Finely divided dust burns rapidly if heated in a flame. Odorless. Insoluble in water.

HEALTH AND SAFETY INFORMATION

Dust, mist, or fumes can cause "metal fume fever," with flu-like symptoms. May cause pulmonary sensitization and lung allergy.

IARC Evaluation: Group 3; unclassifiable substance; inadequate human evidence, inadequate animal evidence.
NTP: Known carcinogen.
IDLH: 250 mg/m^3 as Cr.
OSHA Table Z-1-A Air Contaminant
NIOSH Criteria Document: NIOSH 76-129 (chromium(VI)

Exposure Limits:	**TWA**
ACGIH:	0.5 mg/m^3
NIOSH:*	0.5 mg/m^3
OSHA:*	1 mg/m^3

*Potential occupational carcinogen.

RECOMMENDATIONS FOR RESPIRATOR SELECTION: NIOSH/OSHA

2.5 mg/m^3: DM (any dust and mist respirator).
5 mg/m^3: DMXSQ (any dust and mist respirator except single-use and quarter mask respirators);or SA (any supplied-air respirator).
12.5 mg/m^3: SA:CF (any supplied-air respirator operated in a continuous-flow mode);or PAPRDM (any powered, air-purifying respirator with a dust and mist filter).
25 mg/m^3: HiEF (any air-purifying, full-facepiece respirator with a high-efficiency particulate filter);or PAPRTHiE (any powered, air-purifying respirator with a tight-fitting facepiece and a high-efficiency particulate filter);or SCBAF (any self-contained breathing apparatus with a full facepiece);or SAF (any supplied-air respirator with a full facepiece).
250 mg/m^3: SAF:PD,PP (any supplied-air respirator that has a full facepiece and is operated in a pressure-demand or other positive-pressure mode).

Emergency or Planned Entry into Unknown Concentrations or IDLH Conditions:
SCBAF:PD,PP (any self-contained breathing apparatus that has a full facepiece and is operated in a pressure-demand or other positive-pressure mode);or SAF:PD,PP:ASCBA (any supplied-air respirator that

has a full facepiece and is operated in a pressure-demand or other positive-pressure mode in combination with an auxiliary, self-contained breathing apparatus operated in a pressure-demand or other positive-pressure mode).

Escape: HiEF (any air-purifying, full-facepiece respirator with a high-efficiency particulate filter);or SCBAE (any appropriate escape-type, self-contained breathing apparatus).

CHEMICAL REACTIVITY

Reacts with hydrogen peroxide, nitric oxide, potassium chlorate, sulfur oxide (SO_2). Finely divided chromium forms an explosive mixture with ammonium nitrate. Contact of powder with oxidizers may cause fire and explosions.

FIRE INFORMATION

If material or contaminated runoff enters waterways, notify downstream users of potentially contaminated waters. Notify local health and fire officials and pollution control agencies. Use dry chemicals, alcohol-resistant foam, carbon dioxide. From a secure explosion-proof location, use water spray to cool exposed containers.

FEMA Hazard Classifications: (dust)

Health Hazards (Blue) 0
Flammability (Red) 1
Reactivity (Yellow) 0

FIRST AID

Move victim to fresh air. Call emergency medical care. Apply artificial respiration if victim is not breathing. Administer oxygen if breathing is difficult. Remove and isolate contaminated clothing and shoes. In case of contact with substance, immediately flush skin or eyes with running water for at least 20 minutes. Ensure that medical personnel are aware of the material(s) involved and take precautions to protect themselves.

ENVIRONMENTAL INFORMATION

Clean Air Act: Hazardous Air Pollutants (Title I, Part A, Section 112) as chromium compounds.
Clean Water Act: Section 307 Priority Pollutants; Section 313 Priority Chemicals; Toxic Pollutant (Section 401.15).
EPA Hazardous Waste Number (RCRA No.): D007.
RCRA Section 261 Hazardous Constituents, waste number not listed.
RCRA Toxicity Characteristic (Section 261.24): Maximum Concentration of Contaminants, regulatory level, 5.0 mg/L.
RCRA Land Ban Waste
RCRA Universal Treatment Standards: Wastewater (mg/L), 2.77; Nonwastewater (mg/kg), 0.86 as chromium (total).

RCRA Ground Water Monitoring List: Suggested test method(s) (PQL ug/L): (total) 6010(70); 7190(500); 7191(10).
Safe Drinking Water Act: MCL, 0.1 mg/L; MCLG, 0.1 mg/L; Regulated chemical (47 FR 9352).
EPCRA Section 304: Reportable Quantity (RQ): CERCLA, 5,000 lb (2,270 kg).
EPCRA Section 313: Form R *de minimis* concentration reporting level: 0.1%. Form R Toxic Chemical Category Code: N090.
California's Proposition 65: Carcinogen as chromium (hexavalent compounds).

COBALT C:445

SYNONYMS: AQUACAT; C. I. 77320; COBALT 59; COBALTO (Spanish); EINECS No. 231-158-0; KOBALT (German, Polish); NCI-C60311; SUPER COBALT

IDENTIFICATION

CAS:	7440-48-4
ERG ID:	UN 3077
ERG Guide:	171
ERG Guide Category:	SUBSTANCES (LOW TO MODERATE HAZARD)
Molecular Formula:	Co
RTECS No.:	GF8750000
Properties:	Hard, shiny, silver-gray, magnetic metal. Odorless. Insoluble in water.

HEALTH AND SAFETY INFORMATION

IARC Evaluation: Possible carcinogen, Group 2B; inadequate human evidence.
IDLH: 20 mg/m^3 as cobalt dust and fume.
OSHA Table Z-1-A Air Contaminant
NIOSH Criteria Document: NIOSH 82-107

Exposure Limits:	**TWA**
ACGIH: (1)	0.02 mg/m^3
NIOSH:	0.05 mg/m^3
OSHA:	0.1 mg/m^3

(1) Animal carcinogen

RECOMMENDATIONS FOR RESPIRATOR SELECTION: NIOSH/OSHA

0.25 mg/m^3: DM, *if not present as a fume* (any dust and mist respirator).
0.5 mg/m^3: DMXSQ, *if not present as a fume* (any dust and mist respirator except single-use and quarter mask respirators);or DMFu (any dust, mist, and fume respirator);or SA (any supplied-air respirator).

1.25 mg/m³: SA:CF (any supplied-air respirator operated in a continuous-flow mode);or PAPRDM, *if not present as a fume* (any powered, air-purifying respirator with a dust and mist filter);or PAPRDMFu (any powered, air-purifying respirator with a dust, mist, and fume filter).
2.5 mg/m³: HiEF (any air-purifying, full-facepiece respirator with a high-efficiency particulate filter);or SCBAF (any self-contained breathing apparatus with a full facepiece);or SAF (any supplied-air respirator with a full facepiece).
20 mg/m³: SAF:PD,PP (any supplied-air respirator that has a full facepiece and is operated in a pressure-demand or other positive-pressure mode).

Emergency or Planned Entry into Unknown Concentrations or IDLH Conditions: SCBAF:PD,PP (any self-contained breathing apparatus that has a full facepiece and is operated in a pressure-demand or other positive-pressure mode);or SAF:PD,PP:ASCBA (any supplied-air respirator that has a full facepiece and is operated in a pressure-demand or other positive-pressure mode in combination with an auxiliary, self-contained breathing apparatus operated in a pressure-demand or other positive-pressure mode).

Escape: HiEF (any air-purifying, full-facepiece respirator with a high-efficiency particulate filter);or SCBAE (any appropriate escape-type, self-contained breathing apparatus).

Note: Substance reported to cause eye irritation or damage; may require eye protection.

CHEMICAL REACTIVITY

Reacts with strong oxidizers, ammonium nitrate, sulfuric acid, nitric acid, and nitrates. A noncombustible solid in bulk form, but finely divided dust will burn at high temperatures.

FIRE INFORMATION

Not flammable in bulk form. If material or contaminated runoff enters waterways, notify downstream users of potentially contaminated waters. Notify local health and fire officials and pollution control agencies. Use extinguishing agents suitable for surrounding fire. From a secure explosion-proof location, use water spray to cool exposed containers.

FEMA Hazard Classifications:

Health Hazards (Blue) 1
Flammability (Red) 3
Reactivity (Yellow) 0

FIRST AID

Move victim to fresh air. Call emergency medical care. Apply artificial respiration if victim is not breathing. Administer oxygen if breathing is difficult. Remove and isolate contaminated clothing and shoes. In case of contact with substance, immediately flush skin or eyes with running water for at least 20 minutes. Ensure that medical personnel are aware of the material(s) involved and take precautions to protect themselves.

ENVIRONMENTAL INFORMATION

Clean Air Act: Hazardous Air Pollutants (Title I, Part A, Section 112) as cobalt compounds.
RCRA Ground Water Monitoring List: Suggested test method(s) (PQL ug/L): (total) 6010(70); 7200(500); 7201(10).
EPCRA Section 313: Form R *de minimis* concentration reporting level: 1.0%. Form R Toxic Chemical Category Code: N096.
California's Proposition 65: Carcinogen as cobalt metal powder.
TSCA: 40CFR 716.12(*a*)

COBALTOUS RESINATE, PRECIPITATED C:475

SYNONYMS: COBALT RESINATE, PRECIPITATED; RESINATO COBALTOSO, PRECIPITADO (Spanish)

IDENTIFICATION

CAS:	N/A
ERG ID:	UN 1318
ERG Guide:	133
ERG Guide Category:	FLAMMABLE SOLIDS
Molecular Formula:	$C_{88}H_{124}CoO_8$
RTECS No.:	GG2979800
Properties:	Red-brown powder; insoluble in water. Spontaneously flammable in air; violently reacts with oxidizers.

HEALTH AND SAFETY INFORMATION

IARC Evaluation: Group 2B, possible carcinogen; inadequate human evidence.
IDLH: 20 mg/m^3 as cobalt dust.
OSHA Table Z-1-A Air Contaminant

Exposure Limits:	**TWA**
ACGIH: [1]	0.02 mg/m^3
NIOSH:	0.05 mg/m^3
OSHA:	0.1 mg/m^3

[1] Animal carcinogen

RECOMMENDATIONS FOR RESPIRATOR SELECTION: NIOSH/OSHA

0.25 mg/m^3: DM, *if not present as a fume* (any dust and mist respirator).
0.5 mg/m^3: DMXSQ, *if not present as a fume* (any dust and mist respirator except single-use and quarter mask respirators);or DMFu (any dust, mist, and fume respirator);or SA (any supplied-air respirator).

1.25 mg/m^3: SA:CF (any supplied-air respirator operated in a continuous-flow mode);or PAPRDM, *if not present as a fume* (any powered, air-purifying respirator with a dust and mist filter);or PAPRDMFu (any powered, air-purifying respirator with a dust, mist, and fume filter).
2.5 mg/m^3: HiEF (any air-purifying, full-facepiece respirator with a high-efficiency particulate filter);or SCBAF (any self-contained breathing apparatus with a full facepiece);or SAF (any supplied-air respirator with a full facepiece).
20 mg/m^3: SAF:PD,PP (any supplied-air respirator that has a full facepiece and is operated in a pressure-demand or other positive-pressure mode).

Emergency or Planned Entry into Unknown Concentrations or IDLH Conditions: SCBAF:PD,PP (any self-contained breathing apparatus that has a full facepiece and is operated in a pressure-demand or other positive-pressure mode);or SAF:PD,PP:ASCBA (any supplied-air respirator that has a full facepiece and is operated in a pressure-demand or other positive-pressure mode in combination with an auxiliary, self-contained breathing apparatus operated in a pressure-demand or other positive-pressure mode).

Escape: HiEF (any air-purifying, full-facepiece respirator with a high-efficiency particulate filter);or SCBAE (any appropriate escape-type, self-contained breathing apparatus).

Note: Substance reported to cause eye irritation or damage; may require eye protection.

FIRE INFORMATION

If material or contaminated runoff enters waterways, notify downstream users of potentially contaminated water.

FIRST AID

Move victim to fresh air. Call emergency medical care. Apply artificial respiration if victim is not breathing. Administer oxygen if breathing is difficult. Remove and isolate contaminated clothing and shoes. In case of contact with substance, immediately flush skin or eyes with running water for at least 20 minutes. Removal of solidified molten material from skin requires medical assistance. Keep victim warm and quiet. Ensure that medical personnel are aware of the material(s) involved and take precautions to protect themselves.

ENVIRONMENTAL INFORMATION as cobalt compounds

Clean Air Act: Hazardous Air Pollutants (Title I, Part A, Section 112).

Note: Includes any unique chemical substance that contains cobalt as part of that chemical's infrastructure.

EPCRA Section 313: Includes any unique chemical substance that contains cobalt as part of that chemical's infrastructure. Form R *de minimis* concentration reporting level: 1.0%. Form R Toxic Chemical Category Code: N096.

COLLODION C:495

SYNONYMS: CELLOIDIN; CELLULOSE NITRAT (German); CELLULOSE NITRATE SOLUTION; COLLODION COTTON; COLLODINO-COTONE (Italian); COLLODIONKATOEN (Dutch); COLODION (Spanish); EEC Nos. 603-037-00-6, 603-037-01-3; GUN COTTON; NITRATO de CELULOSA (Spanish); NITROCELLULOSE; NITROCELLULOSE GUM; NITROCELLULOSE SOLUTION; PIROSSILINA (Italian); PIROXILINA (Spanish); PYROXYLIN SOLUTION; BOX TOE GUM; XYLODIDIN

IDENTIFICATION

CAS:	9004-70-0
ERG ID:	UN 2059
ERG Guide:	127
ERG Guide Category:	FLAMMABLE LIQUIDS (POLAR/WATER MISCIBLE)
Molecular Formula:	$C_{25}H_{33}N_7O_{34}$
RTECS No.:	QW09752000
Properties:	Flammable thick liquid or amorphous solid. Colorless to white. Odor depends upon the solvent used; most often ethyl alcohol or etherlike. Floats on water; flammable, irritating vapor is produced. Boiling point is around 94°F.

HEALTH AND SAFETY INFORMATION

May cause dizziness, difficult breathing, loss of consciousness.

RECOMMENDATIONS FOR RESPIRATOR SELECTION

At any concentrations above the NIOSH REL, or where there is no REL, at any detectable concentration: SCBAF:PD,PP (any self-contained breathing apparatus that has a full facepiece and is operated in a pressure-demand or other positive-pressure mode); or SAF:PD,PP:ASCBA (any supplied-air respirator that has a full facepiece and is operated in a pressure-demand or other positive-pressure mode in combination with an auxiliary self-contained breathing apparatus operated in a pressure-demand or other positive-pressure mode).

Escape: GMFOV [any air-purifying, full-facepiece respirator (gas mask) with a chin-style, front- or back-mounted organic vapor canister]; or SCBAE (any appropriate escape-type, self-contained breathing apparatus)

CHEMICAL REACTIVITY

If ether solution evaporates, dry nitrocellulose material is a shock-sensitive explosive and fire hazard. Forms explosive mixture with air. Strong oxidizers may cause fire and explosions. May accumulate static electrical charges, and may cause ignition of its vapors. Attacks some plastics, rubber, and coatings.

FIRE INFORMATION

Poisonous gases may be produced in fire. Vapors are heavier than air and will collect in low areas. Vapors in confined areas may explode when exposed to fire. Vapors may travel long distances to ignition sources and flash back. Storage containers and parts of containers may rocket great distances, in many directions. If material or contaminated runoff enters waterways, notify downstream users of potentially contaminated waters. Notify local health and fire officials and pollution control agencies. Use dry chemicals, foam, carbon dioxide. From a secure explosion-proof location, use water spray to cool exposed containers.

NFPA Hazard Classifications:

Health Hazards (Blue) 1
Flammability (Red) 4
Reactivity (Yellow) 0

Flash Point: -64°F (-53°C) c.c.
Flammable Limits in Air: LEL 1.7%; UEL 48%
Autoignition Temperature: 338°F (206°C)

FIRST AID

Move victim to fresh air. Call emergency medical care. Apply artificial respiration if victim is not breathing. Administer oxygen if breathing is difficult. Remove and isolate contaminated clothing and shoes. In case of contact with substance, immediately flush skin or eyes with running water for at least 20 minutes. Wash skin with soap and water. Keep victim warm and quiet. Ensure that medical personnel are aware of the material(s) involved and take precautions to protect themselves.

COPPER C:500

SYNONYMS: 1721 GOLD; ANAC 110; ARWOOD COPPER; ALLOY 101; ALLOY 102; ALLOY 110; ALLOY 151; ALLOY 155; ALLOY 194; ALLOY 195; ALLOY 210; ALLOY 220; ALLOY 230; ALLOY 240; ALLOY 260; ALLOY 510; ALLOY 511; ALLOY 521; ALLOY 638; ALLOY 725; ALLOY 732; ALLOY 735; ALLOY 762; ALLOY 770; ANAC 110; BRONZE POWDER; CDA 101; CDA 102; CDA 110; CDA 122; C. I. 77400; COBRE (Spanish); COPPER AIRBORNE; COPPER BRONZE; COPPER-MILLED; CUIVRE (French); CUPRUM; EINECS No. 231-159-6; GOLD BRONZE; KAFER COPPER; KOPER (Dutch); KUPFER (German); M1; M2; M3; M4; MI (COPPER); M2 (COPPER); M3 (COPPER); M4 (COPPER); M3R; M35; OFHC Cu; RAME (Italian); RANEY COPPER

IDENTIFICATION

CAS:	7440-50-8
ERG ID:	Not listed
Molecular Formula:	Cu
RTECS No.:	GL5325000

Properties: Lustrous soft, malleable metal. Reddish. Powder or dust are combustible. Odorless. Insoluble in water.

HEALTH AND SAFETY INFORMATION

Long-term exposure may cause liver damage, chronic nose irritation with sores and ulcers, discoloration of the skin and hair, and skin allergy may develop.

IDLH: 100 mg/m^3 as copper dust, mist or fume.
OSHA Table Z-1-A Air Contaminant (copper, dusts and mists).

Exposure Limits:	TWA
ACGIH:	1 mg/m^3 dusts/mists as Cu; 0.2 mg/m^3 fume
NIOSH:	1 mg/m^3 dusts/mists/fumes as Cu
OSHA:	1 mg/m^3 dusts/mists/fumes as Cu

RECOMMENDATIONS FOR RESPIRATOR SELECTION: NIOSH/OSHA

5 mg/m^3: DM (any dust and mist respirator).
10 mg/m^3: DMXSQ, *if not present as a fume* (any dust and mist respirator except single-use and quarter mask respirators);or SA (any supplied-air respirator).
25 mg/m^3: SA:CF (any supplied-air respirator operated in a continuous-flow mode); PAPRDM (any powered, air-purifying respirator with a dust and mist filter).
50 mg/m^3: HiEF (any air-purifying, full-facepiece respirator with a high-efficiency particulate filter);or PAPRTHiE (any powered, air-purifying respirator with a tight-fitting facepiece and a high-efficiency particulate filter);or SCBAF (any self-contained breathing apparatus with a full facepiece);or SAF (any supplied-air respirator with a full facepiece).
100 mg/m^3: SAF:PD,PP (any supplied-air respirator that has a full facepiece and is operated in a pressure-demand or other positive-pressure mode).

Emergency or Planned Entry into Unknown Concentrations or IDLH Conditions:
SCBAF:PD,PP (any self-contained breathing apparatus that has a full facepiece and is operated in a pressure-demand or other positive-pressure mode);or SAF:PD,PP:ASCBA (any supplied-air respirator that has a full facepiece and is operated in a pressure-demand or other positive-pressure mode in combination with an auxiliary, self-contained breathing apparatus operated in a pressure-demand or other positive-pressure mode)

Escape: HiEF (any air-purifying, full-facepiece respirator with a high-efficiency particulate filter);or SCBAE (any appropriate escape-type, self-contained breathing apparatus).
Note: Substance reported to cause eye irritation or damage; may require eye protection.

CHEMICAL REACTIVITY

Forms shock-sensitive compounds with acetylene. Incompatible with acids. Violent reaction if finely divided (powder or dust) copper comes in contact with strong oxidizers.

FIRE INFORMATION

If material or contaminated runoff enters waterways, notify downstream users of potentially contaminated waters. Notify local health and fire officials and pollution control agencies. Use extinguishing agent suitable for surrounding fire.

NFPA Hazard Classifications:

Health Hazards (Blue) 1
Flammability (Red) 0
Reactivity (Yellow) 0

ENVIRONMENTAL INFORMATION

Clean Water Act: Section 307 Priority Pollutants; Section 313 Priority Chemicals; Toxic Pollutant (Section 401.15).
RCRA Ground Water Monitoring List: Suggested test method(s) (PQL ug/L): (total) 6010(60); 7210(200).
Safe Drinking Water Act: MCL, 1.3 mg/L; MCLG, 1 mg/L; SMLC, 1.0 mg/L; Regulated chemical (47 FR 9352).
EPCRA Section 304: Reportable Quantity (RQ): CERCLA, 5,000 lb (2,270 kg).
EPCRA Section 313: Form R *de minimis* concentration reporting level: 1.0%. Form R Toxic Chemical Category Code: N100.

COPPER CYANIDE C:560

SYNONYMS: CIANURO de COBRE (Spanish); CIANURO di RAME (Italian); CUPRICIN; CUPROUS CYANIDE; CYANURE de CUIVRE (French); KUPFERCYANID (German); EINECS No. 208-883-6

IDENTIFICATION

CAS:	544-92-3
ERG ID:	UN 1587
ERG Guide:	151
ERG Guide Category:	TOXIC (NONCOMBUSTIBLE)
Molecular Formula:	CCuN
RTECS No.:	GL7150000
Properties:	White prisms or powder; dark red crystalline solid. Sinks in water.

HEALTH AND SAFETY INFORMATION

IDLH: 25 mg/m^3 as cyanides
OSHA Table Z-1-A Air Contaminant (copper, dusts and mists)
NIOSH Criteria Document: NIOSH 77-108 (cyanide)

Exposure Limits:	**TWA**
ACGIH:	0.2 mg/m^3 fume as Cu; 1 mg/m^3 dust/mist as Cu
NIOSH:	1 mg/m^3 dust/mist as Cu
OSHA:	1 mg/m^3 dust/mist as Cu

RECOMMENDATIONS FOR RESPIRATOR SELECTION: NIOSH as cyanides

50 mg/m³: SA (any supplied-air respirator);or SCBAF (any self-contained breathing apparatus with full facepiece).

Emergency or Planned Entry into Unknown Concentrations or IDLH Conditions: SCBAF:PD,PP (any self-contained breathing apparatus that has a full facepiece and is operated in a pressure-demand or other positive-pressure mode);or SAF:PD,PP:ASCBA (any supplied-air respirator that has a full facepiece and is operated in a pressure-demand or other positive-pressure mode in combination with an auxiliary, self-contained breathing apparatus operated in a pressure-demand or other positive-pressure mode).

Escape: GMFSHiE [any air-purifying, full-facepiece respirator (gas mask) with a chin-style, front- or back-mounted canister providing protection against the compound of concern and having a high-efficiency particulate filter];or SCBAE (any appropriate escape-type, self-contained breathing apparatus).

CHEMICAL REACTIVITY

May form hydrogen cyanide with water. Reacts with chlorine, fluorine, peroxides.

FIRE INFORMATION

Combustion products include toxic cyanide and nitrogen oxides. If material or contaminated runoff enters waterways, notify downstream users of potentially contaminated waters. Notify local health and fire officials and pollution control agencies. Use any extinguishing agent suitable for surrounding fire. From a secure explosion-proof location, use water spray to cool exposed containers.

FIRST AID

Move victim to fresh air. Call emergency medical care. Apply artificial respiration if victim is not breathing. Do not use mouth-to-mouth method if victim ingested or inhaled the substance; induce artificial respiration with the aid of a pocket mask equipped with a one-way valve or other proper respiratory medical device. Administer oxygen if breathing is difficult. Remove and isolate contaminated clothing and shoes. In case of contact with substance, immediately flush skin or eyes with running water for at least 20 minutes. For minor skin contact, avoid spreading material on unaffected skin. Keep victim warm and quiet. Effects of exposure (inhalation, ingestion, or skin contact) may be delayed. Ensure that medical personnel are aware of the material(s) involved and take precautions to protect themselves.

ENVIRONMENTAL INFORMATION

Clean Air Act: Hazardous Air Pollutants (Title I, Part A, Section 112) as cyanide compounds
Clean Water Act: Toxic Pollutant (Section 401.15).
EPA Hazardous Waste Number (RCRA No.): P029.
RCRA Section 261 Hazardous Constituents
RCRA Land Ban Waste
EPCRA Section 304: Reportable Quantity (RQ): CERCLA, 10 lb (4.54 kg).
RCRA Ground Water Monitoring List: Suggested test method(s) (PQL ug/L): 6010(60); 7210(200). *Note:* All species in the ground water that contain copper are included.
EPCRA Section 313: Form R *de minimis* concentration reporting level: 1.0%. Form R Toxic Chemical Category Code: N100 (copper). Form R Toxic Chemical Category Code: N016 (cyanide).
Marine Pollutant (49CFR, Subchapter 172.101, Appendix B).

COPPER OXIDE C:600

SYNONYMS: BLACK COPPER OXIDE; COPPER MONOXIDE; COPPER(II) OXIDE; CUPRIC OXIDE; EINECS No. 215-269-1; KOPER OXIDE (Dutch); KUPFEROXID (German); OSSIDO di COBRE (Italian); OXIDO de COBRE (Spanish); OXYDE de CUIVRE (French); UP13600

IDENTIFICATION

CAS:	1317-39-1
ERG ID:	Not listed
Molecular Formula:	CuO
RTECS No.:	GL8050000
Properties:	Brownish black powder. Very low solubility in water.

HEALTH AND SAFETY INFORMATION

IDLH: 100 mg/m^3 as copper dusts, mists or fume.
OSHA Table Z-1-A Air Contaminant (copper, dusts and mists)

Exposure Limits:	**TWA**
ACGIH:	0.2 mg/m^3 fume as Cu; 1 mg/m^3 dust/mist as Cu
NIOSH:	1 mg/m^3 dust/mist as Cu
OSHA:	1 mg/m^3 dust/mist as Cu

RECOMMENDATIONS FOR RESPIRATOR SELECTION: NIOSH/OSHA as copper

5 mg/m^3: DM (any dust and mist respirator).
10 mg/m^3: DMXSQ, *if not present as a fume* (any dust and mist respirator except single-use and quarter mask respirators);or SA (any supplied-air respirator).

25 mg/m³: SA:CF (any supplied-air respirator operated in a continuous-flow mode); PAPRDM (any powered, air-purifying respirator with a dust and mist filter).
50 mg/m³: HiEF (any air-purifying, full-facepiece respirator with a high-efficiency particulate filter);or PAPRTHiE (any powered, air-purifying respirator with a tight-fitting facepiece and a high-efficiency particulate filter);or SCBAF (any self-contained breathing apparatus with a full facepiece);or SAF (any supplied-air respirator with a full facepiece).
100 mg/m³: SAF:PD,PP (any supplied-air respirator that has a full facepiece and is operated in a pressure-demand or other positive-pressure mode).

Emergency or Planned Entry into Unknown Concentrations or IDLH Conditions: SCBAF:PD,PP (any self-contained breathing apparatus that has a full facepiece and is operated in a pressure-demand or other positive-pressure mode);or SAF:PD,PP:ASCBA (any supplied-air respirator that has a full facepiece and is operated in a pressure-demand or other positive-pressure mode in combination with an auxiliary, self-contained breathing apparatus operated in a pressure-demand or other positive-pressure mode).

Escape: HiEF (any air-purifying, full-facepiece respirator with a high-efficiency particulate filter);or SCBAE (any appropriate escape-type, self-contained breathing apparatus).

Note: Substance reported to cause eye irritation or damage; may require eye protection.

CHEMICAL REACTIVITY

Reacts with acetylene and zirconium.

FIRE INFORMATION

If material or contaminated runoff enters waterways, notify downstream users of potentially contaminated waters. Notify local health and fire officials and pollution control agencies. Use any extinguishing agent suitable for surrounding fire. From a secure explosion-proof location, use water spray to cool exposed containers.

NFPA Hazard Classifications:

Health Hazards (Blue) 2
Flammability (Red) 0
Reactivity (Yellow) 0

FIRST AID

Move victim to fresh air. Call emergency medical care. Apply artificial respiration if victim is not breathing. Administer oxygen if breathing is difficult. Remove and isolate contaminated clothing and shoes. In case of contact with substance, immediately flush skin or eyes with running water for at least 20 minutes. Wash skin with soap and water. Keep victim warm and quiet. Effects of exposure (inhalation, ingestion, or skin contact) to substance may be delayed. Ensure that medical personnel are aware of the material(s) involved, and take precautions to protect themselves.

ENVIRONMENTAL INFORMATION as copper compounds

Clean Water Act: Toxic Pollutant (Section 401.15) as copper and compounds.
RCRA Ground Water Monitoring List: Suggested test method(s) (PQL ug/L): 6010(60); 7210(200). *Note:* All species in the ground water that contain copper are included.
EPCRA Section 313: Includes any unique chemical substance that contains copper as part of that chemical's infrastructure. This category does not include copper phthalocyanide compounds that are substituted with only hydrogen and/or chlorine and/or bromine. Form R *de minimis* concentration reporting level: 1.0%. Form R Toxic Chemical Category Code: N100.

COPPER(I) OXIDE C:605

SYNONYMS: BROWN COPPER OXIDE; C. I. 77402; COPPER HEMIOXIDE; COPPER NORDOX; COPPER PROTOXIDE; COPPER SARDEX; COPPER SUBOXIDE; COPOX; CUPROUS OXIDE; DICOPPER MONOXIDE; DICOPPER OXIDE; DIKOPEROXIDE (Dutch); DIKUPFEROXID (German); EEC No. 029-002-00-X; EINECS No. 215-270-7; FUNGIMAR; KUPFEROXYDUL (German); OLEOCUIVRE (French); OLEO NORDOX; OSSIDO di DIRAME (Italian); OSSIDO di RAME(I) (Italian); OXIDO de DICOBRE(I) (Spanish); OXIDO ROJO de COBRE (Spanish); OXYDE de CUIVRE(I) (French); RED COPPER OXIDE; YELLOW CUPROCIDE

IDENTIFICATION

CAS:	1317-39-1
Molecular Formula:	Cu_2O
RTECS No.:	GL8050000
Properties:	Cubic red or reddish-brown crystals; yellow crystals or powder. Odorless. Insoluble in water.

HEALTH AND SAFETY INFORMATION

IDLH: 100 mg/m^3 as copper dust, mist, or fume.
OSHA Table Z-1-A Air Contaminant (copper, dusts and mists)

Exposure Limits:	**TWA**
ACGIH:	1 mg/m^3 dusts/mists as Cu; 0.2 mg/m^3 fume
NIOSH:	1 mg/m^3 dusts/mists as Cu; 0.1 mg/m^3 fumes as Cu
OSHA:	1 mg/m^3 dusts/mists as Cu; 0.1 mg/m^3 fumes as Cu

RECOMMENDATIONS FOR RESPIRATOR SELECTION: NIOSH/OSHA as copper

5 mg/m^3: DM (any dust and mist respirator).
10 mg/m^3: DMXSQ, *if not present as a fume* (any dust and mist respirator except single-use and quarter mask respirators);or SA (any supplied-air respirator).

25 mg/m³: SA:CF (any supplied-air respirator operated in a continuous-flow mode); PAPRDM (any powered, air-purifying respirator with a dust and mist filter).
50 mg/m³: HiEF (any air-purifying, full-facepiece respirator with a high-efficiency particulate filter);or PAPRTHiE (any powered, air-purifying respirator with a tight-fitting facepiece and a high-efficiency particulate filter);or SCBAF (any self-contained breathing apparatus with a full facepiece);or SAF (any supplied-air respirator with a full facepiece).
100 mg/m³: SAF:PD,PP (any supplied-air respirator that has a full facepiece and is operated in a pressure-demand or other positive-pressure mode).

Emergency or Planned Entry into Unknown Concentrations or IDLH Conditions: SCBAF:PD,PP (any self-contained breathing apparatus that has a full facepiece and is operated in a pressure-demand or other positive-pressure mode);or SAF:PD,PP:ASCBA (any supplied-air respirator that has a full facepiece and is operated in a pressure-demand or other positive-pressure mode in combination with an auxiliary, self-contained breathing apparatus operated in a pressure-demand or other positive-pressure mode).

Escape: HiEF (any air-purifying, full-facepiece respirator with a high-efficiency particulate filter);or SCBAE (any appropriate escape-type, self-contained breathing apparatus).
Note: Substance reported to cause eye irritation or damage; may require eye protection..

CHEMICAL REACTIVITY

Reacts when heated with aluminum. Forms explosive with concentrated peroxyformic acid.

FIRE INFORMATION

If material or contaminated runoff enters waterways, notify downstream users of potentially contaminated waters. Notify local health and fire officials and pollution control agencies. Use any extinguishing agent suitable for surrounding fire. From a secure explosion-proof location, use water spray to cool exposed containers.

FIRST AID

Move victim to fresh air. Call emergency medical care. Apply artificial respiration if victim is not breathing. Do not use mouth-to-mouth method if victim ingested or inhaled the substance; induce artificial respiration with the aid of a pocket mask equipped with a one-way valve or other proper respiratory medical device. Administer oxygen if breathing is difficult. Remove and isolate contaminated clothing and shoes. In case of contact with substance, immediately flush skin or eyes with running water for at least 20 minutes. Wash skin with soap and water. Keep victim warm and quiet. Effects of exposure (inhalation, ingestion, or skin contact) to substance may be delayed. Ensure that medical personnel are aware of the material(s) involved and take precautions to protect themselves.

ENVIRONMENTAL INFORMATION as copper compounds

Clean Water Act: Toxic Pollutant (Section 401.15) as copper and compounds.

RCRA Ground Water Monitoring List: Suggested test method(s) (PQL ug/L): 6010(60); 7210(200) *Note:* All species in the ground water that contain copper are included.
EPCRA Section 313: Includes any unique chemical substance that contains copper as part of that chemical's infrastructure. This category does not include copper phthalocyanide compounds that are substituted with only hydrogen and/or chlorine and/or bromine. Form R *de minimis* concentration reporting level: 1.0%. Form R Toxic Chemical Category Code: N100.

COPPER SULFATE C:620

SYNONYMS: BASICOP; BCS COPPER FUNGICIDE; BLUE COPPER; BLUE STONE; BLUE VITRIOL; COPPER MONOSULFATE; COPPER SULFATE (1:1); COPPER(II) SULFATE; COPPER(2+) SULFATE; COPPER(2+) SULFATE (1:1); COPPER SULFATE PENTAHYDRATE; CP BASIC SULFATE; CUPRIC SULFATE ANHYDROUS; CUPRIC SULPHATE; EINECS No. 231-847-6; GRIFFIN SUPER Cu; KUPFERSULFAT (German); PHYTO-BORDEAUX; ROMAN VITRIOL; SULFATE de CUIVRE (French); SULFURIC ACID, COPPER(2+) SALT (1:1); SULFATE OF COPPER; SULFATO de COBRE (Spanish); TNCS 53; TRIANGLE

IDENTIFICATION

CAS:	7758-98-7
ERG ID:	UN 9109
ERG Guide:	171
ERG Guide Category:	LOW TO MODERATE HAZARD
Molecular Formula:	CuH_2O_4S
RTECS No.:	GL8800000
Properties:	Solid granules, powder, or crystals. Grayish white to greenish-white; in water bright blue. Odorless. Sinks and mixes with water; highly soluble.

HEALTH AND SAFETY INFORMATION

Can cause irritation of the eyes and skin; possible allergen.

IDLH: 100 mg/m^3 as copper dusts, mists, or fume.
OSHA Table Z-1-A Air Contaminant(copper, dusts and mists)

Exposure Limits:	**TWA**
ACGIH:	0.2 mg/m^3 fume as Cu; 1 mg/m^3 dust/mist as Cu
NIOSH:	0.1 mg/m^3 fume as Cu; 1 mg/m^3 dust/mist as Cu
OSHA:	0.1 mg/m^3 fume as Cu; 1 mg/m^3 dust/mist as Cu

RECOMMENDATIONS FOR RESPIRATOR SELECTION: NIOSH/OSHA as copper

5 mg/m^3: DM (any dust and mist respirator).

10 mg/m^3: DMXSQ, *if not present as a fume* (any dust and mist respirator except single-use and quarter mask respirators);or SA (any supplied-air respirator).
25 mg/m^3: SA:CF (any supplied-air respirator operated in a continuous-flow mode); PAPRDM (any powered, air-purifying respirator with a dust and mist filter).
50 mg/m^3: HiEF (any air-purifying, full-facepiece respirator with a high-efficiency particulate filter);or PAPRTHiE (any powered, air-purifying respirator with a tight-fitting facepiece and a high-efficiency particulate filter);or SCBAF (any self-contained breathing apparatus with a full facepiece);or SAF (any supplied-air respirator with a full facepiece).
100 mg/m^3: SAF:PD,PP (any supplied-air respirator that has a full facepiece and is operated in a pressure-demand or other positive-pressure mode).

Emergency or Planned Entry into Unknown Concentrations or IDLH Conditions: SCBAF:PD,PP (any self-contained breathing apparatus that has a full facepiece and is operated in a pressure-demand or other positive-pressure mode);or SAF:PD,PP:ASCBA (any supplied-air respirator that has a full facepiece and is operated in a pressure-demand or other positive-pressure mode in combination with an auxiliary, self-contained breathing apparatus operated in a pressure-demand or other positive-pressure mode).

Escape: HiEF (any air-purifying, full-facepiece respirator with a high-efficiency particulate filter);or SCBAE (any appropriate escape-type, self-contained breathing apparatus).
Note: Substance reported to cause eye irritation or damage; may require eye protection.

CHEMICAL REACTIVITY

Aqueous solution is an acid. Incompatible with strong bases, hydroxylamine, and magnesium.

FIRE INFORMATION

If material or contaminated runoff enters waterways, notify downstream users of potentially contaminated waters. Notify local health and fire officials and pollution control agencies. Use any extinguishing agent suitable for surrounding fire. From a secure explosion-proof location, use water spray to cool exposed containers.

FIRST AID

Move victim to fresh air. Call emergency medical care. Apply artificial respiration if victim is not breathing. Administer oxygen if breathing is difficult. In case of contact with substance, immediately flush skin or eyes with running water for at least 20 minutes. Remove and isolate contaminated clothing and shoes. Ensure that medical personnel are aware of the material(s) involved and take precautions to protect themselves.

ENVIRONMENTAL INFORMATION

Clean Water Act: Section 311 Hazardous Substances/RQ (same as CERCLA); Section 313 Priority Chemicals; Toxic Pollutant (Section 401.15) as copper and compounds.
RCRA Ground Water Monitoring List: Suggested test method(s) (PQL ug/L): 6010(60); 7210(200)

Note: All species in the ground water that contain copper are included.
EPCRA Section 304: Reportable Quantity (RQ): CERCLA, 10 lb (4.54 kg).
EPCRA Section 313: Form R *de minimis* concentration reporting level: 1.0%. Form R Toxic Chemical Category Code: N100 as copper compounds.

CREOSOTE, COAL TAR C:635

SYNONYMS: AWPA #1; BRICK OIL; COAL TAR OIL; COAL TAR CREOSOTE; CREOSOTE FROM COAL TAR; CREOSOTE OIL (NFPA); CREOSOTE P1; CREOSOTUM; CRESYLIC CREOSOTE; DEAD OIL; HEAVY OIL; LIQUID PITCH OIL; NAPHTHALENE OIL; PRESERV-O-SOTE; TAR OIL; WASH OIL

IDENTIFICATION

CAS:	8001-58-9
ERG ID:	UN 1136
ERG Guide:	128
ERG Guide Category:	FLAMMABLE LIQUIDS (NONPOLAR/WATER-IMMISCIBLE)
Molecular Formula:	A mixture
RTECS No.:	GF8655000 (coal tar pitch volatiles)
Properties:	Combustible liquid, yellow to black. Tarry, smoky odor. May float or sink in water; practically insoluble.

HEALTH AND SAFETY INFORMATION

Repeated exposures may cause skin cancer.

IARC Evaluation: Group 2A, animal sufficient evidence; human limited evidence.
NTP: Known carcinogen
IDLH: Potential human carcinogen; 80 mg/m^3 as coal tar products.
OSHA Table Z-1-A Air Contaminant

Exposure Limits:	**TWA**
ACGIH: [1]	0.2 mg/m^3
NIOSH: [2]	0.1 mg/m^3 (cyclohexane-extractable fraction)
OSHA:	0.2 mg/m^3 (benzene-soluble fraction)

[1] Confirmed human carcinogen
[2] Potential human carcinogen. Reduce exposure to lowest feasible level; as coal tar products.

RECOMMENDATIONS FOR RESPIRATOR SELECTION: NIOSH as coal tar products

At any concentrations above the NIOSH REL, or where there is no REL, at any detectable concentration: SCBAF:PD,PP (any self-contained breathing apparatus that has a full facepiece and is operated in a pressure-demand or other positive-pressure mode);or SAF:PD,PP:ASCBA (any supplied-air

respirator that has a full facepiece and is operated in a pressure-demand or other positive-pressure mode in combination with an auxiliary, self-contained breathing apparatus operated in a pressure-demand or other positive pressure mode).

Escape: GMFOVHiE [any air-purifying, full-facepiece respirator (gas mask) with a chin-style, front- or back-mounted organic vapor canister having a high-efficiency particulate filter];or SCBAE (any appropriate escape-type, self-contained breathing apparatus).

CHEMICAL REACTIVITY

Reacts with strong oxidizers and acids.

FIRE INFORMATION

If material or contaminated runoff enters waterways, notify downstream users of potentially contaminated waters. Notify local health and fire officials and pollution control agencies. Use any extinguishing agent suitable for surrounding fire. From a secure explosion-proof location, use water spray to cool exposed containers.

NFPA Hazard Classifications:

Health Hazards (Blue) 2
Flammability (Red) 2
Reactivity (Yellow) 0

Flash Point: 165°F (74°C) c.c.
Autoignition Temperature: 637°F (336°C)

FIRST AID

Move victim to fresh air. Call emergency medical care. Apply artificial respiration if victim is not breathing. If breathing is difficult, give oxygen. Remove and isolate contaminated clothing and shoes. In case of contact with substance, immediately flush skin or eyes with running water for at least 20 minutes. Wash skin with soap and water. Keep victim warm and quiet. Ensure that medical personnel are aware of the material(s) involved and take precautions to protect themselves.

ENVIRONMENTAL INFORMATION

EPA Hazardous Waste Number (RCRA No.): U051
RCRA Toxicity Characteristic (Section 261.24): RCRA Section 261 Hazardous Constituents
RCRA Land Ban Waste
EPCRA Section 304 Reportable Quantity (RQ): CERCLA, 1 lb (0.454 kg)
EPCRA Section 313: Form R *de minimis* concentration reporting level: 0.1%.
California's Proposition 65: Carcinogen as creosote
Marine Pollutant (49CFR, Subchapter 172.101, Appendix B).

CRESOL (MIXED ISOMERS) C:640

SYNONYMS: ACEDE CRESYLIQUE (French); BACILLOL; CRESOLI (Italian); CRESYLIC ACIDS; CRESOLS (ALL ISOMERS); CRESOLS AND CRESYLIC ACIDS, MIXED; CRESOL ISOMERS; CRESOLS (o-; m-; p-); CRESYLIC ACID; EEC No. 604-004-00-9 as ortho-Cresol; HYDROXYMETHYLBENZENES; HYDROXYTOLUOLE (German); HYDROXYTOLUENE; KRESOLE (German); KRESOLEN (Dutch); KREZOL (Polish); MARKEM 501 CLEANER; METHYLPHENOL; OXYTOLUENES; PHENOL, METHYL-; PHENOL, METHYL- (9CI); MICROSTRIP; TAR ACIDS; TEKRESOL; AR-TOLUENOL; TRICRESOL

IDENTIFICATION

CAS:	1319-77-3 (all isomers)
ERG ID:	UN 2076
ERG Guide:	153
ERG Guide Category:	TOXIC AND/OR CORROSIVE (COMBUSTIBLE)
Molecular Formula:	C_7H_8O
RTECS No.:	GO5950000
Properties:	Combustible, watery liquid or solid crystals. Colorless, yellowish, browish-yellow or pinkish. Will turn a dark color with age or after exposure to light. Sweet, tarry, phenolic odor. Slightly soluble in water; what does not dissolve will sink.

HEALTH AND SAFETY INFORMATION

Corrosive to eyes, skin, and respiratory tract. Long-term effects may include chemical burns and rashes from prolonged or repeated contact with exposed areas of the skin; kidney damage. Daily inhalation of airborne cresol may affect the nervous system, including the following symptoms: difficulty in swallowing, vomiting, excess salivation, diarrhea, loss of appetite and a resulting loss of weight, headache, dizziness, and fainting. Most of these effects may go away after exposure has stopped.

IDLH: 250 ppm
Odor Threshold: 0.00005-0.0079 ppm
OSHA Table Z-1-A Air Contaminant (cresol, all isomers)
NIOSH Criteria Document: NIOSH 78-133 CRESOL

Exposure Limits:	**TWA**
ACGIH: (1)	5 ppm; 22 mg/m^3
NIOSH: (1)	2.3 ppm; 10 mg/m^3
OSHA: (1)	5 ppm; 22 mg/m^3

(1) Skin contact contributes to overall exposure

RECOMMENDATIONS FOR RESPIRATOR SELECTION: NIOSH as cresol

23 ppm:CCROVDM (any chemical cartridge respirator with organic vapor cartridge(s) in combination with a dust and mist filter); SA (any supplied-air respirator)
57.5 ppm: SA:CF (any supplied-air respirator operated in a continuous-flow mode);or PAPROVDM (any powered, air-purifying respirator with organic vapor cartridge(s) in combination with a dust and mist filter)
115 ppm: CCRFOVHiE (any chemical cartridge respirator with a full facepiece and organic vapor cartridge(s) in combination with a high-efficiency particulate filter);or GMFOVHiE [any air-purifying, full-facepiece respirator (gas mask) with a chin-style, front- or back-mounted organic vapor canister having a high-efficiency particulate filter];or PAPRTHiE (any powered, air-purifying respirator with a tight-fitting facepiece and a high-efficiency particulate filter);or SAT:CF (any supplied-air respirator that has a tight-fitting facepiece and is operated in a continuous-flow mode);or SCBAF (any self-contained breathing apparatus with a full facepiece);or SAF (any supplied-air respirator with a full facepiece)
250 ppm: SAF:PD,PP (any supplied-air respirator that has a full facepiece and is operated in a pressure-demand or other positive-pressure mode)

Emergency or Planned Entry into Unknown Concentrations or IDLH Conditions:
SCBAF:PD,PP (any self-contained breathing apparatus that has a full facepiece and is operated in a pressure-demand or other positive-pressure mode);or SAF:PD,PP:ASCBA (any supplied-air respirator that has a full facepiece and is operated in a pressure-demand or other positive-pressure mode in combination with an auxiliary, self-contained breathing apparatus operated in a pressure-demand or other positive pressure mode)

Escape: GMFOVHiE [any air-purifying, full-facepiece respirator (gas mask) with a chin-style, front- or back-mounted organic vapor canister having a high-efficiency particulate filter];or SCBAE (any appropriate escape-type, self-contained breathing apparatus).

Note: Substance reported to cause eye irritation or damage; may require eye protection.

CHEMICAL REACTIVITY

Keep away from strong acids, oxidizers (especially chlorine or permanganates), alkalies, aliphatic amines, amides, chlorosulfonic acid, oleum. Liquid attacks some plastics, coatings, and rubber.

FIRE INFORMATION

Poisonous gases may be produced in fire. Vapors are heavier than air and will collect in low areas. Vapors in confined areas may explode when exposed to fire. Vapors may travel long distances to ignition sources and flash back. Storage containers and parts of containers may rocket great distances, in many directions. If material or contaminated runoff enters waterways, notify downstream users of potentially contaminated waters. Notify local health and fire officials and pollution control agencies. Use dry chemicals, foam, carbon dioxide. From a secure explosion-proof location, use water spray to cool exposed containers. If cooling streams are ineffective (venting sound increases in volume and pitch, tank discolors, or shows any signs of deforming), withdraw immediately to a secure position.

NFPA Hazard Classifications:

Health Hazards (Blue) 3
Flammability (Red) 2
Reactivity (Yellow) 0

Flash Point: 175-185°F (79-85°C) o.c.; 178-187°F (81-86°C) c.c.
Flammable Limits in Air: LEL 1.1%-1.4% @ 302°F; UEL NA.
Autoignition Temperature: 1038-1100°F (558-599°C)

FIRST AID

Move victim to fresh air. Call emergency medical care. Apply artificial respiration if victim is not breathing. Do not use mouth-to-mouth method if victim ingested or inhaled the substance; induce artificial respiration with the aid of a pocket mask equipped with a one-way valve or other proper respiratory medical device. Administer oxygen if breathing is difficult. Remove and isolate contaminated clothing and shoes. In case of contact with substance, immediately flush skin or eyes with running water for at least 20 minutes. For minor skin contact, avoid spreading material on unaffected skin. Keep victim warm and quiet. Effects of exposure (inhalation, ingestion, or skin contact) may be delayed. Ensure that medical personnel are aware of the material(s) involved and take precautions to protect themselves.

ENVIRONMENTAL INFORMATION

CLEAN AIR ACT: Hazardous Air Pollutants (Title I, Part A, Section 112)
CLEAN WATER ACT: Section 311 Hazardous Substances/RQ 100 lb (4.54 kg); Section 313 Priority Chemicals.
EPA Hazardous Waste Number (RCRA No.): U052; D026
RCRA Toxicity Characteristic (Section 261.24), Maximum Concentration of Contaminants, regulatory level, 200 mg/L.
Note: If o-, m- and p-Cresol concentrations cannot be differentiated, the total cresol (D026) concentration is used. The regulatory level of total cresol is 200 mg/L.
RCRA Section 261 Hazardous Constituents.
RCRA Land Ban Waste.
EPCRA Section 304 Reportable Quantity (RQ): CERCLA, 1,000 lb (454 kg).
EPCRA Section 313: Form R *de minimus* concentration reporting level: 1.0%.
Marine Pollutant: (49CFR, Subchapter 172.101, Appendix B) as cresols (o-; m-; p- isomers).

m–CRESOL C:660

SYNONYMS: BENZENE, 3-METHYL; 3-CRESOL; meta-CRESOL; m-CRESYLIC ACID; 1-HYDROXY-3-METHYLBENZENE; 3-HYDROXYTOLUENE; m-HYDROXY TOLUENE; m-KRESOL; METACRESOL; m-METHYLPHENOL; 3-METHYLPHENOL; m-OXYTOLUENE; PHENOL, 3-METHYL-; m-TOLUOL

IDENTIFICATION

CAS:	108-39-4
ERG ID:	UN 2076
ERG Guide:	153
ERG Guide Category:	TOXIC AND/OR CORROSIVE (COMBUSTIBLE)
Molecular Formula:	C_7H_8O
RTECS No.:	GO5950000
Properties:	Colorless or yellowish liquid. Will turn a dark color with age or after exposure to light. A solid below 54°F (12°C). Sweet tarry, phenolic odor. Slightly soluble; sinks in water. Poisonous gases may be produced in fire.

HEALTH AND SAFETY INFORMATION

Corrosive to eyes, skin, and respiratory tract. Long-term effects may include chemical burns and rashes from prolonged or repeated contact with exposed areas of the skin; kidney damage. Daily inhalation of airborne cresol may affect the nervous system, including the following symptoms: difficulty in swallowing, vomiting, excess salivation, diarrhea, loss of appetite and a resulting loss of weight, headache, dizziness, and fainting. Most of these effects may go away after exposure has stopped.

IDLH: 250 ppm
Odor Threshold: 0.0005-0.0079 ppm
OSHA Table Z-1-A Air Contaminant (cresol, all isomers)
NIOSH Criteria Document: NIOSH 78-133 CRESOL

Exposure Limits:	**TWA**
ACGIH: [1]	5 ppm; 22 mg/m^3
NIOSH: [1]	2.3 ppm; 10 mg/m^3
OSHA: [1]	5 ppm; 22 mg/m^3

[1] Skin contact contributes to overall exposure

RECOMMENDATIONS FOR RESPIRATOR SELECTION: NIOSH as cresol

23 ppm: CCROVDM (any chemical cartridge respirator with organic vapor cartridge(s) in combination with a dust and mist filter); SA (any supplied-air respirator).
57.5 ppm: SA:CF (any supplied-air respirator operated in a continuous-flow mode);or PAPROVDM (any powered, air-purifying respirator with organic vapor cartridge(s) in combination with a dust and mist filter)
115 ppm: CCRFOVHiE (any chemical cartridge respirator with a full facepiece and organic vapor cartridge(s) in combination with a high-efficiency particulate filter);or GMFOVHiE [any air-purifying, full-facepiece respirator (gas mask) with a chin-style, front- or back-mounted organic vapor canister having a high-efficiency particulate filter];or PAPRTHiE (any powered, air-purifying respirator with a tight-fitting facepiece and a high-efficiency particulate filter);or SAT:CF (any supplied-air respirator that has a tight-fitting facepiece and is operated in a continuous-flow mode);or SCBAF (any self-contained breathing apparatus with a full facepiece);or SAF (any supplied-air respirator with a full facepiece).

250 ppm: SAF:PD,PP (any supplied-air respirator that has a full facepiece and is operated in a pressure-demand or other positive-pressure mode)

Emergency or Planned Entry into Unknown Concentrations or IDLH Conditions: SCBAF:PD,PP (any self-contained breathing apparatus that has a full facepiece and is operated in a pressure-demand or other positive-pressure mode);or SAF:PD,PP:ASCBA (any supplied-air respirator that has a full facepiece and is operated in a pressure-demand or other positive-pressure mode in combination with an auxiliary, self-contained breathing apparatus operated in a pressure-demand or other positive pressure mode)

Escape: GMFOVHiE [any air-purifying, full-facepiece respirator (gas mask) with a chin-style, front- or back-mounted organic vapor canister having a high-efficiency particulate filter];or SCBAE (any appropriate escape-type, self-contained breathing apparatus).
Note: Substance reported to cause eye irritation or damage; may require eye protection.

CHEMICAL REACTIVITY

Forms explosive mixture with air. Strong reducing agent; reacts with strong acids, oxidizers, alkalies, aliphatic amines, amides, chlorosulfonic acid, oleum. Liquid attacks some plastics, coatings, and rubber.

FIRE INFORMATION

Vapors are heavier than air and will collect in low areas. Vapors in confined areas may explode when exposed to fire. Vapors may travel long distances to ignition sources and flash back. Storage containers and parts of containers may rocket great distances, in many directions. If material or contaminated runoff enters waterways, notify downstream users of potentially contaminated waters. Notify local health and fire officials and pollution control agencies. Use dry chemicals, foam, carbon dioxide. From a secure explosion-proof location, use water spray to cool exposed containers. If cooling streams are ineffective (venting sound increases in volume and pitch, tank discolors, or shows any signs of deforming), withdraw immediately to a secure position.

NFPA Hazard Classifications:

Health Hazards (Blue) 3
Flammability (Red) 2
Reactivity (Yellow) 0

Flash Point: 175-185°F (79-85°C) o.c.; 178-187°F (81-86°C) c.c.
Flammable Limits in Air: LEL 1.1%; UEL N/A.
Autoignition Temperature: 1038°F (558°C)

FIRST AID

Move victim to fresh air. Call emergency medical care. Apply artificial respiration if victim is not breathing. Do not use mouth-to-mouth method if victim ingested or inhaled the substance; induce artificial respiration with the aid of a pocket mask equipped with a one-way valve or other proper respiratory

medical device. Administer oxygen if breathing is difficult. Remove and isolate contaminated clothing and shoes. In case of contact with substance, immediately flush skin or eyes with running water for at least 20 minutes. For minor skin contact, avoid spreading material on unaffected skin. Keep victim warm and quiet. Effects of exposure (inhalation, ingestion, or skin contact) may be delayed. Ensure that medical personnel are aware of the material(s) involved and take precautions to protect themselves.

ENVIRONMENTAL INFORMATION

Clean Water Act: Section 311 Hazardous Substances/RQ 100 lb (4.54 kg); Section 313 Priority Chemicals
EPA Hazardous Waste Number (Rcra No.): U052; D024
RCRA Section 261 Hazardous Constituents.
RCRA Toxicity Characteristic (Section 261.24), Maximum Concentration of Contaminants, regulatory level, 200 mg/L.
Note: If o-, m- and p-Cresol concentrations cannot be differentiated, the total cresol (D026) concentration is used. The regulatory level of total cresol is 200 mg/L.
RCRA Land Ban Waste.
RCRA Universal Treatment Standards: Wastewater (mg/L), 0.77; Nonwastewater (mg/kg), 5.6.
Note: Difficult to distinguish from p-cresol.
RCRA Ground Water Monitoring List. Suggested test method(s) (PQL ug/L): 8270(10).
EPCRA Section 304 Reportable Quantity (RQ): CERCLA, 1,000 lb (454 kg).
EPCRA Section 313: Form R *de minimus* concentration reporting level: 1.0%.
Marine Pollutant: (49CFR, Subchapter 172.101, Appendix B).

o–CRESOL C:680

SYNONYMS: BENZENE, 2-METHYL; 2-CRESOL;ortho-CRESOL; o-CRESYLIC ACID; EEC. No. 604-004-00-9; 1-HYDROXY-2-METHYLBENZENE; 2-HYDROXYTOLUENE; o-HYDROXY TOLUENE; o-KRESOL; o-METHYLPHENOL; 2-METHYLPHENOL; o-THOCRESOL; o-OXYTOLUENE; PHENOL, 2-METHYL-; o-TOLUOL

IDENTIFICATION

CAS:	95-48-7
ERG ID:	UN 2076
ERG Guide:	153
ERG Guide Category:	TOXIC AND/OR CORROSIVE (COMBUSTIBLE)
Molecular Formula:	C_7H_8O
RTECS No.:	GO6300000
Properties:	Combustible liquid or crystals. Clear to slightly yellowish liquid above 88°F (31°C). Will trun a dark color with age or after exposure to light. Sweet, tarry, phenolic odor. Slightly soluble in water.

HEALTH AND SAFETY INFORMATION

Corrosive to eyes, skin, and respiratory tract. Long-term effects may include chemical burns and rashes from prolonged or repeated contact with exposed areas of the skin; kidney damage. Daily inhalation of airborne cresol may affect the nervous system, including the following symptoms: difficulty in swallowing, vomiting, excess salivation, diarrhea, loss of appetite and a resulting loss of weight, headache, dizziness, and fainting. Most of these effects may go away after exposure has stopped.

IDLH: 250 ppm
Odor Threshold: 0.00005-0.0079 ppm
OSHA Table Z-1-A Air Contaminant (cresol, all isomers)
NIOSH Criteria Document: NIOSH 78-133 CRESOL

Exposure Limits:	**TWA**
ACGIH: [(1)]	5 ppm; 22 mg/m^3
NIOSH: [(1)]	2.3 ppm; 10 mg/m^3
OSHA: [(1)]	5 ppm; 22 mg/m^3

[(1)] Skin contact contributes to overall exposure

RECOMMENDATIONS FOR RESPIRATOR SELECTION: NIOSH as cresol

23 ppm: CCROVDM (any chemical cartridge respirator with organic vapor cartridge(s) in combination with a dust and mist filter); SA (any supplied-air respirator)
57.5 ppm: SA:CF (any supplied-air respirator operated in a continuous-flow mode);or PAPROVDM (any powered, air-purifying respirator with organic vapor cartridge(s) in combination with a dust and mist filter).
115 ppm: CCRFOVHiE (any chemical cartridge respirator with a full facepiece and organic vapor cartridge(s) in combination with a high-efficiency particulate filter);or GMFOVHiE [any air-purifying, full-facepiece respirator (gas mask) with a chin-style, front- or back-mounted organic vapor canister having a high-efficiency particulate filter];or PAPRTHiE (any powered, air-purifying respirator with a tight-fitting facepiece and a high-efficiency particulate filter);or SAT:CF (any supplied-air respirator that has a tight-fitting facepiece and is operated in a continuous-flow mode);or SCBAF (any self-contained breathing apparatus with a full facepiece);or SAF (any supplied-air respirator with a full facepiece).
250 ppm: SAF:PD,PP (any supplied-air respirator that has a full facepiece and is operated in a pressure-demand or other positive-pressure mode).

Emergency or Planned Entry into Unknown Concentrations or IDLH Conditions:
SCBAF:PD,PP (any self-contained breathing apparatus that has a full facepiece and is operated in a pressure-demand or other positive-pressure mode);or SAF:PD,PP:ASCBA (any supplied-air respirator that has a full facepiece and is operated in a pressure-demand or other positive-pressure mode in combination with an auxiliary, self-contained breathing apparatus operated in a pressure-demand or other positive-pressure mode).

Escape: GMFOVHiE [any air-purifying, full-facepiece respirator (gas mask) with a chin-style, front- or back-mounted organic vapor canister having a high-efficiency particulate filter];or SCBAE (any appropriate escape-type, self-contained breathing apparatus).
Note: Substance reported to cause eye irritation or damage; may require eye protection.

CHEMICAL REACTIVITY

Forms explosive mixture with air. Incompatible with strong acids, oxidizers (especially chlorine or permanganates), alkalies, aliphatic amines, amides, chlorosulfonic acid, oleum. Liquid attacks some plastics and rubber.

FIRE INFORMATION

Poisonous gases may be produced in fire. Vapors are heavier than air and will collect in low areas. Vapors in confined areas may explode when exposed to fire. Vapors may travel long distances to ignition sources and flash back. Storage containers and parts of containers may rocket great distances, in many directions. If material or contaminated runoff enters waterways, notify downstream users of potentially contaminated waters. Notify local health and fire officials and pollution control agencies. Use dry chemicals, foam, carbon dioxide. From a secure explosion-proof location, use water spray to cool exposed containers. If cooling streams are ineffective (venting sound increases in volume and pitch, tank discolors, or shows any signs of deforming), withdraw immediately to a secure position.

NFPA Hazard Classifications:

Health Hazards (Blue) 3
Flammability (Red) 2
Reactivity (Yellow) 0

Flash Point: 175-185°F (79-85°C) o.c.; 178-187°F (81-86°C) c.c.
Flammable Limits in Air: LEL 1.4%; UEL NA.
Autoignition Temperature: 1100°F (599°C)

FIRST AID

Move victim to fresh air. Call emergency medical care. Apply artificial respiration if victim is not breathing. Do not use mouth-to-mouth method if victim ingested or inhaled the substance; induce artificial respiration with the aid of a pocket mask equipped with a one-way valve or other proper respiratory medical device. Administer oxygen if breathing is difficult. Remove and isolate contaminated clothing and shoes. In case of contact with substance, immediately flush skin or eyes with running water for at least 20 minutes. For minor skin contact, avoid spreading material on unaffected skin. Keep victim warm and quiet. Effects of exposure (inhalation, ingestion, or skin contact) may be delayed. Ensure that medical personnel are aware of the material(s) involved and take precautions to protect themselves.

ENVIRONMENTAL INFORMATION

Clean Air Act: Hazardous Air Pollutants (Title I, Part A, Section 112)

Clean Water Act: Section 311 Hazardous Substances/RQ/RQ 100 lb (4.54 kg); Section 313 Priority Chemicals
EPA Hazardous Waste Number (Rcra No.): U052; D023
RCRA Toxicity Characteristic: (Section 261.24), Maximum Concentration of Contaminants, regulatory level, 200.
Note: If o-, m- and p-Cresol concentrations cannot be differentiated, the total cresol (D026) concentration is used. The regulatory level of total cresol is 200 mg/L.
RCRA Section 261 Hazardous Constituents.
RCRA Land Ban Waste.
RCRA Universal Treatment Standards: Wastewater (mg/L), 0.11; Nonwastewater (mg/kg), 5.6
RCRA Ground Water Monitoring List. Suggested test method(s) (PQL ug/L): 8270(10).
EPCRA Section 302 Extremely Hazardous Substances: TPQ = 1,000/10,000 lb (454/4,540 kg).
EPCRA Section 304 Reportable Quantity (RQ): CERCLA, 1,000 lb (454 kg).
EPCRA Section 313: Form R *de minimus* concentration reporting level: 1.0%.
Marine Pollutant: (49CFR, Subchapter 172.101, Appendix B).

p–CRESOL C:700

SYNONYMS: BENZENE, 4-METHYL; 4-CRESOL; para-CRESOL; p-CRESYLIC ACID; 1-HYDROXY-4-METHYLBENZENE; 4-HYDROXYTOLUENE; p-HYDROXY TOLUENE; p-KRESOL; p-METHYLPHENOL; 4-METHYLPHENOL; p-OXYTOLUENE; PARACRESOL; PHENOL, 4-METHYL-; p-TOLUOL

IDENTIFICATION

CAS:	106-44-5
ERG ID:	UN 2076
ERG Guide:	153
ERG Guide Category:	TOXIC AND/OR CORROSIVE (COMBUSTIBLE)
Molecular Formula:	C_7H_8O
RTECS No.:	GO6475000
Properties:	Combustible. Crystalline solid. Colorless or yellow liquid above 95°F (35°C). Will trun a dark color with age or after exposure to light. Sweet, tarry phenolic odor. Slightly soluble; sinks in water. Poisonous gases may be produced in fire.

HEALTH AND SAFETY INFORMATION

Corrosive to eyes, skin, and respiratory tract. Long-term effects may include chemical burns and rashes from prolonged or repeated contact with exposed areas of the skin; kidney damage. Daily inhalation of airborne cresol may affect the nervous system, including the following symptoms: difficulty in swallowing, vomiting, excess salivation, diarrhea, loss of appetite and a resulting loss of weight, headache, dizziness, and fainting. Most of these effects may go away after exposure has stopped.

IDLH: 250 ppm
Odor Threshold: 0.00005-0.0079 ppm
OSHA Table Z-1-A Air Contaminant (cresol, all isomers)
NIOSH Criteria Document: NIOSH 78-133 CRESOL

Exposure Limits:	**TWA**
ACGIH: [1]	5 ppm; 22 mg/m^3
NIOSH: [1]	2.3 ppm; 10 mg/m^3
OSHA: [1]	5 ppm; 22 mg/m^3

[1] Skin contact contributes to overall exposure

RECOMMENDATIONS FOR RESPIRATOR SELECTION: NIOSH as cresol

23 ppm: CCROVDM (any chemical cartridge respirator with organic vapor cartridge(s) in combination with a dust and mist filter); SA (any supplied-air respirator).
57.5 ppm: SA:CF (any supplied-air respirator operated in a continuous-flow mode);or PAPROVDM (any powered, air-purifying respirator with organic vapor cartridge(s) in combination with a dust and mist filter).
115 ppm: CCRFOVHiE (any chemical cartridge respirator with a full facepiece and organic vapor cartridge(s) in combination with a high-efficiency particulate filter);or GMFOVHiE [any air-purifying, full-facepiece respirator (gas mask) with a chin-style, front- or back-mounted organic vapor canister having a high-efficiency particulate filter];or PAPRTHiE (any powered, air-purifying respirator with a tight-fitting facepiece and a high-efficiency particulate filter);or SAT:CF (any supplied-air respirator that has a tight-fitting facepiece and is operated in a continuous-flow mode);or SCBAF (any self-contained breathing apparatus with a full facepiece);or SAF (any supplied-air respirator with a full facepiece).
250 ppm: SAF:PD,PP (any supplied-air respirator that has a full facepiece and is operated in a pressure-demand or other positive-pressure mode).

Emergency or Planned Entry into Unknown Concentrations or IDLH Conditions:
SCBAF:PD,PP (any self-contained breathing apparatus that has a full facepiece and is operated in a pressure-demand or other positive-pressure mode);or SAF:PD,PP:ASCBA (any supplied-air respirator that has a full facepiece and is operated in a pressure-demand or other positive-pressure mode in combination with an auxiliary, self-contained breathing apparatus operated in a pressure-demand or other positive-pressure mode)

Escape: GMFOVHiE [any air-purifying, full-facepiece respirator (gas mask) with a chin-style, front- or back-mounted organic vapor canister having a high-efficiency particulate filter];or SCBAE (any appropriate escape-type, self-contained breathing apparatus).

Note: Substance reported to cause eye irritation or damage; may require eye protection.

CHEMICAL REACTIVITY

Incompatible with strong acids, oxidizers (especially chlorine or permanganates), alkalies, aliphatic amines, amides, chlorosulfonic acid, oleum. Liquid attacks some plastics, coatings, and rubber.

FIRE INFORMATION

Vapors are heavier than air and will collect in low areas. Vapors in confined areas may explode when exposed to fire. Vapors may travel long distances to ignition sources and flash back. Storage containers and parts of containers may rocket great distances, in many directions. If material or contaminated runoff enters waterways, notify downstream users of potentially contaminated waters. Notify local health and fire officials and pollution control agencies. Use dry chemicals, foam, carbon dioxide. From a secure explosion-proof location, use water spray to cool exposed containers. If cooling streams are ineffective (venting sound increases in volume and pitch, tank discolors, or shows any signs of deforming), withdraw immediately to a secure position.

NFPA Hazard Classifications:

Health Hazards (Blue) 3
Flammability (Red) 2
Reactivity (Yellow) 0

Flash Point: 175-185°F (79-85°C) o.c.; 178-187°F (81-86°C) c.c.
Flammable Limits in Air: LEL 1.1%; UEL N/A.
Autoignition Temperature: 1038°F (558°C)

FIRST AID

Move victim to fresh air. Call emergency medical care. Apply artificial respiration if victim is not breathing. Do not use mouth-to-mouth method if victim ingested or inhaled the substance; induce artificial respiration with the aid of a pocket mask equipped with a one-way valve or other proper respiratory medical device. Administer oxygen if breathing is difficult. Remove and isolate contaminated clothing and shoes. In case of contact with substance, immediately flush skin or eyes with running water for at least 20 minutes. For minor skin contact, avoid spreading material on unaffected skin. Keep victim warm and quiet. Effects of exposure (inhalation, ingestion, or skin contact) may be delayed. Ensure that medical personnel are aware of the material(s) involved and take precautions to protect themselves.

ENVIRONMENTAL INFORMATION

Clean Air Act: Hazardous Air Pollutants (Title I, Part A, Section 112)
Clean Water Act: Section 311 Hazardous Substances/RQ/RQ 100 lb (4.54 kg); Section 313 Priority Chemicals
EPA Hazardous Waste Number (Rcra No.): U052; D025
RCRA Toxicity Characteristic. (Section 261.24), MaximumConcentration of Contaminants, regulatory level, 200 mg/L. *Note:* If o-, m- and p-Cresol concentrations cannot be differentiated, the total cresol (D026) concentration is used. The regulatory level of total cresol is 200 mg/L.
RCRA Section 261 Hazardous Constituents.
RCRA Land Ban Waste.
RCRA Universal Treatment Standards: Wastewater (mg/L), 0.77; Nonwastewater (mg/kg), 5.6. *Note:* Difficult to distinguish from m-cresol.
RCRA Ground Water Monitoring List. Suggested test method(s) (PQL ug/L): 8270(10).

EPCRA Section 304 Reportable Quantity (RQ): CERCLA, 1,000 lb (454 kg).
EPCRA Section 313: Form R *de minimus* concentration reporting level: 1.0%.
Marine Pollutant: (49CFR, Subchapter 172.101, Appendix B).

CROTONIC ACID C:703

SYNONYMS: ACIDE CROTONIQUE (French); ACIDO CROTONICO (Spanish); α-BUTENOIC ACID; 2-BUTENOIC ACID; trans-2-BUTENOIC ACID; α-CROTONIC ACID; CROTONSAEURE (German); EINECS No. 223-077-4; 3-METHYLACRYLIC ACID; β-METHYLACRYLIC ACID; 2-PROPNE CARBONIC ACID

IDENTIFICATION

CAS:	3724-65-0
ERG ID:	UN 2823
ERG Guide:	153
ERG Guide Category:	SUBSTANCES-TOXIC AND/OR CORROSIVE (COMBUSTIBLE)
Molecular Formula:	$C_4H_6O_2$
RTECS No.:	GQ2800000
Properties:	Combustible needles or crystals. Colorless to white. Characteristic odor. Soluble in water.

HEALTH AND SAFETY INFORMATION

Corrosive to eyes, skin, and respiratory tract. May cause pulmonary edema, which can be delayed for several hours; there is a risk of death in serious cases.

RECOMMENDATIONS FOR RESPIRATOR SELECTION:

At any concentrations above the NIOSH REL, or where there is no REL, at any detectable concentration: SCBAF:PD,PP (any self-contained breathing apparatus that has a full facepiece and is operated in a pressure-demand or other positive-pressure mode);or SAF:PD,PP:ASCBA (any supplied-air respirator that has a full facepiece and is operated in a pressure-demand or other positive-pressure mode in combination with an auxiliary, self-contained breathing apparatus operated in a pressure-demand or other positive-pressure mode).

Escape: GMFOV [any air-purifying, full-facepiece respirator (gas mask) with a chin-style, front- or back-mounted organic vapor canister];or SCBAE (any appropriate escape-type, self-contained breathing apparatus).

CHEMICAL REACTIVITY

Forms explosive mixture with air. A strong reducing agent. Reacts violently with peroxides, oxidizers, and strong bases. A powerful corrosive. Ultraviolet radiation and/or moisture can cause explosive polymerization.

FIRE INFORMATION

Decomposition products upon heating include toxic gases. Vapors are heavier than air and will collect in low areas. Vapors in confined areas may explode when exposed to fire. Vapors may travel long distances to ignition sources and flash back. Storage containers and parts of containers may rocket great distances, in many directions. If material or contaminated runoff enters waterways, notify downstream users of potentially contaminated waters. Notify local health and fire officials and pollution control agencies. Use dry chemicals, foam, carbon dioxide, water spray. From a secure explosion-proof location, use water spray to cool exposed containers. If cooling streams are ineffective (venting sound increases in volume and pitch, tank discolors, or shows any signs of deforming), withdraw immediately to a secure position.

NFPA Hazard Classifications:

Health Hazards (Blue) 3
Flammability (Red) 2
Reactivity (Yellow) 0

Flash Point: 190°F (88°C) o.c.
Autoignition Temperature: 745°F (396°C)

FIRST AID

Move victim to fresh air and call emergency medical care. Apply artificial respiration if victim is not breathing. Do not use mouth-to-mouth method if victim swallowed or inhaled the substance; induce artificial respiration with the aid of a pocket mask equipped with a one-way valve or other proper respiratory medical device. If breathing is difficult, give oxygen. In case of contact with substance, immediately flush skin or eyes with running water for at least 20 minutes. Remove and isolate contaminated clothing and shoes. Keep victim warm and quiet. Effects of exposure (inhalation, ingestion, or skin contact) to substances may be delayed. Ensure that medical personnel are aware of the material(s) involved and take precautions to protect themselves.

CUMENE C:705

SYNONYMS: BENZENE ISOPROPYL; BENZENE, (1-METHYLETHYL-)-; CUM; CUMEEN (Dutch); CUMENO (Spanish); CUMOL; EEC No. 601-024-00-X; EINECS No. 202-704-5; 2-FENILPROPANO (Italian); 2-FENYL-PROPAN (Dutch); ISOPROPYLBENZEEN (Dutch); ISOPROPILBENZENE (Italian); ISOPROPYL BENZENE; ISOPROPYL BENZENE; ISOPROPYLBENZOL; ISOPROPYL-BENZOL (German); 2-PHENYLPROPANE; (1-METHYLETHYL)BENZENE

IDENTIFICATION

CAS: 98-82-8

ERG ID: UN 1918
ERG Guide: 131
ERG Guide Category: FLAMMABLE LIQUIDS-TOXIC
Molecular Formula: C_9H_{12}
RTECS No.: GR8575000
Properties: Flammable, colorless watery liquid. Gasoline-like odor. Floats on water.

HEALTH AND SAFETY INFORMATION

Irritates eyes, skin, and respiratory tract. Affects the nervous system; may cause dizziness and unconsciousness. Swallowing liquid can cause pneumonia.

IDLH: 900 ppm [LEL]
Odor Threshold: 0.024 ppm
OSHA Table Z-1-A Air Contaminant

Exposure Limits:	TWA
ACGIH: (1)	50 ppm; 246 mg/m^3
NIOSH: (1)	50 ppm; 245 mg/m^3
OSHA: (1)	50 ppm; 245 mg/m^3

(1) Skin contact contributes significantly in overall exposure.

RECOMMENDATIONS FOR RESPIRATOR SELECTION: NIOSH/OSHA

500 ppm: CCROV [any chemical cartridge respirator with organic vapor cartridge(s)];or SA (any supplied-air respirator).
1,000 ppm: SA:CF (any supplied-air respirator operated in a continuous-flow mode);or PAPROV [any powered, air-purifying respirator with organic vapor cartridge(s)];or CCRFOV [any chemical cartridge respirator with a full facepiece and organic vapor cartridge(s)];or GMFOV [any air-purifying, full-facepiece respirator (gas mask) with a chin-style, front- or back-mounted acid gas canister];or SCBAF (any self-contained breathing apparatus with a full facepiece);or SAF (any supplied-air respirator with a full facepiece).

Emergency or Planned Entry into Unknown Concentrations or IDLH Conditions:
SCBAF:PD,PP (any self-contained breathing apparatus that has a full facepiece and is operated in a pressure-demand or other positive-pressure mode);or SAF:PD,PP:ASCBA (any supplied-air respirator that has a full facepiece and is operated in a pressure-demand or other positive-pressure mode in combination with an auxiliary, self-contained breathing apparatus operated in a pressure-demand or other positive-pressure mode).

Escape: GMFOV [any air-purifying, full-facepiece respirator (gas mask) with a chin-style, front- or back-mounted organic vapor canister];or SCBAE (any appropriate escape-type, self-contained breathing apparatus].
Note: Substance reported to cause eye irritation or damage; may require eye protection.

CHEMICAL REACTIVITY

Forms explosive mixture with air. Incompatible with strong acids and strong oxidizers. Air contact forms cumene hydroperoxide. May be able to form unstable peroxides. Attacks rubber. May accumulate static electrical charges and may cause ignition of its vapors.

FIRE INFORMATION

Vapors are heavier than air and will collect in low areas. Vapors in confined areas may explode when exposed to fire. Vapors may travel long distances to ignition sources and flash back. Storage containers and parts of containers may rocket great distances, in many directions. If material or contaminated runoff enters waterways, notify downstream users of potentially contaminated waters. Notify local health and fire officials and pollution control agencies. Use dry chemicals, foam, carbon dioxide. From a secure explosion-proof location, use water spray to cool exposed containers. If cooling streams are ineffective (venting sound increases in volume and pitch, tank discolors, or shows any signs of deforming), withdraw immediately to a secure position.

NFPA Hazard Classifications:

Health Hazards (Blue) 2
Flammability (Red) 3
Reactivity (Yellow) 1

Flash Point: 99°F (37°C) c.c.
Flammable Limits in Air: LEL 0.9%; UEL 6.5%
Autoignition Temperature: 797°F (423°C)

FIRST AID

Move victim to fresh air. Call emergency medical care. Apply artificial respiration if victim is not breathing. Do not use mouth-to-mouth method if victim ingested or inhaled the substance; induce artificial respiration with the aid of a pocket mask equipped with a one-way valve or other proper respiratory medical device. Administer oxygen if breathing is difficult. Remove and isolate contaminated clothing and shoes. In case of contact with substance, immediately flush skin or eyes with running water for at least 20 minutes. Wash skin with soap and water. Keep victim warm and quiet. Effects of exposure (inhalation, ingestion, or skin contact) to substance may be delayed. Ensure that medical personnel are aware of the material(s) involved and take precautions to protect themselves.

ENVIRONMENTAL INFORMATION

Clean Air Act: Hazardous Air Pollutants (Title I, Part A, Section 112).
EPA Hazardous Waste Number (RCRA No.): U055.
RCRA Section 261 Hazardous Constituents.
EPCRA Section 304: Reportable Quantity (RQ): CERCLA, 5,000 lb (2270 kg).
EPCRA Section 313: Form R *de minimis* concentration reporting level: 1.0%.

Marine Pollutant (49CFR, Subchapter 172.101, Appendix B).
TSCA: 40CFR 716.120(*a*); 40CFR 799.18

CUMENE HYDROPEROXIDE C:709

SYNONYMS: α,α-DIMETHYLBENZENE HYDROPEROXIDE; α,α-DIMETHYLBENZYL HYDROPEROXIDE; CHP; CUMEENHYDROPEROXYDE (Dutch); CUMENT HYDROPEROXIDE; α-CUMENYL HYDROPEROXIDE; CUMOLHYDROPEROXID (German); DIMETHYLBENZYL HYDROPEROXIDE; EEC No. 617-002-00-8; EINECS No. 201-254-7; HYDROPEROXIDE, 1-METHYL-1-PHENYLETHYL-; HIDROPEROXIDO de CUMENO (Spanish); HYDROPEROXYDE de CUMYLE (French); IDROPEROSSIDO di CUMOLO (Italian); ISOPROPYLBENZENE HYDROPEROXIDE

IDENTIFICATION

CAS:	80-15-9
ERG ID:	UN 3109
ERG Guide:	145
ERG Guide Category:	ORGANIC PEROXIDES (HEAT AND CONTAMINATION SENSITIVE)
Molecular Formula:	$C_9H_{12}O_2$
RTECS No.:	MX2450000
Properties:	Very flammable liquid. A solid at 16°F (-9°C). Colorless to light yellow. Sharp, irritating, aromatic odor. Sinks in water; soluble.

HEALTH AND SAFETY INFORMATION

Corrosive to eyes, skin, and respiratory tract. Inhalation may cause pulmonary edema, which can be delayed for several hours; there is a risk of death in serious cases.

Odor Threshold: 0.005 ppm
OSHA Process Safety Management of Highly Hazardous Chemicals (29CFR, Part 1910.119, Appendix A): Threshold Quantity: 5,000 pounds.

RECOMMENDATIONS FOR RESPIRATOR SELECTION: NIOSH/OSHA

At any concentrations above the NIOSH REL, or where there is no REL, at any detectable concentration: SCBAF:PD,PP (any self-contained breathing apparatus that has a full facepiece and is operated in a pressure-demand or other positive-pressure mode);or SAF:PD,PP:ASCBA (any supplied-air respirator that has a full facepiece and is operated in a pressure-demand or other positive-pressure mode in

combination with an auxiliary, self-contained breathing apparatus operated in a pressure-demand or other positive-pressure mode).

Escape: GMFOV [any air-purifying, full-facepiece respirator (gas mask) with a chin-style, front- or back-mounted organic vapor canister];or SCBAE (any appropriate escape-type, self-contained breathing apparatus).

CHEMICAL REACTIVITY

A strong oxidizer; violent reaction with reducers, combustible materials, and rust. Explosive decomposition may occur above 120°F (50°C). Forms explosive mixture with air. Decomposition is catalyzed by metals such as aluminum, copper, brass, zinc, and lead. Incompatible with organic and inorganic acids, bases, amines, metallic salts of cobalt, copper, and lead. May accumulate static electrical charges, and may cause ignition of its vapors.

FIRE INFORMATION

Highly explosive. Exposure to heat or fire may cause spontaneous explosion. Increases the intensity of fire upon contact with combustibles. Vapors are heavier than air and will collect in low areas. Vapors in confined areas may explode when exposed to fire. Vapors may travel long distances to ignition sources and flash back. Storage containers and parts of containers may rocket great distances, in many directions. If material or contaminated runoff enters waterways, notify downstream users of potentially contaminated waters. Notify local health and fire officials and pollution control agencies. Use dry chemicals, foam, carbon dioxide. From a secure explosion-proof location, use water spray to cool exposed containers. If cooling streams are ineffective (venting sound increases in volume and pitch, tank discolors, or shows any signs of deforming), withdraw immediately to a secure position.

NFPA Hazard Classifications:

Health Hazards (Blue) 1
Flammability (Red) 2
Reactivity (Yellow) 4
Special situations .... OXY

Flash Point: 147°F (64°C) o.c.; 175°F (79°C) c.c.
Autoignition Temperature: Explodes on heating.

FIRST AID

Move victim to fresh air. Call emergency medical care. Apply artificial respiration if victim is not breathing. Administer oxygen if breathing is difficult. Remove and isolate contaminated clothing and shoes. Remove material from skin immediately. In case of contact with substance, immediately flush skin or eyes with running water for at least 20 minutes. Keep victim warm and quiet. Ensure that medical personnel are aware of the material(s) involved and take precautions to protect themselves.

ENVIRONMENTAL INFORMATION

EPA Hazardous Waste Number (RCRA No.): U096.

RCRA Section 261 Hazardous Constituents
EPCRA Section 304: Reportable Quantity (RQ): CERCLA, 10 lb (4.54 kg).
EPCRA Section 313: Form R *de minimis* concentration reporting level: 1.0%.

CYCLOHEXANE C:760

SYNONYMS: BENZENE HEXAHYDRIDE; BENZENE, HEXAHYDRO-; CICLOESANO (Italian); CICLOHEXANO (Spanish); CYCLOHEXAAN (Dutch); CYCLOHEXAN (German); CYKLOHEKSAN (Polish); EEC No. 601-017-00-1; EINECS No. 203-806-2; HEXAHYDROBENZENE, HEXAMETHYLENE; HEKSAN (Polish); HEXAHYDROBENZENE; HEXAMETHYLENE; HEXANAPHTHENE

IDENTIFICATION

CAS:	110-82-7
ERG ID:	UN 1145
ERG Guide:	128
ERG Guide Category:	FLAMMABLE LIQUIDS (NONPOLAR/WATER-IMMISCIBLE)
Molecular Formula:	C_6H_{12}
RTECS No.:	GU6300000
Properties:	Highly flammable, watery liquid. Colorless. Mild, sweet, gasolinelike odor. May be pungent if impure. Floats on water; insoluble. Flammable irritating vapor is produced in fire. Freezing point is 44°F (7°C).

HEALTH AND SAFETY INFORMATION

Irritating to eyes, skin, and respiratory tract. Prolonged or repeated exposure may cause skin rash and dermatitis; may affect the liver and kidneys.

IDLH: 1,300 ppm [LEL]
Odor Threshold: 83.8 ppm
NIOSH Criteria Document: NIOSH 78-173 KETONES

Exposure Limits:	**TWA**
ACGIH:	300 ppm; 1,030 mg/m^3
NIOSH:	300 ppm; 1,050 mg/m^3
OSHA:	300 ppm; 1,050 mg/m^3

RECOMMENDATIONS FOR RESPIRATOR SELECTION: NIOSH/OSHA

1,000 ppm: SA:CF* (any supplied-air respirator operated in a continuous-flow mode);or PAPROV* [any powered, air-purifying respirator with organic vapor cartridge(s)];or CCRFOV [any chemical cartridge

cartridge respirator with a full facepiece and organic vapor cartridge(s)];or GMFOV [any air-purifying, full-facepiece respirator (gas mask) with a chin-style, front- or back-mounted acid gas canister];or SCBAF (any self-contained breathing apparatus with a full facepiece;or SAF (any supplied-air respirator with a full facepiece)

Emergency or Planned Entry into Unknown Concentrations or IDLH Conditions: SCBAF:PD,PP (any self-contained breathing apparatus that has a full facepiece and is operated in a pressure-demand or other positive-pressure mode);or SAF:PD,PP:ASCBA (any supplied-air respirator that has a full facepiece and is operated in a pressure-demand or other positive-pressure mode in combination with an auxiliary, self-contained breathing apparatus operated in a pressure-demand or other positive-pressure mode).

Escape: GMFOV [any air-purifying, full-facepiece respirator (gas mask) with a chin-style, front- or back-mounted organic vapor canister];or SCBAE (any appropriate escape-type, self-contained breathing apparatus).

[1]Substance causes eye irritation or damage; eye protection needed: shield, hydrocarbon-insoluble rubber, or plastic apron.

CHEMICAL REACTIVITY

Forms explosive mixture with air. Contact with oxidizers such as chromates or permanganates may cause fire and explosions. Explosions occurs when mixed with liquid nitrogen dioxide.

FIRE INFORMATION

Vapors are heavier than air and will collect in low areas. Vapors in confined areas may explode when exposed to fire. Vapors may travel long distances to ignition sources and flash back. Storage containers and parts of containers may rocket great distances, in many directions. If material or contaminated runoff enters waterways, notify downstream users of potentially contaminated waters. Notify local health and fire officials and pollution control agencies. Use dry chemicals, foam, carbon dioxide. From a secure explosion-proof location, use water spray to cool exposed containers. If cooling streams are ineffective (venting sound increases in volume and pitch, tank discolors, or shows any signs of deforming), withdraw immediately to a secure position.

NFPA Hazard Classifications:

Health Hazards (Blue) 1
Flammability (Red) 3
Reactivity (Yellow) 0

Flash Point: -4°F (-20°C) c.c.
Flammable Limits in Air: LEL 1.33%; UEL 8.35%
Autoignition Temperature: 473°F (245°C)

FIRST AID

Move victim to fresh air. Call emergency medical care. Apply artificial respiration if victim is not breathing. If breathing is difficult, give oxygen. Remove and isolate contaminated clothing and shoes. In case of contact with substance, immediately flush skin or eyes with running water for at least 20 minutes. Wash skin with soap and water. Keep victim warm and quiet. Ensure that medical personnel are aware of the material(s) involved and take precautions to protect themselves.

ENVIRONMENTAL INFORMATION

Clean Water Act: Section 311 Hazardous Substances/RQ (same as CERCLA); Section 313 Priority Chemicals.
EPA Hazardous Waste Number (RCRA No.): U056.
RCRA Section 261 Hazardous Constituents
EPCRA Section 304: Reportable Quantity (RQ): CERCLA, 1,000 lb (454 kg).
EPCRA Section 313: Form R *de minimis* concentration reporting level: 1.0%.
TSCA: 40CFR 716.120(*a*); 40 CFR 799.5000

CYCLOHEXANOL C:770

SYNONYMS: ADRONAL; ANOL; CICLOESANOLO (Italian); CICLOHEXANOL (Spanish); 1-CYCLOHEXANOL; CYCLOHEXYL ALCOHOL; CYCLOHEXENE; CYKLOHEKSANOL (Polish); EEC No. 603-009-00-3; EINECS No. 203-630-6; HEXAHYDROPHENOL; HEXALIN; HYDRALIN; HYDROPHENOL; HYDROXYCYCLOHEXANE; NAXOL; PHENOL, HEXAHYDRO-

IDENTIFICATION

CAS:	108-93-0
ERG ID:	UN 1993
ERG Guide:	128
ERG Guide Category:	FLAMMABLE LIQUIDS (NONPOLAR/WATER-IMMISCIBLE)
Molecular Formula:	$C_6H_{12}O$
RTECS No.:	GV7875000
Properties:	Flammable, oily liquid; . Colorless to light yellow. Camphor-like odor. Floats and mixes slowly with water. May solidify. Melting point is 75°F (24°C).

HEALTH AND SAFETY INFORMATION

Irritating to the eyes, skin, and respiratory tract. Has a narcotic effect that can cause drowsiness or unconsciousness.

IDLH: 400 ppm
Odor Threshold: 0.068 ppm
OSHA Table Z-1-A Air Contaminant

Exposure Limits:	**T WA**
ACGIH: [(1)]	50 ppm; 206 mg/m^3
NIOSH: [(1)]	50 ppm; 200 mg/m^3
OSHA: [(1)]	50 ppm; 200mg/m^3

[(1)] Skin contact contributes significantly in overall exposure.

RECOMMENDATIONS FOR RESPIRATOR SELECTION: NIOSH/OSHA

400 ppm: CCROV [any chemical cartridge respirator with organic vapor cartridge(s)];or PAPROV [any powered, air-purifying respirator with organic vapor cartridge(s)];or GMFOV [any air-purifying, full-facepiece respirator (gas mask) with a chin-style, front- or back-mounted organic vapor canister];or SA (any supplied-air respirator);or SCBAF (any self-contained breathing apparatus with a full facepiece).

Emergency or planned entry into unknown concentrations or idlh conditions: SCBAF:PD,PP (any self-contained breathing apparatus that has a full facepiece and is operated in a pressure-demand or other positive-pressure mode);or SAF:PD,PP:ASCBA (any supplied-air respirator that has a full facepiece and is operated in a pressure-demand or other positive-pressure mode in combination with an auxiliary, self-contained breathing apparatus operated in a pressure-demand or other positive pressure mode.

Escape: GMFOV [any air-purifying, full-facepiece respirator (gas mask) with a chin-style, front- or back-mounted organic vapor canister];or SCBAE (any appropriate escape-type, self-contained breathing apparatus).
Note: Substance reported to cause eye irritation or damage; may require eye protection.

CHEMICAL REACTIVITY

Contact with strong oxidizers can cause fire and explosions. Contact with nitric acid at room temperature forms an explosive material.

FIRE INFORMATION

Vapors are heavier than air and will collect in low areas. Vapors in confined areas may explode when exposed to fire. Vapors may travel long distances to ignition sources and flash back. Storage containers and parts of containers may rocket great distances, in many directions. If material or contaminated runoff enters waterways, notify downstream users of potentially contaminated waters. Notify local health and fire officials and pollution control agencies. Use dry chemicals, foam, carbon dioxide. From a secure explosion-proof location, use water spray to cool exposed containers. If cooling streams are ineffective (venting sound increases in volume and pitch, tank discolors, or shows any signs of deforming), withdraw immediately to a secure position.

NFPA Hazard Classifications:

Health Hazards (Blue) 1

Flammability (Red) 2
Reactivity (Yellow) 0

Flash Point: 160°F (71°C) o.c.; 154°F (68°C) c.c.
Autoignition Temperature: 572°F (300°C)

FIRST AID

Move victim to fresh air. Call emergency medical care. Apply artificial respiration if victim is not breathing. If breathing is difficult, give oxygen. Remove and isolate contaminated clothing and shoes. In case of contact with substance, immediately flush skin or eyes with running water for at least 20 minutes. Wash skin with soap and water. Keep victim warm and quiet. Ensure that medical personnel are aware of the material(s) involved and take precautions to protect themselves.

ENVIRONMENTAL INFORMATION

EPCRA Section 313: Form R *de minimis* concentration reporting level: 1.0%.

CYCLOHEXANONE C:780

SYNONYMS: ANOL; ANONE; CICLOESANONE (Italian); CICLOHEXANONA (Spanish); CYCLOHEXANON (Dutch); CYKLOHEKSANON (Polish); CYCLOHEXENE; CYCLOHEXYL ALCOHOL; CYCLOHEXYL KETONE; EEC No. 606-010-00-7; EINECS No. 203-631-1; HEXAHYDROPHENOL; HEXALIN; HEXANON; HYDRALIN; HYTROL O; HYDROOXYCYCLOHEXANE; KETOHEXAMETHYLENE; NADONE; PIMELIC KETON; SEXTONE

IDENTIFICATION

CAS:	108-94-1
ERG ID:	UN 1915
ERG Guide:	127
ERG Guide Category:	FLAMMABLE LIQUIDS (POLAR/WATER MISCIBLE)
Molecular Formula:	$C_6H_{10}O$
RTECS No.:	GW1050000
Properties:	Highly flammable oily liquid. Colorless to light yellow. Sweet, peppermintlike odor. Floats and mixes slowly with water.

HEALTH AND SAFETY INFORMATION

Irritating to skin, eyes, nose, and lungs; prolonged contact will cause skin, eye, and lung burns. Has caused liver, kidney, and vascular injury in experimental animals. May cause liver or kidney damage in humans; skin irritation and rash.

IARC Evaluation: Group 3, unclassifiable; no human data; animal evidence, inadequate.
IDLH: 700 ppm
Odor Threshold: 0.019 ppm
NIOSH Criteria Document: NIOSH 78-173 KETONES

Exposure Limits: TWA

	TWA
ACGIH: [(1)]	25 ppm; 100 mg/m^3
NIOSH: [(1)]	25 ppm; 100 mg/m^3
OSHA: [(1)]	50 ppm; 200 mg/m^3

[(1)] Skin contact contributes significantly to overall exposure

RECOMMENDATIONS FOR RESPIRATOR SELECTION: NIOSH/OSHA

625 ppm: SA:CF (any supplied-air respirator operated in a continuous-flow mode);or PAPROV [any powered, air-purifying respirator with organic vapor cartridge(s)].
700 ppm: CCRFOV [any chemical cartridge respirator with a full facepiece and organic vapor cartridge(s)];or GMFOV [any air-purifying, full-facepiece respirator (gas mask) with a chin-style, front- or back-mounted organic vapor canister];or PAPRTOV [any powered, air-purifying respirator with a tight-fitting facepiece and organic vapor cartridge(s)];or SCBAF (any self-contained breathing apparatus with a full facepiece);or SAF (any supplied-air respirator with a full facepiece).

Emergency or Planned Entry into Unknown Concentrations or IDLH Conditions: SCBAF:PD,PP (any self-contained breathing apparatus that has a full facepiece and is operated in a pressure-demand or other positive-pressure mode);or SAF:PD,PP:ASCBA (any supplied-air respirator that has a full facepiece and is operated in a pressure-demand or other positive-pressure mode in combination with an auxiliary, self-contained breathing apparatus operated in a pressure-demand or other positive-pressure mode).

Escape: GMFOV [any air-purifying, full-facepiece respirator (gas mask) with a chin-style, front- or back-mounted organic vapor canister];or SCBAE (any appropriate escape-type, self-contained breathing apparatus).

Note: Substance causes eye irritation or damage; eye protection needed.

CHEMICAL REACTIVITY

Forms explosive mixture with air. May accumulate static electrical charges, and may cause ignition of its vapors. Attacks red metals and lead. Incompatible with amines, sulfuric acid, nitric acid, oxidizers, strong acids, and aliphatic amines. Dissolves plastics, resins, and rubber.

FIRE INFORMATION

Vapors are heavier than air and will collect in low areas. Vapors in confined areas may explode when exposed to fire. Vapors may travel long distances to ignition sources and flash back. Storage containers and parts of containers may rocket great distances, in many directions. If material or contaminated runoff enters waterways, notify downstream users of potentially contaminated waters. Notify local health and fire officials and pollution control agencies. Use dry chemicals, foam, carbon dioxide. From a secure explosion-proof location, use water spray to cool exposed containers. If cooling streams are ineffective (venting sound increases in volume and pitch, tank discolors, or shows any signs of deforming), withdraw immediately to a secure position.

NFPA Hazard Classifications:

Health Hazards (Blue) 1
Flammability (Red) 2
Reactivity (Yellow) 0

Flash Point: 129°F (54°C) o.c.; 111°F (44°C) c.c.
Flammable Limits in Air: LEL 1.1% @212°F; UEL 9.4%
Autoignition Temperature: 788°F (420°C)

FIRST AID

Move victim to fresh air. Call emergency medical care. Apply artificial respiration if victim is not breathing. If breathing is difficult, give oxygen. Remove and isolate contaminated clothing and shoes. In case of contact with substance, immediately flush skin or eyes with running water for at least 20 minutes. Wash skin with soap and water. Keep victim warm and quiet. Ensure that medical personnel are aware of the material(s) involved and take precautions to protect themselves.

ENVIRONMENTAL INFORMATION

EPA Hazardous Waste Number (RCRA No.): U057.
RCRA Section 261 Hazardous Constituents.
RCRA Land Ban Waste
RCRA Universal Treatment Standards: Wastewater (mg/L), 0.36; Nonwastewater (mg/L), 0.75.
TCLP.
EPCRA Section 304: Reportable Quantity (RQ): CERCLA, 5,000 lb (2,270 kg).
TSCA: 40CFR 716.120(*a*)

CYCLOHEXENYLTRICHLOROSILANE C:830

SYNONYMS: CICLOHEXENILTRICLOROSILANO (Spanish); CYCLOHEXENE, 4-(TRICHLOROSILYL)-; CYCLOHEXENYL TRICHLOROSILANE; TRICHLORO-3-CYCLOHEXENYLSILANE

IDENTIFICATION

CAS:	10137-69-6
ERG ID:	UN 1762

324 Cyclohexenyltrichlorosilane

ERG Guide: 156

ERG Guide Category: SUBSTANCES-TOXIC AND/OR CORROSIVE (COMBUSTIBLE/WATER-SENSITIVE)

Molecular Formula: $C_6H_9Cl_3Si$

RTECS No.: VV2800000

Properties: Combustible liquid. Colorless. Sharp, irritating odor. Reacts with water; fumes in moist air releasing hydrogen chloride.

HEALTH AND SAFETY INFORMATION

Eye and skin contact may cause severe burns and blindness.

IDLH: 50 ppm as hydrogen chloroide

Exposure Limits: **CEILING**

ACGIH:	5 ppm; 7.5 mg/m^3 as hydrogen chloride
NIOSH:	5 ppm; 7 mg/m^3 as hydrogen chloroide
OSHA:	5 ppm; 7 mg/m^3 as hydrogen chloride

RECOMMENDATIONS FOR RESPIRATOR SELECTION: NIOSH/OSHA as hydrogen chloride

50 ppm: CCRS [any chemical cartridge respirator with cartridge(s) providing protection against the compound of concern)];or GMFS [any air-purifying, full-facepiece respirator (gas mask) with a chin-style, front- or back-mounted canister providing protection against the compound of concern];or PAPRS (any powered, air-purifying respirator with cartridge(s) providing protection against the compound of concern);or SA (any supplied-air respirator); SCBAF (any self-contained breathing apparatus with a full facepiece).

Emergency or planned entry into unknown concentrations or IDLH conditions: SCBAF:PD,PP (any self-contained breathing apparatus that has a full facepiece and is operated in a pressure-demand or other positive-pressure mode);or SAF:PD,PP:ASCBA (any supplied-air respirator that has a full facepiece and is operated in a pressure-demand or other positive-pressure mode in combination with an auxiliary, self-contained breathing apparatus operated in a pressure-demand or other positive-pressure mode.

Escape: GMFAG [any air-purifying, full-facepiece respirator (gas mask) with a chin-style, front- or back-mounted organic vapor canister];or SCBAE (any appropriate escape-type, self-contained breathing apparatus).

Note: Substance reported to cause eye irritation or damage; may require eye protection.

CHEMICAL REACTIVITY

Reacts with water to generate hydrogen chloride (hydrochloric acid). Corrodes metals by reacting with surface moisture and generating hydrogen chloride. Reacts with strong bases.

FIRE INFORMATION

Decomposition products upon heating include toxic hydrogen chloride fumes. Vapors are heavier than air and will collect in low areas. Vapors in confined areas may explode when exposed to fire. Vapors may travel long distances to ignition sources and flash back. Storage containers and parts of containers may rocket great distances, in many directions. If material or contaminated runoff enters waterways, notify downstream users of potentially contaminated waters. Notify local health and fire officials and pollution control agencies. Use dry chemicals, foam, carbon dioxide. From a secure explosion-proof location, use water spray to cool exposed containers. If cooling streams are ineffective (venting sound increases in volume and pitch, tank discolors, or shows any signs of deforming), withdraw immediately to a secure position.

FEMA Hazard Classifications:

Health Hazards (Blue) 2
Flammability (Red) 2
Reactivity (Yellow) 1

Flash Point: 200°F (93°C) o.c.

FIRST AID

Move victim to fresh air. Call emergency medical care. Apply artificial respiration if victim is not breathing. Do not use mouth-to-mouth method if victim ingested or inhaled the substance; induce artificial respiration with the aid of a pocket mask equipped with a one-way valve or other proper respiratory medical device. Administer oxygen if breathing is difficult. Remove and isolate contaminated clothing and shoes. In case of contact with substance, immediately flush skin or eyes with running water for at least 20 minutes. For minor skin contact, avoid spreading material on unaffected skin. Keep victim warm and quiet. Effects of exposure (inhalation, ingestion, or skin contact) may be delayed. Ensure that medical personnel are aware of the material(s) involved and take precautions to protect themselves.

ENVIRONMENTAL INFORMATION as hydrogen chloride gas

EPCRA Section 302, Extremely Hazardous Substances: TPQ = 500 lb (227 kg)
EPCRA Section 304 Reportable Quantity (RQ): CERCLA, 5,000 lb (2,270 kg)

CYCLOHEXYLTRICHLOROSILANE C:880

SYNONYMS: CICLOESILTRICLOROSILANO (Spanish); CYCLOHEXANE, 1-(TRICHLOROSILYL)-; EINECS No. 202-639-2; SILANE, TRICHLOROCYCLOHEXYL-; TRICLOROSILANO de CICLOHEXILO (Spanish)

IDENTIFICATION

CAS: 98-12-4

326 Cyclohexyltrichlorosilane

ERG ID: UN 1763
ERG Guide: 156
ERG Guide Category: SUBSTANCES-TOXIC AND/OR CORROSIVE (COMBUSTIBLE/WATER-SENSITIVE)
Molecular Formula: $C_6H_{11}Cl_3Si$
RTECS No.: VV2890000
Properties: Combustible liquid. Colorless to pale yellow. Sharp odor. Insoluble in water; forms hydrogen chloride.

HEALTH AND SAFETY INFORMATION

Corrosive to eyes, skin, and respiratory tract. Poisonous if inhaled or swallowed. May cause pulmonary edema, which can be delayed for several hours; there is a risk of death in serious cases.

RECOMMENDATIONS FOR RESPIRATOR SELECTION:

At any concentrations above the NIOSH REL, or where there is no REL, at any detectable concentration: SCBAF:PD,PP (any self-contained breathing apparatus that has a full facepiece and is operated in a pressure-demand or other positive-pressure mode);or SAF:PD,PP:ASCBA (any supplied-air respirator that has a full facepiece and is operated in a pressure-demand or other positive-pressure mode in combination with an auxiliary, self-contained breathing apparatus operated in a pressure-demand or other positive-pressure mode).

Escape: GMFOV [any air-purifying, full-facepiece respirator (gas mask) with a chin-style, front- or back-mounted organic vapor canister];or SCBAE (any appropriate escape-type, self-contained breathing apparatus).

CHEMICAL REACTIVITY

Reacts with moisture to evolve hydrogen chloride and heat. Attacks metals in the presence of moisture.

FIRE INFORMATION

Decomposition products upon heating include irritating and toxic chlorine and phosgene. Vapors are heavier than air and will collect in low areas. Vapors in confined areas may explode when exposed to fire. Vapors may travel long distances to ignition sources and flash back. Storage containers and parts of containers may rocket great distances, in many directions. If material or contaminated runoff enters waterways, notify downstream users of potentially contaminated waters. Notify local health and fire officials and pollution control agencies. Use dry chemicals, dry sand, or carbon dioxide. From a secure explosion-proof location, use water spray to cool exposed containers. If cooling streams are ineffective (venting sound increases in volume and pitch, tank discolors, or shows any signs of deforming), withdraw immediately to a secure position.

NFPA Hazard Classifications:

Health Hazards (Blue) 2

Flammability (Red) 2
Reactivity (Yellow) 1

Flash Point: 196°F (91°C) o.c.

FIRST AID

Move victim to fresh air. Call emergency medical care. Apply artificial respiration if victim is not breathing. Do not use mouth-to-mouth method if victim ingested or inhaled the substance; induce artificial respiration with the aid of a pocket mask equipped with a one-way valve or other proper respiratory medical device. Administer oxygen if breathing is difficult. Remove and isolate contaminated clothing and shoes. In case of contact with substance, immediately flush skin or eyes with running water for at least 20 minutes. For minor skin contact, avoid spreading material on unaffected skin. Keep victim warm and quiet. Effects of exposure (inhalation, ingestion, or skin contact) may be delayed. Ensure that medical personnel are aware of the material(s) involved and take precautions to protect themselves.

CYCLOPENTENE C:930

SYNONYMS: CICLOPENTENO (Spanish)

IDENTIFICATION

CAS:	142-29-0
ERG ID:	UN 2246
ERG Guide:	128
ERG Guide Category:	FLAMMABLE LIQUIDS (NONPOLAR/WATER-IMMISCIBLE)
Molecular Formula:	C_5H_8
RTECS No.:	GY595000
Properties:	Highly flammable liquid. Colorless. Gasolinelike odor. Floats on the surface of water; insoluble. Boils at 111°F (44°C).

HEALTH AND SAFETY INFORMATION

Irritates the eyes, skin, and respiratory tract. Prolonged contact with skin will cause burns. Swallowing liquid may cause pneumonia. Can have a narcotic effect causing dizziness, drowsiness, and unconsciousness.

RECOMMENDATIONS FOR RESPIRATOR SELECTION:

At any concentrations above the NIOSH REL, or where there is no REL, at any detectable concentration: SCBAF:PD,PP (any self-contained breathing apparatus that has a full facepiece and is operated in a pressure-demand or other positive-pressure mode);or SAF:PD,PP:ASCBA (any supplied-air

respirator that has a full facepiece and is operated in a pressure-demand or other positive-pressure mode in combination with an auxiliary, self-contained breathing apparatus operated in a pressure-demand or other positive-pressure mode).

Escape: GMFOV [any air-purifying, full-facepiece respirator (gas mask) with a chin-style, front- or back-mounted organic vapor canister];or SCBAE (any appropriate escape-type, self-contained breathing apparatus).

CHEMICAL REACTIVITY

Contact with oxidizers can cause fire and explosions. May be able to form peroxides and polymerize.

FIRE INFORMATION

Vapors are heavier than air and will collect in low areas. Vapors in confined areas may explode when exposed to fire. Vapors may travel long distances to ignition sources and flash back. Storage containers and parts of containers may rocket great distances, in many directions. If material or contaminated runoff enters waterways, notify downstream users of potentially contaminated waters. Notify local health and fire officials and pollution control agencies. Use dry chemicals, foam, carbon dioxide. From a secure explosion-proof location, use water spray to cool exposed containers. If cooling streams are ineffective (venting sound increases in volume and pitch, tank discolors, or shows any signs of deforming), withdraw immediately to a secure position.

NFPA Hazard Classifications:

Health Hazards (Blue) 1
Flammability (Red) 3
Reactivity (Yellow) 1

Flash Point: -20°F (-29°C) c.c.
Autoignition Temperature: 743°F (395°C)

FIRST AID

Move victim to fresh air. Call emergency medical care. Apply artificial respiration if victim is not breathing. If breathing is difficult, give oxygen. Remove and isolate contaminated clothing and shoes. In case of contact with substance, immediately flush skin or eyes with running water for at least 20 minutes. Wash skin with soap and water. Keep victim warm and quiet. Ensure that medical personnel are aware of the material(s) involved and take precautions to protect themselves.

p-CYMENE C:980

SYNONYMS: p-CIMENO (Spanish); CYMENE; CYMOL; DOLCYMENE; FEMA No. 2356; 4-ISOPROPYL-1-METHYL BENZENE; ISOPROPYLTOLUOL; p-ISOPROPYLTOLUENE;

4-ISOPROPYL TOLUENE; METHYL PROPYL BENZENE; 1-METHYL-4-ISOPROPYLBENZENE; para-CYMENE; PARA-CYMOL

IDENTIFICATION

CAS:	99-87-6
ERG ID:	UN 2046
ERG Guide:	130
ERG Guide Category:	FLAMMABLE LIQUIDS (NONPOLAR/WATER-IMMISCIBLE/NOXIOUS)
Molecular Formula:	$C_{10}H_{14}$
RTECS No.:	GZ5950000
Properties:	Flammable liquid. Colorless. Mild pleasant odor. Floats on water; insoluble.

HEALTH AND SAFETY INFORMATION

Irritates the eyes, skin, and respiratory tract. May affect the central nervous system. Liquid removes the skin's natural oils. Swallowing liquid can result in pneumonia.

RECOMMENDATIONS FOR RESPIRATOR SELECTION: NIOSH

At any concentrations above the NIOSH REL, or where there is no REL, at any detectable concentration: SCBAF:PD,PP (any self-contained breathing apparatus that has a full facepiece and is operated in a pressure-demand or other positive-pressure mode);or SAF:PD,PP:ASCBA (any supplied-air respirator that has a full facepiece and is operated in a pressure-demand or other positive-pressure mode in combination with an auxiliary, self-contained breathing apparatus operated in a pressure-demand or other positive-pressure mode).

Escape: GMFOV [any air-purifying, full-facepiece respirator (gas mask) with a chin-style, front- or back-mounted organic vapor canister];or SCBAE (any appropriate escape-type, self-contained breathing apparatus).

CHEMICAL REACTIVITY

Light sensitive. Forms an explosive mixture with air. Contact with oxidizers can cause fire and explosions. Will cause rubber to swell and soften.

FIRE INFORMATION

Vapors are heavier than air and will collect in low areas. Vapors in confined areas may explode when exposed to fire. Vapors may travel long distances to ignition sources and flash back. Storage containers and parts of containers may rocket great distances, in many directions. If material or contaminated runoff enters waterways, notify downstream users of potentially contaminated waters. Notify local health and fire officials and pollution control agencies. Use dry chemicals, alcohol-resistant foam, carbon dioxide, AFFF. From a secure explosion-proof location, use water spray to cool exposed containers. If cooling streams are ineffective (venting sound increases in volume and pitch, tank discolors, or shows any signs of deforming), withdraw immediately to a secure position.

NFPA Hazard Classifications:

Health Hazards (Blue) 2
Flammability (Red) 2
Reactivity (Yellow) 0

Flash Point: 140°F (60°C) o.c.; 117°F (47°C) c.c.
Flammable Limits in Air: LEL 0.7% @ 212°F (100°C); UEL 5.6%
Autoignition Temperature: 817°F (436°C)

FIRST AID

Move victim to fresh air. Call emergency medical care. Apply artificial respiration if victim is not breathing. Administer oxygen if breathing is difficult. Remove and isolate contaminated clothing and shoes. In case of contact with substance, immediately flush skin or eyes with running water for at least 20 minutes. Wash skin with soap and water. Keep victim warm and quiet. Effects of exposure (inhalation, ingestion, or skin contact) to substance may be delayed. Ensure that medical personnel are aware of the material(s) involved and take precautions to protect themselves.

DECANOIC ACID — D:010

SYNONYMS: ACIDO DECANOICO (Spanish); CAPRIC ACID; N-CAPRIC ACID; CAPRINIC ACID; CAPRYNIC ACID; N-DECANOIC ACID; N-DECOIC ACID; N-DECYLIC ACID; HEXACID 1095; NEO-FAT 10; 1-NONANECARBOXYLIC ACID

IDENTIFICATION

CAS:	334-48-5
ERG ID:	UN 3261
ERG Guide:	154
ERG Guide Category:	SUBSTANCES-TOXIC and/or CORROSIVE
Molecular Formula:	$C_{10}H_{20}O_2$
Properties:	Combustible, crystalline solid. White. Rancid odor. Insoluble in water.

HEALTH AND SAFETY INFORMATION

Irritating to eyes, skin, and respiratory tract.

Odor Threshold: 0.08-0.09 mg/m^3

RECOMMENDATIONS FOR RESPIRATOR SELECTION:

At any concentrations above the NIOSH REL, or where there is no REL, at any detectable concentration: SCBAF:PD,PP (any self-contained breathing apparatus that has a full facepiece and is operated in a pressure-demand or other positive-pressure mode); or SAF:PD,PP:ASCBA (any supplied-air respirator that has a full facepiece and is operated in a pressure-demand or other positive-pressure mode in combination with an auxiliary, self-contained breathing apparatus operated in a pressure-demand or other positive-pressure mode).

Escape: GMFOV [any air-purifying, full-facepiece respirator (gas mask) with a chin-style, front- or back-mounted organic vapor canister]; or SCBAE (any appropriate escape-type, self-contained breathing apparatus).

CHEMICAL REACTIVITY

Incompatible with sulfuric acid, caustics, ammonia, aliphatic amines, alkanolamines, isocyanates, alkylene oxides, epichlorohydrin. Attacks most common metals.

FIRE INFORMATION

Decomposition products upon heating include toxic gases. Vapors are heavier than air and will collect in low areas. Vapors in confined areas may explode when exposed to fire. Storage containers and parts of containers may rocket great distances, in many directions. If material or contaminated runoff enters waterways, notify downstream users of potentially contaminated waters. Notify local health and fire officials and pollution control agencies. Use dry chemicals, foam, carbon dioxide.

FEMA Hazard Classifications:

Health Hazards (Blue) 0
Flammability (Red) 1
Reactivity (Yellow) 0

Flash Point: More than 230 °F (110 °C) c.c.

FIRST AID

Move victim to fresh air. Call emergency medical care. Apply artificial respiration if victim is not breathing. Do not use mouth-to-mouth method if victim ingested or inhaled the substance; induce artificial respiration with the aid of a pocket mask equipped with a one-way valve or other proper respiratory medical device. Administer oxygen if breathing is difficult. Remove and isolate contaminated clothing and shoes. In case of contact with substance, immediately flush skin or eyes with running water for at least 20 minutes. For minor skin contact, avoid spreading material on unaffected skin. Keep victim warm and quiet. Effects of exposure (inhalation, ingestion, or skin contact) may be delayed. Ensure that medical personnel are aware of the material(s) involved and take precautions to protect themselves.

DIACETONE ALCOHOL D:020

SYNONYMS: ACETONYLDIMETHYLCARBINOL; BLUE EPOXY INK; DIACETON-ALCOHOL (Spanish); DIACETONALCOHOL (Dutch); DIACETONALCOOL (Italian); DIACETONALKOHOL (German); DIACETONE-ALCOOL (French); DIACETONE,4-HYDROXY-4-METHYL-2-PENTATONE,2-METHYL-2-PENTANOL-4-ONE; DIMETHYLACETONYLCARBINOL; DIKETONE ALCOHOL; EEC No. 603-016-00-1; EINECS No. 204-626-7; 4-HYDROXY-2-KETO-4-METHYLPENTANE; 4-HYDROXY-4-METHYL-2-PENTANONE; 4-HYDROXY-4-METHYL-PENTAN-2-ON (German, Dutch); 4-HYDROXY-4-METHYLPENTAN-2-ONE; 4-HYDROXYL-2-KETO-4-METHYLPENTANE; 4-IDROSSI-4-METIL-PENTAN-2-ONE (Italian); 4-METHYL-4-HYDROXY-2-PENTANONE; 2-METHYL-2-PENTANOL-4-ONE; 2-PENTANONE, 4-HYDROXY-4-METHYL-; PERMA DRY INK 555 RED; PERMA DRY INK 555 WHITE; RED EPOXY INK (RINK); TYRANTON

IDENTIFICATION

CAS:	123-42-2
ERG ID:	UN 1148
ERG Guide:	129
ERG Guide Category:	FLAMMABLE LIQUIDS (POLAR/WATER-MIS-CIBLE/NOXIOUS)
Molecular Formula:	$C_6H_{12}O_2$
RTECS No.:	SA9100000
Properties:	Flammable, watery liquid. Colorless to light yellow. Mild, pleasant odor. Floats and mixes with water.

HEALTH AND SAFETY INFORMATION

Irritates eyes, skin, and respiratory tract. Has a narcotic effect on the central nervous system; can cause drowsiness, headache, and unconsciousness. Long-term exposure may cause skin dryness, irritation, and inflammation. May cause liver and kidney damage, anemia, and dermatitis.

IDLH: 1,800 ppm; 8,694 mg/m^3 [LEL].
Odor Threshold: 0.891 ppm
OSHA Table Z-1-A Air Contaminant
NIOSH Criteria Document: NIOSH 78-173 KETONES

Exposure Limits:	**TWA**
ACGIH:	50 ppm; 238 mg/m^3
NIOSH:	50 ppm; 240 mg/m^3
OSHA:	50 ppm; 240 mg/m^3

RECOMMENDATIONS FOR RESPIRATOR SELECTION: NIOSH/OSHA

1,250 ppm: SA:CF (any supplied-air respirator operated in a continuous-flow mode); or PAPROV (any powered, air-purifying respirator with organic vapor cartridge(s)).
1,800 ppm: CCRFOV (any chemical cartridge respirator with a full facepiece and organic vapor cartridge(s)); or GMFOV (any air-purifying, full-facepiece respirator (gas mask) with a chin-style, front- or back-mounted acid gas canister); or PAPRTOV (any powered, air-purifying respirator with a tight-fitting facepiece and organic vapor cartridge(s)); or SCBAF (any self-contained breathing apparatus with a full facepiece); or SAF (any supplied-air respirator with a full facepiece).

Emergency or Planned Entry into Unknown Concentrations or IDLH Conditions: SCBAF:PD,PP (any self-contained breathing apparatus that has a full facepiece and is operated in a pressure-demand or other positive-pressure mode); or SAF:PD,PP:ASCBA (any supplied-air respirator that has a full facepiece and is operated in a pressure-demand; or other positive-pressure mode in combination with an auxiliary, self-contained breathing apparatus operated in a pressure-demand or other positive- pressure mode).

Escape: GMFOV [any air-purifying, full-facepiece respirator (gas mask) with a chin-style, front- or back-mounted organic vapor canister]; or SCBAE (any appropriate escape-type, self-contained breathing apparatus).

Note: Substance causes eye irritation or damage; eye protection needed.

CHEMICAL REACTIVITY

Incompatible with strong acids, strong alkalis (cause formation of flammable acetone vapors), aliphatic amines, isocyanates, oxidizers, and alkali metals. Forms explosive mixture with air. Attacks some forms of plastics, resins, and rubber.

FIRE INFORMATION

Vapors are heavier than air and will collect in low areas. Vapors in confined areas may explode when exposed to fire. Storage containers and parts of containers may rocket great distances, in many directions. If material or contaminated runoff enters waterways, notify downstream users of potentially contaminated waters. Notify local health and fire officials and pollution control agencies. Use dry chemicals, foam, carbon dioxide.

NFPA Hazard Classifications:

Health Hazards (Blue) 1
Flammability (Red) 2
Reactivity (Yellow) 0

Flash Point: 142°F (61°C) o.c.; 125°F (52°C) c.c. (acetone free).
Technical grades containing acetone may have lower flash points varying from 14°F (-20°C) to 320°F (160°C).
Flammable Limits in Air: LEL: 1.8%; UEL 6.9%.
Autoignition Temperature: 118°F (603°C) (acetone free); 1190°F (643°C) (technical grades).

FIRST AID

Move victim to fresh air. Call emergency medical care. Apply artificial respiration if victim is not breathing. Administer oxygen if breathing is difficult. Remove and isolate contaminated clothing and shoes. In case of contact with substance, immediately flush skin or eyes with running water for at least 20 minutes. Wash skin with soap and water. Keep victim warm and quiet. Effects of exposure (inhalation, ingestion, or skin contact) to substance may be delayed. Ensure that medical personnel are aware of the material(s) involved and take precautions to protect themselves.

DIANISIDINE DIISOCYANATE D:022

SYNONYMS: DIISOCIANTO de DIANISIDINA (Spanish); 4,4'-DIISOCYANATO-3,3'-DIMETHOXY-1,1'-BIPHENYL; 3,3'-DIMETHOXYBENZIDINE-4,4'-DIISOCYANATE; 3,3'-DIMETHOXY-4,4'-BIPHENYLENE DIISOCYANATE; ISOCYANIC ACID, 3,3'-DIMETHOXY-4,4'-BIPHENYLENE ESTER; NCI-C02175

IDENTIFICATION

CAS:	91-93-0
ERG ID:	3077, environmentally hazardous substances, solid, n.o.s.
ERG Guide:	171
ERG Guide Category:	SUBSTANCES(LOW TO MODERATE HAZARDS)
Molecular Formula:	$C_{16}H_{12}N_2O_4$
RTECS No.:	NQ8800000
Properties:	Powder. Gray to brown..

HEALTH AND SAFETY INFORMATION

Poisonous. Possible carcinogen. A sensitizer.

IARC Evaluation: Group 3; limited animal evidence

Exposure Limits:	**TWA**	**CEILING** as diisocyanates
NIOSH:	0.005 ppm	0.02 ppm/10M

RECOMMENDATIONS FOR RESPIRATOR SELECTION

At any concentrations above the NIOSH REL, or where there is no REL, at any detectable concentration: SCBAF:PD,PP (any self-contained breathing apparatus that has a full facepiece and is operated in a pressure-demand or other positive-pressure mode); or SAF:PD,PP:ASCBA (any supplied-air respirator that has a full facepiece and is operated in a pressure-demand or other positive-pressure mode in combination with an auxiliary, self-contained breathing apparatus operated in a pressure-demand or other positive-pressure mode).

Escape: GMFOV [any air-purifying, full-facepiece respirator (gas mask) with a chin-style, front- or back-mounted organic vapor canister];or SCBAE (any appropriate escape-type, self-contained breathing apparatus).

CHEMICAL REACTIVITY

Incompatible with acids, acid fumes, caustics, ammonia, aliphatic amines, alkanolamines, aromatic amines, amides, alcohols, glycols, caprolactum solution.

FIRE INFORMATION

Decomposition products upon heating include toxic oxides of nitrogen. If material or contaminated runoff enters waterways, notify downstream users of potentially contaminated waters. Notify local health and fire officials and pollution control agencies.

NFPA Hazard Classifications: Not listed

FIRST AID

Move victim to fresh air; call emergency medical care. Apply artificial respiration if victim is not breathing. Do not use mouth-to-mouth method if victim ingested or inhaled the substance; induce artificial respiration with the aid of a pocket mask equipped with a one-way valve or other proper respiratory medical device. Administer oxygen if breathing is difficult. Remove and isolate contaminated clothing and shoes. In case of contact with substance, immediately flush skin or eyes with running water for at least 20 minutes. For minor skin contact, avoid spreading material on unaffected skin. Keep victim warm and quiet. Effects of exposure (inhalation, ingestion, or skin contact) to substance may be delayed. Ensure that medical personnel are aware of the material(s) involved and take precautions to protect themselves.

ENVIRONMENTAL INFORMATION

EPCRA Section 313: Form R *de minimis* concentration reporting level: 1.0%. Form R Toxic Chemical Category Code: N120.
TSCA: 40CFR 716.120(*d*)8 isocyanates

DIALLYL MALEATE D:024

SYNONYMS: EINECS No. 213-658-0; MALEATE de DIALLYLE (French); MALEATO de DIALILO (Spanish); MALEIC ACID, DIALLYL ESTER; SIPOMER DAM

IDENTIFICATION

CAS:	999-21-3
Molecular Formula:	$C_{10}H_{12}O_4$
RTECS No.:	ON0700000
Properties:	Flammable liquid. Colorless to straw-colored. Insoluble in water.

HEALTH AND SAFETY INFORMATION

Irritates eyes, skin, and respiratory tract.

RECOMMENDATIONS FOR RESPIRATOR SELECTION

At any concentrations above the NIOSH REL, or where there is no REL, at any detectable concentration: SCBAF:PD,PP (any self-contained breathing apparatus that has a full facepiece and is operated in a pressure-demand or other positive-pressure mode); or SAF:PD,PP:ASCBA (any supplied-air respirator that has a full facepiece and is operated in a pressure-demand or other positive-pressure mode in combination with an auxiliary, self-contained breathing apparatus operated in a pressure-demand or other positive-pressure mode).

Escape: GMFOV [any air-purifying, full-facepiece respirator (gas mask) with a chin-style, front- or back-mounted organic vapor canister]; or SCBAE (any appropriate escape-type, self-contained breathing apparatus).

CHEMICAL REACTIVITY

Reacts with sulfuric acid and nitric acid. May form unstable peroxides on storage. Exposure to light and temperatures above 122°F (50°C) may cause polymerization.

FIRE INFORMATION

If material or contaminated runoff enters waterways, notify downstream users of potentially contaminated waters. Notify local health and fire officials and pollution control agencies.

NFPA Hazard Classifications: Not listed

FIRST AID

Move victim to fresh air. Call emergency medical care. Apply artificial respiration if victim is not breathing. Administer oxygen if breathing is difficult. Remove and isolate contaminated clothing and shoes. In case of contact with substance, immediately flush skin or eyes with running water for at least 20 minutes. Wash skin with soap and water. Keep victim warm and quiet. Ensure that medical personnel are aware of the material(s) involved and take precautions to protect themselves.

DIALLYL PHTHALATE D:026

SYNONYMS: DAP; DAPON 35; DAPON R; DI-2-PROPENYL ESTER, 1,2-BENZENEDICARBOXYLIC ACID; EEC No. 607-086-00-4; EINECS No. 205-016-3; FTALATO de DIALILO (Spanish); NCI-C50657; PHTHALLATE de DIALLYLE (French); PHTHALIC ACID, DIALLYL ESTER; o-PHTHALIC ACID, DIALLYL ESTER

IDENTIFICATION

CAS:	131-17-9
ERG ID:	UN 3272
ERG Guide:	127
ERG Guide Category:	FLAMMABLE LIQUIDS (POLAR/WATER- MISCIBLE)
Molecular Formula:	$C_{14}H_{14}O_4$
RTECS No.:	CZ4200000
Properties:	Combustible oily liquid. Colorless. Mild odor. Floats on water; insoluble.

HEALTH AND SAFETY INFORMATION

Irritates eyes, skin, and respiratory tract. Possible carcinogen.

NTP: Evidence of carcinogenicity.

RECOMMENDATIONS FOR RESPIRATOR SELECTION

At any concentrations above the NIOSH REL, or where there is no REL, at any detectable concentration: SCBAF:PD,PP (any self-contained breathing apparatus that has a full facepiece and is operated in a pressure-demand or other positive-pressure mode); or SAF:PD,PP:ASCBA (any supplied-air respirator that has a full facepiece and is operated in a pressure-demand or other positive-pressure mode in combination with an auxiliary, self-contained breathing apparatus operated in a pressure-demand or other positive-pressure mode).

Escape: GMFOV [any air-purifying, full-facepiece respirator (gas mask) with a chin-style, front- or back-mounted organic vapor canister]; or SCBAE (any appropriate escape-type, self-contained breathing apparatus).

CHEMICAL REACTIVITY

Reacts with oxidizers, nitric acid, sulfuric acid. Able to form unstable peroxides. Will polymerize on exposure to peroxides or heat if not inhibited.

FIRE INFORMATION

Storage containers and parts of containers may rocket great distances, in many directions. If material or contaminated runoff enters waterways, notify downstream users of potentially contaminated waters. Notify local health and fire officials and pollution control agencies. Use dry chemicals, alcohol-resistant foam, carbon dioxide.

NFPA Hazard Classifications:

Health Hazards (Blue) 2
Flammability (Red) 1
Reactivity (Yellow) 0

Flash Point: 330°F (165°C) o.c.

FIRST AID

Move victim to fresh air. Call emergency medical care. Apply artificial respiration if victim is not breathing. Administer oxygen if breathing is difficult. Remove and isolate contaminated clothing and shoes. In case of contact with substance, immediately flush skin or eyes with running water for at least 20 minutes. Wash skin with soap and water. Keep victim warm and quiet. Ensure that medical personnel are aware of the material(s) involved and take precautions to protect themselves.

ENVIRONMENTAL INFORMATION

TSCA: 40CFR 716.120(*a*); 40CFR712.30(*w*)

4,4'–DIAMINODIPHENYL METHANE D:028

SYNONYMS: 4-(4-AMINOBENZYL)ANILINE; ANCAMINE TL; ARALDITE HARDENER 972; BENZENAMINE, 4,4'-METHYLENEBIS-; BIS-p-AMINOFENYLMETHAN (German); BIS(4-AMINOPHENYL)METHANE; CURITHANE; DADPM; DDM; DIAMINODIPHENYLMETHANE; p,p'-DIAMINODIPHENYLMETHANE; DI-(4-AMINOPHENYL)METHANE; DIANILINOMETHANE; 4,4'-DIPHENYLMETHANEDIAMINE; DIPHENYLMETHANEDIAMINE; EEC No. 612-051-00-1; EINECS No. 202-974-4; EPICURE DDM; HT 972; JEFFAMINE AP-20; MDA; METHYLENEBIS(ANILINE); 4,4'-METHYLENEBISANILINE; 4,4'-METHYLENEBIS(BENZENEAMINE); METHYLENEDIANILINE; 4,4'-METHYLENEDIANILINE; p,p'-METHYLENEDIANILINE; 4,4-METHYLENEDIANILINE; SUMICURE M; TONOX

IDENTIFICATION

CAS:	101-77-9
ERG ID:	UN 2651
ERG Guide:	153
ERG Guide Category:	SUBSTANCES-TOXIC AND/OR CORROSIVE (COMBUSTIBLE)
Molecular Formula:	$C_{13}H_{14}N_2$
RTECS No.:	BY5425000
Properties:	Combustible flakes, lumps, or leafs. Colorless to yellow; turns light brown on exposure to air. Faint aminelike odor. Slightly soluble in cold water; soluble in hot water.

HEALTH AND SAFETY INFORMATION

Irritates eyes, skin, respiratory tract. Proloned or repeated exposure may affect eyes, liver, central nervous system, spleen. Has caused bladder cancer in animals.

IARC Evaluation: Group 2B; sufficient animal evidence; possible human carcinogen.
NTP: Confirmed carcinogen
IDLH: Potential human carcinogen

OSHA Specifically Regulated Substances: (29 CFR 1910.1050)

Exposure Limits:	**TWA**	**STEL**
ACGIH: (1)	0.1 ppm; 0.81 mg/m^3	
NIOSH: (2)		
OSHA:	0.010 ppm	0.100 ppm

(1) An animal carcinogen. Skin contact contributes significantly to overall exposure.
(2) Potential human carcinogen

RECOMMENDATIONS FOR RESPIRATOR SELECTION:

At any concentrations above the NIOSH REL, or where there is no REL, at any detectable concentration: SCBAF:PD,PP (any self-contained breathing apparatus that has a full facepiece and is operated in a pressure-demand or other positive-pressure mode); or SAF:PD,PP:ASCBA (any supplied-air respirator that has a full facepiece and is operated in a pressure-demand or other positive-pressure mode in combination with an auxiliary, self-contained breathing apparatus operated in a pressure-demand or other positive-pressure mode).

Escape: GMFOV [any air-purifying, full-facepiece respirator (gas mask) with a chin-style, front- or back-mounted organic vapor canister]; or SCBAE (any appropriate escape-type, self-contained breathing apparatus).

CHEMICAL REACTIVITY

Incompatible with strong oxidizers.

FIRE INFORMATION

Combustion products include toxic nitrogen oxides. Firefighting gear (including SCBA) may not provide adequate protection. If exposure occurs, remove and isolate gear immediately and thoroughly decontaminate personnel. Vapors are heavier than air and will collect in low areas. Vapors in confined areas may explode when exposed to fire. Storage containers and parts of containers may rocket great distances, in many directions. If material or contaminated runoff enters waterways, notify downstream users of potentially contaminated waters. Notify local health and fire officials and pollution control agencies. Use dry chemicals, carbon dioxide, foam. Use water spray to cool containers.

NFPA Hazard Classifications:

Health Hazards (Blue) 3
Flammability (Red) 1
Reactivity (Yellow) 0

Flash Point: 440°F (227°C) o.c.

FIRST AID

Move victim to fresh air and call emergency medical care. Apply artificial respiration if victim is not breathing. Do not use mouth-to-mouth method if victim swallowed or inhaled the substance; induce artificial respiration with the aid of a pocket mask equipped with a one-way valve or other proper respiratory medical device. If breathing is difficult, give oxygen. In case of contact with substance, immediately flush skin or eyes with running water for at least 20 minutes. Remove and isolate contaminated clothing and shoes. Keep victim warm and quiet. Effects of exposure (inhalation, ingestion, or skin contact) to substances may be delayed. Ensure that medical personnel are aware of the material(s) involved and take precautions to protect themselves.

ENVIRONMENTAL INFORMATION

Clean Air Act: Hazardous Air Pollutants (Title I, Part A, Section 112)
EPCRA Section 304 Reportable Quantity (RQ): CERCLA, 10 lb (4.54 kg)
EPCRA Section 313: Form R *de minimis* concentration reporting level: 0.1%.

DIBENZOYL PEROXIDE D:030

SYNONYMS: ACETOXYL; BENOXYL; BENZOIC ACID, PEROXIDE; BENZOPEROXIDE; BENZOYL; BENZOYL PEROXIDE (C14H10O4); BENZOYL SUPEROXIDE; BP; BPO; CADOX-40E; CADOX-B 50P; DIBENZOYLPEROXID (German); DIPHENYHL GLYOXAL PEROXIDE; EEC No. 617-008-00-0; EINECS No. 202-327-6; G-20; LUCIDOL; LUCIDOL-40E; LUCIDOL-50P; LUCIDOL-98; LUCIDOL-B 50; LUCIDOL CH 50; LUCIDOL G-20; LUCIDOL KL-50; LUCIDOL S-50; LUPERCO AA; LUPERCO AST; NYPER B; NYPER BO; OXYLITE; PANOXYL; PEROXIDE,

DIBENZOYL; PERSADOX; PEROSSIDO di DIBENZOILE (Italian); PEROXIDO de DIBENZOILO (Spanish); PEROXYDE de DIBENZOYLE (French); SUPEROX 744

IDENTIFICATION

CAS:	94-36-0
ERG ID:	UN 3102 (more than 51–100%); UN 3106 (more than 35–52%)
ERG Guide:	146–(more than 51–100%); 145 (more than 35–52%)
ERG Guide Category:	146–ORGANIC PEROXIDES (HEAT, CONTAMINATION AND FRICTION SENSITIVE); 145-ORGANIC PEROXIDES (HEAT AND CONTAMINATION SENSITIVE).
Molecular Formula:	$C_{14}H_{10}O_4$
RTECS No.:	DM8575000
Properties:	Combustible, solid powder or granules. White. Odorless. Sinks in water. May explode if subjected to heat, shock, contamination, or friction.

HEALTH AND SAFETY INFORMATION

Irritates eyes, skin, and respiratory tract. My affect central nervous system; possible headache, drowsiness, unconsciousness. Long-term effects can cause asthma, eczema.

IARC Evaluation: Unclassifiable substance, Group 3; inadequate human evidence; inadequate animal evidence.
IDLH: 1,500 mg/m^3
OSHA Table Z-1-A Air Contaminant
OSHA Process Safety Management of Highly Hazardous Chemicals (29CFR, Part 1910.119, Appendix A): Threshold Quantity: 7,500 pounds.

Exposure Limits:	**TWA**
ACGIH:	5 mg/m^3
NIOSH:	5 mg/m^3
OSHA:	5 mg/m^3

RECOMMENDATIONS FOR RESPIRATOR SELECTION: NIOSH/OSHA

50 mg/m^3: DMXSQ (any dust and mist respirator except single-use and quarter mask respirators); or SA (any supplied-air respirator).
125 mg/m^3: SA:CF (any supplied-air respirator operated in a continuous-flow mode); or PAPRDM, (any powered, air-purifying respirator with a dust and mist filter).
250 mg/m^3: HiEF (any air-purifying, full-facepiece respirator with a high-efficiency particulate filter); or PAPRTHiE (any powered, air-purifying respirator with a tight-fitting facepiece and a high-efficiency particulate filter); or SCBAF (any self-contained breathing apparatus with a full facepiece); or SAF (any supplied-air respirator with a full facepiece).
1,500 mg/m^3: SAF:PD,PP (any supplied-air respirator that has a full facepiece and is operated in a pressure-demand or other positive-pressure mode).

Emergency or Planned Entry into Unknown Concentrations or IDLH Conditions: SCBAF:PD,PP (any self-contained breathing apparatus that has a full facepiece and is operated in a pressure-demand or other positive-pressure mode); or SAF:PD,PP:ASCBA (any supplied-air respirator that has a full facepiece and is operated in a pressure-demand or other positive-pressure mode in combination with an auxiliary, self-contained breathing apparatus operated in a pressure-demand or other positive-pressure mode).

Escape: HiEF (any air-purifying, full-facepiece respirator with a high-efficiency particulate filter); or SCBAE (any appropriate escape-type, self-contained breathing apparatus).
Note: Substance reported to cause eye irritation or damage; may require eye protection.

CHEMICAL REACTIVITY

Confined storage of dry chemical may lead to decomposition and explosion. A strong oxidant; extremely reactive. Fires and explosion may result from heat, shock, friction, or contamination and from contact with strong acids, combustible materials, oxidizers, acids, bases, alcohols, reducing agents, metals, metal oxides, amines, accelerators, methyl methacrylate, organic matter, lithium aluminum carbide, dimethyl aniline, amines, and metallic naphthenates. May attack some plastics, rubber, and coatings. Protect containers from shock, heat, and friction.

FIRE INFORMATION

Vapors are heavier than air and will collect in low areas. Vapors in confined areas may explode when exposed to fire. Storage containers and parts of containers may rocket great distances, in many directions. If material or contaminated runoff enters waterways, notify downstream users of potentially contaminated waters. Notify local health and fire officials and pollution control agencies. Use dry chemicals, foam, carbon dioxide.

NFPA Hazard Classifications:

Health Hazards (Blue) 1
Flammability (Red) 4
Reactivity (Yellow) 4
Special situations 0XY

Flash Point: 176°F (80°C).

FIRST AID

Move victim to fresh air. Call emergency medical care. Apply artificial respiration if victim is not breathing. Administer oxygen if breathing is difficult. Remove and isolate contaminated clothing and shoes. Remove material from skin immediately. In case of contact with substance, immediately flush skin or eyes with running water for at least 20 minutes. Keep victim warm and quiet. Ensure that medical personnel are aware of the material(s) involved and take precautions to protect themselves.

ENVIRONMENTAL INFORMATION

EPCRA Section 313: Form R *de minimis* concentration reporting level: 1.0%.

DIBENZYL ETHER D:040

SYNONYMS: BENZYL ETHER; 1,1'-[OXYBIS(METHYLENE)] BIS BENZENE; BENZYL OXIDE; EINECS No. 203-118-2; ETER DIBENCILICO (Spanish); FEMA No. 2371; OSSIDO di DIBENZILE (Italian); OXYDE de DIBENZYLE (French)

IDENTIFICATION

CAS:	103-50-4
Molecular Formula:	$C_{14}H_{14}O$
RTECS No.:	DQ6125000
Properties:	Combustible liquid. Colorless to pale yellow. Mild almondlike odor. Floats on water; insoluble.

HEALTH AND SAFETY INFORMATION

Irritating to the eyes, skin, and respiratory tract; affects nervous system. Harmful to skin's natural oils.

RECOMMENDATIONS FOR RESPIRATOR SELECTION:

At any concentrations above the NIOSH REL, or where there is no REL, at any detectable concentration: SCBAF:PD,PP (any self-contained breathing apparatus that has a full facepiece and is operated in a pressure-demand or other

positive-pressure mode); or SAF:PD,PP:ASCBA (any supplied-air respirator that has a full facepiece and is operated in a pressure-demand or other positive-pressure mode in combination with an auxiliary, self-contained breathing apparatus operated in a pressure-demand or other positive-pressure mode).

Escape: GMFOV [any air-purifying, full-facepiece respirator (gas mask) with a chin-style, front- or back-mounted organic vapor canister]; or SCBAE (any appropriate escape-type, self-contained breathing apparatus).

CHEMICAL REACTIVITY

Forms unstable peroxides. An oxidizer; reacts violently with reducing agents, combustible materials. Will also react with strong oxidizers. May accumulate static electricity, and may cause ignition of its vapors.

FIRE INFORMATION

Will increase the intensity of fire. Vapors are heavier than air and will collect in low areas. Vapors in confined areas may explode when exposed to fire. Storage containers and parts of containers may rocket great distances, in many directions. If material or contaminated runoff enters waterways, notify downstream users of potentially contaminated waters. Notify local health and fire officials and pollution control agencies. Use dry chemicals, foam, carbon dioxide, AFFF.

NFPA Hazard Classifications:

Health Hazards (Blue) 0
Flammability (Red) 1
Reactivity (Yellow) 0

Flash Point: 275°F (135°C) c.c.

FIRST AID

Move victim to fresh air. Call emergency medical care. Apply artificial respiration if victim is not breathing. If breathing is difficult, give oxygen. Remove and isolate contaminated clothing and shoes. In case of contact with substance, immediately flush skin or eyes with running water for at least 20 minutes. Wash skin with soap and water. Keep victim warm and quiet. Ensure that medical personnel are aware of the material(s) involved and take precautions to protect themselves.

DIBROMOMETHANE D:050

SYNONYMS: BROMURE de METHYLENE (French); BROMURO di METILENE (Italian); DIBROMMETHAN (German); DIBROMOMETANO (Spanish); EEC No. 602-003-00-8; EINECS No. 200-824-2; METHYLENE BROMIDE; METHYLENE DIBROMIDE; METHANE, DIBROMO-

IDENTIFICATION

CAS:	74-95-3
ERG ID:	UN 2664
ERG Guide:	160
ERG Guide Category:	HALOGENATED SOLVENTS
Molecular Formula:	CH_2Br_2
RTECS No.:	PA7350000
Properties:	Nonflammable, colorless, watery liquid. Sweet, pleasant, chloroformlike odor. Sinks in water; slightly soluble. Vapors are heavier than air; will collect in low areas. Hydrogen bromide and other Toxic gases are produced when heated.

HEALTH AND SAFETY INFORMATION

RECOMMENDATIONS FOR RESPIRATOR SELECTION: NIOSH

At any concentrations above the NIOSH REL, or where there is no REL, at any detectable concentration: SCBAF:PD,PP (any self-contained breathing apparatus that has a full facepiece and is operated in a pressure-demand or other positive-pressure mode); or SAF:PD,PP:ASCBA (any supplied-air respirator that has a full facepiece and is operated in a pressure-demand or other positive-pressure mode in combination with an auxiliary, self-contained breathing apparatus operated in a pressure-demand or other positive pressure mode).

Escape: GMFOV [any air-purifying, full-facepiece respirator (gas mask) with a chin-style, front- or back-mounted organic vapor canister]; or SCBAE (any appropriate escape-type, self-contained breathing apparatus).

CHEMICAL REACTIVITY

Mixture with potassium forms a shock-sensitive explosive.

FIRE INFORMATION

Hydrogen bromide and other toxic gases are produced when heated. Not flammable under normal conditions. If material or contaminated runoff enters waterways, notify downstream users of potentially contaminated waters. Notify local health and fire officials and pollution control agencies. Use any extinguishing agent suitable for surrounding fires.

NFPA Hazard Classifications:

Health Hazards (Blue) 2
Flammability (Red) 0
Reactivity (Yellow) 0

FIRST AID

Move victim to fresh air. Call emergency medical care. Apply artificial respiration if victim is not breathing. Administer oxygen if breathing is difficult. Remove and isolate contaminated clothing and shoes. In case of contact with substance, immediately flush skin or eyes with running water for at least 20 minutes. For minor skin contact, avoid spreading material on unaffected skin. Wash skin with soap and water. Keep victim warm and quiet. Ensure that medical personnel are aware of the material(s) involved and take precautions to protect themselves.

ENVIRONMENTAL INFORMATION

RCRA 40CFR 258, Appendix A: List of hazardous organic constituents.
Safe Drinking Water Act: Priority List (55 FR 1470)
TSCA: 40CFR799.5055 testing of hazardous waste constituents; 40CFR716.120(*d*)4, 40CFR 712.30(*e*)4 brominated flame retardants.

DIBUTOXYETHYL ADIPATE D:053

SYNONYMS: ADIPATO de DIBUTOXIETILO (Spanish); ADITPATO di BIS(2-BUTOSSIETILE (Italian); ADIPIC ACID, DIBUTOXYETHYL ESTER; BIS(2-BUTOXYETHYL) ADIPATE; BUTYL "CELLOSOLVE" ADIPATE; DI(2-BUTOXYETHYL) ADIPATE; DIBUTYL CELLOSOLVE ADIPATE; EINECS No. 205-466-0; HEXANEDIOIC ACID, BIS(2-BUTOXYETHYL) ESTER;

IDENTIFICATION

CAS: 141-18-4
Molecular Formula: $C_{18}H_{34}O_6$
RTECS No.: AU8750000
Properties: Combustible, oily liquid. Colorless. Mild, butyl-like odor.

HEALTH AND SAFETY INFORMATION

RECOMMENDATIONS FOR RESPIRATOR SELECTION

At any concentrations above the NIOSH REL, or where there is no REL, at any detectable concentration: SCBAF:PD,PP (any self-contained breathing apparatus that has a full facepiece and is operated in a pressure-demand or other positive-pressure mode); or SAF:PD,PP:ASCBA (any supplied-air respirator that has a full facepiece and is operated in a pressure-demand or other positive-pressure mode in combination with an auxiliary, self-contained breathing apparatus operated in a pressure-demand or other positive-pressure mode).

Escape: GMFOV [any air-purifying, full-facepiece respirator (gas mask) with a chin-style, front- or back-mounted organic vapor canister]; or SCBAE (any appropriate escape-type, self-contained breathing apparatus).

CHEMICAL REACTIVITY

Incompatible with nitrates, nitric acid, sulfuric acid. Contact with strong oxidizers may cause fire and explosions.

FIRE INFORMATION

If material or contaminated runoff enters waterways, notify downstream users of potentially contaminated waters. Notify local health and fire officials and pollution control agencies.

NFPA Hazard Classifications: Not listed

Flash Point: 370°F (187°C)

FIRST AID

Move victim to fresh air. Call emergency medical care. Apply artificial respiration if victim is not breathing. Administer oxygen if breathing is difficult. Remove and isolate contaminated clothing and shoes. In case of contact with substance, immediately flush skin or eyes with running water for at least 20 minutes. Wash skin with soap and water. Keep victim warm and quiet. Ensure that medical personnel are aware of the material(s) involved and take precautions to protect themselves.

DIBUTOXYETHYL PHTHALATE D:056

SYNONYMS: BIS(2-BUTOXYETHYL) PHTHALATE; 2-BUTOXYETHANOL PHTHALATE (2:1); β-BUTOXYETHYL PHTHALATE; BUTYL "CELLOSOLVE" PHTHALATE; BUTYL GLYCOL PHTHALATE; DIBUTYL CELLOSOLVE PHTHALATE; DIBUTYLGLYCOL PHTHALATE; EINECS No. 204-213-1; FTALATO de DIBUTOXIETILO (Spanish); KESSCOFLEX; KRONISOL

IDENTIFICATION

CAS:	117-83-9
Molecular Formula:	$C_{20}H_{30}O_6$
RTECS No.:	TI0175000
Properties:	Crystalline solid or combustible liquid. White solid or colorless liquid. Insoluble in water.

HEALTH AND SAFETY INFORMATION

RECOMMENDATIONS FOR RESPIRATOR SELECTION

At any concentrations above the NIOSH REL, or where there is no REL, at any detectable concentration: SCBAF:PD,PP (any self-contained breathing apparatus that has a full facepiece and is operated in a pressure-demand or other positive-pressure mode); or SAF:PD,PP:ASCBA (any supplied-air respirator that has a full facepiece and is operated in a pressure-demand or other positive-pressure mode in combination with an auxiliary, self-contained breathing apparatus operated in a pressure-demand or other positive-pressure mode).

Escape: GMFOV [any air-purifying, full-facepiece respirator (gas mask) with a chin-style, front- or back-mounted organic vapor canister]; or SCBAE (any appropriate escape-type, self-contained breathing apparatus).

CHEMICAL REACTIVITY

Incompatible with nitrates, sulfuric acid, nitric acid. Contact with strong oxidizers may cause fire and explsoions.

FIRE INFORMATION

Vapors are heavier than air and will collect in low areas. Vapors in confined areas may explode when exposed to fire. If material or contaminated runoff enters waterways, notify downstream users of potentially contaminated waters. Notify local health and fire officials and pollution control agencies.

NFPA Hazard Classifications:

Health Hazards (Blue) 0
Flammability (Red) 1
Reactivity (Yellow) 0

Flash Point: 407°F (208°C) o.c.

FIRST AID

Move victim to fresh air. Call emergency medical care. Apply artificial respiration if victim is not breathing. Administer oxygen if breathing is difficult. Remove and isolate contaminated clothing and shoes. In case of contact with substance, immediately flush skin or eyes with running water for at least 20 minutes. Wash skin with soap and water. Keep victim warm and quiet. Ensure that medical personnel are aware of the material(s) involved and take precautions to protect themselves.

DI-n-BUTYL ETHER D:060

SYNONYMS: n-DIBUTYL ETHER; n-BUTYL ETHER; BUTYL ETHER; 1-BUTOXY BUTANE; DIBUTYL ETHER; DIBUTYL OXIDE; EEC No. 603-054-00-9; EINECS No. 205-575-3; ETER n-DIBUTILICO (Spanish); ETHER BUTYLIQUE (French); OSSIDO di DIBUTILE (Italian); OXYDE de DIBUTYLE (French); 1,1'-OXYBIS(BUTANE)

IDENTIFICATION

CAS:	142-96-1
ERG ID:	UN 1149
ERG Guide:	127
ERG Guide Category:	FLAMMABLE LIQUIDS (POLAR/WATER- MISCIBLE)
Molecular Formula:	$C_8H_{18}O$
RTECS No.:	EK5425000
Properties:	Flammable liquid. Colorless. Mild pleasant, etherlike odor. Floats on water; soluble. Flammable, irritating vapor is produced.

HEALTH AND SAFETY INFORMATION

Irritates eyes, skin, and respiratory tract.

Odor Threshold: 0.37-2.50 mg/m^3

RECOMMENDATIONS FOR RESPIRATOR SELECTION

At any concentrations above the NIOSH REL, or where there is no REL, at any detectable concentration: SCBAF:PD,PP (any self-contained breathing apparatus that has a full facepiece and is operated in a pressure-demand or other

positive-pressure mode); or SAF:PD,PP:ASCBA (any supplied-air respirator that has a full facepiece and is operated in a pressure-demand or other positive-pressure mode in combination with an auxiliary, self-contained breathing apparatus operated in a pressure-demand or other positive-pressure mode).

Escape: GMFOV [any air-purifying, full-facepiece respirator (gas mask) with a chin-style, front- or back-mounted organic vapor canister]; or SCBAE (any appropriate escape-type, self-contained breathing apparatus).

CHEMICAL REACTIVITY

Forms explosive mixture with air. Violent reaction with strong oxidizers. Incompatible with strong acids. May be able to form unstable peroxides. May accumulate static electricity and cause ignition of its vapors.

FIRE INFORMATION

Vapors are heavier than air and will collect in low areas. Vapors in confined areas may explode when exposed to fire. Vapors may travel long distances to ignition sources and flash back. Storage containers and parts of containers may rocket great distances, in many directions. If material or contaminated runoff enters waterways, notify downstream users of potentially contaminated waters. Notify local health and fire officials and pollution control agencies. Decomposition products upon heating include toxic gases. Use dry chemicals, foam, carbon dioxide, AFFF. Use water to cool containers.

NFPA Hazard Classifications:

Health Hazards (Blue) 2
Flammability (Red) 3
Reactivity (Yellow) 1

Flash Point: 92°F (33°C) o.c.; 77°F (25°C) c.c.
Flammable Limits in Air: LEL 1.5%; UEL 7.6%
Autoignition Temperature: 382°F (194°C)

FIRST AID

Move victim to fresh air. Call emergency medical care. Apply artificial respiration if victim is not breathing. Administer oxygen if breathing is difficult. Remove and isolate contaminated clothing and shoes. In case of contact with substance, immediately flush skin or eyes with running water for at least 20 minutes. Wash skin with soap and water. Keep victim warm and quiet. Ensure that medical personnel are aware of the material(s) involved and take precautions to protect themselves.

DIBUTYL FUMARATE D:064

SYNONYMS: +2-BUTENEDIOIC ACID (E)-, DIBUTYL ESTER; DI-N-BUTYL FUMARATE; DIBUTYLESTER KYSELINY FUMAROVE; EINECS No. 203-327-9; FUMARATO de DIBUTILO (Spanish); FUMARIC ACID, DIBUTYL ESTER

IDENTIFICATION

CAS:	105-75-9
Molecular Formula:	$C_{12}H_{20}O_4$
RTECS No.:	LT1225000
Properties:	Liquid. Colorless and clear. Insoluble in water.

HEALTH AND SAFETY INFORMATION

Irritates eyes, skin , and respiratory tract.

RECOMMENDATIONS FOR RESPIRATOR SELECTION

At any concentrations above the NIOSH REL, or where there is no REL, at any detectable concentration: SCBAF:PD,PP (any self-contained breathing apparatus that has a full facepiece and is operated in a pressure-demand or other positive-pressure mode); or SAF:PD,PP:ASCBA (any supplied-air respirator that has a full facepiece and is operated in a pressure-demand or other positive-pressure mode in combination with an auxiliary, self-contained breathing apparatus operated in a pressure-demand or other positive-pressure mode).

Escape: GMFOV [any air-purifying, full-facepiece respirator (gas mask) with a chin-style, front- or back-mounted organic vapor canister]; or SCBAE (any appropriate escape-type, self-contained breathing apparatus).

CHEMICAL REACTIVITY

Incompatible with nitrates, nitric acid, sulfuric acid. Contact with strong oxidizers may cause fire and explosions.

FIRE INFORMATION

Vapors are heavier than air and will collect in low areas. Vapors in confined areas may explode when exposed to fire. If material or contaminated runoff enters waterways, notify downstream users of potentially contaminated waters. Notify local health and fire officials and pollution control agencies. Use dry chemicals, alcohol-resistant foam, carbon dioxide.

NFPA Hazard Classifications: Not listed

Flash Point: 300°F (149°C) o.c.

FIRST AID

Move victim to fresh air. Call emergency medical care. Apply artificial respiration if victim is not breathing. Administer oxygen if breathing is difficult. Remove and isolate contaminated clothing and shoes. In case of contact with substance, immediately flush skin or eyes with running water for at least 20 minutes. Wash skin with soap and water. Keep victim warm and quiet. Ensure that medical personnel are aware of the material(s) involved and take precautions to protect themselves.

DIBUTYL MALEATE D:068

SYNONYMS: 2-BUTENEDIOIC ACID, DIBUTYL ESTER; DBM; MALEIC ACID, DIBUTYL ESTER; MALEATO de DIBUTILO (Spanish); EINECS No. 203-328-4; RC COMONOMER DBM; STAFLEX DBM

IDENTIFICATION

CAS:	105-76-0
Molecular Formula:	$C_{12}H_{20}O_4$
RTECS No.:	ON0875000
Properties:	Combustible, oily liquid. Colorless. Insoluble in water.

RECOMMENDATIONS FOR RESPIRATOR SELECTION: NIOSH

At any concentrations above the NIOSH REL, or where there is no REL, at any detectable concentration: SCBAF:PD,PP (any self-contained breathing apparatus that has a full facepiece and is operated in a pressure-demand or other positive-pressure mode); or SAF:PD,PP:ASCBA (any supplied-air respirator that has a full facepiece and is operated in a pressure-demand or other positive-pressure mode in combination with an auxiliary, self-contained breathing apparatus operated in a pressure-demand or other positive-pressure mode).

Escape: GMFOV [any air-purifying, full-facepiece respirator (gas mask) with a chin-style, front- or back-mounted organic vapor canister]; or SCBAE (any appropriate escape-type, self-contained breathing apparatus).

CHEMICAL REACTIVITY

Incompatible with nitrates, sulfuric acid, nitric acid. Contact with strong oxidizers may cause fire and explosions.

FIRE INFORMATION

Vapors are heavier than air and will collect in low areas. Vapors in confined areas may explode when exposed to fire. If material or contaminated runoff enters waterways, notify downstream users of potentially contaminated waters. Notify local health and fire officials and pollution control agencies. Use dry chemicals, alcohol-resistant foam, carbon dioxide.

NFPA Hazard Classifications:

Health Hazards (Blue) 1
Flammability (Red) 1
Reactivity (Yellow) 0

Flash Point: 285°F (141°C) o.c.

FIRST AID

Move victim to fresh air. Call emergency medical care. Apply artificial respiration if victim is not breathing. Administer oxygen if breathing is difficult. Remove and isolate contaminated clothing and shoes. In case of contact with substance, immediately flush skin or eyes with running water for at least 20 minutes. Wash skin with soap and water. Keep victim warm and quiet. Ensure that medical personnel are aware of the material(s) involved and take precautions to protect themselves.

DI–tert–BUTYL PEROXIDE D:070

SYNONYMS: CADOX; DIBUTYL PEROXIDE; DI-tert-BUTYLPEROXID (German); DI-tert-BUTYL PEROXYDE (Dutch); DTBP; EEC No. 617-001-00-2; EINECS No. 203-733-6; PEROSSIDO di BUTILE TERZIARIO (Italian); PEROXIDO de terc-BUTILO (Spanish); PEROXYDE de BUTYLE TERTIAIRE (French); (TRIBUTYL)PEROXIDE

IDENTIFICATION

CAS:	110-05-4
ERG ID:	UN 2102
ERG Guide:	145
ERG Guide Category:	ORGANIC PEROXIDES (HEAT AND CONTAMINATION SENSITIVE).
Molecular Formula:	$C_8H_{18}O_2$
RTECS No.:	ER2450000
Properties:	Flammable liquid. Colorless, waterlike. Slightly water soluble.

HEALTH AND SAFETY INFORMATION

Irritates eyes, skin, and respiratory tract.

OSHA Process Safety Management of Highly Hazardous Chemicals (29CFR, Part 1910.119, Appendix A): Threshold Quantity: 5,000 pounds.

RECOMMENDATIONS FOR RESPIRATOR SELECTION

At any concentrations above the NIOSH REL, or where there is no REL, at any detectable concentration: SCBAF:PD,PP (any self-contained breathing apparatus that has a full facepiece and is operated in a pressure-demand or other positive-pressure mode); or SAF:PD,PP:ASCBA (any supplied-air respirator that has a full facepiece and is operated in a pressure-demand or other positive-pressure mode in combination with an auxiliary, self-contained breathing apparatus operated in a pressure-demand or other positive-pressure mode).

Escape: GMFS [any air-purifying, full-facepiece respirator (gas mask) with a chin-style, front- or back-mounted canister providing protection against the compound of concern]; or SCBAE (any appropriate escape-type, self-contained breathing apparatus).

CHEMICAL REACTIVITY

A powerful oxidizer; highly reactive with many other materials. Forms unstable and dangerous peroxides in storage. Heat, friction, shock, contact with reducing agents, thiocyanates, organic matter, combustibles, or contamination may cause explosion. Protect containers from shock, heat and friction. May accumulate static electricity, and cause ignition of its vapors.

FIRE INFORMATION

Fire may produce irritating, corrosive and/or toxic gases. Will increase intensity of fire; may burn rapidly with flare-burning effect. Vapors are heavier than air and will collect in low areas. Vapors in confined areas may explode when exposed to fire. Storage containers and parts of containers may rocket great distances, in many directions. If material or contaminated runoff enters waterways, notify downstream users of potentially contaminated waters. Runoff may create fire or explosion hazard. Notify local health and fire officials and pollution control agencies. Use dry chemicals, foam, carbon dioxide, large quantities of water. Use water spray to cool containers. Material on clothing is a fire hazard; flush immediately with water.

NFPA Hazard Classifications:

Health Hazards (Blue) 3
Flammability (Red) 2
Reactivity (Yellow) 4
Special situations .... OXY

Flash Point: 65°F (18°C) o.c.

FIRST AID

Move victim to fresh air. Call emergency medical care. Apply artificial respiration if victim is not breathing. Administer oxygen if breathing is difficult. Remove and isolate contaminated clothing and shoes. Remove material from skin immediately. In case of contact with substance, immediately flush skin or eyes with running water for at least 20 minutes. Keep victim warm and quiet. Ensure that medical personnel are aware of the material(s) involved and take precautions to protect themselves.

DIBUTYL PHTHALATE D:072

SYNONYMS: o-BENZENEDICARBOXYLIC ACID, DIBUTYL ESTER; BENZENE-o-DICARBOXYLIC ACID DI-N-BUTYL ESTER; 1,2-BENZENEDICARBOXYLIC ACID, DIBUTYL ESTER; DI-N-BUTYL PHTHALATE; N-BUTYLPHTHALATE; PHTHALIC ACID DIBUTYL ESTERCELLUFLEX DBP; DBP; DIBUTYL 1,2-BENZENE DICARBOXYLATE; DIBUTYL-1,2-BENZENEDICARBOXYLATE; DI-N-BUTYL PHTHALATE; ELAOL; EINECS No. 201-557-4; FTALATO de DIBUTILO (Spanish); HEXAPLAS M/B; PALATINOL C; POLYCIZER DBP; PX 104; STAFLEX DBP; WITCIZER 300; BUTYL PHTHALATE; PHTHALIC ACID, DIBUTYL ESTER; RC PLASTICIZER DBP

IDENTIFICATION

CAS:	84-74-2
ERG ID:	UN 9095
ERG Guide:	171
ERG Guide Category:	SUBSTANCES (LOW TO MODERATE HAZARD)
Molecular Formula:	$C_{16}H_{22}O_4$
RTECS No.:	TI0875000
Properties:	Combustible, viscous, oily liquid. Colorless. Odorless. Sinks slowly in water; very slightly soluble (0.001% @ 77 °F).

HEALTH AND SAFETY INFORMATION

Irritates eyes, skin, upper respiratory tract.

IDLH: 4,000 mg/m^3
OSHA Table Z-1-A Air Contaminant

Exposure Limits: TWA

ACGIH:	5 mg/m^3
NIOSH:	5 mg/m^3
OSHA:	5 mg/m^3

RECOMMENDATIONS FOR RESPIRATOR SELECTION: NIOSH/OSHA

50 mg/m^3: DMF (any dust and mist respirator with a full facepiece).
125 mg/m^3: SA:CF (any supplied-air respirator operated in a continuous-flow mode); or PAPRDM (any powered, air-purifying respirator with a dust and mist filter).
250 mg/m^3: HiEF (any air-purifying, full-facepiece respirator with a high-efficiency particulate filter); or SCBAF (any self-contained breathing apparatus with a full facepiece); or SAF (any supplied-air respirator with a full facepiece).
4000 mg/m^3: SAF:PD,PP (any supplied-air respirator that has a full facepiece and is operated in a pressure-demand or other positive-pressure mode).

Emergency or Planned Entry into Unknown Concentrations or Idlh Conditions: SCBAF:PD,PP (any self-contained breathing apparatus that has a full facepiece and is operated in a pressure-demand or other positive-pressure mode); or SAF:PD,PP:ASCBA (any supplied-air respirator that has a full facepiece and is operated in a pressure-demand or other positive-pressure mode in combination with an auxiliary, self-contained breathing apparatus operated in a pressure-demand or other positive-pressure mode.

Escape: HiEF (any air-purifying, full-facepiece respirator with a high-efficiency particulate filter); or SCBAE (any appropriate escape-type, self-contained breathing apparatus).

Note: Substance causes eye irritation or damage; eye protection needed.

CHEMICAL REACTIVITY

Reacts with nitrates, strong oxidizers (violent), alkalis, acids, liquid chlorine.

FIRE INFORMATION

If material or contaminated runoff enters waterways, notify downstream users of potentially contaminated waters. Notify local health and fire officials and pollution control agencies. Use carbon dioxide, dry chemicals, foam.

NFPA Hazard Classifications:

Health Hazards (Blue) 0
Flammability (Red) 1
Reactivity (Yellow) 0

Flash Point: 355°F (179°C) o.c.; 315°F (157°C) c.c.
Flammable Limits in Air: LEL 0.5% @ 456°F (241°C); UEL 2.5% (calculated)
Autoignition Temperature: 757°F (402°C)

FIRST AID

Move victim to fresh air. Call emergency medical care. Apply artificial respiration if victim is not breathing. Administer oxygen if breathing is difficult. In case of contact with substance, immediately flush skin or eyes with running water for at least 20 minutes. Remove and isolate contaminated clothing and shoes. Ensure that medical personnel are aware of the material(s) involved and take precautions to protect themselves.

ENVIRONMENTAL INFORMATION

Clean Air Act: Hazardous Air Pollutants (Title I, Part A, Section 112)
Clean Water Act: Section 307 Priority Pollutants; Section 313 Priority Chemicals
EPA Hazardous Waste Number (RCRA No.): U069, waste stream F039
RCRA Section 261 Hazardous Constituents
RCRA Land Ban Waste
RCRA Universal Treatment Standards: Wastewater (mg/L), 0.057; Nonwastewater (mg/kg), 28.
RCRA Ground Water Monitoring List. Suggested test method(s) (PQL μg/L); 8060 (5); 8270 (10)
EPCRA Section 304 Reportable Quantity (RQ): EHS/CERCLA, 10 lb (4.55 kg)
EPCRA Section 313: Form R *de minimis* concentration reporting level: 1.0%.
Marine Pollutant (49 CFR, Subchapter 172.101, Appendix B)
TSCA: 40CFR 799.5000

DIBUTYLTIN DILAURATE D:075

SYNONYMS: BIS(DODECANOYLOXY)DI-n-BUTYLSTANNANE; BIS(LAUROYLOXY)DIBUTYLSTANNANE; BIS(LAUROYLOXY)DI(n-BUTYL)STANNANE; BUTYNORATE; DBTL; DIBUTYLBIS(LAUROYLOXY)TIN; DI-n-BUTYLTIN DI(DODECANOATE); DIBUTYLTIN LAURATE; DIBUTYL-ZINN-DILAURAT (German); DILAURATO de DIBUTILESTANO (Spanish); EINECS No. 201-039-8; FOMREZ SUL-4; LAUDRAN DI-n-BUTYLCINICITY (Czech); LAURIC ACID, DIBUTYLSTANNYLENE derivative; LAURIC ACID, DIBUTYLSTANNYLENE SALT; STABILIZER D-22; THERM CHEK 820; TIN DIBUTYL DILAURATE; TINOSTAT

IDENTIFICATION

CAS:	77-58-7
ERG ID:	UN 2788 (liquid); UN 3146 (solid)
ERG Guide:	153
ERG Guide Category:	SUBSTANCES-TOXIC AND/OR CORROSIVE (COMBUSTIBLE)
Molecular Formula:	$C_{32}H_{64}O_4Sn$
RTECS No.:	WH7000000
Properties:	Combustible liquid and solid. Pale-yellow liquid, colorless solid.

HEALTH AND SAFETY INFORMATION

Irritates eyes, skin, and respiratory tract.

IDLH: 25 mg/m^3 as tin
OSHA Table Z-1-A Air Contaminant as tin organic compounds

Exposure Limits:	**TWA**	**STEL** as tin
ACGIH: (1)	0.1 mg/m^3	0.2 mg/m^3
NIOSH: (1)	0.1 mg/m^3	
OSHA:	0.1 mg/m^3	

(1) Skin contact contributes significantly to overall exposure.

RECOMMENDATIONS FOR RESPIRATOR SELECTION: NIOSH as tin

10 mg/m^3: DM (any dust and mist respirator)
20 mg/m^3: DMXSQ (any dust and mist respirator except single-use and quarter mask respirators); or SA (any supplied-air respirator).
50 mg/m^3: SA:CF (any supplied-air respirator operated in a continuous-flow mode); or PAPRDM (any powered, air-purifying respirator with a dust and mist filter).
100 mg/m^3: HiEF (Any air-purifying, full-facepiece respirator with a high-efficiency particulate filter); or SCBAF (any self-contained breathing apparatus with a full facepiece); or SAF (any supplied-air respirator with a full facepiece).

Emergency or Planned Entry into Unknown Concentrations or IDLH Conditions: SCBAF:PD,PP (any self-contained breathing apparatus that has a full facepiece and is operated in a pressure-demand or other positive-pressure mode); or SAF:PD,PP:ASCBA (Any supplied-air respirator that has a full facepiece and is operated in a pressure-demand or other positive-pressure mode in combination with an auxiliary, self-contained breathing apparatus operated in a pressure-demand or other positive-pressure mode.

Escape: HiEF (Any air-purifying, full-facepiece respirator with a high-efficiency particulate filter); or SCBAE (Any appropriate escape-type, self-contained breathing apparatus).

CHEMICAL REACTIVITY

Reacts with oxidizers.

FIRE INFORMATION

Vapors are poisonous. Firefighting gear (including SCBA) may not provide adequate protection. If exposure occurs, remove and isolate gear immediately and thoroughly decontaminate personnel. Vapors are heavier than air and will collect in low areas. Vapors in confined areas may explode when exposed to fire. If material or contaminated runoff enters waterways, notify downstream users of potentially contaminated waters. Notify local health and fire officials and pollution control agencies.

NFPA Hazard Classifications: Not listed

Flash Point: 455°F (235°C) o.c.

FIRST AID

Move victim to fresh air and call emergency medical care. Apply artificial respiration if victim is not breathing. Do not use mouth-to-mouth method if victim swallowed or inhaled the substance; induce artificial respiration with the aid of a pocket mask equipped with a one-way valve or other proper respiratory medical device. If breathing is difficult, give oxygen. In case of contact with substance, immediately flush skin or eyes with running water for at least 20 minutes. Remove and isolate contaminated clothing and shoes. Keep victim warm and quiet. Effects of exposure (inhalation, ingestion, or skin contact) to substances may be delayed. Ensure that medical personnel are aware of the material(s) involved and take precautions to protect themselves.

ENVIRONMENTAL INFORMATION as tin

RCRA Ground Water Monitoring List: Suggested test method(s) (PQL ug/L): (total) 7870(8,000).
TSCA: 40CFR716.120(*a*) substance list.; 40CFR712.30(*g*) manufacturers reporting.

DIBUTYLTIN MALEATE D:078

SYNONYMS: DIBUTYLMALOYLOXYSTANNANE; MALEATO de DIBUTILESTANO (Spanish)

IDENTIFICATION

CAS:	15535-69-0
ERG ID:	UN 2788 (liquid); UN 3146 (solid)
ERG Guide:	153
ERG Guide Category:	SUBSTANCES-TOXIC AND/OR CORROSIVE (COMBUSTIBLE)
Molecular Formula:	$C_{12}H_{22}0_5Sn$
RTECS No.:	WH1450000
Properties:	Combustible powder. White. Floats on water.

HEALTH AND SAFETY INFORMATION

Irritates eyes, skin, and respiratory tract.

IDLH: 25 mg/m^3 as tin
OSHA Table Z-1-A Air Contaminant as tin organic compounds

Exposure Limits:	**TWA**	**STEL**
ACGIH: [(1)]	0.1 mg/m^3 as tin	0.2 mg/m^3 as tin
NIOSH: [(1)]	0.1 mg/m^3 as tin	
OSHA:	0.1 mg/m^3 as tin	

[(1)] Skin contact contributes significantly to overall exposure.

RECOMMENDATIONS FOR RESPIRATOR SELECTION: NIOSH as tin

10 mg/m^3: DM (any dust and mist respirator)
20 mg/m^3: DMXSQ (any dust and mist respirator except single-use and quarter mask respirators); or SA (any supplied-air respirator).
50 mg/m^3: SA:CF (any supplied-air respirator operated in a continuous-flow mode); or PAPRDM (any powered, air-purifying respirator with a dust and mist filter).
100 mg/m^3: HiEF (any air-purifying, full-facepiece respirator with a high-efficiency particulate filter); or SCBAF (any self-contained breathing apparatus with a full facepiece); or SAF (any supplied-air respirator with a full facepiece).

Emergency or Planned Entry into Unknown Concentrations or IDLH Conditions: SCBAF:PD,PP (any self-contained breathing apparatus that has a full facepiece and is operated in a pressure-demand or other positive-pressure mode); or SAF:PD,PP:ASCBA (Any supplied-air respirator that has a full facepiece and is operated in a pressure-demand or other positive-pressure mode in combination with an auxiliary, self-contained breathing apparatus operated in a pressure-demand or other positive-pressure mode.

Escape: HiEF (Any air-purifying, full-facepiece respirator with a high-efficiency particulate filter); or SCBAE (Any appropriate escape-type, self-contained breathing apparatus).

CHEMICAL REACTIVITY

Contact with strong oxidizers may cause fire and explosions.

FIRE INFORMATION

Decomposes in heat or fire; vapors are toxic. Firefighting gear (including SCBA) may not provide adequate protection. If exposure occurs, remove and isolate gear immediately and thoroughly decontaminate personnel. Vapors are heavier than air and will collect in low areas. Vapors in confined areas may explode when exposed to fire. If material or contaminated runoff enters waterways, notify downstream users of potentially contaminated waters. Notify local health and fire officials and pollution control agencies.

NFPA Hazard Classifications: Not listed

Flash Point: 400°F (204°C)

FIRST AID

Move victim to fresh air and call emergency medical care. Apply artificial respiration if victim is not breathing. Do not use mouth-to-mouth method if victim swallowed or inhaled the substance; induce artificial respiration with the aid of a pocket mask equipped with a one-way valve or other proper respiratory medical device. If breathing is difficult, give oxygen. In case of contact with substance, immediately flush skin or eyes with running water for at least 20 minutes. Remove and isolate contaminated clothing and shoes. Keep victim warm and quiet. Effects of exposure (inhalation, ingestion, or skin contact) to substances may be delayed. Ensure that medical personnel are aware of the material(s) involved and take precautions to protect themselves.

ENVIRONMENTAL INFORMATION as tin

RCRA Ground Water Monitoring List: Suggested test method(s) (PQL ug/L): (total) 7870(8,000).
TSCA: 40CFR716.120(*a*) substance list.

DIBUTYLTIN SULFIDE D:082

SYNONYMS: DIBUTYLTHIOXOSTANNANE; EINECS No. 224-220-3; SULFURO de DIBUTILESTANO (Spanish); STANNANE, DIBUTYLTHIOXO-; TIN DIBUTYL MERCAPTIDE

IDENTIFICATION

CAS:	4253-22-9
ERG ID:	UN 3282
ERG Guide:	152
ERG Guide Category:	SUBSTANCES-TOXIC (COMBUSTIBLE)
Molecular Formula:	$C_8H_{18}SSn$
RTECS No.:	WH7195000
Properties:	Combustible oily liquid. Colorless.

HEALTH AND SAFETY INFORMATION

Irritates eyes, skin, and respiratory tract.

IDLH: 25 mg/m^3 as tin
OSHA Table Z-1-A Air Contaminant as tin organic compounds

Exposure Limits:	**TWA**	**STEL**
ACGIH: [1]	0.1 mg/m^3 as tin	0.2 mg/m^3 as tin
NIOSH: [1]	0.1 mg/m^3 as tin	
OSHA:	0.1 mg/m^3 as tin	

[1] Skin contact contributes significantly to overall exposure.

RECOMMENDATIONS FOR RESPIRATOR SELECTION: NIOSH as tin

10 mg/m^3: DM (any dust and mist respirator)
20 mg/m^3: DMXSQ (any dust and mist respirator except single-use and quarter mask respirators); or SA (any supplied-air respirator).
50 mg/m^3: SA:CF (any supplied-air respirator operated in a continuous-flow mode); or PAPRDM (any powered, air-purifying respirator with a dust and mist filter).
100 mg/m^3: HiEF (any air-purifying, full-facepiece respirator with a high-efficiency particulate filter); or SCBAF (any self-contained breathing apparatus with a full facepiece); or SAF (any supplied-air respirator with a full facepiece).

Emergency or Planned Entry into Unknown Concentrations or IDLH Conditions: SCBAF:PD,PP (any self-contained breathing apparatus that has a full facepiece and is operated in a pressure-demand or other positive-pressure mode); or SAF:PD,PP:ASCBA (any supplied-air respirator that has a full facepiece and is operated in a pressure-demand or other positive-pressure mode in combination with an auxiliary, self-contained breathing apparatus operated in a pressure-demand or other positive-pressure mode.

Escape: HiEF (any air-purifying, full-facepiece respirator with a high-efficiency particulate filter); or SCBAE (any appropriate escape-type, self-contained breathing apparatus).

CHEMICAL REACTIVITY

Contact with oxidizers may cause fire or explosions.

FIRE INFORMATION

Combustion products include toxic sulfur oxides. Firefighting gear (including SCBA) may not provide adequate protection. If exposure occurs, remove and isolate gear immediately and thoroughly decontaminate personnel. Vapors are heavier than air and will collect in low areas. Vapors in confined areas may explode when exposed to fire. If material or contaminated runoff enters waterways, notify downstream users of potentially contaminated waters. Notify local health and fire officials and pollution control agencies.

NFPA Hazard Classifications: Not listed

FIRST AID

Move victim to fresh air. Call emergency medical care. Apply artificial respiration if victim is not breathing. Do not use mouth-to-mouth method if victim ingested or inhaled the substance; induce artificial respiration with the aid of a pocket mask equipped with a one-way valve or other proper respiratory medical device. Administer oxygen if breathing is difficult. Remove and isolate contaminated clothing and shoes. In case of contact with substance, immediately flush skin or eyes with running water for at least 20 minutes. For minor skin contact, avoid spreading material on unaffected skin. Keep victim warm and quiet. Effects of exposure (inhalation, ingestion, or skin contact) may be delayed. Ensure that medical personnel are aware of the material(s) involved and take precautions to protect themselves.

ENVIRONMENTAL INFORMATION as tin

RCRA Ground Water Monitoring List: Suggested test method(s) (PQL ug/L): (total) 7870(8,000).
TSCA: 40CFR716.120(*a*) substance list.

DICAPRYL PHTHALATE D:086

SYNONYMS: BENZENEDICARBOXYLIC ACID, BIS(1-METHYLHEPTYL) ESTER; BIS(2-OCTYL)PHTHALATE; BIS-(2-OKTYL)ESTER KYSELINY FTALOVE; CAPRYL o-PHTHALATE; DICAPRYL 1,2-BENZENEDICARBOXYLATE; DIOCTANOL-2-PHTHALATE; EINECS No. 205-014-2; FTALATO de DICAPRILO (Spanish); MONOPLEX DCP; PHTHALIC ACID, BIS(2-OCTYL) ESTER; PHTHALIC ACID, DICAPRYL ESTER; PHTHALIC ACID, DI-2-OCTYL ESTER

IDENTIFICATION

CAS:	131-15-7
olecular Formula:	$C_{24}H_{38}O_4$
RTECS No.:	TI0360000
Properties:	Combustible viscous liquid. Insoluble in water.

HEALTH AND SAFETY INFORMATION

RECOMMENDATIONS FOR RESPIRATOR SELECTION

At any concentrations above the NIOSH REL, or where there is no REL, at any detectable concentration: SCBAF:PD,PP (any self-contained breathing apparatus that has a full facepiece and is operated in a pressure-demand or other positive-pressure mode); or SAF:PD,PP:ASCBA (any supplied-air respirator that has a full facepiece and is operated in a pressure-demand or other positive-pressure mode in combination with an auxiliary, self-contained breathing apparatus operated in a pressure-demand or other positive-pressure mode).

Escape: GMFOV [any air-purifying, full-facepiece respirator (gas mask) with a chin-style, front- or back-mounted organic vapor canister]; or SCBAE (any appropriate escape-type, self-contained breathing apparatus).

CHEMICAL REACTIVITY

Incompatible with strong oxidizers, nitrates, strong acids.

FIRE INFORMATION

Vapors are heavier than air and will collect in low areas. Vapors in confined areas may explode when exposed to fire. If material or contaminated runoff enters waterways, notify downstream users of potentially contaminated waters. Notify local health and fire officials and pollution control agencies. Use dry chemicals, alcohol-resistant foam, carbon dioxide.

NFPA Hazard Classifications:

Health Hazards (Blue) 0
Flammability (Red) 1
Reactivity (Yellow) 0

Flash Point: 395°F (201°C)

FIRST AID

Move victim to fresh air. Call emergency medical care. Apply artificial respiration if victim is not breathing. Administer oxygen if breathing is difficult. Remove and isolate contaminated clothing and shoes. In case of contact with substance, immediately flush skin or eyes with running water for at least 20 minutes. Wash skin with soap and water. Keep victim warm and quiet. Ensure that medical personnel are aware of the material(s) involved and take precautions to protect themselves.

3,3'–DICHLOROBENZIDINE D:090

SYNONYMS: BENZIDINE,3,3'-DICHLORO-; C.I. 23060; CURITHANE C126; DCB; 4,4'-DIAMINO-3,3'-DICHLOROBIPHENYL; 4,4'-DIAMINO-3,3'-DICHLORODIPHENYL; 3,3'-DICHLORBENZIDIN (Czech); 3,3'-DICLOROBENZIDINA (Spanish); DICHLOROBENZIDINE; o,o'-DICHLOROBENZIDINE; 3,3'-DICHLOROBIPHENYL-4,4'-DIAMINE; 3,3'-DICHLORO-4,4'-BIPHENYLDIAMINE; 3,3'-DICHLORO-4,4'-DIAMINOBIPHENYL; 3,3'-DICHLORO-4,4'-DIAMINO(1,1-BIPHENYL); EEC No. 612-068-00-4; EINECS No. 202-109-0

IDENTIFICATION

CAS:	91-94-1
ERG ID:	2810
ERG Guide:	153
ERG Guide Category:	SUBSTANCES-TOXIC AND/OR CORROSIVE (COMBUSTIBLE)
Molecular Formula:	$C_{12}H_{10}Cl_2N_2$
RTECS No.:	DD0525000
Properties:	Combustible crystalline solid. Gray to purple in color. Very slightly soluble in water.

HEALTH AND SAFETY INFORMATION

Irritates eyes, skin, respiratory tract. May cause dermatitis; liver, bladder, lung problems.

IARC Evaluation: Group 2B; inadequate human evidence; sufficient animal evidence.
IDLH: Potential human carcinogen
OSHA Table Z-1-A Air Contaminant
OSHA Specifically Regulated Substances: (29 CFR 1910.1007)

Exposure Limits:

ACGIH: Animal carcinogen.
NIOSH: Potential human carcinogen. Reduce exposure to lowest feasible level.
OSHA: Suspected cancer agent. Reduce exposure to lowest feasible level.

+Skin contact contributes significantly to overall exposure.

RECOMMENDATIONS FOR RESPIRATOR SELECTION: NIOSH

At any concentrations above the NIOSH REL, or where there is no REL, at any detectable concentration:
SCBAF:PD,PP (any self-contained breathing apparatus that has a full facepiece and is operated in a pressure-demand or other

positive-pressure mode); or SAF:PD,PP:ASCBA (any supplied-air respirator that has a full facepiece and is operated in a pressure-demand or other positive-pressure mode in combination with an auxiliary, self-contained breathing apparatus operated in a pressure-demand or other positive-pressure mode).

Escape: GMFOV [any air-purifying, full-facepiece respirator (gas mask) with a chin-style, front- or back-mounted organic vapor canister]; or SCBAE (any appropriate escape-type, self-contained breathing apparatus).

CHEMICAL REACTIVITY

An aromatic amine. Incompatible with nonoxidizing mineral acids, sulfuric acid, nitric acid, organic anhydrides, isocyanates, aldehydes.

FIRE INFORMATION

Combustion products include toxic nitrogen oxides and chlorine. If material or contaminated runoff enters waterways, notify downstream users of potentially contaminated waters. Notify local health and fire officials and pollution control agencies.

FEMA Hazard Classifications:

Health Hazards (Blue) 0
Flammability (Red) 1
Reactivity (Yellow) 0

FIRST AID

Move victim to fresh air. Call emergency medical care. Apply artificial respiration if victim is not breathing. Administer oxygen if breathing is difficult. Remove and isolate contaminated clothing and shoes. In case of contact with substance, immediately flush skin or eyes with running water for at least 20 minutes. Ensure that medical personnel are aware of the material(s) involved and take precautions to protect themselves.

ENVIRONMENTAL INFORMATION

Clean Air Act: Hazardous Air Pollutants (Title I, Part A, Section 112) kg)
Clean Water Act: Section 307 Priority Pollutants; Section 313 Priority Chemicals; Toxic Pollutant (Section 401.15)
EPA Hazardous Waste Number (RCRA No.): U073
RCRA Toxicity Characteristic (Section 261.24): RCRA Section 261 Hazardous Constituents
RCRA Ground Water Monitoring List: Suggested test methods(s) (PQL μg/L); 8270 (20)
RCRA Land Ban Waste
EPCRA Section 304 Reportable Quantity (RQ): CERCLA, 1 lb (0.454 kg)
EPCRA Section 313: Form R *de minimis* concentration reporting level: 0.1%.
California's Proposition 65: Carcinogen.

1,1-DICHLOROETHANE D:094

SYNONYMS: AETHYLIDENCHLORID (German); ASYMMETRICAL DICHLOROETHANE; as-DICHLOROETHANE; CHLORINATED HYDROCHLORIC ETHER; CHLORURE D'ETHYLIDENE (French); CLORURO di ETILIDENE (Italian); 1,1-DCE; 1,1-DICHLOORETHAAN (Dutch); 1,1-DICHLORAETHAN (German); 1,1-DICHLORETHANE; 1,1-DICLOROETANO (Italian, Spanish); EEC No. 602-011-00-1; EINECS No. 200-863-5; ETHYLENE DICHLORIDE; ETHYLIDENE CHLORIDE; ETHYLIDENE DICHLORIDE; 1,1-ETHYLIDENE DICHLORIDE; NCI-C04535; ETHANE,1,1-DICHLORO-; DICHLOROMETHYLETHANE

IDENTIFICATION

CAS: 75-34-3

ERG ID:	UN 2362
ERG Guide:	130
ERG Guide Category:	FLAMMABLE LIQUIDS (NONPOLAR/WATER-IMMISCIBLE/NOXIOUS)
Molecular Formula:	$C_2H_4Cl_2$
RTECS No.:	KI0175000
Properties:	Highly flammable, oily liquid. Colorless. Chloroformlike, ethereal odor. Sinks and is almost insoluble in water. Vapors are heavier than air; will collect in low areas. Toxic hydrogen chloride produced when heated.

HEALTH AND SAFETY INFORMATION

Irritates eyes, skin, and respiratory tract. Central nervous system effects. Possible kidney, liver damage.

IDLH: 3,000 ppm
Odor Threshold: 255 ppm
OSHA Table Z-1-A Air Contaminant

Exposure Limits:	TWA	STEL
ACGIH:	100 ppm; 405 mg/m^3	250 ppm
NIOSH:	100 ppm; 400 mg/m^3	None
OSHA:	100 ppm; 400 mg/m^3	None

RECOMMENDATIONS FOR RESPIRATOR SELECTION: NIOSH/OSHA

1,000 ppm: SA (any supplied-air respirator).
2,500 ppm: SA:CF (any supplied-air respirator operated in a continuous-flow mode).
3,000 ppm: SCBAF (any self-contained breathing apparatus with a full facepiece); or SAF (any supplied-air respirator with a full facepiece).

Emergency or Planned Entry into Unknown Concentrations or IDLH Conditions:
SCBAF:PD,PP (any self-contained breathing apparatus that has a full facepiece and is operated in a pressure-demand or other positive-pressure mode); or SAF:PD,PP:ASCBA (any supplied-air respirator that has a full facepiece and is operated in a pressure-demand or other positive-pressure mode in combination with an auxiliary, self-contained breathing apparatus operated in a pressure-demand or other positive-pressure mode).

Escape: GMFOV [any air-purifying, full-facepiece respirator (gas mask) with a chin-style, front- or back-mounted organic vapor canister]; or SCBAE (any appropriate escape-type, self-contained breathing apparatus).

CHEMICAL REACTIVITY

Forms explosive mixture with air. Incompatible with strong oxidizers and strong caustics. Attacks plastics and rubber.

FIRE INFORMATION

Vapors are heavier than air and will collect in low areas. Vapors in confined areas may explode when exposed to fire. Vapors may travel long distances to ignition sources and flash back. Storage containers and parts of containers may rocket great distances, in many directions. If material or contaminated runoff enters waterways, notify downstream users of potentially contaminated waters. Notify local health and fire officials and pollution control agencies. Use dry chemicals, foam, carbon dioxide.

NFPA Hazard Classifications:

Health Hazards (Blue) 2
Flammability (Red) 3
Reactivity (Yellow) 0

Flash Point: 2°F (−17°C) c.c.
Flammable Limits in Air: LEL 5.4%; UEL 11.4%
Autoignition Temperature: 856°F (458°C)

FIRST AID

Move victim to fresh air. Call emergency medical care. Apply artificial respiration if victim is not breathing. Administer oxygen if breathing is difficult. Remove and isolate contaminated clothing and shoes. In case of contact with substance, immediately flush skin or eyes with running water for at least 20 minutes. Wash skin with soap and water. Keep victim warm and quiet. Effects of exposure (inhalation, ingestion, or skin contact) to substance may be delayed. Ensure that medical personnel are aware of the material(s) involved and take precautions to protect themselves.

ENVIRONMENTAL INFORMATION

Clean Air Act: Hazardous Air Pollutants (Title I, Part A, Section 112).
Clean Water Act: Section 307 Priority Pollutants.
EPA Hazardous Waste Number (RCRA No.): U076.
RCRA Section 261 Hazardous Constituents
RCRA Land Ban Waste
RCRA Universal Treatment Standards: Wastewater (mg/L), 0.059; Nonwastewater (mg/kg), 6.0.
RCRA Ground Water Monitoring List: Suggested test method(s) (PQL ug/L): 8010(1); 8240(5).
Safe Drinking Water Act: Priority List (55 FR 1470)
EPCRA Section 304: Reportable Quantity (RQ): CERCLA, 1,000 lb (454 kg).
EPCRA Section 313: Form R *de minimis* concentration reporting level: 1.0%.
Marine Pollutant (49CFR, Subchapter 172.101, Appendix B).
California's Proposition 65: Carcinogen
TSCA: 40CFR712.30(*e*)10; 40CFR799.5055 testing; export notification required.

1,2-DICHLOROETHYLENE — REC. D: 100

SYNONYMS: ACETYLENE DICHLORIDE; cis-ACETYLENE DICHLORIDE; trans-ACETYLENE DICHLORIDE; 1,2-DCE; 1,2-DICHLOR-AETHEN (German); DICHLORO-1,2-ETHYLENE (French); cis-1,2-DICHLOROETHYLENE; sym-DICHLOROETHYLENE; trans-1,2-DICHLOROETHYLENE; 1,2-DICLOROETILENO (Spanish); ETHENE, 1,2-DICHLORO-; DIOFORM; ETHENE, 1,2-DICHLORO-,(E)-; NCI-C56031

IDENTIFICATION

CAS:	156-59-2 (cis- isomer); 156-60-5 (trans- isomer); 540-59-0 (mixed isomers)
ERG ID:	UN 1150
ERG Guide:	132P
ERG Category Guide:	FLAMMABLE LIQUIDS, CORROSIVE
Molecular Formula:	$C_2H_2Cl_2$
RTECS No.:	KV9360000
Properties:	Highly flammable, colorless liquid (usually a mixture of the cis- and transisomers). Pleasant, ether- or chloroform-like. Rapidly evaporates. Insoluble; sinks in water; flammable, irritating vapor is produced. Poisonous gases including hydrogen chloride produced in fire. Vapors are heavier than air; will collect in low areas. Boiling point 119ºF (48ºC).

HEALTH AND SAFETY INFORMATION

Irritates eyes, skin, and respiratory tract. May cause pulmonary edema, which can be delayed for several hours; there is a risk of death in serious cases. May cause kidney, liver damage.

IDLH: 1,000 ppm

Odor Threshold: 19.1 ppm

Note: Long-term, this substance has produced kidney and liver injury at high levels in laboratory animals. May cause lung damage and bronchitis; low blood count.

Exposure Limits:	**TWA**
ACGIH:	200 ppm; 793 mg/m^3
NIOSH:	200 ppm; 790 mg/m^3
OSHA:	200 ppm; 790 mg/m^3

RECOMMENDATIONS FOR RESPIRATOR SELECTION: NIOSH/OSHA

1000 ppm: SA:CF (any supplied-air respirator operated in a continuous-flow mode); or PAPROV (any powered, air-purifying respirator with organic vapor cartridge(s)); or CCRFOV (any chemical cartridge respirator with a full facepiece and organic vapor cartridge(s)); or GMFOV (any air-purifying, full-facepiece respirator (gas mask) with a chin-style, front- or back-mounted acid gas canister); or SCBAF (any self-contained breathing apparatus with a full facepiece); or SAF (any supplied-air respirator with a full facepiece)

Emergency or planned entry into unknown concentrations or idlh conditions: SCBAF:PD,PP (any self-contained breathing apparatus that has a full facepiece and is operated in a pressure-demand or other positive-pressure mode); or SAF:PD,PP:ASCBA (any supplied-air respirator that has a full facepiece and is operated in a pressure-demand or other positive-pressure mode in combination with an auxiliary, self-contained breathing apparatus operated in a pressure-demand or other positive-pressure mode.

Escape: GMFOV [Any air-purifying, full-facepiece respirator (gas mask) with a chin-style, front- or back-mounted organic vapor canister];or SCBAE (any appropriate escape-type, self-contained breathing apparatus).

Note: Substance causes eye irritation or damage; eye protection needed.

CHEMICAL REACTIVITY

Forms explosive mixture with air. Decomposes on contact with acids, acid fumes, hot metal, and ultraviolet light. Reacts violently with sodium, sodium hydroxide, potassium hydroxide, nitrogen peroxide. Contact with strong oxidizers may cause fire and explosions. Incompatible with strong bases, difluoromethylene, dihypofluoride, nitrogen tetroxide (explosive). Attacks some forms of plastics, rubber, and coatings.

FIRE INFORMATION

Vapors are heavier than air and will collect in low areas. Vapors in confined areas may explode when exposed to fire. Vapors may travel long distances to ignition sources and flash back. Storage containers and parts of containers may rocket great distances, in many directions. If material or contaminated runoff enters waterways, notify downstream users of potentially contaminated waters. Notify local health and fire officials and pollution control agencies. Use dry chemicals, foam, carbon dioxide.

NFPA Hazard Classifications:

Health Hazards (Blue) 2
Flammability (Red) 3
Reactivity (Yellow) 2

Flash Point: 37°F (2.5°C) c.c.
Flammable Limits in Air: LEL 9.7%; UEL 12.8%
Autoignition Temperature: 860°F (460°C)

FIRST AID

Move victim to fresh air. Call emergency medical care. Apply artificial respiration if victim is not breathing. Do not use mouth-to-mouth method if victim swallowed or inhaled the substance; induce artifical respiration with the aid of a pocket mask equipped with a one-way valve or other proper respiratory medical device. If breathing is difficult, give oxygen. Remove and isolate contaminated clothing and shoes. In case of contact with substance, immediately flush skin or eyes with running water for at least 20 minutes. Keep victim warm and quiet. Effects of exposure (inhalation, ingestion, or skin contact) to substances may be delayed. Ensure that medical personnel are aware of the material(s) involved and take precautions to protect themselves.

ENVIRONMENTAL INFORMATION

CLEAN WATER ACT: Section 307 Priority Pollutants; Section 313 Priority Chemicals.
EPA HAZARDOUS WASTE NUMBER (RCRA No.): U079 (156-60-5)
RCRA Section 261 Hazardous Constituents.
RCRA Land Ban Waste.
RCRA Universal Treatment Standards: Wastewater (mg/L), 0.054; Nonwastewater (mg/kg), 30
RCRA Ground Water Monitoring List. Suggested test method(s) (PQL ug/L): 8010(1); 8240(5).
SAFE DRINKING WATER ACT: MCL, 0.1 mg/L; MCLG, 0.1 mg/L; Regulated chemical (47 FR 9352).
EPCRA Section 304 Reportable Quantity (RQ): CERCLA, 1,000 lb (454 kg).

2,6–DICHLOROSTYRENE D:103

SYNONYMS: BENZENE,(1,2-DICHLOROETHENYL)-; α,β-DICHLOROSTYRENE; 2,6-DICLOROESTIRENO (Spanish); DWUCHLOROSTYREN (Polish)

IDENTIFICATION

CAS:	6607-45-0
ERG ID:	UN 2810
ERG Guide:	153
ERG Guide Category:	SUBSTANCES-TOXIC AND/OR CORROSIVE (COMBUSTIBLE)
Molecular Formula:	$C_8H_6Cl_2$
RTECS No.:	WL4375000
Properties:	Combustible liquid. Colorless. Insoluble in water.

HEALTH AND SAFETY INFORMATION

Irritates eyes, skin, and respiratory tract.

RECOMMENDATIONS FOR RESPIRATOR SELECTION:

At any concentrations above the NIOSH REL, or where there is no REL, at any detectable concentration: SCBAF:PD,PP (any self-contained breathing apparatus that has a full facepiece and is operated in a pressure-demand or other positive-pressure mode); or SAF:PD,PP:ASCBA (any supplied-air respirator that has a full facepiece and is operated in a pressure-demand or other positive-pressure mode in combination with an auxiliary, self-contained breathing apparatus operated in a pressure-demand or other positive-pressure mode).

Escape: GMFOV [any air-purifying, full-facepiece respirator (gas mask) with a chin-style, front- or back-mounted organic vapor canister]; or SCBAE (any appropriate escape-type, self-contained breathing apparatus).

CHEMICAL REACTIVITY

Contact with oxidizers may cause fire and explosions. May be able to form unstable peroxides.

FIRE INFORMATION

Vapors are heavier than air and will collect in low areas. Vapors in confined areas may explode when exposed to fire. Storage containers and parts of containers may rocket great distances, in many directions. If material or contaminated runoff enters waterways, notify downstream users of potentially contaminated waters. Notify local health and fire officials and pollution control agencies. Use dry chemicals, carbon dioxide, foam. Use water spray to cool containers.

NFPA Hazard Classifications:

Health Hazards (Blue) 2
Flammability (Red) 1
Reactivity (Yellow) 2

Flash Point: 225°F (107°C) o.c.

FIRST AID

Move victim to fresh air. Call emergency medical care. Apply artificial respiration if victim is not breathing. If breathing is difficult, give oxygen. Remove and isolate contaminated clothing and shoes. In case of contact with substance, immediately flush skin or eyes with running water for at least 20 minutes. Wash skin with soap and water. Keep victim warm and quiet. Ensure that medical personnel are aware of the material(s) involved and take precautions to protect themselves.

DICHLORVOS D:106

SYNONYMS: APAVAP; ASTROBOT; ATGARD; BENFOS; BIBESOL; BREVINYL; CANOGARD; CEKUSAN; CYPONA; DDVP; DERABAN; DERRIBANTE; DEVIKOL; DICHLOROPHOS; DICLORVOS (Spanish); DIVIPAN; DUO-KILL; DURAVOS; EEC No. 015-019-00-X; EINECS No. 200-547-7; ENT 20,738; EQUIGEL; ESTROSEL; ESTROSOL; FECAMA; FLY-DIE; FLY FIGHTER; FOSFATO de 2,2-DICLOROVINILO y DIMETILO (Spanish); 2,2-DICHLORVINYLDIMETHYLPHOSPHAT (German); HERKAL; KRECALVIN; MAFU; MARVEX; MOPARI; NCI-C00113; NERKOL; NOGOS; NO-PEST; NO-PEST STRIP; NSC-6738; NUVA; OKO; OMS 14; PHOSPHATE de 2,2-DICHLOROVINYLE et de DIMETHYLE (French); PHOSVIT; SD-1750; TAP 9VP; TASK; TASK TABS; TENAC; TETRAVOS; VAPONA; VAPONITE; VERDICAN; VERDIPOR; VINYLFOS; VINYLOPHOS; 2,2-DICHLOROVINYL O,O-DIMETHYL PHOSPHATE; 2,2-DICHLOROVINYL DIMETHYL PHOSPHATE

IDENTIFICATION

CAS:	62-73-7
ERG ID:	UN 2783
ERG Guide:	152
ERG Guide Category:	SUBSTANCES-TOXIC (COMBUSTIBLE)
Molecular Formula:	$C_4H_7Cl_2O_4P$
RTECS No.:	TC0350000
Properties:	Combustible liquid. Colorless to amber. Pleasant, aromatic, characteristic odor. Sinks and mixes with water.

HEALTH AND SAFETY INFORMATION

Irritates eyes, skin, and respiratory tract. Repeated or prolonged exposure can cause conjunctivitis and dermatitis. Poison by all routes; a cholinesterase inhibitor.

IARC Evaluation: Group 2B; sufficient animal evidence.
NTP: Carcinogen
IDLH: 100 mg/m^3
OSHA Table Z-1-A Air Contaminant

Exposure Limits:	**TWA**
ACGIH:	0.1 ppm; 0.90 mg/m^3
NIOSH:	1 mg/m^3
OSHA:	1 mg/m^3

Note: Skin contact contributes significantly in overall exposure.

RECOMMENDATIONS FOR RESPIRATOR SELECTION: NIOSH/OSHA

10 mg/m³: SA (any supplied-air respirator).
25 mg/m³: SA:CF (any supplied-air respirator operated in a continuous-flow mode).
50 mg/m³: SAT:CF (any supplied-air respirator that has a tight-fitting facepiece and is operated in a continuous-flow mode); or SCBAF (any self-contained breathing apparatus with a full facepiece); or SAF (any supplied-air respirator with a full facepiece)
100 mg/m³: SA:PD,PP (any supplied-air respirator operated in a pressure-demand or other positive-pressure mode).

Emergency or Planned Entry into Unknown Concentrations or IDLH Conditions: SCBAF:PD,PP (any self-contained breathing apparatus that has a full facepiece and is operated in a pressure-demand or other positive-pressure mode); or SAF:PD,PP:ASCBA (any supplied-air respirator that has a full facepiece and is operated in a pressure-demand or other positive-pressure mode in combination with an auxiliary, self-contained breathing apparatus operated in a pressure-demand or other positive-pressure mode.

Escape: GMFOVHiE (any air-purifying, full-facepiece respirator (gas mask) with a chin-style, front- or back-mounted organic vapor canister having a high-efficiency particulate filter); or SCBAE (any appropriate escape-type, self-contained breathing apparatus).

CHEMICAL REACTIVITY

Reacts with strong acids, strong alkalis, oxidizers. Corrosive to iron and mild steel. May attack rubber and coatings. Contact with moisture may cause material to hydrolyze.

FIRE INFORMATION

Decomposition products upon heating include toxic gases, including hydrogen chloride, phosphoric acid, and carbon dioxide. Storage containers and parts of containers may rocket great distances, in many directions. If material or contaminated runoff enters waterways, notify downstream users of potentially contaminated waters. Notify local health and fire officials and pollution control agencies.

FEMA Hazard Classifications:

Health Hazards (Blue) 3
Flammability (Red) 1
Reactivity (Yellow) 0

Flash Point: 350°F (177°C)

FIRST AID

Move victim to fresh air. Call emergency medical care. Apply artificial respiration if victim is not breathing. Do not use mouth-to-mouth method if victim ingested or inhaled the substance; induce artificial respiration with the aid of a pocket mask equipped with a one-way valve or other proper respiratory medical device. Administer oxygen if breathing is difficult. Remove and isolate contaminated clothing and shoes. In case of contact with substance, immediately flush skin or eyes with running water for at least 20 minutes. For minor skin contact, avoid spreading material on unaffected skin. Keep victim warm and quiet. Effects of exposure (inhalation, ingestion, or skin contact) may be delayed. Ensure that medical personnel are aware of the material(s) involved and take precautions to protect themselves.

ENVIRONMENTAL INFORMATION

Clean Air Act: Hazardous Air Pollutants (Title I, Part A, Section 112)
Clean Water Act: Section 311 Hazardous Substances/RQ (same as CERCLA); Section 313 Priority Chemicals
EPCRA Section 302, Extremely Hazardous Substances: TPQ = 1,000 lb (454 kg)
EPCRA Section 304 Reportable Quantity (RQ): EHS/CERCLA, 10 lb (4.54 kg)
EPCRA Section 313: Form R *de minimis* concentration reporting level: 1.0%.
California's Proposition 65: Carcinogen.
Marine Pollutant (49 CFR, Subchapter 172.101, Appendix B).

DICYCLOHEXYLAMINE D:112

SYNONYMS: n-CYCLOHEXYLCYCLOHEXANAMINE; DCHA; DICHA; DICICLOHEXILAMINA (Spanish); DICYKLOHEXYLAMIN (Czech); DODECAHYDRODIPHENYLAMINE; EEC No. 612-066-00-3; EINECS No. 202-980-7

IDENTIFICATION

CAS:	101-83-7
ERG ID:	UN 2565
ERG Guide:	153
ERG Guide Category:	SUBSTANCES-TOXIC AND/OR CORROSIVE (COMBUSTIBLE)
Molecular Formula:	$C_{12}H_{23}N$
RTECS No.:	HY4025000
Properties:	Combustible liquid. Colorless. Faint, fishy odor. Slightly soluble in water.

HEALTH AND SAFETY INFORMATION

Corrosive to eyes, skin, and respiratory tract. Inhalation may cause pulmonary edema, which can be delayed for several hours; there is a risk of death in serious cases.

IARC Evaluation: Group 3; inadequate animal evidence

RECOMMENDATIONS FOR RESPIRATOR SELECTION

At any concentrations above the NIOSH REL, or where there is no REL, at any detectable concentration: SCBAF:PD,PP (any self-contained breathing apparatus that has a full facepiece and is operated in a pressure-demand or other positive-pressure mode); or SAF:PD,PP:ASCBA (any supplied-air respirator that has a full facepiece and is operated in a pressure-demand or other positive-pressure mode in combination with an auxiliary, self-contained breathing apparatus operated in a pressure-demand or other positive-pressure mode).

Escape: GMFOV [any air-purifying, full-facepiece respirator (gas mask) with a chin-style, front- or back-mounted organic vapor canister]; or SCBAE (any appropriate escape-type, self-contained breathing apparatus).

CHEMICAL REACTIVITY

Contact with oxidizers may cause fire and explosions. Incompatible with strong acids. Very corrosive. May attack some plastics and coatings. May ignite on contact with cellulose nitrate of high surface area.

FIRE INFORMATION

Vapors are heavier than air and will collect in low areas. Vapors in confined areas may explode when exposed to fire. If material or contaminated runoff enters waterways, notify downstream users of potentially contaminated waters. Notify local health and fire officials and pollution control agencies. Use dry chemicals, alcohol-resistant foam, carbon dioxide, large quantities of water.

NFPA Hazard Classifications:

Health Hazards (Blue) 3
Flammability (Red) 1
Reactivity (Yellow) 0

Flash Point: 210°F (99°C) o.c.

FIRST AID

Move victim to fresh air and call emergency medical care. Apply artificial respiration if victim is not breathing. Do not use mouth-to-mouth method if victim swallowed or inhaled the substance; induce artificial respiration with the aid of a pocket mask equipped with a one-way valve or other proper respiratory medical device. If breathing is difficult, give oxygen. In case of contact with substance, immediately flush skin or eyes with running water for at least 20 minutes. Remove and isolate contaminated clothing and shoes. Keep victim warm and quiet. Effects of exposure (inhalation, ingestion, or skin contact) to substances may be delayed. Ensure that medical personnel are aware of the material(s) involved and take precautions to protect themselves.

DICYCLOPENTADIENE D:118

SYNONYMS: BICYCLOPENTADIENE; BISCYCLOPENTADIENE; CYCLOPENTADIENE; 1,3-CYCLOPENTADIENE, DIMER; DCPD; 1,3-DICYCLOPENTADIENE DIMER; DICICLOPENTANDIENO (Italian, Spanish); DICYKLOPENTADIEN (Czech); DIMER CYKLOPENTADIENU (Czech); EINECS No. 201-052-9; 4,7-METHANO-1H-INDENE; 3A,4,7,7A-TETRAHYDRO-4,7-METHANOINDENE; TRICYCLO[5.2.1.0 2,6-DECA-3,8-DIENE

IDENTIFICATION

CAS:	77-73-6
ERG ID:	UN 2048
ERG Guide:	129
ERG Guide Category:	FLAMMABLE LIQUIDS (POLAR/WATER-MISCIBLE/NOXIOUS)
Molecular Formula:	$C_{10}H_{12}$
RTECS No.:	PC1050000
Properties:	Flammable liquid, or solid crystals. Colorless. Camphor odor. Floats on water; insoluble. Freezing point is 41°F (5°C) .

HEALTH AND SAFETY INFORMATION

Irritates eyes, skin, and respiratory tract.

Odor Threshold: 0.03 ppm
OSHA Table Z-1-A Air Contaminant

Exposure Limits:	**TWA**
ACGIH:	5 ppm; 27 mg/m^3
NIOSH:	5 ppm; 30 mg/m^3
OSHA:	None

CHEMICAL REACTIVITY

Forms explosive mixture with air. Contact with oxidizers can cause fire and explosions. Incompatible with nitric and organic acids. May be able to form peroxides.

FIRE INFORMATION

Vapors are heavier than air and will collect in low areas. Vapors in confined areas may explode when exposed to fire. Vapors may travel long distances to ignition sources and flash back. Storage containers and parts of containers may rocket great distances, in many directions. If material or contaminated runoff enters waterways, notify downstream users of potentially contaminated waters. Notify local health and fire officials and pollution control agencies.

NFPA Hazard Classifications:

Health Hazards (Blue) 1
Flammability (Red) 3
Reactivity (Yellow) 1

Flash Point: 90°F (32°C) o.c.
Flammable Limits in Air: LEL 0.8%; UEL 6.3%
Autoignition Temperature: 941°F (506°C)

FIRST AID

Move victim to fresh air. Call emergency medical care. Apply artificial respiration if victim is not breathing. Administer oxygen if breathing is difficult. Remove and isolate contaminated clothing and shoes. In case of contact with substance, immediately flush skin or eyes with running water for at least 20 minutes. Wash skin with soap and water. Keep victim warm and quiet. Effects of exposure (inhalation, ingestion, or skin contact) to substance may be delayed. Ensure that medical personnel are aware of the material(s) involved and take precautions to protect themselves.

ENVIRONMENTAL INFORMATION

EPCRA Section 313: Form R *de minimis* concentration reporting level: 1.0%.

DIDECYL PHTHALATE D:124

SYNONYMS: 1,2-BENZENEDICARBOXYLIC ACID, DIDECYL ESTER; DIDECYLPHTHALAT (German); DI-n-DECYL PHTHALATE; EINECS No. 201-561-6; FTALATO de DIDECILO (Spanish)

IDENTIFICATION

CAS:	84-77-5
Molecular Formula:	$C_{28}H_{46}O_4$
RTECS No.:	TI0900000
Properties:	Combustible liquid. Light colored. Insoluble in water.

HEALTH AND SAFETY INFORMATION

Irritates eyes, skin, and respiratory tract.

RECOMMENDATIONS FOR RESPIRATOR SELECTION

At any concentrations above the NIOSH REL, or where there is no REL, at any detectable concentration: SCBAF:PD,PP (any self-contained breathing apparatus that has a full facepiece and is operated in a pressure-demand or other positive-pressure mode); or SAF:PD,PP:ASCBA (any supplied-air respirator that has a full facepiece and is operated in a pressure-demand or other positive-pressure mode in combination with an auxiliary, self-contained breathing apparatus operated in a pressure-demand or other positive-pressure mode).

Escape: GMFOV [any air-purifying, full-facepiece respirator (gas mask) with a chin-style, front- or back-mounted organic vapor canister]; or SCBAE (any appropriate escape-type, self-contained breathing apparatus).

CHEMICAL REACTIVITY

Incompatible with nitrates, nitric acid, sulfuric acid.

FIRE INFORMATION

Vapors are heavier than air and will collect in low areas. Vapors in confined areas may explode when exposed to fire. If material or contaminated runoff enters waterways, notify downstream users of potentially contaminated waters. Notify local health and fire officials and pollution control agencies. Use dry chemicals, alcohol-resistant foam, carbon dioxide.

NFPA Hazard Classifications: Not listed

Flash Point: 445-450°F (229-232°C) o.c.

FIRST AID

Move victim to fresh air. Call emergency medical care. Apply artificial respiration if victim is not breathing. Administer oxygen if breathing is difficult. Remove and isolate contaminated clothing and shoes. In case of contact with substance, immediately flush skin or eyes with running water for at least 20 minutes. Wash skin with soap and water. Keep victim warm and quiet. Ensure that medical personnel are aware of the material(s) involved and take precautions to protect themselves.

DIETHANOLAMINE D:130

SYNONYMS: BIS(2-HYDROXYETHYL)AMINE; N,N-BIS(2-HYDROXYETHYL)AMINE; BUTANE, 1,2,3,4-DIEPOXY; BUTADIENE DIOXIDE; DEA; DIAETHANOLAMIN (German); DIETANOLAMINA (Spanish); DIETHANOLAMIN (Czech); N,N-DIETHANOLAMINE; DIETHYLAMINE, 2,2'-DIHYDROXY-; 2,2'-DIHYDROXYDIETHYLAMINE; DI(2-HYDROXYETHYL)AMINE; DIOLAMINE; EEC No. 603-071-00-1; EINECS No. 203-868-0; ETHANOL, 2,2'-IMINOBIS-; 2,2'IMINOBIS[ETHANOL]; 2,2'-IMINODIETHANOL; 2,2'DIHYDROXYETHYLAMINE; NCI-C55174; DI(2-HYDROXYETHYL) AMINE

IDENTIFICATION

CAS:	111-42-2
ERG ID:	UN 3077 (solid); UN 3082 (liquid)
ERG Guide:	171
ERG Guide Category:	SUBSTANCES (LOW TO MODERATE HAZARD)
Molecular Formula:	$C_4H_{11}NO_2$
RTECS No.:	KL2975000
Properties:	Combustible, viscous, oily liquid or combustible solid crystals. Liquid is colorless; solid is white. Slight dead fish or ammonia odor. Sinks and mixes with water.

HEALTH AND SAFETY INFORMATION

Odor Threshold: 3.06 mg/m^3

Exposure Limits:	**TWA**
ACGIH:	0.46 ppm; 2 mg/m^3
NIOSH:	3 ppm; 15 mg/m^3
OSHA:	None

RECOMMENDATIONS FOR RESPIRATOR SELECTION: NIOSH

At any concentrations above the NIOSH REL, or where there is no REL, at any detectable concentration: SCBAF:PD,PP (any self-contained breathing apparatus that has a full facepiece and is operated in a pressure-demand or other positive-pressure mode); or SAF:PD,PP:ASCBA (any supplied-air respirator that has a full facepiece and is operated in a pressure-demand or other positive-pressure mode in combination with an auxiliary, self-contained breathing apparatus operated in a pressure-demand or other positive-pressure mode).

Escape: GMFOV [any air-purifying, full-facepiece respirator (gas mask) with a chin-style, front- or back-mounted organic vapor canister]; or SCBAE (any appropriate escape-type, self-contained breathing apparatus).

CHEMICAL REACTIVITY

Incompatible with acids, organic anhydrides, isocyanates, vinyl acetate, acrylates, substituted allyls, alkylene oxides, epichlorohydrin, aldehydes, and oxidizers. Corrosive to copper, copper alloys, zinc, and galvanized iron. Reacts with carbon dioxide in air. Hygroscopic.

FIRE INFORMATION

Vapors are heavier than air and will collect in low areas. Vapors in confined areas may explode when exposed to fire. If material or contaminated runoff enters waterways, notify downstream users of potentially contaminated waters. Notify local health and fire officials and pollution control agencies. Use dry chemicals, alcohol-resistant foam, carbon dioxide.

NFPA Hazard Classifications:

Health Hazards (Blue) 1
Flammability (Red) 1
Reactivity (Yellow) 0

Flash Point: 342°F (172°C) o.c.; 297°F (147°C) c.c.
Flammable Limits in Air: LEL 1.6%; UEL 9.8%
Autoignition Temperature: 1224°F (662°C)

FIRST AID

Move victim to fresh air. Call emergency medical care. Apply artificial respiration if victim is not breathing. Administer oxygen if breathing is difficult. Remove and isolate contaminated clothing and shoes. In case of contact with substance, immediately flush skin or eyes with running water for at least 20 minutes. Ensure that medical personnel are aware of the material(s) involved and take precautions to protect themselves.

ENVIRONMENTAL INFORMATION

Clean Air Act: Hazardous Air Pollutants (Title I, Part A, Section 112)
EPCRA Section 304: Reportable Quantity (RQ): CERCLA, 1 lb (0.454 kg)
EPCRA Section 313: Form R *de minimis* concentration reporting level: 1.0%
TSCA: 40CFR 716.120(*a*)

N,N–DIETHYLANILINE D:140

SYNONYMS: BENZENAMINE, N,N-DIETHYL-(9CI); DEA; DIAETHYLANILIN (German); N,N-DIETHYLAMINOBENZENE; DIETHYLANILINE; N,N-DIETHYLBENZENAMINE; DIETHYLPHENYLAMINE; N,N-DIETILANILINA (Spanish); EEC No. 612-054-00-8; EINECS No. 202-088-8

IDENTIFICATION

CAS:	91-66-7
ERG ID:	UN 2432
ERG Guide:	153
ERG Guide Category:	SUBSTANCES-TOXIC AND/OR CORROSIVE (COMBUSTIBLE)
Molecular Formula:	$C_{10}H_{15}N$
RTECS No.:	BX3400000
Properties:	Combustible oily liquid. Colorless to brownish-yellow. Slightly soluble in water.

HEALTH AND SAFETY INFORMATION

Poisonous. Irritates eyes, skin, and respiratory tract. Inhalation can cause dizziness, sleepiness, and unconsciousness. Long-term exposure may cause kidney and liver problems; blood effects due to the formation of methemaglobin.

RECOMMENDATIONS FOR RESPIRATOR SELECTION

At any concentrations above the NIOSH REL, or where there is no REL, at any detectable concentration: SCBAF:PD,PP (any self-contained breathing apparatus that has a full facepiece and is operated in a pressure-demand or other positive-pressure mode); or SAF:PD,PP:ASCBA (any supplied-air respirator that has a full facepiece and is operated in a pressure-demand or other positive-pressure mode in combination with an auxiliary, self-contained breathing apparatus operated in a pressure-demand or other positive-pressure mode).

Escape: GMFOV [any air-purifying, full-facepiece respirator (gas mask) with a chin-style, front- or back-mounted organic vapor canister]; or SCBAE (any appropriate escape-type, self-contained breathing apparatus).

CHEMICAL REACTIVITY

Forms explosive mixture with air. Contact with oxidizers can cause fire and explosions. Incompatible with acids, organic anhydrides, isocyanates, aldehydes.

FIRE INFORMATION

Combustion products include toxic nitrogen oxides. Vapors are heavier than air and will collect in low areas. Vapors in confined areas may explode when exposed to fire. Vapors may travel long distances to ignition sources and flash back. Storage containers and parts of containers may rocket great distances, in many directions. If material or contaminated runoff enters waterways, notify downstream users of potentially contaminated waters. Notify local health and fire officials and pollution control agencies. Use dry chemicals, foam, carbon dioxide.

NFPA Hazard Classifications:

Health Hazards (Blue) 3
Flammability (Red) 2
Reactivity (Yellow) 0

Flash Point: 185°F (85°C)
Autoignition Temperature: 1166°F (630°C)

FIRST AID

Move victim to fresh air and call emergency medical care. Apply artificial respiration if victim is not breathing. Do not use mouth-to-mouth method if victim swallowed or inhaled the substance; induce artificial respiration with the aid of a pocket mask equipped with a one-way valve or other proper respiratory medical device. If breathing is difficult, give oxygen. In case of contact with substance, immediately flush skin or eyes with running water for at least 20 minutes. Remove and isolate contaminated clothing and shoes. Keep victim warm and quiet. Effects of exposure (inhalation, ingestion, or skin contact) to substances may be delayed. Ensure that medical personnel are aware of the material(s) involved and take precautions to protect themselves.

ENVIRONMENTAL INFORMATION

Clean Air Act: Hazardous Air Pollutants (Title I, Part A, Section 112)
EPCRA Section 304 Reportable Quantity (RQ): CERCLA, 1,000 lb (454 kg)

DIETHYL CARBONATE D:150

SYNONYMS: CARBONATE d'ETHYLE (French); CARBONATO de DIETILO (Spanish); CARBONIC ACID, DIETHYL ESTER; DIETHYLCARBONAT (German); EINECS No. 203-311-1; ETHYL CARBONATE; ETHOXYFORMIC ANHYDRIDE; EUFIN; NCI-C60899

IDENTIFICATION

CAS:	105-58-8
ERG ID:	UN 2366
ERG Guide:	127
ERG Guide Category:	FLAMMABLE LIQUIDS (POLAR/WATER- MISCIBLE)
Molecular Formula:	$C_5H_{10}O_3$

RTECS No.: FF9800000
Properties: Highly flammable, watery liquid. Colorless. Mild, sweet odor. Floats on water; insoluble.

HEALTH AND SAFETY INFORMATION

Irritates eyes, skin, and respiratory tract.

RECOMMENDATIONS FOR RESPIRATOR SELECTION:

At any concentrations above the NIOSH REL, or where there is no REL, at any detectable concentration: SCBAF:PD,PP (any self-contained breathing apparatus that has a full facepiece and is operated in a pressure-demand or other positive-pressure mode); or SAF:PD,PP:ASCBA (any supplied-air respirator that has a full facepiece and is operated in a pressure-demand or other positive-pressure mode in combination with an auxiliary, self-contained breathing apparatus operated in a pressure-demand or other positive-pressure mode).

Escape: GMFOV [any air-purifying, full-facepiece respirator (gas mask) with a chin-style, front- or back-mounted organic vapor canister]; or SCBAE (any appropriate escape-type, self-contained breathing apparatus).

CHEMICAL REACTIVITY

Forms explosive mixture with air. Contact with oxidizers can cause fire and explosions.

FIRE INFORMATION

Vapors are heavier than air and will collect in low areas. Vapors in confined areas may explode when exposed to fire. Vapors may travel long distances to ignition sources and flash back. Storage containers and parts of containers may rocket great distances, in many directions. If material or contaminated runoff enters waterways, notify downstream users of potentially contaminated waters. Notify local health and fire officials and pollution control agencies. Use dry chemicals, foam, or carbon dioxide.

NFPA Hazard Classifications:

Health Hazards (Blue) 2
Flammability (Red) 3
Reactivity (Yellow) 1

Flash Point: 115°F (46°C) o.c.; 77°F (25°C) c.c.

FIRST AID

Move victim to fresh air. Call emergency medical care. Apply artificial respiration if victim is not breathing. Administer oxygen if breathing is difficult. Remove and isolate contaminated clothing and shoes. In case of contact with substance, immediately flush skin or eyes with running water for at least 20 minutes. Wash skin with soap and water. Keep victim warm and quiet. Ensure that medical personnel are aware of the material(s) involved and take precautions to protect themselves.

DIETHYLDICHLOROSILANE D:155

SYNONYMS: DICHLOOR(DIETHYL)SILAAN (Dutch); DICHLORODIETHYLSILANE; DIETILDICLOROSILANO (Spanish); EINECS No. 217-005-0; SILANE, DICHLORODIETHYL-

IDENTIFICATION

CAS: 1719-53-5
ERG ID: UN 1767

ERG Guide: 155
ERG Guide Category: SUBSTANCES-TOXIC AND/OR CORROSIVE (FLAMMABLE/WATER-SENSITIVE)
Molecular Formula: $C_4H_{10}Cl_2Si$
RTECS No.: VV3060000
Properties: Flammable liquid. Colorless. Mixes with water; decomposes forming corrosive hydrogen chloride.

HEALTH AND SAFETY INFORMATION

Corrosive to eyes, skin, and respiratory tract. Inhalation may cause pulmonary edema, which can be delayed for several hours; there is a risk of death in serious cases.

RECOMMENDATIONS FOR RESPIRATOR SELECTION

At any concentrations above the NIOSH REL, or where there is no REL, at any detectable concentration: SCBAF:PD,PP (any self-contained breathing apparatus that has a full facepiece and is operated in a pressure-demand or other positive-pressure mode); or SAF:PD,PP:ASCBA (any supplied-air respirator that has a full facepiece and is operated in a pressure-demand or other positive-pressure mode in combination with an auxiliary, self-contained breathing apparatus operated in a pressure-demand or other positive-pressure mode).

Escape: GMFOV [any air-purifying, full-facepiece respirator (gas mask) with a chin-style, front- or back-mounted organic vapor canister]; or SCBAE (any appropriate escape-type, self-contained breathing apparatus).

CHEMICAL REACTIVITY

Contact with water forms heat and toxic and corrosive hydrogen chloride. Contact with oxidizers may cause fire and explosions. Incompatible with ammonia, strong acids, alcohols, glycols.

FIRE INFORMATION

Combustion products include toxic chlorine vapors. Vapors are heavier than air and will collect in low areas. Vapors in confined areas may explode when exposed to fire. Storage containers and parts of containers may rocket great distances, in many directions. If material or contaminated runoff enters waterways, notify downstream users of potentially contaminated waters. Notify local health and fire officials and pollution control agencies. Use dry chemicals, foam, carbon dioxide.

NFPA Hazard Classifications: Not listed

Flash Point: 75°F (24°C) o.c.

FIRST AID

Move victim to fresh air; call emergency medical care. Apply artificial respiration if victim is not breathing. Do not use mouth-to-mouth method if victim ingested or inhaled the substance; induce artificial respiration with the aid of a pocket mask equipped with a one-way valve or other proper respiratory medical device. Administer oxygen if breathing is difficult. Remove and isolate contaminated clothing and shoes. In case of contact with substance, immediately flush skin or eyes with running water for at least 20 minutes. For minor skin contact, avoid spreading material on unaffected skin. Keep victim warm and quiet. Effects of exposure (inhalation, ingestion, or skin contact) to substance may be delayed. Ensure that medical personnel are aware of the material(s) involved and take precautions to protect themselves.

DIETHYLENE GLYCOL D:160

SYNONYMS: BIS(2-HYDROXYETHYL) ETHER; BRECOLANE NDG; CARBITOL; DEACTIVATOR E; DEACTIVATOR H; DEG; DICOL; DIETHYLENE GLYCOL ETHYL ETHER; DIETHYLENE GLYCOL MONOMETHYL ETHER; DIETILENGLICOL (Spanish); DIGLYCOL; DIGGE; 2,2'-DIHYDROXYDIETHYL ETHER; β,β.'-DIHYDROXYDIETHYL ETHER; 2,2'-DIHYDROXYETHYL ETHER; DIOXITOL; DISSOLVANT APV; EINECS No. 203-872-2;

2-(2-ETHOXYETHOXY)ETHANOL; ETHYLENE DIGLYCOL; GLYCOL ETHER; GLYCOL ETHER DE; GLYCOL ETHYL ETHER; HYDRANAL COMPOSITE 5; IBM INK; KODAGRAPH LIQUID DEVELOPER; 3-OXYPENTANE-1,5-DIOL; 2,2'OXYBISETHANOL; 3-OXA-1,5-PENTANEDIOL; 2,2'-OXYDIETHANOL; TL4N

IDENTIFICATION

CAS:	111-46-6
Molecular Formula:	$C_4H_{10}O_3$
RTECS No.:	ID5950000
Properties:	Combustible, thick, oily liquid. Colorless. Practically odorless. Sinks and mixes with water.

HEALTH AND SAFETY INFORMATION

Exposure Limits: TWA

AIHA WEEL:	25 ppm
AIHA WEEL:	50 ppm vapor and aerosols
AIHA WEEL:	10 mg/m^3 aerosols only

RECOMMENDATIONS FOR RESPIRATOR SELECTION: NIOSH

At any concentrations above the NIOSH REL, or where there is no REL, at any detectable concentration: SCBAF:PD,PP (any self-contained breathing apparatus that has a full facepiece and is operated in a pressure-demand or other positive-pressure mode); or SAF:PD,PP:ASCBA (any supplied-air respirator that has a full facepiece and is operated in a pressure-demand or other positive-pressure mode in combination with an auxiliary, self-contained breathing apparatus operated in a pressure-demand or other positive-pressure mode)

ESCAPE: GMFOV [any air-purifying, full-facepiece respirator (gas mask) with a chin-style, front- or back-mounted organic vapor canister]; or SCBAE (any appropriate escape-type, self-contained breathing apparatus).

CHEMICAL REACTIVITY

Incompatible with sulfuric acid, isocyanates, strong oxidizers.

FIRE INFORMATION

Vapors are heavier than air and will collect in low areas. Vapors in confined areas may explode when exposed to fire. If material or contaminated runoff enters waterways, notify downstream users of potentially contaminated waters. Notify local health and fire officials and pollution control agencies. Use dry chemicals, alcohol-resistant foam, carbon dioxide.

NFPA Hazard Classifications:

Health Hazards (Blue) 1
Flammability (Red) 1
Reactivity (Yellow) 0

Flash Point: 255°F (124°C) c.c.
Flammable Limits in Air: LEL 1.6%; UEL 10.8%
Autoignition Temperature: 444°F (229°C)

FIRST AID

Move victim to fresh air. Call emergency medical care. Apply artificial respiration if victim is not breathing. Administer oxygen if breathing is difficult. Remove and isolate contaminated clothing and shoes. In case of contact with substance, immediately flush skin or eyes with running water for at least 20 minutes. Wash skin with soap and water. Keep victim warm and quiet. Effects of exposure (inhalation, ingestion, or skin contact) to substance may be delayed. Ensure that medical personnel are aware of the material(s) involved and take precautions to protect themselves.

ENVIRONMENTAL INFORMATION (as glycol ethers)

Diethylene glycol is among those chemicals released in great quantity by industry and is a water-soluble, volatile organic chemical. It may enter the environment in liquid form via petrochemical plant effluents or as an unburned gas from combustion sources.

CLEAN AIR ACT: Hazardous Air Pollutants (Title I, Part A, Section 112) includes mono- and di-ethers of ethylene glycol, diethyl glycol, and triethylene glycol R-(OCH2CH2)n-OR' where n = 1,2, or 3; R = alkyl or aryl groups; R' = R, H, or groups which when removed, yield glycol ethers with the structure: R-(OCH2CH)n-OH. Polymers are excluded from the glycol category.
EPCRA Section 313: Form R *de minimus* concentration reporting level: 1.0%; Form R Toxic Chemical Category Code: N230
TSCA: 40CFR716.120(*a*); 40CFR799.1560

DIETHYLENE GLYCOL DIBUTYL ETHER D:200

SYNONYMS: BIS(2-BUTOXYETHYL) ETHER; BUTYL DIGLYME; DIBUTILDIGOL (Spanish); 2,2'-DIBUTOXYETHYL ETHER; DIBUTYL CARBITOL; DIETHYLENE GLYCOL DI-N-BUTYL ETHER; DIETILENGLICOL DIBUTIL ETER (Spanish); EINECS No. 204-001-9; OSSIDO di BIS(2-BUTOSSIETILE) (Italian); 5,8,11-TRIOXAPENTADECANE

IDENTIFICATION

CAS:	112-73-2
Molecular Formula:	$C_{12}H_{26}O_3$
RTECS No.:	KN0350000
Properties:	Combustible liquid. Colorless. Floats on water; slightly soluble.

HEALTH AND SAFETY INFORMATION

Irritates eyes and respiratory tract.

IARC Evaluation: Group 3, not classifiable as glycol ethers

RECOMMENDATIONS FOR RESPIRATOR SELECTION

At any concentrations above the NIOSH REL, or where there is no REL, at any detectable concentration: SCBAF:PD,PP (any self-contained breathing apparatus that has a full facepiece and is operated in a pressure-demand or other positive-pressure mode); or SAF:PD,PP:ASCBA (any supplied-air respirator that has a full facepiece and is operated in a pressure-demand or other positive-pressure mode in combination with an auxiliary, self-contained breathing apparatus operated in a pressure-demand or other positive-pressure mode).

Escape: GMFOV [any air-purifying, full-facepiece respirator (gas mask) with a chin-style, front- or back-mounted organic vapor canister]; or SCBAE (any appropriate escape-type, self-contained breathing apparatus).

CHEMICAL REACTIVITY

Reacts with sulfuric acid, nitric acid, caustics, aliphatic amines, amides. Contact with strong oxidizers may cause fire and explosions.

FIRE INFORMATION

Vapors are heavier than air and will collect in low areas. Vapors in confined areas may explode when exposed to fire. Vapors may travel to ignition sources and flash back. Storage containers and parts of containers may rocket great distances, in many directions. If material or contaminated runoff enters waterways, notify downstream users of potentially contaminated waters. Notify local health and fire officials and pollution control agencies.

NFPA Hazard Classifications:

Health Hazards (Blue) 1
Flammability (Red) 1
Reactivity (Yellow) 0

Flash Point: 245°F (118°C) o.c.
Autoignition Temperature: 590°F (310°C)

FIRST AID

Move victim to fresh air. Call emergency medical care. Apply artificial respiration if victim is not breathing. If breathing is difficult, give oxygen. Remove and isolate contaminated clothing and shoes. In case of contact with substance, immediately flush skin or eyes with running water for at least 20 minutes. Wash skin with soap and water. Keep victim warm and quiet. Ensure that medical personnel are aware of the material(s) involved and take precautions to protect themselves.

ENVIRONMENTAL INFORMATION

CLEAN AIR ACT: Hazardous Air Pollutants (Title I, Part A, Section 112) includes mono- and di- ethers of ethylene glycol, diethyl glycol, and triethylene glycol R-(OCH2CH2)n-OR' where n = 1,2, or 3; R = alkyl or aryl groups; R' = R, H, or groups which when removed, yield glycol ethers with the structure: R-(OCH2CH)n-OH. Polymers are excluded from the glycol category.
EPCRA Section 313: Certain glycol ethers are covered. R-(OCH2CH2)n-OR'; Where n = 1,2 or 3; R =alkyl C7 or less; or R = phenyl or alkyl substituted phenyl; R' + H, or alkyl C7 or less; or OR' consisting of carboxylic ester, sulfate, phosphate, nitrate or sulfonate. Form R *de minimus* concentration reporting level: 1.0% Form R Toxic Chemical Category Code: N230
TSCA: 40CFR716.120(*a*); 40CFR799.1560

DIETHYLENE GLYCOL MONOBUTYL ETHER ACETATE D:240

SYNONYMS: ACETATO DEL DIETILENGLICO MONOBUTIL ETER (Spanish); 2-(2-BUTOXYETHOXY) ETHANOL ACETATE; 2-(2-BUTOXYETHOXY) ETHYL ACETATE; BUTYL CARBITOL ACETATE; DIETHYLENE GLYCOL BUTYL ETHER ACETATE; DIGLYCOL MONOBUTYL ETHER ACETATE; EINECS No. 204-685-9; EKTASOLVE DB ACETATE; GLYCOL ETHER DB ACETATE

IDENTIFICATION

CAS:	124-17-4
ERG ID:	UN 3082
ERG Guide:	171
ERG Guide Category:	SUBSTANCES (LOW TO MODERATE HAZARD)
Molecular Formula:	$C_{10}H_{20}O_4$
RTECS No.:	KJ9275000
Properties:	Combustible liquid. Colorless. Mild odor. Floats and mixes slowly with water.

HEALTH AND SAFETY INFORMATION

Can cause irritation of the eyes, skin, and respiratory tract. Absorption through skin can add to exposure.

RECOMMENDATIONS FOR RESPIRATOR SELECTION:

At any concentrations above the NIOSH REL, or where there is no REL, at any detectable concentration: SCBAF:PD,PP (any self-contained breathing apparatus that has a full facepiece and is operated in a pressure-demand or other positive-pressure mode); or SAF:PD,PP:ASCBA (any supplied-air respirator that has a full facepiece and is operated in a pressure-demand or other positive-pressure mode in combination with an auxiliary, self-contained breathing apparatus operated in a pressure-demand or other positive-pressure mode).

Escape: GMFOV [any air-purifying, full-facepiece respirator (gas mask) with a chin-style, front- or back-mounted organic vapor canister]; or SCBAE (any appropriate escape-type, self-contained breathing apparatus).

CHEMICAL REACTIVITY

Reacts with oxidizers, sulfuric acid, isocyanates.

FIRE INFORMATION

Vapors are heavier than air and will collect in low areas. Vapors in confined areas may explode when exposed to fire. Storage containers and parts of containers may rocket great distances, in many directions. If material or contaminated runoff enters waterways, notify downstream users of potentially contaminated waters. Notify local health and fire officials and pollution control agencies. Use alcohol foam, dry chemicals, carbon dioxide, water.

NFPA Hazard Classifications:

Health Hazards (Blue) 1
Flammability (Red) 1
Reactivity (Yellow) 0

Flash Point: 240°F (116°C) o.c.
Flammable Limits in Air: LEL 0.8%; UEL 11.%
Autoignition Temperature: 572°F (300°C)

FIRST AID

Move victim to fresh air. Call emergency medical care. Apply artificial respiration if victim is not breathing. If breathing is difficult, give oxygen. Remove and isolate contaminated clothing and shoes. In case of contact with substance, immediately flush skin or eyes with running water for at least 20 minutes. Wash skin with soap and water. Keep victim warm and quiet. Ensure that medical personnel are aware of the material(s) involved and take precautions to protect themselves.

ENVIRONMENTAL INFORMATION

CLEAN AIR ACT: Hazardous Air Pollutants (Title I, Part A, Section 112) includes mono- and di- ethers of ethylene glycol, diethyl glycol, and triethylene glycol R-(OCH2CH2)n-OR' where n = 1,2, or 3; R = alkyl or aryl groups; R' = R, H, or groups which when removed, yield glycol ethers with the structure: R-(OCH2CH)n-OH. Polymers are excluded from the glycol category.
EPCRA Section 313: Certain glycol ethers are covered. R-(OCH2CH2)n-OR'; Where n = 1,2 or 3; R =alkyl C7 or less; or R = phenyl or alkyl substituted phenyl; R' + H, or alkyl C7 or less; or OR' consisting of carboxylic ester, sulfate, phosphate, nitrate or sulfonate. Form R *de minimus* concentration reporting level: 1.0% Form R Toxic Chemical Category Code: N230
TSCA: 40CFR716.120(*a*); 40CFR799.1560

DIETHYLENETRIAMINE D:250

SYNONYMS: AMINOETHANDIAMINE; AMINOETHYLETHANDIAMINE; 3-AZAPENTANE-1,5-DIAMINE; BIS(2-AMINOETHYL)AMINE; N-(2-AMINOETHYL); DIETHYLENETRIAMINE; DIETILENTRIAMINA (Spanish); EEC No. 612-058-00-X; EINECS No. 203-865-4; 1,2-ETHANEDIAMINE, N-(2-AMINOETHYL)-; ETHYLENEDIAMINE; DETA; BIS(β-AMINOETHYL)AMINE; 2,2'-DIAMINODIETHYLAMINE; 2,2'-IMINOBISETHYLAMINE; 1,4,7-TRIAZAHEPTANE; BIS (2-AMINOETHYL) AMINE

IDENTIFICATION

CAS:	111-40-0
ERG ID:	UN 2079
ERG Guide:	154

ERG Guide Category: SUBSTANCES-TOXIC AND/OR CORROSIVE (NONCOMBUSTIBLE)
Molecular Formula: $C_4H_{13}N_3$
RTECS No.: IE1225000
Properties: Combustible liquid. Colorless to yellow. Ammonialike odor. Floats and mixes with water. Vapors are heavier than air; will collect in low areas.

HEALTH AND SAFETY INFORMATION

Odor Threshold: 35.8 mg/m^3

Exposure Limits:	TWA
ACGIH:	1 ppm; 4.2 mg/m^3
NIOSH:*	1 ppm; 4 mg/m^3
OSHA:	None

*Skin contact contributes significantly in overall exposure.

RECOMMENDATIONS FOR RESPIRATOR SELECTION: NIOSH

At any concentrations above the NIOSH REL, or where there is no REL, at any detectable concentration: CBAF:PD,PP (any self-contained breathing apparatus that has a full facepiece and is operated in a pressure-demand or other positive-pressure mode); or SAF:PD,PP:ASCBA (any supplied-air respirator that has a full facepiece and is operated in a pressure-demand or other positive-pressure mode in combination with an auxiliary, self-contained breathing apparatus operated in a pressure-demand or other positive–pressure mode).

Escape: GMFOV [any air-purifying, full-facepiece respirator (gas mask) with a chin-style, front- or back-mounted organic vapor canister]; or SCBAE (any appropriate escape-type, self-contained breathing apparatus).

CHEMICAL REACTIVITY

Ignites spontaneously with cellulose nitrate. Silver, cobalt, or chromium compounds may cause explosions. Forms explosive mixture with air. Incompatible with acids, organic anhydrides, isocyanates, vinyl acetate, acrylates, substituted allyls, alkylene oxides, epichlorohydrin, ketones, aldehydes, alcohols, glycols, mercury, phenols, cresols, caprolactum solution, and strong oxidizers. Attacks aluminum, copper, lead, tin, zinc, and alloys.

FIRE INFORMATION

Heating or fire produces toxic nitrogen oxides. Firefighting gear, including SCBA, does not provide adequate protection. If exposure occurs, remove and isolate gear immediately and thoroughly decontaminate personnel. Vapors are heavier than air and will collect in low areas. Vapors in confined areas may explode when exposed to fire. Vapors may travel long distances to ignition sources and flash back. Storage containers and parts of containers may rocket great distances, in many directions. If material or contaminated runoff enters waterways, notify downstream users of potentially contaminated waters. Notify local health and fire officials and pollution control agencies. Use dry chemicals, foam, carbon dioxide.

NFPA Hazard Classifications:

Health Hazards (Blue) 3
Flammability (Red) 1
Reactivity (Yellow) 0

Flash Point: 210°F (99°C) o.c.; 208°F (98°C) c.c.
Flammable Limits in Air: LEL 2.0%; UEL 6.7%
Autoignition Temperature: 676°F (358°C)

FIRST AID

Move victim to fresh air. Call emergency medical care. Apply artificial respiration if victim is not breathing. Do not use mouth-to-mouth method if victim ingested or inhaled the substance; induce artificial respiration with the aid of a pocket mask equipped with a one-way valve or other proper respiratory medical device. Administer oxygen if breathing is difficult. Remove and isolate contaminated clothing and shoes. In case of contact with substance, immediately flush skin or eyes with running water for at least 20 minutes. For minor skin contact, avoid spreading material on unaffected skin. Keep victim warm and quiet. Effects of exposure (inhalation, ingestion, or skin contact) may be delayed. Ensure that medical personnel are aware of the material(s) involved and take precautions to protect themselves.

ENVIRONMENTAL INFORMATION

TSCA: 40CFR716.120(*a*); 40CFR799.1575

DI(2-ETHYLHEXYL)FUMARATE D:252

SYNONYMS: BIS(2-ETHYLHEXYL) FUMARATE; 2-BUTENEDIOIC ACID BIS(2-ETHYLHEXYL) ESTER; DI(ETHYLHEXYL)FUMARATE; DIOCTYL FUMARATE; DOF; EINECS No. 205-448-2; 2-ETHYLHEXYL FUMARATE; RC COMONOMER DOF; FUMERATO de DIOCTILO (Spanish)

IDENTIFICATION

CAS:	141-02-6
Molecular Formula:	$C_{20}H_{36}O_4$
RTECS No.:	LT0525000
Properties:	Combustible, mobile liquid. Clear. Mild odor.

HEALTH AND SAFETY INFORMATION

Irritates eyes, skin, and respiratory tract. Inhalation may cause pulmonary edema, which can be delayed for several hours; there is a risk of death in serious cases.

RECOMMENDATIONS FOR RESPIRATOR SELECTION

At any concentrations above the NIOSH REL, or where there is no REL, at any detectable concentration: SCBAF:PD,PP (any self-contained breathing apparatus that has a full facepiece and is operated in a pressure-demand or other positive-pressure mode); or SAF:PD,PP:ASCBA (any supplied-air respirator that has a full facepiece and is operated in a pressure-demand or other positive-pressure mode in combination with an auxiliary, self-contained breathing apparatus operated in a pressure-demand or other positive-pressure mode).

Escape: GMFOV [any air-purifying, full-facepiece respirator (gas mask) with a chin-style, front- or back-mounted organic vapor canister]; or SCBAE (any appropriate escape-type, self-contained breathing apparatus).

CHEMICAL REACTIVITY

Reacts with oxidizers, nitric acid, sulfuric acid.

FIRE INFORMATION

If material or contaminated runoff enters waterways, notify downstream users of potentially contaminated waters. Notify local health and fire officials and pollution control agencies. Use dry chemicals, alcohol-resistant foam, carbon dioxide.

NFPA Hazard Classifications: Not listed

Flash Point: 365°F (185°C) o.c.

FIRST AID

Move victim to fresh air. Call emergency medical care. Apply artificial respiration if victim is not breathing. Administer oxygen if breathing is difficult. In case of contact with substance, immediately flush skin or eyes with running water for at least 20 minutes. Remove and isolate contaminated clothing and shoes. Ensure that medical personnel are aware of the material(s) involved and take precautions to protect themselves.

DI(2-ETHYLHEXYL)MALEATE D:254

SYNONYMS: BIS(2-ETHYLHEXYL)-2-BUTENEDIOATE; BIS(2-ETHYLHEXYL) MALEATE; 2-BUTENEDIOIC ACID (Z)-, BIS(2-ETHYLHEXYL) ESTER;
DIOCTYL MALEATE; DOM; EINECS No. 205-524-5; MALEATO de DI(ETIHEXILO) (Spanish); RC COMONOMER DOM

IDENTIFICATION

CAS:	142-16-5
Molecular Formula:	$C_{20}H_{36}O_4$
RTECS No.:	ON0160000
Properties:	Combustible liquid. Not soluble in water.

RECOMMENDATIONS FOR RESPIRATOR SELECTION

At any concentrations above the NIOSH REL, or where there is no REL, at any detectable concentration: SCBAF:PD,PP (any self-contained breathing apparatus that has a full facepiece and is operated in a pressure-demand or other positive-pressure mode); or SAF:PD,PP:ASCBA (any supplied-air respirator that has a full facepiece and is operated in a pressure-demand or other positive-pressure mode in combination with an auxiliary, self-contained breathing apparatus operated in a pressure-demand or other positive-pressure mode).

Escape: GMFOV [any air-purifying, full-facepiece respirator (gas mask) with a chin-style, front- or back-mounted organic vapor canister]; or SCBAE (any appropriate escape-type, self-contained breathing apparatus).

CHEMICAL REACTIVITY

Reacts with oxidizers.

FIRE INFORMATION

If material or contaminated runoff enters waterways, notify downstream users of potentially contaminated waters. Notify local health and fire officials and pollution control agencies. Use alcohol-resistant foam, carbon dioxide, dry chemicals. Water or foam may cause frothing.

NFPA Hazard Classifications:

Health Hazards (Blue) 0
Flammability (Red) 1
Reactivity (Yellow) 0

Flash Point: 365°F (185°C) o.c.

FIRST AID

Move victim to fresh air. Call emergency medical care. Apply artificial respiration if victim is not breathing. Administer oxygen if breathing is difficult. Remove and isolate contaminated clothing and shoes. In case of contact with substance, immediately flush skin or eyes with running water for at least 20 minutes. Wash skin with soap and water. Keep victim warm and quiet. Ensure that medical personnel are aware of the material(s) involved and take precautions to protect themselves.

ENVIRONMENTAL INFORMATION

TSCA: 40CFR712.30(*e*)13; 40CFR716.120(*d*)13 substantially produced chemicals/need of subchronic tests.

DI-(2-ETHYLHEXYL) PHTHALATE D:256

SYNONYMS: BEHP; 1,2-BENZENEDICARBOXYLIC ACID, BIS(2-ETHYLHEXYL) ESTER;

1,2-BENZENEDICARBOXYLIC ACID, DIOCTYL ESTER; BIS(2-ETILHEXIL)FTALATO (Spanish); BIS(2-ETHYLHEXYL)-1,2-BENZENEDICARBOXYLATE; BIS(2-ETHYLHEXYL)PHTHALATE; BISOFLEX 81; BISOFLEX 82; BISOFLEX DOP; COMPOUND 889; DAF 68; DEHP; DIESTER OF 2-ETHYLHEXYL ALCOHOL AND PHTHALIC ACID; DI(2-ETHYLHEXYL)ORTHOPHTHALATE; DI(2-ETHYLHEXYL)PHTHALATE; DI-s-OCTYL PHTHALATE; DI-sec-OCTYL PHTHALATE (DEHP); DIOCTYL PHTHALATE; DOF; DOF (Russian plasticizer); DOP; EINECS No. 204-211-0; ERGOPLAST FDO; ERGOPLAST FDO-S; ETHYLHEXYL PHTHALATE; 2-ETHYLHEXYL PHTHALATE; EVIPLAST 80; EVIPLAST 81; FLEXIMEL; FLEXOL DOP; FLEXOL PLASTICIZER DOP; FTALATO de(2-ETILHEXILO) (Spanish); GOOD-RITE GP 264; HATCOL DOP; HERCOFLEX 260; KODAFLEX DOP; MITSUBISHI DOP; MOLLANO; NCI-C52733; NUOPLAZ DOP; OCTOIL; OCTYL PHTHALATE; OCTYL PHTHALATE, DI-sec; PALATINOL AH; PHTHALIC ACID DIOCTYL ESTER; PITTSBURGH PX-138; PLATINOL AH; PLATINOL DOP; PLASTHALL DOP; PLASTICIZER 28P; RC PLASTICIZER DOP; POLYCIZER 162; PX-138; REOMOL D 79P; REOMOL DOP; SICOL 150; STAFLEX DOP; TRUFLEX DOP; VESTINOL AH; VINICIZER 80; WITCIZER 312

IDENTIFICATION

CAS:	117-81-7
ERG ID:	UN 3082
ERG Guide:	171
ERG Guide Category:	SUBSTANCES (LOW TO MODERATE HAZARD)
Molecular Formula:	$C_{24}H_{38}O_4$
RTECS No.:	TI0350000; TI1925000
Properties:	Combustible, oily liquid. Colorless. Odorless. Floats on water.

HEALTH AND SAFETY INFORMATION

Irritates eyes, skin, and respiratory tract.

IARC Evaluation: Group 2B, possible carcinogen; no human data; sufficient animal evidence.
NTP: Evidence of carcinogenicity..
IDLH: 5,000 mg/m^3; potential human carcinogen
OSHA Table Z-1-A Air Contaminant

Exposure Limits:	TWA	STEL
ACGIH:*	5 mg/m^3	10 mg/m^3
NIOSH:**	5 mg/m^3	10 mg/m^3
OSHA:	5 mg/m^3	

*Animal carcinogen.
**Potential human carcinogen.

RECOMMENDATIONS FOR RESPIRATOR SELECTION: NIOSH

At any concentrations above the NIOSH REL, or where there is no REL, at any detectable concentration: SCBAF:PD,PP (any self-contained breathing apparatus that has a full facepiece and is operated in a pressure-demand or other positive-pressure mode); or SAF:PD,PP:ASCBA (any supplied-air respirator that has a full facepiece and is operated in a pressure-demand or other positive-pressure mode in combination with an auxiliary, self-contained breathing apparatus operated in a pressure-demand or other positive-pressure mode).

Escape: HiEF (any air-purifying, full-facepiece respirator with a high-efficiency particulate filter); or SCBAE (any appropriate escape-type, self-contained breathing apparatus).

CHEMICAL REACTIVITY

Incompatible with nitrates, strong oxidizers, strong alkalis, or strong acids. Reactions with these compounds may cause fires and explosions.

FIRE INFORMATION

Toxic vapors and gases may be released in fire. Vapors are heavier than air and will collect in low areas. Vapors in confined areas may explode when exposed to fire. If material or contaminated runoff enters waterways, notify downstream users of potentially contaminated waters. Notify local health and fire officials and pollution control agencies. Use dry chemicals, foam, carbon dioxide.

Flash Point: 420°F (215°C) o.c.
Flammable Limits in Air: LEL 0.3% at 474°F (245°C); UEL N/A
Autoignition Temperature: 735°F (390°C)

FIRST AID

Move victim to fresh air. Call emergency medical care. Apply artificial respiration if victim is not breathing. Administer oxygen if breathing is difficult. Remove and isolate contaminated clothing and shoes. In case of contact with substance, immediately flush skin or eyes with running water for at least 20 minutes. Ensure that medical personnel are aware of the material(s) involved and take precautions to protect themselves.

ENVIRONMENTAL INFORMATION

Clean Air Act: Hazardous Air Pollutants (Title I, Part A, Section 112).
Clean Water Act: Section 307 Priority Pollutants.
EPA Hazardous Waste Number (RCRA No.): U028.
RCRA Section 261 Hazardous Constituents
RCRA Land Ban Waste
RCRA Universal Treatment Standards: Wastewater (mg/L), 0.28; Nonwastewater (mg/kg), 28.
RCRA Ground Water Monitoring List: Suggested test method(s) (PQL ug/L): 8060(20); 8270(10).
Safe Drinking Water Act: MCL, 0.006 mg/L; MCLG, zero.
EPCRA Section 304: Reportable Quantity (RQ): CERCLA, 100 lb (45.4 kg).
EPCRA Section 313: Form R *de minimis* concentration reporting level: 0.1%.
California's Proposition 65: Carcinogen.
TSCA: 40CFR799.5025; 40CFR712.30(*d*)

DI(2-ETHYLHEXYL)SEBACATE D:257

SYNONYMS: BISOFLEX DOS; DECANEDIOIC ACID, BIS(2-ETHYLHEXYL) ESTER; DIOCTYL SEBACATE; DOS; EINECS No. 204-558-8; 2-ETHYLHEXYL SEBACATE; MONOPLEX DOS; OCTOIL S; OCTYL SEBACATE; PX 438; SEBACATO de DI(2-ETILHEXILO) (Spanish); SEBACATO de DIOCTILO (Spanish); STALFLEX DOS; UNIFLEX DOS

IDENTIFICATION

CAS: 122-62-3
Molecular Formula: $C_{26}H_{50}O_4$
RTECS No.: VS1000000
Properties: Combustible liquid. Colorless to pale straw-color. Mild odor. Very soluble in water.

HEALTH AND SAFETY INFORMATION

RECOMMENDATIONS FOR RESPIRATOR SELECTION

At any concentrations above the NIOSH REL, or where there is no REL, at any detectable concentration: SCBAF:PD,PP (any self-contained breathing apparatus that has a full facepiece and is operated in a pressure-demand or other positive-pressure mode); or SAF:PD,PP:ASCBA (any supplied-air respirator that has a full facepiece and is operated in a pressure-demand or other positive-pressure mode in combination with an auxiliary, self-contained breathing apparatus operated in a pressure-demand or other positive-pressure mode).

Escape: GMFOV [any air-purifying, full-facepiece respirator (gas mask) with a chin-style, front- or back-mounted organic vapor canister]; or SCBAE (any appropriate escape-type, self-contained breathing apparatus).

CHEMICAL REACTIVITY

Contact with strong oxidizers may cause fire and explosions. Reacts with sulfuric acid, nitric acid.

FIRE INFORMATION

Vapors are heavier than air and will collect in low areas. Vapors in confined areas may explode when exposed to fire.If material or contaminated runoff enters waterways, notify downstream users of potentially contaminated waters. Notify local health and fire officials and pollution control agencies. Use dry chemicals, alcohol-resistant foam, carbon dioxide.

NFPA Hazard Classifications: Not listed

Flash Point: 410°F (210°C) o.c.

FIRST AID

Move victim to fresh air. Call emergency medical care. Apply artificial respiration if victim is not breathing. Administer oxygen if breathing is difficult. Remove and isolate contaminated clothing and shoes. In case of contact with substance, immediately flush skin or eyes with running water for at least 20 minutes. Wash skin with soap and water. Keep victim warm and quiet. Ensure that medical personnel are aware of the material(s) involved and take precautions to protect themselves.

DIETHYL PHTHALATE D:258

SYNONYMS: ANOZOL; 1,2-BENZENEDICARBOXYLIC ACID, DIETHYL ESTER; DEP; DIETHYL ESTER OF PHTHALIC ACID; DIETHYL 1,2-BENZENEDICARBOXYLATE; DIETHYLFTALAAT (Dutch); EINECS No. 201-550-6; ESTOL-1550; ETHYL PHTHALATE; FTALATO de DIETILO (Spanish); FTALATO di DIETILE (Italian); NCI-C60048; NEANTINE; PHTALATE d'ETHYLE (French); PALATINOL A; PHTHALIC ACID, DIETHYL ESTER; PHTHALIC ACID, DIETHYL ESTER; PHTHALOL; PHTHALSAEUREDIAETHYLESTER (German); PLACIDOL E; SOLVANOL

IDENTIFICATION

CAS:	84-66-2
ERG ID:	UN 3082
ERG Guide:	171
ERG Guide Category:	SUBSTANCES (LOW TO MODERATE HAZARD)
Molecular Formula:	$C_{12}H_{12}O_4$
RTECS No.:	TI1050000
Properties:	Combustible liquid. White. Mild chemical odor. Sinks in water. Freezing point is 27°F (13°C).

HEALTH AND SAFETY INFORMATION

Irritates eyes, skin, and respiratory tract. Can affect central nervous system; headache, dizziness, drowsiness.

Exposure Limits: TWA

ACGIH:	5 mg/m^3
NIOSH:	5 mg/m^3
OSHA:	None

CHEMICAL REACTIVITY

Violent reaction with strong acids, strong oxidizers, permanganates, and water. Attacks some forms of plastic.

FIRE INFORMATION

Irritating gases may be produced when heated. Vapors are heavier than air and will collect in low areas. Vapors in confined areas may explode when exposed to fire.If material or contaminated runoff enters waterways, notify downstream users of potentially contaminated waters. Notify local health and fire officials and pollution control agencies. Use dry chemicals, alcohol-resistant foam, carbon dioxide.

NFPA Hazard Classifications:

Health Hazards (Blue) 0
Flammability (Red) 1
Reactivity (Yellow) 0

Flash Point: 322°F (161°C) o.c.
Flammable Limits in Air: LEL 0.75% at 368°F (187°C); UEL N/A
Autoignition Temperature: 855°F (457°C)

FIRST AID

Move victim to fresh air. Call emergency medical care. Apply artificial respiration if victim is not breathing. Administer oxygen if breathing is difficult. Remove and isolate contaminated clothing and shoes. In case of contact with substance, immediately flush skin or eyes with running water for at least 20 minutes. Ensure that medical personnel are aware of the material(s) involved and take precautions to protect themselves.

ENVIRONMENTAL INFORMATION

Clean Water Act: Section 307 Priority Pollutants; Section 313 Priority Chemicals.
EPA Hazardous Waste Number (RCRA No.): U088
RCRA Section 261 Hazardous Constituents
RCRA Land Ban Waste
RCRA Universal Treatment Standards: Wastewater (mg/L), 0.20; Nonwastewater (mg/kg), 28.
RCRA Ground Water Monitoring List: Suggested test method(s) (PQL ug/L): 8060(5); 8270(10).
EPCRA Section 304: Reportable Quantity (RQ): CERCLA, 1,000 lb (454 kg).
EPCRA Section 313: Deleted from EPCRA/SARA 313 July 29, 1996 (FR Vol. 61, No. 146, p. 39356-39357).
TSCA: 40CFR712.30(*d*); 40CFR716.120(*c*)2

DIETHYL SUCCINATE D:260

SYNONYMS: BUTANEDIOIC ACID, DIETHYL ESTER; EINECS No. 204-612-0; SUCCINATE D'ETHYLE (French); SUCCINATO de DIETILO (Spanish); SUCCINIC ACID, DIETHYL ESTER

IDENTIFICATION

CAS: 123-25-1
Molecular Formula: $C_8H_{14}O_4$
RTECS No.: WM7400000
Properties: Combustible liquid. Colorless. Faint, pleasant odor. Mixes with water; slightly soluble.

HEALTH AND SAFETY INFORMATION

Irritates, eyes, skin, and respiratory tract.

RECOMMENDATIONS FOR RESPIRATOR SELECTION

At any concentrations above the NIOSH REL, or where there is no REL, at any detectable concentration: SCBAF:PD,PP (any self-contained breathing apparatus that has a full facepiece and is operated in a pressure-demand or other positive-pressure mode); or SAF:PD,PP:ASCBA (any supplied-air respirator that has a full facepiece and is operated in a pressure-demand or other positive-pressure mode in combination with an auxiliary, self-contained breathing apparatus operated in a pressure-demand or other positive-pressure mode).

Escape: GMFOV [any air-purifying, full-facepiece respirator (gas mask) with a chin-style, front- or back-mounted organic vapor canister]; or SCBAE (any appropriate escape-type, self-contained breathing apparatus).

CHEMICAL REACTIVITY

Incompatible with nitrates, nitric acid, sulfuric acid.

FIRE INFORMATION

Vapors are heavier than air and will collect in low areas. Vapors in confined areas may explode when exposed to fire. If material or contaminated runoff enters waterways, notify downstream users of potentially contaminated waters. Notify local health and fire officials and pollution control agencies. Use dry chemicals, alcohol-resistant foam, carbon dioxide.

NFPA Hazard Classifications: Not listed

Flash Point: 230°F (110°C) o.c.

FIRST AID

Move victim to fresh air. Call emergency medical care. Apply artificial respiration if victim is not breathing. Administer oxygen if breathing is difficult. Remove and isolate contaminated clothing and shoes. In case of contact with substance, immediately flush skin or eyes with running water for at least 20 minutes. Wash skin with soap and water. Keep victim warm and quiet. Ensure that medical personnel are aware of the material(s) involved and take precautions to protect themselves.

DIETHYL SULFATE D:262

SYNONYMS: DIETHYL MONO DIAETHYLSULFAT SULFATE; DIETHYL SULPHATE; DS; DIETHYL ESTER SULFURIC ACID; EEC No. 016-02 7-00-6; EINECS No. 200-589-6; ETHYL SULFATE; SULFATO de DIETILO (Spanish); SULFURIC ACID, DIETHYL ESTER

IDENTIFICATION

CAS:	64-67-5
ERG ID:	UN 1594
ERG Guide:	152
ERG Guide Category:	SUBSTANCES-TOXIC (COMBUSTIBLE)
Molecular Formula:	$C_4H_{10}O_4S$
RTECS No.:	WS7875000
Properties:	Combustible, oily liquid. Colorless but darkens with age. Ethereal, peppermint odor. Sinks and very slowly dissolves in cold water; gradually decomposes to produce sulfuric acid. Contact with hot or warm water may cause a violent reaction.

HEALTH AND SAFETY INFORMATION

Corrosive to the eyes, skin, and respiratory tract. Inhalation may cause pulmonary edema, which can be delayed for several hours; there is a risk of death in serious cases.

IARC Evaluation: Group 2B; limited human evidence; sufficient animal evidence.
OSHA Table Z-1-A Air Contaminant

RECOMMENDATIONS FOR RESPIRATOR SELECTION:

At any concentrations above the NIOSH REL, or where there is no REL, at any detectable concentration: SCBAF:PD,PP (any self-contained breathing apparatus that has a full facepiece and is operated in a pressure-demand or other positive-pressure mode); or SAF:PD,PP:ASCBA (any supplied-air respirator that has a full facepiece and is operated in a pressure-demand or other positive-pressure mode in combination with an auxiliary, self-contained breathing apparatus operated in a pressure-demand or other positive-pressure mode).

Escape: GMFOV [any air-purifying, full-facepiece respirator (gas mask) with a chin-style, front- or back-mounted organic vapor canister]; or SCBAE (any appropriate escape-type, self-contained breathing apparatus).

CHEMICAL REACTIVITY

Reacts slowly with cold water about 0.05 percent per hour at 77°F (25°C), to yield ethyl alcohol and monoethyl sulfate. Reacts vigorously with water at temperature above 122°F (50°C). Sulfuric acid may be produced along with ethyl alcohol and monoethyl sulfate. Avoid contact with aqueous alkali, concentrated nitric acid, and strong oxidizing agents such as peroxides and peracids. Incompatible with aluminum, zinc, galvanized iron, lead, nickel, or copper and its alloys. Violent reaction occurs with potassium and tert-butoxide. It may react with moisture to yield sulfuric acid, which subsequently may react with a metal container to liberate hydrogen gas, resulting in an explosion.

FIRE INFORMATION

Combustion products include toxic sulfur oxides. Vapors in confined areas may explode when exposed to fire. Storage containers and parts of containers may rocket great distances, in many directions. If material or contaminated runoff enters waterways, notify downstream users of potentially contaminated waters. Notify local health and fire officials and pollution control agencies.

NFPA Hazard Classifications:

Health Hazards (Blue) 3
Flammability (Red) 1
Reactivity (Yellow) 1

Flash Point: 235°F (113°C) o.c.; (220°F (104°C) c.c
Flammable Limits in Air: LEL 4.1%; UEL NA
Autoignition Temperature: 817°F (436°C)

FIRST AID

Move victim to fresh air. Call emergency medical care. Apply artificial respiration if victim is not breathing. Do not use mouth-to-mouth method if victim ingested or inhaled the substance; induce artificial respiration with the aid of a pocket mask equipped with a one-way valve or other proper respiratory medical device. Administer oxygen if breathing is difficult. Remove and isolate contaminated clothing and shoes. In case of contact with substance, immediately flush skin or eyes with running water for at least 20 minutes. For minor skin contact, avoid spreading material on unaffected skin. Keep victim warm and quiet. Effects of exposure (inhalation, ingestion, or skin contact) may be delayed. Ensure that medical personnel are aware of the material(s) involved and take precautions to protect themselves.

ENVIRONMENTAL INFORMATION

Clean Air Act: Hazardous Air Pollutants (Title I, Part A, Section 112)
EPCRA Section 304 Reportable Quantity (RQ): CERCLA, 10 lb (4.54 kg)
EPCRA Section 313: Form R *de minimis* concentration reporting level: 0.1%.
California's Proposition 65: Carcinogen.

1,3-DIETHYL THIOUREA D:264

SYNONYMS: N,N'-DIETHYLTHIOCARBAMIDE; N,N'-DIETHYLTHIOUREA; 1,3-DIETHYL-2-THIOUREA; 1,3-DIETILTIOUREA (Spanish); 1,3-DIETHYL-2-THIOHARNSTOFF (German); EINECS No. 203-308-5; NCI-C03816; PENNZONE E; THIATE H; THIOUREA, N,N'-DIETHYL-; U 15030; USAF EK-1803

IDENTIFICATION

CAS:	105-55-5
Molecular Formula:	$C_5H_{12}N_2S$
RTECS No.:	YS9625000; YS9800000
Properties:	Solid. Light tan color. Soluble in water.

HEALTH AND SAFETY INFORMATION

NTP: Evidence of animal carcinogenicity (rat)

RECOMMENDATIONS FOR RESPIRATOR SELECTION:

At any concentrations above the NIOSH REL, or where there is no REL, at any detectable concentration: SCBAF:PD,PP (any self-contained breathing apparatus that has a full facepiece and is operated in a pressure-demand or other positive-pressure mode); or SAF:PD,PP:ASCBA (any supplied-air respirator that has a full facepiece and is operated in a pressure-demand or other positive-pressure mode in combination with an auxiliary, self-contained breathing apparatus operated in a pressure-demand or other positive-pressure mode).

Escape: GMFOV [any air-purifying, full-facepiece respirator (gas mask) with a chin-style, front- or back-mounted organic vapor canister]; or SCBAE (any appropriate escape-type, self-contained breathing apparatus).

FIRE INFORMATION

Combustion products include toxic sulfur and nitrogen oxides. If material or contaminated runoff enters waterways, notify downstream users of potentially contaminated waters. Notify local health and fire officials and pollution control agencies.

NFPA Hazard Classifications: Not listed

FIRST AID

Move victim to fresh air. Call emergency medical care. Apply artificial respiration if victim is not breathing. Administer oxygen if breathing is difficult. Remove and isolate contaminated clothing and shoes. In case of contact with substance, immediately flush skin or eyes with running water for at least 20 minutes. Wash skin with soap and water. Keep victim warm and quiet. Effects of exposure (inhalation, ingestion, or skin contact) to substance may be delayed. Ensure that medical personnel are aware of the material(s) involved and take precautions to protect themselves.

DIETHYLUREA D:268

SYNONYMS: 1,1-DIETHYLHARNSTOFF (German); asym-DIETHYLUREA; 1,1-DIETHYLUREA; n,n-DIETHYLUREA; DIETILUREA (Spanish); EINECS No. 211-220-3; UREA, 1,1-DIETHYL-; UREA,n,n-DIETHYL-

IDENTIFICATION

CAS:	634-95-7
Molecular Formula:	$C_5H_{12}N_2O$
RTECS No.:	YS9353900

HEALTH AND SAFETY INFORMATION

RECOMMENDATIONS FOR RESPIRATOR SELECTION

At any concentrations above the NIOSH REL, or where there is no REL, at any detectable concentration: SCBAF:PD,PP (any self-contained breathing apparatus that has a full facepiece and is operated in a pressure-demand or other positive-pressure mode); or SAF:PD,PP:ASCBA (any supplied-air respirator that has a full facepiece and is operated in a pressure-demand or other positive-pressure mode in combination with an auxiliary, self-contained breathing apparatus operated in a pressure-demand or other positive-pressure mode).

Escape: GMFOV [any air-purifying, full-facepiece respirator (gas mask) with a chin-style, front- or back-mounted organic vapor canister]; or SCBAE (any appropriate escape-type, self-contained breathing apparatus).

CHEMICAL REACTIVITY

Incompatible, and may form explosive mixtures with strong oxidizers, hypochlorites, ammonium nitrate, chromyl chloride, nitrosyl perchlorate, perchloric acid.

FIRE INFORMATION

If material or contaminated runoff enters waterways, notify downstream users of potentially contaminated waters. Notify local health and fire officials and pollution control agencies. Use dry chemicals, alcohol-resistant foam, carbon dioxide.

NFPA Hazard Classifications: Not listed

FIRST AID

Move victim to fresh air. Call emergency medical care. Apply artificial respiration if victim is not breathing. Administer oxygen if breathing is difficult. Remove and isolate contaminated clothing and shoes. In case of contact with substance, immediately flush skin or eyes with running water for at least 20 minutes. Wash skin with soap and water. Keep victim warm and quiet. Effects of exposure (inhalation, ingestion, or skin contact) to substance may be delayed. Ensure that medical personnel are aware of the material(s) involved and take precautions to protect themselves.

DIETHYL ZINC — D:269

SYNONYMS: DIETILZINC (Spanish); EINECS No. 209-161-3; ETHYLZINC; ZINC ETHIDE; ZINC ETHYL; ZINC DIETHYL

IDENTIFICATION

CAS:	557-20-0
ERG ID:	UN 1366
ERG Guide:	135
ERG Guide Category:	SUBSTANCES-SPONTANEOUSLY COMBUSTIBLE
Molecular Formula:	$C_4H_{10}Zn$
RTECS No.:	ZH2070000
Properties:	Flammable, watery, mobile liquid. Colorless. Ignites when exposed to moist air; burns with a blue flame; garliclike vapor is produced. Decomposes in water.

HEALTH AND SAFETY INFORMATION

Corrosive to eyes, skin, and respiratory tract. May cause pulmonary edema, which can be delayed for several hours; there is a risk of death in serious cases.

OSHA Process Safety Management of Highly Hazardous Chemicals (29CFR, Part 1910.119, Appendix A): Threshold Quantity: 10,000 pounds.

RECOMMENDATIONS FOR RESPIRATOR SELECTION

At any concentrations above the NIOSH REL, or where there is no REL, at any detectable concentration: SCBAF:PD,PP (any self-contained breathing apparatus that has a full facepiece and is operated in a pressure-demand or other positive-pressure mode); or SAF:PD,PP:ASCBA (any supplied-air respirator that has a full facepiece and is operated in a pressure-demand or other positive-pressure mode in combination with an auxiliary, self-contained breathing apparatus operated in a pressure-demand or other positive-pressure mode).

Escape: GMFOV [any air-purifying, full-facepiece respirator (gas mask) with a chin-style, front- or back-mounted organic vapor canister]; or SCBAE (any appropriate escape-type, self-contained breathing apparatus).

CHEMICAL REACTIVITY

Reacts violently with water, forming flammable ethane gas and heat. Thermally unstable; spontaneously ignites in moist air or on contact with oxidizers. Reacts, often violently, with many compounds. Will react with surface moisture, generating flammable ethane gas and heat.

FIRE INFORMATION

Do not use water, foam, or halogenatated extinguishing agents. Storage containers and parts of containers may rocket great distances, in many directions. If material or contaminated runoff enters waterways, notify downstream users of potentially contaminated waters. Notify local health and fire officials and pollution control agencies.

NFPA Hazard Classifications:

Health Hazards (Blue) 3
Flammability (Red) 4
Reactivity (Yellow) 3
Special Notice .... Water Reactive

Flash Point: Ignites spontaneously in air. Explosive decomposition above 248°F (120°C)

FIRST AID

Move victim to fresh air. Call emergency medical care. Apply artificial respiration if victim is not breathing. Administer oxygen if breathing is difficult. Remove and isolate contaminated clothing and shoes. In case of contact with substance, immediately flush skin or eyes with running water for at least 20 minutes. Keep victim warm and quiet. Ensure that medical personnel are aware of the material(s) involved and take precautions to protect themselves

DIISOBUTYL ADIPATE D:271

SYNONYMS: ADIPATO de DIISOBUTILO (Spanish); DIBA; EINECS No. 205-450-3; FTAFLEX DIBA; HEXANEDIOIC ACID, BIS(2-METHYLPROPYL) ESTER; ISOBUTYL ADIPATE

IDENTIFICATION

CAS: 141-04-8
Molecular Formula: $C_{14}H_{26}O_4$
Properties: Combustible liquid. Colorless. Odorless. Insoluble in water.

HEALTH AND SAFETY INFORMATION

May irritate eyes, skin, respiratory tract. Moderately toxic.

RECOMMENDATIONS FOR RESPIRATOR SELECTION

At any concentrations above the NIOSH REL, or where there is no REL, at any detectable concentration: SCBAF:PD,PP (any self-contained breathing apparatus that has a full facepiece and is operated in a pressure-demand or other positive-pressure mode); or SAF:PD,PP:ASCBA (any supplied-air respirator that has a full facepiece and is operated in a pressure-demand or other positive-pressure mode in combination with an auxiliary, self-contained breathing apparatus operated in a pressure-demand or other positive-pressure mode).

Escape: GMFOV [any air-purifying, full-facepiece respirator (gas mask) with a chin-style, front- or back-mounted organic vapor canister]; or SCBAE (any appropriate escape-type, self-contained breathing apparatus).

CHEMICAL REACTIVITY

Incompatible with strong oxidizers.

FIRE INFORMATION

Vapors are heavier than air and will collect in low areas. Vapors in confined areas may explode when exposed to fire. If material or contaminated runoff enters waterways, notify downstream users of potentially contaminated waters. Notify local health and fire officials and pollution control agencies. Use dry chemicals, alcohol-resistant foam, carbon dioxide.

NFPA Hazard Classifications: Not listed

FIRST AID

Move victim to fresh air and call emergency medical care. Apply artificial respiration if victim is not breathing. Do not use mouth-to-mouth method if victim swallowed or inhaled the substance; induce artificial respiration with the aid of a pocket mask equipped with a one-way valve or other proper respiratory medical device. If breathing is difficult, give oxygen. In case of contact with substance, immediately flush skin or eyes with running water for at least 20 minutes. Remove and isolate contaminated clothing and shoes. Keep victim warm and quiet. Effects of exposure (inhalation, ingestion, or skin contact) to substances may be delayed. Ensure that medical personnel are aware of the material(s) involved and take precautions to protect themselves.

DIISOBUTYLENE D:272

SYNONYMS: DIISOBUTILENO (Spanish); EEC No. 601-031-00-8; EINECS No. 246-690-9; ISOOCTENE; 2,4,4-TRIMETHYL-1-PENTENE; 2,4,4-TRIMETHYLPENTENE-1; 2,4,4-TRIMETHYLPENTENE-2

IDENTIFICATION

CAS:	25167-70-8
ERG ID:	UN 2050
ERG Guide:	127
ERG Guide Category:	FLAMMABLE LIQUIDS (POLAR/WATER- MISCIBLE)
Molecular Formula:	C_8H_{16}
RTECS No.:	SB2717000
Properties:	Flammable liquid. Colorless. Gasolinelike odor. Floats on water.

HEALTH AND SAFETY INFORMATION

Causes eye irritation. Inhalation can cause coughing, dizziness, headache, sleepiness.

RECOMMENDATIONS FOR RESPIRATOR SELECTION

At any concentrations above the NIOSH REL, or where there is no REL, at any detectable concentration: SCBAF:PD,PP (any self-contained breathing apparatus that has a full facepiece and is operated in a pressure-demand or other positive-pressure mode); or SAF:PD,PP:ASCBA (any supplied-air respirator that has a full facepiece and is operated in a pressure-demand or other positive-pressure mode in combination with an auxiliary, self-contained breathing apparatus operated in a pressure-demand or other positive-pressure mode).

Escape: GMFOV [any air-purifying, full-facepiece respirator (gas mask) with a chin-style, front- or back-mounted organic vapor canister]; or SCBAE (any appropriate escape-type, self-contained breathing apparatus).

CHEMICAL REACTIVITY

Forms explosive mixture with air. Contact with oxidizers may cause fire and explosions. May polymerize; peroxides may accumulate upon prolonged storage in presence of air. May accumulate static electricity and cause ignition of its vapors.

FIRE INFORMATION

Vapors are heavier than air and will collect in low areas. Vapors in confined areas may explode when exposed to fire. Vapors may travel long distances to ignition sources and flash back. Storage containers and parts of containers may rocket great distances, in many directions.If material or contaminated runoff enters waterways, notify downstream users of potentially contaminated waters. Notify local health and fire officials and pollution control agencies. Use dry chemicals, carbon dioxide, dry chemicals, AFFF.

NFPA Hazard Classifications:

Health Hazards (Blue) 1
Flammability (Red) 3
Reactivity (Yellow) 0

Flash Point: 23°F (-5°C) c.c.
Flammable Limits in Air: LEL 0.8%; UEL 4.8%
Autoignition Temperature: 736°F (391°C)

FIRST AID

Move victim to fresh air. Call emergency medical care. Apply artificial respiration if victim is not breathing. Administer oxygen if breathing is difficult. Remove and isolate contaminated clothing and shoes. In case of contact with substance, immediately flush skin or eyes with running water for at least 20 minutes. Wash skin with soap and water. Keep victim warm and quiet. Ensure that medical personnel are aware of the material(s) involved and take precautions to protect themselves.

DIISOBUTYL KETONE D:274

SYNONYMS: DiBk; DIBK; DIISOBUTILCETONA (Spanish); DIISOBUTILCHETONE (Italian); DI-ISOBUTYLCETONE (French); DIISOBUTYLKETON (Dutch, German); DIISIPROPYL-ACETONE; 2,6-DIMETHYL-HEPTAN-4-ON (Dutch); 2,6-DIMETHYL-HEPTAN-4-ON (German); 2,6-DIMETHYLHEPTAN-4-ONE; 2,6-DIMETHYL-4-HEPTANONE; 2,6-DIMETHYLHEPTANONE; 2,6-DIMETIL-EPTAN-4-ONE (Italian); EEC No. 606-005-00-X; EINECS No. 203-620-1; ISOBUTYL KETONE; ISOVALERONE; 4-HEPTANONE,2,6-DIMETHYL-; VALERONE; 2,6-DIMETHYL-4-HEPTANE; 5-DIISOPROPYLACETONE; ISOVALERONE

IDENTIFICATION

CAS:	108-83-8
ERG ID:	UN 1157
ERG Guide:	127
ERG Guide Category:	FLAMMABLE LIQUIDS (POLAR/WATER- MISCIBLE)
Molecular Formula:	$C_9H_{18}O$
RTECS No.:	MJ5775000
Properties:	Combustible oily liquid. Colorless. Mild, sweet, peppermintlike odor. Floats on water; very slightly soluble.

HEALTH AND SAFETY INFORMATION

Irritates eyes, nose, and respiratory tract. Prolonged skin contact can cause dermatitis and other skin disorders.

IDLH: 500 ppm
Odor Threshold: 0.339 ppm
OSHA Table Z-1-A Air Contaminant
NIOSH Criteria Document: No. 78-173 KETONES

Exposure Limits:	**TWA**
ACGIH:	25 ppm; 145 mg/m^3
NIOSH:	25 ppm; 150 mg/m^3
OSHA:	50 ppm; 290 mg/m^3

RECOMMENDATIONS FOR RESPIRATOR SELECTION: NIOSH/OSHA

500 ppm: SA:CF (any supplied-air respirator operated in a continuous-flow mode); or PAPROV (any powered, air-purifying respirator with organic vapor cartridge(s)); or CCRFOV (any air-purifying, full-facepiece respirator (gas mask) with a chin-style, front- or back-mounted acid gas canister); or GMFOV [any air-purifying, full-facepiece respirator (gas mask) with a chin-style, front- or back-mounted organic vapor canister]; or SCBAF (any self-contained breathing apparatus with a full facepiece); or SAF (any supplied-air respirator with a full facepiece)

Emergency or planned entry into unknown concentrations or IDLH conditions: SCBAF:PD,PP (any self-contained breathing apparatus that has a full facepiece and is operated in a pressure-demand or other positive-pressure mode); or SAF:PD,PP:ASCBA (any supplied-air respirator that has a full facepiece and is operated in a pressure-demand or other positive-pressure mode in combination with an auxiliary, self-contained breathing apparatus operated in a pressure-demand or other positive-pressure mode.

Escape: GMFOV [any air-purifying, full-facepiece respirator (gas mask) with a chin-style, front- or back-mounted organic vapor canister]; or SCBAE (any appropriate escape-type, self-contained breathing apparatus).

Note: Substance causes eye irritation or damage; eye protection needed.

CHEMICAL REACTIVITY

Forms explosive mixture with air. Reacts with strong oxidizers, sulfuric acid, nitric acid, alkanolamines. May attack various forms of plastics and coatings. May accumulate static electricity, and may cause ignition of its vapors.

FIRE INFORMATION

Vapors are heavier than air and will collect in low areas. Vapors in confined areas may explode when exposed to fire. Vapors may travel long distances to ignition sources and flash back. Storage containers and parts of containers may rocket great distances, in many directions. If material or contaminated runoff enters waterways, notify downstream users of potentially contaminated waters. Notify local health and fire officials and pollution control agencies. Use dry chemicals, foam, carbon dioxide. Cool containers with water.

NFPA Hazard Classifications:

Health Hazards (Blue) 1
Flammability (Red) 2
Reactivity (Yellow) 0

Flash Point: 131°F (55°C) o.c.; 120°F (49°C) c.c.
Flammable Limits in Air: LEL 0.8%; UEL 7.1%, both @ 200°F (93°C)
Autoignition Temperature: 745°F (396°C)

FIRST AID

Move victim to fresh air. Call emergency medical care. Apply artificial respiration if victim is not breathing. Administer oxygen if breathing is difficult. Remove and isolate contaminated clothing and shoes. In case of contact with substance, immediately flush skin or eyes with running water for at least 20 minutes. Wash skin with soap and water. Keep victim warm and quiet. Ensure that medical personnel are aware of the material(s) involved and take precautions to protect themselves.

DIISOBUTYL PHTHALATE D:276

SYNONYMS: 1,2-BENZENE DICARBOXYLIC ACID, DI-(2-METHYLPROPYL)ESTER; 1,2-BENZENE DICARBOXYLIC ACID, 2-ETHOXY-2-OXOETHYL ETHYL ESTER; DISB; EINECS No. 201-553-2; FTALATO de DIISOBUTIL (Spanish); HEXAPLAS M/B; HEXAPLAS M-1B; ISOBUTYL PHTHALATE; PHTALATE de DIISOBUTYLE (French); PHTHALIC ACID, DI-ISOBUTYL ESTER; PALATINOL-IC

IDENTIFICATION

CAS:	84-69-5
ERG ID:	NA
Molecular Formula:	$C_{16}H_{22}O_4$
RTECS No.:	TI1225000
Properties:	Combustible, oily liquid. Colorless. Slight ester odor. Sinks slowly in water; soluble.

HEALTH AND SAFETY INFORMATION

RECOMMENDATIONS FOR RESPIRATOR SELECTION:

At any concentrations above the NIOSH REL, or where there is no REL, at any detectable concentration: SCBAF:PD,PP (any self-contained breathing apparatus that has a full facepiece and is operated in a pressure-demand or other positive-pressure mode); or SAF:PD,PP:ASCBA (any supplied-air respirator that has a full facepiece and is operated in a pressure-demand or other positive-pressure mode in combination with an auxiliary, self-contained breathing apparatus operated in a pressure-demand or other positive-pressure mode).

Escape: GMFOV [any air-purifying, full-facepiece respirator (gas mask) with a chin-style, front- or back-mounted organic vapor canister]; or SCBAE (any appropriate escape-type, self-contained breathing apparatus).

CHEMICAL REACTIVITY

Reacts with oxidizers, nitrates, strong acids.

FIRE INFORMATION

Decomposition products upon heating include toxic gases. Vapors are heavier than air and will collect in low areas. Vapors in confined areas may explode when exposed to fire. Storage containers and parts of containers may rocket great distances, in many directions. If material or contaminated runoff enters waterways, notify downstream users of potentially contaminated waters. Notify local health and fire officials and pollution control agencies. Use dry chemicals, foam, carbon dioxide. Cool containers with water spray.

NFPA Hazard Classifications:

Health Hazards (Blue) 0
Flammability (Red) 1
Reactivity (Yellow) 0

Flash Point: 365°F (185°C) o.c.; 385°F (196°C) c.c.
Flammable Limits in Air: LEL 0.4%; UEL NA
Autoignition Temperature: 810°F (432°C)

FIRST AID

Move victim to fresh air. Call emergency medical care. Apply artificial respiration if victim is not breathing. Administer oxygen if breathing is difficult. Remove and isolate contaminated clothing and shoes. In case of contact with substance, immediately flush skin or eyes with running water for at least 20 minutes. Wash skin with soap and water. Keep victim warm and quiet. Ensure that medical personnel are aware of the material(s) involved and take precautions to protect themselves.

DIISOOCTYL PHTHALATE D:278

SYNONYMS: 1,2-BENZENEDICARBOXYLIC ACID; 1,2-BENZENEDICARBOXYLIC ACID, DIISOOCTYL ESTER; BIS (6-METHYLHEPTYL) PHTHALATE; BIS (6-METHYLHEPTYL) ESTER OF PHTHALIC ACID; CORFLEX 880; DIOP; DI (6-METHYLHEPTYL) PHTHALATE; EINECS No. 248-523-5; FTALATO de DIOCTILO (Spanish); FLEXOL PLASTICIZER P; HEXAPLAS DIOP; HEXAPLAS M/O; ISOOCTYL PHTHALATE

IDENTIFICATION

CAS:	27554-26-3
ERG ID:	NA
Molecular Formula:	$C_{24}H_{38}O_4$
RTECS No.:	TI1300000
Properties:	Combustible, oily liquid. Colorless. Slight odor. Floats on water; insoluble.

HEALTH AND SAFETY INFORMATION

RECOMMENDATIONS FOR RESPIRATOR SELECTION

At any concentrations above the NIOSH REL, or where there is no REL, at any detectable concentration: SCBAF:PD,PP (any self-contained breathing apparatus that has a full facepiece and is operated in a pressure-demand or other positive-pressure mode); or SAF:PD,PP:ASCBA (any supplied-air respirator that has a full facepiece and is operated in a pressure-demand or other positive-pressure mode in combination with an auxiliary, self-contained breathing apparatus operated in a pressure-demand or other positive-pressure mode).

Escape: GMFOV [any air-purifying, full-facepiece respirator (gas mask) with a chin-style, front- or back-mounted organic vapor canister]; or SCBAE (any appropriate escape-type, self-contained breathing apparatus).

CHEMICAL REACTIVITY

Reacts with strong oxidizers, nitrates ,and strong acids.

FIRE INFORMATION

Vapors are heavier than air and will collect in low areas. Vapors in confined areas may explode when exposed to fire.If material or contaminated runoff enters waterways, notify downstream users of potentially contaminated waters. Notify local health and fire officials and pollution control agencies. Use dry chemicals, alcohol-resistant foam, carbon dioxide.

NFPA Hazard Classifications:

Health Hazards (Blue) 0
Flammability (Red) 1
Reactivity (Yellow) 0

Flash Point: 450°F (232°C) c.c.

FIRST AID

Move victim to fresh air. Call emergency medical care. Apply artificial respiration if victim is not breathing. Administer oxygen if breathing is difficult. Remove and isolate contaminated clothing and shoes. In case of contact with substance, immediately flush skin or eyes with running water for at least 20 minutes. Wash skin with soap and water. Keep victim warm and quiet. Ensure that medical personnel are aware of the material(s) involved and take precautions to protect themselves.

DIMETHYL ACETAMIDE D:280

SYNONYMS: ACETAMIDE N,N-DIMETHYL; ACETDIMETHYLAMIDE; ACETIC ACID DIMETHYLAMIDE; ACETYLDIMETHYLAMINE; N,N-DIMETHYL ACETAMIDE; DIMETHYLACETAMIDE; DIMETHYLACETONE AMIDE; DIMETHYLAMIDE ACETATE; DIMETILACETAMIDA (Spanish); DMA; DMAC; EEC No. 616-011-00-4; NSC 3138; U-5954; PIQ 13

IDENTIFICATION

CAS:	127-19-5
ERG ID:	UN 1993

ERG Guide: 128
ERG Category Guide: FLAMMABLE LIQUIDS (NON-POLAR/WATER-IMMISCIBLE)
Molecular Formula: C_4H_9NO
RTECS No.: AB7700000
Properties: Very flammable, oily liquid. Colorless. Weak ammonia or fishlike odor. Mixes with water (initially floats on surface). Vapors are heavier than air; will collect in low areas. Freezes at -4°F (-20°C).

HEALTH AND SAFETY INFORMATION

Long-term or chronic exposure may cause liver and kidney damage; brain damage, including depression, lethargy, hallucinations, emotionally detached feelings, and personality changes.

IDLH: 300 ppm
Odor Threshold: 47.9 ppm

Limits:	TWA
ACGIH: (1)	10 ppm; 36 mg/m^3
NIOSH: (1)	10 ppm; 35 mg/m^3
OSHA: (1)	10 ppm; 35 mg/m^3

(1) Skin contact contributes significantly to overall exposure.

RECOMMENDATIONS FOR RESPIRATOR SELECTION: NIOSH/OSHA

100 ppm: SA (any supplied-air respirator).
250 ppm: SA:CF (any supplied-air respirator operated in a continuous-flow mode).
400 ppm: SCBAF (any self-contained breathing apparatus with a full facepiece); or SAF (any supplied-air respirator with a full facepiece).

Emergency or Planned Entry into Unknown Concentrations or IDLH Conditions: SCBAF:PD,PP (any self-contained breathing apparatus that has a full facepiece and is operated in a pressure-demand or other positive-pressure mode); or SAF:PD,PP:ASCBA (any supplied-air respirator that has a full facepiece and is operated in a pressure-demand or other positive-pressure mode in combination with an auxiliary, self-contained breathing apparatus operated in a pressure-demand or other positive-pressure mode.

Escape: GMFOV (any air-purifying, full-facepiece respirator (gas mask) with a chin-style, front- or back-mounted organic vapor canister; or SCBAE (any appropriate escape-type, self-contained breathing apparatus).
Note: Substance causes eye irritation or damage; eye protection needed.

CHEMICAL REACTIVITY

Forms explosive mixture with air. Incompatible with nonoxidizing mineral acids, strong acids, ammonia, isocyanates, phenols, cresols, halogenated compounds above 185°F (85°C). Attacks some plastics, rubber, and coatings.

FIRE INFORMATION

Vapors are heavier than air and will collect in low areas. Vapors in confined areas may explode when exposed to fire. Vapors may travel long distances to ignition sources and flash back. Storage containers and parts of containers may rocket great distances, in many directions. If material or contaminated runoff enters waterways, notify downstream users of potentially contaminated waters. Notify local health and fire officials and pollution control agencies. Use dry chemicals, foam, carbon dioxide.

NFPA Hazard Classifications:

Health Hazards (Blue) 2
Flammability (Red) 2
Reactivity (Yellow) 0

Flash Point: 158°F (70°C) o.c.
Flammable Limits in Air: LEL 1.8% @ 212°F (100°C); UEL 11.5% @ 320°F (160°C)
Autoignition Temperature: 914°F (490°C)

FIRST AID

Move victim to fresh air. Call emergency medical care. Apply artificial respiration if victim is not breathing. If breathing is difficult, give oxygen. Remove and isolate contaminated clothing and shoes. In case of contact with substance, immediately flush skin or eyes with running water for at least 20 minutes. Wash skin with soap and water. Keep victim warm and quiet. Ensure that medical personnel are aware of the material(s) involved and take precautions to protect themselves.

3–DIMETHYLAMINOPROPYLAMINE D:281

SYNONYMS: 1-AMINO-3-DIMETHYLAMINOPROPANE; n,n-DIMETHYL-N-(3-AMINOPROPYL)AMINE; n,n-DIMETHYL-1,3-DIAMINOPROPANE; n,n-DIMETHYL-1,3-PROPANEDIAMINE; n,n-DIMETHYL-1,3-PROPYLENEDIAMINE; 3-(DIMETHYLAMINO)PROPYLAMINE; 3-DIMETILAMINOPROPILAMINA (Spanish); EEC No. 612-061-00-6; EINECS No. 203-680-9; 1,3-PROPANEDIAMINE, N,N-DIMETHYL-

IDENTIFICATION

CAS:	109-55-7
ERG ID:	UN 2734 as amines, liquid, corrosive, flammable, n.o.s
ERG Guide:	132
ERG Guide Category:	FLAMMABLE LIQUIDS-CORROSIVE
Molecular Formula:	$C_5H_{14}N_2$
RTECS No.:	TX7525000
Properties:	Liquid. Colorless. Characteristic amine odor. Soluble in water.

HEALTH AND SAFETY INFORMATION

Irritates eyes, skin, and respiratory tract. Inhalation may cause pulmonary edema, which can be delayed for several hours; there is a risk of death in serious cases. Repeated or prolonged contact can cause skin, lung, kidney, and liver damage.

RECOMMENDATIONS FOR RESPIRATOR SELECTION: NIOSH

At any concentrations above the NIOSH REL, or where there is no REL, at any detectable concentration: SCBAF:PD,PP (any self-contained breathing apparatus that has a full facepiece and is operated in a pressure-demand or other positive-pressure mode); or SAF:PD,PP:ASCBA (any supplied-air respirator that has a full facepiece and is operated in a pressure-demand or other positive-pressure mode in combination with an auxiliary, self-contained breathing apparatus operated in a pressure-demand or other positive-pressure mode).

Escape: GMFOV [any air-purifying, full-facepiece respirator (gas mask) with a chin-style, front- or back-mounted organic vapor canister]; or SCBAE (any appropriate escape-type, self-contained breathing apparatus).

CHEMICAL REACTIVITY

Forms explosive mixture with air. Reacts vigorously with oxidizers, organic anhydrides, isocyanates, aldehydes, strong acids, vinyl acetates, acrylates, substituted allyls, alkylene oxides, epichlorohydrin, ketones, alcohols, glycols, phenols, cresols, caprolactum solution. Contact with ethylene dichloride produces explosive acetylene gas. May ignite on contact with cellulose nitrate of high surface area.

FIRE INFORMATION

Vapors are heavier than air and will collect in low areas. Vapors in confined areas may explode when exposed to fire. Vapors may travel long distances to ignition sources and flash back. Storage containers and parts of containers may rocket great distances, in many directions.If material or contaminated runoff enters waterways, notify downstream users of potentially contaminated waters. Notify local health and fire officials and pollution control agencies. Combustion products include toxic nitrogen oxides and combustible gases. Use dry chemicals, alcohol-resistant foam, carbon dioxide. Use water spray to cool containers.

NFPA Hazard Classifications:

Health Hazards (Blue) 3
Flammability (Red) 2
Reactivity (Yellow) 0

Flash Point: 100°F (38°C) o.c.

FIRST AID

Move victim to fresh air; call emergency medical care. Apply artificial respiration if victim is not breathing. Do not use mouth-to-mouth method if victim ingested or inhaled the substance; induce artificial respiration with the aid of a pocket mask equipped with a one-way valve or other proper respiratory medical device. Administer oxygen if breathing is difficult. Remove and isolate contaminated clothing and shoes. In case of contact with substance, immediately flush skin or eyes with running water for at least 20 minutes. For minor skin contact, avoid spreading material on unaffected skin. Keep victim warm and quiet. Effects of exposure (inhalation, ingestion, or skin contact) to substance may be delayed. Ensure that medical ersonnel are aware of the material(s) involved and take precautions to protect themselves.

N,N-DIMETHYLANILINE D:282

SYNONYMS: ANILINE, N,N-DIMETHYL-; BENZENAMINE, N,N-DIMETHYL-; (DIMETHYLAMINO)BENZENE; DIMETHYLANILINE; N,N-DIMETHYLBENZENEAMINE; DIMETHYLPHENYLAMINE; N,N-DIMETHYLPHENYLAMINE; N,N-DIMETILANILINA (Spanish); DWUMETYLOANILINA (Polish); EEC No. 612-016-00-0; EINECS No. 204-493-5; NCI-C56428

IDENTIFICATION

CAS:	121-69-7
ERG ID:	UN 2253
ERG Guide:	152
ERG Guide Category:	SUBSTANCES-TOXIC (COMBUSTIBLE)
Molecular Formula:	$C_8H_{11}N$
RTECS No.:	BX4725000
Properties:	Very flammable, oily liquid. Pale yellow. Aminelike odor. Floats on water; almost insoluble in water. A solid below 36°F (2°C).

HEALTH AND SAFETY INFORMATION

Causes cyanosis, weakness, dizziness, anoxia, ataxia. May interfere with the body's ability to use oxygen.

IDLH: 504 mg/m^3 (100 ppm).

OSHA Table Z-1-A Air Contaminant

Exposure Limits:	**TWA**	**STEL**
ACGIH:*	5 ppm; 25 mg/m^3	10 ppm; 50 mg/m^3
NIOSH:*	5 ppm; 25 mg/m^3	10 ppm; 50 mg/m^3
OSHA:*	5 ppm; 25 mg/m^3	

*Skin contact contributes significantly in overall exposure.

RECOMMENDATIONS FOR RESPIRATOR SELECTION: NIOSH/OSHA

50 ppm: SA (any supplied-air respirator).
100 ppm: SA:CF (any supplied-air respirator operated in a continuous-flow mode); or SCBAF (any self-contained breathing apparatus with a full facepiece); or SAF (any supplied-air respirator with a full facepiece).

Emergency or Planned Entry into Unknown Concentrations or IDLH Conditions: SCBAF:PD,PP (any self-contained breathing apparatus that has a full facepiece and is operated in a pressure-demand or other positive-pressure mode); or SAF:PD,PP:ASCBA (any supplied-air respirator that has a full facepiece and is operated in a pressure-demand or other positive-pressure mode in combination with an auxiliary, self-contained breathing apparatus operated in a pressure-demand or other positive-pressure mode).

Escape: GMFOV [any air-purifying, full-facepiece respirator (gas mask) with a chin-style, front- or back-mounted organic vapor canister]; or SCBAE (any appropriate escape-type, self-contained breathing apparatus).

CHEMICAL REACTIVITY

Forms explosive mixture with air. Contact with strong acids may cause fire and explosions. Contact with hypochlorite bleaches form explosive chloroamines. Incompatible with anhydrides, isocyanates, and aldehydes.

FIRE INFORMATION

Combustion products include toxic nitrogen oxides. Fire fighting gear, including SCBA, does not provide adequate protection. If exposure occurs, remove and isolate gear immediately and thoroughly decontaminate personnel. Vapors are heavier than air and will collect in low areas. Vapors in confined areas may explode when exposed to fire. Vapors may travel long distances to ignition sources and flash back. Storage containers and parts of containers may rocket great distances, in many directions. If material or contaminated runoff enters waterways, notify downstream users of potentially contaminated waters. Notify local health and fire officials and pollution control agencies. Use dry chemicals, foam, carbon dioxide.

NFPA Hazard Classifications:

Health Hazards (Blue) 3
Flammability (Red) 2
Reactivity (Yellow) 0

Flash Point: 142°F (61°C)
Autoignition Temperature: 700°F (371°C)

FIRST AID

Move victim to fresh air. Call emergency medical care. Apply artificial respiration if victim is not breathing. Do not use mouth-to-mouth method if victim ingested or inhaled the substance; induce artificial respiration with the aid of a pocket mask equipped with a one-way valve or other proper respiratory medical device. Administer oxygen if breathing is difficult. Remove and isolate contaminated clothing and shoes. In case of contact with substance, immediately flush skin or eyes with running water for at least 20 minutes. For minor skin contact, avoid spreading material on unaffected skin Keep victim warm and quiet. Effects of exposure (inhalation, ingestion, or skin contact) may be delayed. Ensure that medical personnel are aware of the material(s) involved and take precautions to protect themselves.

ENVIRONMENTAL INFORMATION

Clean Air Act: Hazardous Air Pollutants (Title I, Part A, Section 112)
EPCRA Section 304: Reportable Quantity (RQ): CERCLA, 1 lb (0.454 kg)
EPCRA Section 313: Form R *de minimis* concentration reporting level: 1.0%

3,3'–DIMETHYL–4,4'–BIPHENYL DIISOCYANATE D:283

SYNONYMS: 1,1'-BIPHENYL,4,4'-DIISOCYANATO-3,3'-DIMETHYL-; 4,4'-DIISOCYANATO-3,3'-DIMETHYL-1,1'-BIPHENYL; EINECS No. 202-112-7; ISOCYANIC ACID, 3,3'-DIMETHYL-4,4'-BIPHENYLENE ESTER

IDENTIFICATION

CAS:	91-97-4
ERG ID:	NA
ERG Guide:	171
Molecular Formula:	$C_{16}H_{12}N_2O_2$
RTECS No.:	DV3960000
Properties:	Crystalline solid.

HEALTH AND SAFETY INFORMATION

Exposure Limits:	**TWA**	**CEILING** as diisocyanates
NIOSH:	0.005 ppm	0.02 ppm/10 min.

RECOMMENDATIONS FOR RESPIRATOR SELECTION

At any concentrations above the NIOSH REL, or where there is no REL, at any detectable concentration: SCBAF:PD,PP (any self-contained breathing apparatus that has a full facepiece and is operated in a pressure-demand or other positive-pressure mode); or SAF:PD,PP:ASCBA (any supplied-air respirator that has a full facepiece and is operated in a pressure-demand or other positive-pressure mode in combination with an auxiliary, self-contained breathing apparatus operated in a pressure-demand or other positive-pressure mode).

Escape: GMFOV [any air-purifying, full-facepiece respirator (gas mask) with a chin-style, front- or back-mounted organic vapor canister]; or SCBAE (any appropriate escape-type, self-contained breathing apparatus).

CHEMICAL REACTIVITY

Incompatible with acids, caustics, ammonia, aliphatic amines, alkanolamines, aromatic amines, amides, alcohols, glycols, caprolactum solution.

FIRE INFORMATION

Decomposition products upon heating include toxic gases. If material or contaminated runoff enters waterways, notify downstream users of potentially contaminated waters. Notify local health and fire officials and pollution control agencies.

NFPA Hazard Classifications: Not listed

FIRST AID

Move victim to fresh air. Call emergency medical care. Apply artificial respiration if victim is not breathing. Administer oxygen if breathing is difficult. Remove and isolate contaminated clothing and shoes. In case of contact with substance, immediately flush skin or eyes with running water for at least 20 minutes. Wash skin with soap and water. Keep victim warm and quiet. Ensure that medical personnel are aware of the material(s) involved and take precautions to protect themselves

ENVIRONMENTAL INFORMATION

EPCRA Section 313: Form R *de minimis* concentration reporting level: 1.0%. Form R Toxic Chemical Category Code: N120
TSCA: 40CFR 716.120(*d*)8 isocyanates

N,N–DIMETHYLCYCLOHEXLYAMINE D:285

SYNONYMS: CYCLOHEXYLAMINE, N,N-DIMETHYL; CYCLOHEXYLDIMETHYLAMINE; N-CYCLO-HEXYLDIMETHYLAMINE; N,N-DIMETHYLAMINOCYCLOHEXANE; N-DIMETHYLCYCLOHEXANAMINE; EINECS No. 202-715-5; POLYCAT-8

IDENTIFICATION

CAS:	98-94-2
ERG ID:	UN 2264
ERG Guide:	132
ERG Guide Category:	FLAMMABLE LIQUIDS-CORROSIVE
Molecular Formula:	$C_8H_{17}N$
RTECS No.:	IX0175000
Properties:	Combustible liquid. Colorless. Musky, ammonia odor. Floats and mixes slowly with water.

HEALTH AND SAFETY INFORMATION

Irritates eyes and respiratory tract. Strongly caustic. Severe eye irritant; can cause blindness. Skin contact can cause burns.

RECOMMENDATIONS FOR RESPIRATOR SELECTION

At any concentrations above the NIOSH REL, or where there is no REL, at any detectable concentration: SCBAF:PD,PP (any self-contained breathing apparatus that has a full facepiece and is operated in a pressure-demand or other positive-pressure mode); or SAF:PD,PP:ASCBA (any supplied-air respirator that has a full facepiece and is operated in a pressure-demand or other positive-pressure mode in combination with an auxiliary, self-contained breathing apparatus operated in a pressure-demand or other positive-pressure mode).

Escape: GMFOV [any air-purifying, full-facepiece respirator (gas mask) with a chin-style, front- or back-mounted organic vapor canister]; or SCBAE (any appropriate escape-type, self-contained breathing apparatus).

CHEMICAL REACTIVITY

Forms explosive mixture with air. Incompatible with acids, organic anhydrides, chlorinated hydrocarbons, isocyanates, vinyl acetate, acrylates, substituted allyls, alkylene oxides, epichlorohydrin, ketones, aldehydes, alcohols, glycols, phenols, cresols, caprolactam solution, strong oxidizers. May form explosive compounds with chlorine and hypochlorites, and perchloric acid. compounds. Attacks aluminum, copper, lead, tin, zinc, and alloys.

FIRE INFORMATION

Vapors are heavier than air and will collect in low areas. Vapors in confined areas may explode when exposed to fire. Vapors may travel long distances to ignition sources and flash back. Storage containers and parts of containers may rocket great distances, in many directions. If material or contaminated runoff enters waterways, notify downstream users of potentially contaminated waters. Notify local health and fire officials and pollution control agencies. Use dry chemicals, foam, carbon dioxide. Use water spray to cool containers.

Flash Point: 108°F (42°C) c.c.

FIRST AID

Move victim to fresh air; call emergency medical care. Apply artificial respiration if victim is not breathing. Do not use mouth-to-mouth method if victim ingested or inhaled the substance; induce artificial respiration with the aid of a pocket mask equipped with a one-way valve or other proper respiratory medical device. Administer oxygen if breathing is difficult. Remove and isolate contaminated clothing and shoes. In case of contact with substance, immediately flush skin or eyes with running water for at least 20 minutes. For minor skin contact, avoid spreading material on unaffected skin. Keep victim warm and quiet. Effects of exposure (inhalation, ingestion, or skin contact) to substance may be delayed. Ensure that medical personnel are aware of the material(s) involved and take precautions to protect themselves.

2,5-DIMETHYL-2,5-DI(tert-BUTYLPEROXY)HEXANE D:287

SYNONYMS: AZTEC® 2,5-DI; 2,5-DIMETHYL-2,5-DI(t-BUTYLPEROXY)HEXANE; EINECS No. 201-128-1; ESPERAL®120; LUPERCO 101-P20; LUPERSOL 101; PEROXIDE, (1,1,4,4-TETRAMETHYL-1,4-BUTANEDIYL)BIS(1,1-DIMETHYLETHYL); PEROXIDE, (1,1,4,4-TETRAMETHYLTETRAMETHYLENE)BIS(tert-BUTYL); POLYVEL CR-5F; TRIGONOX® 101-101/45; VAROX

IDENTIFICATION

CAS:	78-63-7
ERG ID:	UN 3105, see organic peroxide table
ERG Guide:	145
ERG Guide Category:	ORGANIC PEROXIDE TYPE D, LIQUID
Molecular Formula:	$C_{16}H_{34}O_4$
RTECS No.:	MO1835000
Properties:	Highly reactive, combustible liquid. Colorless to light yellow. Mixes with water.

HEALTH AND SAFETY INFORMATION

Irritating to the eyes, skin, and respiratory tract.

RECOMMENDATIONS FOR RESPIRATOR SELECTION:

At any concentrations above the NIOSH REL, or where there is no REL, at any detectable concentration: SCBAF:PD,PP (any self-contained breathing apparatus that has a full facepiece and is operated in a pressure-demand or other positive-pressure mode); or SAF:PD,PP:ASCBA (any supplied-air respirator that has a full facepiece and is operated in a pressure-demand or other positive-pressure mode in combination with an auxiliary, self-contained breathing apparatus operated in a pressure-demand or other positive-pressure mode).

Escape: GMFS [any air-purifying, full-facepiece respirator (gas mask) with a chin-style, front- or back-mounted canister providing protection against the compound of concern]; or SCBAE (any appropriate escape-type, self-contained breathing apparatus).

CHEMICAL REACTIVITY

A powerful oxidizer; highly reactive with many other materials. May be able to form unstable and dangerous peroxides in storage. Heat or contact with reducing agents, thiocyanates, organic matter, combustibles, or contamination may cause explosion. May attack some plastics, rubber, and coatings. Protect containers from shock, heat and friction.

FIRE INFORMATION

Firefighting gear (including SCBA) may not provide adequate protection. If exposure occurs, remove and isolate gear immediately and thoroughly decontaminate personnel. Will increase intensity of fire. Vapors are heavier than air and will collect in low areas. Vapors in confined areas may explode when exposed to fire. Vapors may travel long distances to ignition sources and flash back. Storage containers and parts of containers may rocket great distances, in many directions. If material or contaminated runoff enters waterways, notify downstream users of potentially contaminated waters. Runoff may create fire or explosion hazard. Notify local health and fire officials and pollution control agencies.

Flash Point: >180°F (>82°C) o.c.

FIRST AID

Move victim to fresh air. Call emergency medical care. Apply artificial respiration if victim is not breathing. Administer oxygen if breathing is difficult. Remove and isolate contaminated clothing and shoes. Remove material from skin immediately. In case of contact with substance, immediately flush skin or eyes with running water for at least 20 minutes. Keep victim warm and quiet. Ensure that medical personnel are aware of the material(s) involved and take precautions to protect themselves.

DIMETHYLDICHLOROSILANE D:289

SYNONYMS: DICHLORODIMETHYL SILANE; DIMETHYL-DICHLORSILAN (Czech); DICHLOROETHYLSILANE; DIMETILDICLOROSILANO (Spanish); DIMETHYLDICHLOROSILAN; DIMETHYL-DICHLOROSILAN (Czech); EEC No. 014-003-00-X; EINECS No. 200-901-0; INERTON-DMCS; INERTON DW-DMC; SILANE, DICHLORODIMETHYL-

IDENTIFICATION

CAS:	75-78-5
ERG ID:	UN 1162
ERG Guide:	155
ERG Guide Category:	SUBSTANCES-TOXIC AND/OR CORROSIVE (FLAMMABLE/WATER-SENSITIVE)
Molecular Formula:	$C_2H_6Cl_2Si$
RTECS No.:	VV3150000
Properties:	Highly flammable, fuming liquid. Clear, colorless. Sharp, irritating odor, like hydrochloric acid. Sinks in water and decomposes, releasing hydrochloric acid. Produces a large amount of vapor.

HEALTH AND SAFETY INFORMATION

Corrosive and poisonous. Severely irritates eyes, skin, and respiratory tract. Breathing the vapors can cause death, and skin and eye contact can cause burns and blindness.

OSHA Process Safety Management of Highly Hazardous Chemicals (29CFR, Part 1910.119, Appendix A): Threshold Quantity: 1,000 pounds.

RECOMMENDATIONS FOR RESPIRATOR SELECTION: NIOSH/OSHA as hydrogen chloride

50 ppm: CCRS (any chemical cartridge respirator with cartridge(s) providing protection against the compound of concern)); or GMFS [any air-purifying, full-facepiece respirator (gas mask) with a chin-style, front- or back-mounted canister providing protection against the compound of concern]; or PAPRS (any powered, air-purifying respirator with cartridge(s) providing protection against the compound of concern); or SA (any supplied-air respirator); or SCBAF (any self-contained breathing apparatus with a full facepiece).

Emergency or Planned Entry into Unknown Concentrations or IDLH Conditions: SCBAF:PD,PP (any self-contained breathing apparatus that has a full facepiece and is operated in a pressure-demand or other positive-pressure mode); or SAF:PD,PP:ASCBA (any supplied-air respirator that has a full facepiece and is operated in a pressure-demand or other positive-pressure mode in combination with an auxiliary, self-contained breathing apparatus operated in a pressure-demand or other positive-pressure mode).

Escape: GMFAG [any air-purifying, full-facepiece respirator (gas mask) with a chin-style, front- or back-mounted organic vapor canister]; or SCBAE (any appropriate escape-type, self-contained breathing apparatus).

Note: Substance reported to cause eye irritation or damage; may require eye protection.

CHEMICAL REACTIVITY

Forms explosive mixture with air. Reacts violently with acetone, amines, ammonia, alcohols. Contact with water and surface moisture causes vigorous decomposition, generating corrosive hydrogen chloride. Contact with air forms corrosive fumes of hydrochloric acid. Corrosively attacks common metals in the presence of moisture.

FIRE INFORMATION

Combustion products include toxic hydrogen chloride and phosgene. Firefighting gear (including SCBA) does not provide adequate protection. If exposure occurs, remove and isolate gear immediately and thoroughly decontaminate personnel. Vapors are heavier than air and will collect in low areas. Vapors in confined areas may explode when exposed to fire. Vapors may travel long distances to ignition sources and flash back. Storage containers and parts of containers may rocket great distances, in many directions. If material or contaminated runoff enters waterways, notify downstream users of potentially contaminated waters. Notify local health and fire officials and pollution control agencies. DO NOT use water on fire. Use dry chemicals, or low or medium expansion AFFF foam if available in sufficient amounts.

NFPA Hazard Classifications:

Health Hazards (Blue) 3
Flammability (Red) 3
Reactivity (Yellow) 1
Special Situations . . Water (decomposes)

Flash Point: <70°F (<21°C) o.c.
Flammable Limits in Air: LEL 3.4%; UEL >9.5%
Autoignition Temperature: 707°F (375°C)

FIRST AID

Move victim to fresh air; call emergency medical care. Apply artificial respiration if victim is not breathing. Do not use mouth-to-mouth method if victim ingested or inhaled the substance; induce artificial respiration with the aid of a pocket mask equipped with a one-way valve or other proper respiratory medical device. Administer oxygen if breathing is difficult. Remove and isolate contaminated clothing and shoes. In case of contact with substance, immediately flush skin or eyes with running water for at least 20 minutes. For minor skin contact, avoid spreading material on unaffected skin. Keep victim warm and quiet. Effects of exposure (inhalation, ingestion, or skin contact) to substance may be delayed. Ensure that medical personnel are aware of the material(s) involved and take precautions to protect themselves.

ENVIRONMENTAL INFORMATION

Clean Air Act: Section 112[r], Table 3, Accidental Release Prevention/Flammable Substances, TQ = 5,000 lb (2,270 kg)
EPCRA Section 302, Extremely Hazardous Substances: TPQ = 500 lb (227 kg)
EPCRA Section 304 Reportable Quantity (RQ): EHS, 500 lb (227 kg)
EPCRA Section 313: Form R *de minimis* concentration reporting level: 1.0%.

DIMETHYLETHANOLAMINE D:290

SYNONYMS: DEANOL; 2-(DIMETHYLAMINO) ETHANOL; β-DIMETHYLAMINOETHYL ALCOHOL; N,N-DIMETHYLETHANOLAMINE; N,N-DIMETHYL-N-(2-HYDROXYETHYL) AMINE; DIMETILETANOLAMINA (Spanish); DMAE; EEC No. 603-047-00-0; EINECS No. 203-542-8; β-HYDROXYETHYLDIMETHYLAMINE

IDENTIFICATION

CAS:	108-01-0
ERG ID:	UN 2051
ERG Guide:	132
ERG Guide Category:	FLAMMABLE LIQUIDS-CORROSIVE
Molecular Formula:	$C_4H_{11}NO$
RTECS No.:	KK6125000
Properties:	Combustible liquid. Colorless. Amine odor. Floats and mixes with water.

HEALTH AND SAFETY INFORMATION

Odor Threshold: 0.015 ppm detection; 0.045 ppm recognition

RECOMMENDATIONS FOR RESPIRATOR SELECTION

At any concentrations above the NIOSH REL, or where there is no REL, at any detectable concentration: SCBAF:PD,PP (any self-contained breathing apparatus that has a full facepiece and is operated in a pressure-demand or other positive-pressure mode); or SAF:PD,PP:ASCBA (any supplied-air respirator that has a full facepiece and is operated in a pressure-demand or other positive-pressure mode in combination with an auxiliary, self-contained breathing apparatus operated in a pressure-demand or other positive-pressure mode).

Escape: GMFOV [any air-purifying, full-facepiece respirator (gas mask) with a chin-style, front- or back-mounted organic vapor canister]; or SCBAE (any appropriate escape-type, self-contained breathing apparatus).

CHEMICAL REACTIVITY

Forms explosive mixture with air. Violent reaction with oxidizers and acids. Attacks copper and its alloys, galvanized steel, zinc and zinc alloys having more than 10% zinc by weight.

FIRE INFORMATION

Poisonous gases are produced in fire. Vapors are heavier than air and will collect in low areas. Vapors in confined areas may explode when exposed to fire. Vapors may travel long distances to ignition sources and flash back. Storage containers and parts of containers may rocket great distances, in many directions. If material or contaminated runoff enters waterways, notify downstream users of potentially contaminated waters. Notify local health and fire officials and pollution control agencies. Use dry chemicals, foam, carbon dioxide.

NFPA Hazard Classifications:

Health Hazards (Blue) 2
Flammability (Red) 2
Reactivity (Yellow) 0

Flash Point: 105°F (41°C) o.c.
Flammable Limits in Air: LEL 1.6%; UEL 11.9%
Autoignition Temperature: 563°F (295°C)

FIRST AID

Move victim to fresh air. Call emergency medical care. Apply artificial respiration if victim is not breathing. Do not use mouth-to-mouth method if victim swallowed or inhaled the substance; induce artificial respiration with the aid of a pocket mask equipped with a one-way valve or other proper respiratory medical device. Administer oxygen if breathing is difficult. Remove and isolate contaminated clothing and shoes. In case of contact with substance, immediately flush skin or eyes with running water for at least 20 minutes. Keep victim warm and quiet. Effects of exposure (inhalation, ingestion, or skin contact) to substances may be delayed. Ensure that medical personnel are aware of the material(s) involved and take precautions to protect themselves.

DIMETHYL ETHER D:295

SYNONYMS: EEC No. 603-019-00-8; EINECS No. 204-065-8; ETER DIMETILICO (Spanish); METHYL ETHER; OXYBIS METHYANE; DME; WOOD ETHER

IDENTIFICATION

CAS:	115-10-6
ERG ID:	UN 1033
ERG Guide:	115
ERG Guide Category:	GASES-FLAMMABLE (INCLUDING REFRIGERATED LIQUIDS)
Molecular Formula:	C_2H_6O
RTECS No.:	PM4780000
Properties:	Extremely flammable, liquefied compressed gas. Liquid below -12°F (-24°C) Colorless. Sweet, pleasant, etherlike odor. Floats and boils on water; dissolves slowly.

HEALTH AND SAFETY INFORMATION

Liquid may cause frostbite. Irritating to eyes, skin, and respiratory tract.

RECOMMENDATIONS FOR RESPIRATOR SELECTION

At any concentrations above the NIOSH REL, or where there is no REL, at any detectable concentration: SCBAF:PD,PP (any self-contained breathing apparatus that has a full facepiece and is operated in a pressure-demand or other positive-pressure mode); or SAF:PD,PP:ASCBA (any supplied-air respirator that has a full facepiece and is operated in a pressure-demand or other positive-pressure mode in combination with an auxiliary, self-contained breathing apparatus operated in a pressure-demand or other positive-pressure mode).

Escape: GMFOV [any air-purifying, full-facepiece respirator (gas mask) with a chin-style, front- or back-mounted organic vapor canister]; or SCBAE (any appropriate escape-type, self-contained breathing apparatus).

CHEMICAL REACTIVITY

Forms explosive mixture with air. Contact with oxidizers can cause fire and explosions. Technical grade may form unstable peroxides upon prolonged storage in presence of air. Peroxides may be detonated by heating, impact or friction.

FIRE INFORMATION

Vapors are heavier than air and will collect in low areas..Vapors in confined areas may explode when exposed to fire. Vapors may travel long distances to ignition sources and flash back. Cylinders and parts of cylinders may rocket great distances, in many directions. If material or contaminated runoff enters waterways, notify downstream users of potentially contaminated waters. Notify local health and fire officials and pollution control agencies. Stop the release only if it can be done safely from a distance. Do not extinguish fire unless the flow of the gas can be stopped and any remaining gas is out of the line. Specially trained personnel may use fog lines to cool exposures and let fire burn itself out. Cool exposed containers with large quantities of water from unattended equipment. If cylinders are exposed to excessive heat, withdraw immediately to a secure location. Use dry chemicals, carbon dioxide.

NFPA Hazard Classifications:

Health Hazards (Blue) 1
Flammability (Red) 4
Reactivity (Yellow) 1

Flash Point: Flammable gas
Flammable Limits in Air: LEL 3.4%; UEL 27%
Autoignition Temperature: 662°F (350°C)

FIRST AID

Move victim to fresh air. Call emergency medical care. Apply artificial respiration if victim is not breathing. Administer oxygen if breathing is difficult. Remove and isolate contaminated clothing and shoes. Clothing frozen to the skin should be thawed before being removed. In case of contact with liquefied gas, thaw frosted parts with lukewarm water. Keep victim warm and quiet. Ensure that medical personnel are aware of the material(s) involved and take precautions to protect themselves.

ENVIRONMENTAL INFORMATION

Clean Air Act: Section 112[r], Table 3, Accidental Release Prevention/Flammable Substances, TQ = 10,000 lb (4,540 kg)

DIMETHYLFORMAMIDE D:300

SYNONYMS: DIMETHYLFORMAMID (German); DIMETHYL FORMAMIDE; N,N-DIMETHYL FORMAMIDE; N,N-DIMETHYLFORMAMIDE; DIMETILFORMAMIDA (Spanish); DIMETILFORMAMIDE (Italian);

DIMETYLFORMAMIDU (Czech); DMF; DMFA; DWUMETYLOFORMAMID (Polish); DYNASOLVE 100; EEC No. 616-001-00-X; EINECS No. 200-679-5; N-FORMYLDIMETHYLAMINE; FORMAMIDE, N,N-DIMETHYL-; N,N-DIMETHYLMETHANIDE; NCI-C60913; NSC 5356; U-4224; WELD-ON P-70 PRIMER

IDENTIFICATION

CAS:	68-12-2
ERG ID:	UN 2265
ERG Guide:	129
ERG Guide Category:	FLAMMABLE LIQUIDS (POLAR/WATER-MISCIBLE/NOXIOUS)
Molecular Formula:	C_3H_7NO
RTECS No.:	LQ2100000
Properties:	Very flammable, watery liquid. Colorless to very slightly yellow. Faint, ammonia or fishy odor. Floats on water; soluble.

HEALTH AND SAFETY INFORMATION

Irritating to skin, eyes, nose, and lungs. Repeated or prolonged skin contact with the liquid may cause irritation and rash. Prolonged inhalation at 100 ppm has caused liver damage in animals. Kidney damage in animals has also been reported.

IARC Evaluation: Group 2B, possible human carcinogen; human evidence, limited; animal evidence, sufficient.
IDLH: 500 ppm
Odor Threshold: 100 ppm
OSHA Table Z-1-A Air Contaminant

Exposure Limits:	**TWA**
ACGIH:*	10 ppm; 30 mg/m^3
NIOSH:*	10 ppm; 30 mg/m^3
OSHA:*	10 ppm; 30 mg/m^3

*Skin contact contributes significantly to overall exposure.

RECOMMENDATIONS FOR RESPIRATOR SELECTION: NIOSH/OSHA

100 ppm: SA (any supplied-air respirator).
250 ppm: SA:CF (any supplied-air respirator operated in a continuous-flow mode).
500 ppm: SAT:CF (any supplied-air respirator that has a tight-fitting facepiece and is operated in a continuous-flow mode); or SCBAF (any self-contained breathing apparatus with a full facepiece); or SAF (any supplied-air respirator with a full facepiece).

Emergency or Planned Entry into Unknown Concentrations or IDLH Conditions: SCBAF:PD,PP (any self-contained breathing apparatus that has a full facepiece and is operated in a pressure-demand or other positive-pressure mode); or SAF:PD,PP:ASCBA (any supplied-air respirator that has a full facepiece and is operated in a pressure-demand or other positive-pressure mode in combination with an auxiliary, self-contained breathing apparatus operated in a pressure-demand or other positive-pressure mode).

Escape: GMFOV [any air-purifying, full-facepiece respirator (gas mask) with a chin-style, front- or back-mounted organic vapor canister]; or SCBAE (any appropriate escape-type, self-contained breathing apparatus).

Note: Substance reported to cause eye irritation or damage; may require eye protection.

CHEMICAL REACTIVITY

Forms explosive mixture with air. Contact with carbon tetrachloride and other halogenated compounds, particularly in contact with iron or strong oxidizers, may cause fire and explosions. Vigorous reaction with alkylaluminums when they are used as polymerization catalysts. Incompatible with nonoxidizing mineral acids, strong acids, chlorinated hydrocarbons, isocyanates, nitrates, organic nitrates, phenols, cresols, ammonia, bromine, chromic anhydride, magnesium nitrate, methylene diisocyante, phosphorus trioxide, and triethyaluminum. Attacks some plastics, rubber, and coatings. Temperatures above 662°F (350°C) may cause decomposition and increase pressure in closed containers.

FIRE INFORMATION

Toxic gases such as dimethylamine and nitrogen oxides released in a fire. Vapors are heavier than air and will collect in low areas. Vapors in confined areas may explode when exposed to fire. Vapors may travel long distances to ignition sources and flash back. Storage containers and parts of containers may rocket great distances, in many directions. If material or contaminated runoff enters waterways, notify downstream users of potentially contaminated waters. Notify local health and fire officials and pollution control agencies. Use dry chemicals, foam, carbon dioxide.

NFPA Hazard Classifications:

Health Hazards (Blue) 1
Flammability (Red) 2
Reactivity (Yellow) 0

Flash Point: 153°F (67°C) o.c.; 136°F (58°C) c.c.
Flammable Limits in Air: LEL 2.2%; UEL 15.2% both at 212°F (100°C).
Autoignition Temperature: 140°F (60°C).

FIRST AID

Move victim to fresh air. Call emergency medical care. Apply artificial respiration if victim is not breathing. Administer oxygen if breathing is difficult. Remove and isolate contaminated clothing and shoes. In case of contact with substance, immediately flush skin or eyes with running water for at least 20 minutes. Wash skin with soap and water. Keep victim warm and quiet. Effects of exposure (inhalation, ingestion, or skin contact) to substance may be delayed. Ensure that medical personnel are aware of the material(s) involved and take precautions to protect themselves.

Note: Fatal dose for humans has been estimated at 10 grams (about 1/3 ounce).

ENVIRONMENTAL INFORMATION

Clean Air Act: Hazardous Air Pollutants (Title I, Part A, Section 112).
EPCRA Section 304: Reportable Quantity (RQ): CERCLA, 1 lb (0.454 kg).
EPCRA Section 313: Form R *de minimis* concentration reporting level: 1.0%.
TSCA: 716.120(*a*)

2,5–DIMETHYLHEXANE–2,5–DIHYDROPEROXIDE D:303

SYNONYMS: 2,5-DIMETHYL-2,5-DIHYDROPEROXYHEXANE; DIMETHYL-2,5-DIHYDROPEROXYHEXANE(2,5-); DIMETHYLHEXANE.DIHYDROPEROXIDE; EINECS No. 221-184-0; 2,5-HIDROPEROXIDO de 2,5-DIMETILHEXANO (Spanish)

IDENTIFICATION

CAS:	3025-88-5
ERG ID:	UN 3104, see organic peroxide table
ERG Guide:	146
ERG Guide Category:	ORGANIC PEROXIDES (HEAT, CONTAMINATION AND FRICTION SENSITIVE)
Molecular Formula:	$C_8H_{18}O_4$
RTECS No.:	MO1840000
Properties:	Combustible solid.

HEALTH AND SAFETY INFORMATION

Irritating to the eyes, skin, and respiratory tract. Contact with skin or eyes may cause severe injury or burns.

RECOMMENDATIONS FOR RESPIRATOR SELECTION:

At any concentrations above the NIOSH REL, or where there is no REL, at any detectable concentration: SCBAF:PD,PP (any self-contained breathing apparatus that has a full facepiece and is operated in a pressure-demand or other positive-pressure mode); or SAF:PD,PP:ASCBA (any supplied-air respirator that has a full facepiece and is operated in a pressure-demand or other positive-pressure mode in combination with an auxiliary, self-contained breathing apparatus operated in a pressure-demand or other positive-pressure mode).

Escape: GMFOV [any air-purifying, full-facepiece respirator (gas mask) with a chin-style, front- or back-mounted organic vapor canister]; or SCBAE (any appropriate escape-type, self-contained breathing apparatus).

CHEMICAL REACTIVITY

A powerful oxidizer; highly reactive with many other materials. May be able to form unstable and dangerous peroxides in storage. Heat, friction, contact with reducing agents, thiocyanates, organic matter, combustibles, or contamination may cause explosion. May attack some plastics, rubber, and coatings. Protect containers from shock, heat, and friction.

FIRE INFORMATION

Fire may produce irritating, corrosive and/or toxic gases. Will increase intensity of fire; may burn rapidly with flare-burning effect. Vapors are heavier than air and will collect in low areas. Vapors in confined areas may explode when exposed to fire. Storage containers and parts of containers may rocket great distances, in many directions. If material or contaminated runoff enters waterways, notify downstream users of potentially contaminated waters. Runoff may create fire or explosion hazard. Notify local health and fire officials and pollution control agencies. Use dry chemicals, foam, carbon dioxide. Use water spray to cool containers.

FIRST AID

Move victim to fresh air. Call emergency medical care. Apply artificial respiration if victim is not breathing. Administer oxygen if breathing is difficult. Remove and isolate contaminated clothing and shoes. Remove material from skin immediately. In case of contact with substance, immediately flush skin or eyes with running water for at least 20 minutes. Keep victim warm and quiet. Ensure that medical personnel are aware of the material(s) involved and take precautions to protect themselves.

5,5–DIMETHYLHYDANTOIN D:305

SYNONYMS: 5,5-DIMETHYL-2,4-IMIDAZOLIDINEDIONE; 5,5-DIMITLHIDANTOINA (Spanish); DMH; EINECS No. 201-051-3; 2,4-IMIDAZOLIDINEDIONE, 5,5-DIMETHYL-; NSC-8652; T10

IDENTIFICATION

CAS:	77-71-4
Molecular Formula:	$C_6H_8N_2O_2$
RTECS No.:	MU0977000
Properties:	Crystalline solid. White. Soluble in water.

HEALTH AND SAFETY INFORMATION

RECOMMENDATIONS FOR RESPIRATOR SELECTION

At any concentrations above the NIOSH REL, or where there is no REL, at any detectable concentration: SCBAF:PD,PP (any self-contained breathing apparatus that has a full facepiece and is operated in a pressure-demand or other positive-pressure mode); or SAF:PD,PP:ASCBA (any supplied-air respirator that has a full facepiece and is operated in a pressure-demand or other positive-pressure mode in combination with an auxiliary, self-contained breathing apparatus operated in a pressure-demand or other positive-pressure mode).

Escape: GMFOV [any air-purifying, full-facepiece respirator (gas mask) with a chin-style, front- or back-mounted organic vapor canister]; or SCBAE (any appropriate escape-type, self-contained breathing apparatus).

CHEMICAL REACTIVITY

Incompatible with oxidizers.

FIRE INFORMATION

Combustion products include toxic nitrogen oxides. If material or contaminated runoff enters waterways, notify downstream users of potentially contaminated waters. Notify local health and fire officials and pollution control agencies. Use dry chemicals, alcohol-resistant foam, carbon dioxide.

NFPA Hazard Classifications: Not listed

FIRST AID

Move victim to fresh air. Call emergency medical care. Apply artificial respiration if victim is not breathing. Administer oxygen if breathing is difficult. Remove and isolate contaminated clothing and shoes. In case of contact with substance, immediately flush skin or eyes with running water for at least 20 minutes. Wash skin with soap and water. Keep victim warm and quiet. Effects of exposure (inhalation, ingestion, or skin contact) to substance may be delayed. Ensure that medical personnel are aware of the material(s) involved and take precautions to protect themselves.

1,3–DIMETHYLOLUREA D:306

SYNONYMS: n,n'-BIS(HYDROXYMETHYL)UREA; 1,3-BIS(HYDROXYMETHYL)UREA; CAURITE; CSI PASTE; n,n'-DIHYDROXYMETHYLUREA; DIMETHYLOLHARNSTOFF (German); DIMETILOLUREA (Spanish); DMU; EINECS No. 205-444-0; FINISH EN; KAURIT S; KNITTEX ASL; METHURAL; METURAL; OXYMETHUREA; PERMAFRESH 477; PROTESINE DMU; UREOL P

IDENTIFICATION

CAS:	140-95-4
Molecular Formula:	$C_3H_8N_2O_3$
Properties:	Crystalline solid. Soluble in water.

HEALTH AND SAFETY INFORMATION

RECOMMENDATIONS FOR RESPIRATOR SELECTION

At any concentrations above the NIOSH REL, or where there is no REL, at any detectable concentration: SCBAF:PD,PP (any self-contained breathing apparatus that has a full facepiece and is operated in a pressure-demand or other positive-pressure mode); or SAF:PD,PP:ASCBA (any supplied-air respirator that has a full facepiece and is operated in a pressure-demand or other positive-pressure mode in combination with an auxiliary, self-contained breathing apparatus operated in a pressure-demand or other positive-pressure mode).

Escape: GMFOV [any air-purifying, full-facepiece respirator (gas mask) with a chin-style, front- or back-mounted organic vapor canister]; or SCBAE (any appropriate escape-type, self-contained breathing apparatus).

CHEMICAL REACTIVITY

Incompatible, and may form explosive mixtures with strong oxidizers, hypochlorites, ammonium nitrate, chromyl chloride, nitrosyl perchlorate, perchloric acid.

FIRE INFORMATION

If material or contaminated runoff enters waterways, notify downstream users of potentially contaminated waters. Notify local health and fire officials and pollution control agencies. Use dry chemicals, alcohol-resistant foam, carbon dioxide.

NFPA Hazard Classifications: Not listed

FIRST AID

Move victim to fresh air. Call emergency medical care. Apply artificial respiration if victim is not breathing. Administer oxygen if breathing is difficult. Remove and isolate contaminated clothing and shoes. In case of contact with substance, immediately flush skin or eyes with running water for at least 20 minutes. Wash skin with soap and water. Keep victim warm and quiet. Effects of exposure (inhalation, ingestion, or skin contact) to substance may be delayed. Ensure that medical personnel are aware of the material(s) involved and take precautions to protect themselves.

DIMETHYL PHTHALATE D:307

SYNONYMS: AVOLIN; 1,2-BENZENEDICARBOXYLIC ACID, DIMETHYL ESTER; DIMETHYL 1,2-BENZENEDICARBOXYLATE; DIMETHYL BENZENEORTHODICARBOXYLATE; DIMETHYL PHTHALATE; DIMETHYL O-PHTHALATE; N,N-DIMETHYLPHTHALATE; DMP; EINECS No. 205-011-6; ENT 262; FERMINE; FTALATO de DIMETILO (Spanish); METHYL PHTHALATE; MIPAX; NTM; PALATINOL M; PHTHALSAEUREDIMETHYLESTER (German); PHTHALIC ACID DIMETHYL ESTER; SOLVANOM; SOLVARONE

IDENTIFICATION

CAS:	131-11-3
ERG ID:	UN 9188
ERG Guide:	171
ERG Guide Category:	SUBSTANCES (LOW TO MODERATE HAZARD)
Molecular Formula:	$C_{10}H_{10}O_4$
RTECS No.:	TI1575000
Properties:	Combustible oily liquid. A solid below 42○F (6○F). Colorless to pale yellow. Slight aromatic odor. Sinks in water.

HEALTH AND SAFETY INFORMATION

IDLH: 2,000 mg/m^3
OSHA Table Z-1-A Air Contaminant

Exposure Limits:	**TWA**
ACGIH:	5 mg/m^3
NIOSH:	5 mg/m^3
OSHA:	5 mg/m^3

RECOMMENDATIONS FOR RESPIRATOR SELECTION: NIOSH/OSHA

50 mg/m^3: DMF (any dust and mist respirator with a full facepiece).
125 mg/m^3: SA:CF (any supplied-air respirator operated in a continuous-flow mode); or PAPRDM (any powered, air-purifying respirator with a dust and mist filter).
250 mg/m^3: HiEF (any air-purifying, full-facepiece respirator with a high-efficiency particulate filter); or SCBAF (any self-contained breathing apparatus with a full facepiece); or SAF (any supplied-air respirator with a full facepiece).
2,000 mg/m^3: SAF:PD,PP (any supplied-air respirator that has a full facepiece and is operated in a pressure-demand or other positive-pressure mode).

Emergency or Planned Entry into Unknown Concentrations or IDLH Conditions: SCBAF:PD,PP (any self-contained breathing apparatus that has a full facepiece and is operated in a pressure-demand or other positive-pressure mode); or SAF:PD,PP:ASCBA (any supplied-air respirator that has a full facepiece and is operated in a pressure-demand or other positive-pressure mode in combination with an auxiliary, self-contained breathing apparatus operated in a pressure-demand or other positive-pressure mode).

Escape: HiEF (any air-purifying, full-facepiece respirator with a high-efficiency particulate filter); or SCBAE (any appropriate escape-type, self-contained breathing apparatus).
Note: Substance causes eye irritation or damage; eye protection needed.

CHEMICAL REACTIVITY

Incompatible with strong alkalies, strong acids, nitrates, and oxidizers.

FIRE INFORMATION

Vapors are heavier than air and will collect in low areas. Vapors in confined areas may explode when exposed to fire. If material or contaminated runoff enters waterways, notify downstream users of potentially contaminated waters. Notify local health and fire officials and pollution control agencies. Use dry chemicals, alcohol-resistant foam, carbon dioxide.

NFPA Hazard Classifications:

Health Hazards (Blue) 0
Flammability (Red) 1
Reactivity (Yellow) 0

Flash Point: 295°F (146°) c.c.
Flammable Limits in Air: LEL 0.9% at 358°F (181°C); UEL 8.0%
Autoignition Temperature: 915°F (490°C)

FIRST AID

Move victim to fresh air. Call emergency medical care. Apply artificial respiration if victim is not breathing. Administer oxygen if breathing is difficult. Remove and isolate contaminated clothing and shoes. In case of contact with substance, immediately flush skin or eyes with running water for at least 20 minutes. Ensure that medical personnel are aware of the material(s) involved and take precautions to protect themselves.

ENVIRONMENTAL INFORMATION

Clean Air Act: Hazardous Air Pollutants (Title I, Part A, Section 112).
Clean Water Act: Section 307 Priority Pollutants; Section 313 Priority Chemicals; Toxic Pollutant (Section 401.15).
EPA Hazardous Waste Number (RCRA No.): U102.
RCRA Section 261 Hazardous Constituents
RCRA Land Ban Waste
RCRA Universal Treatment Standards: Wastewater (mg/L), 0.047; Nonwastewater (mg/kg), 28.
RCRA Ground Water Monitoring List: Suggested test method(s) (PQL ug/L): 8060(5); 8270(10).
EPCRA Section 304: Reportable Quantity (RQ): CERCLA, 5,000 lb (2,270 kg).
EPCRA Section 313: Form R *de minimis* concentration reporting level: 1.0%.

N,N'–DIMETHYLPIPERAZINE D:313

SYNONYMS: 2,5-DIMETHYLPIPERAZINE; DIMETHYLPIPERAZINE, CIS; EINECS No. 203-409-4; PIPERAZINE, 2,5-DIMETHYL-

IDENTIFICATION

CAS:	106-55-8
Molecular Formula:	$C_6H_{14}N_2$
RTECS No.:	TL5950000
Properties:	Flammable liquid. A heterocylic compound of nitrogen.

HEALTH AND SAFETY INFORMATION

Irritates eyes, skin, and respiratory tract.

RECOMMENDATIONS FOR RESPIRATOR SELECTION

At any concentrations above the NIOSH REL, or where there is no REL, at any detectable concentration: SCBAF:PD,PP (any self-contained breathing apparatus that has a full facepiece and is operated in a pressure-demand or other positive-pressure mode); or SAF:PD,PP:ASCBA (any supplied-air respirator that has a full facepiece and is operated in a pressure-demand or other positive-pressure mode in combination with an auxiliary, self-contained breathing apparatus operated in a pressure-demand or other positive-pressure mode).

Escape: GMFOV [any air-purifying, full-facepiece respirator (gas mask) with a chin-style, front- or back-mounted organic vapor canister]; or SCBAE (any appropriate escape-type, self-contained breathing apparatus).

CHEMICAL REACTIVITY

Contact with oxidizers may cause fire and explosions.

FIRE INFORMATION

Combustion products include toxic nitrogen oxides. Vapors are heavier than air and will collect in low areas. Vapors in confined areas may explode when exposed to fire. Vapors may travel long distances to ignition sources and flash back. Storage containers and parts of containers may rocket great distances, in many directions. If material or contaminated runoff enters waterways, notify downstream users of potentially contaminated waters. Notify local health and fire officials and pollution control agencies. Use dry chemicals, foam, carbon dioxide.

NFPA Hazard Classifications:

Health Hazards (Blue) 2
Flammability (Red) 2
Reactivity (Yellow) 0

Flash Point: 155°F (68°C) o.c.

FIRST AID

Move victim to fresh air. Call emergency medical care. Apply artificial respiration if victim is not breathing. Administer oxygen if breathing is difficult. Remove and isolate contaminated clothing and shoes. In case of contact with substance, immediately flush skin or eyes with running water for at least 20 minutes. Wash skin with soap and water. Keep victim warm and quiet. Effects of exposure (inhalation, ingestion, or skin contact) to substance may be delayed. Ensure that medical personnel are aware of the material(s) involved and take precautions to protect themselves.

DIMETHYL SULFATE D:315

SYNONYMS: DIMETHYLESTER KYSELINY SIROVE (Czech); DIMETHYL ESTER OF SULFURIC ACID; DIMETHYL MONOSULFATE; DIMETHYLSULFAAT (Dutch); DIMETHYLSULFAT (Czech); DIMETHYL SULPHATE; DIMETILSOLFATO (Italian); DMS; DMS (METHYL SULFATE); DWUMETYLOWY SIARCZAN (Polish); EEC No. 016-023-00-4; EINECS No. 201-058-1; METHYL ESTER OF SULFURIC ACID; METHYL SULFATE; METHYLE (SULFATE de) (French); METHYL SULFATE; SULFATE de METHYLE (French); SULFATE DIMETHYLIQUE (French); SULFATO de DIMETILO (Spanish); SULFURIC ACID, DIMETHYL ESTER

IDENTIFICATION

CAS:	77-78-1
ERG ID:	UN 1595
ERG Guide:	156
ERG Guide Category:	SUBSTANCES-TOXIC AND/OR CORROSIVE (COMBUSTIBLE/WATER-SENSITIVE)
Molecular Formula:	$C_2H_6O_4S$

RTECS No.: WS8225000
Properties: Combustible, oily liquid. Colorless. Mild onion odor. Sinks and mixes slowly with water.

HEALTH AND SAFETY INFORMATION

IARC Evaluation: Group 2A, probable carcinogen; inadequate human evidence; sufficient animal evidence.
NTP: Anticipated carcinogen
IDLH: 7 ppm; potential human carcinogen
Odor Threshold: Odor is not a reliable indication of the presence of toxic amounts of gas.
OSHA Table Z-1-A Air Contaminant

Exposure Limits: TWA

ACGIH:(1) (2)	0.1 ppm; 0.52 mg/m^3
NIOSH:(1) (3)	0.1 ppm; 0.5 mg/m^3
OSHA:(1) (4)	1 ppm; 5 mg/m^3

(1)Skin contact contributes significantly in overall exposure.
(2)Animal carcinogen.
(3) Potential carcinogen. Reduce exposure to lowest feasible level.
(4)Suspected carcinogen.

RECOMMENDATIONS FOR RESPIRATOR SELECTION: NIOSH

At any concentrations above the NIOSH REL, or where there is no REL, at any detectable concentration: SCBAF:PD,PP (any self-contained breathing apparatus that has a full facepiece and is operated in a pressure-demand or other positive-pressure mode); or SAF:PD,PP:ASCBA (any supplied-air respirator that has a full facepiece and is operated in a pressure-demand or other positive-pressure mode in combination with an auxiliary, self-contained breathing apparatus operated in a pressure-demand or other positive-pressure mode).

Escape: GMFS [any air-purifying, full-facepiece respirator (gas mask) with a chin-style, front- or back-mounted canister providing protection against the compound of concern]; or SCBAE (any appropriate escape-type, self-contained breathing apparatus).

CHEMICAL REACTIVITY

Forms explosive mixture with air. Incompatible with strong oxidizers, strong acids, strong alkalies, and strong ammonia solutions. Attacks some plastics, rubber, and coatings. Corrodes metals when wet.

FIRE INFORMATION

Firefighting gear (including SCBA) does not provide adequate protection. If exposure occurs, remove and isolate gear immediately and thoroughly decontaminate personnel.Vapors are heavier than air and will collect in low areas. Vapors in confined areas may explode when exposed to fire. Vapors may travel long distances to ignition sources and flash back. Storage containers and parts of containers may rocket great distances, in many directions. If material or contaminated runoff enters waterways, notify downstream users of potentially contaminated waters. Notify local health and fire officials and pollution control agencies. Use dry chemicals, foam, carbon dioxide.

NFPA Hazard Classifications:

Health Hazards (Blue) 4
Flammability (Red) 2
Reactivity (Yellow) 0

Flash Point: 240°F (116°C) o.c.; 180°F (82°C) c.c.
Flammable Limits in Air: LEL 3.6%; UEL 23%
Autoignition Temperature: 370°F (188°C)

FIRST AID

Move victim to fresh air. Call emergency medical care. Apply artificial respiration if victim is not breathing. Do not use mouth-to-mouth method if victim ingested or inhaled the substance; induce artificial respiration with the aid of a pocket mask equipped with a one-way valve or other proper respiratory medical device. Administer oxygen if breathing is difficult. Remove and isolate contaminated clothing and shoes. In case of contact with substance, immediately flush skin or eyes with running water for at least 20 minutes. For minor skin contact, avoid spreading material on unaffected skin. Keep victim warm and quiet. Effects of exposure (inhalation, ingestion, or skin contact) may be delayed. Ensure that medical personnel are aware of the material(s) involved and take precautions to protect themselves.

ENVIRONMENTAL INFORMATION

Clean Air Act: Hazardous Air Pollutants (Title I, Part A, Section 112).
EPA Hazardous Waste Number (RCRA No.): U103.
RCRA Section 261 Hazardous Constituents
EPCRA Section 313: Form R *de minimis* concentration reporting level: 0.1%.
EPCRA Section 302 Extremely Hazardous Substances: TPQ = 500 lb (227 kg).
EPCRA Section 304: Reportable Quantity (RQ): CERCLA, 100 lb (45.4 kg).
California's Proposition 65: Carcinogen.
TSCA: 40CFR 712.30(*e*)10; 40CFR716.120(*d*)10

DIMETHYL SULFOXIDE D:320

SYNONYMS: A-10846; DECAP; DELTAN; DEMASORB; DEMAVET; DEMESO; DEMSODROX; DERMASORB; DIMETHYL SULPHOXIDE; DIMEXIDE; DIPIRATRIL-TROPICO; DMS-70; DMS-90; DMSO; DOLICUR; DOLIGUR; DOMOSO; DURASORB; EINECS No. 200-664-3; GAMASOL-90; HERPID; HYADUR; M-176; INFILTRINA; METHYL SULFOXIDE; NSC-763; RIMSO-50; SOMI-PRONT; SQ 9453; SULFICYL BIS(METHANE); SULFOXIDO de DIMETILO (Spanish); SYNTEXAN; TOPSYM

IDENTIFICATION

CAS:	67-68-5
ERG ID:	N/A
Molecular Formula:	C_3H_6OS
RTECS No.:	PV6210000
Properties:	Combustible liquid. Clear, colorless to pale yellow. Practically odorless to a mild garlic odor. Sinks and mixes with water.

HEALTH AND SAFETY INFORMATION

Irritates eyes and skin. All routes of exposure can produce an intense, garliclike taste and breath odor. Changes in genetic material and birth defects have been seen in laboratory studies. Whether these effects occur in humans is unknown.

RECOMMENDATIONS FOR RESPIRATOR SELECTION: NIOSH

At any concentrations above the NIOSH REL, or where there is no REL, at any detectable concentration: SCBAF:PD,PP (any self-contained breathing apparatus that has a full facepiece and is operated in a pressure-demand or other positive-pressure mode); or SAF:PD,PP:ASCBA (any supplied-air respirator that has a full facepiece and is operated in a pressure-demand or other positive-pressure mode in combination with an auxiliary, self-contained breathing apparatus operated in a pressure-demand or other positive-pressure mode)

Escape: GMFOV [any air-purifying, full-facepiece respirator (gas mask) with a chin-style, front- or back-mounted organic vapor canister]; or SCBAE (any appropriate escape-type, self-contained breathing apparatus).

CHEMICAL REACTIVITY

Forms explosive mixture with air. Contact with acetyl chloride, organic halides, benzenesulfonyl chloride, benzyl chloride, bromobenzoyl acetanilide, carbonyl diisocyanate, cyanuric chloride, decaborane, magnesium perchlorate, iodine pentafluoride, methyl bromide, nitrogen periodate, phosphorous trioxide, silver fluoride and sodium hydride can cause violent reactions. Violent reaction with oxidizers such as permanganates and dichromate. Reacts with ethanoyl chloride, boron compounds, metal alkoxides.

FIRE INFORMATION

Combustion products include toxic sulfur oxides. Vapors are heavier than air and will collect in low areas. Vapors in confined areas may explode when exposed to fire. Vapors may travel long distances to ignition sources and flash back. Storage containers and parts of containers may rocket great distances, in many directions. If material or contaminated runoff enters waterways, notify downstream users of potentially contaminated waters. Notify local health and fire officials and pollution control agencies. Use dry chemicals, foam, carbon dioxide.

NFPA Hazard Classifications:

Health Hazards (Blue) 1
Flammability (Red) 1
Reactivity (Yellow) 0

Flash Point: 203°F (95°C) o.c.; 190°F (102°C) c.c.
Flammable Limits in Air: LEL 2.6%; UEL 41.8%
Autoignition Temperature: 574°F (301°C)

FIRST AID

Move victim to fresh air. Call emergency medical care. Apply artificial respiration if victim is not breathing. Administer oxygen if breathing is difficult. Remove and isolate contaminated clothing and shoes. Clothing frozen to the skin should be thawed before being removed. In case of contact with liquefied gas, thaw frosted parts with lukewarm water. Keep victim warm and quiet. Ensure that medical personnel are aware of the material(s) involved and take precautions to protect themselves.

DIMETHYL TEREPHTHALATE D:323

SYNONYMS: 1,4-BENZENE DICARBOXYLIC ACID METHYL ESTER (9CI); DMT; para-DIMETHYLFTALAAT (Dutch); DIMETHYL p-PHTHALATE; EINECS No. 204-411-8; METHYL-4-CARBOMETHOXY BENZOATE; NCI-C5005; TEREFTALATO de DIMETILO (Spanish); TEREPHTHALIC ACID, DIMETHYL ESTER; TEREPHTHALIC ACID, METHYL ESTER

IDENTIFICATION

CAS:	120-61-6
ERG ID:	Not listed
Molecular Formula:	$C_{10}H_{10}O_4$
RTECS No.:	WZ1225000
Properties:	Combustible solid or heated (molten) liquid. White solid crystals or colorless liquid. Odorless. Solid and liquid sink in water; insoluble in cold water.

HEALTH AND SAFETY INFORMATION

Liquid will irritate and burn eyes, skin.

Exposure Limits: **TWA** as dimethyl phthalate

ACGIH:	5 mg/m^3
NIOSH:	5 mg/m^3
OSHA:	5 mg/m^3

RECOMMENDATIONS FOR RESPIRATOR SELECTION: NIOSH/OSHA as dimethyl phthalate

50 mg/m^3: DMF (any dust and mist respirator with a full facepiece).
125 mg/m^3: SA:CF (any supplied-air respirator operated in a continuous-flow mode); or PAPRDM (any powered, air-purifying respirator with a dust and mist filter).
250 mg/m^3: HiEF (any air-purifying, full-facepiece respirator with a high-efficiency particulate filter); or SCBAF (any self-contained breathing apparatus with a full facepiece); or SAF (any supplied-air respirator with a full facepiece).
2,000 mg/m^3: SAF:PD,PP (any supplied-air respirator that has a full facepiece and is operated in a pressure-demand or other positive-pressure mode).

Emergency or Planned Entry into Unknown Concentrations or IDLH Conditions: SCBAF:PD,PP (any self-contained breathing apparatus that has a full facepiece and is operated in a pressure-demand or other positive-pressure mode); or SAF:PD,PP:ASCBA (any supplied-air respirator that has a full facepiece and is operated in a pressure-demand or other positive-pressure mode in combination with an auxiliary, self-contained breathing apparatus operated in a pressure-demand or other positive-pressure mode).

Escape: HiEF (any air-purifying, full-facepiece respirator with a high-efficiency particulate filter); or SCBAE (any appropriate escape-type, self-contained breathing apparatus).

Note: Substance causes eye irritation or damage; eye protection needed.

CHEMICAL REACTIVITY

Fine particles of material can form explosive mixture with air. Reacts with strong oxidizers. Dry material may accumulate static electricity and cause ignition of its vapors.

FIRE INFORMATION

Decomposition products upon heating include toxic gases. Vapors are heavier than air and will collect in low areas. Vapors in confined areas may explode when exposed to fire. Storage containers and parts of containers may rocket great distances, in many directions. If material or contaminated runoff enters waterways, notify downstream users of potentially contaminated waters. Notify local health and fire officials and pollution control agencies. Use water, dry chemicals, foam, carbon dioxide.

NFPA Hazard Classifications:

Health Hazards (Blue) 1
Flammability (Red) 1
Reactivity (Yellow) 0

Flash Point: 298°F (148°C) o.c. (molten)
Flammable Limits in Air: LEL 0.8%; UEL 11.8%
Autoignition Temperature: 1,058°F (570°C) (dust)

FIRST AID

Move victim to fresh air. Call emergency medical care. Apply artificial respiration if victim is not breathing. Administer oxygen if breathing is difficult. Remove and isolate contaminated clothing and shoes. Clothing frozen to the skin should be thawed before being removed. In case of contact with liquefied gas, thaw frosted parts with lukewarm water. Keep victim warm and quiet. Ensure that medical personnel are aware of the material(s) involved and take precautions to protect themselves.

DIMETHYLTIN DICHLORIDE D:325

SYNONYMS: DICHLORID DIMETHYLCINICITY; DICHLORODIMETHYLSTANNANE; DICHLORODIMETHYLTIN; DICLORURO di DIMETILSTAGNO (Italian); DICLORURO de DIMETILESTANO (Spanish); DIMETHYLDICHLOROSTANNANE; DIMETHYLDICHLOROTIN; DIMETHYLZINNDICHLORID (German); EINECS No. 212-039-2; STANNANE, DICHLORODIMETHYL-; TIN, DIMETHYL-, DICHLORIDE

IDENTIFICATION

CAS:	753-73-1
ERG ID:	UN 3282
ERG Guide:	151
ERG Guide Category:	SUBSTANCES-TOXIC (NONCOMBUSTIBLE)
Molecular Formula:	$C_2H_6Cl_2Sn$
RTECS No.:	WH7245000
Properties:	Crystalline solid. White. Soluble in water.

HEALTH AND SAFETY INFORMATION

Irritates eyes, skin, and respiratory tract.

IDLH: 25 mg/m^3 as tin, organic compounds
OSHA Table Z-1-A Air Contaminant as tin organic compounds

Exposure Limits:	**TWA**	**STEL**
ACGIH: [1]	0.1 mg/m^3 as tin	0.2 mg/m^3 as tin
NIOSH: [1]	0.1 mg/m^3 as tin	
OSHA:	0.1 mg/m^3 as tin	

[1] Skin contact contributes significantly to overall exposure.

RECOMMENDATIONS FOR RESPIRATOR SELECTION: NIOSH as tin

10 mg/m^3: DM (any dust and mist respirator).
20 mg/m^3: DMXSQ (any dust and mist respirator except single-use and quarter mask respirators); or SA (any supplied-air respirator).
50 mg/m^3: SA:CF (any supplied-air respirator operated in a continuous-flow mode); or PAPRDM (any powered, air-purifying respirator with a dust and mist filter).
100 mg/m^3: HiEF (any air-purifying, full-facepiece respirator with a high-efficiency particulate filter); or SCBAF (any self-contained breathing apparatus with a full facepiece); or SAF (any supplied-air respirator with a full facepiece).

Emergency or Planned Entry into Unknown Concentrations or IDLH Conditions: SCBAF:PD,PP (any self-contained breathing apparatus that has a full facepiece and is operated in a pressure-demand or other positive-pressure mode); or SAF:PD,PP:ASCBA (Any supplied-air respirator that has a full facepiece and is operated in a pressure-demand or other positive-pressure mode in combination with an auxiliary, self-contained breathing apparatus operated in a pressure-demand or other positive-pressure mode.

Escape: HiEF (any air-purifying, full-facepiece respirator with a high-efficiency particulate filter); or SCBAE (Any appropriate escape-type, self-contained breathing apparatus).

CHEMICAL REACTIVITY

Incompatible with oxidizers, ethylene oxide, chloric acid, chlorine nitrate, lithium, potassium, sodium acetylide.

FIRE INFORMATION

Combustion products include toxic vapors of tin and chlorine. Firefighting gear (including SCBA) may not provide adequate protection. If exposure occurs, remove and isolate gear immediately and thoroughly decontaminate personnel. Vapors are heavier than air and will collect in low areas. Vapors in confined areas may explode when exposed to fire. If material or contaminated runoff enters waterways, notify downstream users of potentially contaminated waters. Notify local health and fire officials and pollution control agencies.

NFPA Hazard Classifications: Not listed

FIRST AID

Move victim to fresh air and call emergency medical care. Apply artificial respiration if victim is not breathing. Do not use mouth-to-mouth method if victim swallowed or inhaled the substance; induce artificial respiration with the aid of a pocket mask equipped with a one-way valve or other proper respiratory medical device. If breathing is difficult, give oxygen. In case of contact with substance, immediately flush skin or eyes with running water for at least 20 minutes. Remove and isolate contaminated clothing and shoes. Keep victim warm and quiet. Effects of exposure (inhalation, ingestion, or skin contact) to substances may be delayed. Ensure that medical personnel are aware of the material(s) involved and take precautions to protect themselves.

ENVIRONMENTAL INFORMATION as tin

RCRA Ground Water Monitoring List: Suggested test method(s) (PQL ug/L): (total) 7870(8,000).
TSCA: 40CFR716.120(*a*) substance list.; 40CFR712.30(*g*) manufacturers reporting.

DIMETHYLTIN OXIDE D:326

SYNONYMS: DIMETHYLOXOSTANNANE; EINECS No. 218-881-7; OXIDO de DIMETILESTANO (Spanish); STANNANE, DIMETHYLOXO-

IDENTIFICATION

CAS:	2273-45-2
ERG ID:	UN 3282
ERG Guide:	151
ERG Guide Category:	SUBSTANCES-TOXIC (NONCOMBUSTIBLE)
Molecular Formula:	C_2H_6OSn
RTECS No.:	WH7526500
Properties:	Noncombustible powder. White. Sinks in water; insoluble.

HEALTH AND SAFETY INFORMATION

Irritates eyes, skin, and respiratory tract.

IDLH: 25 mg/m^3 as tin, organic compounds
OSHA Table Z-1-A Air Contaminant as tin organic compounds

Exposure Limits:	**TWA**	**STEL**
ACGIH: (1)	0.1 mg/m^3 as tin	0.2 mg/m^3 as tin
NIOSH: (1)	0.1 mg/m^3 as tin	
OSHA:	0.1 mg/m^3 as tin	

(1) Skin contact contributes significantly to overall exposure.

RECOMMENDATIONS FOR RESPIRATOR SELECTION: NIOSH as tin

10 mg/m^3: DM (any dust and mist respirator)
20 mg/m^3: DMXSQ (any dust and mist respirator except single-use and quarter mask respirators); or SA (any supplied-air respirator).
50 mg/m^3: SA:CF (any supplied-air respirator operated in a continuous-flow mode); or PAPRDM (any powered, air-purifying respirator with a dust and mist filter).
100 mg/m^3: HiEF (any air-purifying, full-facepiece respirator with a high-efficiency particulate filter); or SCBAF (any self-contained breathing apparatus with a full facepiece); or SAF (any supplied-air respirator with a full facepiece).

Emergency or Planned Entry into Unknown Concentrations or IDLH Conditions: SCBAF:PD,PP (any self-contained breathing apparatus that has a full facepiece and is operated in a pressure-demand or other positive-pressure mode); or SAF:PD,PP:ASCBA (Any supplied-air respirator that has a full facepiece and is operated in a pressure-demand or other positive-pressure mode in combination with an auxiliary, self-contained breathing apparatus operated in a pressure-demand or other positive-pressure mode.

Escape: HiEF (any air-purifying, full-facepiece respirator with a high-efficiency particulate filter); or SCBAE (any appropriate escape-type, self-contained breathing apparatus).

CHEMICAL REACTIVITY

Incompatible with oxidizers, ammonium nitrate, bromine pentafluoride, chlorine trifluoride, fluorine, hydrazine, hydrogen peroxide, hydrogen trisulfide, lithium, potassium, aluminum + heat, nitroalkanes (nitromethane, nitroethane), 1-nitropropane, selenium oxychloride, silver azide, sulfur dioxide, sulfur trioxide.

FIRE INFORMATION

Combustion products include toxic vapors of tin. Firefighting gear (including SCBA) may not provide adequate protection. If exposure occurs, remove and isolate gear immediately and thoroughly decontaminate personnel. Vapors are heavier than air and will collect in low areas. Vapors in confined areas may explode when exposed to fire. If material or contaminated runoff enters waterways, notify downstream users of potentially contaminated waters. Notify local health and fire officials and pollution control agencies.

NFPA Hazard Classifications: Not listed

FIRST AID

Move victim to fresh air and call emergency medical care. Apply artificial respiration if victim is not breathing. Do not use mouth-to-mouth method if victim swallowed or inhaled the substance; induce artificial respiration with the aid of a pocket mask equipped with a one-way valve or other proper respiratory medical device. If breathing is difficult, give oxygen. In case of contact with substance, immediately flush skin or eyes with running water for at least 20 minutes. Remove and isolate contaminated clothing and shoes. Keep victim warm and quiet. Effects of exposure (inhalation, ingestion, or skin contact) to substances may be delayed. Ensure that medical personnel are aware of the material(s) involved and take precautions to protect themselves.

ENVIRONMENTAL INFORMATION as tin

RCRA Ground Water Monitoring List: Suggested test method(s) (PQL ug/L): (total) 7870(8,000).
TSCA: 40CFR716.120(*a*) substance list.; 40CFR712.30(*g*) manufacturers reporting.

DINITROSOPENTAMETHYLENETETRAMINE D:328

SYNONYMS: ACETO DNPT 40; ACETO DNPT 80; ACETO DNPT 100; CHKHZ 18; N,N-DINITROSOPENTAMETHYLENETETRAMINE; N1,N3-DINITROSOPENTAMETHYLENETETRAMINE; 3,4-DI-N-NITROSOPENTAMETHYLENETETRAMINE; 3,7-DI-N-NITROSOPENTAMETHYLENETETRAMINE; DNPMT; DNPT; EINECS No. 202-928-3; 1,5-METHYLENE-3,7-DINITROSO-1,3,5,7-TETRAAZACYCLOOCTAINE; 1,5-METHYLENE-3,7-DINITROSO-1,3,5,7-TETRAAZACYCLOOCTANE; POROFOR CHKHC-18; POROPHOR B; UNICEL-ND; UNICEL NDX; VULCACEL B-40; VULCACEL BN

IDENTIFICATION

CAS:	101-25-7
ERG ID:	UN 2972
ERG Guide:	149
ERG Guide Category:	SUBSTANCES (SELF-REACTIVE)
Molecular Formula:	$C_5H_{10}N_6O_2$
RTECS No.:	XA5250000
Properties:	Flammable and explosive solid. Light cream-color.

HEALTH AND SAFETY INFORMATION

Irritates eyes and skin.

IARC Evaluation: Group 3; no animal evidence.

RECOMMENDATIONS FOR RESPIRATOR SELECTION

At any concentrations above the NIOSH REL, or where there is no REL, at any detectable concentration: SCBAF:PD,PP (any self-contained breathing apparatus that has a full facepiece and is operated in a pressure-demand or other positive-pressure mode); or SAF:PD,PP:ASCBA (any supplied-air respirator that has a full facepiece and is operated in a pressure-demand or other positive-pressure mode in combination with an auxiliary, self-contained breathing apparatus operated in a pressure-demand or other positive-pressure mode).

Escape: GMFOV [any air-purifying, full-facepiece respirator (gas mask) with a chin-style, front- or back-mounted organic vapor canister]; or SCBAE (any appropriate escape-type, self-contained breathing apparatus).

CHEMICAL REACTIVITY

Burns rapidly and can ignite when handled. Incompatible with oxidizers. May ignite on contact with cellulose nitrate of high surface area. Incompatible with acrolein, acrylonitrile, tert-butyl nitroacetylene, ethylene oxide, isopropyl chlorocarbonate, maleic anhydride, triisobutyl aluminum.

FIRE INFORMATION

NFPA Hazard Classifications: Not listed.

Vapors are heavier than air and will collect in low areas. Vapors in confined areas may explode when exposed to fire. If material or contaminated runoff enters waterways, notify downstream users of potentially contaminated waters. Notify local health and fire officials and pollution control agencies.

Flash Point: None listed; may explode at 390°F (200°C).

FIRST AID

Move victim to fresh air. Call emergency medical care. Apply artificial respiration if victim is not breathing. Administer oxygen if breathing is difficult. Remove and isolate contaminated clothing and shoes. In case of contact with substance, immediately flush skin or eyes with running water for at least 20 minutes. Keep victim warm and quiet. Ensure that medical personnel are aware of the material(s) involved and take precautions to protect themselves.

DIOCTYL ADIPATE D:330

SYNONYMS: ADIPATO de DIOCTILO (Spanish); ADIPATO de DI(2-ETILHEXILO) (Spanish); ADIPATO di BIS(2-ETILESILE) (Italian); ADIPIC ACID BIS (2-ETHYLHEXYL) ESTER; ADIPIC ACID, DIBUTYL ESTER; ADIPOL 2EH; BEHA; BIS(2-ETHYLHEXYL)ADIPATE; BISOFLEX DOA; BIS(2-ETHYLHEXYL)ADIPATE; ADIPOL 2EH; BEHA; BISOFLEX DOA; BUTYL ADIPATE; BIS(2-ETHYLHEXYL) ADIPATE; DEHA; DIBUTYL ADIPINATE; DIBUTYL HEXANEDIOATE; DI(2-ETHYLHEXYL) ADIPATE; DOA; EFFEMOLL DOA; EINECS No. 203-090-1; ERGOPLAST AdDO; FLEXOL A-26; HEXANEDIOIC ACID, DIBUTYL ESTER; KODAFLEX DOA; MONOPLEX DOA; NCI-C5486; OCTYL ADIPATE; PLASTOMOLL DOA; PX-238; REOMOL DOA; RUCOFLEX PLASTICIZER DOA; SICOL 250; TRUFLEX DOA; VESTINOL OA; WICKENOL 158; WITOMOL-320

IDENTIFICATION

CAS:	103-23-1
ERG ID:	UN 3082 ENVIRONMENTALLY HAZARDOUS SUBSTANCES, LIQUID, n.o.s.
ERG Guide:	171

ERG Guide Category: SUBSTANCES (LOW TO MODERATE HAZARD)
Molecular Formula: $C_{22}H_{42}O_4$
Properties: Combustible oily liquid. Clear, straw or pale amber colored. Mild odor. Floats on water; insoluble.

HEALTH AND SAFETY INFORMATION

IARC Evaluation: Group 3, unclassifiable substance; no human data, limited animal evidence.
NTP: Evidence os animal carcinogenicity in mice.

RECOMMENDATIONS FOR RESPIRATOR SELECTION:

At any concentrations above the NIOSH REL, or where there is no REL, at any detectable concentration: SCBAF:PD,PP (any self-contained breathing apparatus that has a full facepiece and is operated in a pressure-demand or other positive-pressure mode); or SAF:PD,PP:ASCBA (any supplied-air respirator that has a full facepiece and is operated in a pressure-demand or other positive-pressure mode in combination with an auxiliary, self-contained breathing apparatus operated in a pressure-demand or other positive–pressure mode).

Escape: GMFOV [any air-purifying, full-facepiece respirator (gas mask) with a chin-style, front- or back-mounted organic vapor canister]; or SCBAE (any appropriate escape-type, self-contained breathing apparatus).

CHEMICAL REACTIVITY

Incompatible with strong acids, nitrates, and oxidizers.

FIRE INFORMATION

Vapors are heavier than air and will collect in low areas. Vapors in confined areas may explode when exposed to fire. Storage containers and parts of containers may rocket great distances, in many directions. If material or contaminated runoff enters waterways, notify downstream users of potentially contaminated waters. Notify local health and fire officials and pollution control agencies.

NFPA Hazard Classifications:

Health Hazards (Blue) 0
Flammability (Red) 1
Reactivity (Yellow) 0

Flash Point: 402°F (206°C)
Flammable Limits in Air: LEL 0.4% at 467°F (242°C); UEL NA.
Autoignition Temperature: 710°F (377°C).

FIRST AID

Move victim to fresh air. Call emergency medical care. Apply artificial respiration if victim is not breathing. Administer oxygen if breathing is difficult. Remove and isolate contaminated clothing and shoes. In case of contact with substance, immediately flush skin or eyes with running water for at least 20 minutes. Ensure that medical personnel are aware of the material(s) involved and take precautions to protect themselves.

ENVIRONMENTAL INFORMATION

Safe Drinking Water Act: MCL, 0.4 mg/L; MCLG, 0.4 mg/L.
EPCRA Section 313: Deleted 7/31/96 (FR Vol. 61, No. 148 p.39891-39892).

DINONYL PHTHALATE D:335

SYNONYMS: BIOFLEX 91; 1,2-BENZENEDICARBOXYLIC ACID, DIONYL ESTER; DINONYL

1,2-BENZENEDICARBOXYLATE; DI-n-NONYL PHTHALATE; EINECS No. 201-560-0; FTALATO de DINONILO (Spanish); PHTHALIC ACID, DINONYL ESTER

IDENTIFICATION

CAS:	84-76-4
ERG ID:	UN 3082
ERG Guide:	171
ERG Guide Category:	SUBSTANCES (LOW TO MODERATE HAZARD)
Molecular Formula:	$C_{26}H_{42}O_4$
RTECS No.:	TI1800000
Properties:	Combustible liquid. Colorless. Odorless. Floats on the surface of water.

HEALTH AND SAFETY INFORMATION

RECOMMENDATIONS FOR RESPIRATOR SELECTION: NIOSH

At any concentrations above the NIOSH REL, or where there is no REL, at any detectable concentration: SCBAF:PD,PP (any self-contained breathing apparatus that has a full facepiece and is operated in a pressure-demand or other positive-pressure mode); or SAF:PD,PP:ASCBA (any supplied-air respirator that has a full facepiece and is operated in a pressure-demand or other positive-pressure mode in combination with an auxiliary, self-contained breathing apparatus operated in a pressure-demand or other positive-pressure mode).

Escape: GMFOV [any air-purifying, full-facepiece respirator (gas mask) with a chin-style, front- or back-mounted organic vapor canister]; or SCBAE (any appropriate escape-type, self-contained breathing apparatus).

CHEMICAL REACTIVITY

Reacts with strong acids, nitrates, and oxidizers. May attack some forms of plastics.

FIRE INFORMATION

Vapors are heavier than air and will collect in low areas. Vapors in confined areas may explode when exposed to fire. Storage containers and parts of containers may rocket great distances, in many directions. If material or contaminated runoff enters waterways, notify downstream users of potentially contaminated waters. Notify local health and fire officials and pollution control agencies. Use dry chemical, foam, carbon dioxide.

FEMA Hazard Classifications:

Health Hazards (Blue) 0
Flammability (Red) 1
Reactivity (Yellow) 0

Flash Point: 420°F (216°C) c.c.

FIRST AID

Move victim to fresh air. Call emergency medical care. Apply artificial respiration if victim is not breathing. Administer oxygen if breathing is difficult. Remove and isolate contaminated clothing and shoes. In case of contact with substance, immediately flush skin or eyes with running water for at least 20 minutes. Wash skin with soap and water. Keep victim warm and quiet. Ensure that medical personnel are aware of the material(s) involved and take precautions to protect themselves.

ENVIRONMENTAL INFORMATION

TSCA: 40CFR716.120(*c*)2, alkyl phthalates

DIOXANE

D:340

SYNONYMS: CHLOROTHENE SM SOLVENT; DIETHYLENE DIOXIDE; DIETHYLENE-1,4-DIOXIDE; 1,4-DIETHYLENEDIOXIDE; DIETHYLENE ETHER; DI(ETHYLENE OXIDE); DIOKAN; DIOKSAN (Polish); DIOSSANO-1,4 (Italian); DIOXAAN-1,4 (Dutch); 1,4-DIOXACYCLOHEXANE; DIOXAN; p-DIOXAN (Czech); DIOXAN-1,4 (German); DIOXANE; p-DIOXANE; 1,4-DIOXANE; DIOXANE-1,4; DIOXANE(1,4-); DIOXANNE (French); DIOXANO (Spanish); p-DIOXAN, TETRAHYDRO-; 1,4-DIOXIN, TETRAHYDRO-; DIOXYETHYLENE ETHER; EEC No. 603-024-00-5; EINECS No. 204-661-8; GLYCOL ETHYLENE ETHER; NCI-C03689; 6200 DRUM CLEANING SOLVENT; 6500 DRUM CLEANING SOLVENT; SOLVENT 111; TETRAHYDRO-p-DIOXIN; TETRAHYDRO-1,4-DIOXIN

IDENTIFICATION

CAS:	123-91-1
ERG ID:	UN 1165
ERG Guide:	127
ERG Guide Category:	FLAMMABLE LIQUIDS (POLAR/WATER MISCIBLE)
Molecular Formula:	$C_4H_8O_2$
RTECS No.:	JG8225000
Properties:	Highly flammable liquid. Colorless. Slight etherlike or alcohol odor. Sinks and mixes with water. Flammable, irritating vapor is produced. Freezing point is 53°F (12°C).

HEALTH AND SAFETY INFORMATION

IARC Evaluation: Group 2B, possible human carcinogen; human evidence, inadequate; animal evidence, sufficient.
NTP: Anticipated human carcinogen.
IDLH: 500 ppm; potential human carcinogen.
Odor Threshold: 7.78 ppm

OSHA Table Z-1-A Air Contaminant

NIOSH Criteria Document: NIOSH 77-226

Prolonged exposure to concentrations above 100 ppm has been shown to lead to severe liver and kidney damage, lung congestion, and death. Drying and cracking of skin has occurred from continuous contact with skin. Repeated skin exposure can contribute to the development of liver and kidney damage.

Exposure Limits:	**TWA**	**CEILING**
ACGIH:(1)	25 ppm; 90 mg/m^3	
NIOSH:(2)		1 ppm; 3.6 mg/m^3 (30 min)
OSHA:(2)	100 ppm; 360 mg/m^3	

(1)Skin contact contributes significantly to overall exposure. ACGIH animal carcinogen.
(2)Potential carcinogen. Reduce exposure to lowest feasible level; skin contact contributes significantly to overall exposure.

RECOMMENDATIONS FOR RESPIRATOR SELECTION: NIOSH

At any concentrations above the NIOSH REL, or where there is no REL, at any detectable concentration: SCBAF:PD,PP (any self-contained breathing apparatus that has a full facepiece and is operated in a pressure-demand or other positive-pressure mode); or SAF:PD,PP:ASCBA (any supplied-air respirator that has a full facepiece and is operated in a pressure-demand or other positive-pressure mode in combination with an auxiliary, self-contained breathing apparatus operated in a pressure-demand or other positive-pressure mode).

Escape: GMFOV [any air-purifying, full-facepiece respirator (gas mask) with a chin-style, front- or back-mounted organic vapor canister]; or SCBAE (any appropriate escape-type, self-contained breathing apparatus).

CHEMICAL REACTIVITY

Anhydrous form forms explosive mixture with air. Air and moisture form unstable peroxides. May accumulate static electrical charges, and may cause ignition of its vapors. Incompatible with strong oxidizers (e.g., permanganates, dichromates, or chlorine), and strong acids. Attacks many plastics. Avoid direct sunlight.

FIRE INFORMATION

Vapors are heavier than air and will collect in low areas. Vapors in confined areas may explode when exposed to fire. Vapors may travel long distances to ignition sources and flash back. Storage containers and parts of containers may rocket great distances, in many directions. If material or contaminated runoff enters waterways, notify downstream users of potentially contaminated waters. Notify local health and fire officials and pollution control agencies. Use dry chemicals, alcohol resistant foam, carbon dioxide, large quantities of water.

NFPA Hazard Classifications:

Health Hazards (Blue) 2
Flammability (Red) 3
Reactivity (Yellow) 1

Flash Point: 74°F (23°C) o.c.; 55°F (13°C) c.c.
Flammable Limits in Air: LEL 1.97%; UEL 22.5%
Autoignition Temperature: 356°F (180°C)

FIRST AID

Move victim to fresh air. Call emergency medical care. Apply artificial respiration if victim is not breathing. If breathing is difficult, give oxygen. Remove and isolate contaminated clothing and shoes. In case of contact with substance, immediately flush skin or eyes with running water for at least 20 minutes. Wash skin with soap and water. Keep victim warm and quiet. Ensure that medical personnel are aware of the material(s) involved and take precautions to protect themselves.

ENVIRONMENTAL INFORMATION

Clean Air Act: Hazardous Air Pollutants (Title I, Part A, Section 112).
EPA Hazardous Waste Number (RCRA No.): U108..
RCRA Section 261 Hazardous Constituents
RCRA Land Ban Waste
RCRA Universal Treatment Standards: Wastewater (mg/L), 12.0; Nonwastewater (mg/kg), 170.
RCRA Ground Water Monitoring List: Suggested test method(s) (PQL ug/L): 8015(150).
EPCRA Section 304: Reportable Quantity (RQ): CERCLA, 100 lb (45.4 kg).
EPCRA Section 313: Form R *de minimis* concentration reporting level: 0.1%.
California's Proposition 65: Carcinogen.

DIPENTENE D:360

SYNONYMS: ACHILLES DIPENTENE; ACINTENE DP; ACINTENE DP DIPENTENE; CAJEPUTENE; CINENE; CYCLOHEXENE, 1-METHYL-4-(1-METHYLETHENYL)-; δ-1,8-TERPODIENE; DIPANOL; DI-p-MENTHA-1,8-DIENE; DIPENTENO (Spanish); EEC No. 601-029-00-7; EINECS No. 205-341-0; 2-HYDROXYMETHYLPHENYL-β-δ-GLUCOPYRANOSIDE; INACTIVE LIMONENE; KAUTSCHIN; LIMONENE; LIMONENE, INACTIVE; DL-LIMONENE; p-MENTHA-1, 8-DIENE; p-MENTHA-1,8-DIENE,DL-; 1,8(9)-p-MENTHADIENE; 1-METHYL-4-ISOPROPENYL-1-CYCLOHEXENE; NESOL; PHELLANDRENE; SALICIN; TERPINENE; δ-1,8-TERPODIENE; UNITENE

IDENTIFICATION

CAS:	138-86-3
ERG ID:	UN 2052
ERG Guide:	128
ERG Guide Category:	FLAMMABLE LIQUIDS (NONPOLAR/WATER-IMMISCIBLE)

Molecular Formula: $C_{10}H_{16}$
Properties: Combustible liquid. Colorless to light yellow. Pleasant, lemonlike odor. Floats on water; insoluble.

HEALTH AND SAFETY INFORMATION

Irritates eyes, skin, and respiratory tract. Prolonged or repeated exposure may cause skin allergy with itching and rash.

RECOMMENDATIONS FOR RESPIRATOR SELECTION

At any concentrations above the NIOSH REL, or where there is no REL, at any detectable concentration: SCBAF:PD,PP (any self-contained breathing apparatus that has a full facepiece and is operated in a pressure-demand or other positive-pressure mode); or SAF:PD,PP:ASCBA (any supplied-air respirator that has a full facepiece and is operated in a pressure-demand or other positive-pressure mode in combination with an auxiliary, self-contained breathing apparatus operated in a pressure-demand or other positive-pressure mode).

Escape: GMFOV [any air-purifying, full-facepiece respirator (gas mask) with a chin-style, front- or back-mounted organic vapor canister]; or SCBAE (any appropriate escape-type, self-contained breathing apparatus).

CHEMICAL REACTIVITY

Forms explosive mixture with air. Contact with strong oxidizers may cause fire and explosions. May form unstable peroxides. May accumulate static electricity and cause ignition of its vapors.

FIRE INFORMATION

Vapors are heavier than air and will collect in low areas. Vapors in confined areas may explode when exposed to fire. Vapors may travel long distances to ignition sources and flash back. Storage containers and parts of containers may rocket great distances, in many directions. If material or contaminated runoff enters waterways, notify downstream users of potentially contaminated waters. Notify local health and fire officials and pollution control agencies. Use dry chemicals, foam, carbon dioxide. Use water spray to cool containers.

NFPA Hazard Classifications:

Health Hazards (Blue) 0
Flammability (Red) 2
Reactivity (Yellow) 0

Flash Point: 115°F (46°C) c.c.
Flammable Limits in Air: LEL 0.7%; UEL 6.1% both at 302°F (150°C)
Autoignition Temperature: 458°F (237°C)

FIRST AID

Move victim to fresh air. Call emergency medical care. Apply artificial respiration if victim is not breathing. If breathing is difficult, give oxygen. Remove and isolate contaminated clothing and shoes. In case of contact with substance, immediately flush skin or eyes with running water for at least 20 minutes. Wash skin with soap and water. Keep victim warm and quiet. Ensure that medical personnel are aware of the material(s) involved and take precautions to protect themselves.

ENVIRONMENTAL INFORMATION

Marine Pollutant (49 CFR, Subchapter 172.101, Appendix B)

DIPHENYL D:390

SYNONYMS: BIPHENYL; BIBENZENE; 1,1'-BIPHENYL; CAROLID AL; DIFENILO (Spanish); DIPHENYLE (French); EINECS No. 202-163-5; PHENADOR-X; LEMONENE; PHPH; PHENYL BENZENE; TETROSIN LY; XENENE

IDENTIFICATION

CAS:	92-52-4
ERG ID:	Not Listed
Molecular Formula:	$C_{12}H_{10}$
RTECS No.:	DU8050000
Properties:	Combustible solid. Colorless to pale yellow. Characteristic aromatic odor. Insoluble in water.

HEALTH AND SAFETY INFORMATION

Irritates eyes, skin, and respiratory tract. Prolonged or repeated exposure can cause skin, kidney, liver, central nervous system problems.

IDLH: 100 mg/m^3
Odor Threshold: 0.0095 mg/m^3
OSHA Table Z-1-A Air Contaminant

Exposure Limits:	**TWA**
ACGIH:	0.2 ppm; 1.3 mg/m^3
NIOSH:	0.2 ppm; 1 mg/m^3
OSHA:	0.2 ppm; 1 mg/m^3

RECOMMENDATIONS FOR RESPIRATOR SELECTION: NIOSH/OSHA

10 mg/m^3: CCROVDM (any chemical cartridge respirator with organic vapor cartridge(s) in combination with a dust and mist filter); SA (any supplied-air respirator)
25 mg/m^3: SA:CF (any supplied-air respirator operated in a continuous-flow mode); or PAPROVDM (any powered, air-purifying respirator with organic vapor cartridge(s) in combination with a dust and mist filter)
50 mg/m^3: CCRFOVHiE (any chemical cartridge respirator with a full facepiece and organic vapor cartridge(s) in combination with a high-efficiency particulate filter); or GMFOVHiE (any air-purifying, full-facepiece respirator (gas mask) with a chin-style, front- or back-mounted organic vapor canister having a high-efficiency particulate filter); or PAPRTOVHiE (any powered, air-purifying respirator with a tight-fitting facepiece and organic vapor cartridge(s) in combination with a high-efficiency particulate filter); or SCBAF (any self-contained breathing apparatus with a full facepiece); or SAF (any supplied-air respirator with a full facepiece)
100 mg/m^3: SAF:PD,PP (any supplied-air respirator that has a full facepiece and is operated in a pressure-demand or other positive-pressure mode)

Emergency or planned entry into unknown concentrations or IDLH conditions: SCBAF:PD,PP (any self-contained breathing apparatus that has a full facepiece and is operated in a pressure-demand or other positive-pressure mode); or SAF:PD,PP:ASCBA (any supplied-air respirator that has a full facepiece and is operated in a pressure-demand or other positive-pressure mode in combination with an auxiliary, self-contained breathing apparatus operated in a pressure-demand or other positive-pressure mode)

Escape: GMFOVHiE (any air-purifying, full-facepiece respirator (gas mask) with a chin-style, front- or back-mounted organic vapor canister having a high-efficiency particulate filter); or SCBAE (any appropriate escape-type, self-contained breathing apparatus).

Note: Substance reported to cause eye irritation or damage; may require eye protection.

CHEMICAL REACTIVITY

Vapor or fine particles form explosive mixture with air. Contact with strong oxidizers may cause fire and explosions.

FIRE INFORMATION

Vapors are heavier than air and will collect in low areas. Vapors in confined areas may explode when exposed to fire. Storage containers and parts of containers may rocket great distances, in many directions. If material or contaminated runoff enters waterways, notify downstream users of potentially contaminated waters. Notify local health and fire officials and pollution control agencies. Use dry chemicals, carbon dioxide, foam , AFFF.

NFPA Hazard Classifications:

Health Hazards (Blue) 2
Flammability (Red) 1
Reactivity (Yellow) 0

Flash Point: 235°F (113°C) c.c.
Flammable Limits in Air: LEL 0.6% @ 232°F (111°C); UEL 5.8% @ 311°F (155°C)
Autoignition Temperature: 100°F (540°C)

FIRST AID

Move victim to fresh air. Call emergency medical care. Apply artificial respiration if victim is not breathing. Administer oxygen if breathing is difficult. In case of contact with substance, immediately flush skin or eyes with running water for at least 20 minutes. Remove and isolate contaminated clothing and shoes. Ensure that medical personnel are aware of the material(s) involved and take precautions to protect themselves.

ENVIRONMENTAL INFORMATION

Clean Air Act: Hazardous Air Pollutants (Title I, Part A, Section 112)
EPCRA Section 304 Reportable Quantity (RQ): CERCLA, 100 lb (45.4 kg)
EPCRA Section 313: Form R *de minimis* concentration reporting level: 1.0%.
Marine Pollutant (49 CFR, Subchapter 172.101, Appendix B)
TSCA: 40CFR716.120(*a*); 40CFR799.925; 40CFR799.18

DIPHENYLDICHLOROSILANE D:420

SYNONYMS: DICHLORODIPHENYLSILANE; DICHLORODIPHENYLSILICANE; DIFENILDICLOROSILANO (Spanish); DIPHENYLSILICON DICHLORIDE; EINECS No. 201-251-0; SILANE, DICHLORODIPHENYL-

IDENTIFICATION

CAS:	80-10-4
ERG ID:	UN 1769
ERG Guide:	156
ERG Guide Category:	SUBSTANCES-TOXIC AND/OR CORROSIVE (COMBUSTIBLE/WATER-SENSITIVE)
Molecular Formula:	$C_{12}H_{10}Cl_2O_4Si$
RTECS No.:	VV3190000
Properties:	Combustible liquid. Colorless. Sharp, irritating odor. Sinks in water; soluble. Reacts with water. Irritating vapor is produced.

HEALTH AND SAFETY INFORMATION

Irritates eyes, skin, and respiratory tract. May cause pulmonary edema, which can be delayed for several hours; there is a risk of death in serious cases.

RECOMMENDATIONS FOR RESPIRATOR SELECTION

At any concentrations above the NIOSH REL, or where there is no REL, at any detectable concentration: SCBAF:PD,PP (any self-contained breathing apparatus that has a full facepiece and is operated in a pressure-demand or other positive-pressure mode); or SAF:PD,PP:ASCBA (any supplied-air respirator that has a full facepiece and is operated in a pressure-demand or other positive-pressure mode in combination with an auxiliary, self-contained breathing apparatus operated in a pressure-demand or other positive-pressure mode).

Escape: GMFOV [any air-purifying, full-facepiece respirator (gas mask) with a chin-style, front- or back-mounted organic vapor canister]; or SCBAE (any appropriate escape-type, self-contained breathing apparatus).

CHEMICAL REACTIVITY

Contact with strong oxidizers may cause fire and explosions. Reacts with water or surface moisture to generate hydrochloric acid. Corrosive to common metals in the presence of moisture.

FIRE INFORMATION

Vapors are heavier than air and will collect in low areas. Vapors in confined areas may explode when exposed to fire. Storage containers and parts of containers may rocket great distances, in many directions. If material or contaminated runoff enters waterways, notify downstream users of potentially contaminated waters. Notify local health and fire officials and pollution control agencies. Decomposition products upon heating includes toxic chlorine.

NFPA Hazard Classifications:

Health Hazards (Blue) 3
Flammability (Red) 1
Reactivity (Yellow) 0

Flash Point: 288°F (142°C) o.c.

FIRST AID

Move victim to fresh air. Call emergency medical care. Apply artificial respiration if victim is not breathing. Do not use mouth-to-mouth method if victim ingested or inhaled the substance; induce artificial respiration with the aid of a pocket mask equipped with a one-way valve or other proper respiratory medical device. Administer oxygen if breathing is difficult. Remove and isolate contaminated clothing and shoes. In case of contact with substance, immediately flush skin or eyes with running water for at least 20 minutes. For minor skin contact, avoid spreading material on unaffected skin. Keep victim warm and quiet. Effects of exposure (inhalation, ingestion, or skin contact) may be delayed. Ensure that medical personnel are aware of the material(s) involved and take precautions to protect themselves.

DIPHENYL ETHER D:450

SYNONYMS: BENZENE, 1,1'-OXYBIS-; BIPHENYL ETHER; BIPHENYL OXIDE; DIPHENYL OXIDE; DOWTHERM; EINECS No. 202-981-2; ETER DIFENILICO (Spanish); GERANIUM CRYSTALS; OSSIDO di DIFENILE (Italian); 1,1-OXYBISBENZENE; OXYDE de DIPHENYLE (French); PHENOXY BENZENE; PHENYL ETHER; PHENYL OXIDE

IDENTIFICATION

CAS:	101-84-8
Molecular Formula:	$C_{12}H_{10}O$
RTECS No.:	KN8970000
Properties:	Combustible liquid [above 81°F (27°C)], or combustible crystalline solid. Colorless. Mild, geraniumlike odor. May float or sink in water; insoluble. Melting point is 81°F (27°C).

HEALTH AND SAFETY INFORMATION

Irritates eyes, skin, and respiratory tract.

IARC Evaluation: Group 3, not classifiable.
IDLH: 100 ppm as phenyl ether vapor
Odor Threshold: 1 ppm
OSHA Table Z-1-A Air Contaminant

Exposure Limits:	TWA	STEL
ACGIH:	1 ppm; 7 mg/m^3 as vapor	2 ppm; 14 mg/m^3 as vapor
NIOSH:	1 ppm; 7 mg/m^3 as vapor	
OSHA:	1 ppm; 7 mg/m^3 as vapor	

RECOMMENDATIONS FOR RESPIRATOR SELECTION: NIOSH/OSHA

25 ppm: SA:CF (any supplied-air respirator operated in a continuous-flow mode); PAPROVDM (any powered, air-purifying respirator with organic vapor cartridge(s) in combination with a dust and mist filter).
50 ppm: CCRFOVHiE (any chemical cartridge respirator with a full facepiece and organic vapor cartridge(s) in combination with a high-efficiency particulate filter); or GMFOVHiE (any air-purifying, full-facepiece respirator (gas mask) with a chin-style, front- or back-mounted organic vapor canister having a high-efficiency particulate filter); or SCBAF (any self-contained breathing apparatus with a full facepiece); or SAF (any supplied-air respirator with a full facepiece).
100 ppm: SAF:PD,PP (any supplied-air respirator that has a full facepiece and is operated in a pressure-demand or other positive-pressure mode).

Emergency or planned entry into unknown concentration or IDLH conditions: SCBAF:PD,PP (any self-contained breathing apparatus that has a full facepiece and is operated in a pressure-demand or other positive-pressure mode); or SAF:PD,PP:ASCBA (any supplied-air respirator that has a full facepiece and is operated in a pressure-demand or other positive-pressure mode in combination with an auxiliary, self-contained breathing apparatus operated in a pressure-demand or other positive-pressure mode).

Escape: GMFOVHiE (any air-purifying, full-facepiece respirator (gas mask) with a chin-style, front- or back-mounted organic vapor canister having a high-efficiency particulate filter);or SCBAE (any appropriate escape-type, self-contained breathing apparatus).

Note: Substance causes eye irritation or damage; eye protection needed.

CHEMICAL REACTIVITY

Incompatible with strong acids, strong oxidizers; may cause fire and explosions. Attacks some plastics, rubber, and coatings. May accumulate static electricity, and cause ignition of its vapors.

FIRE INFORMATION

Vapors are heavier than air and will collect in low areas. Vapors in confined areas may explode when exposed to fire. Vapors may travel long distances to ignition sources and flash back. Storage containers and parts of containers may rocket great distances, in many directions.If material or contaminated runoff enters waterways, notify downstream users of potentially contaminated waters. Notify local health and fire officials and pollution control agencies. Use dry chemicals or carbon dioxide. Cool containers with water spray.

NFPA Hazard Classifications:

Health Hazards (Blue) 1
Flammability (Red) 1
Reactivity (Yellow) 0

Flash Point: 239°F (115°C) c.c.
Flammable Limits in Air: LEL 0.7%; UEL 6.0%
Autoignition Temperature:1144°F (618°C)

FIRST AID

Move victim to fresh air. Call emergency medical care. Apply artificial respiration if victim is not breathing. Administer oxygen if breathing is difficult. Remove and isolate contaminated clothing and shoes. In case of contact with substance, immediately flush skin or eyes with running water for at least 20 minutes. Wash skin with soap and water. Keep victim warm and quiet. Ensure that medical personnel are aware of the material(s) involved and take precautions to protect themselves.

ENVIRONMENTAL INFORMATION

TSCA: 40CFR716.120(*a*); 40CFR799.3300 testing required as unsubstituted phenylenediamines

DIPHENYL KETONE D:480

SYNONYMS: BENZOFENONA (Spanish); BENZOPHENONE; BENZOYL BENZENE; DIPHENYL KETONE; DIPHENYL METHANONE; EINECS No. 204-337-7; FEMA No. 2134; α-OXODIPHENYLMETHANE; α-OXODITANE; PHENYL KETONE

IDENTIFICATION

CAS:	119-61-9
ERG ID:	UN 3276
ERG Guide:	151
ERG Guide Category:	NITRILES, TOXIC, n.o.s.
Molecular Formula:	$C_{13}H_{10}O$
Properties:	Combustible solid. White. Flowery, roselike odor. May float or sink in water.

HEALTH AND SAFETY INFORMATION

Low toxicity. May cause irritation to eyes, skin , and respiratory tract.

NIOSH Criteria Document: NIOSH 78-212 NITRILES

Exposure Limits:	**TWA**
AIHA-WEEL:	5 mg/m^3

At any concentrations above the NIOSH REL, or where there is no REL, at any detectable concentration: SCBAF:PD,PP (any self-contained breathing apparatus that has a full facepiece and is operated in a pressure-demand or other positive-pressure mode); or SAF:PD,PP:ASCBA (any supplied-air respirator that has a full facepiece and is operated in a pressure-demand or other positive-pressure mode in combination with an auxiliary, self-contained breathing apparatus operated in a pressure-demand or other positive-pressure mode).

Escape: GMFOV [any air-purifying, full-facepiece respirator (gas mask) with a chin-style, front- or back-mounted organic vapor canister]; or SCBAE (any appropriate escape-type, self-contained breathing apparatus).

CHEMICAL REACTIVITY

Will attack some plastics. Contact with strong oxidizers may cause fire and explosions.

FIRE INFORMATION

FEMA Hazard Classifications:

Health Hazards (Blue) 1
Flammability (Red) 1
Reactivity (Yellow) 0

FIRST AID

Move victim to fresh air. Call emergency medical care. Apply artificial respiration if victim is not breathing. Do not use mouth-to-mouth method if victim ingested or inhaled the substance; induce artificial respiration with the aid of a pocket mask equipped with a one-way valve or other proper respiratory medical device. Administer oxygen if breathing is difficult. Remove and isolate contaminated clothing and shoes. In case of contact with substance, immediately flush skin or eyes with running water for at least 20 minutes. For minor skin contact, avoid spreading material on unaffected skin. Keep victim warm and quiet. Effects of exposure (inhalation, ingestion, or skin contact) may be delayed. Ensure that medical personnel are aware of the material(s) involved and take precautions to protect themselves.

DIPROPYLENE GLYCOL D:510

SYNONYMS: BIS(2HYDROXYPROPYL) ETHER; 2,2'-DIHYDROXYDIPROPYL ETHER; 1,1'-DIMETHYLDIETHYLENE GLYCOL; DI-α-PROPYLENE GLYCOL; DI-1,2-PROPYLENE GLYCOL; EINECS No. 203-821-4; 1,1'-OXYDI-2-PROPANOL

IDENTIFICATION

CAS: 110-98-5
Molecular Formula: $C_6H_{14}O_3$
RTECS No.: UB8785000
Properties: Combustible, thick liquid. Colorless. Odorless. Sinks and mixes with water.

HEALTH AND SAFETY INFORMATION

Irritates eyes, skin, and respiratory tract.

RECOMMENDATIONS FOR RESPIRATOR SELECTION

At any concentrations above the NIOSH REL, or where there is no REL, at any detectable concentration: SCBAF:PD,PP (any self-contained breathing apparatus that has a full facepiece and is operated in a pressure-demand or other positive-pressure mode); or SAF:PD,PP:ASCBA (any supplied-air respirator that has a full facepiece and is operated in a pressure-demand or other positive-pressure mode in combination with an auxiliary, self-contained breathing apparatus operated in a pressure-demand or other positive-pressure mode).

Escape: GMFOV [any air-purifying, full-facepiece respirator (gas mask) with a chin-style, front- or back-mounted organic vapor canister]; or SCBAE (any appropriate escape-type, self-contained breathing apparatus).

CHEMICAL REACTIVITY

Incompatible with sulfuric acid, perchloric acid, isocyanates, and strong oxidizers.

FIRE INFORMATION

Vapors are heavier than air and will collect in low areas. Vapors in confined areas may explode when exposed to fire. Storage containers and parts of containers may rocket great distances, in many directions. If material or contaminated runoff enters waterways, notify downstream users of potentially contaminated waters. Notify local health and fire officials and pollution control agencies.

NFPA Hazard Classifications:

Health Hazards (Blue) 0
Flammability (Red) 1
Reactivity (Yellow) 0

Flash Point: 250°F (121°C) o.c.
Flammable Limits in Air: LEL 2.9%; UEL 12.6%
Autoignition Temperature: 570°F (299°C)

FIRST AID

Move victim to fresh air. Call emergency medical care. Apply artificial respiration if victim is not breathing. Administer oxygen if breathing is difficult. Remove and isolate contaminated clothing and shoes. In case of contact with substance, immediately flush skin or eyes with running water for at least 20 minutes. Wash skin with soap and water. Keep victim warm and quiet. Ensure that medical personnel are aware of the material(s) involved and take precautions to protect themselves.

ENVIRONMENTAL INFORMATION

TSCA: 40CFR 716.120(*d*)11 propylene glycol ethers and esters.

DIPROPYLENE GLYCOL DIBENZOATE D:540

SYNONYMS: BENZOFLEX 9-88; BENZOFLEX 9-98; DIBENZOATO de DIPOPILENGLICOL (Spanish); DIBENZOL DIPROPYLENE GLYCOL ESTER; DIPROPANEDIOL DIBENZOATE; EINECS No. 202-340-7; K-FLEX DP; 3,3′-OXYDYL-1-PROPANOL DIBENZOATE

IDENTIFICATION

CAS:	94-51-9
Molecular Formula:	$C_{20}H_{22}O_5$
RTECS No.:	UB8787500
Properties:	Combustible, viscous liquid. Straw color. Faint, aromatic odor. Insoluble in water.

HEALTH AND SAFETY INFORMATION

RECOMMENDATIONS FOR RESPIRATOR SELECTION

At any concentrations above the NIOSH REL, or where there is no REL, at any detectable concentration: SCBAF:PD,PP (any self-contained breathing apparatus that has a full facepiece and is operated in a pressure-demand or other positive-pressure mode); or SAF:PD,PP:ASCBA (any supplied-air respirator that has a full facepiece and is operated in a pressure-demand or other positive-pressure mode in combination with an auxiliary, self-contained breathing apparatus operated in a pressure-demand or other positive-pressure mode).

Escape: GMFOV [any air-purifying, full-facepiece respirator (gas mask) with a chin-style, front- or back-mounted organic vapor canister]; or SCBAE (any appropriate escape-type, self-contained breathing apparatus).

CHEMICAL REACTIVITY

Incompatible with strong acids, nitrates, and oxidizers.

FIRE INFORMATION

Vapors are heavier than air and will collect in low areas. Vapors in confined areas may explode when exposed to fire. Storage containers and parts of containers may rocket great distances, in many directions. If material or contaminated runoff enters waterways, notify downstream users of potentially contaminated waters. Notify local health and fire officials and pollution control agencies. Use dry chemicals, alcohol foam, carbon dioxide, water spray.

FEMA Hazard Classifications:

Health Hazards (Blue) 0
Flammability (Red) 1
Reactivity (Yellow) 0

Flash Point: 300°F (149°C) c.c.

FIRST AID

Move victim to fresh air. Call emergency medical care. Apply artificial respiration if victim is not breathing. Administer oxygen if breathing is difficult. Remove and isolate contaminated clothing and shoes. In case of contact with substance, immediately flush skin or eyes with running water for at least 20 minutes. Wash skin with soap and water. Keep victim warm and quiet. Ensure that medical personnel are aware of the material(s) involved and take precautions to protect themselves.

ENVIRONMENTAL INFORMATION

TSCA: 40CFR 716.120(*d*)11 propylene glycol ethers and esters.

DIPROPYL KETONE D:570

SYNONYMS: BUTYRONE; GBL; DIPROPIL CETONA (Spanish); DPK; EEC No. 606-027-00-X; EINECS No. 204-608-9; HEPTAN-4-ONE; 4-HEPTANONE; PROPYL KETONE

IDENTIFICATION

CAS:	123-19-3
ERG ID:	UN 2710
ERG Guide:	127
ERG Guide Category:	FLAMMABLE LIQUIDS (POLAR/WATER- MISCIBLE)
Molecular Formula:	$C_7H_{14}O$
RTECS No.:	MJ5600000
Properties:	Flammable liquid. Colorless. Pleasant odor. Insoluble in water.

HEALTH AND SAFETY INFORMATION

Irritates eyes, skin, respiratory tract.

Exposure Limits:	**TWA**
ACGIH:	50 ppm; 233 mg/m^3
NIOSH:	50 ppm; 235 mg/m^3
OSHA:	None

RECOMMENDATIONS FOR RESPIRATOR SELECTION: NIOSH

At any concentrations above the NIOSH REL, or where there is no REL, at any detectable concentration: SCBAF:PD,PP (any self-contained breathing apparatus that has a full facepiece and is operated in a pressure-demand or other positive-pressure mode); or SAF:PD,PP:ASCBA (any supplied-air respirator that has a full facepiece and is operated in a pressure-demand or other positive-pressure mode in combination with an auxiliary, self-contained breathing apparatus operated in a pressure-demand or other positive-pressure mode).

Escape: GMFOV [any air-purifying, full-facepiece respirator (gas mask) with a chin-style, front- or back-mounted organic vapor canister]; or SCBAE (any appropriate escape-type, self-contained breathing apparatus).

CHEMICAL REACTIVITY

Reacts with oxidizers.

FIRE INFORMATION

Vapors are heavier than air and will collect in low areas. Vapors in confined areas may explode when exposed to fire. Vapors may travel long distances to ignition sources and flash back. Storage containers and parts of containers may rocket great distances, in many directions. If material or contaminated runoff enters waterways, notify downstream users of potentially contaminated waters. Notify local health and fire officials and pollution control agencies. Use dry chemicals, foam, carbon dioxide.

NFPA Hazard Classifications:

Health Hazards (Blue) 2

Flammability (Red) 2
Reactivity (Yellow) 0

Flash Point: 120°F (49°C) c.c.

FIRST AID

Move victim to fresh air. Call emergency medical care. Apply artificial respiration if victim is not breathing. Administer oxygen if breathing is difficult. Remove and isolate contaminated clothing and shoes. In case of contact with substance, immediately flush skin or eyes with running water for at least 20 minutes. Wash skin with soap and water. Keep victim warm and quiet. Ensure that medical personnel are aware of the material(s) involved and take precautions to protect themselves.

DIPROPYL PHTHALATE D:600

SYNONYMS: 1,2-BENZENEDICARBOXYLIC ACID, DIPROPYL ESTER; DI-n-PROPYL PHTHALATE; DPP; EINECS No. 205-015-8; FTALATO de DIPROPILO (Spanish); PHTHALIC ACID, DIPROPYL ESTER

IDENTIFICATION

CAS:	131-16-8
Molecular Formula:	$C_{14}H_{18}O_4$
RTECS No.:	TI1940000
Properties:	Combustible liquid. Colorless. Soluble in water.

HEALTH AND SAFETY INFORMATION

Irritates eyes, skin, and respiratory tract.

RECOMMENDATIONS FOR RESPIRATOR SELECTION

At any concentrations above the NIOSH REL, or where there is no REL, at any detectable concentration: SCBAF:PD,PP (any self-contained breathing apparatus that has a full facepiece and is operated in a pressure-demand or other positive-pressure mode); or SAF:PD,PP:ASCBA (any supplied-air respirator that has a full facepiece and is operated in a pressure-demand or other positive-pressure mode in combination with an auxiliary, self-contained breathing apparatus operated in a pressure-demand or other positive-pressure mode).

Escape: GMFOV [any air-purifying, full-facepiece respirator (gas mask) with a chin-style, front- or back-mounted organic vapor canister]; or SCBAE (any appropriate escape-type, self-contained breathing apparatus).

CHEMICAL REACTIVITY

Contact with oxidizers may cause fire and explosions. Incompatible with nitrates, nitric acid, sulfuric acid.

FIRE INFORMATION

Vapors are heavier than air and will collect in low areas. Vapors in confined areas may explode when exposed to fire. Vapors may travel long distances to ignition sources and flash back. Storage containers and parts of containers may rocket great distances, in many directions. If material or contaminated runoff enters waterways, notify downstream users of potentially contaminated waters. Notify local health and fire officials and pollution control agencies. Use dry chemicals, foam, carbon dioxide.

NFPA Hazard Classifications: Not listed

Flash Point: 230°F (110°C)

FIRST AID

Move victim to fresh air. Call emergency medical care. Apply artificial respiration if victim is not breathing. Administer oxygen if breathing is difficult. Remove and isolate contaminated clothing and shoes. In case of contact with substance, immediately flush skin or eyes with running water for at least 20 minutes. Wash skin with soap and water. Keep victim warm and quiet. Ensure that medical personnel are aware of the material(s) involved and take precautions to protect themselves.

DIPENTENE DIOXIDE D:630

SYNONYMS: 1,2,8,9-DIEPOXYLIMONENE; 1,2:8,9-DIEPOXYMENTHANE; 1,2:8,9-DIEPOXY-p-MENTHANE; DIOXIDO de DIPENTENO (Spanish); EINECS No. 202-475-1; EPOXIDE 269; 4-(1,2-EPOXY-1-METHYLETHYL)-1-METHYL-7-OXABICYCLO(4.1.0)HEPTANE; LIMONENE DIOXIDE; OXYBICLO(4.1.0)-HEPTANE(7-), I-METHYL-4-(2-METHYLOXIRANYL)-; UNOXAT EPOXIDE 269

IDENTIFICATION

CAS:	96-08-2
Molecular Formula:	$C_{10}H_{16}O_2$
RTECS No.:	OS9100000
Properties:	Liquid. Clear. Soluble in water.

HEALTH AND SAFETY INFORMATION

RECOMMENDATIONS FOR RESPIRATOR SELECTION

At any concentrations above the NIOSH REL, or where there is no REL, at any detectable concentration: SCBAF:PD,PP (any self-contained breathing apparatus that has a full facepiece and is operated in a pressure-demand or other positive-pressure mode); or SAF:PD,PP:ASCBA (any supplied-air respirator that has a full facepiece and is operated in a pressure-demand or other positive-pressure mode in combination with an auxiliary, self-contained breathing apparatus operated in a pressure-demand or other positive-pressure mode).

Escape: GMFOV [any air-purifying, full-facepiece respirator (gas mask) with a chin-style, front- or back-mounted organic vapor canister]; or SCBAE (any appropriate escape-type, self-contained breathing apparatus)

CHEMICAL REACTIVITY

Contact with strong oxidizers may cause fire and explosions.

FIRE INFORMATION

If material or contaminated runoff enters waterways, notify downstream users of potentially contaminated waters. Notify local health and fire officials and pollution control agencies. Use dry chemicals, alcohol-resistant foam, carbon dioxide.

NFPA Hazard Classifications: Not listed

FIRST AID

Move victim to fresh air. Call emergency medical care. Apply artificial respiration if victim is not breathing. Administer oxygen if breathing is difficult. Remove and isolate contaminated clothing and shoes. In case of contact with substance, immediately flush skin or eyes with running water for at least 20 minutes. Wash skin with soap and water. Keep victim warm and quiet. Effects of exposure (inhalation, ingestion, or skin contact) to substance may be delayed. Ensure that medical personnel are aware of the material(s) involved and take precautions to protect themselves.

DITHIOBIURET D:660

SYNONYMS: BIURET,2,4-DITHIO-; 2,4-DITHIOBIURET; DTB; EINECS No. 208-784-8; 2-THIO-1-(THIOCARBAMOYL)UREA; UREA, 2-THIO-1-(THIOCARBAMOYL)-; USAF B-44; USAF EK-P-6281

IDENTIFICATION

CAS:	541-53-7
ERG ID:	UN 3077
ERG Guide:	171
ERG Guide Category:	SUBSTANCES (LOW TO MODERATE HAZARD)
Molecular Formula:	$C_2H_5N_3S_2$
RTECS No.:	EC1575000
Properties:	Crystals or needles from water. Soluble in water

HEALTH AND SAFETY INFORMATION

Poisonous if swallowed. Dust may cause irritation to eyes and respiratory tract.

RECOMMENDATIONS FOR RESPIRATOR SELECTION

At any concentrations above the NIOSH REL, or where there is no REL, at any detectable concentration: SCBAF:PD,PP (any self-contained breathing apparatus that has a full facepiece and is operated in a pressure-demand or other positive-pressure mode); or SAF:PD,PP:ASCBA (any supplied-air respirator that has a full facepiece and is operated in a pressure-demand or other positive-pressure mode in combination with an auxiliary, self-contained breathing apparatus operated in a pressure-demand or other positive-pressure mode).

Escape: GMFOV [any air-purifying, full-facepiece respirator (gas mask) with a chin-style, front- or back-mounted organic vapor canister]; or SCBAE (any appropriate escape-type, self-contained breathing apparatus).

CHEMICAL REACTIVITY

Incompatible, and may form explosive mixtures with strong oxidizers, hypochlorites, ammonium nitrate, chromyl chloride, nitrosyl perchlorate, perchloric acid.

FIRE INFORMATION

Combustion products include toxic nitrogen and sulfur oxides. Vapors are heavier than air and will collect in low areas. Vapors in confined areas may explode when exposed to fire. Storage containers and parts of containers may rocket great distances, in many directions. If material or contaminated runoff enters waterways, notify downstream users of potentially contaminated waters. Notify local health and fire officials and pollution control agencies. Use dry chemicals, foam, carbon dioxide.

FEMA Hazard Classifications:

Health Hazards (Blue) 3
Flammability (Red) 1
Reactivity (Yellow) 0

FIRST AID

Move victim to fresh air. Call emergency medical care. Apply artificial respiration if victim is not breathing. Administer oxygen if breathing is difficult. Remove and isolate contaminated clothing and shoes. In case of contact with substance, immediately flush skin or eyes with running water for at least 20 minutes. Wash skin with soap and water. Keep victim warm and quiet. Effects of exposure (inhalation, ingestion, or skin contact) to substance may be delayed. Ensure that medical personnel are aware of the material(s) involved and take precautions to protect themselves.

ENVIRONMENTAL INFORMATION

EPA Hazardous Waste Number (RCRA No.): P049
RCRA Toxicity Characteristic (Section 261.24): RCRA Section 261 Hazardous Constituents
EPCRA Section 302, Extremely Hazardous Substances: TPQ = 100/10,000 lb (45.4/4,450 kg)
EPCRA Section 304 Reportable Quantity (RQ): CERCLA, 100 lb (45.4 kg)
EPCRA Section 313: Form R *de minimis* concentration reporting level: 1.0%.

DIVINYLBENZENE D:690

SYNONYMS: BENZENE, DIVINYL-; 1,4-DIVINYL BENZENE; DIETHYL BENZENE; DVB; DVB-22; DVB-27; DVB-55; EINECS No. 215-325-5; VINYLSTYRENE

IDENTIFICATION

CAS:	1321-74-0
ERG ID:	UN 2049
ERG Guide:	130
ERG Guide Category:	FLAMMABLE LIQUIDS (NONPOLAR/WATER-IMMISCIBLE/NOXIOUS)
Molecular Formula:	$C_{10}H_{10}$
RTECS No.:	CZ9370000
Properties:	Combustible liquid. Pale, straw-color. Floats on water surface; practically insoluble.

HEALTH AND SAFETY INFORMATION

Irritates eyes, skin, and respiratory tract. Can affect nervous system; may cause dizziness and sleepiness. Long-term or prolonged exposure may result in heart, liver, kidney damage.

Odor Threshold: 10-60 ppm

Exposure Limits: TWA

ACGIH:	10 ppm; 53 mg/m^3
NIOSH:	10 ppm; 50 mg/m^3
OSHA:	None

RECOMMENDATIONS FOR RESPIRATOR SELECTION

At any concentrations above the NIOSH REL, or where there is no REL, at any detectable concentration: SCBAF:PD,PP (any self-contained breathing apparatus that has a full facepiece and is operated in a pressure-demand or other positive-pressure mode); or SAF:PD,PP:ASCBA (any supplied-air respirator that has a full facepiece and is operated in a pressure-demand or other positive-pressure mode in combination with an auxiliary, self-contained breathing apparatus operated in a pressure-demand or other positive-pressure mode).

Escape: GMFOV [any air-purifying, full-facepiece respirator (gas mask) with a chin-style, front- or back-mounted organic vapor canister]; or SCBAE (any appropriate escape-type, self-contained breathing apparatus).

CHEMICAL REACTIVITY

Forms explosive mixture with air. Temperatures above 32°F (0°C) can cause polymerization with risk of fire and explosion. Unless inhibited, must be stored and shipped in refrigeration. Commercial product contains mixed isomers, but m-isomer predominates and usually contains an inhibitor to prevent polymerization. Contact with oxidizers can cause fire and explosions. Incompatible with acids, bases, metal salts. May accumulate static electricity, and cause ignition of its vapors.

FIRE INFORMATION

Vapors are heavier than air and will collect in low areas. Vapors in confined areas may explode when exposed to fire. Vapors may travel long distances to ignition sources and flash back. Storage containers and parts of containers may rocket great distances, in many directions. If material or contaminated runoff enters waterways, notify downstream users of potentially contaminated waters. Notify local health and fire officials and pollution control agencies. Use dry chemicals, foam, carbon dioxide.

NFPA Hazard Classifications:

Health Hazards (Blue) 1

Flammability (Red) 2
Reactivity (Yellow) 2

Flash Point: 169°F (76°C) o.c.
Flammable Limits in Air: LEL 0.7%; UEL 6.2%

FIRST AID

Move victim to fresh air. Call emergency medical care. Apply artificial respiration if victim is not breathing. Administer oxygen if breathing is difficult. Remove and isolate contaminated clothing and shoes. In case of contact with substance, immediately flush skin or eyes with running water for at least 20 minutes. Wash skin with soap and water. Keep victim warm and quiet. Effects of exposure (inhalation, ingestion, or skin contact) to substance may be delayed. Ensure that medical personnel are aware of the material(s) involved and take precautions to protect themselves.

ENVIRONMENTAL INFORMATION

TSCA: 40CFR 799.5000; Notification of export required.

DODECENYLSUCCINIC ANHYDRIDE D:720

SYNONYMS: ANHYDRIDE DODECENYLSUCCINIQUE (French); DDS; DDS A; DODECENYLBARNSTEENZUURANHYDRIDE (Dutch); EINECS No. 246-917-1; 2,5-FURANDIONE, 3-(DODECENYL)DIHYDRO-

IDENTIFICATION

CAS:	25377-73-5
Molecular Formula:	$C_{16}H_{27}O_3$
RTECS No.:	WN1225000
Properties:	Combustible oily liquid. Clear to light-yellow.

HEALTH AND SAFETY INFORMATION

Irritates eyes, skin, and respiratory tract. May cause allergic reaction or chronic problems.

RECOMMENDATIONS FOR RESPIRATOR SELECTION

At any concentrations above the NIOSH REL, or where there is no REL, at any detectable concentration: SCBAF:PD,PP (any self-contained breathing apparatus that has a full facepiece and is operated in a pressure-demand or other positive-pressure mode); or SAF:PD,PP:ASCBA (any supplied-air respirator that has a full facepiece and is operated in a pressure-demand or other positive-pressure mode in combination with an auxiliary, self-contained breathing apparatus operated in a pressure-demand or other positive-pressure mode).

Escape: GMFOV [any air-purifying, full-facepiece respirator (gas mask) with a chin-style, front- or back-mounted organic vapor canister]; or SCBAE (any appropriate escape-type, self-contained breathing apparatus).

CHEMICAL REACTIVITY

Reacts with strong oxidizers.

FIRE INFORMATION

Vapors are heavier than air and will collect in low areas. Vapors in confined areas may explode when exposed to fire. If material or contaminated runoff enters waterways, notify downstream users of potentially contaminated waters. Notify local health and fire officials and pollution control agencies. Use dry chemicals, alcohol-resistant foam, carbon dioxide.

NFPA Hazard Classifications: Not listed

Flash Point: 352°F (178°C) o.c.

FIRST AID

Move victim to fresh air. Call emergency medical care. Apply artificial respiration if victim is not breathing. Administer oxygen if breathing is difficult. Remove and isolate contaminated clothing and shoes. In case of contact with substance, immediately flush skin or eyes with running water for at least 20 minutes. Wash skin with soap and water. Keep victim warm and quiet. Effects of exposure (inhalation, ingestion, or skin contact) to substance may be delayed. Ensure that medical personnel are aware of the material(s) involved and take precautions to protect themselves.

DODECYLMETHACRYLATE D:750

SYNONYMS: DODECYL-2-METHYL-2-PROPENOATE; LAURYL METHACRYLATE; METACRILATO de LAURILO (Spanish); METHACRYLIC ACID, DODECYL ESTER; PROPENOIC ACID, 2-METHYL, DODECYL ESTER(2-)

IDENTIFICATION

CAS:	142-90-5
Molecular Formula:	$C_{16}H_{30}O_2$
Properties:	Combustible liquid. Floats on water; insoluble

HEALTH AND SAFETY INFORMATION

Irritates eyes, skin, and respiratory tract.

RECOMMENDATIONS FOR RESPIRATOR SELECTION

At any concentrations above the NIOSH REL, or where there is no REL, at any detectable concentration: SCBAF:PD,PP (any self-contained breathing apparatus that has a full facepiece and is operated in a pressure-demand or other positive-pressure mode); or SAF:PD,PP:ASCBA (any supplied-air respirator that has a full facepiece and is operated in a pressure-demand or other positive-pressure mode in combination with an auxiliary, self-contained breathing apparatus operated in a pressure-demand or other positive-pressure mode).

Escape: GMFOV [any air-purifying, full-facepiece respirator (gas mask) with a chin-style, front- or back-mounted organic vapor canister]; or SCBAE (any appropriate escape-type, self-contained breathing apparatus).

CHEMICAL REACTIVITY

Oxidizers or reducing agents may initiate polymerization.

FIRE INFORMATION

Vapors are heavier than air and will collect in low areas. Vapors in confined areas may explode when exposed to fire. Storage containers and parts of containers may rocket great distances, in many directions. If material or contaminated runoff enters waterways, notify downstream users of potentially contaminated waters. Notify local health and fire officials and pollution control agencies. Use dry chemicals, foam, carbon dioxide. Use water spray to cool containers.

NFPA Hazard Classifications: Not listed

Flash Point: More than 230°F (110°C) c.c.

FIRST AID

Move victim to fresh air. Call emergency medical care. Apply artificial respiration if victim is not breathing. Administer oxygen if breathing is difficult. Remove and isolate contaminated clothing and shoes. In case of contact with substance, immediately flush skin or eyes with running water for at least 20 minutes. Wash skin with soap and water. Keep victim warm and quiet. Effects of exposure (inhalation, ingestion, or skin contact) to substance may be delayed. Ensure that medical personnel are aware of the material(s) involved and take precautions to protect themselves.

DODECYLTRICHLOROSILANE D:780

SYNONYMS: DODECILTRICLOROSILANO (Spanish); DODECYL TRICHLOROSILANE; EINECS No. 224-769-9; SILANE, TRICHLORODODECYL-; TRICHLORODODECYLSILANE; SILANE, DODECYLTRICHLORO-

IDENTIFICATION

CAS:	4484-72-4
ERG ID:	UN 1771
ERG Guide:	156
ERG Guide Category:	SUBSTANCES-TOXIC AND/OR CORROSIVE (COMBUSTIBLE/WATER-SENSITIVE)
Molecular Formula:	$C_{12}H_{25}Cl_3Si$
RTECS No.:	VV3940000
Properties:	Combustible liquid. Colorless to yellow. Sharp, irritating odor. Reacts with water, forming hydrochloric acid. Irritating and corrosive hydrogen chloride gas is produced on contact with water.

HEALTH AND SAFETY INFORMATION

Corrosive to eyes, skin and respiratory tract. Inhalation may cause pulmonary edema, which can be delayed for several hours; there is a risk of death in serious cases.

RECOMMENDATIONS FOR RESPIRATOR SELECTION

At any concentrations above the NIOSH REL, or where there is no REL, at any detectable concentration: SCBAF:PD,PP (any self-contained breathing apparatus that has a full facepiece and is operated in a pressure-demand or other positive-pressure mode); or SAF:PD,PP:ASCBA (any supplied-air respirator that has a full facepiece and is operated in a pressure-demand or other positive-pressure mode in combination with an auxiliary, self-contained breathing apparatus operated in a pressure-demand or other positive-pressure mode).

Escape: GMFOV [any air-purifying, full-facepiece respirator (gas mask) with a chin-style, front- or back-mounted organic vapor canister]; or SCBAE (any appropriate escape-type, self-contained breathing apparatus).

CHEMICAL REACTIVITY

Forms explosive mixture with air. Reacts with strong bases and oxidizers. Hydrolyses with water forming corrosive and toxic hydrogen chloride. Corrosive to common metals in the presence of moisture.

FIRE INFORMATION

Vapors are heavier than air and will collect in low areas. Vapors in confined areas may explode when exposed to fire. Vapors may travel long distances to ignition sources and flash back. Storage containers and parts of containers may rocket great distances, in many directions. If material or contaminated runoff enters waterways, notify downstream users of potentially contaminated waters. Notify local health and fire officials and pollution control agencies. Use dry chemicals, foam, carbon dioxide.

Flash Point: 150°F (66°C) o.c.

FIRST AID

Move victim to fresh air. Call emergency medical care. Apply artificial respiration if victim is not breathing. Do not use mouth-to-mouth method if victim ingested or inhaled the substance; induce artificial respiration with the aid of a pocket mask equipped with a one-way valve or other proper respiratory medical device. Administer oxygen if breathing is difficult. Remove and isolate contaminated clothing and shoes. In case of contact with substance, immediately flush skin or eyes with running water for at least 20 minutes. For minor skin contact, avoid spreading material on unaffected skin. Keep victim warm and quiet. Effects of exposure (inhalation, ingestion, or skin contact) may be delayed. Ensure that medical personnel are aware of the material(s) involved and take precautions to protect themselves.

EPICHLOROHYDRIN E:020

SYNONYMS: 1-CLOOR-2,3-EPOXY-PROPAAN (Dutch); 1-CHLOR-2,3-EPOXY-PROPAN (German); 1-CHLORO-2,3-EPOXYPROPANE; 3-CHLORO-1,2-EPOXYPROPANE; epi-CHLOROHYDRIN; (CHLOROMETHYL)ETHYLENE OXIDE; CHLOROMETHYLOXIRANE; 2-CHLOROPROPYLENE OXIDE; 3-CHLOROPROPYLENE OXIDE; 3-CHLORO-1,2-PROPYLENE OXIDE; CHLOROMETHYL OXIRANE; γ-CHLOROPROPYLENE OXIDE; 1-CHLORO-2,3-EPOXYPROPANE; ECH; EEC No. 603-026-00-6; EPICHLOORHYDRINE (Dutch); EPICHLORHYDRIN (German); EPICHLORHYDRINE (French); α-EPICHLOROHYDRIN; EPICHLOROHYDRYNA (Polish); EPICLORHIDRINA (Spanish); EPICLORIDRINA (Italian); 1,2-EPOXY-3-CHLOROPROPANE; EPOXY RESIN COMPONENT; GLYCEROL EPICHLOROHYDRIN; OXIRANE, 2-(CHLOROMETHYL)-; PHENOXY RESIN COMPONENT; SKEKhg

IDENTIFICATION

CAS:	106-89-8
ERG ID:	UN 2023
ERG Guide:	131
ERG Guide Category:	FLAMMABLE LIQUIDS-TOXIC
Molecular Formula:	C_3H_5ClO
RTECS No.:	TX4900000
Properties:	Highly flammable, watery liquid. Clear, colorless. Irritating, sweet garlic or chloroformlike odor. Sinks and is moderately soluble in water.

HEALTH AND SAFETY INFORMATION

Extremely irritating to skin, eyes, and respiratory tract. Skin and eye contact causes severe burns and blindness.Long–term exposure may cause changes in the eyes and lungs, kidney damage, and chronic asthmatic bronchitis. Epichlorohydrin has been linked to cancers and changes in the genetic material of occupationally exposed humans. May cause cancer of the nose and throat; decreased fertility in males; lung, liver, and adrenal gland damage; antifertility effects, including temporary sterility; skin burns, and other problems.

IARC Evaluation: Group 2A, probable human carcinogen; animal evidence, sufficient.
NTP: Anticipated human carcinogen.
IDLH: 75 ppm; potential carcinogen
Odor Threshold: 0.934 ppm

OSHA Table Z-1-A Air Contaminant
NIOSH Criteria Document: NIOSH 76-206

Exposure Limits:	**TWA**
ACGIH:(1)(2)	0.5 ppm and 1.9 mg/m^3
NIOSH:(3)	
OSHA:(2)	5 ppm; 19 mg/m^3

(1) Skin contact contributes significantly to overall exposure.
(2) Confirmed animal carcinogen.
(3) Potential carcinogen. Limit exposure to lowest feasible concentration

RECOMMENDATIONS FOR RESPIRATOR SELECTION: NIOSH

At any concentrations above the NIOSH REL, or where there is no REL, at any detectable concentration: SCBAF:PD,PP (any self-contained breathing apparatus that has a full facepiece and is operated in a pressure-demand or other positive-pressure mode); or SAF:PD,PP:ASCBA (any supplied-air respirator that has a full facepiece and is operated in a pressure-demand or other positive-pressure mode in combination with an auxiliary, self-contained breathing apparatus operated in a pressure-demand or other positive-pressure mode).

Escape: GMFOVAG [any air-purifying, full-facepiece respirator (gas mask) with a chin-style, front- or back-mounted organic vapor or acid gas canister]; or SCBAE (any appropriate escape-type, self-contained breathing apparatus).

CHEMICAL REACTIVITY

Forms an explosive mixture with air. Heat, acids, alkalies, and metallic halides can cause explosive polymerization. Aliphatic amines, alkaline earths, alkali metals, alkanolamines, powdered metals, and strong oxidizers may cause fire and explosions. Can react violently with nitric acid, sulfuric acid, 2-aminoethanol, chlorosulfonic acid, ethylene diamine, and potassium tert-butoxide. Will pit steel in the presence of moisture. May accumulate static electrical charges and may cause ignition of its vapors. Decomposition forms highly toxic phosgene gas and fumes of chlorides.

FIRE INFORMATION

Combustion products include toxic hydrogen chloride. Firefighting gear, including SCBA, does not provide adequate protection. If exposure occurs, remove and isolate gear immediately and thoroughly decontaminate personnel. May react with itself at high temperatures, blocking relief valves and leading to tank explosion. Vapors may travel long distances to ignition sources and flash back. Vapors in confined areas or containers may explode in fire. If material or contaminated runoff enters waterways, notify downstream users of potentially contaminated water.

NFPA Hazard Classifications:

Health Hazards (Blue) 3
Flammability (Red) 3
Reactivity (Yellow) 2

Flash Point: 88°F (31°C) c.c.
Flammable Limits in Air: LEL 3.7%; UEL 21.4%
Autoignition Temperature: 772°F (411°C)

FIRST AID

Move victim to fresh air. Call emergency medical care. Apply artificial respiration if victim is not breathing. Do not use mouth-to-mouth method if victim ingested or inhaled the substance; induce artificial respiration with the aid of a pocket mask equipped with a one-way valve or other proper respiratory medical device. Administer oxygen if breathing is difficult. Remove and isolate contaminated clothing and shoes. In case of contact with substance, immediately flush skin or eyes with running water for at least 20 minutes. Wash skin with soap and water. Keep victim warm and quiet. Effects of exposure (inhalation, ingestion, or skin contact) to substance may be delayed. Ensure that medical personnel are aware of the material(s) involved and take precautions to protect themselves.

ENVIRONMENTAL INFORMATION

Clean Air Act: Hazardous Air Pollutants (Title I, Part A, Section 112); Accidental Release Prevention/Flammable substances, (Section 112[r], Table 3), TQ = 20,000 lb (9,080 kg).
Clean Water Act: Section 311 Hazardous Substances/RQ (same as CERCLA); Section 313 Priority Chemicals.
EPA Hazardous Waste Number (RCRA No.): U041.
RCRA Section 261 Hazardous Constituents
Safe Drinking Water Act: Regulated chemical (47 FR 9352).
EPCRA Section 302 Extremely Hazardous Substances: TPQ = 1,000 lb (454 kg).
EPCRA Section 304: Reportable Quantity (RQ): CERCLA, 100 lb (45.4 kg).
EPCRA Section 313: Form R *de minimis* concentration reporting level: 0.1%.
California's Proposition 65: Known to cause cancer.

ETHANOLAMINE E:060

SYNONYMS: AETHANOLAMIN (German); 2-AMINOAETHANOL (German); 2-AMINOETHANOL; 2-AMINOETANOLO (Italian); AMINOETHYL ALCOHOL; β-AMINOETHYL ALCOHOL; BACKLAP SLURRY; BAKER PRS-1000R, POSITIVE RESIST STRIPPER; BRAVO EXTRA HEAVY DUTY STRIPPER; COLAMINE; DELTASTRIP 70NS; EEC No. 603-030-00-8; ETANOLAMINA (Italian, Spanish); β-ETHANOLAMINE; ETHYLOLAMINE; GLYCINOL; 2-HYDROXYETHYLAMINE; β-HYDROXYETHYLAMINE; KTI NMD-25(+); LOCOTENGE 520; MARKEM 535 CLEANER; MEA; MICROPOSIT REMOVER 1112A; MONOAETHANOLAMIN (German); MONOETHANOLAMINE;

OLAMINE; PD ACTIVATOR; R-10 STRIPPER; RUSTLICK B-51; SEL-REX CIRCUITPREP SC REPLINISHER/MAKEUP; SEL-REX XR-170A PRETREATMENT; THIOFACO M-50; USAF EK-1597

IDENTIFICATION

CAS:	141-43-5
ERG ID:	UN 2491
ERG Guide:	153
ERG Guide Category:	TOXIC AND/OR CORROSIVE (COMBUSTIBLE)
Molecular Formula:	C_2H_7NO
RTECS No.:	KJ5775000
Properties:	Very flammable, thick, oily liquid; solid below 51°F (11°C). Colorless. Slight ammonia, fishlike odor. Sinks and mixes with water.

HEALTH AND SAFETY INFORMATION

Severely irritating to skin, eyes, nose, and lungs; skin and eye contact causes severe burns and possible blindness. Long–term exposure may cause skin irritation and burns. Continuous exposure to 5 ppm may cause drowsiness and stupor. May cause kidney and liver damage, weight loss, skin irritation, and itching.

IDLH: 76.2 mg/m^3 (30 ppm)
Odor Threshold: 2.59 ppm
OSHA Table Z-1-A Air Contaminant

Exposure Limits:	TWA	STEL
ACGIH:	3 ppm; 7.5 mg/m^3	6 ppm; 15 mg/m^3
NIOSH:	3 ppm; 6 mg/m^{3}y	6 ppm; 15 mg/m^3
OSHA:	3 ppm; 6 mg/m^3	

RECOMMENDATIONS FOR RESPIRATOR SELECTION: NIOSH/OSHA

30 ppm: CCRS [any chemical cartridge respirator with cartridge(s) providing protection against the compound of concern]; or GMFS [any air-purifying, full-facepiece respirator (gas mask) with a chin-style, front- or back-mounted canister providing protection against the compound of concern]; or PAPRS [any powered, air-purifying respirator with cartridge(s) providing protection against the compound of concern];or SA (any supplied-air respirator);or SCBAF (any self-contained breathing apparatus with a full facepiece).

Emergency or Planned Entry into Unknown Concentrations or IDLH Conditions:
SCBAF:PD.PP (any self-contained breathing apparatus that has a full facepiece and is operated in a pressure-demand or other positive-pressure mode);or SAF:PD,PP:ASCBA (any supplied-air respirator that has a full facepiece and is operated in a pressure-demand or other positive-pressure mode in combination with an auxiliary, self-contained breathing apparatus operated in a pressure-demand or other positive-pressure mode).

Escape: GMFS [any air-purifying, full-facepiece respirator (gas mask) with a chin-style, front- or back-mounted canister providing protection against the compound of concern];or SCBAE (any appropriate escape-type, self-contained breathing apparatus).
Note: Substance reported to cause eye irritation or damage; may require eye protection.

CHEMICAL REACTIVITY

Will oxidize in air; often shipped under a pad of inert gas. Forms explosive mixture with air. Contact with strong oxidizers may cause fire and explosions. Can react violently with acetic acid, acetic anhydride, acrolein, acrylic acid, acrylonitrile, chlorosulfonic acid, epichlorohydrin, hydrogen chloride, hydrogen fluoride, mesityl oxide, nitric acid, oleum, sulfuric acid, propiolactone, vinyl acetate, and strong oxidizers such as dichromates, permanganates, and chlorine. A strong base; incompatible with acids, organic anhydrides, isocyanates, alkylene oxides, aldehydes, alcohols, glycols, phenols, cresols, and caprolactum solution. Contact with strong acids may cause spattering. Attacks plastics, rubber, aluminum, copper, tin, and their alloys.

FIRE INFORMATION

Combustion products include toxic nitrogen oxides. Firefighting gear (including SCBA) does not provide adequate protection. If exposure occurs, remove and isolate gear immediately and thoroughly decontaminate personnel. Vapors are heavier than air and will collect in low areas. Vapors in confined areas may explode when exposed to fire. Storage containers and parts of containers may rocket great distances, in many directions. If material or contaminated runoff enters waterways, notify downstream users of potentially contaminated waters. Notify local health and fire officials and pollution control agencies. Use dry chemicals, alcohol-resistant foam, carbon dioxide.

NFPA Hazard Classifications:

Health Hazards (Blue) 3
Flammability (Red) 2
Reactivity (Yellow) 0

Flash Point: 200°F (93°C) o.c.; 186°F (86°C) c.c.
Flammable Limits in Air: LEL 3.0% at 284°F (140°C); UEL 23.5% at 140°F (60°C)
Autoignition Temperature: 766°F (408°C)

FIRST AID

Move victim to fresh air. Call emergency medical care. Apply artificial respiration if victim is not breathing. Do not use mouth-to-mouth method if victim ingested or inhaled the substance; induce artificial respiration with the aid of a pocket mask equipped with a one-way valve or other proper respiratory medical device. Administer oxygen if breathing is difficult. Remove and isolate contaminated clothing and shoes. In case of contact with substance, immediately flush skin or eyes with running water for at least 20 minutes. For minor skin contact, avoid spreading material on unaffected skin. Keep victim warm and quiet. Effects of exposure (inhalation, ingestion, or skin contact) may be delayed. Ensure that medical personnel are aware of the material(s) involved and take precautions to protect themselves.

ENVIRONMENTAL INFORMATION

TSCA: Reporting lists.

2-ETHOXYETHANOL E:080

SYNONYMS: ATHYLENGLYKOL-MONOATHYLATHER (German); CELLOSOLVE; CELLOSOLVE SOLVENT; DAG 154; DEVELOPER 1002; DOWANOL E; DOWANOL EE; DYNASOLVE MP-500; DYNASOLVE MP ALUMINIUM GRADE; EEC No. 603-012-00-x; EGEE; ETHANOL,2-ETHOXY-; ETHER MONOETHYLIQUE de L'ETHYLENE-GLYCOL (French); ETHYL CELLOSOLVE; ETHYLENE GLYCOL ETHYL ETHER; 2-ETOXIETANOL (Spanish); EKTASOLVE EE; GLYCOL MONOETHYL ETHER; ETHYLENE GLYCOL MONOETHYL ETHER; ETOKSYETYLOWY ALKOHOL (Polish); GLYCOL ETHYL ETHER; GLYCOL ETHER; GLYCOL MONOETHYL ETHER; HYDROXY ETHER; JEFFERSOL EE; JUSTRITE THINNER AND CLEANER; MONOETHYL ETHER GLYCOL; NCI-C54853; OXITOL; POLY-SOLV E; POLY-SOLV EE; PYRALIN PI 2563; 2EE; ULTRAMAC 55

IDENTIFICATION

CAS:	110-80-5
ERG ID:	UN 1171
ERG Guide:	127
ERG Guide Category:	FLAMMABLE LIQUIDS (POLAR/WATER MISCIBLE)
Molecular Formula:	$C_4H_{10}O_2$
RTECS No.:	KK8050000
Properties:	Combustible, oily liquid. Colorless. Sweet, etherlike odor. Floats and mixes with water.

HEALTH AND SAFETY INFORMATION

Irritates eyes, skin, and respiratory tract. Long-term exposure may cause damage to the testes, infertility; birth defects, low blood count, liver and kidney damage.

IDLH: 2745 mg/m^3 (500 ppm)
Odor Threshold: 1.22 ppm
OSHA Table Z-1-A Air Contaminant

Exposure Limits:	TWA
ACGIH:*	5 ppm; 18 mg/m^3
NIOSH:*	0.5 ppm; 1.8 mg/m^3, reduce to lowest feasible level as glycol ethers
OSHA:*	200 ppm; 740 mg/m^3

* Skin contact contributes significantly to overall exposure.

RECOMMENDATIONS FOR RESPIRATOR SELECTION: NIOSH

5 ppm: SA (any supplied-air respirator).
12.5 ppm: SA:CF (any supplied-air respirator operated in a continuous-flow mode).
25 ppm: SCBAF (any self-contained breathing apparatus with a full facepiece);or SAF (any supplied-air respirator with a full facepiece).
500 ppm: SA:PD,PP (any supplied-air respirator operated in a pressure-demand or other positive-pressure mode).

Emergency or Planned Entry into Unknown Concentrations or IDLH Conditions: SCBAF:PD,PP (any self-contained breathing apparatus that has a full facepiece and is operated in a pressure-demand or other positive-pressure mode);or SAF:PD,PP:ASCBA (any supplied-air respirator that has a full facepiece and is operated in a pressure-demand or other positive-pressure mode in combination with an auxiliary, self-contained breathing apparatus operated in a pressure-demand or other positive-pressure mode).

Escape: GMFOV[any air-purifying, full-facepiece respirator (gas mask) with a chin-style, front- or back-mounted organic vapor canister];or SCBAE (any appropriate escape-type, self-contained breathing apparatus).

Note: Substance reported to cause eye irritation or damage; may require eye protection.

CHEMICAL REACTIVITY

Forms explosive mixture with air. Strong oxidizers may cause fire and explosions. Attacks some plastics, rubber, and coatings. Able to form unstable peroxides. Incompatible with strong acids and aluminum and its alloys.

FIRE INFORMATION

Vapors are heavier than air and will collect in low areas. Vapors in confined areas may explode when exposed to fire. Vapors may travel long distances to ignition sources and flash back. May react with itself blocking, relief valves and leading to tank explosions. Storage containers and parts of containers may rocket great distances, in many directions. If material or contaminated runoff enters waterways, notify downstream users of potentially contaminated waters. Notify local health and fire officials and pollution control agencies. Use dry chemicals, foam, carbon dioxide.

NFPA Hazard Classifications:

Health Hazards (Blue) 2
Flammability (Red) 2
Reactivity (Yellow) 0

Flash Point: 110°F (43°C) c.c.
Flammable Limits in Air: LEL 1.7% at 200°F (93°C); UEL 15.6% at 200°F (93°C)
Autoignition Temperature: 455°F (235°C)

FIRST AID

Move victim to fresh air. Call emergency medical care. Apply artificial respiration if victim is not breathing. If breathing is difficult, give oxygen. Remove and isolate contaminated clothing and shoes. In case of contact with substance, immediately flush skin or eyes with running water for at least 20 minutes. Wash skin with soap and water. Keep victim warm and quiet. Ensure that medical personnel are aware of the material(s) involved and take precautions to protect themselves.

ENVIRONMENTAL INFORMATION

Glycol ethers are among those chemicals released in the greatest quantity by the electronics/computer industry and are a water-soluble, volatile organic chemicals. They may enter the environment in liquid form via petrochemical plant effluents or as an unburned gas from combustion sources.

EPA Hazardous Waste Number (RCRA No.): U359.
RCRA Section 261 Hazardous Constituents
EPCRA Section 304: Reportable Quantity (RQ): CERCLA, 1,000 lb (454 kg).
EPCRA Section 313: Form R *de minimis* concentration reporting level: 1.0%.
California's Proposition 65: Reproductive toxin (male).

2-ETHOXYETHYL ACETATE E:100

SYNONYMS: ACETATO de 2-ETOXIETILO (Spanish); ACETIC ACID, 2-ETHOXYETHYL ESTER; 1-ACETOXY-2-ETHOXYETHANE; ARISTOLINE(+); AZ 1310-SF(+); AZ 1312-SFD(+); AZ 1318-SFD(+); AZ 1350J(+); AZ 1370(+); AZ 1370-SF(+); AZ 1375(+); AZ 1470(+); AZ 4140(+); AZ 4210(+); AZ 4330(+); AZ 4620(+); AZ PROTECTIVE COATING; AZ THINNER; CELLOSOLVE ACETATE; EE ACETATE; EEC No. 607-037-00-7; EGEEA; EGMEA; ETHANOL, 2-ETHOXY-, ACETATE; 2-ETHOXYETHANOL ACETATE; ETHOXYETHYL ACETATE; β-ETHOXYETHYL ACETATE; ETHYL CELLOSOLVE ACETATE; ETHYLENE GLYCOL ACETATE MONOETTHY ETHER; ETHYLENE GLYCOL ETHYL ETHER ACETATE; ETHYLENE GLYCOL MONOETHYL ETHER ACETATE; ETHYLENE GLYCOL MONOETHYL ETHER MONOACETATE; ETHYLOXITOL ACETATE; GLYCOL MONOETHYL ETHER ACETATE; H.M.D.S. III; HYDROXY ETHER; KODAK MX-936; KODAK PHOTORESIST DEVELOPER; KTI II; KTI 1300 THINNER; KTI 820(+); KTI 820J(+); KTI 1350J(+); KTI 1370; KTI 1375(+); KTI 1470(+); KTI 9000; KTI 9000K; KTI 9010(+); LIQUID ALKALINE STRIP 7463; MARKEM THINNER XF; MICROPOSIT 119 THINNER; MICROPOSIT 111S(+); MICROPOSIT 119S(+); MICROPOSIT 1375(+); MICROPOSIT 1400-33(+); MICROPOSIT 1400S(+); MICROPOSIT 1450J(+); MICROPOSIT 1470(+); MICROPOSIT SAL 601-ER7(+); MICROPOSIT XP-6009(+); MICROPOSIT XP-6012(+); MS-470 URETHANE COATING; OFPR-800 AR-15(+); PR-21 RESIST; PR-55 RESIST; POLY-SOLV EE ACETATE; SELECTILUX P-15(+); SENSOLVE EEA; THINNER E; TSMR 8800(+); ULTAMAC PR-68 RESIN; ULTRAMAC PR-1024 MB-628 RESIN; WAXIVATION COMPOUND; WAYCOAT 204(+); WAYCOAT HPR 205; WAYCOAT 207(+); WAYCOAT RX 507(+); XANTHOCHROME(+)

IDENTIFICATION

CAS:	111-15-9
ERG ID:	UN 1172
ERG Guide:	129
ERG Guide Category:	FLAMMABLE LIQUIDS (POLAR/WATER-MISCIBLE/ NOXIOUS)
Molecular Formula:	$C_6H_{12}O_3$
RTECS No.:	KK8225000
Properties:	Combustible liquid. Colorless. Mild esterlike odor. Floats and slowly mixes with water.

HEALTH AND SAFETY INFORMATION

Irritates eyes, skin, and respiratory tract. Long-term exposure may cause testes damage and infertility; damage to the developing fetus, kidney and liver damage, skin irritation with drying and cracking.

IDLH: 500 ppm
Odor Threshold: 0.182 ppm
OSHA Table Z-1-A Air Contaminant

Exposure Limits:	**TWA**
ACGIH:[(1)]	5 ppm; 27 mg/m^3
NIOSH:[(1)]	0.5 ppm; 2.7 mg/m^3
OSHA:[(1)]	100 ppm; 540 mg/m^3

[(1)] Skin contact contributes significantly to overall exposure.

RECOMMENDATIONS FOR RESPIRATOR SELECTION: NIOSH

5 ppm: CCROV [any chemical cartridge respirator with organic vapor cartridge(s)]; SA (any supplied-air respirator).
12.5 ppm: SA:CF (any supplied-air respirator operated in a continuous-flow mode); PAPROV [any powered, air-purifying respirator with organic vapor cartridge(s)].
25 ppm: CCROV [any chemical cartridge respirator with organic vapor cartridge(s)]; GMFOV [any air-purifying, full-facepiece respirator (gas mask) with a chin-style, front- or back-mounted acid gas canister]; PAPRTOV [any powered, air-purifying respirator with a tight-fitting facepiece and organic vapor cartridge(s)];or SCBAF (any self-contained breathing apparatus with a full facepiece);or SAF (any supplied-air respirator with a full facepiece).
500 ppm: SA:PD,PP (any supplied-air respirator operated in a pressure-demand or other positive-pressure mode).

Emergency or Planned Entry into Unknown Concentrations or IDLH Conditions:
SCBAF:PD,PP (any self-contained breathing apparatus that has a full facepiece and is operated in a pressure-demand or other positive-pressure mode);or SAF:PD,PP:ASCBA (any supplied-air respirator that

has a full facepiece and is operated in a pressure-demand or other positive-pressure mode in combination with an auxiliary, self-contained breathing apparatus operated in a pressure-demand or other positive-pressure mode).

Escape: GMFOV [any air-purifying, full-facepiece respirator (gas mask) with a chin-style, front- or back-mounted organic vapor canister];or SCBAE (any appropriate escape-type, self-contained breathing apparatus).

Note: Substance reported to cause eye irritation or damage; may require eye protection.

CHEMICAL REACTIVITY

Forms explosive mixture with air. Incompatible with strong acids, nitrates. Violent reaction with oxidizers. May form unstable peroxides. Softens many plastics. Attacks some plastics, rubber, and coatings.

FIRE INFORMATION

Vapors are heavier than air and will collect in low areas. Vapors in confined areas may explode when exposed to fire. Vapors may travel long distances to ignition sources and flash back. May react with itself, blocking relief valves and leading to tank explosions. Storage containers and parts of containers may rocket great distances, in many directions. If material or contaminated runoff enters waterways, notify downstream users of potentially contaminated waters. Notify local health and fire officials and pollution control agencies. Use dry chemicals, alcohol-resistant foam, carbon dioxide. From a secure, explosion-proof location, use water spray to cool exposed containers. If cooling streams are ineffective (venting sound increases in volume and pitch, tank discolors, or shows any signs of deforming), withdraw immediately to a secure position.

NFPA Hazard Classifications:

Health Hazards (Blue) 2
Flammability (Red) 2
Reactivity (Yellow) 0

Flash Point: 124°F (53°C) c.c.
Flammable Limits in Air: LEL 1.7%; UEL 8.4%
Autoignition Temperature: 720°F (382°C)

FIRST AID

Move victim to fresh air. Call emergency medical care. Apply artificial respiration if victim is not breathing. Administer oxygen if breathing is difficult. Remove and isolate contaminated clothing and shoes. In case of contact with substance, immediately flush skin or eyes with running water for at least 20 minutes. Wash skin with soap and water. Keep victim warm and quiet. Effects of exposure (inhalation, ingestion, or skin contact) to substance may be delayed. Ensure that medical personnel are aware of the material(s) involved and take precautions to protect themselves.

ENVIRONMENTAL INFORMATION (as glycol ethers)

Glycol ethers are among those chemicals released in the greatest quantity by the electronics/computer industry and is a water-soluble, volatile organic chemicals. They may enter the environment in liquid form via petrochemical plant effluents or as an unburned gas from combustion sources.

Clean Air Act: Hazardous Air Pollutants (Title I, Part A, Section 112)includes mono- and di-ethers of ethylene glycol, diethyl glycol, and triethylene glycol R-(OCH2CH2)n-OR' where n = 1,2, or 3; R = alkyl or aryl groups; R' = R, H, or groups that, when removed, yield glycol ethers with the structure: R-(OCH2CH)n-OH. Polymers are excluded from the glycol category.
EPCRA Section 313: Form R *de minimis* concentration reporting level: 1.0%. Form R Toxic Chemical Category Code: N230.

ETHYL ACETATE E:120

SYNONYMS: ACETATO de ETILO (Spanish); ACETIC ACID, ETHYL ESTER; ACETIC ESTER; ACETIC ETHER; ETHYL ETHANOATE; ACETIDIN; ACETOXYETHANE; AETHYLACETAT (German); ARSENOSILICA FILM 0308; AS-1; AS 1CE; AS 5CE; AS 18CZ5E; AS 18CZ6E; AS 18CZ10A; AS 1400; B446; BORON B-30; BORON B-40; BORON B-5; BORON B-60; CLEANER, INK INDEPENDENT; EEC No. 607-022-00-5; EINECS No. 205-500-4; ESSIGESTER (German); ETHYLACETAAT (Dutch); ETHYL ACETIC ESTER; ETHYLE (ACETATE d') (French); ETHYL ESTER OF ACETIC ACID; ETHYL ETHANOATE; ETILE (ACETATO di) (Italian); FEMA No. 2414; KTI 1470(+); MARKEM 320 CLEANER; OCTAN ETYLU (Polish); VINEGAR NAPHTHA

IDENTIFICATION

CAS:	141-78-6
ERG ID:	UN 1173
ERG Guide:	129
ERG Guide Category:	FLAMMABLE LIQUIDS (POLAR/WATER-MISCIBLE/NOXIOUS)
Molecular Formula:	$C_4H_8O_2$
RTECS No.:	AH5425000
Properties:	Highly flammable, watery liquid. Colorless. Pleasant, fruity odor. Floats and is moderately soluble in water. Flammable, irritating vapor is produced.

HEALTH AND SAFETY INFORMATION

Prolonged or repeated contact may cause irritation of the skin and dermatitis; nervous system changes; liver and kidney damage.

IDLH: 7320 mg/m^3 (2,000 ppm) [10% LEL]
Odor Threshold: 0.61 ppm
OSHA Table Z-1-A Air Contaminant

Exposure Limits:	**TWA**
ACGIH:	400 ppm; 1,440 mg/m^3
NIOSH:	400 ppm; 1,400 mg/m^3
OSHA:	400 ppm; 1,400 mg/m^3

RECOMMENDATIONS FOR RESPIRATOR SELECTION: NIOSH/OSHA

2,000 ppm: SA:CF (any supplied-air respirator operated in a continuous-flow mode);or PAPROV [any powered, air-purifying respirator with organic vapor cartridge(s)];or CCRFOV [any chemical cartridge respirator with a full facepiece and organic vapor cartridge(s)];or GMFOV [any air-purifying, full-facepiece respirator (gas mask) with a chin-style, front- or back-mounted acid gas canister];or SCBAF (any self-contained breathing apparatus with a full facepiece);or SAF (any supplied-air respirator with a full facepiece).

Emergency or Planned Entry into Unknown Concentrations or IDLH Conditions: SCBAF:PD,PP (any self-contained breathing apparatus that has a full facepiece and is operated in a pressure-demand or other positive-pressure mode);or SAF:PD,PP:ASCBA (any supplied-air respirator that has a full facepiece and is operated in a pressure-demand or other positive-pressure mode in combination with an auxiliary, self-contained breathing apparatus operated in a pressure-demand or other positive-pressure mode).

Escape: GMFOV [any air-purifying, full-facepiece respirator (gas mask) with a chin-style, front- or back-mounted organic vapor canister];or SCBAE (any appropriate escape-type, self-contained breathing apparatus).
Note: Substance causes eye irritation or damage; eye protection needed.

CHEMICAL REACTIVITY

Will hydrolyze on standing forming acetic acid and ethyl alcohol. This reaction is greatly accelerated by bases (alkalies). Forms explosive mixture with air. Incompatible with strong acids, nitrates, oxidizers, chlorosulfonic acid, lithium aluminum hydride, and oleum.

FIRE INFORMATION

Vapors are heavier than air and will collect in low areas. Vapors in confined areas may explode when exposed to fire. Vapors may travel long distances to ignition sources and flash back. Storage containers and parts of containers may rocket great distances, in many directions. If material or contaminated runoff enters waterways, notify downstream users of potentially contaminated waters. Notify local health and fire officials and pollution control agencies. Use dry chemicals, foam, carbon dioxide.

NFPA Hazard Classifications:

Health Hazards (Blue) 1
Flammability (Red) 3
Reactivity (Yellow) 0

Flash Point: 55°F (13°C) o.c.; 24°F (-4°C) c.c.
Flammable Limits in Air: LEL 2.0%; UEL 11.5%
Autoignition Temperature: 800°F (426°C)

FIRST AID

Move victim to fresh air. Call emergency medical care. Apply artificial respiration if victim is not breathing. Administer oxygen if breathing is difficult. Remove and isolate contaminated clothing and shoes. In case of contact with substance, immediately flush skin or eyes with running water for at least 20 minutes. Wash skin with soap and water. Keep victim warm and quiet. Effects of exposure (inhalation, ingestion, or skin contact) to substance may be delayed. Ensure that medical personnel are aware of the material(s) involved and take precautions to protect themselves.

ENVIRONMENTAL INFORMATION

EPA Hazardous Waste Number (RCRA No.): U112.
RCRA Section 261 Hazardous Constituents
RCRA Land Ban Waste
RCRA Universal Treatment Standards: Wastewater (mg/L), 0.34; Nonwastewater (mg/kg), 33.
Safe Drinking Water Act: Prohibitions in Injection (40 CFR, Part 148).
EPCRA Section 304: Reportable Quantity (RQ): CERCLA, 5,000 lb (2,270 kg).
TSCA: 40CFR799.5000

ETHYL ACRYLATE, INHIBITED E:140

SYNONYMS: ACRILATO de ETILO (Spanish); ACRYLATE d'ETHYLE (French); ACRYLIC ACID, ETHYL ESTER; ACRYLSAEUREAETHYLESTER (German); AETHYLACRYLAT (German); EEC No. 607-032-00-X; ETHOXYCARBONYL ETHYLENE; ETHYLACRYLAAT (Dutch); ETHYLAKRYLAT (Czech); ETHYL PROPENOATE; ETHYL 2-PROPENOATE; ETIL ACRILATO (Italian); ETILACRILATULUI (Rumanian); FEMA No. 2418; NCI-C50384; 2-PROPENOIC ACID, ETHYL ESTER

IDENTIFICATION

CAS:	140-88-5
ERG ID:	UN 1917
ERG Guide:	129P (inhibited)
ERG Guide Category:	FLAMMABLE LIQUIDS (POLAR/WATER-MISCIBLE/ NOXIOUS)
Molecular Formula:	$C_5H_8O_2$
RTECS No.:	AT0700000
Properties:	Highly flammable liquid. Colorless. Fruity, acrid, penetrating odor. Floats and is slightly soluble in water. Flammable, irritating vapor is produced.

HEALTH AND SAFETY INFORMATION

Very irritating to skin, eyes, nose, and lungs. Repeated or prolonged exposure may cause skin allergy with itching and skin rash; respiratory, liver, and kidney problems.

IARC Evaluation: Group 2B, possible human carcinogen; human data, inadequate; animal evidence, sufficient.
NTP: Anticipated human carcinogen; animal evidence.
IDLH: 300 ppm; potential human carcinogen.
Odor Threshold: 0.0009 ppm
OSHA Table Z-1-A Air Contaminant

Exposure Limits:	TWA	STEL
ACGIH:(1) (2)	5 ppm; 20 mg/m^3	15 ppm; 61 mg/m^3
NIOSH:(3)		
OSHA:(1)	25 ppm; 100 mg/m^3	

(1) Skin contact contributes significantly to overall exposure.
(2) Suspected carcinogen.
(3) Potential human carcinogen. Limit exposure to lowest feasible concentration.

RECOMMENDATIONS FOR RESPIRATOR SELECTION: NIOSH

At any concentrations above the NIOSH REL, or where there is no REL, at any detectable concentration: SCBAF:PD,PP (any self-contained breathing apparatus that has a full facepiece and is operated in a pressure-demand or other positive-pressure mode);or SAF:PD,PP:ASCBA (any supplied-air respirator that has a full facepiece and is operated in a pressure-demand or other positive-pressure mode in combination with an auxiliary, self-contained breathing apparatus operated in a pressure-demand or other positive pressure mode).

Escape: GMFOV [any air-purifying, full-facepiece respirator (gas mask) with a chin-style, front- or back-mounted organic vapor canister];or SCBAE (any appropriate escape-type, self-contained breathing apparatus).

CHEMICAL REACTIVITY

Forms explosive mixture with air. Atmospheric moisture and strong alkalies may cause fire and explosions. Unless properly inhibited (*Note:* Inert gas blanket not recommended), heat, light, or peroxides can cause polymerization. Incompatible with oxidizers (may be violent), strong acids, amines. May accumulate static electricity.

FIRE INFORMATION

Vapors are heavier than air and will collect in low areas. Vapors in confined areas may explode when exposed to fire. Vapors may travel long distances to ignition sources and flash back. Storage containers and parts of containers may rocket great distances, in many directions. If material or contaminated runoff enters waterways, notify downstream users of potentially contaminated waters. Notify local health and fire officials and pollution control agencies. Use dry chemicals, foam, carbon dioxide.

NFPA Hazard Classifications

Health Hazards (Blue) 2
Flammability (Red) 3
Reactivity (Yellow) 2

Flash Point: 44°F (7°C) o.c.; 48°F (9°C) c.c.
Flammable Limits in Air: LEL 1.4%; UEL 14%
Autoignition Temperature: 721°F (383°C)

FIRST AID

Move victim to fresh air. Call emergency medical care. Apply artificial respiration if victim is not breathing. Administer oxygen if breathing is difficult. Remove and isolate contaminated clothing and shoes. In case of contact with substance, immediately flush skin or eyes with running water for at least 20 minutes. Wash skin with soap and water. Keep victim warm and quiet. Effects of exposure (inhalation, ingestion, or skin contact) to substance may be delayed. Ensure that medical personnel are aware of the material(s) involved and take precautions to protect themselves.

ENVIRONMENTAL INFORMATION

Clean Air Act: Hazardous Air Pollutants (Title I, Part A, Section 112).
EPA Hazardous Waste Number (RCRA No.): U113.
RCRA Section 261 Hazardous Constituents
EPCRA Section 304: Reportable Quantity (RQ): CERCLA, 1,000 lb (454 kg).
EPCRA Section 313: Form R *de minimis* concentration reporting level: 0.1%.
Marine Pollutant (49CFR, Subchapter 172.101, Appendix B)
California's Proposition 65: Carcinogen.
TSCA: 40CFR 716.120(*a*) substance list.

ETHYL ALCOHOL E:160

SYNONYMS: ABSOLUTE ALCOHOL; ABSOLUTE ETHANOL; AETHANOL (German); AETHYLDLKOHOL (German); ALCOHOL; ALCOHOL, ANHYDROUS; ALCOHOL C-2; ALCOHOL (CTFA); ALCOHOL, DEHYDRATED; ALCOHOL ETILICO (Spanish); ALCOOL ETHYLIQUE (French); ALCOOL ETILICO (Italian); ALGRAIN; ALKOHOL (German); ALKOHOLU ETYLOWEGO (Polish); ANHYDROL; ARSENOSILICA FILM 0308; AS-1; AS-1CE; AS-5CE; AS-18CZ5E; AS-18CZ6E; AS-18CZ10A; AS-1400; B446; BLANCO-TRON TMS PLUS; COLOGNE SPIRIT; COLOGNE SPIRITS; EEC No. 603-002-00-5; EINECS No. 200-578-6; ENTEC 327; ETAHOLO (Italian); ETHANOL; ETHANOL 200 PROOF; ETHYLALCOHOL (Dutch); ETHYL ALCOHOL ANHYDRO-S; ETHYL HYDRATE; ETHYL HYDROXIDE; ETHYLOL; ETYLOWY ALKOHOL

(Polish); FERMENTATION ALCOHOL; 4282 FLUX; GRAIN ALCOHOL; JAYSOL S; KESTER 2211 ORGANIC FLUX; KESTER 4422 THINNER; KTI COP DEVELOPER I; KTI/MEAD COP I (-); MARKEM 215 CLEANER; MARKEM 452 WASHOUT SOLUTION; MARKEM 501 CLEANER; METHYL CARBINOL; MOLASSES ALCOHOL; MS CLEANER; NCI-CO3134; POTATO ALCOHOL; PURE GRAIN ALCOHOL; PUNCTILIOUS ETHYL ALCOHOL; PYRALIN PI 2563; PYRO; QZ 3289/3290 ADHESION CONCENTRATE; SD ALCOHOL 23-HYDROGEN; SOLDER FLUX; SOLDER FLUX THINNER; SYNASOL SPIRITS OF WINE; SPIRIT; TESCOL

IDENTIFICATION

CAS:	64-17-5
ERG ID:	UN 1170
ERG Guide:	127
ERG Guide Category:	FLAMMABLE LIQUIDS (POLAR/WATER MISCIBLE)
Molecular Formula:	C_2H_6O
RTECS No.:	KQ6300000
Properties:	Highly flammable, watery liquid. Colorless. Pleasant, sweet, fruity; wine or whiskeylike. Mixes with water; soluble. Flammable, irritating vapor is produced.

HEALTH AND SAFETY INFORMATION

Irritating to the skin and eyes. High levels of fumes can cause drowsiness and stupor. Ingestion can cause drunkenness and unconsciousness. Has been linked to birth defects in humans.

IDLH: 3,300 ppm [10% LEL]
Odor Threshold: 0.136 ppm
OSHA Table Z-1-A Air Contaminant

Exposure Limits:	**TWA**
ACGIH:	1,000 ppm; 1,880 mg/m^3
NIOSH:	1,000 ppm; 1,900 mg/m^3
OSHA:	1,000 ppm; 1,900 mg/m^3

RECOMMENDATIONS FOR RESPIRATOR SELECTION: NIOSH/OSHA

3,300 ppm: SA (any supplied-air respirator);or SCBAF (any self-contained breathing apparatus with a full facepiece).

Emergency or Planned Entry into Unknown Concentrations or IDLH Conditions:
SCBAF:PD,PP (any self-contained breathing apparatus that has a full facepiece and is operated in a pressure-demand or other positive-pressure mode);or SAF:PD,PP:ASCBA (any supplied-air respirator that has a full facepiece and is operated in a pressure-demand or other positive-pressure mode in combination with an auxiliary, self-contained breathing apparatus operated in a pressure-demand or other positive–pressure mode).

Escape: SCBAE (any appropriate escape-type, self-contained breathing apparatus).

CHEMICAL REACTIVITY

Forms explosive mixture with air. May accumulate static electrical charges, and may cause ignition of its vapors. Reactions may be violent with oleum, sulfuric acid, nitric acid, bases, aliphatic amines, isocyanates, oxidizers.

FIRE INFORMATION

Vapors are heavier than air and will collect in low areas. Vapors in confined areas may explode when exposed to fire. Vapors may travel long distances to ignition sources and flash back. Storage containers and parts of containers may rocket great distances, in many directions. If material or contaminated runoff enters waterways, notify downstream users of potentially contaminated waters. Notify local health and fire officials and pollution control agencies. Use dry chemicals, foam, carbon dioxide.

NFPA Hazard Classifications:

Health Hazards (Blue) 0
Flammability (Red) 3
Reactivity (Yellow) 0

Flash Point: 64°F (18°C) o.c.; 55°F (13°C) c.c.
Flammable Limits in Air: LEL 3.3%; UEL 19%
Autoignition Temperature: 689°F (365°C)

FIRST AID

Move victim to fresh air. Call emergency medical care. Apply artificial respiration if victim is not breathing. If breathing is difficult, give oxygen. Remove and isolate contaminated clothing and shoes. In case of contact with substance, immediately flush skin or eyes with running water for at least 20 minutes. Wash skin with soap and water. Keep victim warm and quiet. Ensure that medical personnel are aware of the material(s) involved and take precautions to protect themselves.

ENVIRONMENTAL INFORMATION

California's Proposition 65: Reproductive hazard.

ETHYLAMINE E:170

SYNONYMS: AETHYLAMINE (German); AMINOETHANE; 1-AMINOETHANE; EA; EEC No. 612-002-00-4; ETHANAMINE; ETILAMINA (Italian, Spanish); ETN; ETYLOAMINA (Polish); MEA; MONOETHYLAMINE; MONOETHYLAMINE, ANHYDROUS

454 Ethylamine

IDENTIFICATION

CAS:	75-04-7
ERG ID:	UN 1036 (gas); UN 2270 (aqueous solution 50-70% ethylamine)
ERG Guide:	118 (gas); 132 (aqueous solution 50-70% ethylamine)
ERG Guide Category:	118 GASES-FLAMMABLE-CORROSIVE 132 FLAMMABLE LIQUIDS-CORROSIVE
Molecular Formula:	C_2H_7N
RTECS No.:	KH2100000
Properties:	Extremely flammable gas; liquid below 62°F (17°C). Colorless. Strong, pungent, ammonialike odor. Mixes with water. Flammable, irritating vapor is produced.

HEALTH AND SAFETY INFORMATION

Severely irritating to skin, eyes, nose and lungs; can cause severe burns.

IDLH: 1,122 mg/m^3 (600 ppm).
Odor Threshold: 0.324 ppm . Odor is not a reliable indicator of the presence of toxic amounts of gas.
OSHA Table Z-1-A Air Contaminant
OSHA Process Safety Management of Highly Hazardous Chemicals (29 CFR 1910.119, Appendix A): Threshold Quantity: 7,500 pounds.

Exposure Limits:	**TWA**	**STEL**
ACGIH:*	5 ppm; 9.2 mg/m^3	15 ppm; 27.6 mg/m^3
NIOSH:	10 ppm; 18 mg/m^3	
OSHA:	10 ppm; 18 mg/m^3	

* Skin contact contributes significantly in overall exposure.

RECOMMENDATIONS FOR RESPIRATOR SELECTION: NIOSH/OSHA

250 ppm: SA:CF (any supplied-air respirator operated in a continuous-flow mode);or PAPRS (any powered, air-purifying respirator with cartridge(s) providing protection against the compound of concern
500 ppm: CCRFS (any chemical cartridge respirator with a full facepiece and cartridge(s) providing protection against the compound of concern) organic vapor and acid gas cartridge(s);or GMFS [any air-purifying, full-facepiece respirator (gas mask) with a chin-style, front- or back-mounted canister providing protection against the compound of concern];or SCBAF (any self-contained breathing apparatus with a full facepiece);or SAF (any supplied-air respirator with a full facepiece).
600 ppm: SAF:PD,PP (any supplied-air respirator that has a full facepiece and is operated in a pressure-demand or other positive-pressure mode).

Emergency or Planned Entry into Unknown Concentrations or IDLH Conditions:
SCBAF:PD,PP (any self-contained breathing apparatus that has a full facepiece and is operated in a pressure-demand or other positive-pressure mode);or SAF:PD,PP:ASCBA (any supplied-air respirator that

has a full facepiece and is operated in a pressure-demand or other positive-pressure mode in combination with an auxiliary, self-contained breathing apparatus operated in a pressure-demand or other positive-pressure mode).

Escape: GMFS [any air-purifying, full-facepiece respirator (gas mask) with a chin-style, front- or back-mounted canister providing protection against the compound of concern];or SCBAE (any appropriate escape-type, self-contained breathing apparatus).

Note: Substance causes eye irritation or damage; eye protection needed..

CHEMICAL REACTIVITY

Forms explosive mixture with air. Incompatible with acids, organic anhydrides, isocyanates, vinyl acetate, acrylates, substituted allyls, alkylene oxides, epichlorohydrin, ketones, aldehydes, alcohols, glycols, phenols, cresols, caprolactum solution, strong oxidizers. Corrosive to aluminum, copper, lead, tin, zinc and most of their alloys, including brass. Attacks some plastics, rubber, and coatings.

FIRE INFORMATION

Combustion products include toxic nitrogen oxides. Firefighting gear, including SCBA, does not provide adequate protection. If exposure occurs, remove and isolate gear immediately and thoroughly decontaminate personnel. Gases are heavier than air and will collect in low areas. Gases in confined areas may explode when exposed to fire. Gases may travel long distances to ignition sources and flash back. Storage containers and parts of containers may rocket great distances, in many directions. If material or contaminated runoff enters waterways, notify downstream users of potentially contaminated waters. Notify local health and fire officials and pollution control agencies. Do not extinguish the fire unless the flow of the gas can be stopped and any remaining gas is out of the line. Specially trained personnel may use fog lines to cool exposures and let the fire burn itself out. Use dry chemicals, alcohol-resistant foam, carbon dioxide. From a secure explosion-proof location; use water spray to cool exposed containers. If cooling streams are ineffective (venting sound increases in volume and pitch, tank discolors, or shows any signs of deforming), withdraw immediately to a secure position.

NFPA Hazard Classifications:

Health Hazards (Blue) 3
Flammability (Red) 4
Reactivity (Yellow) 0

Flash Point: 0°F (−18°C) o.c.; 1°F (−17°C) c.c.
Flammable Limits in Air: LEL 3.5%; UEL 14%
Autoignition Temperature: 724°F (384°C)

FIRST AID

GASES-FLAMMABLE-CORROSIVE (118): Move victim to fresh air. Call emergency medical care. Apply artificial respiration if victim is not breathing. Do not use mouth-to-mouth method if victim ingested or inhaled the substance; induce artificial respiration with the aid of a pocket mask equipped with a one-way valve or other proper respiratory medical device. Administer oxygen if breathing is difficult. Remove and isolate contaminated clothing and shoes. In case of contact with liquefied gas, thaw frosted parts with lukewarm water. Keep victim warm and quiet. Keep victim under observation. Effects of contact or inhalation may be delayed. Ensure that medical personnel are aware of the material(s) involved and take precautions to protect themselves.

FLAMMABLE LIQUIDS-CORROSIVE (132): Move victim to fresh air. Call emergency medical care. Apply artificial respiration if victim is not breathing. Do not use mouth-to-mouth method if victim swallowed or inhaled the substance; induce artificial respiration with the aid of a pocket mask equipped with a one-way valve or other proper respiratory medical device. Administer oxygen if breathing is difficult. Remove and isolate contaminated clothing and shoes. In case of contact with substance, immediately flush skin or eyes with running water for at least 20 minutes. Keep victim warm and quiet. Effects of exposure (inhalation, ingestion, or skin contact) to substances may be delayed. Ensure that medical personnel are aware of the material(s) involved and take precautions to protect themselves.

ENVIRONMENTAL INFORMATION

Clean Air Act: Accidental Release Prevention/Flammable substances, (Section 112[r], Table 3), TQ = 10,000 lb (4,540 kg).
Clean Water Act: Section 311 Hazardous Substances/RQ (same as CERCLA).
EPCRA Section 304: Reportable Quantity (RQ): CERCLA, 100 lb (45.4 kg).

ETHYL AMYL KETONE E:173

SYNONYMS: AMYL ETHYL KETONE; EAK; EEC No. 606-020-00-1; EINECS No. 208-793-7; 5-METILHEPTANO-3-ONA (Spanish); 5-METHYL-3-HEPTANONE (OSHA); 3-OCTANONE

IDENTIFICATION

CAS:	106-68-3; 541-85-5
ERG ID:	UN 2271
ERG Guide:	127
ERG Guide Category:	FLAMMABLE LIQUIDS (POLAR/WATER- MISCIBLE)
Molecular Formula:	$C_6H_{16}O$
RTECS No.:	MJ3500000
Properties:	Combustible liquid. Colorless. Mild fruity odor. Floats on water; slightly soluble.

HEALTH AND SAFETY INFORMATION

Irritates eyes, skin, and respiratory tract. Can attack the central nervous system; possible headache, drowsiness, unconsciousness.

IDLH: 100 ppm
Odor Threshold: 6 ppm

OSHA Table Z-1-A Air Contaminant
NIOSH Criteria Document: NIOSH 78-173 KETONES

Exposure Limits:	**TWA**
ACGIH:	25 ppm; 131 mg/m^3
NIOSH:	25 ppm; 130 mg/m^3
OSHA:	25 ppm; 130 mg/m^3

RECOMMENDATIONS FOR RESPIRATOR SELECTION: NIOSH/OSHA

100 ppm: CCROV [any chemical cartridge respirator with organic vapor cartridge(s)];or PAPROV [any powered, air-purifying respirator with organic vapor cartridge(s)];or GMFOV[any air-purifying, full-facepiece respirator (gas mask) with a chin-style, front- or back-mounted acid gas canister];or SA (any supplied-air respirator);or SCBAF (any self-contained breathing apparatus with a full facepiece).

Emergency or Planned Entry into Unknown Concentrations or IDLH Conditions: SCBAF:PD,PP (any self-contained breathing apparatus that has a full facepiece and is operated in a pressure-demand or other positive-pressure mode);or SAF:PD,PP:ASCBA (any supplied-air respirator that has a full facepiece and is operated in a pressure-demand or other positive-pressure mode in combination with an auxiliary, self-contained breathing apparatus operated in a pressure-demand or other positive-pressure mode).

Escape: GMFOV [any air-purifying, full-facepiece respirator (gas mask) with a chin-style, front- or back-mounted organic vapor canister];or SCBAE (any appropriate escape-type, self-contained breathing apparatus).

CHEMICAL REACTIVITY

Forms explosive mixture with air. Contact with oxidizers may cause fire and explosions. Incompatible with sulfuric acid, nitric acid, aliphatic amines.

FIRE INFORMATION

Vapors are heavier than air and will collect in low areas. Vapors in confined areas may explode when exposed to fire. Vapors may travel long distances to ignition sources and flash back. Storage containers and parts of containers may rocket great distances, in many directions. If material or contaminated runoff enters waterways, notify downstream users of potentially contaminated waters. Notify local health and fire officials and pollution control agencies. Use dry chemicals, foam, carbon dioxide. Use water to cool containers below the flash point.

FEMA Hazard Classifications:

Health Hazards (Blue) 0
Flammability (Red) 2

Reactivity (Yellow) 0
Flash Point: 115°F (46°C) c.c.

FIRST AID

Move victim to fresh air. Call emergency medical care. Apply artificial respiration if victim is not breathing. Administer oxygen if breathing is difficult. Remove and isolate contaminated clothing and shoes. In case of contact with substance, immediately flush skin or eyes with running water for at least 20 minutes. Wash skin with soap and water. Keep victim warm and quiet. Ensure that medical personnel are aware of the material(s) involved and take precautions to protect themselves.

N–ETHYLANILINE E:177

SYNONYMS: AETHYLANILIN (German); ANILINOETHANE; BENZENAMINE, N-ETHYL-; EINECS No. 203-135-5; N-ETHYLAMINOBENZENE; ETHYLANILINE; N-ETHYLBENZENAMINE; N-ETHYLBENZENAMINO; ETHYLPHENYLAMINE; N-ETILANILINA (Spanish)

IDENTIFICATION

CAS:	103-69-5
ERG ID:	UN 2272
ERG Guide:	153
ERG Guide Category:	SUBSTANCES-TOXIC AND/OR CORROSIVE (COMBUSTIBLE)
Molecular Formula:	$C_8H_{11}N$
RTECS No.:	BX9780000
Properties:	Combustible oily liquid. Clear to yellow-brown. Insoluble in water.

HEALTH AND SAFETY INFORMATION

Poisonous. Serious health hazard. Harmful if inhaled or absorbed through the skin. Can cause blood effects (formation of methemaglobin), skin disorders, liver and kidney damage.

RECOMMENDATIONS FOR RESPIRATOR SELECTION

At any concentrations above the NIOSH REL, or where there is no REL, at any detectable concentration: SCBAF:PD,PP (any self-contained breathing apparatus that has a full facepiece and is operated in a pressure-demand or other positive-pressure mode);or SAF:PD,PP:ASCBA (any supplied-air respirator that has a full facepiece and is operated in a pressure-demand or other positive-pressure mode in combination with an auxiliary, self-contained breathing apparatus operated in a pressure-demand or other positive-pressure mode).

Escape: GMFOV [any air-purifying, full-facepiece respirator (gas mask) with a chin-style, front- or back-mounted organic vapor canister];or SCBAE (any appropriate escape-type, self-contained breathing apparatus).

CHEMICAL REACTIVITY

Decomposes on contact with light or air. Reacts with many materials. Contact with strong oxidizers, strong acids can cause fire, explosions with formation of toxic vapors of aniline and oxides of nitrogen.

FIRE INFORMATION

Combustion products include toxic nitrogen oxides. Vapors are heavier than air and will collect in low areas. Vapors in confined areas may explode when exposed to fire.If material or contaminated runoff enters waterways, notify downstream users of potentially contaminated waters. Notify local health and fire officials and pollution control agencies. Use dry chemicals, alcohol-resistant foam, carbon dioxide. Use water spray to cool containers.

NFPA Hazard Classifications:

Health Hazards (Blue) 3
Flammability (Red) 2
Reactivity (Yellow) 0

Flash Point: 185°F (85°C) o.c.
Flammable Limits in Air: LEL 1.6%; UEL 9.5%

FIRST AID

Move victim to fresh air and call emergency medical care. Apply artificial respiration if victim is not breathing. Do not use mouth-to-mouth method if victim swallowed or inhaled the substance; induce artificial respiration with the aid of a pocket mask equipped with a one-way valve or other proper respiratory medical device. If breathing is difficult, give oxygen. In case of contact with substance, immediately flush skin or eyes with running water for at least 20 minutes. Remove and isolate contaminated clothing and shoes. Keep victim warm and quiet. Effects of exposure (inhalation, ingestion, or skin contact) to substances may be delayed. Ensure that medical personnel are aware of the material(s) involved and take precautions to protect themselves.

ETHYLBENZENE E:180

SYNONYMS: AETHYLBENZOL (German); ARISTOLINE(+); AZ 1470(+); AZ 4210(+); BENCENO ETILICO (Spanish); CEM 388; CEM 420; EB; EEC No. 601-023-00-4; ETHYLBENZEEN (Dutch); ETHYLBENZOL; ETILBENZENE (Italian); ETILBENCENO (Spanish); ETYLOBENZEN (Polish); KTI 1350J(+); KTI PHOTORESIST STANDARD (-); NCI-C56393; PHENYLETHANE; WAYCOAT 28(-); WAYCOAT 43(-); WAYCOAT 59(-); WAYCOAT 204(-); WAYCOAT DEVELOPER(-); WAYCOAT

HNR 80(-); WAYCOAT HNR 120(-); WAYCOAT HPR 205; WAYCOAT 207(+); WAYCOAT HR 200(-); WAYCOAT RX 507(+); WAYCOAT SC100(-); WAYCOAT SC100 CP(+); WAYCOAT SC180(-)

IDENTIFICATION

CAS:	100-41-4
ERG ID:	UN 1175
ERG Guide:	129
ERG Guide Category:	FLAMMABLE LIQUIDS (POLAR/WATER-MISCIBLE/NOXIOUS)
Molecular Formula:	C_8H_{10}
RTECS No.:	DA0700000
Properties:	Highly flammable liquid. Clear, colorless. Sweet, gasolinelike odor. Insoluble; floats on water. Flammable, irritating vapor is produced.

HEALTH AND SAFETY INFORMATION

Irritates eyes, skin, and respiratory tract. Long-term exposure may cause skin rash and irritation of eyes, nose, and throat; brain and nervous system changes.

IDLH: 3526 mg/m^3 (800 ppm)
Odor Threshold: 2.3 ppm
OSHA Table Z-1-A Air Contaminant

Exposure Limits:	**TWA**	**STEL**
ACGIH:	100 ppm; 434 mg/m^3	125 ppm; 543 mg/m^3
NIOSH:	100 ppm; 435 mg/m^3	125 ppm; 545 mg/m^3
OSHA:	100 ppm; 435 mg/m^3	

RECOMMENDATIONS FOR RESPIRATOR SELECTION: NIOSH/OSHA

800 ppm: CCROV [any chemical cartridge respirator with organic vapor cartridge(s)];or GMFOV[any air-purifying, full-facepiece respirator (gas mask) with a chin-style, front- or back-mounted acid gas canister];or PAPROV [any powered, air-purifying respirator with organic vapor cartridge(s)];or SA (any supplied-air respirator);or SCBAF (any self-contained breathing apparatus with a full facepiece).

Emergency or Planned Entry into Unknown Concentrations or IDLH Conditions:
SCBAF:PD,PP (any self-contained breathing apparatus that has a full facepiece and is operated in a pressure-demand or other positive-pressure mode);or SAF:PD,PP:ASCBA (any supplied-air respirator that has a full facepiece and is operated in a pressure-demand or other positive-pressure mode in combination with an auxiliary, self-contained breathing apparatus operated in a pressure-demand or other positive-pressure mode).

Escape: GMFOV [any air-purifying, full-facepiece respirator (gas mask) with a chin-style, front- or back-mounted organic vapor canister];or SCBAE (any appropriate escape-type, self-contained breathing apparatus).
Note: Substance reported to cause eye irritation or damage; may require eye protection.

CHEMICAL REACTIVITY

May accumulate static electrical charges and may cause ignition of its vapors. Forms explosive mixture with air. Incompatible with strong oxidizers and nitric acid. Causes rubber to swell and soften.

FIRE INFORMATION

Vapors are heavier than air and will collect in low areas. Vapors in confined areas may explode when exposed to fire. Vapors may travel long distances to ignition sources and flash back. Storage containers and parts of containers may rocket great distances, in many directions. If material or contaminated runoff enters waterways, notify downstream users of potentially contaminated waters. Notify local health and fire officials and pollution control agencies. Use dry chemicals, foam, carbon dioxide.

NFPA Hazard Classifications:

Health Hazards (Blue) 2
Flammability (Red) 3
Reactivity (Yellow) 0

Flash Point: 59°F (15°C)
Flammable Limits in Air: LEL 0.8%; UEL 6.7%
Autoignition Temperature: 806°F (430°C)

FIRST AID

Move victim to fresh air. Call emergency medical care. Apply artificial respiration if victim is not breathing. Administer oxygen if breathing is difficult. Remove and isolate contaminated clothing and shoes. In case of contact with substance, immediately flush skin or eyes with running water for at least 20 minutes. Wash skin with soap and water. Keep victim warm and quiet. Effects of exposure (inhalation, ingestion, or skin contact) to substance may be delayed. Ensure that medical personnel are aware of the material(s) involved and take precautions to protect themselves.

ENVIRONMENTAL INFORMATION

Clean Air Act: Hazardous Air Pollutants (Title I, Part A, Section 112).
Clean Water Act: Section 311 Hazardous Substances/RQ (same as CERCLA); Section 307 Priority Pollutants; Section 313 Priority Chemicals; Section 307 Toxic Pollutants.
RCRA Universal Treatment Standards: Wastewater (mg/L), 0.057; Nonwastewater (mg/kg), 10.
RCRA Ground Water Monitoring List: Suggested test method(s) (PQL ug/L): 8020(2); 8240(5).
Safe Drinking Water Act: MCL, 0.7 mg/L; MCLG, 0.7 mg/L.
EPCRA Section 304: Reportable Quantity (RQ): CERCLA, 1,000 lb (454 kg).
EPCRA Section 313: Form R *de minimis* concentration reporting level: 1.0%.

ETHYL BENZOATE E:182

SYNONYMS: BENZOATO de ETILO (Spanish); BENZOIC ACID, ETHYL ESTER; BENZOIC ETHER; EINECS No. 202-284-3; ESSENCE of NIOBE; FEMA No. 2422

IDENTIFICATION

CAS:	93-89-0
Molecular Formula:	$C_9H_{10}O_2$
RTECS No.:	DH0200000
Properties:	Combustible liquid. Colorless. Fruity odor. Floats in water; insoluble.

HEALTH AND SAFETY INFORMATION

Irritates eyes, skin, and respiratory tract.

RECOMMENDATIONS FOR RESPIRATOR SELECTION

At any concentrations above the NIOSH REL, or where there is no REL, at any detectable concentration: SCBAF:PD,PP (any self-contained breathing apparatus that has a full facepiece and is operated in a pressure-demand or other positive-pressure mode);or SAF:PD,PP:ASCBA (any supplied-air respirator that has a full facepiece and is operated in a pressure-demand or other positive-pressure mode in combination with an auxiliary, self-contained breathing apparatus operated in a pressure-demand or other positive-pressure mode).

Escape: GMFOV [any air-purifying, full-facepiece respirator (gas mask) with a chin-style, front- or back-mounted organic vapor canister];or SCBAE (any appropriate escape-type, self-contained breathing apparatus).

CHEMICAL REACTIVITY

Forms explosive mixture with air. Contact with oxidizers may cause fire or explosions. Incompatible with sodium hydroxide. May accumulate static electricity and cause ignition of its vapors.

FIRE INFORMATION

Vapors are heavier than air and will collect in low areas. Vapors in confined areas may explode when exposed to fire. If material or contaminated runoff enters waterways, notify downstream users of potentially contaminated waters. Notify local health and fire officials and pollution control agencies. Use dry chemicals, alcohol-resistant foam, carbon dioxide.

NFPA Hazard Classifications:

Health Hazards (Blue) 1

Flammability (Red) 1
Reactivity (Yellow) 0

Flash Point: 190°F (88°C)
Autoignition Temperature: 914°F (490°C)

FIRST AID

Move victim to fresh air. Call emergency medical care. Apply artificial respiration if victim is not breathing. Administer oxygen if breathing is difficult. Remove and isolate contaminated clothing and shoes. In case of contact with substance, immediately flush skin or eyes with running water for at least 20 minutes. Wash skin with soap and water. Keep victim warm and quiet. Ensure that medical personnel are aware of the material(s) involved and take precautions to protect themselves.

ETHYL BUTYL KETONE E:184

SYNONYMS: AETHYLBUTYLKETON (German); BUTYL ETHYL KETONE; n-BUTYL ETHYL KETONE; EEC No. 606-003-00-9; EINECS No. 203-388-1; EPTAN-3-ONE (Italian); HEPTAN-3-ON (Dutch and German); HEPTAN-3-ONE; 3-HEPTANONE; FEMA No. 2545

IDENTIFICATION

CAS:	106-35-4
ERG ID:	UN 1224
ERG Guide:	127
ERG Guide Category:	FLAMMABLE LIQUIDS (POLAR/WATER- MISCIBLE)
Molecular Formula:	$C_7H_{14}O$
RTECS No.:	MJ525000
Properties:	Flammable liquid. Clear, colorless. Mild, fruity odor. Soluble in water.

HEALTH AND SAFETY INFORMATION

Irritates eyes, skin, and respiratory tract. Prolonged or repeated skin contact can cause dermatitis. Exposure to high concentrations may cause headache, dizziness, or unconsciousness.

IDLH: 1,000 ppm
Odor Threshold: 0.1-10 ppm
OSHA Table Z-1-A Air Contaminant
NIOSH Criteria Document: NIOSH 78-173 KETONES

Exposure Limits: TWA

ACGIH: 50 ppm; 234 mg/m^3

NIOSH: 50 ppm; 230 mg/m^3
OSHA: 50 ppm; 230 mg/m^3

RECOMMENDATIONS FOR RESPIRATOR SELECTION: NIOSH/OSHA

500 ppm: CCROV [any chemical cartridge respirator with organic vapor cartridge(s)];or SA (any supplied-air respirator).
1000 ppm: SA:CF (any supplied-air respirator operated in a continuous-flow mode);or PAPROV [any powered, air-purifying respirator with organic vapor cartridge(s)];or CCRFOV [any air-purifying, full-facepiece respirator (gas mask) with a chin-style, front- or back-mounted acid gas canister];or GMFOV [any air-purifying, full-facepiece respirator (gas mask) with a chin-style, front- or back-mounted acid gas canister];or SCBAF (any self-contained breathing apparatus with a full facepiece);or SAF (any supplied-air respirator with a full facepiece).

Emergency or planned entry into unknown concentration or IDLH conditions: SCBAF:PD,PP (any self-contained breathing apparatus that has a full facepiece and is operated in a pressure-demand or other positive-pressure mode);or SAF:PD,PP:ASCBA (any supplied-air respirator that has a full facepiece and is operated in a pressure-demand or other positive-pressure mode in combination with an auxiliary, self-contained breathing apparatus operated in a pressure-demand or other positive pressure mode).

Escape: GMFOV [any air-purifying, full-facepiece respirator (gas mask) with a chin-style, front- or back-mounted organic vapor canister];or SCBAE (any appropriate escape-type, self-contained breathing apparatus).

Note: Substance reported to cause eye irritation or damage; may require eye protection.

CHEMICAL REACTIVITY

Forms explosive mixture with air. Reacts violently with strong oxidizers. Attacks some plastics, rubber, and coatings. Contact with oxidizers can cause fires and explosions. Reacts with acetaldehyde, perchloric acid.

FIRE INFORMATION

Vapors are heavier than air and will collect in low areas. Vapors in confined areas may explode when exposed to fire. If material or contaminated runoff enters waterways, notify downstream users of potentially contaminated waters. Notify local health and fire officials and pollution control agencies. Use dry chemicals, alcohol-resistant foam, carbon dioxide.

NFPA Hazard Classifications:

Health Hazards (Blue) 1
Flammability (Red) 2
Reactivity (Yellow) 0

Flash Point: 115°F (46°C) o.c.
Flammable Limits in Air: LEL 1.4%; UEL 8.8%

FIRST AID

Move victim to fresh air. Call emergency medical care. Apply artificial respiration if victim is not breathing. Administer oxygen if breathing is difficult. Remove and isolate contaminated clothing and shoes. In case of contact with substance, immediately flush skin or eyes with running water for at least 20 minutes. Wash skin with soap and water. Keep victim warm and quiet. Ensure that medical personnel are aware of the material(s) involved and take precautions to protect themselves.

2–ETHYL BUTYRALDEHYDE E:186

SYNONYMS: ALDEHYDE-2-ETHYLBUTYRIQUE (French); BUTANAL, 2-ETHYL-; EINECS No. 202-623-5; 2-ETHYLBUTANAL; ETHYL BUTYRALDEHYDE; α-ETHYLBUTYRALDEHYDE; ETHYL BUTYRALDEHYDE; 2-ETHYLBUTYRIC ALDEHYDE; 2-ETILBUTIRALDEHIDO (Spanish); FEMA No. 2426

IDENTIFICATION

CAS:	97-96-1
ERG ID:	UN 1178
ERG Guide:	129
ERG Guide Category:	FLAMMABLE LIQUIDS (POLAR/WATER-MISCIBLE/NOXIOUS)
Molecular Formula:	$C_6H_{12}O$
RTECS No.:	ES2625000
Properties:	Flammable liquid. Colorless. Pungent odor. Insoluble in water.

HEALTH AND SAFETY INFORMATION

Irritating to eyes, skin, and respiratory tract. May effect the central nervous system causing headache, dizziness, drowsiness, unconsciousness.

RECOMMENDATIONS FOR RESPIRATOR SELECTION

At any concentrations above the NIOSH REL, or where there is no REL, at any detectable concentration: SCBAF:PD,PP (any self-contained breathing apparatus that has a full facepiece and is operated in a pressure-demand or other positive-pressure mode);or SAF:PD,PP:ASCBA (any supplied-air respirator that has a full facepiece and is operated in a pressure-demand or other positive-pressure mode in combination with an auxiliary, self-contained breathing apparatus operated in a pressure-demand or other positive-pressure mode).

Escape: GMFOV [any air-purifying, full-facepiece respirator (gas mask) with a chin-style, front- or back-mounted organic vapor canister];or SCBAE (any appropriate escape-type, self-contained breathing apparatus).

CHEMICAL REACTIVITY

Contact with oxidizers may cause fire and explosions. Incompatible with sulfuric acid, nitric acid, caustics, ammonia, aliphatic amines, alkanolamines, aromatic amines.

FIRE INFORMATION

Vapors are heavier than air and will collect in low areas. Vapors in confined areas may explode when exposed to fire. Vapors may travel long distances to ignition sources and flash back. Storage containers and parts of containers may rocket great distances, in many directions. If material or contaminated runoff enters waterways, notify downstream users of potentially contaminated waters. Notify local health and fire officials and pollution control agencies. Use dry chemicals, foam, carbon dioxide.

NFPA Hazard Classifications:

Health Hazards (Blue) 2
Flammability (Red) 3
Reactivity (Yellow) 1

Flash Point: 70°F (21°C) o.c.
Flammable Limits in Air: LEL 1.2%; UEL 7.7%

FIRST AID

Move victim to fresh air. Call emergency medical care. Apply artificial respiration if victim is not breathing. Administer oxygen if breathing is difficult. Remove and isolate contaminated clothing and shoes. In case of contact with substance, immediately flush skin or eyes with running water for at least 20 minutes. Wash skin with soap and water. Keep victim warm and quiet. Effects of exposure (inhalation, ingestion, or skin contact) to substance may be delayed. Ensure that medical personnel are aware of the material(s) involved and take precautions to protect themselves.

ENVIRONMENTAL INFORMATION

TSCA: 40CFR716.120(*d*)1 as aldehydes

ETHYL CHLOROFORMATE E:188

SYNONYMS: CARBOBOCHLORIDIC ACID, ETHYL ESTER; CHLORAMEISENSAEUREAETHYLESTER (German); CHLOROFORMIC ACID, ETHYL ESTER; ECF; EEC No. 607-020-00-4; EINECS No. 208-778-5; ETHYL CHLOROCARBONATE; ETHYLE, CHLOROFORMIAT d (French); CLOROFORMIATO de ETILO (Spanish); TL423

IDENTIFICATION

CAS:	541-41-3
ERG ID:	UN 1182
ERG Guide:	155
ERG Guide Category:	SUBSTANCES-TOXIC AND/OR CORROSIVE (FLAMMABLE/WATER-SENSITIVE)
Molecular Formula:	$C_3H_5ClO_2$
RTECS No.:	LQ6125000
Properties:	Highly flammable liquid. Colorless to light yellow. Irritating odor, like hydrochloric acid. Slightly soluble in water. Slowly decomposes in water; irritating vapor is produced.

HEALTH AND SAFETY INFORMATION

Poisonous. Corrosive to eyes, skin, and respiratory tract. Inhalation can cause pulmonary edema, which can be delayed for several hours; there is a risk of death in serious cases.

RECOMMENDATIONS FOR RESPIRATOR SELECTION

At any concentrations above the NIOSH REL, or where there is no REL, at any detectable concentration: SCBAF:PD,PP (any self-contained breathing apparatus that has a full facepiece and is operated in a pressure-demand or other positive-pressure mode);or SAF:PD,PP:ASCBA (any supplied-air respirator that has a full facepiece and is operated in a pressure-demand or other positive-pressure mode in combination with an auxiliary, self-contained breathing apparatus operated in a pressure-demand or other positive-pressure mode).

Escape: GMFOV [any air-purifying, full-facepiece respirator (gas mask) with a chin-style, front- or back-mounted organic vapor canister];or SCBAE (any appropriate escape-type, self-contained breathing apparatus).

CHEMICAL REACTIVITY

Slow decomposition in water or alcohol, forming chloroformic acid, ethanol, and carbon dioxide. Reacts violently with strong oxidizers. Slow evolution of hydrogen chloride from surface moisture reaction can cause slow corrosion of metals. Air contact may produce hydrochloric acid fumes. Forms explosive and corrosive mixture with air. Strong oxidizers may cause fire and explosions. Attacks metals in the presence of moisture.

FIRE INFORMATION

Dangerous fire hazard; corrosive, poisonous. Vapors are heavier than air and will collect in low areas. Vapors in confined areas may explode when exposed to fire. Vapors may travel long distances to ignition sources and flash back. Storage containers and parts of containers may rocket great distances, in many directions. If material or contaminated runoff enters waterways, notify downstream users of potentially contaminated waters. Notify local health and fire officials and pollution control agencies. Use dry chemicals, foam, carbon dioxide. Use water spray to cool containers.

NFPA Hazard Classifications:

Health Hazards (Blue) 4
Flammability (Red) 3
Reactivity (Yellow) 1

Flash Point: 82°F (28°C) o.c.; 61°F (16°C) c.c.
Autoignition Temperature: 932°F (500°C)

FIRST AID

Move victim to fresh air; call emergency medical care. Apply artificial respiration if victim is not breathing. Do not use mouth-to-mouth method if victim ingested or inhaled the substance; induce artificial respiration with the aid of a pocket mask equipped with a one-way valve or other proper respiratory medical device. Administer oxygen if breathing is difficult. Remove and isolate contaminated clothing and shoes. In case of contact with substance, immediately flush skin or eyes with running water for at least 20 minutes. For minor skin contact, avoid spreading material on unaffected skin. Keep victim warm and quiet. Effects of exposure (inhalation, ingestion, or skin contact) to substance may be delayed. Ensure that medical personnel are aware of the material(s) involved and take precautions to protect themselves.

ENVIRONMENTAL INFORMATION

EPCRA Section 313: Form R *de minimis* concentration reporting level: 1.0%.

ETHYLENE E:190

SYNONYMS: ACETENE; ATHYLEN (German); BICARBURRETTED HYDROGEN; ATHYLEN (German); DICARBURRETTED HYDROGEN; EEC No. 601-010-00-3; ELAYL; ETHENE; ETILENO (Spanish); ETHYLENE, REFRIGERATED; LIQUID ETHYENE; LIQUID ETHELYNE; NCL-C5066; OLEFIANT GAS

IDENTIFICATION

CAS:	74-85-1
ERG ID:	UN 1962
ERG Guide:	116
ERG Guide Category:	GASES-FLAMMABLE (UNSTABLE)
Molecular Formula:	C_2H_4
RTECS No.:	KU5340000
Properties:	Extremely flammable, compressed gas. Colorless; may appear white. Faint, sweet odor. Floats and boils on water; soluble. Flammable, visible vapor cloud is produced.

HEALTH AND SAFETY INFORMATION

Contact with liquid may cause frostbite. Long-term exposure can cause weight loss, insomnia, irritability, kidney damage, escalated red blood cell count.

IARC Evaluation: Group 3, unclassifiable substance; no human data; no animal data.

Exposure Limits:

Note: Designated by ACGIH as a simple asphyxiant. High concentrations have an anesthetic effect.

RECOMMENDATIONS FOR RESPIRATOR SELECTION: NIOSH

At any concentrations above the NIOSH REL, or where there is no REL, at any detectable concentration: SCBAF:PD,PP (any self-contained breathing apparatus that has a full facepiece and is operated in a pressure-demand or other positive-pressure mode);or SAF:PD,PP:ASCBA (any supplied-air respirator that has a full facepiece and is operated in a pressure-demand or other positive-pressure mode in combination with an auxiliary, self-contained breathing apparatus operated in a pressure-demand or other positive pressure mode).

Escape: GMFOV [any air-purifying, full-facepiece respirator (gas mask) with a chin-style, front- or back-mounted organic vapor canister];or SCBAE (any appropriate escape-type, self-contained breathing apparatus).

CHEMICAL REACTIVITY

A flammable gas. May accumulate static electrical charges and may cause ignition of its vapors. Heat, shock, or friction may cause explosions. Elevated temperatures or oxidizers may cause explosive polymerization and fire. Incompatible with acids, copper and copper compounds, halogens and halogenated hydrocarbons, nitrogen oxides, hydrogen bromide, aluminum chloride, bromotrichloromethane, carbon tetrachloride, chlorine, chlorine dioxide, and nitrogen dioxide.

FIRE INFORMATION

Gases are heavier than air and will collect in low areas. Gases in confined areas may explode when exposed to fire. Gases may travel long distances to ignition sources and flash back. Storage containers and parts of containers may rocket great distances, in many directions. If material or contaminated runoff enters waterways, notify downstream users of potentially contaminated waters. Notify local health and fire officials and pollution control agencies. Do not extinguish the fire unless the flow of the gas can be stopped and any remaining gas is out of the line. Specially trained personnel may use fog lines to cool exposures and let the fire burn itself out. Use dry chemicals, alcohol-resistant foam, carbon dioxide. From a secure explosion-proof location, use water spray to cool exposed containers. If cooling streams are ineffective (venting sound increases in volume and pitch, tank discolors, or shows any signs of deforming), withdraw immediately to a secure position.

NFPA Hazard Classifications:

Health Hazards (Blue) 1
Flammability (Red) 4
Reactivity (Yellow) 2

Flash Point: Gas; -213°F (-136°C) c.c. (approximately)
Flammable Limits in Air: LEL 2.75%; UEL 36%.
Autoignition Temperature: 842°F (450°C).

FIRST AID

Move victim to fresh air. Call emergency medical care. Apply artificial respiration if victim is not breathing. Administer oxygen if breathing is difficult. Remove and isolate contaminated clothing and shoes. In case of contact with liquefied gas, thaw frosted parts with lukewarm water. Keep victim warm and quiet. Ensure that medical personnel are aware of the material(s) involved and take precautions to protect themselves.

ENVIRONMENTAL INFORMATION

Clean Air Act: Accidental Release Prevention/Flammable substances, (Section 112[r], Table 3). TQ = 10,000 lb (4,540 kg).
EPCRA Section 313: Form R *de minimis* concentration reporting level: 1.0%.

ETHYLENE CARBONATE E:194

SYNONYMS: CARBONATO de ETILENO (Spanish); CARBONIC ACID, CYCLIC ETHYLENE ESTER; CYCLIC ETHYLENE CARBONATE; 1,3-DIOXOLAN-2-ONE; DIOXOLONE-2; EINECS No. 202-510-0; ETHYLENE CARBONIC ACID; ETHYLENE GLYCOL CARBONATE; ETHYLENE GLYCOL, CYCLIC CARBONATE; GLYCOL CARBONATE

IDENTIFICATION

CAS:	96-49-1
Molecular Formula:	$C_3H_4O_3$
RTECS No.:	FF9450000; FF9550000
Properties:	Combustible crystalline solid or liquid. Colorless. Soluble in water.

HEALTH AND SAFETY INFORMATION

RECOMMENDATIONS FOR RESPIRATOR SELECTION

At any concentrations above the NIOSH REL, or where there is no REL, at any detectable

concentration: SCBAF:PD,PP (any self-contained breathing apparatus that has a full facepiece and is operated in a pressure-demand or other positive-pressure mode);or SAF:PD,PP:ASCBA (any supplied-air respirator that has a full facepiece and is operated in a pressure-demand or other positive-pressure mode in combination with an auxiliary, self-contained breathing apparatus operated in a pressure-demand or other positive-pressure mode).

Escape: GMFOV [any air-purifying, full-facepiece respirator (gas mask) with a chin-style, front- or back-mounted organic vapor canister];or SCBAE (any appropriate escape-type, self-contained breathing apparatus).

CHEMICAL REACTIVITY

Incompatible with oxidizers, germanium, lead diacetate, magnesium, mercurous chloride, silver nitrate.

FIRE INFORMATION

If material or contaminated runoff enters waterways, notify downstream users of potentially contaminated waters. Notify local health and fire officials and pollution control agencies. Use dry chemicals, alcohol-resistant foam, carbon dioxide.

NFPA Hazard Classifications: Not listed

Flash Point: 290°F (143°C) o.c.

FIRST AID

Move victim to fresh air. Call emergency medical care. Apply artificial respiration if victim is not breathing. Administer oxygen if breathing is difficult. Remove and isolate contaminated clothing and shoes. In case of contact with substance, immediately flush skin or eyes with running water for at least 20 minutes. Wash skin with soap and water. Keep victim warm and quiet. Effects of exposure (inhalation, ingestion, or skin contact) to substance may be delayed. Ensure that medical personnel are aware of the material(s) involved and take precautions to protect themselves.

ETHYLENE CYANOHYDRIN E:197

SYNONYMS: CLORHIDRINA ETILENICA (Spanish); 2-CYANOETHANOL; 2-CYANOETHYL ALCOHOL; EINECS No. 203-704-8; GLYCOL CYANOHYDRIN; β-HPN; HYDRACRYOLNITRILE; 1-HYDROXY-2-CYANOETHANE; 3-HYDROXYPROPANENITRILE; β-HYDROXYPROPIONITRILE; 3-HYDROXYPROPIONITRILE; METHANOLACETONITRILE; USAF RH-7

IDENTIFICATION

CAS: 109-78-4

472 Ethylene Cyanohydrin

ERG ID:	UN 1935
ERG Guide:	157
ERG Guide Category:	SUBSTANCES-TOXIC AND/OR CORROSIVE (NONCOMBUSTIBLE/WATER-SENSITIVE)
Molecular Formula:	C_3H_5NO
RTECS No.:	MU5250000
Properties:	Combustible liquid. Colorless to yellow-brown. Odorless to weak odor. Sinks and mixes with water.

HEALTH AND SAFETY INFORMATION

Irritating to the eyes, skin, and respiratory tract.

IDLH: 25 mg/m^3 as cyanides
OSHA Table Z-1-A Air Contaminant

Exposure Limits:	**TWA**	**CEILING**
ACGIH: [1]	5 mg/m^3 as cyanide	
NIOSH: [1]		4.7 ppm; 5 mg/m^3 per 10 minutes as cyanides
OSHA: [1]	5 mg/m^3 as cyanides	

[1] Skin contact contributes significantly in overall exposure

RECOMMENDATIONS FOR RESPIRATOR SELECTION: NIOSH as cyanides

50 mg/m^3: SA (any supplied-air respirator);or SCBAF (any self-contained breathing apparatus with full facepiece).

Emergency or planned entry into unknown concentration or IDLH conditions:
SCBAF:PD,PP (any self-contained breathing apparatus that has a full facepiece and is operated in a pressure-demand or other positive-pressure mode);or SAF:PD,PP:ASCBA (any supplied-air respirator that has a full facepiece and is operated in a pressure-demand or other positive-pressure mode in combination with an auxiliary, self-contained breathing apparatus operated in a pressure-demand or other positive pressure mode).

Escape: GMFSHiE [any air-purifying, full-facepiece respirator (gas mask) with a chin-style, front- or back-mounted canister providing protection against the compound of concern and having a high-efficiency particulate filter];or SCBAE (any appropriate escape-type, self-contained breathing apparatus).

CHEMICAL REACTIVITY

Forms explosive mixture with air. Unstable: explosive polymerization may occur. A powerful reducing agent; contact with oxidizers may cause fire and explosions. Contact with hot water forms cyanide gas. Reacts violently with sodium hydroxide. Reacts with acids, acid salts, chlorates, nitrates. Attacks mild steel, copper, and copper alloys.

FIRE INFORMATION

Combustion products include hydrogen cyanide and nitrogen oxides. Vapors are heavier than air and will collect in low areas. Vapors in confined areas may explode when exposed to fire. May react with itself when heated, blocking relief valves and leading to container explosion. Storage containers and parts of containers may rocket great distances, in many directions. If material or contaminated runoff enters waterways, notify downstream users of potentially contaminated waters. Notify local health and fire officials and pollution control agencies. Use dry chemicals, alcohol-resistant foam, carbon dioxide. Use water spray to cool containers.

NFPA Hazard Classifications:

Health Hazards (Blue) 1
Flammability (Red) 1
Reactivity (Yellow) 2

Flash Point: 265°F (129°C) o.c.
Flammable Limits in Air: LEL 2.3%; UEL 12.1%
Autoignition Temperature: 922°F (494°C)

FIRST AID

Move victim to fresh air. Call emergency medical care. Apply artificial respiration if victim is not breathing. Do not use mouth-to-mouth method if victim ingested or inhaled the substance; induce artificial respiration with the aid of a pocket mask equipped with a one-way valve or other proper respiratory medical device. Administer oxygen if breathing is difficult. Remove and isolate contaminated clothing and shoes. In case of contact with substance, immediately flush skin or eyes with running water for at least 20 minutes. For minor skin contact, avoid spreading material on unaffected skin. Keep victim warm and quiet. Effects of exposure (inhalation, ingestion, or skin contact) may be delayed. Ensure that medical personnel are aware of the material(s) involved and take precautions to protect themselves.

ENVIRONMENTAL INFORMATION as cyanides (soluble salts)

Clean Air Act: Hazardous Air Pollutants (Title I, Part A, Section 112).
Clean Water Act: Section 307 Priority Pollutants as cyanide, total; Toxic Pollutant (Section 401.15).
EPA Hazardous Waste Number (RCRA No.): P030 as cyanide.
RCRA Toxicity Characteristic (Section 261.24): RCRA Section 261 Hazardous Constituents
RCRA Land Ban Waste
RCRA Universal Treatment Standards: Wastewater (mg/L), 1.2 (total); 0.86 (amenable); Nonwastewater (mg/kg), 590 (total); 30 (amendable).
RCRA Ground Water Monitoring List: Suggested test methods(s) (PQL μg/L); 9010(40).
Safe Drinking Water Act: MCL, 0.2 mg/L; MCLG, 0.2 mg/l; Regulated chemical (47 FR 9352).
EPCRA Section 304 Reportable Quantity (RQ): CERCLA, 10 lb (4.54 kg).
EPCRA Section 313: Form R *de minimis* concentration reporting level: 1.0%. Form R Toxic Chemical Category Code: N016.
Marine Pollutant (49CFR, Subchapter 172.101, Appendix B) as cyanides, inorganic n.o.s.

ETHYLENE DICHLORIDE E:230

SYNONYMS: AETHYLENCHLORID (German); 1,2-BICHLOROETHANE; BICHLORURE-d'ETHYLENE (French); BROCIDE; 1,2-DICHLOROETHANE; DICLORURO de ETILENO (Spanish); DUTCH LIQUID; DUTCH OIL; ETHANE DICHLORIDE; EINECS No. 203-458-1; ETHYLENE CHLORIDE; 1,2-ETHYLENE DICHLORIDE; EDC; EEC No. 602-012-00-7; EINECS No. 203-458-1; ENT 1,656; ETHANE DICHLORIDE; FREON-150; GLYCOL DICHLORIDE; GRANOSAN; NCI-C00511; SYM-DICHLOROETHANE

IDENTIFICATION

CAS:	107-06-2
ERG ID:	UN 1184
ERG Guide:	129
ERG Guide Category:	FLAMMABLE LIQUIDS (POLAR/WATER-MISCIBLE/NOXIOUS)
Molecular Formula:	$C_2H_4Cl_2$
RTECS No.:	KI0525000
Properties:	Highly flammable liquid. Colorless, but decomposes slowly, becoming acidic and darkening in color. Sweet, chloroformlike odor. Sinks in water; practically insoluble.

HEALTH AND SAFETY INFORMATION

Corrosive to the eyes. Irritating to skin and respiratory tract. Attacks the central nervous system; possible headache, nausea, dizziness, sleepiness, unconsciousness, and death. Can cause liver and kidney damage.

IARC Evaluation: Group 2B; possible carcinogen
NTP: Anticipated carcinogen
IDLH: 50 ppm; suspected human carcinogen.
Odor Threshold: 11.2 ppm
OSHA Table Z-1-A Air Contaminant
OSHA Table Z-2 Air Contaminant

Exposure Limits:	**TWA**	**CEILING**	**STEL**
ACGIH:	10 ppm; 40 mg/m^3		
NIOSH: (1)	1 ppm; 4 mg/m^3		2 ppm; 8 mg/m^3
OSHA:	50 ppm	100 ppm; peak 200 ppm(2)	

(1) Potential human carcinogen. Limit exposure to lowest feasible level.
(2) Per 5 minute maximum peak in any 3 hours

RECOMMENDATIONS FOR RESPIRATOR SELECTION: NIOSH

At any concentrations above the NIOSH REL, or where there is no REL, at any detectable concentration: SCBAF:PD,PP (any self-contained breathing apparatus that has a full facepiece and is

operated in a pressure-demand or other positive-pressure mode);or SAF:PD,PP:ASCBA (any supplied-air respirator that has a full facepiece and is operated in a pressure-demand or other positive-pressure mode in combination with an auxiliary, self-contained breathing apparatus operated in a pressure-demand or other positive pressure mode).

Escape: GMFOV[any air-purifying, full-facepiece respirator (gas mask) with a chin-style, front- or back-mounted organic vapor canister];or SCBAE (any appropriate escape-type, self-contained breathing apparatus).

CHEMICAL REACTIVITY

Contact with moisture forms hydrochloric acid. Reacts with strong oxidizers, caustics; chemically active metals such as aluminum (forms dangerous shock-sensitive compounds) or magnesium powder; sodium and potassium; liquid ammonia; alkyl amides. Decomposes to vinyl chloride and hydrochloric acid above 1112°F (600°C). Attacks plastics. Corrodes iron when contaminated with water at elevated temperatures. May accumulate static electricity, and cause ignition of its vapors.

FIRE INFORMATION

Combustion products include toxic hydrogen chloride and phosgene. Vapors are heavier than air and will collect in low areas. Vapors in confined areas may explode when exposed to fire. Vapors may travel long distances to ignition sources and flash back. Storage containers and parts of containers may rocket great distances, in many directions. If material or contaminated runoff enters waterways, notify downstream users of potentially contaminated waters. Notify local health and fire officials and pollution control agencies. Use dry chemicals, foam, carbon dioxide.

NFPA Hazard Classifications:

Health Hazards (Blue) 2
Flammability (Red) 3
Reactivity (Yellow) 0

Flash Point: 60°F (16°C) o.c.; 56°F (13°C) c.c.
Flammable Limits in Air: LEL 6.2%; UEL 16%
Autoignition Temperature: 775°F (413°C)

FIRST AID

Move victim to fresh air. Call emergency medical care. Apply artificial respiration if victim is not breathing. Administer oxygen if breathing is difficult. Remove and isolate contaminated clothing and shoes. In case of contact with substance, immediately flush skin or eyes with running water for at least 20 minutes. Wash skin with soap and water. Keep victim warm and quiet. Effects of exposure (inhalation, ingestion, or skin contact) to substance may be delayed. Ensure that medical personnel are aware of the material(s) involved and take precautions to protect themselves.

ENVIRONMENTAL INFORMATION

Clean Air Act: Hazardous Air Pollutants (Title I, Part A, Section 112)
Clean Water Act: Section 307 Priority Pollutants; Section 311 Hazardous Substances/RQ (same as CERCLA); Section 313 Priority Chemicals; Toxic Pollutant (Section 401.15)
EPA Hazardous Waste Number (RCRA No.): U077, D028
RCRA Section 261 Hazardous Constituents
RCRA Toxicity Characteristic (Section 261.24): Maximum concentration of Contaminants, regulatory level, 0.5 mg/L.
RCRA Land Ban Waste
RCRA Universal Treatment Standards: Wastewater (mg/L), 0.21; Nonwastewater (mg/kg), 6.0.
RCRA Ground Water Monitoring List: Suggested test methods(s) (PQL μg/L); 8010 (0.5); 8240 (5)
Safe Drinking Water Act: MCL, 0.005 mg/L; MCLG, zero; Regulated chemical (47 FR 9352).
EPCRA Section 304 Reportable Quantity (RQ): CERCLA, 100 lb (45.4 kg)
California's Proposition 65: Carcinogen.
Marine Pollutant (49CFR, Subchapter 172.101, Appendix B).

ETHYLENE GLYCOL E:240

SYNONYMS: DOWTHERM SR-1; 1,2-DIHYDROXYETHANE; EEC No. 603-027-00-1; EG; EINECS No. 203-473-3; 1,2-ETHANEDIOL; ETHYLENE ALCOHOL; ETILENGLICOL (Spanish); FAST CURE 45 EPOXY; FRIDEX; GLYCOL; GLYCOL ALCOHOL; 2-HYDROXYETHANOL; KTI NMD-25(+); LUTROL-9; LLEXAN E; LOW DYE-FAST DRY INK; MACROGOL 400 BPC; MEG; ETHYLENE DIHYDRATE; MONOETHYLENE GLYCOL; NCI-C00920; NORKOOL; PD ACTIVATOR; PRO 330 CLEAR THIN SPREAD; 777 ETCH; TESCOL; UCAR-17; VANDALEX 20/24

IDENTIFICATION

CAS:	107-21-1
ERG ID:	UN 3077
ERG Guide:	171
ERG Guide Category:	LOW TO MODERATE HAZARD
Molecular Formula:	$C_2H_6O_2$
RTECS No.:	KW2975000
Properties:	Combustible, slightly syrupy liquid. Clear, colorless. Commercial products often contain colored dyes. Odorless but has a sweet taste. Sinks and mixes with water.

HEALTH AND SAFETY INFORMATION

Occupational exposure to heated ethylene glycol has caused involuntary eye movement that may indicate nerve damage. Some individuals also reported attacks of unconsciousness lasting 5 to 10 minutes, which went away when they stopped working with ethylene glycol. Prolonged or repeated contact may cause irritation of the skin, blistering and dermatitis; brain and central nervous system changes; and liver and kidney damage.

Odor Threshold: more than 0.26 mg/m^3
OSHA Table Z-1-A Air Contaminant

Exposure Limits: **CEILING**

ACGIH: 100 mg/m^3 as vapor or mist

RECOMMENDATIONS FOR RESPIRATOR SELECTION: NIOSH

At any concentrations above the NIOSH REL, or where there is no REL, at any detectable concentration: SCBAF:PD,PP (any self-contained breathing apparatus that has a full facepiece and is operated in a pressure-demand or other positive-pressure mode);or SAF:PD,PP:ASCBA (any supplied-air respirator that has a full facepiece and is operated in a pressure-demand or other positive-pressure mode in combination with an auxiliary, self-contained breathing apparatus operated in a pressure-demand or other positive-pressure mode).

Escape: GMFOV [any air-purifying, full-facepiece respirator (gas mask) with a chin-style, front- or back-mounted organic vapor canister];or SCBAE (any appropriate escape-type, self-contained breathing apparatus).

CHEMICAL REACTIVITY

Reacts with strong oxidizers, strong acids such as sulfuric acid, caustics, aliphatic amines, isocyanates, chlorosulfonic acid, chromium trioxide, potassium permanganate, sodium peroxide, and oleum. Hygroscopic (absorbs moisture from the air). May be explosive when heated.

FIRE INFORMATION

Vapors are heavier than air and will collect in low areas. Vapors in confined areas may explode when exposed to fire. If material or contaminated runoff enters waterways, notify downstream users of potentially contaminated waters. Notify local health and fire officials and pollution control agencies. Use dry chemicals, foam, carbon dioxide.

NFPA Hazard Classifications:

Health Hazards (Blue) 1
Flammability (Red) 1
Reactivity (Yellow) 0

Flash Point: 240°F (116°C) o.c.; 232°F (111°C) c.c.
Flammable Limits in Air: LEL 3.2%; UEL 15.3%.
Autoignition Temperature: 752°F (400°C), pure; 775°F (413°C), antifreeze.

FIRST AID

Move victim to fresh air. Call emergency medical care. Apply artificial respiration if victim is not

breathing. Administer oxygen if breathing is difficult. In case of contact with substance, immediately flush skin or eyes with running water for at least 20 minutes. Remove and isolate contaminated clothing and shoes. Ensure that medical personnel are aware of the material(s) involved and take precautions to protect themselves.

ENVIRONMENTAL INFORMATION

Clean Air Act: Hazardous Air Pollutants (Title I, Part A, Section 112)
EPCRA Section 304: Reportable Quantity (RQ): CERCLA, 1 lb (0.454 kg)
EPCRA Section 313: Form R *de minimis* concentration reporting level: 1.0%

ETHYLENE GLYCOL DIACETATE E:242

SYNONYMS: EINECS No. 203-881-1; 1,2-ETHANEDIOL DIACETATE; ETHYLENE ACETATE; ETHYLENE DIACETATE; ETHYLENE GLYCOL ACETATE; GLYCOL DIACETATE

IDENTIFICATION

CAS:	111-55-7
ERG ID:	NA 1993
ERG Guide:	128
ERG Guide Category:	FLAMMABLE LIQUIDS (NONPOLAR/WATER-IMMISCIBLE)
Molecular Formula:	$C_6H_{10}O_4$
RTECS No.:	KW4025000
Properties:	Combustible liquid. Colorless. Weak fruity odor. Sinks and mixes with water.

HEALTH AND SAFETY INFORMATION

Irritates eyes and respiratory tract.

RECOMMENDATIONS FOR RESPIRATOR SELECTION

At any concentrations above the NIOSH REL, or where there is no REL, at any detectable concentration: SCBAF:PD,PP (any self-contained breathing apparatus that has a full facepiece and is operated in a pressure-demand or other positive-pressure mode);or SAF:PD,PP:ASCBA (any supplied-air respirator that has a full facepiece and is operated in a pressure-demand or other positive-pressure mode in combination with an auxiliary, self-contained breathing apparatus operated in a pressure-demand or other positive-pressure mode).

Escape: GMFOV [any air-purifying, full-facepiece respirator (gas mask) with a chin-style, front- or back-mounted organic vapor canister];or SCBAE (any appropriate escape-type, self-contained breathing apparatus).

CHEMICAL REACTIVITY

Forms explosive mixture with air. Contact with oxidizers may cause fire and explosions. Incompatible with nitrates.

FIRE INFORMATION

Vapors are heavier than air and will collect in low areas. Vapors in confined areas may explode when exposed to fire. Vapors may travel long distances to ignition sources and flash back. Storage containers and parts of containers may rocket great distances, in many directions. If material or contaminated runoff enters waterways, notify downstream users of potentially contaminated waters. Notify local health and fire officials and pollution control agencies. Use dry chemicals, foam, carbon dioxide.

NFPA Hazard Classifications:

Health Hazards (Blue) 1
Flammability (Red) 1
Reactivity (Yellow) 0

Flash Point: 205°F (96°C) o.c.; 191°F (88°C) c.c.
Flammable Limits in Air: LEL 1.6%; UEL 8.4%
Autoignition Temperature: 900°F (482°C)

FIRST AID

Move victim to fresh air. Call emergency medical care. Apply artificial respiration if victim is not breathing. If breathing is difficult, give oxygen. Remove and isolate contaminated clothing and shoes. In case of contact with substance, immediately flush skin or eyes with running water for at least 20 minutes. Wash skin with soap and water. Keep victim warm and quiet. Ensure that medical personnel are aware of the material(s) involved and take precautions to protect themselves.

ETHYLENE GLYCOL SILICATE E:244

SYNONYMS: EINECS No. 241-598-5; SILICATO del ETILENGLICOL (Spanish); SILICIC ACID, (H_4SiO_4), TETRAKIS(2-HYDROXYETHYL) ESTER; TETRAGLYCOL SILICATE; TETRAKIS(β-HYDROXYETHYL)SILICATE; TETRAKIS(2-HYDROXYETHYL)SILICATE; TETRAKIS(2-HYDROXYETHYL) ORTHOSILICATE

IDENTIFICATION

CAS:	17622-94-5
Molecular Formula:	$C_8H_{20}O_8Si$
RTECS No.:	VW2650000
Properties:	Colorless liquid. Mixes with water.

HEALTH AND SAFETY INFORMATION

RECOMMENDATIONS FOR RESPIRATOR SELECTION

At any concentrations above the NIOSH REL, or where there is no REL, at any detectable concentration: SCBAF:PD,PP (any self-contained breathing apparatus that has a full facepiece and is operated in a pressure-demand or other positive-pressure mode);or SAF:PD,PP:ASCBA (any supplied-air respirator that has a full facepiece and is operated in a pressure-demand or other positive-pressure mode in combination with an auxiliary, self-contained breathing apparatus operated in a pressure-demand or other positive-pressure mode).

Escape: GMFOV [any air-purifying, full-facepiece respirator (gas mask) with a chin-style, front- or back-mounted organic vapor canister];or SCBAE (any appropriate escape-type, self-contained breathing apparatus).

CHEMICAL REACTIVITY

Incompatible with oxidizers, sulfuric acid, nitric acid, perchloric acid, caustics, aliphatic amines, amides.

FIRE INFORMATION

If material or contaminated runoff enters waterways, notify downstream users of potentially contaminated waters. Notify local health and fire officials and pollution control agencies. Use dry chemicals, alcohol-resistant foam, carbon dioxide.

NFPA Hazard Classifications: Not listed

FIRST AID

Move victim to fresh air. Call emergency medical care. Apply artificial respiration if victim is not breathing. Administer oxygen if breathing is difficult. Remove and isolate contaminated clothing and shoes. In case of contact with substance, immediately flush skin or eyes with running water for at least 20 minutes. Wash skin with soap and water. Keep victim warm and quiet. Effects of exposure (inhalation, ingestion, or skin contact) to substance may be delayed. Ensure that medical personnel are aware of the material(s) involved and take precautions to protect themselves.

ETHYLENEIMINE, INHIBITED E:246

SYNONYMS: AMINOETHYLENE; AZACYCLOPROPANE; AZIRANE; AZIRIDINE; AZIRINE; 1H-AZIRINE, DIHYDRO-; DIHYDROAZIRINE; DIHYDRO-1-AZIRINE; DIMETHYLENEIMINE; DIMETHYLENIMINE; E-1; EEC No. 613-001-00-1; EINECS No. 205-793-9; ETHYLIMENE; TL 337; ETHYLENE IMINE; ETHYLENIMINE; ETHYLIMINE; ETILENIMIDA (Spanish); ETILENIMINA (Italian); TL 337

IDENTIFICATION

CAS:	151-56-4
ERG ID:	UN 1185
ERG Guide:	131P
ERG Guide Category:	FLAMMABLE LIQUIDS-TOXIC
Molecular Formula:	C_2H_5N
RTECS No.:	KX5075000
Properties:	Flammable, oily liquid. Colorless. Ammonialike odor. Floats and mixes with water. Poisonous, flammable vapor is produced.

HEALTH AND SAFETY INFORMATION

Poison. Breathing the vapors or swallowing the material can cause death. Corrosive; skin and eye contact causes severe burns.

IARC Evaluation: IARC rating 3; limited animal evidence.
IDLH: 100 ppm; uspected human carcinogen.
Odor Threshold: 1.5 ppm
OSHA Table Z-1-A Air Contaminant
OSHA Specifically Regulated Carcinogen (29 CFR 1910.1012)

Exposure Limits: **TWA**

ACGIH: 0.5 ppm; 0.88 mg/m^3
NIOSH: [(1)]
[(1)] Potential human carcinogen; reduce exposure to lowest feasible level; skin contact contributes significantly in overall exposure.

RECOMMENDATIONS FOR RESPIRATOR SELECTION: NIOSH

At any concentrations above the NIOSH REL, or where there is no REL, at any detectable concentration: SCBAF:PD,PP (any self-contained breathing apparatus that has a full facepiece and is operated in a pressure-demand or other positive-pressure mode);or SAF:PD,PP:ASCBA (any supplied-air respirator that has a full facepiece and is operated in a pressure-demand or other positive-pressure mode in combination with an auxiliary, self-contained breathing apparatus operated in a pressure-demand or other positive pressure mode).

Escape: GMFOV [any air-purifying, full-facepiece respirator (gas mask) with a chin-style, front- or back-mounted organic vapor canister];or SCBAE (any appropriate escape-type, self-contained breathing apparatus).

CHEMICAL REACTIVITY

Mild reaction with water; becomes alkaline; reacts with acids. Acids, oxidizers, sunlight, elevated temperatures, silver, silver alloys (e.g., silver solder), or aluminum may cause explosive polymerization.

Attacks rubber, aluminum, zinc and other chemically active metals. May accumulate static electricity and cause ignition of its vapors.

FIRE INFORMATION

Combustion products include toxic nitrogen oxides. Firefighting gear (including SCBA) does not provide adequate protection. If exposure occurs, remove and isolate gear immediately and thoroughly decontaminate personnel. Vapors are heavier than air and will collect in low areas. Vapors in confined areas may explode when exposed to fire. Vapors may travel long distances to ignition sources and flash back. May react with itself, blocking relief valves and leading to tank explosions. Storage containers and parts of containers may rocket great distances, in many directions. If material or contaminated runoff enters waterways, notify downstream users of potentially contaminated waters. Notify local health and fire officials and pollution control agencies. Use dry chemicals, carbon dioxide.

NFPA Hazard Classifications:

Health Hazards (Blue) 4
Flammability (Red) 3
Reactivity (Yellow) 3

Flash Point: 1°F (-17°C) o.c.; 12°F (-11°C) c.c.
Flammable Limits in Air: LEL 3.3%; UEL 54.8%
Autoignition Temperature: 608°F (320°C)

FIRST AID

Move victim to fresh air. Call emergency medical care. Apply artificial respiration if victim is not breathing. Do not use mouth-to-mouth method if victim ingested or inhaled the substance; induce artificial respiration with the aid of a pocket mask equipped with a one-way valve or other proper respiratory medical device. Administer oxygen if breathing is difficult. Remove and isolate contaminated clothing and shoes. In case of contact with substance, immediately flush skin or eyes with running water for at least 20 minutes. Wash skin with soap and water. Keep victim warm and quiet. Effects of exposure (inhalation, ingestion, or skin contact) to substance may be delayed. Ensure that medical personnel are aware of the material(s) involved and take precautions to protect themselves.

ENVIRONMENTAL INFORMATION

Clean Air Act: Hazardous Air Pollutants (Title I, Part A, Section 112); Section 112[r], Accidental Release Prevention/Flammable Substances ((40CFR/68.130; 59FR 4497), TQ = 10,000 lb (4,4550 kg) Reportable Quantity (RQ): CERCLA, 1 lb (0.454 kg).
EPA Hazardous Waste Number (RCRA No.): P054
RCRA Section 261 Hazardous Constituents
EPCRA Section 302, Extremely Hazardous Substances: TPQ = 500 lb (228 kg)
EPCRA Section 304 Reportable Quantity (RQ): CERCLA, 1 lb (0.454 kg)
EPCRA Section 313: Form R *de minimis* concentration reporting level: 0.1%.
California's Proposition 65: Carcinogen.

ETHYLENE OXIDE E:250

SYNONYMS: AETHYLENOXID (German); ANPROLENE; DIHYDROOXIRENE; DIMETHYLENE OXIDE; EEC No. 603-023-00-X; E.O.; 1,2-EPOXYAETHAN (German); EPOXYETHANE (French); 1,2-EPOXYETHANE; ETHYLEENOXIDE (Dutch); ETHYLENE (OXYDE d') (French); ETILENE (OSSIDO DI) (Italian); ETO; ETYLENU TLENEK (Polish); NCI-C50088; OXACYCLOPROPANE; OXANE; OXIDO de ETILENO (Spanish); OXIDOETHANE; αβ-OXIDOETHANE; OXIRAAN (Dutch); OXIRAN; OXIRANE; OXIRENE, DIHYDRO-; 1,2-EPOXY ETHANE

IDENTIFICATION

CAS:	75-21-8
ERG ID:	UN 1040
ERG Guide:	119
ERG Guide Category:	GASES-TOXIC-FLAMMABLE
Molecular Formula:	C_2H_4O
RTECS No.:	KX2450000
Properties:	Extremely flammable liquefied gas; becomes a liquid at 51°F (11°C). Shipped as a liquefied gas under pressure. Colorless. Sweet, etherlike odor. Floats on and mixes with water. Flammable, irritating vapor is produced.

HEALTH AND SAFETY INFORMATION

Very irritating to skin, eyes, nose, and lungs; prolonged contact with skin can cause burns. Contact with liquid may cause frostbite.

IARC Evaluation: Group 2A, probable carcinogen; limited human evidence; sufficient animal evidence.
NTP: Anticipated carcinogen.
IDLH: 800 ppm; potential human carcinogen.
Odor Threshold: 851 ppm
OSHA Table Z-1-A Air Contaminant
OSHA Specifically Regulated Substances (29 CFR 1910.1001-1048).
OSHA Process Safety Management of Highly Hazardous Chemicals (29CFR 1910.119, Appendix A), Threshold Quantity: 5,000 pounds.

Exposure Limits:	**TWA**	**CEILING**
ACGIH:(1)	1 ppm; 1.8 mg/m^3	
NIOSH: (2)	<0.1 ppm; <0.18 mg/m^3	5 ppm; 9 mg/m^3 [10 min/day]
OSHA:	1 ppm; 5 ppm [15 min excursion]	

(1) Suspected carcinogen
(2) Potential carcinogen. Reduce exposure to lowest feasible level.

RECOMMENDATIONS FOR RESPIRATOR SELECTION: NIOSH

5 ppm: GMFS *End of service life indicator (ESLI) required.* [any air-purifying, full-facepiece respirator (gas mask) with a chin-style, front- or back-mounted canister providing protection against the compound of concern];or SCBAF (any self-contained breathing apparatus with a full facepiece);or SAF (any supplied-air respirator with a full facepiece).

Emergency or Planned Entry into Unknown Concentrations or IDLH Conditions: SCBAF:PD,PP (any self-contained breathing apparatus that has a full facepiece and is operated in a pressure-demand or other positive-pressure mode);or SAF:PD,PP:ASCBA (any supplied-air respirator that has a full facepiece and is operated in a pressure-demand or other positive-pressure mode in combination with an auxiliary, self-contained breathing apparatus operated in a pressure-demand or other positive-pressure mode).

Escape: GMFS *End of service life indicator (ESLI) required.* [any air-purifying, full-facepiece respirator (gas mask) with a chin-style, front- or back-mounted canister providing protection against the compound of concern];or SCBAE (any appropriate escape-type, self-contained breathing apparatus).

CHEMICAL REACTIVITY

Forms explosive mixture with air. Reacts slowly with water, forming hydrate; not hazardous at normal temperatures. Reacts, possibly violently, with many substances, including strong acids, alcohols, alkalis, amines, and oxidizers; chlorides/oxides of iron, aluminum and tin; mercury and magnesium. May polymerize violently when in contact with highly active catalytic surfaces such as anhydrous iron, tin and aluminum chlorides, pure iron and aluminum oxides, and alkali metal hydroxides. Attacks copper, silver, and their alloys. Protect container from physical damage, sun, and heat. Attacks some plastics, rubber, and coatings.

FIRE INFORMATION

Firefighting gear (including SCBA) does not provide adequate protection. If exposure occurs, remove and isolate gear immediately and thoroughly decontaminate personnel. Gases are heavier than air and will collect in low areas. Gases in confined areas may explode when exposed to fire. Gases may travel long distances to ignition sources and flash back. Storage containers and parts of containers may rocket great distances, in many directions. If material or contaminated runoff enters waterways, notify downstream users of potentially contaminated waters. Notify local health and fire officials and pollution control agencies. Do not extinguish the fire unless the flow of the gas can be stopped and any remaining gas is out of the line. Specially trained personnel may use fog lines to cool exposures and let the fire burn itself out. Use dry chemicals, foam, carbon dioxide. From a secure explosion-proof location, use water spray to cool exposed containers. If cooling streams are ineffective (venting sound increases in volume and pitch, tank discolors, or shows any signs of deforming), withdraw immediately to a secure position.

NFPA Hazard Classifications:

Health Hazards (Blue) 3
Flammability (Red) 4
Reactivity (Yellow) 3

Flash Point: -20°F (-29°C) c.c. (liquid)
Flammable Limits in Air: LEL 2.5%; UEL 100%
Autoignition Temperature: 804°F (430°C)

FIRST AID

Move victim to fresh air. Call emergency medical care. Apply artificial respiration if victim is not breathing. Do not use mouth-to-mouth method if victim ingested or inhaled the substance; induce artificial respiration with the aid of a pocket mask equipped with a one-way valve or other proper respiratory medical device. Administer oxygen if breathing is difficult. Remove and isolate contaminated clothing and shoes. In case of contact with substance, immediately flush skin or eyes with running water for at least 20 minutes. In case of contact with liquefied gas, thaw frosted parts with lukewarm water. Keep victim warm and quiet. Keep victim under observation. Effects of contact or inhalation may be delayed. Ensure that medical personnel are aware of the material(s) involved and take precautions to protect themselves.

ENVIRONMENTAL INFORMATION

Clean Air Act: Hazardous Air Pollutants (Title I, Part A, Section 112); Accidental Release Prevention/Flammable substances, (Section 112[r], Table 3), TQ = 10,000 lb (4,540 kg).
EPA Hazardous Waste Number (RCRA No.): U115.
RCRA Section 261 Hazardous Constituents
RCRA Land Ban Waste
RCRA Universal Treatment Standards: Wastewater (mg/L), 0.12; Nonwastewater (mg/kg), N/A.
EPCRA Section 302 Extremely Hazardous Substances: TPQ = 1,000 lb (454 kg).
EPCRA Section 304: Reportable Quantity (RQ): CERCLA, 10 lb (4.54 kg).
EPCRA Section 313: Form R *de minimis* concentration reporting level: 0.1%.
California's Proposition 65: Carcinogen; Reproductive toxin (female).

ETHYL ETHER — E:260

SYNONYMS: AETHER; ANAESTHETIC ETHER; ANESTHESIA ETHER; ANESTHETIC ETHER; DIAETHYLAETHER (German); DIETHYL ETHER; DIETHYL OXIDE; DWUETYLOWYETER (Polish); EEC No. 603-022-00-4; EINECS No. 200-467-2; ETER ETILICO (Spanish); ETHANE, 1,1'-OXYBIS-; ETERE ETILICO (Italian); ETHER; ETHER, ETHYL; ETHER ETHYLIQUE (French); ETHOXYETHANE; OXYDE d'ETHYLE (French); ETHYL ETHER UPS/ACS; SOLVENT ETHER; SULFURIC ETHER

IDENTIFICATION

CAS:	60-29-7
ERG ID:	UN 1155
ERG Guide:	127
ERG Guide Category:	FLAMMABLE LIQUIDS (POLAR/WATER MISCIBLE)
Molecular Formula:	$C_4H_{10}O$

RTECS No.: KI5775000

Properties: Extremely flammable, watery liquid. Clear, colorless. Sweet, pungent odor. Floats on water. Flammable, irritating vapor is produced. Boiling point is 94°F (34°C).

HEALTH AND SAFETY INFORMATION

Mildly irritating to the eyes, skin, nose and lungs. Has anesthetic or narcotic properties; high levels can cause drowsiness, unconsciousness, and possible death. Skin contact may cause drying and cracking.

IARC Evaluation: Overall evaluation not assigned; animal data, not adequate.
IDLH: 1,900 ppm [LEL]
Odor Threshold: 2.29 ppm
OSHA Table Z-1-A Air Contaminant

Exposure Limits:	TWA	STEL
ACGIH:	400 ppm; 1210 mg/m^3	500 ppm; 1520 mg/m^3
NIOSH:		
OSHA:	400 ppm; 1200 mg/m^3	

RECOMMENDATIONS FOR RESPIRATOR SELECTION: OSHA

1,900 ppm: CCROV [any chemical cartridge respirator with organic vapor cartridge(s)];or GMFOV[any air-purifying, full-facepiece respirator (gas mask) with a chin-style, front- or back-mounted acid gas canister];or PAPROV [any powered, air-purifying respirator with organic vapor cartridge(s)];or SA (any supplied-air respirator);or SCBAF (any self-contained breathing apparatus with a full facepiece

Emergency or Planned Entry into Unknown Concentrations or IDLH Conditions: SCBAF:PD,PP (any self-contained breathing apparatus that has a full facepiece and is operated in a pressure-demand or other positive-pressure mode);or SAF:PD,PP:ASCBA (any supplied-air respirator that has a full facepiece and is operated in a pressure-demand or other positive-pressure mode in combination with an auxiliary, self-contained breathing apparatus operated in a pressure-demand or other positive-pressure mode).

Escape: GMFOV [any air-purifying, full-facepiece respirator (gas mask) with a chin-style, front- or back-mounted organic vapor canister];or SCBAE (any appropriate escape-type, self-contained breathing apparatus).
Note: Substance reported to cause eye irritation or damage; may require eye protection.

CHEMICAL REACTIVITY

Forms explosive mixture with air. Incompatible with strong acids, strong oxidizers. Can form peroxides from air or light; may explode when container is unstoppered or otherwise opened. Attacks some plastics, rubber, and coatings. Being a nonconductor, chemical may accumulate static electric charges that may result in ignition of vapor.

FIRE INFORMATION

Vapors are heavier than air and will collect in low areas. Vapors in confined areas may explode when exposed to fire. Vapors may travel long distances to ignition sources and flash back. Storage containers and parts of containers may rocket great distances, in many directions. If material or contaminated runoff enters waterways, notify downstream users of potentially contaminated waters. Notify local health and fire officials and pollution control agencies. Use dry chemicals, foam, carbon dioxide.

NFPA Hazard Classifications:

Health Hazards (Blue) 1
Flammability (Red) 4
Reactivity (Yellow) 1

Flash Point: −40°F (−40°C) o.c.: −49°F (−45°C) c.c.
Flammable Limits in Air: LEL 1.85%; UEL 36.0%
Autoignition Temperature: 356°F (180°C)

FIRST AID

Move victim to fresh air. Call emergency medical care. Apply artificial respiration if victim is not breathing. If breathing is difficult, give oxygen. Remove and isolate contaminated clothing and shoes. In case of contact with substance, immediately flush skin or eyes with running water for at least 20 minutes. Wash skin with soap and water. Keep victim warm and quiet. Ensure that medical personnel are aware of the material(s) involved and take precautions to protect themselves.

ENVIRONMENTAL INFORMATION

Clean Air Act: Accidental Release Prevention/Flammable substances, (Section 112[r], Table 3), TQ = 10,000 lb (4,540 kg).
EPA Hazardous Waste Number (RCRA No.): U117.
RCRA Section 261 Hazardous Constituents
RCRA Land Ban Waste
EPCRA Section 304: Reportable Quantity (RQ): CERCLA, 100 lb (45.4 kg).
RCRA Universal Treatment Standards: Wastewater (mg/L), 0.12; Nonwastewater (mg/kg), 160.
TSCA:40CFR712.30(*e*)10; 40CFR716.120(*d*)10 dermal absorption testing.

2–ETHYL HEXANOL E:265

SYNONYMS: 1-AETHYLHEXANOL (German); EINECS No. 203-234-3; 2-ETHYL-1-HEXANOL; 2-ETHYLHEXYL ALCOHOL; 2-ETILHEXANOL (Spanish); OCTYL ALCOHOL

IDENTIFICATION

CAS:	104-76-7

488 2-Ethyl Hexanol

ERG ID:	UN 1987
ERG Guide:	127
ERG Guide Category:	FLAMMABLE LIQUIDS (POLAR/WATER- MISCIBLE)
Molecular Formula:	$C_8H_{18}O$
RTECS No.:	MP0350000
Properties:	Flammable oily liquid. Colorless. Faint odor. Floats on water; slightly soluble.

HEALTH AND SAFETY INFORMATION

RECOMMENDATIONS FOR RESPIRATOR SELECTION

At any concentrations above the NIOSH REL, or where there is no REL, at any detectable concentration: SCBAF:PD,PP (any self-contained breathing apparatus that has a full facepiece and is operated in a pressure-demand or other positive-pressure mode);or SAF:PD,PP:ASCBA (any supplied-air respirator that has a full facepiece and is operated in a pressure-demand or other positive-pressure mode in combination with an auxiliary, self-contained breathing apparatus operated in a pressure-demand or other positive-pressure mode).

Escape: GMFOV [any air-purifying, full-facepiece respirator (gas mask) with a chin-style, front- or back-mounted organic vapor canister];or SCBAE (any appropriate escape-type, self-contained breathing apparatus).

CHEMICAL REACTIVITY

Forms explosive mixture with air. Incompatible with strong acids, caustics, aliphatic amines, isocyanates, oxidizers. May accumulate static electrical charges and may cause ignition of its vapors.

FIRE INFORMATION

Vapors are heavier than air and will collect in low areas. Vapors in confined areas may explode when exposed to fire. Vapors may travel long distances to ignition sources and flash back. Storage containers and parts of containers may rocket great distances, in many directions. If material or contaminated runoff enters waterways, notify downstream users of potentially contaminated waters. Notify local health and fire officials and pollution control agencies. Use dry chemicals, foam, carbon dioxide.

NFPA Hazard Classifications:

Health Hazards (Blue) 2
Flammability (Red) 2
Reactivity (Yellow) 0

Flash Point: 164°F (73°C) c.c.
Flammable Limits in Air: LEL 0.88%; UEL 9.7%
Autoignition Temperature: 448°F (231°C)

FIRST AID

Move victim to fresh air. Call emergency medical care. Apply artificial respiration if victim is not breathing. Administer oxygen if breathing is difficult. Remove and isolate contaminated clothing and shoes. In case of contact with substance, immediately flush skin or eyes with running water for at least 20 minutes. Wash skin with soap and water. Keep victim warm and quiet. Ensure that medical personnel are aware of the material(s) involved and take precautions to protect themselves.

2–ETHYLHEXYL ACRYLATE E:270

SYNONYMS: ACRILATO DE 2-ETILHEXILO (Spanish); ACRYLIC ACID, 2-ETHYLHEXYL ESTER; EEC No. 607-107-00-7; EINECS No. 203-080-7; 2-ETHYLHEXYL, 2-PROPENOATE; OCTYL ACRYLATE

IDENTIFICATION

CAS:	103-11-7
ERG ID:	NA 1993, combustible liquids, n.o.s.
ERG Guide:	128
ERG Guide Category:	FLAMMABLE LIQUIDS (NONPOLAR/WATER-IMMISCIBLE)
Molecular Formula:	$C_{13}H_{10}Cl_2S$
RTECS No.:	AT08550000
Properties:	Combustible liquid. Colorless. Sharp odor. Floats on water; insoluble.

HEALTH AND SAFETY INFORMATION

Corrosive to eyes, skin, and respiratory tract. Inhalation can cause pulmonary edema, which can be delayed for several hours; there is a risk of death in serious cases.

RECOMMENDATIONS FOR RESPIRATOR SELECTION

At any concentrations above the NIOSH REL, or where there is no REL, at any detectable concentration: SCBAF:PD,PP (any self-contained breathing apparatus that has a full facepiece and is operated in a pressure-demand or other positive-pressure mode);or SAF:PD,PP:ASCBA (any supplied-air respirator that has a full facepiece and is operated in a pressure-demand or other positive-pressure mode in combination with an auxiliary, self-contained breathing apparatus operated in a pressure-demand or other positive-pressure mode).

Escape: GMFOV [any air-purifying, full-facepiece respirator (gas mask) with a chin-style, front- or back-mounted organic vapor canister];or SCBAE (any appropriate escape-type, self-contained breathing apparatus).

CHEMICAL REACTIVITY

Sunlight, heat, contaminants, or peroxides can cause polymerization. Reacts violently with strong oxidizers, with risk of fire and explosions. Forms explosive mixture with air. Incompatible with strong acids, aliphatic amines, alkanolamines.

FIRE INFORMATION

Vapors are heavier than air and will collect in low areas. Vapors in confined areas may explode when exposed to fire. Vapors may travel long distances to ignition sources and flash back. Storage containers and parts of containers may rocket great distances, in many directions. If material or contaminated runoff enters waterways, notify downstream users of potentially contaminated waters. Notify local health and fire officials and pollution control agencies. Use dry chemicals, alcohol resistant foam, carbon dioxide.

NFPA Hazard Classifications:
Health Hazards (Blue) 2
Flammability (Red) 2
Reactivity (Yellow) 2

Flash Point: 180°F (82°C) o.c.
Flammable Limits in Air: LEL 0.8%; UEL 6.4%
Autoignition Temperature: 485°F (252°C)

FIRST AID

Move victim to fresh air. Call emergency medical care. Apply artificial respiration if victim is not breathing. If breathing is difficult, give oxygen. Remove and isolate contaminated clothing and shoes. In case of contact with substance, immediately flush skin or eyes with running water for at least 20 minutes. Wash skin with soap and water. Keep victim warm and quiet. Ensure that medical personnel are aware of the material(s) involved and take precautions to protect themselves.

2–ETHYLHEXYL CHLORIDE E:275

SYNONYMS: 1-CHLORO-2-ETHYLHEXANE; 3-CHLOROMETHYLHEPTANE; EINECS No. 204-594-4; 2-ETHYLHEXYL-1-CHLORIDE; HEPTANE, 3-(CHLOROMETHYL)-

IDENTIFICATION

CAS:	123-04-6
ERG ID:	UN 2810 as toxic liquid, organic, n.o.s.
ERG Guide:	153
ERG Guide Category:	SUBSTANCES-TOXIC AND/OR CORROSIVE (COMBUSTIBLE)
Molecular Formula:	$C_8H_{17}Cl$

RTECS No.: MI8760000
Properties: Flammable liquid. Colorless. Floats in water; insoluble.

HEALTH AND SAFETY INFORMATION

Irritates eyes, skin, and respiratory tract.

RECOMMENDATIONS FOR RESPIRATOR SELECTION

At any concentrations above the NIOSH REL, or where there is no REL, at any detectable concentration: SCBAF:PD,PP (any self-contained breathing apparatus that has a full facepiece and is operated in a pressure-demand or other positive-pressure mode);or SAF:PD,PP:ASCBA (any supplied-air respirator that has a full facepiece and is operated in a pressure-demand or other positive-pressure mode in combination with an auxiliary, self-contained breathing apparatus operated in a pressure-demand or other positive-pressure mode).

Escape: GMFOV [any air-purifying, full-facepiece respirator (gas mask) with a chin-style, front- or back-mounted organic vapor canister];or SCBAE (any appropriate escape-type, self-contained breathing apparatus).

CHEMICAL REACTIVITY

Reacts vigorously with oxidizers.

FIRE INFORMATION

Decomposition products upon heating include toxic phosgene and chlorine. Vapors are heavier than air and will collect in low areas. Vapors in confined areas may explode when exposed to fire. Vapors may travel long distances to ignition sources and flash back. Storage containers and parts of containers may rocket great distances, in many directions. If material or contaminated runoff enters waterways, notify downstream users of potentially contaminated waters. Notify local health and fire officials and pollution control agencies. Use dry chemicals, foam, carbon dioxide.

NFPA Hazard Classifications:

Health Hazards (Blue) 2
Flammability (Red) 2
Reactivity (Yellow) 0

Flash Point: 140°F (60°C) o.c.

FIRST AID

Move victim to fresh air. Call emergency medical care. Apply artificial respiration if victim is not breathing. Do not use mouth-to-mouth method if victim ingested or inhaled the substance; induce artificial respiration with the aid of a pocket mask equipped with a one-way valve or other proper respiratory

medical device. Administer oxygen if breathing is difficult. Remove and isolate contaminated clothing and shoes. In case of contact with substance, immediately flush skin or eyes with running water for at least 20 minutes. Wash skin with soap and water. Keep victim warm and quiet. Effects of exposure (inhalation, ingestion, or skin contact) to substance may be delayed. Ensure that medical personnel are aware of the material(s) involved and take precautions to protect themselves.

ETHYL METHACRYLATE E:294

SYNONYMS: EEC No. 607-071-00-2; EINECS No. 202-597-5; ETHYL 1-2-METHACRYLATE; ETHYL-α-METHYLACRYLATE; 1-2-METHACRYLIC ACID, ETHYL ESTER; 2-METHYLE-2-PROPANOIC ACID, ETHYL ESTER; ETHYL METHYL ACRYLATE; METACRILATO de ETILO (Spanish); 2-PROPANOIC ACID, 1-METHYL-, ETHYL ESTER; RHOPLEX AC-33; ETHYL METHACRYLATE-INHIBITED; ETHYL 2-METHACRYLATE; ETHYL 2-METHYL-2-PROPENOATE

IDENTIFICATION

CAS:	97-63-2
ERG ID:	UN 2277
ERG Guide:	129P
ERG Guide Category:	FLAMMABLE LIQUIDS (POLAR/WATER-MISCIBLE/NOXIOUS)
Molecular Formula:	$C_6H_{10}O$
RTECS No.:	OZ4550000
Properties:	Flammable liquid. Colorless. Sharp unpleasant odor. Floats on water; insoluble.

HEALTH AND SAFETY INFORMATION

RECOMMENDATIONS FOR RESPIRATOR SELECTION

At any concentrations above the NIOSH REL, or where there is no REL, at any detectable concentration: SCBAF:PD,PP (any self-contained breathing apparatus that has a full facepiece and is operated in a pressure-demand or other positive-pressure mode);or SAF:PD,PP:ASCBA (any supplied-air respirator that has a full facepiece and is operated in a pressure-demand or other positive-pressure mode in combination with an auxiliary, self-contained breathing apparatus operated in a pressure-demand or other positive-pressure mode).

Escape: GMFOV [any air-purifying, full-facepiece respirator (gas mask) with a chin-style, front- or back-mounted organic vapor canister];or SCBAE (any appropriate escape-type, self-contained breathing apparatus).

CHEMICAL REACTIVITY

Easily forms explosive mixture with air. Contact with oxidizers may cause fire and explosions. Incompatible with oxidizers, strong acids, amines. Corrosive to mild steel.

FIRE INFORMATION

Vapors are heavier than air and will collect in low areas. Vapors in confined areas may explode when exposed to fire. Vapors may travel long distances to ignition sources and flash back. Storage containers and parts of containers may rocket great distances, in many directions. If material or contaminated runoff enters waterways, notify downstream users of potentially contaminated waters. Notify local health and fire officials and pollution control agencies. Use dry chemicals, foam, carbon dioxide, large quantities of water. Use water spray to cool containers.

NFPA Hazard Classifications:

Health Hazards (Blue) 2
Flammability (Red) 3
Reactivity (Yellow) 0

Flash Point: 68°F (20°C) o.c.
Flammable Limits in Air: LEL 1.8% to saturation.
Autoignition Temperature: 734°F (390°C)

FIRST AID

Move victim to fresh air. Call emergency medical care. Apply artificial respiration if victim is not breathing. Administer oxygen if breathing is difficult. Remove and isolate contaminated clothing and shoes. In case of contact with substance, immediately flush skin or eyes with running water for at least 20 minutes. Wash skin with soap and water. Keep victim warm and quiet. Effects of exposure (inhalation, ingestion, or skin contact) to substance may be delayed. Ensure that medical personnel are aware of the material(s) involved and take precautions to protect themselves.

ENVIRONMENTAL INFORMATION

EPA Hazardous Waste Number (RCRA No.): U118
RCRA Toxicity Characteristic (Section 261.24): RCRA Section 261 Hazardous Constituents
RCRA Land Ban Waste
RCRA Universal Treatment Standards: Wastewater (mg/L), 0.14; Nonwastewater (mg/kg), 160.
RCRA Ground Water Monitoring List: Suggested test methods(s) (PQL μg/L); 8015(10); 8240(5).
EPCRA Section 304 Reportable Quantity (RQ): CERCLA, 1,000 lb (454 kg)
TSCA: 40CFR799.5055 subject to testing.

ETHYL METHYL ETHER E:296

SYNONYMS: EEC No. 603-020-00-3; ETER ETILMETILICO (Spanish); ETHANE, METHOXY-; ETHOXYMETHANE; METHOXYETHANE; METHYL ETHYL ETHER; OXYDE d'ETHYLE et de METHYLE (French)

IDENTIFICATION

CAS:	540-67-0
ERG ID:	UN 1039
ERG Guide:	115
ERG Guide Category:	GASES-FLAMMABLE (INCLUDING REFRIGERATED LIQUIDS)
Molecular Formula:	C_3H_8O
RTECS No.:	KO0260000
Properties:	Highly flammable liquid or gas [liquifies at 51°F (11°C)]. Colorless. Soluble in water.

HEALTH AND SAFETY INFORMATION

Can act as an anesthetic, causing dizziness, tiredness, unconsciousness.

RECOMMENDATIONS FOR RESPIRATOR SELECTION

At any concentrations above the NIOSH REL, or where there is no REL, at any detectable concentration: SCBAF:PD,PP (any self-contained breathing apparatus that has a full facepiece and is operated in a pressure-demand or other positive-pressure mode); or SAF:PD,PP:ASCBA (any supplied-air respirator that has a full facepiece and is operated in a pressure-demand or other positive-pressure mode in combination with an auxiliary, self-contained breathing apparatus operated in a pressure-demand or other positive-pressure mode).

Escape: GMFOV [any air-purifying, full-facepiece respirator (gas mask) with a chin-style, front- or back-mounted organic vapor canister]; or SCBAE (any appropriate escape-type, self-contained breathing apparatus).

CHEMICAL REACTIVITY

Forms explosive mixture with air. Contact with oxidizers may cause fire and explosions. May form unstable peroxides in storage, in the presence of O_2 or sunlight; may explode spontaneously, or when heated.

FIRE INFORMATION

Gases are heavier than air and will collect in low areas. Gases in confined areas may explode when exposed to fire. Gases may travel long distances to ignition sources and flash back. Storage containers and parts of containers may rocket great distances, in many directions. If material or contaminated runoff enters waterways, notify downstream users of potentially contaminated waters. Notify local health and fire officials and pollution control agencies. Do not extinguish the fire unless the flow of the gas can be stopped and any remaining gas is out of the line. Specially trained personnel may use fog lines to cool exposures and let the fire burn itself out. Use dry chemicals, foam, carbon dioxide. From a secure explosion-proof location, use water spray to cool exposed containers. If cooling streams are ineffective (venting sound increases in volume and pitch, tank discolors, or shows any signs of deforming), withdraw immediately to a secure position.

NFPA Hazard Classifications:

Health Hazards (Blue) 1
Flammability (Red) 4
Reactivity (Yellow) 1

Flash Point: -35°F (-37°C) c.c.
Flammable Limits in Air: LEL 2.0%; UEL 10.1%
Autoignition Temperature: 374°F (190°C)

FIRST AID

Move victim to fresh air. Call emergency medical care. Apply artificial respiration if victim is not breathing. Administer oxygen if breathing is difficult. Remove and isolate contaminated clothing and shoes. Clothing frozen to the skin should be thawed before being removed. In case of contact with liquefied gas, thaw frosted parts with lukewarm water. Keep victim warm and quiet. Ensure that medical personnel are aware of the material(s) involved and take precautions to protect themselves.

4–ETHYLMORPHOLINE E:298

SYNONYMS: EINECS No. 202-885-0; n-ETHYLMORPHOLINE; NEM

IDENTIFICATION

CAS:	100-74-3
ERG ID:	UN 1993
ERG Guide:	128
ERG Guide Category:	FLAMMABLE LIQUIDS (NONPOLAR/WATER-IMMISCIBLE)
Molecular Formula:	$C_6H_{13}NO$
RTECS No.:	QE4025000
Properties:	Highly flammable liquid. Colorless to slight yellow. Ammonialike odor. Soluble in water.

HEALTH AND SAFETY INFORMATION

Irritates eyes, skin, and respiratory tract.

IDLH: 100 ppm
Odor Threshold: 0.275 ppm
OSHA Table Z-1-A Air Contaminant

Exposure Limits:	**TWA**
ACGIH: [1]	5 ppm; 24 mg/m^3
NIOSH: [1]	5 ppm; 23 mg/m^3
OSHA: [1]	20 ppm; 94 mg/m^3

[1] Skin contact contributes significantly in overall exposure.

RECOMMENDATIONS FOR RESPIRATOR SELECTION: NIOSH/OSHA

50 ppm: CCRFOV [any chemical cartridge respirator with a full facepiece and organic vapor cartridge(s)];or SA (any supplied-air respirator).
100 ppm: SA:CF (any supplied-air respirator operated in a continuous-flow mode);or PAPROV [any powered, air-purifying respirator with organic vapor cartridge(s)];or CCRFOV [any chemical cartridge respirator with a full facepiece and organic vapor cartridge(s)];or GMFOV [any air-purifying, full-facepiece respirator (gas mask) with a chin-style, front- or back-mounted acid gas canister];or SCBAF (any self-contained breathing apparatus with a full facepiece);or SAF (any supplied-air respirator with a full facepiece).

Emergency or planned entry into unknown concentration or IDLH conditions: SCBAF:PD,PP (any self-contained breathing apparatus that has a full facepiece and is operated in a pressure-demand or other positive-pressure mode);or SAF:PD,PP:ASCBA (any supplied-air respirator that has a full facepiece and is operated in a pressure-demand or other positive-pressure mode in combination with an auxiliary, self-contained breathing apparatus operated in a pressure-demand or other positive pressure mode).

Escape: GMFOV [any air-purifying, full-facepiece respirator (gas mask) with a chin-style, front- or back-mounted organic vapor canister];or SCBAE (any appropriate escape-type, self-contained breathing apparatus).

Note: Substance reported to cause eye irritation or damage; may require eye protection.

CHEMICAL REACTIVITY

Oxidizing materials can cause a vigorous reaction. Reacts with strong acids. Attacks some plastics, rubber, and coatings.

FIRE INFORMATION

Combustion products may include toxic nitrogen oxides ammonia. Vapors are heavier than air and will collect in low areas. Vapors in confined areas may explode when exposed to fire. Vapors may travel long distances to ignition sources and flash back. Storage containers and parts of containers may rocket great distances, in many directions. If material or contaminated runoff enters waterways, notify downstream users of potentially contaminated waters. Notify local health and fire officials and pollution control agencies. Use dry chemicals, foam, carbon dioxide.

NFPA Hazard Classifications:

Health Hazards (Blue) 2
Flammability (Red) 3
Reactivity (Yellow) 0

Flash Point: 90 °F (32 °C) c.c.

FIRST AID

Move victim to fresh air. Call emergency medical care. Apply artificial respiration if victim is not breathing. If breathing is difficult, give oxygen. Remove and isolate contaminated clothing and shoes. In case of contact with substance, immediately flush skin or eyes with running water for at least 20 minutes. Wash skin with soap and water. Keep victim warm and quiet. Ensure that medical personnel are aware of the material(s) involved and take precautions to protect themselves.

ETHYLTRICHLOROSILANE E:500

SYNONYMS: EINECS No. 204-072-6; ETHYL SILICON TRICHLORIDE; ETILTRICLOROSILANO (Spanish); SILANE, TRICHLOROETHYL- SILANE; SILICANE,TRICHLOROETHYL- SILICANE; TRICHLOROETHYLSILANE; TRICHLOROETHYLSILICANE

IDENTIFICATION

CAS:	115-21-9
ERG ID:	UN 1196
ERG Guide:	155
ERG Guide Category:	SUBSTANCES-TOXIC AND/OR CORROSIVE (FLAMMABLE/WATER-SENSITIVE)
Molecular Formula:	$C_2H_5Cl_3Si$
RTECS No.:	VV4200000
Properties:	Highly flammable, fuming liquid. Colorless. Sharp, intolerable, irritating odor. Reacts violently with water. Irritating hydrogen chloride gas is produced on contact with water.

HEALTH AND SAFETY INFORMATION

Poison. Breathing vapors can cause death. Skin and eye contact can cause severe burns and blindness.

RECOMMENDATIONS FOR RESPIRATOR SELECTION

At any concentrations above the NIOSH REL, or where there is no REL, at any detectable concentration: SCBAF:PD,PP (any self-contained breathing apparatus that has a full facepiece and is operated in a pressure-demand or other positive-pressure mode);or SAF:PD,PP:ASCBA (any supplied-air respirator that has a full facepiece and is operated in a pressure-demand or other positive-pressure mode in combination with an auxiliary, self-contained breathing apparatus operated in a pressure-demand or other positive-pressure mode).

Escape: GMFOV [any air-purifying, full-facepiece respirator (gas mask) with a chin-style, front- or back-mounted organic vapor canister];or SCBAE (any appropriate escape-type, self-contained breathing apparatus).

CHEMICAL REACTIVITY

Forms explosive mixture with air. Reacts vigorously with water, forming heat and hydrogen chloride (hydrochloric acid). Reacts with surface moisture to form hydrogen chloride, which is corrosive to common metals.

FIRE INFORMATION

Do not use water. Decomposition products upon heating include hydrogen chloride, phosgene, and other toxic gases. Vapors in confined areas may explode when exposed to fire. Vapors may travel long distances to ignition sources and flash back. Storage containers and parts of containers may rocket great distances, in many directions. If material or contaminated runoff enters waterways, notify downstream users of potentially contaminated waters. Notify local health and fire officials and pollution control agencies. Use dry chemicals, foam, carbon dioxide.

NFPA Hazard Classifications:

Health Hazards (Blue) 3
Flammability (Red) 3
Reactivity (Yellow) 2
Special Notice .... Water

Flash Point: 72°F (22°C) o.c.

FIRST AID

Move victim to fresh air; call emergency medical care. Apply artificial respiration if victim is not breathing. Do not use mouth-to-mouth method if victim ingested or inhaled the substance; induce artificial respiration with the aid of a pocket mask equipped with a one-way valve or other proper respiratory medical device. Administer oxygen if breathing is difficult. Remove and isolate contaminated clothing and shoes. In case of contact with substance, immediately flush skin or eyes with running water for at least 20 minutes. For minor skin contact, avoid spreading material on unaffected skin. Keep victim warm and quiet.

Effects of exposure (inhalation, ingestion, or skin contact) to substance may be delayed. Ensure that medical personnel are aware of the material(s) involved and take precautions to protect themselves.

ENVIRONMENTAL INFORMATION

EPCRA Section 302, Extremely Hazardous Substances: TPQ = 500 lb (227 kg)
EPCRA Section 304 Reportable Quantity (RQ): EHS, 500 lb (227 kg)

FERRIC CHLORIDE F:020

SYNONYMS: CLORURO FERRICO (Spanish); EINECS No. 231-729-4; EISENTRICHLORID (German); FERRIC CHLORIDE, ANHYDROUS; FERRIC CHLORIDE, HEXAHYDRATE; FERRIC CHLORIDE, SOLID; FERROUS(III) CHLORIDE; FLORES MARTIS; JERNTRICHLORIDE (Dutch); IRON(III) CHLORIDE; IRON PERCHLORIDE; IRON TRICHLORIDE; PERCHLORURE de FER (French); TRICLORURO di FERRO (Italian)

IDENTIFICATION

CAS:	7705-08-0
ERG ID:	UN 1773; UN 2582 (solution)
ERG Guide:	157; 154 (solution)
ERG Guide Category:	TOXIC AND/OR CORROSIVE (NONCOMBUSTIBLE/ WATER SENSITIVE)
Molecular Formula:	Cl_3Fe
RTECS No.:	LJ9100000
Properties:	Brown, corrosive solid. Black or greenish-black. Odorless. Sinks and mixes with water.

HEALTH AND SAFETY INFORMATION

Exposure Limits: TWA

ACGIH:	1 mg/m^3 as soluble iron salts
NIOSH:	1 mg/m^3 as soluble iron salts
OSHA:	None

RECOMMENDATIONS FOR RESPIRATOR SELECTION: NIOSH

At any concentrations above the NIOSH REL, or where there is no REL, at any detectable concentration: SCBAF:PD,PP (any self-contained breathing apparatus that has a full facepiece and is operated in a pressure-demand or other positive-pressure mode);or SAF:PD,PP:ASCBA (any supplied-air respirator that has a full facepiece and is operated in a pressure-demand or other positive-pressure mode in combination with an auxiliary, self-contained breathing apparatus operated in a pressure-demand or other positive-pressure mode).

Escape: GMFOV [any air-purifying, full-facepiece respirator (gas mask) with a chin-style, front- or back-mounted organic vapor canister];or SCBAE (any appropriate escape-type, self-contained breathing apparatus).

CHEMICAL REACTIVITY

Incompatible with water. Solution is very acidic; violent reaction with bases. Shock- and friction-sensitive explosive material formed with potassium, sodium, and other active metals. Reacts with allyl chloride, allyl alcohol, ethylene oxide. Attacks metals when wet.

FIRE INFORMATION

If material or contaminated runoff enters waterways, notify downstream users of potentially contaminated waters. Notify local health and fire officials and pollution control agencies. Use dry chemicals, alcohol-resistant foam, carbon dioxide.

FEMA Hazard Classifications:

Health Hazards (Blue) 1
Flammability (Red) 0
Reactivity (Yellow) 0

FIRST AID

Solid: Move victim to fresh air. Call emergency medical care. Apply artificial respiration if victim is not breathing. Do not use mouth-to-mouth method if victim ingested or inhaled the substance; induce artificial respiration with the aid of a pocket mask equipped with a one-way valve or other proper respiratory medical device. Administer oxygen if breathing is difficult. Remove and isolate contaminated clothing and shoes. In case of contact with substance, immediately flush skin or eyes with running water for at least 20 minutes. For minor skin contact, avoid spreading material on unaffected skin. Keep victim warm and quiet. Effects of exposure (inhalation, ingestion, or skin contact) may be delayed. Ensure that medical personnel are aware of the material(s) involved and take precautions to protect themselves.

Solution: Move victim to fresh air. Call emergency medical care. Apply artificial respiration if victim is not breathing. Do not use mouth-to-mouth method if victim ingested or inhaled the substance; induce artificial respiration with the aid of a pocket mask equipped with a one-way valve or other proper respiratory medical device. Administer oxygen if breathing is difficult. Remove and isolate contaminated clothing and shoes. In case of contact with substance, immediately flush skin or eyes with running water for at least 20 minutes. For minor skin contact, avoid spreading material on unaffected skin. Keep victim warm and quiet. Effects of exposure (inhalation, ingestion, or skin contact) may be delayed. Ensure that medical personnel are aware of the material(s) involved and take precautions to protect themselves.

ENVIRONMENTAL INFORMATION

Clean Water Act: Section 311 Hazardous Substances/RQ (same as CERCLA).
EPCRA Section 304: Reportable Quantity (RQ): CERCLA, 1,000 lb (454 kg).

FERRIC NITRATE F:030

SYNONYMS: EINECS No. 233-899-5; FERRIC NITRATE, NONHYDRATE; IJZERTRINITRAAT (Dutch); IRON NITRATE; IRON(III) NITRATE, ANHYDROUS; IRON TRINITRATE; NITRATO FERRICO (Spanish); NITRIC ACID, IRON(3+) SALT; NITRIC ACID, IRON(III) SALT

IDENTIFICATION

CAS:	10421-48-4
ERG ID:	UN 1466
ERG Guide:	140
ERG Guide Category:	OXIDIZERS
Molecular Formula:	$N_3O_9{\bullet}Fe$
Properties:	Noncombustible crystalline solid. Greyish-white to pale violet or green. Odorless. Sinks and mixes with water; the solution is acidic.

HEALTH AND SAFETY INFORMATION

Dust may cause eye and respiratory tract irritation. Liquid can be corrosive to eyes, skin, and respiratory tract.

Exposure Limits:	**TWA**
ACGIH:	1 mg/m^3 as iron salts
NIOSH:	1 mg/m^3 as soluble iron salts
OSHA:	None

RECOMMENDATIONS FOR RESPIRATOR SELECTION

At any concentrations above the NIOSH REL, or where there is no REL, at any detectable concentration: SCBAF:PD,PP (any self-contained breathing apparatus that has a full facepiece and is operated in a pressure-demand or other positive-pressure mode);or SAF:PD,PP:ASCBA (any supplied-air respirator that has a full facepiece and is operated in a pressure-demand or other positive-pressure mode in combination with an auxiliary, self-contained breathing apparatus operated in a pressure-demand or other positive-pressure mode).

Escape: GMFOV [any air-purifying, full-facepiece respirator (gas mask) with a chin-style, front- or back-mounted organic vapor canister];or SCBAE (any appropriate escape-type, self-contained breathing apparatus).

CHEMICAL REACTIVITY

An oxidizer; contact of solid with combustible materials (such as wood, paper) or reducing agents may cause fire and explosions. Incompatible with sodium salicylate, 2-naphthol, resorcinol, sulfur. Solutions are corrosive to most metals. At the melting point [95°F (35°C)] this material decomposes.

FIRE INFORMATION

Combustion products include toxic nitrogen oxides and nitric acid. If material or contaminated runoff enters waterways, notify downstream users of potentially contaminated waters. Notify local health and fire officials and pollution control agencies.

NFPA Hazard Classifications: Not listed

FIRST AID

Move victim to fresh air. Call emergency medical care. Apply artificial respiration if victim is not breathing. Administer oxygen if breathing is difficult. Remove and isolate contaminated clothing and shoes. In case of contact with substance, immediately flush skin or eyes with running water for at least 20 minutes. Keep victim warm and quiet. Ensure that medical personnel are aware of the material(s) involved and take precautions to protect themselves.

ENVIRONMENTAL INFORMATION

Clean Water Act: Section 311 Hazardous Substances/RQ (same as CERCLA)
EPCRA Section 304 Reportable Quantity (RQ): CERCLA, 1,000 lb (4.54 kg)
Marine Pollutant (49CFR, Subchapter 172.101, Appendix B).

FIBROUS GLASS DUST F:065

SYNONYMS: FIBERGLASS; FIBRA de VIDRIO (Spanish); GLAS; GLASS FIBERS; GLASS WOOL

IDENTIFICATION

CAS:	14808-60-7
RTECS No.:	LK3651000
Properties:	Solid. Fibrous glass is a man-made fiber in which the fiber-forming substance is a class of materials made form silicon dioxide with other elements, including oxides of various metals. Glasses solidify from the molten state without crystallization.

HEALTH AND SAFETY INFORMATION as respirable dust.

May cause irritation to the eyes and skin.

IARC Evaluation: Group 2A, sufficient animal evidence as silica, crystalline quartz
OSHA Table Z-1-A Air Contaminant as silica, crystalline quartz.
OSHA Table Z-3-A Air Contaminant as mineral dusts

NIOSH Criteria Document: Publication No. 77-152

Exposure Limits:	**TWA**
ACGIH: [(1)]	10 mg/m^3 (dust)
NIOSH:	3 fibers/cm^3 (fibers$\leq$ 3.5 μm in diameter & $\geq$ 10 μm in length) 5 mg/m^3 (total fibrous glass)
OSHA:	15 mg/m^3 (total dust); 5 mg/m^3 (respiratory dust)

[(1)] Notice of intended change as animal carcinogen, not classifiable as a human carcinogen; 1 f/cc (fibers longer than 5 μm; diameter > 3μm; aspect ration > 5:1 as determined by the membrane filer method at 400-450X magnification (4-mm objective) phase contrast illumination.

RECOMMENDATIONS FOR RESPIRATOR SELECTION: NIOSH as silica crystalline quartz

0.5 mg/m^3: HiE (any air-purifying respirator with a high-efficiency particulate filter.
1.25 mg/m^3: PAPRHiE (any powered, air-purifying respirator with a high-efficiency particulate filter);or SA:CF (any supplied-air respirator operated in a continuous-flow mode).
2.5 mg/m^3: HiEF (any air-purifying, full-facepiece respirator with a high-efficiency particulate filter);or PAPRTHiE (any powered, air-purifying respirator with a tight-fitting facepiece and a high-efficiency particulate filter).
25 mg/m^3: SA:PD,PP (any supplied-air respirator operated in a pressure-demand or other positive-pressure mode).

Emergency or Planned Entry into Unknown Concentrations or IDLH Conditions: SCBAF:PD,PP (any self-contained breathing apparatus that has a full facepiece and is operated in a pressure-demand or other positive-pressure mode);or SAF:PD,PP:ASCBA (any supplied-air respirator that has a full facepiece and is operated in a pressure-demand or other positive-pressure mode in combination with an auxiliary, self-contained breathing apparatus operated in a pressure-demand or other positive-pressure mode).

Escape: HiEF (any air-purifying, full-facepiece respirator with a high-efficiency particulate filter);or SCBAE (any appropriate escape-type, self-contained breathing apparatus).

CHEMICAL REACTIVITY

Incompatible with strong oxidizers.

FIRE INFORMATION

If material or contaminated runoff enters waterways, notify downstream users of potentially contaminated waters. Notify local health and fire officials and pollution control agencies. Use any extinguishing agent suitable for surrounding fire.

NFPA Hazard Classifications: Noncombustible

FIRST AID

Move victim to fresh air. Call emergency medical care. Apply artificial respiration if victim is not

breathing. Administer oxygen if breathing is difficult. Remove and isolate contaminated clothing and shoes. In case of contact with substance, immediately flush skin or eyes with running water for at least 20 minutes. Removal of solidified molten material from skin requires medical assistance. Keep victim warm and quiet. Ensure that medical personnel are aware of the material(s) involved and take precautions to protect themselves.

FLUOROBENZENE F:105

SYNONYMS: BENZENE, FLUORO; EINECS No. 207-321-7; PHENYL FLUORIDE; MONOFLUOROBENZENE; PHENYL FLUORIDE; BENZENE FLUORIDE; MFB

IDENTIFICATION

CAS:	462-06-6
ERG ID:	UN 2387
ERG Guide:	130
ERG Guide Category:	FLAMMABLE LIQUIDS (NONPOLAR/WATER-IMMISCIBLE/NOXIOUS)
Molecular Formula:	C_6H_5F
RTECS No.:	DA0800000
Properties:	Flammable, watery liquid. Colorless. Benzene odor. May float or sink in water. Flammable vapor is produced

RECOMMENDATIONS FOR RESPIRATOR SELECTION:

Repeated exposure may result in liver and kidney problems.

At any concentrations above the NIOSH REL, or where there is no REL, at any detectable concentration: SCBAF:PD,PP (any self-contained breathing apparatus that has a full facepiece and is operated in a pressure-demand or other positive-pressure mode);or SAF:PD,PP:ASCBA (any supplied-air respirator that has a full facepiece and is operated in a pressure-demand or other positive-pressure mode in combination with an auxiliary, self-contained breathing apparatus operated in a pressure-demand or other positive-pressure mode).

Escape: GMFOV [any air-purifying, full-facepiece respirator (gas mask) with a chin-style, front- or back-mounted organic vapor canister];or SCBAE (any appropriate escape-type, self-contained breathing apparatus)

CHEMICAL REACTIVITY

A halogenated hydrocarbon. Reacts with alkali metals (such as sodium or potassium and their alloys), ammonium nitrate, chromic acid, halogens, hydrogen peroxide, nitric acid, finely divided aluminum, calcium, magnesium. Contact with strong oxidizers and heat may cause explosions.

FIRE INFORMATION

Vapors are heavier than air and will collect in low areas. Vapors in confined areas may explode when exposed to fire. Vapors may travel long distances to ignition sources and flash back. Storage containers and parts of containers may rocket great distances, in many directions. If material or contaminated runoff enters waterways, notify downstream users of potentially contaminated waters. Notify local health and fire officials and pollution control agencies. Use dry chemicals, foam, carbon dioxide.

NFPA Hazard Classifications:

Health Hazards (Blue) . . . . -
Flammability (Red) 3
Reactivity (Yellow) 0

Flash Point: 5°F (-15°C) c.c.

FIRST AID

Move victim to fresh air. Call emergency medical care. Apply artificial respiration if victim is not breathing. Administer oxygen if breathing is difficult. Remove and isolate contaminated clothing and shoes. In case of contact with substance, immediately flush skin or eyes with running water for at least 20 minutes. Wash skin with soap and water. Keep victim warm and quiet. Effects of exposure (inhalation, ingestion, or skin contact) to substance may be delayed. Ensure that medical personnel are aware of the material(s) involved and take precautions to protect themselves.

FOLPET F:115

SYNONYMS: EEC No. 613-045-00-1; EINECS No. 205-088-6; EPA PESTICIDE CHEMICAL CODE 081601; FOLPAN; FTALAN; ORTHOPHALTAN; PHALTAN; PHTHALTAN; THIOPHAL; n-(TRICHLOR-METHYLTHIO)-PHTHALAMID (German); n-(TRICHLOROMETHYLMERCAPTO)-PHTHALIMIDE; 2-((TRICHLOROMETHYL)THIO)-1H-ISOINDOLE-1,3(2H)-DIONE; TROYSAN ANTI-MILDEW O; n-(TRICHLOROMETHYLTHIO)PHTHALIMIDE; TRIFOL; VINICOIL

IDENTIFICATION

CAS:	133-07-3
ERG ID:	UN 2811 as toxic solid, organic, n.o.s.
ERG Guide:	154
ERG Guide Category:	SUBSTANCES-TOXIC AND/OR CORROSIVE (NONCOMBUSTIBLE)
Molecular Formula:	$C_9H_4Cl_3NO_2S$
RTECS No.:	TI5685000
Properties:	Crystalline or powder solid. White crystals; light colored powder. Floats on water; very slightly soluble.

HEALTH AND SAFETY INFORMATION

May cause irritation of the eyes, skin, and respiratory tract.

RECOMMENDATIONS FOR RESPIRATOR SELECTION

At any concentrations above the NIOSH REL, or where there is no REL, at any detectable concentration: SCBAF:PD,PP (any self-contained breathing apparatus that has a full facepiece and is operated in a pressure-demand or other positive-pressure mode);or SAF:PD,PP:ASCBA (any supplied-air respirator that has a full facepiece and is operated in a pressure-demand or other positive-pressure mode in combination with an auxiliary, self-contained breathing apparatus operated in a pressure-demand or other positive-pressure mode).

Escape: GMFOV [any air-purifying, full-facepiece respirator (gas mask) with a chin-style, front- or back-mounted organic vapor canister];or SCBAE (any appropriate escape-type, self-contained breathing apparatus).

CHEMICAL REACTIVITY

Aromatic halogen compound. Incompatible with alkalies, phosphine, tetraselenium tetranitride.

FIRE INFORMATION

Combustion products include toxic nitrogen oxides and sulfur. If material or contaminated runoff enters waterways, notify downstream users of potentially contaminated waters. Notify local health and fire officials and pollution control agencies. Use dry chemicals, alcohol-resistant foam, carbon dioxide.

NFPA Hazard Classifications:

Health Hazards (Blue) 1
Flammability (Red) .... -
Reactivity (Yellow) 0

FIRST AID

Move victim to fresh air; call emergency medical care. Apply artificial respiration if victim is not breathing. Do not use mouth-to-mouth method if victim ingested or inhaled the substance; induce artificial respiration with the aid of a pocket mask equipped with a one-way valve or other proper respiratory medical device. Administer oxygen if breathing is difficult. Remove and isolate contaminated clothing and shoes. In case of contact with substance, immediately flush skin or eyes with running water for at least 20 minutes. For minor skin contact, avoid spreading material on unaffected skin. Keep victim warm and quiet. Effects of exposure (inhalation, ingestion, or skin contact) to substance may be delayed. Ensure that medical personnel are aware of the material(s) involved and take precautions to protect themselves.

ENVIRONMENTAL INFORMATION

EPCRA Section 313: Form R *de minimis* concentration reporting level: 1.0%.
California's Proposition 65: Carcinogen.

FORMALDEHYDE F:120

SYNONYMS: ALDEHYDE FORMIQUE (French); ALDEIDE FORMICA (Italian); BFV; CUPOSIT CP-74M ELECTROLESS COPPER; CUPOSIT CP-74R COPPER REPLINISHER; DEHIDO (Spanish); EEC No. 605-001-00-5; EINECS No. 200-001-8; FORMALDEHYD (Czech, Polish); FORMALIN; FORMALIN 40; FORMALINA (Italian); FORMALINE (German); FORMALIN-LOESUNGEN (German); FORMALITH; FORMIC ALDEHYDE; FORMOL; FYDE; HERCULES 37M6-8; IVALON; KARSAN; LOW DYE-FAST DRY INK; LYSOFORM; MAGNIFLOC 156C FLOCCULANT; METHANAL; METHANAL SOLUTION; METHYL ALDEHYDE; METHYLENE GLYCOL; METHYLENE OXIDE; MORBICID; NCI-C02799; OPLOSSINGEN (Dutch); OXOMETHANE; OXYMETHYLENE; PARAFORM; POLYOXYMETHYLENE GLYCOLS; SUPERLYSOFORM; TETRAOXYMETHYLENE; TRIOXANE

CAS:	50-00-0
ERG ID:	UN 2209 (solution); UN 1198 (flammable solutions)
ERG Guide:	132
ERG Guide Category:	FLAMMABLE LIQUIDS, CORROSIVE
Molecular Formula:	CH_2O
RTECS No.:	LP8925000
Properties:	Extremely flammable watery liquid. Usually shipped as a 30 -40% solution in water or 15% or less solution in methanol. Colorless or cloudy. Pungent, irritating odor. Sinks and mixes with water.

HEALTH AND SAFETY INFORMATION

The major concerns of repeated exposure are sensitization and cancer. In sensitized persons, formaldehyde can cause asthma and contact dermatitis. In persons who are not sensitized, prolonged inhalation of formaldehyde at low levels is likely to result in chronic pulmonary injury. The U.S. Department of Health and Human Services has determined that formaldehyde may reasonably be anticipated to be a carcinogen. In humans, formaldehyde exposure has been associated with increased risk of leukemia and nasal, mouth, throat, lung, and brain cancer. There is limited evidence that formaldehyde causes adverse reproductive effects. Also, long–term inhalation can result in respiratory congestion with associated coughing and shortness of breath. Daily skin contact can lead to drying and scaling. Subsequent contact may cause skin rashes and asthma, and reactions may become more severe if exposure persists.

IARC Evaluation: Group 2A, probable human carcinogen; human evidence, limited; animal evidence, sufficient.
NTP: Anticipated human carcinogen.

IDLH: 20 ppm.; potential human carcinogen
Odor Threshold: 0.871 ppm. Odor is not a reliable indicator of the presence of toxic amounts of vapor.
OSHA Table Z-1-A Air Contaminant
OSHA Table Z-2 Air Contaminant
OSHA Specifically Regulated Substances: (29 CFR 1910.1048).
OSHA Process Safety Management of Highly Hazardous Chemicals (29 CFR, 1910.119):
Threshold Quantity: 1,000 pounds.

Exposure Limits:	**TWA**	**CEILING**	**STEL**
ACGIH:(1)		0.3 ppm; 0.37 mg/m^3	
NIOSH:(2)	0.016 ppm	0.1 ppm [15 min]	
OSHA:	0.75 ppm		2 ppm

(1)Suspected carcinogen.
(2)Potential carcinogen. Reduce exposure to lowest feasible level.

RECOMMENDATIONS FOR RESPIRATOR SELECTION: NIOSH

At any concentrations above the NIOSH REL, or where there is no REL, at any detectable concentration: SCBAF:PD,PP (any self-contained breathing apparatus that has a full facepiece and is operated in a pressure-demand or other positive-pressure mode);or SAF:PD,PP:ASCBA (any supplied-air respirator that has a full facepiece and is operated in a pressure-demand or other positive-pressure mode in combination with an auxiliary, self-contained breathing apparatus operated in a pressure-demand or other positive-pressure mode).

Escape: GMFS [any air-purifying, full-facepiece respirator (gas mask) with a chin-style, front- or back-mounted canister providing protection against the compound of concern];or SCBAE (any appropriate escape-type, self-contained breathing apparatus).

CHEMICAL REACTIVITY

May polymerize unless properly inhibited (usually with methanol). Forms explosive mixture with air. Incompatible with strong acids, amines, strong oxidizers, alkaline materials, nitrogen dioxide, and performic acid. Reaction with hydrochloric acid forms bis-chloromethyl ether, a carcinogen.

FIRE INFORMATION

Firefighting gear (including SCBA) does not provide adequate protection. If exposure occurs, remove and isolate gear immediately and thoroughly decontaminate personnel. Vapors are heavier than air and will collect in low areas. Vapors in confined areas may explode when exposed to fire. Vapors may travel long distances to ignition sources and flash back.Storage containers and parts of containers may rocket great distances, in many directions. If material or contaminated runoff enters waterways, notify downstream users of potentially contaminated waters. Notify local health and fire officials and pollution control agencies. Use dry chemicals, foam, carbon dioxide.

NFPA Hazard Classifications:

Health Hazards (Blue) 3
Flammability (Red) 4
Reactivity (Yellow) 0

Solution (37% methanol free; 37%, 15% methanol)

Health Hazards (Blue) 3
Flammability (Red) 2
Reactivity (Yellow) 0

Flash Point: (37% methanol-free) 185°F (85°C) c.c.; (15% methanol) 122°F (50°C) c.c.
Flammable Limits in Air: LEL 7.0%; UEL 73% (gas)
Autoignition Temperature: 806°F (430°C) (solution); 572°F (300°C) (gas)

FIRST AID

Move victim to fresh air. Call emergency medical care. Apply artificial respiration if victim is not breathing. Do not use mouth-to-mouth method if victim swallowed or inhaled the substance; induce artificial respiration with the aid of a pocket mask equipped with a one-way valve or other proper respiratory medical device. If breathing is difficult, give oxygen. Remove and isolate contaminated clothing and shoes. In case of contact with substance, immediately flush skin or eyes with running water for at least 20 minutes. Keep victim warm and quiet. Effects of exposure (inhalation, ingestion, or skin contact) to substances may be delayed. Ensure that medical personnel are aware of the material(s) involved and take precautions to protect themselves.

ENVIRONMENTAL INFORMATION

Clean Air Act: Hazardous Air Pollutants (Title I, Part A, Section 112); Accidental Release Prevention/Flammable substances, (Section 112[r], Table 3), TQ = 15,000 lb (6,810 kg).
Clean Water Act: Section 311 Hazardous Substances/RQ (same as CERCLA); Section 313 Priority Chemicals.
EPA Hazardous Waste Number (RCRA No.): U122.
RCRA Section 261 Hazardous Constituents
EPCRA Section 302 Extremely Hazardous Substances: TPQ = 500 lb (227 kg).
EPCRA Section 304: Reportable Quantity (RQ): CERCLA, 100 lb (45.4 kg).
EPCRA Section 313: Form R *de minimis* concentration reporting level: 0.1%.
California's Proposition 65: Carcinogen as formaldehyde (gas).
TSCA: 40CFR716.120(*d*)1 as aldehydes

FORMIC ACID F:140

SYNONYMS: ACIDE FORMIQUE (French); ACIDO FORMICO (Italian, Spanish); ADD F; AMASIL; AMEISENSAEURE (German); AMINIC ACID; BILORIN; EEC No. 607-001-00-1; EINECS No.

200-579-1; FIDELITY 1215A; FIDELITY 1215B SOLDER STRIPPER; FORMIC ACID, 85-95 % IN AQUEOUS SOLUTION; FORMYLIC ACID; HYDROGEN CARBOXYLIC ACID; METHANOIC ACID; MIERENZUUR (Dutch); MS-111 EPOXY STRIPPER; 911 STRIPPER; 1030 ACID CLEANER

IDENTIFICATION

CAS:	64-18-6
ERG ID:	UN 1779
ERG Guide:	153
ERG Guide Category:	TOXIC AND/OR CORROSIVE (COMBUSTIBLE)
Molecular Formula:	CH_2O_2
RTECS No.:	LQ4900000
Properties:	Very flammable liquid; solid below 47°F (8°C). Colorless, may be fuming. Penetrating, irritating odor. Soluble; sinks and mixes with water.

HEALTH AND SAFETY INFORMATION

Very irritating to skin, eyes, nose, and lungs; skin and eye contact can cause severe burns and possible blindness. Prolonged or repeated exposure may cause irritation of the skin and changes in the kidney, indicated by albumin and blood in the urine. Possible allergy or lung injury.

IDLH: 30 ppm
Odor Threshold: 28.2 ppm
OSHA Table Z-1-A Air Contaminant

Exposure Limits:	**TWA**	**STEL**
ACGIH:	5 ppm; 9.4 mg/m^3	10 ppm; 19 mg/m^3
NIOSH:	5 ppm; 9 mg/m^3	
OSHA:	5 ppm; 9 mg/m^3	

RECOMMENDATIONS FOR RESPIRATOR SELECTION: NIOSH/OSHA

30 ppm: SA (any supplied-air respirator);or SCBAF (any self-contained breathing apparatus with a full facepiece).

Emergency or Planned Entry into Unknown Concentrations or IDLH Conditions:
SCBAF:PD,PP (any self-contained breathing apparatus that has a full facepiece and is operated in a pressure-demand or other positive-pressure mode);or SAF:PD,PP:ASCBA (any supplied-air respirator that has a full facepiece and is operated in a pressure-demand or other positive-pressure mode in combination with an auxiliary, self-contained breathing apparatus operated in a pressure-demand or other positive-pressure mode).

Escape: GMFOVHiE [any air-purifying, full-facepiece respirator (gas mask) with a chin-style, front- or back-mounted organic vapor canister having a high-efficiency particulate filter];or SCBAE (any appropriate escape-type, self-contained breathing apparatus).

Note: Substance reported to cause eye irritation or damage; may require eye protection.

CHEMICAL REACTIVITY

Forms explosive mixture with air. A strong reducing agent. Violent reaction with oxidizers, strong caustics such as lye, furfuryl alcohol, hydrogen peroxide, hydrated thallium nitrate, nitromethane, powdered metals. Contact with concentrated sulfuric acid causes formation of toxic and flammable carbon monoxide gas. Incompatible with bases, ammonia, aliphatic amines, alkanolamines, isocyanates, alkylene oxides, and epichlorohydrin. Attacks aluminum, cast iron, steel, some plastics, rubber, and coatings. May deteriorate upon normal storage causing pressure buildup and container failure.

FIRE INFORMATION

Firefighting gear (including SCBA) does not provide adequate protection. If exposure occurs, remove and isolate gear immediately and thoroughly decontaminate personnel.Vapors are heavier than air and will collect in low areas. Vapors in confined areas may explode when exposed to fire. Vapors may travel long distances to ignition sources and flash back.Storage containers and parts of containers may rocket great distances, in many directions. If material or contaminated runoff enters waterways, notify downstream users of potentially contaminated waters. Notify local health and fire officials and pollution control agencies. Use dry chemicals, foam, carbon dioxide.

NFPA Hazard Classifications:

Health Hazards (Blue).......... 3
Flammability (Red) 2
Reactivity (Yellow)............. 0

Flash Point: 156°F (69°C) o.c.; 122°F (50°C), o.c. (90% solution)
Flammable Limits in Air: LEL 4%; UEL 33%; LEL 18%; UEL 57% (90% solution)
Autoignition Temperature: 1004°F (540°C); 813°F (434°C) (90% solution)

FIRST AID

Move victim to fresh air. Call emergency medical care. Apply artificial respiration if victim is not breathing. Do not use mouth-to-mouth method if victim ingested or inhaled the substance; induce artificial respiration with the aid of a pocket mask equipped with a one-way valve or other proper respiratory medical device. Administer oxygen if breathing is difficult. Remove and isolate contaminated clothing and shoes. In case of contact with substance, immediately flush skin or eyes with running water for at least 20 minutes. For minor skin contact, avoid spreading material on unaffected skin. Keep victim warm and quiet. Effects of exposure (inhalation, ingestion, or skin contact) may be delayed. Ensure that medical personnel are aware of the material(s) involved and take precautions to protect themselves.

ENVIRONMENTAL INFORMATION

Clean Water Act: Section 311 Hazardous Substances/RQ (same as CERCLA).
EPCRA Section 304: Reportable Quantity (RQ): CERCLA, 5,000 lb (2,270 kg).
EPA Hazardous Waste Number (RCRA No.): U123.
RCRA Section 261 Hazardous Constituents
EPCRA Section 313: Form R *de minimis* concentration reporting level: 1.0%.

FREON 11 F:160

SYNONYMS: ALGOFRENE TYPE 1; ARCTON 9; DISTILLEX DS6; EINECS No. 200-892-3; ELECTRO-CF 11; ESKIMON 11; FC-11; FLUOROCARBON-11; FLUOROTRICHLOROMETHANE; FLUOTRICLOROMETANO (Spanish); FLUOROTROJCHLOROMETAN (Polish); FREON MF; FRIGEN 11; GENESOLV A; GENETRON 11; HALOCARBON 11; ISCEON 131; ISOTRON 11; LEDON 11; METHANE, TRICHLOROFLUORO-; MONOCHLOROTRICHLOROMETHANE; REFRIGERANT 11; R-11; TRICHLOROMONOFLUOROMETHANE; UCON 11

IDENTIFICATION

CAS:	75-69-4
ERG ID:	UN 3082
ERG Guide:	171
ERG Guide Category:	LOW TO MODERATE HAZARD
Molecular Formula:	CCl_3F
RTECS No.:	PB6125000
Properties:	Colorless to water-white liquid. Gas above 75°F (24°C). Nearly odorless.

HEALTH AND SAFETY INFORMATION

IDLH: 2,000 ppm
Odor Threshold: 16.3 ppm

Exposure Limits:	**TWA**	**CEILING**
ACGIH:		1,000 ppm; 5,620 mg/m^3
NIOSH:		1,000 ppm; 5,600 mg/m^3
OSHA:	1,000 ppm; 5,600 mg/m^3	

RECOMMENDATIONS FOR RESPIRATOR SELECTION: NIOSH/OSHA

2,000 ppm: SA (any supplied-air respirator);or SCBAF (any self-contained breathing apparatus with a full facepiece).

Emergency or Planned Entry into Unknown Concentrations or IDLH Conditions: SCBAF:PD,PP (any self-contained breathing apparatus that has a full facepiece and is operated in a pressure-demand or other positive-pressure mode);or SAF:PD,PP:ASCBA (any supplied-air respirator that has a full facepiece and is operated in a pressure-demand or other positive-pressure mode in combination with an auxiliary, self-contained breathing apparatus operated in a pressure-demand or other positive-pressure mode).

Escape: GMFOV [any air-purifying, full-facepiece respirator (gas mask) with a chin-style, front- or back-mounted organic vapor canister];or SCBAE (any appropriate escape-type, self-contained breathing apparatus).

CHEMICAL REACTIVITY

Reacts, possibly violently, with chemically active metals such as sodium, potassium, calcium, magnesium, zinc, alkali amides, powdered aluminum, granular barium, and lithium shavings. Heat causes decomposition and formation of toxic gas.

FIRE INFORMATION

Noncombustible liquid; nonflammable gas.

FIRST AID

Move victim to fresh air. Call emergency medical care. Apply artificial respiration if victim is not breathing. Administer oxygen if breathing is difficult. In case of contact with substance, immediately flush skin or eyes with running water for at least 20 minutes. Remove and isolate contaminated clothing and shoes. Ensure that medical personnel are aware of the material(s) involved and take precautions to protect themselves.

ENVIRONMENTAL INFORMATION

Clean Air Act: Stratospheric ozone protection (Title VI, Subpart A, Appendix A), Class I, Ozone Depletion Potential = 1.0.
EPA Hazardous Waste Number (RCRA No.): U121.
RCRA Section 261 Hazardous Constituents
RCRA Land Ban Waste
RCRA Universal Treatment Standards: Wastewater (mg/L), 0.020; Nonwastewater (mg/kg), 30.
RCRA Ground Water Monitoring List: Suggested test method(s) (PQL ug/L): 8010(10); 8240(5).
EPCRA Section 304: Reportable Quantity (RQ): CERCLA, 5,000 lb (2,270 kg).
EPCRA Section 313: Form R *de minimis* concentration reporting level: 1.0%.

FREON 12 F:180

SYNONYMS: ALGOFRENE TYPE 2; ARCTON 6; DICHLOORDIFLUORMETHAAN (Dutch);

DICHLORODIFLUOROMETHANE; DICLORODIFLUOMETANO (Spanish); DIFLUORODICHLOROMETHANE; EINCES No. 200-893-9; ELECTRO-CF 12; ESKIMON 12; F 12; FC 12; FLUOROCARBON 12; FREON F 12; FRIGEN 12; GENETRON 12; HALON 122; ISCON 122; ISOTRON 12; KAISER CHEMICALS 12; LEDON 12; PROPELLANT 12; R-12; REFRIGERANT 12; STERETHOX; UCON 12

IDENTIFICATION

CAS:	75-71-8
ERG ID:	UN 1028
ERG Guide:	126
ERG Guide Category:	GASES-COMPRESSED OR LIQUEFIED (INCLUDING REFRIGERANT GASES)
Molecular Formula:	CCl_2F_2
RTECS No.:	PA8200000
Properties:	Colorless, noncombustible gas. Etherlike odor at extremely high concentrations. Shipped as a liquefied compressed gas.

HEALTH AND SAFETY INFORMATION

IDLH: 15,000 ppm

Exposure Limits:	**TWA**
ACGIH:	1,000 ppm; 4,950 mg/m^3
NIOSH:	1,000 ppm; 4,950 mg/m^3
OSHA:	1,000 ppm; 4,950 mg/m^3

RECOMMENDATIONS FOR RESPIRATOR SELECTION: NIOSH/OSHA

10,000 ppm: SA (any supplied-air respirator).
15,000 ppm: SA:CF (any supplied-air respirator operated in a continuous-flow mode); SCBAF (any self-contained breathing apparatus with a full facepiece); SAF (any supplied-air respirator with a full facepiece).

Emergency or Planned Entry into Unknown Concentrations or IDLH Conditions: SCBAF:PD,PP (any self-contained breathing apparatus that has a full facepiece and is operated in a pressure-demand or other positive-pressure mode);or SAF:PD,PP:ASCBA (any supplied-air respirator that has a full facepiece and is operated in a pressure-demand or other positive-pressure mode in combination with an auxiliary, self-contained breathing apparatus operated in a pressure-demand or other positive-pressure mode).

Escape: GMFOV [any air-purifying, full-facepiece respirator (gas mask) with a chin-style, front- or back-mounted organic vapor canister] or SCBAE (any appropriate escape-type, self-contained breathing apparatus).

CHEMICAL REACTIVITY

Reacts with chemically active metals such as beryllium, powdered aluminum, zinc, magnesium, calcium, and sodium. Attacks some plastics, rubber, and coatings.

FIRE INFORMATION

Nonflammable gas. Storage containers and parts of containers may rocket great distances, in many directions. Notify local health and fire officials and pollution control agencies. Use any extinguishing agent suitable for surrounding fires. From a secure, explosion-proof location, use water spray to cool exposed containers. If cooling streams are ineffective (venting sound increases in volume and pitch, tank discolors, or shows any signs of deforming), withdraw immediately to a secure position.

FEMA Hazard Classifications:

If material or contaminated runoff enters waterways, notify downstream users of potentially contaminated waters. Notify local health and fire officials and pollution control agencies. Use dry chemicals, foam, carbon dioxide.

Health Hazards (Blue) 1
Flammability (Red) 0
Reactivity (Yellow) 0

FIRST AID

Move victim to fresh air. Call emergency medical care. Apply artificial respiration if victim is not breathing. Administer oxygen if breathing is difficult. Remove and isolate contaminated clothing and shoes. In case of contact with liquefied gas, thaw frosted parts with lukewarm water. Keep victim warm and quiet. Keep victim under observation. Ensure that medical personnel are aware of the material(s) involved and take precautions to protect themselves.

ENVIRONMENTAL INFORMATION

Clean Air Act: Stratospheric ozone protection (Title VI, Subpart A, Appendix A), Class I, Ozone Depletion Potential = 1.0.
EPA Hazardous Waste Number (RCRA No.): U075.
RCRA Section 261 Hazardous Constituents
RCRA Land Ban Waste
RCRA Universal Treatment Standards: Wastewater (mg/L), 0.23; Nonwastewater (mg/kg), 7.2.
RCRA Ground Water Monitoring List: Suggested test method(s) (PQL ug/L): 8010(10); 8240(5).
Safe Drinking Water Act: Priority List (55 FR 1470).
EPCRA Section 304: Reportable Quantity (RQ): CERCLA, 5,000 lb (2,270 kg).
EPCRA Section 313: Form R *de minimis* concentration reporting level: 1.0%

FREON 113 F:300

SYNONYMS: ARCTON 63; ARKLONE; ARKLONE P; CFC 113; CHLOROFLUOROCARBON-113; DIAFLON S-3; DISTILLEX DS5; EINECS No. 200-939-1; F-13; FLUOROCARBON 113; FREON 113TR-T; FREON TF SOLV; FRIGEN 113a; GENESOLV D SOLVENT; GENETRON 113; HALOCARBON 113; ISCEON 113; KAISER CHEMICAL 11; KHLADON 113; R-13; REFRIGERANT 113; TRICHLOROTRIFLUOROETHANE; 1,1,2-TRICHLORO-1,2,2-TRIFLUOROETHANE; TRICLOROTRIFLUOETANO (Spanish); TTE; UCON 113; UCON FLUOROCARBON 113

IDENTIFICATION

CAS:	76-13-1
ERG ID:	UN 3082
ERG Guide:	171
ERG Guide Category:	LOW TO MODERATE HAZARD
Molecular Formula:	$C_2Cl_3F_3$
RTECS No.:	KJ4000000
Properties:	Colorless to water-white liquid. Carbon tetrachloridelike odor. A gas above 118°F (46°C). Insoluble in water; sinks. High temperatures will break it down to toxic gases.

HEALTH AND SAFETY INFORMATION

Can accumulate in the brain and kidney from exposure below 2,000 ppm. Irregular heartbeat may occur. Levels above 2,500 ppm may cause loss of concentration, tiredness, and a feeling of heaviness of the head. Prolonged exposure can cause narcotic effect or suffocation.

IDLH: 2,000 ppm
Odor Threshold: 487 ppm

Exposure Limits:	**TWA**	**STEL**
ACGIH:	1,000 ppm; 7,670 mg/m^3	1,250 ppm; 9,590 mg/m^3
NIOSH:	1,000 ppm; 7,600 mg/m^3	1,250 ppm; 9,500 mg/m^3
OSHA:	1,000 ppm; 7,600 mg/m^3	

RECOMMENDATIONS FOR RESPIRATOR SELECTION: NIOSH/OSHA

2,000 ppm: SA (any supplied-air respirator);or SCBAF (any self-contained breathing apparatus with a full facepiece).

Emergency or Planned Entry into Unknown Concentrations or IDLH Conditions:
SCBAF:PD,PP (any self-contained breathing apparatus that has a full facepiece and is operated in a pressure-demand or other positive-pressure mode);or SAF:PD,PP:ASCBA (any supplied-air respirator that has a full facepiece and is operated in a pressure-demand or other positive-pressure mode in combination

with an auxiliary, self-contained breathing apparatus operated in a pressure-demand or other positive-pressure mode).

Escape: GMFOV [any air-purifying, full-facepiece respirator (gas mask) with a chin-style, front- or back-mounted organic vapor canister];or SCBAE (any appropriate escape-type, self-contained breathing apparatus).

CHEMICAL REACTIVITY

Reacts with chemically active metals such as barium, calcium, powdered aluminum, lithium, zinc, magnesium, titanium, and beryllium. Contact with alloys containing more than 2% magnesium may cause decomposition (hydrogen chloride, hydrogen fluoride, and carbon monoxide released). Attacks some plastics, rubber, and coatings.

FIRE INFORMATION

Noncombustible liquid at ordinary temperatures but will ignite and burn weakly at 1256°F (678°C). If material or contaminated runoff enters waterways, notify downstream users of potentially contaminated waters. Notify local health and fire officials and pollution control agencies. Use dry chemicals, foam, carbon dioxide.

FIRST AID

Move victim to fresh air. Call emergency medical care. Apply artificial respiration if victim is not breathing. Administer oxygen if breathing is difficult. Remove and isolate contaminated clothing and shoes. In case of contact with substance, immediately flush skin or eyes with running water for at least 20 minutes. Wash skin with soap and water. Keep victim warm and quiet. Effects of exposure (inhalation, ingestion, or skin contact) to substance may be delayed. Ensure that medical personnel are aware of the material(s) involved and take precautions to protect themselves.

ENVIRONMENTAL INFORMATION

Freon 113 is a potent ozone depleter. All of the Freon 113 produced is eventually lost as air emissions and builds up in the atmosphere. Stratospheric ozone depletion causes an increase in the levels of ultraviolet solar radiation reaching the earth's surface, which in turn is linked to increased incidence of skin cancers, immune system depression, cataracts, and disruptions in terrestrial and aquatic ecosystems. In addition, increased UV-B radiation is expected to increase photochemical smog and aggravating related health problems in urban and industrialized areas.

If released on land, Freon 113 will leach into the ground and volatilize from the soil surface. No degradative processes are known to occur in the soil. Freon 113 is not very water soluble and is removed rapidly from water via volatilization. Chemical hydrolysis, bioaccumulation, and adsorption to sediments are not significant fate processed in water.

Freon 113 is extremely stable in the lower atmosphere and will disperse over the globe and diffuse slowly into the stratosphere where it will be lost by photolysis. In this process, chlorine atoms are released that attack ozone.

Clean Air Act: Stratospheric ozone protection (Title VI, Subpart A, Appendix A), Class I, Ozone Depletion Potential = 0.8.
RCRA Land Ban Waste
RCRA Universal Treatment Standards: Wastewater (mg/L), 0.057; Nonwastewater (mg/kg), 30.
EPCRA Section 313: Form R *de minimis* concentration reporting level: 1.0%.

FUMERIC ACID F:347

SYNONYMS: ACIDO FUMARICO (Spanish); ALLOMALEIC ACID; BOLETIC ACID; 2-BUTENEDIOIC ACID (E); trans-BUTENEDIOIC ACID; (E)-BUTENEDIOIC ACID; BUTENEDIOIC ACID, (E)-; EEC No. 607-146-00-X; EINECS No. 203-743-0; 1,2-ETHENEDICARBOXYLIC ACID, trans-; FUMARSYRE (French); FUMARSAEURE (German); trans-1,2-ETHYLENEDICARBOXYLIC ACID; 1,2-ETHYLENEDICARBOXYLIC ACID, (E); KYSELINA FUMAROVA (Czech); LICHENIC ACID; NSC-2752; U-1149; USAF EK-P-583

IDENTIFICATION

CAS:	110-17-8
ERG ID:	UN 9126
ERG Guide:	171
ERG Guide Category:	SUBSTANCES (LOW TO MODERATE HAZARD)
Molecular Formula:	$C_4H_4O_4$
RTECS No.:	LS9625000
Properties:	Combustible, crystalline solid. White. Odorless. Fruity-acidic taste. Sinks and mixes with water; slightly soluble.

HEALTH AND SAFETY INFORMATION

Can irritate eyes, skin, and respiratory tract.

RECOMMENDATIONS FOR RESPIRATOR SELECTION

At any concentrations above the NIOSH REL, or where there is no REL, at any detectable concentration: SCBAF:PD,PP (any self-contained breathing apparatus that has a full facepiece and is operated in a pressure-demand or other positive-pressure mode);or SAF:PD,PP:ASCBA (any supplied-air respirator that has a full facepiece and is operated in a pressure-demand or other positive-pressure mode in combination with an auxiliary, self-contained breathing apparatus operated in a pressure-demand or other positive-pressure mode).

Escape: GMFOV [any air-purifying, full-facepiece respirator (gas mask) with a chin-style, front- or back-mounted organic vapor canister];or SCBAE (any appropriate escape-type, self-contained breathing apparatus).

CHEMICAL REACTIVITY

Reacts with strong oxidizers. Incompatible with sulfuric acid, caustics, ammonia, amines, isocyanates, alkylene oxides, epichlorohydrin.

FIRE INFORMATION

Combustion product includes toxic maleic anhydride. If material or contaminated runoff enters waterways,

notify downstream users of potentially contaminated waters. Notify local health and fire officials and pollution control agencies. Use dry chemicals, water spray, foam, carbon dioxide.

FEMA Hazard Classifications:

Health Hazards (Blue) 0
Flammability (Red) 1
Reactivity (Yellow) 0

Flash Point: >156°F (>69°C)
Autoignition Temperature: 1364°F (740°C) as powder

FIRST AID

Move victim to fresh air. Call emergency medical care. Apply artificial respiration if victim is not breathing. Administer oxygen if breathing is difficult. In case of contact with substance, immediately flush skin or eyes with running water for at least 20 minutes. Remove and isolate contaminated clothing and shoes. Ensure that medical personnel are aware of the material(s) involved and take precautions to protect themselves.

ENVIRONMENTAL INFORMATION

Clean Water Act: Section 311 Hazardous Substances/RQ (same as CERCLA)
EPCRA Section 304 Reportable Quantity (RQ): CERCLA, 5,000 lb (2,270 kg.)

FURFURAL F:354

SYNONYMS: ARTIFICIAL ANT OIL; EEC No. 605-010-00-4; EINECS No. 202-627-7; 2-FURANCARBOXALDEHYDE; FURAL; FURALE; 2-FURALDEHYDE; 2-FURANALDEHYDE; 2-FURANCARBONAL; FURFURALDEHYDE; FURFUROL; 2-FURYL-METHANAL; FUROLE; α-FUROLE; NCI-C56177; PYROMUCIC ALDEHYDE; FURFUROLE; FURAL/PYROMUCIC ALDEHYDE; QUAKERAL

IDENTIFICATION

CAS: 98-01-1

520 Furfural

ERG ID: UN 1199
ERG Guide: 132P
ERG Guide Category: FLAMMABLE LIQUIDS-CORROSIVE
Molecular Formula: $C_5H_4O_2$
RTECS No.: LT7000000
Properties: Combustible oily liquid. New and pure material is colorless; turns yellow to amber-brown on contact with air. Almond odor. Caramel taste. Sinks in water; soluble

HEALTH AND SAFETY INFORMATION

Corrosive to eyes, skin, and respiratory tract. Inhalation may cause pulmonary edema, which can be delayed for several hours; there is a risk of death in serious cases.

IDLH: 100 ppm
Odor Threshold: 0.058 ppm
OSHA Table Z-1-A Air Contaminant

Exposure Limits:	TWA
ACGIH: (1)	2 ppm; 7.9 mg/m^3
OSHA: (1)	5 ppm; 20 mg/m^3

(1) Skin contact contributes significantly in overall exposure.

RECOMMENDATIONS FOR RESPIRATOR SELECTION: OSHA

50 ppm: CCROV [any chemical cartridge respirator with organic vapor cartridge(s)];or SA (any supplied-air respirator).
100 ppm: SA:CF (any supplied-air respirator operated in a continuous-flow mode);or CCRFOV [any chemical cartridge respirator with a full facepiece and organic vapor cartridge(s)];or PAPROV [any powered, air-purifying respirator with organic vapor cartridge(s)];or GMFOV[any air-purifying, full-facepiece respirator (gas mask) with a chin-style, front- or back-mounted acid gas canister];or SCBAF (any self-contained breathing apparatus with a full facepiece);or SAF (any supplied-air respirator with a full facepiece).

Emergency or planned entry into unknown concentrations or IDLH conditions:
SCBAF:PD,PP (any self-contained breathing apparatus that has a full facepiece and is operated in a pressure-demand or other positive-pressure mode);or SAF:PD,PP:ASCBA (any supplied-air respirator that has a full facepiece and is operated in a pressure-demand or other positive-pressure mode in combination with an auxiliary, self-contained breathing apparatus operated in a pressure-demand or other positive pressure mode.

Escape: GMFOV[any air-purifying, full-facepiece respirator (gas mask) with a chin-style, front- or back-mounted organic vapor canister];or SCBAE (any appropriate escape-type, self-contained breathing apparatus).
Note: Substance reported to cause eye irritation or damage; may require eye protection.

CHEMICAL REACTIVITY

Forms explosive mixture with air. Violent reaction with strong acids, oxidizers, strong bases. Attacks many plastic materials.

FIRE INFORMATION

Vapors are heavier than air and will collect in low areas. Vapors in confined areas may explode when exposed to fire. If material or contaminated runoff enters waterways, notify downstream users of potentially contaminated waters. Notify local health and fire officials and pollution control agencies. Use dry chemicals, alcohol-resistant foam, carbon dioxide. Use water spray to cool exposed containers.

NFPA Hazard Classifications:

Health Hazards (Blue) 3
Flammability (Red) 2
Reactivity (Yellow) 0

Flash Point: 153°F (67°C) o.c.; 140°F (60°C) c.c.
Flammable Limits in Air: LEL: 2.1%; UEL: 19.3%
Autoignition Temperature: 597°F (314°C)

FIRST AID

Move victim to fresh air; call emergency medical care. Apply artificial respiration if victim is not breathing. Do not use mouth-to-mouth method if victim ingested or inhaled the substance; induce artificial respiration with the aid of a pocket mask equipped with a one-way valve or other proper respiratory medical device. Administer oxygen if breathing is difficult. Remove and isolate contaminated clothing and shoes. In case of contact with substance, immediately flush skin or eyes with running water for at least 20 minutes. For minor skin contact, avoid spreading material on unaffected skin. Keep victim warm and quiet. Effects of exposure (inhalation, ingestion, or skin contact) to substance may be delayed. Ensure that medical personnel are aware of the material(s) involved and take precautions to protect themselves.

ENVIRONMENTAL INFORMATION

Clean Water Act: Section 311 Hazardous Substances/RQ (same as CERCLA)
EPA Hazardous Waste Number (RCRA No.): U125
RCRA Section 261 Hazardous Constituents
RCRA Land Ban Waste
EPCRA Section 304 Reportable Quantity (RQ): CERCLA, 5,000 lb (2,270 kg.)
TSCA: 40CFR716.120(*d*)1 as aldehydes

FURFURYL ALCOHOL F:360

SYNONYMS: ALCOHOL FURFURILICO (Spanish); ALCOOL FURFURYLIQUE (French); EEC No.

603-018-00-2; EINECS No. 202-626-1; 2-FURANCARBINOL; 2-FURANMETHANOL; FURFURAL ALCOHOL; FURFURALCOHOL; FURFURYLALKOHOL (German); FURYLALCOHOL; 2-FURYLCARBINOL; 2-HYDROXYMETHYLFURAN; MICROPOSIT REMOVER 1112A; NCI-C56224; QO FURFURYL ALCOHOL (FA)

IDENTIFICATION

CAS:	98-00-0
ERG ID:	UN 2874
ERG Guide:	153
ERG Guide Category:	TOXIC AND/OR CORROSIVE (COMBUSTIBLE)
Molecular Formula:	$C_5H_6O_2$
RTECS No.:	LU9100000
Properties:	Combustible liquid. Colorless to amber. Mild irritating odor. Mixes with water.

HEALTH AND SAFETY INFORMATION

Prolonged or repeated contact may cause headache, nausea, and irritation of eyes, mouth and stomach; possible dermatitis and skin rash.

IDLH: 75 ppm
Odor Threshold: 7.83 ppm
OSHA Table Z-1-A Air Contaminant
NIOSH Criteria Document: NIOSH 79-133

Exposure Limits:	**TWA**	**STEL**
ACGIH:(1)	10 ppm; 40 mg/m^3	15 ppm; 60 mg/m^3
NIOSH:(1)	10 ppm; 40 mg/m^3	15 ppm; 60 mg/m^3
OSHA:	50 ppm; 200 mg/m^3	

(1)Skin contact contributes significantly to overall exposure.

RECOMMENDATIONS FOR RESPIRATOR SELECTION: NIOSH/OSHA

75 ppm: CCROV [any chemical cartridge respirator with organic vapor cartridge(s)];or GMFOV[any air-purifying, full-facepiece respirator (gas mask) with a chin-style, front- or back-mounted acid gas canister];or PAPROV [any powered, air-purifying respirator with organic vapor cartridge(s)];or SA (any supplied-air respirator);or SCBAF (any self-contained breathing apparatus with a full facepiece).

Emergency or Planned Entry into Unknown Concentrations or IDLH Conditions: SCBAF:PD,PP (any self-contained breathing apparatus that has a full facepiece and is operated in a pressure-demand or other positive-pressure mode);or SAF:PD,PP:ASCBA (any supplied-air respirator that has a full facepiece and is operated in a pressure-demand or other positive-pressure mode in combination with an auxiliary, self-contained breathing apparatus operated in a pressure-demand or other positive-pressure mode).

Escape: GMFOV [any air-purifying, full-facepiece respirator (gas mask) with a chin-style, front- or back-mounted organic vapor canister];or SCBAE (any appropriate escape-type, self-contained breathing apparatus).
Note: Substance reported to cause eye irritation or damage; may require eye protection.

CHEMICAL REACTIVITY

Forms explosive mixture with air. Strong acids (including some organic acids) or acid catalysts may cause polymerization. Strong reaction with oxidizers and acids. Incompatible with caustics, aliphatic amines, and isocyanates. Attacks some plastics, coatings, and rubber.

FIRE INFORMATION

NFPA Hazard Classifications:

Health Hazards (Blue) 1
Flammability (Red) 2
Reactivity (Yellow) 1

Flash Point: 167°F (75°C) o.c.; 149°F (65°C) c.c.
Flammable Limits in Air: LEL 1.8%; UE: 16.3%
Autoignition Temperature: 915°F (491°C)

FIRST AID

Move victim to fresh air. Call emergency medical care. Apply artificial respiration if victim is not breathing. Do not use mouth-to-mouth method if victim ingested or inhaled the substance; induce artificial respiration with the aid of a pocket mask equipped with a one-way valve or other proper respiratory medical device. Administer oxygen if breathing is difficult. Remove and isolate contaminated clothing and shoes. In case of contact with substance, immediately flush skin or eyes with running water for at least 20 minutes. For minor skin contact, avoid spreading material on unaffected skin. Keep victim warm and quiet. Effects of exposure (inhalation, ingestion, or skin contact) may be delayed. Ensure that medical personnel are aware of the material(s).

FUSEL OIL — F:390

SYNONYMS: EINECS No. 232-395-2; FEMA No. 2497; FUSELOEL (German); FUSEL OIL, REFINED; HUILE de FUSEL (French)

IDENTIFICATION

CAS:	8013-75-0
ERG ID:	UN 1201
ERG Guide:	127

524 Fusel Oil

ERG Guide Category: FLAMMABLE LIQUIDS (POLAR/WATER MISCIBLE)

Molecular Formula: Combination of amyl alcohols, primarily isoamyl alcohol, and 2-methyl-1-butanol. $C_5H_{12}O$ as isoamyl alcohol.

Properties: Flammable liquid. Colorless to pale yellow. Odorless. Slightly soluble in water.

HEALTH AND SAFETY INFORMATION

RECOMMENDATIONS FOR RESPIRATOR SELECTION: NIOSH

At any concentrations above the NIOSH REL, or where there is no REL, at any detectable concentration: SCBAF:PD,PP (any self-contained breathing apparatus that has a full facepiece and is operated in a pressure-demand or other positive-pressure mode);or SAF:PD,PP:ASCBA (any supplied-air respirator that has a full facepiece and is operated in a pressure-demand or other positive-pressure mode in combination with an auxiliary, self-contained breathing apparatus operated in a pressure-demand or other positive-pressure mode).

Escape: GMFOV [any air-purifying, full-facepiece respirator (gas mask) with a chin-style, front- or back-mounted organic vapor canister];or SCBAE (any appropriate escape-type, self-contained breathing apparatus).

CHEMICAL REACTIVITY

Reacts vigorously with oxidizing material.

FIRE INFORMATION

Vapors are heavier than air and will collect in low areas. Vapors in confined areas may explode when exposed to fire. Vapors may travel long distances to ignition sources and flash back. Storage containers and parts of containers may rocket great distances, in many directions. If material or contaminated runoff enters waterways, notify downstream users of potentially contaminated waters. Notify local health and fire officials and pollution control agencies. Use dry chemicals, foam, carbon dioxide.

NFPA Hazard Classifications:

Health Hazards (Blue) 1
Flammability (Red) 2
Reactivity (Yellow) 0

Flash Point: 109°F (43°C)
Flammable Limits in Air: LEL 1.2%; UEL 9.0% @ 212°F (100°C)
Autoignition Temperature: 662°F (350°C)

FIRST AID

Move victim to fresh air. Call emergency medical care. Apply artificial respiration if victim is not

breathing. Administer oxygen if breathing is difficult. Remove and isolate contaminated clothing and shoes. In case of contact with substance, immediately flush skin or eyes with running water for at least 20 minutes. Wash skin with soap and water. Keep victim warm and quiet. Ensure that medical personnel are aware of the material(s) involved and take precautions to protect themselves.

GLYCERIN G:160

SYNONYMS: CP 290B ACTIVATOR; CRODEROL G7000; EINECS No. 200-289-5; EMERY 912; GLICERINA (Spanish); GLYCERIN, ANHYDROUS; GLYCERIN, SYNTHETIC; GLYCERITOL; GLYCEROL; GLYCON G 100; GLYCON G 300; GLYCYL ALCOHOL; GLYROL; GROCOLENE; KEMSTRENE; KEMSTRENE 96%; MAGIC GLASS CLEANING AND ANTI-FOGGING FLUID; MOON; OSMOGLYN; PRICERINE 9071; 1,2,3-PROPANETRIOL; SHUR-COAL FCA; STAR; SUPEROL; SYNTHETIC GLYCERIN; 90 TECHNICAL GLYCERIN; TRIHYDROXYPROPANE; 1,2,3-TRIHYDROXYPROPANE

IDENTIFICATION

CAS: 56-81-5
ERG ID: UN 1993
ERG Guide: 128
ERG Guide Category: FLAMMABLE LIQUIDS (NON-POLAR/WATER-IMMISCIBLE)
Molecular Formula: $C_3H_8O_3$
RTECS No.: MA8050000
Properties: Combustible, oily liquid. Clear, colorless. Odorless. Sinks and mixes with water. Syrupy liquid or solid below 64°F (18°C). The solid form melts above 64°F (18°C) but the liquid freezes at a much lower temperature.

HEALTH AND SAFETY INFORMATION

Irritates eyes and skin. Can cause kidney damage.

OSHA Table Z-1-A Air Contaminant as mist

Exposure Limits:	**TWA**
ACGIH:	10 mg/m^3 (mist)
NIOSH:	No established REL
OSHA:	15 mg/m^3 (total) 5 mg/m^3 (repiratory)

RECOMMENDATIONS FOR RESPIRATOR SELECTION: NIOSH

At any concentrations above the NIOSH REL, or where there is no REL, at any detectable

concentration: SCBAF:PD,PP (any self-contained breathing apparatus that has a full facepiece and is operated in a pressure-demand or other positive-pressure mode);or SAF:PD,PP:ASCBA (any supplied-air respirator that has a full facepiece and is operated in a pressure-demand or other positive-pressure mode in combination with an auxiliary, self-contained breathing apparatus operated in a pressure-demand or other positive pressure mode).

ESCAPE: GMFOV [any air-purifying, full-facepiece respirator (gas mask) with a chin-style, front- or back-mounted organic vapor canister];or SCBAE (any appropriate escape-type, self-contained breathing apparatus).

CHEMICAL REACTIVITY

Polymerizes at approximately 300°F (150°C). Incompatible with acetic anhydride, potassium permanganate, strong acids, caustics, aliphatic amines, isocyanates, oxidizers.

FIRE INFORMATION

NFPA Hazard Classifications:

Health Hazards (Blue) 1
Flammability (Red) 1
Reactivity (Yellow) 0

Flash Point: 350°F (177°C) o.c.; 320°F (160°C) c.c.
Flammable Limits in Air: LEL 0.9%
Autoignition Temperature: 698°F (370°C)

FIRST AID

Move victim to fresh air. Call emergency medical care. Apply artificial respiration if victim is not breathing. If breathing is difficult, give oxygen. Remove and isolate contaminated clothing and shoes. In case of contact with substance, immediately flush skin or eyes with running water for at least 20 minutes. Wash skin with soap and water. Keep victim warm and quiet. Ensure that medical personnel are aware of the material(s) involved and take precautions to protect themselves.

GOSSYPOL G:186

SYNONYMS:
2,2'-BIS(1,6,7-TRIHYDROXY-3-METHYL-5-ISOPROPYL)-8-ALDEHYDONAPHTHALENE; 8-FORMYL-1,6,7-TRIHYDROXY-5-ISOPROPYL-3-METHYL-2,2'-BISNAPHTHALENE

IDENTIFICATION

CAS: 303-45-7

ERG ID:	UN 2811 as toxic solid, organic, n.o.s.
ERG Guide:	154
ERG Guide Category:	SUBSTANCES-TOXIC AND/OR CORROSIVE (NONCOMBUSTIBLE)
Molecular Formula:	$C_{30}H_{30}O_8$
RTECS No.:	DU3100000
Properties:	Crystalline solid. Yellow. Floats in water; insoluble.

HEALTH AND SAFETY INFORMATION

May be irritating to the eyes and respiratory tract. Inhalation may cause pulmonary edema, which can be delayed for several hours; there is a risk of death in serious cases.

RECOMMENDATIONS FOR RESPIRATOR SELECTION

At any concentrations above the NIOSH REL, or where there is no REL, at any detectable concentration: SCBAF:PD,PP (any self-contained breathing apparatus that has a full facepiece and is operated in a pressure-demand or other positive-pressure mode);or SAF:PD,PP:ASCBA (any supplied-air respirator that has a full facepiece and is operated in a pressure-demand or other positive-pressure mode in combination with an auxiliary, self-contained breathing apparatus operated in a pressure-demand or other positive-pressure mode).

Escape: GMFOV [any air-purifying, full-facepiece respirator (gas mask) with a chin-style, front- or back-mounted organic vapor canister];or SCBAE (any appropriate escape-type, self-contained breathing apparatus).

CHEMICAL REACTIVITY

A natural polyphenol. Phenols are incompatible with sulfuric acid, nitric acid, caustics, aliphatic amines, amides.

FIRE INFORMATION

If material or contaminated runoff enters waterways, notify downstream users of potentially contaminated waters. Notify local health and fire officials and pollution control agencies. Use dry chemicals, alcohol-resistant foam, carbon dioxide.

NFPA Hazard Classifications: Not listed

FIRST AID

Move victim to fresh air; call emergency medical care. Apply artificial respiration if victim is not breathing. Do not use mouth-to-mouth method if victim ingested or inhaled the substance; induce artificial respiration with the aid of a pocket mask equipped with a one-way valve or other proper respiratory medical device. Administer oxygen if breathing is difficult. Remove and isolate contaminated clothing and shoes. In case of contact with substance, immediately flush skin or eyes with running water for at least 20 minutes. For minor skin contact, avoid spreading material on unaffected skin. Keep victim warm and quiet. Effects of exposure (inhalation, ingestion, or skin contact) to substance may be delayed. Ensure that medical personnel are aware of the material(s) involved and take precautions to protect themselves.

GLUTARIC ACID G:192

SYNONYMS: ACIDO GLUTARICO (Spanish); EINECS No. 203-817-2; GLUTARSAEURE (German); PENTANDIOIC ACID; PENTANEDIOIC ACID; 1,5-PENTANEDIOIC ACID; 1,3-PROPANEDICARBOXYLIC ACID

IDENTIFICATION

CAS:	110-94-1
Molecular Formula:	$C_5H_8O_4$
RTECS No.:	MA3740000
Properties:	Solid crystals or needles. Colorless. Soluble in water. May be closely related to adipic acid.

HEALTH AND SAFETY INFORMATION

Solution is a medium strong acid. Corrosive to eyes, skin and respiratory tract. Inhalation may cause pulmonary edema, which can be delayed for several hours; there is a risk of death in serious cases.

RECOMMENDATIONS FOR RESPIRATOR SELECTION

At any concentrations above the NIOSH REL, or where there is no REL, at any detectable concentration: SCBAF:PD,PP (any self-contained breathing apparatus that has a full facepiece and is operated in a pressure-demand or other positive-pressure mode);or SAF:PD,PP:ASCBA (any supplied-air respirator that has a full facepiece and is operated in a pressure-demand or other positive-pressure mode in combination with an auxiliary, self-contained breathing apparatus operated in a pressure-demand or other positive-pressure mode).

Escape: GMFOV [any air-purifying, full-facepiece respirator (gas mask) with a chin-style, front- or back-mounted organic vapor canister];or SCBAE (any appropriate escape-type, self-contained breathing apparatus).

CHEMICAL REACTIVITY

Incompatible with oxidizers, sulfuric acid, caustics, ammonia, aliphatic amines, alkanolamines, isocyanates, alkylene oxides, epichlorohydrin.

FIRE INFORMATION

Decomposition products upon heating includes irritating and toxic gases. If material or contaminated runoff enters waterways, notify downstream users of potentially contaminated waters. Notify local health and fire officials and pollution control agencies. Use dry chemicals, water spray, carbon dioxide.

NFPA Hazard Classifications: Not listed

FIRST AID

Move victim to fresh air. Call emergency medical care. Apply artificial respiration if victim is not breathing. Administer oxygen if breathing is difficult. Remove and isolate contaminated clothing and shoes. In case of contact with substance, immediately flush skin or eyes with running water for at least 20 minutes. Wash skin with soap and water. Keep victim warm and quiet. Effects of exposure (inhalation, ingestion, or skin contact) to substance may be delayed. Ensure that medical personnel are aware of the material(s) involved and take precautions to protect themselves.

GLYCIDOL G:200

SYNONYMS: EINECS No. 209-128-3; EPIHYDRIN ALCOHOL; 2,3-EPOSSI-L-PROPANOLO (Italian); 2,3-EPOXYPROPANOL; 2,3-EPOXY-1-PROPANOL; EPOXYPROPYL ALCOHOL; GLYCIDE; GLYCIDYL ALCOHOL; 3-HYDROXY-1,2-EPOXYPROPANE; HYDROXYMETHYL ETHYLENE OXIDE; 2-HYDROXYMETHYL OXIRAN; 3-HYDROXYPROPYLENE OXIDE; METHANOL, OXIRANYL-; NCI-C55549; OXYRANYL METHANOL

IDENTIFICATION

CAS:	556-52-5
ERG ID:	UN 2810 as toxic liquid, organic
ERG Guide:	153
ERG Guide Category:	SUBSTANCES-TOXIC AND/OR CORROSIVE (COMBUSTIBLE)
Molecular Formula:	$C_3H_6O_2$
RTECS No.:	UB4375000
Properties:	Combustible liquid. Colorless. Extremely soluble in water.

HEALTH AND SAFETY INFORMATION

Irritates eyes, skin, and respiratory tract. May affect the central nervous system; with risk of unconsciousness. Mutagenic. May cause dryness, scaling, and other skin disorders.

IDLH: 150 ppm
OSHA Table Z-1-A Air Contaminant

Exposure Limits:	**TWA**
ACGIH: (1)	2 ppm; 6.1 mg/m^3

NIOSH: 25 ppm; 75 mg/m³
OSHA: 50 ppm; 150 mg/m³

(1) Animal carcinogen

RECOMMENDATIONS FOR RESPIRATOR SELECTION

At any concentrations above the NIOSH REL, or where there is no REL, at any detectable concentration: SCBAF:PD,PP (any self-contained breathing apparatus that has a full facepiece and is operated in a pressure-demand or other positive-pressure mode);or SAF:PD,PP:ASCBA (any supplied-air respirator that has a full facepiece and is operated in a pressure-demand or other positive-pressure mode in combination with an auxiliary, self-contained breathing apparatus operated in a pressure-demand or other positive-pressure mode).

Escape: GMFOV [any air-purifying, full-facepiece respirator (gas mask) with a chin-style, front- or back-mounted organic vapor canister];or SCBAE (any appropriate escape-type, self-contained breathing apparatus).

CHEMICAL REACTIVITY

forms explosive mixture with air. Explodes or explosive polymerization when heated or in the presence of strong acids, bases, metals (e.g., copper, zinc), and metal salts (e.g., aluminum chloride, iron(III) chloride, tin(IV) chloride). Reacts with strong oxidizers (may be violent) and nitrates.

FIRE INFORMATION

Vapors are heavier than air and will collect in low areas. Vapors in confined areas may explode when exposed to fire.If material or contaminated runoff enters waterways, notify downstream users of potentially contaminated waters. Storage containers and parts of containers may rocket great distances, in many directions. Notify local health and fire officials and pollution control agencies. Use dry chemicals, alcohol-resistant foam, carbon dioxide. Use water spray to cool exposed containers.

NFPA Hazard Classifications: Not listed

Flash Point: 178°F (81°C)

FIRST AID

Move victim to fresh air. Call emergency medical care. Apply artificial respiration if victim is not breathing. Do not use mouth-to-mouth method if victim ingested or inhaled the substance; induce artificial respiration with the aid of a pocket mask equipped with a one-way valve or other proper respiratory medical device. Administer oxygen if breathing is difficult. Remove and isolate contaminated clothing and shoes. In case of contact with substance, immediately flush skin or eyes with running water for at least 20 minutes. Wash skin with soap and water. Keep victim warm and quiet. Effects of exposure (inhalation, ingestion, or skin contact) to substance may be delayed. Ensure that medical personnel are aware of the material(s) involved and take precautions to protect themselves.

ENVIRONMENTAL INFORMATION

TSCA: 40CFR716.120(*c*)13 as glycidol (oxiranemethanol) and its derivitives.

GRAPHITE, NATURAL G:210

SYNONYMS: BLACK LEAD; EINECS No. 231-955-3; MINERAL CARBON; PLUMBAGO; SILVER GRAPHITE; STOVE BLACK

IDENTIFICATION

CAS:	7782-42-5
ERG ID:	Not listed
Molecular Formula:	C
RTECS No.:	MD9659600
Properties:	Combustible steel gray to black solid. Greasy feeling. Odorless. Insoluble in water.

HEALTH AND SAFETY INFORMATION

IDLH: 1,250 mg/m^3
OSHA Table Z-1-A Air Contaminant
OSHA Table Z-3 Mineral Dusts: 15 mppcf

Exposure Limits:	**TWA**
ACGIH:	2 mg/m^3 (respirable fraction)
NIOSH:	2.5 mg/m^3 (respirable fraction)
OSHA:	15 mppcf (respirable fraction)

RECOMMENDATIONS FOR RESPIRATOR SELECTION: NIOSH

12.5 mg/m^3: D (any dust respirator).
25 mg/m^3: DXSQ (any dust respirator except single-use and quarter-mask respirators;or SA (any supplied respirator).
62.5 mg/m^3: PAPRD (any powered, air-purifying respirator with a dust filter);or SA:CF (any supplied-air respirator operated in a continuous-flow mode.
125 mg/m^3: HiEF (any air-purifying, full-facepiece respirator with a high-efficiency particulate filter;or PAPRTHiE (any powered, air-purifying respirator with a tight-fitting facepiece and a high-efficiency particulate filter;or, SAT:CF (any supplied-air respirator that has a tight-fitting facepiece and is operated in a continuous-flow mode; SCBAF (any self-contained breathing apparatus with a full facepiece).
1250 mg/m^3: SAF:PD,PP (any supplied-air respirator that has a full facepiece and is operated in a pressure-demand or other positive-pressure mode).

Emergency or Planned Entry into Unknown Concentrations or IDLH Conditions: SCBAF:PD,PP (any self-contained breathing apparatus that has a full faceplate and is operated in a pressure-demand or other positive-pressure mode);or SAF:PD,PP:ASCBA (any supplied-air respirator that has a full facepiece and is operated in a pressure-demand or other positive-pressure mode in combination with an auxiliary, self-contained breathing apparatus operated in a pressure-demand or other positive-pressure mode).

Escape: HiEF (any air-purifying, full-facepiece respirator with a high-efficiency particulate filter);or SCBAE (any appropriate escape-type, self-contained breathing apparatus).

CHEMICAL REACTIVITY

Reacts with very strong oxidizers (e.g., fluorine, chlorine trifluoride, and potassium peroxide).

FIRE INFORMATION

Combustible solid. If material or contaminated runoff enters waterways, notify downstream users of potentially contaminated waters. Notify local health and fire officials and pollution control agencies. Use dry chemicals, foam, carbon dioxide.

FIRST AID

Move victim to fresh air. Call emergency medical care. Apply artificial respiration if victim is not breathing. Administer oxygen if breathing is difficult. Remove and isolate contaminated clothing and shoes. In case of contact with substance, immediately flush skin or eyes with running water for at least 20 minutes. Wash skin with soap and water. Keep victim warm and quiet. Effects of exposure (inhalation, ingestion, or skin contact) to substance may be delayed. Ensure that medical personnel are aware of the material(s) involved and take precautions to protect themselves.

HELIUM H:020

SYNONYMS: EINECS No. 231-168-5; ELIO (Italian); HELIO (Spanish); HELIUM, COMPRESSED; HELIUM GAS; HELIUM, REFRIGERATED LIQUID; HELIUM CRYOGENIC LIQUID

IDENTIFICATION

CAS: 7440-59-7
ERG ID: UN 1046 (compressed); UN 1963 (liquid, refrigerated)
ERG Guide: 121 (compressed); 120 (liquid, refrigerated)
ERG Guide Category: GASES- INERT
Molecular Formula: He
RTECS No.: MH6520000
Properties: Noncombustible, inert gas. Colorless, Odorless. Insoluble in water

HEALTH AND SAFETY INFORMATION

A simple asphyxiant. Gas is lighter than air, but cold vapor is heavier than air and can accumulate at ground level, causing oxygen deficiency if atmosphere becomes saturated. Contact with liquid can cause frostbite.

RECOMMENDATIONS FOR RESPIRATOR SELECTION

At any concentrations above the NIOSH REL, or where there is no REL, at any detectable concentration: SCBAF:PD,PP (any self-contained breathing apparatus that has a full facepiece and is operated in a pressure-demand or other positive-pressure mode);or SAF:PD,PP:ASCBA (any supplied-air respirator that has a full facepiece and is operated in a pressure-demand or other positive-pressure mode in combination with an auxiliary, self-contained breathing apparatus operated in a pressure-demand or other positive-pressure mode).

Escape: GMFOV [any air-purifying, full-facepiece respirator (gas mask) with a chin-style, front- or back-mounted organic vapor canister];or SCBAE (any appropriate escape-type, self-contained breathing apparatus).

FIRE INFORMATION

Storage containers and parts of containers may rocket great distances, in many directions. Notify local health and fire officials and pollution control agencies. Do not spray liquid helium with water; otherwise, any extinguishing agent suitable for surrounding fire can be used. From a secure, explosion-proof location, use water spray to cool exposed containers. If cooling streams are ineffective (venting sound increases in volume and pitch, tank discolors, or shows any signs of deforming), withdraw immediately to a secure position.

FIRST AID

Move victim to fresh air. Call emergency medical care. Apply artificial respiration if victim is not breathing. Administer oxygen if breathing is difficult. Clothing frozen to the skin should be thawed before being removed. In case of contact with liquefied gas, thaw frosted parts with lukewarm water. Keep victim warm and quiet. Ensure that medical personnel are aware of the material(s) involved and take precautions to protect themselves.

HEPTANE H:030

SYNONYMS: DIPROPYL METHANE; DIPROPYLMETHANE; EEC NO. 601-008-00-2; EINECS No. 205-563-8; EPTANI (Italian); HEPTAN (Polish); n-HEPTANE; normal-HEPTANE; HEPTANEN (Dutch); HEPTANO (Spanish); HEPTYL HYDRIDE; DIPROPAL METHANE; SKELLY-SOLVE C

IDENTIFICATION

CAS: 142-82-5

534 Heptane

ERG ID:	UN 1206
ERG Guide:	128
ERG Guide Category:	FLAMMABLE LIQUIDS (NONPOLAR/WATER-IMMISCIBLE)
Molecular Formula:	C_7H_{16}
RTECS No.:	MI7700000
Properties:	Highly flammable, colorless, watery liquid. Gasolinelike odor. Floats on water; insoluble.

HEALTH AND SAFETY INFORMATION

IDLH: 750 ppm
Odor Threshold: 9.77 ppm
OSHA Table z-1-A Air Contaminant

Irritating to the skin, eyes, nose and lungs.

Exposure Limits:	**TWA**	**CEILING**	**STEL**
ACGIH:	400 ppm; 1,650 mg/m³		500 ppm; 2,050 mg/m³
NIOSH:	85 ppm; 350 mg/m³	440 ppm; 1,800 mg/m³/15 min	
OSHA:	500 ppm; 2,000 mg/m³		

RECOMMENDATIONS FOR RESPIRATOR SELECTION: NIOSH/OSHA

750 ppm: CCROV [any chemical cartridge respirator with organic vapor cartridge(s)];or GMFOV [any air-purifying, full-facepiece respirator (gas mask) with a chin-style, front- or back-mounted acid gas canister];or PAPROV [any powered, air-purifying respirator with organic vapor cartridge(s)];or SA (any supplied-air respirator);or SCBAF (any self-contained breathing apparatus with a full facepiece).

Emergency or Planned Entry into Unknown Concentrations or IDLH Conditions: SCBAF:PD,PP (any self-contained breathing apparatus that has a full facepiece and is operated in a pressure-demand or other positive-pressure mode);or SAF:PD,PP:ASCBA (any supplied-air respirator that has a full facepiece and is operated in a pressure-demand or other positive-pressure mode in combination with an auxiliary, self-contained breathing apparatus operated in a pressure-demand or other positive-pressure mode).

Escape: GMFOV [any air-purifying, full-facepiece respirator (gas mask) with a chin-style, front- or back-mounted organic vapor canister];or SCBAE (any appropriate escape-type, self-contained breathing apparatus).

CHEMICAL REACTIVITY

Forms explosive mixture with air. Contact with strong oxidizers may cause fire and explosions. Attacks some plastics, rubber, and coatings. May accumulate static electric charges that can ignite its vapors.

FIRE INFORMATION

Vapors are heavier than air and will collect in low areas. Vapors in confined areas may explode when exposed to fire. Vapors may travel long distances to ignition sources and flash back.Storage containers and parts of containers may rocket great distances, in many directions.If material or contaminated runoff enters waterways, notify downstream users of potentially contaminated waters. Notify local health and fire officials and pollution control agencies. Use dry chemicals, foam, carbon dioxide.

NFPA Hazard Classifications:

Health Hazards (Blue) 1
Flammability (Red) 3
Reactivity (Yellow) 0

Flash Point: 25°F (-4°C) c.c.
Flammable Limits in Air: LEL 1.0%; UEL 6.7%
Autoignition Temperature: 392°F (200°C)

FIRST AID

Move victim to fresh air. Call emergency medical care. Apply artificial respiration if victim is not breathing. Administer oxygen if breathing is difficult. Remove and isolate contaminated clothing and shoes. In case of contact with substance, immediately flush skin or eyes with running water for at least 20 minutes. Wash skin with soap and water. Keep victim warm and quiet. Ensure that medical personnel are aware of the material(s) involved and take precautions to protect themselves.

ENVIRONMENTAL INFORMATION

TSCA: 40CFR712.30(*e*)10; 40CFR716.120(*d*)10 re dermal absorption testing

HEXACHLOROBENZENE H:035

SYNONYMS: AMATIN; ANTICARIE; BENZENE, HEXACHLORO-; BENZENE HEXACHLORIDE; BUNT-CURE; BUNT-NO-MORE; CO-OP HEXA; EEC No. 602-065-00-6; EINECS No. 204-273-9; GRANOX MN; HCB; HEXA C. B.; HEXACHLOROBENZOL (German); HEXACLOROBENCENO (Spanish); JULIN'S CARBON CHLORIDE; NO BUNT LIQUID; PENTACHLOROPHENYL CHLORIDE; PERCHLOROBENZENE; PHENYL PERCHLORYL; SANOCIDE; SMUT-GO; SNIECIOTOX; BENZENE, HEXACHLORO-

IDENTIFICATION

CAS:	118-74-1
ERG ID:	UN 2729

536 Hexachlorobenzene

ERG Guide:	152
ERG Guide Category:	SUBSTANCES- TOXIC (COMBUSTIBLE)
Molecular Formula:	C_6Cl_6
RTECS No.:	DA2975000
Properties:	Combustible solid, needlelike crystals. White. Faint, not unpleasant odor. Sinks in water; insoluble.

HEALTH AND SAFETY INFORMATION

Poisonous. Irritates eyes, skin, and respiratory tract. May cause blood, liver, kidney disorders.

IARC Evaluation: Group 2B, sufficient animal evidence; limited human evidence.

Exposure Limits: **TWA**

ACGIH: [1] 0.025 mg/m^3

[1] Animal carcinogen. Notice in 1996 of intended change to 0.002 mg/m^3. Skin contact contributes significantly in overall exposure.

RECOMMENDATIONS FOR RESPIRATOR SELECTION

At any concentrations above the NIOSH REL, or where there is no REL, at any detectable concentration: SCBAF:PD,PP (any self-contained breathing apparatus that has a full facepiece and is operated in a pressure-demand or other positive-pressure mode);or SAF:PD,PP:ASCBA (any supplied-air respirator that has a full facepiece and is operated in a pressure-demand or other positive-pressure mode in combination with an auxiliary, self-contained breathing apparatus operated in a pressure-demand or other positive-pressure mode).

Escape: GMFOV [any air-purifying, full-facepiece respirator (gas mask) with a chin-style, front- or back-mounted organic vapor canister];or SCBAE (any appropriate escape-type, self-contained breathing apparatus).

CHEMICAL REACTIVITY

A halogenated hydrocarbon. Incompatible with oxidizers, N,N'-dimethylformamide.

FIRE INFORMATION

Combustion products include toxic fumes of chlorides. Vapors are heavier than air and will collect in low areas. Vapors in confined areas may explode when exposed to fire. Storage containers and parts of containers may rocket great distances, in many directions. If material or contaminated runoff enters waterways, notify downstream users of potentially contaminated waters. Notify local health and fire officials and pollution control agencies. Use dry chemicals, foam, carbon dioxide. Use water spray to cool exposed containers.

FEMA Hazard Classifications:

Health Hazards (Blue) 1
Flammability (Red) 1
Reactivity (Yellow) 0

Flash Point: 468°F (242°C) c.c.

FIRST AID

Move victim to fresh air. Call emergency medical care. Apply artificial respiration if victim is not breathing. Do not use mouth-to-mouth method if victim ingested or inhaled the substance; induce artificial respiration with the aid of a pocket mask equipped with a one-way valve or other proper respiratory medical device. Administer oxygen if breathing is difficult. Remove and isolate contaminated clothing and shoes. In case of contact with substance, immediately flush skin or eyes with running water for at least 20 minutes. For minor skin contact, avoid spreading material on unaffected skin. Keep victim warm and quiet. Effects of exposure (inhalation, ingestion, or skin contact) may be delayed. Ensure that medical personnel are aware of the material(s) involved and take precautions to protect themselves.

ENVIRONMENTAL INFORMATION

Clean Air Act: Hazardous Air Pollutants (Title I, Part A, Section 112)
Clean Water Act: Section 313 Priority Chemicals
EPA Hazardous Waste Number (RCRA No.): U127; D032
RCRA Section 261 Hazardous Constituents
RCRA Toxicity Characteristic (Section 261.24): Maximum Concentration of Contaminants, regulatory level, 0.13 mg/L.
RCRA Land Ban Waste
RCRA Universal Treatment Standards: Wastewater (mg/L), 0.055; Nonwastewater (mg/kg), 10.
RCRA Ground Water Monitoring List: Suggested test methods(s) (PQL μg/L); 8120 (0.05); 8270 (10)
Safe Drinking Water Act: MCL, 0.001 mg/L; MCLG, zero
EPCRA Section 304 Reportable Quantity (RQ): CERCLA, 10 lb (4.54 kg)
EPCRA Section 313: Form R *de minimis* concentration reporting level: 0.1%.
California's Proposition 65: Carcinogen, reproductive toxin

HEXACHLOROPROPYLENE H:064

SYNONYMS: EINECS No. 217-560-9; HEXACHLOROPROPENE; HEXACLOROPROPILENO (Spanish); 1-PROPENE,1,1,2,3,3-HEXACHLORO-

IDENTIFICATION

CAS: 1888-71-7

538 Hexachloropropylene

ERG ID:	UN 3082 as environmentally hazardous substances, liquid, n.o.s..
ERG Guide:	171
ERG Guide Category:	SUBSTANCES (LOW TO MODERATE HAZARD)
Molecular Formula:	C_3Cl_6
RTECS No.:	UD0175000
Properties:	Liquid. Poisonous gases produced when heated.

HEALTH AND SAFETY INFORMATION

Toxic; a suspected carcinogen. Long-term effects may include liver, blood, skin, lung problems.

RECOMMENDATIONS FOR RESPIRATOR SELECTION

At any concentrations above the NIOSH REL, or where there is no REL, at any detectable concentration: SCBAF:PD,PP (any self-contained breathing apparatus that has a full facepiece and is operated in a pressure-demand or other positive-pressure mode);or SAF:PD,PP:ASCBA (any supplied-air respirator that has a full facepiece and is operated in a pressure-demand or other positive-pressure mode in combination with an auxiliary, self-contained breathing apparatus operated in a pressure-demand or other positive-pressure mode).

Escape: GMFOV [any air-purifying, full-facepiece respirator (gas mask) with a chin-style, front- or back-mounted organic vapor canister];or SCBAE (any appropriate escape-type, self-contained breathing apparatus).

CHEMICAL REACTIVITY

Incompatible with oxidizers, acrolein, acrylonitrile, tert-butyl nitroacetylene, ethylene oxide, isopropyl chlorocarbonate, maleic anhydride, liquid or gaseous oxygen or fluorine, potassium, triisobutylaluminum, s-trioxane, dinitrogen tetroxide. May ignite on contact with cellulose nitrate of high surface area.

FIRE INFORMATION

Combustion products include toxic fumes of chlorides. If material or contaminated runoff enters waterways, notify downstream users of potentially contaminated waters. Notify local health and fire officials and pollution control agencies. Use dry chemicals, alcohol-resistant foam, carbon dioxide.

NFPA Hazard Classifications: Not listed

FIRST AID

Move victim to fresh air; call emergency medical care. Apply artificial respiration if victim is not breathing. Do not use mouth-to-mouth method if victim ingested or inhaled the substance; induce artificial respiration with the aid of a pocket mask equipped with a one-way valve or other proper respiratory medical device. Administer oxygen if breathing is difficult. Remove and isolate contaminated clothing and shoes. In case of contact with substance, immediately flush skin or eyes with running water for at least 20 minutes. For minor skin contact, avoid spreading material on unaffected skin. Keep victim warm and quiet. Effects of exposure (inhalation, ingestion, or skin contact) to substance may be delayed. Ensure that medical personnel are aware of the material(s) involved and take precautions to protect themselves.

ENVIRONMENTAL INFORMATION

EPA Hazardous Waste Number (RCRA No.): U243
RCRA Land Ban Waste
RCRA Universal Treatment Standards: Wastewater (mg/L), 0.035; Nonwastewater (mg/kg), 30.
RCRA Ground Water Monitoring List: Suggested test methods(s) (PQL μg/L): 8270(10)
EPCRA Section 304 Reportable Quantity (RQ): CERCLA, 1,000 lb (454 kg)
TSCA: 40CFR721.4145

HEXADECYLTRICHLOROSILANE H:068

SYNONYMS: EINECS No. 227-575-2; SILANE, TRICHLOROHEXADECYL-; TRICHLOROHEXADECYLSILANE

IDENTIFICATION

CAS:	5894-60-0
ERG ID:	UN 1781
ERG Guide:	156
ERG Guide Category:	SUBSTANCES- TOXIC AND/OR CORROSIVE (COMBUSTIBLE/WATER-SENSITIVE)
Molecular Formula:	$C_{16}H_{33}Cl_3Si$
RTECS No.:	VV4260000
Properties:	Combustible liquid. Colorless to yellow. Soluble in water; decomposes forming hydrochloric acid and heat.

HEALTH AND SAFETY INFORMATION

Poisonous. Corrosive to the eyes, skin, and respiratory tract. Inhalation may cause pulmonary edema, which can be delayed for several hours; there is a risk of death in serious cases.

RECOMMENDATIONS FOR RESPIRATOR SELECTION

At any concentrations above the NIOSH REL, or where there is no REL, at any detectable concentration: SCBAF:PD,PP (any self-contained breathing apparatus that has a full facepiece and is operated in a pressure-demand or other positive-pressure mode);or SAF:PD,PP:ASCBA (any supplied-air respirator that has a full facepiece and is operated in a pressure-demand or other positive-pressure mode in combination with an auxiliary, self-contained breathing apparatus operated in a pressure-demand or other positive-pressure mode).

Escape: GMFOV [any air-purifying, full-facepiece respirator (gas mask) with a chin-style, front- or back-mounted organic vapor canister];or SCBAE (any appropriate escape-type, self-contained breathing apparatus).

CHEMICAL REACTIVITY

Contact with water forms toxic and corrosive hydrogen chloride. Incompatible with ammonia; may form an explosive or self-igniting material. Attacks most metals in the presence of moisture.

FIRE INFORMATION

Combustion products include toxic and corrosive chlorides and hydrogen chloride. Vapors are heavier than air and will collect in low areas. Vapors in confined areas may explode when exposed to fire. If material or contaminated runoff enters waterways, notify downstream users of potentially contaminated waters. Notify local health and fire officials and pollution control agencies. Use dry chemicals, alcohol-resistant foam, carbon dioxide.

NFPA Hazard Classifications:

Health Hazards (Blue) 3
Flammability (Red) 1
Reactivity (Yellow) 0

Flash Point: 295°F (146°C) o.c.

FIRST AID

Move victim to fresh air. Call emergency medical care. Apply artificial respiration if victim is not breathing. Do not use mouth-to-mouth method if victim ingested or inhaled the substance; induce artificial respiration with the aid of a pocket mask equipped with a one-way valve or other proper respiratory medical device. Administer oxygen if breathing is difficult. Remove and isolate contaminated clothing and shoes. In case of contact with substance, immediately flush skin or eyes with running water for at least 20 minutes. For minor skin contact, avoid spreading material on unaffected skin. Keep victim warm and quiet. Effects of exposure (inhalation, ingestion, or skin contact) may be delayed. Ensure that medical personnel are aware of the material(s) involved and take precautions to protect themselves.

HEXADIENE H:072

SYNONYMS: ALLYLPROPENYL; 1,4-HEXADIENE; HEXA-1,4-DIENE

IDENTIFICATION

CAS: 592-45-0; 42296-74-2

ERG ID: UN 2458
ERG Guide: 130
ERG Guide Category: FLAMMABLE LIQUIDS (NONPOLAR/WATER-IMMISCIBLE/NOXIOUS)
Molecular Formula: C_6H_{10}
RTECS No.: MM1996000
Properties: Flammable liquid. Colorless. Floats on water; insoluble.

HEALTH AND SAFETY INFORMATION

Allyl compounds, in general, are toxic.

RECOMMENDATIONS FOR RESPIRATOR SELECTION

At any concentrations above the NIOSH REL, or where there is no REL, at any detectable concentration: SCBAF:PD,PP (any self-contained breathing apparatus that has a full facepiece and is operated in a pressure-demand or other positive-pressure mode);or SAF:PD,PP:ASCBA (any supplied-air respirator that has a full facepiece and is operated in a pressure-demand or other positive-pressure mode in combination with an auxiliary, self-contained breathing apparatus operated in a pressure-demand or other positive-pressure mode).

Escape: GMFOV [any air-purifying, full-facepiece respirator (gas mask) with a chin-style, front- or back-mounted organic vapor canister];or SCBAE (any appropriate escape-type, self-contained breathing apparatus).

CHEMICAL REACTIVITY

Contact with oxidizers may cause fire and explosions.

FIRE INFORMATION

Vapors are heavier than air and will collect in low areas. Vapors in confined areas may explode when exposed to fire. Vapors may travel long distances to ignition sources and flash back. Storage containers and parts of containers may rocket great distances, in many directions. If material or contaminated runoff enters waterways, notify downstream users of potentially contaminated waters. Notify local health and fire officials and pollution control agencies. Use dry chemicals, foam, carbon dioxide.

NFPA Hazard Classifications:

Health Hazards (Blue) 0
Flammability (Red) 3
Reactivity (Yellow) 0

Flash Point:-6°F (-21°C)
Flammable Limits in Air: LEL 2%; UEL 6.1%

FIRST AID

Move victim to fresh air. Call emergency medical care. Apply artificial respiration if victim is not breathing. Administer oxygen if breathing is difficult. Remove and isolate contaminated clothing and shoes. In case of contact with substance, immediately flush skin or eyes with running water for at least 20 minutes. Wash skin with soap and water. Keep victim warm and quiet. Effects of exposure (inhalation, ingestion, or skin contact) to substance may be delayed. Ensure that medical personnel are aware of the material(s) involved and take precautions to protect themselves.

HEXAFLUOROETHANE H:076

SYNONYMS: ETHANE, HEXAFLUORO-; F 116; FREON 116; HEXAFLUORETANO (Spanish); PERFLUOROETHANE; R 116

IDENTIFICATION

CAS:	76-16-4
ERG ID:	UN 2193
ERG Guide:	126
ERG Guide Category:	GASES- COMPRESSED OR LIQUEFIED (INCLUDING REFRIGERANT GASES)
Molecular Formula:	$C_2H_3F_3$
RTECS No.:	KI4110000
Properties:	Nonflammable gas. Insoluble in water.

RECOMMENDATIONS FOR RESPIRATOR SELECTION

At any concentrations above the NIOSH REL, or where there is no REL, at any detectable concentration: SCBAF:PD,PP (any self-contained breathing apparatus that has a full facepiece and is operated in a pressure-demand or other positive-pressure mode);or SAF:PD,PP:ASCBA (any supplied-air respirator that has a full facepiece and is operated in a pressure-demand or other positive-pressure mode in combination with an auxiliary, self-contained breathing apparatus operated in a pressure-demand or other positive-pressure mode).

Escape: GMFOV [any air-purifying, full-facepiece respirator (gas mask) with a chin-style, front- or back-mounted organic vapor canister];or SCBAE (any appropriate escape-type, self-contained breathing apparatus).

CHEMICAL REACTIVITY

Generally stable.

FIRE INFORMATION

Storage containers and parts of containers may rocket great distances, in many directions. Notify local health and fire officials and pollution control agencies. Use any extinguishing agent suitable for surrounding fires. From a secure, explosion-proof location, use water spray to cool exposed containers. If cooling streams are ineffective (venting sound increases in volume and pitch, tank discolors, or shows any signs of deforming), withdraw immediately to a secure position.

NFPA Hazard Classifications

Health Hazards (Blue) 1
Flammability (Red) 0
Reactivity (Yellow) 0

FIRST AID

Move victim to fresh air. Call emergency medical care. Apply artificial respiration if victim is not breathing. Administer oxygen if breathing is difficult. Remove and isolate contaminated clothing and shoes. In case of contact with liquefied gas, thaw frosted parts with lukewarm water. Keep victim warm and quiet. Keep victim under observation. Ensure that medical personnel are aware of the material(s) involved and take precautions to protect themselves.

HEXAMETHYLENEDIAMINE H:085

SYNONYMS: 1,6-DIAMINOHEXANE; EINECS No. 204-679-6; 1,6-HEXANEDIAMINE; HEXAMETILENDIAMINA (Spanish); 1,6-HEXANEDIAMINE, SOLUTION; HEXAMETHYLENEDIAMINE, SOLID; HEXAMETHYLENEDIAMINE, SOLUTION; HMDA; NCI-C61405

IDENTIFICATION

CAS:	124-09-4
ERG ID:	UN 2280 (solid); UN 1783 (solution)
ERG Guide:	153
ERG Guide Category:	SUBSTANCES- TOXIC AND/OR CORROSIVE (COMBUSTIBLE)
Molecular Formula:	$C_4H_{16}N_2$
RTECS No.:	MO1180000
Properties:	Combustible solid. Colorless. Weak ammonia, or fishy odor. Floats in water; soluble.

HEALTH AND SAFETY INFORMATION

Corrosive to eyes, skin, and respiratory tract. Inahalation can cause pulmonary edema, which can be delayed for several hours; there is a risk of death in serious cases. Long-term exposure can result in skin, blood, kidney and liver problems.

Odor Threshold: 0.0041 mg/m^3

RECOMMENDATIONS FOR RESPIRATOR SELECTION

At any concentrations above the NIOSH REL, or where there is no REL, at any detectable concentration: SCBAF:PD,PP (any self-contained breathing apparatus that has a full facepiece and is operated in a pressure-demand or other positive-pressure mode);or SAF:PD,PP:ASCBA (any supplied-air respirator that has a full facepiece and is operated in a pressure-demand or other positive-pressure mode in combination with an auxiliary, self-contained breathing apparatus operated in a pressure-demand or other positive-pressure mode).

Escape: GMFOV [any air-purifying, full-facepiece respirator (gas mask) with a chin-style, front- or back-mounted organic vapor canister];or SCBAE (any appropriate escape-type, self-contained breathing apparatus).

CHEMICAL REACTIVITY

A reducing agent; incompatible with oxidizers. A strong base when dissolved in water; reacts with acids, organic anhydrides, isocyanates, alkylene oxides, epichlorohydrin, aldehydes, alcohols, glycols, phenols, cresols, caprolactum solution. In the presence of moisture corrodes chemically active metals such as aluminum and zinc.

FIRE INFORMATION

Vapors are heavier than air and will collect in low areas. Vapors in confined areas may explode when exposed to fire. If material or contaminated runoff enters waterways, notify downstream users of potentially contaminated waters. Notify local health and fire officials and pollution control agencies. Use dry chemicals, alcohol-resistant foam, carbon dioxide, water. Use water to cool exposed containers.

FEMA Hazard Classifications:

Health Hazards (Blue) 1
Flammability (Red) 2
Reactivity (Yellow) 0

Flash Point: 160°F (71°C) o.c.
Flammable Limits in Air: LEL 0.7%; UEL 6.3%
Autoignition Temperature: 585°F (307°C)

FIRST AID

Move victim to fresh air and call emergency medical care. Apply artificial respiration if victim is not breathing. Do not use mouth-to-mouth method if victim swallowed or inhaled the substance; induce artificial respiration with the aid of a pocket mask equipped with a one-way valve or other proper respiratory medical device. If breathing is difficult, give oxygen. In case of contact with substance, immediately flush skin or eyes with running water for at least 20 minutes. Remove and isolate contaminated clothing and shoes. Keep victim warm and quiet. Effects of exposure (inhalation, ingestion, or skin contact) to substances may be delayed. Ensure that medical personnel are aware of the material(s) involved and take precautions to protect themselves.

HEXAMETHYLENE GLYCOL H:093

SYNONYMS: 1,6-DIHYDROXYHEXANE; EINECS No. 211-074-0; HDO; HEXAMETHYLENEDIOL; a-w-HEXANEDIOL; w-HEXANEDIOL; HEXANEDIOL; 1,6-HEXANEDIOL; HEXANE-1,6-DIOL; HEXAMETILENGLICOL (Spanish); 1,6-HEXYLENE GLYCOL

IDENTIFICATION

CAS:	629-11-8
Molecular Formula:	$C_6H_{14}O_2$
RTECS No.:	MO2100000
Properties:	Combustible needlelike solid.

HEALTH AND SAFETY INFORMATION

Irritates eyes, skin, and respiratory tract. Due to the availability of insufficient data on human effects, caution should be exercised.

RECOMMENDATIONS FOR RESPIRATOR SELECTION

At any concentrations above the NIOSH REL, or where there is no REL, at any detectable concentration: SCBAF:PD,PP (any self-contained breathing apparatus that has a full facepiece and is operated in a pressure-demand or other positive-pressure mode);or SAF:PD,PP:ASCBA (any supplied-air respirator that has a full facepiece and is operated in a pressure-demand or other positive-pressure mode in combination with an auxiliary, self-contained breathing apparatus operated in a pressure-demand or other positive-pressure mode).

Escape: GMFOV [any air-purifying, full-facepiece respirator (gas mask) with a chin-style, front- or back-mounted organic vapor canister];or SCBAE (any appropriate escape-type, self-contained breathing apparatus).

CHEMICAL REACTIVITY

Reacts with oxidizers. Incompatible with sulfuric acid, nitric acid, caustics, aliphatic amines, isocyanates.

FIRE INFORMATION

Vapors are heavier than air and will collect in low areas. Vapors in confined areas may explode when exposed to fire. If material or contaminated runoff enters waterways, notify downstream users of potentially contaminated waters. Notify local health and fire officials and pollution control agencies. Use dry chemicals, alcohol-resistant foam, carbon dioxide.

NFPA Hazard Classifications: Not listed

Flash Point: 266°F (130°C)

FIRST AID

Move victim to fresh air. Call emergency medical care. Apply artificial respiration if victim is not breathing. Do not use mouth-to-mouth method if victim swallowed or inhaled the substance; induce artificial respiration with the aid of a pocket mask equipped with a one-way valve or other proper respiratory medical device. Administer oxygen if breathing is difficult. Remove and isolate contaminated clothing and shoes. In case of contact with substance, immediately flush skin or eyes with running water for at least 20 minutes. For minor skin contact, avoid spreading material on unaffected skin. Keep victim warm and quiet. Effects of exposure (inhalation, ingestion, or skin contact) to substances may be delayed. Ensure that medical personnel are aware of the material(s) involved and take precautions to protect themselves.

HEXAMETHYLENETETRAMINE H:095

SYNONYMS: AMINOFORM; AMMONIOFORMALDEHYDE; EINECS No. 202-905-8; HEXA; HEXAMETILENTETRAMINA (Spanish); HEXAMINE; METHENEAMINE; UROTROPIN;

IDENTIFICATION

CAS:	100-97-0
ERG ID:	UN 1328
ERG Guide:	133
ERG Guide Category:	FLAMMABLE SOLIDS
Molecular Formula:	$C_6H_{12}N_4$
RTECS No.:	MN4725000
Properties:	Combustible solid crystals or powder. White. Mild ammonia odor. Sinks and mixes with water.

HEALTH AND SAFETY INFORMATION

Irritates eyes, skin, and respiratory tract. Inhalation may cause pulmonary edema, which can be delayed for several hours; there is a risk of death in serious cases. Long-term: mutagenic data reported by EPA; possible dermatitis.

RECOMMENDATIONS FOR RESPIRATOR SELECTION

At any concentrations above the NIOSH REL, or where there is no REL, at any detectable concentration: SCBAF:PD,PP (any self-contained breathing apparatus that has a full facepiece and is operated in a pressure-demand or other positive-pressure mode);or SAF:PD,PP:ASCBA (any supplied-air respirator that has a full facepiece and is operated in a pressure-demand or other positive-pressure mode in

combination with an auxiliary, self-contained breathing apparatus operated in a pressure-demand or other positive-pressure mode).

Escape: GMFOV [any air-purifying, full-facepiece respirator (gas mask) with a chin-style, front- or back-mounted organic vapor canister];or SCBAE (any appropriate escape-type, self-contained breathing apparatus).

CHEMICAL REACTIVITY

May ignite on contact with cellulose nitrate of high surface area. Incompatible with acrolein, acrylonitrile, tert-butyl nitroacetylene, ethylene oxide, isopropyl chlorocarbonate, maleic anhydride, triisobutylaluminum. Reacts with oxidizers; violent reaction with sodium peroxide.

FIRE INFORMATION

Combustion products include toxic nitrogen oxides. Vapors are heavier than air and will collect in low areas. Vapors in confined areas may explode when exposed to fire.If material or contaminated runoff enters waterways, notify downstream users of potentially contaminated waters. Notify local health and fire officials and pollution control agencies. Use dry chemicals, alcohol-resistant foam, carbon dioxide.

FEMA Hazard Classifications:

Health Hazards (Blue) 2
Flammability (Red) 1
Reactivity (Yellow) 0

Flash Point: 482°F (250°C) c.c.
Autoignition Temperature: 482°F (250°C)

FIRST AID

Move victim to fresh air. Call emergency medical care. Apply artificial respiration if victim is not breathing. Administer oxygen if breathing is difficult. Remove and isolate contaminated clothing and shoes. In case of contact with substance, immediately flush skin or eyes with running water for at least 20 minutes. Removal of solidified molten material from skin requires medical assistance. Keep victim warm and quiet. Ensure that medical personnel are aware of the material(s) involved and take precautions to protect themselves.

ENVIRONMENTAL INFORMATION

TSCA: 40CFR716.120(*d*)1 as aldehydes

HEXANAL H:098

SYNONYMS: ALDEHYDE C-6; CAPRONIC ALDEHYDE; CAPROALDEHYDE;

548 Hexanal

CAPRONALDEHYDE; N-CAPROYLALDEHYDE; EINECS No. 200-624-5; FEMA NO. 2557; n-HEXALDEHIDO (Spanish); n-HEXALDEHYDE; 1-HEXANAL

IDENTIFICATION

CAS:	66-25-1
ERG ID:	UN 1207
ERG Guide:	129
ERG Guide Category:	FLAMMABLE LIQUIDS (POLAR/WATER-MISCIBLE/NOXIOUS)
Molecular Formula:	$C_6H_{11}O$
RTECS No.:	MN7175000
Properties:	Liquid. Colorless. Sharp unpleasant odor. Floats on water; insoluble. Flammable, irritating vapor is produced.

HEALTH AND SAFETY INFORMATION

Irritates eyes, skin, and respiratory tract.

RECOMMENDATIONS FOR RESPIRATOR SELECTION

At any concentrations above the NIOSH REL, or where there is no REL, at any detectable concentration: SCBAF:PD,PP (any self-contained breathing apparatus that has a full facepiece and is operated in a pressure-demand or other positive-pressure mode);or SAF:PD,PP:ASCBA (any supplied-air respirator that has a full facepiece and is operated in a pressure-demand or other positive-pressure mode in combination with an auxiliary, self-contained breathing apparatus operated in a pressure-demand or other positive-pressure mode).

Escape: GMFOV [any air-purifying, full-facepiece respirator (gas mask) with a chin-style, front- or back-mounted organic vapor canister];or SCBAE (any appropriate escape-type, self-contained breathing apparatus).

CHEMICAL REACTIVITY

Forms explosive mixture with air. Contact with oxidizers may cause fire and explosions. May attack some forms of plastics.

FIRE INFORMATION

Vapors are heavier than air and will collect in low areas. Vapors in confined areas may explode when exposed to fire. Vapors may travel long distances to ignition sources and flash back. Storage containers and parts of containers may rocket great distances, in many directions. If material or contaminated runoff enters waterways, notify downstream users of potentially contaminated waters. Notify local health and fire officials and pollution control agencies. Use dry chemicals, foam, carbon dioxide.

NFPA Hazard Classifications:

Health Hazards (Blue) 2
Flammability (Red) 3
Reactivity (Yellow) 1

Flash Point: 90°F (32°C) o.c.

FIRST AID

Move victim to fresh air. Call emergency medical care. Apply artificial respiration if victim is not breathing. Administer oxygen if breathing is difficult. Remove and isolate contaminated clothing and shoes. In case of contact with substance, immediately flush skin or eyes with running water for at least 20 minutes. Wash skin with soap and water. Keep victim warm and quiet. Effects of exposure (inhalation, ingestion, or skin contact) to substance may be delayed. Ensure that medical personnel are aware of the material(s) involved and take precautions to protect themselves.

ENVIRONMENTAL INFORMATION

TSCA: 40CFR716.120(*d*)1 as aldehydes

HEXANE H:100

SYNONYMS: EEC No. 601-037-00-0; EINECS No. 203-777-6; ESANI (Italian); EXXSOL HEXANE; GENESOLV 404 AZEOTROPE; HEKSAN (Polish); normal-HEXANE; n-HEXANE; HEXANEN (Dutch); n-HEXANO (Spanish); NCI-C60571; HEXYL HYDRIDE; SKELLYSOLVE-B

IDENTIFICATION

CAS:	110-54-3; 64742-49-0
ERG ID:	UN 1208
ERG Guide:	128
ERG Guide Category:	FLAMMABLE LIQUIDS (NONPOLAR/WATER-IMMISCIBLE)
Molecular Formula:	C_6H_{14}
RTECS No.:	MN9275000
Properties:	Highly flammable, watery liquid. Clear, colorless. Gasolinelike odor. Floats on water; insoluble.

HEALTH AND SAFETY INFORMATION

Irritating to skin, eyes, nose and lungs. Long-term exposure may cause damage to male reproductive glands, central nervous system, and brain; dry and cracked skin with possible rash.

550 Hexane

IDLH: 1,100 ppm
Odor Threshold: 21.9 ppm
OSHA Table Z-1-A Air Contaminant
NIOSH Criteria Document: NIOSH 77-151 ALKANES

Exposure Limits:	**TWA**
ACGIH:	50 ppm; 176 mg/m^3
NIOSH:	50 ppm; 180 mg/m^3
OSHA:	500 ppm; 1,800 mg/m^3

RECOMMENDATIONS FOR RESPIRATOR SELECTION: NIOSH/OSHA

500 ppm: SA (any supplied-air respirator)
1,100 ppm: SA:CF (any supplied-air respirator operated in a continuous-flow mode);or SCBAF (any self-contained breathing apparatus with a full facepiece);or SAF (any supplied-air respirator with a full facepiece).

Emergency or Planned Entry into Unknown Concentrations or IDLH Conditions: SCBAF:PD,PP (any self-contained breathing apparatus that has a full facepiece and is operated in a pressure-demand or other positive-pressure mode);or SAF:PD,PP:ASCBA (any supplied-air respirator that has a full facepiece and is operated in a pressure-demand or other positive-pressure mode in combination with an auxiliary, self-contained breathing apparatus operated in a pressure-demand or other positive-pressure mode).

Escape: GMFOV [any air-purifying, full-facepiece respirator (gas mask) with a chin-style, front- or back-mounted organic vapor canister];or SCBAE (any appropriate escape-type, self-contained breathing apparatus).
Note: Substance reported to cause eye irritation or damage; may require eye protection.

CHEMICAL REACTIVITY

Forms explosive mixture with air. Strong oxidizers such as permanganates, chromates, or chlorine may cause fire and explosions. Attacks some plastics, rubber, and coatings. May accumulate static electrical charges, and may cause ignition of its vapors.

FIRE INFORMATION

Vapors are heavier than air and will collect in low areas. Vapors in confined areas may explode when exposed to fire. Vapors may travel long distances to ignition sources and flash back. Storage containers and parts of containers may rocket great distances, in many directions. If material or contaminated runoff enters waterways, notify downstream users of potentially contaminated waters. Notify local health and fire officials and pollution control agencies. Use dry chemicals, foam, carbon dioxide.

NFPA Hazard Classifications:

Health Hazards (Blue) 1
Flammability (Red) 3
Reactivity (Yellow) 0

Flash Point: –7°F (–22°C) c.c.
Flammable Limits in Air: LEL 1.1%; UEL 7.5%
Autoignition Temperature: 437°F (225°C)

FIRST AID

Move victim to fresh air. Call emergency medical care. Apply artificial respiration if victim is not breathing. If breathing is difficult, give oxygen. Remove and isolate contaminated clothing and shoes. In case of contact with substance, immediately flush skin or eyes with running water for at least 20 minutes. Wash skin with soap and water. Keep victim warm and quiet. Ensure that medical personnel are aware of the material(s) involved and take precautions to protect themselves.

ENVIRONMENTAL INFORMATION

Clean Air Act: Hazardous Air Pollutants (Title I, Part A, Section 112)
EPCRA Section 304: Reportable Quantity (RQ): CERCLA, 1 lb (0.454 kg)
EPCRA Section 313: Form R *de minimis* concentration reporting level: 1.0%

1,2,6–HEXANETRIOL H:108

SYNONYMS: EINECS No. 203-424-6; HEXANETRIOL-1,2,6; HEXANE-1,2,6-TRIOL

IDENTIFICATION

CAS:	106-69-4
ERG ID:	UN 1208 as hexanes
ERG Guide:	128
ERG Guide Category:	FLAMMABLE LIQUIDS (NONPOLAR/WATER-IMMISCIBLE)
Molecular Formula:	$C_6H_{14}O_3$
RTECS No.:	MO4650000
Properties:	Combustible liquid. Colorless. Soluble in water.

HEALTH AND SAFETY INFORMATION

Irritates eyes and respiratory tract.

RECOMMENDATIONS FOR RESPIRATOR SELECTION

At any concentrations above the NIOSH REL, or where there is no REL, at any detectable

concentration: SCBAF:PD,PP (any self-contained breathing apparatus that has a full facepiece and is operated in a pressure-demand or other positive-pressure mode);or SAF:PD,PP:ASCBA (any supplied-air respirator that has a full facepiece and is operated in a pressure-demand or other positive-pressure mode in combination with an auxiliary, self-contained breathing apparatus operated in a pressure-demand or other positive-pressure mode).

Escape: GMFOV [any air-purifying, full-facepiece respirator (gas mask) with a chin-style, front- or back-mounted organic vapor canister];or SCBAE (any appropriate escape-type, self-contained breathing apparatus).

CHEMICAL REACTIVITY

Reacts with oxidizing materials.

FIRE INFORMATION

Vapors are heavier than air and will collect in low areas. Vapors in confined areas may explode when exposed to fire. If material or contaminated runoff enters waterways, notify downstream users of potentially contaminated waters. Notify local health and fire officials and pollution control agencies. Use dry chemicals, alcohol-resistant foam, carbon dioxide.

NFPA Hazard Classifications:

Health Hazards (Blue) 1
Flammability (Red) 1
Reactivity (Yellow) 0

Flash Point: 375°F (191°C) o.c.

FIRST AID

Move victim to fresh air. Call emergency medical care. Apply artificial respiration if victim is not breathing. If breathing is difficult, give oxygen. Remove and isolate contaminated clothing and shoes. In case of contact with substance, immediately flush skin or eyes with running water for at least 20 minutes. Wash skin with soap and water. Keep victim warm and quiet. Ensure that medical personnel are aware of the material(s) involved and take precautions to protect themselves.

HEXANOL H:115

SYNONYMS: ALCOHOL C-6; ALCOOL HEXYLIQUE (French); n-AMYL CARBINOL; AMYL CARBINOL; CAPROYL ALCOHOL; EEC No. 603-059-00-6; EINECS No. 203-852-3; EPAL-6; FEMA No. 2567; HEXANOL; 1-HEXANOL; n-HEXANOL; HEXYL ALCOHOL; n-HEXYL ALCOHOL; 1-HYDROXYHEXANE; PENTYLCARBINOL

IDENTIFICATION

CAS:	111-27-3
ERG ID:	UN 2282
ERG Guide:	129
ERG Guide Category:	FLAMMABLE LIQUIDS (POLAR/WATER-MISCIBLE/NOXIOUS)
Molecular Formula:	$C_6H_{14}O$
RTECS No.:	MQ4025000
Properties:	Flammable, clear liquid. Colorless. Sweet, fruity odor. Floats on water; slightly soluble.

HEALTH AND SAFETY INFORMATION

Irritates eyes, skin, and respiratory tract.

RECOMMENDATIONS FOR RESPIRATOR SELECTION

At any concentrations above the NIOSH REL, or where there is no REL, at any detectable concentration: SCBAF:PD,PP (any self-contained breathing apparatus that has a full facepiece and is operated in a pressure-demand or other positive-pressure mode);or SAF:PD,PP:ASCBA (any supplied-air respirator that has a full facepiece and is operated in a pressure-demand or other positive-pressure mode in combination with an auxiliary, self-contained breathing apparatus operated in a pressure-demand or other positive-pressure mode).

Escape: GMFOV [any air-purifying, full-facepiece respirator (gas mask) with a chin-style, front- or back-mounted organic vapor canister];or SCBAE (any appropriate escape-type, self-contained breathing apparatus).

CHEMICAL REACTIVITY

Forms explosive mixture with air. Contact with strong oxidizers may cause fire and explosions. Incompatible with sulfuric acid, nitric acid, caustics, aliphatic amines, isocyanates.

FIRE INFORMATION

Vapors are heavier than air and will collect in low areas. Vapors in confined areas may explode when exposed to fire. Vapors may travel long distances to ignition sources and flash back. Storage containers and parts of containers may rocket great distances, in many directions. If material or contaminated runoff enters waterways, notify downstream users of potentially contaminated waters. Notify local health and fire officials and pollution control agencies. Use dry chemicals, foam, carbon dioxide.

NFPA Hazard Classifications:

Health Hazards (Blue) 1

Flammability (Red) 2
Reactivity (Yellow) 0

Flash Point: 149°F (65°C) o.c.; 145°F (63°C) c.c.
Flammable Limits in Air: LEL 1.2%; UEL 7.7% (calculated)
Autoignition Temperature: 554°F (290°C)

FIRST AID

Move victim to fresh air. Call emergency medical care. Apply artificial respiration if victim is not breathing. Administer oxygen if breathing is difficult. Remove and isolate contaminated clothing and shoes. In case of contact with substance, immediately flush skin or eyes with running water for at least 20 minutes. Wash skin with soap and water. Keep victim warm and quiet. Effects of exposure (inhalation, ingestion, or skin contact) to substance may be delayed. Ensure that medical personnel are aware of the material(s) involved and take precautions to protect themselves.

2-HEXANONE H:120

SYNONYMS: BUTYL METHYL KETONE; n-BUTYL METHYL KETONE; EEC No. 606-030-00-6; 2-HEXANONE; KETONE, BUTYL METHYL; MBK; METHYL BUTYL KETONE; METHYL n-BUTYL KETONE; METIL n-BUTIL CETONA (Spanish); MNBK; PROPYLACETONE

IDENTIFICATION

CAS:	591-78-6
ERG ID:	UN 1993
ERG Guide:	128
ERG Guide Category:	FLAMMABLE LIQUIDS (NON-POLAR/WATER-IMMISCIBLE)
Molecular Formula:	$C_6H_{12}O$
RTECS No.:	MP1400000
Properties:	Flammable liquid. Clear. Disagreeable acetonelike odor. Floats and mixes with water.

HEALTH AND SAFETY INFORMATION

IDLH: 1,600 ppm
Odor Threshold: 0.166 ppm

NIOSH Criteria Document: NIOSH 78-173 KETONES

Exposure Limits:	**TWA**
ACGIH: [(1)]	5 ppm; 20 mg/m^3

NIOSH: 1 ppm; 4 mg/m^3
OSHA: 100 ppm; 410 mg/m^3

(1) Skin contact contributes significantly to overall exposure.

RECOMMENDATIONS FOR RESPIRATOR SELECTION: NIOSH

10 ppm: SA (any supplied-air respirator).
25 ppm: SA:CF (any supplied-air respirator operated in a continuous-flow mode).
50 ppm: SAT:CF (any supplied-air respirator that has a tight-fitting facepiece and is operated in a continuous-flow mode);or SCBAF (any self-contained breathing apparatus with a full facepiece);or SAF (any supplied-air respirator with a full facepiece).
1600 ppm: SAF:PD,PP (any supplied-air respirator that has a full facepiece and is operated in a pressure-demand or other positive-pressure mode).

Emergency or Planned Entry Into Unknown Concentrations or Idlh Conditions:
SCBAF:PD,PP (any self-contained breathing apparatus that has a full facepiece and is operated in a pressure-demand or other positive-pressure mode);or SAF:PD,PP:ASCBA (any supplied-air respirator that has a full facepiece and is operated in a pressure-demand or other positive-pressure mode in combination with an auxiliary, self-contained breathing apparatus operated in a pressure-demand or other positive pressure mode.

Escape: GMFOV (any air-purifying, full-facepiece respirator (gas mask) with a chin-style, front- or back-mounted acid gas canister);or SCBAE (any appropriate escape-type, self-contained breathing apparatus).

CHEMICAL REACTIVITY

Forms explosive mixture with air. Contact with strong oxidizers may cause fire and explosions. Dissolves some plastics, resins, and rubbers.

FIRE INFORMATION

NFPA Hazard Classifications:

Health Hazards (Blue) 2
Flammability (Red) 3
Reactivity (Yellow) 0

Flash Point: 83°F (28°C) o.c.; 77°F (25°C) c.c.
Flammable Limits in Air: LEL 1.3%; UEL 8.0%
Autoignition Temperature: 795°F (423°C)

FIRST AID

Move victim to fresh air. Call emergency medical care. Apply artificial respiration if victim is not

breathing. If breathing is difficult, give oxygen. Remove and isolate contaminated clothing and shoes. In case of contact with material, immediately flush skin or eyes with running water for at least 20 minutes. Wash skin with soap and water. Keep victim warm and quiet. Ensure that medical personnel are aware of the material(s) involved and take precautions to protect themselves.

HEXENE H:130

SYNONYMS: BUTYL ETHYLENE; EINECS No. 209-753-1; 1-HEXENE; α-HEXENE; HEXYLENE

IDENTIFICATION

CAS:	592-41-6
ERG ID:	UN 2370
ERG Guide:	128
ERG Guide Category:	FLAMMABLE LIQUIDS (NONPOLAR/WATER-IMMISCIBLE)
Molecular Formula:	C_6H_{12}
RTECS No.:	MP6600100
Properties:	Highly flammable liquid. Colorless. Mild, gasolinelike odor. Floats on water; insoluble. Produces a large amount of vapor.

HEALTH AND SAFETY INFORMATION

Irritates eyes, skin, and respiratory tract.

RECOMMENDATIONS FOR RESPIRATOR SELECTION

At any concentrations above the NIOSH REL, or where there is no REL, at any detectable concentration: SCBAF:PD,PP (any self-contained breathing apparatus that has a full facepiece and is operated in a pressure-demand or other positive-pressure mode);or SAF:PD,PP:ASCBA (any supplied-air respirator that has a full facepiece and is operated in a pressure-demand or other positive-pressure mode in combination with an auxiliary, self-contained breathing apparatus operated in a pressure-demand or other positive-pressure mode).

Escape: GMFOV [any air-purifying, full-facepiece respirator (gas mask) with a chin-style, front- or back-mounted organic vapor canister];or SCBAE (any appropriate escape-type, self-contained breathing apparatus).

CHEMICAL REACTIVITY

Contact with oxidizers can cause fire and explosions.

FIRE INFORMATION

Vapors are heavier than air and will collect in low areas. Vapors in confined areas may explode when

exposed to fire. Vapors may travel long distances to ignition sources and flash back. Storage containers and parts of containers may rocket great distances, in many directions. If material or contaminated runoff enters waterways, notify downstream users of potentially contaminated waters. Notify local health and fire officials and pollution control agencies. Use dry chemicals, foam, carbon dioxide. Use water spray to cool exposed containers and to control vapor.

NFPA Hazard Classifications:

Health Hazards (Blue) 1
Flammability (Red) 3
Reactivity (Yellow) 0

Flash Point: –15°F (–26°C) c.c.
Flammable Limits in Air: LEL 1.2%; UEL 6.9%
Autoignition Temperature: 487°F (253°C)

FIRST AID

Move victim to fresh air. Call emergency medical care. Apply artificial respiration if victim is not breathing. If breathing is difficult, give oxygen. Remove and isolate contaminated clothing and shoes. In case of contact with substance, immediately flush skin or eyes with running water for at least 20 minutes. Wash skin with soap and water. Keep victim warm and quiet. Ensure that medical personnel are aware of the material(s) involved and take precautions to protect themselves.

HYDRAZINE H:160

SYNONYMS: AMERZINE; DIAMIDE; DIAMINE; DIAMINE, HYDRAZINE BASE; EEC No. 007-008-01-0 (solution); HIDRAZINA (Spanish); HYDRAZINE, ANHYDROUS; HYDRAZINE BASE; HYDRAZYNA (Polish); SCAV-OX; SCAV-OX II; ULTRA PURE

IDENTIFICATION

CAS:	302-01-2
ERG ID:	UN 2029 (anhydrous or aqueous solution >64%) UN 2030 (hydrate or solutions with not less than 37%, but not >64%) UN 3293 (solution not >37% hydrazine)
ERG Guide:	132 (UN 2029) 153 (UN 2030) 152 (UN 3293)
ERG Guide Category:	132 FLAMMABLE LIQUIDS, CORROSIVE 153 TOXIC AND/OR CORROSIVE (COMBUSTIBLE) 152 TOXIC (COMBUSTIBLE)
Molecular Formula:	H_4N_2

RTECS No.:	MU7175000
Properties:	Highly flammable, oily , fuming liquid. Colorless. Ammonialike, fishy odor. Soluble in water. Freezing point is 36°F (2°C).

HEALTH AND SAFETY INFORMATION

Poison. Breathing the vapors can cause death. Skin contact or swallowing the liquid can cause death. Skin and eye contact causes severe burns and blindness. Repeated or prolonged exposure may cause lethargy, vomiting, tremors; itching and burning eyes and skin, conjuntivitis, contact dermatitis; cancer; liver, kidney, nervous system damag; red cell damage and anemia; bronchitis; skin allergy.

IARC Evaluation: Group 2B, possible human carcinogen; human evidence, inadequate; animal evidence, sufficient.
NTP anticipated human carcinogen.
IDLH: 50 ppm; potential human carcinogen
Odor Threshold: 3.6 ppm
OSHA Table Z-1-A Air Contaminant
NIOSH Criteria Document: NIOSH 78-172

Exposure Limits:	**TWA**	**Ceiling**
ACGIH:[(1)]	0.01 ppm; 0.013 mg/m^3	
NIOSH:[(2)]		0.03 ppm; 0.04 mg/m^3/2 hrs
OSHA: [(3)]	1 ppm; 1.3 mg/m^3	

[(1)] animal carcinogen
[(2)] potential carcinogen, reduce exposure to lowest feasible level.
[(3)] Skin contact contributes significantly in overall exposure.

RECOMMENDATIONS FOR RESPIRATOR SELECTION: NIOSH

At any concentrations above the NIOSH REL, or where there is no REL, at any detectable concentration: SCBAF:PD,PP (any self-contained breathing apparatus that has a full facepiece and is operated in a pressure-demand or other positive-pressure mode);or SAF:PD,PP:ASCBA (any supplied-air respirator that has a full facepiece and is operated in a pressure-demand or other positive-pressure mode in combination with an auxiliary, self-contained breathing apparatus operated in a pressure-demand or other positive pressure mode)

Escape: SCBAE (any appropriate escape-type, self-contained breathing apparatus).

CHEMICAL REACTIVITY

Forms explosive mixture with air. A highly reactive reducing agent and a strong base; ignites spontaneously upon contact with porous materials such as wood, cloth, rusting metal, or soil. Oxides of iron or copper, manganese, lead, copper, or their alloys can cause fire and explosions. Violent reaction with acids and halogens. Very corrosive to glass and rubber. Attacks cork, some plastics, and coatings.

FIRE INFORMATION

Combustion products include toxic nitrogen oxides. Highly flammable; may continue to burn in the absence of air. Firefighting gear (including SCBA) does not provide adequate protection. If exposure occurs, remove and isolate gear immediately and thoroughly decontaminate personnel. Vapors are heavier than air and will collect in low areas. Vapors in confined areas may explode when exposed to fire. Vapors may travel long distances to ignition sources and flash back. Storage containers and parts of containers may rocket great distances, in many directions. If material or contaminated runoff enters waterways, notify downstream users of potentially contaminated waters. Notify local health and fire officials and pollution control agencies. Use dry chemicals, foam, carbon dioxide.

NFPA Hazard Classifications:

Health Hazards (Blue) 3
Flammability (Red) 3
Reactivity (Yellow) 2

Flash Point: 100°F (38°C) o.c.
Flammable Limits in Air: LEL 2.9%; UEL 98%
Autoignition Temperature: 518°F (270°C)

FIRST AID

FLAMMABLE LIQUIDS, CORROSIVE: Move victim to fresh air. Call emergency medical care. Apply artificial respiration if victim is not breathing. Do not use mouth-to-mouth method if victim swallowed or inhaled the substance; induce artifical respiration with the aid of a pocket mask equipped with a one-way valve or other proper respiratory medical device. If breathing is difficult, give oxygen. Remove and isolate contaminated clothing and shoes. In case of contact with substance, immediately flush skin or eyes with running water for at least 20 minutes. Keep victim warm and quiet. Effects of exposure (inhalation, ingestion, or skin contact) to substances may be delayed. Ensure that medical personnel are aware of the material(s) involved and take precautions to protect themselves.

TOXIC AND/OR CORROSIVE (COMBUSTIBLE): Move victim to fresh air. Call emergency medical care. Apply artificial respiration if victim is not breathing. Do not use mouth-to-mouth method if victim ingested or inhaled the substance; induce artificial respiration with the aid of a pocket mask equipped with a one-way valve or other proper respiratory medical device. Administer oxygen if breathing is difficult. Remove and isolate contaminated clothing and shoes. In case of contact with substance, immediately flush skin or eyes with running water for at least 20 minutes. For minor skin contact, avoid spreading material on unaffected skin. Keep victim warm and quiet. Effects of exposure (inhalation, ingestion, or skin contact) may be delayed. Ensure that medical personnel are aware of the material(s) involved and take precautions to protect themselves.

TOXIC (COMBUSTIBLE): Move victim to fresh air. Call emergency medical care. Apply artificial respiration if victim is not breathing. Do not use mouth-to-mouth method if victim ingested or inhaled the substance; induce artificial respiration with the aid of a pocket mask equipped with a one-way valve or other proper respiratory medical device. Administer oxygen if breathing is difficult. Remove and isolate contaminated clothing and shoes. In case of contact with substance, immediately flush skin or eyes with running water for at least 20 minutes. For minor skin contact, avoid spreading material on unaffected skin. Keep victim warm and quiet. Effects of exposure (inhalation, ingestion, or skin contact) may be delayed. Ensure that medical personnel are aware of the material(s) involved and take precautions to protect themselves.

ENVIRONMENTAL INFORMATION

CLEAN AIR ACT: Hazardous Air Pollutants (Title I, Part A, Section 112); Accidental Release Prevention/Flammable substances, (Section 112[r], Table 3), TQ = 15,000 lb (6,810 kg).
EPA HAZARDOUS WASTE NUMBER (RCRA No.): U133
RCRA Section 261 Hazardous Constituents.
EPCRA Section 302 Extremely Hazardous Substances: TPQ = 1,000 lb (454 kg).
EPCRA Section 304 Reportable Quantity (RQ): EHS, CERCLA, 1 lb. (0.454 kg.).
EPCRA Section 313: Form R *de minimus* concentration reporting level: 0.1%.
CALIFORNIA'S PROPOSITION 65: Carcinogen.

HYDROBENZOIC ACID — H:195

SYNONYMS: ACIDO-m-IDROSSIBENZOICO (Italian); 3-CARBOXYPHENOL; EINECS No. 202-726-5; M-HBA; 3-HYDROXYBENZOESAEURE (German); m-HYDROXYBENZOIC ACID; m-SALICYLIC ACID

IDENTIFICATION

CAS:	99-06-9
Molecular Formula:	$C_7H_6O_3$
RTECS No.:	DH1750000
Properties:	Powdery solid. White. Very soluble in water.

HEALTH AND SAFETY INFORMATION

RECOMMENDATIONS FOR RESPIRATOR SELECTION

At any concentrations above the NIOSH REL, or where there is no REL, at any detectable concentration: SCBAF:PD,PP (any self-contained breathing apparatus that has a full facepiece and is operated in a pressure-demand or other positive-pressure mode);or SAF:PD,PP:ASCBA (any supplied-air respirator that has a full facepiece and is operated in a pressure-demand or other positive-pressure mode in combination with an auxiliary, self-contained breathing apparatus operated in a pressure-demand or other positive-pressure mode).

Escape: GMFOV [any air-purifying, full-facepiece respirator (gas mask) with a chin-style, front- or back-mounted organic vapor canister];or SCBAE (any appropriate escape-type, self-contained breathing apparatus).

CHEMICAL REACTIVITY

Incompatible with oxidizers, fluorine, lead diacetate.

FIRE INFORMATION

If material or contaminated runoff enters waterways, notify downstream users of potentially contaminated waters. Notify local health and fire officials and pollution control agencies. Use dry chemicals, alcohol-resistant foam, carbon dioxide.

NFPA Hazard Classifications: Not listed

FIRST AID

Move victim to fresh air. Call emergency medical care. Apply artificial respiration if victim is not breathing. Administer oxygen if breathing is difficult. Remove and isolate contaminated clothing and shoes. In case of contact with substance, immediately flush skin or eyes with running water for at least 20 minutes. Wash skin with soap and water. Keep victim warm and quiet. Effects of exposure (inhalation, ingestion, or skin contact) to substance may be delayed. Ensure that medical personnel are aware of the material(s) involved and take precautions to protect themselves.

HYDROBROMIC ACID H:200

SYNONYMS: ACIDE BROMHYDRIQUE (French); ACIDO BROMIDRICO (Italian); ANHYDROUS HYDROBROMIC ACID; HYDROGEN BROMIDE; BROMOWODOR (Polish); BROMURO de HIDROGENO (Spanish); BROMWASSERSTOFF (German); BROOMWATERSTOF (Dutch); EEC No. 035-002-00-0; HYDROBROMIC ACID, ANHYDROUS; HYDROGEN BROMIDE, ANHYDROUS

IDENTIFICATION

CAS:	10035-10-6
ERG ID:	UN 1048 (anhydrous) UN 1788 (solution)
ERG Guide:	125 (UN 1048, anhydrous) 154 (UN 1788, solution)
ERG Guide Category:	154 TOXIC AND/OR CORROSIVE (NON-COMBUSTIBLE) 125 GASES- CORROSIVE
Molecular Formula:	HBr
RTECS No.:	MW3850000

Properties: Nonflammable gas. Colorless; but may be shipped as a pale yellow liquid under pressure. Irritating odor. Sinks and mixes with water; heat and poisonous white vapor cloud is produced.

HEALTH AND SAFETY INFORMATION

Poison. Breathing the gas can kill you. Skin and eye contact causes severe burns and blindness. Can cause pulmonary edema, which can be delayed for several hours; there is a risk of death in serious cases. Prolonged or repeated exposure may cause irritation of the nose and throat with mucous production; chronic indigestion, lung damage; loss of the sense of smell.

IDLH: 30 ppm
Odor Threshold: 1.99 ppm
OSHA Table Z-1-A Air Contaminant as hydrobromic acid
OSHA Process Safety Management of Highly Hazardous Chemicals (29 CFR 1910.119, Appendix A), Threshold Quantity: 5,000 pounds as hydrogen bromide.

Exposure Limits:	TWA	Ceiling
ACGIH:		3 ppm; 9.9 mg/m^3
NIOSH:		3 ppm; 10 mg/m^3
OSHA:	3 ppm; 10 mg/m^3	

RECOMMENDATIONS FOR RESPIRATOR SELECTION: NIOSH/OSHA

30 ppm: SA:CF (any supplied-air respirator operated in a continuous-flow mode);or PAPRAG (any powered, air-purifying respirator with acid gas cartridge(s).);or GMFAG (any air-purifying, full-facepiece respirator (gas mask) with a chin-style, front- or back-mounted organic vapor canister);or SCBAF (any self-contained breathing apparatus with a full facepiece);or SAF (any supplied-air respirator with a full facepiece)

Emergency or Planned Entry into Unknown Concentrations or Idlh Conditions:
SCBAF:PD,PP (any self-contained breathing apparatus that has a full facepiece and is operated in a pressure-demand or other positive-pressure mode);or SAF:PD,PP:ASCBA (any supplied-air respirator that has a full facepiece and is operated in a pressure-demand or other positive-pressure mode in combination with an auxiliary, self-contained breathing apparatus operated in a pressure-demand or other positive pressure mode.)

Escape: GMFAG (any air-purifying, full-facepiece respirator (gas mask) with a chin-style, front- or back-mounted organic vapor canister);or SCBAE (any appropriate escape-type, self-contained breathing apparatus).
Note: Substance causes eye irritation or damage; eye protection needed.

CHEMICAL REACTIVITY

Incompatible with aliphatic amines, alkanolamines, alkylene oxides, aromatic amines, amides, ammonia, ammonium hydroxide, bases, calcium oxide, epichlorohydrin, fluorine, isocyanates, oleum, organic anhydrides, sulfuric acid, sodium tetrahydroborate, strong oxidizers, vinyl acetate, water. Attacks most metals with the formation of highly flammable hydrogen gas.

FIRE INFORMATION

Firefighting gear (including SCBA) does not provide adequate protection. If exposure occurs, remove and isolate gear immediately and thoroughly decontaminate personnel.Gases are heavier than air and will collect in low areas. Gases in confined areas may explode when exposed to fire. Gases may travel long distances to ignition sources and flash back. Storage containers and parts of containers may rocket great distances, in many directions. If material or contaminated runoff enters waterways, notify downstream users of potentially contaminated waters. Notify local health and fire officials and pollution control agencies. Do not extinguish the fire unless the flow of the gas can be stopped and any remaining gas is out of the line. Specially trained personnel may use fog lines to cool exposures and let the fire burn itself out. Use dry chemicals, foam, carbon dioxide. From a secure, explosion-proof location, use water spray to cool exposed containers. If cooling streams are ineffective (venting sound increases in volume and pitch, tank discolors, or shows any signs of deforming), withdraw immediately to a secure position.

NFPA Hazard Classifications:

Health Hazards (Blue) 3
Flammability (Red) 0
Reactivity (Yellow) 0

FIRST AID

GASES- CORROSIVE: Move victim to fresh air. Call emergency medical care. Apply artificial respiration if victim is not breathing. Do not use mouth-to-mouth method if victim ingested or inhaled the substance; induce artificial respiration with the aid of a pocket mask equipped with a one-way valve or other proper respiratory medical device. Administer oxygen if breathing is difficult. Remove and isolate contaminated clothing and shoes. In case of contact with liquefied gas, thaw frosted parts with lukewarm water. In case of contact with substance, immediately flush skin or eyes with running water for at least 20 minutes. Keep victim warm and quiet. Keep victim under observation. Effects of contact or inhalation may be delayed. Ensure that medical personnel are aware of the material(s) involved and take precautions to protect themselves.

TOXIC AND/OR CORROSIVE (NON-COMBUSTIBLE): Move victim to fresh air. Call emergency medical care. Apply artificial respiration if victim is not breathing. Do not use mouth-to-mouth method if victim ingested or inhaled the substance; induce artificial respiration with the aid of a pocket mask equipped with a one-way valve or other proper respiratory medical device. Administer oxygen if breathing is difficult. Remove and isolate contaminated clothing and shoes. In case of contact with substance, immediately flush skin or eyes with running water for at least 20 minutes. For minor skin contact, avoid spreading material on unaffected skin. Keep victim warm and quiet. Effects of exposure (inhalation, ingestion, or skin contact) may be delayed. Ensure that medical personnel are aware of the material(s) involved and take precautions to protect themselves. Move victim to fresh air. Call emergency medical care. Apply artificial respiration if victim is not breathing. Do not use mouth-to-mouth method if victim ingested or inhaled the substance; induce artificial respiration with the aid of a pocket mask equipped with a one-way valve or other proper respiratory medical device. Administer oxygen if breathing is difficult. Remove and isolate contaminated clothing and shoes. In case of contact with substance, immediately flush skin or eyes with running water for at least 20 minutes. For minor skin contact, avoid spreading material on unaffected skin. Keep victim warm and quiet. Effects of exposure (inhalation, ingestion, or skin contact) may be delayed. Ensure that medical personnel are aware of the material(s) involved and take precautions to protect themselves.

HYDROCHLORIC ACID H:220

SYNONYMS: ACIDE CHLORHYDRIQUE (French); ACIDO CLORIDRICO (Italian); ANHYDROUS HYDROCHLORIC ACID; AQUEOUS HYDROGEN CHLORIDE; CHLOORWATERSTOF (Dutch); CHLOROHYDRIC ACID; CHLOROWODOR (Polish); CHLORWASSERSTOFF (German); CLORURO de HIDROGENO (Spanish); EEC No. 017-002-00-2; EINECS No. 231-595-7; HYDROCHLORIC ACID, ANHYDROUS; HYDROCHLORIDE; HYDROGEN CHLORIDE; MAGNIFLOC 509C FLOCCULANT; MURIATIC ACID; SPIRITS OF SALT

IDENTIFICATION

CAS:	7647-01-0
ERG ID:	UN 1050 (anhydrous); UN 1789 (solution); UN 2186 (refrigerated)
ERG Guide:	125 (anhydrous); 157 (solution); 125(refrigerated)
ERG Guide Category:	125 GASES-CORROSIVE 157 TOXIC AND/OR CORROSIVE (NONCOMBUSTIBLE/ WATER SENSITIVE)
Molecular Formula:	HCl
RTECS No.:	MW4025000
Properties:	Nonflammable, watery liquid or gas. Colorless to yellow. Sharp, pungent, irritating odor. Sinks and mixes with water. Irritating vapor is produced. Shipped as a liquefied compressed gas often used in an aqueous solution. May fume upon release, depending on concentration.

HEALTH AND SAFETY INFORMATION

Poison. Breathing the gas can cause death. Skin and eye contact causes severe burns and blindness. A primary concern of hydrochloric acid is its aerosol form. Chronic or prolonged exposure to gas (above the OSHA PEL) or to mist has been associated with changes in pulmonary function, chronic inflammation of the bronchi, nasal ulceration and symptoms resembling acute viral infection of the upper respiratory tract as well as inflammation of the skin, discoloration and erosion of dental enamel, and inflammation of the eye membrane. Also, Long-term exposure to levels greater than 5 ppm of hydrogen chloride can cause some dental erosion.

IDLH: 50 ppm
Odor Threshold: 6.31 ppm; sharp, choking odor at 1 to 5 ppm.
OSHA Table Z-1-A Air Contaminant
OSHA Process Safety Management of Highly Hazardous Chemicals (29 CFR 1910.119, Appendix A), Threshold Quantity: 5,000 lb as hydrochloric acid, anhydrous, or hydrogen chloride.

Exposure Limits:	**CEILING**
ACGIH:	5 ppm; 7.5 mg/m^3
NIOSH:	5 ppm; 7 mg/m^3
OSHA:	5 ppm; 7 mg/m^3

RECOMMENDATIONS FOR RESPIRATOR SELECTION: NIOSH/OSHA

50 ppm: CCRS [any chemical cartridge respirator with cartridge(s) providing protection against the compound of concern)];or GMFS [any air-purifying, full-facepiece respirator (gas mask) with a chin-style, front- or back-mounted canister providing protection against the compound of concern];or PAPRS (any powered, air-purifying respirator with cartridge(s) providing protection against the compound of concern);or SA (any supplied-air respirator);or SCBAF (any self-contained breathing apparatus with a full facepiece).

Emergency or Planned Entry into Unknown Concentrations or IDLH Conditions: SCBAF:PD,PP (any self-contained breathing apparatus that has a full facepiece and is operated in a pressure-demand or other positive-pressure mode);or SAF:PD,PP:ASCBA (any supplied-air respirator that has a full facepiece and is operated in a pressure-demand or other positive-pressure mode in combination with an auxiliary, self-contained breathing apparatus operated in a pressure-demand or other positive-pressure mode).

Escape: GMFAG [any air-purifying, full-facepiece respirator (gas mask) with a chin-style, front- or back-mounted organic vapor canister];or SCBAE (any appropriate escape-type, self-contained breathing apparatus).
Note: Substance reported to cause eye irritation or damage; may require eye protection.

CHEMICAL REACTIVITY

Reacts with acetic anhydride, aliphatic amines, alkanolamines, alkylene oxides, aromatic amines, amides, 2-aminoethanol, ammonia, ammonium hydroxide, calcium phosphide, chlorosulfonic acid, ethylene diamine, ethyleneimine, epichlorohydrin, isocyanates, metal acetylides, oleum, organic anhydrides, perchloric acid, 3-propiolactone, uranium phosphide, sulfuric acid, sodium hydroxide and other bases, strong oxidizers, vinyl acetate, and vinylidene fluoride. Forms flammable hydrogen gas on contact with most common metals. Attacks some plastics, rubber, and coatings.

FIRE INFORMATION

Firefighting gear (including SCBA) does not provide adequate protection. If exposure occurs, remove and isolate gear immediately and thoroughly decontaminate personnel. Gases are heavier than air and will collect in low areas. Gases in confined areas may explode when exposed to fire. Gases may travel long distances to ignition sources and flash back. Storage containers and parts of containers may rocket great distances, in many directions. If material or contaminated runoff enters waterways, notify downstream users of potentially contaminated waters. Notify local health and fire officials and pollution control agencies. Do not extinguish the fire unless the flow of the gas can be stopped and any remaining gas is out of the line. Specially trained personnel may use fog lines to cool exposures and let the fire burn itself out. Use dry chemicals, foam, carbon dioxide. DO NOT apply water to cryogenic liquid containers. If cooling streams are ineffective (venting sound increases in volume and pitch, tank discolors, or shows any signs of deforming), withdraw immediately to a secure position.

NFPA Hazard Classifications:

Health Hazards (Blue) 3
Flammability (Red) 0
Reactivity (Yellow) 0

FIRST AID

GASES- CORROSIVE (125): Move victim to fresh air. Call emergency medical care. Apply artificial respiration if victim is not breathing. Do not use mouth-to-mouth method if victim ingested or inhaled the substance; induce artificial respiration with the aid of a pocket mask equipped with a one-way valve or other proper respiratory medical device. Administer oxygen if breathing is difficult. Remove and isolate contaminated clothing and shoes. In case of contact with liquefied gas, thaw frosted parts with lukewarm water. In case of contact with substance, immediately flush skin or eyes with running water for at least 20 minutes. Keep victim warm and quiet. Keep victim under observation. Effects of contact or inhalation may be delayed. Ensure that medical personnel are aware of the material(s) involved and take precautions to protect themselves.

TOXIC AND/OR CORROSIVE (NONCOMBUSTIBLE/WATER SENSITIVE) (157): Move victim to fresh air. Call emergency medical care. Apply artificial respiration if victim is not breathing. Do not use mouth-to-mouth method if victim ingested or inhaled the substance; induce artificial respiration with the aid of a pocket mask equipped with a one-way valve or other proper respiratory medical device. Administer oxygen if breathing is difficult. Remove and isolate contaminated clothing and shoes. In case of contact with substance, immediately flush skin or eyes with running water for at least 20 minutes. For minor skin contact, avoid spreading material on unaffected skin. Keep victim warm and quiet. Effects of exposure (inhalation, ingestion, or skin contact) may be delayed. Ensure that medical personnel are aware of the material(s) involved and take precautions to protect themselves.

ENVIRONMENTAL INFORMATION

Ecologically, accidental release of solution forms of hydrochloric acid may adversely affect aquatic life by including a transient lowering of the pH of surface waters. Releases of hydrochloric acid to surface waters and soils will be neutralized to an extent due to the buffering capacities of both systems. The extent of these reactions will depend on the characteristics of the specific environment.

Clean Air Act: Hazardous Air Pollutants (Title I, Part A, Section 112); (anhydrous or gas) Accidental Release Prevention/Flammable substances, (Section 112[r], Table 3), TQ = 5,000 lb (2,270 kg).

Clean Water Act: Section 311 Hazardous Substances/RQ (same as CERCLA); Section 313 Priority Chemicals.
EPCRA Section 304: Reportable Quantity (RQ): CERCLA, 5,000 lb (2,270 kg).
EPCRA Section 313: Form R *de minimis* concentration reporting level: 1.0%.

Note: Nonaerosol forms of hydrochloric acid have been deleted from EPCRA/SARA 313 reporting, 7/29/96 (FR vol. 61, No. 146, p. 39356-39357).

HYDROCYANIC ACID H:240

SYNONYMS: ACIDE CYANHYDRIQUE (French); ACIDO CIANIDRICO (Italian); AERO LIQUID HCN; BLAUSAEURE (German); BLAUWZUUR (Dutch); CIANURO de HIDROGENO (Spanish); CYAANWATERSTOF (Dutch); CYANWASSERSTOFF (German); CYCLON; CYCLONE B; CYJANOWODOR (Polish); EEC No. 006-006-00-X; EINECS No. 200-821-6; FORMONITRILE; HYDROGEN CYANIDE; PRUSSIC ACID; ZACLON DISCOIDS; ZYKLON

IDENTIFICATION

CAS:	74-90-8
ERG ID:	UN 1051 (anhydrous); UN 1613 (solution)
ERG Guide:	117 (anhydrous); 154 (solution)
ERG Guide Category:	117 GASES-TOXIC-FLAMMABLE (EXTREME HAZARD) 154 TOXIC AND/OR CORROSIVE (NON-COMBUSTIBLE)
Molecular Formula:	HCN
RTECS No.:	MW6825000
Properties:	Extremely flammable gas or liquid. Because its boiling point 78°F (26°C) is near room temperature, hydrogen cyanide may be a colorless gas or colorless liquid. Transported in red and white candy-striped containers. It is also sold as a water solution at a strength of 2% (USP) and commercial strengths of 5% and 10%. Colorless or pale blue. Bitter almond odor; some people cannot smell it. Sinks and mixes with water. Poisonous, flammable vapor is produced and rises.

HEALTH AND SAFETY INFORMATION

Poison. Breathing the vapors or skin contact can cause death. Chronically exposed workers may complain of headache, eye irritation, easy fatigue, chest discomfort, palpitations, loss of appetite. Itching scarlet rash, red bumps, and severe nose itch leading to bleeding and possibly holes in the nose may result from long-term exposure to hydrogen cyanide. Headache, nausea, vomiting, weakness, and enlarged thyroid gland have also been reported at exposures from 4 to 12 ppm.

IDLH: 50 ppm.

Odor Threshold: 0.063 ppm. Odor is not a reliable indicator of the presence of toxic amounts of vapor. Almond odor reported at >1 ppm; inadequate warning for acute or chronic exposure. Perception of odor is a genetic trait (20% to 40% of the general population cannot detect hydrogen cyanide); also, rapid olfactory fatigue can occur.

OSHA Table Z-1-A Air Contaminant

OSHA Process Safety Management of Highly Hazardous Chemicals (29CFR 1910.119, Appendix A), Threshold Quantity: 1,000 pounds (anhydrous).

NIOSH Criteria Document: NIOSH 78-212 NITRILES

Exposure Limits:	TWA	CEILING	STEL
ACGIH:[(1)]		4.7 ppm; 5 mg/m^3	
NIOSH:[(1)]			4.7 ppm; 5 mg/m^3
OSHA:[(1)]		10 ppm;11 mg/m^3	

[(1)]Skin contact contributes significantly to overall exposure.

RECOMMENDATIONS FOR RESPIRATOR SELECTION: NIOSH/OSHA

47 ppm: SA (any supplied-air respirator).
50 ppm: SA:CF (any supplied-air respirator operated in a continuous-flow mode);or SCBAF (any self-contained breathing apparatus with a full facepiece);or SAF (any supplied-air respirator with a full facepiece).

Emergency or Planned Entry into Unknown Concentrations or IDLH Conditions: SCBAF:PD,PP (any self-contained breathing apparatus that has a full facepiece and is operated in a pressure-demand or other positive-pressure mode);or SAF:PD,PP:ASCBA (any supplied-air respirator that has a full facepiece and is operated in a pressure-demand or other positive-pressure mode in combination with an auxiliary, self-contained breathing apparatus operated in a pressure-demand or other positive-pressure mode).

Escape: GMFS [any air-purifying, full-facepiece respirator (gas mask) with a chin-style, front- or back-mounted canister providing protection against the compound of concern];or SCBAE (any appropriate escape-type, self-contained breathing apparatus).

CHEMICAL REACTIVITY

Unless stabilized and maintained, samples stored more than 90 days are hazardous; may polymerize to form a dark brown explosive powder. May polymerize at 122-140°F (50-60°C). Samples containing more than 2-5% water are less stable than dry material. Can be self-reactive, forming an explosive mixture with air. Heat, moisture, or contact with amines or strong bases can cause, or speed up, polymerization. Incompatible with acetaldehyde. Oxidizers may cause fire and explosions. Attacks some plastics, rubber, and coatings.

FIRE INFORMATION

Combustion products are less toxic than the material itself. May react with itself without warning with explosive violence.
Firefighting gear (including SCBA) does not provide adequate protection. If exposure occurs, remove and isolate gear immediately and thoroughly decontaminate personnel. Gases are heavier than air and will collect in low areas. Gases in confined areas may explode when exposed to fire. Gases may travel long distances to ignition sources and flash back. Storage containers and parts of containers may rocket great distances, in many directions. If material or contaminated runoff enters waterways, notify downstream users of potentially contaminated waters. Notify local health and fire officials and pollution control agencies. Do not extinguish the fire unless the flow of the gas can be stopped and any remaining gas is out of the line. Specially trained personnel may use fog lines to cool exposures and let the fire burn itself out. Use dry chemicals, foam, carbon dioxide. From a secure, explosion-proof location, use water spray to cool exposed containers. If cooling streams are ineffective (venting sound increases in volume and pitch, tank discolors, or shows any signs of deforming), withdraw immediately to a secure position.

NFPA Hazard Classifications:

Health Hazards (Blue) 4
Flammability (Red) 4
Reactivity (Yellow) 2

Flash Point: 0°F (-18°C), c.c (96%)
Flammable Limits in Air: LEL 5.6%; UEL 40.0%
Autoignition Temperature: 1,004°F (540°C)

FIRST AID

GASES-TOXIC-FLAMMABLE (EXTREME HAZARD) (117): Move victim to fresh air. Call emergency medical care. Apply artificial respiration if victim is not breathing. Do not use mouth-to-mouth method if victim ingested or inhaled the substance; induce artificial respiration with the aid of a pocket mask equipped with a one-way valve or other proper respiratory medical device. Administer oxygen if breathing is difficult. Remove and isolate contaminated clothing and shoes. In case of contact with substance, immediately flush skin or eyes with running water for at least 20 minutes. In case of contact with liquefied gas, thaw frosted parts with lukewarm water. Keep victim warm and quiet. Keep victim under observation. Effects of contact or inhalation may be delayed. Ensure that medical personnel are aware of the material(s) involved and take precautions to protect themselves.

TOXIC AND/OR CORROSIVE (NONCOMBUSTIBLE) (154): Move victim to fresh air. Call emergency medical care. Apply artificial respiration if victim is not breathing. Do not use mouth-to-mouth method if victim ingested or inhaled the substance; induce artificial respiration with the aid of a pocket mask equipped with a one-way valve or other proper respiratory medical device. Administer oxygen if breathing is difficult. Remove and isolate contaminated clothing and shoes. In case of contact with substance, immediately flush skin or eyes with running water for at least 20 minutes. For minor skin contact, avoid spreading material on unaffected skin. Keep victim warm and quiet. Effects of exposure (inhalation, ingestion, or skin contact) may be delayed. Ensure that medical personnel are aware of the material(s) involved and take precautions to protect themselves.

ENVIRONMENTAL INFORMATION as hydrogen cyanide

Clean Air Act: Accidental Release Prevention/Flammable substances, (Section 112[r], Table 3), TQ = 2,500 lb (1,135 kg).
Clean Water Act: Section 311 Hazardous Substances/RQ (same as CERCLA); Section 313 Priority Chemicals.
EPA Hazardous Waste Number (RCRA No.): P063.
Marine Pollutant (49CFR, Subchapter 172.101, Appendix B).
EPCRA Section 302 Extremely Hazardous Substances: TPQ = 100 lb (45.4 kg).
EPCRA Section 304: Reportable Quantity (RQ): CERCLA, 10 lb (4.54 kg).
EPCRA Section 313: Form R *de minimis* concentration reporting level: 1.0%.
Marine Pollutant (49CFR, Subchapter 172.101, Appendix B).

HYDROFLUORIC ACID H:260

SYNONYMS: ANHYDROUS HYDROGEN FLUORIDE; ANHYDROFLUORIC ACID; AQUEOUS HYDROGEN FLUORIDE; BOE (BUFFERED OXIDE ETCH); C-P 8 SOLUTION; DOPED POLY ETCH; EEC No. 009-002-00-6; 231-634-8; FLUERWATERSTOF (Dutch); FLUORURO de HIDROGENO (Spanish); FLUORURO di IDROGENO (Italian); FLUSSAEURE (German); FLUSSYRE (French); FRECKLE ETCH; HYDROFLUORIDE; HF-A; HYDROGEN FLUORIDE; IMPLANTER FUMER; KTI BUFFERED OXIDE ETCH 6:1; KTI BUFFERED OXIDE ETCH 50:1; KTI OXIDE ETCH 5:1; KTI OXIDE ETCH 10:1; KTI OXIDE ETCH 50:1; MAE ETCHANTS; MIXED ACID ETCH; POLY ETCH 95%; RDX-221 ALUMETCH LK; RUBIGINE; SILICON ETCH SOLUTION; WRIGHT ETCH

IDENTIFICATION

CAS:	7664-39-3
ERG ID:	UN 1052 (anhydrous); UN 1790 (solution)
ERG Guide:	125 (anhydrous); 157 (solution)
ERG Guide Category:	125 GASES– CORROSIVE 157 TOXIC AND/OR CORROSIVE (NONCOMBUSTIBLE/ WATER SENSITIVE)
Molecular Formula:	HF
RTECS No.:	MW7875000
Properties:	Nonflammable gas or fuming, watery liquid [below 67°F (19°C)]. Colorless to green; weak solutions have appearance of water. Shipped as a 40% to 70% solution of hydrogen fluoride in water. Sharp, irritating odor. Sinks and mixes with water, with release of heat and toxic hydrofluoric acid and heat. The pure compound is a crystalline solid below 12°F (-11°C). Often used in aqueous solution. Gas is slightly lighter than air.

HEALTH AND SAFETY INFORMATION

Poison. Breathing the vapor can cause death. Skin and eye contact causes blindness and severe burns, which may not be immediately painful. Long-term hydrogen fluoride exposure has been reported to damage the kidneys and liver. Repeated exposure to fluoride chemicals may cause stiffness in muscles or ligaments, and even crippling.

IDLH: 30 ppm as F
Odor Threshold: 0.036 ppm . Disagreeable, pungent odor at <1 ppm; irritation of eyes and throat at 3 ppm.
OSHA Table Z-1-A and Z-2 Air Contaminant
OSHA Process Safety Management of Highly Hazardous Chemicals (29 CFR1910.119, Appendix A), Threshold Quantity: 1,000 pounds as hydrofluoric acid, anhydrous or hydrogen fluoride.
NIOSH Criteria Document: NIOSH 76-143

Exposure Limits:	**TWA**	**CEILING**
ACGIH:		3 ppm; 2.3 mg/m^3 as F
NIOSH:	3 ppm; 2.5 mg/m^3 as F	6 ppm; 5 mg/m^3/15 min as F
OSHA:	3 ppm; 2.5 mg/m^3 as F	

RECOMMENDATIONS FOR RESPIRATOR SELECTION: NIOSH/OSHA

30 ppm: CCRS (any chemical cartridge respirator with cartridge(s) providing protection against the compound of concern);or PAPRS (any powered, air-purifying respirator with cartridge(s) providing protection against the compound of concern);or GMFS [any air-purifying, full-facepiece respirator (gas mask) with a chin-style, front- or back-mounted canister providing protection against the compound of concern];or SA (any supplied-air respirator);or SCBA (any self-contained breathing apparatus).

Emergency or Planned Entry into Unknown Concentrations or IDLH Conditions:
SCBAF:PD,PP (any self-contained breathing apparatus that has a full facepiece and is operated in a pressure-demand or other positive-pressure mode);or SAF:PD,PP:ASCBA (any supplied-air respirator that has a full facepiece and is operated in a pressure-demand or other positive-pressure mode in combination with an auxiliary, self-contained breathing apparatus operated in a pressure-demand or other positive-pressure mode).

Escape: GMFS [any air-purifying, full-facepiece respirator (gas mask) with a chin-style, front- or back-mounted canister providing protection against the compound of concern];or SCBAE (any appropriate escape-type, self-contained breathing apparatus).
Note: Substance reported to cause eye irritation or damage; may require eye protection.

CHEMICAL REACTIVITY

A strong acid. Reacts–possibly with violence–with acetic anhydride, aliphatic amines, alcohols, alkanolamines, alkylene oxides, aromatic amines, amides, 2-aminoethanol, ammonia, ammonium hydroxide, arsenic trioxide, bismuthic acid, calcium oxide, ethylene diamine, ethyleneimine, epichlorohydrin, isocyanates, metal acetylides, nitrogen trifluoride, oleum, organic anhydrides, oxygen difluoride, phosphorus pentoxide, sulfuric acid, sodium hydroxide and other bases, strong oxidizers, vinyl acetate, and vinylidene fluoride. Attacks glass, concrete, ceramics, some metals (forming flammable hydrogen gas), rubber, leather, plastics, and coatings; may cause some plastics to become brittle.

FIRE INFORMATION

Firefighting gear (including SCBA) does not provide adequate protection. If exposure occurs, remove and isolate gear immediately and thoroughly decontaminate personnel. Gases are heavier than air and will collect in low areas. Gases in confined areas may explode when exposed to fire. Gases may travel long distances to ignition sources and flash back. Storage containers and parts of containers may rocket great distances, in many directions. If material or contaminated runoff enters waterways, notify downstream users of potentially contaminated waters. Notify local health and fire officials and pollution control agencies. Do not extinguish the fire unless the flow of the gas can be stopped and any remaining gas is out of the line. Specially trained personnel may use fog lines to cool exposures and let the fire burn itself out. Use dry chemicals, foam, carbon dioxide. From a secure, explosion-proof location, use water spray to cool exposed containers. If cooling streams are ineffective (venting sound increases in volume and pitch, tank discolors, or shows any signs of deforming), withdraw immediately to a secure position.

NFPA Hazard Classifications:

Health Hazards (Blue) 4
Flammability (Red) 0
Reactivity (Yellow) 0

FIRST AID

GASES-CORROSIVE (125): Move victim to fresh air. Call emergency medical care. Apply artificial respiration if victim is not breathing. Do not use mouth-to-mouth method if victim ingested or inhaled the substance; induce artificial respiration with the aid of a pocket mask equipped with a one-way valve or other proper respiratory medical device. Administer oxygen if breathing is difficult. Remove and isolate contaminated clothing and shoes. In case of contact with liquefied gas, thaw frosted parts with lukewarm water. In case of contact with substance, immediately flush skin or eyes with running water for at least 20 minutes. Keep victim warm and quiet. Keep victim under observation. Effects of contact or inhalation may be delayed. Ensure that medical personnel are aware of the material(s) involved and take precautions to protect themselves.

TOXIC AND/OR CORROSIVE (NONCOMBUSTIBLE/WATER SENSITIVE) (157): Move victim to fresh air. Call emergency medical care. Apply artificial respiration if victim is not breathing. Do not use mouth-to-mouth method if victim ingested or inhaled the substance; induce artificial respiration with the aid of a pocket mask equipped with a one-way valve or other proper respiratory medical device. Administer oxygen if breathing is difficult. Remove and isolate contaminated clothing and shoes. In case of contact with substance, immediately flush skin or eyes with running water for at least 20 minutes. For minor skin contact, avoid spreading material on unaffected skin. Keep victim warm and quiet. Effects of exposure (inhalation, ingestion, or skin contact) may be delayed. Ensure that medical personnel are aware of the material(s) involved and take precautions to protect themselves.

ENVIRONMENTAL INFORMATION

Clean Air Act: Hazardous Air Pollutants (Title I, Part A, Section 112).
Clean Water Act: Section 311 Hazardous Substances/RQ (same as CERCLA); Section 313 Priority Chemicals.
EPA Hazardous Waste Number (RCRA No.): U134.
EPCRA Section 302 Extremely Hazardous Substances: TPQ = 100 lb (45.4 kg).
EPCRA Section 304: Reportable Quantity (RQ): CERCLA, 100 lb (45.4 kg).
EPCRA Section 313: Form R *de minimis* concentration reporting level: 1.0%.

HYDROFLUOROSILICIC ACID (25% OR LESS) H:270

SYNONYMS: ACIDO HIDROFLUOSILICICO (Spanish); EINECS No. 241-034-8; FLUOROSILICIC ACID; FLUOSILICIC ACID; HEXAFLUOROKISELSYRE (French); HEXAFLUOSILICIC ACID; SAND ACID

IDENTIFICATION

CAS:	16961-83-4
ERG ID:	UN 1778
ERG Guide:	154
ERG Guide Category:	SUBSTANCES-TOXIC AND/OR CORROSIVE (NONCOMBUSTIBLE)
Molecular Formula:	H_2F_6Si
RTECS No.:	VV8225000
Properties:	Liquid. Colorless to straw yellow. Unpleasant, sour odor. Soluble in water; releases heat. Freezes at 63°F (17°C).

HEALTH AND SAFETY INFORMATION

Poison. Breathing the vapors can cause death. Skin and eye contact can cause severe burns and blindness.

Exposure Limits:	**TWA**	**STEL**
ACGIH:	1 ppm; 1.6 mg/m^3 as fluorine	2 ppm; 3.1 mg/m^3 as fluorine
NIOSH:	0.1 ppm; 0.2 mg/m^3 as fluorine	
OSHA:	0.1 ppm; 0.2 mg/m^3 as fluorine	

RECOMMENDATIONS FOR RESPIRATOR SELECTION: NIOSH/OSHA as fluorine

Exposure Limits:	**TWA**	**CEILING**
ACGIH:		3 ppm; 2.3 mg/m^3 as F
NIOSH:	3 ppm; 2.5 mg/m^3 as F	6 ppm; 5 mg/m^3/15 min as F
OSHA:	3 ppm; 2.5 mg/m^3 as F	

RECOMMENDATIONS FOR RESPIRATOR SELECTION: as hydrogen fluoride

30 ppm: CCRS (any chemical cartridge respirator with cartridge(s) providing protection against the compound of concern);or PAPRS (any powered, air-purifying respirator with cartridge(s) providing protection against the compound of concern);or GMFS [any air-purifying, full-facepiece respirator (gas mask) with a chin-style, front- or back-mounted canister providing protection against the compound of concern];or SA (any supplied-air respirator);or SCBA (any self-contained breathing apparatus).

Emergency or Planned Entry into Unknown Concentrations or IDLH Conditions: SCBAF:PD,PP (any self-contained breathing apparatus that has a full facepiece and is operated in a pressure-demand or other positive-pressure mode);or SAF:PD,PP:ASCBA (any supplied-air respirator that has a full facepiece and is operated in a pressure-demand or other positive-pressure mode in combination with an auxiliary, self-contained breathing apparatus operated in a pressure-demand or other positive-pressure mode).

Escape: GMFS [any air-purifying, full-facepiece respirator (gas mask) with a chin-style, front- or back-mounted canister providing protection against the compound of concern];or SCBAE (any appropriate escape-type, self-contained breathing apparatus).
Note: Substance reported to cause eye irritation or damage; may require eye protection.

CHEMICAL REACTIVITY

Reacts with water and generates heat. Can react with strong acids to release hydrogen fluoride fumes. Will react with metals to release hydrogen gas. Will attack glass and materials containing silica; stoneware.

FIRE INFORMATION

Combustion products include corrosive and toxic hydrogen fluoride. Firefighting gear (including SCBA) does not provide adequate protection. If exposure occurs, remove and isolate gear immediately and thoroughly decontaminate personnel. Vapors are heavier than air and will collect in low areas. Storage containers and parts of containers may rocket great distances, in many directions.If material or contaminated runoff enters waterways, notify downstream users of potentially contaminated waters. Notify local health and fire officials and pollution control agencies. Use dry chemicals, foam, carbon dioxide.

NFPA Hazard Classifications:

Health Hazards (Blue) 3
Flammability (Red) 0
Reactivity (Yellow) 0

Flash Point: Not flammable

FIRST AID

Move victim to fresh air; call emergency medical care. Apply artificial respiration if victim is not breathing. Do not use mouth-to-mouth method if victim ingested or inhaled the substance; induce artificial respiration with the aid of a pocket mask equipped with a one-way valve or other proper respiratory medical device. Administer oxygen if breathing is difficult. Remove and isolate contaminated clothing and shoes. In case of contact with substance, immediately flush skin or eyes with running water for at least 20 minutes. For minor skin contact, avoid spreading material on unaffected skin. Keep victim warm and quiet. Effects of exposure (inhalation, ingestion, or skin contact) to substance may be delayed. Ensure that medical personnel are aware of the material(s) involved and take precautions to protect themselves.

HYDROGEN H:280

SYNONYMS: EEC No. 001-001-00-9; FORMING GAS; HYDROGEN, COMPRESSED; HIDROGENO (Spanish); HYDROGEN, REFRIGERATED LIQUID; LIQUID HYDROGEN; PARA HYDROGEN

IDENTIFICATION

CAS:	1333-74-0
ERG ID:	UN 1966 (Refrigerated) UN 1049 (Compressed)
ERG Guide:	115
ERG Guide Category:	GASES- FLAMMABLE (INCLUDING REFRIGERATED LIQUIDS)
Molecular Formula:	H_2
RTECS No.:	MW8900000
Properties:	Extremely flammable, compressed gas. Colorless. Odorless. Floats and boils on water; insoluble. Lighter than air.

HEALTH AND SAFETY INFORMATION

Contact with liquid may cause frostbite.

CHEMICAL REACTIVITY

Vapors form explosive or combustible mixture with air over a wide range of concentrations. Ignites easily with oxygen. Violent reaction with oxidizers, halogens, acetylene, bromine, chlorine, fluorine, nitrous oxide, and other gases. Mild steel and most iron alloys become brittle at liquid hydrogen temperatures.

FIRE INFORMATION

Burns with an invisible flame (even in the dark). High-pressure releases often ignite without any apparent source of ignition. Gases are heavier than air and will collect in low areas. Gases in confined areas may explode when exposed to fire. Gases may travel long distances to ignition sources and flash back. Storage containers and parts of containers may rocket great distances, in many directions. If material or contaminated runoff enters waterways, notify downstream users of potentially contaminated waters. Notify local health and fire officials and pollution control agencies. Do not extinguish the fire unless the flow of the gas can be stopped and any remaining gas is out of the line. Specially trained personnel may use fog lines to cool exposures and let the fire burn itself out. Use dry chemicals, foam, carbon dioxide. From a secure, explosion-proof location, use water spray to cool exposed containers. If cooling streams are ineffective (venting sound increases in volume and pitch, tank discolors, or shows any signs of deforming), withdraw immediately to a secure position.

NFPA Hazard Classifications

Health Hazards (Blue) 0
Flammability (Red) 4
Reactivity (Yellow) 0

Flammable Limits in Air: LEL 4.1%; UEL 74.2%
Autoignition Temperature: 752°F (400°C)

FIRST AID

Move victim to fresh air. Call emergency medical care. Apply artificial respiration if victim is not breathing. Administer oxygen if breathing is difficult. Remove and isolate contaminated clothing and shoes. Clothing frozen to the skin should be thawed before being removed. In case of contact with liquefied gas, thaw frosted parts with lukewarm water. Keep victim warm and quiet. Keep victim under observation. Ensure that medical personnel are aware of the material(s) involved and take precautions to protect themselves.

ENVIRONMENTAL INFORMATION

CLEAN AIR ACT: Accidental Release Prevention/Flammable substances, Section 112[r], Table 3), TQ = 10,000 lb (4,540 kg).

HYDROGEN PEROXIDE H:300

SYNONYMS: ALBONE; CARRO'S ACID; EINECS No. 231-765-0; HIGH STRENGTH HYDROGEN PEROXIDE; HOOH; HYOXYL; HYDROGEN DIOXIDE; HYDROPEROXIDE; INHIBINE; LEA RONAL NP-A/NP-B SOLDER STRIPPER; NANOSTRIP; OXYDOL; PATCLIN 948 SOLDER STRIPPER; PERHYDROL; PERONE; PEROXAN; PEROXIDE; PEROXIDO de HIDROGENO (Spanish); PIRANHA ETCH; RCA CLEAN (STEPS 1 AND 2); SUPEROXOL; SUPER SOLDER STRIP 1805; T-STUFF; WASSERSTOSSOXIDE (German); WATERSTOFPEROXYDE (Dutch); WCD 10BL CORROSION INHIBITOR

IDENTIFICATION

CAS:	7722-84-1
ERG ID:	UN 2984 (8%–20% solution)
	UN 2014 (20%–52% solution)
	UN 2015 (>52% solution)

ERG Guide:	140 (8%–20% solution) 140 (20%–52% solution) 143 (>52% solution)
ERG Guide Category:	140 OXIDIZERS 143 OXIDIZERS (UNSTABLE)
Molecular Formula:	H_2O_2
RTECS No.:	MX0900000
Properties:	Noncombustible watery liquid. Colorless. Shipped in water solution. Slightly sharp, irritating odor. Sinks and mixes with water. Nonflammable, but contact with combustible materials may cause ignition.

HEALTH AND SAFETY INFORMATION

Corrosive. May cause severe eye and skin burns. Repeated exposure to vapor may cause chronic irritation of the respiratory tract and partial or complete lung collapse. Repeated contact with vapor or solution may result in bleaching of the skin or hair. Also, long-term exposure may cause irritation to eyes, nose, and throat.

IARC Evaluation: Group 3, unclassifiable; no human data; animal evidence, limited.
IDLH: 75 ppm
Odor Threshold: Odor is inadequate as index of exposure; irritating only at high concentrations.
OSHA Table Z-1-A Air Contaminant (conc. >52%)
OSHA Process Safety Management of Highly Hazardous Chemicals (29 CFR 1910.119, Appendix A), Threshold Quantity: 7,500 pounds (52% by weight or greater).

Exposure Limits:	**TWA**
ACGIH:*	1 ppm; 1.4 mg/m^3
NIOSH:	1 ppm; 1.4 mg/m^3
OSHA:	1 ppm; 1.4 mg/m^3

*Animal carcinogen

RECOMMENDATIONS FOR RESPIRATOR SELECTION: NIOSH/OSHA

10 ppm: SA (any supplied-air respirator).
25 ppm: SA:CF (any supplied-air respirator operated in a continuous-flow mode).
50 ppm: SCBAF (any self-contained breathing apparatus with a full facepiece);or SAF (any supplied-air respirator with a full facepiece).
75 ppm: SAF:PD,PP (any supplied-air respirator that has a full facepiece and is operated in a pressure-demand or other positive-pressure mode).

Emergency or Planned Entry into Unknown Concentrations or IDLH Conditions:
SCBAF:PD,PP (any self-contained breathing apparatus that has a full facepiece and is operated in a pressure-demand or other positive-pressure mode);or SAF:PD,PP:ASCBA (any supplied-air respirator that

has a full facepiece and is operated in a pressure-demand or other positive-pressure mode in combination with an auxiliary, self-contained breathing apparatus operated in a pressure-demand or other positive-pressure mode).

Escape: GMFS [any air-purifying, full-facepiece respirator (gas mask) with a chin-style, front- or back-mounted canister providing protection against the compound of concern];or SCBAE (any appropriate escape-type, self-contained breathing apparatus).

Note: Substance reported to cause eye irritation or damage; may require eye protection.

CHEMICAL REACTIVITY

A powerful oxidizer; attacks many substances. Contact with most organic, readily oxidizable materials, reducing agents, and combustibles causes fire and explosions. Contact with iron, copper, brass, bronze, chromium, zinc, lead, manganese, silver, and other catalytic metals (and their salts), especially in a basic (pH 7 or above) environment, cause rapid decomposition with evolution and release of oxygen gas. Attacks, and may ignite, some plastics, rubber, and coatings. Decomposes slowly at ordinary temperatures and builds up pressure in a closed container. The rate of decomposition doubles for each 50°F (10°C) rise (1.5 times 10°C rise) in temperature and becomes self-sustaining at 285°F (141°C). Reacts with iron, copper, brass, and many other metals.

FIRE INFORMATION

Noncombustible but will increase intensity of a fire. Firefighting gear, including SCBA, does not provide adequate protection. If exposure occurs, remove and isolate gear immediately and thoroughly decontaminate personnel. Storage containers and parts of containers may rocket great distances, in many directions. If material or contaminated runoff enters waterways, notify downstream users of potentially contaminated waters. Notify local health and fire officials and pollution control agencies. Use flooding amounts of water; dry chemicals, foam, carbon dioxide. Use water spray to cool exposed containers.

NFPA Hazard Classifications:

Health Hazards (Blue) 2
Flammability (Red) 0
Reactivity (Yellow) 3
Special situations 0XY

Flash Point: Not flammable but may cause fire and react violently on contact with combustibles and metals.

FIRST AID

OXIDIZERS (140): Move victim to fresh air. Call emergency medical care. Apply artificial respiration if victim is not breathing. Administer oxygen if breathing is difficult. Remove and isolate contaminated clothing and shoes. In case of contact with substance, immediately flush skin or eyes with running water for at least 20 minutes. Keep victim warm and quiet. Ensure that medical personnel are aware of the material(s) involved and take precautions to protect themselves.

OXIDIZERS (UNSTABLE) (143): Move victim to fresh air. Call emergency medical care. Apply artificial respiration if victim is not breathing. Administer oxygen if breathing is difficult. Remove and isolate contaminated clothing and shoes. In case of contact with substance, immediately flush skin or eyes with running water for at least 20 minutes. Keep victim warm and quiet. Ensure that medical personnel are aware of the material(s) involved and take precautions to protect themselves.

ENVIRONMENTAL INFORMATION (for concentrations >52%).
EPCRA Section 302 Extremely Hazardous Substances: TPQ = 1,000 lb (454 kg).
EPCRA Section 304: Reportable Quantity (RQ): EHS, 1000 lb (454 kg).

HYDROGEN SULFIDE H:340

SYNONYMS: ACIDE SULFHYDRIQUE (French); EEC No. 016-001-00-4; HEPATIC GAS; HYDROGENE SULFURE (French); HYDROSULFURIC ACID; IDROGENOSOLFORATO (Italian); ROTTEN EGG GAS; SCHWEFELWASSERSTOFF (German); SEWER GAS; SIARKOWODOR (Polish); STINK DAMP; SULFURETTED HYDROGEN; SULFUR HYDRIDE; SULFUR HYDROXIDE; SULFURO de HIDROGENO (Spanish); ZWAVELWATERSTOF (Dutch); SULPHURETTED HYDROGEN

IDENTIFICATION

CAS:	7783-06-4
ERG ID:	UN 1053
ERG Guide:	117
ERG Guide Category:	GASES- TOXIC- FLAMMABLE (EXTREME HAZARD)
Molecular Formula:	H_2S
RTECS No.:	MX1225000
Properties:	Extremely flammable, poisonous, liquified compressed gas. Colorless; forms a visible cloud. Rotten egg odor, at concentrations below 30 ppm; from 30-100 ppm, sickeningly sweet; many people rapidly lose the ability to detect the smell, so odor is not a good warning of excess levels. Sinks and "boils" in water; insoluble.

HEALTH AND SAFETY INFORMATION

Poison. Breathing the gas can cause death. Contact with the liquid can cause frostbite. Repeated or prolonged exposure has been reported to cause low blood pressure, blurred vision, headache, nausea, loss of appetite, weight loss, ataxia, eye-membrane inflammation, and chronic cough. Neurologic symptoms, including psychological disorders, have been associated with chronic exposure.

IDLH: 100 ppm
Odor Threshold: 0.0005 ppm. Warning properties may not be dependable; odor is not a reliable indicator of the presence of toxic amounts of gas. Characteristic rotten-egg odor detectable at about 0.02 ppm, but olfactory nerve fatigue occurs in 2 to 15 minutes at 100 to 150 ppm.

OSHA Table Z-1-A Air Contaminant
OSHA Table Z-2 Air Contaminant
OSHA Process Safety Management of Highly Hazardous Chemicals (29 CFR 1910.119, Appendix A), Threshold Quantity: 1,500 pounds
NIOSH Criteria Document: NIOSH 77-158

Exposure Limits:	TWA	Ceiling	STEL
ACGIH:	10 ppm; 14 mg/m^3		15 ppm; 21 mg/m^3
NIOSH:		10 ppm; 15 mg/m^3 10 min	
OSHA:		20 ppm; 50 ppm 10 min. max peak	

RECOMMENDATIONS FOR RESPIRATOR SELECTION: NIOSH/OSHA

100 ppm: PAPRS (any powered, air-purifying respirator with cartridge(s) providing protection against the compound of concern);or GMFS (any air-purifying, full-facepiece respirator (gas mask) with a chin-style, front- or back-mounted canister providing protection against the compound of concern);or SA (any supplied-air respirator);or SCBAF (any self-contained breathing apparatus with a full facepiece)

Emergency or planned entry into unknown concentrations or idlh conditions: SCBAF:PD,PP (any self-contained breathing apparatus that has a full facepiece and is operated in a pressure-demand or other positive-pressure mode);or SAF:PD,PP:ASCBA (any supplied-air respirator that has a full facepiece and is operated in a pressure-demand or other positive-pressure mode in combination with an auxiliary, self-contained breathing apparatus operated in a pressure-demand or other positive pressure mode.

Escape: GMFS (any air-purifying, full-facepiece respirator (gas mask) with a chin-style, front- or back-mounted canister providing protection against the compound of concern);or SCBAE (any appropriate escape-type, self-contained breathing apparatus).

Note: Substance reported to cause eye irritation or damage; may require eye protection.

CHEMICAL REACTIVITY

A highly flammable and reactive gas. Reacts explosively with fuming or concentrated nitric acid. Strong reaction with most oxidizers such as permanganates, chlorine, dichromates. Incompatible with acetaldehyde, barium pentafluoride, chlorine monoxide, chlorine trifluoride, chromic anhydride, copper, lead dioxide, nitrogen iodide, nitrogen trichloride, nitrogen trifluoride, oxygen difluoride, phenyl diazonium chloride, sodium, sodium peroxide. Will burn or explode in the presence of metal oxides. Reacts with alkali metals and alkali earth metals. Attacks or tarnishes most metals.

FIRE INFORMATION

Combustion products include toxic sulfur oxides. Firefighting gear (including SCBA) does not provide adequate protection. If exposure occurs, remove and isolate gear immediately and thoroughly decontaminate personnel. Gases are heavier than air and will collect in low areas. Gases in confined areas may explode when exposed to fire. Gases may travel long distances to ignition sources and flash back. Storage containers and parts of containers may rocket great distances, in many directions. If material or contaminated runoff enters waterways, notify downstream users of potentially contaminated waters. Notify local health and fire officials and pollution control agencies. Do not extinguish the fire unless the flow of the gas can be stopped and any remaining gas is out of the line. Specially trained personnel may use fog lines to cool exposures and let the fire burn itself out. Use dry chemicals, foam, carbon dioxide. From a secure, explosion-proof location, use water spray to cool exposed containers. If cooling streams are ineffective (venting sound increases in volume and pitch, tank discolors, or shows any signs of deforming), withdraw immediately to a secure position.

NFPA Hazard Classifications:

Health Hazards (Blue) 3
Flammability (Red) 4
Reactivity (Yellow) 0

Flash Point: Flammable gas.
Flammable Limits in Air: LEL 4.3%; UEL 45%
Autoignition Temperature: 500°F (260°C).

FIRST AID

Move victim to fresh air. Call emergency medical care. Apply artificial respiration if victim is not breathing. Do not use mouth-to-mouth method if victim ingested or inhaled the substance; induce artificial respiration with the aid of a pocket mask equipped with a one-way valve or other proper respiratory medical device. Administer oxygen if breathing is difficult. Remove and isolate contaminated clothing and shoes. In case of contact with substance, immediately flush skin or eyes with running water for at least 20 minutes. In case of contact with liquefied gas, thaw frosted parts with lukewarm water. Keep victim warm and quiet. Keep victim under observation. Effects of contact or inhalation may be delayed. Ensure that medical personnel are aware of the material(s) involved and take precautions to protect themselves.

ENVIRONMENTAL INFORMATION

CLEAN AIR ACT: Accidental Release Prevention/Flammable substances, (Section 112[r], Table 3), TQ = 10,000 lb (4,540 kg).
CLEAN WATER ACT: Section 311 Hazardous Substances/RQ (same as CERCLA).
EPA HAZARDOUS WASTE NUMBER (RCRA No.): U135
RCRA Section 261 Hazardous Constituents.
EPCRA Section 302 Extremely Hazardous Substances: TPQ = 500 lb (227 kg).
EPCRA Section 304 Reportable Quantity (RQ): EHS, CERCLA, 100 lb (45.4 kg).
EPCRA Section 313: Subject to an administrative stay until further notice. See 59 FR 43048, August 22, 1994.

HYDROQUINONE H:360

SYNONYMS: ARCTUVIN; p-BENZENEDIOL; 1,4-BENZENEDIOL; BENZOHYDROQUINONE; BENZOQUINOL; BLACK AND WHITE BLEACHING CREAM; BOYDES PTS DEVELOPER; CRONAFLEX PDC DEVELOPER; 1,4-DIHYDROXY-BENZEEN (Dutch); 1,4-DIHYDROXYBENZEN (Czech); DIHYDROXYBENZENE; p-DIHYDROXYBENZENE; 1,4-DIHYDROXYBENZENE; 1,4-DIHYDROXY-BENZOL (German); 1,4-DIIDROBENZENE (Italian); p-DIOXOBENZENE; EEC No. 604-005-00-4; EINECS No. 204-617-8; ELDOPAQUE; ELDOQUIN; HIDROQUINONA (Spanish); HYDROCHINON (Czech); HYDROCHINON (Polish); HYDROQUINOL; HYDROQUINOLE; α-HYDROQUINONE; p-HYDROQUINONE; para-HYDROQUINONE; para-HYDROXYPHENOL; IDROCHINONE (Italian); KODAGRAPH LIQUID DEVELOPER; KODAK 55/66 DEVELOPER; NCI-C55834; PD-86 DEVELOPER; QUINOL; b-QUINOL; USAF EK-356; PYROGENTISIC ACID; SR-201: TECQUINOL; TENOX HQ

IDENTIFICATION

CAS:	123-31-9
ERG ID:	UN 2662
ERG Guide:	153
ERG Guide Category:	TOXIC AND/OR CORROSIVE (COMBUSTIBLE)
Molecular Formula:	$C_6H_6O_2$
RTECS No.:	MX3500000
Properties:	Combustible crystals. Colorless, white, light tan to gray; air causes darkening. Sinks and mixes with water.

HEALTH AND SAFETY INFORMATION

Corrosive to the eyes, skin, and respiratory tract. May cause pulmonary edema, which can be delayed for several hours; there is a risk of death in serious cases. An allergen. Prolonged or repeated exposures may cause skin and eye discoloration; eye clouding, blurred vision ,and permanent vision damage; skin rash; lung and blood problems.

IARC Evaluation: Group 3, unclassifiable; no human data; animal evidence, inadequate.
IDLH: 50 mg/m^3.
OSHA Table Z-1-A Air Contaminant
NIOSH Criteria Document: NIOSH 78-155

Exposure Limits:	**TWA**	**CEILING**
ACGIH:[1]	2 mg/m^3	
NIOSH:		2 mg/m^3 [15 min.]
OSHA:	2 mg/m^3	

[1]Animal carcinogen.

RECOMMENDATIONS FOR RESPIRATOR SELECTION: NIOSH/OSHA

50 mg/m³: PAPRD (any powered, air-purifying respirator with a dust filter);or HiEF (any air-purifying, full-facepiece respirator with a high-efficiency particulate filter);or SAT:CF (any supplied-air respirator that has a tight-fitting facepiece and is operated in a continuous-flow mode);or SCBAF (any self-contained breathing apparatus with a full facepiece);or SAF (any supplied-air respirator with a full facepiece).

Emergency or Planned Entry into Unknown Concentrations or IDLH Conditions: SCBAF:PD,PP (any self-contained breathing apparatus that has a full faceplate and is operated in a pressure-demand or other positive-pressure mode);or SAF:PD,PP:ASCBA (any supplied-air respirator that has a full facepiece and is operated in a pressure-demand or other positive-pressure mode in combination with an auxiliary, self-contained breathing apparatus operated in a pressure-demand or other positive-pressure mode).

Escape: HiEF (any air-purifying, full-facepiece respirator with a high-efficiency particulate filter);or SCBAE (any appropriate escape-type, self-contained breathing apparatus).

Note: Substance causes eye irritation or damage; eye protection needed.

CHEMICAL REACTIVITY

Incompatible with strong oxidizers and caustics. May be oxidized to quinone at room temperatures in the presence of moisture. May explode on contact with oxygen.

FIRE INFORMATION

Vapors are heavier than air and will collect in low areas. Vapors in confined areas may explode when exposed to fire. Vapors may travel long distances to ignition sources and flash back. Storage containers and parts of containers may rocket great distances, in many directions. If material or contaminated runoff enters waterways, notify downstream users of potentially contaminated waters. Notify local health and fire officials and pollution control agencies. Use dry chemicals, water spray.

NFPA Hazard Classifications:

Health Hazards (Blue) 2
Flammability (Red) 1
Reactivity (Yellow) 0

Flash Point: 329°F (165°C) o.c. (molten)
Autoignition Temperature: 960°F (516°C)

FIRST AID

Move victim to fresh air. Call emergency medical care. Apply artificial respiration if victim is not breathing. Do not use mouth-to-mouth method if victim ingested or inhaled the substance; induce artificial respiration with the aid of a pocket mask equipped with a one-way valve or other proper respiratory

medical device. Administer oxygen if breathing is difficult. Remove and isolate contaminated clothing and shoes. In case of contact with substance, immediately flush skin or eyes with running water for at least 20 minutes. For minor skin contact, avoid spreading material on unaffected skin. Keep victim warm and quiet. Effects of exposure (inhalation, ingestion, or skin contact) may be delayed. Ensure that medical personnel are aware of the material(s) involved and take precautions to protect themselves.

ENVIRONMENTAL INFORMATION

Clean Air Act: Hazardous Air Pollutants (Title I, Part A, Section 112)
EPCRA Section 302 Extremely Hazardous Substances: TPQ = 500/10,000 lb (227/4,540 kg)
EPCRA Section 304: Reportable Quantity (RQ): EHS, CERCLA, 100 lb (045.4 kg)
EPCRA Section 313: Form R *de minimis* concentration reporting level: 1.0%
TSCA:40CFR716.120(*a*); 40CFR799.18; Section 8(*d*) reporting requirements.

HYDROQUINONE MONOMETHYL ETHER H:366

SYNONYMS: EINECS No. 205-769-8; HIDROQUINONA-MONOMETIL ETER (Spanish); HQMME; MEQUINOL; para-HYDROXYANISOLE; 4-METHOXY PHENOL; P-METHOXY PHENOL; MME; MONOMETHYL ETHER HYDROQUINONE; PHENOL, 4-METHOXY-; USAF AN-7

IDENTIFICATION

CAS: 150-76-5
Molecular Formula: $C_7H_8O_2$
RTECS No.: SL7700000
Properties: Combustible solid or leaflets. Colorless to white. Caramel and phenol odor. Slightly soluble in water.

HEALTH AND SAFETY INFORMATION

Irritates eyes, skin, and respiratory tract. May affect central nervous system; possible headaches, dizziness, unconsciousness.

Exposure Limits:	TWA
NIOSH:	5 mg/m^3
OSHA:	None

RECOMMENDATIONS FOR RESPIRATOR SELECTION

At any concentrations above the NIOSH REL, or where there is no REL, at any detectable concentration: SCBAF:PD,PP (any self-contained breathing apparatus that has a full facepiece and is operated in a pressure-demand or other positive-pressure mode);or SAF:PD,PP:ASCBA (any supplied-air

respirator that has a full facepiece and is operated in a pressure-demand or other positive-pressure mode in combination with an auxiliary, self-contained breathing apparatus operated in a pressure-demand or other positive-pressure mode).

Escape: GMFOV [any air-purifying, full-facepiece respirator (gas mask) with a chin-style, front- or back-mounted organic vapor canister];or SCBAE (any appropriate escape-type, self-contained breathing apparatus).

CHEMICAL REACTIVITY

Reacts with strong oxidizers, strong bases, acid chlorides, and acid aldehydes. Incompatible with sulfuric acid, nitric acid.

FIRE INFORMATION

If material or contaminated runoff enters waterways, notify downstream users of potentially contaminated waters. Notify local health and fire officials and pollution control agencies. Use dry chemicals, alcohol-resistant foam, carbon dioxide.

NFPA Hazard Classifications:

Health Hazards (Blue) . . . . -
Flammability (Red) 1
Reactivity (Yellow) 0

Flash Point: 270°F (132°C) o.c.
Flammable Limits in Air: LEL 3.1%; UEL 16%
Autoignition Temperature: 421°F (790°C)

FIRST AID

Move victim to fresh air. Call emergency medical care. Apply artificial respiration if victim is not breathing. Administer oxygen if breathing is difficult. Remove and isolate contaminated clothing and shoes. In case of contact with substance, immediately flush skin or eyes with running water for at least 20 minutes. Removal of solidified molten material from skin requires medical assistance. Keep victim warm and quiet. Ensure that medical personnel are aware of the material(s) involved and take precautions to protect themselves.

HYDROXYADIPALDEHYDE H:372

SYNONYMS: EINECS No. 205-479-1; HEXANEDIAL, 2-HYDROXY-; 2-HIDROXIADIPALDEHIDO (Spanish); α-HYDROXYADIPALDEHYDE; 2-HYDROXYADIPALDEHYDE; 2-HYDROXYHEXANEDIAL

586 Hydroxyadipaldehyde

IDENTIFICATION

CAS:	141-31-1
Molecular Formula:	$C_6H_{10}O_3$
RTECS No.:	MO1130000
Properties:	Combustible liquid. Soluble in water.

HEALTH AND SAFETY INFORMATION

Moderately toxic. Irritates eyes, skin, and respiratory tract. Many aldehydes are mutagens.

RECOMMENDATIONS FOR RESPIRATOR SELECTION

At any concentrations above the NIOSH REL, or where there is no REL, at any detectable concentration: SCBAF:PD,PP (any self-contained breathing apparatus that has a full facepiece and is operated in a pressure-demand or other positive-pressure mode);or SAF:PD,PP:ASCBA (any supplied-air respirator that has a full facepiece and is operated in a pressure-demand or other positive-pressure mode in combination with an auxiliary, self-contained breathing apparatus operated in a pressure-demand or other positive-pressure mode).

Escape: GMFOV [any air-purifying, full-facepiece respirator (gas mask) with a chin-style, front- or back-mounted organic vapor canister];or SCBAE (any appropriate escape-type, self-contained breathing apparatus).

CHEMICAL REACTIVITY

Incompatible with oxidizers.

FIRE INFORMATION

Vapors are heavier than air and will collect in low areas. Vapors in confined areas may explode when exposed to fire. If material or contaminated runoff enters waterways, notify downstream users of potentially contaminated waters. Notify local health and fire officials and pollution control agencies. Use dry chemicals, alcohol-resistant foam, carbon dioxide.

NFPA Hazard Classifications: Not listed

FIRST AID

Move victim to fresh air. Call emergency medical care. Apply artificial respiration if victim is not breathing. Administer oxygen if breathing is difficult. Remove and isolate contaminated clothing and shoes. In case of contact with substance, immediately flush skin or eyes with running water for at least 20 minutes. Wash skin with soap and water. Keep victim warm and quiet. Effects of exposure (inhalation, ingestion, or skin contact) to substance may be delayed. Ensure that medical personnel are aware of the material(s) involved and take precautions to protect themselves.

ENVIRONMENTAL INFORMATION

TSCA: 40CFR716.120(*d*)1 as aldehydes

2–HYDROXYETHYL METHACRYLATE H:378

SYNONYMS: EEC No. 607-124-00-X; EINECS No. 212-782-2; ETHYLENE GLYCOL METHACRYLATE; ETHYLENE GLYCOL, MONOMETHACRYLATE; GLYCOL METHACRYLATE; GLYCOL MONOMETHACRYLATE; HEMA; 2-HYDROXYETHYL ESTER METHACRYLIC ACID; β-HYDROXYETHYL METHACRYLATE; 2-IDROSSIETILE METACRILATO (Italian); MHOROMER; MONOMER MG-1; MONOMETHACRYLIC ETHER of ETHYLENE GLYCOL; 2-PROPENOIC ACID, 2-METHYL-, 2-HYDROXYETHYL ESTER

IDENTIFICATION

CAS:	868-77-9
Molecular Formula:	$C_6H_{10}O_3$
RTECS No.:	OZ4725000
Properties:	Combustible liquid. Water soluble.

HEALTH AND SAFETY INFORMATION

RECOMMENDATIONS FOR RESPIRATOR SELECTION

At any concentrations above the NIOSH REL, or where there is no REL, at any detectable concentration: SCBAF:PD,PP (any self-contained breathing apparatus that has a full facepiece and is operated in a pressure-demand or other positive-pressure mode);or SAF:PD,PP:ASCBA (any supplied-air respirator that has a full facepiece and is operated in a pressure-demand or other positive-pressure mode in combination with an auxiliary, self-contained breathing apparatus operated in a pressure-demand or other positive-pressure mode).

Escape: GMFOV [any air-purifying, full-facepiece respirator (gas mask) with a chin-style, front- or back-mounted organic vapor canister];or SCBAE (any appropriate escape-type, self-contained breathing apparatus).

CHEMICAL REACTIVITY

Incompatible with oxidizers, sulfuric acid, nitric acid, caustics, aliphatic amines, isocyanates.

FIRE INFORMATION

Vapors are heavier than air and will collect in low areas. Vapors in confined areas may explode when exposed to fire. If material or contaminated runoff enters waterways, notify downstream users of potentially contaminated waters. Notify local health and fire officials and pollution control agencies. Use dry chemicals, alcohol-resistant foam, carbon dioxide.

NFPA Hazard Classifications: Not listed

Flash Point: 207°F (97°C)

FIRST AID

Move victim to fresh air. Call emergency medical care. Apply artificial respiration if victim is not breathing. Administer oxygen if breathing is difficult. Remove and isolate contaminated clothing and shoes. In case of contact with substance, immediately flush skin or eyes with running water for at least 20 minutes. Wash skin with soap and water. Keep victim warm and quiet. Effects of exposure (inhalation, ingestion, or skin contact) to substance may be delayed. Ensure that medical personnel are aware of the material(s) involved and take precautions to protect themselves.

HYDROXYLAMINE HYDROCHLORIDE H:384

SYNONYMS: CLORHIDRATO de HIDROXILAMINA (Spanish); EINECS No. 226-798-2; HYDROXYAMMONIUM CHLORIDE; HYDROXYLAMINE CHLORIDE; HYDROXYLAMINE CHLORIDE (1:1); HYDROXYLAMMONIUM CHLORIDE; OXAMMONIUM HYDROCHLORIDE

IDENTIFICATION

CAS:	5470-11-1
ERG ID:	UN 2811 as toxic solid, organic, n.o.s.
ERG Guide:	154
ERG Guide Category:	SUBSTANCES- TOXIC AND/OR CORROSIVE (NONCOMBUSTIBLE)
Molecular Formula:	ClH_4NO
RTECS No.:	NC3675000
Properties:	Crystalline solid. Colorless. Soluble in water; decomposes, forming hydrochloric acid.

HEALTH AND SAFETY INFORMATION

May cause severe irritation to eyes, skin, and respiratory tract. Due to the availability of sufficient data on human effects, caution should be exercised.

RECOMMENDATIONS FOR RESPIRATOR SELECTION

At any concentrations above the NIOSH REL, or where there is no REL, at any detectable concentration: SCBAF:PD,PP (any self-contained breathing apparatus that has a full facepiece and is operated in a pressure-demand or other positive-pressure mode);or SAF:PD,PP:ASCBA (any supplied-air

respirator that has a full facepiece and is operated in a pressure-demand or other positive-pressure mode in combination with an auxiliary, self-contained breathing apparatus operated in a pressure-demand or other positive-pressure mode).

Escape: GMFOV [any air-purifying, full-facepiece respirator (gas mask) with a chin-style, front- or back-mounted organic vapor canister];or SCBAE (any appropriate escape-type, self-contained breathing apparatus).

CHEMICAL REACTIVITY

Powerful reducing agent; violent reaction with oxidizers. Amines are incompatible with acrolein, acrylonitrile, tert-butyl nitroacetylene, ethylene oxide, isopropyl chlorocarbonate, maleic anhydride, triisobutylaluminum; may ignite on contact with cellulose nitrate of high surface area. May corrode metals in the presence of moisture.

FIRE INFORMATION

Combustion products include toxic nitrogen oxides. If material or contaminated runoff enters waterways, notify downstream users of potentially contaminated waters. Notify local health and fire officials and pollution control agencies. Use dry chemicals, alcohol-resistant foam, carbon dioxide.

NFPA Hazard Classifications: Not listed

FIRST AID

Move victim to fresh air; call emergency medical care. Apply artificial respiration if victim is not breathing. Do not use mouth-to-mouth method if victim ingested or inhaled the substance; induce artificial respiration with the aid of a pocket mask equipped with a one-way valve or other proper respiratory medical device. Administer oxygen if breathing is difficult. Remove and isolate contaminated clothing and shoes. In case of contact with substance, immediately flush skin or eyes with running water for at least 20 minutes. For minor skin contact, avoid spreading material on unaffected skin. Keep victim warm and quiet. Effects of exposure (inhalation, ingestion, or skin contact) to substance may be delayed. Ensure that medical personnel are aware of the material(s) involved and take precautions to protect themselves.

ENVIRONMENTAL INFORMATION

TSCA: 40CFR704.225(*a*)

HYDROXYLAMINE SULFATE (2:1) H:390

SYNONYMS: BIS(HYDROXYLAMINE) SULFATE; EINECS 233-118-8; HYDROXYLAMINE NEUTRAL; SULFATE; HS; HYDROXYLAMINE SULPHATE (2:1); HYDROXYLAMMONIUM SULFATE; OXAMMONIUM SULFATE; SULFATE de HIDROXILAMINA (Spanish)

590 Hydroxylamine Sulfate

IDENTIFICATION

CAS:	10039-54-0 (2:1); 10046-01-0 (1:1)
ERG ID:	UN 2865
ERG Guide:	154
ERG Guide Category:	SUBSTANCES- TOXIC AND/OR CORROSIVE (NON-COMBUSTIBLE)
Molecular Formula:	$H_8N_2O_6S$
RTECS No.:	NC5425000
Properties:	Noncombustible solid. White. Odorless. Sinks and mixes with water. Poisonous gases may be produced in fire.

HEALTH AND SAFETY INFORMATION

RECOMMENDATIONS FOR RESPIRATOR SELECTION

At any concentrations above the NIOSH REL, or where there is no REL, at any detectable concentration: SCBAF:PD,PP (any self-contained breathing apparatus that has a full facepiece and is operated in a pressure-demand or other positive-pressure mode);or SAF:PD,PP:ASCBA (any supplied-air respirator that has a full facepiece and is operated in a pressure-demand or other positive-pressure mode in combination with an auxiliary, self-contained breathing apparatus operated in a pressure-demand or other positive-pressure mode).

Escape: GMFOV [any air-purifying, full-facepiece respirator (gas mask) with a chin-style, front- or back-mounted organic vapor canister];or SCBAE (any appropriate escape-type, self-contained breathing apparatus).

CHEMICAL REACTIVITY

A strong oxidizer. Heat or friction may cause explosions. Violent, possibly explosive, reaction with caustics, reducing agents, and combustibles. Incompatible with strong acids, organic anhydrides, isocyanates, and aldehydes. Magnesium or aluminum contact may be explosive.

FIRE INFORMATION

Combustible solid. Dust cloud may explode if ignited in an enclosed area. Poisonous. If material or contaminated runoff enters waterways, notify downstream users of potentially contaminated water.

FIRST AID

Move victim to fresh air. Call emergency medical care. Apply artificial respiration if victim is not breathing. Do not use mouth-to-mouth method if victim ingested or inhaled the substance; induce artificial respiration with the aid of a pocket mask equipped with a one-way valve or other proper respiratory medical device. Administer oxygen if breathing is difficult. Remove and isolate contaminated clothing and shoes. In case of contact with substance, immediately flush skin or eyes with running water for at least 20 minutes. For minor skin contact, avoid spreading material on unaffected skin. Keep victim warm and quiet.

Effects of exposure (inhalation, ingestion, or skin contact) may be delayed. Ensure that medical personnel are aware of the material(s) involved and take precautions to protect themselves.

ENVIRONMENTAL INFORMATION

TSCA: 40CFR704.225(*a*)

HYDROXY–2–PROPANONE H:460

SYNONYMS: ACETONE ALCOHOL; ALCOOL ACETONYLIQUE (French); ACETOL (1); EINECS No. 204-124-8; HIDROXIACETONA (Spanish); HYDROXYACETONE; 1-HYDROXY-2-PROPANONE; IDROSSIACETONE (Italian); PYRUVIC ALCOHOL

IDENTIFICATION

CAS:	116-09-6
Molecular Formula:	$C_3H_6O_2$
RTECS No.:	UC2800000
Properties:	Combustible liquid. Colorless. Water soluble.

HEALTH AND SAFETY INFORMATION

Toxic and an allergen.

RECOMMENDATIONS FOR RESPIRATOR SELECTION

At any concentrations above the NIOSH REL, or where there is no REL, at any detectable concentration: SCBAF:PD,PP (any self-contained breathing apparatus that has a full facepiece and is operated in a pressure-demand or other positive-pressure mode);or SAF:PD,PP:ASCBA (any supplied-air respirator that has a full facepiece and is operated in a pressure-demand or other positive-pressure mode in combination with an auxiliary, self-contained breathing apparatus operated in a pressure-demand or other positive-pressure mode).

Escape: GMFOV [any air-purifying, full-facepiece respirator (gas mask) with a chin-style, front- or back-mounted organic vapor canister];or SCBAE (any appropriate escape-type, self-contained breathing apparatus).

CHEMICAL REACTIVITY

Incompatible with oxidizers, sulfuric acid, nitric acid, caustics, aliphatic amines, isocyanates.

FIRE INFORMATION

Vapors are heavier than air and will collect in low areas. Vapors in confined areas may explode when exposed to fire. If material or contaminated runoff enters waterways, notify downstream users of potentially contaminated waters. Notify local health and fire officials and pollution control agencies. Use dry chemicals, alcohol-resistant foam, carbon dioxide.

NFPA Hazard Classifications: Not listed

FIRST AID

Move victim to fresh air. Call emergency medical care. Apply artificial respiration if victim is not breathing. Do not use mouth-to-mouth method if victim ingested or inhaled the substance; induce artificial respiration with the aid of a pocket mask equipped with a one-way valve or other proper respiratory medical device. Administer oxygen if breathing is difficult. Remove and isolate contaminated clothing and shoes. In case of contact with substance, immediately flush skin or eyes with running water for at least 20 minutes. Wash skin with soap and water. Keep victim warm and quiet. Effects of exposure (inhalation, ingestion, or skin contact) to substance may be delayed. Ensure that medical personnel are aware of the material(s) involved and take precautions to protect themselves.

HYDROXYPROPYL ACRYLATE H:510

SYNONYMS: ACRILATO de 2-HIDROXIPROPILO (Spanish); ACRYLIC ACID-2-HYDROXYPROPYL ESTER; EEC No. 607-108-00-2; EINECS No. 213-663-8; β-HYDROXYPROPYL ACRYLATE; HPA; 1,2-PROPANEDIOL-1-ACRYLATE; PROPYLENE GLYCOL MONOACRYLATE

IDENTIFICATION

CAS:	999-61-1
ERG ID:	UN 2810
ERG Guide:	153
ERG Guide Category:	SUBSTANCES- TOXIC AND/OR CORROSIVE (COMBUSTIBLE)
Molecular Formula:	$C_6H_{10}O_3$
RTECS No.:	AT1925000
Properties:	Combustible liquid. Clear to light yellow. Sweetish odor. Water soluble.

HEALTH AND SAFETY INFORMATION

Irritates eyes, skin, and respiratory tract.

OSHA Table Z-1-A Air Contaminant

Exposure Limits:	**TWA**
ACGIH: (1)	0.5 ppm; 2.8 mg/m^3

NIOSH: [(1)] 0.5 ppm; 3 mg/m^3
OSHA: None

[(1)] Skin contact contributes significantly to overall exposure.

RECOMMENDATIONS FOR RESPIRATOR SELECTION

At any concentrations above the NIOSH REL, or where there is no REL, at any detectable concentration: SCBAF:PD,PP (any self-contained breathing apparatus that has a full facepiece and is operated in a pressure-demand or other positive-pressure mode);or SAF:PD,PP:ASCBA (any supplied-air respirator that has a full facepiece and is operated in a pressure-demand or other positive-pressure mode in combination with an auxiliary, self-contained breathing apparatus operated in a pressure-demand or other positive-pressure mode).

Escape: GMFOV [any air-purifying, full-facepiece respirator (gas mask) with a chin-style, front- or back-mounted organic vapor canister];or SCBAE (any appropriate escape-type, self-contained breathing apparatus).

CHEMICAL REACTIVITY

Incompatible with strong acids and nitrates Can become unstable at high temperatures and pressures or may react with water with some release of energy, but not violently.

FIRE INFORMATION

Vapors are heavier than air and will collect in low areas. Vapors in confined areas may explode when exposed to fire. Storage containers and parts of containers may rocket great distances, in many directions. If material or contaminated runoff enters waterways, notify downstream users of potentially contaminated waters. Notify local health and fire officials and pollution control agencies. Use dry chemicals, foam, carbon dioxide.

NFPA Hazard Classifications:

Health Hazards (Blue) 3
Flammability (Red) 1
Reactivity (Yellow) 2

Flash Point: 212°F (100°C) o.c.; 207°F (97°C) c.c.
Flammable Limits in Air: LEL 1.4%

FIRST AID

Move victim to fresh air and call emergency medical care. Apply artificial respiration if victim is not breathing. Do not use mouth-to-mouth method if victim swallowed or inhaled the substance; induce artificial respiration with the aid of a pocket mask equipped with a one-way valve or other proper respiratory medical device. If breathing is difficult, give oxygen. In case of contact with substance, immediately flush skin or eyes with running water for at least 20 minutes. Remove and isolate contaminated clothing and shoes. Keep victim warm and quiet. Effects of exposure (inhalation, ingestion, or skin contact) to substances may be delayed. Ensure that medical personnel are aware of the material(s) involved and take precautions to protect themselves.

IODINE MONOCHLORIDE I:070

SYNONYMS: EINECS No. 232-236-7; IODINE CHLORIDE; MONOCHLORURE D'IODE (French); MONOCLORURO de YODO (Spanish); PROTOCHLORURE d'IODE (French); WIJS CHLORIDE

IDENTIFICATION

CAS:	7790-99-0
ERG ID:	UN 1792
ERG Guide:	157
ERG Guide Category:	SUBSTANCES- TOXIC AND/OR CORROSIVE (NONCOMBUSTIBLE/WATER-SENSITIVE)
Molecular Formula:	ClI
RTECS No.:	NN1650000; NN1655000
Properties:	Potentially explosive crystalline solid or oily liquid. Brownish-black crystals or reddish-brown liquid. Odorless. Mixes with water.

HEALTH AND SAFETY INFORMATION

Highly corrosive to the eyes and skin. Vapors are irritating to eyes, skin, and respiratory tract

RECOMMENDATIONS FOR RESPIRATOR SELECTION

At any concentrations above the NIOSH REL, or where there is no REL, at any detectable concentration: SCBAF:PD,PP (any self-contained breathing apparatus that has a full facepiece and is operated in a pressure-demand or other positive-pressure mode);or SAF:PD,PP:ASCBA (any supplied-air respirator that has a full facepiece and is operated in a pressure-demand or other positive-pressure mode in combination with an auxiliary, self-contained breathing apparatus operated in a pressure-demand or other positive-pressure mode).

Escape: GMFOV [any air-purifying, full-facepiece respirator (gas mask) with a chin-style, front- or back-mounted organic vapor canister];or SCBAE (any appropriate escape-type, self-contained breathing apparatus).

CHEMICAL REACTIVITY

Reacts with water or steam to produce toxic and corrosive fumes. Violent reactions with metals, e.g., sodium (mixture explodes on impact), potassium (explodes on contact), and aluminum (ignition after a delay period). Reacts violently with aluminum foil, cadmium, lead, organic matter, phosphorous, phosphorus trichloride, rubber, and metal sulfides, e.g., silver sulfide and zinc sulfide. Hydrogen chloride and hydrogen iodide may form if exposed to heat, moisture, water, or air.

FIRE INFORMATION

Combustion products include toxic fumes of chlorides and Iodides. If material or contaminated runoff enters waterways, notify downstream users of potentially contaminated waters. Notify local health and fire officials and pollution control agencies.

Hazard Classifications: based on NFPA system

Health Hazards (Blue) 3
Flammability (Red) 0
Reactivity (Yellow) 3

FIRST AID

Move victim to fresh air. Call emergency medical care. Apply artificial respiration if victim is not breathing. Do not use mouth-to-mouth method if victim ingested or inhaled the substance; induce artificial respiration with the aid of a pocket mask equipped with a one-way valve or other proper respiratory medical device. Administer oxygen if breathing is difficult. Remove and isolate contaminated clothing and shoes. In case of contact with substance, immediately flush skin or eyes with running water for at least 20 minutes. For minor skin contact, avoid spreading material on unaffected skin. Keep victim warm and quiet. Effects of exposure (inhalation, ingestion, or skin contact) may be delayed. Ensure that medical personnel are aware of the material(s) involved and take precautions to protect themselves.

IRON OXIDE, RED I:081

SYNONYMS: ANHYDROUS IRON OXIDE; ANHYDROUS OXIDE of IRON; ARMENIAN BOLE; BAUXITE RESIDUE; BLACK OXIDE of IRON; BLENDED RED OXIDES of IRON; BURNT ISLAND RED; BURNT SIENNA; BURNT UMBER; ALCOTONE RED; CAPUT MORTUUM; C.I. 77491; C.I. PIGMENT RED 101; COLCOTHAR;; COLLOIDAL FERRIC OXIDE; CROCUS MARTIS ADSTRINGENS; DEANOX; EINECS No. 215-570-8; EISENOXYD; ENGLISH RED; FERRIC OXIDE; FERROUS FERRITE; FERRUGO; INDIAN RED; IRON(III) OXIDE; IRON OXIDE RED 130B; IRON OXIDE, SPENT; IRON SESQUIOXIDE; IRON SPONGE; JEWELERS ROUGE; LEVANOX RED 130A; LIGHT RED; MANUFACTURED IRON OXIDES; MARS BROWN; MARS RED; MIO 40GN; NATURAL IRON OXIDES; NATURAL RED OXIDE; OCHRE; PRUSSIAN BROWN; RADDLE; 11554 RED; RED IRON OXIDE; RED OCHRE; RED OXIDE; ROUGE; RUBIGO; SIENNA; SIFERRIT; SPECULAR IRON; STONE RED; SUPRA; SYNTHETIC IRON OXIDE; TURKEY RED; VENETIAN RED; VITRIOL RED; VOGELS IRON RED; YELLOW FERRIC OXIDE; YELLOW OXIDE of IRON; ZELAZA THLENK (Polish)

IDENTIFICATION

CAS:	1309-37-1; 1332-37-2 (spent)
ERG ID:	UN 1376
ERG Guide:	135
ERG Guide Category:	SUBSTANCES - SPONTANEOUSLY COMBUSTIBLE
Molecular Formula:	Fe_2O_3; FeO_2
RTECS No.:	NO7400000
Properties:	Flammable fine powder. Dark red or bluish. Usually used in cake form or impregnated in paper or cloth. Insoluble in water.

HEALTH AND SAFETY INFORMATION

May cause metal fume fever with flulike symptoms, including fever, chills, chest tightness, cough, and metallic taste in mouth. May cause irritation of the eyelids, permanent staining of the eyes. May cause bronchitis and chronic cough.

IARC Evaluation: Group 3, human evidence limited
IDLH: 2,500 mg/m^3 as Iron oxide dust and fume as Fe
OSHA Table Z-1-A Air Contaminant as fume

Exposure Limits:	**TWA**
ACGIH:[(1)]	5 mg/m^3 vapor, fume as Fe; 10 ppm; 5 mg/m^3 (respirable, as rouge)
NIOSH:	5 mg/m^3 dust and fume as Fe
OSHA:	(total)10 mg/m^3, dust and fume as Fe; 10 mg/m^3; 5 mg/m^3, (respirable as rouge).

[(1)]Inhalable particulate matter containing no asbestos and <1% crystalline silica; not classifiable as a human carcinogen.

RECOMMENDATIONS FOR RESPIRATOR SELECTION: NIOSH as iron oxide dust and fume as Fe

50 mg/m^3: DMFu (any dust, mist, and fume respirator);or SA (any supplied-air respirator)
125 mg/m^3: SA:CF (any supplied-air respirator operated in a continuous-flow mode);or PAPRDMFu (any powered, air-purifying respirator with a dust, mist, and fume filter).
250 mg/m^3: HiEF (any air-purifying, full-facepiece respirator with a high-efficiency particulate filter);or PAPRTHiE (any powered, air-purifying respirator with a tight-fitting facepiece and a high-efficiency particulate filter);or SCBAF (any self-contained breathing apparatus with a full facepiece);or SAF (any supplied-air respirator with a full facepiece).
2,500 mg/m^3: SA:PD,PP (any supplied-air respirator operated in a pressure-demand or other positive-pressure mode).

Emergency or Planned Entry into Unknown Concentrations or IDLH Conditions:
SCBAF:PD,PP (any self-contained breathing apparatus that has a full facepiece and is operated in a pressure-demand or other positive-pressure mode);or SAF:PD,PP:ASCBA (any supplied-air respirator that has a full facepiece and is operated in a pressure-demand or other positive-pressure mode in combination

with an auxiliary, self-contained breathing apparatus operated in a pressure-demand or other positive-pressure mode).

Escape: HiEF (any air-purifying, full-facepiece respirator with a high-efficiency particulate filter);or SCBAE (any appropriate escape-type, self-contained breathing apparatus).

CHEMICAL REACTIVITY

Contact with hydrogen peroxide, ethylene oxide, and calcium hypochlorite will cause explosions. Reacts violently with powdered aluminum, hydrazine, and hydrogen trisulfide. Contact with carbon monoxide may form an explosive material.

FIRE INFORMATION

Poisonous gases produced in fire. If material or contaminated runoff enters waterways, notify downstream users of potentially contaminated waters. Notify local health and fire officials and pollution control agencies.

NFPA Hazard Classifications: Not listed

FIRST AID

Move victim to fresh air. Call emergency medical care. Apply artificial respiration if victim is not breathing. Administer oxygen if breathing is difficult. Remove and isolate contaminated clothing and shoes. In case of contact with substance, immediately flush skin or eyes with running water for at least 20 minutes. Keep victim warm and quiet. Ensure that medical personnel are aware of the material(s) involved and take precautions to protect themselves.

ISOAMYLACETATE I:083

SYNONYMS: ACETATE d'ISOAMYLE (French); ACETATO de ISOAMILO (Spanish); ACETIC ACID, ISOPENTYL ESTER; AMYL ACETATE; sec-AMYL ACETATE; AMYLACETIC ESTER; BANANA OIL; 1-BUTANOL,3-METHYL-,ACETATE; EINECS No. 204-662-3; ISOAMYL ACETATE; ISOAMYL ETHANOATE; ISOPENTYL ACETATE; 3-METHYL-1-BUTYL ACETATE; 3-METHYL-1-BUTANOL ACETATE; 3-METHYLBUTYL ESTER OF ACETIC ACID; 3-METHYLBUTYL ETHANOATE; METHYLBUTYL ETHANOATE; PEAR OIL

IDENTIFICATION

CAS:	123-92-2
ERG ID:	UN 1104
ERG Guide:	129
ERG Guide Category:	FLAMMABLE LIQUIDS (POLAR/WATER-MISCIBLE/NOXIOUS)

Molecular Formula: $C_7H_{14}O_2$
RTECS No.: NS9800000
Properties: Flammable, oily liquid. Colorless. Pure: banana-pear odor; impure: penetrating, irritating and slightly applelike odor. Floats on water; slightly soluble. Flammable, irritating vapor is produced.

HEALTH AND SAFETY INFORMATION

Irritates eyes, skin, and respiratory tract.

IDLH: 1,000 ppm
Odor Threshold: 0.004 ppm
OSHA Table Z-1-A Air Contaminant

Exposure Limits:	**TWA**
ACGIH:	100 ppm; 532 mg/m^3
NIOSH:	100 ppm; 525 mg/m^3
OSHA:	100 ppm; 525 mg/m^3

RECOMMENDATIONS FOR RESPIRATOR SELECTION: NIOSH

1000 ppm: CCROV [any chemical cartridge respirator with organic vapor cartridge(s)];or PAPROV [any powered, air-purifying respirator with organic vapor cartridge(s)];or GMFOV[any air-purifying, full-facepiece respirator (gas mask) with a chin-style, front- or back-mounted organic vapor canister];or SA (any supplied-air respirator);or SCBAF (any self-contained breathing apparatus with a full facepiece)

Emergency or planned entry into unknown concentrations or idlh conditions: SCBAF:PD,PP (any self-contained breathing apparatus that has a full facepiece and is operated in a pressure-demand or other positive-pressure mode);or SAF:PD,PP:ASCBA (any supplied-air respirator that has a full facepiece and is operated in a pressure-demand or other positive-pressure mode in combination with an auxiliary, self-contained breathing apparatus operated in a pressure-demand or other positive pressure mode.

Escape: GMFOV(any air-purifying, full-facepiece respirator (gas mask) with a chin-style, front- or back-mounted organic vapor canister);or SCBAE (any appropriate escape-type, self-contained breathing apparatus).

CHEMICAL REACTIVITY

Reacts with nitrates, strong oxidizers, alkalis, and acids. Attacks asbestos. Softens, then dissolves many plastic materials and rubber.

FIRE INFORMATION

Vapors are heavier than air and will collect in low areas. Vapors in confined areas may explode when exposed to fire. Vapors may travel long distances to ignition sources and flash back. Storage containers and parts of containers may rocket great distances, in many directions. If material or contaminated runoff enters waterways, notify downstream users of potentially contaminated waters. Notify local health and fire officials and pollution control agencies. Use dry chemicals, foam, carbon dioxide.

NFPA Hazard Classifications:

Health Hazards (Blue) 1
Flammability (Red) 3
Reactivity (Yellow) 0

Flash Point: 77°F (25°C) c.c.
Flammable Limits in Air: LEL 1.0% @ 212°F (100°C); UEL 7.5%
Autoignition Temperature: 680°F (360°C)

FIRST AID

Move victim to fresh air. Call emergency medical care. Apply artificial respiration if victim is not breathing. Administer oxygen if breathing is difficult. Remove and isolate contaminated clothing and shoes. In case of contact with substance, immediately flush skin or eyes with running water for at least 20 minutes. Wash skin with soap and water. Keep victim warm and quiet. Effects of exposure (inhalation, ingestion, or skin contact) to substance may be delayed. Ensure that medical personnel are aware of the material(s) involved and take precautions to protect themselves.

ENVIRONMENTAL INFORMATION

Clean Water Act: Section 311 Hazardous Substances/RQ (same as CERCLA)
EPCRA Section 304 Reportable Quantity (RQ): CERCLA, 5,000 lb (2,270 kg)
TSCA: 40CFR712.(*e*)10; 40CFR716.120(*d*)10

ISOAMYL BUTYRATE I:084

SYNONYMS: BUTANOIC ACID, 3-METHYLBUTYL ESTER; EINECS No. 203-380-8; ISOAMYL BUTANOATE; ISOPENTYL BUTANOATE; ISOPENTYLBUTYRATE; 3-METHYLBUTYL BUTYRATE

IDENTIFICATION

CAS:	106-27-4
ERG ID:	UN 1993 as combustible liquid, n.o.s.
ERG Guide:	128
ERG Guide Category:	FLAMMABLE LIQUIDS (NON-POLAR/WATER-IMMISICIBLE)
Molecular Formula:	$C_9H_{18}O_2$
RTECS No.:	ET5034000
Properties:	Combustible liquid. Colorless. Fruitlike odor. Insoluble in water.

HEALTH AND SAFETY INFORMATION

Irritates eyes, skin, and respiratory tract.

RECOMMENDATIONS FOR RESPIRATOR SELECTION

At any concentrations above the NIOSH REL, or where there is no REL, at any detectable concentration: SCBAF:PD,PP (any self-contained breathing apparatus that has a full facepiece and is operated in a pressure-demand or other positive-pressure mode);or SAF:PD,PP:ASCBA (any supplied-air respirator that has a full facepiece and is operated in a pressure-demand or other positive-pressure mode in combination with an auxiliary, self-contained breathing apparatus operated in a pressure-demand or other positive-pressure mode).

Escape: GMFOV [any air-purifying, full-facepiece respirator (gas mask) with a chin-style, front- or back-mounted organic vapor canister];or SCBAE (any appropriate escape-type, self-contained breathing apparatus).

CHEMICAL REACTIVITY

Contact with oxidizers may cause fire and explosions.

FIRE INFORMATION

NFPA Hazard Classifications:

Health Hazards (Blue) . . . . -
Flammability (Red) 2
Reactivity (Yellow) . . . . -

Flash Point: 138°F (59°C)

FIRST AID

Move victim to fresh air. Call emergency medical care. Apply artificial respiration if victim is not breathing. Administer oxygen if breathing is difficult. Remove and isolate contaminated clothing and shoes. In case of contact with substance, immediately flush skin or eyes with running water for at least 20 minutes. Wash skin with soap and water. Keep victim warm and quiet. Effects of exposure (inhalation, ingestion, or skin contact) to substance may be delayed. Ensure that medical personnel are aware of the material(s) involved and take precautions to protect themselves.

ISOAMYL CHLORIDE I:085

SYNONYMS: BUTANE, 1-CHLORO-3-METHYL-; 1-CHLORO-3-METHYLBUTANE; EINECS No. 203-525-5

IDENTIFICATION

CAS:	107-84-6
ERG ID:	UN 1992 as flammable liquid, toxic, n.o.s.
ERG Guide:	131
ERG Guide Category:	FLAMMABLE LIQUIDS-TOXIC
Molecular Formula:	$C_5H_{11}Cl$ (various isomers have the same name)
RTECS No.:	NA
Properties:	Highly flammable liquid. Colorless to light yellow. Slightly soluble in water.

HEALTH AND SAFETY INFORMATION

A toxic chlorinated hydrocarbon (aliphatic). Suspected carcinogen; may cause blood, and liver, skin, kidney problems including tumors.

RECOMMENDATIONS FOR RESPIRATOR SELECTION

At any concentrations above the NIOSH REL, or where there is no REL, at any detectable concentration: SCBAF:PD,PP (any self-contained breathing apparatus that has a full facepiece and is operated in a pressure-demand or other positive-pressure mode);or SAF:PD,PP:ASCBA (any supplied-air respirator that has a full facepiece and is operated in a pressure-demand or other positive-pressure mode in combination with an auxiliary, self-contained breathing apparatus operated in a pressure-demand or other positive-pressure mode).

Escape: GMFOV [any air-purifying, full-facepiece respirator (gas mask) with a chin-style, front- or back-mounted organic vapor canister];or SCBAE (any appropriate escape-type, self-contained breathing apparatus).

CHEMICAL REACTIVITY

Forms explosive mixture with air. Contact with oxidizers may cause fire and explosions. Aliphatic chlorinated hydrocarbons react with aluminum, potassium, sodium, liquid oxygen.

FIRE INFORMATION

Combustion products include toxic fumes of chlorides. Vapors are heavier than air and will collect in low areas. Vapors in confined areas may explode when exposed to fire. Vapors may travel long distances to ignition sources and flash back. Storage containers and parts of containers may rocket great distances, in many directions. If material or contaminated runoff enters waterways, notify downstream users of potentially contaminated waters. Notify local health and fire officials and pollution control agencies. Use dry chemicals, foam, carbon dioxide. Use water spray to cool exposed containers.

NFPA Hazard Classifications:

Health Hazards (Blue) . . . . -

Flammability (Red) 3
Reactivity (Yellow) .... -

Flash Point: <70°F (<21°C)
Flammable Limits in Air: LEL 1.5%; UEL 7.7%
Autoignition Temperature:

FIRST AID

Move victim to fresh air. Call emergency medical care. Apply artificial respiration if victim is not breathing. Administer oxygen if breathing is difficult. Remove and isolate contaminated clothing and shoes. In case of contact with substance, immediately flush skin or eyes with running water for at least 20 minutes. Wash skin with soap and water. Keep victim warm and quiet. Effects of exposure (inhalation, ingestion, or skin contact) to substance may be delayed. Ensure that medical personnel are aware of the material(s) involved and take precautions to protect themselves.

ISOBUTANE I:087

SYNONYMS: 1,1-DIMETHYLETHANE; EINECS No. 200-857-2; ISOBUTANO (Spanish); 2-METILPROPANO (Spanish); 2-METHYLPROPANE; PROPANE, 2-METHYL; STCC 4905747; TRIMETHYLMETHANE

IDENTIFICATION

CAS:	75-28-5
ERG ID:	UN 1969; UN 1011 (butane mixture)
ERG Guide:	115
ERG Guide Category:	GASES- FLAMMABLE (INCLUDING REFRIGERATED LIQUIDS)
Molecular Formula:	C_4H_{10}
RTECS No.:	TZ4300000
Properties:	Extremely flammable, colorless gas or liquid (below 11°F; −12°C). Shipped as a liquefied, compressed gas. Gasolinelike odor. Liquid floats on the surface of water and "boils," forming a visible vapor cloud; slightly soluble in water.

HEALTH AND SAFETY INFORMATION

Contact with liquid may cause frostbite.

Exposure Limits: **TWA**

ACGIH: None

NIOSH: 800 ppm; 1,900 mg/m^3
OSHA: None

RECOMMENDATIONS FOR RESPIRATOR SELECTION: NIOSH

At any concentrations above the NIOSH REL, or where there is no REL, at any detectable concentration: SCBAF:PD,PP (any self-contained breathing apparatus that has a full facepiece and is operated in a pressure-demand or other positive-pressure mode);or SAF:PD,PP:ASCBA (any supplied-air respirator that has a full facepiece and is operated in a pressure-demand or other positive-pressure mode in combination with an auxiliary, self-contained breathing apparatus operated in a pressure-demand or other positive-pressure mode).

Escape: GMFOV [any air-purifying, full-facepiece respirator (gas mask) with a chin-style, front- or back-mounted organic vapor canister];or SCBAE (any appropriate escape-type, self-contained breathing apparatus).

CHEMICAL REACTIVITY

Forms explosive mixture with air. Contact with strong oxidizers (e.g., nitrates and perchlorates, chlorine, fluorine) may cause fire and explosions. May accumulate static electrical charges and may cause ignition of its vapors.

FIRE INFORMATION

Firefighting gear (including SCBA) does not provide adequate protection. If exposure occurs, remove and isolate gear immediately and thoroughly decontaminate personnel. Gases are heavier than air and will collect in low areas. Gases in confined areas may explode when exposed to fire. Gases may travel long distances to ignition sources and flash back. Storage containers and parts of containers may rocket great distances, in many directions. If material or contaminated runoff enters waterways, notify downstream users of potentially contaminated waters. Notify local health and fire officials and pollution control agencies. Do not extinguish the fire unless the flow of the gas can be stopped and any remaining gas is out of the line. Specially trained personnel may use fog lines to cool exposures and let the fire burn itself out. Use dry chemicals, foam, carbon dioxide. From a secure, explosion-proof location, use water spray to cool exposed containers. If cooling streams are ineffective (venting sound increases in volume and pitch, tank discolors, or shows any signs of deforming), withdraw immediately to a secure position.

NFPA Hazard Classifications:

Health Hazards (Blue) 1
Flammability (Red) 4
Reactivity (Yellow) 0

Flash Point: Flammable gas.
Flammable Limits in Air: LEL 1.6%; UEL 8.4%
Autoignition Temperature: 860°F (460°C)

FIRST AID

Move victim to fresh air. Call emergency medical care. Apply artificial respiration if victim is not breathing. Administer oxygen if breathing is difficult. Remove and isolate contaminated clothing and shoes. Clothing frozen to the skin should be thawed before being removed. In case of contact with liquefied gas, thaw frosted parts with lukewarm water. Keep victim warm and quiet. Ensure that medical personnel are aware of the material(s) involved and take precautions to protect themselves.

ENVIRONMENTAL INFORMATION

Clean Air Act: Accidental Release Prevention/Flammable substances, (Section 112[r], Table 3), TQ = 10,000 lb (4,540 kg).

ISOBUTYL ACETATE I:093

SYNONYMS: ACETATE d'ISOBUTYLE (French); ACETATO de ISOBUTILO (Spanish); ACETATO di ISOBUTILE (Italian); ACETIC ACID, ISOBUTYL ESTER; ACETIC ACID-2-METHYLPROPYL ESTER; EEC No. 607-026-00-7; EINECS No. 203-745-1; FEMA No. 2175; ISOBUTYLESTER KYSELINY OCTOVE (Czech); ISOBUTYL ESTER of ACETIC ACID; 2-METHYLPROPYL ACETATE; 2-METHYLPROPYL ESTER of ACETIC ACID; 2-METHYL-1-PROPYL ACETATE; β-METHYLPROPYL ETHANOATE

IDENTIFICATION

CAS:	110-19-0
ERG ID:	UN 1213
ERG Guide:	129
ERG Guide Category:	GASES-COMPRESSED OR LIQUEFIED (INCLUDING REFRIGERANT GASES)
Molecular Formula:	$C_6H_{12}O_2$
RTECS No.:	AI4025000
Properties:	Flammable, colorless liquid. Fruity, floral odor. Slightly soluble in water.

HEALTH AND SAFETY INFORMATION

IDLH: 1,300 ppm; 6,279 mg/m^3 [LEL]
Odor Threshold: 0.479 ppm
OSHA Table Z-1-A Air Contaminant

Exposure Limits:	**TWA**
ACGIH:	150 ppm; 713 mg/m^3

NIOSH: 150 ppm; 700 mg/m^3
OSHA: 150 ppm; 700 mg/m^3

RECOMMENDATIONS FOR RESPIRATOR SELECTION: NIOSH/OSHA

1,300 ppm: SA:CF (any supplied-air respirator operated in a continuous-flow mode);or CCRFOV [any chemical cartridge respirator with a full facepiece and organic vapor cartridge(s)];or GMFOV [any air-purifying, full-facepiece respirator (gas mask) with a chin-style, front- or back-mounted acid gas canister);or PAPROV [any powered, air-purifying respirator with organic vapor cartridge(s)]; SCBAF (any self-contained breathing apparatus with a full facepiece);or SAF (any supplied-air respirator with a full facepiece).

Emergency or Planned Entry into Unknown Concentrations or IDLH Conditions: SCBAF:PD,PP (any self-contained breathing apparatus that has a full facepiece and is operated in a pressure-demand or other positive-pressure mode);or SAF:PD,PP:ASCBA (any supplied-air respirator that has a full facepiece and is operated in a pressure-demand or other positive-pressure mode in combination with an auxiliary, self-contained breathing apparatus operated in a pressure-demand or other positive-pressure mode).

Escape: GMFOV [any air-purifying, full-facepiece respirator (gas mask) with a chin-style, front- or back-mounted organic vapor canister];or SCBAE (any appropriate escape-type, self-contained breathing apparatus).

Note: Substance causes eye irritation or damage; eye protection needed.

CHEMICAL REACTIVITY

Forms explosive mixture with air. Reacts with water on standing, producing acetic acid and n-butyl alcohol. Reacts violently with strong oxidizers. Reacts with caustics, strong acids, and nitrates. Dissolves rubber, many plastics, resins, and some coatings. May accumulate static electrical charges and cause ignition of its vapors.

FIRE INFORMATION

Combustion may produce irritants and toxic vapors. Gases are heavier than air and will collect in low areas. Gases in confined areas may explode when exposed to fire. Gases may travel long distances to ignition sources and flash back. Storage containers and parts of containers may rocket great distances, in many directions. If material or contaminated runoff enters waterways, notify downstream users of potentially contaminated waters. Notify local health and fire officials and pollution control agencies. Do not extinguish the fire unless the flow of the gas can be stopped and any remaining gas is out of the line. Specially trained personnel may use fog lines to cool exposures and let the fire burn itself out. Use dry chemicals, foam, carbon dioxide. From a secure, explosion-proof location, use water spray to cool exposed containers. If cooling streams are ineffective (venting sound increases in volume and pitch, tank discolors, or shows any signs of deforming), withdraw immediately to a secure position.

NFPA Hazard Classifications:

Health Hazards (Blue) 1
Flammability (Red) 3
Reactivity (Yellow) 0

Flash Point: 64°F (18°C)
Flammable Limits in Air: LEL 1.3%; UEL 10.5%
Autoignition Temperature: 790°F (421°C)

FIRST AID

Move victim to fresh air. Call emergency medical care. Apply artificial respiration if victim is not breathing. Administer oxygen if breathing is difficult. Remove and isolate contaminated clothing and shoes. In case of contact with substance, immediately flush skin or eyes with running water for at least 20 minutes. Wash skin with soap and water. Keep victim warm and quiet. Effects of exposure (inhalation, ingestion, or skin contact) to substance may be delayed. Ensure that medical personnel are aware of the material(s) involved and take precautions to protect themselves.

ISOBUTYL ALCOHOL I:096

SYNONYMS: ALCOHOL ISOBUTILICO (Spanish); ALCOOL ISOBUTYLIQUE (French); ALCOWIPE; ALPHA 100 FLUX; ALPHA 850-33 FLUX; AQUA-SOL FLUX; AVANTINE; BURMAR LAB CLEAN; C-589; CHEMTRANIC FLUX STRIPPER; COPPER 2 REAGENT; CP 290B ACTIVATOR; DAG 154; DAZZLENS CLEANER; EINECS No. 201-148-0; ENTAC 349 BIOCIDE; ENTEC 327 SURFACTANT; EPOXY CURE AGENT; FC-95; FEMA No. 2179; FERMENTATION BUTYL ALCOHOL; FILM REMOVER; 4282 FLUX; GLID-GUARD EPOXY SAFETY BLUE; HARDNESS 2 TEST SOLUTION; HIGH GRADE 1086; 1-HYDROXYMETHYLPROPANE; IBA; ISOBUTANOL; ISOBUTANOL (Spanish); ISOBUTYLALKOHOL (Czech); ISOPROPYLCARBINOL; KESTER 1585 ROSIN FLUX; KTI COP RINSE I/II; KTI MASK PROTECTIVE COATING; KTI NMD-25(+); KTI PBS RINSE; KTI PMMA RINSE; LENS CLEANER M6015; MARKEM 320 CLEANER; MAGIC GLASS CLEANER AND ANTIFOGGING FLUID; 2-METHYL-1-PROPANOL; 2-METHYLPROPYL ALCOHOL; MICROPOSIT NPE-210 SOLUTION; OMEGA METER SOLUTION; OPTI SKAN SCAN CLEANER; ORGANO FLUX 3355-11; PBS DEVELOPER; PBS RINSE; PC-96 SOLVENT SOLUBLE RESIST; PRIMER 910-S; 1-PROPANOL, 2-METHYL-; RN-10 E-BEAM NEGATIVE RESIST RINSE; RN-11 DEVELOPER; RN-11 E-BEAM NEGATIVE RESIST RINSE; 197 ROSIN FLUX; ROSIN FLUX KESTER 135/1544 MIL; ROSIN FLUX; RP-10 E-BEAM POSITIVE RESIST RINSE; SCAN KLEEN; SOLDER FLUX; SOLDER FLUX 2163 ORGANIC; SOLDER FLUX THINNER; STCC 4909131; STERETS PRE-INJECTION SWABS; SURFYNOL 104PA SURFACTANT; TRUE BLUE GLASS CLEANER; UVEX PRIMER 910S; VANDALEX 20; VANDALEX 124; WRS 200S SOLUTION; XEROX FILM REMOVER, TIP WIPES; XEROX CLEANER, FORMULA A

IDENTIFICATION

CAS: 78-83-1

ERG ID: UN 1212
ERG Guide: 129
ERG Guide Category: FLAMMABLE LIQUIDS (POLAR/WATER-MISCIBLE/NOXIOUS)
Molecular Formula: $C_4H_{10}O$
RTECS No.: NP9625000
Properties: Flammable, colorless, oily liquid. Sweet, musty odor. Slightly soluble in water.

HEALTH AND SAFETY INFORMATION

IDLH: 1,600 ppm; 4,928 mg/m^3
Odor Threshold: 0.832 ppm
OSHA Table Z-1-A Air Contaminant

Exposure Limits:	**TWA**
ACGIH:	50 ppm; 152 mg/m^3
NIOSH:	50 ppm; 150 mg/m^3
OSHA:	100 ppm; 300 mg/m^3

RECOMMENDATIONS FOR RESPIRATOR SELECTION: NIOSH

500 ppm: CCROV [any chemical cartridge respirator with organic vapor cartridge(s)];or SA (any supplied-air respirator).
1,250 ppm: SA:CF (any supplied-air respirator operated in a continuous-flow mode);or PAPROV [any powered, air-purifying respirator with organic vapor cartridge(s)].
1,600 ppm: CCRFOV [any chemical cartridge respirator with a full facepiece and organic vapor cartridge(s)];or GMFOV [any air-purifying, full-facepiece respirator (gas mask) with a chin-style, front- or back-mounted acid gas canister];or PAPRTOV [any powered, air-purifying respirator with a tight-fitting facepiece and organic vapor cartridge(s)];or SCBAF (any self-contained breathing apparatus with a full facepiece);or SAF (any supplied-air respirator with a full facepiece).

Emergency or Planned Entry into Unknown Concentrations or IDLH Conditions:
SCBAF:PD;PP (any self-contained breathing apparatus that has a full facepiece and is operated in a pressure-demand or other positive-pressure mode);or SAF:PD,PP:ASCBA (any supplied-air respirator that has a full facepiece and is operated in a pressure-demand or other positive-pressure mode in combination with an auxiliary, self-contained breathing apparatus operated in a pressure-demand or other positive-pressure mode).

Escape:GMFOV [any air-purifying, full-facepiece respirator (gas mask) with a chin-style, front- or back-mounted organic vapor canister];or SCBAE (any appropriate escape-type, self-contained breathing apparatus).

Note: Substance reported to cause eye irritation or damage; may require eye protection.

CHEMICAL REACTIVITY

Forms explosive mixture with air. Incompatible with strong acids, mineral acids, strong oxidizers, caustics, aliphatic amines, isocyanates, alkali metals, and alkaline earth. Attacks some plastics, rubber, and coatings. May react with aluminum at high temperatures.

FIRE INFORMATION

Vapors are heavier than air and will collect in low areas. Vapors in confined areas may explode when exposed to fire. Vapors may travel long distances to ignition sources and flash back. Storage containers and parts of containers may rocket great distances, in many directions. If material or contaminated runoff enters waterways, notify downstream users of potentially contaminated waters. Notify local health and fire officials and pollution control agencies. Use dry chemicals, foam, carbon dioxide.

NFPA Hazard Classifications:

Health Hazards (Blue) 1
Flammability (Red) 3
Reactivity (Yellow) 0

Flash Point: 82°F (28°C)
Flammable Limits in Air: LEL at (123°F; 51°C) 1.7%; UEL at (202°F; 94°C) 10.6%
Autoignition Temperature: 780°F (415°C)

FIRST AID

Move victim to fresh air. Call emergency medical care. Apply artificial respiration if victim is not breathing. Administer oxygen if breathing is difficult. Remove and isolate contaminated clothing and shoes. In case of contact with substance, immediately flush skin or eyes with running water for at least 20 minutes. Wash skin with soap and water. Keep victim warm and quiet. Effects of exposure (inhalation, ingestion, or skin contact) to substance may be delayed. Ensure that medical personnel are aware of the material(s) involved and take precautions to protect themselves.

ENVIRONMENTAL INFORMATION

EPA Hazardous Waste Number (RCRA No.): U140.
RCRA Section 261 Hazardous Constituents
RCRA Land Ban Waste
RCRA Universal Treatment Standards: Wastewater (mg/L), 5.6; Nonwastewater (mg/kg), 170.
RCRA Ground Water Monitoring List: Suggested test method(s) (PQL ug/L): 8015(50).
EPCRA Section 304: Reportable Quantity (RQ): CERCLA, 5,000 lb (2,270 kg).
TSCA: 40CFR799.5000 re testing consent orders.

ISOBUTYLENE I:100

SYNONYMS: γ-BUTYLENE; EEC No. 601-012-00-4; ISOBUTENE; ISOBUTILAMINA (Spanish); LIQUIFIED PETROLEUM GAS; 2-METHYLPROPENE

IDENTIFICATION

CAS:	115-11-7
ERG ID:	UN 1055
ERG Guide:	115 (22)
ERG Guide Category:	GASES- FLAMMABLE (INCLUDING REFRIGERATED LIQUIDS)
Molecular Formula:	C_4H_8
RTECS No.:	UD0890000
Properties:	Extremely flammable, liquefied, compressed gas. Becomes liquid below 20°F (7°C). Colorless. Sweet, gasolinelike odor. Floats and "boils" on water; insoluble. Flammable, visible vapor cloud is produced.

RECOMMENDATIONS FOR RESPIRATOR SELECTION

At any concentrations above the NIOSH REL, or where there is no REL, at any detectable concentration: SCBAF:PD,PP (any self-contained breathing apparatus that has a full facepiece and is operated in a pressure-demand or other positive-pressure mode);or SAF:PD,PP:ASCBA (any supplied-air respirator that has a full facepiece and is operated in a pressure-demand or other positive-pressure mode in combination with an auxiliary, self-contained breathing apparatus operated in a pressure-demand or other positive pressure mode)

Escape: GMFOV (any air-purifying, full-facepiece respirator (gas mask) with a chin-style, front- or back-mounted organic vapor canister;or SCBAE (any appropriate escape-type, self-contained breathing apparatus).

CHEMICAL REACTIVITY

Flammable gas; Forms explosive mixture with air. May be able to form unstable peroxides; may cause polymerization. Violent reaction with strong oxidizers, strong acids, oxides of nitrogen. May accumulate static electrical charges and may cause ignition of its vapors.

FIRE INFORMATION

Gases are heavier than air and will collect in low areas. Gases in confined areas may explode when exposed to fire. Gases may travel long distances to ignition sources and flash back. Storage containers and parts of containers may rocket great distances, in many directions. If material or contaminated runoff enters waterways, notify downstream users of potentially contaminated waters. Notify local health and fire officials and pollution control agencies. Do not extinguish the fire unless the flow of the gas can be stopped and any remaining gas is out of the line. Specially trained personnel may use fog lines to cool exposures and let the fire burn itself out. Use dry chemicals, foam, carbon dioxide. From a secure, explosion-proof location, use water spray to cool exposed containers. If cooling streams are ineffective (venting sound increases in volume and pitch, tank discolors, or shows any signs of deforming), withdraw immediately to a secure position.

NFPA Hazard Classifications:

Health Hazards (Blue) 1
Flammability (Red) 4
Reactivity (Yellow) 0

Flash Point: Flammable gas
Flammable Limits in Air: LEL 1.8%; UEL 9.6%
Autoignition Temperature: 869°F (465°C).

FIRST AID

Move victim to fresh air. Call emergency medical care. Apply artificial respiration if victim is not breathing. Administer oxygen if breathing is difficult. Remove and isolate contaminated clothing and shoes. Clothing frozen to the skin should be thawed before being removed. In case of contact with liquefied gas, thaw frosted parts with lukewarm water. Keep victim warm and quiet. Keep victim under observation. Ensure that medical personnel are aware of the material(s) involved and take precautions to protect themselves.

ISOOCTYL ALCOHOL I:111

SYNONYMS: ALCOHOL OCTILICO (Spanish); DIMETHYL-1-HEXANOLS; EINECS No. 248-133-5; ISOOCTANOL; 6-METHYL-1-HEPTANOL; OXO OCTYL ALCOHOL

IDENTIFICATION

CAS:	26952-21-6
ERG ID:	UN 1993
ERG Guide:	128
ERG Guide Category:	FLAMMABLE LIQUIDS (NON-POLAR/WATER-IMMISCIBLE)
Molecular Formula:	$C_8H_{18}O$
RTECS No.:	NS7700000
Properties:	Combustible liquid. Colorless. Mild odor. Floats on water; insoluble

HEALTH AND SAFETY INFORMATION

Irritating to eyes, skin, and respiratory tract. Absorbed through the skin.

OSHA Table Z-1-A Air Contaminant

Exposure Limits: **TWA**

ACGIH: (1)	50 ppm; 266 mg/m^3
NIOSH: (1)	50 ppm; 270 mg/m^3
OSHA:	None

(1) Skin contact contributes to overall exposure.

RECOMMENDATIONS FOR RESPIRATOR SELECTION

At any concentrations above the NIOSH REL, or where there is no REL, at any detectable concentration: SCBAF:PD,PP (any self-contained breathing apparatus that has a full facepiece and is operated in a pressure-demand or other positive-pressure mode);or SAF:PD,PP:ASCBA (any supplied-air respirator that has a full facepiece and is operated in a pressure-demand or other positive-pressure mode in combination with an auxiliary, self-contained breathing apparatus operated in a pressure-demand or other positive-pressure mode).

Escape: GMFOV [any air-purifying, full-facepiece respirator (gas mask) with a chin-style, front- or back-mounted organic vapor canister];or SCBAE (any appropriate escape-type, self-contained breathing apparatus).

FIRE INFORMATION

Vapors are heavier than air and will collect in low areas. Vapors in confined areas may explode when exposed to fire. Vapors may travel long distances to ignition sources and flash back. Storage containers and parts of containers may rocket great distances, in many directions. If material or contaminated runoff enters waterways, notify downstream users of potentially contaminated waters. Notify local health and fire officials and pollution control agencies. Use dry chemicals, foam, carbon dioxide.

NFPA Hazard Classifications:

Health Hazards (Blue) 0
Flammability (Red) 2
Reactivity (Yellow) 0

Flash Point: 180°F (82°C) o.c.
Flammable Limits in Air: LEL 0.9% (calc.); UEL 5.7% (est.)
Autoignition Temperature: 530°F (277°C) (est.)

FIRST AID

Move victim to fresh air. Call emergency medical care. Apply artificial respiration if victim is not breathing. Administer oxygen if breathing is difficult. Remove and isolate contaminated clothing and shoes. In case of contact with substance, immediately flush skin or eyes with running water for at least 20 minutes. Wash skin with soap and water. Keep victim warm and quiet. Effects of exposure (inhalation, ingestion, or skin contact) to substance may be delayed. Ensure that medical personnel are aware of the material(s) involved and take precautions to protect themselves.

ISOPENTANE I:112

SYNONYMS: BUTANE, 2-METHYL-; EEC No. 601-006-00-1; EINECS No. 201-142-8; ETHYL DIMETHYL METHANE; ISOAMYL HYDRIDE; ISOPENTANO (Spanish); 2-METHYLBUTANE

IDENTIFICATION

CAS:	78-78-4
ERG ID:	UN 1265
ERG Guide:	128
ERG Guide Category:	FLAMMABLE LIQUIDS (NONPOLAR/WATER-IMMISCIBLE)
Molecular Formula:	C_5H_{12}
RTECS No.:	EK4430000
Properties:	Highly flammable, watery liquid. Colorless. Gasolinelike odor. Floats on water; insoluble. Flammable, irritating vapor is produced. Boiling point is 82°F (28°C).

HEALTH AND SAFETY INFORMATION

Irritates eyes, skin, and respiratory tract. May cause lung problems, including pneumonia and pulmonary edema, which can be delayed for several hours; there is a risk of death in serious cases.

Exposure Limits:	**TWA**	**CEILING**
ACGIH:	600 ppm; 1,770 mg/m^3 as pentane	
NIOSH:	120 ppm; 350 mg/m^3	610 ppm; 1,800 mg/m^3 15-min. as n-pentane
OSHA:	1,000 ppm; 2,950 mg/m^3	

RECOMMENDATIONS FOR RESPIRATOR SELECTION: NIOSH as n-pentane

1200 ppm: SA (any supplied-air respirator)
1500 ppm: SA:CF (any supplied-air respirator operated in a continuous-flow mode);or SCBAF (any self-contained breathing apparatus with a full facepiece);or SAF (any supplied-air respirator with a full facepiece)

Emergency or planned entry into unknown concentrations or IDLH conditions: SCBAF:PD,PP (any self-contained breathing apparatus that has a full facepiece and is operated in a pressure-demand or other positive-pressure mode);or SAF:PD,PP:ASCBA (any supplied-air respirator that has a full facepiece and is operated in a pressure-demand or other positive-pressure mode in combination with an auxiliary, self-contained breathing apparatus operated in a pressure-demand or other positive pressure mode.

Escape: GMFOV(any air-purifying, full-facepiece respirator (gas mask) with a chin-style, front- or back-mounted organic vapor canister;or SCBAE (any appropriate escape-type, self-contained breathing apparatus).

CHEMICAL REACTIVITY

Dangerous reaction with oxidizers. Highly volatile liquid; vapors may explode when mixed with air. May react with some plastics. May accumulate static electricity and cause ignition of its vapors.

FIRE INFORMATION

Vapors are heavier than air and will collect in low areas. Vapors in confined areas may explode when exposed to fire. Vapors may travel long distances to ignition sources and flash back. Storage containers and parts of containers may rocket great distances, in many directions. If material or contaminated runoff enters waterways, notify downstream users of potentially contaminated waters. Notify local health and fire officials and pollution control agencies. Use dry chemicals, foam, carbon dioxide.

NFPA Hazard Classifications:

Health Hazards (Blue) 1
Flammability (Red) 4
Reactivity (Yellow) 0

Flash Point: less than-60°F (less than-51°C) c.c.
Flammable Limits in Air: LEL 1.4%; UEL 7.6%
Autoignition Temperature: 788°F (420°C)

FIRST AID

Move victim to fresh air. Call emergency medical care. Apply artificial respiration if victim is not breathing. If breathing is difficult, give oxygen. Remove and isolate contaminated clothing and shoes. In case of contact with substance, immediately flush skin or eyes with running water for at least 20 minutes. Wash skin with soap and water. Keep victim warm and quiet. Ensure that medical personnel are aware of the material(s) involved and take precautions to protect themselves.

ENVIRONMENTAL INFORMATION

Clean Air Act: Accidental Release Prevention/Flammable Substances (Section 112[r] Table 3), TQ = 10,000 lb (4,540 kg)

ISOPENTANOIC ACID I:113

SYNONYMS: ACIDO ISOVALERICO (Spanish); DELPHINIC ACID; EINECS No. 207-975-3; FEMA No. 3102; ISOPROPYLACETIC ACID; ISOVALERIANIC ACID; ISOVALERIC ACID; 3-METHYLBUTANOIC ACID; b-METHYLBUTYRIC ACID; 3-METHYLBUTYRIC ACID

IDENTIFICATION

CAS: 503-74-2

614 Isopentanoic Acid

ERG ID: UN 1993 as combustible liquid, n.o.s.
ERG Guide: 128
ERG Guide Category: FLAMMABLE LIQUIDS (NON-POLAR/WATER-IMMISCIBLE)
Molecular Formula: $C_5H_{10}O_2$
RTECS No.: NY1400000
Properties: Combustible liquid. Colorless, oily. Penetrating, strong cheeselike odor. Acid taste. Insoluble in water.

HEALTH AND SAFETY INFORMATION

Corrosive to eyes, skin, and respiratory tract. Inhalation of fumes may cause pulmonary edema, which can be delayed for several hours; there is a risk of death in serious cases.

RECOMMENDATIONS FOR RESPIRATOR SELECTION

At any concentrations above the NIOSH REL, or where there is no REL, at any detectable concentration: SCBAF:PD,PP (any self-contained breathing apparatus that has a full facepiece and is operated in a pressure-demand or other positive-pressure mode);or SAF:PD,PP:ASCBA (any supplied-air respirator that has a full facepiece and is operated in a pressure-demand or other positive-pressure mode in combination with an auxiliary, self-contained breathing apparatus operated in a pressure-demand or other positive-pressure mode).

Escape: GMFOV [any air-purifying, full-facepiece respirator (gas mask) with a chin-style, front- or back-mounted organic vapor canister];or SCBAE (any appropriate escape-type, self-contained breathing apparatus).

CHEMICAL REACTIVITY

Contact with oxidizers may cause fire and explosion. An organic acid; incompatible with sulfuric acid, caustics, ammonia, aliphatic amines, alkanolamines, isocyanates, alkylene oxides, epichlorihydrin..

FIRE INFORMATION

Vapors are heavier than air and will collect in low areas. Vapors in confined areas may explode when exposed to fire. If material or contaminated runoff enters waterways, notify downstream users of potentially contaminated waters. Notify local health and fire officials and pollution control agencies. Use dry chemicals, alcohol-resistant foam, carbon dioxide.

NFPA Hazard Classifications

Health Hazards (Blue) 1
Flammability (Red) . . . . -
Reactivity (Yellow) 0

Flash Point: 158°F (70°C)

Autoignition Temperature: 781°F (416°C)

FIRST AID

Move victim to fresh air. Call emergency medical care. Apply artificial respiration if victim is not breathing. Do not use mouth-to-mouth method if victim ingested or inhaled the substance; induce artificial respiration with the aid of a pocket mask equipped with a one-way valve or other proper respiratory medical device. Administer oxygen if breathing is difficult. Remove and isolate contaminated clothing and shoes. In case of contact with substance, immediately flush skin or eyes with running water for at least 20 minutes. Wash skin with soap and water. Keep victim warm and quiet. Effects of exposure (inhalation, ingestion, or skin contact) to substance may be delayed. Ensure that medical personnel are aware of the material(s) involved and take precautions to protect themselves.

ISOPHORONE I:114

SYNONYMS: EEC No. 606-012-00-8; EINECS No. 201-126-0; ISOACETOPHORONE; ISOFORONA (Spanish); 1,1,3-TRIMETHYL-3-CYCLOHEXEN-5-ONE; 3,5,5-TRIMETHYL-2-CYCLOHEXANE-1-ONE; 3,5,5-TRIMETHYL-2-CYCLOHENONE

IDENTIFICATION

CAS:	78-59-1
ERG ID:	UN 1993
ERG Guide:	128
ERG Guide Category:	FLAMMABLE LIQUIDS (NONPOLAR/WATER-IMMISCIBLE)
Molecular Formula:	$C_9H_{14}O$
RTECS No.:	GW7700000
Properties:	Combustible liquid. Colorless. Camphorlike; peppermint odor. Floats on water and mixes slowly; slightly soluble.

HEALTH AND SAFETY INFORMATION

A very dangerous ketone. May have a narcotic effect. Breathing vapors can cause death. Irritates eyes, skin, and respiratory tract. Possible skin, cornea, liver, and kidney damage.

IDLH: 200 ppm
Odor Threshold: 0.631 ppm
OSHA Table Z-1-A Air Contaminant
NIOSH Criteria Document: NIOSH 78-173 KETONES

Exposure Limits:	**TWA**	**CEILING**
ACGIH: (1)		5 ppm; 28 mg/m^3

NIOSH: 4 ppm; 23 mg/m^3
OSHA: 25 ppm; 140 mg/m^3

$^{(1)}$ Animal carcinogen

CHEMICAL REACTIVITY

Forms explosive mixture with air. Reacts with oxidizers, strong alkalis, amines.

FIRE INFORMATION

Vapors are heavier than air and will collect in low areas. Vapors in confined areas may explode when exposed to fire. Storage containers and parts of containers may rocket great distances, in many directions. If material or contaminated runoff enters waterways, notify downstream users of potentially contaminated waters. Notify local health and fire officials and pollution control agencies. Stop flow if possible; otherwise do not extinguish. Use dry chemicals, alcohol-resistant foam, carbon dioxide.
Use water spray to cool exposed containers.

NFPA Hazard Classifications:

Health Hazards (Blue) 2
Flammability (Red) 0
Reactivity (Yellow) 0

Flash Point: 205°F (96°C) o.c.; 184°F (84°C) c.c.
Flammable Limits in Air: LEL 0.84%; UEL 3.8%
Autoignition Temperature: 860°F (460°C)

FIRST AID

Move victim to fresh air. Call emergency medical care. Apply artificial respiration if victim is not breathing. If breathing is difficult, give oxygen. Remove and isolate contaminated clothing and shoes. In case of contact with substance, immediately flush skin or eyes with running water for at least 20 minutes. Wash skin with soap and water. Keep victim warm and quiet. Ensure that medical personnel are aware of the material(s) involved and take precautions to protect themselves.

ENVIRONMENTAL INFORMATION

Clean Air Act: Hazardous Air Pollutants (Title I, Part A, Section 112)
Clean Water Act: Section 307 Toxic Pollutants; Section 307 Priority Pollutants
RCRA Ground Water Monitoring List: Suggested test methods(s) (PQL μg/L); 8090 (60); 8270 (10)
Safe Drinking Water Act: Priority List (55FR 1470)
EPCRA Section 304 Reportable Quantity (RQ): CERCLA, 5,000 lb (2,270 kg)
TSCA: 40CFR716.120(*a*)

ISOPHORONE DIAMINE I:115

SYNONYMS: AMINO-1,3,3-TRIMETHYLCYCLOHEXANEMETHANAMINE(5-); 3-AMINOMETHYL-3,5,5-TRIMETHYLCYCLO-HEXYLAMINE; CYCLOHEXANEMETHANAMINE, 5-AMINO-1,3,3-TRIMETHYL-; EEC No. 612-067-00-9; EINECS No. 220-666-8; VESTAMIN® IPD

IDENTIFICATION

CAS:	2855-13-2
ERG ID:	UN 2289
ERG Guide:	153
ERG Guide Category:	SUBSTANCES- TOXIC AND/OR CORROSIVE (COMBUSTIBLE)
Molecular Formula:	$C_{10}H_{22}N_2$
Properties:	Combustible liquid. Colorless. Faint amine odor. Floats and mixes with water. Freezing point is 50°F (10°C).

HEALTH AND SAFETY INFORMATION

Corrosive to eyes, skin, and respiratory tract. Inhalation may cause pulmonary edema, which can be delayed or several hours; there is a risk of death in serious cases.

RECOMMENDATIONS FOR RESPIRATOR SELECTION

At any concentrations above the NIOSH REL, or where there is no REL, at any detectable concentration: SCBAF:PD,PP (any self-contained breathing apparatus that has a full facepiece and is operated in a pressure-demand or other positive-pressure mode);or SAF:PD,PP:ASCBA (any supplied-air respirator that has a full facepiece and is operated in a pressure-demand or other positive-pressure mode in combination with an auxiliary, self-contained breathing apparatus operated in a pressure-demand or other positive-pressure mode).

Escape: GMFOV [any air-purifying, full-facepiece respirator (gas mask) with a chin-style, front- or back-mounted organic vapor canister];or SCBAE (any appropriate escape-type, self-contained breathing apparatus).

CHEMICAL REACTIVITY

Reacts with acids, alcohols, glycols. May ignite on contact with cellulose nitrate of high surface area. Incompatible with acrolein, acrylonitrile, tert-butyl nitroacetylene, ethylene oxide, isopropyl chlorocarbonate, maleic anhydride, triisobutylaluminum. Corrodes aluminum and steel in the presence of moisture and carbon dioxide.

FIRE INFORMATION

Vapors are heavier than air and will collect in low areas. Vapors in confined areas may explode when exposed to fire. If material or contaminated runoff enters waterways, notify downstream users of potentially contaminated waters. Notify local health and fire officials and pollution control agencies. Use dry chemicals, alcohol-resistant foam, carbon dioxide.

FEMA Hazard Classifications

Health Hazards (Blue) 1
Flammability (Red) 1
Reactivity (Yellow) 0

Flash Point: 230°F (110°C) o.c.

FIRST AID

Move victim to fresh air and call emergency medical care. Apply artificial respiration if victim is not breathing. Do not use mouth-to-mouth method if victim swallowed or inhaled the substance; induce artificial respiration with the aid of a pocket mask equipped with a one-way valve or other proper respiratory medical device. If breathing is difficult, give oxygen. In case of contact with substance, immediately flush skin or eyes with running water for at least 20 minutes. Remove and isolate contaminated clothing and shoes. Keep victim warm and quiet. Effects of exposure (inhalation, ingestion, or skin contact) to substances may be delayed. Ensure that medical personnel are aware of the material(s) involved and take precautions to protect themselves.

ISOPHORONE DIISOCYANATE I:116:

SYNONYMS: DIISOCIANATO de ISOFORONA (Spanish); EEC No. 615-008-00-5; EINECS No. 223-861-6; IPDI; ISOPHORONE DIAMINE DIISOCYANATE; 3-ISOCYANATOMETHYL-3,5,5-TRIMETHYLCYCLOHEXYL-ISOCYATE; ISOPHORONE DIAMINE DIISOCYANATE

IDENTIFICATION

CAS:	4098-71-9
ERG ID:	UN 2290
ERG Guide:	156
ERG Guide Category:	SUBSTANCES- TOXIC AND/OR CORROSIVE (COMBUSTIBLE/WATER-SENSITIVE)
Molecular Formula:	$C_{12}H_{18}N_2O_2$
RTECS No.:	NQ9370000
Properties:	Combustible liquid. Colorless or yellowish. Sinks and reacts with water to produce carbon dioxide and toxic isophorone diamine.

HEALTH AND SAFETY INFORMATION

Poisonous. Irritating to eyes, skin, and respiratory tract. May be absorbed through the skin.

OSHA Table Z-1-A Air Contaminant

Exposure Limits:	**TWA**	**STEL**
ACGIH:	0.005 ppm; 0.045 mg/m^3	
NIOSH: [(1)]	0.005 ppm; 0.045 mg/m^3	0.02 ppm; 0.180 mg/m^3

[(1)] Skin contact contributes significantly to overall exposure.

RECOMMENDATIONS FOR RESPIRATOR SELECTION: NIOSH

0.05 ppm: SA (any supplied-air respirator)
0.125 ppm: SA:CF (any supplied-air respirator operated in a continuous-flow mode)
0.25 ppm: SCBAF (any self-contained breathing apparatus with a full facepiece);or SAF (any supplied-air respirator with a full facepiece)

Emergency or planned entry into unknown concentrations or idlh conditions: SCBAF:PD,PP (any self-contained breathing apparatus that has a full facepiece and is operated in a pressure-demand or other positive-pressure mode);or SAF:PD,PP:ASCBA (any supplied-air respirator that has a full facepiece and is operated in a pressure-demand or other positive-pressure mode in combination with an auxiliary, self-contained breathing apparatus operated in a pressure-demand or other positive pressure mode.

Escape: GMFOV (any air-purifying, full-facepiece respirator (gas mask) with a chin-style, front- or back-mounted organic vapor canister;or SCBAE (any appropriate escape-type, self-contained breathing apparatus).

Note: Substance reported to cause eye irritation or damage; may require eye protection.

CHEMICAL REACTIVITY

Reacts with water to produce water-soluble isophorone diamine (a toxic and corrosive compound) and carbon dioxide. Reacts with alcohols, amines, mercaptans, amides, urethanes, ureas. Contact with aluminum, aluminum alloys, copper, or copper alloys is prohibited.

FIRE INFORMATION

Vapors are heavier than air and will collect in low areas. Vapors in confined areas may explode when exposed to fire.If material or contaminated runoff enters waterways, notify downstream users of potentially contaminated waters. Notify local health and fire officials and pollution control agencies. Use dry chemicals, alcohol-resistant foam, carbon dioxide. Use water spray to cool exposed containers.

NFPA Hazard Classifications:

Health Hazards (Blue) 2
Flammability (Red) 1
Reactivity (Yellow) 1
Special Notice . . . . Water reactive

Flash Point: 212°F (100°C)

FIRST AID

Move victim to fresh air. Call emergency medical care. Apply artificial respiration if victim is not breathing. Do not use mouth-to-mouth method if victim ingested or inhaled the substance; induce artificial respiration with the aid of a pocket mask equipped with a one-way valve or other proper respiratory medical device. Administer oxygen if breathing is difficult. Remove and isolate contaminated clothing and shoes. In case of contact with substance, immediately flush skin or eyes with running water for at least 20 minutes. For minor skin contact, avoid spreading material on unaffected skin. Keep victim warm and quiet. Effects of exposure (inhalation, ingestion, or skin contact) may be delayed. Ensure that medical personnel are aware of the material(s) involved and take precautions to protect themselves.

ENVIRONMENTAL INFORMATION

EPCRA Section 302, Extremely Hazardous Substances: TPQ = 100 lb (45.4 kg)
EPCRA Section 304 Reportable Quantity (RQ): EHS, 100 lb (45.4 kg)
EPCRA Section 313: Form R *de minimis* concentration reporting level: 1.0%. Form R Toxic Chemical Category Code: N120
TSCA: 40CFR 712.30(*e*)8; 40CFR716.120(*d*)8

ISOPHTHALIC ACID I:117

SYNONYMS: 1,3-BENZENEDICARBOXYLIC ACID; m-PHTHALIC ACID; ACIDO ISOFTALICO (Spanish)

IDENTIFICATION

CAS:	121-91-5
Molecular Formula:	$C_8H_6O_4$
RTECS No.:	NT2007000
Properties:	Combustible crystalline powder. Colorless. Slightly soluble in water.

HEALTH AND SAFETY INFORMATION

Irritates eyes and skin.

RECOMMENDATIONS FOR RESPIRATOR SELECTION

At any concentrations above the NIOSH REL, or where there is no REL, at any detectable concentration: SCBAF:PD,PP (any self-contained breathing apparatus that has a full facepiece and is operated in a pressure-demand or other positive-pressure mode);or SAF:PD,PP:ASCBA (any supplied-air respirator that has a full facepiece and is operated in a pressure-demand or other positive-pressure mode in

combination with an auxiliary, self-contained breathing apparatus operated in a pressure-demand or other positive-pressure mode).

Escape: GMFOV [any air-purifying, full-facepiece respirator (gas mask) with a chin-style, front- or back-mounted organic vapor canister];or SCBAE (any appropriate escape-type, self-contained breathing apparatus).

CHEMICAL REACTIVITY

Incompatible with sulfuric acid, caustics, ammonia, aliphatic amines, alkanolamines, isocyanates, alkylene oxides, epichlorohydrin..

FIRE INFORMATION

If material or contaminated runoff enters waterways, notify downstream users of potentially contaminated waters. Notify local health and fire officials and pollution control agencies. Use dry chemicals, alcohol-resistant foam, carbon dioxide.

FEMA Hazard Classifications:

Health Hazards (Blue) 1
Flammability (Red) 1
Reactivity (Yellow) 0

Autoignition Temperature: > 1202°F (>650°C)

FIRST AID

Move victim to fresh air. Call emergency medical care. Apply artificial respiration if victim is not breathing. Administer oxygen if breathing is difficult. Remove and isolate contaminated clothing and shoes. In case of contact with substance, immediately flush skin or eyes with running water for at least 20 minutes. Wash skin with soap and water. Keep victim warm and quiet. Effects of exposure (inhalation, ingestion, or skin contact) to substance may be delayed. Ensure that medical personnel are aware of the material(s) involved and take precautions to protect themselves.

ISOPHTHALOYL CHLORIDE I:118

SYNONYMS: 1,3-BENZENEDICARBONYL CHLORIDE; EINECS No. 202-774-7; ISOPHTHALIC ACID CHLORIDE; ISOPHTHALIC ACID DICHLORIDE; ISOPHTHALOYL DICHLORIDE; ISOPHTHALYL DICHLORIDE; m-PHTHALIC DICHLORIDE; m-PHTHALOYL CHLORIDE

IDENTIFICATION

CAS: 99-63-8

622 Isophthloyl Chloride

Molecular Formula: $C_8H_4Cl_2O_2$
RTECS No.: NT2625000
Properties: Combustible, crystalline solid. White. Insoluble in water.

HEALTH AND SAFETY INFORMATION

Irritates eyes, skin, and respiratory tract.

RECOMMENDATIONS FOR RESPIRATOR SELECTION

At any concentrations above the NIOSH REL, or where there is no REL, at any detectable concentration: SCBAF:PD,PP (any self-contained breathing apparatus that has a full facepiece and is operated in a pressure-demand or other positive-pressure mode);or SAF:PD,PP:ASCBA (any supplied-air respirator that has a full facepiece and is operated in a pressure-demand or other positive-pressure mode in combination with an auxiliary, self-contained breathing apparatus operated in a pressure-demand or other positive-pressure mode).

Escape: GMFOV [any air-purifying, full-facepiece respirator (gas mask) with a chin-style, front- or back-mounted organic vapor canister];or SCBAE (any appropriate escape-type, self-contained breathing apparatus).

CHEMICAL REACTIVITY

May react with water and alcohol. Incompatible with strong oxidizers.

FIRE INFORMATION

Combustion products include toxic fumes of chlorides and may include phosgenes. If material or contaminated runoff enters waterways, notify downstream users of potentially contaminated waters. Notify local health and fire officials and pollution control agencies. Use dry chemicals, alcohol-resistant foam, carbon dioxide, water spray.

NFPA Hazard Classifications

Health Hazards (Blue) .... -
Flammability (Red) 1
Reactivity (Yellow) 0

Flash Point: 365°F (185°C) o.c.

FIRST AID

Move victim to fresh air. Call emergency medical care. Apply artificial respiration if victim is not breathing. Do not use mouth-to-mouth method if victim swallowed or inhaled the substance; induce artificial respiration with the aid of a pocket mask equipped with a one-way valve or other proper respiratory medical device. Administer oxygen if breathing is difficult. Remove and isolate contaminated clothing and shoes. In case of contact with substance, immediately flush skin or eyes with running water for at least 20 minutes. For minor skin contact, avoid spreading material on unaffected skin. Keep victim warm and quiet. Effects of exposure (inhalation, ingestion, or skin contact) to substances may be delayed. Ensure that medical personnel are aware of the material(s) involved and take precautions to protect themselves.

ENVIRONMENTAL INFORMATION

TSCA: 40CFR716.120(*d*)13

ISOPRENE I:120

SYNONYMS: 1,3-BUTADIENE, 2-METHYL; EEC No. 601-014-00-5; EINECS No. 201-143-3; ISOPENTADIENE; ISOPRENO (Spanish); 2-METILBUTA-1,3-DIENE (Italian); β-METHYLBIVINYL; 2-METHYLBUTADIENE; 2-METHYL-1,3-BUTADIENE; 3-METHYL-1,3-BUTADIENE

IDENTIFICATION

CAS:	78-79-5
ERG ID:	UN 1218
ERG Guide:	130P
ERG Guide Category:	FLAMMABLE LIQUIDS (NONPOLAR/WATER IMMISCIBLE/NOXIOUS)
Molecular Formula:	C_5H_8
RTECS No.:	NT4037000
Properties:	Extremely flammable, watery liquid. Colorless. Mild petroleumlike odor. Floats on water; insoluble. Liquid boils at 93°F (35°C). Vapor is heavier than air; will collect in low areas.

HEALTH AND SAFETY INFORMATION

Odor Threshold: 0.005 ppm

Irritating to skin, eyes, nose, and lungs.

RECOMMENDATIONS FOR RESPIRATOR SELECTION

At any concentrations above the NIOSH REL, or where there is no REL, at any detectable concentration: SCBAF:PD,PP (any self-contained breathing apparatus that has a full facepiece and is operated in a pressure-demand or other positive-pressure mode);or SAF:PD,PP:ASCBA (any supplied-air respirator that has a full facepiece and is operated in a pressure-demand or other positive-pressure mode in combination with an auxiliary, self-contained breathing apparatus operated in a pressure-demand or other positive pressure mode).

Escape: GMFOV [any air-purifying, full-facepiece respirator (gas mask) with a chin-style, front- or back-mounted organic vapor canister];or SCBAE (any appropriate escape-type, self-contained breathing apparatus).

CHEMICAL REACTIVITY

Store under inert atmosphere (preferably nitrogen) with at least 50 ppm tert-butylcatechol present as an inhibitor. Forms explosive mixture with air. Air contact forms unstable peroxides; heat or buildup of peroxides may cause violent polymerization. Violent reaction with oxidizers, acids, and reducing agents. Attacks some paints, rubber, and coatings. May accumulate static electrical charges and may cause ignition of its vapors.

FIRE INFORMATION

Vapors may travel long distances to ignition sources and flash back. Vapors in confined areas or containers may explode in fire. If material or contaminated runoff enters waterways, notify downstream users of potentially contaminated water.

NFPA Hazard Classifications:

Health Hazards (Blue) 1
Flammability (Red) 4
Reactivity (Yellow) 2

Flash Point: −65°F (−54°C) c.c.
Flammable Limits in Air: LEL 2.0%; UEL 9.0%
Autoignition Temperature: 743°F (395°C)

FIRST AID

Move victim to fresh air. Call emergency medical care. Apply artificial respiration if victim is not breathing. Administer oxygen if breathing is difficult. Remove and isolate contaminated clothing and shoes. In case of contact with substance, immediately flush skin or eyes with running water for at least 20 minutes. Wash skin with soap and water. Keep victim warm and quiet. Effects of exposure (inhalation, ingestion, or skin contact) to substance may be delayed. Ensure that medical personnel are aware of the material(s) involved and take precautions to protect themselves.

ENVIRONMENTAL INFORMATION

Clean Air Act: Hazardous Air Pollutants (Title I, Part A, Section 112); Accidental Release Prevention/Flammable substances, (Section 112[r], Table 3), TQ = 10,000 lb (4,540 kg).
Clean Water Act: Section 311 Hazardous Substances/RQ (same as CERCLA); Section 307 Priority Pollutants; Toxic Pollutant (Section 401.15).
RCRA Ground Water Monitoring List: Suggested test method(s) (PQL ug/L): 8090(60); 8270(10).
EPCRA Section 304: Reportable Quantity (RQ): CERCLA, 100 lb (45.4 kg).

ISOPROPYL ACETATE

I:150

SYNONYMS: ACETATE d'ISOPROPYLE (French); ACETATO de ISOPROPILO (Spanish); ACETIC ACID ISOPROPYL ESTER; ACETIC ACID-1-METHYL ETHYL ESTER; 2-ACETOXYPROPANE; FEMA No. 2926; ISOPROPILE (ACETO di) (Italian); ISOPROPYLACETAAT (Dutch); ISOPROPYL ESTER of ACETIC ACID; ISOPROPYLACETAT (German); ISOPROPYL (ACETATE d') (French); ISOPROPYLESTER KYSELINY OCTOVE (Czech); 1-METHYLETHYL ESTER of ACETIC ACID; 2-PROPYL ACETATE

IDENTIFICATION

CAS:	108-21-4
ERG ID:	UN 1220
ERG Guide:	129
ERG Guide Category:	FLAMMABLE LIQUID (POLAR/WATER-MISCIBLE/ NOXIOUS)
Molecular Formula:	$C_5H_{10}O_2$
RTECS No.:	AI4930000
Properties:	Flammable, colorless liquid. Fruity odor. Slightly soluble in water.

HEALTH AND SAFETY INFORMATION

IDLH: 1,800 ppm
Odor Threshold: 2.4 ppm
OSHA Table Z-1-A Air Contaminant

Exposure Limits:	**TWA**	**STEL**
ACGIH:	250 ppm; 1,040 mg/m^3	310 ppm; 1,290 mg/m^3
NIOSH:	No established REL	
OSHA:	250 ppm; 950 mg/m^3	

RECOMMENDATIONS FOR RESPIRATOR SELECTION: NIOSH

1,800 ppm: SA:CF (any supplied-air respirator operated in a continuous-flow mode);or SCBAF (any self-contained breathing apparatus with a full facepiece);or SAF (any supplied-air respirator with a full facepiece).

Emergency or Planned Entry into Unknown Concentrations or IDLH Conditions:
SCBAF:PD,PP (any self-contained breathing apparatus that has a full facepiece and is operated in a pressure-demand or other positive-pressure mode);or SAF:PD,PP:ASCBA (any supplied-air respirator that has a full facepiece and is operated in a pressure-demand or other positive-pressure mode in combination with an auxiliary, self-contained breathing apparatus operated in a pressure-demand or other positive-pressure mode).

Escape: GMFOV [any air-purifying, full-facepiece respirator (gas mask) with a chin-style, front- or back-mounted organic vapor canister];or SCBAE (any appropriate escape-type, self-contained breathing apparatus).

Note: Substance reported to cause eye irritation or damage; may require eye protection.

CHEMICAL REACTIVITY

Forms explosive mixture with air. Contact with strong oxidizers may cause fire and explosions. Incompatible with strong acids, caustics, nitrates. Dissolves rubber and many plastic materials. Contact with steel may cause slow decomposition. May accumulate static electrical charges, and may cause ignition of its vapors.

FIRE INFORMATION

Vapors may travel long distances to ignition sources and flash back. Vapors in confined areas may explode in fire.

NFPA Hazard Classifications:

Health Hazards (Blue) 1
Flammability (Red) 3
Reactivity (Yellow) 0

Flash Point: 36°F (33°C)
Flammable Limits in Air: LEL at 100°F (38°C) 1.8%; UEL 8%
Autoignition Temperature: 860°F (460°C)

FIRST AID

Move victim to fresh air. Call emergency medical care. Apply artificial respiration if victim is not breathing. Administer oxygen if breathing is difficult. Remove and isolate contaminated clothing and shoes. In case of contact with substance, immediately flush skin or eyes with running water for at least 20 minutes. Wash skin with soap and water. Keep victim warm and quiet. Effects of exposure (inhalation, ingestion, or skin contact) to substance may be delayed. Ensure that medical personnel are aware of the material(s) involved and take precautions to protect themselves.

ISOPROPYL ALCOHOL I:160

SYNONYMS: ALCOHOL ISOPROPILICO (Spanish); ALCOOL ISOPROPYLIQUE (French); ALCOWIPE; AVANTINE; C-589; BURMAR LAB CLEAN; CHEMTRANIC FLUX STRIPPER; COPPER 2 REAGENT; CP 290B ACTIVATOR; DAG 154; DAZZLENS CLEANER; DIMETHYLCARBINOL; EEC No. 603-003-00-0; EINECS No. 200-661-7; ENTEC 327 SURFACTANT; ENTAC 349 BIOCIDE; EPOXY CURE AGENT; FC-95; FILM REMOVER; GLID-GUARD EPOXY SAFETY BLUE; HARDNESS 2 TEST SOLUTION; HIGH GRADE 1086; IPA; ISOPROPANOL;

KESTER 103 THINNER; KESTER 108 THINNER; KTI COP RINSE I/II; KTI MASK PROTECTIVE COATING; KTI NMD-25(+); KTI PBS RINSE; KTI PMMA RINSE; LENS CLEANER M6015; MAGIC GLASS CLEANER AND ANTIFOGGING FLUID; MARKEM 320 CLEANER; MICROPOSIT NPE-210 SOLUTION; OMEGA METER SOLUTION; 197 ROSIN FLUX; OPTI SKAN SCAN CLEANER; ORGANO FLUX 3355-11; PBS DEVELOPER; PC-96 SOLVENT SOLUBLE RESISPETROHOL; PRIMER 910-S; PROPAAN-2-OL (Dutch); 2-PROPANOL; sec-PROPYL ALCOHOL; RN-10 E-BEAM NEGATIVE RESIST RINSE; RN-11 E-BEAM NEGATIVE RESIST RINSE; RN-11 DEVELOPER; RP-10 E-BEAM POSITIVE RESIST RINSE; RUBBING ALCOHOL; SCAN KLEEN; SOLDER FLUX THINNER; SPINRITE ARSENIC; STERETS PRE-INJECTION SWABS; SURFYNOL 104PA SURFACTANT; TRUE BLUE GLASS CLEANER; UVEX PRIMER 910S; VANDALEX 20; VANDALEX 124; VWR GLASS CLEANER; WHIRLWIND GLASS CLEANER; WRS 200S SOLUTION

IDENTIFICATION

CAS:	67-63-0
ERG ID:	UN 1219
ERG Guide:	129
ERG Guide Category:	FLAMMABLE LIQUIDS (POLAR/WATER-MISCIBLE/ NOX IOUS)
Molecular Formula:	C_3H_8O
RTECS No.:	NT8050000
Properties:	Extremely flammable, clear, colorless, watery liquid. Odor like rubbing alcohol or acetone. Water soluble.

HEALTH AND SAFETY INFORMATION

Irritates eyes (corrosive), skin and respiratory tract. Destroys skins natural oils. Long-term exposure may cause dryness and cracking of skin; sinus cancer.

IARC Evaluation: Group 3, unclassifiable substance; human evidence, inadequate; animal evidence, inadequate.
IDLH: 2,000 ppm [10% LEL]
Odor Threshold: 0.442 ppm
OSHA Table Z-1-A Air Contaminant
NIOSH Criteria Document: NIOSH 76-142

Exposure Limits:	**TWA**	**STEL**
ACGIH:	400 ppm; 983 mg/m^3	500 ppm; 1,230 mg/m^3
NIOSH:	400 ppm; 980 mg/m^3	500 ppm; 1,225 mg/m^3
OSHA:	400 ppm; 980 mg/m^3	

RECOMMENDATIONS FOR RESPIRATOR SELECTION: NIOSH/OSHA

2,000 ppm: SA:CF (any supplied-air respirator operated in a continuous-flow mode);or CCRFOV [any chemical cartridge respirator wit a full-facepiece and organic vapor cartridge(s)];or GMFOV [any air-purifying, full-facepiece respirator (gas mask) with a chin-style, front- or back-mounted organic vapor canister];or PAPROV [any powered, air-purifying respirator with organic vapor cartridge(s)];or SCBAF (any self-contained breathing apparatus with a full facepiece);or SAF (any supplied-air respirator with a full facepiece).

Emergency or Planned Entry into Unknown Concentrations or IDLH Conditions: SCBAF:PD,PP (any self-contained breathing apparatus that has a full facepiece and is operated in a pressure-demand or other positive-pressure mode);or SAF:PD,PP:ASCBA (any supplied-air respirator that has a full facepiece and is operated in a pressure-demand or other positive-pressure mode in combination with an auxiliary, self-contained breathing apparatus operated in a pressure-demand or other positive-pressure mode).

Escape: GMFOV [any air-purifying, full-facepiece respirator (gas mask) with a chin-style, front- or back-mounted organic vapor canister];or SCBAE (any appropriate escape-type, self-contained breathing apparatus).
Note: Substance causes eye irritation or damage; eye protection needed.

CHEMICAL REACTIVITY

Forms explosive mixture with air. Violent reaction with phosgene, potassium tert-butoxide, oleum, and nitroform and hydrogen with palladium catalyst. Incompatible with alkaline earth and alkali metals, crotonaldehyde, strong acids, amines, ammonia, caustics, and strong oxidizers. Attacks some plastics, rubber, and coatings. Reacts with metallic aluminum at high temperatures.

FIRE INFORMATION

Vapors are heavier than air and will collect in low areas. Vapors in confined areas may explode when exposed to fire. Vapors may travel long distances to ignition sources and flash back. Storage containers and parts of containers may rocket great distances, in many directions. If material or contaminated runoff enters waterways, notify downstream users of potentially contaminated waters. Notify local health and fire officials and pollution control agencies. Use dry chemicals, alcohol-resistant foam, carbon dioxide, large quantities of water. Use water to cool exposed containers.

NFPA Hazard Classifications:

Health Hazards (Blue) 1
Flammability (Red) 3
Reactivity (Yellow) 0

Flash Point: 65°F (18°C) o.c.; 53°F (12°C) c.c
Flammable Limits in Air: LEL 2.0%; UEL 12.7% @ 200°F (93°C)
Autoignition Temperature: 750°F (399°C)

FIRST AID

Move victim to fresh air. Call emergency medical care. Apply artificial respiration if victim is not breathing. Administer oxygen if breathing is difficult. Remove and isolate contaminated clothing and shoes. In case of contact with substance, immediately flush skin or eyes with running water for at least 20 minutes. Wash skin with soap and water. Keep victim warm and quiet. Effects of exposure (inhalation, ingestion, or skin contact) to substance may be delayed. Ensure that medical personnel are aware of the material(s) involved and take precautions to protect themselves.

ENVIRONMENTAL INFORMATION

EPCRA Section 313: (reportable if being manufactured by the strong acid process only; no supplier notification); Form R *de minimis* concentration reporting level: 1.0%.
TSCA: 40CFR799.2325

ISOPROPYLAMINE I:200

SYNONYMS: 1-AMINO-2-HYDROXYPROPANE; 2-AMINOPROPANE; 1-AMINO-PROPANOL-2; EEC No. 612-007-00-1; EINECS No. 200-860-9; 2-HYDROXYPROPYLAMINE; ISOPROPILAMINA (Spanish); MONOISOPROPYLAMINE; 2-PROPANAMINE; sec-PROPYLAMINE

IDENTIFICATION

CAS:	75-31-0
ERG ID:	UN 1221
ERG Guide:	132
ERG Guide Category:	FLAMMABLE LIQUIDS- CORROSIVE
Molecular Formula:	C_3H_9N
RTECS No.:	NT8400000
Properties:	Extremely flammable liquid. Colorless. Ammonialike odor. Floats and mixes with water; large amounts of irritating vapor is produced. Boiling point is 91°F (33°C).

HEALTH AND SAFETY INFORMATION

Severely irritating to the eyes, skin, and respiratory tract. Skin and eye contact can cause severe burns and blindness.

IDLH: 750 ppm
Odor Threshold: 0.6 ppm
OSHA Table Z-1-A Air Contaminant
OSHA Process Safety Management of Highly Hazardous Chemicals (29CFR, Part 1910.119, Appendix A): Threshold Quantity: 5,000 pounds.

Exposure Limits:	TWA	STEL
ACGIH:	5 ppm; 12 mg/m^3	10 ppm; 24 mg/m^3
OSHA:	5 ppm; 12 mg/m^3	

RECOMMENDATIONS FOR RESPIRATOR SELECTION: OSHA

125 ppm: SA:CF (any supplied-air respirator operated in a continuous-flow mode);or PAPRS (any powered, air-purifying respirator with cartridge(s) providing protection against the compound of concern.
250 ppm: CCRFS(any chemical cartridge respirator with a full facepiece and cartridge(s) providing protection against the compound of concern)organic vapor and acid gas cartridge(s);or GMFS (any air-purifying, full-facepiece respirator (gas mask) with a chin-style, front- or back-mounted canister protection against the compound of concern);or PAPRTS (any powered, air-purifying respirator with a tight-fitting facepiece and cartridge(s) providing protection against the compound of concern);or SCBAF (any self-contained breathing apparatus with a full facepiece);or SAF (any supplied-air respirator with a full facepiece).
750 ppm: SAF:PD,PP (any supplied-air respirator that has a full facepiece and is operated in a pressure-demand or other positive-pressure mode).

Emergency or planned entry into unknown concentrations or idlh conditions: SCBAF:PD,PP (any self-contained breathing apparatus that has a full facepiece and is operated in a pressure-demand or other positive-pressure mode);or SAF:PD,PP:ASCBA (any supplied-air respirator that has a full facepiece and is operated in a pressure-demand or other positive-pressure mode in combination with an auxiliary, self-contained breathing apparatus operated in a pressure-demand or other positive pressure mode.

Escape: GMFS (any air-purifying, full-facepiece respirator (gas mask) with a chin-style, front- or back-mounted canister protection against the compound of concern);or SCBAE (any appropriate escape-type, self-contained breathing apparatus).

Note: Substance causes eye irritation or damage; eye protection needed.

CHEMICAL REACTIVITY

Contact with strong oxidizers may cause fire and explosions. Reacts with strong acids, aldehydes, ketones, epoxides. Severely corrodes aluminum, copper, and copper-based alloys (except monel).

FIRE INFORMATION

Combustion products include toxic nitrogen oxides. Firefighting gear (including SCBA) does not provide adequate protection. If exposure occurs, remove and isolate gear immediately and thoroughly decontaminate personnel. Vapors are heavier than air and will collect in low areas. Vapors in confined areas may explode when exposed to fire. Vapors may travel long distances to ignition sources and flash back. Storage containers and parts of containers may rocket great distances, in many directions.If material or contaminated runoff enters waterways, notify downstream users of potentially contaminated waters. Notify local health and fire officials and pollution control agencies. Use dry chemicals, foam, carbon dioxide.

NFPA Hazard Classifications:

Health Hazards (Blue) 3
Flammability (Red) 4
Reactivity (Yellow) 0

Flash Point:-35°F (-37°C) o.c.
Flammable Limits in Air: LEL 2.3%; UEL 12%
Autoignition Temperature: 756°F (402°C)

FIRST AID

Move victim to fresh air; call emergency medical care. Apply artificial respiration if victim is not breathing. Do not use mouth-to-mouth method if victim ingested or inhaled the substance; induce artificial respiration with the aid of a pocket mask equipped with a one-way valve or other proper respiratory medical device. Administer oxygen if breathing is difficult. Remove and isolate contaminated clothing and shoes. In case of contact with substance, immediately flush skin or eyes with running water for at least 20 minutes. For minor skin contact, avoid spreading material on unaffected skin. Keep victim warm and quiet. Effects of exposure (inhalation, ingestion, or skin contact) to substance may be delayed. Ensure that medical personnel are aware of the material(s) involved and take precautions to protect themselves.

ENVIRONMENTAL INFORMATION

Clean Air Act: Accidental Release Prevention/Flammable Substances (Section 112[r], Table 3), TQ = 10,000 lb (4,540 kg)
Marine Pollutant (49CFR, Subchapter 172.101, Appendix B).
TSCA: 40CFR721.4500 isopropyl amine, distillation residues (CAS 79771-08-7)

ISOPROPYLBENZENE PEROXIDE I:250

SYNONYMS: ACTIVE DICUMYL PEROXIDE; BIS(α,α-DIMETHYLBENZYL)PEROXIDE; CUMENE PEROXIDE; CUMYL PEROXIDE; DOLPHON, CC-130; DI-α-CUMYL PEROXIDE; DICUMYL PEROXIDE (DOT); DI-CUP; DI-CUP 40 KF; DI-CUPR; DIISOPROPYLBENZENE PEROXIDE; IMPCO HI-TEMP CATALYST; LUPERCO; LUPEROX; LUPEROX 500R; LUPEROX 500T; VAROX DCP-R; VAROX DCP-T

IDENTIFICATION

CAS:	80-43-3
ERG ID:	UN 3110 as organic peroxide type F, solid (exempt at ≤42%) UN 3109 as organic peroxide type F, liquid (exempt at ≤42%)
ERG Guide:	145
ERG Guide Category:	ORGANIC PEROXIDES (HEAT, CONTAMINATION AND FRICTION SENSITIVE)
Molecular Formula:	$C_{18}H_{22}O_2$
RTECS No.:	SD8150000; SD7885000
Properties:	Combustible, granular solid. Clear, amber, or pale yellow-white. Fruity odor. Practically insoluble in water.

HEALTH AND SAFETY INFORMATION

Irritates eyes, skin, and respiratory tract. Skin contact may contribute significantly to overall exposure. Reported to have low toxicity. Prolonged or repeated exposure can cause liver and lung damage.

RECOMMENDATIONS FOR RESPIRATOR SELECTION

At any concentrations above the NIOSH REL, or where there is no REL, at any detectable concentration: SCBAF:PD,PP (any self-contained breathing apparatus that has a full facepiece and is operated in a pressure-demand or other positive-pressure mode);or SAF:PD,PP:ASCBA (any supplied-air respirator that has a full facepiece and is operated in a pressure-demand or other positive-pressure mode in combination with an auxiliary, self-contained breathing apparatus operated in a pressure-demand or other positive-pressure mode).

Escape: GMFOV [any air-purifying, full-facepiece respirator (gas mask) with a chin-style, front- or back-mounted organic vapor canister];or SCBAE (any appropriate escape-type, self-contained breathing apparatus).

CHEMICAL REACTIVITY

A powerful oxidizer; reacts with strong acids, amines, reducing agents, combustibles, organic compounds, copper alloys, lead, iron, and many other substances and materials. May explode from heat or contamination.

FIRE INFORMATION

Combustion products include carbon dioxide, carbon monoxide, phenols, and organic acids. Vapors are heavier than air and will collect in low areas. Vapors in confined areas may explode when exposed to fire. Vapors may travel long distances to ignition sources and flash back. Storage containers and parts of containers may rocket great distances, in many directions. If material or contaminated runoff enters waterways, notify downstream users of potentially contaminated waters. Notify local health and fire officials and pollution control agencies. Use dry chemicals, carbon dioxide, water fog. From a secure, explosion-proof location, use water spray to cool exposed containers.

NFPA Hazard Classifications: Not listed
Flash Point: 260°F (127°C)

FIRST AID

Move victim to fresh air. Call emergency medical care. Apply artificial respiration if victim is not breathing. Administer oxygen if breathing is difficult. Remove and isolate contaminated clothing and shoes. Remove material from skin immediately. In case of contact with substance, immediately flush skin or eyes with running water for at least 20 minutes. Keep victim warm and quiet. Ensure that medical personnel are aware of the material(s) involved and take precautions to protect themselves.

ISOPROPYL ETHER I:300

SYNONYMS: DIISOPROPYL ETHER; DIISOPROPYL OXIDE; DI-(1-METHYLETHYL)ETHER; EEC No. 603-045-00-X; EINECS No. 203-560-6; ETER ISOPROPILICO (Spanish); 2-ISOPROPOXY PROPANE; OSSIDO di DIISOPROPILE (Italian); 2,2'-OXYBISPROPANE; OXYDE de DIISOPROPYLE (French); PROPANE,2,2'-OXYBIS-

IDENTIFICATION

CAS:	108-20-1
ERG ID:	UN 1159
ERG Guide:	127
ERG Guide Category:	FLAMMABLE LIQUIDS (POLAR/WATER- MISCIBLE)
Molecular Formula:	$C_6H_{14}O$
RTECS No.:	TZ5425000
Properties:	Flammable liquid. Colorless. Sweet, camphorlike odor. Floats and mixes slowly with water. Flammable, irritating vapor is produced.

HEALTH AND SAFETY INFORMATION

Irritates eyes and respiratory tract. Can have a narcotic effect with headache, drowsiness, unconsciousness.

IDLH: 1400 ppm [10% LEL]
Odor Threshold: 0.055 ppm

Exposure Limits:	**TWA**	**STEL**
ACGIH:	250 ppm; 1040 mg/m^3	310 ppm; 1300 mg/m^3
NIOSH:	500 ppm; 2100 mg/m^3	
OSHA:	500 ppm; 2100 mg/m^3	

RECOMMENDATIONS FOR RESPIRATOR SELECTION: NIOSH/OSHA

1400 ppm: CCRFOV (any air-purifying, full-facepiece respirator (gas mask) with a chin-style, front- or back-mounted acid gas canister);or PAPROV [any powered, air-purifying respirator with organic vapor cartridge(s)];or GMFOV(any air-purifying, full-facepiece respirator (gas mask) with a chin-style, front- or back-mounted organic vapor canister;or SA (any supplied-air respirator);or SCBAF (any self-contained breathing apparatus with a full facepiece)

Emergency or planned entry into unknown concentrations or IDLH conditions:
SCBAF:PD,PP (any self-contained breathing apparatus that has a full facepiece and is operated in a pressure-demand or other positive-pressure mode);or SAF:PD,PP:ASCBA (any supplied-air respirator that has a full facepiece and is operated in a pressure-demand or other positive-pressure mode in combination with an auxiliary, self-contained breathing apparatus operated in a pressure-demand or other positive pressure mode.

Escape: GMFOV(any air-purifying, full-facepiece respirator (gas mask) with a chin-style, front- or back-mounted organic vapor canister;or SCBAE (any appropriate escape-type, self-contained breathing apparatus).

Note: Substance reported to cause eye irritation or damage; may require eye protection.

CHEMICAL REACTIVITY

Contact with oxidizers, acids may cause fire and explosions. Contact with air may form unstable peroxides.

FIRE INFORMATION

Vapors are heavier than air and will collect in low areas. Vapors in confined areas may explode when exposed to fire. Vapors may travel long distances to ignition sources and flash back. Storage containers and parts of containers may rocket great distances, in many directions. If material or contaminated runoff enters waterways, notify downstream users of potentially contaminated waters. Notify local health and fire officials and pollution control agencies. Use water, dry chemicals, foam, carbon dioxide. Use water spray to cool exposed containers.

NFPA Hazard Classifications:

Health Hazards (Blue) 1
Flammability (Red) 3
Reactivity (Yellow) 1

Flash Point:-15°F (-26°C) o.c.;-18°F (-28°C) c.c.
Flammable Limits in Air: LEL 1.4%; UEL 7.9%
Autoignition Temperature: 830°F (443°C)

FIRST AID

Move victim to fresh air. Call emergency medical care. Apply artificial respiration if victim is not breathing. Administer oxygen if breathing is difficult. Remove and isolate contaminated clothing and shoes. In case of contact with substance, immediately flush skin or eyes with running water for at least 20 minutes. Wash skin with soap and water. Keep victim warm and quiet. Ensure that medical personnel are aware of the material(s) involved and take precautions to protect themselves.

ISOPROPYL GLYCIDYL ETHER I:350

SYNONYMS: EINECS No. 223-672-9; GLYCIDYL ISOPROPYL ETHER; 1,2-EPOXY-3-ISOPROPOXYPROPANE; IGE; ISOPROPIL GLICIDIL ETER (Spanish); ISOPROPYL EPOXYPROPYL ETHER; (ISOPROXYMETHYL)OXIRANE; NCI-C56439; OSSIDO di 2,3-EPOSSIPROPILE E ISOPROPILE (Italian); OXYDE de 2,3-EPOXYPROPYLE ET d' ISOPROPYLE

IDENTIFICATION

CAS:	4016-14-2
ERG ID:	UN 1993
ERG Guide:	128
ERG Guide Category:	FLAMMABLE LIQUIDS (NONPOLAR/WATER-IMMISCIBLE)
Molecular Formula:	$C_6H_{12}O_2$
RTECS No.:	TZ3500000
Properties:	Flammable liquid. Colorless. Water soluble.

HEALTH AND SAFETY INFORMATION

Irritates eyes, skin, and respiratory tract.

IDLH: 400 ppm
Odor Threshold: 297 ppm
OSHA Table Z-1-A Air Contaminant

Exposure Limits:	TWA	CEILING	STEL
ACGIH:	50 ppm; 238 mg/m^3		75 ppm; 356 mg/m^3
NIOSH:		50 ppm; 240 mg/m^3 [15-min]	
OSHA:	50 ppm; 240 mg/m^3		

RECOMMENDATIONS FOR RESPIRATOR SELECTION: NIOSH/OSHA

1000 ppm: SA:CF (any supplied-air respirator operated in a continuous-flow mode);or SCBAF (any self-contained breathing apparatus with a full facepiece).

Emergency or planned entry into unknown concentrations or IDLH conditions: SCBAF:PD,PP (any self-contained breathing apparatus that has a full facepiece and is operated in a pressure-demand or other positive-pressure mode);or SAF:PD,PP:ASCBA (any supplied-air respirator that has a full facepiece and is operated in a pressure-demand or other positive-pressure mode in combination with an auxiliary, self-contained breathing apparatus operated in a pressure-demand or other positive pressure mode).

Escape: GMFOV(any air-purifying, full-facepiece respirator (gas mask) with a chin-style, front- or back-mounted organic vapor canister;or SCBAE (any appropriate escape-type, self-contained breathing apparatus).

Note: Substance causes eye irritation or damage; eye protection needed.

CHEMICAL REACTIVITY

Contact with strong oxidizers can cause fires and explosions. Contact with strong caustics may cause polymerization. Exposure to air or light may cause formation of explosive peroxides.

FIRE INFORMATION

Vapors are heavier than air and will collect in low areas. Vapors in confined areas may explode when exposed to fire. Vapors may travel long distances to ignition sources and flash back. Storage containers and parts of containers may rocket great distances, in many directions. If material or contaminated runoff enters waterways, notify downstream users of potentially contaminated waters. Notify local health and fire officials and pollution control agencies. Use dry chemicals, foam, carbon dioxide.

NFPA Hazard Classifications: Not listed

Flash Point: 92°F (33°C) c.c.

FIRST AID

Move victim to fresh air. Call emergency medical care. Apply artificial respiration if victim is not breathing. If breathing is difficult, give oxygen. Remove and isolate contaminated clothing and shoes. In case of contact with substance, immediately flush skin or eyes with running water for at least 20 minutes. Wash skin with soap and water. Keep victim warm and quiet. Ensure that medical personnel are aware of the material(s) involved and take precautions to protect themselves.

o–ISOPROPYLPHENOL I:400

SYNONYMS: EINECS No. 201-852-8; 2-ISOPROPILFENOLO (Spanish); 2-ISOPROPYLPHENOL; 2-(1-METHYLETHYL)PHENOL; PHENOL, o-ISOPROPYL-; PHENOL, 2-(1-METHYLETHYL)-(9CI); PRODOX 131

IDENTIFICATION

CAS:	88-69-7
ERG ID:	UN 1993
ERG Guide:	128
ERG Guide Category:	FLAMMABLE LIQUIDS (NONPOLAR/WATER-IMMISCIBLE)
Molecular Formula:	$C_9H_{12}O$
RTECS No.:	SL5900000
Properties:	Combustible liquid. Light yellow. Floats on water.

HEALTH AND SAFETY INFORMATION

RECOMMENDATIONS FOR RESPIRATOR SELECTION

At any concentrations above the NIOSH REL, or where there is no REL, at any detectable concentration: SCBAF:PD,PP (any self-contained breathing apparatus that has a full facepiece and is operated in a pressure-demand or other positive-pressure mode);or SAF:PD,PP:ASCBA (any supplied-air

respirator that has a full facepiece and is operated in a pressure-demand or other positive-pressure mode in combination with an auxiliary, self-contained breathing apparatus operated in a pressure-demand or other positive-pressure mode).

Escape: GMFOV [any air-purifying, full-facepiece respirator (gas mask) with a chin-style, front- or back-mounted organic vapor canister];or SCBAE (any appropriate escape-type, self-contained breathing apparatus).

CHEMICAL REACTIVITY

Contact with strong oxidizers may cause fire and explosions. Incompatible with sulfuric acid, nitric acid, caustics, aliphatic amines, amides.

FIRE INFORMATION

Vapors are heavier than air and will collect in low areas. Vapors in confined areas may explode when exposed to fire. If material or contaminated runoff enters waterways, notify downstream users of potentially contaminated waters. Notify local health and fire officials and pollution control agencies. Use dry chemicals, alcohol-resistant foam, carbon dioxide.

NFPA Hazard Classifications: Not listed
Flash Point: 220°F (104°C) o.c.

FIRST AID

Move victim to fresh air. Call emergency medical care. Apply artificial respiration if victim is not breathing. If breathing is difficult, give oxygen. Remove and isolate contaminated clothing and shoes. In case of contact with substance, immediately flush skin or eyes with running water for at least 20 minutes. Wash skin with soap and water. Keep victim warm and quiet. Ensure that medical personnel are aware of the material(s) involved and take precautions to protect themselves.

ENVIRONMENTAL INFORMATION

TSCA: 490CFR716.120(*a*)

ISOVALERALDEHYDE I:450

SYNONYMS: ALDEHIDO ISOVALERIANICO (Spanish); BUTANAL, 3-METHYL-; EINECS No. 209-691-5; ISOPENTALDEHYDE; ISOVALERIC ALDEHYDE; ISOVALERAL; 3-METHYLBUTANAL; 3-METHYLBUTYRALDEHYDE

IDENTIFICATION

CAS:	590-86-3
ERG ID:	UN 1989

638 Isovaleraldehyde

ERG Guide: 129
ERG Guide Category: FLAMMABLE LIQUIDS (POLAR/WATER-MISCIBLE/NOXIOUS)
Molecular Formula: $C_5H_{10}O$
RTECS No.: ES3450000
Properties: Flammable liquid. Colorless. Strong, fruity odor. Soluble in water.

HEALTH AND SAFETY INFORMATION

Odor Threshold: 0.00015 ppm (detection); 13.0 ppm (recognition)

Exposure Limits: TWA

ACGIH: 50 ppm; 176 mg/m^3 as n-valeraldehyde
NIOSH: 50 ppm; 176 mg/m^3 as n-valeraldehyde
OSHA: None

RECOMMENDATIONS FOR RESPIRATOR SELECTION

At any concentrations above the NIOSH REL, or where there is no REL, at any detectable concentration: SCBAF:PD,PP (any self-contained breathing apparatus that has a full facepiece and is operated in a pressure-demand or other positive-pressure mode);or SAF:PD,PP:ASCBA (any supplied-air respirator that has a full facepiece and is operated in a pressure-demand or other positive-pressure mode in combination with an auxiliary, self-contained breathing apparatus operated in a pressure-demand or other positive-pressure mode).

Escape: GMFOV [any air-purifying, full-facepiece respirator (gas mask) with a chin-style, front- or back-mounted organic vapor canister];or SCBAE (any appropriate escape-type, self-contained breathing apparatus).

CHEMICAL REACTIVITY

Forms explosive mixture with air. Reacts violently with strong oxidizers. May accumulate static electrical charges, and may cause ignition of its vapors.

FIRE INFORMATION

Vapors are heavier than air and will collect in low areas. Vapors in confined areas may explode when exposed to fire. If material or contaminated runoff enters waterways, notify downstream users of potentially contaminated waters. Notify local health and fire officials and pollution control agencies. Use dry chemicals, alcohol-resistant foam, carbon dioxide.

NFPA Hazard Classifications as valeraldehyde

Health Hazards (Blue) 1

Flammability (Red) 3
Reactivity (Yellow) 0

Flash Point: 23°F (-5°C) o.c.

FIRST AID

Move victim to fresh air. Call emergency medical care. Apply artificial respiration if victim is not breathing. Administer oxygen if breathing is difficult. Remove and isolate contaminated clothing and shoes. In case of contact with substance, immediately flush skin or eyes with running water for at least 20 minutes. Wash skin with soap and water. Keep victim warm and quiet. Effects of exposure (inhalation, ingestion, or skin contact) to substance may be delayed. Ensure that medical personnel are aware of the material(s) involved and take precautions to protect themselves.

ENVIRONMENTAL INFORMATION

TSCA: 40CFR716.120(*d*)1 as aldehydes; listed as butanal, 3-methyl.

KEROSENE K:020

SYNONYMS: ASTRAL OIL; COAL OIL; DEOBASE; 6200 DRUM CLEANING SOLVENT; 6500 DRUM CLEANING SOLVENT; EEC No. 650-001-02-5; EINECS No. 232-366-4; ILLUMINATING OIL; 11-1193 ISOCUT CUTTING FLUID; FUEL OIL No. 1; KEROSINE; KERSOSENO (Spanish); JET A; JET FUEL, JP-1; JP-4; PEARL; RANGE OIL; STRIGHT RUN KEROSENE

IDENTIFICATION

CAS:	8008-20-6
ERG ID:	UN 1223
ERG Guide:	128
ERG Guide Category:	FLAMMABLE LIQUIDS (NONPOLAR/WATER-IMMISCIBLE)
RTECS No.:	OA5500000
Properties:	Very flammable, watery liquid. Clear, colorless to yellowish, or light brown-amber. Petroleumlike odor. Floats on and insoluble in water.

Note: A petroleum distillate with a typical boiling point range of 347–617°F (175–325°C) and a typical composition of 25% normal paraffins, 11% branched paraffins, 30% monocycloparaffins, 12% dicycloparaffins, 1% tricycloparaffins, 16% mononuclear aromatics and 5% dinuclear aromatics. Deororized kerosene has a boiling range of 408.2–525.2°F (209–274°C) and a typical composition of 55.2% paraffins, 40.9% naphthenes, and 3.9% aromatics.

HEALTH AND SAFETY INFORMATION

Irritating to the eyes and skin. Absorption through the skin is slow but repeated skin contact over many years has caused muscular weakness, anemia, changes in white blood cells, fever, and death.

Odor Threshold: 7.07 mg/m^3 (1 ppm)

Exposure Limits:	**TWA**
ACGIH:	None
NIOSH:	100 mg/m^3
OSHA:	None

RECOMMENDATIONS FOR RESPIRATOR SELECTION: NIOSH as kerosene

1,000 ppm: CCROV [any chemical cartridge respirator with organic vapor cartridge(s)];or SA (any supplied-air respirator).
2,500 ppm: SA:CF (any supplied-air respirator operated in a continuous-flow mode);or PAPROV [any powered, air-purifying respirator with organic vapor cartridge(s)].
5,000 ppm: CCRFOV [any chemical cartridge respirator with a full facepiece and organic vapor cartridge(s)];or GMFOV [any air-purifying, full-facepiece respirator (gas mask) with a chin-style, front- or back-mounted acid gas canister);or PAPRTOV [any powered, air-purifying respirator with a tight-fitting facepiece and organic vapor cartridge(s)];or SCBAF (any self-contained breathing apparatus with a full facepiece);or SAF (any supplied-air respirator with a full facepiece).

Emergency or Planned Entry into Unknown Concentrations or IDLH Conditions: SCBAF:PD,PP (any self-contained breathing apparatus that has a full facepiece and is operated in a pressure-demand or other positive-pressure mode);or SAF:PD,PP:ASCBA (any supplied-air respirator that has a full facepiece and is operated in a pressure-demand or other positive-pressure mode in combination with an auxiliary, self-contained breathing apparatus operated in a pressure-demand or other positive-pressure mode).

Escape: GMFOV [any air-purifying, full-facepiece respirator (gas mask) with a chin-style, front- or back-mounted organic vapor canister];or SCBAE (any appropriate escape-type, self-contained breathing apparatus).

CHEMICAL REACTIVITY

Forms explosive mixture with air (flash point 100°F/43°C). Incompatible with nitric acid; oxidizers may cause fire and explosions. May rust steel. May accumulate static electrical charges, and may cause ignition of its vapors.

FIRE INFORMATION

Vapors are heavier than air and will collect in low areas. Storage containers and parts of containers may rocket great distances, in many directions. If material or contaminated runoff enters waterways, notify downstream users of potentially contaminated waters. Notify local health and fire officials and pollution control agencies. Use dry chemicals, alcohol-resistant foam, carbon dioxide. Use water spray to cool exposed containers.

NFPA Hazard Classifications:

Health Hazards (Blue) 0
Flammability (Red) 2
Reactivity (Yellow) 0

Flash Point: 100–162°F (38–72°C) c.c.
Flammable Limits in Air: LEL 0.7%; UEL 5%
Autoignition Temperature: 444°F (229°C)

FIRST AID

Move victim to fresh air. Call emergency medical care. Apply artificial respiration if victim is not breathing. If breathing is difficult, give oxygen. Remove and isolate contaminated clothing and shoes. In case of contact with substance, immediately flush skin or eyes with running water for at least 20 minutes. Wash skin with soap and water. Keep victim warm and quiet. Ensure that medical personnel are aware of the material(s) involved and take precautions to protect themselves.

KETENE K:030

SYNONYMS: CARBOMETHENE; CETENE; CETENA (Spanish); ETHENONE; KETEN; KETOETHYLENE

IDENTIFICATION

CAS:	463-51-4
ERG ID:	Cannot be shipped in gaseous state.
Molecular Formula:	C_2H_2O
RTECS No.:	OA7700000
Properties:	Flammable gas. Colorless. Penetrating odor. Soluble in water; decomposes. Melting point-240°F (-151°C). Boiling Point-69°F (-56°C).

HEALTH AND SAFETY INFORMATION

Highly toxic. Breathing vapors can cause death. Severe irritant to eyes, skin, and respiratory tract; possible severe burns and blindness. Can cause pulmonary edema, which can be delayed for several hours; there is a risk of death in serious cases.

IDLH: 5 ppm

Odor Threshold: Odor is not a reliable indicator of the presence of toxic amounts of gas.
OSHA Table Z-1-A Air Contaminant
OSHA Process Safety Management of Highly Hazardous Chemicals (29CFR, Part 1910.119, Appendix A): Threshold Quantity: 100 pounds.
NIOSH Criteria Document: NIOSH 78-173 KETONES

Exposure Limits:	**TWA**	**STEL**
ACGIH:	0.5 ppm; 0.86 mg/m^3	1.5 ppm; 2.6 mg/m^3
NIOSH:	0.5 ppm; 0.9 mg/m^3	1.5 ppm; 3.0 mg/m^3
OSHA:	0.5 ppm; 0.9 mg/m^3	

RECOMMENDATIONS FOR RESPIRATOR SELECTION

At any concentrations above the NIOSH REL, or where there is no REL, at any detectable concentration: SCBAF:PD,PP (any self-contained breathing apparatus that has a full facepiece and is operated in a pressure-demand or other positive-pressure mode);or SAF:PD,PP:ASCBA (any supplied-air respirator that has a full facepiece and is operated in a pressure-demand or other positive-pressure mode in combination with an auxiliary, self-contained breathing apparatus operated in a pressure-demand or other positive-pressure mode).

Escape: GMFOV [any air-purifying, full-facepiece respirator (gas mask) with a chin-style, front- or back-mounted organic vapor canister];or SCBAE (any appropriate escape-type, self-contained breathing apparatus).

CHEMICAL REACTIVITY

Forms explosive mixture with air. Violent reaction with oxidizers. Reacts with hydrogen peroxide to form explosive diacetyl peroxide. Reacts with water, alcohols, and ammonia. Readily polymerizes. Reacts with water to form acetic acid.

FIRE INFORMATION

Extreme health hazard. Firefighting gear (including SCBA) does not provide adequate protection. If exposure occurs, remove and isolate gear immediately and thoroughly decontaminate personnel. Gas in confined areas is highly toxic and may explode when exposed to fire. Gas may travel long distances to ignition sources and flash back. Storage containers and parts of containers may rocket great distances, in many directions. Notify local health and fire officials and pollution control agencies.

FEMA Hazard Classifications:

Health Hazards (Blue) 4
Flammability (Red) 2
Reactivity (Yellow) 3
Special Notice .... Water

Flash Point: Not reported. Flammable gas.
Flammable Limits in Air: Not reported, but is flammable and explosive in air.
Autoignition Temperature: Not reported.

FIRST AID

Do not put yourself in danger by entering a contaminated area to rescue a victim. Move victim to fresh air. Call emergency medical care. Provide basic life support/CPR as needed. Decontaminate the victim as follows: Inhalation–remove victim to fresh air and give oxygen if available; Skin–remove and isolate contaminated clothing (including shoes) and wash skin with soap and large volumes of water for at least 20 minutes. Eye–rinse eyes with large volumes of water or saline for at least 20 minutes. Swallowed–do not make victim vomit. Victims should be examined by a physician as soon as possible. Ensure that medical personnel are aware of the material(s) involved and take precautions to protect themselves. Toxic effects may be delayed. For skin burns, decontaminate with water and apply a clean dressing.

LAUROYL PEROXIDE L:010

SYNONYMS: ALPEROX C; DECANOYL PEROXIDE; DILAUROYL PEROXIDE; DIAURYL PEROXIDE; DPY-97 F; EEC No. 617-003-00-3; EINECS No. 203-326-3; LAUROX; LAURYDOL; LYP; LYP 97; DODECANOYL PEROXIDE; PEROXIDO de DILAUROILO (Spanish)

IDENTIFICATION

CAS:	105-74-8
ERG ID:	UN 2124 (pure); UN 2893 (< 42% solution)
ERG Guide:	145
ERG Guide Category:	ORGANIC PEROXIDES (HEAT AND CONTAMINATION SENSITIVE).
Molecular Formula:	$C_{24}H_{46}O_4$
RTECS No.:	OF2625000
Properties:	Combustible, solid powder or flakes. White. Faint soapy odor. Insoluble in water.

HEALTH AND SAFETY INFORMATION

May cause eye, skin, and respiratory tract irritation. A questionable carcinogen.

IARC Evaluation: IARC Group 3; weak carcinogen in mice.

RECOMMENDATIONS FOR RESPIRATOR SELECTION

At any concentrations above the NIOSH REL, or where there is no REL, at any detectable concentration: SCBAF:PD,PP (any self-contained breathing apparatus that has a full facepiece and is operated in a pressure-demand or other positive-pressure mode);or SAF:PD,PP:ASCBA (any supplied-air

respirator that has a full facepiece and is operated in a pressure-demand or other positive-pressure mode in combination with an auxiliary, self-contained breathing apparatus operated in a pressure-demand or other positive-pressure mode).

Escape: GMFOV [any air-purifying, full-facepiece respirator (gas mask) with a chin-style, front- or back-mounted organic vapor canister];or SCBAE (any appropriate escape-type, self-contained breathing apparatus).

CHEMICAL REACTIVITY

An organic peroxide. A powerful oxidizer. Heat or direct sunlight may cause explosive decomposition. May ignite or explode spontaneously when mixed with chemical accelerators; contact with combustible, readily oxidizable, or organic compounds.

FIRE INFORMATION

Material on clothing is a fire hazard; flush immediately with water. Will increase the intensity of an existing fire. Vapors are heavier than air and will collect in low areas. Vapors in confined areas may explode when exposed to fire. Vapors may travel long distances to ignition sources and flash back. Storage containers and parts of containers may rocket great distances, in many directions. If material or contaminated runoff enters waterways, notify downstream users of potentially contaminated waters. Notify local health and fire officials and pollution control agencies. Use water spray, dry chemical, alcohol-resistant foam, carbon dioxide, from a safe location.

NFPA Hazard Classifications:

Health Hazards (Blue) 0
Flammability (Red) 2
Reactivity (Yellow) 3
Special Notice . . . . OXY

Flash Point: Oxidizing combustible solid; self-accelerating ; decomposes at 120°F (49°C).

FIRST AID

Move victim to fresh air. Call emergency medical care. Apply artificial respiration if victim is not breathing. Administer oxygen if breathing is difficult. Remove and isolate contaminated clothing and shoes. Remove material from skin immediately. In case of contact with substance, immediately flush skin or eyes with running water for at least 20 minutes. Keep victim warm and quiet. Ensure that medical personnel are aware of the material(s) involved and take precautions to protect themselves.

LAURYL MERCAPTAN L:015

SYNONYMS: 1-DODECANETHIOL; DODECYL MERCAPTAN; EINECS No. 203-984-1;

LAURILMERCAPITANO (Spanish); 1-MERCAPTODODECANE; NCI-C60935; PENNFLOAT M; PENNFLOAT S

IDENTIFICATION

CAS:	112-55-0
ERG ID:	UN 3071
ERG Guide:	131
ERG Guide Category:	FLAMMABLE LIQUIDS- TOXIC
Molecular Formula:	$C_{12}H_{26}S$
RTECS No.:	JR3155000
Properties:	Combustible, oily liquid. Colorless to pale yellow. Mild skunklike odor. Insoluble in water; may produce toxic and flammable vapors. Freezing point is 15°F (-9°C).

HEALTH AND SAFETY INFORMATION

Irritates eyes, skin, respiratory tract. Central nervous system and blood effects.

Odor Threshold: 4 mg/m^3

Exposure Limits: **CEILING**

NIOSH:	0.5 ppm; 4.1 mg/m^3 [15 min]
OSHA:	None

RECOMMENDATIONS FOR RESPIRATOR SELECTION: NIOSH

50 ppm: CCROV [any chemicall cartridge respirator with organic vapor cartridge(s)];or PAPROV [any powered, air-purifying respirator with organic vapor cartridge(s)];or GMFOV [any air-purifying, full-facepiece respirator (gas mask) with a chin-style, front- or back-mounted acid gas canister];or SA (any supplied-air respirator);or SCBAF (any self-contained breathing apparatus with a full facepiece).

12.5 ppm: SA:CF (any supplied-air respirator operated in a continuous-flow mode); ;or PAPROV [any powered, air-purifying respirator with organic vapor cartridge(s)].

25 ppm:: CCRFOV [any air-purifying, full-facepiece respirator (gas mask) with a chin-style, front-or-back-mounted acid gas canister];or GMFOV [any air-purifying, full-facepiece respirator (gas mask) with a chin-style, front- or back-mounted organic vapor canister];or PAPRTOV [any powered, air- purifying respirator with a tight-fitting facepiece and organic vapor cartridge(s)];or SCBAF (any self- contained breathing apparatus with a full facepiece);or SAF (any supplied- air respirator with a full facepiece).

At any concentrations above the NIOSH REL, or where there is no REL, at any detectable concentration: SCBAF:PD,PP (any self-contained breathing apparatus that has a full facepiece and is operated in a pressure-demand or other positive-pressure mode);or SAF:PD,PP:ASCBA (any supplied-air

respirator that has a full facepiece and is operated in a pressure-demand or other positive-pressure mode in combination with an auxiliary, self-contained breathing apparatus operated in a pressure-demand or other positive-pressure mode).

Escape: GMFOV [any air-purifying, full-facepiece respirator (gas mask) with a chin-style, front- or back-mounted organic vapor canister];or SCBAE (any appropriate escape-type, self-contained breathing apparatus).

CHEMICAL REACTIVITY

Reacts with strong oxidizers and acids, strong bases, reducing agents, alkali metals, water, steam..

FIRE INFORMATION

Combustion products include toxic sulfur oxides. Vapors are heavier than air and will collect in low areas. Vapors in confined areas may explode when exposed to fire. Vapors may travel long distances to ignition sources and flash back. Storage containers and parts of containers may rocket great distances, in many directions. If material or contaminated runoff enters waterways, notify downstream users of potentially contaminated waters. Notify local health and fire officials and pollution control agencies. Use dry chemicals, alcohol-resistant foam (may cause frothing), carbon dioxide from a safe location. Use water spray to cool exposed containers.

NFPA Hazard Classifications:

Health Hazards (Blue) 2
Flammability (Red) 1
Reactivity (Yellow) 0

Flash Point: 262°F (128°C) o.c.

FIRST AID

Move victim to fresh air. Call emergency medical care. Apply artificial respiration if victim is not breathing. Do not use mouth-to-mouth method if victim ingested or inhaled the substance; induce artificial respiration with the aid of a pocket mask equipped with a one-way valve or other proper respiratory medical device. Administer oxygen if breathing is difficult. Remove and isolate contaminated clothing and shoes. In case of contact with substance, immediately flush skin or eyes with running water for at least 20 minutes. Wash skin with soap and water. Keep victim warm and quiet. Effects of exposure (inhalation, ingestion, or skin contact) to substance may be delayed. Ensure that medical personnel are aware of the material(s) involved and take precautions to protect themselves.

LEAD L:020

SYNONYMS: BLEI (German); CHUMBO (Portugese); C.I. 77575; C.I. PIGMENT METAL 4; EINECS No. 231-100-4; GLOVER; HARO MIX CE-701; HARO MIX CK-711; HARO MIX MH-204; JMI

SLOOP; LEAD FLAKE; LEAD METAL; LEAD S2; LOOD (Dutch); OLOW (Polish); OMAHA; OMAHA & GRANT; PIOMBO (Italian); PLOMB (French); PLOMO (Spanish); PLUMBUM; SI; SO

IDENTIFICATION

CAS:	7439-92-1
Molecular Formula:	Pb
RTECS No.:	OF7525000
Properties:	Heavy, soft metal. Bluish white to silvery grey. Odorless. Insoluble in water.

HEALTH AND SAFETY INFORMATION

Lead can accumulate in the body over a period of time. Therefore, long–term exposures to lower levels can result in a buildup of lead in the body and more severe symptoms. These may include anemia, pale skin, a blue line at the gum margin, decreased hand grip strength, abdominal pain, severe constipation, nausea, vomiting, and paralysis of the wrist joint. Prolonged exposure may also result in kidney damage. If the central nervous system is affected, usually due to very high exposures, the resulting effects include severe headache, convulsions, coma, delirium, and death. In nonfatal cases, recovery is slow and not always complete. Alcohol ingestion and physical exertion may bring on symptoms. Continuous exposure can result in decreased fertility. Elevated lead exposure of either parent before pregnancy can increase the chances of miscarriage or birth defects. Exposure of the mother during pregnancy can cause birth defects.

IARC Evaluation: Group 2B, possible human carcinogen; human evidence, inadequate; animal evidence, sufficient.
IDLH: 100 mg/m^3 as lead.
OSHA Specifically Regulated Substances (29 CFR 1910.1025).
NIOSH Criteria Document: NIOSH 78-158 LEAD, inorganic dusts and fumes.

Exposure Limits:	**TWA**
ACGIH:[(1)]	0.05 mg/m^3
NIOSH:[(2)]	0.100 mg/m^3 (10-hour)
OSHA:[(2)]	0.050 mg/m^3 (8-hour)

[(1)]Animal carcinogen.
[(2)]Suspected carcinogen.

RECOMMENDATIONS FOR RESPIRATOR SELECTION: OSHA as lead

0.5 mg/m^3: HiE (any air-purifying, respirator with a high-efficiency particulate filter); SA (any supplied-air respirator).
1.25 mg/m^3: SA:CF (any supplied-air respirator operated in a continuous-flow mode);or PAPRHiE (any powered, air-purifying respirator with a high-efficiency particulate filter).
2.5 mg/m^3: HiEF (any air-purifying, full-facepiece respirator with a high-efficiency particulate filter);or SAT:CF (any supplied-air respirator that has a tight-fitting facepiece and is operated in a continuous-flow

mode);or PAPRTHiE (any powered, air-purifying respirator with a tight-fitting facepiece and a high-efficiency particulate filter);or SCBAF (any self-contained breathing apparatus with a full facepiece);or SAF (any supplied-air respirator with a full facepiece).
50 mg/m³: SA:PD,PP (any supplied-air respirator operated in a pressure-demand or other positive-pressure mode).
100 mg/m³: SAF:PD,PP (any supplied-air respirator that has a full facepiece and is operated in a pressure-demand or other positive-pressure mode).

Emergency or Planned Entry into Unknown Concentrations or IDLH Conditions:
SCBAF:PD,PP (any self-contained breathing apparatus that has a full facepiece and is operated in a pressure-demand or other positive-pressure mode);or SAF:PD,PP:ASCBA (any supplied-air respirator that has a full facepiece and is operated in a pressure-demand or other positive-pressure mode in combination with an auxiliary, self-contained breathing apparatus operated in a pressure-demand or other positive-pressure mode).

Escape: HiEF (any air-purifying, full-facepiece respirator with a high-efficiency particulate filter);or SCBAE (any appropriate escape-type, self-contained breathing apparatus).

CHEMICAL REACTIVITY

Violent reaction with potassium. Reacts with strong oxidizers, hydrogen peroxide, and acids. Powder is incompatible with ammonium nitrate.

FIRE INFORMATION:

If material or contaminated runoff enters waterways, notify downstream users of potentially contaminated water. Use any extinguishing agent.

FIRST AID

EYE: If this chemical contacts the eyes, immediately wash them with large amounts of water, occasionally lifting the lower and upper lids. Get medical attention immediately. Contact lenses should not be worn when working with this chemical.
SKIN: If this chemical contacts the skin, immediately wash the contaminated skin with soap and water. If this chemical penetrates the clothing, immediately remove clothing, wash the skin with soap and water, and get medical attention promptly.
BREATH: If a person breathes large amounts of this chemical, move exposed person to fresh air at once. If breathing has stopped, perform mouth-to-mouth resuscitation. Keep the affected person warm and at rest. Get medical attention as soon as possible.
SWALLOW: If this chemical has been swallowed, get medical attention immediately.

ENVIRONMENTAL INFORMATION

Clean Water Act: Section 307 Priority Pollutants; Section 307 Toxic Pollutants as lead and compounds; Section 313 Priority Chemicals.
EPA Hazardous Waste Number (RCRA No.): D008.

RCRA Toxicity Characteristic (Section 261.24): Maximum Concentration of Contaminants, regulatory level, 5.0 mg/L.
RCRA Section 261 Hazardous Constituents, as lead compounds n.o.s., waste number not listed.
RCRA Universal Treatment Standards: Wastewater (mg/L), 0.69; Nonwastewater (mg/L), 0.37 TCLP.
RCRA Ground Water Monitoring List: Suggested test method(s) (PQL ug/L): 6010(40); 7420(1,000); 7421(10).
Safe Drinking Water Act: MCL, zero; MCLG, zero; Regulated chemical (47 FR 9352).
EPCRA Section 304: Reportable Quantity (RQ): CERCLA, 10 lb (4.54 kg).
EPCRA Section 313: Form R *de minimis* concentration reporting level: 0.1%.
California's Proposition 65: Carcinogen; Reproductive toxin (male, female).

LEAD ACETATE L:040

SYNONYMS: ACETATE de PLOMB (French); ACETATO de PLOMO (Spanish); ACETIC ACID LEAD(2+) SALT; BLACK MARKING INK, 105E; BLEIACETAT (German); DI(ACETATO) de CHUMBO (Portugese); DI(ACETATO) di PIOMBO (Italian); DIBASIC LEAD ACETATE; EEC No. 082-005-00-8; EINECS No. 206-104-4; LEADAC; LEAD(2+) ACETATE; LEAD ACETATE TRIHYDRATE; LEAD ACETATE(II), TRIHYDRATE; LEAD DIACETATE; LEAD DIBASIC ACETATE; LEAD(II) ACETATE; LOODDI(ACETAAT) (Dutch); MULTILAYER DIELECTRIC INK HD; NEUTRAL LEAD ACETATE; NORMAL LEAD ACETATE; PLUMBOUS ACETATE; SALT OF SATURN; SUGAR OF LEAD

IDENTIFICATION

CAS:	301-04-2; 6080-56-4 (lead (2+), trihydrate
ERG ID:	UN 1616
ERG Guide:	151
ERG Guide Category:	TOXIC (NONCOMBUSTIBLE)
Molecular Formula:	$C_4H_6O_4Pb \cdot 3H_2O$
RTECS No.:	AI5250000
Properties:	Noncombustible crystalline or powdery solid. Colorless or white. Odorless. Sinks and mixes with water.

HEALTH AND SAFETY INFORMATION

IARC Evaluation: Group 3, unclassifiable; human evidence, limited; animal evidence, sufficient.
NTP: Anticipated human carcinogen; possible mutagen.
IDLH: 100 mg/m^3 as lead.

Exposure Limits:	**TWA**
ACGIH:[(1)]	0.05 mg/m^3 as lead

NIOSH:[2] 0.100 mg/m^3 as lead
OSHA:[2] 0.050 mg/m^3

[1]Animal carcinogen.
[2]Potential carcinogen. Reduce exposure to lowest feasible level.

RECOMMENDATIONS FOR RESPIRATOR SELECTION: OSHA as lead

0.5 mg/m^3: HiE (any air-purifying respirator with a high-efficiency particulate filter); SA (any supplied-air respirator).
1.25 mg/m^3: SA:CF (any supplied-air respirator operated in a continuous-flow mode);or PAPRHiE (any powered, air-purifying respirator with a high-efficiency particulate filter).
2.5 mg/m^3: HiEF (any air-purifying, full-facepiece respirator with a high-efficiency particulate filter);or SAT:CF (any supplied-air respirator that has a tight-fitting facepiece and is operated in a continuous-flow mode);or PAPRTHiE (any powered, air-purifying respirator with a tight-fitting facepiece and a high-efficiency particulate filter);or SCBAF (any self-contained breathing apparatus with a full facepiece);or SAF (any supplied-air respirator with a full facepiece).
50 mg/m^3: SA:PD,PP (any supplied-air respirator operated in a pressure-demand or other positive-pressure mode).
100 mg/m^3: SAF:PD,PP (any supplied-air respirator that has a full facepiece and is operated in a pressure-demand or other positive-pressure mode).

Emergency or Planned Entry into Unknown Concentrations or IDLH Conditions:
SCBAF:PD,PP (any self-contained breathing apparatus that has a full facepiece and is operated in a pressure-demand or other positive-pressure mode);or SAF:PD,PP:ASCBA (any supplied-air respirator that has a full facepiece and is operated in a pressure-demand or other positive-pressure mode in combination with an auxiliary, self-contained breathing apparatus operated in a pressure-demand or other positive-pressure mode).

Escape: HiEF (any air-purifying, full-facepiece respirator with a high-efficiency particulate filter);or SCBAE (any appropriate escape-type, self-contained breathing apparatus).

CHEMICAL REACTIVITY

Contact with strong acids form acetic acid. Incompatible with bases, ammonia, amines, cresols, isocyanates, alkylene oxides, epichlorohydrin, phenols, and sulfites. Violent reaction with bromates and strong oxidizers.

FIRE INFORMATION

Combustion products include toxic lead oxides. If material or contaminated runoff enters waterways, notify downstream users of potentially contaminated waters. Notify local health and fire officials and pollution control agencies.

FEMA Hazard Classifications:

Health Hazards (Blue) 1

Flammability (Red) 0
Reactivity (Yellow) 0

FIRST AID

Move victim to fresh air. Call emergency medical care. Apply artificial respiration if victim is not breathing. Do not use mouth-to-mouth method if victim ingested or inhaled the substance; induce artificial respiration with the aid of a pocket mask equipped with a one–way valve or other proper respiratory medical device. Administer oxygen if breathing is difficult. Remove and isolate contaminated clothing and shoes. In case of contact with substance, immediately flush skin or eyes with running water for at least 20 minutes. For minor skin contact, avoid spreading material on unaffected skin. Keep victim warm and quiet. Effects of exposure (inhalation, ingestion, or skin contact) may be delayed. Ensure that medical personnel are aware of the material(s) involved and take precautions to protect themselves.

ENVIRONMENTAL INFORMATION

Clean Water Act: Section 311 Hazardous Substances/RQ (same as CERCLA); Section 313 Priority Chemicals; Section 307 Toxic Pollutants as lead and compounds.
EPA Hazardous Waste Number (RCRA No.): U144
RCRA Section 261 Hazardous Constituents
EPCRA Section 304: Reportable Quantity (RQ): CERCLA, 10 lb (4.54 kg).
EPCRA Section 313: Form R *de minimis* concentration reporting level: 1.0%. Form R Toxic Chemical Category Code: N420.
Marine Pollutant (49CFR, Subchapter 172.101, Appendix B).
California's Proposition 65: Carcinogen.

LEAD DIOXIDE L:045

SYNONYMS: BIOXYDE de PLOMB (French); BLEIDIOXID (German); C.I. 77580; DIOXIDO de PLOMO (Spanish); EINECS No. 215-174-5; LEAD BROWN; LEAD(4+) OXIDE; LEAD BROWN; LEAD(IV) OXIDE; LEAD OXIDE BROWN; LEAD PEROXIDE; LEAD SUPEROXIDE; LOODDIOXIDE (Dutch); PEROSSIDO di PIOMBO (Italian); PEROXIDO de CHUMBO (Portugese); PEROXYDE de PLOMB (French)

IDENTIFICATION

CAS:	1309-60-0
ERG ID:	UN 1872
ERG Guide:	141
ERG Guide Category:	OXIDIZERS–TOXIC (SOLID)
Molecular Formula:	O_2Pb
RTECS No.:	OG0700000
Properties:	Dark brown powder or crystalline solid. Insoluble in water.

HEALTH AND SAFETY INFORMATION

IDLH: 100 mg/m^3 as lead
OSHA Specifically Regulated Substances (29 CFR 1910.1025): as inorganic lead.

Exposure Limits:	**TWA**
ACGIH:[1]	0.05 mg/m^3 as lead
NIOSH:[2]	0.100 mg/m^3 as lead
OSHA:[2]	0.050 mg/m^3

[1]Animal carcinogen.
[2]Potential carcinogen. Reduce exposure to lowest feasible level.

RECOMMENDATIONS FOR RESPIRATOR SELECTION: OSHA as lead

0.5 mg/m^3: HiE (any air-purifying respirator with a high-efficiency particulate filter); SA (any supplied-air respirator).
1.25 mg/m^3: SA:CF (any supplied-air respirator operated in a continuous-flow mode);or PAPRHiE (any powered, air-purifying respirator with a high-efficiency particulate filter).
2.5 mg/m^3: HiEF (any air-purifying, full-facepiece respirator with a high-efficiency particulate filter);or SAT:CF (any supplied-air respirator that has a tight-fitting facepiece and is operated in a continuous-flow mode);or PAPRTHiE (any powered, air-purifying respirator with a tight-fitting facepiece and a high-efficiency particulate filter);or SCBAF (any self-contained breathing apparatus with a full facepiece);or SAF (any supplied-air respirator with a full facepiece).
50 mg/m^3: SA:PD,PP (any supplied-air respirator operated in a pressure-demand or other positive-pressure mode).
100 mg/m^3: SAF:PD,PP (any supplied-air respirator that has a full facepiece and is operated in a pressure-demand or other positive-pressure mode).

Emergency or Planned Entry into Unknown Concentrations or IDLH Conditions:
SCBAF:PD,PP (any self-contained breathing apparatus that has a full facepiece and is operated in a pressure-demand or other positive-pressure mode);or SAF:PD,PP:ASCBA (any supplied-air respirator that has a full facepiece and is operated in a pressure-demand or other positive-pressure mode in combination with an auxiliary, self-contained breathing apparatus operated in a pressure-demand or other positive-pressure mode).

Escape: HiEF (any air-purifying, full-facepiece respirator with a high-efficiency particulate filter);or SCBAE (any appropriate escape-type, self-contained breathing apparatus).

CHEMICAL REACTIVITY

A strong oxidizer; violent reaction with aluminum carbide, barium sulfide (possibly explosive), calcium sulfide, chlorine, hydrazine, hydrogen peroxide, hydrogen sulfide, hydroxylamine, magnesium, molybdenum, phosphorus, rubidium acetylide (explosive), sodium, sulfuric oxychloride, reducing agents and combustible materials. Incompatible with acids.

FIRE INFORMATION

Combustion products include toxic lead oxides. Combustion and decomposition products upon heating include lead. If material or contaminated runoff enters waterways, notify downstream users of potentially contaminated water.

FIRST AID

Move victim to fresh air. Call emergency medical care. Apply artificial respiration if victim is not breathing. Administer oxygen if breathing is difficult. Remove and isolate contaminated clothing and shoes. In case of contact with substance, immediately flush skin or eyes with running water for at least 20 minutes. Keep victim warm and quiet. Ensure that medical personnel are aware of the material(s) involved and take precautions to protect themselves.

ENVIRONMENTAL INFORMATION as lead compounds

Clean Water Act: Section 307 Toxic Pollutants as lead and compounds.
RCRA Section 261 Hazardous Constituents, waste number not listed, as lead compounds, n.o.s.
EPCRA Section 313: Includes any unique chemical substance that contains lead as part of that chemical's infrastructure. Form R *de minimis* concentration reporting level: inorganic compounds 0.1%;organic compounds 1.0%. Form R Toxic Chemical Category Code: N420.
Marine Pollutant (49CFR, Subchapter 172.101, Appendix B) as lead compounds, soluble, n.o.s.
California's Proposition 65: Carcinogen.

LEAD NITRATE L:060

SYNONYMS: EEC No. 082-001-00-6; EINECS No. 233-245-9; LEAD DINITRATE; LEADNI; LEAD(2+) NITRATE(1:2); NITRATE de PLOMB (French); NITRATO de PLOMO (Spanish); NITRIC ACID, LEAD(2+) SALT; NITRIC ACID, LEAD(II) SALT; STCC 4918726

IDENTIFICATION

CAS:	10099-74-8
ERG ID:	UN 1469
ERG Guide:	141
ERG Guide Category:	OXIDIZERS–TOXIC (SOLID)
Molecular Formula:	N_2O_6Pb
RTECS No.:	OG2100000
Properties:	Noncombustible crystalline or powdery solid. White. Odorless. Sinks and mixes with water.

HEALTH AND SAFETY INFORMATION

IDLH: 100 mg/m^3 as lead
OSHA Specifically Regulated Substances (29 CFR 1910.1025): as inorganic lead.

Exposure Limits: TWA

ACGIH:(1)	0.05 mg/m^3 as lead
NIOSH:(2)	0.100 mg/m^3 as lead
OSHA:(2)	0.050 mg/m^3

(1)Animal carcinogen.
(2)Potential carcinogen. Reduce exposure to lowest feasible level.

RECOMMENDATIONS FOR RESPIRATOR SELECTION: OSHA as lead

0.5 mg/m^3: HiE (any air-purifying respirator with a high-efficiency particulate filter); SA (any supplied-air respirator).
1.25 mg/m^3: SA:CF (any supplied-air respirator operated in a continuous-flow mode);or PAPRHiE (any powered, air-purifying respirator with a high-efficiency particulate filter).
2.5 mg/m^3: HiEF (any air-purifying, full-facepiece respirator with a high-efficiency particulate filter);or SAT:CF (any supplied-air respirator that has a tight-fitting facepiece and is operated in a continuous-flow mode);or PAPRTHiE (any powered, air-purifying respirator with a tight-fitting facepiece and a high-efficiency particulate filter);or SCBAF (any self-contained breathing apparatus with a full facepiece);or SAF (any supplied-air respirator with a full facepiece).
50 mg/m^3: SA:PD,PP (any supplied-air respirator operated in a pressure-demand or other positive-pressure mode).
100 mg/m^3: SAF:PD,PP (any supplied-air respirator that has a full facepiece and is operated in a pressure-demand or other positive-pressure mode).

Emergency or Planned Entry into Unknown Concentrations or IDLH Conditions:
SCBAF:PD,PP (any self-contained breathing apparatus that has a full facepiece and is operated in a pressure-demand or other positive-pressure mode);or SAF:PD,PP:ASCBA (any supplied-air respirator that has a full facepiece and is operated in a pressure-demand or other positive-pressure mode in combination with an auxiliary, self-contained breathing apparatus operated in a pressure-demand or other positive-pressure mode).

Escape: HiEF (any air-purifying, full-facepiece respirator with a high-efficiency particulate filter);or SCBAE (any appropriate escape-type, self-contained breathing apparatus).

CHEMICAL REACTIVITY

A strong oxidizer; contact with wood, paper, and combustibles may cause fire. Violent reaction with reducing materials, carbon, lead hypophosphite, ammonium sulfocyanate, cyclopentadienyl (may for an explosive), sodium, potassium acetate, and many other substances.

FIRE INFORMATION

Combustion products include toxic nitrogen oxides and oxygen. Will increase the activity of fire and other substances. If material or contaminated runoff enters waterways, notify downstream users of potentially contaminated waters. Notify local health and fire officials and pollution control agencies.

FEMA Hazard Classifications:

Health Hazards (Blue) 1
Flammability (Red) 0
Reactivity (Yellow) 0
Special situations . . . . OXY

FIRST AID

Move victim to fresh air. Call emergency medical care. Apply artificial respiration if victim is not breathing. Administer oxygen if breathing is difficult. Remove and isolate contaminated clothing and shoes. In case of contact with substance, immediately flush skin or eyes with running water for at least 20 minutes. Keep victim warm and quiet. Ensure that medical personnel are aware of the material(s) involved and take precautions to protect themselves.

ENVIRONMENTAL INFORMATION

Clean Water Act: Section 307 Toxic Pollutants as lead and compounds; Section 311 Hazardous Substances/RQ (same as CERCLA); Section 313 Priority Chemicals.
RCRA Section 261 Hazardous Constituents, waste number not listed, as lead compounds, n.o.s.
EPCRA Section 304: Reportable Quantity (RQ): CERCLA, 10 lb (4.54 kg).
EPCRA Section 313: Form R *de minimis* concentration reporting level: 0.1%; Form R Toxic Chemical Category Code: N420.
Marine Pollutant (49CFR, Subchapter 172.101, Appendix B).
California's Proposition 65: Carcinogen.

LEAD PHOSPHATE L:066

SYNONYMS: BIS (ORTOFOSFATO) DE TRIPLOMO (Spanish); BLEIPHOSPHAT (German); C.I. 77622; EEC No. 082-006-00-3; EINECS No. 231-205-5; LEAD ORTHOPHOSPHATE; LEAD(II) PHOSPHATE (3:2); LEAD PHOSPHATE (3:2); LEAD(2+) PHOSPHATE; NORMAL LEAD ORTHOPHOSPHATE; PHOSPHORIC ACID, LEAD(2+) SALT (2:3); PLUMBOUS PHOSPHATE; TRILEAD BIS(ORTHOPHOSPHATE); TRILEAD PHOSPHATE

IDENTIFICATION

CAS:	7446-27-7
ERG ID:	UN 2811
ERG Guide:	154
ERG Guide Category:	SUBSTANCES- TOXIC AND/OR CORROSIVE (NONCOMBUSTIBLE)
Molecular Formula:	$C_8P_2Pb_3$
RTECS No.:	OG3675000
Properties:	Crystalline solid or powder. Colorless (crystals) or white (powder). Insoluble in water.

HEALTH AND SAFETY INFORMATION

May be corrosive to the eyes. Irritates skin and respiratory tract.

IARC Evaluation: Group 2B rating; possible carcinogen, limited human evidence; sufficient animal evidence.
NTP: Carcinogen
IDLH: 100 mg/m^3 as lead
OSHA Table Z-1-A Air Contaminant as lead
OSHA Specifically Regulated Substances: (29CFR Part 1910.1025, Appendix C)

Exposure Limits:	TWA
ACGIH: [(1)]	0.05 mg/m^3 as lead
NIOSH:[(2)]	0.100 mg/m^3 as lead
OSHA:[(2)]	0.050 mg/m^3 as lead

[(1)] Animal carcinogen
[(2)] Potential carcinogen. Reduce exposure to lowest feasible level.

RECOMMENDATIONS FOR RESPIRATOR SELECTION: OSHA as lead

0.5 mg/m^3: HiE (any air-purifying respirator with a high-efficiency particulate filter); SA (any supplied-air respirator).
1.25 mg/m^3: SA:CF (any supplied-air respirator operated in a continuous-flow mode);or PAPRHiE (any powered, air-purifying respirator with a high-efficiency particulate filter).
2.5 mg/m^3: HiEF (any air-purifying, full-facepiece respirator with a high-efficiency particulate filter);or SAT:CF (any supplied-air respirator that has a tight-fitting facepiece and is operated in a continuous-flow mode);or PAPRTHiE (any powered, air-purifying respirator with a tight-fitting facepiece and a high-efficiency particulate filter);or SCBAF (any self-contained breathing apparatus with a full facepiece);or SAF (any supplied-air respirator with a full facepiece).
50 mg/m^3: SA:PD,PP (any supplied-air respirator operated in a pressure-demand or other positive-pressure mode).
100 mg/m^3: SAF:PD,PP (any supplied-air respirator that has a full facepiece and is operated in a pressure-demand or other positive-pressure mode).

Emergency or Planned Entry into Unknown Concentrations or IDLH Conditions:
SCBAF:PD,PP (any self-contained breathing apparatus that has a full facepiece and is operated in a pressure-demand or other positive-pressure mode);or SAF:PD,PP:ASCBA (any supplied-air respirator that has a full facepiece and is operated in a pressure-demand or other positive-pressure mode in combination with an auxiliary, self-contained breathing apparatus operated in a pressure-demand or other positive-pressure mode).

Escape: HiEF (any air-purifying, full-facepiece respirator with a high-efficiency particulate filter);or SCBAE (any appropriate escape-type, self-contained breathing apparatus).

CHEMICAL REACTIVITY

Incompatible with acids, antimony petachloride, magnesium, silver nitrate, zinc acetate.

FIRE INFORMATION

Combustion products include toxic lead and phosphorus fumes. Firefighting gear (including SCBA) may not provide adequate protection. If exposure occurs, remove and isolate gear immediately and thoroughly decontaminate personnel. If material or contaminated runoff enters waterways, notify downstream users of potentially contaminated waters. Notify local health and fire officials and pollution control agencies.

FIRST AID

Move victim to fresh air; call emergency medical care. Apply artificial respiration if victim is not breathing. Do not use mouth-to-mouth method if victim ingested or inhaled the substance; induce artificial respiration with the aid of a pocket mask equipped with a one-way valve or other proper respiratory medical device. Administer oxygen if breathing is difficult. Remove and isolate contaminated clothing and shoes. In case of contact with substance, immediately flush skin or eyes with running water for at least 20 minutes. For minor skin contact, avoid spreading material on unaffected skin. Keep victim warm and quiet. Effects of exposure (inhalation, ingestion, or skin contact) to substance may be delayed. Ensure that medical personnel are aware of the material(s) involved and take precautions to protect themselves.

ENVIRONMENTAL INFORMATION as lead compounds

Clean Water Act: Section 307 Toxic Pollutants as lead and compounds.
RCRA Section 261 Hazardous Constituents, waste number not listed, as lead compounds, n.o.s.
EPCRA Section 313: Includes any unique chemical substance that contains lead as part of that chemical's infrastructure. Form R *de minimis* concentration reporting level: inorganic compounds 0.1%;organic compounds 1.0%. Form R Toxic. Chemical Category Code: N420.
Marine Pollutant (49CFR, Subchapter 172.101, Appendix B) as lead compounds, soluble, n.o.s.
California's Proposition 65: Carcinogen.

LEAD PHOSPHITE, DIBASIC L:068

SYNONYMS: C.I. 77620; DIBASIC LEAD METAPHOSPHATE; DIBASIC LEAD PHOSPHITE; FOSFITO DIBASICO de PLOMO (Spanish); LEAD DIBASIC PHOSPHITE; LEAD OXIDE PHOSPHONATE, HEMIHYDRATE

IDENTIFICATION

CAS:	1344-40-7
ERG ID:	UN 2989
ERG Guide:	133
ERG Guide Category:	FLAMMABLE SOLIDS

658 Lead Phosphite, Dibasic

Molecular Formula: $HO_5PPb_3.1/2\ H_2O$
RTECS No.: OG2880000
Properties: Crystalline solid. White. Insoluble in water.

HEALTH AND SAFETY INFORMATION

IARC Evaluation: Group 2B rating; possible carcinogen, limited human evidence, sufficient animal evidence.
NTP: Carcinogen
IDLH: 100 mg/m^3 as lead
OSHA Table Z-1-A Air Contaminant as lead
OSHA Specifically Regulated Substances: (29CFR Part 1910.1025, Appendix C)

Exposure Limits: TWA

	TWA
ACGIH:(1)	0.05 mg/m^3 as lead
NIOSH:(2)	0.100 mg/m^3 as lead
OSHA:(2)	0.050 mg/m^3 as lead

(1) Animal carcinogen.
(2) Potential carcinogen. Reduce exposure to lowest feasible level.

RECOMMENDATIONS FOR RESPIRATOR SELECTION: OSHA as lead

0.5 mg/m^3: HiE (any air-purifying respirator with a high-efficiency particulate filter); SA (any supplied-air respirator).
1.25 mg/m^3: SA:CF (any supplied-air respirator operated in a continuous-flow mode);or PAPRHiE (any powered, air-purifying respirator with a high-efficiency particulate filter).
2.5 mg/m^3: HiEF (any air-purifying, full-facepiece respirator with a high-efficiency particulate filter);or SAT:CF (any supplied-air respirator that has a tight-fitting facepiece and is operated in a continuous-flow mode);or PAPRTHiE (any powered, air-purifying respirator with a tight-fitting facepiece and a high-efficiency particulate filter);or SCBAF (any self-contained breathing apparatus with a full facepiece);or SAF (any supplied-air respirator with a full facepiece).
50 mg/m^3: SA:PD,PP (any supplied-air respirator operated in a pressure-demand or other positive-pressure mode).
100 mg/m^3: SAF:PD,PP (any supplied-air respirator that has a full facepiece and is operated in a pressure-demand or other positive-pressure mode).

Emergency or Planned Entry into Unknown Concentrations or IDLH Conditions:
SCBAF:PD,PP (any self-contained breathing apparatus that has a full facepiece and is operated in a pressure-demand or other positive-pressure mode);or SAF:PD,PP:ASCBA (any supplied-air respirator that has a full facepiece and is operated in a pressure-demand or other positive-pressure mode in combination with an auxiliary, self-contained breathing apparatus operated in a pressure-demand or other positive-pressure mode).

Escape: HiEF (any air-purifying, full-facepiece respirator with a high-efficiency particulate filter);or SCBAE (any appropriate escape-type, self-contained breathing apparatus).

FIRE INFORMATION

Combustion products include toxic lead and phosphorus fumes. Firefighting gear (including SCBA) may not provide adequate protection. If exposure occurs, remove and isolate gear immediately and thoroughly decontaminate personnel. If material or contaminated runoff enters waterways, notify downstream users of potentially contaminated waters. Notify local health and fire officials and pollution control agencies.

NFPA Hazard Classifications: Not listed

FIRST AID

Move victim to fresh air. Call emergency medical care. Apply artificial respiration if victim is not breathing. Administer oxygen if breathing is difficult. Remove and isolate contaminated clothing and shoes. In case of contact with substance, immediately flush skin or eyes with running water for at least 20 minutes. Removal of solidified molten material from skin requires medical assistance. Keep victim warm and quiet. Ensure that medical personnel are aware of the material(s) involved and take precautions to protect themselves.

ENVIRONMENTAL INFORMATION as lead compounds

Clean Water Act: Section 307 Toxic Pollutants as lead and compounds.
RCRA Section 261 Hazardous Constituents, waste number not listed, as lead compounds, n.o.s.
EPCRA Section 313: Includes any unique chemical substance that contains lead as part of that chemical's infrastructure. Form R *de minimis* concentration reporting level: inorganic compounds 0.1%;organic compounds 1.0%. Form R Toxic **Chemical Category Code:** N420.
Marine Pollutant (49CFR, Subchapter 172.101, Appendix B) as lead compounds, soluble, n.o.s.
California's Proposition 65: Carcinogen.

LEAD STEARATE L:073

SYNONYMS: BLEISTEARAT (German); EEC No. 082-001-00-6; EINECS No. 231-068-1; ESTEARATO de PLOMO (Spanish); OCTADECANOIC ACID, LEAD SALT; STEARIC ACID, LEAD SALT; NEUTRAL LEAD STEARATE

IDENTIFICATION

CAS:	7428-48-0
ERG ID:	UN 3077
ERG Guide:	171
ERG Guide Category:	SUBSTANCES (LOW TO MODERATE HAZARD)
Molecular Formula:	$C_{18}H_{36}0_{2\bullet x}Pb$
Properties:	Solid powder. White. Slight fatty odor. Sinks in water; mixes in hot water.

HEALTH AND SAFETY INFORMATION

Corrosive to the eyes. May be absorbed through the skin. May cause lung or skin cancer.

IARC Evaluation: IARC possible carcinogen, rating 2B
IDLH: 100 mg/m^3, OSHA as lead
OSHA Table Z-1-A Air Contaminant
OSHA Specifically Regulated Substances: 29CFR 1910.1025 as lead

Exposure Limits:	**TWA**
ACGIH:(1)	0.05 mg/m^3 as lead
NIOSH:(2)	0.10 mg/m^3 as lead
OSHA:(2)	0.05 mg/m^3 as lead

(1) Animal carcinogen.
(2) Potential carcinogen. Reduce exposure to lowest feasible level.

RECOMMENDATIONS FOR RESPIRATOR SELECTION: OSHA as lead

0.5 mg/m^3: HiE (any air-purifying respirator with a high-efficiency particulate filter); SA (any supplied-air respirator)
1.25 mg/m^3: SA:CF (any supplied-air respirator operated in a continuous-flow mode);or PAPRHiE (any powered, air-purifying respirator with a high-efficiency particulate filter)
2.5 mg/m^3: HiEF (any air-purifying, full-facepiece respirator with a high-efficiency particulate filter);or SAT:CF (any supplied-air respirator that has a tight-fitting facepiece and is operated in a continuous-flow mode);or PAPRTHiE (any powered, air-purifying respirator with a tight-fitting facepiece and a high-efficiency particulate filter);or SCBAF (any self-contained breathing apparatus with a full facepiece);or SAF (any supplied-air respirator with a full facepiece)
50 mg/m: SA:PD,PP (any supplied-air respirator operated in a pressure-demand or other positive-pressure mode)
100 mg/m: SAF:PD,PP (any supplied-air respirator that has a full facepiece and is operated in a pressure-demand or other positive-pressure mode)

Emergency or planned entry into unknown concentrations or IDLH conditions:
SCBAF:PD,PP (any self-contained breathing apparatus that has a full facepiece and is operated in a pressure-demand or other positive-pressure mode);or SAF:PD,PP:ASCBA (any supplied-air respirator that has a full facepiece and is operated in a pressure-demand or other positive-pressure mode in combination with an auxiliary, self-contained breathing apparatus operated in a pressure-demand or other positive pressure mode.

Escape: HiEF (any air-purifying, full-facepiece respirator with a high-efficiency particulate filter);or SCBAE (any appropriate escape-type, self-contained breathing apparatus).

CHEMICAL REACTIVITY

Reacts with strong oxidizers. Water or steam may intensify toxicity.

FIRE INFORMATION

Combustion products include toxic fumes of lead. If material or contaminated runoff enters waterways, notify downstream users of potentially contaminated waters. Notify local health and fire officials and pollution control agencies. Use foam, carbon dioxide from a safe location.

FEMA Hazard Classifications

Health Hazards (Blue) 1
Flammability (Red) 0
Reactivity (Yellow) 0

Flash Point: More than 450°F (232°C) o.c.

FIRST AID

Move victim to fresh air. Call emergency medical care. Apply artificial respiration if victim is not breathing. Administer oxygen if breathing is difficult. In case of contact with substance, immediately flush skin or eyes with running water for at least 20 minutes. Remove and isolate contaminated clothing and shoes. Ensure that medical personnel are aware of the material(s) involved and take precautions to protect themselves.

ENVIRONMENTAL INFORMATION

Clean Water Act: Section 307 Toxic Pollutants as lead and compounds; Section 311 Hazardous Substances/RQ (same as CERCLA); Section 313 Priority Chemicals
RCRA Section 261 Hazardous Constituents, waste number not listed, as lead compounds, n.o.s.
EPCRA Section 304 Reportable Quantity (RQ): CERCLA, 10 lb (4.54 kg)
EPCRA Section 313: Form R *de minimis* concentration reporting level: 1.0%. Form R Toxic Chemical Category Code: N420
California's Proposition 65: Carcinogen.
Marine Pollutant (49CFR, Subchapter 172.101, Appendix B)

LEAD SULFATE L:076

SYNONYMS: ANGLISITE; BLEISULFAT (German); C.I. 77630; C.I. PIGMENT WHITE 3; EEC No. 082-001-00-6; EINECS No. 231-198-9; FAST WHITE; FREEMANS WHITE LEAD; LANARKITE; LEAD BOTTOMS; LEAD (+2) SULFATE; LEAD(II) SULFATE (1:1); LOODSULFAAT (Dutch); MILK WHITE; MULHOUSE WHITE; SULFATE de PLOMB (French); SULFATO de PLOMO (Spanish); SULFURIC ACID, LEAD (2+) SALT (1:1); WHITE LEAD

662 Lead Sulfate

IDENTIFICATION

CAS:	7446-14-2; 15739-80-7
ERG ID:	UN 1794
ERG Guide:	154
ERG Guide Category:	SUBSTANCES- TOXIC AND/OR CORROSIVE (NONCOMBUSTIBLE)
Molecular Formula:	O_4PbS
RTECS No.:	OG4375000
Properties:	Crystalline solid or powder. White to yellow-green. Odorless. Soluble in water.

HEALTH AND SAFETY INFORMATION

Corrosive to eyes, skin, and respiratory tract. In addition to its toxicity, fumes may cause pulmonary edema, which can be delayed for several hours; there is a risk of death in serious cases.

IARC Evaluation: IARC possible carcinogen, rating 2B
IDLH: 100 mg/m^3, OSHA as lead
OSHA Table Z-1-A Air Contaminant

OSHA Specifically Regulated Substances: 29CFR 1910.1025 as lead

Exposure Limits:	**TWA**
ACGIH:[(1)]	0.15 mg/m^3 as lead
NIOSH:[(2)]	0.100 mg/m^3 as lead
OSHA:[(2)]	0.05 mg/m^3 as lead

[(1)] Animal carcinogen.
[(2)] Potential carcinogen. Reduce exposure to lowest feasible level.

RECOMMENDATIONS FOR RESPIRATOR SELECTION: OSHA as lead

0.5 mg/m^3: HiE (any air-purifying respirator with a high-efficiency particulate filter); SA (any supplied-air respirator).
1.25 mg/m^3: SA:CF (any supplied-air respirator operated in a continuous-flow mode);or PAPRHiE (any powered, air-purifying respirator with a high-efficiency particulate filter).
2.5 mg/m^3: HiEF (any air-purifying, full-facepiece respirator with a high-efficiency particulate filter);or SAT:CF (any supplied-air respirator that has a tight-fitting facepiece and is operated in a continuous-flow mode);or PAPRTHiE (any powered, air-purifying respirator with a tight-fitting facepiece and a high-efficiency particulate filter);or SCBAF (any self-contained breathing apparatus with a full facepiece);or SAF (any supplied-air respirator with a full facepiece).
50 mg/m: SA:PD,PP (any supplied-air respirator operated in a pressure-demand or other positive-pressure mode).
100 mg/m: SAF:PD,PP (any supplied-air respirator that has a full facepiece and is operated in a pressure-demand or other positive-pressure mode).

Emergency or planned entry into unknown concentrations or idlh conditions: SCBAF:PD,PP (any self-contained breathing apparatus that has a full facepiece and is operated in a pressure-demand or other positive-pressure mode);or SAF:PD,PP:ASCBA (any supplied-air respirator that has a full facepiece and is operated in a pressure-demand or other positive-pressure mode in combination with an auxiliary, self-contained breathing apparatus operated in a pressure-demand or other positive pressure mode.

Escape: HiEF (any air-purifying, full-facepiece respirator with a high-efficiency particulate filter);or SCBAE (any appropriate escape-type, self-contained breathing apparatus).

CHEMICAL REACTIVITY

Violent reaction with potassium, aluminum, magnesium.

FIRE INFORMATION

Combustion products include toxic lead fumes. Firefighting gear (including SCBA) may not provide adequate protection. If exposure occurs, remove and isolate gear immediately and thoroughly decontaminate personnel. If material or contaminated runoff enters waterways, notify downstream users of potentially contaminated waters. Notify local health and fire officials and pollution control agencies. Use lots of water from a safe distance and location. Use water spray to cool exposed containers.

FIRST AID

Move victim to fresh air; call emergency medical care. Apply artificial respiration if victim is not breathing. Do not use mouth-to-mouth method if victim ingested or inhaled the substance; induce artificial respiration with the aid of a pocket mask equipped with a one-way valve or other proper respiratory medical device. Administer oxygen if breathing is difficult. Remove and isolate contaminated clothing and shoes. In case of contact with substance, immediately flush skin or eyes with running water for at least 20 minutes. For minor skin contact, avoid spreading material on unaffected skin. Keep victim warm and quiet. Effects of exposure (inhalation, ingestion, or skin contact) to substance may be delayed. Ensure that medical personnel are aware of the material(s) involved and take precautions to protect themselves.

ENVIRONMENTAL INFORMATION

Clean Water Act: Section 307 Toxic Pollutants as lead and compounds; Section 311 Hazardous Substances/RQ (same as CERCLA); Section 313 Priority Chemicals
RCRA Section 261 Hazardous Constituents waste number not listed, as lead compounds, n.o.s.
RCRA Land Ban Waste
EPCRA Section 304 Reportable Quantity (RQ): CERCLA, 10 lb (4.54 kg)
EPCRA Section 313: Form R *de minimis* concentration reporting level: 0.1%. Form R Toxic Chemical Category Code: N420.
California's Proposition 65: Carcinogen.
Marine Pollutant (49CFR, Subchapter 172.101, Appendix B)

LIQUEFIED PETROLEUM GAS L:080

SYNONYMS: BOTTLED GAS; COMPRESSED PETROLEUM GAS; GAS de PETROLEO LICUADO (Spanish); LIQUEFIED HYDROCARBON GAS; L.P.G.; PETROLEUM GAS, LIQUEFIED; PROPANE-BUTANE-(PROPYLENE); PYROFAX

IDENTIFICATION

CAS:	68476-85-7
ERG ID:	UN 1075
ERG Guide:	115
ERG Guide Category:	GASES-FLAMMABLE (INCLUDING REFRIGERATED LIQUIDS)
Molecular Formula:	C_3H_8-C_3H_6-C_4H_{10}-C_4 H_8 (mixture)
RTECS No.:	SE7545000
Properties:	Extremely flammable gas. May be shipped as a liquefied gas. Colorless. Odorless gas when pure; a foul-smelling odorant is usually added. Floats and "boils" on water; insoluble. Flammable vapor cloud is produced.

HEALTH AND SAFETY INFORMATION

Contact with liquid may cause frostbite. High concentrations can cause headaches, unconsciousness.

IDLH: 2,000 ppm
Odor Threshold: 5,000-18,000 ppm. Odor is not a reliable indicator of the presence of toxic amounts of gas.
OSHA Table Z-1-A Air Contaminant

Exposure Limits:	**TWA**
ACGIH:	1000 ppm; 1800 mg/m^3
NIOSH:	1000 ppm; 1800 mg/m^3
OSHA:	1000 ppm; 1800 mg/m^3

RECOMMENDATIONS FOR RESPIRATOR SELECTION: NIOSH/OSHA

2000 ppm: SA (any supplied-air respirator);or SCBAF (any self-contained breathing apparatus with a full facepiece).

Emergency or Planned Entry into Unknown Concentrations or Idlh Conditions:

SCBAF:PD,PP (any self-contained breathing apparatus that has a full facepiece and is operated in a pressure-demand or other positive-pressure mode);or SAF:PD,PP:ASCBA (any supplied-air respirator that has a full facepiece and is operated in a pressure-demand or other positive-pressure mode in combination with an auxiliary, self-contained breathing apparatus operated in a pressure-demand or other positive pressure mode.

Escape: SCBAE (any appropriate escape-type, self-contained breathing apparatus).

CHEMICAL REACTIVITY

Forms explosive mixture with air. Strong oxidizers may cause fire and explosions. Attacks some plastics, rubber and coatings.

FIRE INFORMATION

Gases are heavier than air and will collect in low areas. Gases in confined areas may explode when exposed to fire. Gases may travel long distances to ignition sources and flash back. Storage containers and parts of containers may rocket great distances, in many directions. If material or contaminated runoff enters waterways, notify downstream users of potentially contaminated waters. Notify local health and fire officials and pollution control agencies. Do not extinguish the fire unless the flow of the gas can be stopped and any remaining gas is out of the line. Specially trained personnel may use fog lines to cool exposures and let the fire burn itself out. Use dry chemicals, alcohol-resistant foam, carbon dioxide. From a secure, explosion-proof location, use water spray to cool exposed containers. If cooling streams are ineffective (venting sound increases in volume and pitch, tank discolors, or shows any signs of deforming), withdraw immediately to a secure position.

NFPA Hazard Classifications:

Health Hazards (Blue) 1
Flammability (Red) 4
Reactivity (Yellow) 0

Flash Point: Flammable gas. Propane:-156°F (-104°C) c.c. Butane:-176°F (-60°C) c.c.
Flammable Limits in Air: LEL 2.1% (Propane), LEL 1.9% (Butane); UEL 9.5% (Propane), UEL 8.5% (Butane).
Autoignition Temperature: About 752°F (400°C). Propane: 871°F (466°C); Butane: 761°F (405°C)

FIRST AID

Move victim to fresh air. Call emergency medical care. Apply artificial respiration if victim is not breathing. Administer oxygen if breathing is difficult. Remove and isolate contaminated clothing and shoes. Clothing frozen to the skin should be thawed before being removed. In case of contact with liquefied gas, thaw frosted parts with lukewarm water. Keep victim warm and quiet. Keep victim under observation. Ensure that medical personnel are aware of the material(s) involved and take precautions to protect themselves.

MAGNESIUM CARBONATE M:050

SYNONYMS: CARBONATE MAGNESIUM; CARBONIC ACID, MAGNESIUM SALT; C.I. 77713;

666 Magnesium Carbonate

DCI LIGHT MAGNESIUM CARBONATE; EINECS No. 208-915-9; HYDROMAGNESITE; MAGMASTER; MAGNESIA ALBA; MAGNESIUM(II) CARBONATE (1:1); MAGNESIUM CARBONATE, PRECIPITATED; MAGNESITE; MARINCO; STAN-MAG MAGNESIUM CARBONATE

IDENTIFICATION

CAS:	546-93-0
Molecular Formula:	$CMgO_3$
RTECS No.:	OM2470000
Properties:	Crystalline powder. White, grayish-white, yellow, or brownish yellow. Odorless. Insoluble in water.

HEALTH AND SAFETY INFORMATION

Exposure Limits:	**TWA**
ACGIH:[(1)]	10 mg/m^3 total dust; 3 mg/m^3 respiratory fraction
OSHA:[(1)]	15 mg/m^3 total dust; 5 mg/m^3 respiratory fraction

[(1)] a nuisance particulate

RECOMMENDATIONS FOR RESPIRATOR SELECTION

At any concentrations above the NIOSH REL, or where there is no REL, at any detectable concentration: SCBAF:PD,PP (any self-contained breathing apparatus that has a full facepiece and is operated in a pressure-demand or other positive-pressure mode);or SAF:PD,PP:ASCBA (any supplied-air respirator that has a full facepiece and is operated in a pressure-demand or other positive-pressure mode in combination with an auxiliary, self-contained breathing apparatus operated in a pressure-demand or other positive-pressure mode).

Escape: GMFOV [any air-purifying, full-facepiece respirator (gas mask) with a chin-style, front- or back-mounted organic vapor canister];or SCBAE (any appropriate escape-type, self-contained breathing apparatus).

CHEMICAL REACTIVITY

Reacts with formaldehyde and acids. Carbonates are incompatible with calcium, lead diacetate, mercurous chloride, silicon, silver nitrate, titanium, zinc acetate.

FIRE INFORMATION

Noncombustible. If material or contaminated runoff enters waterways, notify downstream users of potentially contaminated waters. Notify local health and fire officials and pollution control agencies. Use water, dry chemicals, foam, carbon dioxide.

FIRST AID

Move victim to fresh air. Call emergency medical care. Apply artificial respiration if victim is not breathing. Do not use mouth-to-mouth method if victim ingested or inhaled the substance; induce artificial respiration with the aid of a pocket mask equipped with a one-way valve or other proper respiratory medical device. Administer oxygen if breathing is difficult. Remove and isolate contaminated clothing and shoes. In case of contact with substance, immediately flush skin or eyes with running water for at least 20 minutes. For minor skin contact, avoid spreading material on unaffected skin. Keep victim warm and quiet. Effects of exposure (inhalation, ingestion, or skin contact) may be delayed. Ensure that medical personnel are aware of the material(s) involved and take precautions to protect themselves.

MAGNESIUM FLUOROSILICATE M:085

SYNONYMS: EINECS No. 241-022-2; FLUOSILICATE de MAGNESIUM (French); FLUOSILICATO MAGNESICO (Spanish); HEXAFLUOROSILICATE(2-) MAGNESIUM (1:1); MAGNESIUM FLUOSILICATE; MAGNESIUM SILICOFLUORIDE; SILICATE(2-), HEXAFLUORO-, MAGNESIUM (1:1)

IDENTIFICATION

CAS:	18972-56-0
ERG ID:	UN 2853
ERG Guide:	151
ERG Guide Category:	TOXIC (NONCOMBUSTIBLE)
Molecular Formula:	$F_6Si{\bullet}Mg$
RTECS No.:	VV8575000
Properties:	Crystalline powder. White, efflorescent. Soluble in water.

HEALTH AND SAFETY INFORMATION

Repeated exposure to fluoride chemicals may cause stiffness in muscles or ligaments, and even crippling.

IDLH: 250 mg/m³ as F
NIOSH Criteria Document: NIOSH 76-103 INORGANIC FLUORIDE

Exposure Limits:	**TWA**
ACGIH:	2.5 mg/m^3 as F
NIOSH:	2.5 mg/m^3 as F
OSHA:	2.5 mg/m^3 as F

RECOMMENDATIONS FOR RESPIRATOR SELECTION: NIOSH/OSHA as fluorine

1 ppm: SA (any supplied-air respirator).

2.5 ppm: SA:CF (any supplied-air respirator operated in a continuous-flow mode).
5 ppm: SCBAF (any self-contained breathing apparatus with a full facepiece);or SAF (any supplied-air respirator with a full facepiece).
25 ppm: SAF:PD,PP (any supplied-air respirator that has a full facepiece and is operated in a pressure-demand or other positive-pressure mode).

Emergency or Planned Entry into Unknown Concentrations or IDLH Conditions: SCBAF:PD,PP (any self-contained breathing apparatus that has a full facepiece and is operated in a pressure-demand or other positive-pressure mode);or SAF:PD,PP:ASCBA (any supplied-air respirator that has a full facepiece and is operated in a pressure-demand or other positive-pressure mode in combination with an auxiliary, self-contained breathing apparatus operated in a pressure-demand or other positive-pressure mode).

Escape: GMFS end of service life indicator (ESLI) required (any air-purifying, full-facepiece respirator (gas mask) with a chin-style, front- or back-mounted canister protection against the compound of concern);or SCBAE (any appropriate escape-type, self-contained breathing apparatus).

Note: Substance reported to cause eye irritation or damage; may require eye protection.

FIRE INFORMATION

Combustion products include toxic fumes of fluorides. If material or contaminated runoff enters waterways, notify downstream users of potentially contaminated waters. Notify local health and fire officials and pollution control agencies. Use dry chemicals, alcohol-resistant foam, carbon dioxide.

NFPA Hazard Classifications: Not listed

FIRST AID

Move victim to fresh air. Call emergency medical care. Apply artificial respiration if victim is not breathing. Do not use mouth-to-mouth method if victim ingested or inhaled the substance; induce artificial respiration with the aid of a pocket mask equipped with a one-way valve or other proper respiratory medical device. Administer oxygen if breathing is difficult. Remove and isolate contaminated clothing and shoes. In case of contact with substance, immediately flush skin or eyes with running water for at least 20 minutes. For minor skin contact, avoid spreading material on unaffected skin. Keep victim warm and quiet. Effects of exposure (inhalation, ingestion, or skin contact) may be delayed. Ensure that medical personnel are aware of the material(s) involved and take precautions to protect themselves.

ENVIRONMENTAL INFORMATION

Safe Drinking Water Act: 40CFR141.82 as silicate.

MAGNESIUM HEXAFLUOROSILICATE M:090

SYNONYMS: EINECS No. 241-022-2; EULAVA SM; HEXAFLUOSILICATE de MAGNESIUM

(French); HEXAFLUOSILICATO MAGNESICO (Spanish); MAGNESIUM FLUOSILICATE; MAGNESIUM SILICOFLUORIDE; SILICOFLUORURO MAGNESICO (Spanish)

IDENTIFICATION

CAS:	18972-56-0
ERG ID:	UN 2853
ERG Guide:	151
ERG Guide Category:	TOXIC (NONCOMBUSTIBLE)
Molecular Formula:	$F_6Si•Mg•6H_2O$
RTECS No.:	VV8575000
Properties:	Crystalline powder. White. Highly soluble in water.

HEALTH AND SAFETY INFORMATION

Repeated exposure to fluoride chemicals may cause stiffness in muscles or ligaments, and even crippling.

IDLH: 250 mg/m^3 as F
OSHA Table Z-1-A Air Contaminant as fluoride compounds
OSHA Table Z-2 Air Contaminant as fluoride compounds
NIOSH Criteria Document: NIOSH 76-103 INORGANIC FLUORIDE

Exposure Limits:	**TWA**
ACGIH:	2.5 mg/m^3 as F
NIOSH:	2.5 mg/m^3 as F
OSHA:	2.5 mg/m^3 as F

RECOMMENDATIONS FOR RESPIRATOR SELECTION: NIOSH/OSHA as fluorine

1 ppm: SA (any supplied-air respirator).
2.5 ppm: SA:CF (any supplied-air respirator operated in a continuous-flow mode).
5 ppm: SCBAF (any self-contained breathing apparatus with a full facepiece);or SAF (any supplied-air respirator with a full facepiece).
25 ppm: SAF:PD,PP (any supplied-air respirator that has a full facepiece and is operated in a pressure-demand or other positive-pressure mode).

Emergency or Planned Entry into Unknown Concentrations or IDLH Conditions:
SCBAF:PD,PP (any self-contained breathing apparatus that has a full facepiece and is operated in a pressure-demand or other positive-pressure mode);or SAF:PD,PP:ASCBA (any supplied-air respirator that has a full facepiece and is operated in a pressure-demand or other positive-pressure mode in combination with an auxiliary, self-contained breathing apparatus operated in a pressure-demand or other positive-pressure mode).

Escape: GMFS end of service life indicator (ESLI) required [any air-purifying, full-facepiece respirator (gas mask) with a chin-style, front- or back-mounted canister protection against the compound of concern];or SCBAE (any appropriate escape-type, self-contained breathing apparatus).

Note: Substance reported to cause eye irritation or damage; may require eye protection.

FIRST AID

Move victim to fresh air. Call emergency medical care. Apply artificial respiration if victim is not breathing. Do not use mouth-to-mouth method if victim ingested or inhaled the substance; induce artificial respiration with the aid of a pocket mask equipped with a one-way valve or other proper respiratory medical device. Administer oxygen if breathing is difficult. Remove and isolate contaminated clothing and shoes. In case of contact with substance, immediately flush skin or eyes with running water for at least 20 minutes. For minor skin contact, avoid spreading material on unaffected skin. Keep victim warm and quiet. Effects of exposure (inhalation, ingestion, or skin contact) may be delayed. Ensure that medical personnel are aware of the material(s) involved and take precautions to protect themselves.

ENVIRONMENTAL INFORMATION

Safe Drinking Water Act: 40CFR141.82 as silicate.

MAGNESIUM NITRATE M:095

SYNONYMS: EINECS No. 233-826-7; MAGNESIUM(II) NITRATE(1:2); MAGNESIUM(2+) NITRATE(1:2); HEXAHYDRATE; NITRIC ACID, MAGNESIUM SALT; NITRATO MAGNESICO (Spanish); NITROMAGNESITE

IDENTIFICATION

CAS:	10377-60-3; 13446-18-9
ERG ID:	UN 1474
ERG Guide:	140
ERG Guide Category:	OXIDIZERS
Molecular Formula:	$MgN_2O_6 \cdot 6H_2O$
RTECS No.:	OM3756000
Properties:	Solid deliquescent crystals. White. Odorless. Soluble in water.

HEALTH AND SAFETY INFORMATION

Irritates eyes, skin, and respiratory tract. May cause metal fume fever with flulike symptoms; risk of seizures.

At any concentrations above the NIOSH REL, or where there is no REL, at any detectable concentration: SCBAF:PD,PP (any self-contained breathing apparatus that has a full facepiece and is operated in a pressure-demand or other positive-pressure mode);or SAF:PD,PP:ASCBA (any supplied-air

respirator that has a full facepiece and is operated in a pressure-demand or other positive-pressure mode in combination with an auxiliary, self-contained breathing apparatus operated in a pressure-demand or other positive-pressure mode).

Escape: GMFOV [any air-purifying, full-facepiece respirator (gas mask) with a chin-style, front- or back-mounted organic vapor canister];or SCBAE (any appropriate escape-type, self-contained breathing apparatus).

CHEMICAL REACTIVITY

A strong oxidizer. Contact with dimethylformamide; combustible, organic, reducing materials and oxidizable materials can generate heat, perhaps causing ignition and causing violent combustion. Nitrates are incompatible with esters.

FIRE INFORMATION

Decomposition products include oxygen, which will increase fire activity and toxic oxides of nitrogen. If material or contaminated runoff enters waterways, notify downstream users of potentially contaminated waters. Notify local health and fire officials and pollution control agencies. Use water, dry chemicals, alcohol-resistant foam, carbon dioxide.

NFPA Hazard Classifications:

Health Hazards (Blue) 1
Flammability (Red) 0
Reactivity (Yellow) 0
Special Notice . . . . OXY

Flash Point: Not combustible.

FIRST AID

Move victim to fresh air. Call emergency medical care. Apply artificial respiration if victim is not breathing. Administer oxygen if breathing is difficult. Remove and isolate contaminated clothing and shoes. In case of contact with substance, immediately flush skin or eyes with running water for at least 20 minutes. Keep victim warm and quiet. Ensure that medical personnel are aware of the material(s) involved and take precautions to protect themselves.

MAGNESIUM OXIDE M:100

SYNONYMS: BISLUMINA; CALCINED BRUCITE; CALCINED MAGNESIA; CALCINED MAGNESITE; COROX; DYNAMAG; EINECS No. 215-171-9; ELASTOMAG; FLAMARRET; INSULMAG; KEN-MAG; MAGCHEM; MAGCHEM 10; MAGCHEM 20; MAGCHEM 60; MAGLITE; MAG; MAGNESIA; MAGNESIA FUME; MAGNESIA USTA; MAGNEZU (Polish); OXIDO MAGNESICO (Spanish); SEAWATER MAGNESIA

672 Magnesium Oxide

IDENTIFICATION

CAS:	1309-48-4
Molecular Formula:	MgO
RTECS No.:	OM3850000
Properties:	Very fine powder. White. Odorless. Slightly soluble in water

HEALTH AND SAFETY INFORMATION

Irritates eyes and respiratory tract. Can cause metal fume fever, with flulike symptoms.

IDLH: 750 mg/m^3 fume
OSHA Table Z-1-A Air Contaminant

Exposure Limits: TWA

ACGIH:	10 mg/m^3 as fume
NIOSH:	No established REL
OSHA:	15 mg/m^3 as fume

RECOMMENDATIONS FOR RESPIRATOR SELECTION: OSHA
150 mg/m^3: DMFu (any dust, mist, and fume respirator);or SA (any supplied-air respirator)
375 mg/m^3: SA:CF (any supplied-air respirator operated in a continuous-flow mode);or PAPRDMFu (any powered, air-purifying respirator with a dust, mist, and fume filter).
750 mg/m^3: HiEF (any air-purifying, full-facepiece respirator with a high-efficiency particulate filter);or PAPRTHiE (any powered, air-purifying respirator with a tight-fitting facepiece and a high-efficiency particulate filter);or SCBAF (any self-contained breathing apparatus with a full facepiece);or SAF (any supplied-air respirator with a full facepiece).

Emergency or Planned Entry into Unknown Concentrations or IDLH Conditions:
SCBAF:PD,PP (any self-contained breathing apparatus that has a full facepiece and is operated in a pressure-demand or other positive-pressure mode);or SAF:PD,PP:ASCBA (any supplied-air respirator that has a full facepiece and is operated in a pressure-demand or other positive-pressure mode in combination with an auxiliary, self-contained breathing apparatus operated in a pressure-demand or other positive-pressure mode).

Escape: HiEF (any air-purifying, full-facepiece respirator with a high-efficiency particulate filter);or SCBAE (any appropriate escape-type, self-contained breathing apparatus).

CHEMICAL REACTIVITY

Violent reaction with chlorine trifluoride, bromine pentalfluoride, phosphorous pentachloride. Incompatible with acids.

FIRE INFORMATION

If material or contaminated runoff enters waterways, notify downstream users of potentially contaminated waters. Notify local health and fire officials and pollution control agencies.

NFPA Hazard Classifications:

Health Hazards (Blue) 1
Flammability (Red) 0
Reactivity (Yellow) 0

FIRST AID

Move victim to fresh air. Call emergency medical care. Apply artificial respiration if victim is not breathing. Administer oxygen if breathing is difficult. Remove and isolate contaminated clothing and shoes. In case of contact with substance, immediately flush skin or eyes with running water for at least 20 minutes. Wash skin with soap and water. Keep victim warm and quiet. Effects of exposure (inhalation, ingestion, or skin contact) to substance may be delayed. Ensure that medical personnel are aware of the material(s) involved and take precautions to protect themselves.

MALEIC ANHYDRIDE M:102

SYNONYMS: ANHIDRIDO MALEICO (Spanish); ANHYDRIDE MALEIQUE (French); BM 10; cis-BUTENEDIOIC ANHYDRIDE; DIHYDRO-2,5-DIOXOFURAN; 2,5-DIHYDROFURAN-2,5-DIONE; EEC No. 607-096-00-9; EINECS No. 203-571-6; 2,5-FURANDIONE; 2,5-FURANEDIONE; MALEIC ACID ANHYDRIDE; MALEINSAEUREANHYDRID (German); TOXILIC ANHYDRIDE

IDENTIFICATION

CAS:	108-31-6
ERG ID:	UN 2215
ERG Guide:	156
ERG Guide Category:	SUBSTANCES-TOXIC AND/OR CORROSIVE (COMBUSTIBLE/WATER-SENSITIVE)
Molecular Formula:	$C_4H_2O_3$
RTECS No.:	ON3675000
Properties:	Combustible liquid or solid (solid difficult to ignite). Colorless liquid, or white solid as needles, crystals, lumps or pellets with an irritating, acrid, choking odor. Mixes with water forming a strong acid.

HEALTH AND SAFETY INFORMATION

Irritates eyes, skin, and respiratory tract. Inhalation can cause pulmonary edema, which can be delayed for several hours; there is a risk of death in serious cases.

IDLH: 10 mg/m^3
Odor Threshold: 0.318 ppm
OSHA Table Z-1-A Air Contaminant

Exposure Limits:	**TWA**
ACGIH:	0.25 ppm; 1.0 mg/m^3
NIOSH:	1 mg/m^3; 0.25 ppm
OSHA:	1 mg/m^3; 0.25 ppm

RECOMMENDATIONS FOR RESPIRATOR SELECTION: NIOSH/OSHA

10 mg/m^3: SA:CF (any supplied-air respirator operated in a continuous-flow mode);or SCBAF (any self-contained breathing apparatus with a full facepiece);or SAF (any supplied-air respirator with a full facepiece).

Emergency or Planned Entry into Unknown Concentrations or IDLH Conditions: SCBAF:PD,PP (any self-contained breathing apparatus that has a full facepiece and is operated in a pressure-demand or other positive-pressure mode);or SAF:PD,PP:ASCBA (any supplied-air respirator that has a full facepiece and is operated in a pressure-demand or other positive-pressure mode in combination with an auxiliary, self-contained breathing apparatus operated in a pressure-demand or other positive-pressure mode).

Escape: GMFOVHiE [any air-purifying, full-facepiece respirator (gas mask) with a chin-style, front- or back-mounted canister having a high-efficiency particulate filter] or SCBAE (any appropriate escape-type, self-contained breathing apparatus).

Note: Substance causes eye irritation or damage; eye protection needed.

CHEMICAL REACTIVITY

Reacts (hydrolyses) slowly with cold water to form maleic acid. Contact with strong oxidizers may cause fire and explosions. Contact with amines or alkali metals (even in low concentrations of 200 ppm) cause polymerization if temperature is greater than 150°F(66°C). Attacks some plastics, rubber, and coatings. Corrosive to metals except in the presence of water.

FIRE INFORMATION

Vapors are heavier than air and will collect in low areas.Vapors in confined areas may explode when exposed to fire. Vapors may travel long distances to ignition sources and flash back.Storage containers and parts of containers may rocket great distances, in many directions.If material or contaminated runoff enters waterways, notify downstream users of potentially contaminated waters. Notify local health and fire officials and pollution control agencies. Use dry chemicals, foam, carbon dioxide. From a secure explosion-proof location, use water spray to cool exposed containers.

NFPA Hazard Classifications:

Health Hazards (Blue) 3
Flammability (Red) 1
Reactivity (Yellow) 1

Flash Point: 218°F (103°C)
Flammable Limits in Air: LEL 1.4%; UEL 7.1%
Autoignition Temperature: 878°F (470°C)

FIRST AID

Move victim to fresh air. Call emergency medical care. Apply artificial respiration if victim is not breathing. Do not use mouth-to-mouth method if victim ingested or inhaled the substance; induce artificial respiration with the aid of a pocket mask equipped with a one-way valve or other proper respiratory medical device. Administer oxygen if breathing is difficult. Remove and isolate contaminated clothing and shoes. In case of contact with material, immediately flush skin or eyes with running water for at least 20 minutes. For minor skin contact, avoid spreading material on unaffected skin. Keep victim warm and quiet. Effects of exposure (inhalation, ingestion, or skin contact) may be delayed. Ensure that medical personnel are aware of the material(s) involved and take precautions to protect themselves.

ENVIRONMENTAL INFORMATION

Clean Air Act: Hazardous Air Pollutants (Title I, Part A, Section 112).
Clean Water Act: Section 311 Hazardous Substances/RQ (same as CERCLA); Section 313 Priority Chemicals.
EPA Hazardous Waste Number (RCRA No.): U147.
RCRA Section 261 Hazardous Constituents
EPCRA Section 304: Reportable Quantity (RQ): CERCLA, 5,000 lb (2,270 kg).
EPCRA Section 313: Form R *de minimis* concentration reporting level: 1.0%.
TSCA: 40CFR716.120(*d*)

MANGANESE M:104

SYNONYMS: EINECS No. 231-105-1; MANGAN (Polish); MANGANAT; MANGANESE-55; MANGANESO (Spanish); MANGAN NITRIDOVANY (Czech); TRIPART LIQUID MANGANESE; TRONAMANG

IDENTIFICATION

CAS:	7439-96-5
Molecular Formula:	Mn
RTECS No.:	OO9275000 (metal)
Properties:	Combustible. A lustrous, brittle, silvery, or reddish gray solid.

HEALTH AND SAFETY INFORMATION

Irritates eyes and skin. Can cause brain damage.

IDLH: 500 mg/m^3
OSHA Table Z-1-A Air Contaminant as manganese and compounds

Exposure Limits:	TWA	CEILING	STEL
ACGIH:	0.2 mg/m^3		
NIOSH:	1 mg/m^3		3 mg/m^3
OSHA:		5 mg/m^3	

RECOMMENDATIONS FOR RESPIRATOR SELECTION: NIOSH/OSHA

10 mg/m^3: DMXSQ, if not present as a fume (any dust and mist respirator except single-use and quarter-mask respirators);or SA (any supplied-air respirator).
25 mg/m^3: SA:CF (any supplied-air respirator operated in a continuous-flow mode);or PAPRDM, if not present as a fume (any powered, air-purifying respirator with a dust and mist filter).
50 mg/m^3: HiEF (any air-purifying, full-facepiece respirator with a high-efficiency particulate filter);or SAT:CF (any supplied-air respirator that has a tight-fitting facepiece and is operated in a continuous-flow mode);or PAPRTHiE (any powered, air-purifying respirator with a tight-fitting facepiece and a high-efficiency particulate filter);or SCBAF (any self-contained breathing apparatus with a full facepiece);or SAF (any supplied-air respirator with a full facepiece).
500 mg/m^3: SA:PD,PP (any supplied-air respirator operated in a pressure-demand or other positive-pressure mode).

Emergency or Planned Entry into Unknown Concentrations or IDLH Conditions: SCBAF:PD,PP (any self-contained breathing apparatus that has a full facepiece and is operated in a pressure-demand or other positive-pressure mode);or SAF:PD,PP:ASCBA (any supplied-air respirator that has a full facepiece and is operated in a pressure-demand or other positive-pressure mode in combination with an auxiliary, self-contained breathing apparatus operated in a pressure-demand or other positive-pressure mode).

Escape: HiEF (any air-purifying, full-facepiece respirator with a high-efficiency particulate filter);or SCBAE (any appropriate escape-type, self-contained breathing apparatus).

CHEMICAL REACTIVITY

Contact with water forms explosive hydrogen gas. Contact with oxidizers, nitric acid, finely divided metals, or sulfur dioxide may cause fire and explosions.

FIRE INFORMATION

If material or contaminated runoff enters waterways, notify downstream users of potentially contaminated waters. Notify local health and fire officials and pollution control agencies. Use any extinguishing agent.

FEMA Hazard Classifications: (dust)

Health Hazards (Blue) 0
Flammability (Red) 2
Reactivity (Yellow) 2

FIRST AID

Move victim to fresh air. Call emergency medical care. Apply artificial respiration if victim is not breathing. Administer oxygen if breathing is difficult. Remove and isolate contaminated clothing and shoes. In case of contact with substance, immediately flush skin or eyes with running water for at least 20 minutes. Wash skin with soap and water. Keep victim warm and quiet. Effects of exposure (inhalation, ingestion, or skin contact) to substance may be delayed. Ensure that medical personnel are aware of the material(s) involved and take precautions to protect themselves.

ENVIRONMENTAL INFORMATION

Safe Drinking Water Act: Priority List; SMCL, 0.05 mg/L.
EPCRA Section 313: Form R *de minimis* concentration reporting level: 1.0%; Form R Toxic Chemical Category Code: N450.

MELAMINE M:140

SYNONYMS: AERO; AMMELIDE; CYANURAMIDE; CYANURIC TRIAMIDE; CYANUROTRIAMIDE; CYANUROTRIAMINE; CYMEL; EINECS No. 203-615-4; HICOPHOR PR; ISOMELAMINE; METALAMINE. NCI-C50715; PLURAGARD; PLURAGARD C 133; TEOHARN; THEOHARN; 2,4,6-TRIAMINO-s-TRIAZINE; 2,4,6-TRIAMINE-1,3,5-TRIAZINE; 1,3,5-TRIAZINE-2,4,6-TRIAMINE; s-TRIAZINE, 2,4,6-TRIAMINO-; VIRSET 656-4; TRIAMINO-1,3,5-TRIAZINE(2,4,6-)

IDENTIFICATION

CAS:	108-78-1
ERG ID:	UN 2811 as toxic solid, organic, n.o.s.
ERG Guide:	154
ERG Guide Category:	SUBSTANCES-TOXIC AND/OR CORROSIVE (NONCOMBUSTIBLE)
Molecular Formula:	$C_3H_6N_6$
RTECS No.:	OS0700000
Properties:	Prism or crystalline solid. Colorless to white. Very slightly soluble in water.

HEALTH AND SAFETY INFORMATION

Irritates eyes, skin, and respiratory tract. Repeated or prolonged exposure can cause kidney damage and skin problems.

IARC Evaluation: Group 3; unclassifiable; no human data; inadequate animal evidence.
NTP: carcinogenic in animals.

RECOMMENDATIONS FOR RESPIRATOR SELECTION

At any concentrations above the NIOSH REL, or where there is no REL, at any detectable concentration: SCBAF:PD,PP (any self-contained breathing apparatus that has a full facepiece and is operated in a pressure-demand or other positive-pressure mode);or SAF:PD,PP:ASCBA (any supplied-air respirator that has a full facepiece and is operated in a pressure-demand or other positive-pressure mode in combination with an auxiliary, self-contained breathing apparatus operated in a pressure-demand or other positive-pressure mode).

Escape: GMFOV [any air-purifying, full-facepiece respirator (gas mask) with a chin-style, front- or back-mounted organic vapor canister];or SCBAE (any appropriate escape-type, self-contained breathing apparatus).

CHEMICAL REACTIVITY

Incompatible with acrolein, acrylonitrile, tert-butyl nitroacetylene, ethylene oxide, isopropyl chlorocarbonate, maleic anhydride, triisobutylaluminum, strong oxidizers. Amines can ignite on contact with cellulose nitrate of high surface area.

FIRE INFORMATION

Decomposes at less than 525°F (273°C); combustion products include toxic ammonia, hydrocyanic acid, and nitrogen oxides. Vapors are heavier than air and will collect in low areas. Vapors in confined areas may explode when exposed to fire.If material or contaminated runoff enters waterways, notify downstream users of potentially contaminated waters. Notify local health and fire officials and pollution control agencies. Use water, dry chemicals, alcohol-resistant foam, carbon dioxide. Use water to cool exposed containers.

FEMA Hazard Classifications:

Health Hazards (Blue) 0
Flammability (Red) 2
Reactivity (Yellow) 0

Flash Point: 572°F (300°C)
Autoignition Temperature: > 932°F (>498°C)

FIRST AID

Move victim to fresh air; call emergency medical care. Apply artificial respiration if victim is not breathing. Do not use mouth-to-mouth method if victim ingested or inhaled the substance; induce artificial respiration with the aid of a pocket mask equipped with a one-way valve or other proper respiratory medical device. Administer oxygen if breathing is difficult. Remove and isolate contaminated clothing and shoes. In case of contact with substance, immediately flush skin or eyes with running water for at least 20 minutes. For minor skin contact, avoid spreading material on unaffected skin. Keep victim warm and quiet. Effects of exposure (inhalation, ingestion, or skin contact) to substance may be delayed. Ensure that medical personnel are aware of the material(s) involved and take precautions to protect themselves.

ENVIRONMENTAL INFORMATION

TSCA: CHIP available.

MENTHANEDIAMINE M:147

SYNONYMS: 4-AMINO-a,a,4-TRIMETHYLCYCLOHEXANEMETHAMINE; 1,8-DIAMINO-p-MENTHANE; EINECS No. 201-287-7; p-MENTHANE-1,8-DIAMINE; p-MENTHANE-1,8-DIYLDIAMINE; USAF RH-4

IDENTIFICATION

CAS:	80-52-4
ERG ID:	UN 2735 as amines, liquid, corrosive, n.o.s.
ERG Guide:	153
ERG Guide Category:	SUBSTANCES-TOXIC AND/OR CORROSIVE (COMBUSTIBLE)
Molecular Formula:	$C_{10}H_{22}N_2$
RTECS No.:	OS8925000
Properties:	Liquid. Clear. Paint-thinnerlike odor. Soluble in water.

HEALTH AND SAFETY INFORMATION

Irritates eyes and skin.

RECOMMENDATIONS FOR RESPIRATOR SELECTION

At any concentrations above the NIOSH REL, or where there is no REL, at any detectable concentration: SCBAF:PD,PP (any self-contained breathing apparatus that has a full facepiece and is operated in a pressure-demand or other positive-pressure mode);or SAF:PD,PP:ASCBA (any supplied-air respirator that has a full facepiece and is operated in a pressure-demand or other positive-pressure mode in combination with an auxiliary, self-contained breathing apparatus operated in a pressure-demand or other positive-pressure mode).

Escape: GMFOV [any air-purifying, full-facepiece respirator (gas mask) with a chin-style, front- or back-mounted organic vapor canister];or SCBAE (any appropriate escape-type, self-contained breathing apparatus).

CHEMICAL REACTIVITY

Incompatible with acrolein, acrylonitrile, tert-butyl nitroacetylene, ethylene oxide, isopropyl chlorocarbonate, maleic anhydride, triisobutylaluminum. Amines may ignite on contact with cellulose nitrate.

FIRE INFORMATION

Combustion products include toxic nitrogen oxides. Vapors are heavier than air and will collect in low areas. Vapors in confined areas may explode when exposed to fire. Vapors may travel long distances to ignition sources and flash back. Storage containers and parts of containers may rocket great distances, in many directions. If material or contaminated runoff enters waterways, notify downstream users of potentially contaminated waters. Notify local health and fire officials and pollution control agencies. Use dry chemicals, foam, carbon dioxide. From a secure explosion-proof location, use water spray to cool exposed containers.

NFPA Hazard Classifications: Not listed

Flash Point: 199°F (93°C)

FIRST AID

Move victim to fresh air and call emergency medical care. Apply artificial respiration if victim is not breathing. Do not use mouth-to-mouth method if victim swallowed or inhaled the substance; induce artificial respiration with the aid of a pocket mask equipped with a one-way valve or other proper respiratory medical device. If breathing is difficult, give oxygen. In case of contact with substance, immediately flush skin or eyes with running water for at least 20 minutes. Remove and isolate contaminated clothing and shoes. Keep victim warm and quiet. Effects of exposure (inhalation, ingestion, or skin contact) to substances may be delayed. Ensure that medical personnel are aware of the material(s) involved and take precautions to protect themselves.

2–MERCAPTOETHANOL M:154

SYNONYMS: EINECS No. 200-464-6; EMERY 5791; 1-ETHANOL-2-THIOL; 2-HYDROXY-1-ETHANETHIOL; 2-HYDROXYETHYLMERCAPTAN; 1-HYDROXY-2-MERCATOETHANE; 2-ME; MERCAPTOETHANOL; b-MERCAPTOETHANOL; MONOTHIOETHYLENEGLYCOL; 2-THIOETHANOL; THIOETHYLENEGLYCOL; THIOGLYCOL; THIOMONOGLYCOL; USAF EK-4196

IDENTIFICATION

CAS:	60-24-2
ERG ID:	UN 2966
ERG Guide:	153
ERG Guide Category:	SUBSTANCES-TOXIC AND/OR CORROSIVE (COMBUSTIBLE)
Molecular Formula:	C_2H_6OS
RTECS No.:	KL5600000
Properties:	Combustible liquid. Clear to white. Faint, disagreeable odor. Soluble in water.

HEALTH AND SAFETY INFORMATION

Irritates eyes, skin, and respiratory tract.

Odor Threshold: 0.1-1.0 ppm

Exposure Limits:	**TWA**
AIHAWEEL (1)	0.2 ppm

(1)Skin contact contributes significantly to overall exposure.

RECOMMENDATIONS FOR RESPIRATOR SELECTION

At any concentrations above the NIOSH REL, or where there is no REL, at any detectable concentration: SCBAF:PD,PP (any self-contained breathing apparatus that has a full facepiece and is operated in a pressure-demand or other positive-pressure mode);or SAF:PD,PP:ASCBA (any supplied-air respirator that has a full facepiece and is operated in a pressure-demand or other positive-pressure mode in combination with an auxiliary, self-contained breathing apparatus operated in a pressure-demand or other positive-pressure mode).

Escape: GMFOV [any air-purifying, full-facepiece respirator (gas mask) with a chin-style, front- or back-mounted organic vapor canister];or SCBAE (any appropriate escape-type, self-contained breathing apparatus).

CHEMICAL REACTIVITY

Forms explosive mixture with air. Contact with oxidizers may cause fire and explosions.

FIRE INFORMATION

Combustion products include toxic oxides of sulfur. Vapors are heavier than air and will collect in low areas. Vapors in confined areas may explode when exposed to fire. Vapors may travel long distances to ignition sources and flash back. Storage containers and parts of containers may rocket great distances, in many directions. If material or contaminated runoff enters waterways, notify downstream users of potentially contaminated waters. Notify local health and fire officials and pollution control agencies. Use dry chemicals, alcohol-resistant foam, carbon dioxide, water spray. From a secure explosion-proof location, use water spray to cool exposed containers.

NFPA Hazard Classifications:

Health Hazards (Blue) 2
Flammability (Red) 2
Reactivity (Yellow) . . . . -

Flash Point: 165°F (74°C) o.c.

FIRST AID

Move victim to fresh air and call emergency medical care. Apply artificial respiration if victim is not breathing. Do not use mouth-to-mouth method if victim swallowed or inhaled the substance; induce artificial respiration with the aid of a pocket mask equipped with a one-way valve or other proper respiratory medical device. If breathing is difficult, give oxygen. In case of contact with substance, immediately flush skin or eyes with running water for at least 20 minutes. Remove and isolate contaminated clothing and shoes. Keep victim warm and quiet. Effects of exposure (inhalation, ingestion, or skin contact) to substances may be delayed. Ensure that medical personnel are aware of the material(s) involved and take precautions to protect themselves.

MERCURIC CHLORIDE M:166

SYNONYMS: BICHLORIDE OF MERCURY; BICHLORURE de MERCURE (French); CALOCHLOR; CHLORID RTUTNATY (Czech); CHLORURE MERCURIQUE (French); CLORURO MERURICO (Spanish); CLORURO di MERCURIO (Italian); CORROSIVE MERCURY CHLORIDE; EEC No. 080-002-00-6; EINECS No. 231-299-8; FUNGCHEX; MC; MERCURIC BICHLORIDE; MERCURY BICHLORIDE; MERCURY(II) CHLORIDE; MERCURY(2+) CHLORIDE; MERCURY PERCHLORIDE; NCI-C60173; QUECKSILBER CHLORID (German); PERCHLORIDE OF MERCURY; SULEMA (Russian); SUBLIMAT (Czech); TL 898

IDENTIFICATION

CAS:	7487-94-7
ERG ID:	UN 1624
ERG Guide:	154
ERG Guide Category:	SUBSTANCES-TOXIC AND/OR CORROSIVE (NONCOMBUSTIBLE)
Molecular Formula:	Cl_2Hg
RTECS No.:	OV910000
Properties:	Noncombustible solid. White. Sinks and mixes slowly with water.

HEALTH AND SAFETY INFORMATION

IARC Evaluation: Group 3; not classifiable.
IDLH: 10 mg/m^3 as mercury
OSHA Table Z-1-A Air Contaminant

Exposure Limits:	TWA	CEILING
ACGIH: [1]	0.025 mg/m^3 as Hg	
NIOSH: [1]	0.05 mg/m^3 as Hg vapor	0.1 mg/m^3 as Hg
OSHA:		0.1 mg/m^3 as Hg

[1] Skin contact contributes significantly in overall exposure

RECOMMENDATIONS FOR RESPIRATOR SELECTION: NIOSH/OSHA as mercury compounds

0.5 mg/m^3: CCRS *End of service life indicator (ESLI) required.* [any chemical cartridge respirator with cartridge(s) providing protection against the compound of concern)];or SA (any supplied-air respirator)
1.25 mg/m^3: SA:CF (any supplied-air respirator operated in a continuous-flow mode);or PAPRS *End of service life indicator (ESLI) required.* (any powered, air-purifying respirator with cartridge(s) providing protection against the compound of concern)
2.5 mg/m^3: CCRFS *End of service life indicator (ESLI) required.* (any chemical cartridge respirator with a full facepiece and cartridge(s) providing protection against the compound of concern);or GMFS *End of service life indicator (ESLI) required.* (any air-purifying, full-facepiece respirator (gas mask) with a chin-style, front- or back-mounted canister providing protection against the compound of concern);or SAT:CF (any supplied-air respirator that has a tight-fitting facepiece and is operated in a continuous-flow mode);or PAPRTS (any powered, air-purifying respirator with a tight-fitting facepiece and cartridge(s) providing protection against the compound of concern) ;or SCBAF (any self-contained breathing apparatus with a full facepiece);or SAF (any supplied-air respirator with a full facepiece)
10 mg/m^3: SA:PD,PP (any supplied-air respirator operated in a pressure-demand or other positive-pressure mode)

Emergency or planned entry into unknown concentrations or IDLH conditions:
SCBAF:PD,PP (any self-contained breathing apparatus that has a full facepiece and is operated in a pressure-demand or other positive-pressure mode);or SAF:PD,PP:ASCBA (any supplied-air respirator that has a full facepiece and is operated in a pressure-demand or other positive-pressure mode in combination with an auxiliary, self-contained breathing apparatus operated in a pressure-demand or other positive pressure mode.

Escape: GMFS (any air-purifying, full-facepiece respirator (gas mask) with a chin-style, front- or back-mounted canister providing protection against the compound of concern);or SCBAE (any appropriate escape-type, self-contained breathing apparatus).

CHEMICAL REACTIVITY

Violent reaction with potassium and sodium. Incompatible with light metals.

FIRE INFORMATION

Combustion products include toxic mercury fumes. If material or contaminated runoff enters waterways, notify downstream users of potentially contaminated waters. Notify local health and fire officials and pollution control agencies. Use any extinguishing agent.

NFPA Hazard Classifications:

Health Hazards (Blue) 3
Flammability (Red) 0
Reactivity (Yellow) 0

Flash Point: Not flammable

FIRST AID

Move victim to fresh air; call emergency medical care. Apply artificial respiration if victim is not breathing. Do not use mouth-to-mouth method if victim ingested or inhaled the substance; induce artificial respiration with the aid of a pocket mask equipped with a one-way valve or other proper respiratory medical device. Administer oxygen if breathing is difficult. Remove and isolate contaminated clothing and shoes. In case of contact with substance, immediately flush skin or eyes with running water for at least 20 minutes. For minor skin contact, avoid spreading material on unaffected skin. Keep victim warm and quiet. Effects of exposure (inhalation, ingestion, or skin contact) to substance may be delayed. Ensure that medical personnel are aware of the material(s) involved and take precautions to protect themselves.

ENVIRONMENTAL INFORMATION

Clean Water Act: Section 307 Toxic Pollutants as mercury and compounds
RCRA Section 261 Hazardous Constituents, waste number not listed, as mercury compounds, n.o.s.
EPCRA Section 302, Extremely Hazardous Substances: TPQ = 500/10,000 lb (227/4,540 kg)
EPCRA Section 304 Reportable Quantity (RQ): EHS, 500 lb (227 kg)
EPCRA Section 313 (as mercury compound): Form R *de minimis* concentration reporting level: 1.0%. Form R Toxic Chemical Category Code: N458.
California's Proposition 65: Reproductive toxin.
Marine Pollutant (49CFR, Subchapter 172.101, Appendix B); Severe Pollutant

MERCURIC IODIDE M:174

SYNONYMS: DIIODURE de MERCURE (French); EINECS No. 231-873-8; HYDRARGYRUM BIJODATUM (German); MERCURIC IODIDE, RED; MERCURY BINIODIDE; MERCURY(II) IODIDE; MERCURY(2+) IODIDE; QUECKSILBERDIIODID (German); RED MERCURIC IODIDE; YODURO MERCURICO (Spanish)

IDENTIFICATION

CAS:	7774-29-0
ERG ID:	UN 1638
ERG Guide:	151
ERG Guide Category:	SUBSTANCES-TOXIC (NONCOMBUSTIBLE)
Molecular Formula:	HgI_2
RTECS No.:	OW5250000; OW5300000
Properties:	Noncombustible, powdery solid. Scarlet red. Odorless. Sinks in water and is slightly soluble.

HEALTH AND SAFETY INFORMATION

IARC Evaluation: Group 3; not classifiable.
IDLH: 10mg/m^3 as mercury
OSHA Table Z-1-A Air Contaminant

Exposure Limits:	**TWA**	**CEILING**
ACGIH: [1]	0.025 mg/m^3 as Hg	
NIOSH: [1]	0.05 mg/m^3 as Hg vapor	0.1 mg/m^3 as Hg
OSHA:		0.1 mg/m^3 as Hg

[1] Skin contact contributes significantly to overall exposure.

RECOMMENDATIONS FOR RESPIRATOR SELECTION: NIOSH/OSHA as mercury compounds

0.5 mg/m^3: CCRS *End of service life indicator (ESLI) required.* [any chemical cartridge respirator with cartridge(s) providing protection against the compound of concern)];or SA (any supplied-air respirator).
1.25 mg/m^3: SA:CF (any supplied-air respirator operated in a continuous-flow mode);or PAPRS *End of service life indicator (ESLI) required.* (any powered, air-purifying respirator with cartridge(s) providing protection against the compound of concern).
2.5 mg/m^3: CCRFS *End of service life indicator (ESLI) required.* [(any chemical cartridge respirator with a full facepiece and cartridge(s) providing protection against the compound of concern];or GMFS *End of service life indicator (ESLI) required.* [(any air-purifying, full-facepiece respirator (gas mask) with a chin-style, front- or back-mounted canister providing protection against the compound of concern];or SAT:CF (any supplied-air respirator that has a tight-fitting facepiece and is operated in a continuous-flow mode);or PAPRTS [any powered, air-purifying respirator with a tight-fitting facepiece and cartridge(s) providing protection against the compound of concern];or SCBAF (any self-contained breathing apparatus with a full facepiece);or SAF (any supplied-air respirator with a full facepiece).
10 mg/m^3: SA:PD,PP (any supplied-air respirator operated in a pressure-demand or other positive-pressure mode).

Emergency or planned entry into unknown concentrations or IDLH conditions:

SCBAF:PD,PP (any self-contained breathing apparatus that has a full facepiece and is operated in a pressure-demand or other positive-pressure mode);or SAF:PD,PP:ASCBA (any supplied-air respirator that

has a full facepiece and is operated in a pressure-demand or other positive-pressure mode in combination with an auxiliary, self-contained breathing apparatus operated in a pressure-demand or other positive pressure mode.

Escape: GMFS (any air-purifying, full-facepiece respirator (gas mask) with a chin-style, front- or back-mounted canister providing protection against the compound of concern);or SCBAE (any appropriate escape-type, self-contained breathing apparatus).

CHEMICAL REACTIVITY

Violent reaction with chlorine, sodium, and fluorine.

FIRE INFORMATION

Combustion products include toxic mercury vapors. If material or contaminated runoff enters waterways, notify downstream users of potentially contaminated waters. Notify local health and fire officials and pollution control agencies. Use lots of water, dry chemicals, foam, carbon dioxide.

NFPA Hazard Classifications: Not listed

FIRST AID

Move victim to fresh air. Call emergency medical care. Apply artificial respiration if victim is not breathing. Do not use mouth-to-mouth method if victim ingested or inhaled the substance; induce artificial respiration with the aid of a pocket mask equipped with a one-way valve or other proper respiratory medical device. Administer oxygen if breathing is difficult. Remove and isolate contaminated clothing and shoes. In case of contact with substance, immediately flush skin or eyes with running water for at least 20 minutes. For minor skin contact, avoid spreading material on unaffected skin. Keep victim warm and quiet. Effects of exposure (inhalation, ingestion, or skin contact) may be delayed. Ensure that medical personnel are aware of the material(s) involved and take precautions to protect themselves.

ENVIRONMENTAL INFORMATION

Clean Air Act: Hazardous Air Pollutants (Title I, Part A, Section 112)
Clean Water Act: Section 307 Toxic Pollutants as mercury and compounds
RCRA Section 261 Hazardous Constituents, waste number not listed, as mercury compounds, n.o.s.
EPCRA Section 313 (as mercury compound): Form R *de minimis* concentration reporting level: 1.0%. Form R Toxic Chemical Category Code: N458.
California's Proposition 65: Reproductive toxin.
Marine Pollutant (49CFR, Subchapter 172.101, Appendix B); Severe Pollutant as mercury

MERCURY M:200

SYNONYMS: COLLOIDAL MERCURY; EEC No. 080-001-00-0; EINECS No. 231-106-7; KWIK

(Dutch); MERCURE (French); MERCURIO (Italian, Spanish); MERCURY, ELEMENTAL; MERCURY, METALLIC; METALLIC MERCURY; NCI-C60399; QUECKSILBER (German); QUICK SILVER; RTEC (Polish)

IDENTIFICATION

CAS:	7439-97-6
ERG ID:	UN 2809
ERG Guide:	172
ERG Guide Category:	GALLIUM AND MERCURY
Molecular Formula:	Hg
RTECS No.:	OV4550000
Properties:	Heavy liquid. Silverywhite. Odorless. Sinks in water.

HEALTH AND SAFETY INFORMATION

Mercury accumulates in the brain quickly during exposure but is released from the brain very slowly. This may result in a buildup in brain tissue over a long time. The liver and kidneys may also be damaged by mercury accumulation. Long-term exposure may cause headache, dizziness, restlessness, irritability, sleepiness, tremors, defective muscle control, increased salivation, loose teeth, irritation of the gums with blue line between teeth and gums, loss of appetite, nausea, vomiting, diarrhea, liver damage, changes in urine, raised red areas and blisters on skin, impaired memory, and possibly permanent brain damage. Frequency of complaints and severity of symptoms increase with levels of exposure, most notably above 0.1 mg/m^3. However, many of these symptoms have been reported at levels below recommended limits due to the accumulation of mercury over long- term exposure.

IDLH: 10 mg/m^3
OSHA Table Z-1-A Air Contaminant
OSHA Table Z-2 Air Contaminant
NIOSH Criteria Document: NIOSH 73-11024 INORGANIC MERCURY

Exposure Limits:	**TWA**	**CEILING**
ACGIH:[(1)]	0.025 mg/m^3	
NIOSH:[(1)]	0.05 mg/m^3 as vapor	0.1 mg/m^3
OSHA:		0.1 mg/m^3

[(1)]Skin contact contributes significantly to overall exposure.

RECOMMENDATIONS FOR RESPIRATOR SELECTION: NIOSH/OSHA as mercury

0.5 mg/m^3: CCRS end of service life indicator (ESLI) required (any chemical cartridge respirator with cartridge(s) providing protection against the compound of concern);or SA (any supplied-air respirator).
1.25 mg/m^3: SA:CF (any supplied-air respirator operated in a continuous-flow mode);or PAPRS* (any powered, air-purifying respirator with cartridge(s) providing protection against the compound of concern).
2.5 mg/m^3: CCRFS end of service life indicator (ESLI) required (any chemical cartridge respirator with a

full facepiece and cartridge(s) providing protection against the compound of concern);or GMFS end of service life indicator (ESLI) required [any air-purifying, full-facepiece respirator (gas mask) with a chin-style, front- or back-mounted canister providing protection against the compound of concern];or SAT:CF (any supplied-air respirator that has a tight-fitting facepiece and is operated in a continuous-flow mode);or PAPRTS (any powered, air-purifying respirator with a tight-fitting facepiece and cartridge(s) providing protection against the compound of concern);or SCBAF (any self-contained breathing apparatus with a full facepiece);or SAF (any supplied-air respirator with a full facepiece).
10 mg/m^3: SA:PD,PP (any supplied-air respirator operated in a pressure-demand or other positive-pressure mode).

Emergency or Planned Entry into Unknown Concentrations or IDLH Conditions: SCBAF:PD,PP (any self-contained breathing apparatus that has a full facepiece and is operated in a pressure-demand or other positive-pressure mode);or SAF:PD,PP:ASCBA (any supplied-air respirator that has a full facepiece and is operated in a pressure-demand or other positive-pressure mode in combination with an auxiliary, self-contained breathing apparatus operated in a pressure-demand or other positive-pressure mode).

Escape: GMFS [any air-purifying, full-facepiece respirator (gas mask) with a chin-style, front- or back-mounted canister providing protection against the compound of concern];or SCBAE (any appropriate escape-type, self-contained breathing apparatus).

CHEMICAL REACTIVITY

Reaction with acetylene, acetylene products, and ammonia gases can form shock-sensitive solids that can initiate fires of combustibles. Violent reactions with boron phosphodiiodide, chlorine, chlorine dioxide, and methyl azide. Reacts with nitrates, chlorates, hot sulfuric acid, alkalies, and dry bromine. Attacks copper and copper alloys.

FIRE INFORMATION

If material or contaminated runoff enters waterways, notify downstream users of potentially contaminated waters. Notify local health and fire officials and pollution control agencies.

FIRST AID

Move victim to fresh air. Call emergency medical care. Apply artificial respiration if victim is not breathing. Administer oxygen if breathing is difficult. Remove and isolate contaminated clothing and shoes. In case of contact with material, immediately flush skin or eyes with running water for at least 20 minutes. Keep victim warm and quiet. Ensure that medical personnel are aware of the material(s) involved and take precautions to protect themselves.

ENVIRONMENTAL INFORMATION

Clean Water Act: Section 307 Toxic Pollutants as mercury and compounds; Section 313 Priority Chemicals.
EPA Hazardous Waste Number (RCRA No.): U151.
RCRA Section 261 Hazardous Constituents

RCRA Toxicity Characteristic (Section 261.24): Maximum Concentration of Contaminants, regulatory level, 0.2 mg/L.
RCRA Land Ban Waste
RCRA Universal Treatment Standards: Wastewater (mg/L), 0.15; Nonwastewater (mg/L), 0.25 TCLP; Wastewater from retort, N/A; Nonwastewater from retort (mg/L), 0.20 TCLP.
RCRA Ground Water Monitoring List: Suggested test method(s) (PQL ug/L): 7420(2) as mercury (total).
Safe Drinking Water Act: MCL, 0.002 mg/L; MCLG, 0.002 mg/L.
EPCRA Section 304: Reportable Quantity (RQ): CERCLA, 1 lb (0.454 kg).
EPCRA Section 313: Form R *de minimis* concentration reporting level: 1.0%. Form R Toxic Chemical Category Code: N458.
California's Proposition 65: Reproductive toxin.

MESITYLENE M:212

SYNONYMS: BENZENE, 1,3,5-TRIMETHYL-; EEC No. 601-025-005; EINECS No. 203-604-4; FLEET-X; TMB; TRIMETHYL BENZENE; 1,3,5-TRIMETHYL BENZENE; sym-TRIMETHYLBENZENE; TRIMETHYL BEN; MESITILENO (Spanish); ZOL

IDENTIFICATION

CAS:	108-67-8
ERG ID:	UN 2325
ERG Guide:	129
ERG Guide Category:	FLAMMABLE LIQUIDS (POLAR/WATER-MISCIBLE/NOXIOUS)
Molecular Formula:	C_9H_{12}
RTECS No.:	OX6825000
Properties:	Flammable liquid. Clear, colorless. Aromatic odor. Insoluble in water.

HEALTH AND SAFETY INFORMATION

Irritates eyes, nose and respiratory tract. Affects central nervous system with possible headache, dizziness, fatigue, nausea; and blood (hypochromic anemia). May cause chemical pneumonia.

Exposure Limits:	**TWA**
ACGIH:	25 ppm; 123 mg/m^3
NIOSH:	25 ppm; 125 mg/m^3
OSHA:	None

RECOMMENDATIONS FOR RESPIRATOR SELECTION

At any concentrations above the NIOSH REL, or where there is no REL, at any detectable

concentration: SCBAF:PD,PP (any self-contained breathing apparatus that has a full facepiece and is operated in a pressure-demand or other positive-pressure mode);or SAF:PD,PP:ASCBA (any supplied-air respirator that has a full facepiece and is operated in a pressure-demand or other positive-pressure mode in combination with an auxiliary, self-contained breathing apparatus operated in a pressure-demand or other positive-pressure mode).

Escape: GMFOV [any air-purifying, full-facepiece respirator (gas mask) with a chin-style, front- or back-mounted organic vapor canister];or SCBAE (any appropriate escape-type, self-contained breathing apparatus).

CHEMICAL REACTIVITY

Reacts vigorously with oxidizers and nitric acid.

FIRE INFORMATION

Vapors are heavier than air and will collect in low areas. Vapors in confined areas may explode when exposed to fire. Vapors may travel long distances to ignition sources and flash back. Storage containers and parts of containers may rocket great distances, in many directions. If material or contaminated runoff enters waterways, notify downstream users of potentially contaminated waters. Notify local health and fire officials and pollution control agencies. Use dry chemicals, foam, carbon dioxide. From a secure, explosion-proof location, use water spray to cool exposed containers.

NFPA Hazard Classifications:

Health Hazards (Blue) 0
Flammability (Red) 2
Reactivity (Yellow) 0

Flash Point: 122°F (50°C)
Autoignition Temperature: 1039°F (559°C)

FIRST AID

Move victim to fresh air. Call emergency medical care. Apply artificial respiration if victim is not breathing. Administer oxygen if breathing is difficult. Remove and isolate contaminated clothing and shoes. In case of contact with substance, immediately flush skin or eyes with running water for at least 20 minutes. Wash skin with soap and water. Keep victim warm and quiet. Effects of exposure (inhalation, ingestion, or skin contact) to substance may be delayed. Ensure that medical personnel are aware of the material(s) involved and take precautions to protect themselves.

ENVIRONMENTAL INFORMATION

Safe Drinking Water Act: Monitoring is at the discretion of the state.
TSCA: 40CFR716.120(*a*); 40CFR799.2175 testing requirements.

MESITYL OXIDE **M:213**

SYNONYMS: EEC No. 606-009-00-1; ISOBUTENYL METHYL KETONE; ISOPROPYLIDENEACETONE; METHYL ISOBUTENYL KETONE; 4-METHYL-3-PENTENE-2-ONE; OXIDO de MESITILO (Spanish); OXYDE de MESITYLE (French)

IDENTIFICATION

CAS:	141-79-7
ERG ID:	UN 1229
ERG Guide:	129
ERG Guide Category:	FLAMMABLE LIQUIDS (POLAR/WATER-MISCIBLE/NOXIOUS)
Molecular Formula:	$C_6H_{10}O$
RTECS No.:	SB4200000
Properties:	Highly flammable, oily liquid. Colorless to light yellow. Strong, sweet mintlike or honeylike odor. Floats on water; moderately soluble. Flammable, irritating vapor is produced.

HEALTH AND SAFETY INFORMATION

Irritates eyes, skin, and respiratory tract. Can cause burns to eyes. Inhalation may cause headache, drowsiness, unconsciousness.

IDLH: 1,400 ppm [LEL]
Odor Threshold: 0.056 ppm
OSHA Table Z-1-A Air Contaminant
NIOSH Criteria Document: NIOSH 78-173 KETONES

Exposure Limits:	**TWA**	**STEL**
ACGIH:	15 ppm; 60 mg/m^3	25 ppm; 100 mg/m^3
NIOSH:	10 ppm; 40 mg/m^3	
OSHA:	25 ppm; 100 mg/m^3	

RECOMMENDATIONS FOR RESPIRATOR SELECTION: NIOSH

250 ppm: SA:CF (any supplied-air respirator operated in a continuous-flow mode);or PAPROV [any powered, air-purifying respirator with organic vapor cartridge(s)].
500 ppm: CCRFOV [any chemical cartridge respirator with a full facepiece and organic vapor cartridge(s)];or GMFOV [any air-purifying, full-facepiece respirator (gas mask) with a chin-style, front- or back-mounted acid gas canister];or PAPRTOV [any powered, air-purifying respirator with a tight-fitting facepiece and organic vapor cartridge(s)];or SCBAF (any self-contained breathing apparatus with a full facepiece);or SAF (any supplied-air respirator with a full facepiece).
1400 ppm: SAF:PD,PP (any supplied-air respirator that has a full facepiece and is operated in a pressure-demand or other positive-pressure mode).

Emergency or planned entry into unknown concentrations or idlh conditions: SCBAF:PD,PP (any self-contained breathing apparatus that has a full facepiece and is operated in a pressure-demand or other positive-pressure mode);or SAF:PD,PP:ASCBA (any supplied-air respirator that has a full facepiece and is operated in a pressure-demand or other positive-pressure mode in combination with an auxiliary, self-contained breathing apparatus operated in a pressure-demand or other positive pressure mode.

Escape: GMFOV(any air-purifying, full-facepiece respirator (gas mask) with a chin-style, front- or back-mounted organic vapor canister or SCBAE (any appropriate escape-type, self-contained breathing apparatus).

Note: Substance causes eye irritation or damage; eye protection needed.

CHEMICAL REACTIVITY

May form unstable peroxides. Forms explosive mixture with air. Contact with oxidizers may cause fire and explosions. Reacts with sulfuric acid, nitric acid, chlorosulfonic acid, ethylene diamine. Attacks copper, plastics, and some coatings.

FIRE INFORMATION

Vapors are heavier than air and will collect in low areas. Vapors in confined areas may explode when exposed to fire. Vapors may travel long distances to ignition sources and flash back. Storage containers and parts of containers may rocket great distances, in many directions. If material or contaminated runoff enters waterways, notify downstream users of potentially contaminated waters. Notify local health and fire officials and pollution control agencies. Use dry chemicals, foam, carbon dioxide. From a secure, explosion-proof location, use water spray to cool exposed containers.

NFPA Hazard Classifications:

Health Hazards (Blue) 2
Flammability (Red) 3
Reactivity (Yellow) 1

Flash Point: 84°F (29°C) o.c.; 87°F (31°C) c.c.
Flammable Limits in Air: LEL 1.4%; UEL 7.2%
Autoignition Temperature: 652°F (344°C)

FIRST AID

Move victim to fresh air. Call emergency medical care. Apply artificial respiration if victim is not breathing. Administer oxygen if breathing is difficult. Remove and isolate contaminated clothing and shoes. In case of contact with substance, immediately flush skin or eyes with running water for at least 20 minutes. Wash skin with soap and water. Keep victim warm and quiet. Effects of exposure (inhalation, ingestion, or skin contact) to substance may be delayed. Ensure that medical personnel are aware of the material(s) involved and take precautions to protect themselves.

ENVIRONMENTAL INFORMATION

TSCA: 40CFR716.120(*a*); 40CFR799, subpart B, Section 8(*d*) [799.2500; 799.5000] manufacturers, importers, processors are required to submit unpublished health studies to the U.S. EPA.

METHACRYLIC ACID M:215

SYNONYMS: ACIDE METHACRYLIQUE (French); ACIDO METACRILICO (Spanish); ACIDO α-METACRILICO (Spanish); ACRYLIC ACID, 2-METHYL-; EEC No. 607-088-00-5; EINECS No. 201-204-4; GLACIAL METHACRYLIC ACID; KYSELINA METHAKRYLOVA (Czech); METHACRYLIC ACID, INHIBITED; α-METHYL-ACRYLIC ACID; METHACRYLSAEURE (German); METHACRYLZUUR (Dutch); 2-METHYLPROPENOIC ACID; 2-METHYL-2-PROPENOIC ACID; 2-PROPENOIC ACID, 2-METHYL-; PROPIONIC ACID, 2-METHYLENE

IDENTIFICATION

CAS:	79-41-4
ERG ID:	UN 2531
ERG Guide:	153P
ERG Guide Category:	SUBSTANCES-TOXIC AND/OR CORROSIVE (COMBUSTIBLE)
Molecular Formula:	$C_4H_6O_2$
RTECS No.:	OZ2975000
Properties:	Highly flammable, colorless liquid or solid [below 61°F (16°C)]. Acrid, repulsive odor. Sinks slowly in water and is moderately soluble.

HEALTH AND SAFETY INFORMATION

OSHA Table Z-1-A Air Contaminant

Exposure Limits:	**TWA**
ACGIH:	20 ppm; 70 mg/m^3
NIOSH:[(1)]	20 ppm; 70 mg/m^3
OSHA:	None

[(1)]Skin contact contributes significantly to overall exposure.

RECOMMENDATIONS FOR RESPIRATOR SELECTION: NIOSH

At any concentrations above the NIOSH REL, or where there is no REL, at any detectable concentration: SCBAF:PD,PP (any self-contained breathing apparatus that has a full facepiece and is operated in a pressure-demand or other positive-pressure mode);or SAF:PD,PP:ASCBA (any supplied-air

respirator that has a full facepiece and is operated in a pressure-demand or other positive-pressure mode in combination with an auxiliary, self-contained breathing apparatus operated in a pressure-demand or other positive-pressure mode).

Escape: GMFOV [any air-purifying, full-facepiece respirator (gas mask) with a chin-style, front- or back-mounted organic vapor canister];or SCBAE (any appropriate escape-type, self-contained breathing apparatus).

CHEMICAL REACTIVITY

A reducing agent; reaction with oxidizers may be violent. Forms unstable peroxides; can polymerize violently, especially above temperatures above 120°F (49°C). Incompatible with strong acids, caustics, ammonia, amines, isocyanates, alkylene oxides, and epichlorohydrin. Corrosive to metals. Typically contains 100 ppm of the monomethyl ether of hydroquinone to prevent polymerization.

FIRE INFORMATION

Firefighting gear (including SCBA) does not provide adequate protection. If exposure occurs, remove and isolate gear immediately and thoroughly decontaminate personnel. Vapors are heavier than air and will collect in low areas. Vapors in confined areas may explode when exposed to fire. Vapors may travel long distances to ignition sources and flash back. Storage containers and parts of containers may rocket great distances, in many directions. If material or contaminated runoff enters waterways, notify downstream users of potentially contaminated waters. Notify local health and fire officials and pollution control agencies. Use dry chemicals, foam, carbon dioxide. From a secure, explosion-proof location, use water spray to cool exposed containers.

NFPA Hazard Classifications:

Health Hazards (Blue) 3
Flammability (Red) 2
Reactivity (Yellow) 2

Flash Point: 152°F (67°C)
Flammable Limits in Air: LEL 1.6%; UEL 8.8%
Autoignition Temperature: 154°F (68°C)

FIRST AID

Move victim to fresh air. Call emergency medical care. Apply artificial respiration if victim is not breathing. Do not use mouth-to-mouth method if victim ingested or inhaled the substance; induce artificial respiration with the aid of a pocket mask equipped with a one-way valve or other proper respiratory medical device. Administer oxygen if breathing is difficult. Remove and isolate contaminated clothing and shoes. In case of contact with substance, immediately flush skin or eyes with running water for at least 20 minutes. For minor skin contact, avoid spreading material on unaffected skin. Keep victim warm and quiet. Effects of exposure (inhalation, ingestion, or skin contact) may be delayed. Ensure that medical personnel are aware of the material(s) involved and take precautions to protect themselves.

METHACRYLONITRILE M:217

SYNONYMS: 2-CYANOPROPENE-1; 2-CYANO-1-PROPENE; EEC No. 608-010-00-2; EINECS No. 204-817-5; ISOPROPENE CYANIDE; ISOPROPENYLNITRILE; α-METHYLACRYLONITRILE; 2-METHYLACRYLONITRILE; 2-METHYLPROPENENITRILE; 2-METHYL-2-PROPENENITRILE; 2-PROPENENITRILE, 2-METHYL-; USAF ST40

IDENTIFICATION

CAS:	126-98-7
ERG ID:	UN 3079
ERG Guide:	131P
ERG Guide Category:	FLAMMABLE LIQUIDS-TOXIC
Molecular Formula:	C_4H_5N
RTECS No.:	UD1400000
Properties:	Extremely flammable liquid. Colorless. Bitter almond odor. Practically insoluble in water.

HEALTH AND SAFETY INFORMATION

Poisonous; a cyanide compound. Breathing the vapors can cause death. Swallowing the material or skin contact can cause serious injury or death. Irritates eyes and can be absorbed through the skin.

Odor Threshold: 6.9 ppm. Odor is not a reliable indicator of the presence of toxic amounts of material.
NIOSH Criteria Document: NIOSH 78-212 NITRILES

Exposure Limits: TWA

ACGIH: [(1)]	1 ppm; 2.7 mg/m^3
NIOSH: [(1)]	1 ppm; 3 mg/m^3
OSHA:	None

[(1)] Skin contact contributes significantly in overall exposure.

RECOMMENDATIONS FOR RESPIRATOR SELECTION: NIOSH as cyanides

50 mg/m^3: SA (any supplied-air respirator);or SCBAF (any self-contained breathing apparatus with full facepiece).

Emergency or Planned Entry into Unknown Concentrations or IDLH Conditions:
SCBAF:PD,PP (any self-contained breathing apparatus that has a full facepiece and is operated in a pressure-demand or other positive-pressure mode);or SAF:PD,PP:ASCBA (any supplied-air respirator that has a full facepiece and is operated in a pressure-demand or other positive-pressure mode in combination with an auxiliary, self-contained breathing apparatus operated in a pressure-demand or other positive-pressure mode).

Escape: GMFSHiE [any air-purifying, full-facepiece respirator (gas mask) with a chin-style, front- or back-mounted canister providing protection against the compound of concern and having a high-efficiency particulate filter];or SCBAE (any appropriate escape-type, self-contained breathing apparatus).

CHEMICAL REACTIVITY

May react with itself without warning with explosive violence. Forms explosive mixture with air. Incompatible with aliphatic amines, alkanolamines. Reacts violently with oxidizers. Incompatible with strong acids, strong bases, or exposure to light; may cause polymerization.

FIRE INFORMATION

Combustion products include toxic nitrogen oxides and cyanides. Firefighting gear (including SCBA) does not provide adequate protection. If exposure occurs, remove and isolate gear immediately and thoroughly decontaminate personnel. Vapors are heavier than air and will collect in low areas. Vapors in confined areas may explode when exposed to fire. Vapors may travel long distances to ignition sources and flash back. Storage containers and parts of containers may rocket great distances, in many directions. If material or contaminated runoff enters waterways, notify downstream users of potentially contaminated waters. Notify local health and fire officials and pollution control agencies. Use dry chemicals, foam, carbon dioxide. From a secure, explosion-proof location, use water spray to cool exposed containers.

NFPA Hazard Classifications:

Health Hazards (Blue) 2
Flammability (Red) 3
Reactivity (Yellow) 2

Flash Point: 34°F (1°C) c.c.
Flammable Limits in Air: LEL 2%; UEL 6.8%

FIRST AID

Move victim to fresh air. Call emergency medical care. Apply artificial respiration if victim is not breathing. Do not use mouth-to-mouth method if victim ingested or inhaled the substance; induce artificial respiration with the aid of a pocket mask equipped with a one-way valve or other proper respiratory medical device. Administer oxygen if breathing is difficult. Remove and isolate contaminated clothing and shoes. In case of contact with substance, immediately flush skin or eyes with running water for at least 20 minutes. Wash skin with soap and water. Keep victim warm and quiet. Effects of exposure (inhalation, ingestion, or skin contact) to substance may be delayed. Ensure that medical personnel are aware of the material(s) involved and take precautions to protect themselves.

ENVIRONMENTAL INFORMATION

Clean Air Act: Accidental Release Prevention/Flammable Substances (Section 112[r], Table 3), TQ = 10,000 lb (4,540 kg).
EPA Hazardous Waste Number (RCRA No.): U152, D003, P030

RCRA Toxicity Characteristic (Section 261.24): RCRA Section 261 Hazardous Constituents
RCRA Land Ban Waste
RCRA Universal Treatment Standards: Wastewater (mg/L), 0.24; Nonwastewater (mg/kg), 84.
RCRA Ground Water Monitoring List: Suggested test methods(s) (PQL μg/L); 8015(5); 8240(5)
EPCRA Section 302, Extremely Hazardous Substances: TPQ = 500 lb (227 kg)
EPCRA Section 304 Reportable Quantity (RQ): CERCLA, 1,000 lb (454 kg)
EPCRA Section 313: Form R *de minimis* concentration reporting level: 1.0%.

METHANE M:220

SYNONYMS: BIOGAS; EEC No. 601-001-00-4; EINECS No. 200-812-7; FIRE DAMP; MARSH GAS; METHANE GAS; METANO (Italian, Spanish); METHYL HYDRIDE; NATURAL GAS; P-10 GAS; R50

IDENTIFICATION

CAS:	74-82-8
ERG ID:	UN 1971; UN 1972 (cryogenic gas)
ERG Guide:	115
ERG Guide Category:	GASES–FLAMMABLE (INCLUDING REFRIGERATED LIQUIDS)
Molecular Formula:	CH_4
RTECS No.:	PA1490000
Properties:	Extremely flammable compressed gas, or cryogenic liquid. Colorless. No odor, or a weak skunklike odor. Liquid floats and "boils" on water; insoluble. Flammable visible vapor cloud is produced.

HEALTH AND SAFETY INFORMATION

Methane is a simple asphyxiant. Contact with liquid may cause frostbite.

RECOMMENDATIONS FOR RESPIRATOR SELECTION: NIOSH

At any concentrations above the NIOSH REL, or where there is no REL, at any detectable concentration: SCBAF:PD,PP (any self-contained breathing apparatus that has a full facepiece and is operated in a pressure-demand or other positive-pressure mode);or SAF:PD,PP:ASCBA (any supplied-air respirator that has a full facepiece and is operated in a pressure-demand or other positive-pressure mode in combination with an auxiliary, self-contained breathing apparatus operated in a pressure-demand or other positive-pressure mode).

Escape: GMFOV [any air-purifying, full-facepiece respirator (gas mask) with a chin-style, front- or back-mounted organic vapor canister];or SCBAE (any appropriate escape-type, self-contained breathing apparatus).

CHEMICAL REACTIVITY

Forms an explosive mixture with air. Violent reaction with strong oxidizers, bromine pentafluoride, chlorine dioxide, nitrogen trifluoride, liquid oxygen, and oxygen difluoride.

FIRE INFORMATION

Gases are heavier than air and will collect in low areas. Gases in confined areas may explode when exposed to fire. Gases may travel long distances to ignition sources and flash back. Storage containers and parts of containers may rocket great distances, in many directions. If material or contaminated runoff enters waterways, notify downstream users of potentially contaminated waters. Notify local health and fire officials and pollution control agencies. Do not extinguish the fire unless the flow of the gas can be stopped and any remaining gas is out of the line. Specially trained personnel may use fog lines to cool exposures and let the fire burn itself out. Use dry chemicals, alcohol-resistant foam, carbon dioxide. From a secure, explosion-proof location, use water spray to cool exposed containers. If cooling streams are ineffective (venting sound increases in volume and pitch, tank discolors, or shows any signs of deforming), withdraw immediately to a secure position.

NFPA Hazard Classifications:

Health Hazards (Blue) 1
Flammability (Red) 4
Reactivity (Yellow) 0

Flash Point: Flammable gas
Flammable Limits in Air: LEL 5.0%; UEL 15.0%
Autoignition Temperature: 1004°F (540°C)

FIRST AID

Move victim to fresh air. Call emergency medical care. Apply artificial respiration if victim is not breathing. Administer oxygen if breathing is difficult. Remove and isolate contaminated clothing and shoes. Clothing frozen to the skin should be thawed before being removed. In case of contact with liquefied gas, thaw frosted parts with lukewarm water. Keep victim warm and quiet. Keep victim under observation. Ensure that medical personnel are aware of the material(s) involved and take precautions to protect themselves.

ENVIRONMENTAL INFORMATION

Clean Air Act: Accidental Release Prevention/Flammable substances, (Section 112[r], Table 3), TQ = 10,000 lb (4,540 kg).

p–MENTHANE HYDROPEROXIDE M:224

SYNONYMS: 1-ISOPROPYL-4-METHYLCYCLOHANE HYDROPEROXIDE; p-METHANE

HYDROPEROXIDE; p-METHYL HYDROPEROXIDE; PARAMENTHANE HYDROPEROXIDE; PHMP

IDENTIFICATION

CAS:	52061-60-6; 80-47-7
ERG ID:	UN 3109; UN 3105 (>56%);
ERG Guide:	ORGANIC PEROXIDE TYPE F, LIQUID (3109) ORGANIC PEROXIDE TYPE D, LIQUID (3105)
ERG Guide Category:	145
Molecular Formula:	$C_{10}H_{20}O_2$
RTECS No.:	MX2500000
Properties:	Liquid. Clear to pale yellow.

HEALTH AND SAFETY INFORMATION

Severely irritates eyes, skin, and respiratory tract.

RECOMMENDATIONS FOR RESPIRATOR SELECTION

At any concentrations above the NIOSH REL, or where there is no REL, at any detectable concentration: SCBAF:PD,PP (any self-contained breathing apparatus that has a full facepiece and is operated in a pressure-demand or other positive-pressure mode);or SAF:PD,PP:ASCBA (any supplied-air respirator that has a full facepiece and is operated in a pressure-demand or other positive-pressure mode in combination with an auxiliary, self-contained breathing apparatus operated in a pressure-demand or other positive-pressure mode).

Escape: GMFOV [any air-purifying, full-facepiece respirator (gas mask) with a chin-style, front- or back-mounted organic vapor canister];or SCBAE (any appropriate escape-type, self-contained breathing apparatus).

CHEMICAL REACTIVITY

Unstable; an organic peroxide. A strong oxidizing agent; may ignite combustibles (wood, paper, oil, clothing, etc.). May explode from heat or contamination. Reacts vigorously with organic materials.

FIRE INFORMATION

Wash material off clothing; may be a fire hazard. Will increase the activity of fire. May burn rapidly with a flare-burning effect. Vapors are heavier than air and will collect in low areas. Vapors in confined areas may explode when exposed to fire. Vapors may travel long distances to ignition sources and flash back. Storage containers and parts of containers may rocket great distances, in many directions. If material or contaminated runoff enters waterways, notify downstream users of potentially contaminated waters. Notify local health and fire officials and pollution control agencies. Runoff may created a fire or explosion hazard. Use dry chemicals, foam, carbon dioxide. For large fires consider evacuation for at least 800 feet (250 meters); tank car or truck: isolate for at least ½-mile (800 meters) in all directions.

NFPA Hazard Classifications: Not listed

FIRST AID

Move victim to fresh air. Call emergency medical care. Apply artificial respiration if victim is not breathing. Administer oxygen if breathing is difficult. Remove and isolate contaminated clothing and shoes. In case of contact with substance, immediately flush skin or eyes with running water for at least 20 minutes. Wash skin with soap and water. Keep victim warm and quiet. Effects of exposure (inhalation, ingestion, or skin contact) to substance may be delayed. Ensure that medical personnel are aware of the material(s) involved and take precautions to protect themselves.

METHANESULFONIC ACID M:227

SYNONYMS: ACIDE METHANESULFONIQUE (Spanish); ACIDO de METANOSULFONICO (Spanish); ALKANE SULFONIC ACID; EEC No. 607-145-00-4; EINECS No. 200-898-6; KYSELENA METHANSULFONOVA (Czech); METHAANSULFONZUUR (Dutch)

IDENTIFICATION

CAS:	75-75-2
ERG ID:	UN 3265 (corrosive liquid, acidic, organic, n.o.s.); UN 3261 (corrosive solid, acidic, organic, n.o.s.)
ERG Guide:	153 (liquid); 154 (solid)
ERG Guide Category:	SUBSTANCES-TOXIC AND/OR CORROSIVE (COMBUSTIBLE)(liquid) SUBSTANCES-TOXIC AND/OR CORROSIVE (NON-COMBUSTIBLE)(solid)
Molecular Formula:	CH_4O_3S
RTECS No.:	PB1140000
Properties:	Combustible liquid (at room temperature) or solid. Soluble in water.

HEALTH AND SAFETY INFORMATION

Corrosive to eyes, skin, and respiratory tract. Inhalation of dust or liquid may cause pulmonary edema, which can be delayed for several hours; there is a risk of death in serious cases.

RECOMMENDATIONS FOR RESPIRATOR SELECTION

At any concentrations above the NIOSH REL, or where there is no REL, at any detectable concentration: SCBAF:PD,PP (any self-contained breathing apparatus that has a full facepiece and is operated in a pressure-demand or other positive-pressure mode);or SAF:PD,PP:ASCBA (any supplied-air respirator that has a full facepiece and is operated in a pressure-demand or other positive-pressure mode in

combination with an auxiliary, self-contained breathing apparatus operated in a pressure-demand or other positive-pressure mode).

Escape: GMFOV [any air-purifying, full-facepiece respirator (gas mask) with a chin-style, front- or back-mounted organic vapor canister];or SCBAE (any appropriate escape-type, self-contained breathing apparatus).

CHEMICAL REACTIVITY

Contact with acids form toxic sulfur oxides. Explosive reaction with ethyl vinyl ether. Incompatible with hydrogen fluoride, and oxidizers. Organic acids are incompatible with sulfuric acid, caustics, ammonia, aliphatic amines, alkanolamines, isocyanates, alkylene oxides, epichlorohydrin. Corrosive to most common metals.

FIRE INFORMATION

Combustion products include toxic sulfur oxides. If material or contaminated runoff enters waterways, notify downstream users of potentially contaminated waters. Notify local health and fire officials and pollution control agencies. Use dry chemicals, alcohol-resistant foam, carbon dioxide. From a secure, explosion-proof location, use water spray to cool exposed containers.

NFPA Hazard Classifications: Not listed

Flash Point: 234°F (112°C)

FIRST AID

Move victim to fresh air. Call emergency medical care. Apply artificial respiration if victim is not breathing. Administer oxygen if breathing is difficult. Remove and isolate contaminated clothing and shoes. In case of contact with substance, immediately flush skin or eyes with running water for at least 20 minutes. Wash skin with soap and water. Keep victim warm and quiet. Effects of exposure (inhalation, ingestion, or skin contact) to substance may be delayed. Ensure that medical personnel are aware of the material(s) involved and take precautions to protect themselves.

2-METHOXYETHANOL M:240

SYNONYMS: ACCUSPIN SPIN-N DOPANT; AETHYLENGYKOL-MONOMETHYLAETHER (German); DOWANOL EM; EEC No. 603-011-00-4; EGM; EGME; EINECS No. 203-713-7; ENTEC 763; ETHER MONOMETHYLIQUE de l'ETHYLENE-GLYCIL (French); ETHYLENE GLYCOL METHYL ETHER; ETHYLENE GLYCOLMONOMETHYL ETHER; GLYCOL ETHER EM; GLYCOL METHYL ETHER; GLYCOL MONOMETHYL ETHER; JEFFERSOL EM; KTI 732; KTI 732R; KTI 732-A313A(-); KTI 747; KTI 747R; KTI 747-A313A(-); KTI 752(-); MECS; METHYL CELLOSOLVE; METHYL ETHOXOL; METHYL GLYCOL; METHYL ICINOL; METHYL OXITOL; METIL CELLOSOLVE (Italian); METILGLICOL (Spanish); METOKSYETYLOWY ALCOHOL (Polish); METHOXY ETHANOL(2-); POLY-SOLV EM; PRIST; URESOLVE PLUS AND SG

702 2-Methoxyethanol

IDENTIFICATION

CAS:	109-86-4
ERG ID:	UN 1188
ERG Guide:	127
ERG Guide Category:	FLAMMABLE LIQUIDS (POLAR/WATER MISCIBLE)
Molecular Formula:	$C_3H_8O_2$
RTECS No.:	KL5775000
Properties:	Very flammable, colorless liquid. Mild, sweet, etherlike odor. Floats on surface, and mixes with water.

HEALTH AND SAFETY INFORMATION

Irritatates skin, eyes, and respiratory tract.

IDLH: 200 ppm
Odor Threshold: 0.11 ppm
OSHA Table Z-1-A Air Contaminant

Exposure Limits:	**TWA**
ACGIH:[1]	5 ppm; 16 mg/m^3
NIOSH:[1]	0.1 ppm; 0.3 mg/m^3
OSHA:[1]	25 ppm; 80 mg/m^3

[1]Skin contact contributes significantly to overall exposure.

RECOMMENDATIONS FOR RESPIRATOR SELECTION: NIOSH

1 ppm: SA (any supplied-air respirator).
2.5 ppm SA:CF (any supplied-air respirator operated in a continuous-flow mode).
5 ppm: SCBAF (any self-contained breathing apparatus with a full facepiece);SAF (any supplied-air respirator with a full facepiece).
100 ppm: SA:PD,PP (any supplied-air respirator operated in a pressure-demand or other positive-pressure mode).
200 ppm: SAF:PD,PP (any supplied-air respirator that has a full facepiece and is operated in a pressure-demand or other positive-pressure mode).

Emergency or Planned Entry into Unknown Concentrations or IDLH Conditions:

SCBAF:PD,PP (any self-contained breathing apparatus that has a full facepiece and is operated in a pressure-demand or other positive-pressure mode);or SAF:PD,PP:ASCBA (any supplied-air respirator that has a full facepiece and is operated in a pressure-demand or other positive-pressure mode in combination with an auxiliary, self-contained breathing apparatus operated in a pressure-demand or other positive-pressure mode).

Escape: GMFOV [any air-purifying, full-facepiece respirator (gas mask) with a chin-style, front- or back-mounted organic vapor canister];or SCBAE (any appropriate escape-type, self-contained breathing apparatus).

Note: Substance reported to cause eye irritation or damage; may require eye protection.

CHEMICAL REACTIVITY

Forms explosive mixture with air. Heat or oxidizers may form unstable peroxides. Attacks many metals. Strong oxidizers cause fire and explosions. Strong bases cause decomposition. Attacks some plastics, rubber, and coatings. May accumulate static electrical charges and may cause ignition of its vapors.

FIRE INFORMATION

Vapors are heavier than air and will collect in low areas. Vapors in confined areas may explode when exposed to fire. Vapors may travel long distances to ignition sources and flash back. Storage containers and parts of containers may rocket great distances, in many directions. If material or contaminated runoff enters waterways, notify downstream users of potentially contaminated waters. Notify local health and fire officials and pollution control agencies. Use dry chemicals, foam, carbon dioxide. From a secure, explosion-proof location, use water spray to cool exposed containers.

NFPA Hazard Classifications:

Health Hazards (Blue) 2
Flammability (Red) 2
Reactivity (Yellow) 0

Flash Point: 120°F (49°C) o.c.; 102°F (39°C) c.c.
Flammable Limits in Air: LEL 1.8%; UEL 14%
Autoignition Temperature: 551°F (288°C)

FIRST AID

Move victim to fresh air. Call emergency medical care. Apply artificial respiration if victim is not breathing. If breathing is difficult, give oxygen. Remove and isolate contaminated clothing and shoes. In case of contact with material, immediately flush skin or eyes with running water for at least 20 minutes. Wash skin with soap and water. Keep victim warm and quiet. Ensure that medical personnel are aware of the material(s) involved and take precautions to protect themselves.

ENVIRONMENTAL INFORMATION

EPCRA Section 313: Form R *de minimis* concentration reporting level: 1.0%
California's Proposition 65: Reproductive toxin (male)

2-METHOXYETHYL ACETATE M:260

SYNONYMS: ACETATO de 2-METOXIETILO (Spanish); ACETATO di 2-METOSSIETILE (Italian); ACETIC ACID, 2-METHOXYETHYL ESTER; ACETYL METHYL CELLOSOLVE; EEC No. 607-036-00-1; EINECS No. 203-772-9; EGMEA; ETHANOL, 2-METHOXY-, ACETATE; ETHYLENE GLYCOL METHYL ETHER ACETATE; ETHYLENE GLYCOL MONOMETHYL ETHER ACETATE; GLYCOL MONOMETHYL ETHER ACETATE; KTI KPR NEGATIVE DEVELOPER; 2-METHOXYETHYL ACRYLATE; METHYL CELLOSOLVE ACETATE; METHYL GLYCOL ACETATE; METHYL GLYCOL MONOACETATE

IDENTIFICATION

CAS:	110-49-6
ERG ID:	UN 1189
ERG Guide:	129
ERG Guide Category:	FLAMMABLE LIQUIDS (POLAR/WATER-MISCIBLE/NOXIOUS)
Molecular Formula:	$C_5H_{10}O_3$
RTECS No.:	KL5950000
Properties:	Combustible liquid. Colorless. Mild, etherlike odor. Soluble in water.

HEALTH AND SAFETY INFORMATION

Irritates eyes, skin, and respiratory tract. May cause skin problems, central nervous system effects, kidney damage.

IDLH: 200 ppm; potential human carcinogen
Odor Threshold: 1.07 ppm
OSHA Table Z-1-A Air Contaminant

Exposure Limits:	**TWA**
ACGIH:+	5 ppm; 24 mg/m^3
NIOSH:+•	0.1 ppm; 0.5 mg/m^3
OSHA:+	25 ppm; 120 mg/m^3

+Skin contact contributes significantly to overall exposure.
•Potential carcinogen. Reduce exposure to lowest feasible level.

RECOMMENDATIONS FOR RESPIRATOR SELECTION: NIOSH

1 ppm: SA (any supplied-air respirator).
2.5 ppm: SA:CF (any supplied-air respirator operated in a continuous-flow mode).
5 ppm: SCBAF (any self-contained breathing apparatus with a full facepiece);SAF (any supplied-air respirator with a full facepiece).

100 ppm: SA:PD,PP (any supplied-air respirator operated in a pressure-demand or other positive-pressure mode).
200 ppm: SAF:PD,PP (any supplied-air respirator that has a full facepiece and is operated in a pressure-demand or other positive-pressure mode).

Emergency or Planned Entry into Unknown Concentrations or IDLH Conditions: SCBAF:PD,PP (any self-contained breathing apparatus that has a full facepiece and is operated in a pressure-demand or other positive-pressure mode);or SAF:PD,PP:ASCBA (any supplied-air respirator that has a full facepiece and is operated in a pressure-demand or other positive-pressure mode in combination with an auxiliary, self-contained breathing apparatus operated in a pressure-demand or other positive-pressure mode).

Escape: GMFOV[any air-purifying, full-facepiece respirator (gas mask) with a chin-style, front- or back-mounted organic vapor canister];or SCBAE (any appropriate escape-type, self-contained breathing apparatus).

Note: Substance reported to cause eye irritation or damage; may require eye protection.

CHEMICAL REACTIVITY

Forms explosive mixture with air. Contact with air may form unstable and explosive peroxides. Incompatible with nitrates, strong oxidizers, strong alkalies, and strong acids.

FIRE INFORMATION

Vapors are heavier than air and will collect in low areas. Vapors in confined areas may explode when exposed to fire. Vapors may travel long distances to ignition sources and flash back. Storage containers and parts of containers may rocket great distances, in many directions. If material or contaminated runoff enters waterways, notify downstream users of potentially contaminated waters. Notify local health and fire officials and pollution control agencies. Use dry chemicals, foam, carbon dioxide. From a secure, explosion-proof location, use water spray to cool exposed containers.

NFPA Hazard Classifications:

Health Hazards (Blue) 1
Flammability (Red) 2
Reactivity (Yellow) .... -

Flash Point: 120°F (49°C) c.c.
Flammable Limits in Air: LEL 1.7%; UEL 8.2%
Autoignition Temperature: 740°C (392°C)

FIRST AID

Move victim to fresh air. Call emergency medical care. Apply artificial respiration if victim is not breathing. Administer oxygen if breathing is difficult. Remove and isolate contaminated clothing and shoes. In case of contact with substance, immediately flush skin or eyes with running water for at least 20 minutes. Wash skin with soap and water. Keep victim warm and quiet. Effects of exposure (inhalation, ingestion, or skin contact) to substance may be delayed. Ensure that medical personnel are aware of the material(s) involved and take precautions to protect themselves.

ENVIRONMENTAL INFORMATION (as glycol ethers)

Clean Air Act: Hazardous Air Pollutants (Title I, Part A, Section 112) includes mono- and di- ethers of ethylene glycol, diethyl glycol, and triethylene glycol R-(OCH2CH2)n-OR' where n = 1,2, or 3; R = alkyl or aryl groups; R' = R, H, or groups which when removed, yield glycol ethers with the structure: R-(OCH2CH)n-OH. Polymers are excluded from the glycol category.
EPCRA Section 313: Form R *de minimis* concentration reporting level: 1.0%. Form R Toxic Chemical Category Code: N230.

METHYL ABIETATE M:265

SYNONYMS: ABALYN; ABIETATE de METHYLE, TECHNIQUE (French); ABIETATO de METILO (Spanish); ABIETIC ACID, METHYL ESTER; EINECS No. 204-832-7; METHYL ESTER of WOOD ROSIN; METHYL ESTER of WOOD ROSIN, partially hydrogenated

IDENTIFICATION

CAS:	127-25-3
Molecular Formula:	$C_{21}H_{32}O_2$
RTECS No.:	TP8590000
Properties:	Combustible, thick liquid. Colorless to yellow. Little or no odor. Insoluble in water.

HEALTH AND SAFETY INFORMATION

May cause irritation to eyes and skin.

RECOMMENDATIONS FOR RESPIRATOR SELECTION

At any concentrations above the NIOSH REL, or where there is no REL, at any detectable concentration: SCBAF:PD,PP (any self-contained breathing apparatus that has a full facepiece and is operated in a pressure-demand or other positive-pressure mode);or SAF:PD,PP:ASCBA (any supplied-air respirator that has a full facepiece and is operated in a pressure-demand or other positive-pressure mode in combination with an auxiliary, self-contained breathing apparatus operated in a pressure-demand or other positive-pressure mode).

Escape: GMFOV [any air-purifying, full-facepiece respirator (gas mask) with a chin-style, front- or back-mounted organic vapor canister];or SCBAE (any appropriate escape-type, self-contained breathing apparatus).

CHEMICAL REACTIVITY

Contact with strong oxidizers may cause fire and explosions.

FIRE INFORMATION

Vapors in confined areas may explode when exposed to fire. If material or contaminated runoff enters waterways, notify downstream users of potentially contaminated waters. Notify local health and fire officials and pollution control agencies. Use dry chemicals, alcohol-resistant foam, carbon dioxide. From a secure, explosion-proof location, use water spray to cool exposed containers.

NFPA Hazard Classifications

Health Hazards (Blue) 0
Flammability (Red) 1
Reactivity (Yellow) 0

Flash Point: 356°F (180°C) o.c.

FIRST AID

Move victim to fresh air. Call emergency medical care. Apply artificial respiration if victim is not breathing. Administer oxygen if breathing is difficult. Remove and isolate contaminated clothing and shoes. In case of contact with substance, immediately flush skin or eyes with running water for at least 20 minutes. Wash skin with soap and water. Keep victim warm and quiet. Ensure that medical personnel are aware of the material(s) involved and take precautions to protect themselves.

METHYL ACETYLENE AND PROPADIENE MIXTURE M:270

SYNONYMS: ALLENE-METHYL ACETYLENE MIXTURE; MAPP GAS; METHYL ACETYLENE-ALLENE MIXTURE; PROPADIENE AND METHYLACETYLENE MIXTURE; PROPYNE-ALLENE MIXTURE; PROPYNE-PROPADIENE MIXTURE

IDENTIFICATION

CAS:	59355-75-8
ERG ID:	UN 1060
ERG Guide:	116P
ERG Guide Category:	GASES–FLAMMABLE (UNSTABLE)
Molecular Formula:	C_3H_4
RTECS No.:	UK4920000
Properties:	Extremely flammable gas. Shipped as a liquified compressed gas. Colorless. Strong, characteristic, garliclike odor. Poor solubility in water; floats and "boils" on water surface.

HEALTH AND SAFETY INFORMATION

Irritating to eyes, skin, and respiratory tract.. Contact with liquid can cause frostbite.

IDLH: 3400 ppm [10% LEL]
Odor Threshold: 100 ppm
OSHA Table Z-1-A Air Contaminant

Exposure Limits:	**TWA**	**STEL**
ACGIH:	1,000 ppm; 1,800 mg/m^3	1,250 ppm; 2,250 mg/m^3
NIOSH:	1,000 ppm; 1,800 mg/m^3	1,250 ppm; 2,250 mg/m^3
OSHA:	1,000 ppm; 1,800 mg/m^3	

RECOMMENDATIONS FOR RESPIRATOR SELECTION: NIOSH

At any concentrations above the NIOSH REL, or where there is no REL, at any detectable concentration: SCBAF:PD,PP (any self-contained breathing apparatus that has a full facepiece and is operated in a pressure-demand or other positive-pressure mode);or SAF:PD,PP:ASCBA (any supplied-air respirator that has a full facepiece and is operated in a pressure-demand or other positive-pressure mode in combination with an auxiliary, self-contained breathing apparatus operated in a pressure-demand or other positive-pressure mode).

Escape: GMFOV [any air-purifying, full-facepiece respirator (gas mask) with a chin-style, front- or back-mounted organic vapor canister];or SCBAE (any appropriate escape-type, self-contained breathing apparatus).

CHEMICAL REACTIVITY

Forms explosive mixture with air. Incompatible with strong oxidizers. Contact with copper alloys may form explosive compounds. Attacks some plastics, rubber, and coatings. May accumulate static electrical charges and may cause ignition of its vapors.

FIRE INFORMATION

Gases are heavier than air and will collect in low areas. Gases in confined areas may explode when exposed to fire. Gases may travel long distances to ignition sources and flash back. Storage containers and parts of containers may rocket great distances, in many directions. If material or contaminated runoff enters waterways, notify downstream users of potentially contaminated waters. Notify local health and fire officials and pollution control agencies. Do not extinguish the fire unless the flow of the gas can be stopped and any remaining gas is out of the line. Specially trained personnel may use fog lines to cool exposures and let the fire burn itself out. Use dry chemicals, foam, carbon dioxide. From a secure, explosion-proof location, use water spray to cool exposed containers. If cooling streams are ineffective (venting sound increases in volume and pitch, tank discolors, or shows any signs of deforming), withdraw immediately to a secure position.

NFPA Hazard Classifications:

Health Hazards (Blue) 2
Flammability (Red) 4
Reactivity (Yellow) 2

Flash Point: NA (gas)
Flammable Limits in Air: LEL 3.4%; UEL 10.8%

FIRST AID

Move victim to fresh air. Call emergency medical care. Apply artificial respiration if victim is not breathing. If breathing is difficult, give oxygen. Remove and isolate contaminated clothing and shoes. In case of contact with liquefied gas, thaw frosted parts with lukewarm water. Keep victim warm and quiet. Ensure that medical personnel are aware of the material(s) involved and take precautions to protect themselves.

METHYL ACRYLATE, INHIBITED M:290

SYNONYMS: ACRILATO de METILO (Spanish); ACRYLATE de METHYLE (French); ACRYLIC ACID, METHYL ESTER; ACRYLSAEUREMETHYLESTER (German); CURITHANE 103; EINECS No. 202-500-6; METHOXYCARBONYLETHYLENE; METHYLACRYLATE; METHYLACRYLAAT (Dutch); METHYL-ACRYLAT (German); METHYL PROPENATE; METHYL PROPENOATE; METHYL-2-PROPENOATE; METILACRILATO (Italian); PROPENOIC ACID, METHYL ESTER; 2-PROPENOIC ACID, METHYL ESTER; STCC 4907245

IDENTIFICATION

CAS:	96-33-3
ERG ID:	UN 1919
ERG Guide:	129P
ERG Guide Category:	FLAMMABLE LIQUIDS (POLAR/WATER-MISCIBLE/NOXIOUS)
Molecular Formula:	$C_4H_6O_2$
RTECS No.:	AT280000
Properties:	Highly flammable, colorless liquid. Acrid odor. Floats on the surface of water; slightly soluble in water. Vapors are heavier than air; will collect in low areas.

HEALTH AND SAFETY INFORMATION

Highly irritating to eyes, skin, and respiratory tract. Eye contact can cause severe burns and blindness.

IARC Evaluation: Group 3, unclassifiable substance; no human data; inadequate animal evidence.
IDLH: 250 ppm
Odor Threshold: 0.263 ppm
OSHA Table Z-1-A Air Contaminant

Exposure Limits:	**TWA**
ACGIH:[1]	2 ppm; 7 mg/m^3
NIOSH:[1]	0 ppm; 35 mg/m^3
OSHA:[1]	0 ppm; 35 mg/m^3

[1]Skin contact contributes significantly to overall exposure

RECOMMENDATIONS FOR RESPIRATOR SELECTION: NIOSH

100 ppm: SA (any supplied-air respirator).
250 ppm: SA:CF (any supplied-air respirator operated in a continuous-flow mode);or SCBAF (any self-contained breathing apparatus with a full facepiece);or SAF (any supplied-air respirator with a full facepiece).

Emergency or Planned Entry into Unknown Concentrations or IDLH Conditions: SCBAF:PD,PP (any self-contained breathing apparatus that has a full facepiece and is operated in a pressure-demand or other positive-pressure mode);or SAF:PD,PP:ASCBA (any supplied-air respirator that has a full facepiece and is operated in a pressure-demand or other positive-pressure mode in combination with an auxiliary, self-contained breathing apparatus operated in a pressure-demand or other positive-pressure mode).

Escape: GMFOV [any air-purifying, full-facepiece respirator (gas mask) with a chin-style, front- or back-mounted organic vapor canister];or SCBAE (any appropriate escape-type, self-contained breathing apparatus).

CHEMICAL REACTIVITY

Heat, light, and/or lack of appropriate inhibitor (such as hydroquinone) concentration can result in explosive polymerization. Polymerization may cause relief valves to be blocked leading to tank explosions. Forms explosive mixture with air. Incompatible with strong acids, strong alkalis, oxidizers, peroxides, aliphatic amines, and alkanolamines.

FIRE INFORMATION

Firefighting gear (including SCBA) does not provide adequate protection. If exposure occurs, remove and isolate gear immediately and thoroughly decontaminate personnel. Vapors are heavier than air and will collect in low areas. Vapors in confined areas may explode when exposed to fire. Vapors may travel long distances to ignition sources and flash back. Storage containers and parts of containers may rocket great distances, in many directions. If material or contaminated runoff enters waterways, notify downstream users of potentially contaminated waters. Notify local health and fire officials and pollution control agencies. Use dry chemicals, foam, carbon dioxide. From a secure, explosion-proof location, use water spray to cool exposed containers.

NFPA Hazard Classifications:

Health Hazards (Blue) 3
Flammability (Red) 3
Reactivity (Yellow) 2

Flash Point: 27°F (-3°C)
Flammable Limits in Air: LEL 2.8%; UEL 25%
Autoignition Temperature: 875°F (468°C)

FIRST AID

Move victim to fresh air. Call emergency medical care. Apply artificial respiration if victim is not breathing. Administer oxygen if breathing is difficult. Remove and isolate contaminated clothing and shoes. In case of contact with substance, immediately flush skin or eyes with running water for at least 20 minutes. Wash skin with soap and water. Keep victim warm and quiet. Effects of exposure

(inhalation, ingestion, or skin contact) to substance may be delayed. Ensure that medical personnel are aware of the material(s) involved and take precautions to protect themselves.

ENVIRONMENTAL INFORMATION

EPA Hazardous Waste Number (RCRA No.): U328.
RCRA Section 261 Hazardous Constituents
EPCRA Section 304: Reportable Quantity (RQ): CERCLA, 100 lb (45.4 kg).
EPCRA Section 313: Form R *de minimis* concentration reporting level: 1.0%

METHYL ALCOHOL M:300

SYNONYMS: ALCOHOL C-1; ALCOHOL METILICO (Spanish); ALCOOL METHYLIQUE (French); ALCOOL METILICO (Italian); ARSENOSILICA FILM 0308; BLANCO-TRON TMS PLUS; BORON B-30; BORON B-40; BORON B-50; BORON B-60; CARBINOL; COLONIAL SPIRIT; COLUMBIAN SPIRITS; DYNASOLVE 190; DYNASOLVE MP-710; DYNASOLVE MP-750; EEC No. 603-001-00-X; EINECS No. 200-659-6; 4282 FLUX; FREON TMS; GENESOLV DFX; METHANOL; METANOLO (Italian); METHYLOL; METHYLALKOHOL (German); METHYL HYDROXIDE; METHYL HYDRATE; METYLOWY ALKOHOL (Polish); MONOHYDROXYMETHANE; MU-9; P-12571D; P-12572; PYRALIN PI 2563; PYROXYLIC SPIRIT; PYROLIGNEOUS SPIRIT; SLOTELET ADDITIVE K; SOLDERON SG ADDITIVE; REMOVER; WOOD ALCOHOL; WOOD NAPHTHA; WOOD SPIRIT

IDENTIFICATION

CAS:	67-56-1
ERG ID:	UN 1230
ERG Guide:	131
ERG Guide Category:	FLAMMABLE LIQUIDS-TOXIC
Molecular Formula:	CH_4O
RTECS No.:	PC1400000
Properties:	Highly flammable, watery liquid. Clear, colorless. Sweet, alcohol odor. Floats and mixes with water. Flammable, irritating vapor is produced.

HEALTH AND SAFETY INFORMATION

Irritating to eyes and skin. A poison which is eliminated very slowly from the body. Below 500 ppm symptoms are rarely felt. Can cause headache, vomiting, irritation of the nose and throat, dilation of the pupils, feeling of intoxication, loss of muscle coordination, excessive sweating, bronchitis, and convulsions. Very high exposures may result in stupor, cramps, and visual difficulties such as spotted vision, sensitivity to light, eye tenderness, and blindness. Because methyl alcohol is slowly eliminated from the body, repeated low exposures may build up to high levels causing severe symptoms. Recovery is not always complete.

IDLH: 6,000 ppm
Odor Threshold: 141 ppm
OSHA Table Z-1-A Air Contaminant
NIOSH Criteria Document: NIOSH 76-148

Exposure Limits:	**TWA**	**STEL**
ACGIH:*	200 ppm; 262 mg/m^3	250 ppm; 328 mg/m^3
NIOSH:*	200 ppm; 260 mg/m^3	250 ppm; 325 mg/m^3
OSHA:*	200 ppm; 260 mg/m^3	

*Skin contact contributes significantly to overall exposure.

RECOMMENDATIONS FOR RESPIRATOR SELECTION: NIOSH/OSHA

2,000 ppm: SA (any supplied-air respirator).
5,000 ppm: SA:CF (any supplied-air respirator operated in a continuous-flow mode).
6,000 ppm: SAT:CF (any supplied-air respirator that has a tight-fitting facepiece and is operated in a continuous-flow mode);or SCBAF (any self-contained breathing apparatus with a full facepiece);or SAF (any supplied-air respirator with a full facepiece).

Emergency or Planned Entry into Unknown Concentrations or IDLH Conditions:

SCBAF:PD,PP (any self-contained breathing apparatus that has a full facepiece and is operated in a pressure-demand or other positive-pressure mode);or SAF:PD,PP:ASCBA (any supplied-air respirator that

has a full facepiece and is operated in a pressure-demand or other positive-pressure mode in combination with an auxiliary, self-contained breathing apparatus operated in a pressure-demand or other positive-pressure mode).

Escape: SCBAE (any appropriate escape-type, self-contained breathing apparatus).

CHEMICAL REACTIVITY

May accumulate static electrical charges and may cause ignition of its vapors. Forms explosive mixture with air. Strong reaction with magnesium, bromine, and nitric acid. Incompatible with strong acids, strong oxidizers (such as chromates, permanganates, hypochlorites, and peroxides), caustics, aliphatic amines, isocyanates, chromic anhydride, lead perchlorate, perchloric acid, and phosphorus trioxide. May react with metallic aluminum at high temperature. Attacks some plastics, rubber, and coatings.

FIRE INFORMATION

Burns with an invisible flame. Vapors are heavier than air and will collect in low areas. Vapors in confined areas may explode when exposed to fire. Vapors may travel long distances to ignition sources and flash back. Storage containers and parts of containers may rocket great distances, in many directions. If material or contaminated runoff enters waterways, notify downstream users of potentially contaminated waters. Notify local health and fire officials and pollution control agencies. Use dry chemicals, foam, carbon dioxide. From a secure, explosion-proof location, use water spray to cool exposed containers.

NFPA Hazard Classifications:

Health Hazards (Blue) 1
Flammability (Red) 3
Reactivity (Yellow) 0

Flash Point: 61°F (16°C) o.c.; 52°F (11°C) c.c.
Flammable Limits in Air: LEL 6.0%; UEL 36.5%
Autoignition Temperature: 867°F (464°C)

FIRST AID

Move victim to fresh air. Call emergency medical care. Apply artificial respiration if victim is not breathing. Do not use mouth-to-mouth method if victim ingested or inhaled the substance; induce artificial respiration with the aid of a pocket mask equipped with a one-way valve or other proper respiratory medical device. Administer oxygen if breathing is difficult. Remove and isolate contaminated clothing and shoes. In case of contact with substance, immediately flush skin or eyes with running water for at least 20 minutes. Wash skin with soap and water. Keep victim warm and quiet. Effects of exposure (inhalation, ingestion, or skin contact) to substance may be delayed. Ensure that medical personnel are aware of the material(s) involved and take precautions to protect themselves.

ENVIRONMENTAL INFORMATION

Ecologically, methanol is expected to have low toxicity to aquatic organisms. Concentrations lethal to half the organisms of a test population are expected to exceed 1 mg methanol per liter water. Methanol is not likely to persist in water or to bioaccumulate in aquatic organisms. Liquid methanol is likely to evaporate when left exposed. Methanol reacts in air to produce formaldehyde which contributes to the formation of air pollutants. In the atmosphere it can react with other atmospheric chemicals or be washed out by rain. Methanol is readily degraded by microorganisms in soils and surface waters.

Clean Air Act: Hazardous Air Pollutants (Title I, Part A, Section 112); Accidental Release Prevention/Flammable substances, (Section 112[r], Table 3), TQ = 5,000 lb (2,270 kg).
EPA Hazardous Waste Number (RCRA No.): U154.
RCRA Section 261 Hazardous Constituents
RCRA Land Ban Waste
RCRA Universal Treatment Standards: Wastewater (mg/L), 5.6; Nonwastewater (mg/L/TCLP), 0.75.
EPCRA Section 304: Reportable Quantity (RQ): CERCLA, 5,000 lb (2,270 kg).
EPCRA Section 313: Form R *de minimis* concentration reporting level: 1.0%.

METHYL ALLYL ALCOHOL M:304

SYNONYMS: ALCOHOL METILALILICO (Spanish); METHALLYL ALCOHOL; ISOPROPENYL CARBINOL; 2-METHYL-2-PROPEN-1-OL

IDENTIFICATION

CAS:	513-42-8
ERG ID:	UN 2614
ERG Guide:	129
ERG Guide Category:	FLAMMABLE LIQUIDS (POLAR/WATER-MISCIBLE/NOXIOUS)
Molecular Formula:	C_4H_8O
Properties:	Flammable liquid. Colorless. Pungent odor. Soluble in water.

HEALTH AND SAFETY INFORMATION

Irritates eyes, skin, and respiratory tract.

RECOMMENDATIONS FOR RESPIRATOR SELECTION

At any concentrations above the NIOSH REL, or where there is no REL, at any detectable concentration: SCBAF:PD,PP (any self-contained breathing apparatus that has a full facepiece and is operated in a pressure-demand or other positive-pressure mode);or SAF:PD,PP:ASCBA (any supplied-air respirator that has a full facepiece and is operated in a pressure-demand or other positive-pressure mode in

combination with an auxiliary, self-contained breathing apparatus operated in a pressure-demand or other positive-pressure mode).

Escape: GMFOV [any air-purifying, full-facepiece respirator (gas mask) with a chin-style, front- or back-mounted organic vapor canister];or SCBAE (any appropriate escape-type, self-contained breathing apparatus).

CHEMICAL REACTIVITY

Forms explosive mixture with air. Contact with oxidizers may cause fire and explosions. Incompatible with sulfuric acid, nitric acid, caustics, aliphatic amines, isocyanates.

FIRE INFORMATION

Vapors are heavier than air and will collect in low areas. Vapors in confined areas may explode when exposed to fire. Vapors may travel long distances to ignition sources and flash back. Storage containers and parts of containers may rocket great distances, in many directions. If material or contaminated runoff enters waterways, notify downstream users of potentially contaminated waters. Notify local health and fire officials and pollution control agencies. Use dry chemicals, foam, carbon dioxide. From a secure, explosion-proof location, use water spray to cool exposed containers.

NFPA Hazard Classifications: Not listed

Flash Point: 92°F (33°C) c.c.

FIRST AID

Move victim to fresh air. Call emergency medical care. Apply artificial respiration if victim is not breathing. Administer oxygen if breathing is difficult. Remove and isolate contaminated clothing and shoes. In case of contact with substance, immediately flush skin or eyes with running water for at least 20 minutes. Wash skin with soap and water. Keep victim warm and quiet. Effects of exposure (inhalation, ingestion, or skin contact) to substance may be delayed. Ensure that medical personnel are aware of the material(s) involved and take precautions to protect themselves.

METHYL–n–AMYL CARBINOL M:307

SYNONYMS: ALCOOL HEPTYLIQUE, SECONDAIRE (French); AMYL METHYL CARBINOL; EINECS No. 208-844-3; HEPTANE-2-OL; HEPTANOL-2; 2-HEPTANOL; HEPTANOL-2,2-METHYL-1-PENTANOL; 2-HEPTYL ALCOHOL; 2-HYDROXYHEPTANE; METIL-n-AMILCETONA (Spanish)

IDENTIFICATION

CAS: 543-49-7

716 Methyl-n-amyl Carbinol

ERG ID:	UN 1105 as amyl alcohols
ERG Guide:	129
ERG Guide Category:	FLAMMABLE LIQUIDS (POLAR/WATER-MISCIBLE/NOXIOUS)
Molecular Formula:	$C_7H_{16}O$
RTECS No.:	MJ2975000
Properties:	Combustible liquid. Colorless. Mild odor. Floats on water; insoluble.

HEALTH AND SAFETY INFORMATION

Irritates eyes, skin, and respiratory tract. Skin contact contributes significantly to overall exposure. May affect the central nervous system; headache, dizziness, nausea, unconsciousness.

RECOMMENDATIONS FOR RESPIRATOR SELECTION

At any concentrations above the NIOSH REL, or where there is no REL, at any detectable concentration: SCBAF:PD,PP (any self-contained breathing apparatus that has a full facepiece and is operated in a pressure-demand or other positive-pressure mode);or SAF:PD,PP:ASCBA (any supplied-air respirator that has a full facepiece and is operated in a pressure-demand or other positive-pressure mode in combination with an auxiliary, self-contained breathing apparatus operated in a pressure-demand or other positive-pressure mode).

Escape: GMFOV [any air-purifying, full-facepiece respirator (gas mask) with a chin-style, front- or back-mounted organic vapor canister];or SCBAE (any appropriate escape-type, self-contained breathing apparatus).

CHEMICAL REACTIVITY

Forms explosive mixture with air. Contact with oxidizers may cause fires and explosions. Alcohols are incompatible with sulfuric acid, nitric acid, caustics, aliphatic amines, isocyanates..

FIRE INFORMATION

Vapors are heavier than air and will collect in low areas. Vapors in confined areas may explode when exposed to fire. Vapors may travel long distances to ignition sources and flash back. Storage containers and parts of containers may rocket great distances, in many directions. If material or contaminated runoff enters waterways, notify downstream users of potentially contaminated waters. Notify local health and fire officials and pollution control agencies. Use dry chemicals, foam, carbon dioxide. From a secure, explosion-proof location, use water spray to cool exposed containers.

NFPA Hazard Classifications: (2-heptanol)

Health Hazards (Blue) 0
Flammability (Red) 2
Reactivity (Yellow) 0

Flash Point: 160°F (71°C) c.c.

FIRST AID

Move victim to fresh air. Call emergency medical care. Apply artificial respiration if victim is not breathing. Administer oxygen if breathing is difficult. Remove and isolate contaminated clothing and shoes. In case of contact with substance, immediately flush skin or eyes with running water for at least 20 minutes. Wash skin with soap and water. Keep victim warm and quiet. Effects of exposure (inhalation, ingestion, or skin contact) to substance may be delayed. Ensure that medical personnel are aware of the material(s) involved and take precautions to protect themselves.

METHYL BENZOATE M:313

SYNONYMS: BENZOATO de METILO (Spanish); BENZOIC ACID, METHYL ESTER; EINECS No. 202-259-7; METHYL BENZENECARBOXYLATE; OXIDATE LE; NIOBE OIL; OIL OF NIOBE; ESSENCE OF NIOBE; FEMA No..2683

IDENTIFICATION

CAS:	93-58-3
ERG ID:	UN 2938
ERG Guide:	152
ERG Guide Category:	SUBSTANCES-TOXIC (COMBUSTIBLE)
Molecular Formula:	$C_8H_8O_2$
RTECS No.:	DH3850000
Properties:	Combustible, oily liquid. Colorless. Pleasant odor. Insoluble in water.

HEALTH AND SAFETY INFORMATION

Poisonous. Irritates eyes, skin, and respiratory tract.

NIOSH Criteria Document: NIOSH 78-173 KETONES

RECOMMENDATIONS FOR RESPIRATOR SELECTION

At any concentrations above the NIOSH REL, or where there is no REL, at any detectable concentration: SCBAF:PD,PP (any self-contained breathing apparatus that has a full facepiece and is operated in a pressure-demand or other positive-pressure mode);or SAF:PD,PP:ASCBA (any supplied-air respirator that has a full facepiece and is operated in a pressure-demand or other positive-pressure mode in combination with an auxiliary, self-contained breathing apparatus operated in a pressure-demand or other positive-pressure mode).

Escape: GMFOV [any air-purifying, full-facepiece respirator (gas mask) with a chin-style, front- or back-mounted organic vapor canister];or SCBAE (any appropriate escape-type, self-contained breathing apparatus).

CHEMICAL REACTIVITY

Forms explosive mixture with air. Contact with strong oxidizers may cause fire and explosions. Incompatible with strong acids such as nitric acid and sulfuric acid, nitrates.

FIRE INFORMATION

Vapors are heavier than air and will collect in low areas. Vapors in confined areas may explode when exposed to fire.If material or contaminated runoff enters waterways, notify downstream users of potentially contaminated waters. Notify local health and fire officials and pollution control agencies. Use dry chemicals, alcohol-resistant foam, carbon dioxide. From a secure, explosion-proof location, use water spray to cool exposed containers.

NFPA Hazard Classifications:

Health Hazards (Blue) 0
Flammability (Red) 2
Reactivity (Yellow) 0

Flash Point: 181°F (83°C) c.c.

FIRST AID

Move victim to fresh air. Call emergency medical care. Apply artificial respiration if victim is not breathing. Do not use mouth-to-mouth method if victim ingested or inhaled the substance; induce artificial respiration with the aid of a pocket mask equipped with a one-way valve or other proper respiratory medical device. Administer oxygen if breathing is difficult. Remove and isolate contaminated clothing and shoes. In case of contact with substance, immediately flush skin or eyes with running water for at least 20 minutes. For minor skin contact, avoid spreading material on unaffected skin. Keep victim warm and quiet. Effects of exposure (inhalation, ingestion, or skin contact) may be delayed. Ensure that medical personnel are aware of the material(s) involved and take precautions to protect themselves.

α–METHYLBENZYL ETHER — M:316

SYNONYMS: BENZENE, 1,1'-(OXYDIETHYLIDENE)BIS-; EINECS No. 202-290-6; ETHER-BIS(α-METHYLBENZYL); OSSIDO DI BIS(1-FENILETILE) (Italian); 1,1'-(OXYDIETHYLIDENE)BISBENZENE

IDENTIFICATION

CAS: 93-96-9

ERG ID: UN 3271 as ethers, n.o.s.
ERG Guide: 127
ERG Guide Category: FLAMMABLE LIQUIDS (POLAR/WATER- MISCIBLE)
Molecular Formula: $C_{16}H_{18}O$
RTECS No.: KN3350000
Properties: Combustible liquid. Straw yellow. Faint odor. Practically insoluble in water.

HEALTH AND SAFETY INFORMATION

Irritates eyes, skin, and respiratory tract.

RECOMMENDATIONS FOR RESPIRATOR SELECTION

At any concentrations above the NIOSH REL, or where there is no REL, at any detectable concentration: SCBAF:PD,PP (any self-contained breathing apparatus that has a full facepiece and is operated in a pressure-demand or other positive-pressure mode);or SAF:PD,PP:ASCBA (any supplied-air respirator that has a full facepiece and is operated in a pressure-demand or other positive-pressure mode in combination with an auxiliary, self-contained breathing apparatus operated in a pressure-demand or other positive-pressure mode).

Escape: GMFOV [any air-purifying, full-facepiece respirator (gas mask) with a chin-style, front- or back-mounted organic vapor canister];or SCBAE (any appropriate escape-type, self-contained breathing apparatus)

CHEMICAL REACTIVITY

Contact with strong oxidizers may cause fire and explosions. May form unstable peroxides.

FIRE INFORMATION

Vapors are heavier than air and will collect in low areas. Vapors in confined areas may explode when exposed to fire.If material or contaminated runoff enters waterways, notify downstream users of potentially contaminated waters. Notify local health and fire officials and pollution control agencies. Use dry chemicals, alcohol-resistant foam, carbon dioxide. Use water spray to cool exposed containers. From a secure, explosion-proof location, use water spray to cool exposed containers.

NFPA Hazard Classifications

Health Hazards (Blue) 2
Flammability (Red) 1
Reactivity (Yellow) 0

Flash Point: 275°F (135°C) o.c.

FIRST AID

Move victim to fresh air. Call emergency medical care. Apply artificial respiration if victim is not breathing. Administer oxygen if breathing is difficult. Remove and isolate contaminated clothing and shoes. In case of contact with substance, immediately flush skin or eyes with running water for at least 20 minutes. Wash skin with soap and water. Keep victim warm and quiet. Ensure that medical personnel are aware of the material(s) involved and take precautions to protect themselves.

2–METHYL–1–BUTANOL M:318

SYNONYMS: AMYL ALCOHOL, PRIMARY; 1-BUTANOL, 2-METHYL-; dl-sec-BUTYLCARBINOL; EEC No. 603-006-00-7; EINECS No. 205-289-9; 2-METHYLBUTANOL; 2-METHYL-1-BUTANOL; PRIMARY PENTYL ALCOHOL

IDENTIFICATION

CAS:	137-32-6
ERG ID:	UN 1986 as alcohols, flammable, toxic, n.o.s.
ERG Guide:	131
ERG Guide Category:	FLAMMABLE LIQUIDS-TOXIC
Molecular Formula:	$C_5H_{12}O$
RTECS No.:	EL5250000
Properties:	Flammable liquid. Colorless. Mixes slightly in water.

HEALTH AND SAFETY INFORMATION

Irritates eyes, skin, and respiratory tract. Affects central nervous system; blood (may form methemaglobin). Toxic by all routes; skin contact contributes significantly to overall exposure. Consumption of alcohol may contribute to overall exposure.

RECOMMENDATIONS FOR RESPIRATOR SELECTION

At any concentrations above the NIOSH REL, or where there is no REL, at any detectable concentration: SCBAF:PD,PP (any self-contained breathing apparatus that has a full facepiece and is operated in a pressure-demand or other positive-pressure mode);or SAF:PD,PP:ASCBA (any supplied-air respirator that has a full facepiece and is operated in a pressure-demand or other positive-pressure mode in combination with an auxiliary, self-contained breathing apparatus operated in a pressure-demand or other positive-pressure mode).

Escape: GMFOV [any air-purifying, full-facepiece respirator (gas mask) with a chin-style, front- or back-mounted organic vapor canister];or SCBAE (any appropriate escape-type, self-contained breathing apparatus).

CHEMICAL REACTIVITY

Forms explosive mixture with air. Contact with oxidizers may cause fire and explosions. Reacts with alkali metals, hydrogen trisulfide, sulfuric acid, nitric acid, caustics, aliphatic amines, isocyanates. May accumulate static electricity, and cause ignition of its vapors.

FIRE INFORMATION

Vapors are heavier than air and will collect in low areas. Vapors in confined areas may explode when exposed to fire. Vapors may travel long distances to ignition sources and flash back. Storage containers and parts of containers may rocket great distances, in many directions. If material or contaminated runoff enters waterways, notify downstream users of potentially contaminated waters. Notify local health and fire officials and pollution control agencies. Use dry chemicals, foam, carbon dioxide. From a secure, explosion-proof location, use water spray to cool exposed containers.

FEMA Hazard Classifications:

Health Hazards (Blue) 2
Flammability (Red) 2
Reactivity (Yellow) 0

Flash Point: 122°F (50°C) o.c
Flammable Limits in Air: LEL 1.4%; UEL 9.0%
Autoignition Temperature: 725°F (385°C)

FIRST AID

Move victim to fresh air. Call emergency medical care. Apply artificial respiration if victim is not breathing. Do not use mouth-to-mouth method if victim ingested or inhaled the substance; induce artificial respiration with the aid of a pocket mask equipped with a one-way valve or other proper respiratory medical device. Administer oxygen if breathing is difficult. Remove and isolate contaminated clothing and shoes. In case of contact with substance, immediately flush skin or eyes with running water for at least 20 minutes. Wash skin with soap and water. Keep victim warm and quiet. Effects of exposure (inhalation, ingestion, or skin contact) to substance may be delayed. Ensure that medical personnel are aware of the material(s) involved and take precautions to protect themselves.

METHYL CHLORIDE M:320

SYNONYMS: ARTIC; CHLOOR-METHAAN (Dutch); CHLOR-METHAN (German); CHLOROMETHANE; CHLORURE de METHYLE (French); CLOROMETANO (Italian); CLORURO de METILO (Spanish); CLORURO di METILE (Italian); EEC No. 602-001-00-7; EINECS No. 200-817-4; METHYLCHLORID (German); METYLU CHLOREK (Polish); MONOCHLOROMETHANE; METHANE, CHLORO-

IDENTIFICATION

CAS:	74-87-3
ERG ID:	UN 1063
ERG Guide:	115
ERG Guide Category:	GASES-FLAMMABLE (INCLUDING REFRIGERATED LIQUIDS)
Molecular Formula:	CH_3Cl
RTECS No.:	PA6300000
Properties:	Extremely flammable gas. A liquid below −11°F (−24°C). May be shipped as a liquified compressed gas. Colorless. Odorless or faint sweet etherlike odor, which may not be noticable at dangerous concentrations. Floats and boils on water. Flammable, visible vapor cloud is formed.

HEALTH AND SAFETY INFORMATION

Irritating to the skin, eyes nose, and lungs. Contact with liquid may cause frostbite. Some health effects may last months or longer. Overexposure may be delayed (24 to 48 hours) with onset of dizziness, headache, mental confusion, slurred speech, double vision and sleepiness. Recovery from effects may be slow. May cause central nervous system damage; damage to the testes with decreased male hormone and sperm production; liver, kidney cardiovascular, and bone marrow damage; brain damage; and possible lung damage.

IARC Evaluation: Group 3, unclassifiable substance: human evidence, inadequate; animal evidence, inadequate.
IDLH: 2,000 ppm; Potential human carcinogen.
Odor Threshold: 10.2 ppm
OSHA Table Z-1-A Air Contaminant
OSHA Table Z-2 Air Contaminant
OSHA Process Safety Management of Highly Hazardous Chemicals (29CFR 1910.119, Appendix A), Threshold Quantity: 15,000 pounds.

Exposure Limits:	**TWA**	**CEILING**	**STEL**
ACGIH:[1]		50 ppm; 103 mg/m³	100 ppm; 207 mg/m³
NIOSH:[2]			
OSHA:	100 ppm	200 ppm; 300 ppm; (5 min max peak in any 3 hrs)	

[1]Skin contact contributes significantly to overall exposure.
[2]Potential carcinogen. Reduce exposure to lowest feasible level.

RECOMMENDATIONS FOR RESPIRATOR SELECTION: NIOSH

At any concentrations above the NIOSH REL, or where there is no REL, at any detectable concentration: SCBAF:PD,PP (any self-contained breathing apparatus that has a full facepiece and is

operated in a pressure-demand or other positive-pressure mode);or SAF:PD,PP:ASCBA (any supplied-air respirator that has a full facepiece and is operated in a pressure-demand or other positive-pressure mode in combination with an auxiliary, self-contained breathing apparatus operated in a pressure-demand or other positive-pressure mode).

Escape: SCBAE (any appropriate escape-type, self-contained breathing apparatus).

CHEMICAL REACTIVITY

Moisture causes decomposition. Contact with aluminum forms a product which may ignite spontaneously in air. Reacts violently with magnesium, sodium and potassium metals. Reacts with barium, lithium, titanium. Attacks plastics, rubber and coatings.

FIRE INFORMATION

Gases are heavier than air and will collect in low areas. Gases in confined areas may explode when exposed to fire. Gases may travel long distances to ignition sources and flash back. Storage containers and parts of containers may rocket great distances, in many directions. If material or contaminated runoff enters waterways, notify downstream users of potentially contaminated waters. Notify local health and fire officials and pollution control agencies. Do not extinguish the fire unless the flow of the gas can be stopped and any remaining gas is out of the line. Specially trained personnel may use fog lines to cool exposures and let the fire burn itself out. Use dry chemicals, foam, carbon dioxide. From a secure, explosion-proof location, use water spray to cool exposed containers. If cooling streams are ineffective (venting sound increases in volume and pitch, tank discolors, or shows any signs of deforming), withdraw immediately to a secure position.

NFPA Hazard Classifications:

Health Hazards (Blue) 1
Flammability (Red) 4
Reactivity (Yellow) 0

Flash Point: -50°F (-46°C) c.c.
Flammable Limits in Air: LEL 8.1%; UEL 17.4%
Autoignition Temperature: 1170°F (632°C)

FIRST AID

Move victim to fresh air. Call emergency medical care. Apply artificial respiration if victim is not breathing. Administer oxygen if breathing is difficult. Remove and isolate contaminated clothing and shoes. Clothing frozen to the skin should be thawed before being removed. In case of contact with liquefied gas, thaw frosted parts with lukewarm water. Keep victim warm and quiet. Keep victim under observation. Ensure that medical personnel are aware of the material(s) involved and take precautions to protect themselves.

ENVIRONMENTAL INFORMATION

Clean Air Act: Hazardous Air Pollutants (Title I, Part A, Section 112); Accidental Release Prevention/Flammable substances, (Section 112[r], Table 3), TQ = 10,000 lb (4,540 kg).
Clean Water Act: Section 307 Priority Pollutants; Section 313 Priority Chemicals.
EPA Hazardous Waste Number (RCRA No.): U045.
RCRA Section 261 Hazardous Constituents
RCRA Land Ban Waste
Safe Drinking Water Act: Priority List (55 FR 1470)
RCRA Universal Treatment Standards: Wastewater (mg/L), 0.19; Nonwastewater (mg/kg), 30.
RCRA Ground Water Monitoring List: Suggested test method(s) (PQL ug/L): 8010(1); 8240(10).
EPCRA Section 304: Reportable Quantity (RQ): CERCLA, 100 lb (45.4 kg).
EPCRA Section 313: Form R *de minimis* concentration reporting level: 1.0%.
TSCA: 40CFR799.5025 hazardous waste constituents, testing requirements.

METHYL CYCLOHEXANE — M:325

SYNONYMS: CYCLOHEXANE, METHYL-; CYCLOHEXYLMETHANE; EEC No. 601-018-00-7; EINECS No. 203-624-3; HEPTANAPHTHENE; HEXAHYDROTOLUENE; METILCICLOHEXANO (Spanish); METHYLOCYKLOHEKSAN (Polish); METILCICLOESANO (Italian); SEXTONE B; TOLUENE HEXAHYDRIDE

IDENTIFICATION

CAS:	108-87-2
ERG ID:	UN 2298
ERG Guide:	128
ERG Guide Category:	FLAMMABLE LIQUIDS (NONPOLAR/WATER-IMMISCIBLE)
Molecular Formula:	C_7H_{14}
RTECS No.:	GV6125000
Properties:	Flammable liquid. Colorless. Faint benzenelike odor. Floats on water; insoluble.

HEALTH AND SAFETY INFORMATION

Low toxicity. Irritates eyes, skin, and respiratory system.

IDLH: 1,200 ppm [10% LEL]
Odor Threshold: 550-630 ppm
OSHA Table Z-1-A Air Contaminant

Exposure Limits:	**TWA**
ACGIH:	400 ppm; 1,610 mg/m^3

NIOSH: 400 ppm; 1,600 mg/m^3
OSHA: 500 ppm; 2,000 mg/m^3

RECOMMENDATIONS FOR RESPIRATOR SELECTION: NIOSH/OSHA

1,200 ppm: SA (any supplied-air respirator);or SCBAF (any self-contained breathing apparatus with a full facepiece)

Emergency or planned entry into unknown concentrations or IDLH conditions:
SCBAF:PD,PP (any self-contained breathing apparatus that has a full facepiece and is operated in a pressure-demand or other positive-pressure mode);or SAF:PD,PP:ASCBA (any supplied-air respirator that has a full facepiece and is operated in a pressure-demand or other positive-pressure mode in combination with an auxiliary, self-contained breathing apparatus operated in a pressure-demand or other positive pressure mode.

Escape: GMFOV (any air-purifying, full-facepiece respirator (gas mask) with a chin-style, front- or back-mounted acid gas canister);or SCBAE (any appropriate escape-type, self-contained breathing apparatus).

CHEMICAL REACTIVITY

Contact with oxidizers may cause fire and explosions. May accumulate static electricity and cause ignition of its vapors.

FIRE INFORMATION

Vapors are heavier than air and will collect in low areas. Vapors in confined areas may explode when exposed to fire. Vapors may travel long distances to ignition sources and flash back. Storage containers and parts of containers may rocket great distances, in many directions. If material or contaminated runoff enters waterways, notify downstream users of potentially contaminated waters. Notify local health and fire officials and pollution control agencies. Use dry chemicals, foam, carbon dioxide. Use water spray to cool exposed containers. Fight fire from a secure area.

NFPA Hazard Classifications:

Health Hazards (Blue) 2
Flammability (Red) 3
Reactivity (Yellow) 0

Flash Point: 25°F (-4°C) c.c.
Flammable Limits in Air: LEL 1.2%; UEL 6.7%
Autoignition Temperature: 482°F (250°C)

FIRST AID

Move victim to fresh air. Call emergency medical care. Apply artificial respiration if victim is not

breathing. If breathing is difficult, give oxygen. Remove and isolate contaminated clothing and shoes. In case of contact with substance, immediately flush skin or eyes with running water for at least 20 minutes. Wash skin with soap and water. Keep victim warm and quiet. Ensure that medical personnel are aware of the material(s) involved and take precautions to protect themselves.

4,4'-METHYLENEBIS(2-CHLOROANILINE) M:330

SYNONYMS: BIS AMINE; CURALIN M; CURENE 442; CYANASET; DI(-4-AMINO-3-CHLOROPHENYL)METHANE; DI-(4-AMINO-3-CLOROFENIL)METANO (Italian); 4,4'-DIAMINO-3,3'-DICHLORODIPHENYLMETHANE; 3,3'-DICHLORO-4,4'-DIAMINODIPHENYLMETHANE; EEC No. 612-078-00-9; EINECS No. 202-918-9; MBOCA; METHYLENE-4,4'-BIS(o-CHLOROANILINE); p,p'-METHYLENEBIS(a-CHLOROANILINE); p,p'-METHYLENEBIS(o-CHLOROANILINE); 4,4'-METHYLENE(BIS)-CHLOROANILINE; 4,4'-METHYLENEBIS(o-CHLOROANILINE); 4,4'-METHYLENEBIS-2-CHLOROBENZENAMINE; METHYLENE-BIS-ORTHOCHLOROANILINE; 4,4-METILENE-BIS-o-CLOROANILINA (Italian); MOCA

IDENTIFICATION

CAS:	101-14-4
ERG ID:	UN 2018 (Solid); UN 2019 (Liquid) as chloroanilines
ERG Guide:	152
ERG Guide Category:	SUBSTANCES-TOXIC (COMBUSTIBLE)
Molecular Formula:	$C_{13}H_{12}Cl_2N_2$
RTECS No.:	CY1050000
Properties:	Solid pellets or flakes. Tan. Faint aminelike odor. Slightly soluble in water.

HEALTH AND SAFETY INFORMATION

Highly toxic. Avoid any skin contact. Effects of contact or inhalation may be delayed. Confirmed carcinogen. Irritates eyes, skin, and respiratory tract. May affect liver, kidneys, blood (forms methemaglobin).

IARC Evaluation: Group 2A, sufficient animal evidence. Suspected human carcinogen.
NTP: Potential carcinogen
OSHA Table Z-1-A Air Contaminant

Exposure Limits:	**TWA**
ACGIH: (1)	0.01 ppm; 0.11 mg/m^3
NIOSH: (2)	0.003 mg/m^3
OSHA:	None

(1) Suspected human carcinogen. Skin contact contributes significantly to overall exposure.
(2) Potential carcinogen. Reduce exposure to lowest feasible level. Skin contact contributes significantly to overall exposure.

RECOMMENDATIONS FOR RESPIRATOR SELECTION:

At any concentrations above the NIOSH REL, or where there is no REL, at any detectable concentration: SCBAF:PD,PP (any self-contained breathing apparatus that has a full facepiece and is operated in a pressure-demand or other positive-pressure mode);or SAF:PD,PP:ASCBA (any supplied-air respirator that has a full facepiece and is operated in a pressure-demand or other positive-pressure mode in combination with an auxiliary, self-contained breathing apparatus operated in a pressure-demand or other positive-pressure mode).

Escape: GMFOV [any air-purifying, full-facepiece respirator (gas mask) with a chin-style, front- or back-mounted organic vapor canister];or SCBAE (any appropriate escape-type, self-contained breathing apparatus).

CHEMICAL REACTIVITY

Reacts with chemicallyactive metals (e.g., potassium, sodium, magnesium, and zinc). May ignite on contact with cellulose nitrate of high surface area. Incompatible with acrolein, acrylonitrile, tert-butyl nitroacetylene, ethylene oxide, isopropyl chlorocarbonate, maleic anhydride, triisobutylaluminum.

FIRE INFORMATION

Combustion products include toxic chlorides and nitrogen oxides. Firefighting gear (including SCBA) may not provide adequate protection. If exposure occurs, remove and isolate gear immediately and thoroughly decontaminate personnel. Vapors in confined areas may explode when exposed to fire. Storage containers and parts of containers may rocket great distances, in many directions. If material or contaminated runoff enters waterways, notify downstream users of potentially contaminated waters. Notify local health and fire officials and pollution control agencies. Use dry chemicals, foam, carbon dioxide. From a secure, explosion-proof location, use water spray to cool exposed containers.

FEMA Hazard Classifications:

Health Hazards (Blue) 1
Flammability (Red) 1
Reactivity (Yellow) 0

FIRST AID

Move victim to fresh air. Call emergency medical care. Apply artificial respiration if victim is not breathing. Administer oxygen if breathing is difficult. Remove and isolate contaminated clothing and shoes. In case of contact with substance, immediately flush skin or eyes with running water for at least 20 minutes. Wash skin with soap and water. Keep victim warm and quiet. Effects of exposure (inhalation,

ingestion, or skin contact) to substance may be delayed. Ensure that medical personnel are aware of the material(s) involved and take precautions to protect themselves.

ENVIRONMENTAL INFORMATION

Clean Air Act: Hazardous Air Pollutants (Title I, Part A, Section 112)
EPA Hazardous Waste Number (RCRA No.): U158
RCRA Toxicity Characteristic (Section 261.24)
RCRA Land Ban Waste
EPCRA Section 304 Reportable Quantity (RQ): CERCLA, 10 lb (4.54 kg)
EPCRA Section 313: Form R *de minimis* concentration reporting level: 0.1%.
California's Proposition 65: Carcinogen.
TSCA: 40CFR 704.175 as MBOCA; TSCA CHIP available.

METHYLENE BIS(4-CYCLOHEXYLISOCYANATE) M:335

SYNONYMS: BIS(4-ISOCYANATOCYCLOHEXYL)METHANE; CYCLOHEXANE, 1,1'-METHYLENEBIS(4-ISOCYANATO-)DICYCLOHEXYLMETHANE-4,4'-DIISOCYANATE; DIISOCIANATO de 4,4'-DICICLOHEXILMETANO (Spanish); DMDI; EEC No. 615-009-00-0; EINECS No. 225-863-2; HMDI; HYDROGENATED MDI; METHYLENE BIS(4-CYCLOHEXYLISOCYANATE); NACCONATE H 12; REDUCED MDI; SATURATED MDI

IDENTIFICATION

CAS:	5124-30-1
ERG ID:	NA 2206 isocyanate solution, toxic, n.o.s.
ERG Guide:	155
ERG Guide Category:	SUBSTANCES-TOXIC AND/OR CORROSIVE
Molecular Formula:	$C_{15}H_{22}N_2O_2$
RTECS No.:	NQ9250000
Properties:	Combustible liquid. Colorless to light yellow. Poisonous gases emitted when heated.

HEALTH AND SAFETY INFORMATION

Poisonous. Irritates eyes, skin, and respiratory tract.

OSHA Table Z-1-A Air Contaminant

Exposure Limits:	TWA	CEILING
ACGIH:	0.005 ppm; 0.054 mg/m^3	
NIOSH:		0.01 ppm; 0.11 mg/m^3
OSHA:	None	

RECOMMENDATIONS FOR RESPIRATOR SELECTION

At any concentrations above the NIOSH REL, or where there is no REL, at any detectable concentration: SCBAF:PD,PP (any self-contained breathing apparatus that has a full facepiece and is operated in a pressure-demand or other positive-pressure mode);or SAF:PD,PP:ASCBA (any supplied-air respirator that has a full facepiece and is operated in a pressure-demand or other positive-pressure mode in combination with an auxiliary, self-contained breathing apparatus operated in a pressure-demand or other positive-pressure mode).

Escape: GMFOV [any air-purifying, full-facepiece respirator (gas mask) with a chin-style, front- or back-mounted organic vapor canister];or SCBAE (any appropriate escape-type, self-contained breathing apparatus)

CHEMICAL REACTIVITY

Reacts with water, acids, and organotin catalysts, caustics, ammonia, aliphatic amines, alkanolamines, amides, alcohols, glycols, caprolactum solution. May slowly polymerize if heated above 122°F.

FIRE INFORMATION

Combustion products include toxic nitrogen oxides and cyanide. Vapors in confined areas may explode when exposed to fire. If material or contaminated runoff enters waterways, notify downstream users of potentially contaminated waters. Notify local health and fire officials and pollution control agencies. Use dry chemicals, alcohol-resistant foam, carbon dioxide. From a secure, explosion-proof location, use water spray to cool exposed containers.

NFPA Hazard Classifications: Not listed

Flash Point: > 395°F (>202°C)

FIRST AID

Move victim to fresh air. Call emergency medical care. Apply artificial respiration if victim is not breathing. Administer oxygen if breathing is difficult. Remove and isolate contaminated clothing and shoes. In case of contact with substance, immediately flush skin or eyes with running water for at least 20 minutes. Wash skin with soap and water. Keep victim warm and quiet. Effects of exposure (inhalation, ingestion, or skin contact) to substance may be delayed. Ensure that medical personnel are aware of the material(s) involved and take precautions to protect themselves.

ENVIRONMENTAL INFORMATION

EPCRA Section 313: Form R *de minimis* concentration reporting level: 1.0%.
TSCA: 40CFR716.120(*a*); 40CFR712.30(*e*)8 isocyanates.

METHYLENE BISPHENYL ISOCYANATE M:340

SYNONYMS: CARADATE 30; CARWINATE 125 M; DESMODUR-44; 4,4'-DIPHENYLMETHANE DIISOCYANATE; DIPHENYLMETHANE-4,4'-DIISOCYANATE; EEC No. 615-005-01-6; EINECS No. 202-966-0; HYLENE M50; INSTAFILL COMPONENT A; INSTAPAK 40 COMPONENT A; ISONATE; MDI; MBI; METHYLENE BIS(PHENYLISOCYANATE); METHYLENE BIS(4-PHENYL ISOCYANATE); 4,4-METHYLENEDIPHENYL DIISOCYANATE; METHYLENE DI-p-PHENYLENE ESTER OF ISOCYANIC ACID; METILENBIS(FENILISOCIANATO) (Spanish); MULTRATHANE M; NACCONATE 300; NCI-C50668; RITCHPAK-40, A-SIDE; RITCHPAK-42, A-SIDE; RITCHPAK-50, A-SIDE; RITCHPAK-70, A-SIDE; RUBINATE; TR-24A; TR-240IMA; TR-250A; VILRATHANE 4300

IDENTIFICATION

CAS:	101-68-8
ERG ID:	UN 2489
ERG Guide:	156
ERG Guide Category:	TOXIC AND/OR CORROSIVE (COMBUSTIBLE/WATER-SENSITIVE)
Molecular Formula:	$C_{15}H_{10}N_2O_2$
RTECS No.:	NQ9350000
Properties:	Combustible, solid flakes. White to light yellow. Sinks in water.

HEALTH AND SAFETY INFORMATION

Poisonous. Irritates eyes and skin. An allergen.

IARC Evaluation: Group 3, unclassifiable; no human data; no animal data.
IDLH: 75 mg/m^3
Odor Threshold: 0.384 ppm
OSHA Table Z-1-A Air Contaminant

Exposure Limits:	**TWA**	**CEILING**
ACGIH:	0.005 ppm; 0.051 mg/m^3	
NIOSH:	0.005 ppm; 0.05 mg/m^3	0.020 ppm; 0.2 mg/m^3/10 min
OSHA:		0.02 ppm; 0.2 mg/m^3

RECOMMENDATIONS FOR RESPIRATOR SELECTION: NIOSH

0.5 mg/m^3: SA (any supplied-air respirator);or SCBA (any self-contained breathing apparatus).
1.25 mg/m^3: SA:CF (any supplied-air respirator operated in a continuous-flow mode).
2.5 mg/m^3: SCBAF (any self-contained breathing apparatus with a full facepiece);or SAF (any supplied-air respirator with a full facepiece).
75 mg/m^3: SAF:PD,PP (any supplied-air respirator that has a full facepiece and is operated in a pressure-demand or other positive-pressure mode).

Emergency or Planned Entry into Unknown Concentrations or IDLH Conditions: SCBAF:PD,PP (any self-contained breathing apparatus that has a full facepiece and is operated in a pressure-demand or other positive-pressure mode);or SAF:PD,PP:ASCBA (any supplied-air respirator that has a full facepiece and is operated in a pressure-demand or other positive-pressure mode in combination with an auxiliary, self-contained breathing apparatus operated in a pressure-demand or other positive-pressure mode).

Escape: GMFOVHiE [any air-purifying, full-facepiece respirator (gas mask) with a chin-style, front- or back-mounted organic vapor canister having a high-efficiency particulate filter];or SCBAE (any appropriate escape-type, self-contained breathing apparatus).

Note: Substance reported to cause eye irritation or damage; may require eye protection.

CHEMICAL REACTIVITY

Incompatible with strong alkalies, acids, alcohols, ammonia, amines, amides, glycols, and caprolactum. Unstable above 100°F (38°C). Attacks some plastics, rubber, and coatings.

FIRE INFORMATION

Combustion products include toxic nitrogen oxides, sulfur oxides, and cyanides. Firefighting gear (including SCBA) may not provide adequate protection. If exposure occurs, remove and isolate gear immediately and thoroughly decontaminate personnel. Vapors in confined areas may explode when exposed to fire. Storage containers and parts of containers may rocket great distances, in many directions. If material or contaminated runoff enters waterways, notify downstream users of potentially contaminated waters. Notify local health and fire officials and pollution control agencies. Use dry chemicals, foam, carbon dioxide. From a safe location, use water spray to cool exposed containers.

FEMA Hazard Classifications:

Health Hazards (Blue) 2
Flammability (Red) 1
Reactivity (Yellow) 1

Flash Point: 425°F (218°C) o.c.; 390°F (199°C) c.c.

FIRST AID

Move victim to fresh air. Call emergency medical care. Apply artificial respiration if victim is not breathing. Do not use mouth-to-mouth method if victim ingested or inhaled the substance; induce artificial respiration with the aid of a pocket mask equipped with a one-way valve or other proper respiratory medical device. Administer oxygen if breathing is difficult. Remove and isolate contaminated clothing and shoes. In case of contact with material, immediately flush skin or eyes with running water for at least 20 minutes. For minor skin contact, avoid spreading material on unaffected skin. Keep victim warm and quiet. Effects of exposure (inhalation, ingestion, or skin contact) may be delayed. Ensure that medical personnel are aware of the material(s) involved and take precautions to protect themselves.

ENVIRONMENTAL INFORMATION

Clean Air Act: Hazardous Air Pollutants (Title I, Part A, Section 112).
EPCRA Section 304: Reportable Quantity (RQ): CERCLA, 5,000 lb (2,270 kg).
EPCRA Section 313: Form R *de minimis* concentration reporting level: 1.0%.
TSCA: 40CFR716.120(*a*); 40CFR712.30(*e*)8 isocyanates.

METHYLENE CHLORIDE M:360

SYNONYMS: AEROTHENE; BLANCO-TRON TMC; CHEMTRANIC FLUX OFF; CHEMTRANIC FLUX STRIPPER; CLORURO de METILENO (Spanish); CLORURO di METILENE (Italian); DICHLOROMETHANE; DRIVERIT; EEC No. 602-004-00-3; EINECS No. 200-838-9; FREON TMC; GENESOLV DM AZEOTROPE; JASCO INK REMOVER; M-CLEAN D; MERKEM 531 GRAVURE PLATE STRIPPER; METHYLENE DICHLORIDE; MS-111 EPOXY STRIPPER; MS-190 FLUX REMOVER; MS-190 HD FLUX REMOVER; MS-470 URETHANE COATING; METHOKLONE; MU-9; METHYLENE BICHLORIDE; METHYLENE DICHLORIDE; NEVOLIN; SISULFASOL; STRIP L; STRIPTEX MU-7 RESIST REMOVER; TEFLON DRY LUBE 708; TPD THINNER

IDENTIFICATION

CAS:	75-09-2
ERG ID:	UN 1593
ERG Guide:	160
ERG Guide Category:	HALOGENATED SOLVENTS
Molecular Formula:	CH_2Cl_2
RTECS No.:	PA8050000
Properties:	Flammable, watery liquid. Clear, colorless. Sweet, pleasant (like chloroform or ether) odor. Sinks in water and is soluble in water.

HEALTH AND SAFETY INFORMATION

Harmful or fatal if swallowed; produces carbon monoxide in the body. Chronic occupational exposure to various solvents, including methylene chloride, have been associated with chronic neurological effects such as memory loss with intellectual impairment and balance disturbances. Cardiovascular effects have been documented in case reports (ATSDR) but have not been demonstrated in epidemiologic studies of workers exposed to methylene chloride. Irritant contact dermatitis manifested by inflammation and hives has been noted in workers who have chronic skin exposure. The New York State Department of Health reports that prolonged exposure can cause changes in the blood, hallucinations, and decreased response to visual and auditory stimulation. Long–term exposures have also resulted in damage to the liver.

IARC Evaluation: Group 2B, possible human carcinogen; human evidence, inadequate; animal evidence, sufficient.
NTP: Anticipated human carcinogen; animal carcinogen.

EPA probable human carcinogen.
IDLH: 2,300 ppm; potential human carcinogen.
Odor Threshold: 0.912 ppm ; may be inadequate for exposure; olfactory fatigue may occur.
OSHA Table Z-1-A and Z-2 Air Contaminant
NIOSH Criteria Document: NIOSH 76-138

Exposure Limits:	**TWA**	**STEL**
ACGIH:[1]	50 ppm; 174 mg/m^3	
NIOSH:[2]		
OSHA:[3]	25 ppm	125 ppm per 15 min

[1]Suspected carcinogen.
[2]Potential carcinogen. Reduce exposure to lowest feasible level.
[3] PEL effective 4/10/97

RECOMMENDATIONS FOR RESPIRATOR SELECTION: NIOSH

At any concentrations above the NIOSH REL, or where there is no REL, at any detectable concentration: SCBAF:PD,PP (any self-contained breathing apparatus that has a full facepiece and is operated in a pressure-demand or other positive-pressure mode);or SAF:PD,PP:ASCBA (any supplied-air respirator that has a full facepiece and is operated in a pressure-demand or other positive-pressure mode in combination with an auxiliary, self-contained breathing apparatus operated in a pressure-demand or other positive-pressure mode).

Escape: GMFOV [any air-purifying, full-facepiece respirator (gas mask) with a chin-style, front- or back-mounted organic vapor canister];or SCBAE (any appropriate escape-type, self-contained breathing apparatus).

CHEMICAL REACTIVITY

Contact with alkali metals (sodium, potassium) may cause explosions. Incompatible with strong oxidizers, caustic substances, chemically active metals such as aluminum and magnesium powders, potassium, sodium, and concentrated nitric acid. Will corrode iron, stainless steel, and copper, especially at high temperatures and if metals are wet. Attacks some plastics, rubber, and coatings. May accumulate static electrical charges and cause ignition of its vapors. Can produce phosgene in the presence of space heaters, wood stoves, or when smoking.

FIRE INFORMATION

Combustion products include toxic phosgene gases and hydrogen chloride; these decomposition products are considered more toxic than methylene chloride. Vapors are heavier than air and will collect in low areas. Vapors in confined areas may explode when exposed to fire. Storage containers and parts of containers may rocket great distances, in many directions. If material or contaminated runoff enters waterways, notify downstream users of potentially contaminated waters. Notify local health and fire officials and pollution control agencies. Use any extinguishing agent. From a safe location, use water spray to cool exposed containers.

NFPA Hazard Classifications:

Health Hazards (Blue) 2
Flammability (Red) 1
Reactivity (Yellow) 0

Flash Point: No flash point by conventional methods, but forms flammable vapor-air mixtures at 212°F (100°C) and higher.
Flammable Limits in Air: LEL 13%; UEL 23%
Autoignition Temperature: 1033°F (556°C)

FIRST AID

Move victim to fresh air. Call emergency medical care. Apply artificial respiration if victim is not breathing. Administer oxygen if breathing is difficult. Remove and isolate contaminated clothing and shoes. In case of contact with material, immediately flush skin or eyes with running water for at least 20 minutes. For minor skin contact, avoid spreading material on unaffected skin. Wash skin with soap and water. Keep victim warm and quiet. Ensure that medical personnel are aware of the material(s) involved and take precautions to protect themselves.

ENVIRONMENTAL INFORMATION

Methylene chloride is among those chemicals released in the greatest quantity by the fabricated metal products industry. When spilled on land, methylene chloride is rapidly lost from the soil surface through volatilization. The remainder leaches through the subsoil into the groundwater. Biodegradation is possible in natural waters but will probably be very slow compared with evaporation. Little is known about bioconcentration in aquatic organisms or adsorption to sediments, but these are not likely to be significant processes. Hydrolysis is not an important process under normal environmental conditions. Methylene chloride released into the atmosphere degrades via contact with other gases with a half-life of several months. A small fraction of the chemical diffuses to the stratosphere, where it rapidly degrades through exposure to ultraviolet radiation and contact with chlorine ions. Being a moderately soluble chemical, methylene chloride is expected to partially return to earth in rain.

Clean Air Act: Hazardous Air Pollutants (Title I, Part A, Section 112).
Clean Water Act: Section 307 Priority Pollutants.
EPA Hazardous Waste Number (RCRA No.): U080..
RCRA Section 261 Hazardous Constituents
RCRA Land Ban Waste
Safe Drinking Water Act: Regulated chemical (47 FR 9352); MCL, 0.005 mg/L; MCLG, zero.
RCRA Universal Treatment Standards: Wastewater (mg/L), 0.089; Nonwastewater (mg/kg), 30.
RCRA Ground Water Monitoring List: Suggested test method(s) (PQL ug/L): 8010(5); 8240(5).
Safe Drinking Water Act: MCL, 0.005mg/L; MCLG, zero; Regulated chemical (47 FR 9352).
EPCRA Section 304: Reportable Quantity (RQ): CERCLA, 1,000 lb (454 kg).
EPCRA Section 313: Form R *de minimis* concentration reporting level: 1.0%.
California's Proposition 65: Carcinogen.
TSCA: 40CFR716.120(*a*)

METHYL ETHYL KETONE M:380

SYNONYMS: ACETONE, METHYL-; AETHYLMETHYLKETON (German); BUTANONE; 2-BUTANONE; BUTANONE 2 (French); EEC No. 606-002-00-3; EINECS No. 201-159-0; EPOXY CHROMATE METAL PRIMER; EPOXY SOLVENT CURE AGENT; ETHYL METHYL CETONE (French); ETHYLMETHYLKETON (Dutch); ETHYL METHYL KETONE; GLID-GUARD EPOXY-SAFETY BLUE; HUMISEAL 503; HUMISEAL 1B31; MEK; METHYL ACETONE; METHYL KETONE; KETONE, ETHYL METHYL; MEETCO; METIL ETIL CETONA (Spanish); METILETILCHETONE (Italian); METYLOETYLOKETON (Polish); VANDELEX 20; VANDELEX 24; WELD-ON P-70 PRIMER

IDENTIFICATION

CAS:	78-93-3
ERG ID:	UN 1193; UN 1232
ERG Guide:	127
ERG Guide Category:	FLAMMABLE LIQUIDS (POLAR/WATER MISCIBLE)
Molecular Formula:	C_4H_8O
RTECS No.:	EL6475000
Properties:	Highly flammable, colorless liquid. Sweet odor, like acetone or fingernail polish remover. Floats on and mixes with water. Large amounts of flammable, irritating vapor is produced.

HEALTH AND SAFETY INFORMATION

Irritating to eyes, skin, and respiratory tract. Can cause eye damage. Breathing moderate amounts of MEK for short periods of time can cause adverse effects on the nervous system, ranging from headaches, dizziness, nausea, and numbness in the fingers and toes to unconsciousness. Prolonged contact or repeated exposures to moderate to high amounts may cause liver and kidney effects, irritation of the skin, dermatitis, central nervous system changes, disturbed sleep pattern, birth defects, and miscarriage. No agreement exists over the carcinogenicity of MEK although limited animal evidence exists.

IDLH: 3,000 ppm
Odor Threshold: 0.27 ppm
OSHA Table Z-1-A Air Contaminant as 2-butanone
NIOSH Criteria Document: NIOSH-78-173 KETONES

Exposure Limits:	**TWA**	**STEL**
ACGIH:	200 ppm; 590 mg/m^3	300 ppm; 885 mg/m^3
NIOSH:	200 ppm; 590 mg/m^3	300 ppm; 885 mg/m^3
OSHA:	200 ppm; 590 mg/m^3	

RECOMMENDATIONS FOR RESPIRATOR SELECTION: NIOSH/OSHA

3,000 ppm: SA:CF (any supplied-air respirator operated in a continuous-flow mode);or PAPROV [any powered, air-purifying respirator with organic vapor cartridge(s)];or CCRFOV (any chemical cartridge respirator with a full facepiece and organic vapor cartridge);or GMFOV [any air-purifying, full-facepiece respirator (gas mask) with a chin-style, front- or back-mounted acid gas canister]; SCBAF (any self-contained breathing apparatus with a full facepiece);or SAF (any supplied-air respirator with a full facepiece).

Emergency or Planned Entry into Unknown Concentrations or IDLH Conditions: SCBAF:PD,PP (any self-contained breathing apparatus that has a full facepiece and is operated in a pressure-demand or other positive-pressure mode);or SAF:PD,PP:ASCBA (any supplied-air respirator that has a full facepiece and is operated in a pressure-demand or other positive-pressure mode in combination with an auxiliary, self-contained breathing apparatus operated in a pressure-demand or other positive-pressure mode).

Escape: GMFOV [any air-purifying, full-facepiece respirator (gas mask) with a chin-style, front- or back-mounted organic vapor canister];or SCBAE (any appropriate escape-type, self-contained breathing apparatus).
Note: Substance causes eye irritation or damage; eye protection needed.

CHEMICAL REACTIVITY

Forms explosive mixture with air. Extremely flammable. Incompatible with sulfuric acid, nitric acid, aliphatic amines, strong oxidizers, potassium tert-butoxide, 2-propanol, chlorosulfonic acid, and oleum.

FIRE INFORMATION

Vapors are heavier than air and will collect in low areas. Vapors in confined areas may explode when exposed to fire. Vapors may travel long distances to ignition sources and flash back. Storage containers and parts of containers may rocket great distances, in many directions. If material or contaminated runoff enters waterways, notify downstream users of potentially contaminated waters. Notify local health and fire officials and pollution control agencies. Use dry chemicals, foam, carbon dioxide. From a safe location, use water spray to cool exposed containers.

NFPA Hazard Classifications:

Health Hazards (Blue) 1
Flammability (Red) 3
Reactivity (Yellow) 0

Flash Point: 22°F (-6°C) o.c.; 16°F (-9°C) c.c.
Flammable Limits in Air: LEL 1.8%; UEL 11.5%
Autoignition Temperature: 960°F (515°C)

FIRST AID

Move victim to fresh air. Call emergency medical care. Apply artificial respiration if victim is not breathing. If breathing is difficult, give oxygen. Remove and isolate contaminated clothing and shoes. In case of contact with material, immediately flush skin or eyes with running water for at least 20 minutes.

Wash skin with soap and water. Keep victim warm and quiet. Ensure that medical personnel are aware of the material(s) involved and take precautions to protect themselves.

ENVIRONMENTAL INFORMATION

Methyl ethyl ketone (MEK) is among those chemicals released in the greatest quantity by the fabricated metal products industry. Most of the MEK released to the environment ends up in the atmosphere. MEK can contribute to the formation of air pollutants in the lower atmosphere. It can be degraded by microorganisms living in water and soil.

Clean Air Act: Hazardous Air Pollutants (Title I, Part A, Section 112).
EPA Hazardous Waste Number (RCRA No.): U159.
RCRA Section 261 Hazardous Constituents
RCRA Land Ban Waste
RCRA Toxicity Characteristic (Section 261.24): Maximum Concentration of Contaminants, regulatory level, 200.0 mg/L.
RCRA Universal Treatment Standards: Wastewater (mg/L), 0.28; Nonwastewater (mg/kg), 36.
RCRA Ground Water Monitoring List: Suggested test method(s) (PQL ug/L): 8015(10); 8240(100).
Safe Drinking Water Act: Priority List (55 FR 1470).
EPCRA Section 304: Reportable Quantity (RQ): CERCLA, 5,000 lb (2,270 kg).
EPCRA Section 313: Form R *de minimis* concentration reporting level: 1.0%.
TSCA: 40CFR716.120(*a*)

METHYL ETHYL KETONE PEROXIDE M:390

SYNONYMS: 2-BUTANONE PEROXIDE; BUTANOX M50; CHALOXYD; EINECS No. 215-661-2; ETHYL METHYL KETONE PEROXIDE; HI-POINT 90; KETONOX; LUPERSOL; LPT; M105; MEKP; MEKP-HA 1; MEKP-LA 1; MEK PEROXIDE; METHYL ETHYL KETONE HYDROPEROXIDE; NCI-C55447; PEROXIDO de 2-BUTANONA (Spanish); QUICKSET EXTRA; SPRAYSET MEKP; THERMACURE

IDENTIFICATION

CAS:	1338-23-4
ERG ID:	UN 2550; UN 2127 (60% MEKP); UN 3101
ERG Guide:	147 (60% MEPK); 146 (UN 3101, organic peroxide, type B, liquid)
ERG Guide Category:	147-ORGANIC PEROXIDES (HEAT AND CONTAMINATION SENSITIVE/SEVERE IRRITANTS) 146-ORGANIC PEROXIDES (HEAT, CONTAMINATION AND FRICTION SENSITIVE)
Molecular Formula:	$C_8H_{16}O_4$
RTECS No.:	EL9450000
Properties:	Flammable liquid. Colorless. Acetonelike odor. Not soluble in water.

HEALTH AND SAFETY INFORMATION

Corrosive to eyes; irritates skin and respiratory tract. Inhalation may cause pulmonary edema, which can be delayed for several hours; there is a risk of death in serious cases. May cause kidney and liver damage.

OSHA Table Z-1-A Air Contaminant
OSHA Process Safety Management of Highly Hazardous Chemicals (29CFR, Part 1910.119, Appendix A): Threshold Quantity: 5,000 pounds. (Concentration > 60%)

Exposure Limits: CEILING

ACGIH:	0.2 ppm; 1.5 mg/m^3
NIOSH:	0.2 ppm; 1.5 mg/m^3
OSHA:	None

RECOMMENDATIONS FOR RESPIRATOR SELECTION: NIOSH

At any concentrations above the NIOSH REL, or where there is no REL, at any detectable concentration: SCBAF:PD,PP (any self-contained breathing apparatus that has a full facepiece and is operated in a pressure-demand or other positive-pressure mode);or SAF:PD,PP:ASCBA (any supplied-air respirator that has a full facepiece and is operated in a pressure-demand or other positive-pressure mode in combination with an auxiliary, self-contained breathing apparatus operated in a pressure-demand or other positive-pressure mode).

Escape: GMFOV [any air-purifying, full-facepiece respirator (gas mask) with a chin-style, front- or back-mounted organic vapor canister];or SCBAE (any appropriate escape-type, self-contained breathing apparatus).

CHEMICAL REACTIVITY

Forms explosive mixture with air. Explosive decomposition above 200°F (93°C). Pure substance is shock-sensitive. Strong oxidizer. Violent reaction with chemical accelerants; strong acids, strong bases, reducing agents, combustible substances, organic materials, oxides of heavy metals, salts, trace contaminants, and amines. May accumulate static electrical charges; may cause ignition of its vapors.

FIRE INFORMATION

Vapors are heavier than air and will collect in low areas. Vapors in confined areas may explode when exposed to fire. Vapors may travel long distances to ignition sources and flash back. Storage containers and parts of containers may rocket great distances, in many directions. If material or contaminated runoff enters waterways, notify downstream users of potentially contaminated waters. Notify local health and fire officials and pollution control agencies. Use water spray, dry chemicals, carbon dioxide. From a safe location, use water spray to cool exposed containers.

Flash Point: 125–200°F (52–93°C) o.c.; 180°F (82°C) c.c. (60% MEKP).

FIRST AID

Move victim to fresh air. Call emergency medical care. Apply artificial respiration if victim is not breathing. Administer oxygen if breathing is difficult. Remove and isolate contaminated clothing and shoes. Remove material from skin immediately. In case of contact with substance, immediately flush skin or eyes with running water for at least 20 minutes. Keep victim warm and quiet. Ensure that medical personnel are aware of the material(s) involved and take precautions to protect themselves.

ENVIRONMENTAL INFORMATION

EPA Hazardous Waste Number (RCRA No.): U160.
RCRA Section 261 Hazardous Constituents.
EPCRA Section 304: Reportable Quantity (RQ): CERCLA, 10 lb (4.54 kg).

METHYL FORMATE M:393

SYNONYMS: EEC No. 607-014-00-1; EINECS No. 203-481-7; FORMIATE de METHYLE (French); FORMIATO de METILO (Spanish); FORMIC ACID, METHYL ESTER; METHYL ESTER OF FORMIC ACID; METHYLE FORMIATE de (French); METHYLFORMIAAT (Dutch); METHYLFORMIAT (German); METHYL METHANOATE; METIL (FORMIATO di) (Italian);

IDENTIFICATION

CAS:	107-31-3
ERG ID:	UN 1243
ERG Guide:	129
ERG Guide Category:	FLAMMABLE LIQUIDS (POLAR/WATER-MISCIBLE/NOXIOUS)
Molecular Formula:	$C_2H_4O_2$
RTECS No.:	LQ8925000
Properties:	Extremely flammable liquid. Colorless. Pleasant, fruity odor. Initially floats on water; soluble. Flammable, irritating vapor is produced. Boiling point is 88°F (31°C).

HEALTH AND SAFETY INFORMATION

Irritates eyes, skin, and respiratory tract. Inhalation may cause pulmonary edema, which can be delayed for several hours; there is a risk of death in serious cases. May affect the central nervous system and cause optic nerve damage.

IDLH: 4,500 ppm
Odor Threshold: 93.3 ppm
OSHA Table Z-1-A Air Contaminant

740 Methyl Formate

Exposure Limits:	TWA	STEL
ACGIH:	100 ppm; 246 mg/m^3	150 ppm; 368 mg/m^3
NIOSH:	100 ppm; 250 mg/m^3	150 ppm; 375 mg/m^3
OSHA:	100 ppm; 250 mg/m^3	

RECOMMENDATIONS FOR RESPIRATOR SELECTION: NIOSH/OSHA

1000 ppm: SA (any supplied-air respirator).
2500 ppm: SA:CF (any supplied-air respirator operated in a continuous-flow mode).
4500 ppm: SCBAF (any self-contained breathing apparatus with a full facepiece);or SAF (any supplied-air respirator with a full facepiece).

Emergency or planned entry into unknown concentration or IDLH conditions: SCBAF:PD,PP (any self-contained breathing apparatus that has a full facepiece and is operated in a pressure-demand or other positive-pressure mode);or SAF:PD,PP:ASCBA (any supplied-air respirator that has a full facepiece and is operated in a pressure-demand or other positive-pressure mode in combination with an auxiliary, self-contained breathing apparatus operated in a pressure-demand or other positive pressure mode).

Escape: GMFOV (any air-purifying, full-facepiece respirator (gas mask) with a chin-style, front- or back-mounted acid gas canister);or SCBAE (any appropriate escape-type, self-contained breathing apparatus).

Note: Substance reported to cause eye irritation or damage; may require eye protection.

CHEMICAL REACTIVITY

Slow reaction with water to form formic acid and methyl alcohol. Forms explosive mixture with air. Incompatible with strong acids, caustics, water, nitrates. Reacts violently with strong oxidizers. Attacks some plastics and coatings. May accumulate static electrical charges, and may cause ignition of its vapors.

FIRE INFORMATION

Vapors are heavier than air and will collect in low areas. Vapors in confined areas may explode when exposed to fire. Vapors may travel long distances to ignition sources and flash back. Storage containers and parts of containers may rocket great distances, in many directions. If material or contaminated runoff enters waterways, notify downstream users of potentially contaminated waters. Notify local health and fire officials and pollution control agencies. Use dry chemicals, foam, carbon dioxide.

NFPA Hazard Classifications:

Health Hazards (Blue) 2
Flammability (Red) 4
Reactivity (Yellow) 0

Flash Point: -2°F (-19°C) c.c.
Flammable Limits in Air: LEL 4.5%; UEL 23%
Autoignition Temperature: 840°F (449°C)

FIRST AID

Move victim to fresh air. Call emergency medical care. Apply artificial respiration if victim is not breathing. Administer oxygen if breathing is difficult. Remove and isolate contaminated clothing and shoes. In case of contact with substance, immediately flush skin or eyes with running water for at least 20 minutes. Wash skin with soap and water. Keep victim warm and quiet. Effects of exposure (inhalation, ingestion, or skin contact) to substance may be delayed. Ensure that medical personnel are aware of the material(s) involved and take precautions to protect themselves.

ENVIRONMENTAL INFORMATION

Clean Air Act: Accidental Release Prevention/Flammable Substances (Section 112[r], Table 3), TQ = 10,000 lb (4,540 kg)
TSCA: 40CFR712.30(*e*)10; 40CFR716.120(*d*)10 OSHA chemicals; need for dermal testing

METHYL HEXYL KETONE M:396

SYNONYMS: EINECS No. 203-837-1; ESILMETILCHETONE (Italian); FEMA No. 2802; HEXYL METHYL KETONE; OCTAN-2-ONE; 2-0CTANONE

IDENTIFICATION

CAS:	111-13-7
ERG ID:	UN 1224 as ketones, liquids, n.o.s.
ERG Guide:	127
ERG Guide Category:	FLAMMABLE LIQUIDS (POLAR/WATER- MISCIBLE)
Molecular Formula:	$C_8H_{16}O$
RTECS No.:	RH1484
Properties:	Flammable liquid. Colorless. Fruity, applelike odor. Practically insoluble in water.

HEALTH AND SAFETY INFORMATION

May cause irritation of eyes, skin, and respiratory tract.

NIOSH Criteria Document: NIOSH 78-173 KETONES

RECOMMENDATIONS FOR RESPIRATOR SELECTION

At any concentrations above the NIOSH REL, or where there is no REL, at any detectable concentration: SCBAF:PD,PP (any self-contained breathing apparatus that has a full facepiece and is

operated in a pressure-demand or other positive-pressure mode);or SAF:PD,PP:ASCBA (any supplied-air respirator that has a full facepiece and is operated in a pressure-demand or other positive-pressure mode in combination with an auxiliary, self-contained breathing apparatus operated in a pressure-demand or other positive-pressure mode).

Escape: GMFOV [any air-purifying, full-facepiece respirator (gas mask) with a chin-style, front- or back-mounted organic vapor canister];or SCBAE (any appropriate escape-type, self-contained breathing apparatus)

CHEMICAL REACTIVITY

Contact with oxidizers may cause fire and explosions. Ketones react violently with aldehydes (e.g., acetaldehyde), bromine, nitric acid, perchloric acid; forms explosive mixture with 4,6-dinitrobenzofurazan-n-oxide.

FIRE INFORMATION

Vapors are heavier than air and will collect in low areas. Vapors in confined areas may explode when exposed to fire. Vapors may travel long distances to ignition sources and flash back. Storage containers and parts of containers may rocket great distances, in many directions.If material or contaminated runoff enters waterways, notify downstream users of potentially contaminated waters. Notify local health and fire officials and pollution control agencies. Use dry chemicals, foam, carbon dioxide.

NFPA Hazard Classifications

Health Hazards (Blue) 0
Flammability (Red) 2
Reactivity (Yellow) 0

Flash Point: 125°F (52°C)

FIRST AID

Move victim to fresh air. Call emergency medical care. Apply artificial respiration if victim is not breathing. Administer oxygen if breathing is difficult. Remove and isolate contaminated clothing and shoes. In case of contact with substance, immediately flush skin or eyes with running water for at least 20 minutes. Wash skin with soap and water. Keep victim warm and quiet. Ensure that medical personnel are aware of the material(s) involved and take precautions to protect themselves.

2–METHYLIMIDAZOLE M:398

SYNONYMS: EINECS No. 211-765-7; 2-METILIMIDAZOLE (Spanish)

IDENTIFICATION

CAS: 693-98-1
Molecular Formula: $C_4H_6N_2$
RTECS No.: NI7175000
Properties: Needles. Mixes with water.

HEALTH AND SAFETY INFORMATION

Due to the availability of insufficient data on human effects, caution should be exercised.

RECOMMENDATIONS FOR RESPIRATOR SELECTION

At any concentrations above the NIOSH REL, or where there is no REL, at any detectable concentration: SCBAF:PD,PP (any self-contained breathing apparatus that has a full facepiece and is operated in a pressure-demand or other positive-pressure mode);or SAF:PD,PP:ASCBA (any supplied-air respirator that has a full facepiece and is operated in a pressure-demand or other positive-pressure mode in combination with an auxiliary, self-contained breathing apparatus operated in a pressure-demand or other positive-pressure mode).

Escape: GMFOV [any air-purifying, full-facepiece respirator (gas mask) with a chin-style, front- or back-mounted organic vapor canister];or SCBAE (any appropriate escape-type, self-contained breathing apparatus).

FIRE INFORMATION

Combustion products include toxic nitrogen oxides. If material or contaminated runoff enters waterways, notify downstream users of potentially contaminated waters. Notify local health and fire officials and pollution control agencies.

NFPA Hazard Classifications: Not listed

FIRST AID

Move victim to fresh air. Call emergency medical care. Apply artificial respiration if victim is not breathing. Administer oxygen if breathing is difficult. Remove and isolate contaminated clothing and shoes. In case of contact with substance, immediately flush skin or eyes with running water for at least 20 minutes. Wash skin with soap and water. Keep victim warm and quiet. Effects of exposure (inhalation, ingestion, or skin contact) to substance may be delayed. Ensure that medical personnel are aware of the material(s) involved and take precautions to protect themselves.

METHYL ISOAMYL KETONE M:400

SYNONYMS: DP-13 E BEAM POSITIVE RESIST DEVELOPER; DP-18 E BEAM POSITIVE RESIST

744 Methy Isoamyl Ketone

DEVELOPER; ISOAMYL METHYL KETONE; ISOPENTYL METHYL KETONE; KTI PBS DEVELOPER II; KTI PBS RINSE; 2-METHYL-5-HEXANONE; 5-METHYL-2-HEXANONE; METIL ISOAMIL CETONA (Spanish); MIAK; PBS DEVELOPER; PBS RINSE; RP-10 E BEAM POSITIVE RESIST RINSE

IDENTIFICATION

CAS: 110-12-3
ERG ID: UN 2302
ERG Guide: 127
ERG Guide Category: FLAMMABLE LIQUIDS (POLAR/WATER MISCIBLE)
Molecular Formula: $C_7H_{14}O$
RTECS No.: MP3850000
Properties: Flammable liquid. Colorless. Pleasant odor.

HEALTH AND SAFETY INFORMATION

Prolonged or repeated contact may cause irritation of the skin and dermatitis with dryness and cracking; nervous system changes; liver damage.

Odor Threshold: 0.013 ppm
OSHA Table Z-1-A Air Contaminant
NIOSH Criteria Document: NIOSH-78-173 KETONES

Exposure Limits: TWA

ACGIH: 50 ppm; 234 mg/m^3
NIOSH: 50 ppm; 240 mg/m^3
OSHA: 100 ppm; 475 mg/m^3

RECOMMENDATIONS FOR RESPIRATOR SELECTION

At any concentrations above the NIOSH REL, or where there is no REL, at any detectable concentration: SCBAF:PD,PP (any self-contained breathing apparatus that has a full facepiece and is operated in a pressure-demand or other positive-pressure mode);or SAF:PD,PP:ASCBA (any supplied-air respirator that has a full facepiece and is operated in a pressure-demand or other positive-pressure mode in combination with an auxiliary, self-contained breathing apparatus operated in a pressure-demand or other positive-pressure mode).

Escape: GMFOV [any air-purifying, full-facepiece respirator (gas mask) with a chin-style, front- or back-mounted organic vapor canister];or SCBAE (any appropriate escape-type, self-contained breathing apparatus).

CHEMICAL REACTIVITY

Forms explosive mixture with air. Contact with oxidizers may cause fire and explosions. Ketones can react violently with aldehydes (acetaldehyde), bromine, nitric acid, perchloric acid. Forms explosive mixture with 4,6-dinitrobenzofurazan-N-oxide.

FIRE INFORMATION

Vapors are heavier than air and will collect in low areas. Vapors in confined areas may explode when exposed to fire. Vapors may travel long distances to ignition sources and flash back. Storage containers and parts of containers may rocket great distances, in many directions. If material or contaminated runoff enters waterways, notify downstream users of potentially contaminated waters. Notify local health and fire officials and pollution control agencies. Use dry chemicals, foam, carbon dioxide. From a secure, explosion-proof location, use water spray to cool exposed containers.

NFPA Hazard Classifications:

Health Hazards (Blue) 1
Flammability (Red) 2
Reactivity (Yellow) 0

Flash Point: 110°F (43°C) o.c.
Flammable Limits in Air: LEL 1.0%; UEL 8.2% both @ 200°F (93°C)
Autoignition Temperature: 375°F (191°C)

FIRST AID

Move victim to fresh air. Call emergency medical care. Apply artificial respiration if victim is not breathing. If breathing is difficult, give oxygen. Remove and isolate contaminated clothing and shoes. In case of contact with material, immediately flush skin or eyes with running water for at least 20 minutes. Wash skin with soap and water. Keep victim warm and quiet. Ensure that medical personnel are aware of the material(s) involved and take precautions to protect themselves.

METHYL ISOBUTYL CARBINOL M:410

SYNONYMS: ALCOOL METHYL AMYLIQUE (French); EEC No. 603-008-00-8; EINECS No. 203-551-7; ISOBUTYLMETHYLCARBINOL; ISOBUTYLMETHYLMETHANOL; MAOH; METHYL AMYL ALCOHOL; METHYLISOBUTYL CARBINOL; 2-METHYL-4-PENTANOL; 4-METHYLPENTANOL-2; 4-METHYL-2-PENTANOL; METILAMIL ALCOHOL (Italian); METILISOBUTILCARBINOL (Spanish); 4-METILPENTAN-2-OLO (Italian); MIBC; MIC; 3-MIC; 2-PENTANOL, 4-METHYL-

IDENTIFICATION

CAS: 108-11-2

746 Methy Isobutyl Carbinol

ERG ID:	UN 2053
ERG Guide:	129
ERG Guide Category:	FLAMMABLE LIQUIDS (POLAR/WATER-MISCIBLE/ NOXIOUS)
Molecular Formula:	$C_6H_{14}O$
RTECS No.:	SA7350000
Properties:	Flammable, colorless liquid. Sharp, irritating odor. Floats on water; slightly soluble.

HEALTH AND SAFETY INFORMATION

Irritatates eyes, skin, and respiratory tract.

IDLH: 400 ppm
Odor Threshold: 0.043–4.25 mg/m^3
OSHA Table Z-1-A Air Contaminant

Exposure Limits:	**TWA**	**STEL**
ACGIH:[(1)]	25 ppm; 104 mg/m^3	40 ppm; 167 mg/m^3
NIOSH:[(1)]	25 ppm; 100 mg/m^3	40 ppm; 165 mg/m^3
OSHA:[(1)]	25 ppm; 100 mg/m^3	

[(1)]Skin contact contributes significantly to overall exposure.

RECOMMENDATIONS FOR RESPIRATOR SELECTION: NIOSH

250 ppm: SA (any supplied-air respirator).
400 ppm: SA:CF (any supplied-air respirator operated in a continuous-flow mode);or SCBAF (any self-contained breathing apparatus with a full facepiece);or SAF (any supplied-air respirator with a full facepiece).

Emergency or Planned Entry into Unknown Concentrations or IDLH Conditions:
SCBAF:PD,PP (any self-contained breathing apparatus that has a full facepiece and is operated in a pressure-demand or other positive-pressure mode);or SAF:PD,PP:ASCBA (any supplied-air respirator that has a full facepiece and is operated in a pressure-demand or other positive-pressure mode in combination with an auxiliary, self-contained breathing apparatus operated in a pressure-demand or other positive-pressure mode).

Escape: GMFOV [any air-purifying, full-facepiece respirator (gas mask) with a chin-style, front- or back-mounted organic vapor canister];or SCBAE (any appropriate escape-type, self-contained breathing apparatus).
Note: Substance causes eye irritation or damage; eye protection needed.

CHEMICAL REACTIVITY

Forms explosive mixture with air. Contact with alkali metals forms explosive hydrogen gas. Incompatible with strong acids, caustics, aliphatic amines, and isocyanates. Attacks some plastics, rubber, and coatings. May accumulate static electrical charges and may cause ignition of its vapors.

FIRE INFORMATION

Vapors may travel long distances to ignition sources and flash back. Vapors in confined areas or containers may explode in fire.

NFPA Hazard Classifications:

Vapors are heavier than air and will collect in low areas. Vapors in confined areas may explode when exposed to fire. Vapors may travel long distances to ignition sources and flash back. Storage containers and parts of containers may rocket great distances, in many directions. If material or contaminated runoff enters waterways, notify downstream users of potentially contaminated waters. Notify local health and fire officials and pollution control agencies. Use dry chemicals, foam, carbon dioxide. From a secure, explosion-proof location, use water spray to cool exposed containers.

Health Hazards (Blue) 2
Flammability (Red) 2
Reactivity (Yellow) 0

Flash Point: 106°F (41°C)
Flammable Limits in Air: LE: 1.0%; UEL 5.5%

FIRST AID

Move victim to fresh air. Call emergency medical care. Apply artificial respiration if victim is not breathing. Administer oxygen if breathing is difficult. Remove and isolate contaminated clothing and shoes. In case of contact with substance, immediately flush skin or eyes with running water for at least 20 minutes. Wash skin with soap and water. Keep victim warm and quiet. Effects of exposure (inhalation, ingestion, or skin contact) to substance may be delayed. Ensure that medical personnel are aware of the material(s) involved and take precautions to protect themselves.

METHYL ISOBUTYL KETONE M:420

SYNONYMS: EEC No. 606-004-00-4; EINECS No. 203-550-1; FEMA No. 2731; HEXON (Czech); HEXONE; ISOBUTYL METHYL KETONE; ISOPROPYLACETONE; KTI COP RINSE I; KTI PMMA RINSE; METHYL-ISOBUTYL-CETONE (French); 4-METHYL-2-PENTANON (Czech); 2-METHYL-4-PENTANONE; 4-METHYL-2-PENTANONE; MIBK; MIK; 2-PENTANONE, 4-METHYL-; METIL ISOBUTIL CETONA (Spanish); RN-10 E BEAM NEGATIVE RESIST RINSE; SHELL MIBK

IDENTIFICATION

CAS: 108-10-1

748 Methy Isobutyl Ketone

ERG ID:	UN 1245
ERG Guide:	127
ERG Guide Category:	FLAMMABLE LIQUIDS (POLAR/WATER MISCIBLE)
Molecular Formula:	$C_6H_{12}O$
RTECS No.:	SA9275000
Properties:	Highly flammable, watery liquid. Colorless. Mild, pleasant odor. Floats and mixes slowly with water. Flammable, irritating vapor is produced.

HEALTH AND SAFETY INFORMATION

Extremely irritating to eyes, skin, and respiratory tract. Poisonous by various routes. Prolonged or repeated contact may cause irritation of the skin, dermatitis with flaking and blisters, liver, and kidney damage.

IDLH: 500 ppm
Odor Threshold: 0.121 ppm
OSHA Table Z-1-A Air Contaminant as hexone
NIOSH Criteria Document: NIOSH 78-173 KETONES

Exposure Limits:	**TWA**	**STEL**
ACGIH:	50 ppm; 205 mg/m^3	75 ppm; 307 mg/m^3
NIOSH:	50 ppm; 205 mg/m^3	75 ppm; 300 mg/m^3
OSHA:	100 ppm: 410 mg/m^3	

RECOMMENDATIONS FOR RESPIRATOR SELECTION: NIOSH/OSHA

500 ppm: CCROV [any chemical cartridge respirator with organic vapor cartridge(s)];or GMFOV [any air-purifying, full-facepiece respirator (gas mask) with a chin-style, front- or back-mounted organic vapor canister];or PAPROV [any powered, air-purifying respirator with organic vapor cartridge(s)];or SA (any supplied-air respirator);or SCBAF (any self-contained breathing apparatus with a full facepiece)

Emergency or Planned Entry into Unknown Concentrations or IDLH Conditions:
SCBAF:PD,PP (any self-contained breathing apparatus that has a full facepiece and is operated in a pressure-demand or other positive-pressure mode);or SAF:PD,PP:ASCBA (any supplied-air respirator that has a full facepiece and is operated in a pressure-demand or other positive-pressure mode in combination with an auxiliary, self-contained breathing apparatus operated in a pressure-demand or other positive-pressure mode)

Escape: GMFOV [any air-purifying, full-facepiece respirator (gas mask) with a chin-style, front- or back-mounted organic vapor canister];or SCBAE (any appropriate escape-type, self-contained breathing apparatus).

Note: Substance reported to cause eye irritation or damage; may require eye protection.

CHEMICAL REACTIVITY

Forms explosive mixture with air. Incompatible with strong oxidizers, strong acids, and aliphatic amines. Dissolves some plastics, resins, and rubber.

FIRE INFORMATION

Vapors are heavier than air and will collect in low areas. Vapors in confined areas may explode when exposed to fire. Vapors may travel long distances to ignition sources and flash back. Storage containers and parts of containers may rocket great distances, in many directions. If material or contaminated runoff enters waterways, notify downstream users of potentially contaminated waters. Notify local health and fire officials and pollution control agencies. Use dry chemicals, foam, carbon dioxide. From a secure, explosion-proof location, use water spray to cool exposed containers.

NFPA Hazard Classifications:

Health Hazards (Blue) 2
Flammability (Red) 3
Reactivity (Yellow) 1

Flash Point: 75°F (24°C) o.c.; 64°F (18°C) c.c.
Flammable Limits in Air: LEL 1.2%; UEL 8.0% both @ 200°F (93°C)
Autoignition Temperature: 846°F (452°C)

FIRST AID

Move victim to fresh air. Call emergency medical care. Apply artificial respiration if victim is not breathing. If breathing is difficult, give oxygen. Remove and isolate contaminated clothing and shoes. In case of contact with material, immediately flush skin or eyes with running water for at least 20 minutes. Wash skin with soap and water. Keep victim warm and quiet. Ensure that medical personnel are aware of the material(s) involved and take precautions to protect themselves.

ENVIRONMENTAL INFORMATION

Clean Air Act: Hazardous Air Pollutants (Title I, Part A, Section 112).
EPA Hazardous Waste Number (RCRA No.): U161.
RCRA Section 261 Hazardous Constituents
RCRA Land Ban Waste
RCRA Universal Treatment Standards: Wastewater (mg/L), 0.14; Nonwastewater (mg/kg), 33.
Safe Drinking Water Act: Priority List (55 FR 1470)
EPCRA Section 304: Reportable Quantity (RQ): CERCLA, 5,000 lb (2,270 kg).
EPCRA Section 313: Form R *de minimis* concentration reporting level: 1.0%.
TSCA: 40CFR716.120(*a*) 40CFR799.5000 test consenting orders.

METHYL ISOPROPYL KETONE M:440

SYNONYMS: 2-ACETYL PROPANE; DP-13 E BEAM POSITIVE RESIST DEVELOPER; DP-18 E BEAM POSITIVE RESIST DEVELOPER; ISOPROPYL METHYL KETONE; 3-METHYL-2-BUTANONE; 3-METHYL BUTAN-2-ONE; METIL ISOPROPIL CETONA (Spanish); MIPK

IDENTIFICATION

CAS:	563-80-4
ERG ID:	UN 2397
ERG Guide:	127
ERG Guide Category:	FLAMMABLE LIQUIDS (POLAR/WATER MISCIBLE)
Molecular Formula:	$C_5H_{10}O$
RTECS No.:	EL9100000
Properties:	Combustible liquid. Colorless. Acetonelike odor. Floats on water.

HEALTH AND SAFETY INFORMATION

Prolonged or repeated contact may cause irritation of the skin and dermatitis with dryness and cracking; nervous system changes.

Odor Threshold: 4.47 ppm
OSHA Table Z-1-A Air Contaminant
NIOSH Criteria Document: NIOSH-78-173 KETONES

Exposure Limits:	**TWA**
ACGIH:	200 ppm; 705 mg/m^3
NIOSH:	200 ppm; 705 mg/m^3
OSHA:	None

RECOMMENDATIONS FOR RESPIRATOR SELECTION: NIOSH

At any concentrations above the NIOSH REL, or where there is no REL, at any detectable concentration: SCBAF:PD,PP (any self-contained breathing apparatus that has a full facepiece and is operated in a pressure-demand or other positive-pressure mode);or SAF:PD,PP:ASCBA (any supplied-air respirator that has a full facepiece and is operated in a pressure-demand or other positive-pressure mode in combination with an auxiliary, self-contained breathing apparatus operated in a pressure-demand or other positive pressure mode).

Escape: GMFOV (any air-purifying, full-facepiece respirator (gas mask) with a chin-style, front- or back-mounted organic vapor canister;or SCBAE (any appropriate escape-type, self-contained breathing apparatus).

CHEMICAL REACTIVITY

Forms explosive mixture with air. Violent reaction with strong oxidizers. Ketones can react violently with aldehydes (acetaldehyde), bromine, nitric acid, perchloric acid. Forms explosive mixture with 4,6-dinitrobenzofurazan-N-oxide.

FIRE INFORMATION

FEMA Hazard Classifications:

Health Hazards (Blue) 1
Flammability (Red) 3
Reactivity (Yellow) 0

Flash Point: 43°F (6°C) c.c.

FIRST AID

Move victim to fresh air. Call emergency medical care. Apply artificial respiration if victim is not breathing. If breathing is difficult, give oxygen. Remove and isolate contaminated clothing and shoes. In case of contact with material, immediately flush skin or eyes with running water for at least 20 minutes. Wash skin with soap and water. Keep victim warm and quiet. Ensure that medical personnel are aware of the material(s) involved and take precautions to protect themselves.

METHYL ISOPROPROPENYL KETONE, INHIBITED M:445

SYNONYMS: 3-BUTEN-2-ONE-3-METHYL-; 3-METHYL-3-BUTEN-2-0N (German);2-METHYL-1-BUTENE-3-ONE; 2-METHYL-3-BUTENE-2-ONE; 2-METHYL-1-BUTANONE; METIL ISOPROPENIL CETONA (Spanish); ISOPROPENYL METHYL KETONE

IDENTIFICATION

CAS:	814-78-8
ERG ID:	UN 1246
ERG Guide:	127P
ERG Guide Category:	FLAMMABLE LIQUIDS (POLAR/WATER- MISCIBLE)
Molecular Formula:	C_5H_8O
RTECS No.:	EN0175000
Properties:	Highly flammable liquid. Clear, colorless. Sweet, pleasant odor. Floats on water; insoluble. Flammable, irritating vapor is produced.

752 Methy Isoproprenyl Ketone, Inhibited

HEALTH AND SAFETY INFORMATION

Extremely irritating to eyes, skin, and respiratory tract.

NIOSH Criteria Document: NIOSH 78-173 KETONES

RECOMMENDATIONS FOR RESPIRATOR SELECTION

At any concentrations above the NIOSH REL, or where there is no REL, at any detectable concentration: SCBAF:PD,PP (any self-contained breathing apparatus that has a full facepiece and is operated in a pressure-demand or other positive-pressure mode);or SAF:PD,PP:ASCBA (any supplied-air respirator that has a full facepiece and is operated in a pressure-demand or other positive-pressure mode in combination with an auxiliary, self-contained breathing apparatus operated in a pressure-demand or other positive-pressure mode).

Escape: GMFOV [any air-purifying, full-facepiece respirator (gas mask) with a chin-style, front- or back-mounted organic vapor canister];or SCBAE (any appropriate escape-type, self-contained breathing apparatus)

CHEMICAL REACTIVITY

Forms explosive mixture with air. Contact with strong oxidizers may cause fire and explosions. Heat and/or inappropriate level of inhibitor may cause polymerization. May react with itself blocking relief valves and leading to tank explosions. Ketones can react violently with aldehydes (acetaldehyde), bromine, nitric acid, perchloric acid; form explosive mixture with 4,6-dinitrobenzofurazan-N-oxide.

FIRE INFORMATION

Vapors are heavier than air and will collect in low areas. Vapors in confined areas may explode when exposed to fire. Vapors may travel long distances to ignition sources and flash back. Storage containers and parts of containers may rocket great distances, in many directions. If material or contaminated runoff enters waterways, notify downstream users of potentially contaminated waters. Notify local health and fire officials and pollution control agencies. Use dry chemicals, foam, carbon dioxide.

NFPA Hazard Classifications:

Health Hazards (Blue) 2
Flammability (Red) . . . . -
Reactivity (Yellow) 0

Flash Point: 73°F (23°C) c.c.
Flammable Limits in Air: LEL 1.8%; UEL 9.0%

FIRST AID

Move victim to fresh air. Call emergency medical care. Apply artificial respiration if victim is not

breathing. Administer oxygen if breathing is difficult. Remove and isolate contaminated clothing and shoes. In case of contact with substance, immediately flush skin or eyes with running water for at least 20 minutes. Wash skin with soap and water. Keep victim warm and quiet. Ensure that medical personnel are aware of the material(s) involved and take precautions to protect themselves.

METHYL METHACRYLATE, INHIBITED M:450

SYNONYMS: ACRYLIC ACID, 2-METHYL-, METHYL ESTER; DIAKON; EINECS No. 201-297-1; METACRILATO de METILO (Spanish); METAKRYLAN METYLU (Polish); METHACRYLATE de METHYLE (French); METHACRYLIC ACID MET; METHACRYLIC ACID, METHYL ESTER; METHACRYLSAEUREMETHYL ESTER (German); METHYL ESTER of METHACRYLIC ACID; METHYLMETHACRYLAAT (Dutch); METHYL-METHACRYLAT (German); METHYL METHACRYLATE MONOMER; METHYL-α-METHYLACRYLATE; METHYL-2-METHYLPROPENOATE; METHYL-2-METHYL-2-PROPENOATE; 2-METHYL PROPENOIC ACID, METHYL ESTER; METIL METACRILATO (Italian); MMA; MONOCITE METHACRYLATE MONOMER; MER; NCI-C50680; PEGALAN; 2-PROPENOIC ACID, 2-METHYL-, METHYL ESTER

IDENTIFICATION

CAS:	80-62-6
ERG ID:	UN 1247
ERG Guide:	129P (inhibited)
ERG Guide Category:	FLAMMABLE LIQUIDS (POLAR/WATER-MISCIBLE/ NOXIOUS)
Molecular Formula:	$C_5H_8O_2$
RTECS No.:	OZ5075000
Properties:	Highly flammable, colorless liquid. Acrid, fruity odor. Floats on and is slightly soluble in water.

HEALTH AND SAFETY INFORMATION

Very irritating to eyes, skin, and respiratory tract.

IARC Evaluation: Group 3, unclassifiable substance; no human data; inadequate animal evidence.
IDLH: 1,000 ppm
Odor Threshold: 0.085 ppm
OSHA Table Z-1-A Air Contaminant

Exposure Limits:	**TWA**
ACGIH:	100 ppm; 410 mg/m^3
NIOSH:	100 ppm; 410 mg/m^3
OSHA:	100 ppm; 410 mg/m^3

RECOMMENDATIONS FOR RESPIRATOR SELECTION: NIOSH

1,000 ppm: SA:CF (any supplied-air respirator operated in a continuous-flow mode);or CCRFOV [any chemical cartridge respirator with a full facepiece and organic vapor cartridge (s)];or GMFOV [any air-purifying, full-facepiece respirator (gas mask) with a chin-style, front- or back-mounted acid gas canister];or PAPROV [any powered, air-purifying respirator with organic vapor cartridge(s)];or SCBAF (any self-contained breathing apparatus with a full facepiece);or SAF (any supplied-air respirator with a full facepiece).

Emergency or Planned Entry into Unknown Concentrations or IDLH Conditions: SCBAF:PD,PP (any self-contained breathing apparatus that has a full facepiece and is operated in a pressure-demand or other positive-pressure mode);or SAF:PD,PP:ASCBA (any supplied-air respirator that has a full facepiece and is operated in a pressure-demand or other positive-pressure mode in combination with an auxiliary, self-contained breathing apparatus operated in a pressure-demand or other positive-pressure mode).

Escape: GMFOV [any air-purifying, full-facepiece respirator (gas mask) with a chin-style, front- or back-mounted organic vapor canister];or SCBAE (any appropriate escape-type, self-contained breathing apparatus).

Note: Substance causes eye irritation or damage; eye protection needed.

CHEMICAL REACTIVITY

Forms explosive mixture with air. Incompatible with caustics, nitrates, strong acids, aliphatic amines, alkanolamines, peroxides, and strong oxidizers. Heat and/or lack of appropriate level of inhibitor can cause polymerization. May accumulate static electrical charges and may cause ignition of its vapors.

FIRE INFORMATION

Vapors are heavier than air and will collect in low areas. Vapors in confined areas may explode when exposed to fire. Vapors may travel long distances to ignition sources and flash back. Storage containers and parts of containers may rocket great distances, in many directions. If material or contaminated runoff enters waterways, notify downstream users of potentially contaminated waters. Notify local health and fire officials and pollution control agencies. Use dry chemicals, foam, carbon dioxide. From a secure, explosion-proof location, use water spray to cool exposed containers.

NFPA Hazard Classifications:

Health Hazards (Blue) 2
Flammability (Red) 3
Reactivity (Yellow) 2

Flash Point: 50°F (10°C) o.c.
Flammable Limits in Air: LEL 1.7%; UEL 8.2%

FIRST AID

Move victim to fresh air. Call emergency medical care. Apply artificial respiration if victim is not breathing. Administer oxygen if breathing is difficult. Remove and isolate contaminated clothing and shoes. In case of contact with substance, immediately flush skin or eyes with running water for at least 20 minutes. Wash skin with soap and water. Keep victim warm and quiet. Effects of exposure (inhalation, ingestion, or skin contact) to substance may be delayed. Ensure that medical personnel are aware of the material(s) involved and take precautions to protect themselves.

ENVIRONMENTAL INFORMATION

Clean Air Act: Hazardous Air Pollutants (Title I, Part A, Section 112).
Clean Water Act: Section 311 Hazardous Substances/RQ (same as CERCLA); Section 313 Priority Chemicals.
EPA Hazardous Waste Number (RCRA No.): U162.
RCRA Section 261 Hazardous Constituents
RCRA Land Ban Waste
RCRA Universal Treatment Standards: Wastewater (mg/L), 0.14; Nonwastewater (mg/kg), 160.
RCRA Ground Water Monitoring List: Suggested test method(s) (PQL ug/L): 8015(2); 8240(5).
EPCRA Section 304: Reportable Quantity (RQ): CERCLA, 1,000 lb (454 kg)
EPCRA Section 313: Form R *de minimis* concentration reporting level: 1.0%.
Marine Pollutant (49CFR, Subchapter 172.101, Appendix B).
TSCA: 40CFR716.120(*a*)

n–METHYL MORPHOLINE M:452

SYNONYMS: METHYLMORPHOLINE; 4-METHYLMORPHOLINE; MORPHOLINE, n-METHYL-

IDENTIFICATION

CAS:	109-02-4
ERG ID:	UN 2535
ERG Guide:	132
ERG Guide Category:	FLAMMABLE LIQUIDS-CORROSIVE
Molecular Formula:	$C_5H_{11}NO$
RTECS No.:	QE5775000
Properties:	Flammable liquid. Clear. Ammonia odor. Soluble in water.

HEALTH AND SAFETY INFORMATION

Moderately toxic by skin contact. Irritates eyes, skin, and respiratory tract.

RECOMMENDATIONS FOR RESPIRATOR SELECTION

At any concentrations above the NIOSH REL, or where there is no REL, at any detectable concentration: SCBAF:PD,PP (any self-contained breathing apparatus that has a full facepiece and is operated in a pressure-demand or other positive-pressure mode);or SAF:PD,PP:ASCBA (any supplied-air respirator that has a full facepiece and is operated in a pressure-demand or other positive-pressure mode in combination with an auxiliary, self-contained breathing apparatus operated in a pressure-demand or other positive-pressure mode).

Escape: GMFOV [any air-purifying, full-facepiece respirator (gas mask) with a chin-style, front- or back-mounted organic vapor canister];or SCBAE (any appropriate escape-type, self-contained breathing apparatus).

CHEMICAL REACTIVITY

Forms explosive mixture with air. Contact with oxidizers can cause fire and explosions.

FIRE INFORMATION

Combustion products include toxic nitrogen oxides. Vapors are heavier than air and will collect in low areas. Vapors in confined areas may explode when exposed to fire. Vapors may travel long distances to ignition sources and flash back. Storage containers and parts of containers may rocket great distances, in many directions. If material or contaminated runoff enters waterways, notify downstream users of potentially contaminated waters. Notify local health and fire officials and pollution control agencies. Use dry chemicals, foam, carbon dioxide.

NFPA Hazard Classifications

Health Hazards (Blue) 2
Flammability (Red) 3
Reactivity (Yellow) 0

Flash Point: 75°F (24°C) c.c.

FIRST AID

Move victim to fresh air; call emergency medical care. Apply artificial respiration if victim is not breathing. Do not use mouth-to-mouth method if victim ingested or inhaled the substance; induce artificial respiration with the aid of a pocket mask equipped with a one-way valve or other proper respiratory medical device. Administer oxygen if breathing is difficult. Remove and isolate contaminated clothing and shoes. In case of contact with substance, immediately flush skin or eyes with running water for at least 20 minutes. For minor skin contact, avoid spreading material on unaffected skin. Keep victim warm and quiet. Effects of exposure (inhalation, ingestion, or skin contact) to substance may be delayed. Ensure that medical personnel are aware of the material(s) involved and take precautions to protect themselves.

2-METHYL-1,3-PENTADIENE M:454

SYNONYMS: 2- METIL-1,3-PENTADIENO (Spanish); METHYLPENTADIENE; 1,3-PENTADIENE, 2-METHYL

IDENTIFICATION

CAS:	1118-58-7; 54363-49-4; 39382-31-5 (methylpentadienes)
ERG ID:	UN 2461
ERG Guide:	127
ERG Guide Category:	FLAMMABLE LIQUIDS (POLAR/WATER MISCIBLE)
Molecular Formula:	C_6H_{10}
RTECS No.:	RZ2473000
Properties:	Highly flammable liquid. Soluble in water; may be slight.

HEALTH AND SAFETY INFORMATION

Toxic. Irritates eyes, skin, and respiratory tract. Vapors may affect the central nervous system; headache, dizziness; high concentrations may have a narcotic effect.

RECOMMENDATIONS FOR RESPIRATOR SELECTION

At any concentrations above the NIOSH REL, or where there is no REL, at any detectable concentration: SCBAF:PD,PP (any self-contained breathing apparatus that has a full facepiece and is operated in a pressure-demand or other positive-pressure mode);or SAF:PD,PP:ASCBA (any supplied-air respirator that has a full facepiece and is operated in a pressure-demand or other positive-pressure mode in combination with an auxiliary, self-contained breathing apparatus operated in a pressure-demand or other positive-pressure mode).

Escape: GMFOV [any air-purifying, full-facepiece respirator (gas mask) with a chin-style, front- or back-mounted organic vapor canister];or SCBAE (any appropriate escape-type, self-contained breathing apparatus).

CHEMICAL REACTIVITY

Contact with oxidizers may cause fire or explosions. May polymerize when heated or exposed to fire.

FIRE INFORMATION

Vapors are heavier than air and will collect in low areas. Vapors in confined areas may explode when exposed to fire. Vapors may travel long distances to ignition sources and flash back. Storage containers and parts of containers may rocket great distances, in many directions. If material or contaminated runoff enters waterways, notify downstream users of potentially contaminated waters. Notify local health and fire officials and pollution control agencies. Use dry chemicals, foam, carbon dioxide. Use water spray to cool exposed containers.

NFPA Hazard Classifications

Health Hazards (Blue) 0
Flammability (Red) 3
Reactivity (Yellow) 0

Flash Point: < -4°F (< 120°C)

FIRST AID

Move victim to fresh air. Call emergency medical care. Apply artificial respiration if victim is not breathing. Administer oxygen if breathing is difficult. Remove and isolate contaminated clothing and shoes. In case of contact with substance, immediately flush skin or eyes with running water for at least 20 minutes. Wash skin with soap and water. Keep victim warm and quiet. Effects of exposure (inhalation, ingestion, or skin contact) to substance may be delayed. Ensure that medical personnel are aware of the material(s) involved and take precautions to protect themselves.

4-METHYL-1,3-PENTADIENE M:456

SYNONYMS: 4- METIL-1,3-PENTADIENO (Spanish); METHYLPENTADIENE; 1,3-PENTADIENE, 4-METHYL

IDENTIFICATION

CAS:	926-56-7; 54363-49-4; 39382-31-5 (methylpentadienes)
ERG ID:	UN 2461
ERG Guide:	127
ERG Guide Category:	FLAMMABLE LIQUIDS (POLAR/WATER MISCIBLE)
Molecular Formula:	C_6H_{10}
RTECS No.:	RZ2473000
Properties:	Highly flammable liquid. Insoluble in water.

HEALTH AND SAFETY INFORMATION

Irritates eyes, skin, and respiratory tract. Vapors may affect the central nervous system; headache, dizziness.

RECOMMENDATIONS FOR RESPIRATOR SELECTION

At any concentrations above the NIOSH REL, or where there is no REL, at any detectable concentration: SCBAF:PD,PP (any self-contained breathing apparatus that has a full facepiece and is operated in a pressure-demand or other positive-pressure mode);or SAF:PD,PP:ASCBA (any supplied-air

respirator that has a full facepiece and is operated in a pressure-demand or other positive-pressure mode in combination with an auxiliary, self-contained breathing apparatus operated in a pressure-demand or other positive-pressure mode).

Escape: GMFOV [any air-purifying, full-facepiece respirator (gas mask) with a chin-style, front- or back-mounted organic vapor canister];or SCBAE (any appropriate escape-type, self-contained breathing apparatus).

CHEMICAL REACTIVITY

Contact with oxidizers may cause fire or explosions. May polymerize when heated or exposed to fire.

FIRE INFORMATION

Vapors are heavier than air and will collect in low areas.Vapors in confined areas may explode when exposed to fire. Vapors may travel long distances to ignition sources and flash back.Storage containers and parts of containers may rocket great distances, in many directions.If material or contaminated runoff enters waterways, notify downstream users of potentially contaminated waters. Notify local health and fire officials and pollution control agencies. Use dry chemicals, foam, carbon dioxide. Use water spray to cool exposed containers.

NFPA Hazard Classifications

Health Hazards (Blue) 0
Flammability (Red) 3
Reactivity (Yellow) 1

Flash Point: -30°F (-34°C)

FIRST AID

Move victim to fresh air. Call emergency medical care. Apply artificial respiration if victim is not breathing. Administer oxygen if breathing is difficult. Remove and isolate contaminated clothing and shoes. In case of contact with substance, immediately flush skin or eyes with running water for at least 20 minutes. Wash skin with soap and water. Keep victim warm and quiet. Effects of exposure (inhalation, ingestion, or skin contact) to substance may be delayed. Ensure that medical personnel are aware of the material(s) involved and take precautions to protect themselves.

4–METHYL–1–PENTENE M:458

SYNONYMS:4- METHYLPENTENE; METHYL-1-PENTENE(4-); 4-METIL-1-PENTENO (Spanish)

IDENTIFICATION

CAS: 691-37-2

760 4-Methyl-1-Pentene

ERG ID: UN 1993
ERG Guide: 128
ERG Guide Category: FLAMMABLE LIQUIDS (NONPOLAR/WATER-IMMISCIBLE)
Molecular Formula: C_6H_{12}
Properties: Flammable liquid. Clear.

HEALTH AND SAFETY INFORMATION

Irritates eyes and respiratory tract. May have a narcotic effect in high concentrations.

RECOMMENDATIONS FOR RESPIRATOR SELECTION

At any concentrations above the NIOSH REL, or where there is no REL, at any detectable concentration: SCBAF:PD,PP (any self-contained breathing apparatus that has a full facepiece and is operated in a pressure-demand or other positive-pressure mode);or SAF:PD,PP:ASCBA (any supplied-air respirator that has a full facepiece and is operated in a pressure-demand or other positive-pressure mode in combination with an auxiliary, self-contained breathing apparatus operated in a pressure-demand or other positive-pressure mode).

Escape: GMFOV [any air-purifying, full-facepiece respirator (gas mask) with a chin-style, front- or back-mounted organic vapor canister];or SCBAE (any appropriate escape-type, self-contained breathing apparatus).

CHEMICAL REACTIVITY

Forms explosive mixture with air. Contact with oxidizers may cause fire and explosions. May attack plastics and coatings.

FIRE INFORMATION

Vapors are heavier than air and will collect in low areas. Vapors in confined areas may explode when exposed to fire. Vapors may travel long distances to ignition sources and flash back. Storage containers and parts of containers may rocket great distances, in many directions. If material or contaminated runoff enters waterways, notify downstream users of potentially contaminated waters. Notify local health and fire officials and pollution control agencies. Use dry chemicals, foam, carbon dioxide. Use water spray to cool exposed containers.

NFPA Hazard Classifications:

Health Hazards (Blue) 1
Flammability (Red) 3
Reactivity (Yellow) 0

Flash Point: <20°F (<-7°C)
Autoignition Temperature: 572°F (300°C)

FIRST AID

Move victim to fresh air. Call emergency medical care. Apply artificial respiration if victim is not breathing. If breathing is difficult, give oxygen. Remove and isolate contaminated clothing and shoes. In case of contact with substance, immediately flush skin or eyes with running water for at least 20 minutes. Wash skin with soap and water. Keep victim warm and quiet. Ensure that medical personnel are aware of the material(s) involved and take precautions to protect themselves.

METHYL PROPIONATE M:467

SYNONYMS: EEC No. 607-027-2; EINECS No. 209-060-4; METHYL PROPANOATE; METHYL PROPYLATE; PROPANOIC ACID, METHYL ESTER; PROPIONATE de METHYLE (French); PROPIONATO de METILO (Spanish)

IDENTIFICATION

CAS:	554-12-1
ERG ID:	UN 1248
ERG Guide:	129
ERG Guide Category:	FLAMMABLE LIQUIDS (POLAR/WATER-MISCIBLE/NOXIOUS)
Molecular Formula:	$C_4H_8O_2$
RTECS No.:	UF5970000
Properties:	Extremely flammable liquid. Colorless. Perceptible, characteristic odor. Mixes with water.

HEALTH AND SAFETY INFORMATION

Toxic. Irritates eyes, skin, and respiratory tract.

Odor Threshold: 10-12 mg/m^3 (recognition)

RECOMMENDATIONS FOR RESPIRATOR SELECTION

At any concentrations above the NIOSH REL, or where there is no REL, at any detectable concentration: SCBAF:PD,PP (any self-contained breathing apparatus that has a full facepiece and is operated in a pressure-demand or other positive-pressure mode);or SAF:PD,PP:ASCBA (any supplied-air respirator that has a full facepiece and is operated in a pressure-demand or other positive-pressure mode in combination with an auxiliary, self-contained breathing apparatus operated in a pressure-demand or other positive-pressure mode).

Escape: GMFOV [any air-purifying, full-facepiece respirator (gas mask) with a chin-style, front- or back-mounted organic vapor canister];or SCBAE (any appropriate escape-type, self-contained breathing apparatus).

CHEMICAL REACTIVITY

Contact with oxidizers may cause fire and explosions. May accumulate static electricity, and cause ignition of its vapors.

FIRE INFORMATION

Vapors are heavier than air and will collect in low areas. Vapors in confined areas may explode when exposed to fire. Vapors may travel long distances to ignition sources and flash back. Storage containers and parts of containers may rocket great distances, in many directions. If material or contaminated runoff enters waterways, notify downstream users of potentially contaminated waters. Notify local health and fire officials and pollution control agencies. Use dry chemicals, foam, carbon dioxide.

NFPA Hazard Classifications:

Health Hazards (Blue) 1
Flammability (Red) 3
Reactivity (Yellow) 0

Flash Point: 28°F (-2°C) o.c.
Flammable Limits in Air: LEL 2.5%; UEL 13%
Autoignition Temperature: 876°F (469°C)

FIRST AID

Move victim to fresh air. Call emergency medical care. Apply artificial respiration if victim is not breathing. Administer oxygen if breathing is difficult. Remove and isolate contaminated clothing and shoes. In case of contact with substance, immediately flush skin or eyes with running water for at least 20 minutes. Wash skin with soap and water. Keep victim warm and quiet. Effects of exposure (inhalation, ingestion, or skin contact) to substance may be delayed. Ensure that medical personnel are aware of the material(s) involved and take precautions to protect themselves.

2–METHYLPYRIDINE M:474

SYNONYMS: EEC No. 613-036-00-2; EINECS No. 203-643-7; α-METHYLPYRIDINE; 2-METILPIRIDINA (Spanish); α-PICOLINE; 2-PICOLINE; 2-PYRIDINE,2-METHYL-

IDENTIFICATION

CAS:	109-06-8
ERG ID:	UN 2313
ERG Guide:	130
ERG Guide Category:	FLAMMABLE LIQUIDS (NONPOLAR/WATER-IMMISCIBLE/NOXIOUS)

Molecular Formula:	C_6H_7N
RTECS No.:	TJ4900000
Properties:	Flammable liquid. Colorless. Strong, unpleasant odor. Insoluble in water. Poisonous, flammable vapor is produced.

HEALTH AND SAFETY INFORMATION

Irritates eyes, skin, and respiratory tract. Skin contact contributes significantly in overall exposure. Causes dry skin.

Odor Threshold: 0.0003 ppm

RECOMMENDATIONS FOR RESPIRATOR SELECTION

At any concentrations above the NIOSH REL, or where there is no REL, at any detectable concentration: SCBAF:PD,PP (any self-contained breathing apparatus that has a full facepiece and is operated in a pressure-demand or other positive-pressure mode); or SAF:PD,PP:ASCBA (any supplied-air respirator that has a full facepiece and is operated in a pressure-demand or other positive-pressure mode in combination with an auxiliary, self-contained breathing apparatus operated in a pressure-demand or other positive-pressure mode).

Escape: GMFOV [any air-purifying, full-facepiece respirator (gas mask) with a chin-style, front- or back-mounted organic vapor canister]; or SCBAE (any appropriate escape-type, self-contained breathing apparatus).

CHEMICAL REACTIVITY

Forms explosive mixture with air. Contact with strong oxidizers may cause fire and explosions. Attacks copper and its alloys.

FIRE INFORMATION

Combustion products include toxic nitrogen oxides. Vapors are heavier than air and will collect in low areas. Vapors in confined areas may explode when exposed to fire. If material or contaminated runoff enters waterways, notify downstream users of potentially contaminated waters. Notify local health and fire officials and pollution control agencies. Use dry chemicals, alcohol-resistant foam, carbon dioxide. From a secure, explosion-proof location, use water spray to cool exposed containers.

NFPA Hazard Classifications:

Health Hazards (Blue) 2
Flammability (Red) 2
Reactivity (Yellow) 0

Flash Point: 102°F (39°C) o.c.; 79°F (26°C) c.c.

Flammable Limits in Air: LEL 1.4%; UEL 8.6%
Autoignition Temperature: 1000°F (538°C)

FIRST AID

Move victim to fresh air. Call emergency medical care. Apply artificial respiration if victim is not breathing. Administer oxygen if breathing is difficult. Remove and isolate contaminated clothing and shoes. In case of contact with substance, immediately flush skin or eyes with running water for at least 20 minutes. Wash skin with soap and water. Keep victim warm and quiet. Effects of exposure (inhalation, ingestion, or skin contact) to substance may be delayed. Ensure that medical personnel are aware of the material(s) involved and take precautions to protect themselves.

ENVIRONMENTAL INFORMATION

EPA Hazardous Waste Number (RCRA No.): U191
RCRA Section 261 Hazardous Constitutents.
EPCRA Section 304 Reportable Quantity (RQ): CERCLA, 5,000 lb (2,270 kg)
EPCRA Section 313: Form R *de minimis* concentration reporting level: 1.0%.
California's Proposition 65: Carcinogen.
Marine Pollutant (49CFR, Subchapter 172.101, Appendix B).
TSCA: 40CFR716.120(*a*)

1–METHYLPYRROLIDONE M:480

SYNONYMS: ACCUSTRIP P-300; ARC CLEANER; BURMAR LAB CLEAN; BURMAR POSISTRIP 830; BURMAR POSISTRIP LE; EEC No. 606-021-00-7; EMT-130 POSITIVE PHOTORESIST STRIPPER; EMT-500; IBM INK; 1-METHYL-2-PYRROLIDINONE; n-METHYL-2-PYRROLIDINONE; n-METHYLPYRROLIDINONE; 1-METHYLPYRROLIDINONE; n-METHYL-α-PYRROLIDONE; 1-METILPIRROLIDONA (Spanish); NMP; P-12571D; P-12572; PIQ-13; PIQ COUPLER 3; PIX-1400X-1; PIX-3500; POLY-6; POLYMIDE COATING PI 2545; POLYMIDE COATING PI 2561; POSISTRIP 830 DEVELOPER; PRS-2000 CLEANER; PRS-3000 CLEANER; PYRALIN PI-2563; PYRALIN PI 2570; m-PYROL

IDENTIFICATION

CAS:	872-50-4
ERG ID:	UN 1760
ERG Guide:	154
ERG Guide Category:	TOXIC AND/OR CORROSIVE (NON-COMBUSTIBLE)
Molecular Formula:	C_5H_9NO
RTECS No.:	UY5970000
Properties:	Flammable liquid. White. Mild fishy odor. May float or sink in water. Poisonous gases may be produced when heated.

RECOMMENDATIONS FOR RESPIRATOR SELECTION

At any concentrations above the NIOSH REL, or where there is no REL, at any detectable concentration: SCBAF:PD,PP (any self-contained breathing apparatus that has a full facepiece and is operated in a pressure-demand or other positive-pressure mode);or SAF:PD,PP:ASCBA (any supplied-air respirator that has a full facepiece and is operated in a pressure-demand or other positive-pressure mode in combination with an auxiliary, self-contained breathing apparatus operated in a pressure-demand or other positive pressure mode)

Escape: GMFOV (any air-purifying, full-facepiece respirator (gas mask) with a chin-style, front- or back-mounted organic vapor canister;or SCBAE (any appropriate escape-type, self-contained breathing apparatus).

CHEMICAL REACTIVITY

Forms explosive mixture with air. A strong base. Incompatible with strong acids, organic anhydrides, isocyanates, aldehydes, light metals, flammable and porous materials. Attacks some plastics, rubber and coatings.

FIRE INFORMATION

Vapors are heavier than air and will collect in low areas. Vapors in confined areas may explode when exposed to fire. Vapors may travel long distances to ignition sources and flash back. Storage containers and parts of containers may rocket great distances, in many directions. If material or contaminated runoff enters waterways, notify downstream users of potentially contaminated waters. Notify local health and fire officials and pollution control agencies. Use dry chemicals, foam, carbon dioxide. From a secure, explosion-proof location, use water spray to cool exposed containers.

NFPA Hazard Classifications:

Health Hazards (Blue) 2
Flammability (Red) 3
Reactivity (Yellow) 1

Flash Point: 7°F (-14°C) o.c.

FIRST AID

Move victim to fresh air. Call emergency medical care. Apply artificial respiration if victim is not breathing. Do not use mouth-to-mouth method if victim ingested or inhaled the substance; induce artificial respiration with the aid of a pocket mask equipped with a one-way valve or other proper respiratory medical device. Administer oxygen if breathing is difficult. Remove and isolate contaminated clothing and shoes. In case of contact with material, immediately flush skin or eyes with running water for at least 20 minutes. For minor skin contact, avoid spreading material on unaffected skin. Keep victim warm and quiet. Effects of exposure (inhalation, ingestion, or skin contact) may be delayed. Ensure that medical personnel are aware of the material(s) involved and take precautions to protect themselves.

ENVIRONMENTAL INFORMATION

EPCRA Section 313: Form R *de minimus* concentration reporting level: 1.0%.
TSCA: 40CFR799.5000 testing consent orders.

METHYL STEARATE M:484

SYNONYMS: EINECS No. 203-990-4; EMERY 2218; ESTEARATO de METILO (Spanish); METHOLENE 2218; METHYL ESTER STEARIC ACID; METHYL OCTADECANOATE; OCTADECANOIC ACID, METHYL ESTER; STEARATO di METILE (Italian)

IDENTIFICATION

CAS:	112-61-8
Molecular Formula:	$C_{19}H_{38}O_2$
RTECS No.:	WI4460000
Properties:	Combustible liquid or semisolid. Clear to white. Insoluble in water.

HEALTH AND SAFETY INFORMATION

May cause tumors.

RECOMMENDATIONS FOR RESPIRATOR SELECTION

At any concentrations above the NIOSH REL, or where there is no REL, at any detectable concentration: SCBAF:PD,PP (any self-contained breathing apparatus that has a full facepiece and is operated in a pressure-demand or other positive-pressure mode);or SAF:PD,PP:ASCBA (any supplied-air respirator that has a full facepiece and is operated in a pressure-demand or other positive-pressure mode in combination with an auxiliary, self-contained breathing apparatus operated in a pressure-demand or other positive-pressure mode).

Escape: GMFOV [any air-purifying, full-facepiece respirator (gas mask) with a chin-style, front- or back-mounted organic vapor canister];or SCBAE (any appropriate escape-type, self-contained breathing apparatus).

CHEMICAL REACTIVITY

Contact with strong oxidizers may cause fire and explosions. Esters are incompatible with acrolein, acrylonitrile, tert-butyl nitroacetylene, ethylene oxide, isopropyl chlorocarbonate, maleic anhydride, triisobutylaluminum; may ignite on contact with cellulose nitrate.

FIRE INFORMATION

Vapors are heavier than air and will collect in low areas. Vapors in confined areas may explode when

exposed to fire. If material or contaminated runoff enters waterways, notify downstream users of potentially contaminated waters. Notify local health and fire officials and pollution control agencies. Use dry chemicals, alcohol-resistant foam, carbon dioxide.

NFPA Hazard Classifications

Health Hazards (Blue) 0
Flammability (Red) 1
Reactivity (Yellow) 0

Flash Point: 307°F (153°C) c.c.

FIRST AID

Move victim to fresh air. Call emergency medical care. Apply artificial respiration if victim is not breathing. Administer oxygen if breathing is difficult. Remove and isolate contaminated clothing and shoes. In case of contact with substance, immediately flush skin or eyes with running water for at least 20 minutes. Wash skin with soap and water. Keep victim warm and quiet. Ensure that medical personnel are aware of the material(s) involved and take precautions to protect themselves.

METHYL–p–TOLUENESULFONATE M:487

SYNONYMS: BENZENESULFONIC ACID, 4-METHYL-, METHYL ESTER; EINECS No. 201-283-5; METHYLESTER KYSELINY p-TOLUENSULFONOVE (Czech); METHYLBENZENESULFONIC ACID, METHYL ESTER(4-); METHYL-p-METHYLBENZENESULFONATE; METHYL-4-METHYLBENZENESULFONATE; METHYL TOLUENE-4-SULFONATE; METHYL TOSYLATE; METHYL-p-TOSYLATE; METHYL TOLUENE SULFONATE; METHYL TOLUENE-4-SULFONATE; 4-TOLUENE SULFONIC ACID, METHYL ESTER; P-TOLUENE SULFONIC ACID, METHYL ESTER; p-TOLUOLSULFONSAEURE METHYL ESTER (German); p-TOLUENSULFONATO de METILO (Spanish)

IDENTIFICATION

CAS:	80-48-8
ERG ID:	UN 2811 as toxic solid, organic, n.o.s.
ERG Guide:	154
ERG Guide Category:	SUBSTANCES -TOXIC AND/OR CORROSIVE (NONCOMBUSTIBLE)
Molecular Formula:	$C_8H_{10}O_3S$
RTECS No.:	XT7000000
Properties:	Crystalline solid. White to light brown. Melting point 75°F (24°C). Insoluble in water.

HEALTH AND SAFETY INFORMATION

Irritates eyes and skin. Dermatitis or allergy may develop. May cause tumors.

RECOMMENDATIONS FOR RESPIRATOR SELECTION

At any concentrations above the NIOSH REL, or where there is no REL, at any detectable concentration: SCBAF:PD,PP (any self-contained breathing apparatus that has a full facepiece and is operated in a pressure-demand or other positive-pressure mode);or SAF:PD,PP:ASCBA (any supplied-air respirator that has a full facepiece and is operated in a pressure-demand or other positive-pressure mode in combination with an auxiliary, self-contained breathing apparatus operated in a pressure-demand or other positive-pressure mode).

Escape: GMFOV [any air-purifying, full-facepiece respirator (gas mask) with a chin-style, front- or back-mounted organic vapor canister];or SCBAE (any appropriate escape-type, self-contained breathing apparatus)

CHEMICAL REACTIVITY

Incompatible with strong oxidizers. Esters are incompatible with nitrates, nitric acid, sulfuric acid.

FIRE INFORMATION

Combustion products include toxic sulfur oxides. If material or contaminated runoff enters waterways, notify downstream users of potentially contaminated waters. Notify local health and fire officials and pollution control agencies. Use dry chemicals, alcohol-resistant foam, carbon dioxide.

NFPA Hazard Classifications

Health Hazards (Blue) 2
Flammability (Red) 1
Reactivity (Yellow) 0

Flash Point: 306°F (152°C) o.c.

FIRST AID

Move victim to fresh air; call emergency medical care. Apply artificial respiration if victim is not breathing. Do not use mouth-to-mouth method if victim ingested or inhaled the substance; induce artificial respiration with the aid of a pocket mask equipped with a one-way valve or other proper respiratory medical device. Administer oxygen if breathing is difficult. Remove and isolate contaminated clothing and shoes. In case of contact with substance, immediately flush skin or eyes with running water for at least 20 minutes. For minor skin contact, avoid spreading material on unaffected skin. Keep victim warm and quiet. Effects of exposure (inhalation, ingestion, or skin contact) to substance may be delayed. Ensure that medical personnel are aware of the material(s) involved and take precautions to protect themselves.

ENVIRONMENTAL INFORMATION

TSCA: 40CFR716.120(*d*); 40CFR712.30(*e*)14 as sulphones

METHYL TRICHLOROSILANE M:490

SYNONYMS: EEC No. 014-004-00-5; EINECS No. 200-902-6; METHYL-TRICHLORSILAN (Czech); SILANE, TRICHLOROMETHYL-; SILANE,METHYLTRICHLORO-; METHYL CHLOROSILANE; TRICHLOROMETHYLSILANE

IDENTIFICATION

CAS:	75-79-6
ERG ID:	UN 1250
ERG Guide:	155
ERG Guide Category:	SUBSTANCES-TOXIC AND/OR CORROSIVE (FLAMMABLE/WATER-SENSITIVE)
Molecular Formula:	$C_2H_3Cl_3Si$
RTECS No.:	VV4550000
Properties:	Extremely flammable liquid. Colorless, fuming. Sharp, acrid, irritating odor. Reacts violently with water to form hydrochloric acid.

HEALTH AND SAFETY INFORMATION

Poisonous. Breathing the vapor can cause death. Corrosive to eyes, skin, and respiratory tract; skin and eye contact causes sever burns and blindness. Inhalation may cause pulmonary edema, which can be delayed for several hours; there is a risk of death in serious cases.

IDLH: 100 ppm as HCl.
Odor Threshold: Decomposes in moist air, creating HCl with odor threshold of 0.255-10.06 ppm
OSHA Process Safety Management of Highly Hazardous Chemicals (29CFR, Part 1910.119, Appendix A): Threshold Quantity: 500 pounds.

Exposure Limits:	**TWA**
AIHAWEEL	1 ppm

RECOMMENDATIONS FOR RESPIRATOR SELECTION: as hydrochloric acid

50 ppm: CCRS [any chemical cartridge respirator with cartridge(s) providing protection against the compound of concern)];or GMFS [any air-purifying, full-facepiece respirator (gas mask) with a chin-style, front- or back-mounted canister providing protection against the compound of concern];or PAPRS [any powered, air-purifying respirator with cartridge(s) providing protection against the compound of concern];or SA (any supplied-air respirator);or SCBAF (any self-contained breathing apparatus with a full facepiece).

Emergency or Planned Entry into Unknown Concentrations or IDLH Conditions: SCBAF:PD,PP (any self-contained breathing apparatus that has a full facepiece and is operated in a pressure-demand or other positive-pressure mode);or SAF:PD,PP:ASCBA (any supplied-air respirator that has a full facepiece and is operated in a pressure-demand or other positive-pressure mode in combination with an auxiliary, self-contained breathing apparatus operated in a pressure-demand or other positive-pressure mode).

Escape: GMFAG [any air-purifying, full-facepiece respirator (gas mask) with a chin-style, front- or back-mounted organic vapor canister];or SCBAE (any appropriate escape-type, self-contained breathing apparatus).
Note: Substance reported to cause eye irritation or damage; may require eye protection.

CHEMICAL REACTIVITY

Forms explosive gas mixture with air. Contact with water or moist air forms hydrochloric acid (hydrogen chloride). Contact with oxidizers can cause fire and explosions. Incompatible with bases. Contact with ammonia may form a self-igniting compound. Corrodes most metals. Attacks some plastics, rubber, and coatings.

FIRE INFORMATION

Do not use water. Combustion products include toxic hydrogen chloride and phosgene. Firefighting gear (including SCBA) does not provide adequate protection. If exposure occurs, remove and isolate gear immediately and thoroughly decontaminate personnel. Vapors are heavier than air and will collect in low areas. Vapors in confined areas may explode when exposed to fire. Vapors may travel long distances to ignition sources and flash back. Storage containers and parts of containers may rocket great distances, in many directions. If material or contaminated runoff enters waterways, notify downstream users of potentially contaminated waters. Notify local health and fire officials and pollution control agencies. If material is on fire and conditions permit, do not extinguish. Use dry chemicals, low or medium expansion AFFF foam, carbon dioxide. Use large amounts of water well away from the material to disperse vapors. If possible contain runoff.

NFPA Hazard Classifications:

Health Hazards (Blue) 3
Flammability (Red) 3
Reactivity (Yellow) 2
Special Notice ..... WATER REACTIVE

Flash Point: 15°F (-9°C) c.c.
Flammable Limits in Air: LEL 7.6%; UEL more than 20%
Autoignition Temperature: > 760°F (404°C)

FIRST AID

Do not put yourself in danger by entering a contaminated area to rescue a victim. Move victim to fresh air; call emergency medical care. Apply artificial respiration if victim is not breathing. Do not use mouth-to-mouth method if victim ingested or inhaled the substance; induce artificial respiration with the aid of a pocket mask equipped with a one-way valve or other proper respiratory medical device. Administer oxygen if breathing is difficult. Remove and isolate contaminated clothing and shoes. In case of contact with substance, immediately flush skin or eyes with running water for at least 20 minutes. For minor skin contact, avoid spreading material on unaffected skin. Keep victim warm and quiet. Effects of exposure (inhalation, ingestion, or skin contact) to substance may be delayed. Ensure that medical personnel are aware of the material(s) involved and take precautions to protect themselves.

ENVIRONMENTAL INFORMATION

Clean Air Act: Accidental Release Prevention/Flammable Substances (Section 112[r], Table 3), TQ = 5,000 lb (2.270 kg)
EPCRA Section 302, Extremely Hazardous Substances: TPQ = 500 lb (227 kg)
EPCRA Section 304 Reportable Quantity (RQ): EHA, 500 lb (227 kg)
EPCRA Section 313: Form R *de minimis* concentration reporting level: 1.0%.

2–METHYLVALERIC ACID M:492

SYNONYMS: ACIDO 2-METILVALERICO (Spanish); EINECS No. 202-594-9; 2-METHYLPENTANOIC ACID; METHYLPROPYLACETIC ACID; α-METHYLVALERIC ACID; 2-PENTANECARBOXYLIC ACID; PENTANOIC ACID, 2-METHYL-

IDENTIFICATION

CAS:	97-61-0
ERG ID:	UN 2922 as corrosive liquid, toxic, n.o.s.
ERG Guide:	154
ERG Guide Category:	SUBSTANCES-TOXIC AND/OR CORROSIVE
Molecular Formula:	$C_6H_{12}O_2$
RTECS No.:	YV7700000
Properties:	Combustible liquid. Water-white. Insoluble in water.

HEALTH AND SAFETY INFORMATION

Corrosive to eyes, skin, and respiratory tract. Inhalation may cause pulmonary edema, which can be delayed for several hours; there is a risk of death in serious cases.

RECOMMENDATIONS FOR RESPIRATOR SELECTION

At any concentrations above the NIOSH REL, or where there is no REL, at any detectable concentration: SCBAF:PD,PP (any self-contained breathing apparatus that has a full facepiece and is

operated in a pressure-demand or other positive-pressure mode);or SAF:PD,PP:ASCBA (any supplied-air respirator that has a full facepiece and is operated in a pressure-demand or other positive-pressure mode in combination with an auxiliary, self-contained breathing apparatus operated in a pressure-demand or other positive-pressure mode).

Escape: GMFOV [any air-purifying, full-facepiece respirator (gas mask) with a chin-style, front- or back-mounted organic vapor canister];or SCBAE (any appropriate escape-type, self-contained breathing apparatus).

CHEMICAL REACTIVITY

Incompatible with sulfuric acid, caustics, ammonia, aliphatic amines, alkanolamines, isocyanates, alkylene oxides, epichlorohydrin., oxidizers.

FIRE INFORMATION

Vapors are heavier than air and will collect in low areas. Vapors in confined areas may explode when exposed to fire.If material or contaminated runoff enters waterways, notify downstream users of potentially contaminated waters. Notify local health and fire officials and pollution control agencies. Use dry chemicals, alcohol-resistant foam, carbon dioxide.

NFPA Hazard Classifications

Health Hazards (Blue) 0
Flammability (Red) 1
Reactivity (Yellow) 0

Flash Point: 225°F (107°C) o.c.
Autoignition Temperature: 712°F (378°C)

FIRST AID

Move victim to fresh air. Call emergency medical care. Apply artificial respiration if victim is not breathing. Administer oxygen if breathing is difficult. Remove and isolate contaminated clothing and shoes. In case of contact with substance, immediately flush skin or eyes with running water for at least 20 minutes. Wash skin with soap and water. Keep victim warm and quiet. Effects of exposure (inhalation, ingestion, or skin contact) to substance may be delayed. Ensure that medical personnel are aware of the material(s) involved and take precautions to protect themselves.

METHYL VINYL KETONE M:494

SYNONYMS: ACETONE, METHYLENE; ACETYL ETHYLENE; BUTENONE; 1-BUTEN-3-ONE; 3-BUTEN-2-ONE; 2-BUTENONE; Δ (3)-2-BUTENONE; γ-OXO-α-BUTYLENE; EINECS No. 201-160-6; KETONE, METHYL VINYL; METHYLENE ACETONE; VINYL METHYL KETONE

IDENTIFICATION

CAS:	78-94-4
ERG ID:	UN 1251
ERG Guide:	131P
ERG Guide Category:	FLAMMABLE LIQUIDS - TOXIC
Molecular Formula:	C_4H_6O
RTECS No.:	EM9800000
Properties:	Highly flammable liquid. Solid at 20°F (-7°C) Colorless to light yellow. Strong, irritating odor. Floats on water; soluble.

HEALTH AND SAFETY INFORMATION

Poisonous; breathing the vapors can cause death; can be absorbed through the skin. Corrosive to eyes, skin, and respiratory tract; can cause severe burns and blindness. Inhalation may cause pulmonary edema, which can be delayed for several hours; there is a risk of death in serious cases.

Odor Threshold: 0.5 mg/m^3

RECOMMENDATIONS FOR RESPIRATOR SELECTION

At any concentrations above the NIOSH REL, or where there is no REL, at any detectable concentration: SCBAF:PD,PP (any self-contained breathing apparatus that has a full facepiece and is operated in a pressure-demand or other positive-pressure mode); or SAF:PD,PP:ASCBA (any supplied-air respirator that has a full facepiece and is operated in a pressure-demand or other positive-pressure mode in combination with an auxiliary, self-contained breathing apparatus operated in a pressure-demand or other positive-pressure mode).

Escape: GMFOV [any air-purifying, full-facepiece respirator (gas mask) with a chin-style, front- or back-mounted organic vapor canister]; or SCBAE (any appropriate escape-type, self-contained breathing apparatus)

CHEMICAL REACTIVITY

Forms explosive mixture with air. May accumulate static electrical charges, and may cause ignition of its vapors. Heat can cause polymerization. May react with itself, blocking relief valves and leading to tank explosion. Reacts violently with strong oxidizers.

FIRE INFORMATION

Firefighting gear (including SCBA) does not provide adequate protection. If exposure occurs, remove and isolate gear immediately and thoroughly decontaminate personnel. Vapors are heavier than air and will collect in low areas. Vapors in confined areas may explode when exposed to fire. Vapors may travel long distances to ignition sources and flash back. Storage containers and parts of containers may rocket great distances, in many directions. If material or contaminated runoff enters waterways, notify downstream users of potentially contaminated waters. Notify local health and fire officials and pollution control agencies. For large fires, specially trained personnel operating from a safe distance can fight fires using alcohol resistant (AFFF) foam or dry chemical if available in sufficient amounts. Under favorable conditions, experienced crews can use coordinating fog streams to sweep the flames of the surface of the burning liquid. Keep exposures cool to protect against re-ignition. Do not direct streams into the liquid. From a secure, explosion-proof location, use water spray to cool exposed containers. If cooling streams are ineffective (venting sound increases in volume and pitch, tank discolors, or shows any signs of deforming), withdraw immediately to a secure position. For smaller fires, use dry chemicals, alcohol-resistant foam, carbon dioxide, large quantities of water.

NFPA Hazard Classifications:

Health Hazards (Blue) 4
Flammability (Red) 3
Reactivity (Yellow) 2

Flash Point: 30°F (–1°C) o.c.; 20°F (–7°C) c.c.
Flammable Limits in Air: LEL 2.1%; UEL 15.6%
Autoignition Temperature: 915°F (491°C)

FIRST AID

Move victim to fresh air. Call emergency medical care. Apply artificial respiration if victim is not breathing. Do not use mouth-to-mouth method if victim ingested or inhaled the substance; induce artificial respiration with the aid of a pocket mask equipped with a one-way valve or other proper respiratory medical device. Administer oxygen if breathing is difficult. Remove and isolate contaminated clothing and shoes. In case of contact with substance, immediately flush skin or eyes with running water for at least 20 minutes. Wash skin with soap and water. Keep victim warm and quiet. Effects of exposure (inhalation, ingestion, or skin contact) to substance may be delayed. Ensure that medical personnel are aware of the material(s) involved and take precautions to protect themselves.

ENVIRONMENTAL INFORMATION

EPCRA Section 302, Extremely Hazardous Substances: TPQ = 10 lb (4.54 kg)
EPCRA Section 304 Reportable Quantity (RQ): EHS, 10 lb (4.54 kg)

2–METHYL–5–VINYLPYRIDINE M:497

SYNONYMS: EINECS No. 205-432-5; 5-ETHENYL-2-METHYLPYRIDINE; 2-METIL-5-VINILPIRIDINA (Spanish); PYRIDINE, 5-ETHENYL-2-METHYL-; PYRIDINE, 2-METHYL-5-VINYL; 5-VINYL-2-PICOLINE

IDENTIFICATION

CAS: 140-76-1

Molecular Formula: C_8H_9N
RTECS No.: UT2975000

HEALTH AND SAFETY INFORMATION

May be poisonous by inhalation and intravenous routes. Due to the availability of insufficient data on human effects, caution should be exercised.

RECOMMENDATIONS FOR RESPIRATOR SELECTION

Use any supplied-air respirator.

CHEMICAL REACTIVITY

Contact with oxidizers may cause fire and explosions.

FIRE INFORMATION

Combustion products include toxic nitrogen oxides. Vapors are heavier than air and will collect in low areas. Vapors in confined areas may explode when exposed to fire.If material or contaminated runoff enters waterways, notify downstream users of potentially contaminated waters. Notify local health and fire officials and pollution control agencies. Use dry chemicals, alcohol-resistant foam, carbon dioxide.

NFPA Hazard Classifications: Not listed

Flash Point: 149°F (65°C)

FIRST AID

Move victim to fresh air. Call emergency medical care. Apply artificial respiration if victim is not breathing. Administer oxygen if breathing is difficult. Remove and isolate contaminated clothing and shoes. In case of contact with substance, immediately flush skin or eyes with running water for at least 20 minutes. Wash skin with soap and water. Keep victim warm and quiet. Effects of exposure (inhalation, ingestion, or skin contact) to substance may be delayed. Ensure that medical personnel are aware of the material(s) involved and take precautions to protect themselves.

ENVIRONMENTAL INFORMATION as pyridine, 2-methyl-5-vinyl

EPCRA Section 302 Extremely Hazardous Substances: TPQ = 500 lb (227kg).
EPCRA Section 304 Reportable Quantity (RQ): EHS, 1 lb (0.454 kg).

MINERAL SPIRITS M:505

SYNONYMS: NAPHTHA SOLVENT; PETROLEUM SPIRITS; SOLTROL ISOPARAFFINIC SOLVENTS; STODDARD SOLVENT; TERPENTINE SUBSTITUTE; VARNOLINE; WHITE SPIRITS

776 Mineral Spirits

IDENTIFICATION

CAS:	8052-41-3 (stoddard solvent); 8032-32-4; 64742-47-7; 64475-85-0
ERG ID:	UN1268 (petroleum distillate); UN 1300
ERG Guide:	128
ERG Guide Category:	FLAMMABLE LIQUIDS (NONPOLAR/WATER-IMMISCIBLE)
Molecular Formula:	C_9H_{20} (stoddard solvent)
RTECS No.:	OI6180000; WJ8925000
Properties:	Flammable, watery liquid. Colorless. Gasoline or kerosenelike odor. Floats on water; insoluble.

HEALTH AND SAFETY INFORMATION

Irritates eyes, skin, and respiratory tract. Affects central nervous system. Skin contact can cause defatting, irritation, and redness. May cause kidney problems.

IDLH: 5,150 ppm as stoddard solvent
Odor Threshold: 1-30 ppm as stoddard solvent.

Exposure Limits:	**TWA**	**CEILING** as stoddard solvent
ACGIH: [(1)]	100 ppm	
NIOSH:	350 mg/m^3	1800 mg/m^3 [15 min]
OSHA:	500 ppm (2,900 mg/m^3)	

[(1)] *Note*: TLV lower than PEL

RECOMMENDATIONS FOR RESPIRATOR SELECTION: NIOSH as stoddard solvent, mineral spirits

3,500 ppm: CCROV* [any chemical cartridge respirator with organic vapor cartridge(s)]; or SA* (any supplied-air respirator).
8,750 ppm: SA:CF* [any supplied-air respirator operated in a continuous-flow mode); or PAPROV* (any powered, air-purifying respirator with organic vapor cartridge(s)]
17,500 ppm: CCRFOV [any chemical cartridge respirator with a full facepiece and organic vapor cartridge(s)]; or GMFOV [any air-purifying, full-facepiece respirator (gas mask) with a chin-style, front- or back-mounted acid gas canister]; or PAPRTOV [any powered, air-purifying respirator with a tight-fitting facepiece and organic vapor cartridge(s)]; or SCBAF (any self-contained breathing apparatus with a full facepiece); or SAF (any supplied-air respirator with a full facepiece).
20,000 ppm: SAF:PD,PP:ASCBA (any supplied-air respirator that has a full facepiece and is operated in a pressure-demand or other positive-pressure mode in combination with an auxiliary, self-contained breathing apparatus operated in a pressure-demand or other positive- pressure mode).
**Note:* Substance causes eye irritation or damage; eye protection needed.

Emergency or Planned Entry into Unknown Concentrations or IDLH Conditions:
SCBAF:PD,PP (any self-contained breathing apparatus that has a full facepiece and is operated in a pressure-demand or other positive-pressure mode);or SAF:PD,PP:ASCBA (any supplied-air respirator that

has a full facepiece and is operated in a pressure-demand or other positive-pressure mode in combination with an auxiliary, self-contained breathing apparatus operated in a pressure-demand or other positive-pressure mode).

Escape: GMFOV [any air-purifying, full-facepiece respirator (gas mask) with a chin-style, front- or back-mounted organic vapor canister]; or SCBAE (any appropriate escape-type, self-contained breathing apparatus].

CHEMICAL REACTIVITY

Forms explosive mixture with air. Incompatible with nitric acid, strong oxidizers, perchlorates, nitrogen tetroxide.. Attacks some plastics, rubber, and coatings.

FIRE INFORMATION

NFPA Hazard Classifications:

Health Hazards (Blue) 0
Flammability (Red) 2
Reactivity (Yellow) 0

Flash Point: 105–140°F (41–60°C) c.c., depending on grade
Flammable Limits in Air: LEL 0.8% @ 212°F (100°C); UEL 5.0%
Autoignition Temperature: 470°F (244°C)

FIRST AID

Move victim to fresh air. Call emergency medical care. Apply artificial respiration if victim is not breathing. If breathing is difficult, give oxygen. Remove and isolate contaminated clothing and shoes. In case of contact with substance, immediately flush skin or eyes with running water for at least 20 minutes. Wash skin with soap and water. Keep victim warm and quiet. Ensure that medical personnel are aware of the material(s) involved and take precautions to protect themselves.

MORPHOLINE M:520

SYNONYMS: BURMAR POSISTRIP LE; DIETHYLENEIMIDE OXIDE; DIETHYLENE IMIDOXIDE; DIETHYLENE OXIMIDE; N,N-DIMETHYLACETAMIDE; DIETHYLENIMIDE OXIDE; EEC No. 613-028-00-9; EINECS No. 203-815-1; P-ISOXAZINE,TETRAHYDRO-; MORFOLINA (Spanish); 1-OXA-4-AZACYCLOHEXANE; 2H-1,4-OXAZINE, TETRAHYDRO-; TETRAHYDRO-1,4-ISOXAZINE; TETRAHYDRO-1,4-OXAZINE; TETRAHYDRO-2H-1,4-OXAZINE; TETRAHYDRO-P-OXAZINE

IDENTIFICATION

CAS: 110-91-8

ERG ID: UN 2054; UN 1760 (aqueous mixtures)
ERG Guide: 132; 154 (aqueous mixtures)
ERG Guide Category: 132 FLAMMABLE LIQUIDS, CORROSIVE
154 TOXIC AND/OR CORROSIVE (NONCOM- BUSTIBLE)
Molecular Formula: C_4H_9NO
RTECS No.: QD6475000
Properties: Highly flammable, oily, liquid. Colorless. Fishy, or ammonialike odor. Freezes at 23°F (-5°C). Floats and mixes with water. Irritating vapor is produced.

HEALTH AND SAFETY INFORMATION

Severely irritating to skin, eyes, nose and lungs; skin and eye contact causes severe burns and blindness. Repeated exposure may cause liver and kidney damage, chronic respiratory disease, skin and eye problems, birth defects, and cancer.

IARC Evaluation: Group 3, unclassifiable; no human data; animal evidence, inadequate.
IDLH: 1,400 ppm [10% LEL]
Odor Threshold: 0.036 mg/m^3
OSHA Table Z-1-A Air Contaminant

Exposure Limits:	TWA	STEL
ACGIH:(1)	20 ppm; 71 mg/m^3	
NIOSH:(1)	20 ppm; 70 mg/m^3	30 ppm; 105 mg/m^3
OSHA:(1)	20 ppm; 70 mg/m^3	

(1)Skin contact contributes significantly to overall exposure.

RECOMMENDATIONS FOR RESPIRATOR SELECTION: NIOSH/OSHA

550 ppm: SA:CF (any supplied-air respirator operated in a continuous-flow mode);or PAPROV [any powered, air-purifying respirator with organic vapor cartridge(s)].
1,000 ppm: CCRFOV [any chemical cartridge respirator with a full facepiece and organic vapor cartridge(s)];or GMFOV [any air-purifying, full-facepiece respirator (gas mask) with a chin-style, front- or back-mounted acid gas canister];or PAPRTOV [any powered, air-purifying respirator with a tight-fitting facepiece and organic vapor cartridge(s)];or SCBAF (any self-contained breathing apparatus with a full facepiece);or SAF (any supplied-air respirator with a full facepiece).
1,400 ppm: SAF:PD,PP (any supplied-air respirator that has a full facepiece and is operated in a pressure-demand or other positive-pressure mode).

Emergency or Planned Entry into Unknown Concentrations or IDLH Conditions:
SCBAF:PD,PP (any self-contained breathing apparatus that has a full facepiece and is operated in a pressure-demand or other positive-pressure mode);or SAF:PD,PP:ASCBA (any supplied-air respirator that has a full facepiece and is operated in a pressure-demand or other positive-pressure mode in combination with an auxiliary, self-contained breathing apparatus operated in a pressure-demand or other positive-pressure mode).

Escape: GMFOV [any air-purifying, full-facepiece respirator (gas mask) with a chin-style, front- or back-mounted organic vapor canister];or SCBAE (any appropriate escape-type, self-contained breathing apparatus).

Note: Substance causes eye irritation or damage; eye protection needed.

CHEMICAL REACTIVITY

Forms explosive mixture with air. Incompatible with acids, organic anhydrides, isocyanates, vinyl acetate, acrylates, substituted allyls, alkylene oxides, epichlorohydrin, ketones, aldehydes, alcohols, glycols, phenols, cresols, caprolactum solution, and strong oxidizers. Attacks aluminum, copper, lead, tin, zinc and alloys; some plastics, rubber, and coatings.

FIRE INFORMATION

Combustion products include toxic nitrogen oxides. Firefighting gear (including SCBA) does not provide adequate protection. If exposure occurs, remove and isolate gear immediately and thoroughly decontaminate personnel.Vapors are heavier than air and will collect in low areas. Vapors in confined areas may explode when exposed to fire. Vapors may travel long distances to ignition sources and flash back. Storage containers and parts of containers may rocket great distances, in many directions. If material or contaminated runoff enters waterways, notify downstream users of potentially contaminated waters. Notify local health and fire officials and pollution control agencies. Use dry chemicals, foam, carbon dioxide. From a secure, explosion-proof location, use water spray to cool exposed containers.

NFPA Hazard Classifications:

Health Hazards (Blue) 2
Flammability (Red) 3
Reactivity (Yellow) 0

Flash Point: 98°F (37°C) o.c.
Flammable Limits in Air: LEL 1.4%; UEL 11.2%
Autoignition Temperature: 590°F (310°C)

FIRST AID

FLAMMABLE LIQUIDS, CORROSIVE (132): Move victim to fresh air. Call emergency medical care. Apply artificial respiration if victim is not breathing. Do not use mouth-to-mouth method if victim swallowed or inhaled the substance; induce artificial respiration with the aid of a pocket mask equipped with a one-way valve or other proper respiratory medical device. If breathing is difficult, give oxygen. Remove and isolate contaminated clothing and shoes. In case of contact with material, immediately flush skin or eyes with running water for at least 20 minutes. Keep victim warm and quiet. Effects of exposure (inhalation, ingestion, or skin contact) to substances may be delayed. Ensure that medical personnel are aware of the material(s) involved and take precautions to protect themselves.

SUBSTANCES-TOXIC AND/OR CORROSIVE (NONCOMBUSTIBLE) (154): Move victim to fresh air. Call emergency medical care. Apply artificial respiration if victim is not breathing. Do not use mouth-to-mouth method if victim ingested or inhaled the substance; induce artificial respiration with the aid of a pocket mask equipped with a one-way valve or other proper respiratory medical device. Administer oxygen if breathing is difficult. Remove and isolate contaminated clothing and shoes. In case of contact with material, immediately flush skin or eyes with running water for at least 20 minutes. For minor skin contact, avoid spreading material on unaffected skin. Keep victim warm and quiet. Effects of exposure (inhalation, ingestion, or skin contact) may be delayed. Ensure that medical personnel are aware of the material(s) involved and take precautions to protect themselves.

1,4–NAPTHOQUINONE N:045

SYNONYMS: 1,4-DIHYDRO-1,4-DIKETONAPHTHALENE; EINECS No. 204-977-6; 1,4-NAPHTHALENEDIONE; α-NAPHTHOQUINONE; USAF CY-10

IDENTIFICATION

CAS:	130-15-4
ERG ID:	UN 1325
ERG Guide:	133
ERG Guide Category:	FLAMMABLE SOLIDS
Molecular Formula:	$C_{10}H_6O_2$
RTECS No.:	QL6475000; QL7175000
Properties:	Combustible solid, powder. Yellow-tan to greenishyellow. A pungent, chlorinelike odor; acrid, irritating, penetrating. Mixes slightly in water.

HEALTH AND SAFETY INFORMATION

Irritant, hematological, allergenic activity. May affect nerves.

RECOMMENDATIONS FOR RESPIRATOR SELECTION

At any concentrations above the NIOSH REL, or where there is no REL, at any detectable concentration: SCBAF:PD,PP (any self-contained breathing apparatus that has a full facepiece and is operated in a pressure-demand or other positive-pressure mode); or SAF:PD,PP:ASCBA (any supplied-air respirator that has a full facepiece and is operated in a pressure-demand or other positive-pressure mode in combination with an auxiliary, self-contained breathing apparatus operated in a pressure-demand or other positive-pressure mode).

Escape: GMFOV [any air-purifying, full-facepiece respirator (gas mask) with a chin-style, front- or back-mounted organic vapor canister];or SCBAE (any appropriate escape-type, self-contained breathing apparatus).

CHEMICAL REACTIVITY

Reacts with strong oxidizing agents and strong reducing agents. Decomposes to toxic fumes of carbon monoxide and carbon dioxide.

FIRE INFORMATION

If material or contaminated runoff enters waterways, notify downstream users of potentially contaminated waters. Notify local health and fire officials and pollution control agencies. Use dry chemicals, alcohol-resistant foam, carbon dioxide. From a secure, explosion-proof location, use water spray to cool exposed containers.

NFPA Hazard Classifications: Not listed

FIRST AID

Move victim to fresh air. Call emergency medical care. Apply artificial respiration if victim is not breathing. Administer oxygen if breathing is difficult. Remove and isolate contaminated clothing and shoes. In case of contact with substance, immediately flush skin or eyes with running water for at least 20 minutes. Removal of solidified molten material from skin requires medical assistance. Keep victim warm and quiet. Ensure that medical personnel are aware of the material(s) involved and take precautions to protect themselves.

ENVIRONMENTAL INFORMATION

EPA Hazardous Waste Number (RCRA No.): U166
RCRA Section 261 Hazardous Constituents
RCRA Ground Water Monitoring List: Suggested test methods(s) (PQL μg/L); 8270 (10)
EPCRA Section 304 Reportable Quantity (RQ): CERCLA, 5,000 lb (2,270 kg)

NEOPENTANE N:050

SYNONYMS: 2,2-DIMETHYLPROPANE; 2,2-DIMETHYLPROPANE, other than pentane and isopentane; EEC No. 601-005-00-6; EINECS No. 207-343-7; tert-PENTANE; PROPANE, 2,2-DIMETHYL-

IDENTIFICATION

CAS:	463-82-1
ERG ID:	UN 2044
ERG Guide:	115
ERG Guide Category:	GASES-FLAMMABLE (INCLUDING REFRIGERATED LIQUIDS)
Molecular Formula:	C_5H_{12}
RTECS No.:	TY1190000

782 Neopentane

Properties: Highly flammable gas, or compressed liquid. Colorless. Gasolinelike odor. Insoluble in water.

HEALTH AND SAFETY INFORMATION

Can cause suffocation and is a narcotic at high concentrations. Liquid can cause frostbite. Irritates skin, with possible blisters, itching and pain.. Long-term exposure may cause nervous system damage with numbness, and weakness in arms and legs; dry, cracked skin; liver and kidney disease; chronic respiratory problems.

NIOSH Criteria Document: NIOSH 77-151 ALKANES

Exposure Limits:	TWA	CEILING
ACGIH:	600 ppm; 1770 mg/m^3 (1)	750 ppm; 2210 mg/m^3
NIOSH:	120 ppm; 350 mg/m^3	610 ppm; 1800 mg/m^3 [15 min]
OSHA:	1000 ppm; 2950 mg/m^3	

(1) ACGIH notice of intended change eliminates STEL for all isomers of pentane.

RECOMMENDATIONS FOR RESPIRATOR SELECTION: NIOSH

1200 ppm: SA (any supplied-air respirator).
1500 ppm: SA:CF (any supplied-air respirator operated in a continuous-flow mode); or SCBAF (any self-contained breathing apparatus with a full facepiece); or SAF (any supplied-air respirator with a full facepiece).

Emergency or planned entry into unknown concentrations or IDLH conditions: SCBAF:PD,PP (any self-contained breathing apparatus that has a full facepiece and is operated in a pressure-demand or other positive-pressure mode); or SAF:PD,PP:ASCBA (any supplied-air respirator that has a full facepiece and is operated in a pressure-demand or other positive-pressure mode in combination with an auxiliary, self-contained breathing apparatus operated in a pressure-demand or other positive pressure mode.

Escape: GMFOV (any air-purifying, full-facepiece respirator (gas mask) with a chin-style, front-or back-, mounted organic vapor canister; or SCBAE (any appropriate escape-type, self-contained breathing apparatus).

CHEMICAL REACTIVITY

Forms explosive mixture with air. May accumulate static electricity, and cause ignition of its vapors.

FIRE INFORMATION

Gases are heavier than air and will collect in low areas. Gases in confined areas may explode when exposed to fire. Gases may travel long distances to ignition sources and flash back. Storage containers and parts of containers may rocket great distances, in many directions. If material or contaminated runoff enters waterways, notify downstream users of potentially contaminated waters. Notify local health and fire officials and pollution control agencies. Do not extinguish the fire unless the flow of the gas can be stopped and any remaining gas is out of the line. Specially trained personnel may use fog lines to cool exposures and let the fire burn itself out. Use dry chemicals, foam, carbon dioxide. From a secure, explosion-proof location, use water spray to cool exposed containers. If cooling streams are ineffective (venting sound increases in volume and pitch, tank discolors, or shows any signs of deforming), withdraw immediately to a secure position.

NFPA Hazard Classifications:

Health Hazards (Blue) 0
Flammability (Red) 4
Reactivity (Yellow) 0

Flash Point: Gas
Flammable Limits in Air: LEL 1.4%; UEL 7.5%
Autoignition Temperature: 842°F (450°C)

FIRST AID

Move victim to fresh air. Call emergency medical care. Apply artificial respiration if victim is not breathing. Administer oxygen if breathing is difficult. Remove and isolate contaminated clothing and shoes. Clothing frozen to the skin should be thawed before being removed. In case of contact with liquefied gas, thaw frosted parts with lukewarm water. Keep victim warm and quiet. Ensure that medical personnel are aware of the material(s) involved and take precautions to protect themselves.

ENVIRONMENTAL INFORMATION as 2,2-dimethylpropane

Clean Air Act: Accidental Release Prevention/Flammable Substances (Section 112[r], Table 3), TQ = 10,000 lb (4,540 kg)

NEOPENTY GLYCOL N:055

SYNONYMS: 2,2–DIMETHYLPROPANE–1,3–DIOL; 2,2-DIMETHYL-1,3-PROPANEDIOL; EINECS No. 204-781-0; 1,3-PROPANEDIOL, 2,2-DIMETHYL; DIMETHYLOL PROPANE; DIMETHYLTRIMETHYLENE GLYCOL; NEOL; NEOPENTYLENE GLYCOL; 1,3-PROPANEDIOL, 2,2-DIMETHYL-

IDENTIFICATION

CAS:	126-30-7
Molecular Formula:	$C_5H_{12}O_2$
RTECS No.:	TY5775000
Properties:	Combustible hygroscopic crystalline solid. Colorless to white. Odorless. Soluble in water.

HEALTH AND SAFETY INFORMATION

Irritates eyes, skin, and respiratory tract. Due to the availability of insufficient data on human effects, caution should be exercised.

RECOMMENDATIONS FOR RESPIRATOR SELECTION

At any concentrations above the NIOSH REL, or where there is no REL, at any detectable concentration: SCBAF:PD,PP (any self-contained breathing apparatus that has a full facepiece and is operated in a pressure-demand or other positive-pressure mode); or SAF:PD,PP:ASCBA (any supplied-air respirator that has a full facepiece and is operated in a pressure-demand or other positive-pressure mode in combination with an auxiliary, self-contained breathing apparatus operated in a pressure-demand or other positive-pressure mode).

Escape: GMFOV [any air-purifying, full-facepiece respirator (gas mask) with a chin-style, front- or back-mounted organic vapor canister]; or SCBAE (any appropriate escape-type, self-contained breathing apparatus).

CHEMICAL REACTIVITY

Dust or finely dispersed particles can forms explosive mixture with air. Incompatible with strong acids, caustics, aliphatic amines, isocyanates, and oxidizers.

FIRE INFORMATION

Vapors in confined areas may explode when exposed to fire. If material or contaminated runoff enters waterways, notify downstream users of potentially contaminated waters. Notify local health and fire officials and pollution control agencies. Use dry chemicals, alcohol-resistant foam, carbon dioxide. From a secure, explosion-proof location, use water spray to cool exposed containers.

NFPA Hazard Classifications

Health Hazards (Blue) 1
Flammability (Red) 1
Reactivity (Yellow) 0

Flash Point: 225°F (107°C) c.c.
Flammable Limits in Air: LEL 1.37% @ 300°F (149°C); UEL 18.8 @ 351°F (177°C)
Autoignition Temperature: 7500°F (399°C)

FIRST AID

Move victim to fresh air. Call emergency medical care. Apply artificial respiration if victim is not breathing. Administer oxygen if breathing is difficult. Remove and isolate contaminated clothing and shoes. In case of contact with substance, immediately flush skin or eyes with running water for at least 20 minutes. Wash skin with soap and water. Keep victim warm and quiet. Effects of exposure (inhalation,

ingestion, or skin contact) to substance may be delayed. Ensure that medical personnel are aware of the material(s) involved and take precautions to protect themselves.

NICKEL N:060

SYNONYMS: ALLOY 725; ALLOY 732; ALLOY 735; ALLOY 762; ALLOY 770; C.I.77775; EINECS No. 231-111-4; NICHEL (Italian); NICKEL, ELEMENTAL; NICKEL 200; NICKEL 201; NICKEL 204; NICKEL 205; NICKEL 211; NICKEL 212; NICKEL 213; NICKEL 222; NICKEL 223; NICKEL 229; Ni 270; NICKEL 270; NICKEL 0901-S; NICKEL 4303T; NIKLAD 794-A; NIKKEL (Dutch); NIQUEL (Spanish); NP2; RANEY ALLOY; RANEY NICKEL

IDENTIFICATION

CAS:	7440-02-0
ERG ID:	UN 2881
ERG Guide:	135
ERG Guide Category:	SUBSTANCES - SPONTANEOUSLY COMBUSTIBLE
Molecular Formula:	Ni
RTECS No.:	QR5950000 (metal)
Properties:	Silvery white, workable metal. Odorless. Insoluble in water.

HEALTH AND SAFETY INFORMATION

Long-term exposure may cause impairment of sense of smell, chest pain, destruction of nasal tissues, and asthmatic lung disease; allergic sensitivity may also develop. Inhalation of nickel dust has been associated with an increased risk of lung, throat, and nasal cancer.

IARC Carcinogen Evaluation: Group 1, known human carcinogen; human evidence, sufficient; animal evidence, sufficient
NTP: Anticipated human carcinogen
IDLH: 10 mg/m^3; potential occupational carcinogen.
OSHA Table Z-1-A Air Contaminant
NIOSH Criteria Document: NIOSH 77-164

Exposure Limits:	**TWA**
ACGIH: (1)	1.5 mg/m^3
NIOSH: (2)	0.015 mg/m^3
OSHA:	1 mg/m^3

(1) Not suspected as a human carcinogen.
(2) Potential carcinogen. Reduce exposure to lowest feasible level.

CHEMICAL REACTIVITY

Powder may self-ignite in air. Contact with strong mineral acids form flammable and explosive hydrogen gas. Sulfur contact evolves heat. Violent reaction with strong oxidizers, fluorine, ammonia nitrate, hydrazine, performic acid, phosphorus, sulfur, selenium, nitric acid, carbon monoxide, and many other chemicals.

FIRE INFORMATION

Fine nickel dust or fumes are flammable. If material or contaminated runoff enters waterways, notify downstream users of potentially contaminated waters. Notify local health and fire officials and pollution control agencies.

NFPA Hazard Classifications: (NFPA49 or NFPA325)

Health Hazards (Blue) 2
Flammability (Red) 4
Reactivity (Yellow) 1

FIRST AID

Move victim to fresh air. Call emergency medical care. Apply artificial respiration if victim is not breathing. Administer oxygen if breathing is difficult. Remove and isolate contaminated clothing and shoes. In case of contact with substance, immediately flush skin or eyes with running water for at least 20 minutes. Wash skin with soap and water. Keep victim warm and quiet. Effects of exposure (inhalation, ingestion, or skin contact) to substance may be delayed. Ensure that medical personnel are aware of the material(s) involved and take precautions to protect themselves.

ENVIRONMENTAL INFORMATION

Clean Air Act: Hazardous Air Pollutants (Title I, Part A, Section 112) as nickel compounds.
Clean Water Act: Section 307 Toxic Pollutants as nickel and compounds; Section 313 Priority Chemicals.
RCRA Section 261 Hazardous Constituents, waste number not listed.
RCRA Land Ban Waste
RCRA Universal Treatment Standards: Wastewater (mg/L), 3.98; Nonwastewater (mg/L), 5.0 TCLP.
RCRA Ground Water Monitoring List: Suggested test method(s) (PQL ug/L). (total) 6010(50); 7520(400).
Safe Drinking Water Act: Regulated chemical (47 FR 9352).
EPCRA Section 304: Reportable Quantity (RQ): CERCLA, 100 lb (45.4 kg).
EPCRA Section 313: Form R *de minimis* concentration reporting level: 0.1%; Form R Toxic Chemical Category Code: N495.
California's Proposition 65: Carcinogen as nickel and certain nickel compounds.

NITRIC ACID N:120

SYNONYMS: ACIDE NITRIQUE (French); ACIDO NITRICO (Italian, Spanish); ALUMINUM ETCH

16-1-1-2; ALUMINUM ETCH 82-3-15-0; AQUA FORTIS; AQUA REGIA; AZOTIC ACID; AZOTOWY KWAS (Polish); C-P 8 SOLUTION; CHROME ETCH KTI; COPPER, BRASS BRITE DIP 127; COPPER, BRASS BRITE DIP 1127; COPPERLITE RD-25; DOPED POLY ETCH; EEC No. 007-004-01-9; EINECS No. 231-714-2; ENGRAVERS ACID; FRECKLE ETCH; HYDROGEN NITRATE; KOVAR BRIGHT DIP (412X); KOVAR BRIGHT DIP (RDX-555); KTI ALUMINUM ETCH I; KTI ALUMINUM ETCH II; KTI ALUMINUM ETCH III; KTI CHROME ETCH; MAE ETCHANTS; MIXED ACID ETCH (5-2-2); MIXED ACID ETCH (6-1-1-); NITAL; NITRALINE; NF SOLDER STRIPPER 3114-B; NITRIC ACID, RED FUMING; NITRIC ACID, WHITE FUMING; NITROUS FUMES; NITRYL HYDROXIDE; PASSIVATION SOLUTION; PATCLIN 958; POLY ETCH 95%; RED FUMING NITRIC ACID; RFNA; RT-2 STRIPPING SOLUTION; SALPETERSAURE (German); SALPETERZUUROPLOSSINGEN (Dutch); SILICON ETCH SOLUTION; SOLDER STRIP NP-A; STRESS RELIEF ETCH; WET K-ETCH; WFNA; WHITE FUMING NITRIC ACID

IDENTIFICATION

CAS:	7697-37-2
ERG ID:	UN 1760 (40% acid); UN 2031 (>40% acid); UN 2032 (red fuming)
ERG Guide:	154 (40% acid); 157 (>40% acid and fuming)
ERG Guide Category:	TOXIC AND/OR CORROSIVE (NONCOMBUSTIBLE/ WATER SENSITIVE)
Molecular Formula:	HNO_3
RTECS No.:	QU5775000 (up to 70%); QU5900000 (more than 70%)
Properties:	Pale yellow, reddish brown liquid; or red fuming liquid when concentrated. Acrid, suffocating odor. Sinks and mixes (violently) with water; giving off heat.

HEALTH AND SAFETY INFORMATION

Poisonous. Inhaling the vapor can cause death. Skin and eye contact causes severe burns and blindness. Long–term inhalation of mists or vapors can cause tooth erosion, bronchitis, lung inflammation, and skin irritation.

IDLH: 25 ppm; 65.5 mg/m^3
Odor Threshold: 0.267 ppm
OSHA Table Z-1-A Air Contaminant
OSHA Process Safety Management of Highly Hazardous Chemicals (29CFR, Part 1910.119, Appendix A): Threshold Quantity (94.5% by weight or greater): 500 pounds.
NIOSH Criteria Document: NIOSH 76-141

Exposure Limits:	TWA	STEL
ACGIH:	2 ppm; 5.2 mg/m^3	4 ppm; 10 mg/m^3
NIOSH:	2 ppm; 5 mg/m^3	4 ppm; 10 mg/m^3
OSHA:	2 ppm; 5 mg/m^3	4 ppm; 10 mg/m^3

RECOMMENDATIONS FOR RESPIRATOR SELECTION: NIOSH/OSHA

25 ppm: SA:CF (any supplied-air respirator operated in a continuous-flow mode); or CCRFS (any chemical cartridge respirator with a full facepiece and cartridge(s) providing protection against the compound of concern); or GMFS [any air-purifying, full-facepiece respirator (gas mask) with a chin-style, front- or back-mounted canister providing protection against the compound of concern]; or SCBAF (any self-contained breathing apparatus with a full facepiece); or SAF (any supplied-air respirator with a full facepiece).

Emergency or Planned Entry into Unknown Concentrations or IDLH Conditions: SCBAF:PD,PP (any self-contained breathing apparatus that has a full facepiece and is operated in a pressure-demand or other positive-pressure mode); or SAF:PD,PP:ASCBA (any supplied-air respirator that has a full facepiece and is operated in a pressure-demand or other positive-pressure mode in combination with an auxiliary, self-contained breathing apparatus operated in a pressure-demand or other positive pressure mode.

Escape: GMFS [any air-purifying, full-facepiece respirator (gas mask) with a chin-style, front- or back-mounted canister providing protection against the compound of concern]; or SCBAE (any appropriate escape-type, self-contained breathing apparatus).
Note: Substance reported to cause eye irritation or damage; may require eye protection.
Note: Only nonoxidizable sorbents are allowed (*not charcoal*).

CHEMICAL REACTIVITY

A strong oxidizer that can react violently with reducing agents, strong bases, or combustible materials. Incompatible with many substances including acrylates, aliphatic amines, alcohols, aldehydes, alkanolamines, alkylene oxides, anion exchange resins, aromatic amines, amides, cresols, cyanides, cyclic ketones, epichlorohydrin, glycols, isocyanates, ketones, oleum, organic anhydrides, phenols, substituted allyls, sulfuric acid, strong oxidizers, and terpenes. Attacks some plastics, rubber, and coatings. Corrosive to most metals releasing highly flammable hydrogen gas.

FIRE INFORMATION

Highly toxic nitrogen oxides produced when heated. Will increase the intensity of fire. Firefighting gear (including SCBA) does not provide adequate protection. If exposure occurs, remove and isolate gear immediately and thoroughly decontaminate personnel. Vapors are heavier than air and will collect in low areas. Vapors in confined areas may explode when exposed to fire. Storage containers and parts of containers may rocket great distances, in many directions. If material or contaminated runoff enters waterways, notify downstream users of potentially contaminated waters. Notify local health and fire officials and pollution control agencies. Use dry chemicals, foam, carbon dioxide. From a secure, explosion-proof location, use water spray to cool exposed containers.

NFPA Hazard Classifications:

Fuming (UN 2032)

Health Hazards (Blue) 4

Flammability (Red) 0
Reactivity (Yellow) 1
Special situations . . . . OXY

More than 40% (UN 031)

Health Hazards (Blue) 4
Flammability (Red) 0
Reactivity (Yellow) 0
Special situations . . . . OXY

40% or less (NA 1760)

Health Hazards (Blue) 3
Flammability (Red) 0
Reactivity (Yellow) 0

FIRST AID

Move victim to fresh air. Call emergency medical care. Apply artificial respiration if victim is not breathing. Do not use mouth-to-mouth method if victim ingested or inhaled the substance; induce artificial respiration with the aid of a pocket mask equipped with a one-way valve or other proper respiratory medical device. Administer oxygen if breathing is difficult. Remove and isolate contaminated clothing and shoes. In case of contact with substance, immediately flush skin or eyes with running water for at least 20 minutes. For minor skin contact, avoid spreading material on unaffected skin. Keep victim warm and quiet. Effects of exposure (inhalation, ingestion, or skin contact) may be delayed. Ensure that medical personnel are aware of the material(s) involved and take precautions to protect themselves.

ENVIRONMENTAL INFORMATION

Clean Water Act: Section 311 Hazardous Substances/RQ (same as CERCLA);Section 313 Priority Chemicals.
EPCRA Section 302 Extremely Hazardous Substances: TPQ = 1,000 lb (454 kg)
EPCRA Section 304: Reportable Quantity (RQ): CERCLA, 1,000 lb (454 kg)
EPCRA Section 313: Form R *de minimis* concentration reporting level: 1.0%

ENVIRONMENTAL INFORMATION - NITRIC ACID, CONCENTRATION > 80%:

Clean Air Act: Accidental Release Prevention/Flammable substances, (Section 112[r], Table 3), TQ = 15,000 lb (1,362 kg).
EPCRA Section 302 Extremely Hazardous Substances: TPQ = 1,000 lb (454 kg)
EPCRA Section 304: Reportable Quantity (RQ): CERCLA, 1,000 lb (454 kg)
EPCRA Section 313: Form R *de minimis* concentration reporting level: 1.0%

NITROBENZENE N:145

SYNONYMS: BENZENE, NITRO-; EEC No. 609-003-00-7; EINECS No. 202-716-0; ESSENCE OF MIRBANE; ESSENCE OF MYRBANE; MIRBANE OIL; NCI-C60082; NITROBENCENO (Spanish); NITROBENZEEN (Dutch); NITROBENZEN (Polish); NITROBENZENO (Italian); NITROBENZOL (French); NITROBENZOL, L; NITRO, LIQUID; OIL OF MIRBANE; OIL OF MYRBANE; STCC 4921455

IDENTIFICATION

CAS:	98-95-3
ERG ID:	UN 1662
ERG Guide:	152
ERG Guide Category:	SUBSTANCES - TOXIC (COMBUSTIBLE)
Molecular Formula:	$C_6H_5NO_2$
RTECS No.:	DA6475000
Properties:	Very flammable, light yellow-green to brown, oily liquid; solid below 42°F (6°C). Pungent odor like bitter almonds or paste shoe polish. Some may not be able to smell this substance. Sinks in water; insoluble.

HEALTH AND SAFETY INFORMATION

Poisonous. Breathing the vapor, swallowing the material, or absorption through the skin can cause death. Irritates eyes, skin, and respiratory tract. May interfere with the body's ability to use oxygen. A blood poison and a nerve poison which is easily absorbed by the body.

IDLH: 200 ppm; 1,024 mg/m^3
Odor Threshold: 0.044 ppm . Odor is not a reliable indicator of the presence of toxic amounts of vapor.
OSHA Table Z-1-A Air Contaminant

Exposure Limits:	**TWA**
ACGIH: [1)]	1 ppm; 5 mg/m^3
NIOSH: [(1)]	1 ppm; 5 mg/m^3
OSHA: [(1)]	1 ppm; 5 mg/m^3

[(1)] Skin contact contributes significantly to overall exposure.

RECOMMENDATIONS FOR RESPIRATOR SELECTION: NIOSH

At any concentrations above the NIOSH REL, or where there is no REL, at any detectable concentration: SCBAF:PD,PP (any self-contained breathing apparatus that has a full facepiece and is operated in a pressure-demand or other positive-pressure mode); or SAF:PD,PP:ASCBA (any supplied-air respirator that has a full facepiece and is operated in a pressure-demand or other positive-pressure mode in combination with an auxiliary, self-contained breathing apparatus operated in a pressure-demand or other positive-pressure mode).

Escape: GMFOV [any air-purifying, full-facepiece respirator (gas mask) with a chin-style, front- or back-mounted organic vapor canister]; or SCBAE (any appropriate escape-type, self-contained breathing apparatus).

CHEMICAL REACTIVITY

Forms explosive mixture with air. Incompatible with nitric acid, nitrogen tetroxide, caustics, ammonia, amines, phosphorus pentachloride, and chemically active metals such as tin or zinc. Attacks some plastics, rubber, and coatings.

FIRE INFORMATION

Combustion products include toxic nitrogen oxides. Firefighting gear (including SCBA) does not provide adequate protection. If exposure occurs, remove and isolate gear immediately and thoroughly decontaminate personnel. Vapors are heavier than air and will collect in low areas. Storage containers and parts of containers may rocket great distances, in many directions. If material or contaminated runoff enters waterways, notify downstream users of potentially contaminated waters. Notify local health and fire officials and pollution control agencies. Use dry chemicals, foam, carbon dioxide. From a secure, explosion-proof location, use water spray to cool exposed containers.

NFPA Hazard Classifications:

Health Hazards (Blue) 3
Flammability (Red) 2
Reactivity (Yellow) 1

Flash Point: 190°F (88°C)
Flammable Limits in Air: LEL 1.8%; UEL NA%
Autoignition Temperature: 900°F (482°C)

FIRST AID

Move victim to fresh air. Call emergency medical care. Apply artificial respiration if victim is not breathing. Do not use mouth-to-mouth method if victim ingested or inhaled the substance; induce artificial respiration with the aid of a pocket mask equipped with a one–way valve or other proper respiratory medical device. Administer oxygen if breathing is difficult. Remove and isolate contaminated clothing and shoes. In case of contact with substance, immediately flush skin or eyes with running water for at least 20 minutes. For minor skin contact, avoid spreading material on unaffected skin. Keep victim warm and quiet. Effects of exposure (inhalation, ingestion, or skin contact) may be delayed. Ensure that medical personnel are aware of the material(s) involved and take precautions to protect themselves.

ENVIRONMENTAL INFORMATION

Clean Air Act: Hazardous Air Pollutants (Title I, Part A, Section 112).
Clean Water Act: Section 307 Toxic Pollutants; Section 311 Hazardous Substances/RQ (same as CERCLA); Section 313 Priority Chemicals.
EPA Hazardous Waste Number (RCRA No.): U169.

RCRA Section 261 Hazardous Constituents
RCRA Land Ban Waste
RCRA Universal Treatment Standards: Wastewater (mg/L), 0.068; nonwastewater (mg/kg), 14.
RCRA Ground Water Monitoring List: Suggested test method(s) (PQL ug/L): 8090(40); 8270(10).
Safe Drinking Water Act: Regulated chemical (47 FR 9352).
EPCRA Section 302 Extremely Hazardous Substances: TPQ = 10,000 lb (4,540 kg).
EPCRA Section 304: Reportable Quantity (RQ): CERCLA, 1,000 lb (454 kg).
EPCRA Section 313: Form R *de minimis* concentration reporting level: 1.0%.
TSCA: 40CFR716.120(*a*)

NITROBIPHENYL(4-) N:149

SYNONYMS: BIPHENYL (1,1'-), 4-NITRO-; EEC No. 609-039-00-3; EINECS No. 202-204-7; 4-NITROBIPHENYL; NITRO DIPHENYL(4-); o-NITROPHENYL; o-NITRODIPHENYL; ONB; PNB

IDENTIFICATION

CAS:	92-93-3
ERG ID:	UN 2811 as toxic solid, organic, n.o.s.
ERG Guide:	154
ERG Guide Category:	SUBSTANCES-TOXIC AND/OR CORROSIVE
Molecular Formula:	$C_{12}H_9NO_2$
RTECS No.:	DV5600000
Properties:	Combustible plates, needles, or liquid. Light yellow to reddish. Floats on water; practically insoluble.

HEALTH AND SAFETY INFORMATION

A confirmed carcinogen. All contact should be avoided. May cause severe irritation to eyes, skin, and respiratory tract.

IARC Evaluation: Group 3, limited animal evidence.
NTP: Confirmed human carcinogen
OSHA Table Z-1-A Air Contaminant
OSHA Specifically Regulated Substances

Exposure Limits:	**TWA**
ACGIH:	Confirmed human carcinogen, no assigned TWA value.
NIOSH:	Potential Occupational Carcinogen, limit contact to lowest feasible concentration.
OSHA:	OSHA regulated carcinogen use 29CFR1910.1003

RECOMMENDATIONS FOR RESPIRATOR SELECTION: NIOSH

At any concentrations above the NIOSH REL, or where there is no REL, at any detectable concentration: SCBAF:PD,PP (any self-contained breathing apparatus that has a full facepiece and is operated in a pressure-demand or other positive-pressure mode); or SAF:PD,PP:ASCBA (any supplied-air respirator that has a full facepiece and is operated in a pressure-demand or other positive-pressure mode in combination with an auxiliary, self-contained breathing apparatus operated in a pressure-demand or other positive-pressure mode).

Escape: HiEF (Any air-purifying, full-facepiece respirator with a high-efficiency particulate filter);or SCBAE (any appropriate escape-type, self-contained breathing apparatus).

CHEMICAL REACTIVITY

Contact with strong oxidizers may cause fire and explosions. Incompatible with alkalis.

FIRE INFORMATION

Combustion products include toxic nitrogen oxides. Firefighting gear (including SCBA) may not provide adequate protection. If exposure occurs, remove and isolate gear immediately and thoroughly decontaminate personnel. Vapors in confined areas may explode when exposed to fire. Vapors may travel long distances to ignition sources and flash back. Storage containers and parts of containers may rocket great distances, in many directions. If material or contaminated runoff enters waterways, notify downstream users of potentially contaminated waters. Notify local health and fire officials and pollution control agencies. Use dry chemicals, foam, carbon dioxide. From a secure, explosion-proof location, use water spray to cool exposed containers.

NFPA Hazard Classifications

Health Hazards (Blue) 2
Flammability (Red) 1
Reactivity (Yellow) 0

Flash Point: 290°F (143°C)
Autoignition Temperature: 356°F (180°C)

FIRST AID

Move victim to fresh air; call emergency medical care. Apply artificial respiration if victim is not breathing. *Do not use mouth-to-mouth method if victim ingested or inhaled the substance; induce artificial respiration with the aid of a pocket mask equipped with a one-way valve or other proper respiratory medical device.* Administer oxygen if breathing is difficult. Remove and isolate contaminated clothing and shoes. In case of contact with material, immediately flush skin or eyes with running water for at least 20 minutes. For minor skin contact, avoid spreading material on unaffected skin. Keep victim warm and quiet. Effects of exposure (inhalation, ingestion, or skin contact) to substance may be delayed. Ensure that medical personnel are aware of the material(s) involved and take precautions to protect themselves.

ENVIRONMENTAL INFORMATION

CLEAN AIR ACT: Hazardous Air Pollutants (Title I, Part A, Section 112)
EPCRA Section 304 Reportable Quantity (RQ): CERCLA, 1 lb (0.454 kg).
EPCRA Section 313: Form R deminimus concentration reporting level: 0.1%.
CALIFORNIA'S PROPOSITION 65: Carcinogen.

NITROGEN N:160

SYNONYMS: AZOTE (French); AZOTO (Italian); EINECS No. 231-783-9; NITROGEN, COMPRESSED; NITROGENO (Spanish); STIKSTOFF (German)

IDENTIFICATION

CAS:	7727-37-9
ERG ID:	UN 1066 (compressed); UN 1977 (liquid, refrigerated)
ERG Guide:	121 (compressed); 120 (liquid, refrigerated)
ERG Guide Category:	GASES - INERT
Molecular Formula:	N_2
RTECS No.:	QW9700000
Properties:	Nonflammable, inert compressed gas or cryogenic liquid. Colorless gas; faintly yellow liquid. Odorless. Gas is insoluble in water; liquid floats on water surface and is slightly soluble. Vapors (cold) are heavier than air; will collect in low areas.

HEALTH AND SAFETY INFORMATION

A simple asphyxiant. Regulate level to maintain 18% oxygen in working spaces. Cold vapor is heavier than air and can accumulate at ground level, causing oxygen deficiency and risk of suffocation. Contact with liquid can cause frostbite. Under certain conditions caisson disease or the "bends" may develop.

Odor Threshold: Odor is not a reliable indicator of the presence of toxic amounts of gas.

RECOMMENDATIONS FOR RESPIRATOR SELECTION

At any concentrations above the NIOSH REL, or where there is no REL, at any detectable concentration: SCBAF:PD,PP (any self-contained breathing apparatus that has a full facepiece and is operated in a pressure-demand or other positive-pressure mode); or SAF:PD,PP:ASCBA (any supplied-air respirator that has a full facepiece and is operated in a pressure-demand or other positive-pressure mode in combination with an auxiliary, self-contained breathing apparatus operated in a pressure-demand or other positive pressure mode).

Escape: GMFOV [any air-purifying, full-facepiece respirator (gas mask) with a chin-style, front- or back-mounted organic vapor canister]; or SCBAE (any appropriate escape-type, self-contained breathing apparatus).

FIRE INFORMATION

Storage containers and parts of containers may rocket great distances, in many directions. Notify local health and fire officials and pollution control agencies. Use any extinguishing agent suitable for surrounding fires. DO NOT apply water to cryogenic liquid containers. If cryogenic liquid containers are exposed to direct flame or elevated temperatures for prolonged times, withdraw immediately to a secure location.

NFPA Hazard Classifications:

Health Hazards (Blue) 3
Flammability (Red) 0
Reactivity (Yellow) 0

FIRST AID

Move victim to fresh air. Call emergency medical care. Apply artificial respiration if victim is not breathing. Administer oxygen if breathing is difficult. Remove and isolate contaminated clothing and shoes. Clothing frozen to the skin should be thawed before being removed. In case of contact with liquefied gas, thaw frosted parts with lukewarm water. Keep victim warm and quiet. Keep victim under observation. Ensure that medical personnel are aware of the material(s) involved and take precautions to protect themselves.

NITROGEN DIOXIDE N:180

SYNONYMS: AZOTE (French); AZOTO (Italian); DINITROGEN TETROXIDE (N2O4); DIOXIDO de NITROGENO (Spanish); EEC No. 007-002-00-0; EINECS No. 233-272-6; NITRITO; NITROGEN PEROXIDE; STICKSTOFFDIOXIDE (German); STIKSTOFDIOXYDE (Dutch).

IDENTIFICATION

CAS:	10102-44-0
ERG ID:	UN 1067
ERG Guide:	124
ERG Guide Category:	GASES - TOXIC AND/OR CORROSIVE-OXIDIZING
Molecular Formula:	NO_2
RTECS No:	QW9800000
Properties:	Nonflammable liquid or gas. Boils at 70°F (21°C). Yellow-brown liquid or reddishbrown to dark brown gas. When frozen, [freezes at 17°F (–8°C)] may appear as blue-white snow. Acrid, pungent odor; not a reliable indicator of the presence of toxic amounts of gas. Dissolves in water forming corrosive nitric acid solution.

HEALTH AND SAFETY INFORMATION

Corrosive to eyes, skin, and respiratory tract.Inhalation may cause pulmonary edema, which can be delayed for several hours; there is a risk of death in serious cases. Long-term exposure can cause headache, weakness, loss of sleep and appetite, sores in nose and mouth, nausea, and erosion of teeth. Exposure to 0.4-2,7 ppm for 4-6 years has been associated with emphysema and bronchitis. Genetic changes have been found in experimental animals.

IDLH: 20 ppm
Odor Threshold: 0.186 ppm
OSHA Table Z-1-A Air Contaminant
OSHA Process Safety Management of Highly Hazardous Chemicals (29CFR, Part 1910.119, Appendix A): Threshold Quantity: 250 pounds.
NIOSH Criteria Document: NIOSH 76-149 NITROGEN OXIDES

Exposure Limits:	TWA	Ceiling	STEL
ACGIH:	3 ppm; 5.6 mg/m^3		5 ppm; 9.4 mg/m^3
NIOSH:			1 ppm; 1.8 mg/m^3 per 15 minutes
OSHA:		5 ppm; 9 mg/m^3	

CHEMICAL REACTIVITY

A powerful oxidizer. Water forms nitric acid and nitric oxide. Incompatible with nitric oxide reacts with air forming more nitrogen tetroxide. Explosive reaction with alcohols, boron trichloride, carbon disulfide, cyclohexane, fluorine, formaldehyde, nitrobenzene, propylene, petroleum, reducing agents, toluene, tungsten carbide, trichloroethylene, fuels and petroleum, organic matter, anhydrous ammonia, chlorinated hydrocarbons, toluene and other combustible and organic substances. Attacks metals in presence of moisture.

FIRE INFORMATION

Combustion products include oxygen and toxic nitrogen oxides. Noncombustible but will intensify combustion of other substances. Firefighting gear (including SCBA) does not provide adequate protection. If exposure occurs, remove and isolate gear immediately and thoroughly decontaminate personnel. Gases are heavier than air and will collect in low areas. Gases in confined areas may explode when exposed to fire. Gases may travel long distances to ignition sources and flash back. Storage containers and parts of containers may rocket great distances, in many directions. If material or contaminated runoff enters waterways, notify downstream users of potentially contaminated waters. Notify local health and fire officials and pollution control agencies. Do not extinguish the fire unless the flow of the gas can be stopped and any remaining gas is out of the line. Specially trained personnel may use fog lines to cool exposures and let the fire burn itself out. Use dry chemicals, foam, carbon dioxide. From a secure, explosion-proof location, use water spray to cool exposed containers. If cooling streams are ineffective (venting sound increases in volume and pitch, tank discolors, or shows any signs of deforming), withdraw immediately to a secure position.

NFPA Hazard Classifications:

Health Hazards (Blue) 3
Flammability (Red) 0
Reactivity (Yellow) 0
Special Notice .... OXY

Flash Point: Not flammable; but will accelerate and intensify the burning of combustible materials.

FIRST AID

Move victim to fresh air. Call emergency medical care. Apply artificial respiration if victim is not breathing. Do not use mouth-to-mouth method if victim ingested or inhaled the substance; induce artificial respiration with the aid of a pocket mask equipped with a one-way valve or other proper respiratory medical device. Administer oxygen if breathing is difficult. Clothing frozen to the skin should be thawed before being removed. Remove and isolate contaminated clothing and shoes. In case of contact with substance, immediately flush skin or eyes with running water for at least 20 minutes. Keep victim warm and quiet. Keep victim under observation. Effects of contact or inhalation may be delayed. Ensure that medical personnel are aware of the material(s) involved and take precautions to protect themselves.

ENVIRONMENTAL INFORMATION

CLEAN WATER ACT: Section 311 Hazardous Substances/RQ (same as CERCLA)
EPA HAZARDOUS WASTE NUMBER (RCRA No.): P078
RCRA Section 261 Hazardous Constituents.
EPCRA Section 302 Extremely Hazardous Substances: TPQ = 100 lb (45.4 kg).
EPCRA Section 304 Reportable Quantity (RQ): CERCLA, 10 lb (4.54 kg).

NITROMETHANE N:220

SYNONYMS: BLANCO-TRON TMS PLUS; CHLOROTHENE SM SOLVENT; EEC No. 609-036-00-7; EINECS No. 200-876-6; FREON TMS; GENESOLV DFX; METHANE, NITRO-; NITROCARBOL; NITROFUEL; NITROMETAN (Polish); NITROMETANO (Spanish); NM; NM-55; 6200 DRUM CLEANING SOLVENT; 6500 DRUM CLEANING SOLVENT

IDENTIFICATION

CAS:	75-52-5
ERG ID:	UN 1261
ERG Guide:	129
ERG Guide Category:	FLAMMABLE LIQUIDS (POLAR/WATER-MISCIBLE/NOXIOUS)
Molecular Formula:	CH_3NO_2
RTECS No:	PA9800000

798 Nitromethane

Properties: Highly flammable, oily liquid. Clear, colorless. Strong, disagreeable odor. Sinks and mixes slowly with water; moderately soluble. Irritating vapor is produced.

HEALTH AND SAFETY INFORMATION

Irritates eyes, skin, and respiratory tract. Long-term exposure includes kidney and liver damage; chronic respiratory disease. Following exposure regular medical monitoring and checkups are advisable.

IDLH: 750 ppm
Odor Threshold: 3.5 ppm
OSHA Table Z-1-A Air Contaminant
OSHA Process Safety Management of Highly Hazardous Chemicals (29CFR, Part 1910.119, Appendix A): Threshold Quantity: 2,500 pounds.

Exposure Limits:	TWA
ACGIH:	20 ppm; 50 mg/m^3
OSHA:	100 ppm; 250 mg/m^3

RECOMMENDATIONS FOR RESPIRATOR SELECTION: OSHA

750 ppm: SA:CF (any supplied-air respirator operated in a continuous-flow mode); or SCBAF (any self-contained breathing apparatus with a full facepiece); or SAF (any supplied-air respirator with a full facepiece).

Emergency or Planned Entry Into Unknown Concentrations or IDLH Conditions: SCBAF:PD,PP (any self-contained breathing apparatus that has a full facepiece and is operated in a pressure-demand or other positive-pressure mode); or SAF:PD,PP:ASCBA (any supplied-air respirator that has a full facepiece and is operated in a pressure-demand or other positive-pressure mode in combination with an auxiliary, self-contained breathing apparatus operated in a pressure-demand or other positive pressure mode.

Escape: SCBAE (any appropriate escape-type, self-contained breathing apparatus).
Note: Substance causes eye irritation or damage; eye protection needed.

CHEMICAL REACTIVITY

May explode when heated above 446°F (230°C) or shocked. Forms explosive mixture with air. Caustics, ammonia, strong acids, amines can cause sensitization which may result in explosions. Contact with strong oxidizers may cause fire and explosions. Mixtures with hydrocarbons (and other combustible materials) are highly flammable. Contact with metallic oxides may cause decomposition and development of pressure. Attacks some plastics, rubber, and coatings.

FIRE INFORMATION

Combustion products include toxic nitrogen oxides. Vapors are heavier than air and will collect in low areas. Vapors in confined areas may explode when exposed to fire. Vapors may travel long distances to ignition sources and flash back.Storage containers and parts of containers may rocket great distances, in many directions. If material or contaminated runoff enters waterways, notify downstream users of potentially contaminated waters. Notify local health and fire officials and pollution control agencies. Use alcohol-resistant foam, carbon dioxide. From a secure, explosion-proof location, use water spray to cool exposed containers.

NFPA Hazard Classifications:

Health Hazards (Blue) 1
Flammability (Red) 3
Reactivity (Yellow) 4

Flash Point: 110°F (43°C) o.c.; 95°F (35°C) c.c.
Flammable Limits in Air: LEL 7.3%; UEL 62%
Autoignition Temperature: 785°F (418°C)

FIRST AID

Move victim to fresh air. Call emergency medical care. Apply artificial respiration if victim is not breathing. Administer oxygen if breathing is difficult. Remove and isolate contaminated clothing and shoes. In case of contact with substance, immediately flush skin or eyes with running water for at least 20 minutes. Wash skin with soap and water. Keep victim warm and quiet. Effects of exposure (inhalation, ingestion, or skin contact) to substance may be delayed. Ensure that medical personnel are aware of the material(s) involved and take precautions to protect themselves.

ENVIRONMENTAL INFORMATION

TSCA: 40CFR716.120(*a*)

2–NITROPROPANE N:237

SYNONYMS: DIMETHYLNITROMETHANE; EEC NO. 609-002-00-1; EINECS No. 201-209-1; ISONITROPROPANE; NIPAR S-20; NIPARS-20 SOLVENT; NIPAR S-30 SOLVENT; NITROISOPROPANE; β-NITROPROPANE; 2-NITROPROPANO (Spanish); 2-NP; PROPANE, 2-NITRO; sec-NITROPROPANE

IDENTIFICATION

CAS: 79-46-9; 25322-01-4 (nitropropane, nitrosamines)
ERG ID: UN 2608
ERG Guide: 129
ERG Guide Category: FLAMMABLE LIQUIDS (POLAR/WATER-MISCIBLE/NOXIOUS)
Molecular Formula: $C_3H_7NO_2$

800 2-Nitropropane

RTECS No.: TZ5250000

Properties: Highly flammable, colorless liquid. Mild, fruity odor. May float or sink in water; slightly soluble.

HEALTH AND SAFETY INFORMATION

Irritates eyes, skin, and respiratory tract. Long-term problems include blood, liver, and kidney damage.

IARC Evaluation: Group 2B, possible carcinogen; No human data; sufficient animal evidence.
NTP: Anticipated carcinogen.
IDLH: 1000 ppm; suspected human carcinogen.
Odor Threshold: 4.85 ppm
OSHA Table Z-1-A Air Contaminant

Exposure Limits:	TWA
ACGIH: (1)	10 ppm; 36 mg/m^3
NIOSH: (2)	
OSHA:	25 ppm; 90 mg/m^3

(1) Animal carcinogen.
(2) Suspected human carcinogen. Reduce exposure to lowest feasible level.

RECOMMENDATIONS FOR RESPIRATOR SELECTION: NIOSH

Any detectable concentration: SCBAF:PD,PP (any self-contained breathing apparatus that has a full facepiece and is operated in a pressure-demand or other positive-pressure mode); or SAF:PD,PP:ASCBA (any supplied-air respirator that has a full facepiece and is operated in a pressure-demand or other positive-pressure mode in combination with an auxiliary, self-contained breathing apparatus operated in a pressure-demand or other positive-pressure mode).

Escape: SCBAE (any appropriate escape-type, self-contained breathing apparatus).

CHEMICAL REACTIVITY

Heat may cause explosive decomposition. Reacts with amines, strong acids, alkalis, oxidizers, chlorosulfonic acid, metal oxides, oleum, and combustible materials. Nitric acid and bases form shock-sensitive materials. May attack some plastics.

FIRE INFORMATION

Combustion products include toxic nitrogen oxides. Vapors are heavier than air and will collect in low areas. Vapors in confined areas may explode when exposed to fire. Vapors may travel long distances to ignition sources and flash back. Storage containers and parts of containers may rocket great distances, in many directions. If material or contaminated runoff enters waterways, notify downstream users of potentially contaminated waters. Notify local health and fire officials and pollution control agencies. Use dry chemicals, foam, carbon dioxide, water spray. From a secure, explosion-proof location, use water spray to cool exposed containers.

NFPA Hazard Classifications:

Health Hazards (Blue) 1
Flammability (Red) 3
Reactivity (Yellow) 2

Flash Point: 100°F (38°C) o.c.; 82°F (28°C) c.c.
Flammable Limits in Air: LEL 2.6%; UEL 11.0%
Autoignition Temperature: 802°F (428°C)

FIRST AID

Move victim to fresh air. Call emergency medical care. Apply artificial respiration if victim is not breathing. Administer oxygen if breathing is difficult. Remove and isolate contaminated clothing and shoes. In case of contact with substance, immediately flush skin or eyes with running water for at least 20 minutes. Wash skin with soap and water. Keep victim warm and quiet. Effects of exposure (inhalation, ingestion, or skin contact) to substance may be delayed. Ensure that medical personnel are aware of the material(s) involved and take precautions to protect themselves.

ENVIRONMENTAL INFORMATION

Clean Air Act: Hazardous Air Pollutants (Title I, Part A, Section 112).
EPA Hazardous Waste Number (RCRA No.): U171.
RCRA Section 261 Hazardous Constituents
EPCRA Section 304: Reportable Quantity (RQ): CERCLA, 10 lb (4.54 kg).
EPCRA Section 313: Form R *de minimis* concentration reporting level: 0.1%.
California's Proposition 65: Carcinogen.
TSCA: 40CFR716.120(*d*)10; 40CFR712.30(*e*)10 re dermal absorption testing.

NONYL PHENOL N:360

SYNONYMS: 2,6-DIMETHYL-4-HEPTYLPHENOL, (o- and p-); EINECS No. 246-672-0; HYDROXY No. 253; NONYL PHENOL (MIXED ISOMERS); PHENOL, NONYL; PHENOL, 4-NONYL, BRANCHED; PREVOSTSEL VON

IDENTIFICATION

CAS:	104-40-5; 25154-52-3 (mixed isomers); 84852-15-3 (branched)
ERG ID:	UN 2904
ERG Guide:	154
ERG Guide Category:	SUBSTANCES-TOXIC AN/OR CORROSIVE (COMBUSTIBLE)
Molecular Formula:	$C_{15}H_{24}O$
RTECS No.:	SM5600000

Properties: Flammable, thick liquid. Light yellow, straw color. Medicinal odor. Practically insoluble in water.

HEALTH AND SAFETY INFORMATION

Corrosive to eyes, skin, and respiratory tract. Inhalation may cause pulmonary edema, which can be delayed for several hours; there is a risk of death in serious cases. Skin contact contributes significantly to overall exposure

RECOMMENDATIONS FOR RESPIRATOR SELECTION

At any concentrations above the NIOSH REL, or where there is no REL, at any detectable concentration: SCBAF:PD,PP (any self-contained breathing apparatus that has a full facepiece and is operated in a pressure-demand or other positive-pressure mode); or SAF:PD,PP:ASCBA (any supplied-air respirator that has a full facepiece and is operated in a pressure-demand or other positive-pressure mode in combination with an auxiliary, self-contained breathing apparatus operated in a pressure-demand or other positive-pressure mode).

Escape: GMFOV [any air-purifying, full-facepiece respirator (gas mask) with a chin-style, front- or back-mounted organic vapor canister]; or SCBAE (any appropriate escape-type, self-contained breathing apparatus).

CHEMICAL REACTIVITY

Contact with strong oxidizers may cause fire and explosions. Incompatible with strong acids, caustics, isocyanates, aliphatic amines, amides. Copper and its alloys promote color formation.

FIRE INFORMATION

Vapors in confined areas may explode when exposed to fire. If material or contaminated runoff enters waterways, notify downstream users of potentially contaminated waters. Notify local health and fire officials and pollution control agencies. Use dry chemicals, alcohol-resistant foam, carbon dioxide. From a secure, explosion-proof location, use water spray to cool exposed containers.

NFPA Hazard Classifications:

Health Hazards (Blue) 2
Flammability (Red) 1
Reactivity (Yellow) 0

Flash Point: 300°F (149°C) o.c.; 285°F (141°C) c.c.

FIRST AID

Move victim to fresh air. Call emergency medical care. Apply artificial respiration if victim is not breathing. Administer oxygen if breathing is difficult. In case of contact with substance, immediately flush skin or eyes with running water for at least 20 minutes. Remove and isolate contaminated clothing and shoes. Ensure that medical personnel are aware of the material(s) involved and take precautions to protect themselves.

ENVIRONMENTAL INFORMATION

TSCA: 40CFR799.5000 as 4-nonylphenol, branched.

OCTADECYLTRICHLOROSILANE O:007

SYNONYMS: EINECS No. 203-930-7; 1,6-HEXANEDIOIC ACID; SILANE, TRICHLOROOCTADECYL-; TRICHLOROOCTADECYLSILANE; SILANE, OCTADECYLTRICHLORO-; SILANE, TRICHLOROOCTADECYL-

IDENTIFICATION

CAS:	112-04-9
ERG ID:	UN 1800
ERG Guide:	156
ERG Guide Category:	SUBSTANCES - TOXIC AND/OR CORROSIVE (COMBUSTIBLE/WATER-SENSITIVE)
Molecular Formula:	$C_{18}H_{37}Cl_3Si$
RTECS No.:	VV4680000
Properties:	Combustible and corrosive liquid. Soluble in water. May react with water or steam to form corrosive and toxic hydrochloric acid.

HEALTH AND SAFETY INFORMATION

Toxic by ingestion. Corrosive to eyes, skin, and respiratory tract. Inhalation may cause pulmonary edema, which can be delayed for several hours; there is a risk of death in serious cases.

RECOMMENDATIONS FOR RESPIRATOR SELECTION

At any concentrations above the NIOSH REL, or where there is no REL, at any detectable concentration: SCBAF:PD,PP (any self-contained breathing apparatus that has a full facepiece and is operated in a pressure-demand or other positive-pressure mode); or SAF:PD,PP:ASCBA (any supplied-air respirator that has a full facepiece and is operated in a pressure-demand or other positive-pressure mode in combination with an auxiliary, self-contained breathing apparatus operated in a pressure-demand or other positive-pressure mode).

Escape: GMFOV [any air-purifying, full-facepiece respirator (gas mask) with a chin-style, front- or back-mounted organic vapor canister]; or SCBAE (any appropriate escape-type, self-contained breathing apparatus).

CHEMICAL REACTIVITY

Chlorosilanes are incompatible with ammonia, steam, and water. Contact with metals in the presence of moisture can cause corrosion and form explosive hydrogen gas.

FIRE INFORMATION

When heated may form explosive mixture with air. Combustion products include toxic fumes of chlorides. Vapors are heavier than air and will collect in low areas. Vapors in confined areas may explode when exposed to fire. Storage containers and parts of containers may rocket great distances, in many directions. If material or contaminated runoff enters waterways, notify downstream users of potentially contaminated waters. Notify local health and fire officials and pollution control agencies. Use AFFF, dry chemicals, alcohol-resistant medium expansion foam, carbon dioxide. From a secure, explosion-proof location, use water spray to cool exposed containers.

NFPA Hazard Classifications

Health Hazards (Blue) 3
Flammability (Red) 2
Reactivity (Yellow) 2

Flash Point: 193°F (89°C)

FIRST AID

Move victim to fresh air. Call emergency medical care. Apply artificial respiration if victim is not breathing. Do not use mouth-to-mouth method if victim ingested or inhaled the substance; induce artificial respiration with the aid of a pocket mask equipped with a one-way valve or other proper respiratory medical device. Administer oxygen if breathing is difficult. Remove and isolate contaminated clothing and shoes. In case of contact with substance, immediately flush skin or eyes with running water for at least 20 minutes. For minor skin contact, avoid spreading material on unaffected skin. Keep victim warm and quiet. Effects of exposure (inhalation, ingestion, or skin contact) may be delayed. Ensure that medical personnel are aware of the material(s) involved and take precautions to protect themselves.

OCTAFLUORO–2–BUTENE O:014

SYNONYMS: EINECS No. 206-640-9; FC-1318; OCTAFLUO-2-BUTENO (Spanish); OCTAFLUOROBUTENE-2; OCTAFLUOROBUT-2-ENE; 1,1,1,2,3,4,4,4-OCTAFLUORO-2-BUTENE; PERFLUOROBUT-2-ENE; PERFLUORO-2-BUTENE

IDENTIFICATION

CAS:	360-89-4
ERG ID:	UN 2422
ERG Guide:	126

ERG Guide Category: GASES - COMPRESSED OR LIQUEFIED (INCLUDING REFRIGERANT GASES)

Molecular Formula: C_4F_8

RTECS No.: EM8980000

Properties: Nonflammable liquefied gas. Colorless.

HEALTH AND SAFETY INFORMATION

Contact with liquid may cause frostbite. Repeated exposure to fluoride chemicals may cause stiffness in muscles or ligaments and even crippling.

IDLH: 25 ppm as fluorine
Odor Threshold: 0.126 ppm as fluorine

Exposure Limits:	**TWA**
ACGIH:	2.5 mg/m^3 as F
NIOSH:	2.5 mg/m^3 as F
OSHA:	2.5 mg/m^3 as F

RECOMMENDATIONS FOR RESPIRATOR SELECTION: NIOSH/OSHA as fluorine

1 ppm: SA[1] (any supplied-air respirator).
2.5 ppm: SA:CF[1] (any supplied-air respirator operated in a continuous-flow mode).
5 ppm: SCBAF (any self-contained breathing apparatus with a full facepiece); or SAF (any supplied-air respirator with a full facepiece).
25 ppm: SAF:PD,PP (any supplied-air respirator that has a full facepiece and is operated in a pressure-demand or other positive-pressure mode).

Emergency or Planned Entry into Unknown Concentrations or IDLH Conditions:
SCBAF:PD,PP (any self-contained breathing apparatus that has a full facepiece and is operated in a pressure-demand or other positive-pressure mode); or SAF:PD,PP:ASCBA (any supplied-air respirator that has a full facepiece and is operated in a pressure-demand or other positive-pressure mode in combination with an auxiliary, self-contained breathing apparatus operated in a pressure-demand or other positive-pressure mode).

Escape: GMFS *end of service life indicator (ESLI) required* (any air-purifying, full-facepiece respirator (gas mask) with a chin-style, front- or back-mounted canister providing protection against the compound of concern); or SCBAE (any appropriate escape-type, self-contained breathing apparatus).

[1]Substance reported to cause eye irritation or damage; may require eye protection.

CHEMICAL REACTIVITY

Incompatible with barium, sodium, potassium.

FIRE INFORMATION

Not flammable. Combustion products include toxic fumes of fluorides. Storage containers and parts of containers may rocket great distances, in many directions. Notify local health and fire officials and pollution control agencies. Use any extinguishing agent suitable for surrounding fires. From a secure, explosion-proof location, use water spray to cool exposed containers. If cooling streams are ineffective (venting sound increases in volume and pitch, tank discolors, or shows any signs of deforming), withdraw immediately to a secure position.

FIRST AID

Move victim to fresh air. Call emergency medical care. Apply artificial respiration if victim is not breathing. Administer oxygen if breathing is difficult. Remove and isolate contaminated clothing and shoes. In case of contact with liquefied gas, thaw frosted parts with lukewarm water. Keep victim warm and quiet. Keep victim under observation. Ensure that medical personnel are aware of the material(s) involved and take precautions to protect themselves.

n–OCTYL BROMIDE O:063

SYNONYMS: BROMO OCTANE; 1-BROMOOCTANE; BROMURE D'octyle (French); BROMURO di OTTILE (Italian); CAPRYL BROMIDE; CAPRYLIC BROMIDE; EINECS No. 203-912-9; OCTANE,1-BROMO-

IDENTIFICATION

CAS:	111-83-1
Molecular Formula:	$C_8H_{17}Br$
RTECS No.:	RG8575000
Properties:	Liquid. Colorless. Floats on water; insoluble.

HEALTH AND SAFETY INFORMATION

Organic bromides are toxic. Due to the availability of insufficient data on human effects, caution should be exercised.

IDLH: 3 ppm as bromine

Exposure Limits:	**TWA**	**STEL**
ACGIH:	0.1 ppm; 0.66 mg/m^3 as Br	0.2 ppm; 1.3 mg/m^3 as Br
NIOSH:	0.1 ppm; 0.7 mg/m^3 as Br	0.3 ppm; 2.0 mg/m^3 as Br
OSHA:	0.1 ppm; 0.7 mg/m^3 as Br	

RECOMMENDATIONS FOR RESPIRATOR SELECTION

At any concentrations above the NIOSH REL, or where there is no REL, at any detectable concentration: SCBAF:PD,PP (any self-contained breathing apparatus that has a full facepiece and is operated in a pressure-demand or other positive-pressure mode); or SAF:PD,PP:ASCBA (any supplied-air respirator that has a full facepiece and is operated in a pressure-demand or other positive-pressure mode in combination with an auxiliary, self-contained breathing apparatus operated in a pressure-demand or other positive-pressure mode).

Escape: GMFOV [any air-purifying, full-facepiece respirator (gas mask) with a chin-style, front- or back-mounted organic vapor canister]; or SCBAE (any appropriate escape-type, self-contained breathing apparatus).

CHEMICAL REACTIVITY

Contact with strong oxidizers may cause a violent reaction.

FIRE INFORMATION

Combustion products include toxic bromides. Vapors are heavier than air and will collect in low areas. Vapors in confined areas may explode when exposed to fire. Vapors may travel long distances to ignition sources and flash back. Storage containers and parts of containers may rocket great distances, in many directions. If material or contaminated runoff enters waterways, notify downstream users of potentially contaminated waters. Notify local health and fire officials and pollution control agencies. Use dry chemicals, foam, carbon dioxide. From a secure, explosion-proof location, use water spray to cool exposed containers.

NFPA Hazard Classifications: Not listed

Flash Point: 172°F (78°C)

FIRST AID

Move victim to fresh air; call emergency medical care. Apply artificial respiration if victim is not breathing. Do not use mouth-to-mouth method if victim ingested or inhaled the substance; induce artificial respiration with the aid of a pocket mask equipped with a one-way valve or other proper respiratory medical device. Administer oxygen if breathing is difficult. Remove and isolate contaminated clothing and shoes. In case of contact with substance, immediately flush skin or eyes with running water for at least 20 minutes. For minor skin contact, avoid spreading material on unaffected skin. Keep victim warm and quiet. Effects of exposure (inhalation, ingestion, or skin contact) to substance may be delayed. Ensure that medical personnel are aware of the material(s) involved and take precautions to protect themselves.

OCTYL DECYL PHTHALATE O:066

SYNONYMS: 1,2-BENZENEDICARBOXYLIC ACID, DECYL OCTYL ESTER; N-DECYL n-OCTYL

PHTHALATE; DINOPOL 235; EINECS No. 204-295-9; FTALATO de N-OCTIL-n-DECILO (Spanish); N-OCTYL n-DECYL PHTHALATE; POLYCIZER 532; POLYCIZER 562; STAFLEX 500

IDENTIFICATION

CAS:	119-07-3
Molecular Formula:	$C_{26}H_{42}O_4$
RTECS No.:	TI0550000
Properties:	Combustible liquid. Colorless. Mild, characteristic odor. Floats on water.

HEALTH AND SAFETY INFORMATION

RECOMMENDATIONS FOR RESPIRATOR SELECTION

At any concentrations above the NIOSH REL, or where there is no REL, at any detectable concentration: SCBAF:PD,PP (any self-contained breathing apparatus that has a full facepiece and is operated in a pressure-demand or other positive-pressure mode); or SAF:PD,PP:ASCBA (any supplied-air respirator that has a full facepiece and is operated in a pressure-demand or other positive-pressure mode in combination with an auxiliary, self-contained breathing apparatus operated in a pressure-demand or other positive-pressure mode).

Escape: GMFOV [any air-purifying, full-facepiece respirator (gas mask) with a chin-style, front- or back-mounted organic vapor canister]; or SCBAE (any appropriate escape-type, self-contained breathing apparatus).

CHEMICAL REACTIVITY

Contact with strong oxidizers may cause fire and explosions. Incompatible with nitrates, nitric acid, sulfuric acid.

FIRE INFORMATION

Vapors in confined areas may explode when exposed to fire. Containers may explode in fire. If material or contaminated runoff enters waterways, notify downstream users of potentially contaminated waters. Notify local health and fire officials and pollution control agencies. Use dry chemicals, alcohol-resistant foam, carbon dioxide. From a secure, explosion-proof location, use water spray to cool exposed containers.

NFPA Hazard Classifications: Not listed

Flash Point: 455°F (235°C) c.c.

FIRST AID

Move victim to fresh air. Call emergency medical care. Apply artificial respiration if victim is not breathing. Administer oxygen if breathing is difficult. Remove and isolate contaminated clothing and shoes. In case of contact with substance, immediately flush skin or eyes with running water for at least 20

minutes. Wash skin with soap and water. Keep victim warm and quiet. Effects of exposure (inhalation, ingestion, or skin contact) to substance may be delayed. Ensure that medical personnel are aware of the material(s) involved and take precautions to protect themselves.

ENVIRONMENTAL INFORMATION

TSCA: 40CFR712.30(*d*); 40CFR716.120(*c*)2 alkyl esters of 1,2-benzenedicarboxylic acid; Health and Safety Reporting List

tert–OCTYL MERCAPTAN O:069

SYNONYMS: EINECS No. 205-490-1; METHYL HEPTANETHIOL; tert-OCTANETHIOL; terc-OCTILMERCAPTANO (Spanish); t-OCTYL MERCAPTAN; terc.OKTANTHIOL (German); 2-PENTANETHIOL, 2,4,4-TRIMETHYL-; 2,4,4-TRIMETHYL-2-PENTANETHIOL

IDENTIFICATION

CAS:	141-59-3
ERG ID:	UN 3023
ERG Guide:	131
ERG Guide Category:	FLAMMABLE LIQUIDS - TOXIC
Molecular Formula:	$C_8H_{18}S$
Properties:	Flammable liquid. Mercaptans have an offensive odor. Floats on water; insoluble; may react forming toxic, corrosive and flammable vapors.

HEALTH AND SAFETY INFORMATION

Poisonous by ingestion. Irritates eyes, skin, and respiratory tract.

RECOMMENDATIONS FOR RESPIRATOR SELECTION

At any concentrations above the NIOSH REL, or where there is no REL, at any detectable concentration: SCBAF:PD,PP (any self-contained breathing apparatus that has a full facepiece and is operated in a pressure-demand or other positive-pressure mode); or SAF:PD,PP:ASCBA (any supplied-air respirator that has a full facepiece and is operated in a pressure-demand or other positive-pressure mode in combination with an auxiliary, self-contained breathing apparatus operated in a pressure-demand or other positive-pressure mode).

Escape: GMFOV [any air-purifying, full-facepiece respirator (gas mask) with a chin-style, front- or back-mounted organic vapor canister]; or SCBAE (any appropriate escape-type, self-contained breathing apparatus).

CHEMICAL REACTIVITY

Forms explosive mixture with air. Contact with oxidizers may cause fire and explosions. Contact with water or acids may form toxic, corrosive and flammable vapors.

FIRE INFORMATION

Combustion products include toxic fumes of sulfides. Vapors are initially heavier than air and will collect in low areas. Vapors in confined areas may explode when exposed to fire. If material or contaminated runoff enters waterways, notify downstream users of potentially contaminated waters. Notify local health and fire officials and pollution control agencies. Use dry chemicals, alcohol-resistant foam, carbon dioxide. From a secure, explosion-proof location, use water spray to cool exposed containers.

NFPA Hazard Classifications:

Health Hazards (Blue) 2
Flammability (Red) 2
Reactivity (Yellow) 0

Flash Point: 115°F (46°C) o.c.

FIRST AID

Move victim to fresh air. Call emergency medical care. Apply artificial respiration if victim is not breathing. Do not use mouth-to-mouth method if victim ingested or inhaled the substance; induce artificial respiration with the aid of a pocket mask equipped with a one-way valve or other proper respiratory medical device. Administer oxygen if breathing is difficult. Remove and isolate contaminated clothing and shoes. In case of contact with substance, immediately flush skin or eyes with running water for at least 20 minutes. Wash skin with soap and water. Keep victim warm and quiet. Effects of exposure (inhalation, ingestion, or skin contact) to substance may be delayed. Ensure that medical personnel are aware of the material(s) involved and take precautions to protect themselves.

OCTYL PHENOL — O:072

SYNONYMS: DIISOBUTYL PHENOL; EINECS No. 248-310-7; para-OCTYL PHENOL; PHENOL, (1,1,3,3-TETRAMETHYLBUTYL)-; (1,1,3,3-TETRAMETHYLBUTYL)PHENOL; USAF RH-6

IDENTIFICATION

CAS:	1806-26-4; 27193-28-8 (mixed isomers)
ERG ID:	UN 2928 as toxic solid, corrosive, organic, n.o.s.
ERG Guide:	154
ERG Guide Category:	SUBSTANCES-TOXIC AND/OR CORROSIVE
Molecular Formula:	$C_{14}H_{22}O$
RTECS No.:	SM5775000
Properties:	Combustible flaky solid or powder. White to light pink. Insoluble in water.

HEALTH AND SAFETY INFORMATION

Poisonous. Corrosive to eyes, skin, and respiratory tract. Skin contact contributes significantly to overall exposure. Long-term exposure may include skin discoloration. Inhalation may cause pulmonary edema, which can be delayed for several hours; there is a risk of death in serious cases.

RECOMMENDATIONS FOR RESPIRATOR SELECTION

At any concentrations above the NIOSH REL, or where there is no REL, at any detectable concentration: SCBAF:PD,PP (any self-contained breathing apparatus that has a full facepiece and is operated in a pressure-demand or other positive-pressure mode); or SAF:PD,PP:ASCBA (any supplied-air respirator that has a full facepiece and is operated in a pressure-demand or other positive-pressure mode in combination with an auxiliary, self-contained breathing apparatus operated in a pressure-demand or other positive-pressure mode).

Escape: GMFOV [any air-purifying, full-facepiece respirator (gas mask) with a chin-style, front- or back-mounted organic vapor canister]; or SCBAE (any appropriate escape-type, self-contained breathing apparatus).

CHEMICAL REACTIVITY

Reacts vigorously with oxidizing materials. Finely dispersed particles may form an explosive mixture with air.

FIRE INFORMATION

Combustion products include toxic vapors of phenol. If material or contaminated runoff enters waterways, notify downstream users of potentially contaminated waters. Notify local health and fire officials and pollution control agencies. Use dry chemicals, water spray. From a secure, explosion-proof location, use water spray to cool exposed containers.

NFPA Hazard Classifications: Not listed

FIRST AID

Move victim to fresh air. Call emergency medical care. Apply artificial respiration if victim is not breathing. Do not use mouth-to-mouth method if victim ingested or inhaled the substance; induce artificial respiration with the aid of a pocket mask equipped with a one-way valve or other proper respiratory medical device. Administer oxygen if breathing is difficult. Remove and isolate contaminated clothing and shoes. In case of contact with substance, immediately flush skin or eyes with running water for at least 20 minutes. For minor skin contact, avoid spreading material on unaffected skin. Keep victim warm and quiet. Effects of exposure (inhalation, ingestion, or skin contact) may be delayed. Ensure that medical personnel are aware of the material(s) involved and take precautions to protect themselves.

OCTYLTRICHLOROSILANE O:075

SYNONYMS: EINECS No. 226-112-1; SILANE, TRICHLOROOCTYL-; TRICHLOROOCTYLSILANE

812 Octlytrichlorosilane

IDENTIFICATION

CAS:	5283-66-9
ERG ID:	UN 1801
ERG Guide:	156
ERG Guide Category:	SUBSTANCES - TOXIC AND/OR CORROSIVE (COMBUSTIBLE/WATER-SENSITIVE)
Molecular Formula:	$C_8H_{17}Cl_3Si$
RTECS No.:	VV4700000
Properties:	Fuming liquid. Water-white. Pungent, irritating odor. Contact with moisture forms irritating fumes.

HEALTH AND SAFETY INFORMATION

Corrosive to eyes, skin, and respiratory tract. Inhalation may cause pulmonary edema, which can be delayed for several hours; there is a risk of death in serious cases.

Exposure Limits:	**TWA**	**CEILING**	**STEL**
ACGIH:	0.5 ppm; 1.5 (Cl)mg/m^3		1.0 ppm; 2.9(Cl) mg/m^3
NIOSH:		0.5 ppm; 1.45(Cl) mg/m^3 [15 min]	
OSHA:		1 ppm; 3 (Cl) mg/m^3	

RECOMMENDATIONS FOR RESPIRATOR SELECTION

At any concentrations above the NIOSH REL, or where there is no REL, at any detectable concentration: SCBAF:PD,PP (any self-contained breathing apparatus that has a full facepiece and is operated in a pressure-demand or other positive-pressure mode); or SAF:PD,PP:ASCBA (any supplied-air respirator that has a full facepiece and is operated in a pressure-demand or other positive-pressure mode in combination with an auxiliary, self-contained breathing apparatus operated in a pressure-demand or other positive-pressure mode).

Escape: GMFOV [any air-purifying, full-facepiece respirator (gas mask) with a chin-style, front- or back-mounted organic vapor canister]; or SCBAE (any appropriate escape-type, self-contained breathing apparatus).

CHEMICAL REACTIVITY

Incompatible with ammonia. Contact with strong oxidizers may cause fire and explosions. Reacts with water or steam to produce toxic and corrosive fumes.

FIRE INFORMATION

Combustion products include toxic fumes of chlorides. If material or contaminated runoff enters waterways, notify downstream users of potentially contaminated waters. Notify local health and fire officials and pollution control agencies. Use dry chemicals, alcohol-resistant foam, carbon dioxide. From a secure, explosion-proof location, use water spray to cool exposed containers.

FIRST AID

Move victim to fresh air. Call emergency medical care. Apply artificial respiration if victim is not breathing. Do not use mouth-to-mouth method if victim ingested or inhaled the substance; induce artificial respiration with the aid of a pocket mask equipped with a one-way valve or other proper respiratory medical device. Administer oxygen if breathing is difficult. Remove and isolate contaminated clothing and shoes. In case of contact with substance, immediately flush skin or eyes with running water for at least 20 minutes. For minor skin contact, avoid spreading material on unaffected skin. Keep victim warm and quiet. Effects of exposure (inhalation, ingestion, or skin contact) may be delayed. Ensure that medical personnel are aware of the material(s) involved and take precautions to protect themselves.

OLEYL ALCOHOL O:078

SYNONYMS: ADOL; ADOL 34; ADOL 80; ADOL 85; ADOL 90; ADOL 320; ADOL 330; ADOL 340; ATALCO O; CACHALOT O-1; CACHALOT O-3; CACHALOT O-8; CACHALOT O-15; CONDITIONER 1; CRODACOL A.10; CRODACOL-O; DASCOOL; DERMAFFINE; EINECS No. 205-597-3; H.D. EUTANOL; HD OLEYL ALCOHOL 70/75; HD OLEYL ALCOHOL 80/85; HD OLEYL ALCOHOL 90/95; HD OLEYL ALCOHOL CG; LANCOL; LOXANOL 95; LOXANOL M; NOVOL; OCENOL; OCEOL; cis-9-OCTADECEN-1-OL; (Z)-9-OCTADECEN-1-OL; OLEOL; SATOL; SIPOL O; UNJECOL 50; UNJECOL 70; UNJECOL 90; UNJECOL 110

IDENTIFICATION

CAS:	143-28-2
Molecular Formula:	$C_{18}H_{36}O$
RTECS No.:	RG4120000
Properties:	Combustible, thick, oily liquid. Clear to pale yellow. Insoluble in water.

HEALTH AND SAFETY INFORMATION

Irritates skin and eyes.

RECOMMENDATIONS FOR RESPIRATOR SELECTION

At any concentrations above the NIOSH REL, or where there is no REL, at any detectable concentration: SCBAF:PD,PP (any self-contained breathing apparatus that has a full facepiece and is operated in a pressure-demand or other positive-pressure mode); or SAF:PD,PP:ASCBA (any supplied-air respirator that has a full facepiece and is operated in a pressure-demand or other positive-pressure mode in combination with an auxiliary, self-contained breathing apparatus operated in a pressure-demand or other positive-pressure mode).

Escape: GMFOV [any air-purifying, full-facepiece respirator (gas mask) with a chin-style, front- or back-mounted organic vapor canister]; or SCBAE (any appropriate escape-type, self-contained breathing apparatus).

CHEMICAL REACTIVITY

Contact with strong oxidizers may cause fire and explosions. Alcohols are incompatible with sulfuric acid, nitric acid, caustics, aliphatic amines, isocyanates.

FIRE INFORMATION

Decomposition products upon heating include irritating and toxic gases. If material or contaminated runoff enters waterways, notify downstream users of potentially contaminated waters. Notify local health and fire officials and pollution control agencies. Use dry chemicals, alcohol-resistant foam, carbon dioxide. From a secure, explosion-proof location, use water spray to cool exposed containers.

NFPA Hazard Classifications: Not listed

Flash Point: 234°F (112°C) c.c.

FIRST AID

Move victim to fresh air. Call emergency medical care. Apply artificial respiration if victim is not breathing. Administer oxygen if breathing is difficult. Remove and isolate contaminated clothing and shoes. In case of contact with substance, immediately flush skin or eyes with running water for at least 20 minutes. Wash skin with soap and water. Keep victim warm and quiet. Effects of exposure (inhalation, ingestion, or skin contact) to substance may be delayed. Ensure that medical personnel are aware of the material(s) involved and take precautions to protect themselves.

OXYGEN O:100

SYNONYMS: EEC No. 008-001-00-8; EINECS No. 231-956-9; LIQUID OXYGEN; LOX; OSSIGENO (Italian); OXIGENO (Spanish); OXYGEN, ELEMENTAL; OXYGEN, LIQUID; PDE-100; POLY I GAS; SAUERSTOFF (German); ZUURSTOF (Dutch)

IDENTIFICATION

CAS:	7782-44-7
ERG ID:	UN 1072 (compressed gas); UN 1073 (cryogenic liquid)
ERG Guide:	122 (cryogenic liquid; compressed gas)
ERG Guide Category:	GASES–OXIDIZING (INCLUDING REFRIGERATED LIQUIDS)
Molecular Formula:	O_2
RTECS No.:	RS2060000
Properties:	Nonflammable (but may cause combustible materials to ignite) compressed gas, or cryogenic liquid. Colorless gas; blue liquid. Odorless. Sinks and "boils" in water.

HEALTH AND SAFETY INFORMATION

Breathing 50-100% oxygen at normal pressure, even intermittently over a prolonged period, may cause lung damage. Contact with liquid may cause frostbite.

CHEMICAL REACTIVITY

Heat of water will vigorously vaporize liquid oxygen. A strong oxidizer; violent reaction with reducing agents, combustibles, and organic and easily oxidizable materials; contact may cause fire and explosions. The low temperature may cause brittleness on contact with some materials.

FIRE INFORMATION

A strong oxidizer. Nonflammable, but will increase the intensity of a fire. Gases are heavier than air and will collect in low areas. Storage containers and parts of containers may rocket great distances, in many directions. Notify local health and fire officials and pollution control agencies. Material does not burn; use extinguishing agent appropriate for surrounding fire. Do not apply water to cryogenic liquid containers. If cryogenic liquid containers are exposed to direct flames or elevated temperatures for prolonged times, withdraw immediately to secure location.

NFPA Hazard Classifications:

Health Hazards (Blue) 3
Flammability (Red) 0
Reactivity (Yellow) 0
Special situations . . . . OXY

Flash Point: Not flammable but supports combustion, and may cause combustible to ignite.

FIRST AID

Move victims to fresh air. Call emergency medical care. Apply artificial respiration if victim is not breathing. Administer oxygen if breathing is difficult. Remove and isolate contaminated clothing and shoes. Clothing frozen to the skin should be thawed before being removed. In case of contact with liquefied gas, thaw frosted parts with lukewarm water. Keep victim warm and quiet. Ensure that medical personnel are aware of the material(s) involved and take precautions to protect themselves.

OXYGEN DIFLUORIDE O:110

SYNONYMS: DIFLUORINE MONOXIDE; DIFLUORURO de OXIGENO (Spanish); EINECS No. 231-996-7; FLUORINE MONOXIDE; FLUORINE OXIDE; OXYGEN FLUORIDE

IDENTIFICATION

CAS: 7783-41-7

816 Oxygen Difluoride

ERG ID:	UN 2190
ERG Guide:	124
ERG Guide Category:	GASES - TOXIC AND/OR CORROSIVE - OXIDIZING
Molecular Formula:	F_2O
RTECS No.:	RS2100000
Properties:	Nonflammable colorless gas. A peculiar, foul odor. Shipped as a nonliquified compressed gas. Explode on contact with air and water. A powerful oxidizer.

HEALTH AND SAFETY INFORMATION

IDLH: 0.5 ppm; 1.12 mg/m^3
Odor Threshold: 0.098 ppm
OSHA Table Z-1-A Air Contaminant
OSHA Process Safety Management of Highly Hazardous Chemicals (29CFR 1910.119, Appendix A), Threshold Quantity: 100 pounds

Exposure Limits:	**CEILING**
ACGIH:	0.05 ppm; 0.11 mg/m^3
NIOSH:	0.05 ppm; 0.10 mg/m^3
OSHA:	0.05 ppm; 0.10 mg/m^3

Poisonous. Inhaling gas can cause death. Skin and eye contact causes severe burns and blindness. contact with liquid can cause frostbite.

RECOMMENDATIONS FOR RESPIRATOR SELECTION: NIOSH

0.5 ppm: SA (any supplied-air respirator); or SCBAF (any self-contained breathing apparatus with a full facepiece)

Emergency or Planned Entry into Unknown Concentrations or IDLH Conditions: SCBAF:PD,PP (any self-contained breathing apparatus that has a full facepiece and is operated in a pressure-demand or other positive-pressure mode); or SAF:PD,PP:ASCBA (any supplied-air respirator that has a full facepiece and is operated in a pressure-demand or other positive-pressure mode in combination with an auxiliary, self-contained breathing apparatus operated in a pressure-demand or other positive-pressure mode)

Escape: GMFS *Only nonoxidizable sorbents are allowed (not charcoal).* (any air-purifying, full-facepiece respirator (gas mask) with a chin-style, front- or back-mounted canister providing protection against the compound of concern); or SCBAE (any appropriate escape-type, self-contained breathing apparatus).

CHEMICAL REACTIVITY

A strong oxidizer. Reacts with water, forming hydrofluoric acid. Forms explosive reactions with absorbents like silica gel, alumina, and molecular sieve. Reacts with combustible materials, chlorine, bromine, iodine, platinum, metal oxides, moist air, hydrogen sulfide, and hydrocarbons.

FIRE INFORMATION

Combustion products include toxic fluorine gas. Water contact may cause explosions. Firefighting gear, including SCBA, does not provide adequate protection. If exposure occurs, remove and isolate gear immediately and thoroughly decontaminate personnel. Vapors are heavier than air and will collect in low areas. Nonflammable gas, but a strong oxidizer; will increase intensity of fire.

NFPA Hazard Classifications:

Material does not burn; use extinguishing agents appropriate for surrounding fires. If water must be used, use it in flooding quantities. From a secure, explosion-proof location, use water spray to cool exposed containers. If cooling streams are ineffective (venting sound increases in volume and pitch, tank discolors, or shows any signs of deforming), withdraw immediately to a secure position. If material or contaminated runoff enters waterways, notify downstream users of potentially contaminated waters. Notify local health and fire officials and pollution control agencies.

Health Hazards (Blue) 4
Flammability (Red) 0
Reactivity (Yellow) 4
Special situations . . . . OXY

FIRST AID

Do not put yourself in danger by entering a contaminated area to rescue a victim. Move victim to fresh air. Call emergency medical care. Provide basic life support/CPR as needed. Decontaminate the victim as follows: Inhalation–remove victim to fresh air and give oxygen if available; Skin–remove and isolate contaminated clothing (including shoes) and wash skin with soap and large volumes of water for at least 20 minutes. Eye–rinse eyes with large volumes of water or saline for at least 20 minutes. Swallowed–do not make victim vomit. Victims should be examined by a physician as soon as possible. Ensure that medical personnel are aware of the material(s) involved and take precautions to protect themselves. Toxic effects may be delayed. For skin burns, decontaminate with water and apply a clean dressing.

PARALDEHYDE P:023

SYNONYMS: p-ACETALDEHYDE; ACETALDEHYDE, TRIMER; EEC No. 605-004-00-1; EINECS No. 204-639-8; ELALDEHYDE; PARAACETALDEHYDE; PARACETALDEHYDE; PARAL; PARALDEHIDO (Spanish); PARALDEHYD (German); PARALDEIDE (Italian); PCHO; PORAL; TRIACETALDEHYDE (French); 2,4,6-TRIMETHYL-1,3,5-TRIOXAAN (Dutch); 2,4,6-TRIMETHYL-1,3,5-TRIOXANE; 2,4,6-TRIMETHYL-S-TRIOXANE; s-TRIMETHYL-TRIOXYMETHYLENE; TRIMETHYLTRIOXYMETHYLENE; 2,4,6-TRIMETIL-1,3,5-TRIOSSANO (Italian); 1,3,5-TRIOXANE,2,4,6-TRIMETHYL-

818 Paraldehyde

IDENTIFICATION

CAS:	123-63-7
ERG ID:	UN 1264
ERG Guide:	129
ERG Guide Category:	FLAMMABLE LIQUIDS (POLAR/WATER-MISCIBLE/NOXIOUS)
Molecular Formula:	$C_6H_{12}O_3$
RTECS No.:	YK0525000
Properties:	Highly flammable liquid. Colorless. Characteristic aromatic odor. Floats on water; soluble.

HEALTH AND SAFETY INFORMATION

Irritates eyes, skin, and respiratory tract. High exposure may affect central nervous system; narcotic effects.

Odor Threshold: 0.023 mg/m^3 (recognition).

RECOMMENDATIONS FOR RESPIRATOR SELECTION

At any concentrations above the NIOSH REL, or where there is no REL, at any detectable concentration: SCBAF:PD,PP (any self-contained breathing apparatus that has a full facepiece and is operated in a pressure-demand or other positive-pressure mode); or SAF:PD,PP:ASCBA (any supplied-air respirator that has a full facepiece and is operated in a pressure-demand or other positive-pressure mode in combination with an auxiliary, self-contained breathing apparatus operated in a pressure-demand or other positive-pressure mode).

Escape: GMFOV [any air-purifying, full-facepiece respirator (gas mask) with a chin-style, front- or back-mounted organic vapor canister]; or SCBAE (any appropriate escape-type, self-contained breathing apparatus).

CHEMICAL REACTIVITY

Forms explosive mixture with air. Reacts with strong acids, caustics, ammonia, amines, and oxidizers. Contact with acids produces acetaldehyde. Decomposes on standing; exposure to air and light causes decomposition to acetaldehyde; oxidizes to acetic acid. Attacks rubber and some plastics (e.g., polystyrene, styrene-acrylonitrile copolymer).

FIRE INFORMATION

Vapors are heavier than air and will collect in low areas. Vapors in confined areas may explode when exposed to fire. Vapors may travel long distances to ignition sources and flash back. Storage containers and parts of containers may rocket great distances, in many directions. If material or contaminated runoff enters waterways, notify downstream users of potentially contaminated waters. Notify local health and fire officials and pollution control agencies. Use dry chemicals, foam, carbon dioxide. From a secure, explosion-proof location, use water spray to cool exposed containers.

NFPA Hazard Classifications:

Health Hazards (Blue) 2
Flammability (Red) 3
Reactivity (Yellow) 1

Flash Point: 96°F (36°C) o.c.
Flammable Limits in Air: LEL 1.3%; UEL NA
Autoignition Temperature: 460°F (238°C)

FIRST AID

Move victim to fresh air. Call emergency medical care. Apply artificial respiration if victim is not breathing. Administer oxygen if breathing is difficult. Remove and isolate contaminated clothing and shoes. In case of contact with substance, immediately flush skin or eyes with running water for at least 20 minutes. Wash skin with soap and water. Keep victim warm and quiet. Effects of exposure (inhalation, ingestion, or skin contact) to substance may be delayed. Ensure that medical personnel are aware of the material(s) involved and take precautions to protect themselves.

ENVIRONMENTAL INFORMATION

EPA Hazardous Waste Number (RCRA No.): U182
RCRA Section 261 Hazardous Constituents.
EPCRA Section 304 Reportable Quantity (RQ): CERCLA, 1,000 lb (454 kg)
EPCRA Section 313: Form R *de minimis* concentration reporting level: 1.0%.

PELARGONIC ACID P:026

SYNONYMS: ACIDE PELARGONIQUE (French); ACIDO PELARGONICO (Spanish); CIRRASOL 185A; EINECS No. 203-931-2; EMFAC 1202; HEXACID C-9; NONANOIC ACID; n-NONOIC ACID; n-NONYLIC ACID; 1-OCTANECARBOXYLIC ACID; PELARGIC ACID; PELARGONSAEURE (German) PELARGOONZUUR (Dutch)

IDENTIFICATION

CAS:	112-05-0
Molecular Formula:	$C_9H_{18}O_2$
RTECS No.:	RA6650000
Properties:	Combustible, oily liquid. Yellowish to colorless. Slight odor. Practically insoluble in water.

HEALTH AND SAFETY INFORMATION

Corrosive to the eyes. Irritates skin and respiratory tract. Breathing can become extremely difficult. Inhalation may cause pulmonary edema, which can be delayed for several hours; there is a risk of death in serious cases. Long-term exposure can result in skin problems.

RECOMMENDATIONS FOR RESPIRATOR SELECTION

At any concentrations above the NIOSH REL, or where there is no REL, at any detectable concentration: SCBAF:PD,PP (any self-contained breathing apparatus that has a full facepiece and is operated in a pressure-demand or other positive-pressure mode); or SAF:PD,PP:ASCBA (any supplied-air respirator that has a full facepiece and is operated in a pressure-demand or other positive-pressure mode in combination with an auxiliary, self-contained breathing apparatus operated in a pressure-demand or other positive-pressure mode).

Escape: GMFOV [any air-purifying, full-facepiece respirator (gas mask) with a chin-style, front- or back-mounted organic vapor canister]; or SCBAE (any appropriate escape-type, self-contained breathing apparatus).

CHEMICAL REACTIVITY

Contact with strong oxidizers may cause fire and explosions. Organic acids are incompatible with sulfuric acid, caustics, ammonia, aliphatic amines, alkanolamines, isocyanates, alkylene oxides, epichlorohydrin..

FIRE INFORMATION

Combustion may produce irritants and toxic vapors. Vapors in confined areas may explode when exposed to fire. If material or contaminated runoff enters waterways, notify downstream users of potentially contaminated waters.Notify local health and fire officials and pollution control agencies. Use dry chemicals, alcohol-resistant foam, carbon dioxide. From a secure, explosion-proof location, use water spray to cool exposed containers.

NFPA Hazard Classifications: Not listed

Flash Point: 291°F (114°C)
Flammable Limits in Air: LEL 1.2%; UEL NA
Autoignition Temperature: 761°F (405°C)

FIRST AID

Move victim to fresh air. Call emergency medical care. Apply artificial respiration if victim is not breathing. Administer oxygen if breathing is difficult. Remove and isolate contaminated clothing and shoes. In case of contact with substance, immediately flush skin or eyes with running water for at least 20 minutes. Wash skin with soap and water. Keep victim warm and quiet. Effects of exposure (inhalation, ingestion, or skin contact) to substance may be delayed. Ensure that medical personnel are aware of the material(s) involved and take precautions to protect themselves.

PELARGONYL PEROXIDE P:029

SYNONYMS: BIS(1-OXONONYL) PEROXIDE; DIPELARGONYL PEROXIDE; EINECS No. 212-093-7; NONANOYL PEROXIDE; PEROXIDE, BIS(1-OXONONYL); PEROXIDO de PELARGONILO (Spanish)

IDENTIFICATION

CAS:	762-13-0
ERG ID:	UN 2130
ERG Guide:	148
ERG Guide Category:	ORGANIC PEROXIDES (HEAT AND CONTAMINATION SENSITIVE/TEMPERATURE CONTROLLED)
Molecular Formula:	$C_{18}H_{34}O_4$
RTECS No.:	RA8435000
Properties:	Flammable liquid. Water-white. Faint odor. Floats on water; soluble.

HEALTH AND SAFETY INFORMATION

Severe eye irritant. Irritates skin and respiratory tract. Inhalation may cause pulmonary edema, which can be delayed for several hours; there is a risk of death in serious cases. Due to the availability of insufficient data on human effects, caution should be exercised.

RECOMMENDATIONS FOR RESPIRATOR SELECTION

At any concentrations above the NIOSH REL, or where there is no REL, at any detectable concentration: SCBAF:PD,PP (any self-contained breathing apparatus that has a full facepiece and is operated in a pressure-demand or other positive-pressure mode); or SAF:PD,PP:ASCBA (any supplied-air respirator that has a full facepiece and is operated in a pressure-demand or other positive-pressure mode in combination with an auxiliary, self-contained breathing apparatus operated in a pressure-demand or other positive-pressure mode).

Escape: GMFOV [any air-purifying, full-facepiece respirator (gas mask) with a chin-style, front- or back-mounted organic vapor canister]; or SCBAE (any appropriate escape-type, self-contained breathing apparatus).

CHEMICAL REACTIVITY

A strong oxidizer; violent reaction with combustibles, reducing agents, with risk of fire and explosions.

FIRE INFORMATION

Will increase the intensity of fire. Storage containers and parts of containers may rocket great distances, in many directions. If material or contaminated runoff enters waterways, notify downstream users of potentially contaminated waters. Notify local health and fire officials and pollution control agencies. Use dry chemicals, foam, carbon dioxide, water spray. From a secure, explosion-proof location, use water spray to cool exposed containers.

NFPA Hazard Classifications: Not listed.

FIRST AID

Move victim to fresh air. Call emergency medical care. Apply artificial respiration if victim is not breathing. Administer oxygen if breathing is difficult. Remove and isolate contaminated clothing and shoes. Remove material from skin immediately. In case of contact with substance, immediately flush skin or eyes with running water for at least 20 minutes. Keep victim warm and quiet. Ensure that medical personnel are aware of the material(s) involved and take precautions to protect themselves.

1,3–PENTADIENE P:032

SYNONYMS: EINECS No. 207-995-2; α-METHYLBIVINYL; 1-METHYLBUTADIENE; cis-PENTADIENE-1,3; PIPERYLENE; trans-PENTADIENE-1,3; 1,3-PENTADIENO (Spanish)

IDENTIFICATION

CAS:	504-60-9
ERG ID:	UN 1993
ERG Guide:	128
ERG Guide Category:	FLAMMABLE LIQUIDS (NONPOLAR/WATER-IMMISCIBLE)
Molecular Formula:	C_5H_8
RTECS No.:	RZ2465000
Properties:	Extremely flammable liquid. Colorless. Floats on water; practically insoluble. Boils at 108°F (42°C). Produces a large amount of vapor.

HEALTH AND SAFETY INFORMATION

Irritates eyes, skin, and respiratory tract. Poisonous if ingested.

RECOMMENDATIONS FOR RESPIRATOR SELECTION

At any concentrations above the NIOSH REL, or where there is no REL, at any detectable concentration: SCBAF:PD,PP (any self-contained breathing apparatus that has a full facepiece and is operated in a pressure-demand or other positive-pressure mode); or SAF:PD,PP:ASCBA (any supplied-air respirator that has a full facepiece and is operated in a pressure-demand or other positive-pressure mode in combination with an auxiliary, self-contained breathing apparatus operated in a pressure-demand or other positive-pressure mode).

Escape: GMFOV [any air-purifying, full-facepiece respirator (gas mask) with a chin-style, front- or back-mounted organic vapor canister]; or SCBAE (any appropriate escape-type, self-contained breathing apparatus).

CHEMICAL REACTIVITY

Forms explosive mixture with air. Reacts violently with oxidizers. Attacks some plastics, rubber, and coatings. May accumulate static electrical charges, and may cause ignition of its vapors.

FIRE INFORMATION

Vapors are heavier than air and will collect in low areas. Vapors in confined areas may explode when exposed to fire. Vapors may travel long distances to ignition sources and flash back. Storage containers and parts of containers may rocket great distances, in many directions. If material or contaminated runoff enters waterways, notify downstream users of potentially contaminated waters.Notify local health and fire officials and pollution control agencies. Use dry chemicals, foam, carbon dioxide. From a secure, explosion-proof location, use water spray to cool exposed containers.

NFPA Hazard Classifications:

Health Hazards (Blue) 0
Flammability (Red) 4
Reactivity (Yellow) 2

Flash Point: –20°F (–29°C) c.c. (cis and trans mixture)
Flammable Limits in Air: LEL 1.5%; UEL 8.3%

FIRST AID

Move victim to fresh air. Call emergency medical care. Apply artificial respiration if victim is not breathing. If breathing is difficult, give oxygen. Remove and isolate contaminated clothing and shoes. In case of contact with substance, immediately flush skin or eyes with running water for at least 20 minutes. Wash skin with soap and water. Keep victim warm and quiet. Ensure that medical personnel are aware of the material(s) involved and take precautions to protect themselves.

ENVIRONMENTAL INFORMATION

Clean Air Act: Accidental Release Prevention/Flammable Substances (Section 112[r], Table 3), TQ = 10,000 lb (4,540 kg)
EPA Hazardous Waste Number (RCRA No.): U186
RCRA Toxicity Characteristic (Section 261.24) Section 261 Hazardous Constitutents.
EPCRA Section 304 Reportable Quantity (RQ): CERCLA, 100 lb (45.4 kg)

PENTAERYTHRITOL P:037

SYNONYMS: 2,2-BIS(HYDROXYMETHYL)-1,3-PROPANEDIOL; EINECS No. 204-104-9; PENTAERYTHRITE; TETRAHYDROXYMETHYLMETHANE; PE; MONO PE; PENTEK

IDENTIFICATION

CAS:	115-77-5
Molecular Formula:	$C_5H_{12}O_4$
RTECS No.:	RZ2490000
Properties:	Combustible, crystalline solid. Colorless to white. Odorless. Sinks and mixes slowly with water.

HEALTH AND SAFETY INFORMATION

Nuisance dust or particulate.

OSHA Table Z-1-A Air Contaminant

Exposure Limits: TWA

ACGIH:	10 mg/m^3
NIOSH:	10 mg/m^3 (total); 5 mg/m^3 (respirable fraction)
OSHA:	15 mg/m^3 (total); 5 mg/m^3 (respirable fraction)

RECOMMENDATIONS FOR RESPIRATOR SELECTION

At any concentrations above the NIOSH REL, or where there is no REL, at any detectable concentration: SCBAF:PD,PP (any self-contained breathing apparatus that has a full facepiece and is operated in a pressure-demand or other positive-pressure mode); or SAF:PD,PP:ASCBA (any supplied-air respirator that has a full facepiece and is operated in a pressure-demand or other positive-pressure mode in combination with an auxiliary, self-contained breathing apparatus operated in a pressure-demand or other positive-pressure mode).

Escape: GMFOV [any air-purifying, full-facepiece respirator (gas mask) with a chin-style, front- or back-mounted organic vapor canister]; or SCBAE (any appropriate escape-type, self-contained breathing apparatus).

CHEMICAL REACTIVITY

Reacts with organic acids and oxidizers. Explosive compound is formed when mixture of PE and thiophosphoryl chloride is heated.

FIRE INFORMATION

Storage containers and parts of containers may rocket great distances, in many directions. If material or contaminated runoff enters waterways, notify downstream users of potentially contaminated waters. Notify local health and fire officials and pollution control agencies. Use dry chemicals, foam, carbon dioxide. From a secure, explosion-proof location, use water spray to cool exposed containers.

FEMA Hazard Classifications

Health Hazards (Blue) 0
Flammability (Red) 2
Reactivity (Yellow) 0

Autoignition Temperature: 842°F (450°C) as dust cloud

FIRST AID

Move victim to fresh air. Call emergency medical care. Apply artificial respiration if victim is not breathing. Administer oxygen if breathing is difficult. Remove and isolate contaminated clothing and shoes. In case of contact with substance, immediately flush skin or eyes with running water for at least 20 minutes. Wash skin with soap and water. Keep victim warm and quiet. Effects of exposure (inhalation, ingestion, or skin contact) to substance may be delayed. Ensure that medical personnel are aware of the material(s) involved and take precautions to protect themselves.

PENTANE P:040

SYNONYMS: AMYL HYDRATE; EEC No. 601-006-00-1; EINECS No. 203-692-4; PENTAN; n-PENTANE; normal-PENTANE; PENTANO (Spanish)

IDENTIFICATION

CAS:	109-66-0
ERG ID:	UN 1265
ERG Guide:	128
ERG Guide Category:	FLAMMABLE LIQUIDS (NONPOLAR/WATER-IMMISCIBLE)
Molecular Formula:	C_5H_{12}
RTECS No.:	RZ9450000
Properties:	Extremely flammable liquid. Colorless. Gasolinelike odor. Floats on water; insoluble. Flammable vapor is produced. Boiling point is 97°F (36°C).

HEALTH AND SAFETY INFORMATION

Irritates eyes, skin, and respiratory tract. Swallowing liquid can cause chemical pneumonia. Prolonged or long-term exposure may cause central nervous system damage with numbness and weakness in arms and legs; dry, cracked skin; liver and kidney disease; and chronic respiratory problems.

IDLH: 1,500 ppm; 4,500 mg/m^3 [10% LEL]
Odor Threshold: 31.6 ppm
OSHA Table Z-1-A Air Contaminant
NIOSH Criteria Document: NIOSH 77-151 ALKANES

Exposure Limits:	**TWA**	**CEILING**
ACGIH:	600 ppm; 1,770 mg/m^3	
NIOSH:	120 ppm; 350 mg/m^3	610 ppm; 1,800 mg/m^3 [5 min]
OSHA:	1,000 ppm; 2,950 mg/m^3	

RECOMMENDATIONS FOR RESPIRATOR SELECTION: NIOSH

1,200 ppm: SA (any supplied-air respirator).
1,500 ppm: SA:CF (any supplied-air respirator operated in a continuous-flow mode); or SCBAF (any self-contained breathing apparatus with a full facepiece); or SAF (any supplied-air respirator with a full facepiece).

Emergency or Planned Entry into Unknown Concentrations or IDLH Conditions: SCBAF:PD,PP (any self-contained breathing apparatus that has a full facepiece and is operated in a pressure-demand or other positive-pressure mode); or SAF:PD,PP:ASCBA (any supplied-air respirator that has a full facepiece and is operated in a pressure-demand or other positive-pressure mode in combination with an auxiliary, self-contained breathing apparatus operated in a pressure-demand or other positive-pressure mode).

Escape: GMFOV [any air-purifying, full-facepiece respirator (gas mask) with a chin-style, front- or back-mounted organic vapor canister]; or SCBAE (any appropriate escape-type, self-contained breathing apparatus).

CHEMICAL REACTIVITY

Forms explosive mixture with air. Contact with oxidizers can cause fire and explosions. Attacks some plastics, rubber, and coatings.

FIRE INFORMATION

Vapors are heavier than air and will collect in low areas. Vapors in confined areas may explode when exposed to fire. Vapors may travel long distances to ignition sources and flash back. Storage containers and parts of containers may rocket great distances, in many directions. If material or contaminated runoff enters waterways, notify downstream users of potentially contaminated waters. Notify local health and fire officials and pollution control agencies. Use dry chemicals, foam, carbon dioxide. From a secure, explosion-proof location, use water spray to cool exposed containers.

NFPA Hazard Classifications:

Health Hazards (Blue) 1
Flammability (Red) 4
Reactivity (Yellow) 0

Flash Point: –57°F (–49°C) c.c.
Flammable Limits in Air: LEL 1.5%; UEL 7.8%
Autoignition Temperature: 544°F (284°C)

FIRST AID

Move victim to fresh air. Call emergency medical care. Apply artificial respiration if victim is not breathing. If breathing is difficult, give oxygen. Remove and isolate contaminated clothing and shoes. In case of contact with substance, immediately flush skin or eyes with running water for at least 20 minutes.

Wash skin with soap and water. Keep victim warm and quiet. Ensure that medical personnel are aware of the material(s) involved and take precautions to protect themselves.

ENVIRONMENTAL INFORMATION

Clean Air Act: Accidental Release Prevention/Flammable substances, (Section 112[r], Table 3), TQ = 10,000 lb (4,540 kg).
TSCA: 40CFR716.120(*d*)10; 40CFR712.30(*e*)10 re dermal absorption testing.

1,5–PENTANEDIOL P:150

SYNONYMS: 1,5-DIHYDROXYPENTANE; EINECS No. 203-854-4; PENTAMETHYLENE GLYCOL; PENTANE-1,5-DIOL; 1,5-PENTYLENE GLYCO

IDENTIFICATION

CAS:	111-29-5
Molecular Formula:	$C_5H_{12}O_2$
RTECS No.:	SA0480000
Properties:	Combustible liquid. Colorless. Odorless. Mixes with water. Decomposes to toxic fumes of carbon monoxide and carbon dioxide.

HEALTH AND SAFETY INFORMATION

Irritates eyes, skin, and respiratory tract.

RECOMMENDATIONS FOR RESPIRATOR SELECTION

At any concentrations above the NIOSH REL, or where there is no REL, at any detectable concentration: SCBAF:PD,PP (any self-contained breathing apparatus that has a full facepiece and is operated in a pressure-demand or other positive-pressure mode); or SAF:PD,PP:ASCBA (any supplied-air respirator that has a full facepiece and is operated in a pressure-demand or other positive-pressure mode in combination with an auxiliary, self-contained breathing apparatus operated in a pressure-demand or other positive-pressure mode).

Escape: GMFOV [any air-purifying, full-facepiece respirator (gas mask) with a chin-style, front- or back-mounted organic vapor canister]; or SCBAE (any appropriate escape-type, self-contained breathing apparatus).

CHEMICAL REACTIVITY

Contact with strong oxidizers may cause fire and explosions. Reacts with acid chlorides, acid anhydrides, chloroformates, and reducing agents. Glycols are incompatible with sulfuric acid, nitric acid, caustics, aliphatic amines, isocyanates.

FIRE INFORMATION

Combustion may produce irritants and toxic vapors. Vapors in confined areas may explode when exposed to fire. If material or contaminated runoff enters waterways, notify downstream users of potentially contaminated waters. Notify local health and fire officials and pollution control agencies. Use dry chemicals, alcohol-resistant foam, carbon dioxide. From a secure, explosion-proof location, use water spray to cool exposed containers.

NFPA Hazard Classifications

Health Hazards (Blue) 1
Flammability (Red) 1
Reactivity (Yellow) 0

Flash Point: 265°F (129°C)
Flammable Limits in Air: LEL 1.4%; UEL 13.2%
Autoignition Temperature: 635°F (335°C)

FIRST AID

Move victim to fresh air. Call emergency medical care. Apply artificial respiration if victim is not breathing. If breathing is difficult, give oxygen. Remove and isolate contaminated clothing and shoes. In case of contact with substance, immediately flush skin or eyes with running water for at least 20 minutes. Wash skin with soap and water. Keep victim warm and quiet. Ensure that medical personnel are aware of the material(s) involved and take precautions to protect themselves.

PERACETIC ACID P:170

SYNONYMS: ACETIC PEROXIDE; ACETYL HYDROPEROXIDE; ACIDE PERACETIQUE (French); ACIDE PEROXYACETIQUE (French); EEC No. 607-094-00-8; EINECS No. 201-186-8; HYDROPEROXIDE, ACETYL; ETHANEPEROXIC ACID; PAA; PERACETIC ACID SOLUTION; PEROXYACETIC ACID; PEROXYAZIJNZUUR (Dutch); PEROXYESSIGSAEURE (German)

IDENTIFICATION

CAS:	79-21-0
ERG ID:	UN 2131
ERG Guide:	147
ERG Guide Category:	ORGANIC PEROXIDES (HEAT AND CONTAMINATION SENSITIVE/SEVERE IRRITANTS)
Molecular Formula:	$C_2H_4O_3$
RTECS No.:	SD8927000
Properties:	Combustible liquid. Colorless. Strong acrid odor. Water soluble. Flammable, irritating vapor is produced.

HEALTH AND SAFETY INFORMATION

Corrosive to eyes, skin, and respiratory tract. May be fatal if inhaled or absorbed through the skin. Inhalation may cause pulmonary edema, which can be delayed for several hours; there is a risk of death in serious cases.

OSHA Process Safety Management of Highly Hazardous Chemicals (29CFR, Part 1910.119, Appendix A), Threshold Quantity: (concentration >60% acetic acid) 1,000 pounds.

RECOMMENDATIONS FOR RESPIRATOR SELECTION

At any concentrations above the NIOSH REL, or where there is no REL, at any detectable concentration: SCBAF:PD,PP (any self-contained breathing apparatus that has a full facepiece and is operated in a pressure-demand or other positive-pressure mode); or SAF:PD,PP:ASCBA (any supplied-air respirator that has a full facepiece and is operated in a pressure-demand or other positive-pressure mode in combination with an auxiliary, self-contained breathing apparatus operated in a pressure-demand or other positive-pressure mode).

Escape: GMFOV [any air-purifying, full-facepiece respirator (gas mask) with a chin-style, front- or back-mounted organic vapor canister]; or SCBAE (any appropriate escape-type, self-contained breathing apparatus).

CHEMICAL REACTIVITY

A powerful oxidizer and thermally explosive: decomposes violently at 230°F (110°C). Reacts violently with many substances, including acetic anhydride, combustibles, strong bases, ether solvents, olefins, organic matter, magnesium, metal oxides, heavy metals, metal chloride solutions, reducing agents, nickel, phosphorus, sodium nitride, and zinc. May cause fire in contact with organic materials such as wood, cotton, or straw. Solution reacts with bases and thiocyanates. Corrosive to most metals, including aluminum. Stability during transport: Stable if kept cool and out of contact with most metals. At 30°C concentration decreases about 0.4% each month.

FIRE INFORMATION

Fight fire from protected location or maximum possible distance. Vapors are heavier than air and will collect in low areas. Vapors in confined areas may explode when exposed to fire. Vapors may travel long distances to ignition sources and flash back. Storage containers and parts of containers may rocket great distances, in many directions. If material or contaminated runoff enters waterways, notify downstream users of potentially contaminated waters. Notify local health and fire officials and pollution control agencies. Use flooding amounts of water. From a secure, explosion-proof location, use water spray to cool exposed containers.

NFPA Hazard Classifications:

Health Hazards (Blue) 3
Flammability (Red) 2

Reactivity (Yellow) 4
Special Notice . . . . OXY

Flash Point: 105°F (41°C) o.c.
Autoignition Temperature: 392°C (200°C)

FIRST AID

Move victim to fresh air. Call emergency medical care. Apply artificial respiration if victim is not breathing. Administer oxygen if breathing is difficult. Remove and isolate contaminated clothing and shoes. Remove material from skin immediately. In case of contact with substance, immediately flush skin or eyes with running water for at least 20 minutes. Keep victim warm and quiet. Ensure that medical personnel are aware of the material(s) involved and take precautions to protect themselves.

ENVIRONMENTAL INFORMATION

Clean Air Act: Accidental Release Prevention/Flammable Substances (Section 112[r], Table 3), TQ = 10,000 lb (4,540 kg)
EPCRA Section 304 Reportable Quantity (RQ): CERCLA, 500 lb (227 kg)
EPCRA Section 313: Form R *de minimis* concentration reporting level: 1.0%.

PHENOL P:120

SYNONYMS: A20; A30; A40; ACIDE CARBOLIQUE (French); BAKER'S P AND S LIQUID; BAKER'S P AND S OINTMENT; BENZENE,HYDROXY-; BURMAR 712D; CARBOLIC ACID; CARBOLSAURE (German); EEC No. 604-001-00-2; EINECS No. 203-632-7; FENOL (Dutch, Polish, Spanish); FENOLO (Italian); HYDROXYBENZENE; J100; MONOHYDROXY BENZENE; MS-111 EPOXY STRIPPER; NCI-C50124; OXYBENZENE; PHENIC ACID; PHENOLE (German); PHENYL ALCOHOL; PHENYL HYDRATE; PHENYL HYDROXIDE; PHENYLIC ACID; STRIP L; ULTRAMAC S-40

IDENTIFICATION

CAS:	108-95-2
ERG ID:	UN 1671 (solid); UN 2312 (molten); UN 2821 (solution)
ERG Guide:	153
ERG Guide Category:	TOXIC AND/OR CORROSIVE (COMBUSTIBLE)
Molecular Formula:	C_6H_6O
RTECS No.:	SJ3325000
Properties:	Very flammable solid crystals or syrupy liquid solution. White solid that may redden on exposure to light and heat. Sweet, tarry, medicinal odor. May float or sink, and mixes slowly with water.

HEALTH AND SAFETY INFORMATION

Swallowing the material or inhaling the vapors can cause death. Skin and eye contact causes severe burns and blindness. Repeated phenol exposure in the workplace has caused renal damage, including kidney inflammation, swelling in the kidney tubules and cells, and degenerative changes in glomeruli. Liver damage and pigment changes of the skin have been noted in some workers. Chronic exposure has been correlated with an increased risk of coronary artery disease and insufficient blood supply to the heart in workers. Prolonged exposure by any route to levels above the standard may cause difficulty swallowing, loss of appetite, headache, dizziness, fainting, mental disturbances, dark coloration of urine, and skin rash. Phenol has been reported to be embryotoxic and fetotoxic, but may not be teratogenic.

IARC Evaluation: Group 3, unclassifiable; human evidence, inadequate; animal evidence, inadequate.
IDLH: 250 ppm; 998 mg/m^3
Odor Threshold: 0.011 ppm Odor is not a reliable indicator of toxic amounts of vapor.
OSHA Table Z-1-A Air Contaminant
NIOSH Criteria Document: NIOSH 76-196

Exposure Limits:	TWA	CEILING
ACGIH: (1)	5 ppm; 19 mg/m^3	
NIOSH: (1)	5 ppm; 19 mg/m^3	15.6 ppm; 60 mg/m^3/15 minutes
OSHA: (1)	5 ppm; 19 mg/m^3	

(1) Skin contact contributes significantly to overall exposure.

RECOMMENDATIONS FOR RESPIRATOR SELECTION: NIOSH/OSHA

50 ppm: CCROVDM (any chemical cartridge respirator with organic vapor cartridge(s) in combination with a dust and mist filter); SA (any supplied-air respirator).
125 ppm: SA:CF (any supplied-air respirator operated in a continuous-flow mode); or PAPROVDM (any powered, air-purifying respirator with organic vapor cartridge(s) in combination with a dust and mist filter).
250 ppm: CCRFOVHiE (any chemical cartridge respirator with a full facepiece and organic vapor cartridge(s) in combination with a high-efficiency particulate filter); or GMFOVHiE [any air-purifying, full-facepiece respirator (gas mask) with a chin-style, front- or back-mounted organic vapor canister having a high-efficiency particulate filter]; or PAPRTOVHiE (any powered, air-purifying respirator with a tight-fitting facepiece and organic vapor cartridge(s) in combination with a high-efficiency particulate filter); or SCBAF (any self-contained breathing apparatus with a full facepiece); or SAF (any supplied-air respirator with a full facepiece).

Emergency or Planned Entry into Unknown Concentrations or IDLH Conditions:
SCBAF:PD,PP (any self-contained breathing apparatus that has a full facepiece and is operated in a pressure-demand or other positive-pressure mode); or SAF:PD,PP:ASCBA (any supplied-air respirator that has a full facepiece and is operated in a pressure-demand or other positive-pressure mode in combination with an auxiliary, self-contained breathing apparatus operated in a pressure-demand or other positive-pressure mode).

Escape: GMFOVHiE [any air-purifying, full-facepiece respirator (gas mask) with a chin-style, front- or back-mounted organic vapor canister having a high-efficiency particulate filter]; or SCBAE (any appropriate escape-type, self-contained breathing apparatus).

CHEMICAL REACTIVITY

Forms explosive mixture with air. Can react violently with butadiene and mixtures of aluminum chloride and nitrobenzene. Vigorous reaction with strong oxidizers such as permanganates or chlorine. Incompatible with acids, aluminum chloride, caustics, aliphatic amines, amides, oxidizers, formaldehyde, and calcium hypochlorite. Liquid attacks some plastics, rubber, and coatings; hot liquid attacks aluminum, magnesium, lead, and zinc metals.

FIRE INFORMATION

Combustion products include toxic carbon monoxide. Firefighting gear (including SCBA) does not provide adequate protection. If exposure occurs, remove and isolate gear immediately and thoroughly decontaminate personnel. Vapors are heavier than air and will collect in low areas. Vapors in confined areas may explode when exposed to fire. Vapors may travel long distances to ignition sources and flash back. Storage containers and parts of containers may rocket great distances, in many directions. If material or contaminated runoff enters waterways, notify downstream users of potentially contaminated waters. Notify local health and fire officials and pollution control agencies. Use dry chemicals, foam, carbon dioxide. From a secure, explosion-proof location, use water spray to cool exposed containers.

NFPA Hazard Classifications:

Health Hazards (Blue) 3
Flammability (Red) 2
Reactivity (Yellow) 0

Flash Point: 185°F (85°C) o.c.; 175°F (79°C) c.c.
Flammable Limits in Air: LEL 1.7%; UEL 8.6%
Autoignition Temperature: 1319°F (715°C)

FIRST AID

Move victim to fresh air. Call emergency medical care. Apply artificial respiration if victim is not breathing. Do not use mouth-to-mouth method if victim ingested or inhaled the substance; induce artificial respiration with the aid of a pocket mask equipped with a one-way valve or other proper respiratory medical device. Administer oxygen if breathing is difficult. Remove and isolate contaminated clothing and shoes. In case of contact with substance, immediately flush skin or eyes with running water for at least 20 minutes. For minor skin contact, avoid spreading material on unaffected skin. Keep victim warm and quiet. Effects of exposure (inhalation, ingestion, or skin contact) may be delayed. Ensure that medical personnel are aware of the material(s) involved and take precautions to protect themselves.

ENVIRONMENTAL INFORMATION

Clean Air Act: Hazardous Air Pollutants (Title I, Part A, Section 112).

Clean Water Act: Section 307 Toxic Pollutants; Section 307 Priority Pollutants; Section 311 Hazardous Substances/RQ (same as CERCLA); Section 313 Priority Chemicals.
EPA Hazardous Waste Number (RCRA No.): U188.
RCRA Section 261 Hazardous Constituents
RCRA Land Ban Waste
RCRA Universal Treatment Standards: Wastewater (mg/L), 0.039; Nonwastewater (mg/kg), 6.2.
RCRA Ground Water Monitoring List: Suggested test method(s) (PQL ug/L): 8040(1); 8270(10).
EPCRA Section 302 Extremely Hazardous Substances: TPQ = 500/10,000 lb (227/4,540 kg).
EPCRA Section 304: Reportable Quantity (RQ): EHS, CERCLA, 1,000 lb (454 kg).
EPCRA Section 313: Form R *de minimis* concentration reporting level: 1.0%.
TSCA: 40CFR716.120(*a*)

PHENOLSULFONIC ACID P:130

SYNONYMS: ACIDO FENOLSULFONICO (Spanish); ELTESOL PSA 65; HYDROXYBENZENE-SULFONIC ACID; PHENOLSULPHONIC ACID; SULFPCARBOLIC ACID

IDENTIFICATION

CAS:	1333-39-7
ERG ID:	UN 1803
ERG Guide:	153
ERG Guide Category:	SUBSTANCES-TOXIC AND/OR CORROSIVE (COMBUSTIBLE)
Molecular Formula:	$C_6H_6O_4S$
RTECS No.:	DB6950000
Properties:	Combustible, yellow liquid; turns brown on contact with air. Soluble in water.

HEALTH AND SAFETY INFORMATION

Corrosive to eyes, skin, and respiratory tract. Inhalation may cause pulmonary edema, which can be delayed for several hours; there is a risk of death in serious cases.

RECOMMENDATIONS FOR RESPIRATOR SELECTION

At any concentrations above the NIOSH REL, or where there is no REL, at any detectable concentration: SCBAF:PD,PP (any self-contained breathing apparatus that has a full facepiece and is operated in a pressure-demand or other positive-pressure mode); or SAF:PD,PP:ASCBA (any supplied-air respirator that has a full facepiece and is operated in a pressure-demand or other positive-pressure mode in combination with an auxiliary, self-contained breathing apparatus operated in a pressure-demand or other positive-pressure mode).

Escape: GMFOV [any air-purifying, full-facepiece respirator (gas mask) with a chin-style, front- or back-mounted organic vapor canister]; or SCBAE (any appropriate escape-type, self-contained breathing apparatus).

CHEMICAL REACTIVITY

Decomposes above 122°F (50°C) forming sulfuric acid fumes. Contact with strong oxidizers may cause fire and explosions. Organic acids are incompatible with sulfuric acid, caustics, ammonia, aliphatic amines, alkanolamines, isocyanates, alkylene oxides, epichlorohydrin..

FIRE INFORMATION

Combustible liquid. Vapors are heavier than air and will collect in low areas. Vapors in confined areas may explode when exposed to fire. Vapors may travel long distances to ignition sources and flash back. Storage containers and parts of containers may rocket great distances, in many directions. If material or contaminated runoff enters waterways, notify downstream users of potentially contaminated waters. Notify local health and fire officials and pollution control agencies. Use dry chemicals, foam, carbon dioxide. From a secure, explosion-proof location, use water spray to cool exposed containers.

FIRST AID

Move victim to fresh air; call emergency medical care. Apply artificial respiration if victim is not breathing. Do not use mouth-to-mouth method if victim ingested or inhaled the substance; induce artificial respiration with the aid of a pocket mask equipped with a one-way valve or other proper respiratory medical device. Administer oxygen if breathing is difficult. Remove and isolate contaminated clothing and shoes. In case of contact with substance, immediately flush skin or eyes with running water for at least 20 minutes. For minor skin contact, avoid spreading material on unaffected skin. Keep victim warm and quiet. Effects of exposure (inhalation, ingestion, or skin contact) to substance may be delayed. Ensure that medical personnel are aware of the material(s) involved and take precautions to protect themselves.

PHENYL CYANIDE P:133

SYNONYMS: BENZENE, CYANO-; BENZENENITRILE; BENZOIC ACID NITRILE; BENZONITRILE; BENZONITRILO (Spanish); CYANOBENZENE; EEC No. 608-012-00-3; EINECS No. 202-855-7; PHENYL CYANIDE

IDENTIFICATION

CAS:	100-47-0
ERG ID:	UN 2224
ERG Guide:	152
ERG Guide Category:	SUBSTANCES-TOXIC (COMBUSTIBLE)
Molecular Formula:	C_7H_5N
RTECS No.:	DI2450000
Properties:	Combustible liquid. Colorless. Almondlike odor. May float or sink in water; slightly soluble.

HEALTH AND SAFETY INFORMATION

Poisonous. Prolonged skin contact may cause burns. Breathing the vapors, swallowing the liquid, or skin contact can cause death. Phenly cyanide (benzonitrile) passes through the skin; converted to cyanide in the body. Long-term or repeated exposure can cause liver and nervous system damage.

IDLH: 25 mg/m^3 as cyanides

Exposure Limits:	**TWA**	**CEILING**
ACGIH: [(1)]	5 mg/m^3 as cyanide	
NIOSH: [(1)]		4.7 ppm; 5 mg/m^3 per 10 minutes as cyanides
OSHA: [(1)]	5 mg/m^3 as cyanides	

[(1)] Skin contact contributes significantly in overall exposure.

RECOMMENDATIONS FOR RESPIRATOR SELECTION: NIOSH as cyanides

50 mg/m^3: SA (any supplied-air respirator); or SCBAF (any self-contained breathing apparatus with full facepiece).

Emergency or Planned Entry into Unknown Concentrations or Idlh Conditions: SCBAF:PD,PP (any self-contained breathing apparatus that has a full facepiece and is operated in a pressure-demand or other positive-pressure mode); or SAF:PD,PP:ASCBA (any supplied-air respirator that has a full facepiece and is operated in a pressure-demand or other positive-pressure mode in combination with an auxiliary, self-contained breathing apparatus operated in a pressure-demand or other positive pressure mode.

Escape: GMFSHiE [any air-purifying, full-facepiece respirator (gas mask) with a chin-style, front-or back-, mounted canister providing protection against the compound of concern and having a high-efficiency particulate filter]; or SCBAE (any appropriate escape-type, self-contained breathing apparatus).

CHEMICAL REACTIVITY

Forms explosive mixture with air. Contact with strong acids forms hydrocyanic acid. Contact with oxidizers may cause fire and explosions. Attacks some plastics.

FIRE INFORMATION

Combustion products include toxic cyanide and nitrogen oxide fumes. Firefighting gear (including SCBA) does not provide adequate protection. If exposure occurs, remove and isolate gear immediately and thoroughly decontaminate personnel. Vapors are heavier than air and will collect in low areas. Vapors in confined areas may explode when exposed to fire. Vapors may travel long distances to ignition sources and flashback. Storage containers and parts of containers may rocket great distances, in many directions. If material or contaminated runoff enters waterways, notify downstream users of potentially contaminated waters. Notify local health and fire officials and pollution control agencies. Use dry chemicals, foam, carbon dioxide. From a secure, explosion-proof location use water spray to cool exposed containers. If cooling streams are ineffective (venting sound increases in volume and pitch, tank discolors, or shows any signs of deforming), withdraw immediately to a secure position.

FEMA Hazard Classifications:

Health Hazards (Blue) 4
Flammability (Red) 2
Reactivity (Yellow) 0

Flash Point: 167°F (75°C) c.c.
Flammable Limits in Air: LEL 1.4%; UEL 7.2%
Autoignition Temperature: 1022°F (550°C)

FIRST AID

Move victim to fresh air. Call emergency medical care. Apply artificial respiration if victim is not breathing. Do not use mouth-to-mouth method if victim ingested or inhaled the substance; induce artificial respiration with the aid of a pocket mask equipped with a one-way valve or other proper respiratory medical device. Administer oxygen if breathing is difficult. Remove and isolate contaminated clothing and shoes. In case of contact with substance, immediately flush skin or eyes with running water for at least 20 minutes. For minor skin contact, avoid spreading material on unaffected skin. Keep victim warm and quiet. Effects of exposure (inhalation, ingestion, or skin contact) may be delayed. Ensure that medical personnel are aware of the material(s) involved and take precautions to protect themselves.

ENVIRONMENTAL INFORMATION

Clean Water Act: Section 311 Hazardous Substances/RQ (same as CERCLA); Section 313 Priority Chemical
EPCRA Section 304 Reportable Quantity (RQ): CERCLA, 5,000 lb (2,270 kg)

p–PHENYLENEDIAMINE P:135

SYNONYMS: 4-AMINOANILINE; p-AMINOANILINE; BASF URSOL D; 1,4-BENZENEDIAMINE; p-BENZENEDIAMINE; BENZOFUR D; C.I. 76060; C.I. DEVELOPER 13; C.I. OXIDATION BASE 10; DEVELOPER 13; DEVELOPER PF; p-DIAMINOBENZENE; 1,4-DIAMINOBENZENE; DURAFUR BLACK R; EEC No. 612-028-00-6; EINECS No. 203-404-7; p-FENILENDIAMINA (Spanish); FENYLENODWUAMINA (Polish); FOURAMINE D; FOURRINE D; FOURRINE 1; FUR BLACK 41867; FUR BROWN 41866; FURRO D; FUR YELLOW; FUTRAMINE D; NAKO H; ORSIN; PARA; PARAPHENYLEN-DIAMINE; PELAGOL GREY D; PELTOL D; PHENYLENEDIAMINE, PARA, solid; 1,4-PHENYLENEDIAMINE; PPD; RENAL PF; SANTOFLEX IC; TERTRAL D; URSOL D; USAF EK-394; VULKANOX 4020; ZOBA BLACK D

IDENTIFICATION

CAS: 106-50-3

ERG ID:	UN 1673
ERG Guide:	153
ERG Guide Category:	SUBSTANCES-TOXIC AND/OR CORROSIVE (COMBUSTIBLE)
Molecular Formula:	$C_6H_8N_2$
RTECS No.:	SS8050000
Properties:	Combustible, crystalline solid. White to slightly red; oxidizes in air to purple and black. Soluble in water.

HEALTH AND SAFETY INFORMATION

Corrosive to eyes, skin, and respiratory tract. Inhalation may cause pulmonary edema, which can be delayed for several hours; there is a risk of death in serious cases. Can cause eczema, kidney damage, neurological disorders.

IARC Evaluation: Group 3; unclassifiable; insufficient human evidence.
IDLH: 25 mg/m^3
OSHA Table Z-1-A Air Contaminant

Exposure Limits:	**TWA**
ACGIH:	0.1 mg/m^3
NIOSH: (1)	0.1 mg/m^3
OSHA: (1)	0.1 mg/m^3

(1) Skin contact contributes significantly to overall exposure.

RECOMMENDATIONS FOR RESPIRATOR SELECTION: NIOSH/OSHA

2.5 mg/m^3: SA:CF * (any supplied-air respirator operated in a continuous-flow mode)
5 mg/m^3: SCBAF (any self-contained breathing apparatus with a full facepiece); or SAF (any supplied-air respirator with a full facepiece).
25 mg/m^3:SAF:PD,PP (Any supplied-air respirator that has a full facepiece and is operated in a pressure-demand or other positive-pressure mode) .

EMERGENCY OR PLANNED ENTRY INTO UNKNOWN CONCENTRATIONS OR IDLH CONDITIONS: SCBAF:PD,PP (any self-contained breathing apparatus that has a full facepiece and is operated in a pressure-demand or other positive-pressure mode); or SAF:PD,PP:ASCBA (Any supplied-air respirator that has a full facepiece and is operated in a pressure-demand or other positive-pressure mode in combination with an auxiliary, self-contained breathing apparatus operated in a pressure-demand or other positive pressure mode.

Escape: GMFSHiE [any air-purifying, full-facepiece respirator (gas mask) with a chin-style, front- or back-mounted organic vapor canister providing protection against the compound of concern and having a high-efficiency particulate filter.]; or SCBAE (any appropriate escape-type, self-contained breathing apparatus).

*Substance causes eye irritation or damage; may require eye protection.

CHEMICAL REACTIVITY

A strong reducing agent; violent reaction with oxidizers, chlorates, nitrates, permanganates, peroxides, aluminum chlorate, aluminum bichlorate, benzoyl peroxide, bromine dioxide, cadmium chlorate, red lead oxide, magnesium peroxide, mercuric oxide, nitrogen trifluoride, oxygen difluoride, ozone, potassium chlorate, potassium nitrate, tetrafluoro hydrazine, zinc permanganate. Incompatible with ammonium chromate, ammonium magnesium chromate, ammonium peroxychromate, ammonium persulfate. Aromatic amines are incompatible with nitric acid, sulfuric acid, nonoxidizing mineral acids, organic anhydrides, isocyanates, aldehydes.

FIRE INFORMATION

Combustion products include toxic nitrogen oxides. Storage containers and parts of containers may rocket great distances, in many directions. If material or contaminated runoff enters waterways, notify downstream users of potentially contaminated waters. Notify local health and fire officials and pollution control agencies. Use dry chemicals, foam, carbon dioxide. From a secure, explosion-proof location, use water spray to cool exposed containers.

NFPA Hazard Classifications: as phenylenediamine

Health Hazards (Blue) 2
Flammability (Red) 1
Reactivity (Yellow) 0

Flash Point: 313°F (156°C)

FIRST AID

Move victim to fresh air and call emergency medical care. Apply artificial respiration if victim is not breathing. Do not use mouth-to-mouth method if victim swallowed or inhaled the substance; induce artificial respiration with the aid of a pocket mask equipped with a one-way valve or other proper respiratory medical device. If breathing is difficult, give oxygen. In case of contact with substance, immediately flush skin or eyes with running water for at least 20 minutes. Remove and isolate contaminated clothing and shoes. Keep victim warm and quiet. Effects of exposure (inhalation, ingestion, or skin contact) to substances may be delayed. Ensure that medical personnel are aware of the material(s) involved and take precautions to protect themselves.

ENVIRONMENTAL INFORMATION

Clean Air Act: Hazardous Air Pollutants (Title I, Part A, Section 112)
RCRA Ground Water Monitoring List: Suggested test methods(s) (PQL μg/L); 8270 (10)
EPCRA Section 304 Reportable Quantity (RQ): CERCLA, 5,000 lb (2,270 kg)
EPCRA Section 313: Form R *de minimis* concentration reporting level: 1.0%.
TSCA: 40CFR716.120(*a*); 40CFR 799.3300

PHENYLMERCURIC ACETATE P:143

SYNONYMS: ACETATE PHENYLMERCURIQUE (French); ACETATO FENILMERCURICO (Spanish); ACETIC ACID, PHENYLMERCURY DERIVATIVE; AGROSAN; ALGIMYCIN; ANTIMUCIN WDR; BUFEN; CEKUSIL; CELMER; CERESAN; CONTRA CREME CYANACIDE; EINECS No. 200-532-5; FEMMA; FENYLMERCURIACETAT (Czech); FMA; FUNGITOX OR; GALLOTOX; HL-331; HONG KIEN; HOSTAQUICK; KWIKSAN; LEYTOSAN; LIQUIPHENE; MERCURY, (ACETO-O)PHENYL-; NORFORMS; NYMERATE; PAMISAN; PHENMAD; PHENOMERCURIC ACETATE; PHENOMERCURY ACETATE; PHENYLQUECKSILBERACETAT (German); PMA; PMAC; PMACETATE; PMAL; PMAS; PURASAN-SC-10; PURATURF 10; QUICKSAN; SANITIZED SPG; SC-110; SEEDTOX; SPOR-KIL; TAG; TAG FUNGICIDE; ZIARNIK; ACETATOPHENYLMERCURY; (ACETATO-O) PHENYL MERCURY; COCURE 26; COSAN -100

IDENTIFICATION

CAS:	62-38-4
ERG ID:	UN 1674
ERG Guide:	151
ERG Guide Category:	SUBSTANCES-TOXIC (NONCOMBUSTIBLE)
Molecular Formula:	$C_8H_8HgO_2$
RTECS No.:	0V6475000
Properties:	Noncombustible, crystalline solid or powder. White powder or lustrous white crystals. Slight vinegarlike odor and taste. Floats on water; soluble.

HEALTH AND SAFETY INFORMATION

Poisonous. Extremely irritating to the eyes, skin, and respiratory tract. Passes through the unbroken skin. Inhalation may cause pulmonary edema, which can be delayed for several hours; there is a risk of death in serious cases.

IDLH: 10 mg/m^3 as mercury

Exposure Limits:	**TWA**	**CEILING**
ACGIH: (1)	0.1 ppm	
NIOSH: (1)	0.05 mg/m^3	0.1 mg/m^3
OSHA:		0.1 mg/m^3

(1) Skin contact contributes significantly in overall exposure.

RECOMMENDATIONS FOR RESPIRATOR SELECTION: NIOSH/OSHA as mercury

0.5 mg/m^3: CRS *End of service life indicator (ESLI) required.* [any chemical cartridge respirator with cartridge(s) providing protection against the compound of concern)]; or SA (any supplied-air respirator).

1.25 mg/m³: SA:CF (any supplied-air respirator operated in a continuous-flow mode); or PAPRS *End of service life indicator (ESLI) required.* (any powered, air-purifying respirator with cartridge(s) providing protection against the compound of concern).
2.5 mg/m³: CCRFS *End of service life indicator (ESLI) required.* [any chemical cartridge respirator with a full facepiece and cartridge(s) providing protection against the compound of concern]; or GMFS *End of service life indicator (ESLI) required.* [any air-purifying, full-facepiece respirator (gas mask) with a chin-style, front- or back-mounted canister providing protection against the compound of concern]; or SAT:CF (any supplied-air respirator that has a tight-fitting facepiece and is operated in a continuous-flow mode); or PAPRTS (any powered, air-purifying respirator with a tight-fitting facepiece and cartridge(s) providing protection against the compound of concern); or SCBAF (any self-contained breathing apparatus with a full facepiece); or SAF (any supplied-air respirator with a full facepiece).
10 mg/m³: SA:PD,PP (any supplied-air respirator operated in a pressure-demand or other positive-pressure mode).

Emergency or planned entry into unknown concentrations or IDLH conditions: SCBAF:PD,PP (any self-contained breathing apparatus that has a full facepiece and is operated in a pressure-demand or other positive-pressure mode); or SAF:PD,PP:ASCBA (any supplied-air respirator that has a full facepiece and is operated in a pressure-demand or other positive-pressure mode in combination with an auxiliary, self-contained breathing apparatus operated in a pressure-demand or other positive pressure mode.

Escape: GMFS [any air-purifying, full-facepiece respirator (gas mask) with a chin-style, front- or back-mounted canister providing protection against the compound of concern]; or SCBAE (any appropriate escape-type, self-contained breathing apparatus).

CHEMICAL REACTIVITY

Contact with oxidizers can cause fire and explosions. Reacts with halides: boron tribromide, arsenic trichloride, potassium trichloride, acetyl chloride, benzenesulfonyl chloride, cyanuric chloride, iodine pentafluoride, silver fluoride, sodium hydride, sulfur dichloride, thionyl chloride, iron oxide, rust, aluminum chloride, silicon tetrachloride, methyl bromide, hydrogen chloride, hydrogen bromide, sulfuryl chloride, boron triiodide, and others.

FIRE INFORMATION

Combustion products include toxic fumes of mercury. Storage containers and parts of containers may rocket great distances, in many directions. If material or contaminated runoff enters waterways, notify downstream users of potentially contaminated waters.Notify local health and fire officials and pollution control agencies. If in organic solution, treat fire according to the characteristics of the solvent; otherwise, use dry chemicals, alcohol-resistant foam, carbon dioxide. From a secure, explosion-proof location, use water spray to cool exposed containers.

NFPA Hazard Classifications:

Dry

Health Hazards (Blue) 3

Flammability (Red) 1
Reactivity (Yellow) 0

Organic Solution

Health Hazards (Blue) 3
Flammability (Red) 2
Reactivity (Yellow) 0

Flash Point: >100°F (38°C)

FIRST AID

Move victim to fresh air. Call emergency medical care. Apply artificial respiration if victim is not breathing. Do not use mouth-to-mouth method if victim ingested or inhaled the substance; induce artificial respiration with the aid of a pocket mask equipped with a one-way valve or other proper respiratory medical device. Administer oxygen if breathing is difficult. Remove and isolate contaminated clothing and shoes. In case of contact with substance, immediately flush skin or eyes with running water for at least 20 minutes. For minor skin contact, avoid spreading material on unaffected skin. Keep victim warm and quiet. Effects of exposure (inhalation, ingestion, or skin contact) may be delayed. Ensure that medical personnel are aware of the material(s) involved and take precautions to protect themselves.

ENVIRONMENTAL INFORMATION

Clean Air Act: Hazardous Air Pollutants (Title I, Part A, Section 112)
Clean Water Act: Section 307 Toxic Pollutants as mercury and compounds
RCRA Section 261 Hazardous Constituents, waste number not listed, as mercury compounds, n.o.s.
EPCRA Section 313: Form R *de minimis* concentration reporting level: 1.0%. Form R Toxic Chemical Category Code: N458.
California's Proposition 65: Reproductive toxin
Marine Pollutant (49CFR, Subchapter 172.101, Appendix B)

PHENYLMETHYL ACETATE P:147

SYNONYMS: ACETATO de BENCILO (Spanish); ACETIC ACID, PHENYLMETHYL ESTER; ACETIC ACID, BENZYL ESTER; ACETOMETHYLBENZENE; α-ACETOXYTOLUENE; BENZYL ACETATE; BENZYL ETHANOATE; EINECS No. 205-399-7; FEMA No. 2135; NCI-C06508

IDENTIFICATION

CAS:	140-11-4
ERG ID:	UN 3272
ERG Guide:	127
ERG Guide Category:	FLAMMABLE LIQUIDS (POLAR/WATER MISCIBLE)

842 Phenymethyl Acetate

Molecular Formula: $C_9H_{10}O_2$
RTECS No.: AF5075000
Properties: Combustible liquid. Colorless. Pearlike odor. Slightly soluble in water.

HEALTH AND SAFETY INFORMATION

Causes irritation of the eyes, skin, and respiratory tract. Will dry skin. Affects nervous system.

At any concentrations above the NIOSH REL, or where there is no REL, at any detectable concentration: SCBAF:PD,PP (any self-contained breathing apparatus that has a full facepiece and is operated in a pressure-demand or other positive-pressure mode); or SAF:PD,PP:ASCBA (any supplied-air respirator that has a full facepiece and is operated in a pressure-demand or other positive-pressure mode in combination with an auxiliary, self-contained breathing apparatus operated in a pressure-demand or other positive-pressure mode).

Escape: GMFOV (any air-purifying, full-facepiece respirator (gas mask) with a chin-style, front- or back-mounted organic vapor canister); or SCBAE (any appropriate escape-type, self-contained breathing apparatus).

CHEMICAL REACTIVITY

Forms explosive mixture with air. Contact with oxidizers may cause fire and explosions.

FIRE INFORMATION

If material or contaminated runoff enters waterways, notify downstream users of potentially contaminated waters. Notify local health and fire officials and pollution control agencies. Use dry chemicals. Foam, carbon dioxide, AFFF. From a secure, explosion-proof location, use water spray to cool exposed containers. If cooling streams are ineffective (venting sound increases in volume and pitch, tank discolors, or shows any signs of deforming), withdraw immediately to a secure position.

NFPA Hazard Classifications:

Health Hazards (Blue) 1
Flammability (Red) 1
Reactivity (Yellow) 0

Flash Point: 195°F (91°C) c.c.
Autoignition Temperature: 860°F (460°C)

FIRST AID

Move victim to fresh air. Call emergency medical care. Apply artificial respiration if victim is not breathing. Administer oxygen if breathing is difficult. Remove and isolate contaminated clothing and shoes. In case of contact with substance, immediately flush skin or eyes with running water for at least 20

minutes. Wash skin with soap and water. Keep victim warm and quiet. Ensure that medical personnel are aware of the material(s) involved and take precautions to protect themselves.

n–PHENYLPIPERAZINE P:161

SYNONYMS: EINECS No. 202-165-6; 1-FENYLPIPERAZIN; 1-PHENYLPIPERAZINE

IDENTIFICATION

CAS:	92-54-6
Molecular Formula:	$C_{10}H_{14}N_2$
RTECS No.:	TM2625000
Properties:	Combustible, oily liquid. Pale yellow. Floats on water; insoluble. Decomposes to produce poisonous gases.

HEALTH AND SAFETY INFORMATION

Irritates eyes (severe), skin, and respiratory tract.

RECOMMENDATIONS FOR RESPIRATOR SELECTION

At any concentrations above the NIOSH REL, or where there is no REL, at any detectable concentration: SCBAF:PD,PP (any self-contained breathing apparatus that has a full facepiece and is operated in a pressure-demand or other positive-pressure mode); or SAF:PD,PP:ASCBA (any supplied-air respirator that has a full facepiece and is operated in a pressure-demand or other positive-pressure mode in combination with an auxiliary, self-contained breathing apparatus operated in a pressure-demand or other positive-pressure mode).

Escape: GMFOV [any air-purifying, full-facepiece respirator (gas mask) with a chin-style, front- or back-mounted organic vapor canister]; or SCBAE (any appropriate escape-type, self-contained breathing apparatus).

CHEMICAL REACTIVITY

Contact with strong oxidizers may cause fire and explosions. Supports combustion.

FIRE INFORMATION

Combustion products include toxic nitrogen oxides. Vapors in confined areas may explode when exposed to fire. If material or contaminated runoff enters waterways, notify downstream users of potentially contaminated waters. Notify local health and fire officials and pollution control agencies. Use dry chemicals, alcohol-resistant foam, carbon dioxide. From a secure, explosion-proof location, use water spray to cool exposed containers.

NFPA Hazard Classifications: Not listed
Flash Point: 285°F (141°C) o.c.

FIRST AID

Move victim to fresh air. Call emergency medical care. Apply artificial respiration if victim is not breathing. If breathing is difficult, give oxygen. Remove and isolate contaminated clothing and shoes. In case of contact with substance, immediately flush skin or eyes with running water for at least 20 minutes. Wash skin with soap and water. Keep victim warm and quiet. Ensure that medical personnel are aware of the material(s) involved and take precautions to protect themselves.

PHOSGENE P:165

SYNONYMS: CARBONE (OXYCHLORURE de) (French); CARBON OXYCHLORIDE; CARBONIC DICHLORIDE; CARBONIO (OSSICLORURO di) (Italian); CARBONYLCHLORID (German); CARBONYL CHLORIDE; CG; CHLOROFORMYL CHLORIDE; CLORURO de CARBONILO (Spanish); DIPHOSGENE; EEC No. 006-002-00-8; EINECS No. 200-870-3; FOSGEEN (Dutch); FOSGEN (Polish); FOSGENE (Italian); FOSGENO (Spanish); KOOLSTOFOXYCHLORIDE (Dutch); NCI-C60219; PHOSGEN (German)

IDENTIFICATION

CAS:	75-44-5
ERG ID:	UN 1076
ERG Guide:	125
ERG Guide Category:	GASES-CORROSIVE
Molecular Formula:	CCl_2O
RTECS No.:	SY5600000
Properties:	Noncombustible, liquefied compressed gas or fuming liquid below 47°F (8°C). Colorless to light yellow. Sweet or sharp odor of new-mown hay, but may not be detectable at higher concentrations. Liquid sinks in water; reacts to form hydrochloric acid and carbon dioxide.

HEALTH AND SAFETY INFORMATION

Poisonous and corrosive gas. Breathing gas can cause death. Skin and eye contact causes severe burns and blindness. Inhalation may cause pulmonary edema, which can be delayed for several hours; there is a risk of death in serious cases.

IDLH: 2 ppm
Odor Threshold: 0.55 ppm Poor warning: odor is not a reliable indicator of the presence of toxic amounts of gas.
OSHA Table Z-1-A Air Contaminant

OSHA Process Safety Management of Highly Hazardous Chemicals (29CFR, Part 1910.119, Appendix A): Threshold Quantity: 100 pounds.

Exposure Limits:	**TWA**	**CEILING**
ACGIH:	0.1 ppm; 0.40 mg/m^3	
NIOSH:	0.1 ppm; 0.40 mg/m^3	0.2 ppm; 0.8 mg/m^3 [15 min]
OSHA:	0.1 ppm; 0.40 mg/m^3	

RECOMMENDATIONS FOR RESPIRATOR SELECTION: NIOSH/OSHA

1 ppm: SA (any supplied-air respirator).
2 ppm: SCBAF (any self-contained breathing apparatus with a full facepiece); or SAF (any supplied-air respirator with a full facepiece).

Emergency or planned entry into unknown concentration or IDLH conditions:
SCBAF:PD,PP (any self-contained breathing apparatus that has a full facepiece and is operated in a pressure-demand or other positive-pressure mode); or SAF:PD,PP:ASCBA (any supplied-air respirator that has a full facepiece and is operated in a pressure-demand or other positive-pressure mode in combination with an auxiliary, self-contained breathing apparatus operated in a pressure-demand or other positive pressure mode).

Escape: GMFS [any air-purifying, full-facepiece respirator (gas mask) with a chin-style, front- or back-mounted canister providing protection against the compound of concern]; or SCBAE (any appropriate escape-type, self-contained breathing apparatus).

Note: Substance reported to cause eye irritation or damage; may require eye protection.

CHEMICAL REACTIVITY

Incompatible with water, producing hydrochloric acid. Reacts violently with strong oxidizers, anhydrous ammonia, isopropyl alcohol, and chemically active metals. Forms shock-sensitive material with potassium. Attacks most metals in moist conditions. Attacks some plastics and coatings.

FIRE INFORMATION

Not flammable, but decomposition products include toxic chlorine gas and hydrogen chloride. Vapors are initially heavier than air and will collect in low areas. If material or contaminated runoff enters waterways, notify downstream users of potentially contaminated waters. Notify local health and fire officials and pollution control agencies. Use remote equipment wherever possible. Extinguish fire using appropriate agent suitable for surrounding fire. From a secure, explosion-proof location, use water spray to cool exposed containers. If containers are exposed to direct flame or elevated temperatures for prolonged times, withdraw immediately to a secure location

NFPA Hazard Classifications:

Health Hazards (Blue) 4

Flammability (Red) 0
Reactivity (Yellow) 1

FIRST AID

Move victim to fresh air. Call emergency medical care. Apply artificial respiration if victim is not breathing. Do not use mouth-to-mouth method if victim ingested or inhaled the substance; induce artificial respiration with the aid of a pocket mask equipped with a one-way valve or other proper respiratory medical device. Administer oxygen if breathing is difficult. Remove and isolate contaminated clothing and shoes. In case of contact with liquefied gas, thaw frosted parts with lukewarm water. Keep victim warm and quiet. Keep victim under observation. Effects of contact or inhalation may be delayed. Ensure that medical personnel are aware of the material(s) involved and take precautions to protect themselves.

ENVIRONMENTAL INFORMATION

Clean Air Act: Hazardous Air Pollutants (Title I, Part A, Section 112); List of high risk pollutants (Section 63.74); Accidental Release Prevention/Flammable Substances (Section 112[r], Table 3), TQ = 500 lb (227 kg)
Clean Water Act: Section 311 Hazardous Substances/RQ (same as CERCLA); Section 313 Priority Chemicals
EPA Hazardous Waste Number (RCRA No.): P095
RCRA Section 261 Hazardous Constituents.
EPCRA Section 302, Extremely Hazardous Substances: TPQ = 10 lb (4.54 kg)
EPCRA Section 304 Reportable Quantity (RQ): CERCLA, 10 lb (4.54 kg)
EPCRA Section 313: Form R *de minimis* concentration reporting level: 1.0%.
California's Proposition 65: Carcinogen.

PHOSPHORIC ACID P:180

SYNONYMS: ACIDE PHOSPHORIQUE (French); ACIDO FOSFORICO (Spanish); ALUMINUM ETCH 16-1-1-2; ALUMINUM ETCH 82-3-15-0; COPPER, BRASS BRITE DIP 127; COPPER, BRASS BRITE DIP 1127; EEC No. 015-011-00-6; EINECS No. 231-633-2; ENPLATE CONDITIONER 470; ENTEC 310A; ENTEC 313; ENTEC 318; ENTEC 353; ENTEC 356; ENTEC 436; FRECKLE ETCH; KESTER 5765 SCALE-OFF; KTI ALUMINUM ETCH I; KTI ALUMINUM ETCH II; KTI ALUMINUM ETCH III; MARPHOS; METAPHOSPHORIC ACID; NFB; ORTHOPHOSPHORIC ACID; PHOSPHORIC ACID, AQUEOUS; RDH LIME SOLVENT; SONAC; SUPER Q ETCH; SUPERPHOS-105; TOP DIP; WC-REINIGER; WET K-ETCH; WHITE PHOSPHORIC ACID

IDENTIFICATION

CAS:	7664-38-2
ERG ID:	UN 1805
ERG Guide:	154
ERG Guide Category:	TOXIC AND/OR CORROSIVE (NONCOMBUSTIBLE)

Molecular Formula:	H_3O_4P
RTECS No.:	TB63000000
Properties:	Nonflammable, colorless, thick, syruplike liquid, or solid crystals below 70°F (21°C). Odorless. Corrosive. Sinks and mixes with water. When heated will generate toxic gases and phosphoric acid fumes.

HEALTH AND SAFETY INFORMATION

Corrosive to eyes, skin, and respiratory tract. Inhalation may cause pulmonary edema, which can be delayed for several hours; there is a risk of death in serious cases. Repeated or prolonged skin exposure may cause irritation; bronchitis may also result from prolonged exposure.

IDLH: 1,000 mg/m^3
OSHA Table Z-1-A Air Contaminant

Exposure Limits:	**TWA**	**STEL**
ACGIH:	1 mg/m^3	3 mg/m^3
NIOSH:	1 mg/m^3	3 mg/m^3
OSHA:	1 mg/m^3	

RECOMMENDATIONS FOR RESPIRATOR SELECTION: NIOSH/OSHA

25 mg/m^3: SA:CF (any supplied-air respirator operated in a continuous-flow mode).
50 mg/m^3: HiEF (any air-purifying, full-facepiece respirator with a high-efficiency particulate filter); or SCBAF (any self-contained breathing apparatus with a full facepiece); or SAF (any supplied-air respirator with a full facepiece).
1,000 mg/m^3: SAF:PD,PP (any supplied-air respirator that has a full facepiece and is operated in a pressure-demand or other positive-pressure mode).

Emergency or Planned Entry into Unknown Concentrations or IDLH Conditions:
SCBAF:PD,PP (any self-contained breathing apparatus that has a full facepiece and is operated in a pressure-demand or other positive-pressure mode); or SAF:PD,PP:ASCBA (any supplied-air respirator that has a full facepiece and is operated in a pressure-demand or other positive-pressure mode in combination with an auxiliary, self-contained breathing apparatus operated in a pressure-demand or other positive-pressure mode).

Escape: HiEF (any air-purifying, full-facepiece respirator with a high-efficiency particulate filter); or SCBAE (any appropriate escape-type, self-contained breathing apparatus).
Note: Substance reported to cause eye irritation or damage; may require eye protection.

CHEMICAL REACTIVITY

A strong acid; contact with strong caustic can generate a lot of heat and violent spattering. Incompatible with aliphatic amines, alkanolamines, alkylene oxides, aromatic amines, amides, ammonia, ammonium hydroxide, bases, calcium oxide, epichlorohydrin, isocyanates, explosive with nitromethane, oleum, organic anhydrides, sulfuric acid, sodium tetrahydroborate, strong oxidizers, vinyl acetate, and water. Contact with most metals forms flammable hydrogen gas. If used as a metal cleaner, impurities in the metals may lead to the formation of extremely poisonous phosphine gas. Attacks some plastics, rubber, coatings, glass, and ceramics.

FIRE INFORMATION

Combustion products include toxic phosphoric oxides. If material or contaminated runoff enters waterways, notify downstream users of potentially contaminated waters. Notify local health and fire officials and pollution control agencies. Use water, dry chemicals, alcohol-resistant foam, carbon dioxide. From a secure, explosion-proof location, use water spray to cool exposed containers.

NFPA Hazard Classifications:

Health Hazards (Blue) 3
Flammability (Red) 0
Reactivity (Yellow) 0

FIRST AID

Move victim to fresh air. Call emergency medical care. Apply artificial respiration if victim is not breathing. Do not use mouth-to-mouth method if victim ingested or inhaled the substance; induce artificial respiration with the aid of a pocket mask equipped with a one-way valve or other proper respiratory medical device. Administer oxygen if breathing is difficult. Remove and isolate contaminated clothing and shoes. In case of contact with substance, immediately flush skin or eyes with running water for at least 20 minutes. For minor skin contact, avoid spreading material on unaffected skin. Keep victim warm and quiet. Effects of exposure (inhalation, ingestion, or skin contact) may be delayed. Ensure that medical personnel are aware of the material(s) involved and take precautions to protect themselves.

ENVIRONMENTAL INFORMATION

Clean Water Act: Section 311 Hazardous Substances/RQ (same as CERCLA); Section 313 Priority Chemicals.
EPCRA Section 304: Reportable Quantity (RQ): CERCLA, 5,000 lb (2,270 kg).
EPCRA Section 313: Form R *de minimis* concentration reporting level: 1.0%.

PHOSPHORUS, YELLOW OR WHITE — P:210

SYNONYMS: BONIDE BLUE DEATH RAT KILLER; COMMON SENSE COCKROACH AND RAT PREPARATIONS; EEC No. 015-001-00-1; EINECS No. 231-768-7; ELEMENTAL PHOSPHORUS; EXOLITE 405; EXOLIT LPKN 275; EXOLIT VPK-N 361; FOSFORO BLANCO (Spanish); FOSFORO BIANCO (Italian); GELBER PHOSPHOR (German); MOLTEN PHOSPHORUS; PHOSPHORE BLANC (French); PHOSPHORUS ELEMENTAL, WHITE; PHOSPHOROUS YELLOW; PHOSPHORUS-31; STCC 4916140; TETRAFOSFOR (Dutch); TETRAPHOSPHOR (German); WEISS PHOSPHOR (German); WHITE PHOSPHORUS; WP; YELLOW PHOSPHORUS

IDENTIFICATION

CAS:	7723-14-0
ERG ID:	UN 1381(dry, under water or in water solution); UN 2447 (white, molten)
ERG Guide:	136
ERG Guide Category:	SUBSTANCES-SPONTANEOUSLY COMBUSTIBLE-TOXIC (AIR-REACTIVE)
Molecular Formula:	P_4
RTECS No.:	TH3500000
Properties:	Extremely flammable, soft, waxy solid. White to yellow; darkens on contact with light. Pyrophoric; burns spontaneously on contact with air. Sharp, acrid, garliclike smell. Sinks in water; slightly soluble. Stable when excluded from air; usually shipped or stored in water or inert gas. Melts at 111°F (44°C) to a yellow or white liquid.

HEALTH AND SAFETY INFORMATION

Poisonous. Breathing the vapor or swallowing the material can cause death. Corrosive to eyes, skin, and respiratory tract. Inhalation may cause pulmonary edema, which can be delayed for several hours; there is a risk of death in serious cases. Skin contact causes deep burns which are slow-healing. Prolonged or repeated contact can cause skin, bone, liver, and kidney disorders.

IDLH: 5 mg/m^3
OSHA Table Z-1-A Air Contaminant

Exposure Limits:	**TWA**
ACGIH:	0.02 ppm; 0.1 mg/m^3
NIOSH:	0.1 mg/m^3
OSHA:	0.1 mg/m^3

RECOMMENDATIONS FOR RESPIRATOR SELECTION: NIOSH

1 mg/m^3: SA (any supplied-air respirator); or SCBA (any self-contained breathing apparatus).
2.5 mg/m^3: SA:CF* (any supplied-air respirator operated in a continuous-flow mode).
5 mg/m^3: SCBAF (any self-contained breathing apparatus with a full facepiece); or SAF (any supplied-air respirator with a full facepiece).

Emergency or Planned Entry into Unknown Concentrations or IDLH Conditions:
SCBAF:PD,PP (any self-contained breathing apparatus that has a full facepiece and is operated in a pressure-demand or other positive-pressure mode); or SAF:PD,PP:ASCBA (any supplied-air respirator that has a full facepiece and is operated in a pressure-demand or other positive-pressure mode in combination with an auxiliary, self-contained breathing apparatus operated in a pressure-demand or other positive-pressure mode).

Escape: GMFOV [any air-purifying, full-facepiece respirator (gas mask) with a chin-style, front- or back-mounted organic vapor canister]; or SCBAE (any appropriate escape-type, self-contained breathing apparatus).

*Substance causes eye irritation or damage; eye protection needed.

CHEMICAL REACTIVITY

A strong reducing agent and fire hazard. Ignites spontaneously on contact with air above 86°F (30°C). Contact with oxidizers (including sulfur) will cause fire and explosions. Reacts with halogens and halides. Contact with strong caustics forms poisonous and flammable gas. Melts at 111°F (44°C) to a yellow or white liquid; liquid attacks some plastics, rubber, and coatings. Contact with metals form highly reactive phosphide materials.

FIRE INFORMATION

If material or contaminated runoff enters waterways, notify downstream users of potentially contaminated waters. Notify local health and fire officials and pollution control agencies. Use dry chemicals, water spray, fog, earth, sand. From a secure, explosion-proof location, use water spray to cool exposed containers.

NFPA Hazard Classifications:

Health Hazards (Blue) 4
Flammability (Red) 4
Reactivity (Yellow) 2

FIRST AID

Move victim to fresh air. Call emergency medical care. Apply artificial respiration if victim is not breathing. Administer oxygen if breathing is difficult. In case of contact with substance, keep exposed skin areas immersed in water or covered with wet bandages until medical attention is received. Removal of solidified molten material from skin requires medical assistance. Remove and isolate contaminated clothing and shoes at the site and place in metal container filled with water. Fire hazard if allowed to dry. Effects of exposure (inhalation, ingestion, or skin contact) to substance may be delayed. Keep victim warm and quiet. Ensure that medical personnel are aware of the material(s) involved and take precautions to protect themselves.

ENVIRONMENTAL INFORMATION

Clean Air Act: Hazardous Air Pollutants (Title I, Part A, Section 112).
Clean Water Act: Section 311 Hazardous Substances/RQ (same as CERCLA); Section 313 Priority Chemicals.
EPCRA Section 302 Extremely Hazardous Substances: TPQ = 100 lb (45.4 kg).
EPCRA Section 304: Reportable Quantity (RQ): CERCLA, 1 lb (0.454 kg).
EPCRA Section 313: (yellow or white) Form R *de minimis* concentration reporting level: 1.0%

Marine Pollutant (49CFR, Subchapter 172.101, Appendix B), severe pollutant, white, yellow dry, molten or in solution.
TSCA: 40CFR716.120(*a*).

PHOSPHORUS OXYCHLORIDE P:220

SYNONYMS: EEC No. 015-009-00-5; EINECS No. 233-046-7; OXICLORURO de FOSFORO (Spanish); PHOSPHORUS CHLORIDE; PHOSPHOROUS OXYCHLORIDE SILICON DOPANT; PHOSPHORUS OXYTRICHLORIDE; PHOSPHORYL CHLORIDE

IDENTIFICATION

CAS:	10025-87-3
ERG ID:	UN 1810
ERG Guide:	137
ERG Guide Category:	WATER REACTIVE-CORROSIVE
Molecular Formula:	Cl_3OP
RTECS No:	TH4897000
Properties:	Nonflammable, oily liquid; fumes in air. Colorless to light yellow. Musty odor. Sinks and reacts violently with water, forming toxic and corrosive phosphoric and hydrochloric acid. Freezing point is 34°F (1°C).

HEALTH AND SAFETY INFORMATION

Poison. Breathing the vapors can cause death. Corrosive to eyes, skin, and respiratory tract; can cause severe burns and blindness. Inhalation may cause pulmonary edema, which can be delayed for several hours; there is a risk of death in serious cases. Repeated or prolonged exposures may cause kidney and liver damage; lung problems.

OSHA Table Z-1-A Air Contaminant
OSHA Process Safety Management of Highly Hazardous Chemicals (29CFR, Part 1910.119, Appendix A): Threshold Quantity: 1,000 pounds.

Exposure Limits:	**TWA**	**STE**
ACGIH:	0.1 ppm; 0.63 mg/m3	
NIOSH:	0.1 ppm; 0.6 mg/m^3	0.5 ppm; 3 mg/m^3
OSHA:	None	

CHEMICAL REACTIVITY

Water contact forms hydrochloric and phosphoric acids, phosphine, and may self-ignite. Air contact forms corrosive fumes. Reacts with acids, alcohols, alkali metals, caustics, combustible materials, carbon disulfide, dimethyl formamide, organic matter, strong bases, zinc powder. Rapid corrosion of steel and most metals, except lead, in the presence of moisture.

FIRE INFORMATION

Combustion products include toxic hydrogen chloride and phosphoric acid. Firefighting gear (including SCBA) does not provide adequate protection. If exposure occurs, remove and isolate gear immediately and thoroughly decontaminate personnel. Vapors are heavier than air and will collect in low areas. Vapors in confined areas may explode when exposed to fire. Storage containers and parts of containers may rocket great distances, in many directions. If material or contaminated runoff enters waterways, notify downstream users of potentially contaminated waters. Notify local health and fire officials and pollution control agencies. Do not use water. Use dry chemicals, earth, or sand. From a secure, explosion-proof location, or using unattended equipment, cool exposed, closed containers with water spray.

NFPA Hazard Classifications:

Health Hazards (Blue) 3
Flammability (Red) 0
Reactivity (Yellow) 2
Special Notice .... WATER REACTIVE

FIRST AID

Move victim to fresh air. Call emergency medical care. Apply artificial respiration if victim is not breathing. Do not use mouth-to-mouth method if victim ingested or inhaled the substance; induce artificial respiration with the aid of a pocket mask equipped with a one-way valve or other proper respiratory medical device. Administer oxygen if breathing is difficult. Remove and isolate contaminated clothing and shoes. In case of contact with substance, immediately flush skin or eyes with running water for at least 20 minutes. For minor skin contact, avoid spreading material on unaffected skin. Removal of solidified molten material from skin requires medical assistance. Keep victim warm and quiet. Effects of exposure (inhalation, ingestion, or skin contact) to substance may be delayed. Ensure that medical personnel are aware of the material(s) involved and take precautions to protect themselves.

ENVIRONMENTAL INFORMATION

Clean Air Act: Accidental Release Prevention/Flammable substances, (Section 112[r], Table 3), TQ = 5,000 lb (2,270 kg).
Clean Water Act: Section 311 Hazardous Substances/RQ (same as CERCLA).
EPCRA Section 302 Extremely Hazardous Substances: TPQ = 500 lb (227 kg).
EPCRA Section 304 Reportable Quantity (RQ): EHS, CERCLA, 1,000 lb (454 kg).

PHOSPHORUS PENTOXIDE P:260

SYNONYMS: ANHIDRIDO FOSFORICO (Spanish); ANHYDRIDE PHOSPHORIQUE (French); DIPHOSPHORUS PENTOXIDE; EEC No. 015-010-00-0; EINECS No. 215-236-1; PHOSPHORIC

ANHYDRIDE; PHOSPHORIC OXIDE; PHOSPHORUS PENTOXIDE; PHOSPHORUS(V) OXIDE; PHOSPHORIC ACID, ANHYDROUS; POX

IDENTIFICATION

CAS:	1314-56-3
ERG ID:	UN 1807
ERG Guide:	137
ERG Guide Category:	WATER REACTIVE-CORROSIVE
Molecular Formula:	P_2O_5
RTECS No.:	TH3945000
Properties:	Crystalline solid or powder. White, deliquescent. Very soluble in water; dangerously reactive.

HEALTH AND SAFETY INFORMATION

Fumes and dust from powder are extremely corrosive to eyes, skin, and respiratory tract. Eye contact can cause blindness. Skin contact can cause severe burns. Inhalation may cause pulmonary edema, which can be delayed for several hours; there is a risk of death in serious cases.

Exposure Limits:	**TWA**
DFG	1.0 mg/m^3

RECOMMENDATIONS FOR RESPIRATOR SELECTION

At any concentrations above the NIOSH REL, or where there is no REL, at any detectable concentration: SCBAF:PD,PP (any self-contained breathing apparatus that has a full facepiece and is operated in a pressure-demand or other positive-pressure mode); or SAF:PD,PP:ASCBA (any supplied-air respirator that has a full facepiece and is operated in a pressure-demand or other positive-pressure mode in combination with an auxiliary, self-contained breathing apparatus operated in a pressure-demand or other positive-pressure mode).

Escape: GMFOV [any air-purifying, full-facepiece respirator (gas mask) with a chin-style, front- or back-mounted organic vapor canister]; or SCBAE (any appropriate escape-type, self-contained breathing apparatus).

CHEMICAL REACTIVITY

Reacts violently with water to form phosphoric acid; dangerous fire risk. With the appropriate conditions it undergoes hazardous reactions with ammonia, calcium oxide, chlorine trifluoride, formic acid, hydrogen fluoride, inorganic bases, hydrofluoric acid, iodides, metals (aluminum, magnesium), methyl hydroperoxide, oxididizers (e.g., bromine, pentafluoride, chlorine trifluoride, perchloric acid, oxygen, oxygen difluoride, hydrogen peroxide), potassium, propargyl alcohol, 3-propynol, sodium hydroxide. Contact with perchloric acid in chloroform causes a violent explosion.

FIRE INFORMATION

Combustion products include toxic phosphorus oxides. If material or contaminated runoff enters waterways, notify downstream users of potentially contaminated waters. Notify local health and fire officials and pollution control agencies. Use dry chemicals, alcohol-resistant foam, carbon dioxide. Do not get water inside containers. From a secure, explosion-proof location, use water spray to cool exposed containers.

FEMA Hazard Classifications:

Health Hazards (Blue) 3
Flammability (Red) 0
Reactivity (Yellow) 2
Special Situations . WATER REACTIVE

FIRST AID

Move victim to fresh air. Call emergency medical care. Apply artificial respiration if victim is not breathing. Do not use mouth-to-mouth method if victim ingested or inhaled the substance; induce artificial respiration with the aid of a pocket mask equipped with a one-way valve or other proper respiratory medical device. Administer oxygen if breathing is difficult. Remove and isolate contaminated clothing and shoes. In case of contact with substance, immediately flush skin or eyes with running water for at least 20 minutes. For minor skin contact, avoid spreading material on unaffected skin. Removal of solidified molten material from skin requires medical assistance. Keep victim warm and quiet. Effects of exposure (inhalation, ingestion, or skin contact) to substance may be delayed. Ensure that medical personnel areaware of te material(s) involved and take precautions to protect themselves.

ENVIRONMENTAL INFORMATION

Removed from EHS List (FR VOL. 61, No. 89, Page 20477

PHOSPHOTUNGSTIC ACID P:282

SYNONYMS: EINECS No. 235-087-6; PHOSPHO-12-TUNGSTIC ACID; PHOSPHOTUNGSTIC ACID, Nhydrate; PHOSPHOWOLFRAMIC ACID; TUNGSTEN HYDROXIDE OXIDE PHOSPHATE; TUNGSTOPHOSPHORIC ACID (8CI); 12-TUNGSTOPHOSPHORIC ACID

IDENTIFICATION

CAS:	12067-99-1
Molecular Formula:	$H_3O_{40}PW_{12} \bullet xH_2O$
RTECS No.:	TH5650000
Properties:	Yellow crystals or white powder. Odorless. Mixes with water.

HEALTH AND SAFETY INFORMATION

Liquid and dust from powder are corrosive to eyes, skin, and respiratory tract. Inhalation may cause pulmonary edema, which can be delayed for several hours; there is a risk of death in serious cases.

Exposure Limits:	TWA	STEL
ACGIH:	1.0 mg/m^3 as tungsten	3.0 mg/m^3 as tungsten
NIOSH:	5.0 mg/m^3 as tungsten	10 mg/m^3 as tungsten

RECOMMENDATIONS FOR RESPIRATOR SELECTION

At any concentrations above the NIOSH REL, or where there is no REL, at any detectable concentration: SCBAF:PD,PP (any self-contained breathing apparatus that has a full facepiece and is operated in a pressure-demand or other positive-pressure mode); or SAF:PD,PP:ASCBA (any supplied-air respirator that has a full facepiece and is operated in a pressure-demand or other positive-pressure mode in combination with an auxiliary, self-contained breathing apparatus operated in a pressure-demand or other positive-pressure mode).

Escape: GMFOV [any air-purifying, full-facepiece respirator (gas mask) with a chin-style, front- or back-mounted organic vapor canister]; or SCBAE (any appropriate escape-type, self-contained breathing apparatus).

CHEMICAL REACTIVITY

A strong oxidizing agent in aqueous solution; a strong acid in free acid form. Unstable with alkali metals, most common metals, metal oxides, strong reducing agents, carbonates, and cyanides. Incompatible with sulfuric acid, caustics, ammonia, aliphatic amines, alkanolamines, aromatic amines, amides, organic anhydrides, isocyanates, alkylene oxides, epichlorohydrin..

FIRE INFORMATION

Combustion products include toxic phosphorus oxides. If material or contaminated runoff enters waterways, notify downstream users of potentially contaminated waters. Notify local health and fire officials and pollution control agencies. Use dry chemicals, alcohol-resistant foam, carbon dioxide, water spray. From a secure, explosion-proof location, use water spray to cool exposed containers.

FIRST AID

Move victim to fresh air. Call emergency medical care. Apply artificial respiration if victim is not breathing. Do not use mouth-to-mouth method if victim ingested or inhaled the substance; induce artificial respiration with the aid of a pocket mask equipped with a one-way valve or other proper respiratory medical device. Administer oxygen if breathing is difficult. Remove and isolate contaminated clothing and shoes. In case of contact with substance, immediately flush skin or eyes with running water for at least 20 minutes. For minor skin contact, avoid spreading material on unaffected skin. Removal of solidified molten material from skin requires medical assistance. Keep victim warm and quiet. Effects of exposure (inhalation, ingestion, or skin contact) to substance may be delayed. Ensure that medical personnel areaware of te material(s) involved and take precautions to protect themselves.

PHTHALIC ACID, DIISODECYL ESTER P:285

SYNONYMS: 1,2-BENZENEDICARBOXYLIC ACID, DIISODECYL ESTER; BIS (ISODECYL) PHTHALATE; DIDP; DIISODECYL PHTHALATE; EINECS No. 247-977-1; PHTHALIC ACID, BIS (8-METHYL-NONYL) ESTER; PLASTICIZED DDP; PX-120; SICOL-184; SICOL 184; VESTINOL-DZ

IDENTIFICATION

CAS:	26761-40-0; 6851-49-1 (mixed isomers)
Molecular Formula:	$C_{28}H_{46}O_4$
RTECS No.:	TI1270000
Properties:	Combustible, oily liquid. Colorless. Mild odor. May float or sink in water; practically insoluble.

HEALTH AND SAFETY INFORMATION

Poisonous by inhalation. Due to the availability of insufficient data on human effects, caution should be exercised.

RECOMMENDATIONS FOR RESPIRATOR SELECTION

At any concentrations above the NIOSH REL, or where there is no REL, at any detectable concentration: SCBAF:PD,PP (any self-contained breathing apparatus that has a full facepiece and is operated in a pressure-demand or other positive-pressure mode); or SAF:PD,PP:ASCBA (any supplied-air respirator that has a full facepiece and is operated in a pressure-demand or other positive-pressure mode in combination with an auxiliary, self-contained breathing apparatus operated in a pressure-demand or other positive-pressure mode).

Escape: GMFOV [any air-purifying, full-facepiece respirator (gas mask) with a chin-style, front- or back-mounted organic vapor canister]; or SCBAE (any appropriate escape-type, self-contained breathing apparatus).

CHEMICAL REACTIVITY

Incompatible with strong acids, strong alkalies, nitrates, and oxidizers. May attack some forms of plastics.

FIRE INFORMATION

If material or contaminated runoff enters waterways, notify downstream users of potentially contaminated waters. Notify local health and fire officials and pollution control agencies. Use dry chemicals, alcohol-resistant foam, carbon dioxide; water may be ineffective on fire. From a secure, explosion-proof location, use water spray to cool exposed containers.

NFPA Hazard Classifications:

Health Hazards (Blue) 0
Flammability (Red) 1
Reactivity (Yellow) 0

Flash Point: 450°F (232°C) o.c.
Flammable Limits in Air: LEL 0.3% @ 508°F (264°C); UEL NA
Autoignition Temperature: 755°F (402°C)

FIRST AID

Move victim to fresh air. Call emergency medical care. Apply artificial respiration if victim is not breathing. Administer oxygen if breathing is difficult. Remove and isolate contaminated clothing and shoes. In case of contact with substance, immediately flush skin or eyes with running water for at least 20 minutes. Wash skin with soap and water. Keep victim warm and quiet. Effects of exposure (inhalation, ingestion, or skin contact) to substance may be delayed. Ensure that medical personnel are aware of the material(s) involved and take precautions to protect themselves.

ENVIRONMENTAL INFORMATION

Clean Water Act: Section 307 Toxic Pollutants (40CFR 403, Appendix B; 40CFR 401.15) as phthalate esters.
RCRA Section 261 Hazardous Constituents as phthalic acid esters, n.o.s.
EPCRA 40CFR Section 302.4: Reportable Quantity (RQ): EHS/CERCLA, 1 lb (0.454 kg)
TSCA: 40CFR799.5000 as diisodecyl phthalate

PHTHALIC ANHYDRIDE, SOLID P:287

SYNONYMS: ANHIDRIDO FTALICO (Spanish); ANHYDRIDE PHTALIQUE (French); ANIDRIDE FTALICA (Italian); 1,2-BENZENEDICARBOXYLIC ANHYDRIDE; 1,3-DIOXOPHTHALAN; EEC No. 607-009-00-4; EINECS No. 201-607-5; ESEN; FTALOWY BEZWODNIK (Polish); FTAALZUURANHYDRIDE (Dutch); ISOBENZOFURAN,1,3-DIHYDRO-1,3-DIOXO-; 1,3-ISOBENZOFURANDIONE; NCI-C03601; PAN; PHTHALANDIONE; 1,3-PHTHALANDIONE; PHTHALIC ACID ANHYDRIDE; PHTHALSAEUREANHYDRID (German); RETARDER AK; RETARDER ESEN; RETARDER PD

IDENTIFICATION

CAS:	85-44-9
ERG ID:	UN 2214
ERG Guide:	156
ERG Guide Category:	SUBSTANCES-TOXIC AND/OR CORROSIVE (COMBUSTIBLE/WATER-SENSITIVE)

858 Phthalic Anhydride, Solid

Molecular Formula: $C_8H_4O_3$
RTECS No.: TI3150000
Properties: Combustible solid flakes or liquid (heated). Colorless or pale yellow. Acrid, choking odor. Solid sinks in water. Liquid solidifies and sinks in water; very slightly soluble; decomposes slowly, forming phthalic acid.

HEALTH AND SAFETY INFORMATION

Corrosive to eyes, skin, and respiratory tract. Inhalation may cause pulmonary edema, which can be delayed for several hours; there is a risk of death in serious cases. Repeated or prolonged exposure to skin can cause skin disorders including eczema.

IDLH: 60 mg/m^3
Odor Threshold: 0.052 ppm
OSHA Table Z-1-A Air Contaminant

Exposure Limits:	TWA
ACGIH:	1 ppm; 6.1 mg/m^3
NIOSH:	1 ppm; 6 mg/m^3
OSHA:	2 ppm; 12 mg/m^3

RECOMMENDATIONS FOR RESPIRATOR SELECTION: NIOSH

30 mg/m^3: DM (any dust and mist respirator).
60 mg/m^3: DMXSQ (any dust and mist respirator except single-use and quarter mask respirators); or HiEF (any air-purifying, full-facepiece respirator with a high-efficiency particulate filter); or PAPRDM (any powered, air-purifying respirator with a dust and mist filter); or SA (any supplied-air respirator); or SCBAF (any self-contained breathing apparatus with a full facepiece).

Emergency or planned entry into unknown concentration or IDLH conditions: SCBAF: PD, PP (any self-contained breathing apparatus that has a full facepiece and is operated in a pressure-demand or other positive-pressure mode); or SAF: PD, PP: ASCBA (any supplied-air respirator that has a full facepiece and is operated in a pressure-demand or other positive-pressure mode in combination with an auxiliary, self-contained breathing apparatus operated in a pressure-demand or other positive pressure mode).

Escape: HiEF (any air-purifying, full-facepiece respirator with a high-efficiency particulate filter); or SCBAE (any appropriate escape-type, self-contained breathing apparatus).

Note: Substance reported to cause eye irritation or damage; may require eye protection.

CHEMICAL REACTIVITY

Dust forms explosive mixture with air; can ignite spontaneously. Solid reacts very slowly in water. Liquid spatters when in contact with water. Slowly converted to phthalic acid in cold water; faster in hot water.

Incompatible with strong acids (violent reaction with nitric acid), caustics, ammonia, amines, and strong oxidizers. Attacks some plastics, rubber, and coatings. Mixture with air can be self-reactive. Solution attacks ordinary iron and mild steel.

FIRE INFORMATION

If material or contaminated runoff enters waterways, notify downstream users of potentially contaminated waters. Notify local health and fire officials and pollution control agencies. Use dry chemicals, carbon dioxide, water spray. From a secure, explosion-proof location, use water spray to cool exposed containers.

NFPA Hazard Classifications:

Health Hazards (Blue) 3
Flammability (Red) 1
Reactivity (Yellow) 0

Flash Point: 329°F (165°C) o.c.; 305°F (152°C) c.c.
Flammable Limits in Air: LEL 1.7%; UEL 10.5%
Autoignition Temperature: 1058°F (570°C)

FIRST AID

Move victim to fresh air. Call emergency medical care. Apply artificial respiration if victim is not breathing. Do not use mouth-to-mouth method if victim ingested or inhaled the substance; induce artificial respiration with the aid of a pocket mask equipped with a one-way valve or other proper respiratory medical device. Administer oxygen if breathing is difficult. Remove and isolate contaminated clothing and shoes. In case of contact with substance, immediately flush skin or eyes with running water for at least 20 minutes. For minor skin contact, avoid spreading material on unaffected skin. Keep victim warm and quiet. Effects of exposure (inhalation, ingestion, or skin contact) may be delayed. Ensure that medical personnel are aware of the material(s) involved and take precautions to protect themselves.

ENVIRONMENTAL INFORMATION

Clean Air Act: Hazardous Air Pollutants (Title I, Part A, Section 112)
EPA Hazardous Waste Number (RCRA No.): U190
RCRA Section 261 Hazardous Constituents.
RCRA Land Ban Waste
RCRA Universal Treatment Standards: Wastewater (mg/L), 0.055; Nonwastewater (mg/kg), 28.
EPCRA Section 304 Reportable Quantity (RQ): CERCLA, 5,000 lb (2,270 kg)
EPCRA Section 313: Form R *de minimis* concentration reporting level: 1.0%.

PINE OIL P:310

SYNONYMS: ACEITE de PINO (Spanish); ARIZOLE; OIL OF PINE; OLEUM ABIETIS; TERPENTINOEL (German); UNIPINE; YARMORE; YARMORE PINE OIL

860 Pine Oil

IDENTIFICATION

CAS:	8002-09-3; 80000-26-8 (scotch pine); 8021-29-2 (siberian pine); 8021-27-0 (silver pine)
ERG ID:	UN 1272
ERG Guide:	129
ERG Guide Category:	FLAMMABLE LIQUIDS (POLAR/WATER-MISCIBLE/ NOXIOUS)
RTECS No.:	TK5100000
Properties:	Flammable, liquid. Pale yellow. Penetrating, disinfectant odor. Insoluble in water.

HEALTH AND SAFETY INFORMATION

Skin irritant.

RECOMMENDATIONS FOR RESPIRATOR SELECTION

At any concentrations above the NIOSH REL, or where there is no REL, at any detectable concentration: SCBAF:PD,PP (any self-contained breathing apparatus that has a full facepiece and is operated in a pressure-demand or other positive-pressure mode); or SAF:PD,PP:ASCBA (any supplied-air respirator that has a full facepiece and is operated in a pressure-demand or other positive-pressure mode in combination with an auxiliary, self-contained breathing apparatus operated in a pressure-demand or other positive-pressure mode).

Escape: GMFOV [any air-purifying, full-facepiece respirator (gas mask) with a chin-style, front- or back-mounted organic vapor canister]; or SCBAE (any appropriate escape-type, self-contained breathing apparatus).

CHEMICAL REACTIVITY

Forms explosive mixture with air. Incompatible with strong acids and oxidizers.

FIRE INFORMATION

Vapors are heavier than air and will collect in low areas. Vapors in confined areas may explode when exposed to fire. Vapors may travel long distances to ignition sources and flash back. Storage containers and parts of containers may rocket great distances, in many directions. If material or contaminated runoff enters waterways, notify downstream users of potentially contaminated waters. Notify local health and fire officials and pollution control agencies. Use dry chemicals, foam, carbon dioxide. From a secure, explosion-proof location, use water spray to cool exposed containers.

NFPA Hazard Classifications:

Health Hazards (Blue) 0
Flammability (Red) 2
Reactivity (Yellow) 0

Flash Point:138°F (59°C); 172°F (78°C) steam distilled

FIRST AID

Move victim to fresh air. Call emergency medical care. Apply artificial respiration if victim is not breathing. Administer oxygen if breathing is difficult. Remove and isolate contaminated clothing and shoes. In case of contact with substance, immediately flush skin or eyes with running water for at least 20 minutes. Wash skin with soap and water. Keep victim warm and quiet. Effects of exposure (inhalation, ingestion, or skin contact) to substance may be delayed. Ensure that medical personnel are aware of the material(s) involved and take precautions to protect themselves.

PINE TAR P:312

SYNONYMS: EINECS No. 232-374-8; TAR, PINE

IDENTIFICATION

CAS:	8011-48-1
ERG ID:	UN 1993
ERG Guide:	128
ERG Guide Category:	FLAMMABLE LIQUIDS (NONPOLAR/WATER-IMMISCIBLE)
RTECS No.:	WW7530000
Properties:	Sticky, thick ,flammable liquid. Dark brown to black liquid or semisolid. Hardens with aging. Strong odor.

CHEMICAL REACTIVITY

Contact with oxidizers may cause fire and explosions.

FIRE INFORMATION

Use any extinguishing material. From a secure, explosion-proof location, use water spray to cool exposed containers.

NFPA Hazard Classifications:

Health Hazards (Blue) 0
Flammability (Red) 2
Reactivity (Yellow) 0

Flash Point:130°F (54°C) c.c. (pine tar); 144°F (62°C) c.c. (pine tar oil)
Autoignition Temperature: 130°F (54°C) c.c.;

FIRST AID

Move victim to fresh air. Call emergency medical care. Apply artificial respiration if victim is not breathing. If breathing is difficult, give oxygen. Remove and isolate contaminated clothing and shoes. In case of contact with substance, immediately flush skin or eyes with running water for at least 20 minutes. Wash skin with soap and water. Keep victim warm and quiet. Ensure that medical personnel are aware of the material(s) involved and take precautions to protect themselves.

PIPERIDINE P:314

SYNONYMS: AZACYCLOHEXANE; CYCLOPENTIMINE; CYPENTIL; EEC No. 613-027-00-3; EINECS No. 203-813-0; HEXAHYDROPYRIDINE; HEXAZANE; PENTAMETHYLENEIMINE; PIPERIDIN (German); PIPERIDINA (Spanish)

IDENTIFICATION

CAS:	110-89-4
ERG ID:	UN 2401
ERG Guide:	132
ERG Guide Category:	FLAMMABLE LIQUIDS-CORROSIVE
Molecular Formula:	$C_5H_{11}N$
RTECS No.:	TM3500000
Properties:	Highly flammable liquid. Clear, colorless. Fishy ammonia or pepperlike odor. Salty taste. Has a soapy feel. Initially floats on water and is soluble. Freezes at 19°F (–7°C).

HEALTH AND SAFETY INFORMATION

Poisonous. Corrosive to eyes, skin, and respiratory tract. Eye contact can cause blindness; prolonged contact with the skin causes burns. Inhalation may cause pulmonary edema, which can be delayed for several hours; there is a risk of death in serious cases. Affects the central nervous system.

RECOMMENDATIONS FOR RESPIRATOR SELECTION

At any concentrations above the NIOSH REL, or where there is no REL, at any detectable concentration: SCBAF:PD,PP (any self-contained breathing apparatus that has a full facepiece and is operated in a pressure-demand or other positive-pressure mode); or SAF:PD,PP:ASCBA (any supplied-air respirator that has a full facepiece and is operated in a pressure-demand or other positive-pressure mode in combination with an auxiliary, self-contained breathing apparatus operated in a pressure-demand or other positive-pressure mode).

Escape: GMFOV [any air-purifying, full-facepiece respirator (gas mask) with a chin-style, front- or back-mounted organic vapor canister]; or SCBAE (any appropriate escape-type, self-contained breathing apparatus).

CHEMICAL REACTIVITY

A strong reducing agent; contact with oxidizers can cause fire and explosions. Liquid is a strong acid. Reacts vigorously with acids, acid chlorides, acid anhydrides, and carbon dioxide. Explodes on contact with 1-perchloryl-piperidine, dicyanofurazan, or n-nitrosoacetanilide. Liquid corrodes aluminum, zinc, and other metals. May attack some plastics and coatings.

FIRE INFORMATION

Thermal decomposition may produce carbon monoxide, carbon dioxide, and nitrogen oxides. Vapors are heavier than air and will collect in low areas. Vapors in confined areas may explode when exposed to fire. Vapors may travel long distances to ignition sources and flash back. Storage containers and parts of containers may rocket great distances, in many directions. If material or contaminated runoff enters waterways, notify downstream users of potentially contaminated waters. Notify local health and fire officials and pollution control agencies. Use dry chemicals, foam, carbon dioxide. From a secure, explosion-proof location, use water spray to cool exposed containers.

NFPA Hazard Classifications

Health Hazards (Blue) 3
Flammability (Red) 3
Reactivity (Yellow) 0

Flash Point: 61°F (16°C)

FIRST AID

Move victim to fresh air; call emergency medical care. Apply artificial respiration if victim is not breathing. Do not use mouth-to-mouth method if victim ingested or inhaled the substance; induce artificial respiration with the aid of a pocket mask equipped with a one-way valve or other proper respiratory medical device. Administer oxygen if breathing is difficult. Remove and isolate contaminated clothing and shoes. In case of contact with substance, immediately flush skin or eyes with running water for at least 20 minutes. For minor skin contact, avoid spreading material on unaffected skin. Keep victim warm and quiet. Effects of exposure (inhalation, ingestion, or skin contact) to substance may be delayed. Ensure that medical personnel are aware of the material(s) involved and take precautions to protect themselves.

ENVIRONMENTAL INFORMATION

Clean Air Act: Accidental Release Prevention/Flammable Substances (Section 112[r], Table 3), TQ = 15,000 lb (6,810 kg)
EPCRA Section 302, Extremely Hazardous Substances: TPQ = 1,000 lb (454 kg)
EPCRA Section 304 Reportable Quantity (RQ): EHS, 1,000 lb (454 kg)

PIPERAZINE P:317

SYNONYMS: ANTIREN; 1,4-DIAZACYCLOHEXANE; 1,4-DIETHYLENEDIAMINE;

864 Piperazine

DIETHYLENEIMINE; DISPERMINE; EEC No. 612-057-00-4; EINECS No. 203-808-3; HEXAHYDRO-1,4-DIAZINE; HEXAHYDROPYRAZINE; LUMBRICAL; LUMBUCAL; PIPERAZIDINE; PIPERAZIN (German); PIPERAZINA (Spanish); PIPERAZINE, ANHYDROUS; PYRAZINE HEXAHYDRIDE; PYRAZINE, HEXAHYDRO-

IDENTIFICATION

CAS:	110-85-0 (base)
ERG ID:	UN 2579
ERG Guide:	153
ERG Guide Category:	SUBSTANCES-TOXIC AND/OR CORROSIVE (COMBUSTIBLE)
Molecular Formula:	$C_4H_{10}N_2$ (base)
RTECS No.:	TK7800000 (base)
Properties:	Combustible solid. Transparent to white. Mild, ammonialike, fishy odor. Sinks in water; slightly soluble.

HEALTH AND SAFETY INFORMATION

Severe irritant to skin and respiratory tract (dust). Corrosive to the eyes. May affect central nervous system. May cause lung allergy; asthma; skin allergies. Can pass through unbroken skin; can increase the severity of the symptoms listed.

OSHA Table Z-1-A Air Contaminant as piperazine dihydrochloride

Exposure Limits: **TWA** as piperazine dihydrochloride

ACGIH:	5 mg/m^3
NIOSH:	5 mg/m^3
OSHA:	None

RECOMMENDATIONS FOR RESPIRATOR SELECTION

At any concentrations above the NIOSH REL, or where there is no REL, at any detectable concentration: SCBAF:PD,PP (any self-contained breathing apparatus that has a full facepiece and is operated in a pressure-demand or other positive-pressure mode); or SAF:PD,PP:ASCBA (any supplied-air respirator that has a full facepiece and is operated in a pressure-demand or other positive-pressure mode in combination with an auxiliary, self-contained breathing apparatus operated in a pressure-demand or other positive-pressure mode).

Escape: GMFOV [any air-purifying, full-facepiece respirator (gas mask) with a chin-style, front- or back-mounted organic vapor canister];or SCBAE (any appropriate escape-type, self-contained breathing apparatus).

CHEMICAL REACTIVITY

A medium strong base. Reacts on contact with acids, carbon tetrachloride, nitrogen compounds.

Incompatible with amides, isocyanates, alkylene oxides, epichlorohydrin, aldehydes, alcohols, glycols, phenols, cresols, caprolactum solution. Contact with strong oxidizers may cause fire or explosions. Attacks nickel, copper, and cobalt. Corrosive to aluminum, magnesium, and zinc

FIRE INFORMATION

Combustion products include toxic nitrogen oxides. If material or contaminated runoff enters waterways, notify downstream users of potentially contaminated waters. Notify local health and fire officials and pollution control agencies. Use dry chemicals, alcohol-resistant foam, carbon dioxide, water spray. Use water spray to cool exposed containers.

NFPA Hazard Classifications:

Health Hazards (Blue) 2
Flammability (Red) 2
Reactivity (Yellow) 0

Flash Point: 178°F (81°C) o.c. (molten solid)
Autoignition Temperature: 851°F (455°C)

FIRST AID

Move victim to fresh air and call emergency medical care. Apply artificial respiration if victim is not breathing. Do not use mouth-to-mouth method if victim swallowed or inhaled the substance; induce artificial respiration with the aid of a pocket mask equipped with a one-way valve or other proper respiratory medical device. If breathing is difficult, give oxygen. In case of contact with substance, immediately flush skin or eyes with running water for at least 20 minutes. Remove and isolate contaminated clothing and shoes. Keep victim warm and quiet. Effects of exposure (inhalation, ingestion, or skin contact) to substances may be delayed. Ensure that medical personnel are aware of the material(s) involved and take precautions to protect themselves.

POLYBUTENES P:325

SYNONYMS: 1-BUTENE, HOMOPOLYMER; BUTENE, POLYMERS; BUTENE RESINS; INDOPOL L 50; PIB; POLYISOBUTYLENE WAXES; POLYISOBUTYLENE PLASTICS; POLYISOBUTYLENE RESINS

IDENTIFICATION

CAS:	9003-29-6; 9003-28-5
ERG ID:	UN 1866
ERG Guide:	127
ERG Guide Category:	FLAMMABLE LIQUIDS (POLAR/WATER- MISCIBLE)
Molecular Formula:	$(C_4H_8)_N$
Properties:	Combustible, oily liquid. Colorless. Odorless. Floats on water.

HEALTH AND SAFETY INFORMATION

Low toxicity by inhalation (SAX). Due to the availability of insufficient data on human effects, caution should be exercised

RECOMMENDATIONS FOR RESPIRATOR SELECTION

At any concentrations above the NIOSH REL, or where there is no REL, at any detectable concentration: SCBAF:PD,PP (any self-contained breathing apparatus that has a full facepiece and is operated in a pressure-demand or other positive-pressure mode); or SAF:PD,PP:ASCBA (any supplied-air respirator that has a full facepiece and is operated in a pressure-demand or other positive-pressure mode in combination with an auxiliary, self-contained breathing apparatus operated in a pressure-demand or other positive-pressure mode).

Escape: GMFOV [any air-purifying, full-facepiece respirator (gas mask) with a chin-style, front- or back-mounted organic vapor canister]; or SCBAE (any appropriate escape-type, self-contained breathing apparatus).

CHEMICAL REACTIVITY

Contact with strong oxidizers may cause fire and explosions. Incompatible with sulfuric acid and nitric acid

FIRE INFORMATION

Vapors are initially heavier than air and will collect in low areas. Notify local health and fire officials and pollution control agencies. Use dry chemicals, alcohol-resistant foam, carbon dioxide. From a secure, explosion-proof location, use water spray to cool exposed containers.

FEMA Hazard Classifications

Health Hazards (Blue) 0
Flammability (Red) 1
Reactivity (Yellow) 0

Flash Point: 215-470°F (102-243°C) o.c.; 280°F (138°C) c.c.

FIRST AID

Move victim to fresh air. Call emergency medical care. Apply artificial respiration if victim is not breathing. Administer oxygen if breathing is difficult. Remove and isolate contaminated clothing and shoes. In case of contact with substance, immediately flush skin or eyes with running water for at least 20 minutes. Wash skin with soap and water. Keep victim warm and quiet. Ensure that medical personnel are aware of the material(s) involved and take precautions to protect themselves.

POLYCHLORINATED BIPHENYLS P:330

SYNONYMS: AROCLOR®; AROCLOR® 1221; AROCLOR® 1232; AROCLOR® 1242; AROCLOR ® 1248; AROCLOR® 1254; AROCLOR® 1260; AROCLOR® 1262; AROCLOR® 1268; AROCLOR® 2565; AROCLOR® 4465; BIPHENYL, POLYCHLORO-; CHLOPHEN; CHLOREXTOL; CHLORINATED BIPHENYL; CHLORINATED DIPHENYL; CHLORINATED DIPHENYLENE; CHLORO BIPHENYL; CHLORODIPHENYL (54% CHLORINE); CHLORO 1,1-BIPHENYL; CLOPHEN; DYKANOL; EEC No. 602-039-00-4; EINECS No. 215-648-1; FENCLOR; INERTEEN; KANECHLOR®; KANECHLOR® 300; KANECHLOR® 400; KANECHLOR® 500; MONTAR; NOFLAMOL; PCB; PHENOCHLOR; PHENOCLOR; BIFENILOS POLICLORADOS (Spanish); POLYCHLOROBIPHENYL; PYRALENE; PYRANOL; SANTOTHERM; SANTOTHERM FR; SOVOL; THERMINOL FR-1; HALOGENATED WAXES

IDENTIFICATION

CAS:	1336-36-3 (generic CAS); 12672-29-6 (aroclor 1248); 53469-21-9 (aroclor 1242); 11097-69-1 (aroclor 1254); 11096-82-5 (aroclor 1260); 11141-16-5 (aroclor 1232); 12674-11-2 (aroclor 1016); 11104-28-2 (aroclor 1221)
ERG ID:	UN 2315
ERG Guide:	171
ERG Guide Category:	SUBSTANCES (LOW TO MODERATE HAZARD)
Molecular Formula:	$C_{12}H_7Cl_3$ (approximately)
RTECS No.:	TQ1350000; TQ 1352000 (aroclor 1221); TQ1354000 (aroclor 1232); TQ1356000 (aroclor 1242); TQ1358000 (aroclor 1248); TQ1360000 (aroclor 1254); TQ1362000 (aroclor 1260)
Properties:	Nonflammable, oily liquid, solid, or powder. Aroclors 1221, 1232, and 1248 are colorless oily liquids. Aroclor 1254 is a light yellow, thick oil; solid below 50°F (10°C). Aroclor 1260 and 1262 are sticky resins. Aroclor is a white powder. Mild hydrocarbon odor. Sinks in water; practically insoluble. Severely restricted or banned in many countries.

HEALTH AND SAFETY INFORMATION

Poisonous and carcinogenic. Irritates skin (strong) and respiratory tract. Skin contact can cause irritation. Skin contact contributes significantly to overall exposure. Toxic skin and liver effects.

IARC Evaluation: Group 2A; probable carcinogen; limited human evidence; sufficient animal evidence.
NTP: Anticipated carcinogen.
IDLH: 5 mg/m^3; potential carcinogen.
NIOSH Criteria Document: No. 77-225 PCBs.

Exposure Limits:	**TWA**
ACGIH:(1) (2)	1 mg/m^3 (42% chlorine); 0.5 mg/m^3 (54% chlorine)
NIOSH:(1)	0.001 mg/m^3 (applies to all PCBs)

OSHA:[1] 1 mg/m^3 (42% chlorine); 0.5 mg/m^3 (54% chlorine)

[1] Skin contact contributes significantly to overall exposure.
[2] ACGIH animal carcinogen.

RECOMMENDATIONS FOR RESPIRATOR SELECTION: NIOSH

At any concentrations above the NIOSH REL, or where there is no REL, at any detectable concentration: SCBAF:PD,PP (any self-contained breathing apparatus that has a full facepiece and is operated in a pressure-demand or other positive-pressure mode); or SAF:PD,PP:ASCBA (any supplied-air respirator that has a full facepiece and is operated in a pressure-demand or other positive-pressure mode in combination with an auxiliary, self-contained breathing apparatus operated in a pressure-demand or other positive pressure mode).

Escape: GMFOV [any air-purifying, full-facepiece respirator (gas mask) with a chin-style, front- or back-mounted organic vapor canister]; or SCBAE (any appropriate escape-type, self-contained breathing apparatus).

CHEMICAL REACTIVITY

Reacts with strong oxidizers.

FIRE INFORMATION

Nonflammable, but exposure in a fire results in the formation of a black soot containing PCBs, polychlorinated dibenzofurans and chlorinated dibenzo-p-dioxins. Firefighting gear (including SCBA) does not provide adequate protection. If exposure occurs, remove and isolate gear immediately and thoroughly decontaminate personnel. Storage containers and parts of containers may rocket great distances, in many directions. If material or contaminated runoff enters waterways, notify downstream users of potentially contaminated waters. Notify local health and fire officials and pollution control agencies. Use dry chemicals, foam, carbon dioxide. From a secure, explosion-proof location, use water spray to cool exposed containers.

NFPA Hazard Classifications:

Health Hazards (Blue) 2
Flammability (Red) 1
Reactivity (Yellow) 0

Flash Point: More than 286-385°F (141-196°C)

FIRST AID

Move victim to fresh air. Call emergency medical care. Apply artificial respiration if victim is not breathing. Administer oxygen if breathing is difficult. Remove and isolate contaminated clothing and shoes. In case of contact with substance, immediately flush skin or eyes with running water for at least 20 minutes. Ensure that medical personnel are aware of the material(s) involved and take precautions to protect themselves.

ENVIRONMENTAL INFORMATION as PCBs

Clean Air Act: Hazardous Air Pollutants (Title I, Part A, Section 112).
Clean Water Act: Section 307 Toxic Pollutants; Section 307 Priority Pollutants; Section 311 Hazardous Substances/RQ (same as CERCLA); Section 313 Priority Chemicals.
RCRA Land Ban Waste
RCRA Universal Treatment Standards: Wastewater (mg/L), 0.10; Nonwastewater (mg/kg), 10, total PCBs, sum of all PCB isomers, or all AROCLORS.
RCRA Ground Water Monitoring List: Suggested test method(s) (PQL ug/L): 8080(50); 8250(100).
Safe Drinking Water Act: MCL, 0.0005 mg/L; MCLG, zero; Regulated chemical (47 FR 9352).
EPCRA Section 304: Reportable Quantity (RQ): CERCLA, 1 lb (0.454 kg).
EPCRA Section 313: Form R *de minimis* concentration reporting level: 0.1%.
Marine Pollutant (49CFR, Subchapter 172.101, Appendix B), severe pollutant.
California's Proposition 65: Carcinogen; Reproductive toxin.

POLYETHYLENE P:335

SYNONYMS: AC 8; AC 394; AC 680; AC 1220; AC GA; ACP 6; AC 8 (POLYMER); ACROART; AGILENE; ALATHON;
ALATHON 14; ALATHON 15; ALATHON 1560; ALATHON 6600; ALATHON 7026; ALATHON 7040; ALATHON 7050;
ALATHON 7140; ALATHON 7511; ALATHON 5B; ALATHON 71XHN; ALCOWAX 6; ALDYL A; ALITHON 7050; ALKATHENE; ALKATHENE 17/04/00; ALKATHENE 22 300; ALKATHENE 200; ALKATHENE ARN 60; ALKATHENE WJG 11; ALKATHENE WNG 14; ALKATHENE XDG 33; ALKATHENE XJK 25; ALLIED PE 617; ALPHEX FIT 221;
AMBYTHENE; AMOCO 610A4; A 60-20R; A 60-70R; BAKELITE DFD 330; BAKELITE DHDA 4080; BAKELITE DYNH; BARECO POLYWAX 2000; BARECO WAX C 7500; BICOLENE C; BPE-I; BRALEN KB 2-11; BRALEN RB 03-23; BULEN A; BULEN A 30; CARLONA 58-030; CARLONA 900; CARLONA 18020 FA; CARLONA PXB; CHEMCOR; CHEMPLEX 3006; CIPE; COATHYLENE HA 1671; COURLENE-X3; CPE; CPE 16; CPE 25; CRYOPOLYTHENE; CRY-O-VAC L; DAISOLAC; DAPLEN; DAPLEN 1810 H; DFD 0173; DFD 0188; DFD 2005; DFD 6005; DFD 6032; DFD 6040; DFDJ 5505; DGNB 3825; DIOTHENE; DIXOPAK; DMDJ 4309; DMDJ 5140; DMDJ 7008; DOWLEX FILM; DQDA 1868; DQWA 0355; DXM 100; DYALL; DYLAN; DYLAN SUPER; DYLAN WPD 205; DYNH; DYNK 2; EINECS No. 200-815-3; ELTEX; ELTEX 6037; ELTEX A 1050; EPOLENE C; EPOLENE C 10; EPOLENE C 11; EPOLENE E; EPOLENE E 10; EPOLENE E 12; EPOLENE N: ETHENE POLYMER; ETHERIN; ETHEROL E; ETHYLENE HOMOPOLYMER; ETHYLENE POLYMER; ETHYLENE POLYMERS (8CI); FABRITONE PE; FB 217; FERTENE; FLAMOLIN MF 15711; FLOTHENE; FM 510; FORTIFLEX 6015; FORTIFLEX A 60/500; FP 4; 2100 GP; G-RESINS; GREX; GREX PP 60-002; GRISOLEN; HFDB 4201; HI-FAX; HI-FAX 1900; HI-FAX 4401; HI-FAX 4601; HIZEX; HIZEX 5000; HIZEX 5100; HIZEX 3000B; HIZEX 3300F; HIZEX 7000F; HIZEX 7300F; HIZEX 1091J; HIZEX 1291J; HIZEX 1300J; HIZEX 2100J; HIZEX 2200J; HIZEX 2100LP; HIZEX 5100LP; HIZEX 6100P; HIZEX 3000S; HIZEX 3300S; HIZEX 5000S; HOECHST PA 190; HOECHST WAX PA 520; HOSTALEN; HOSTALEN GD 620; HOSTALEN GD 6250; HOSTALEN GF 4760; HOSTALEN GF 5750; HOSTALEN GM 5010; HOSTALEN GUR; HOSTALEN HDPE; INTERFLO; IRAX; IRRATHENE R; LACQTEN 1020; LD 400; LD 600; LDPE 4; LUPOLEN 4261A; LUPOLEN

6042D; LUPOLEN 1010H; LUPOLEN 1800H; LUPOLEN 1810H; LUPOLEN 6011H; LUPOLEN KR 1032; LUPOLEN KR 1051; LUPOLEN KR 1257; LUPOLEN 6011L; LUPOLEN L 6041D; LUPOLEN N; LUPOLEN 1800S; MANOLENE 6050; MARLEX 9; MARLEX 50; MARLEX 60; MARLEX 960; MARLEX 6003; MARLEX 6009; MARLEX 6015; MARLEX 6050; MARLEX 6060; MARLEX EHM 6001; MARLEX M 309; MARLEX TR 704; MARLEX TR 880; MARLEX TR 885; MARLEX TR 906; MICROTHENE; MICROTHENE 510; MICROTHENE 704; MICROTHENE 710; MICROTHENE F; MICROTHENE FN 500; MICROTHENE FN 510; MICROTHENE MN 754-18; MIKROLOUR; MIRASON 9; MIRASON 16; MIRASON M 15; MIRASON M 50; MIRASON M 68; MIRASON NEO 23H; MIRATHEN; MIRATHEN 1313; MIRATHEN 1350; MOPLEN RO-QG 6015; NEOPOLEN; NEOPOLEN 30N; NEOZEX 45150; NEOZEX 4010B; NOPOL (POLYMER); NOVATEC JUO 80; NOVATEC JVO 80; NVC 9025; OKITEN G 23; ORIZON; ORIZON 805; 6020P; PA 130; PA 190; PA 520; PA 560; PAD 522; P 2010B; PE 512; PE 617; PEN 100; PEP 211; PES 100; PES 200; PETROTHENE; PETROTHENE LB 861; PETROTHENE LC 731; PETROTHENE LC 941; PETROTHENE NA 219; PETROTHENE NA 227; PETROTHENE XL 6301; P 4007EU; P 4070L; PLANIUM; PLASKON PP 60-002; PLASTAZOTE X 1016; PLASTRONGA; PLASTYLENE MA 2003; PLASTYLENE MA 7007; POLITEN; POLITEN I 020; POLYAETHYLEN (German); POLY-EM 12; POLY-EM 40; POLY-EM 41; POLYETHYLENE AS; POLYETHYLENE RESINS; POLYMIST A12; POLYMUL CS 81; POLYSION N 22; POLYTHENE; POLYWAX 1000; POROLEN; P 2070P; PPE 2; PROCENE UF 1.5; PROFAX A 60-008; P 2020T; P 2050T; P 4007T; PTS 2; PVP 8T; PY 100; RCH 1000; REPOC; RIGIDEX; RIGIDEX 35; RIGIDEX 50; RIGIDEX TYPE 2; ROPOL; ROPOTHENE OB.03-110; SANWAX 161P; SCLAIR 59; SCLAIR 2911; SCLAIR 19A; SCLAIR 96A; SCLAIR 59C; SCLAIR 79D; SCLAIR 11K; SCLAIR 19X6; SDP 640; SHOLEX 5003; SHOLEX 5100; SHOLEX 6000; SHOLEX 6002; SHOLEX F 171; SHOLEX F 6050C; SHOLEX F 6080C; SHOLEX 4250HM; SHOLEX L 131; SHOLEX S 6008; SHOLEX SUPER; SHOLEX XMO 314; SOCAREX; SRM 1475; SRM 1476; STAFLEN E 650; STAMYLAN 900; STAMYLAN 1000; STAMYLAN 1700; STAMYLAN 8200; STAMYLAN 8400; SUMIKATHENE; SUMIKATHENE F 01-1; SUMIKATHENE F 210-3; SUMIKATHENE F 702; SUMIKATHENE G 201; SMIKATHENE G 202; SUMIKATHENE G 701; SUMIKATHENE G 801; SUMIKATHENE G 806; SUMIKATHENE HARD 2052; SUNWAX 151; SUPER DYLAN; SUPRATHEN; SUPRATHEN C 100; SUPRATHEN C 100; TAKATHENE; TAKATHENE P 3; TAKATHENE P 12; TELCOTENE; TELECOTHENE; TENAPLAS; TENITE 800; TENITE 1811; TENITE 2910; TENITE 2918; TENITE 3300; TENITE 3340; TROVIDUR PE; 23F203; TYRIN; TYVEK; UNIFOS DYOB S; UNIFOS EFD 0118; VALERON; VALSPEX 155-53; VELUSTRAL KPA; VESTOLEN; VESTOLEN A 616; VESTOLEN A 6016; WAX LE; WJG 11; WNF 15; WVG 23; XL 335-1; XL 1246; XNM 68; XO 440; YUKALON EH 30; YUKALON HE 60; YUKALON K 3212; YUKALON LK 30; YUKALON MS 30; YUKALON PS 30; YUKALON YK 30; ZF 36; ZINPO

IDENTIFICATION

CAS:	9002-88-4
ERG ID:	UN 3314 as plastic molding compound
ERG Guide:	171
ERG Guide Category:	SUBSTANCES (LOW TO MODERATE HAZARD)
Molecular Formula:	$(C_2H_4)_N$
RTECS No.:	TQ3325000
Properties:	Combustible, leathery, resinous, pellets or powder. White. Waxy odor. Floats on water; insoluble.

HEALTH AND SAFETY INFORMATION

IARC Evaluation: Group 3; sufficient animal evidence; inadequate human evidence.

CHEMICAL REACTIVITY

Contact with strong oxidizers may cause fire and explosions. Reacts vigorously with flourine, xenon tetrafluoride. Incompatible with nitric acid, sodium chlorate, trinitromethane.

FIRE INFORMATION

NFPA Hazard Classifications: Not listed

Flash Point: 448°F (231°C)

FIRST AID

Move victim to fresh air. Call emergency medical care. Apply artificial respiration if victim is not breathing. Administer oxygen if breathing is difficult. In case of contact with substance, immediately flush skin or eyes with running water for at least 20 minutes. Remove and isolate contaminated clothing and shoes. Ensure that medical personnel are aware of the material(s) involved and take precautions to protect themselves.

POLYETHYLENE GLYCOL P:340

SYNONYMS: ALKAPOL PEG-200; ALKAPOL PEG-300; ALKAPOL PEG-600; ALKAPOL PEG-6000; ALKAPOL PEG-8000; CARBOWAX; EINECS No. 203-473-3; ETHYLENE GLYCOL POLYMER; ETHYLENE GLYCOL HOMOPOLYMER; ETHYLENE OXIDE, HOMOPOLYMER; ETHYLENE OXIDE POLYMER; α-HYDRO-ω-HYDROXYPOLY(OXY-1,2-ETHANEDIYL); JEFFOX; JORCHEM 400 ML; LUTROL; OXYGUARD GEL; OXIRANE, HOMOPOLYMER; OXIRANE POLYMER; OXYETHYLENE POLYMER; PEG; PLURACOL E-200; PLURACOL E-300; PLURACOL E-400; PLURACOL E-600; PLURACOL E-1500; PLURACOL E 1500; PLURACOL E-4000; PLURACOL E-6000; PLURACOL P-410; PLURACOL P-710; PLURACOL P-1010; PLURACOL P-2010; PLURACOL P-3010; PLURACOL P-4010; POLYETHER GLYCOL; POLYETHYLENE DIOL; POLY(ETHYLENE OXIDE); POLY-G SERIES; POLYOX; POLY(OXY-1,2-ETHANEDIYL), a-HYDRO-ω-HYDROXY-; POLY(OXYETHYLENE); POLYOXYETHYLENEDIO

IDENTIFICATION

CAS: 25322-68-3
Molecular Formula: $(C_2H_4O)_N \cdot H_2O$
RTECS No.: TQ3500000; TQ4025000; TQ4100000; TQ4950000
Properties: Combustible liquid or solid. Clear liquid; white solid. Completely soluble in water.

HEALTH AND SAFETY INFORMATION

Irritates eyes and skin.

RECOMMENDATIONS FOR RESPIRATOR SELECTION

At any concentrations above the NIOSH REL, or where there is no REL, at any detectable concentration: SCBAF:PD,PP (any self-contained breathing apparatus that has a full facepiece and is operated in a pressure-demand or other positive-pressure mode); or SAF:PD,PP:ASCBA (any supplied-air respirator that has a full facepiece and is operated in a pressure-demand or other positive-pressure mode in combination with an auxiliary, self-contained breathing apparatus operated in a pressure-demand or other positive-pressure mode).

Escape: GMFOV [any air-purifying, full-facepiece respirator (gas mask) with a chin-style, front- or back-mounted organic vapor canister]; or SCBAE (any appropriate escape-type, self-contained breathing apparatus).

CHEMICAL REACTIVITY

Contact with strong oxidizers may cause fire and explosions. Incompatible with sulfuric acid, nitric acid, caustics, aliphatic amines, isocyanates.

FIRE INFORMATION

Vapors in confined areas may explode when exposed to fire. If material or contaminated runoff enters waterways, notify downstream users of potentially contaminated waters. Notify local health and fire officials and pollution control agencies. Use dry chemicals, alcohol-resistant foam, carbon dioxide. From a secure, explosion-proof location, use water spray to cool exposed containers.

NFPA Hazard Classifications as polyethylene glycols

Health Hazards (Blue) 0
Flammability (Red) 1
Reactivity (Yellow) 0

Flash Point: 471°F (244°C); 360-550°F (182-287°C) o.c. as polyethylene glycols

FIRST AID

Move victim to fresh air. Call emergency medical care. Apply artificial respiration if victim is not breathing. Administer oxygen if breathing is difficult. In case of contact with substance, immediately flush skin or eyes with running water for at least 20 minutes. Remove and isolate contaminated clothing and shoes. Ensure that medical personnel are aware of the material(s) involved and take precautions to protect themselves.

POLYETHYLENE POLYAMINES P:342

SYNONYMS: CORCAT; POLIETILENIMINA (Spanish); POLYETHYLENEIMIN (Czech); POLY(ETHYLENEIMINE); POLYETHYLENEIMINE

IDENTIFICATION

CAS:	26913-06-4
ERG ID:	UN 2735
ERG Guide:	153
ERG Guide Category:	SUBSTANCES-TOXIC AND/OR CORROSIVE (COMBUSTIBLE)
Molecular Formula:	$(C_2H_5N)_N$
RTECS No.:	TQ7700000
Properties:	Combustible liquid. Yellowish to reddish brown. Amine, ammonialike odor. Soluble in water.

HEALTH AND SAFETY INFORMATION

Irritates eyes. Toxic by unspecified routes. Due to the availability of insufficient data on human effects, caution should be exercised.

RECOMMENDATIONS FOR RESPIRATOR SELECTION

At any concentrations above the NIOSH REL, or where there is no REL, at any detectable concentration: SCBAF:PD,PP (any self-contained breathing apparatus that has a full facepiece and is operated in a pressure-demand or other positive-pressure mode); or SAF:PD,PP:ASCBA (any supplied-air respirator that has a full facepiece and is operated in a pressure-demand or other positive-pressure mode in combination with an auxiliary, self-contained breathing apparatus operated in a pressure-demand or other positive-pressure mode).

Escape: GMFOV [any air-purifying, full-facepiece respirator (gas mask) with a chin-style, front- or back-mounted organic vapor canister]; or SCBAE (any appropriate escape-type, self-contained breathing apparatus).

CHEMICAL REACTIVITY

Contact with strong oxidizers may cause fire and explosions. Forms explosive mixture with air. Incompatible with acids, organic anhydrides, isocyanates, vinyl acetate, acrylates, substituted allyls, alkylene oxides, epichlorohydrin, ketones, aldehydes, alcohols, glycols, phenols, cresols, captrolactam solution. Attacks aluminum, copper, lead, tin, zinc, and alloys.

FIRE INFORMATION

Combustion products include toxic nitrogen oxides. Vapors are heavier than air and will collect in low areas. Vapors in confined areas may explode when exposed to fire. If material or contaminated runoff enters waterways, notify downstream users of potentially contaminated waters. Notify local health and fire officials and pollution control agencies. Use dry chemicals, alcohol-resistant foam, carbon dioxide. From a secure, explosion-proof location, use water spray to cool exposed containers.

FEMA Hazard Classifications

Health Hazards (Blue) 0
Flammability (Red) 1
Reactivity (Yellow) 0

Flash Point: 207°F (98°C)
Autoignition Temperature: 743°F (395°C)

FIRST AID

Move victim to fresh air and call emergency medical care. Apply artificial respiration if victim is not breathing. Do not use mouth-to-mouth method if victim swallowed or inhaled the substance; induce artificial respiration with the aid of a pocket mask equipped with a one-way valve or other proper respiratory medical device. If breathing is difficult, give oxygen. In case of contact with substance, immediately flush skin or eyes with running water for at least 20 minutes. Remove and isolate contaminated clothing and shoes. Keep victim warm and quiet. Effects of exposure (inhalation, ingestion, or skin contact) to substances may be delayed. Ensure that medical personnel are aware of the material(s) involved and take precautions to protect themselves.

POLYGLYCOL DISTEARATE P:343

SYNONYMS: CARBOWAX 1000 DISTEARATE; EMALEX 200 DI-S; EMALEX 6300 DI-S; GENAPOL® TS POWDER; HELOXAMATE® 400 DS; HODAG® 602-S; KESSCO® PEG 200DS; LIPOEG® 4-DS; LIPOPEG® 6000-S; MAPEG® 200 DS; MAPEG® 1540 DS; MAPEG® 6000 DS; MARLOSOL® BS; PEG; POLYETHYLENE GLYCOL DISTEARATE 300; POLYETHYLENE GLYCOL DISTEARATE; POLYETHYLENE GLYCOL DISTEARATE 400; POLYETHYLENE GLYCOL DISTEARATE 600; POLYETHYLENE GLYCOL DISTEARATE 1000; REWOPAL® PEG 6000 DS; RITAPEG® 400 DS; WITCONOL® 2642

IDENTIFICATION

CAS:	9005-08-7
Molecular Formula:	$C_{36}H_{70}O_3(C_2H_4O)_n$
RTECS No.:	TQ4375000
Properties:	Combustible, soft, waxy solid. Off-white. Insoluble in water.

HEALTH AND SAFETY INFORMATION

May cause eye and skin irritation. Due to the availability of insufficient data on human effects, caution should be exercised.

CHEMICAL REACTIVITY

Reacts with strong oxidizing agents.

FIRE INFORMATION

Combustion products include toxic fumes of carbon monoxide, carbon dioxide. If material or contaminated runoff enters waterways, notify downstream users of potentially contaminated waters. Notify local health and fire officials and pollution control agencies. Use dry chemicals, alcohol-resistant foam, carbon dioxide. From a secure, explosion-proof location, use water spray to cool exposed containers.

NFPA Hazard Classifications: Not listed

Flash Point: 230°F (110°C)

FIRST AID

Move victim to fresh air. Call emergency medical care. Apply artificial respiration if victim is not breathing. Administer oxygen if breathing is difficult. In case of contact with substance, immediately flush skin or eyes with running water for at least 20 minutes. Remove and isolate contaminated clothing and shoes. Ensure that medical personnel are aware of the material(s) involved and take precautions to protect themselves.

POLYPROPYLENE P:344

SYNONYMS: AMCO; AZDEL; CLYSAR; ELPON; GERFIL; LAMBETH; NOBLEN; POLY PROPENE; PROFAX; REXENE; SHELL 5520; PROPENE POLYMER; TENITE 423; TUFFLITE; ULSTRON; VISCOL 350P; W 101; WEX 1242

IDENTIFICATION

CAS:	9003-07-0; 9010-79-1 (nucleated)
Molecular Formula:	$(C_3H_6)_x$
RTECS No.:	UD1842000
Properties:	Combustible; slow-burning, crystalline solid. Tan to white. Odorless. Floats on water.

HEALTH AND SAFETY INFORMATION

Causes bronchitis in laboratory animals. May be toxic by ingestion and other routes. Due to the availability of insufficient data on human effects, caution should be exercised.

Exposure Limits: **TWA** as inert dusts, particulates not otherwise regulated (PNOR)

OSHA: 15mg/m^3 (total); 5mg/m^3 (resp.)

RECOMMENDATIONS FOR RESPIRATOR SELECTION

At any concentrations above the NIOSH REL, or where there is no REL, at any detectable concentration: SCBAF:PD,PP (any self-contained breathing apparatus that has a full facepiece and is operated in a pressure-demand or other positive-pressure mode); or SAF:PD,PP:ASCBA (any supplied-air respirator that has a full facepiece and is operated in a pressure-demand or other positive-pressure mode in combination with an auxiliary, self-contained breathing apparatus operated in a pressure-demand or other positive-pressure mode).

Escape: GMFOV [any air-purifying, full-facepiece respirator (gas mask) with a chin-style, front- or back-mounted organic vapor canister]; or SCBAE (any appropriate escape-type, self-contained breathing apparatus).

CHEMICAL REACTIVITY

A polyalkene. Incompatible with strong acids, chlorine and other strong oxidizers, potassium permanganate, isopropyl oils.

FIRE INFORMATION

If material or contaminated runoff enters waterways, notify downstream users of potentially contaminated waters.Notify local health and fire officials and pollution control agencies. Use dry chemicals, alcohol-resistant foam, carbon dioxide. From a secure, explosion-proof location, use water spray to cool exposed containers.

NFPA Hazard Classifications: Not listed

Flash Point: NA (a combustible solid)

FIRST AID

Move victim to fresh air. Call emergency medical care. Apply artificial respiration if victim is not breathing. Administer oxygen if breathing is difficult. Remove and isolate contaminated clothing and shoes. In case of contact with substance, immediately flush skin or eyes with running water for at least 20 minutes. Wash skin with soap and water. Keep victim warm and quiet. Effects of exposure (inhalation, ingestion, or skin contact) to substance may be delayed. Ensure that medical personnel are aware of the material(s) involved and take precautions to protect themselves.

POLYPROPYLENE GLYCOL P:345

SYNONYMS: ALKAPOL PPG-1200; JEFFOX; POLYOXYPROPYLENE GLYCOL; POLYOXPROPYLENE GLYCOL; PLURACOL POLYOL; POLYPROPYLENE GLYCOL 750; POLYPROPYLENE GLYCOLS P400 to P4000; POLYPROPYLENE OXIDE; PROPYLENE OXIDE HOMOPOLYMER; THANOL PPG

IDENTIFICATION

CAS:	25322-69-4
ERG ID:	UN 3271
ERG Guide:	127
ERG Guide Category:	FLAMMABLE LIQUIDS (POLAR/WATER- MISCIBLE)
Molecular Formula:	$(C_3H_8O_2)_N$ where n averages 2–34
RTECS No.:	TR5250000; TR5425000; TR5600000; TR5775000; TR5950000; TR6125000; MC9800000; MD0175000; MD0350000; MD0525000; MD0700000; TR5300000; TR6200000; TR6210000; TR6220000
Properties:	Combustible liquid. Colorless to lightly colored clear. Odorless or mild, sweet odor. Soluble in water.

HEALTH AND SAFETY INFORMATION

Irritates eyes. May be toxic by unspecified routes. Due to the availability of insufficient data on human effects, caution should be exercised.

RECOMMENDATIONS FOR RESPIRATOR SELECTION

At any concentrations above the NIOSH REL, or where there is no REL, at any detectable concentration: SCBAF:PD,PP (any self-contained breathing apparatus that has a full facepiece and is operated in a pressure-demand or other positive-pressure mode); or SAF:PD,PP:ASCBA (any supplied-air respirator that has a full facepiece and is operated in a pressure-demand or other positive-pressure mode in combination with an auxiliary, self-contained breathing apparatus operated in a pressure-demand or other positive-pressure mode).

Escape: GMFOV [any air-purifying, full-facepiece respirator (gas mask) with a chin-style, front- or back-mounted organic vapor canister]; or SCBAE (any appropriate escape-type, self-contained breathing apparatus).

CHEMICAL REACTIVITY

Contact with strong oxidizers may cause fire and explosions. Incompatible with sulfuric acid, nitric acid, caustics, aliphatic amines, isocyanates, perchloric acid.

FIRE INFORMATION

Vapors are heavier than air and will collect in low areas. Vapors in confined areas may explode when exposed to fire. If material or contaminated runoff enters waterways, notify downstream users of potentially contaminated waters. Notify local health and fire officials and pollution control agencies. Use dry chemicals, alcohol-resistant foam, carbon dioxide. From a secure, explosion-proof location, use water spray to cool exposed containers.

NFPA Hazard Classifications:

Health Hazards (Blue) 0
Flammability (Red) 1
Reactivity (Yellow) 0

Flash Point: 365–495°F (185–257°C) o.c.

FIRST AID

Move victim to fresh air. Call emergency medical care. Apply artificial respiration if victim is not breathing. Administer oxygen if breathing is difficult. Remove and isolate contaminated clothing and shoes. In case of contact with substance, immediately flush skin or eyes with running water for at least 20 minutes. Wash skin with soap and water. Keep victim warm and quiet. Ensure that medical personnel are aware of the material(s) involved and take precautions to protect themselves.

POLYSTYRENE POLYMER P:347

SYNONYMS: 3A; A 3-80; AFCOLENE; AFCOLENE 666; AFCOLENE S 100; AMOCO® H2R; AMOCO® H3E; AMOCO® R1; AMOCO® R5; ATACTIC POLYSTYRENE; BACTOLATEX; BAKELITE SMD 3500; BASF III; BDH 29-790; BENZENE, ETHENYL-, HOMOPOLYMER (9CI); BEXTRENE XL 750; BICOLASTIC A 75; BICOLENE H; BIO-BEADS S-S 2; BP-KLP; BSB-S 40; BSB-S-E; BUSTREN; BUSTREN K 500; BUSTREN K 525-19; BUSTREN U 825; BUSTREN U 825E11; BUSTREN Y 825; BUSTREN Y 3532; CADCO 0115; CARINEX GP; CARINEX HR; CARINEX HRM; CARINEX SB 59; CARINEX SB 61; CARINEX SL 273; CARINEX TGX/MF; COPAL Z; COSDEN 550; COSDEN 945E; DENKA QP3; DIAREX 43G; DIAREX HF 55; DIAREX HF 77; DIAREX HF 55-247; DIAREX HS 77; DIAREX HT 88; DIAREX HT 88A; DIAREX HT 90; DIAREX HT 190; DIAREX HT 500; DIAREX YH 476; DORVON; DORVON FR 100; DOW 360; DOW 456; DOW 665; DOW 860; DOW 1683; DOW MX 5514; DOW MX 5516; DYLARK 250; DYLENE; DYLENE 8; DYLENE 8G; DYLENE 9; DYLITE® D195B; DYLITE® R295B ESP; DYLITE® F 40; DYLITE® F 40L; EDISTIR® FA; EDISTIR® N 1280; EDISTIR® N 1281; EDISTIR® RC; EDISTIR® RB; EDISTIR® RK; EDISTIR® RKV; EDISTIR® RV 8 EDISTIR® SR 550; EDISTIR® SRL 550; EDISTIR® UT/1; EDISTIR® UT/SF; ELECTRAFIL® J-30/CF/20; EMPEE® PS-921; ESBRITE; ESBRITE 2; ESBRITE 4; ESBRITE 4-62; ESBRITE 8; ESBRITE G 10; ESBRITE G-P 2; ESBRITE 500HM; ESBRITE LBL; ESCOREZ 7404' ESTYRENE 4-62; ESTYRENE G 15; ESTYRENE G 20; ESTYRENE G-P 4; ESTYRENE H 61; ESTYRENE 500SH; ETHENYLBENZENE HOMOPOLYMER; FC-MY 5450; FG 834; FOSTER GRANT 834; GEDEX; 454H; HF 10; HF 55; HF 77; HH 102; HHI 11; HI-STYROL; HOSTYREN N; HOSTYREN N 4000; HOSTYREN N 7001; HOSTYREN N 4000V; HOSTYREN S; HT 88; HT 88A; HT 91-1; HT-F 76;HUNTSMAN 201; HUNTSMAN 240; HUNTSMAN 312; HUNTSMAN 331; HUNTSMAN 351; HUNTSMAN 474; HUNTSMAN 765; IT 40; K 525; KB (POLYMER); KM; KM (POLYMER); KOPLEN 2; KR 2537; KRASTEN 1.4; KRASTEN 052; KRASTEN SB; LACQREN 506; LACQREN 550; LS 061A; LS 1028E; LUSTREX; LUSTREX H 77; LUSTREX HH 101; LUSTREX HP 77; LUSTREX HT 88; MOBIL 1240; MOBIL 2120; MOBIL 5350; MOBIL 5600; MOBIL 8020; MOBIL MX 4354; MX 4500; MX 5514; MX 5516; MX 5517-02; 168N15; NaPSt; NBS 706; N 4000V; OWISPOL GF; PELASPAN 333; PELASPAN ESP 109s; PICCOLASTIC; PICCOLASTIC A; PICCOLASTIC A 5; PICCOLASTIC A

25; PICCOLASTIC A 50; PICCOLASTIC A 75; PICCOLASTIC C 125; PICCOLASTIC D; PICCOLASTIC D-100; PICCOLASTIC D 125; PICCOLASTIC D 150; PICCOLASTIC E 75; PICCOLASTIC E 100; PICCOLASTIC E 200; POLIGOSTYRENE; POLYCO 220NS; POLYFLEX; POLYPENCO® Q200.5; POLYSPHERE 3000 SP; POLYSTROL D; POLYSTYRENE 101; POLYSTYRENE 220; POLYSTYRENE 410; POLYSTYRENE 2122; POLYSTYRENE BW; POLYSTYRENE RESIN; POLYSTYROL; PRINTELS; PRX 1195; PS 1; PS 2; PS 200; PS 209; PS-B; PSB-C; PSB-S; PSB-S 40; PSB-S-E; PS 454H; PS 2 (POLYMER); PS 5 (POLYMER); PSV-L; PSV-L 1; PSV-L 2; PSV-L 1S; PY2763; R 3; R 3612; PEPLAY RP 2177; REPLAY RP 2236; RETAIN PS-4000; REXOLITE 1422; RHODOLNE; RTP 401; S 173; SB 475K; SD 188; SHELL 300; 686E; SMD 3500; SPS 600; SRM 705; SRM 706; ST 90; STERNITE 30; STERNITE ST 30VL; ST 30UL; STYRAFOIL; STYRAGEL; STYRENE POLYMER; STYREX C; STYROCELL PM; STRODUR®; STYROFAN 2D; STYROFLEX; STYROFOAM; STYROLUX; STYRON; STYRON 475; STYRON 492; STYRON 666; STYRON 678; STYRON 679; STYRON 683; STYRON 685; STRYON 686; STYRON 690; STYRON 69021; STYRON 440A; STYRON 470A; STYRON 475D; STYRON GP; STYRON 666K27; STYRON PS 3; STYRON T 679; STYRON 666U; STYRON 666V; STYROPIAN; STYROPIAN FH 105; STYROPOL HT 500; STYROPOL IBE; STYROPOL JQ 300; STYROPOL KA; STYROPOR®; STYVEX 22000 NA; TC 3-30; TGD 5161; TMDE 6500; TOPOREX 500; TOPOREX 830; TOPOREX 550-02; TOPOREX 850-51; TOPOREX 855-51; TROLITUL; TRYCITE 1000; 825TV-PS; 475U; U625; UBATOL U 2001; UCC 6863; UP 1; UP 2; UP 27; UPM; UPM703; UPM508L; VESTOLEN P 5232G; VESTYRON; VESTYRON 512; VESTYRON 114-12; VESTYRON MB; VESTYRON N; VINAMUL N 710; VINAMUL N 7700; VINYLBENZENE POLYMER; VINYL PRODUCTS R 3612; X 600

IDENTIFICATION

CAS:	9003-53-6
ERG ID:	UN 2211 as polystyrene beads, expandable
ERG Guide:	133
ERG Guide Category:	FLAMMABLE SOLIDS
Molecular Formula:	$(C_8H_8)_N$
RTECS No.:	WL6475000
Properties:	Combustible solid. Transparent. Insoluble in water. May be mixed with a flammable liquid, or as a resin.

HEALTH AND SAFETY INFORMATION

IARC Evaluation: Group 3; limited animal evidence.

Exposure Limits: **TWA** as inert dusts, particulates not otherwise regulated (PNOR)

OSHA: 15mg/m^3 (total); 5mg/m^3 (resp.)

RECOMMENDATIONS FOR RESPIRATOR SELECTION

At any concentrations above the NIOSH REL, or where there is no REL, at any detectable concentration: SCBAF:PD,PP (any self-contained breathing apparatus that has a full facepiece and is operated in a pressure-demand or other positive-pressure mode); or SAF:PD,PP:ASCBA (any supplied-air respirator that has a full facepiece and is operated in a pressure-demand or other positive-pressure mode in

combination with an auxiliary, self-contained breathing apparatus operated in a pressure-demand or other positive-pressure mode).

Escape: GMFOVHiE [any air-purifying, full-facepiece respirator (gas mask) with a chin-style, front- or back-mounted organic vapor canister having a high-efficiency particulate filter]; or SCBAE (any appropriate escape-type, self-contained breathing apparatus).

CHEMICAL REACTIVITY

Incompatible with strong oxidizers, hydrocarbon solvents.

FIRE INFORMATION

Fire may produce irritating and/or toxic vapors. If material or contaminated runoff enters waterways, notify downstream users of potentially contaminated waters. Notify local health and fire officials and pollution control agencies. Use dry chemicals, alcohol-resistant foam, carbon dioxide. Use water spray to cool exposed containers.

NFPA Hazard Classifications: Not listed

Autoignition Temperature: 800°F (427°C)

FIRST AID

Move victim to fresh air. Call emergency medical care. Apply artificial respiration if victim is not breathing. Administer oxygen if breathing is difficult. Remove and isolate contaminated clothing and shoes. In case of contact with substance, immediately flush skin or eyes with running water for at least 20 minutes. Removal of solidified molten material from skin requires medical assistance. Keep victim warm and quiet. Ensure that medical personnel are aware of the material(s) involved and take precautions to protect themselves.

POLYTETRAFLUOROETHYLENE P:348

SYNONYMS: AFLON; ALGLOFLON; ALGOFLON SV; ALKATHENE RXDG33; AMIP 15m; BALFON 7000; BDH 29-801; CHROMOSORB T; DIXON 164; DLX-6000; DUROID 5870; EK 1108GY-A; ETHICON PTFE; FLUO-KEM; FLUON; FLUOROFLEX; FLUOROLON 4; FLUOROPAK 80; FLUORPLAST 4; FTORLON 4; FTOROPLAST 4; GORE-TEX; HALON TFEG 180; HEYDEFLON; HOSTAFLON; MOLYKOTE 522; POLIFEN; POLITEF; POLY(ETHYLENE TETRAFLUORIDE); POLYFENE; POLYFLON; POLYTEF; POLYTETRAFLUOROETHENE; PTFE; SOREFLON 604; TARFLEN; TEFLON; TETRAFLUOROETHENE HOMOPOLYMER; TETRAFLUOROETHENE POLYMER; TETRAFLUOROETHYLENE HOMOPOLYMER; TETRAFLUOROETHYLENE POLYMERS; TETRAN PTFE; TFE; VALFLON; VELFLON; ZITEX H 662-124

IDENTIFICATION

CAS: 9002-84-0

Molecular Formula: $(C_2F_4)_n$
RTECS No.: KX4025000
Properties: Soft, waxy solid. Milky-white.

HEALTH AND SAFETY INFORMATION

Dust may cause polymer fume fever.

IARC Evaluation: Group 3; sufficient animal evidence; inadequate human evidence.

Exposure Limits: **TWA** as inert dusts, particulates not otherwise regulated (PNOR)

OSHA: 15mg/m^3 (total); 5mg/m^3 (resp.)

RECOMMENDATIONS FOR RESPIRATOR SELECTION

At any concentrations above the NIOSH REL, or where there is no REL, at any detectable concentration: SCBAF:PD,PP (any self-contained breathing apparatus that has a full facepiece and is operated in a pressure-demand or other positive-pressure mode); or SAF:PD,PP:ASCBA (any supplied-air respirator that has a full facepiece and is operated in a pressure-demand or other positive-pressure mode in combination with an auxiliary, self-contained breathing apparatus operated in a pressure-demand or other positive-pressure mode).

Escape: GMFOV [any air-purifying, full-facepiece respirator (gas mask) with a chin-style, front- or back-mounted organic vapor canister]; or SCBAE (any appropriate escape-type, self-contained breathing apparatus).

CHEMICAL REACTIVITY

Reacts with fluorine and sodium potassium alloy. Under the proper conditions, it undergoes hazardous reactions with boron, magnesium, or titanium.

FIRE INFORMATION

Nonflammable. Poisonous fluorine gases are produced when heated to above 750°F (399°C). If material or contaminated runoff enters waterways, notify downstream users of potentially contaminated waters. Notify local health and fire officials and pollution control agencies. Use dry chemicals, alcohol-resistant foam, carbon dioxide.

FIRST AID

Move victim to fresh air. Call emergency medical care. Apply artificial respiration if victim is not breathing. Administer oxygen if breathing is difficult. In case of contact with substance, immediately flush skin or eyes with running water for at least 20 minutes. Remove and isolate contaminated clothing and shoes. Ensure that medical personnel are aware of the material(s) involved and take precautions to protect themselves.

POLYURETHANE P:349

SYNONYMS: ANDUR; ASPECT® TPPE; AUTOFROTH; AUTOPAK; AUTOPOUR; CURENE; DESMOPAN® 150; DESMOPAN® 385; DESMOPAN® 585; DESMOPAN® 786; DESMOPA® KA 8333; ELECTRAFIL® J-100/CF/30; ESTANE® 5701 F1; ESTANE® 58092; ESTANE® 58300; ETHERON; ETHERON SPONGE; ISOPLAST 101; ISOPLAST 302; ISOURETHANE; NCI-C56451; PEARLSTICK 46-10/06; PLIOGRIP; POLIURETANO (Spanish); POLYFOAM PLASTIC SPONGE; POLYFOAM SPONGE; POLYURETHANE A; POLYURETHANE ESTER FOAM; POLYURETHANE ETHER FOAM; POLYURETHANE SPONGE; RESIN 164; RTP 2301A; RTP 2381A; SPENKEL; SPENLITE; TEXIN 480-A; TEXIN 985-A; TEXIN 5286; THERMOCOMP® TF-1004; URETHANE POLYMERS

IDENTIFICATION

CAS:	9009-54-5
RTECS No.:	TR7875000
Properties:	Combustible fibers, coatings, rigid and flexible foams. Insoluble in water

HEALTH AND SAFETY INFORMATION

IARC Evaluation: Group 3; sufficient animal evidence.

Exposure Limits: **TWA** as inert dusts, particulates not otherwise regulated (PNOR)

OSHA: 15mg/m^3 (total); 5mg/m^3 (resp.)

RECOMMENDATIONS FOR RESPIRATOR SELECTION

At any concentrations above the NIOSH REL, or where there is no REL, at any detectable concentration: SCBAF:PD,PP (any self-contained breathing apparatus that has a full facepiece and is operated in a pressure-demand or other positive-pressure mode); or SAF:PD,PP:ASCBA (any supplied-air respirator that has a full facepiece and is operated in a pressure-demand or other positive-pressure mode in combination with an auxiliary, self-contained breathing apparatus operated in a pressure-demand or other positive-pressure mode).

Escape: GMFOV [any air-purifying, full-facepiece respirator (gas mask) with a chin-style, front- or back-mounted organic vapor canister]; or SCBAE (any appropriate escape-type, self-contained breathing apparatus).

CHEMICAL REACTIVITY

Attacked by chlorinated and aromatic solvents.

FIRE INFORMATION

Combustion products include toxic vapors of cyanide and nitrogen. If material or contaminated runoff enters

waterways, notify downstream users of potentially contaminated waters. Notify local health and fire officials and pollution control agencies. Use dry chemicals, alcohol-resistant foam, carbon dioxide. From a secure, explosion-proof location, use water spray to cool exposed containers.

NFPA Hazard Classifications: Not listed

FIRST AID

Move victim to fresh air. Call emergency medical care. Apply artificial respiration if victim is not breathing. Administer oxygen if breathing is difficult. In case of contact with substance, immediately flush skin or eyes with running water for at least 20 minutes. Remove and isolate contaminated clothing and shoes. Ensure that medical personnel are aware of the material(s) involved and take precautions to protect themselves.

POLYVINYL ACETATE P:350

SYNONYMS: ACETIC ACID ETHENYL ESTER HOMOPOLYMER; ACETIC ACID VINYL ESTER POLYMERS; ASAHISOL 1527; ASB 516; AYAA; AYAF; BAKELITE AYAA; BAKELITE LP 90; BASCOREZ; BOND CH 18; BOOKSAVER; BORDEN 2123; CATOMER VA; CEVIAN A 678; D 50; DANFIRM; DARATAK; DCA 70; DUVILAX BD 20; ELMERS GLUE ALL; EP 1463; EVERFLEX® SP-1084; FORMVAR 1285; GELVA CSV 16; GOHSENYL E 50 Y; KURARE OM 100; LEMAC 1000; LIQUID LATEX; MERCKOGEN 6000; MOVINYL 114; NATIONAL 120-1207; PROTEX (POLYMER); QUABOND®; RHODOPAS M; ROVACE 2113; SOVIOL; SP 60 ESTER; TOABOND 40H; UCAR 130; VA 0112;VINAC® B 7; VINAC® XX-210; VINAC® XX-220; VINAC® 230; VINAC® 240; VINYL ACETATE HOMOPOLYMER; VINYL ACETATE POLYMER; VINYL ACETATE RESIN; VINYL PRODUCTS R 10688; WINACET D

IDENTIFICATION

CAS:	9003-20-7
Molecular Formula:	$(C_4H_6O_2)_n$
RTECS No.:	AK0920000
Properties:	Combustible, solid resin. Clear, colorless. Mild, characteristic odor. Floats on water; insoluble.

HEALTH AND SAFETY INFORMATION

IARC Evaluation: Inadequate animal evidence.

Exposure Limits: **TWA** as inert dusts, particulates not otherwise regulated (PNOR)

OSHA: 15mg/m^3 (total); 5mg/m^3 (resp.)

CHEMICAL REACTIVITY

Esters are incompatible with nitrates, nitric acid, sulfuric acid.

FIRE INFORMATION

Combustion products include carbon monoxide. If material or contaminated runoff enters waterways, notify downstream users of potentially contaminated waters. Notify local health and fire officials and pollution control agencies. Use dry chemicals, alcohol-resistant foam, carbon dioxide. From a secure, explosion-proof location, use water spray to cool exposed containers.

NFPA Hazard Classifications: Not listed

FIRST AID

Move victim to fresh air. Call emergency medical care. Apply artificial respiration if victim is not breathing. Administer oxygen if breathing is difficult. Remove and isolate contaminated clothing and shoes. In case of contact with substance, immediately flush skin or eyes with running water for at least 20 minutes. Wash skin with soap and water. Keep victim warm and quiet. Effects of exposure (inhalation, ingestion, or skin contact) to substance may be delayed. Ensure that medical personnel are aware of the material(s) involved and take precautions to protect themselves.

POLYVINYL ACETATE CHLORIDE P:351

SYNONYMS: ACETIC ACID, VINYL ESTER, CHLOROETHYLENE COPOLYMER; POLYVINYLCHLORIDE ACETATE; VINYL CHLORIDE ACETATE COPOLYMER; VINYL CHLORIDE VINYL ACETATE COPOLYMER

IDENTIFICATION

CAS:	34149-92-3
ERG ID:	UN 3314
ERG Guide:	171
ERG Guide Category:	SUBSTANCES (LOW TO MODERATE HAZARD)
Molecular Formula:	$(C_4H_7ClO)_N$
RTECS No.:	TR8090000
Properties:	Solid powder or granules.

HEALTH AND SAFETY INFORMATION

IARC Evaluation: Group 3; inadequate animal evidence; no adequate human data.

Exposure Limits: **TWA** as inert dusts, particulates not otherwise regulated (PNOR)

OSHA: 15mg/m^3 (total); 5mg/m^3 (resp.)

CHEMICAL REACTIVITY

Incompatible with nitrates, nitric acid, sulfuric acid.

FIRE INFORMATION

Combustion products include toxic fumes of chlorides. If material or contaminated runoff enters waterways, notify downstream users of potentially contaminated waters. Notify local health and fire officials and pollution control agencies. Use dry chemicals, alcohol-resistant foam, carbon dioxide. From a secure, explosion-proof location, use water spray to cool exposed containers.

NFPA Hazard Classifications: Not listed

FIRST AID

Move victim to fresh air. Call emergency medical care. Apply artificial respiration if victim is not breathing. Administer oxygen if breathing is difficult. In case of contact with substance, immediately flush skin or eyes with running water for at least 20 minutes. Remove and isolate contaminated clothing and shoes. Ensure that medical personnel are aware of the material(s) involved and take precautions to protect themselves.

POLYVINYL ALCOHOL P:352

SYNONYMS: ALCOTEX 88/05; ALCOTEX 88/10; ALKOTEX; ALVYL; ARACET APV; CIPOVIOL W 72; COVOL; COVOL 971; EINECS No. 209-183-3; ELVANOL; ELVANOL 50-42; ELVANOL 52-22; ELVANOL 70-05; ELVANOL 71-30; ELVANOL 90-50; ELVANOL 522-22; ELVANOL 73125G; EP 160 EP 160; ETHENOL HOMOPOLYMER (9CI); GALVATOL 1-60; GELVATOL; GELVATOL 1-30; GELVATOL 1-90; GELVATOL 3-91; GELVATOL 20-30; GELVATOL 2090; GH 20; GL 02; GL 03; GLO 5; GM 14; GOHSENOL; GOHSENOL AH 22; GOHSENOL GH; GOHSENOL GH 17; GOHSENOL GH 20; GOHSENOL GH 23; GOHSENOL GL 03; GOHSENOL GL 05; GOHSENOL GL 08; GOHSENOL GM 14; GOHSENOL GM 94; GOHSENOL GM 14L; GOHSENOL KH 17; GOHSENOL NH 05; GOHSENOL NH 17; GOHSENOL NH 18; GOHSENOL NH 20; GOHSENOL NH 26; GOHSENOL NK 114; GOHSENOL NL 05; GOHSENOL NM 14; IVALON; KURALON VP; KURARE POVAL 1700; KURARE PVA 205; KURATE POVAL 120; LEMOL; LEMOL 5-88; LEMOL 5-98; LEMOL 12-88; LEMOL 16-98; LEMOL 24-98; LEMOL 30-98; LEMOL 51-98; LEMOL 60-98; LEMOL 75-98; M 13/20; MOWIOL; MOWIOL N 30-88; MOWIOL N 50-98; MOWIOL N 70-98; NH 18; NM 11; NM 14; POLYDESIS; POLYSIZER 173; POLYVINOL; POLYVIOL; POLYVIOL M 13/140; POLYVIOL MO 5/140; POLYVIOL W 25/140; POLYVIOL W 40/140; POVAL 117; POVAL 120; POVAL 203; POVAL 205; POVAL 217; POVAL 1700; POVAL C 17; PVA; PVA 008; PVOH; PVS 4; RESISTOFLEX; RHODOVIOL; RHODOVIOL 4/125; RHODOVIOL 16/200; RHODOVIOL 4-125P; RHODOVIOL R 16/20; SOLVAR; SUMITEX H 10; VIBATEX S; VINACOL MH; VINALAK; VINAROL; VINAROL DT; VINAROLE; VINAROL ST; VINAVILOL 2-98; VINNAROL; VINOL; VINOL 125; VINOL 205; VINOL UNISIZE; VINYL ALCOHOL POLYMER; VINYLON FILM 2000

IDENTIFICATION

CAS:	9002-89-5
Molecular Formula:	$(C_2H_4O)_N$
RTECS No.:	TR8100000
Properties:	Flammable powder. Colorless, white, or cream. Mixes with water.

HEALTH AND SAFETY INFORMATION

IARC Evaluation: Group 3; limited animal evidence; inadequate human evidence.

Exposure Limits: **TWA** as inert dusts, particulates not otherwise regulated (PNOR)

OSHA: 15mg/m^3 (total); 5mg/m^3 (resp.)

CHEMICAL REACTIVITY

Contact with strong oxidizers may cause fire and explosions. Dust forms an explosive mixture with air. Incompatible with sulfuric acid, nitric acid, caustics, aliphatic amines, isocyanates.

FIRE INFORMATION

Combustion products include carbon monoxide. Vapors are heavier than air and will collect in low areas. Vapors in confined areas may explode when exposed to fire. If material or contaminated runoff enters waterways, notify downstream users of potentially contaminated waters. Notify local health and fire officials and pollution control agencies. Use dry chemicals, alcohol-resistant foam, carbon dioxide. From a secure, explosion-proof location, use water spray to cool exposed containers.

NFPA Hazard Classifications

Health Hazards (Blue) 0
Flammability (Red) 1
Reactivity (Yellow) 0

Flash Point: 175°F (79°C) o.c.

FIRST AID

Move victim to fresh air. Call emergency medical care. Apply artificial respiration if victim is not breathing. Administer oxygen if breathing is difficult. Remove and isolate contaminated clothing and shoes. In case of contact with substance, immediately flush skin or eyes with running water for at least 20 minutes. Wash skin with soap and water. Keep victim warm and quiet. Effects of exposure (inhalation, ingestion, or skin contact) to substance may be delayed. Ensure that medical personnel are aware of the material(s) involved and take precautions to protect themselves.

POLYVINYL BUTYRAL P:353

SYNONYMS: BAKELITE XYHL; BUTACITE® (DuPont); BUTVAR;POLIVINIL BUTIRAL (Spanish); ESLEX W-201; MOWITAL B60H; POLY(2-PROPYL-m-DIOXANE-4,6-DIYLENE); POLYVINYLBUTYRAL (Czech); POLYVINYL BUTYRAL RESINS; PVB; SAFLEX®; VINYL ACETAL POLYMERS, BUTYRALS; VL-023

IDENTIFICATION

CAS: 63148-65-2
Molecular Formula: $H_2{\cdot}(C_8H_{14}O_2)_n$
RTECS No.: TR4955000
Properties: Granular solid. White. Insoluble in water.

HEALTH AND SAFETY INFORMATION

Irritates eyes.

Exposure Limits: **TWA** as inert dusts, particulates not otherwise regulated (PNOR)

OSHA: 15mg/m^3 (total); 5mg/m^3 (resp.)

CHEMICAL REACTIVITY

Reacts with strong oxidizing agents. Decomposes to butyraldehyde, butyric acid, and carbon monoxide.

FIRE INFORMATION

NFPA Hazard Classifications: Not listed

FIRST AID

Move victim to fresh air. Call emergency medical care. Apply artificial respiration if victim is not breathing. Administer oxygen if breathing is difficult. In case of contact with substance, immediately flush skin or eyes with running water for at least 20 minutes. Remove and isolate contaminated clothing and shoes. Ensure that medical personnel are aware of the material(s) involved and take precautions to protect themselves.

POLYVINYL CHLORIDE P:354

SYNONYMS: ARMODOUR; ARON COMPOUND HW; ASTRALON; ATACTIC POLY(VINYL CHLORIDE); BLACAR 1716; BOLATRON; BONLOID; BREON; CHLOROETHENE HOMOPOLYMER; CHLOROETHYLENE POLYMER; CHLOROSTOP; COBEX (polymer); CONTIZELL; CORVIC 55/9; DACOVIN; DANUVIL 70; DARVIC 110; DARVIS CLEAR 025; DECELITH H; DENKA VINYL SS 80; DIAMOND SHAMROCK 40; DORLYL; DUROFOL P; DYNADUR; E 62; E 66P; EINECS No. 208-750-2; EKAVYL SD 2; E-PVC; ESCAMBIA 2160; ETHENE, CHOLOR-, HOMOPOLYMERS; ETHYLENE, CHLORO-, POLYMERS; EUROPHAN; EXON 605; FC 4648; FLOCOR; FLOCOR; GAFCOTE; GENOTHERM; GEON®; GEON® LATEX 151; GUTTAGENA; HALVIC 223; HISHIREX 502; HISPAVIC 229; HOSTALIT; IGELITE F; IMPROVED WILT PRUF; KAYLITE; KLEGECELL; KOROSEAL; LONZA G; LUCOFLEX; LUCOVYL PE; LUTOFAN; MARVINAL; MIRREX MCFD 1025; MOVINYL 100; MYRAFORM; NCI-C60797; NIKA-TEMP; NIKAVINYL SG 700; NIPEON A 21; NIPOL 576; NORVINYL; NOVON 712; ONGROVIL S 165; OPALON; ORTUDUR; PANTASOTE R 873; PARCLOID; PATTINA V 82;

PEVIKON D 61; PLIOVIC; POLIVINIT; POLY(CHLOROETHYLENE); POLYTHERM; PROTOTYPE III SOFT; PVC; QSAH 7; QUIRVIL; QYSA; RAVINYL; RUCON B 20; S 65 (polymer); SCON 5300; SICRON; S-LON; SOLVIC; SP 60 (CHLOROCARBON); SUMILIT EXA 13; SUMITOMO PX 11; TAKILON; TECHNOPOR; TENNECO 1742; TK 1000; TROVIDUR; TROVITHERN HTL; U 1 (polymer); ULTRON; UNICHEM; VERON P 130/1; VESTOLIT B 7021; VINIKA KR 600; VINIKULON; VINIPLAST; VINIPLEN P 73; VINNOL E 75; VINOFLEX; VINYLCHLON 4000LL; VINYL CHLORIDE HOMOPOLYMER; VINYL CHLORIDE POLYMER; VYGEN 85; WELVIC G 2/5; WILT PRUF; WINIDUR; X-AB

IDENTIFICATION

CAS:	9002-86-2
ERG ID:	UN 3314 as plastic molding compound
ERG Guide:	171
ERG Guide Category:	SUBSTANCES (LOW TO MODERATE HAZARD)
Molecular Formula:	$(C_2H_3Cl)_N$
RTECS No.:	KV0350000
Properties:	Combustible powder or granules. White powder or colorless granules. Insoluble in water.

HEALTH AND SAFETY INFORMATION

Prolonged or repeated contact with PVC dust can cause lung irritation, skin irritation, liver and blood damage.

IARC Evaluation: Group 3; inadequate human evidence; inadequate animal evidence.

Exposure Limits: **TWA** as inert dusts, particulates not otherwise regulated (PNOR)

OSHA: 15mg/m^3 (total); 5mg/m^3 (resp.)

Note: Various countries including United Kingdom, Germany, Switzerland, and the Commonwealth of Independent States (former Soviet Union countries) have established workplace limits for respirable and inhalable dusts.

CHEMICAL REACTIVITY

Reacts violently with flourine. Incompatible with liquid oxygen.

FIRE INFORMATION

Combustion products include toxic fumes of chlorides. If material or contaminated runoff enters waterways, notify downstream users of potentially contaminated waters. Notify local health and fire officials and pollution control agencies. Use dry chemicals, alcohol-resistant foam, carbon dioxide.

NFPA Hazard Classifications: Not listed but ignites at high temperature.

FIRST AID

Move victim to fresh air. Call emergency medical care. Apply artificial respiration if victim is not breathing. Administer oxygen if breathing is difficult. In case of contact with substance, immediately flush skin or eyes with running water for at least 20 minutes. Remove and isolate contaminated clothing and shoes. Ensure that medical personnel are aware of the material(s) involved and take precautions to protect themselves.

POLYVINYLPYRROLIDONE P:356

SYNONYMS: AGENT AT 717; ALBIGEN A; ALDACOL Q; AT 717; BOLINAN; EINECS No. 201-800-4; 1-ETHENYL-2-PYRROLIDINONE HOMOPOLYMER; 1-ETHENYL-2-PYRROLIDINONE POLYMERS; GANEX P 804; HEMODESIS; HEMODEZ; K25 (polymer); KOLLIDON; LUVISKOL; MPK 90; NCI-C60582; NEOCOMPENSAN; PERAGAL ST; PERISTON; PLASDONE; POLYCLAR L; POLY(1-(2-OXO-1-PYRROLIDINYL)ETHYLENE); POLY(1-VINYL-2-PYRROLIDINONE) HOMOPOLYMER; POLYVIDONE; POLY(n-VINYLBUTYROLACTAM); POLYVINYLPYRROLIDONE; POVIDONE (USP XIX); PROTAGENT; PVP; SUBTOSAN; VINISIL; n-VINYLBUTYROLACTAM POLYMER; n-VINYLPYRROLIDONE POLYMER

IDENTIFICATION

CAS:	9003-39-8
ERG ID:	UN 3314 as plastic molding compound
ERG Guide:	171
ERG Guide Category:	SUBSTANCES (LOW TO MODERATE HAZARD)
Molecular Formula:	$(C_6H_9ON)_N$
RTECS No.:	TR8370000
Properties:	Powder, solid, or aqueous solution. White or light yellow. Mixes with water.

HEALTH AND SAFETY INFORMATION

IARC Evaluation: Group 3; animal carcinogen

Exposure Limits: **TWA** as inert dusts, particulates not otherwise regulated (PNOR)

OSHA: 15mg/m^3 (total); 5mg/m^3 (resp.)

RECOMMENDATIONS FOR RESPIRATOR SELECTION

At any concentrations above the NIOSH REL, or where there is no REL, at any detectable concentration: SCBAF:PD,PP (any self-contained breathing apparatus that has a full facepiece and is operated in a pressure-demand or other positive-pressure mode); or SAF:PD,PP:ASCBA (any supplied-air respirator that has a full facepiece and is operated in a pressure-demand or other positive-pressure mode in combination with an auxiliary, self-contained breathing apparatus operated in a pressure-demand or other positive-pressure mode).

Escape: GMFOV [any air-purifying, full-facepiece respirator (gas mask) with a chin-style, front- or back-mounted organic vapor canister]; or SCBAE (any appropriate escape-type, self-contained breathing apparatus).

CHEMICAL REACTIVITY

Reacts with strong oxidizing agents. Decomposes to carbon monoxide, carbon dioxide, and nitrogen oxides.

FIRE INFORMATION

Combustion products include toxic nitrogen oxides. If material or contaminated runoff enters waterways, notify downstream users of potentially contaminated waters. Notify local health and fire officials and pollution control agencies. Use dry chemicals, alcohol-resistant foam, carbon dioxide. From a secure, explosion-proof location, use water spray to cool exposed containers.

NFPA Hazard Classifications: Not listed

FIRST AID

Move victim to fresh air. Call emergency medical care. Apply artificial respiration if victim is not breathing. Administer oxygen if breathing is difficult. In case of contact with substance, immediately flush skin or eyes with running water for at least 20 minutes. Remove and isolate contaminated clothing and shoes. Ensure that medical personnel are aware of the material(s) involved and take precautions to protect themselves.

POTASSIUM BOROHYDRIDE P:357

SYNONYMS: BORATE(1-), TETRAHYDRO-, POTASSIUM; BOROHIDRURO POTASICO (Spanish); BOROHYDRURE de POTASSIUM (French); EINECS No. 237-360-5; KALIUMBORHYDRID (German); KBH: POTASSIUM BOROHYDRATE; TETRAHYDROBORATE(1-) POTASSIUM

IDENTIFICATION

CAS:	13762-51-1
ERG ID:	UN 1870
ERG Guide:	138
ERG Guide Category:	WATER-REACTIVE (EMITTING FLAMMABLE GASES)
Molecular Formula:	BH_4K
RTECS No.:	TS7525000
Properties:	Flammable, crystalline solid. White. Soluble in water.

HEALTH AND SAFETY INFORMATION

Corrosive to eyes. Irritates skin and respiratory tract. Inhalation may cause pulmonary edema, which can be delayed for several hours; there is a risk of death in serious cases. Boron compounds are poisonous and may affect the central nervous system.

Exposure Limits: **TWA** as inert dusts, particulates not otherwise regulated (PNOR)

OSHA: 15mg/m^3 (total); 5mg/m^3 (resp.)

RECOMMENDATIONS FOR RESPIRATOR SELECTION

At any concentrations above the NIOSH REL, or where there is no REL, at any detectable concentration: SCBAF:PD,PP (any self-contained breathing apparatus that has a full facepiece and is operated in a pressure-demand or other positive-pressure mode); or SAF:PD,PP:ASCBA (any supplied-air respirator that has a full facepiece and is operated in a pressure-demand or other positive-pressure mode in combination with an auxiliary, self-contained breathing apparatus operated in a pressure-demand or other positive-pressure mode).

Escape: GMFOV [any air-purifying, full-facepiece respirator (gas mask) with a chin-style, front- or back-mounted organic vapor canister]; or SCBAE (any appropriate escape-type, self-contained breathing apparatus).

CHEMICAL REACTIVITY

Incompatible with moisture and acids; decomposes, forming explosive hydrogen. May form explosive mixture with air; may be volatile and explosive. Violent reaction with strong oxidizers. Dust clouds are explosive. Keep away from heat, sparks, and flame. Obtain specific handling instructions from the manufacturer.

FIRE INFORMATION

Combustion products include toxic fumes of potassium oxides. Dusts and vapors in confined areas may explode when exposed to fire. If material or contaminated runoff enters waterways, notify downstream users of potentially contaminated waters. Notify local health and fire officials and pollution control agencies. Use dry chemicals, alcohol-resistant foam, carbon dioxide. From a secure, explosion-proof location, use water spray to cool exposed containers. If containers are exposed to direct flame or elevated temperatures for prolonged times, withdraw immediately to a secure location

FIRST AID

Move victim to fresh air. Call emergency medical care. Apply artificial respiration if victim is not breathing. Administer oxygen if breathing is difficult. Remove and isolate contaminated clothing and shoes. In case of contact with substance, immediately flush skin or eyes with running water for at least 20 minutes. Keep victim warm and quiet. Ensure that medical personnel are aware of the material(s) involved and take precautions to protect themselves.

POTASSIUM BROMATE P:358

SYNONYMS: BROMATO POTASICO (Spanish); BROMIC ACID, POTASSIUM SALT; EEC No. 035-003-00-6; EINECS No. 231-829-8; NSC 215200

892 Potassium Bromate

IDENTIFICATION

CAS:	7758-01-2
ERG ID:	UN 1484
ERG Guide:	140
ERG Guide Category:	OXIDIZERS
Molecular Formula:	$BrKO_3$
RTECS No.:	EF8725000
Properties:	Noncombustible, colorless, or white crystalline solid. Soluble in water.

HEALTH AND SAFETY INFORMATION

Long-term effects include brain and kidney damage.

IARC Evaluation: Group 2B; no human data; sufficient animal evidence.

Exposure Limits: **TWA** as inert dusts, particulates not otherwise regulated (PNOR)

OSHA: 15mg/m^3 (total); 5mg/m^3 (resp.)

RECOMMENDATIONS FOR RESPIRATOR SELECTION

At any concentrations above the NIOSH REL, or where there is no REL, at any detectable concentration: SCBAF:PD,PP (any self-contained breathing apparatus that has a full facepiece and is operated in a pressure-demand or other positive-pressure mode); or SAF:PD,PP:ASCBA (any supplied-air respirator that has a full facepiece and is operated in a pressure-demand or other positive-pressure mode in combination with an auxiliary, self-contained breathing apparatus operated in a pressure-demand or other positive pressure mode).

Escape: GMFOV [any air-purifying, full-facepiece respirator (gas mask) with a chin-style, front- or back-mounted organic vapor canister]; or SCBAE (any appropriate escape-type, self-contained breathing apparatus).

CHEMICAL REACTIVITY

A powerful oxidizer. Violent reaction with reducing agents, combustible materials, and organic substances. Incompatible with aluminum, ammonium salts, powdered metals, phosphorus, sulfur, and many other substances. Forms friction- and shock-sensitive materials with organic materials or some metals.

FIRE INFORMATION

Noncombustible, but will intensify an existing fire. If material or contaminated runoff enters waterways, notify downstream users of potentially contaminated waters. Notify local health and fire officials and pollution control agencies. From a secure, explosion-proof location, use water spray to cool exposed containers.

FIRST AID

Move victim to fresh air. Call emergency medical care. Apply artificial respiration if victim is not breathing. Administer oxygen if breathing is difficult. Remove and isolate contaminated clothing and shoes. In case of contact with substance, immediately flush skin or eyes with running water for at least 20 minutes. Keep victim warm and quiet. Ensure that medical personnel are aware of the material(s) involved and take precautions to protect themselves.

ENVIRONMENTAL INFORMATION

EPCRA Section 313: Form R *de minimis* concentration reporting level: 0.1%
California's Proposition 65: Carcinogen

POTASSIUM CYANIDE P:460

SYNONYMS: CIANURO POTASICO (Spanish); CYANIDE; CYANIDE OF POTASSIUM; CYANURE de POTASSIUM (French); EEC No. 006-007-00-5; EINECS No. 205-729-3; ENSTRIP AU-78 GOLD ETCH; 1500 REPLENISHING SALTS; GOLD TECHNISTRIP; HYDROCYANIC ACID, POTASSIUM SALT; KALIUM-CYANID (German); KCN; POTASSIUM CYANIDE SOLUTION; POTASSIUM SALT OF HYDROCYANIC ACID; Q-LEVEL BRIGHTENER; SILVER GLO 3K; SILVER PLATING SOLUTION; TECHNISTRIP AU

IDENTIFICATION

CAS:	151-50-8
ERG ID:	UN 1680
ERG Guide:	157
ERG Guide Category:	TOXIC AND/OR CORROSIVE (NONCOMBUSTIBLE/WATER-SENSITIVE)
Molecular Formula:	CKN
RTECS No.:	TS8750000
Properties:	Poisonous, solid crystals or solution. White; colorless solution. Bitter almond odor. Sinks and mixes with water.

HEALTH AND SAFETY INFORMATION

Poisonous by all routes. Wear protection and approved respirator. Long-term exposure may cause enlarged thyroid, nose sores, and bleeding. High doses may decrease fertility in women and men.

IDLH: 25 mg/m^3 as CN
OSHA Table Z-1-A Air Contaminant as CN

Exposure Limits:	**TWA**	**CEILING**
ACGIH:[(1)]	4.7 ppm; 5 mg/m^3	

NIOSH:		4.7 ppm; 5 mg/m^3 (10 min)
OSHA:	5 mg/m^3	

[1] Skin contact contributes significantly in overall exposure.

RECOMMENDATIONS FOR RESPIRATOR SELECTION: NIOSH/OSHA

25 mg/m^3: SA (any supplied-air respirator); or SCBAF (any self-contained breathing apparatus with a full facepiece).

Emergency or Planned Entry into Unknown Concentrations or IDLH Conditions: SCBAF:PD,PP (any self-contained breathing apparatus that has a full facepiece and is operated in a pressure-demand or other positive-pressure mode); or SAF:PD,PP:ASCBA (any supplied-air respirator that has a full facepiece and is operated in a pressure-demand or other positive-pressure mode in combination with an auxiliary, self-contained breathing apparatus operated in a pressure-demand or other positive-pressure mode)

Escape: GMFOVHiE [any air-purifying, full-facepiece respirator (gas mask) with a chin-style, front- or back-mounted canister having a high-efficiency particulate filter]; or SCBAE (any appropriate escape-type, self-contained breathing apparatus).

CHEMICAL REACTIVITY

Incompatible with acids, organic anhydrides, isocyanates, alkylene oxides, epichlorohydrin, aldehydes, alcohols, glycols, phenols, cresols, caprolactum, strong oxidizers, and sodium chlorate. Attacks aluminum, copper, and zinc in the presence of moisture. Absorbs moisture from the air, forming a syrup.

FIRE INFORMATION

Combustion products include toxic oxides of nitrogen, potassium, and cyanide. If material or contaminated runoff enters waterways, notify downstream users of potentially contaminated waters. Notify local health and fire officials and pollution control agencies. Use extinguishing agent suitable for surrounding fire. Use water spray to cool exposed containers.

NFPA Hazard Classifications:

Health Hazards (Blue) 3
Flammability (Red) 0
Reactivity (Yellow) 0

FIRST AID

Move victim to fresh air. Call emergency medical care. Apply artificial respiration if victim is not breathing. Do not use mouth-to-mouth method if victim ingested or inhaled the substance; induce artificial respiration with the aid of a pocket mask equipped with a one-way valve or other proper respiratory medical device. Administer oxygen if breathing is difficult. Remove and isolate contaminated clothing and shoes. In case of contact with substance, immediately flush skin or eyes with running water for at least 20 minutes. For minor skin contact, avoid spreading material on unaffected skin. Keep victim warm and quiet. Effects of exposure (inhalation, ingestion, or skin contact) may be delayed. Ensure that medical personnel are aware of the material(s) involved and take precautions to protect themselves.

ENVIRONMENTAL INFORMATION

Clean Water Act: Section 311 Hazardous Substances/RQ (same as CERCLA); Section 313 Priority Chemicals.
EPA Hazardous Waste Number (RCRA No.): P098.
RCRA Section 261 Hazardous Constituents
EPCRA Section 302 Extremely Hazardous Substances: TPQ = 100 lb (45.4 kg).
EPCRA Section 304: Reportable Quantity (RQ): EHS, CERCLA, 10 lb (4.54 kg).
Marine Pollutant (49CFR, Subchapter 172.101, Appendix B).

POTASSIUM FLUORIDE P:490

SYNONYMS: EEC No. 009-005-00-2; EINECS No. 232-151-5; FLUORURO POTASICO (Spanish); FLUORURE de POTASSIUM (French); KALIUMFLUORID (German); POTASSIO FLUORURO (Italian); POTASSIUM FLUORURE (French)

IDENTIFICATION

CAS:	7789-23-3
ERG ID:	UN 1812
ERG Guide:	154
ERG Guide Category:	SUBSTANCES-TOXIC AND/OR CORROSIVE (NONCOMBUSTIBLE)
Molecular Formula:	FK
RTECS No.:	TT0700000
Properties:	Noncombustible crystalline solid. White. Sharp salty taste. Soluble in water.

HEALTH AND SAFETY INFORMATION

Corrosive to eyes, skin, and respiratory tract. Inhalation may cause pulmonary edema, which can be delayed for several hours; there is a risk of death in serious cases. Repeated exposure to fluoride chemicals may cause stiffness in muscles or ligaments, and even crippling; bone and tooth disorders.

IDLH: 250 mg/m^3 as F
OSHA Table Z-1-A Air Contaminant as F
NIOSH Criteria Document: NIOSH 76-103 INORGANIC FLUORIDE

Exposure Limits:	**TWA**
ACGIH:	2.5 mg/m^3 as fluorides

NIOSH: 2.5 mg/m^3 as fluorides
OSHA: 2.5 mg/m^3 as fluorides

RECOMMENDATIONS FOR RESPIRATOR SELECTION: NIOSH/OSHA as fluorides
12.5 mg/m^3: DM (any dust and mist respirator).
25 mg/m^3: DMXSQ * (any dust and mist respirator except single-use and quarter-mask respirators); or SA * (any supplied-air respirator).
62.5 mg/m^3: SA:CF * [any supplied-air respirator operated in a continuous-flow mode)]; or PAPRDM *+ *if not present as a fume* (any powered, air-purifying respirator with a dust and mist filter).
125 mg/m^3: HiEF + (any air-purifying, full-facepiece respirator with a high-efficiency particulate filter); or SCBAF (any self-contained breathing apparatus with a full facepiece); or SAF (any supplied-air respirator with a full facepiece).
250 mg/m^3: SA:PD,PP (any supplied-air respirator operated in a pressure-demand or other positive-pressure mode).

Emergency or Planned Entry into Unknown Concentrations or IDLH Conditions:
SCBAF:PD,PP (any self-contained breathing apparatus that has a full faceplate and is operated in a pressure-demand or other positive-pressure mode); or SAF:PD,PP:ASCBA (any supplied-air respirator that has a full facepiece and is operated in a pressure-demand or other positive-pressure mode in combination with an auxiliary, self-contained breathing apparatus operated in a pressure-demand or other positive-pressure mode).

Escape: HiEF + (any air-purifying, full-facepiece respirator with a high-efficiency particulate filter); or SCBAE (any appropriate escape-type, self-contained breathing apparatus).
*Substance reported to cause eye irritation or damage; may require eye protection.
+ May need acid gas sorbent.

CHEMICAL REACTIVITY

Incompatible with acids.

FIRE INFORMATION

Noncombustible solid. If material or contaminated runoff enters waterways, notify downstream users of potentially contaminated waters. Notify local health and fire officials and pollution control agencies. Use dry chemicals, foam, carbon dioxide.

FIRST AID

Move victim to fresh air; call emergency medical care. Apply artificial respiration if victim is not breathing. Do not use mouth-to-mouth method if victim ingested or inhaled the substance; induce artificial respiration with the aid of a pocket mask equipped with a one-way valve or other proper respiratory medical device. Administer oxygen if breathing is difficult. Remove and isolate contaminated clothing and shoes. In case of contact with substance, immediately flush skin or eyes with running water for at least 20 minutes. For minor skin contact, avoid spreading material on unaffected skin. Keep victim warm and quiet. Effects of exposure (inhalation, ingestion, or skin contact) to substance may be delayed. Ensure that medical personnel are aware of the material(s) involved and take precautions to protect themselves.

POTASSIUM HYDROXIDE P:500

SYNONYMS: AZ 400K DEVELOPER; BOYDES PTS DEVELOPER; C-585; CAUSTIC LYE; CAUSTIC POTASH; COLEC 302; CRONAFLEX PDC DEVELOPER; DYNASOLVE 190; DYNASOLVE MP-500; DYNASOLVE MP-710; DYNASOLVE MP-750; DYNASOLVE MP ALUMINUM GRADE; EEC No. 019-002-00-8; EINECS No. 215-181-3; ENTEC 310A; ENTEC 353; ENTEC 356; HIDROXIDO POTASICO (Spanish); HYDROXIDE de POTASSIUM (French); KALIUMHYDROXID (German); KALIUMHYDROXYDE (Dutch); KODAK 66 DEVELOPER; K-O-H; LYE; LYE WATER; METEX COPPER ADDITION AGENT S-1; MICROPOSIT 452; MICROPOSIT 452 DEVELOPER; MICROPOSIT 454 DEVELOPER; MICROPOSIT 455 DEVELOPER; POTASSA; POTASSE CAUSTIQUE (French); POTASSIO (IDROSSIDO di) (Italian); POTASSIUM HYDRATE; POTASSIUM (HYDRIXYDE de) (French); Q-LEVEL BRIGHTENER; ROCHELTEX

IDENTIFICATION

CAS:	1310-58-3
ERG ID:	UN 1813 (solid); UN 1814 (solution)
ERG Guide:	154
ERG Guide Category:	TOXIC AND/OR CORROSIVE (NONCOMBUSTIBLE)
Molecular Formula:	KOH
RTECS No.:	TT2100000
Properties:	Solid crystals, pellets, or watery liquid. White solid; colorless cloudy liquid. Odorless. Solid sinks and dissolves mixes slowly with water; liquid mixes readily with water, forming heat and corrosive and toxic fumes.

HEALTH AND SAFETY INFORMATION

Corrosive to eyes, skin, and respiratory tract. Inhalation may cause pulmonary edema, which can be delayed for several hours; there is a risk of death in serious cases. Long-term exposure may cause sores in the lining of the nose and lung problems.

OSHA Table Z-1-A Air Contaminant

Exposure Limits:	TWA	CEILING
ACGIH:		2 mg/m^3
NIOSH:	2 mg/m^3	
OSHA:	None	

RECOMMENDATIONS FOR RESPIRATOR SELECTION

At any concentrations above the NIOSH REL, or where there is no REL, at any detectable concentration: SCBAF:PD,PP (any self-contained breathing apparatus that has a full facepiece and is operated in a pressure-demand or other positive-pressure mode); or SAF:PD,PP:ASCBA (any supplied-air respirator that has a full facepiece and is operated in a pressure-demand or other positive-pressure mode in

combination with an auxiliary, self-contained breathing apparatus operated in a pressure-demand or other positive-pressure mode).

Escape: GMFOV [any air-purifying, full-facepiece respirator (gas mask) with a chin-style, front- or back-mounted organic vapor canister]; or SCBAE (any appropriate escape-type, self-contained breathing apparatus).

CHEMICAL REACTIVITY

Contact of solid material with water produces a vigorous, exothermic reaction. Aqueous solution is a strong base. Reacts violently with acids, organic anhydrides, isocyanates, alkylene oxides, epichlorohydrin, aldehydes, alcohols, glycols, phenols, cresols, and caprolactum solution. Dissolves in water forming high heat, caustic fumes, and flammable hydrogen gas. Also reacts with halogenated hydrocarbons and maleic anhydride. Attacks metals such as aluminum, tin, lead, and zinc, producing flammable hydrogen gas.

FIRE INFORMATION

Noncombustible. Vapors are heavier than air and will collect in low areas. Vapors in confined areas may explode when exposed to fire. If material or contaminated runoff enters waterways, notify downstream users of potentially contaminated waters. Notify local health and fire officials and pollution control agencies. Potassium hydroxide may ignite nearby material if it absorbs water. Use extinguishing agent suitable for surrounding fire.

NFPA Hazard Classifications:

Health Hazards (Blue) 3
Flammability (Red) 0
Reactivity (Yellow) 1

FIRST AID

Move victim to fresh air. Call emergency medical care. Apply artificial respiration if victim is not breathing. Do not use mouth-to-mouth method if victim ingested or inhaled the substance; induce artificial respiration with the aid of a pocket mask equipped with a one-way valve or other proper respiratory medical device. Administer oxygen if breathing is difficult. Remove and isolate contaminated clothing and shoes. In case of contact with substance, immediately flush skin or eyes with running water for at least 20 minutes. For minor skin contact, avoid spreading material on unaffected skin. Keep victim warm and quiet. Effects of exposure (inhalation, ingestion, or skin contact) may be delayed. Ensure that medical personnel are aware of the material(s) involved and take precautions to protect themselves.

ENVIRONMENTAL INFORMATION

Clean Water Act: Section 311 Hazardous Substances/RQ (same as CERCLA).
EPCRA Section 304: Reportable Quantity (RQ): CERCLA, 1,000 lb (454 kg).

POTASSIUM PERMANGANATE P:580

SYNONYMS: CAIROX; CHAMELEON MINERAL; C.I. 77755; CONDY'S CRYSTALS; EEC No. 025-002-00-9; EINECS No. 231-760-3; KALIUMPERMANGANAAT (Dutch); KALIUMPERMANGANAT (German); PERMANGANATE de POTASSIUM (French); PERMANGANATE OF POTASH; PERMANGANATO POTASICO (Spanish); POTASSIO (PERMANGANATO di) (Italian); POTASSIUM (PERMANGANATE de) (French); PURPLE SALT

IDENTIFICATION

CAS:	7722-64-7
ERG ID:	UN 1490
ERG Guide:	140
ERG Guide Category:	OXIDIZERS
Molecular Formula:	$KMnO_4$
RTECS No.:	SD6475000
Properties:	Dark purple, bronzelike, blue-metallic crystalline solid; solution may be various shades of purple. Odorless. Sinks and mixes slowly with water.

HEALTH AND SAFETY INFORMATION

Corrosive to eyes, skin, and respiratory tract. Inhalation may cause pulmonary edema, which can be delayed for several hours; there is a risk of death in serious cases. Long-term or repeated exposures may cause genetic changes in living cells; lung problems.

IDLH: 500 mg/m^3 as manganese

Exposure Limits:	**TWA**	**CEILING**	**STEL**
ACGIH:	0.2 mg/m^3 as Mn		
NIOSH:	1 mg/m^3 as Mn		3 mg/m^3 as Mn
OSHA:		5 mg/m^3 as Mn	

RECOMMENDATIONS FOR RESPIRATOR SELECTION: NIOSH/OSHA as manganese

10 mg/m^3: DMXSQ *if not present as a fume* (any dust and mist respirator except single-use and quarter mask respirators); or SA (any supplied-air respirator).
25 mg/m^3: SA:CF (any supplied-air respirator operated in a continuous-flow mode.); or PAPRDM *if not present as a fume* (any powered, air-purifying respirator with a dust and mist filter).
50 mg/m^3: HiEF (any air-purifying, full-facepiece respirator with a high-efficiency particulate filter); or SAT:CF (any supplied-air respirator that has a tight-fitting facepiece and is operated in a continuous-flow mode); or PAPRTHiE (any powered, air-purifying respirator with a tight-fitting facepiece and a high-efficiency particulate filter); or SCBAF (any self-contained breathing apparatus with a full facepiece); or SAF (any supplied-air respirator with a full facepiece).

500 mg/m³: SA:PD:PP (any supplied-air respirator operated in a pressure-demand or other positive-pressure mode).

Emergency or Planned Entry into Unknown Concentrations or IDLH Conditions: SCBAF:PD,PP (any self-contained breathing apparatus that has a full facepiece and is operated in a pressure-demand or other positive-pressure mode); or SAF:PD,PP:ASCBA (any supplied-air respirator that has a full facepiece and is operated in a pressure-demand or other positive-pressure mode in combination with an auxiliary, self-contained breathing apparatus operated in a pressure-demand or other positive-pressure mode.

Escape: HiEF (any air-purifying, full-facepiece respirator with a high-efficiency particulate filter); or SCBAE (any appropriate escape-type, self-contained breathing apparatus).

Solution

At any concentrations above the NIOSH REL, or where there is no REL, at any detectable concentration: SCBAF:PD,PP (any self-contained breathing apparatus that has a full facepiece and is operated in a pressure-demand or other positive-pressure mode); or SAF:PD,PP:ASCBA (any supplied-air respirator that has a full facepiece and is operated in a pressure-demand or other positive-pressure mode in combination with an auxiliary, self-contained breathing apparatus operated in a pressure-demand or other positive-pressure mode).

Escape: GMFOV [any air-purifying, full-facepiece respirator (gas mask) with a chin-style, front- or back-mounted organic vapor canister]; or SCBAE (any appropriate escape-type, self-contained breathing apparatus).

CHEMICAL REACTIVITY

A powerful oxidizer. Forms heat- and shock-sensitive compound with sulfuric acid. Reacts violently with combustibles or reducing agents. It will decompose and release oxygen if brought into contact with alcohol acids, ferrous salts, iodides, and oxalates. Reacts explosively or initiates fire if brought into contact with acetic acid, acetic anhydride, anhydrous ammonia, antimony, arsenic, hydrogen peroxide, hydroxylamine, hydrogen trisulfide, glycerine, ethylene glycol, hydroxylamine, organic matter, polypropylene, phosphorus, powdered sulfur, and finely divided combustible material, etc. Incompatible with nitric acid, producing toxic chlorine fumes. A dangerous fire and explosion hazard; isolate from all other materials. Slowly decomposes with time.

FIRE INFORMATION

Combustion products include irritants and toxic vapors. Not combustible, but it is powerful oxidizer and a fire hazard. Decomposes at 464°F (240°C) and releases oxygen, which will greatly intensify an ongoing fire. Vapors are heavier than air and will collect in low areas. Vapors in confined areas may explode when exposed to fire. Vapors may travel long distances to ignition sources and flash back. Storage containers and parts of containers may rocket great distances, in many directions. If material or contaminated runoff enters waterways, notify downstream users of potentially contaminated waters. Notify local health and fire officials and pollution control agencies. Use any extinguishing medium suitable for surrounding fires. From a secure, explosion-proof location, use water spray to cool exposed containers.

NFPA Hazard Classifications:

Health Hazards (Blue) 1
Flammability (Red) 0
Reactivity (Yellow) 0
Special Situations .... .OXY

FIRST AID

Move victim to fresh air. Call emergency medical care. Apply artificial respiration if victim is not breathing. Administer oxygen if breathing is difficult. Remove and isolate contaminated clothing and shoes. In case of contact with substance, immediately flush skin or eyes with running water for at least 20 minutes. Keep victim warm and quiet. Ensure that medical personnel are aware of the material(s) involved and take precautions to protect themselves.

ENVIRONMENTAL INFORMATION

Clean Water Act: Section 311 Hazardous Substances/RQ (same as CERCLA).
EPCRA Section 304: Reportable Quantity (RQ): CERCLA, 100 lb (45.4 kg).

POTASSIUM PERSULFATE P:590

SYNONYMS: ANTHION; DIPOTASSIUM PEROXODISULPHATE; EINECS No. 231-781-8; KALIUMPERSULFAT (German); PEROXYDISULFURIC ACID DIPOTASSIUM SALT; PERSULFATO POTASICO (Spanish); POTASSIUM PEROXYDISULFATE; POTASSIUM PEROXYDISULPHATE

IDENTIFICATION

CAS:	7727-21-1
ERG ID:	UN 1492
ERG Guide:	140
ERG Guide Category:	OXIDIZERS
Molecular Formula:	$K_2O_8S_2$
RTECS No.:	SE0400000
Properties:	Crystalline solid. White. Odorless. Moderately soluble in water. Flammable when exposed to heat of chemical reaction. Poisonous gases emitted when heated.

HEALTH AND SAFETY INFORMATION

Corrosive to eyes, skin, and respiratory tract. Inhalation may cause pulmonary edema, which can be delayed for several hours; there is a risk of death in serious cases. Skin contact can cause irritation, redness, and eczema.

902 Potassium Persulfate

Exposure Limits: **TWA**

ACGIH: 0.1 mg/m^3 as S_2O_8

RECOMMENDATIONS FOR RESPIRATOR SELECTION

At any concentrations above the NIOSH REL, or where there is no REL, at any detectable concentration: SCBAF:PD,PP (any self-contained breathing apparatus that has a full facepiece and is operated in a pressure-demand or other positive-pressure mode); or SAF:PD,PP:ASCBA (any supplied-air respirator that has a full facepiece and is operated in a pressure-demand or other positive-pressure mode in combination with an auxiliary, self-contained breathing apparatus operated in a pressure-demand or other positive-pressure mode).

Escape: GMFOV [any air-purifying, full-facepiece respirator (gas mask) with a chin-style, front- or back-mounted organic vapor canister]; or SCBAE (any appropriate escape-type, self-contained breathing apparatus).

CHEMICAL REACTIVITY

A strong oxidizer. Reacts violently with reducing agents, combustible materials, organic substances. Strong reaction with strong caustics such as caustic potash or oxidizable materials such as sodium, sulfur, chemically active metals, metallic dusts (e.g., aluminum), chlorates, perchlorates. Contact with potassium hydroxide in the presence of moisture may cause fire and explosions. Aqueous solution decomposes at room temperature. Material with react with moisture in a closed container, producing a mixture of ozone and sulfuric acid.

FIRE INFORMATION

Combustion products include irritants and toxic vapors. Noncombustible, but causes combustibles to burn. Cover exposed material from wind, rain, or water spray. Decomposes when heated above 212°F (100°C) or in solution above122 °F (50°C) forming corrosive sulfuric and hydrochloric acids. Flush contaminated clothing with water. Combustion products include toxic oxides of sulfur and potassium. Containers may explode when exposed to fire. If material or contaminated runoff enters waterways, notify downstream users of potentially contaminated waters. Notify local health and fire officials and pollution control agencies. Use dry chemicals, alcohol-resistant foam, carbon dioxide. From a secure, explosion-proof location, use water spray to cool exposed containers.

NFPA Hazard Classifications: Not listed

FIRST AID

Move victim to fresh air. Call emergency medical care. Apply artificial respiration if victim is not breathing. Administer oxygen if breathing is difficult. Remove and isolate contaminated clothing and shoes. In case of contact with substance, immediately flush skin or eyes with running water for at least 20 minutes. Keep victim warm and quiet. Ensure that medical personnel are aware of the material(s) involved and take precautions to protect themselves.

PROPADIENE

P:630

SYNONYMS: ALENO (Spanish); ALLENE; sym-ALLENE; sym-ALLYENE; DIMETHYLENE-METANE; EINECS No. 207-335-3; 1,2-PROPADIENE

IDENTIFICATION

CAS:	463-49-0
ERG ID:	UN 2200
ERG Guide:	116P
ERG Guide Category:	GASES-FLAMMABLE (UNSTABLE)
Molecular Formula:	C_2H_4
RTECS No.:	BA0400000
Properties:	Extremely flammable, unstable, compressed liquefied gas. Colorless. Mild, sweet odor. Insoluble in water.

HEALTH AND SAFETY INFORMATION

Can saturate the air and cause suffocation. Irritates eyes, skin, and respiratory tract. Contact with liquid can cause frostbite. May affect central nervous system.

RECOMMENDATIONS FOR RESPIRATOR SELECTION

At any concentrations above the NIOSH REL, or where there is no REL, at any detectable concentration: SCBAF:PD,PP (any self-contained breathing apparatus that has a full facepiece and is operated in a pressure-demand or other positive-pressure mode); or SAF:PD,PP:ASCBA (any supplied-air respirator that has a full facepiece and is operated in a pressure-demand or other positive-pressure mode in combination with an auxiliary, self-contained breathing apparatus operated in a pressure-demand or other positive pressure mode).

Escape: GMFOV [any air-purifying, full-facepiece respirator (gas mask) with a chin-style, front- or back-mounted organic vapor canister]; or SCBAE (any appropriate escape-type, self-contained breathing apparatus).

CHEMICAL REACTIVITY

Dangerous fire and explosion hazard. Contact with heat or oxidizers can cause fire and explosions. Contact with materials containing 63% or more copper will form methyl acetylene [propyne (OSHA)], a dangerous fire hazard. Contact with nitric oxide can form an explosive compound.

FIRE INFORMATION

Vapors are heavier than air and will collect in low areas. Vapors in confined areas may explode when exposed to fire. Vapors may travel long distances to ignition sources and flash back. Storage containers and parts of containers may rocket great distances, in many directions. If material or contaminated runoff enters waterways, notify downstream users of potentially contaminated waters. Notify local health and fire officials and pollution control agencies. Shut off valve if possible; if not possible and there is no danger to surroundings, allow fire to burn itself out. Otherwise, use dry chemicals, foam, carbon dioxide. From a secure, explosion-proof location, use water spray to cool exposed containers.

FEMA Hazard Classifications:

Health Hazards (Blue) 2
Flammability (Red) 4
Reactivity (Yellow) 3

Flash Point: Flammable gas.
Flammable Limits in Air: LEL 3.1%; UEL 12%

FIRST AID

Move victim to fresh air. Call emergency medical care. Apply artificial respiration if victim is not breathing. Administer oxygen if breathing is difficult. Remove and isolate contaminated clothing and shoes. In case of contact with liquefied gas, thaw frosted parts with lukewarm water. Keep victim warm and quiet. Ensure that medical personnel are aware of the material(s) involved and take precautions to protect themselves.

PROPANE P:640

SYNONYMS: A-108; BOTTLED GAS; DIMETHYLMETHANE; EEC No. 601-003-00-5; EINECS No. 200-827-9; LIQUIFIED PETROLEUM GAS; PROPAN VERFLUSSIGT (German); n-PROPANE; PROPANO (Spanish); PROPANO LICUADO (Spanish); PROPYL HYDRATE; PROPYL HYDRIDE; TEFLON DRY LUBE 708; TWINKLE STAINLESS STEEL CLEANER

IDENTIFICATION

CAS:	74-98-6
ERG ID:	UN 1978
ERG Guide:	115
ERG Guide Category:	GASES-FLAMMABLE (INCLUDING REFRIGERATED LIQUIDS)
Molecular Formula:	C_3H_8
RTECS No.:	TX2275000
Properties:	Extremely flammable, liquefied gas; may be a liquid under pressure or refrigeration. Colorless. Faint petroleumlike odor; may contain additives whose odor will warn of leaking gas; has been described as a "skunk" odor. Liquid floats and "boils" on water; practically insoluble. Flammable visible vapor cloud is produced

HEALTH AND SAFETY INFORMATION

Can saturate air and cause suffocation. Skin contact with liquid may cause frostbite.

IDLH: 2100 ppm [10% LEL]
Odor Threshold: 2690 ppm
OSHA Table Z-1-A Air Contaminant

Exposure Limits: TWA

ACGIH:	2500 ppm; 4508 mg/m^3.
NIOSH:	1,000 ppm; 1,800 mg/m^3
OSHA:	1,000 ppm; 1,800 mg/m^3

RECOMMENDATIONS FOR RESPIRATOR SELECTION: NIOSH/OSHA

2,100 ppm: SA (any supplied-air respirator); or SCBAF (any self-contained breathing apparatus with a full facepiece).

Emergency or Planned Entry into Unknown Concentrations or IDLH Conditions:
SCBAF:PD,PP (any self-contained breathing apparatus that has a full facepiece and is operated in a pressure-demand or other positive-pressure mode); or SAF:PD,PP:ASCBA (any supplied-air respirator that has a full facepiece and is operated in a pressure-demand or other positive-pressure mode in combination with an auxiliary, self-contained breathing apparatus operated in a pressure-demand or other positive-pressure mode).

Escape: SCBAE (any appropriate escape-type, self-contained breathing apparatus).

CHEMICAL REACTIVITY

Forms explosive gas mixture with air or chlorine dioxide. Strong oxidizers may cause fire and explosions. Liquid attacks some plastics, rubber, and coatings. May accumulate static electrical charges and may cause ignition of its vapors.

FIRE INFORMATION

Gas is heavier than air and will collect in low areas. Gas in confined areas may explode when exposed to fire. Vapors may travel long distances to ignition sources and flash back. Storage containers and parts of containers may rocket great distances, in many directions. Notify local health and fire officials and pollution control agencies. Shut off valve if possible; if not possible and there is no danger to surroundings, allow fire to burn itself out. Otherwise, use dry chemicals, foam, carbon dioxide. From a secure, explosion-proof location, use water spray to cool exposed containers. If containers are exposed to direct flame or elevated temperatures for prolonged times, withdraw immediately to a secure location.

NFPA Hazard Classifications:

Health Hazards (Blue) 1
Flammability (Red) 4
Reactivity (Yellow) 0

Flash Point: -156°F (-104°C) c.c.

Flammable Limits in Air: LEL 2.1%; UEL 9.5%
Autoignition Temperature: 842°F (450°C)

FIRST AID

Move victim to fresh air. Call emergency medical care. Apply artificial respiration if victim is not breathing. Administer oxygen if breathing is difficult. Remove and isolate contaminated clothing and shoes. Clothing frozen to the skin should be thawed before being removed. In case of contact with liquefied gas, thaw frosted parts with lukewarm water. Keep victim warm and quiet. Ensure that medical personnel are aware of the material(s) involved and take precautions to protect themselves.

ENVIRONMENTAL INFORMATION

Clean Air Act: Accidental Release Prevention/Flammable substances, (Section 112[r], Table 3), TQ = 10,000 lb (4,540 kg).

PROPANETHIOL P:645

SYNONYMS: EINECS No.. 203-455-5; 3-MERCAPTOPROPANE; 3-MERCAPTOPROPANOL; n-PROPILMERCAPTANO (Spanish); n-PROPYL MERCAPTAN; 1-PROPANETHIOL; PROPANE-1-THIO

IDENTIFICATION

CAS:	107-03-9
ERG ID:	UN 2402
ERG Guide:	130
ERG Guide Category:	FLAMMABLE LIQUIDS (NONPOLAR/WATER-IMMISCIBLE/NOXIOUS)
Molecular Formula:	C_3H_8S
RTECS No.:	TZ7300000
Properties:	Extremely flammable liquid. Colorless. Offensive, skunklike, rotten cabbage odor. Floats on water; slightly soluble.

HEALTH AND SAFETY INFORMATION

Odor Threshold: 0.00075 ppm

RECOMMENDATIONS FOR RESPIRATOR SELECTION: NIOSH

5 ppm: CCROV [any chemical cartridge respirator with organic vapor cartridge(s)]; or SA (any supplied-air respirator).
12.5 ppm: SA:CF (any supplied-air respirator operated in a continuous-flow mode); or PAPROV [any powered, air-purifying respirator with organic vapor cartridge(s)].
25 ppm: CCROV [any chemical cartridge respirator with organic vapor cartridge(s)]; or GMFOV [any

air-purifying, full-facepiece respirator (gas mask) with a chin-style, front- or back-mounted acid gas canister]; or PAPRTOV [any powered, air-purifying respirator with a tight-fitting facepiece and organic vapor cartridge(s)]; or SCBAF (any self-contained breathing apparatus with a full facepiece) or SAF (any supplied-air respirator with a full facepiece).

Emergency or planned entry into unknown concentrations or IDLH conditions: SCBAF:PD,PP (any self-contained breathing apparatus that has a full facepiece and is operated in a pressure-demand or other positive-pressure mode); or SAF:PD,PP:ASCBA (any supplied-air respirator that has a full facepiece and is operated in a pressure-demand or other positive-pressure mode in combination with an auxiliary, self-contained breathing apparatus operated in a pressure-demand or other positive pressure mode.

Escape: GMFOV [any air-purifying, full-facepiece respirator (gas mask) with a chin-style, front- or back-mounted organic vapor canister]; or SCBAE (any appropriate escape-type, self-contained breathing apparatus).

CHEMICAL REACTIVITY

Contact with oxidizers may cause fire and explosions. Reacts with reducing agents, strong acids and bases, alkali metals, calcium hypochlorite. May react with steam, water, or acids to produce toxic and flammable vapors.

FIRE INFORMATION

Combustion products include toxic fumes of sulfides. Vapors are heavier than air and will collect in low areas. Vapors in confined areas may explode when exposed to fire. Vapors may travel long distances to ignition sources and flash back. Storage containers and parts of containers may rocket great distances, in many directions. If material or contaminated runoff enters waterways, notify downstream users of potentially contaminated waters. Notify local health and fire officials and pollution control agencies. Use dry chemicals, foam, carbon dioxide. Water may be ineffective. From a secure, explosion-proof location, use water spray to cool exposed containers.

NFPA Hazard Classifications: Not listed

Flash Point: -5°F (-21°C) o.c.

FIRST AID

Move victim to fresh air. Call emergency medical care. Apply artificial respiration if victim is not breathing. Administer oxygen if breathing is difficult. Remove and isolate contaminated clothing and shoes. In case of contact with substance, immediately flush skin or eyes with running water for at least 20 minutes. Wash skin with soap and water. Keep victim warm and quiet. Effects of exposure (inhalation, ingestion, or skin contact) to substance may be delayed. Ensure that medical personnel are aware of the material(s) involved and take precautions to protect themselves.

PROPIONALDEHYDE P:655

SYNONYMS: ALDEHYDE PROPIONIQUE (French); EEC No. 605-018-00-8; EINECS No. 204-623-0; FEMA No. 2923; METHYLACETALDEHYDE; NCI-C61029; PROPALDEHYDE; PROPANAL; PROPANALDEHYDE; PROPIONAL; PROPIONIC-ALDEHYDE; PROPYL-ALDEHYDE; PROPYLALDEHYDE; PROPYLIC-ALDEHYDE

IDENTIFICATION

CAS:	123-38-6
ERG ID:	UN 1275
ERG Guide:	129
ERG Guide Category:	FLAMMABLE LIQUIDS (POLAR/WATER-MISCIBLE/NOXIOUS)
Molecular Formula:	C_3H_6O
RTECS No.:	UE0350000
Properties:	Flammable liquid. Colorless. Suffocating, fruity, unpleasant, choking odor. Floats and mixes slowly with water.

HEALTH AND SAFETY INFORMATION

Moderate health hazard. Irritates eyes, skin, and respiratory tract. High concentrations of vapor can cause narcosis, nausea, and loss of consciousness. Inhalation may cause pulmonary edema, which can be delayed for several hours; there is a risk of death in serious cases. Long-term exposure may result in liver problems.

Odor Threshold: 1 ppm

RECOMMENDATIONS FOR RESPIRATOR SELECTION

At any concentrations above the NIOSH REL, or where there is no REL, at any detectable concentration: SCBAF:PD,PP (any self-contained breathing apparatus that has a full facepiece and is operated in a pressure-demand or other positive-pressure mode); or SAF:PD,PP:ASCBA (any supplied-air respirator that has a full facepiece and is operated in a pressure-demand or other positive-pressure mode in combination with an auxiliary, self-contained breathing apparatus operated in a pressure-demand or other positive-pressure mode).

Escape: GMFOV [any air-purifying, full-facepiece respirator (gas mask) with a chin-style, front- or back-mounted organic vapor canister]; or SCBAE (any appropriate escape-type, self-contained breathing apparatus).

CHEMICAL REACTIVITY

Forms explosive mixture with air. Incompatible with strong acids, caustics, and amines. Reacts violently with strong oxidizers, methyl methacrylate. Can self-ignite if finely dispersed on porous or combustible material. Hazardous polymerization may occur. Heat or ultraviolet can cause decomposition. May accumulate static electrical charges, and may cause ignition of its vapors.

FIRE INFORMATION

Vapors are heavier than air and will collect in low areas. Vapors in confined areas may explode when exposed to fire. Vapors may travel long distances to ignition sources and flash back. Storage containers and parts of containers may rocket great distances, in many directions. If material or contaminated runoff enters waterways, notify downstream users of potentially contaminated waters. Notify local health and fire officials and pollution control agencies. Use dry chemicals, foam, carbon dioxide. Water may be ineffective. From a secure, explosion-proof location, use water spray to cool exposed containers.

NFPA Hazard Classifications:

Health Hazards (Blue) 2
Flammability (Red) 3
Reactivity (Yellow) 2

Flash Point: –22°F (–30°C) o.c.
Flammable Limits in Air: LEL 2.6%; UEL 17%
Autoignition Temperature: 405°F (207°C)

FIRST AID

Move victim to fresh air. Call emergency medical care. Apply artificial respiration if victim is not breathing. Administer oxygen if breathing is difficult. Remove and isolate contaminated clothing and shoes. In case of contact with substance, immediately flush skin or eyes with running water for at least 20 minutes. Wash skin with soap and water. Keep victim warm and quiet. Effects of exposure (inhalation, ingestion, or skin contact) to substance may be delayed. Ensure that medical personnel are aware of the material(s) involved and take precautions to protect themselves.

ENVIRONMENTAL INFORMATION

Clean Air Act: Hazardous Air Pollutants (Title I, Part A, Section 112)
EPCRA Section 304 Reportable Quantity (RQ): CERCLA, 1,000 lb (454 kg)
EPCRA Section 313: Form R *de minimis* concentration reporting level: 1.0%.

PROPIONIC ACID P:660

SYNONYMS: ACIDE PROPIONIQUE (French); ACIDO PROPIONICO (Spanish); CARBONYETHANE; CARBOXYETHANE; EEC No. 607-089-00-0; EINECS No. 201-176-3; ETHANECARBOXYLIC ACID; ETHYLFORMIC ACID; LUPROSIL; METACETONIC ACID; METHYLACETIC ACID; PROPANOIC ACID; PROZOIN; PSEUDOACETIC ACID

IDENTIFICATION

CAS: 79-09-4

910 Propionic Acid

ERG ID:	UN 1848
ERG Guide:	132
ERG Guide Category:	FLAMMABLE LIQUIDS-CORROSIVE
Molecular Formula:	$C_3H_6O_2$
RTECS No.:	UE5950000
Properties:	Very flammable, colorless, oily liquid. Sharp, pungent, rancid odor. Folats on water initially; soluble. Irritating vapor is produced. Freezes at -6°F (-21°C).

HEALTH AND SAFETY INFORMATION

Corrosive to eyes, skin, and respiratory tract. Inhalation may cause pulmonary edema, which can be delayed for several hours; there is a risk of death in serious cases. Prolonged or repeated exposures may cause asthmalike allergy; lung problems.

IDLH: 0.06–0.6 mg/m^3; 0.16 ul/L
Odor Threshold: 0.037 ppm

Exposure Limits:	**TWA**	**STE**
ACGIH:	10 ppm; 30 mg/m^3	
NIOSH:	10 ppm; 30 mg/m^3	15 ppm; 45 mg/m^3
OSHA:	None	

RECOMMENDATIONS FOR RESPIRATOR SELECTION

At any concentrations above the NIOSH REL, or where there is no REL, at any detectable concentration: SCBAF:PD,PP (any self-contained breathing apparatus that has a full facepiece and is operated in a pressure-demand or other positive-pressure mode); or SAF:PD,PP:ASCBA (any supplied-air respirator that has a full facepiece and is operated in a pressure-demand or other positive-pressure mode in combination with an auxiliary, self-contained breathing apparatus operated in a pressure-demand or other positive-pressure mode).

Escape: GMFOV [any air-purifying, full-facepiece respirator (gas mask) with a chin-style, front- or back-mounted organic vapor canister]; or SCBAE (any appropriate escape-type, self-contained breathing apparatus).

CHEMICAL REACTIVITY

Forms explosive mixture with air. Incompatible with sulfuric acid, strong bases (reaction evolves heat), ammonia, aliphatic amines, alkanolamines, isocyanates, alkylene oxides, epichlorohydrin, and oxidizers. Attacks metals such as steel, forming flammable hydrogen gas.

FIRE INFORMATION

Combustion products include carbon monoxide and carbon dioxide. Vapors are heavier than air and will collect in low areas. Vapors in confined areas may explode when exposed to fire. Vapors may travel long distances to ignition sources and flash back. Storage containers and parts of containers may rocket great distances, in many directions. If material or contaminated runoff enters waterways, notify downstream users of potentially contaminated waters. Notify local health and fire officials and pollution control agencies. Use dry chemicals, foam, carbon dioxide. From a secure, explosion-proof location, use water spray to cool exposed containers.

NFPA Hazard Classifications:

Health Hazards (Blue) 3
Flammability (Red) 2
Reactivity (Yellow) 0

Flash Point: 132°F (57°C) o.c.; 126°F (52°C) c.c.
Flammable Limits in Air: LEL 2.9%; UEL 12.1%
Autoignition Temperature: 1105°F (462°C)

FIRST AID

Move victim to fresh air. Call emergency medical care. Apply artificial respiration if victim is not breathing. Do not use mouth-to-mouth method if victim swallowed or inhaled the substance; induce artificial respiration with the aid of a pocket mask equipped with a one-way valve or other proper respiratory medical device. Administer oxygen if breathing is difficult. Remove and isolate contaminated clothing and shoes. In case of contact with substance, immediately flush skin or eyes with running water for at least 20 minutes. Keep victim warm and quiet. Effects of exposure (inhalation, ingestion, or skin contact) to substances may be delayed. Ensure that medical personnel are aware of the material(s) involved and take precautions to protect themselves.

ENVIRONMENTAL INFORMATION

Clean Water Act: Section 311 Hazardous Substances/RQ (same as CERCLA)
EPCRA Section 304: Reportable Quantity (RQ): CERCLA, 5,000 lb (2,270 kg)

PROPIONIC ANHYDRIDE P:662

SYNONYMS: ANHIDRIDO PROPIONICO (Spanish); ANHYDRIDE PROPIONIQUE (French); EEC No. 607-010-00-X; EINECS No. 204-638-2; METHYLACETIC ACID ANHYDRIDE; PROPIONIC ACID ANHYDRIDE; PROPIONYL OXIDE

IDENTIFICATION

CAS:	123-62-6
ERG ID:	UN 2496
ERG Guide:	156
ERG Guide Category:	SUBSTANCES-TOXIC AND/OR CORROSIVE (COMBUSTIBLE/WATER-SENSITIVE)

912 Propionic Anhydride

Molecular Formula: $C_6H_{10}O_3$
RTECS No.: UF9100000
Properties: Combustible liquid. Colorless. Sharp odor. Sinks and mixes slowly with water.

HEALTH AND SAFETY INFORMATION

Corrosive to eyes, skin, and respiratory tract. Inhalation may cause pulmonary edema, which can be delayed for several hours; there is a risk of death in serious cases.

Exposure Limits:	TWA	STE
ACGIH:	10 ppm; 30 mg/m^3 as propionic acid	
NIOSH:	10 ppm; 30 mg/m^3 as propionic acid	15 ppm; 45 mg/m^3 as propionic acid
OSHA:	None	

RECOMMENDATIONS FOR RESPIRATOR SELECTION

At any concentrations above the NIOSH REL, or where there is no REL, at any detectable concentration: SCBAF:PD,PP (any self-contained breathing apparatus that has a full facepiece and is operated in a pressure-demand or other positive-pressure mode); or SAF:PD,PP:ASCBA (any supplied-air respirator that has a full facepiece and is operated in a pressure-demand or other positive-pressure mode in combination with an auxiliary, self-contained breathing apparatus operated in a pressure-demand or other positive-pressure mode).

Escape: GMFOV [any air-purifying, full-facepiece respirator (gas mask) with a chin-style, front- or back-mounted organic vapor canister]; or SCBAE (any appropriate escape-type, self-contained breathing apparatus).

CHEMICAL REACTIVITY

Reacts with water to form heat and weak propionic acid. May accumulate static electricity and cause ignition of its vapors. Contact with oxidizers, acids, bases may cause fire and explosions. Slowly corrosive to metals in the presence of moisture.

FIRE INFORMATION

Combustion may produce irritants and toxic vapors. Vapors are heavier than air and will collect in low areas. Vapors in confined areas may explode when exposed to fire.If material or contaminated runoff enters waterways, notify downstream users of potentially contaminated waters. Notify local health and fire officials and pollution control agencies. Use dry chemicals, alcohol-resistant foam, carbon dioxide. From a secure, explosion-proof location, use water spray to cool exposed containers.

NFPA Hazard Classifications:

Health Hazards (Blue) 3

Flammability (Red) 2
Reactivity (Yellow) 1

Flash Point: 156°F (69°C) o.c.; 145°F (63°C) c.c.
Flammable Limits in Air: LEL 1.5% @ 165°F (74°C); UEL 11.9% @ 261°F (127°C)
Autoignition Temperature: 540°F (286°C)

FIRST AID

Move victim to fresh air. Call emergency medical care. Apply artificial respiration if victim is not breathing. Do not use mouth-to-mouth method if victim ingested or inhaled the substance; induce artificial respiration with the aid of a pocket mask equipped with a one-way valve or other proper respiratory medical device. Administer oxygen if breathing is difficult. Remove and isolate contaminated clothing and shoes. In case of contact with substance, immediately flush skin or eyes with running water for at least 20 minutes. For minor skin contact, avoid spreading material on unaffected skin. Keep victim warm and quiet. Effects of exposure (inhalation, ingestion, or skin contact) may be delayed. Ensure that medical personnel are aware of the material(s) involved and take precautions to protect themselves.

ENVIRONMENTAL INFORMATION

Clean Water Act: Section 311 Hazardous Substances/RQ (same as CERCLA)
EPCRA Section 304 Reportable Quantity (RQ): CERCLA, 5,000 lb (2,270 kg)
TSCA: 40CFR712.30(*e*)13; 40CFR716.120(*d*)13 chemicals in need of subchronic tests.

n–PROPYL ACETATE P:663

SYNONYMS: ACETATE de PROPYLE NORMAL (French); ACETATO de n-PROPILO (Spanish); ACETIC ACID, PROPYL ESTER; ACETIC ACID, n-PROPYL ESTER; 1-ACETOXYPROPANE; EEC No. 607-024-00-6; EINECS No. 203-686-1; OCTAN PROPYLU (Polish); PROPYL ACETATE; n-PROPYL ESTER of ACETIC ACID; 1-PROPYL ACETATE

IDENTIFICATION

CAS:	109-60-4
ERG ID:	UN 1276
ERG Guide:	129
ERG Guide Category:	FLAMMABLE LIQUIDS (POLAR/WATER-MISCIBLE/NOXIOUS)
Molecular Formula:	$C_5H_{10}O_2$
RTECS No.:	AJ3675000
Properties:	Highly flammable liquid. Clear, colorless. Pleasant, sweet odor. Floats on water; slightly soluble.

HEALTH AND SAFETY INFORMATION

Irritating to eyes, skin, and respiratory tract. High levels of exposure can affect the central nervous system.

914 n-Propyl Acetate

IDLH: 1,700 ppm
Odor Threshold: 0.575 ppm
OSHA Table Z-1-A Air Contaminant

Exposure Limits:	TWA	STEL
ACGIH:	200 ppm; 835 mg/m^3	250 ppm; 1040 mg/m^3
NIOSH:	200 ppm; 840 mg/m^3	250 ppm; 1050 mg/m^3
OSHA:	200 ppm; 840 mg/m^3	

RECOMMENDATIONS FOR RESPIRATOR SELECTION: NIOSH

1700 ppm: SA:CF (any supplied-air respirator operated in a continuous-flow mode); or CCRFOV [any air-purifying, full-facepiece respirator (gas mask) with a chin-style, front- or back-mounted acid gas canister]; or GMFOV [any air-purifying, full-facepiece respirator (gas mask) with a chin-style, front- or back-mounted organic vapor canister]; or PAPROV [any powered, air-purifying respirator with organic vapor cartridge(s)]; or SCBAF (any self-contained breathing apparatus with a full facepiece); or SAF (any supplied-air respirator with a full facepiece)

Emergency or planned entry into unknown concentrations or IDLH conditions: SCBAF:PD,PP (any self-contained breathing apparatus that has a full facepiece and is operated in a pressure-demand or other positive-pressure mode); or SAF:PD,PP:ASCBA (any supplied-air respirator that has a full facepiece and is operated in a pressure-demand or other positive-pressure mode in combination with an auxiliary, self-contained breathing apparatus operated in a pressure-demand or other positive pressure mode.

Escape: GMFOV [any air-purifying, full-facepiece respirator (gas mask) with a chin-style, front- or back-mounted organic vapor canister]; or SCBAE (any appropriate escape-type, self-contained breathing apparatus).

CHEMICAL REACTIVITY

Hydrolyzes on standing to form acetic acid and n-propyl alcohol. The presence of bases speeds up the reaction. Reacts vigorously with oxidizers, nitrates, strong alkalis, and strong acids. Softens or dissolves many plastics.

FIRE INFORMATION

Vapors are heavier than air and will collect in low areas. Vapors in confined areas may explode when exposed to fire. Vapors may travel long distances to ignition sources and flash back. Storage containers and parts of containers may rocket great distances, in many directions. If material or contaminated runoff enters waterways, notify downstream users of potentially contaminated waters. Notify local health and fire officials and pollution control agencies. Use dry chemicals, foam, carbon dioxide. From a secure, explosion-proof location, use water spray to cool exposed containers.

NFPA Hazard Classifications:

Health Hazards (Blue) 1
Flammability (Red) 3
Reactivity (Yellow) 0

Flash Point: 65°F (18°C) o.c.; 58°F (14°C) c.c.
Flammable Limits in Air: LEL 2.0%; UEL 8.0%
Autoignition Temperature: 842°F (450°C)

FIRST AID

Move victim to fresh air. Call emergency medical care. Apply artificial respiration if victim is not breathing. Administer oxygen if breathing is difficult. Remove and isolate contaminated clothing and shoes. In case of contact with substance, immediately flush skin or eyes with running water for at least 20 minutes. Wash skin with soap and water. Keep victim warm and quiet. Effects of exposure (inhalation, ingestion, or skin contact) to substance may be delayed. Ensure that medical personnel are aware of the material(s) involved and take precautions to protect themselves.

n–PROPYL ALCOHOL P:665

SYNONYMS: ALCOOL PROPILICO (Italian); ALCOOL PROPYLIQUE (French); EEC No. 603-003-00-0; EINECS No. 200-746-9; ETHYL CARBINOL; 1-HYDROXYPROPANE; NORMAL PROPYL ALCOHOL; OPTAL; OSMOSOL EXTRA; n-PROPANOL; PROPANOL-1; 1-PROPANOL (Spanish); PROPANOLE (German); PROPANOLEN (Dutch); PROPANOLI (Italian); PROPYL ALCOHOL; 1-PROPYL ALCOHOL; PROPYL ALCOHOL, NORMAL; n-PROPYL ALKOHOL (German); PROPYLIC ALCOHOL; PROPYLOWY ALKOHOL (Polish)

IDENTIFICATION

CAS:	71-23-8
ERG ID:	UN 1274
ERG Guide:	129
ERG Guide Category:	FLAMMABLE LIQUIDS(POLAR/WATER-MISCIBLE/ NOXIOUS)
Molecular Formula:	C_3H_8O
RTECS No.:	UH8225000
Properties:	Flammable, clear liquid. Mild, alcohollike odor. Mixes with water.

HEALTH AND SAFETY INFORMATION

IDLH: 800 ppm
Odor Threshold: 2.4 ppm
OSHA Table Z-1-A Air Contaminant

916 n-Propyl Alcohol

Exposure Limits:	**TWA**	**STEL**
ACGIH: [(1)]	200 ppm; 492 mg/m^3	250 ppm; 614 mg/m^3
NIOSH: [(1)]	200 ppm; 500 mg/m^3	250 ppm; 625 mg/m^3
OSHA:	200 ppm; 500 mg/m^3	

[(1)] Skin contact contributes significantly to overall exposure.

RECOMMENDATIONS FOR RESPIRATOR SELECTION: NIOSH

800 ppm: CCROV [any chemical cartridge respirator with organic vapor cartridge(s)]; or PAPROV [any powered, air-purifying respirator with organic vapor cartridge(s)]; or GMFOV [any air-purifying, full-facepiece respirator (gas mask) with a chin-style, front- or back-mounted acid gas canister]; or SA (any supplied-air respirator); or SCBAF (any self-contained breathing apparatus with a full facepiece)

Emergency or Planned Entry into Unknown Concentrations or IDLH Conditions: SCBAF:PD,PP (any self-contained breathing apparatus that has a full facepiece and is operated in a pressure-demand or other positive-pressure mode); or SAF:PD,PP:ASCBA (any supplied-air respirator that has a full facepiece and is operated in a pressure-demand or other positive-pressure mode in combination with an auxiliary, self-contained breathing apparatus operated in a pressure-demand or other positive-pressure mode)

Escape: GMFOV [any air-purifying, full-facepiece respirator (gas mask) with a chin-style, front- or back-mounted organic vapor canister]; or SCBAE (any appropriate escape-type, self-contained breathing apparatus).
Note: Substance reported to cause eye irritation or damage; eye protection needed.

CHEMICAL REACTIVITY

Forms explosive mixture with air. Contact with strong oxidizers may cause fire and explosions. Attacks some plastics, rubber, and coatings. Incompatible with strong acids, caustics, aliphatic amines, and isocyanates.

FIRE INFORMATION

Vapors are heavier than air and will collect in low areas. Vapors in confined areas may explode when exposed to fire. Vapors may travel long distances to ignition sources and flash back. Storage containers and parts of containers may rocket great distances, in many directions. If material or contaminated runoff enters waterways, notify downstream users of potentially contaminated waters. Notify local health and fire officials and pollution control agencies. Use dry chemicals, alcohol foam, carbon dioxide; water may be ineffective. From a secure, explosion-proof location, use water spray to cool exposed containers.

NFPA Hazard Classifications:

Health Hazards (Blue) 1
Flammability (Red) 3
Reactivity (Yellow) 0

Flash Point: 72°F (22°C)
Flammable Limits in Air: LEL 2.2%; UEL 13.7%
Autoignition Temperature: 775°F (412°C)

FIRST AID

Move victim to fresh air. Call emergency medical care. Apply artificial respiration if victim is not breathing. Administer oxygen if breathing is difficult. Remove and isolate contaminated clothing and shoes. In case of contact with substance, immediately flush skin or eyes with running water for at least 20 minutes. Wash skin with soap and water. Keep victim warm and quiet. Effects of exposure (inhalation, ingestion, or skin contact) to substance may be delayed. Ensure that medical personnel are aware of the material(s) involved and take precautions to protect themselves.

PROPYLENE P:670

SYNONYMS: EINECS No. 204-062-1; ISOBUTYLENE; METHYLETHENE; METHYLETHYLENE; NCI-C50077; PROPENE; 1-PROPENE; PROPILENO (Spanish); 1-PROPYLENE; R 1270; STCC 4905782

IDENTIFICATION

CAS:	115-07-1
ERG ID:	UN 1077; UN 1075
ERG Guide:	115
ERG Guide Category:	GASES-FLAMMABLE (INCLUDING REFRIGERATED LIQUIDS)
Molecular Formula:	C_3H_6
RTECS No.:	UD0890000
Properties:	Extremely flammable, colorless gas. Faint petroleum odor. Shipped and stored as a liquefied, compressed gas. Floats and "boils" on water surface; insoluble in water.

HEALTH AND SAFETY INFORMATION

Can saturate the air and cause suffocation. ACGIH simple asphyxiant, not suspected as a human carcinogen. Contact with liquid may cause frostbite. Can affect the central nervous system.

IARC Evaluation: Group 3, unclassifiable substance; no human data; no animal data.

RECOMMENDATIONS FOR RESPIRATOR SELECTION: NIOSH

At any concentrations above the NIOSH REL, or where there is no REL, at any detectable concentration: SCBAF:PD,PP (any self-contained breathing apparatus that has a full facepiece and is operated in a pressure-demand or other positive-pressure mode); or SAF:PD,PP:ASCBA (any supplied-air respirator that has a full facepiece and is operated in a pressure-demand or other positive-pressure mode in

combination with an auxiliary, self-contained breathing apparatus operated in a pressure-demand or other positive-pressure mode).

Escape: GMFOV [any air-purifying, full-facepiece respirator (gas mask) with a chin-style, front- or back-mounted organic vapor canister]; or SCBAE (any appropriate escape-type, self-contained breathing apparatus).

CHEMICAL REACTIVITY

Forms explosive mixture with air. Able to form dangerous peroxides; may polymerize. Contact with strong oxidizers or strong acids may cause fire and explosions. May accumulate electrical charges and may cause ignition of its vapors.

FIRE INFORMATION

Vapors are heavier than air and will collect in low areas. Vapors in confined areas may explode when exposed to fire. Vapors may travel long distances to ignition sources and flash back. Storage containers and parts of containers may rocket great distances, in many directions. If material or contaminated runoff enters waterways, notify downstream users of potentially contaminated waters. Notify local health and fire officials and pollution control agencies. Shut off valve if possible; if not possible and there is no danger to surroundings, allow fire to burn itself out. Use dry chemicals, foam, carbon dioxide. From a secure, explosion-proof location, use water spray to cool exposed containers. If containers are exposed to direct flame or elevated temperatures for prolonged times, withdraw immediately to a secure location

NFPA Hazard Classifications:

Health Hazards (Blue) 1
Flammability (Red) 4
Reactivity (Yellow) 1

Flash Point: –162°F (–108°C)
Flammable Limits in Air: LEL 2.0%; UEL 11.1%
Autoignition Temperature: 851°F (455°C)

FIRST AID

Move victim to fresh air. Call emergency medical care. Apply artificial respiration if victim is not breathing. Administer oxygen if breathing is difficult. Remove and isolate contaminated clothing and shoes. Clothing frozen to the skin should be thawed before being removed. In case of contact with liquefied gas, thaw frosted parts with lukewarm water. Keep victim warm and quiet. Keep victim under observation. Ensure that medical personnel are aware of the material(s) involved and take precautions to protect themselves.

ENVIRONMENTAL INFORMATION

Clean Air Act: Accidental Release Prevention/Flammable substances, (Section 112[r], Table 3), TQ = 10,000 lb (4,540 kg).
EPCRA Section 313: Form R *de minimis* concentration reporting level: 1.0%.

PROPYLENE GLYCOL

P:680

SYNONYMS: CHILLSA FE; 1,2-DIHYDROXYPROPANE; DOWFROST; EINECS No. 200-338-0; FIDELITY 1215B SOLDER STRIPPER; HARDNESS 2 TEST SOLUTION; 2-HYDROXYPROPANOL; ILEXAN; ILEXAN P; METHYLETHYLENEGLYCOL; METHYLETHYLENE GLYCOL; METHYL GLYCOL; MONOPROPYLENE GLYCOL; MONAWET MO-84R2W; PG12; PROPANE-1,2-DIOL; 1,2-PROPANEDIOL; PROPILENGLICOL (Spanish); 1,2-PROPYLENE GLYCOL; SIRLENE; SOLAR WINTER BAN; TRIMETHYL GLYCO

IDENTIFICATION

CAS:	57-55-6
ERG ID:	UN 1993
ERG Guide:	128
ERG Guide Category:	FLAMMABLE LIQUIDS (NON-POLAR/WATER-IMMISCIBLE)
Molecular Formula:	$C_3H_8O_2$
RTECS No:	TY2000000
Properties:	Thick, syrupy, combustible liquid. Colorless. Odorless. Mixes with water.

HEALTH AND SAFETY INFORMATION

Irritates eyes, skin, and respiratory tract. Prolonged or repeated contact may cause mutations; skin effects.

RECOMMENDATIONS FOR RESPIRATOR SELECTION

At any concentrations above the NIOSH REL, or where there is no REL, at any detectable concentration: SCBAF:PD,PP (any self-contained breathing apparatus that has a full facepiece and is operated in a pressure-demand or other positive-pressure mode); or SAF:PD,PP:ASCBA (any supplied-air respirator that has a full facepiece and is operated in a pressure-demand or other positive-pressure mode in combination with an auxiliary, self-contained breathing apparatus operated in a pressure-demand or other positive pressure mode).

Escape: GMFOV (any air-purifying, full-facepiece respirator (gas mask) with a chin-style, front- or back-mounted organic vapor canister; or SCBAE (any appropriate escape-type, self-contained breathing apparatus).

CHEMICAL REACTIVITY

Contact with oxidizers such as permanganates, chlorine, or dichromates can cause a violent reaction. Incompatible with strong acids, caustics, aliphatic amines, isocyanates.

FIRE INFORMATION

If material or contaminated runoff enters waterways, notify downstream users of potentially contaminated waters. Notify local health and fire officials and pollution control agencies. Use dry chemicals, foam, carbon dioxide. From a secure, explosion-proof location, use water spray to cool exposed containers.

NFPA Hazard Classifications:

Health Hazards (Blue) 0
Flammability (Red) 1
Reactivity (Yellow) 0

Flash Point: 225°F (107°C) o.c.; 210°F (99°C) c.c.
Flammable Limits in Air: LEL 2.6%; UEL 12.5%
Autoignition Temperature: 790°F (375°C)

FIRST AID

Move victim to fresh air. Call emergency medical care. Apply artificial respiration if victim is not breathing. Administer oxygen if breathing is difficult. Remove and isolate contaminated clothing and shoes. In case of contact with substance, immediately flush skin or eyes with running water for at least 20 minutes. Wash skin with soap and water. Keep victim warm and quiet. Ensure that medical personnel are aware of the material(s) involved and take precautions to protect themselves.

PROPYLENE GLYCOL PHENYL ETHER P:690

SYNONYMS: DOWANOL PPH GLYCOL ETHER; EINECS No. 224-027-4; 2-PHENOXYPROPANOL; 2-PHENOXYPROPYL ALCOHOL; 1-PROPANOL, 2-PHENOXY-; PROPILENGLICOL FENIL ETER (Spanish)

IDENTIFICATION

CAS:	4169-04-4
ERG ID:	UN 3271 as ethers, n.o.s.
ERG Guide:	127
ERG Guide Category:	FLAMMABLE LIQUIDS (POLAR/WATER- MISCIBLE)
Molecular Formula:	$C_9H_{12}O_2$
RTECS No.:	UB8886000
Properties:	Combustible liquid. Colorless to faint yellow. Soluble in water.

HEALTH AND SAFETY INFORMATION

Irritates eyes, skin, and respiratory tract. Prolonged exposure may cause personality changes. Skin contact may cause inflammation, redness, and contributes significantly to overall exposure.

RECOMMENDATIONS FOR RESPIRATOR SELECTION

At any concentrations above the NIOSH REL, or where there is no REL, at any detectable concentration: SCBAF:PD,PP (any self-contained breathing apparatus that has a full facepiece and is operated in a pressure-demand or other positive-pressure mode); or SAF:PD,PP:ASCBA (any supplied-air

respirator that has a full facepiece and is operated in a pressure-demand or other positive-pressure mode in combination with an auxiliary, self-contained breathing apparatus operated in a pressure-demand or other positive-pressure mode).

Escape: GMFOV [any air-purifying, full-facepiece respirator (gas mask) with a chin-style, front- or back-mounted organic vapor canister]; or SCBAE (any appropriate escape-type, self-contained breathing apparatus).

CHEMICAL REACTIVITY

Contact with oxidizers may cause fire and explosions. May be able to form an explosive mixture with ai and unstable peroxides. Keep away from heat, sparks and flame. Incompatible with sulfuric acid, nitric acid, caustics, aliphatic amines, isocyanates.

FIRE INFORMATION

Combustion products include irritating and toxic gases. Vapors are heavier than air and will collect in low areas. Vapors in confined areas may explode when exposed to fire. If material or contaminated runoff enters waterways, notify downstream users of potentially contaminated waters. Notify local health and fire officials and pollution control agencies. Use dry chemicals, alcohol-resistant foam, carbon dioxide, water spray. From a secure, explosion-proof location, use water spray to cool exposed containers.

NFPA Hazard Classifications: Not listed

FIRST AID

Move victim to fresh air. Call emergency medical care. Apply artificial respiration if victim is not breathing. Administer oxygen if breathing is difficult. Remove and isolate contaminated clothing and shoes. In case of contact with substance, immediately flush skin or eyes with running water for at least 20 minutes. Wash skin with soap and water. Keep victim warm and quiet. Ensure that medical personnel are aware of the material(s) involved and take precautions to protect themselves.

ENVIRONMENTAL INFORMATION

Although "glycol ethers" are on the Community Right-to-Know List (EPCRA Section 313), this category does *not* contain glycol ethers based on propylene glycol, dipropylene glycol, or tripropylene glycol.

TSCA: 40CFR712.30(e)11, glycol ethers and esters; 40CFR716.120(*d*)10, dermal absorption testing.

PROPYLENE OXIDE P:730

SYNONYMS: AD 6; EINECS No. 200-879-2; EPOXYPROPANE; 1,2-EPOXYPROPANE; 2,3-EPOXYPROPANE; METHYL ETHYLENE OXIDE; METHYL OXIRANE; NCIC50099; OXIDO de PROPILENO (Spanish); OXYDE de PROPYLENE (French); OXIRANE, METHYL-; PROPANE, 1,2-EPOXY-; PROPENE OXIDE; PROPYLENE EPOXIDE; 1,2-PROPYLENE OXIDE; STCC 4906620

922 Propylene Oxide

IDENTIFICATION

CAS:	75-56-9
ERG ID:	UN 1280
ERG Guide:	127P
ERG Guide Category:	FLAMMABLE LIQUIDS (POLAR/WATER-MISCIBLE)
Molecular Formula:	C_3H_6O
RTECS No.:	TZ2975000
Properties:	Extremely flammable, clear, colorless liquid; a gas above 94° (34°C). Sweet, benzene or etherlike odor. Soluble in water.

HEALTH AND SAFETY INFORMATION

Corrosive to eyes, skin, and respiratory tract; skin and eye contact can cause severe burns and blindness. Inhalation may cause pulmonary edema, which can be delayed for several hours; there is a risk of death in serious cases. Skin disorders may be caused by prolonged or repeated skin contact.

IARC Evaluation: Group 2A, probable carcinogen; inadequate human evidence; sufficient animal evidence.
NTP: Anticipated carcinogen.
IDLH: 400 ppm; carcinogen.
Odor Threshold: 33.1 ppm
OSHA Table Z-1-A Air Contaminant

Exposure Limits:	**TWA**
ACGIH: (1)	20 ppm; 48 mg/m^3
NIOSH: (2)	
OSHA:	100 ppm; 240 mg/m^3

(1) Animal carcinogen.
(2) Potential carcinogen. Reduce occupational exposures to the lowest feasible concentrations.

RECOMMENDATIONS FOR RESPIRATOR SELECTION

At any concentrations above the NIOSH REL, or where there is no REL, at any detectable concentration: SCBAF:PD,PP (any self-contained breathing apparatus that has a full facepiece and is operated in a pressure-demand or other positive-pressure mode); or SAF:PD,PP:ASCBA (any supplied-air respirator that has a full facepiece and is operated in a pressure-demand or other positive-pressure mode in combination with an auxiliary, self-contained breathing apparatus operated in a pressure-demand or other positive pressure mode).

Escape: GMFOV [any air-purifying, full-facepiece respirator (gas mask) with a chin-style, front- or back-mounted organic vapor canister]; or SCBAE (any appropriate escape-type, self-contained breathing apparatus).

CHEMICAL REACTIVITY

Reacts with water. Polymerization may occur due to high temperatures or contamination with alkalis, aqueous acids, amines, and acidic alcohols. Incompatible with anhydrous metal chlorides, iron, strong acids, caustics, peroxides, salts, and combustible materials.

FIRE INFORMATION

Firefighting gear (including SCBA) does not provide adequate protection. If exposure occurs, remove and isolate gear immediately and thoroughly decontaminate personnel. Vapors are heavier than air and will collect in low areas. Vapors in confined areas may explode when exposed to fire. Vapors may travel long distances to ignition sources and flash back. Storage containers and parts of containers may rocket great distances, in many directions. If material or contaminated runoff enters waterways, notify downstream users of potentially contaminated waters. Notify local health and fire officials and pollution control agencies. Use dry chemicals, foam, carbon dioxide. From a secure, explosion-proof location, use water spray to cool exposed containers. May self-react, blocking relief valves and leading to tank explosion.

NFPA Hazard Classifications:

Health Hazards (Blue) 3
Flammability (Red) 4
Reactivity (Yellow) 2

Flash Point: −35°F (−37°C); 73°F (23°C) 1% aqueous solution.
Flammable Limits in Air: LEL 2.3%; UEL 36%
Autoignition Temperature: 840°F (449°C)

FIRST AID

Move victim to fresh air. Call emergency medical care. Apply artificial respiration if victim is not breathing. If breathing is difficult, give oxygen. Remove and isolate contaminated clothing and shoes. In case of contact with substance, immediately flush skin or eyes with running water for at least 20 minutes. Wash skin with soap and water. Keep victim warm and quiet. Ensure that medical personnel are aware of the material(s) involved and take precautions to protect themselves.

ENVIRONMENTAL INFORMATION

Clean Air Act: Hazardous Air Pollutants (Title I, Part A, Section 112); Accidental Release Prevention/Flammable substances, (Section 112[r], Table 3), TQ = 10,000 lb (4,540 kg).
Clean Water Act: Section 311 Hazardous Substances/RQ (same as CERCLA); Section 313 Priority Chemicals.
EPCRA Section 302 Extremely Hazardous Substances: TPQ = 10,000 lb (4,540 kg).
EPCRA Section 304: Reportable Quantity (RQ): CERCLA, 100 lb (45.4 kg).
EPCRA Section 313: Form R *de minimis* concentration reporting level: 0.1%.
California's Proposition 65: Carcinogen.
TSCA: 40CFR799.18

n–PROPYLTRICHLOROSILANE P:733

SYNONYMS: EINECS No. 205-489-6; PROPYLTRICHLOROSILANE; SILANE, TRICHLOROPROPYL-; TRICHLOROPROPYLSILANE

IDENTIFICATION

CAS:	141-57-1
ERG ID:	UN 1816
ERG Guide:	155
ERG Guide Category:	SUBSTANCES-TOXIC AND/OR CORROSIVE (FLAMMABLE/WATER-SENSITIVE)
Molecular Formula:	$C_3H_7Cl_3Si$
RTECS No.:	VV5300000
Properties:	Flammable and corrosive liquid. Colorless. Insoluble in water; decomposes, forming toxic and corrosive hydrogen chloride fumes. Shipped packaged under nitrogen gas blanket.

HEALTH AND SAFETY INFORMATION

Corrosive to eyes, skin, and respiratory tract; skin and eye contact causes severe burns and blindness. May be harmful if absorbed through the skin. Inhalation may cause pulmonary edema, which can be delayed for several hours; there is a risk of death in serious cases.

RECOMMENDATIONS FOR RESPIRATOR SELECTION

At any concentrations above the NIOSH REL, or where there is no REL, at any detectable concentration: SCBAF:PD,PP (any self-contained breathing apparatus that has a full facepiece and is operated in a pressure-demand or other positive-pressure mode); or SAF:PD,PP:ASCBA (any supplied-air respirator that has a full facepiece and is operated in a pressure-demand or other positive-pressure mode in combination with an auxiliary, self-contained breathing apparatus operated in a pressure-demand or other positive-pressure mode).

Escape: GMFOV [any air-purifying, full-facepiece respirator (gas mask) with a chin-style, front- or back-mounted organic vapor canister];or SCBAE (any appropriate escape-type, self-contained breathing apparatus).

CHEMICAL REACTIVITY

Reacts with water, forming hydrogen chloride. Contact with oxidizers can cause fire and explosions.

FIRE INFORMATION

Combustion products include hydrogen chloride and other toxic fumes of chlorides. Vapors are heavier than air and will collect in low areas. Vapors in confined areas may explode when exposed to fire. Vapors may travel long distances to ignition sources and flash back. Storage containers and parts of containers may rocket great distances, in many directions. If material or contaminated runoff enters waterways, notify downstream users of potentially contaminated waters. Notify local health and fire officials and pollution control agencies. Use dry chemicals, foam, carbon dioxide, or water spray. From a secure, explosion-proof location, use water spray to cool exposed containers.

NFPA Hazard Classifications

Health Hazards (Blue) 3
Flammability (Red) 3
Reactivity (Yellow) 1

Flash Point: 98°F (37°C) o.c.

FIRST AID

Move victim to fresh air; call emergency medical care. Apply artificial respiration if victim is not breathing. Do not use mouth-to-mouth method if victim ingested or inhaled the substance; induce artificial respiration with the aid of a pocket mask equipped with a one-way valve or other proper respiratory medical device. Administer oxygen if breathing is difficult. Remove and isolate contaminated clothing and shoes. In case of contact with substance, immediately flush skin or eyes with running water for at least 20 minutes. For minor skin contact, avoid spreading material on unaffected skin. Keep victim warm and quiet. Effects of exposure (inhalation, ingestion, or skin contact) to substance may be delayed. Ensure that medical personnel are aware of the material(s) involved and take precautions to protect themselves.

PSEUDOPINENE P:737

SYNONYMS: 6,6-DIMETHYL-2-METHYLENEBICYCLO(3.1.1)HEPTANE; EINECS No. 204-872-5; FEMA No. 2903; NOPINEN; NOPINENE; β-PINENE; 2(10)-PINENE; PSEUDOPINEN

IDENTIFICATION

CAS:	127-91-3
ERG ID:	UN 1993
ERG Guide:	128
ERG Guide Category:	FLAMMABLE LIQUIDS (NONPOLAR/WATER-IMMISCIBLE)
Molecular Formula:	$C_{10}H_{16}$
RTECS No.:	DT5077000
Properties:	Flammable liquid. Colorless. Pine odor. Floats on water.

RECOMMENDATIONS FOR RESPIRATOR SELECTION

Irritates eyes; moderate skin irritant.

CHEMICAL REACTIVITY

Forms explosive mixture with air. Keep away from heat, sparks ,and open flame. Contact with oxidizers can cause fire and explosions. Incompatible with nitric acid.

FIRE INFORMATION

Vapors are heavier than air and will collect in low areas. Vapors in confined areas may explode when exposed to fire. Vapors may travel long distances to ignition sources and flash back. Storage containers and parts of containers may rocket great distances, in many directions. If material or contaminated runoff enters waterways, notify downstream users of potentially contaminated waters. Notify local health and fire officials and pollution control agencies. Use dry chemicals, foam, carbon dioxide. From a secure, explosion-proof location, use water spray to cool exposed containers.

NFPA Hazard Classifications: Not listed

Flash Point: 88°F (31°C)

FIRST AID

Move victim to fresh air. Call emergency medical care. Apply artificial respiration if victim is not breathing. If breathing is difficult, give oxygen. Remove and isolate contaminated clothing and shoes. In case of contact with substance, immediately flush skin or eyes with running water for at least 20 minutes. Wash skin with soap and water. Keep victim warm and quiet. Ensure that medical personnel are aware of the material(s) involved and take precautions to protect themselves.

PYRIDINE P:740

SYNONYMS: AZABENZENE; AZINE; EEC No. 613-002-00-7; EINECS No. 203-809-9; NCI-C55301; PIRIDINA (Spanish); PYRIDIN (German); PRINDINA (Italian); PIRYDYNA (Polish)

IDENTIFICATION

CAS:	110-86-1
ERG ID:	UN 1281
ERG Guide:	129
ERG Guide Category:	FLAMMABLE LIQUIDS (POLAR/WATER-MISCIBLE/NOXIOUS)
Molecular Formula:	C_5H_5N
RTECS No.:	UR8400000
Properties:	Highly flammable, colorless to light yellow liquid. Sharp, nauseating, fishlike odor. Mixes with water; initially floats. Poisonous, flammable vapors of cyanides, nitrogen oxides, carbon dioxide, and carbon monoxide may be produced.

HEALTH AND SAFETY INFORMATION

Irritates eyes, skin, and respiratory tract. High concentrations can cause unconsciousness. Repeated exposure, even to low levels, may cause liver and brain damage. Long-term exposure may cause loss of appetite; nausea; fatigue; depression; bluish tint to the skin; liver, kidney, central nervous system damage; and death. Can cause the skin of exposed persons to become hypersensitive to sunlight.

IDLH: 1,000 ppm
Odor Threshold: 0.085 ppm
OSHA Table Z-1-A Air Contaminant

Exposure Limits: TWA

ACGIH:	5 ppm; 16 mg/m^3
NIOSH:	5 ppm; 15 mg/m^3
OSHA:	5 ppm; 15 mg/m^3

RECOMMENDATIONS FOR RESPIRATOR SELECTION: NIOSH/OSHA

125 ppm: SA:CF (any supplied-air respirator operated in a continuous-flow mode); or PAPROV [any powered, air-purifying respirator with organic vapor cartridge(s)]
250 ppm: CCRFOV [any air-purifying, full-facepiece respirator (gas mask) with a chin-style, front- or back-mounted acid gas canister]; or GMFOV [any air-purifying, full-facepiece respirator (gas mask) with a chin-style, front- or back-mounted organic vapor canister]; or PAPRTOV [any powered, air-purifying respirator with a tight-fitting facepiece and organic vapor cartridge(s)]; or SCBAF (any self-contained breathing apparatus with a full facepiece); or SAF (any supplied-air respirator with a full facepiece)
1,000 ppm: SAF:PD,PP (any supplied-air respirator that has a full facepiece and is operated in a pressure-demand or other positive-pressure mode).

Emergency or Planned Entry into Unknown Concentrations or IDLH Conditions:
SCBAF:PD,PP (any self-contained breathing apparatus that has a full facepiece and is operated in a pressure-demand or other positive-pressure mode); or SAF:PD,PP:ASCBA (any supplied-air respirator that has a full facepiece and is operated in a pressure-demand or other positive-pressure mode in combination with an auxiliary, self-contained breathing apparatus operated in a pressure-demand or other positive-pressure mode).

Escape: GMFOV[any air-purifying, full-facepiece respirator (gas mask) with a chin-style, front- or back-mounted organic vapor canister]; or SCBAE (any appropriate escape-type, self-contained breathing apparatus).

Note: Substance causes eye irritation or damage; eye protection needed.

CHEMICAL REACTIVITY

Forms explosive mixture with air. May react violently with chlorosulfonic acid, chromium oxide, maleic anhydride, nitric acid, oleum, a-propiolactone, sulfuric acid, silver perchlorate, and oxidizers such as permanganates. Strong acids may cause violent spattering. Attacks some plastics, rubber, and coatings.

FIRE INFORMATION

Combustion products include toxic nitrogen oxides. Vapors are heavier than air and will collect in low areas. Vapors in confined areas may explode when exposed to fire. Vapors may travel long distances to ignition sources and flash back. Storage containers and parts of containers may rocket great distances, in many directions. If material or contaminated runoff enters waterways, notify downstream users of potentially contaminated waters. Notify local health and fire officials and pollution control agencies. Use dry chemicals, alcohol resistant (AFFF) foam, carbon dioxide, large quantities of water to disperse vapors [contain runoff]. From a secure, explosion-proof location, use water spray to cool exposed containers.

NFPA Hazard Classifications:

Health Hazards (Blue) 3
Flammability (Red) 3
Reactivity (Yellow) 0

Flash Point: Liquid ignites at 68°F (20°C) c.c.
Flammable Limits in Air: LEL 1.8%; UEL 12.4%
Autoignition Temperature: 900°F (482°C)

FIRST AID

Move victim to fresh air. Call emergency medical care. Apply artificial respiration if victim is not breathing. Administer oxygen if breathing is difficult. Remove and isolate contaminated clothing and shoes. In case of contact with substance, immediately flush skin or eyes with running water for at least 20 minutes. Wash skin with soap and water. Keep victim warm and quiet. Effects of exposure (inhalation, ingestion, or skin contact) to substance may be delayed. Ensure that medical personnel are aware of the material(s) involved and take precautions to protect themselves.

ENVIRONMENTAL INFORMATION

EPA Hazardous Waste Number (RCRA No.): U196, DO38.
RCRA Section 261 Hazardous Constituents
RCRA Toxicity Characteristic (Section 261.24): Maximum Concentration of Contaminants, regulatory level, 5.0 mg/L.
RCRA Land Ban Waste
RCRA Universal Treatment Standards: Wastewater (mg/L), 0.014; Nonwastewater (mg/kg), 16.
RCRA Ground Water Monitoring List: Suggested test method(s) (PQL ug/L): 8240(5); 8270(10).
EPCRA Section 304: Reportable Quantity (RQ): CERCLA, 1,000 lb (454 kg).
EPCRA Section 313: Form R *de minimis* concentration reporting level: 1.0%.
TSCA: 40CFR716.120(*a*)

QUINOLINE Q:500

SYNONYMS: 1-AZANAPHTHALENE; 1-BENZAZINE; 1-BENZINE; BENZO(B) PYRIDINE;

CHINOLEINE; CHINOLIN (Czech); CHINOLINE; EINECS No. 202-051-6; LEUCOL; LEUCOLINE; LEUKOL; QUINOLEINA (Spanish); USAF EK-218

IDENTIFICATION

CAS:	91-22-5
ERG ID:	UN 2656
ERG Guide:	154
ERG Guide Category:	SUBSTANCES-TOXIC AND/OR CORROSIVE (NONCOMBUSTIBLE)
Molecular Formula:	C_9H_7N
RTECS No.:	VA9275000
Properties:	Combustible liquid. Colorless to brown. Strong, peculiar, unpleasant odor. Sinks in water; insoluble

HEALTH AND SAFETY INFORMATION

Irritates eyes, skin, and respiratory tract. Passes through the skin; can increase severity of exposure. Single high eye exposure can cause permanent eye damage. Inhalation may cause pulmonary edema, which can be delayed for several hours; there is a risk of death in serious cases.

Odor Threshold: 0.015 ppm

RECOMMENDATIONS FOR RESPIRATOR SELECTION

At any concentrations above the NIOSH REL, or where there is no REL, at any detectable concentration: SCBAF:PD,PP (any self-contained breathing apparatus that has a full facepiece and is operated in a pressure-demand or other positive-pressure mode); or SAF:PD,PP:ASCBA (any supplied-air respirator that has a full facepiece and is operated in a pressure-demand or other positive-pressure mode in combination with an auxiliary, self-contained breathing apparatus operated in a pressure-demand or other positive-pressure mode).

Escape: GMFOV [any air-purifying, full-facepiece respirator (gas mask) with a chin-style, front- or back-mounted organic vapor canister]; or SCBAE (any appropriate escape-type, self-contained breathing apparatus).

CHEMICAL REACTIVITY

Contact with strong oxidizers may cause fire and explosions. May attack some forms of plastics.

FIRE INFORMATION

Combustion products include toxic nitrogen oxides. Vapors are heavier than air and will collect in low areas. Vapors in confined areas may explode when exposed to fire. If material or contaminated runoff enters waterways, notify downstream users of potentially contaminated waters. Notify local health and fire officials and pollution control agencies. Use dry chemicals, alcohol-resistant foam, carbon dioxide. From a secure, explosion-proof location, use water spray to cool exposed containers.

NFPA Hazard Classifications:

Health Hazards (Blue) 2
Flammability (Red) 1
Reactivity (Yellow) 0

Flash Point: 225°F (107°C) c.c.
Autoignition Temperature: 896°F (480°C)

FIRST AID

Move victim to fresh air; call emergency medical care. Apply artificial respiration if victim is not breathing. Do not use mouth-to-mouth method if victim ingested or inhaled the substance; induce artificial respiration with the aid of a pocket mask equipped with a one-way valve or other proper respiratory medical device. Administer oxygen if breathing is difficult. Remove and isolate contaminated clothing and shoes. In case of contact with substance, immediately flush skin or eyes with running water for at least 20 minutes. For minor skin contact, avoid spreading material on unaffected skin. Keep victim warm and quiet. Effects of exposure (inhalation, ingestion, or skin contact) to substance may be delayed. Ensure that medical personnel are aware of the material(s) involved and take precautions to protect themselves.

ENVIRONMENTAL INFORMATION

Clean Air Act: Hazardous Air Pollutants (Title I, Part A, Section 112)
Clean Water Act: Section 311 Hazardous Substances/RQ (same as CERCLA); Section 313 Priority Chemicals
EPCRA Section 304 Reportable Quantity (RQ): CERCLA, 5,000 lb (2,270 kg)
EPCRA Section 313: Form R *de minimis* concentration reporting level: 1.0%.

RESORCINOL R:100

SYNONYMS: BENZENE, M-DIHYDROXY-; m-BENZENEDIOL; 1,3-BENZENEDIOL; C.I. 76505; C.I. DEVELOPER 4; C.I. OXIDATION BASE 31; DEVELOPER O; DEVELOPER R; DEVELOPER RS; m-DIHYDROXYBENZENE; 1,3-DIHYDROXYBENZENE; m-DIOXYBENZENE; DURAFUR DEVELOPER G; EINECS No. 203-585-2; FOURAMINE RS; FOURRINE 79; FOURRINE EW; m-HYDROQUINONE; 3-HYDROXYCYCLOHEXADIEN-1-ONE; m-HYDROXYPHENOL; 3-HYDROXYPHENOL; NAKO TGG; NCI-C05970; PELAGOL GREY RS; PELAGOL RS; PHENOL, M-HYDROXY-; RESORCIN; RESORCINA (Spanish); RESORCINE; STCC 4966774

IDENTIFICATION

CAS:	108-46-3
ERG ID:	UN 2876
ERG Guide:	153
ERG Guide Category:	SUBSTANCES-TOXIC AND/OR CORROSIVE

	(COMBUSTIBLE)
Molecular Formula:	$C_6H_6O_2$
RTECS No.:	VG9625000
Properties:	Combustible liquid. Hygroscopic white needles, plates, crystals, flakes, or powder; turns pink on exposure to air or light, or contact with iron. Faint, characteristic odor. Water soluble.

HEALTH AND SAFETY INFORMATION

Irritates eyes, skin, and respiratory tract. Skin contact contributes significantly to overall exposure; can affect the blood's ability to carry oxygen. Long-term or repeated exposure can lead to skin, blood, spleen, liver, kidney, and heart damage.

IARC Evaluation: Group 3, unclassifiable substance; no human data; inadequate animal evidence.
OSHA Table Z-1-A Air Contaminant

Exposure Limits:	**TWA**	**STEL**
ACGIH: (1)	10 ppm; 45 mg/m^3	20 ppm; 90 mg/m^3
NIOSH:	10 ppm; 45 mg/m^3	20 ppm; 90 mg/m^3
OSHA:	None	

(1) Not suspected as a human carcinogen.

RECOMMENDATIONS FOR RESPIRATOR SELECTION

At any concentrations above the NIOSH REL, or where there is no REL, at any detectable concentration: SCBAF:PD,PP (any self-contained breathing apparatus that has a full facepiece and is operated in a pressure-demand or other positive-pressure mode); or SAF:PD,PP:ASCBA (any supplied-air respirator that has a full facepiece and is operated in a pressure-demand or other positive-pressure mode in combination with an auxiliary, self-contained breathing apparatus operated in a pressure-demand or other positive-pressure mode).

Escape: GMFOV [any air-purifying, full-facepiece respirator (gas mask) with a chin-style, front- or back-mounted organic vapor canister]; or SCBAE (any appropriate escape-type, self-contained breathing apparatus).

CHEMICAL REACTIVITY

Hygroscopic; absorbs moisture from air. Reacts violently with strong oxidizers or nitric acid. Incompatible with acetanilide, albumin, alkalis, antipyrine, camphor, ferric salts, menthol, spirit nitrous ether, acids, and bases. May accumulate static electrical charges, and may cause ignition of its finely dispersed particles.

FIRE INFORMATION

Combustible liquid; but difficult to ignite. Vapors are heavier than air and will collect in low areas. Vapors in confined areas may explode when exposed to fire. Vapors may travel long distances to ignition sources and flash back. Storage containers and parts of containers may rocket great distances, in many directions. If material or contaminated runoff enters waterways, notify downstream users of potentially contaminated waters. Notify local health and fire officials and pollution control agencies. Use dry chemicals, water, carbon dioxide. From a secure, explosion-proof location, use water spray to cool exposed containers.

NFPA Hazard Classifications:

Health Hazards (Blue) 2
Flammability (Red) 1
Reactivity (Yellow) 0

Flash Point: 261°F (127°C)
Flammable Limits in Air: LEL 1.4% at 392°F (200°C); UEL N/A
Autoignition Temperature: 1,126°F (608°C)

FIRST AID

Move victim to fresh air; call emergency medical care. Apply artificial respiration if victim is not breathing. Do not use mouth-to-mouth method if victim ingested or inhaled the substance; induce artificial respiration with the aid of a pocket mask equipped with a one-way valve or other proper respiratory medical device. Administer oxygen if breathing is difficult. Remove and isolate contaminated clothing and shoes. In case of contact with substance, immediately flush skin or eyes with running water for at least 20 minutes. For minor skin contact, avoid spreading material on unaffected skin. Keep victim warm and quiet. Effects of exposure (inhalation, ingestion, or skin contact) to substance may be delayed. Ensure that medical personnel are aware of the material(s) involved and take precautions to protect themselves.

ENVIRONMENTAL INFORMATION

Clean Water Act: Section 311 Hazardous Substances/RQ (same as CERCLA).
EPA Hazardous Waste Number (RCRA No.): U201.
RCRA Section 261 Hazardous Constituents
EPCRA Section 304: Reportable Quantity (RQ): CERCLA, 5,000 lb (2,270 kg).

RESORCINOL DIGLYCIDYL ETHER — R:150

SYNONYMS: ARALDITE ERE 1359; m-BIS(2,3-EPOXYPROPOXY)BENZENE; 1,3-BIS(2,3-EPOXYPROPOXY)BENZENE; m-BIS(GLYCIDYLOXY)BENZENE; DGRE; 1,3-DIGLYCIDYLOXYBENZENE; DIGLYCIDYLOXYBENZENE; DIGLYCIDYL RESORCINOL ETHER; ERE 1359; NCI-C54966; OXIRANE, 2,2'-(1,3-PHENYLENEBIS(OXYMETHYLENE)BIS-; 2,2'-[1,3-PHENYLENEBIS(OXYMETHYLENE)]BISOXIRANE; RDGE; RESORCINOL BIS(2,3-EPOXYPROPYL)ETHER

IDENTIFICATION

CAS: 101-90-6

ERG ID: UN 2929 as toxic liquid, flammable, organic, n.o.s.
ERG Guide: 131
ERG Guide Category: FLAMMABLE LIQUIDS-TOXIC
Molecular Formula: $C_{12}H_{14}O_4$
RTECS No.: VH1050000
Properties: Combustible liquid. Straw yellow.

HEALTH AND SAFETY INFORMATION

Irritates skin and eyes. A carcinogen; limit exposure to lowest feasible concentration.

IARC Evaluation: Group 2B; sufficient animal evidence.
NTP: Confirmed carcinogen

RECOMMENDATIONS FOR RESPIRATOR SELECTION

At any concentrations above the NIOSH REL, or where there is no REL, at any detectable concentration: SCBAF:PD,PP (any self-contained breathing apparatus that has a full facepiece and is operated in a pressure-demand or other positive-pressure mode); or SAF:PD,PP:ASCBA (any supplied-air respirator that has a full facepiece and is operated in a pressure-demand or other positive-pressure mode in combination with an auxiliary, self-contained breathing apparatus operated in a pressure-demand or other positive-pressure mode).

Escape: GMFOV [any air-purifying, full-facepiece respirator (gas mask) with a chin-style, front- or back-mounted organic vapor canister]; or SCBAE (any appropriate escape-type, self-contained breathing apparatus).

CHEMICAL REACTIVITY

Contact with strong oxidizers may cause fire and explosions. Incompatible with sulfuric and nitric acid. May form unstable and potentially explosive peroxides when exposed to air.

FIRE INFORMATION

If material or contaminated runoff enters waterways, notify downstream users of potentially contaminated waters. Notify local health and fire officials and pollution control agencies. Use dry chemicals, foam, carbon dioxide. From a secure, explosion-proof location, use water spray to cool exposed containers. If cooling streams are ineffective (venting sound increases in volume and pitch, tank discolors, or shows any signs of deforming), withdraw immediately to a secure position.

NFPA Hazard Classifications: Not listed

Flash Point: 350°F (176°C) o.c.

FIRST AID

Move victim to fresh air. Call emergency medical care. Apply artificial respiration if victim is not breathing.

Do not use mouth-to-mouth method if victim ingested or inhaled the substance; induce artificial respiration with the aid of a pocket mask equipped with a one-way valve or other proper respiratory medical device. Administer oxygen if breathing is difficult. Remove and isolate contaminated clothing and shoes. In case of contact with substance, immediately flush skin or eyes with running water for at least 20 minutes. Wash skin with soap and water. Keep victim warm and quiet. Effects of exposure (inhalation, ingestion, or skin contact) to substance may be delayed. Ensure that medical personnel are aware of the material(s) involved and take precautions to protect themselves.

ENVIRONMENTAL INFORMATION

EPCRA Section 313: Form R *de minimis* concentration reporting level: 0.1%.
California's Proposition 65: Carcinogen.

SEBACIC ACID S:030

SYNONYMS: ACIDO SEBACICO (Spanish); DECANEDIOIC ACID; EINECS No. 203-845-5; 1,8-OCTANEDICARBOXYLIC ACID; SEBACINSAEURE (German); SEBACINEZUUR (Dutch); SEBACYLIC ACID; USAF HC-1

IDENTIFICATION

CAS:	111-20-6
Molecular Formula:	$C_{10}H_{18}O_4$
RTECS No.:	VS0875000
Properties:	Combustible powder, tablets, or leaflets. White. Mild, fatty-acid odor. Slightly soluble in water.

HEALTH AND SAFETY INFORMATION

Mildly toxic if swallowed. Due to the availability of insufficient data on human effects, caution should be exercised.

CHEMICAL REACTIVITY

Reacts with oxidizing agents, bases, and reducing agents. Organic acids are incompatible with sulfuric acid, caustics, ammonia, aliphatic amines, alkanolamines, isocyanates, alkylene oxides, epichlorohydrin.

FIRE INFORMATION

Combustion may produce irritants and toxic vapors. Vapors in confined areas may explode when exposed to fire. If material or contaminated runoff enters waterways, notify downstream users of potentially contaminated waters. Notify local health and fire officials and pollution control agencies. Use dry chemicals, alcohol-resistant foam, carbon dioxide. From a secure, explosion-proof location, use water spray to cool exposed containers.

NFPA Hazard Classifications: Not listed

FIRST AID

Move victim to fresh air. Call emergency medical care. Apply artificial respiration if victim is not breathing. Administer oxygen if breathing is difficult. Remove and isolate contaminated clothing and shoes. In case of contact with substance, immediately flush skin or eyes with running water for at least 20 minutes. Removal of solidified molten material from skin requires medical assistance. Keep victim warm and quiet. Ensure that medical personnel are aware of the material(s) involved and take precautions to protect themselves.

SEBACONITRILE S:037

SYNONYMS: DECANEDINITRILE; 1,8-DICYANOOCTANE; EINECS No. 217-494-0; OCTAMETHYLENE DICYANIDE; OKTAMETHYLENDIKYANID (German)

IDENTIFICATION

CAS:	1871-96-1
ERG ID:	UN 3276 as nitriles, toxic, n.o.s.
ERG Guide:	151
ERG Guide Category:	SUBSTANCES-TOXIC (NONCOMBUSTIBLE)
Molecular Formula:	$C_{10}H_{16}N_2$
RTECS No.:	VS1500000
Properties:	Oily liquid. Light yellow.

HEALTH AND SAFETY INFORMATION

Many nitriles have a cyanide-like effects; can be extremely poisonous to humans by most routes of entry, especially by ingestion and inhalation.

IDLH: 25 mg/m^3 as cyanide
NIOSH Criteria Document: NIOSH 78-212 NITRILES

Exposure Limits:	**TWA**	**CEILING**
ACGIH: [1]	5 mg/m^3 as cyanide	
NIOSH: [1]		4.7 ppm; 5 mg/m^3 per 10 minutes as cyanides
OSHA: [1]	5 mg/m^3 as cyanides	

[1] Skin contact contributes significantly in overall exposure

RECOMMENDATIONS FOR RESPIRATOR SELECTION

At any concentrations above the NIOSH REL, or where there is no REL, at any detectable concentration: SCBAF:PD,PP (any self-contained breathing apparatus that has a full facepiece and is operated in a pressure-demand or other positive-pressure mode); or SAF:PD,PP:ASCBA (any supplied-air respirator that has a full facepiece and is operated in a pressure-demand or other positive-pressure mode in combination with an auxiliary, self-contained breathing apparatus operated in a pressure-demand or other positive-pressure mode).

Escape: GMFOV [any air-purifying, full-facepiece respirator (gas mask) with a chin-style, front- or back-mounted organic vapor canister]; or SCBAE (any appropriate escape-type, self-contained breathing apparatus).

CHEMICAL REACTIVITY

Nitriles may react with acids, acid fumes emitting cyanide fumes. Incompatible with chlorates (potassium chlorate, sodium chlorate), fluorine, mercurous chloride, sodium nitrite.

FIRE INFORMATION

Combustion products include toxic nitrogen oxides and cyanide. If material or contaminated runoff enters waterways, notify downstream users of potentially contaminated waters. Notify local health and fire officials and pollution control agencies. Use dry chemicals, alcohol-resistant foam, carbon dioxide.

NFPA Hazard Classifications: Not listed

FIRST AID

Move victim to fresh air. Call emergency medical care. Apply artificial respiration if victim is not breathing. Do not use mouth-to-mouth method if victim ingested or inhaled the substance; induce artificial respiration with the aid of a pocket mask equipped with a one-way valve or other proper respiratory medical device. Administer oxygen if breathing is difficult. Remove and isolate contaminated clothing and shoes. In case of contact with substance, immediately flush skin or eyes with running water for at least 20 minutes. For minor skin contact, avoid spreading material on unaffected skin. Keep victim warm and quiet. Effects of exposure (inhalation, ingestion, or skin contact) may be delayed. Ensure that medical personnel are aware of the material(s) involved and take precautions to protect themselves.

ENVIRONMENTAL INFORMATION as cyanide compounds

CLEAN AIR ACT: Hazardous Air Pollutants (Title I, Part A, Section 112)
CLEAN WATER ACT: Section 307 Priority Pollutants as cyanide, total.
EPA HAZARDOUS WASTE NUMBER (RCRA No.): P030 as cyanides soluble salts and complexes, n.o.s.
RCRA Section 261 Hazardous Constituents. as cyanides, soluble salts and complexes, n.o.s.

EPCRA (Section 313): X+CN- where X = H+ or any other group where a formal dissociation may occur. For example, KCN or $Ca(CN)_2$. Form R deminimus concentration reporting level: 1.0%. Form R Toxic Chemical Category Code: N016
MARINE POLLUTANT (49CFR, Subchapter 172.101, Appendix B) as cyanide mixtures

SELENIUM S:020

SYNONYMS: C.I. 77805; COLLOIDAL SELENIUM; ELEMENTAL SELENIUM; SELEN (Polish); SELENATE; SELENIO (Spanish); SELENIUM ALLOY; SELENIUM BASE; SELENIUM (COLLODIAL); SELENIUM DUST; SELENIUM ELEMENT; SELENIUM HOMOPOLYMER; SELENIUM POWDER; VANDEX

IDENTIFICATION

CAS:	7782-49-2
ERG ID:	UN 2658 (powder)
ERG Guide:	152
ERG Guide Category:	SUBSTANCES-TOXIC (COMBUSTIBLE)
Molecular Formula:	Se
RTECS No.:	VS7700000
Properties:	Combustible amorphous powder or crystalline solid. Red to brownish red (powder); grey (crystals). Insoluble in water.

HEALTH AND SAFETY INFORMATION

Dust and concentrated mist irritates eyes, skin, and respiratory system. May cause liver, kidney spleen, blood problems.

IARC Evaluation: Group 3, unclassifiable substance; inadequate human evidence; inadequate animal evidence.
IDLH: 1 mg/m^3
OSHA Table Z-1-A Air Contaminant

Exposure Limits:	**TWA**
ACGIH:	0.05 ppm; 0.16 mg/m^3
NIOSH:	0.05 ppm; 0.2 mg/m^3
OSHA:	0.05 ppm; 0.2 mg/m^3

RECOMMENDATIONS FOR RESPIRATOR SELECTION: NIOSH/OSHA

1 mg/m^3: DM if not present as a fume (any dust and mist respirator); or DMFu (any dust, mist, and fume respirator); or HiEF (any air-purifying, full-facepiece respirator with a high-efficiency particulate filter); or PAPRDM if not present as a fume (any powered, air-purifying respirator with a dust and mist filter); or

PAPRDMFu (any powered, air-purifying respirator with a dust, mist, and fume filter); or SA (any supplied-air respirator); or SCBAF (any self-contained breathing apparatus with a full facepiece); or SAF (any supplied-air respirator with a full facepiece).

Emergency or Planned Entry into Unknown Concentrations or IDLH Conditions: SCBAF:PD,PP (any self-contained breathing apparatus that has a full facepiece and is operated in a pressure-demand or other positive-pressure mode); or SAF:PD,PP:ASCBA (any supplied-air respirator that has a full facepiece and is operated in a pressure-demand or other positive-pressure mode in combination with an auxiliary, self-contained breathing apparatus operated in a pressure-demand or other positive-pressure mode).

Escape: HiEF (any air-purifying, full-facepiece respirator with a high-efficiency particulate filter); or SCBAE (any appropriate escape-type, self-contained breathing apparatus).

Note: Substance reported to cause eye irritation or damage; may require eye protection.

CHEMICAL REACTIVITY

Reacts violently with many materials including strong acids such as nitric acid, chlorates, metals, barium carbide, calcium carbide, cadmium, chromium trioxide, fluorine, lithium silicide, nickel, potassium, potassium bromate, sodium, strontium carbide, uranium, and zinc. Contact with strong oxidizers may cause fire and explosions.

FIRE INFORMATION

Combustible solid. If material or contaminated runoff enters waterways, notify downstream users of potentially contaminated waters. Notify local health and fire officials and pollution control agencies. Use dry chemicals, alcohol-resistant foam, carbon dioxide. From a secure, explosion-proof location, use water spray to cool exposed containers.

FEMA Hazard Classifications: (powder)

Health Hazards (Blue) 1
Flammability (Red) 1
Reactivity (Yellow) 1

FIRST AID

Move victim to fresh air. Call emergency medical care. Apply artificial respiration if victim is not breathing. Do not use mouth-to-mouth method if victim ingested or inhaled the substance; induce artificial respiration with the aid of a pocket mask equipped with a one-way valve or other proper respiratory medical device. Administer oxygen if breathing is difficult. Remove and isolate contaminated clothing and shoes. In case of contact with substance, immediately flush skin or eyes with running water for at least 20 minutes. For minor skin contact, avoid spreading material on unaffected skin. Keep victim warm and quiet. Effects of exposure (inhalation, ingestion, or skin contact) may be delayed. Ensure that medical personnel are aware of the material(s) involved and take precautions to protect themselves.

ENVIRONMENTAL INFORMATION

Clean Water Act: Section 307 Priority Pollutants; Section 307 Toxic Pollutants; Section 313 Priority Chemicals.
EPA Hazardous Waste Number (RCRA No.): D010.
RCRA Section 261 Hazardous Constituents, waste number not listed.
RCRA Land Ban Waste
RCRA Toxicity Characteristic (Section 261.24): Maximum Concentration of Contaminants, regulatory level, 1.0 mg/L.
RCRA Universal Treatment Standards: Wastewater (mg/L), 0.82; Nonwastewater (mg/L), 0.16 TCLP.
RCRA Ground Water Monitoring List: Suggested test method(s) (PQL ug/L): (total) 6010(750); 7740(20); 7741(20).
Safe Drinking Water Act: MCL, 0.05 mg/L; MCLG, 0.05 mg/L; Regulated chemical (47 FR 9352).
EPCRA Section 304: Reportable Quantity (RQ): CERCLA, 100 lb (45.4 kg).
EPCRA Section 313: Form R *de minimis* concentration reporting level: 1.0%. Form R Toxic Chemical Category Code: N725.

SILICA, AMORPHOUS S:050

SYNONYMS: AMORPHOUS FUSED SILICA; DIATOMACEOUS EARTH; DIATOMACEOUS SILICA; DIATOMITE; EINECS No. 231-545-4 (fused); EINECS No. 262-373-8 (amorphous); QUARTZ GLASS; QUARTZ POWDER; SILICA AEROGEL; SILICA AMORPHOUS FUSED; SILICA GEL; SILICA, VITREOUS; SILICA XEROGEL; SILICIC ACID; SILICON DIOXIDE (AMORPHOUS); TIERRA de DIATOMEAS (Spanish); VITREOUS QUARTZ

IDENTIFICATION

CAS:	7631-86-9; 60676-86-0 (fused); 61790-53-2 (amorphous)
Molecular Formula:	O_2Si
RTECS No.:	VV7310000; VV7320000 (fused)
Properties:	Noncombustible, white to gray, powder. Odorless. Slightly soluble in water.

HEALTH AND SAFETY INFORMATION

Prolonged or repeated exposures to dust can cause silicosis, an irreversible lung disease that can cause death.

IRAC: Group 3, insufficient animal and human evidence.
IDLH: 3,000 mg/m^3
OSHA Table Z-1-A Air Contaminant
OSHA Specifically Regulated Substances (29 CFR 1910.1001-1048)
NIOSH Criteria Document available

Exposure Limits:	**TWA**
ACGIH:	0.1 mg/m^3 (silica fused) 10 mg/m^3 (diatomaceous earth, inhalable particulate) 3 mg/m^3 (diatomaceous earth, respirable particulate) 10 mg/m^3 (silica gel)
NIOSH:	6 mg/m^3
OSHA:	either 80 mg/m^3 divided by the value "%SiO2," or 20 mppcf.*

*Millions of particles per cubic foot of air, based on impinger samples counted by light-field techniques.

RECOMMENDATIONS FOR RESPIRATOR SELECTION: NIOSH

30 mg/m^3: DM (any dust and mist respirator).
60 mg/m^3: XSQ (any dust and mist respirator except single-use and quarter mask respirators); or SA (any supplied-air respirator).
150 mg/m^3: SA:CF (any supplied-air respirator operated in a continuous-flow mode); or PAPRDM (any powered, air-purifying respirator with a dust and mist filter).
300 mg/m^3: HiEF (any air-purifying, full-facepiece respirator with a high-efficiency particulate filter); or SAT:CF (any supplied-air respirator that has a tight-fitting facepiece and is operated in a continuous-flow mode); or PAPRTHiE (any powered, air-purifying respirator with a tight-fitting facepiece and a high-efficiency particulate filter); or SCBAF (any self-contained breathing apparatus with a full facepiece); or SAF (any supplied-air respirator with a full facepiece).
3,000 mg/m^3: SAF:PD,PP (any supplied-air respirator that has a full facepiece and is operated in a pressure-demand or other positive-pressure mode).

Emergency or Planned Entry into Unknown Concentrations or IDLH Conditions:
SCBAF:PD,PP (any self-contained breathing apparatus that has a full facepiece and is operated in a pressure-demand or other positive-pressure mode); or SAF:PD,PP:ASCBA (any supplied-air respirator that has a full facepiece and is operated in a pressure-demand or other positive-pressure mode in combination with an auxiliary, self-contained breathing apparatus operated in a pressure-demand or other positive-pressure mode).

Escape: HiEF (any air-purifying, full-facepiece respirator with a high-efficiency particulate filter); or SCBAE (any appropriate escape-type, self-contained breathing apparatus).

CHEMICAL REACTIVITY

Incompatible with strong oxidizers: fluorine, oxygen difluoride, manganese trioxide, chlorine trifluoride, etc.

FIRE INFORMATION

Noncombustible solid.

FIRST AID

EYE: If this chemical contacts the eyes, immediately wash them with large amounts of water, occasionally lifting the lower and upper lids. Get medical attention immediately. Contact lenses should not be worn when working with this chemical.
SKIN: If this chemical contacts the skin, immediately wash the contaminated skin with soap and water. If this chemical penetrates the clothing, immediately remove clothing, wash the skin with soap and water, and get medical attention promptly.
BREATH: If a person breathes large amounts of this chemical, move exposed person to fresh air at once. If breathing has stopped, perform mouth-to-mouth resuscitation. Keep the affected person warm and at rest. Get medical attention as soon as possible.
SWALLOW: If this chemical has been swallowed, get medical attention immediately.

ENVIRONMENTAL INFORMATION

Safe Drinking Water Act: 40CFR141.87 water quality parameter requirements.

SILICON S:080

SYNONYMS: EINECS No. 231-130-8; ELEMENTAL SILICON; LAMINATIONS, TRANSFORMER; SILICIO (Italian, Spanish); SILICIUM (Dutch, French, German)

IDENTIFICATION

CAS:	7440-21-3
ERG ID:	UN 1346
ERG Guide:	170 (amorphous powder)
ERG Guide Category:	METALS (POWDERS, DUST, SHAVINGS, BORINGS, TURNINGS, CUTTINGS, ETC.)
Molecular Formula:	Si
RTECS No.:	VW0400000
Properties:	Combustible in powder form. Lustrous black to gray needlelike crystals; amorphous form is a dark brown powder.

HEALTH AND SAFETY INFORMATION

Powder can cause skin and respiratory tract irritation. Silicosis may develop over many years; disability may occur.

OSHA Table Z-1-A Air Contaminant
OSHA Specifically Regulated Substances (29 CFR 1910.1001-1048)
NIOSH Criteria Document: NIOSH 75-120 (crystalline silica)

Exposure Limits: **TWA**

ACGIH: [1]	10 mg/m^3 of total dust
NIOSH:	10 mg/m^3 total; 5 mg/m^3 respirable fraction
OSHA:	15 mg/m^3 total; 5 mg/m^3 respirable fraction

[1] Total dust containing no asbestos and less than 1% crystalline silica.

RECOMMENDATIONS FOR RESPIRATOR SELECTION: NIOSH

At any concentrations above the NIOSH REL, or where there is no REL, at any detectable concentration: SCBAF:PD,PP (any self-contained breathing apparatus that has a full facepiece and is operated in a pressure-demand or other positive-pressure mode); or SAF:PD,PP:ASCBA (any supplied-air respirator that has a full facepiece and is operated in a pressure-demand or other positive-pressure mode in combination with an auxiliary, self-contained breathing apparatus operated in a pressure-demand or other positive-pressure mode).

Escape: GMFOV [any air-purifying, full-facepiece respirator (gas mask) with a chin-style, front- or back-mounted organic vapor canister]; or SCBAE (any appropriate escape-type, self-contained breathing apparatus).

CHEMICAL REACTIVITY

Contact with powerful oxidizers may cause fire and explosions. Reacts with chlorine, chlorine trifluoride, fluorine, calcium, cesium carbide, hydrogen peroxide, acetylene, alkaline carbonates.

FIRE INFORMATION

Flammable solid in powder form. If material or contaminated runoff enters waterways, notify downstream users of potentially contaminated waters. Notify local health and fire officials and pollution control agencies. Use water, dry chemicals, foam, carbon dioxide.

FEMA Hazard Classifications:

Health Hazards (Blue) 1
Flammability (Red) 2
Reactivity (Yellow) 0

FIRST AID

Move victim to fresh air. Call emergency medical care. Apply artificial respiration if victim is not breathing. Administer oxygen if breathing is difficult. Remove and isolate contaminated clothing and shoes. In case of contact with substance, immediately flush skin or eyes with running water for at least 20 minutes. Keep victim warm and quiet. Ensure that medical personnel are aware of the material(s) involved and take precautions to protect themselves.

SILVER NITRATE S:200

SYNONYMS: ARGENTI NITRAS; EINECS No. 231-853-9; LUNAR CAUSTIC; NITRATE d'ARGENT (French); NITRATO de PLATA (Spanish); NITRIC ACID, SILVER(1+) SALT; NITRIC ACID, SILVER(I) SALT; SILBERNITRAT (German); SILVER(1+) NITRATE; SILVER(I) NITRATE; SILBERNITRAT; STCC 4918742

IDENTIFICATION

CAS:	7761-88-8
ERG ID:	UN 1493
ERG Guide:	140
ERG Guide Category:	OXIDIZERS
Molecular Formula:	$AgNO_3$
RTECS No.:	VW4725000
Properties:	Noncombustible, colorless crystalline solid; exposure to light may cause material to become dark gray. Soluble in water.

HEALTH AND SAFETY INFORMATION

Irritates eyes, skin, and respiratory tract. Swallowing may cause convulsions, and may affect the blood. Silver accumulates in the body and may cause permanent skin discoloration.

IDLH: 10 mg/m^3 as soluble silver compounds.
OSHA Table Z-1-A Air Contaminant

Exposure Limits:	**TWA**
ACGIH:	0.01 mg/m^3 soluble compounds as Ag
NIOSH:	0.01 mg/m^3 soluble compounds as Ag
OSHA:	0.01 mg/m^3 soluble compounds as Ag

RECOMMENDATIONS FOR RESPIRATOR SELECTION: NIOSH

0.25 mg/m^3: SA:CF (any supplied-air respirator operated in a continuous-flow mode); or PAPRHiE (any powered, air-purifying respirator with a high-efficiency particulate filter).
0.5 mg/m^3: HiEF (any air-purifying, full-facepiece respirator with a high-efficiency particulate filter); or SCBAF (any self-contained breathing apparatus with a full facepiece); or SAF (any supplied-air respirator with a full facepiece).
10 mg/m^3: SAF:PD,PP (any supplied-air respirator that has a full facepiece and is operated in a pressure-demand or other positive-pressure mode).

Note: Substance causes eye irritation or damage; eye protection needed.

Emergency or Planned Entry into Unknown Concentrations or IDLH Conditions:
SCBAF:PD,PP (any self-contained breathing apparatus that has a full facepiece and is operated in a

pressure-demand or other positive-pressure mode); or SAF:PD,PP:ASCBA (any supplied-air respirator that has a full facepiece and is operated in a pressure-demand or other positive-pressure mode in combination with an auxiliary, self-contained breathing apparatus operated in a pressure-demand or other positive-pressure mode).

Escape: HiEF (any air-purifying, full-facepiece respirator with a high-efficiency particulate filter); or SCBAE (any appropriate escape-type, self-contained breathing apparatus).

CHEMICAL REACTIVITY

An oxidizer; violent reaction with reducing agents or combustible materials. Contact with acetylene forms shock-sensitive materials. Contact with ammonia forms compounds that are explosive when dry. Hydrogen peroxide causes violent decomposition to oxygen gas. Attacks some plastics, rubber, and coatings.

FIRE INFORMATION

Noncombustible solid. Combustion products include toxic oxides of nitrogen. If material or contaminated runoff enters waterways, notify downstream users of potentially contaminated waters. Notify local health and fire officials and pollution control agencies. Use water, dry chemicals, foam, carbon dioxide.

FEMA Hazard Classifications:

Health Hazards (Blue) 2
Flammability (Red) 0
Reactivity (Yellow) 0
Special Situations . . . . OXY

FIRST AID

Move victim to fresh air. Call emergency medical care. Apply artificial respiration if victim is not breathing. Administer oxygen if breathing is difficult. Remove and isolate contaminated clothing and shoes. In case of contact with substance, immediately flush skin or eyes with running water for at least 20 minutes. Keep victim warm and quiet. Ensure that medical personnel are aware of the material(s) involved and take precautions to protect themselves.

ENVIRONMENTAL INFORMATION as silver compounds

Clean Water Act: Section 307 Toxic Pollutants as silver and compounds.
RCRA Section 261 Hazardous Constituents, as silver compounds, n.o.s., waste number not listed.
Land Ban Chemical
EPCRA Section 313: Includes any unique chemical substance that contains silver as part of that chemical's infrastructure. Form R *de minimis* concentration reporting level: 1.0%. Form R Toxic Chemical Category Code: N740

SODIUM S:210

SYNONYMS: EEC NO. 011-001-00-0; NATRIUM (German); SODIO (Italian, Spanish); SODIUM, METAL LIQUID ALLOY; SODIUM METAL

IDENTIFICATION

CAS:	7440-23-5
ERG ID:	UN 1428
ERG Guide:	138
ERG Guide Category:	SUBSTANCES–WATER-REACTIVE (EMITTING FLAMMABLE GASES)
Molecular Formula:	Na
RTECS No.:	VY0686000
Properties:	Highly flammable, soft silvery to grayish-white solid. Shipped or stored under inert gas or mineral oil. Odorless. Floats and reacts violently with water forming flammable and explosive hydrogen gas.

HEALTH AND SAFETY INFORMATION

Skin and eye contact causes severe burns and possible blindness. Fumes from burning metals are extremely irritating to skin, eyes, nose, and lungs.

RECOMMENDATIONS FOR RESPIRATOR SELECTION: NIOSH

At any concentrations above the NIOSH REL, or where there is no REL, at any detectable concentration: SCBAF:PD,PP (any self-contained breathing apparatus that has a full facepiece and is operated in a pressure-demand or other positive-pressure mode); or SAF:PD,PP:ASCBA (any supplied-air respirator that has a full facepiece and is operated in a pressure-demand or other positive-pressure mode in combination with an auxiliary, self-contained breathing apparatus operated in a pressure-demand or other positive pressure mode).

Escape: GMFOV [any air-purifying, full-facepiece respirator (gas mask) with a chin-style, front- or back-mounted organic vapor canister]; or SCBAE (any appropriate escape-type, self-contained breathing apparatus).

CHEMICAL REACTIVITY

Reacts violently with water, with formation of flammable hydrogen gas and caustic soda solution; a fire often occurs. Ignites spontaneously in moist air. Strong reaction with oxidizers, halogens, carbon dioxide, and many hydrocarbon substances (especially halogenated).

FIRE INFORMATION

Highly flammable when heated; may ignite spontaneously in air. Do not use water; reacts violently to form highly flammable and explosive hydrogen. May react violently with carbon dioxide or halogenated extinguishing agents. Containers may explode when exposed to fire. If material or contaminated runoff enters waterways, notify downstream users of potentially contaminated waters. Notify local health and fire officials and pollution control agencies. Use graphite, Class D extinguisher, soda ash or inert powder. From a secure, explosion-proof location, use water spray to cool exposed containers. If containers are exposed to direct flame or elevated temperatures for prolonged times, withdraw immediately to a secure location.

NFPA Hazard Classifications:

Health Hazards (Blue) 3
Flammability (Red) 1
Reactivity (Yellow) 2
Special Situations . . . . WATER

Autoignition Temperature: 250°F (121°C)

FIRST AID

Move victim to fresh air. Call emergency medical care. Apply artificial respiration if victim is not breathing. Administer oxygen if breathing is difficult. Remove and isolate contaminated clothing and shoes. In case of contact with substance, wipe from skin immediately; flush skin or eyes with running water for at least 20 minutes. Keep victim warm and quiet. Ensure that medical personnel are aware of the material(s) involved and take precautions to protect themselves.

ENVIRONMENTAL INFORMATION

Clean Water Act: Section 311 Hazardous Substances/RQ (same as CERCLA).
EPCRA Section 304: Reportable Quantity (RQ): CERCLA, 10 lb (4.54 kg).

SODIUM AZIDE S:225

SYNONYMS: AI3-50436; AXIUM; AZIDA SODICO (Spanish); AZIDE; AZIUM; AZOTURE de SODIUM (French); AZOTURO di SODIO (Italian); CASWELL No. 744A; DAZOE; EEC No. 011-004-00-7; EINECS No. 247-852-1; EPA PESTICIDE CHEMICAL CODE 107701; HYDRAZOIC ACID, SODIUM SALT; KAZOE; NATRIUMAZID (German); NATRIUMAZIDE (Dutch); NCI-C06462; NSC 3072; SMITE; SODIUM ARDE; SODIUM AZIDE, SALMONELLA TYHIMURIUM METABOLITE; SODIUM, AZOTURE de (French); SODIUM, AZOTURO di (Italian); STCC 4923465; U-3886

IDENTIFICATION

CAS: 26628-22-8
ERG ID: UN 1687
ERG Guide: 153

ERG Guide Category: SUBSTANCES-TOXIC AND/OR CORROSIVE (COMBUSTIBLE)

Molecular Formula: N_3Na

RTECS No.: VY8050000

Properties: Crystalline solid or powder. Colorless to white (powder). Combustible if heated above 572°F (300°C). Odorless. Initially sinks in water and is soluble.

HEALTH AND SAFETY INFORMATION

Irritates eyes, respiratory tract, mucous membranes. A metabolic poison; affects the central nervous system; can cause seizures, unconsciousness and death. Breathing in the dust, swallowing the material, or absorption through the skin can cause decreased blood pressure, serious illness. Repeated or prolonged exposure can cause blood vessel effects. Following exposure regular medical checkups are advisable.

OSHA Table Z-1-A Air Contaminant

Exposure Limits: CEILING

ACGIH:	0.3 mg/m^3 as sodium azide 0.11 mg/m^3 as hydrazoic acid vapor
NIOSH: [(1)]	0.1 ppm as NH_3 0.3 mg/m^3 as NaH_3
OSHA:	None

[(1)] Skin contact contributes significantly to overall exposure.

RECOMMENDATIONS FOR RESPIRATOR SELECTION: NIOSH

At any concentrations above the NIOSH REL, or where there is no REL, at any detectable concentration: SCBAF:PD,PP (any self-contained breathing apparatus that has a full facepiece and is operated in a pressure-demand or other positive-pressure mode); or SAF:PD,PP:ASCBA (any supplied-air respirator that has a full facepiece and is operated in a pressure-demand or other positive-pressure mode in combination with an auxiliary, self-contained breathing apparatus operated in a pressure-demand or other positive-pressure mode).

Escape: GMFOV [any air-purifying, full-facepiece respirator (gas mask) with a chin-style, front- or back-mounted organic vapor canister]; or SCBAE (any appropriate escape-type, self-contained breathing apparatus).

CHEMICAL REACTIVITY

Water contact forms hydrazoic acid. Reacts with acids and metals. Over a period of time sodium azide may react with copper, lead, brass, or solder in plumbing systems, forming an accumulation of the *highly explosive* compounds of lead azide and copper azide. Rapid heating or mechanical shock may cause material to explode. Attacks aluminum.

FIRE INFORMATION

Firefighting gear, including SCBA, does not provide adequate protection. If exposure occurs, remove and isolate gear immediately and thoroughly decontaminate personnel. Decomposition products from heating include toxic hydrazoic acid fumes and nitrogen oxides. Shock or heat may cause material to explode. Containers may explode in fire. If material or contaminated runoff enters waterways, notify downstream users of potentially contaminated water. Notify local health and fire officials and pollution control agencies. Use dry sand, dry chemicals, foam or carbon dioxide; flooding quantities of water. Do not use halons. From a secure, explosion-proof location, use water spray to cool exposed containers.

FEMA Hazard Classifications:

Health Hazards (Blue) 3
Flammability (Red) 0
Reactivity (Yellow) 3

FIRST AID

Move victim to fresh air; call emergency medical care. Apply artificial respiration if victim is not breathing. Do not use mouth-to-mouth method if victim ingested or inhaled the substance; induce artificial respiration with the aid of a pocket mask equipped with a one-way valve or other proper respiratory medical device. Administer oxygen if breathing is difficult. Remove and isolate contaminated clothing and shoes. In case of contact with substance, immediately flush skin or eyes with running water for at least 20 minutes. For minor skin contact, avoid spreading material on unaffected skin. Keep victim warm and quiet. Effects of exposure (inhalation, ingestion, or skin contact) to substance may be delayed. Ensure that medical personnel are aware of the material(s) involved and take precautions to protect themselves.

ENVIRONMENTAL INFORMATION

EPA Hazardous Waste Number (RCRA No.): P105
RCRA Section 261 Hazardous Constituents
EPCRA Section 302 Extremely Hazardous Substances: TPQ = 500 lb (227 kg)
EPCRA Section 304: Reportable Quantity (RQ): CERCLA, 1,000 lb (454 kg)
EPCRA Section 313: Form R *de minimis* concentration reporting level: 1.0%

SODIUM BIFLUORIDE S:255

SYNONYMS: BIFLUORURO SODICO (Spanish); EEC No. 009-007-00-3; EINECS No. 215-608-3; HYDROFLUORIC ACID, SODIUM SALT (2:1); NATRIUMHYDROGENFLUORID (German); SODIUM ACID FLUORIDE; SODIUM DIFLUORIDE; SODIUM HYDROGEN FLUORIDE; SODIUM HYDROGEN DIFLUORIDE

IDENTIFICATION

CAS: 1333-83-1

ERG ID: UN 2439
ERG Guide: 154
ERG Guide Category: SUBSTANCES-TOXIC AND/OR CORROSIVE (NONCOMBUSTIBLE)
Molecular Formula: HF_2Na
RTECS No.: WB03550010
Properties: Noncombustible solid crystalline powder. Colorless to white. Mild odor. Sinks in water; soluble.

HEALTH AND SAFETY INFORMATION

Highly toxic if swallowed. Corrosive to eyes, skin, and respiratory tract. Inhalation may cause pulmonary edema, which can be delayed for several hours; there is a risk of death in serious cases. Repeated exposure to fluoride chemicals may cause stiffness in muscles or ligaments, and even crippling; bone and tooth disorders. Can cause kidney damage.

IDLH: 250 mg/m^3 as fluorine

Exposure Limits:	**TWA**
ACGIH:	2.5 mg/m^3 as fluorides
NIOSH:	2.5 mg/m^3 as fluorides
OSHA:	2.5 mg/m^3 as fluorides

RECOMMENDATIONS FOR RESPIRATOR SELECTION: NIOSH/OSHA as fluorides

12.5 mg/m^3: DM (any dust and mist respirator).
25 mg/m^3: DMXSQ * (any dust and mist respirator except single-use and quarter mask respirators); or SA * (any supplied-air respirator).
62.5 mg/m^3: SA:CF * [any supplied-air respirator operated in a continuous-flow mode)]; or PAPRDM *+ *if not present as a fume* (any powered, air-purifying respirator with a dust and mist filter).
125 mg/m^3: HiEF + (any air-purifying, full-facepiece respirator with a high-efficiency particulate filter); or SCBAF (any self-contained breathing apparatus with a full facepiece); or SAF (any supplied-air respirator with a full facepiece).
250 mg/m^3: SA:PD,PP (any supplied-air respirator operated in a pressure-demand or other positive-pressure mode).

Emergency or Planned Entry into Unknown Concentrations or IDLH Conditions:

SCBAF:PD,PP (any self-contained breathing apparatus that has a full faceplate and is operated in a pressure-demand or other positive-pressure mode); or SAF:PD,PP:ASCBA (any supplied-air respirator that has a full facepiece and is operated in a pressure-demand or other positive-pressure mode in combination with an auxiliary, self-contained breathing apparatus operated in a pressure-demand or other positive-pressure mode).

Escape: HiEF [+] (any air-purifying, full-facepiece respirator with a high-efficiency particulate filter); or SCBAE (any appropriate escape-type, self-contained breathing apparatus).

[*] Substance reported to cause eye irritation or damage; may require eye protection.
[+] *Note:* May need acid gas sorbent.

CHEMICAL REACTIVITY

Reacts with water, acids liberating heat and forming corrosive and flammable hydrogen cyanide and hydrogen fluoride solution. Incompatible with acid salts, chlorates, nitrates. Absorbs water from the air forming a corrosive syrup. Aqueous solution corrodes glass, concrete, and certain metals, especially those containing silica such as cast iron. Will attack natural rubber, leather, and many organic materials. May generate hydrogen gas on contact with some metals.

FIRE INFORMATION

FEMA Hazard Classifications

Combustion products include highly toxic sodium oxides and fluorides. If material or contaminated runoff enters waterways, notify downstream users of potentially contaminated waters. Notify local health and fire officials and pollution control agencies. Use dry chemicals, foam, carbon dioxide.

Health Hazards (Blue) 3
Flammability (Red) 0
Reactivity (Yellow) 1

FIRST AID

Move victim to fresh air; call emergency medical care. Apply artificial respiration if victim is not breathing. Do not use mouth-to-mouth method if victim ingested or inhaled the substance; induce artificial respiration with the aid of a pocket mask equipped with a one-way valve or other proper respiratory medical device. Administer oxygen if breathing is difficult. Remove and isolate contaminated clothing and shoes. In case of contact with substance, immediately flush skin or eyes with running water for at least 20 minutes. For minor skin contact, avoid spreading material on unaffected skin. Keep victim warm and quiet. Effects of exposure (inhalation, ingestion, or skin contact) to substance may be delayed. Ensure that medical personnel are aware of the material(s) involved and take precautions to protect themselves.

ENVIRONMENTAL INFORMATION

Clean Water Act: Section 311 Hazardous Substances/RQ (same as CERCLA)
EPCRA Section 304 Reportable Quantity (RQ): CERCLA, 100 lb (45.4 kg)

SODIUM BISULFATE S:260

SYNONYMS: BISULFATO SODICO (Spanish); EINECS No. 231-665-7; GBS; IDROGENOSOFLATO di SODIO (Italian); NATRIUMHYDROGENSULFAT (German); NITRE CAKE; SODIUM ACID

SULFATE; SODIUM BISULFATE, FUSED; SODIUM BISULFATE, SOLID; SODIUM HYDROGEN SULFATE; SODIUM PYROSULFATE; SULFURIC ACID, MONOSODIUM SALT

IDENTIFICATION

CAS:	7681-38-1
ERG ID:	UN 1821 (solid); UN 2837 (solution)
ERG Guide:	154
ERG Guide Category:	TOXIC AND/OR CORROSIVE (NONCOMBUSTIBLE)
Molecular Formula:	$HNaO_4S$
RTECS No.:	VZ1860000
Properties:	Noncombustible, corrosive solid. Colorless. Poisonous gas produced in fire.

HEALTH AND SAFETY INFORMATION

Corrosive to eyes, skin, and respiratory tract. Inhalation may cause pulmonary edema, which can be delayed for several hours; there is a risk of death in serious cases.

RECOMMENDATIONS FOR RESPIRATOR SELECTION

At any concentrations above the NIOSH REL, or where there is no REL, at any detectable concentration: SCBAF:PD,PP (any self-contained breathing apparatus that has a full facepiece and is operated in a pressure-demand or other positive-pressure mode); or SAF:PD,PP:ASCBA (any supplied-air respirator that has a full facepiece and is operated in a pressure-demand or other positive-pressure mode in combination with an auxiliary, self-contained breathing apparatus operated in a pressure-demand or other positive-pressure mode).

Escape: HiEF (any air-purifying, full-facepiece respirator with a high-efficiency particulate filter); or SCBAE (any appropriate escape-type, self-contained breathing apparatus).

CHEMICAL REACTIVITY

Contact with water forms sulfuric acid. Reacts with calcium hypochlorite. Attacks metals in the presence of moisture.

FIRE INFORMATION

Noncombustible solid. Combustion products include toxic sulfur and sodium oxides. If material or contaminated runoff enters waterways, notify downstream users of potentially contaminated waters. Notify local health and fire officials and pollution control agencies. Use water, dry chemicals, alcohol-resistant foam, carbon dioxide.

FIRST AID

Move victim to fresh air. Call emergency medical care. Apply artificial respiration if victim is not breathing. Do not use mouth-to-mouth method if victim ingested or inhaled the substance; induce artificial respiration with the aid of a pocket mask equipped with a one–way valve or other proper respiratory medical device. Administer oxygen if breathing is difficult. Remove and isolate contaminated clothing and shoes. In case of contact with substance, immediately flush skin or eyes with running water for at least 20 minutes. For minor skin contact, avoid spreading material on unaffected skin. Keep victim warm and quiet. Effects of exposure (inhalation, ingestion, or skin contact) may be delayed. Ensure that medical personnel are aware of the material(s) involved and take precautions to protect themselves.

SODIUM BISULFITE (SOLID) S:280

SYNONYMS: AMERSITE; AMERSITE 2; BISULFITE de SODIUM (French); BISULFITO SIDICO SOLIDO (Spanish); CB-16 CLEARING BATH; CRONAFLEX PFC DEVELOPER; EINECS No. 231-548-0; HYDROGEN SULFITE SODIUM; KODAK CHROMIUM INTENSIFIER, PART B; MONOSODIUM SALT of SULFUROUS ACID; SODIUM ACID BISULFITE; SODIUM ACID SULFITE; SODIUM BISULFITE, SOLUTION; SODIUM BISULFITE, SOLID; SODIUM HYDROGEN SULFITE; SODIUM HYDROGEN SULFITE, SOLID; SODIUM SULFHYDRATE; SODIUM METABISULFITE; SODIUM PYROSULFITE; SULFUROUS ACID, MONOSODIUM SALT

IDENTIFICATION

CAS:	7631-90-5
ERG ID:	UN 2693
ERG Guide:	154 (solution)
ERG Guide Category:	TOXIC AND/OR CORROSIVE (NON-COMBUSTIBLE)
Molecular Formula:	$HNaO_3S$
RTECS No:	VZ2000000
Properties:	Solid powder or granules. White. Faint, irritating, sulfurous odor. Sinks and mixes with water.

HEALTH AND SAFETY INFORMATION

Corrosive to eyes, skin, and respiratory tract. Inhalation may cause pulmonary edema, which can be delayed for several hours; there is a risk of death in serious cases. May cause asthmalike allergy; lung damage.

Exposure Limits:	**TWA**
ACGIH:	5 mg/m^3
NIOSH:	5 mg/m^3
OSHA:	None

RECOMMENDATIONS FOR RESPIRATOR SELECTION: NIOSH

Where there is no REL, at any detectable concentration: SCBAF:PD,PP (any self-contained breathing apparatus that has a full facepiece and is operated in a pressure-demand or other positive-pressure mode);or SAF:PD,PP:ASCBA (any supplied-air respirator that has a full facepiece and is operated in a pressure-demand or other positive-pressure mode in combination with an auxiliary, self-contained breathing apparatus operated in a pressure-demand or other positive pressure mode).

Escape: GMFOV [any air-purifying, full-facepiece respirator (gas mask) with a chin-style, front- or back-mounted organic vapor canister]; or SCBAE (any appropriate escape-type, self-contained breathing apparatus).

CHEMICAL REACTIVITY

Slowly oxidized to the sulfate on contact with air. Reaction with heat, oxidizers, or acids forms corrosive and toxic dioxides including sulfur dioxide gas. Attacks metals.

FIRE INFORMATION

Noncombustible solid. If material or contaminated runoff enters waterways, notify downstream users of potentially contaminated waters. Notify local health and fire officials and pollution control agencies. Use dry chemicals, alcohol-resistant foam, carbon dioxide. Use water spray to cool exposed containers.

FEMA Hazard Classifications:

Health Hazards (Blue) 1
Flammability (Red) 1
Reactivity (Yellow) 0

FIRST AID

Move victim to fresh air. Call emergency medical care. Apply artificial respiration if victim is not breathing. Do not use mouth-to-mouth method if victim ingested or inhaled the substance; induce artificial respiration with the aid of a pocket mask equipped with a one-way valve or other proper respiratory medical device. Administer oxygen if breathing is difficult. Remove and isolate contaminated clothing and shoes. In case of contact with substance, immediately flush skin or eyes with running water for at least 20 minutes. For minor skin contact, avoid spreading material on unaffected skin. Keep victim warm and quiet. Effects of exposure (inhalation, ingestion, or skin contact) may be delayed. Ensure that medical personnel are aware of the material(s) involved and take precautions to protect themselves.

ENVIRONMENTAL INFORMATION

CLEAN WATER ACT: Section 311 Hazardous Substances/RQ (same as CERCLA).
EPCRA Section 304 Reportable Quantity (RQ): CERCLA, 5,000 lb (2,270 kg).

SODIUM CHLORATE S:310

SYNONYMS: ASEX; ATLACIDE; ATRATOL B-HERBATOX; CHLORATE OF SODA; CHLORATE SALT OF SODIUM; CHLORAX; CHLORIC ACID, SODIUM SALT; CHLORSAURE (German); CLORATO di SODIO (Italian); CLORATO SODICO (Spanish); DE-FOL-ATE; DESOLET; DREXEL

DEFOL; DROP LEAF; EEC No. 017-005-00-9; EINECS No. 231-887-4; EVAU-SUPERFALL; GRAIN SORGHUM HARVEST AID; GRANEX OK; HARVEST-AID; KLOREX; KUSA-TOHRU; KUSATOL; LOREX; NATRIUMCHLORAAT (Dutch); NATRIUMCHLORAT (German); ORTHO C-1 DEFOLIANT & WEED KILLER; OXYCIL; RASIKAL; SHED-A-LEAF; SHED-A-LEAF "L"; SODA CHLORATE; SODIO (CLORATO di) (Italian); SODIUM (CHLORATE de) (French); TRAVEX; TUMBLEAF; UNITED CHEMICAL DEFOLIANT NO. 1; VAL-DROP

IDENTIFICATION

CAS:	7775-09-9
ERG ID:	UN 1495; UN 2428 (aqueous solution)
ERG Guide:	140
ERG Guide Category:	OXIDIZERS
Molecular Formula:	$ClNaO_3$
RTECS No.:	FO0525000
Properties:	Noncombustible, colorless or white crystalline solid. Odorless. Soluble in water.

HEALTH AND SAFETY INFORMATION

Poisonous. Irritates eyes, skin, and respiratory tract; eye and skin contact can cause burns and blindness. Long-term exposure may cause blood changes; kidney and liver damage.

RECOMMENDATIONS FOR RESPIRATOR SELECTION

At any concentrations above the NIOSH REL, or where there is no REL, at any detectable concentration: SCBAF:PD,PP (any self-contained breathing apparatus that has a full facepiece and is operated in a pressure-demand or other positive-pressure mode); or SAF:PD,PP:ASCBA (any supplied-air respirator that has a full facepiece and is operated in a pressure-demand or other positive-pressure mode in combination with an auxiliary, self-contained breathing apparatus operated in a pressure-demand or other positive-pressure mode).

Escape: GMFOV [any air-purifying, full-facepiece respirator (gas mask) with a chin-style, front- or back-mounted organic vapor canister]; or SCBAE (any appropriate escape-type, self-contained breathing apparatus).

CHEMICAL REACTIVITY

A powerful oxidizer; reacts violently with reducing agents and combustible substances. Explosions may result from contact with ammonia salts, carbon, oils, metal sulfides, nitrobenzene, powdered metals, sugar. Contact with strong acids gives off carbon dioxide. Forms shock-sensitive mixture with some organic materials. Solution (50%) decomposes at 300°F (149°C) liberating oxygen and increasing fire risk. Do not use combustible absorbents on spills. Incompatible with calcium, lead diacetate, germanium, magnesium, mercurous chloride, silicon, silver nitrate, titanium, zinc acetate.

FIRE INFORMATION

Noncombustible solid but may increase intensity of fire or cause combustible materials to ignite or explode. Material on clothing is a fire hazard; flush immediately with water. If material or contaminated runoff enters waterways, notify downstream users of potentially contaminated waters. Notify local health and fire officials and pollution control agencies. Use dry chemicals, dry sand, carbon dioxide. From a secure, explosion-proof location, use water spray to cool exposed containers.

FEMA Hazard Classifications:

Solid

Health Hazards (Blue) 1
Flammability (Red) 0
Reactivity (Yellow) 2
Special Situations . . . . OXY

Solution

Health Hazards (Blue) 1
Flammability (Red) 0
Reactivity (Yellow) 0
Special Situations . . . . OXY

FIRST AID

Move victim to fresh air. Call emergency medical care. Apply artificial respiration if victim is not breathing. Administer oxygen if breathing is difficult. Remove and isolate contaminated clothing and shoes. In case of contact with substance, immediately flush skin or eyes with running water for at least 20 minutes. Keep victim warm and quiet. Ensure that medical personnel are aware of the material(s) involved and take precautions to protect themselves.

SODIUM CYANIDE S:360

SYNONYMS: CIANETO de SODIO (Portugese); CIANURO di SODIO (Italian); CIANURO SODICO (Spanish); CYANIDE OF SODIUM; CYANOBRIK; CYANOGRAN; CYANURE de SODIUM (French); CYMAG; EEC No. 006-007-00-5; HYDROCYANIC ACID, SODIUM SALT; KYANID SODNY (Czech); NATRIUMCYANID (German); SODIUM CYANIDE, SOLID; SODIUM CYANIDE, SOLUTION

IDENTIFICATION

CAS:	143-33-9
ERG ID:	UN 1689 (solid); UN 1935 (solution)
ERG Guide:	157

ERG Guide Category: TOXIC AND/OR CORROSIVE (NONCOMBUSTIBLE/WATER-SENSITIVE)
Molecular Formula: CNNa
RTECS No.: VZ7530000
Properties: Nonflammable, colorless, cubes, granules, flakes or egglike lumps. Odorless to almondlike odor when damp; odorless when completely dry. Sinks and mixes with water.

HEALTH AND SAFETY INFORMATION

Poison. Breathing or swallowing the dust, or skin contact can cause death.

IDLH: 25 mg/m^3 as cyanide
OSHA Table Z-1-A Air Contaminant as CN

Poison. Inhaling or swallowing the dust, or skin contact may cause death. Repeated or prolonged exposure may cause loss of appetite, headache, nausea, and respiratory irritation; skin rash; enlargement of the thyroid gland and thyroid problems; nose irritation with sores and nose bleed; central nervous system damage; changes to blood cell count.

Exposure Limits:	TWA	CEILING
NIOSH:(1)		4.7 ppm; 5 mg/m^3 [10 min] as CN
OSHA:(1)	5 mg/m^3 as CN	

(1) Skin contact contributes significantly to overall exposure.

RECOMMENDATIONS FOR RESPIRATOR SELECTION: NIOSH/OSHA

25 mg/m^3: SA (any supplied-air respirator); or SCBA (any self-contained breathing apparatus).

Emergency or Planned Entry into Unknown Concentrations or IDLH Conditions:
SCBAF:PD,PP (any self-contained breathing apparatus that has a full facepiece and is operated in a pressure-demand or other positive-pressure mode); or SAF:PD,PP:ASCBA (any supplied-air respirator that has a full facepiece and is operated in a pressure-demand or other positive-pressure mode in combination with an auxiliary, self-contained breathing apparatus operated in a pressure-demand or other positive-pressure mode).

Escape: GMFSHiE [any air-purifying, full-facepiece respirator (gas mask) with a chin-style, front- or back-mounted canister providing protection against the compound of concern and having a high-efficiency particulate filter]; or SCBAE (any appropriate escape-type, self-contained breathing apparatus).

CHEMICAL REACTIVITY

Aqueous solution is caustic. Contact with acids or carbon dioxide releases poisonous hydrogen cyanide. Incompatible with organic anhydrides, isocyanates, alkylene oxides, epichlorohydrin, aldehydes, alcohols, glycols, phenols, cresols, caprolactum solution, strong oxidizers, nitrates, nitrites. Attacks aluminum, copper, and zinc.

FIRE INFORMATION

Combustion products include toxic cyanide and nitrogen oxides. Firefighting gear, including SCBA, does not provide adequate protection. Do not use carbon dioxide. If exposure occurs, remove and isolate gear immediately and thoroughly decontaminate personnel. If material or contaminated runoff enters waterways, notify downstream users of potentially contaminated waters. Notify local health and fire officials and pollution control agencies. Use extinguisher suitable for surrounding material; water, dry chemicals, alcohol-resistant foam. Do not use carbon dioxide; releases cyanide gas. From a secure, explosion-proof location, use water spray to cool exposed containers.

NFPA Hazard Classifications:

Health Hazards (Blue) 3
Flammability (Red) 0
Reactivity (Yellow) 0

FIRST AID

Move victim to fresh air. Call emergency medical care. Apply artificial respiration if victim is not breathing. Do not use mouth-to-mouth method if victim ingested or inhaled the substance; induce artificial respiration with the aid of a pocket mask equipped with a one–way valve or other proper respiratory medical device. Administer oxygen if breathing is difficult. Remove and isolate contaminated clothing and shoes. In case of contact with substance, immediately flush skin or eyes with running water for at least 20 minutes. For minor skin contact, avoid spreading material on unaffected skin. Keep victim warm and quiet. Effects of exposure (inhalation, ingestion, or skin contact) may be delayed. Ensure that medical personnel are aware of the material(s) involved and take precautions to protect themselves.

ENVIRONMENTAL INFORMATION

Clean Water Act: Section 311 Hazardous Substances/RQ (same as CERCLA); Section 313 Priority Chemicals.
EPA Hazardous Waste Number (RCRA No.): P106.
RCRA Section 261 Hazardous Constituents
EPCRA Section 302 Extremely Hazardous Substances: TPQ = 100 lb (45.4 kg).
EPCRA Section 304: Reportable Quantity (RQ): EHS, CERCLA, 10 lb (4.54 kg).
EPCRA Section 313: See cyanide compounds.
Marine Pollutant (49CFR, Subchapter 172.101, Appendix B).
TSCA: 40CFR712.30(*d*); 40CFR716.120(*a*); 40CFR799.5000

SODIUM HYDROSULFITE S:395

SYNONYMS: BLANKIT; BURMOL; DISODIUM-DITHIONITE; DISODIUM HYDROSULFITE; DITHIONOUS ACID, DISODIUM SALT; D-OX; EEC No. 016-028-00-1; EINECS No. 231-890-0; HIDROSULFITO SODICO(Spanish); HYDROLIN; HYDROS; K-BRITE; NATRIUMDITHIONIT

(German); REDUCTONE; SODIUM DITHIONITE; SODIUM HYDROSULPHITE; SODIUM HYPOSULFITE; SODIUM SULFOXYLATE; VATROLITE; V-BRITE B; VIRCHEM; VIRTEX CC; VIRTEX D; VIRTEX L; VIRTEX RD

IDENTIFICATION

CAS:	7775-14-6
ERG ID:	UN 1384
ERG Guide:	135
ERG Guide Category:	SUBSTANCES-SPONTANEOUSLY COMBUSTIBLE
Molecular Formula:	$N_2O_4S_2$
RTECS No.:	JP2100000
Properties:	Sontaneously combustible crystalline powder. White or grayish white. Faint sulfurous odor. Soluble in water; decomposes.

HEALTH AND SAFETY INFORMATION

Irritates eyes, skin, and respiratory tract. May be harmful if inhaled. Absorbed through the skin; contact may contribute to overall exposure.

CHEMICAL REACTIVITY

May ignite with moisture and air. May ignite in the presence of combustible materials. Heat above 122°F (50°C) can initiate thermal decomposition with release of sulfur dioxides. A strong reducing agent; contact with oxidizers can cause fire and explosions. Reacts with strong acids (forming toxic sulfur dioxides), and sodium chlorite.

FIRE INFORMATION

Combustion products include toxic sulfur dioxides. Vapors in confined areas may explode when exposed to fire. If material or contaminated runoff enters waterways, notify downstream users of potentially contaminated waters. Notify local health and fire officials and pollution control agencies. Keep water away from any release. On fire, use flooding quantities of water, dry chemicals, dry sand, carbon dioxide. Foam may be ineffective. Monitor temperatures of containers exposed to fire for at least 48 hours to verify decomposition has not begun in the sealed containers (NFPA).

FEMA Hazard Classifications:

Health Hazards (Blue) 2
Flammability (Red) 1
Reactivity (Yellow) 2

FIRST AID

Move victim to fresh air. Call emergency medical care. Apply artificial respiration if victim is not breathing. Administer oxygen if breathing is difficult. Remove and isolate contaminated clothing and shoes. In case of contact with substance, immediately flush skin or eyes with running water for at least 20

minutes. Keep victim warm and quiet. Ensure that medical personnel are aware of the material(s) involved and take precautions to protect themselves.

SODIUM HYDROXIDE S:400

SYNONYMS: AZ 303-A DEVELOPER; AZ 351 DEVELOPER; CAUSTIC SODA; CAUSTIC SODA, BEAD; CAUSTIC SODA, DRY; CAUSTIC SODA, FLAKE; CAUSTIC SODA, GRANULAR; CAUSTIC SODA, SOLID; COLEC 302; CUPOSIT CP-74B ELECTROLESS COPPER; EEC No. 011-002-00-6 (solid); EEC No. 011-002-01-3 (solution); EINECS No. 215-185-5; ELECTROCLEANER 111; ENTEC 313 CORROSION INHIBITOR/DISPERSANT; ENTEC 318 CORROSION INHIBITOR/DISPERSANT; FIDELEX 218 METAL STRIPPER; G-182 ACTIVATOR; HIDROXIDO SODICO (Spanish); HYDROXYDE de SODIUM (French); KTI DE-3(+); KTI DE-5; KTI SB-351(+); LEWIS-RED DEVIL LYE; LYE; LYE SOLUTION; METALEX TIN STRIPPER L-2; METALEX W SPECIAL; MICROPOSIT 303A; MICROPOSIT 351; MICROPOSIT 352; MICROPOSIT 353 DEVELOPER; MICROPOSIT 354 DEVELOPER; NATRIUMHYDROXID (German); NATRIUMHYDROXYDE (Dutch); PELS; ROSTRIP M-12; RS-4 STRIPPER, POSITIVE; S-1634; SODA LYE; SODIO(IDROSSIDO di) (Italian); SODIUM HYDRATE; SODIUM HYDROXIDE, BEAD; SODIUM HYDROXIDE, DRY; SODIUM HYDROXIDE, FLAKE; SODIUM HYDROXIDE, GRANULAR; SODIUM HYDROXIDE, SOLID; SODIUM(HYDROXYDE de) (French); SODIUM HYDROXIDE CAUSTIC SODA SOLUTION; SODIUM HYDRATE SOLUTION; SODIUM HYDROXIDE LIQUID; SODIUM HYDROXIDE SOLUTION; SUNNY SOL 100; SUNNY SOL 150; ULTRAMAC D-89; WHITE CAUSTIC; WHITE CAUSTIC, SOLUTION

IDENTIFICATION

CAS:	1310-73-2
ERG ID:	UN 1823 (solid); UN 1824 (solution)
ERG Guide:	154
ERG Guide Category:	TOXIC AND/OR CORROSIVE (NONCOMBUSTIBLE)
Molecular Formula:	HNaO
RTECS No.:	WB4900000
Properties:	Nonflammable solid flakes, pellets, sticks, cakes; or solution in water. White to colorless solid; milky solution. Odorless. Sinks and mixes with water; liberates sufficient heat to ignite combustible materials. Absorbs moisture from the air.

HEALTH AND SAFETY INFORMATION

Corrosive to eyes, skin, and respiratory tract. Inhalation may cause pulmonary edema, which can be delayed for several hours; there is a risk of death in serious cases. Repeated exposure to mists or aerosols can produce a tolerance for its irritant effects. Low concentrations can defat the skin, causing an irritant dermatitis with redness, drying, fissuring, and scaling. Cancer of the esophagus has been reported in patients who have corrosion-induced strictures. However, it is believed that these cancers were the result of tissue destruction and scar formation rather than a direct cancer-causing action of sodium hydroxide.

960 Sodium Hydroxide

IDLH: 10 mg/m^3
Odor Threshold: Not listed; mucous membrane irritation at 2 mg/m^3
OSHA Table Z-1-A Air Contaminant
NIOSH Criteria Document: NIOSH 76-105

Exposure Limits:	TWA	CEILING
ACGIH:		2 mg/m^3
NIOSH:		2 mg/m^3
OSHA:	2 mg/m^3	

RECOMMENDATIONS FOR RESPIRATOR SELECTION: NIOSH/OSHA

10 mg/m^3: SA:CF (any supplied-air respirator operated in a continuous-flow mode.); or HiEF (any air-purifying, full-facepiece respirator with a high-efficiency particulate filter); PAPRDM (any powered, air-purifying respirator with a dust and mist filter); or SCBAF (any self-contained breathing apparatus with a full facepiece); or SAF (any supplied-air respirator with a full facepiece).

Emergency or Planned Entry into Unknown Concentrations or IDLH Conditions: SCBAF:PD,PP (any self-contained breathing apparatus that has a full facepiece and is operated in a pressure-demand or other positive-pressure mode); or SAF:PD,PP:ASCBA (any supplied-air respirator that has a full facepiece and is operated in a pressure-demand or other positive-pressure mode in combination with an auxiliary, self-contained breathing apparatus operated in a pressure-demand or other positive-pressure mode).

Escape: HiEF (any air-purifying, full-facepiece respirator with a high-efficiency particulate filter); or SCBAE (any appropriate escape-type, self-contained breathing apparatus).

Note: Substance causes eye irritation or damage; eye protection needed.

CHEMICAL REACTIVITY

Contact with water, acids, flammable liquids, and organic halogens, especially trichloroethylene, may cause fires and explosions. Contact with nitromethane and similar nitro compounds form shock-sensitive salts. Water contact forms a large amount of heat and corrosive fumes; may splatter. Attacks some plastics, rubber, and coatings. Contact with metals such as aluminum, tin, lead, and zinc causes corrosion and the formation of flammable hydrogen gas; corrosive to most metals.

FIRE INFORMATION

Combustion products include toxic sodium oxides. Firefighting gear, including SCBA, does not provide adequate protection. If exposure occurs, remove and isolate gear immediately and thoroughly decontaminate personnel. Material does not burn; fight surrounding fire with an agent appropriate for the burning material. If material or contaminated runoff enters waterways, notify downstream users of potentially contaminated waters. Notify local health and fire officials and pollution control agencies. Use any extinguishing material: water, dry chemicals, foam, carbon dioxide. From a secure, explosion-proof location, use water spray to cool exposed containers.

NFPA Hazard Classifications:

Health Hazards (Blue) 3
Flammability (Red) 0
Reactivity (Yellow) 1

FIRST AID

Move victim to fresh air. Call emergency medical care. Apply artificial respiration if victim is not breathing. Do not use mouth-to-mouth method if victim ingested or inhaled the substance; induce artificial respiration with the aid of a pocket mask equipped with a one–way valve or other proper respiratory medical device. Administer oxygen if breathing is difficult. Remove and isolate contaminated clothing and shoes. In case of contact with substance, immediately flush skin or eyes with running water for at least 20 minutes. For minor skin contact, avoid spreading material on unaffected skin. Keep victim warm and quiet. Effects of exposure (inhalation, ingestion, or skin contact) may be delayed. Ensure that medical personnel are aware of the material(s) involved and take precautions to protect themselves.

ENVIRONMENTAL INFORMATION

Clean Water Act: Section 311 Hazardous Substances/RQ (same as CERCLA).
EPCRA Section 304: Reportable Quantity (RQ): CERCLA, 1,000 lb (454 kg).

SODIUM METHYLATE S:455

SYNONYMS: EEC NO. 603-040-00-2; EINECS No. 204-699-5; METILATO SODICO (Spanish); METHANOL, SODIUM SALT; METHYL ALCOHOL, SODIUM SALT; NATRIUMMETHANOLAT (German); SODIUM METHANOLATE; SODIUM METHOXIDE

IDENTIFICATION

CAS:	124-41-4
ERG ID:	UN 1431
ERG Guide:	138
ERG Guide Category:	WATER-REACTIVE (EMITTING FLAMMABLE GASES)
Molecular Formula:	CH_3NaO
RTECS No.:	PC3570000
Properties:	Flammable powdery solid. White. Odorless. Mixes with water; reacts violently. Combustible liquid (solution in methyl alcohol).

HEALTH AND SAFETY INFORMATION

Corrosive to eyes, skin, and respiratory tract. Skin contact in the presence of moisture can cause serious burns. Inhalation may cause pulmonary edema, which can be delayed for several hours; there is a risk of death in serious cases.

RECOMMENDATIONS FOR RESPIRATOR SELECTION

At any concentrations above the NIOSH REL, or where there is no REL, at any detectable concentration: SCBAF:PD,PP (any self-contained breathing apparatus that has a full facepiece and is operated in a pressure-demand or other positive-pressure mode); or SAF:PD,PP:ASCBA (any supplied-air respirator that has a full facepiece and is operated in a pressure-demand or other positive-pressure mode in combination with an auxiliary, self-contained breathing apparatus operated in a pressure-demand or other positive-pressure mode).

Escape: GMFOV [any air-purifying, full-facepiece respirator (gas mask) with a chin-style, front- or back-mounted organic vapor canister]; or SCBAE (any appropriate escape-type, self-contained breathing apparatus).

CHEMICAL REACTIVITY

Spontaneously combustible in air above 259°F (126°C). Ignites on contact with water or steam; produces a caustic sodium hydroxide solution and a solution of methyl alcohol. A strong reducing agent; reacts violently with oxidizers. Solution reacts with acids, mixtures of chloroform and methanol. Attacks certain plastics such as nylon and polyesters. Reacts with metals producing flammable and potentially explosive hydrogen gas.

FIRE INFORMATION

Combustion products include highly toxic sodium oxides. May be corrosive to equipment. If material or contaminated runoff enters waterways, notify downstream users of potentially contaminated waters. Notify local health and fire officials and pollution control agencies. Use dry chemicals, sand, carbon dioxide. Do not use water-based extinguishing agents. If containers are exposed to direct flame or elevated temperatures for prolonged times, withdraw immediately to a secure location.

FEMA Hazard Classifications

Health Hazards (Blue) 2
Flammability (Red) 4
Reactivity (Yellow) 1
Special Situations . WATER REACTIVE

Flash Point: Flammable solid that may ignite in the presence of moisture. Solution, in methyl alcohol, approximately 70°F (24°C).
Autoignition Temperature: 157°F (70°C)

FIRST AID

Move victim to fresh air. Call emergency medical care. Apply artificial respiration if victim is not breathing. Administer oxygen if breathing is difficult. Remove and isolate contaminated clothing and shoes. In case of contact with substance, immediately flush skin or eyes with running water for at least 20 minutes. Keep victim warm and quiet. Ensure that medical personnel are aware of the material(s) involved and take precautions to protect themselves.

ENVIRONMENTAL INFORMATION

Clean Water Act: Section 311 Hazardous Substances/RQ (same as CERCLA)
EPCRA Section 304 Reportable Quantity (RQ): CERCLA, 1,000 lb (454 kg)

SODIUM MONOXIDE S:472

SYNONYMS: CALCINED SODA; DISODIUM MONOXIDE; DISODIUM OXIDE; EINECS No. 235-641-7; SODIUM OXIDE

IDENTIFICATION

CAS:	12401-86-4
ERG ID:	UN 1825
ERG Guide:	157
ERG Guide Category:	SUBSTANCES-TOXIC AND/OR CORROSIVE (NONCOMBUSTIBLE/WATER-SENSITIVE)
Molecular Formula:	Na_2O
RTECS No.:	WC4800000
Properties:	Noncombustible powder. White to grey. Odorless. Mixes with water forming corrosive sodium hydroxide.

HEALTH AND SAFETY INFORMATION

Corrosive to eyes, skin, and respiratory tract. Inhalation may cause pulmonary edema, which can be delayed for several hours; there is a risk of death in serious cases.

RECOMMENDATIONS FOR RESPIRATOR SELECTION

At any concentrations above the NIOSH REL, or where there is no REL, at any detectable concentration: SCBAF:PD,PP (any self-contained breathing apparatus that has a full facepiece and is operated in a pressure-demand or other positive-pressure mode); or SAF:PD,PP:ASCBA (any supplied-air respirator that has a full facepiece and is operated in a pressure-demand or other positive-pressure mode in combination with an auxiliary, self-contained breathing apparatus operated in a pressure-demand or other positive-pressure mode).

Escape: GMFOV [any air-purifying, full-facepiece respirator (gas mask) with a chin-style, front- or back-mounted organic vapor canister]; or SCBAE (any appropriate escape-type, self-contained breathing apparatus.

CHEMICAL REACTIVITY

Reacts with water; decomposes to sodium hydroxide. Reacts with nitric oxide above 212°F (100°C),

2,4-dinitrotoluene, phosphorus(VI) oxide in the presence of heat and/or water.. A strong caustic when wet. Caustics react on contact with acids, and are incompatible with amides, isocyanates, alkylene oxides, epichlorohydrin, aldehydes, alcohols, glycols, phenols, cresols, caprolactum solution.

FIRE INFORMATION

Combustion products include toxic sodium oxides. If material or contaminated runoff enters waterways, notify downstream users of potentially contaminated waters. Notify local health and fire officials and pollution control agencies. Use dry chemicals, alcohol-resistant foam, carbon dioxide.

NFPA Hazard Classifications: Not listed

FIRST AID

Move victim to fresh air. Call emergency medical care. Apply artificial respiration if victim is not breathing. Do not use mouth-to-mouth method if victim ingested or inhaled the substance; induce artificial respiration with the aid of a pocket mask equipped with a one-way valve or other proper respiratory medical device. Administer oxygen if breathing is difficult. Remove and isolate contaminated clothing and shoes. In case of contact with substance, immediately flush skin or eyes with running water for at least 20 minutes. For minor skin contact, avoid spreading material on unaffected skin. Keep victim warm and quiet. Effects of exposure (inhalation, ingestion, or skin contact) may be delayed. Ensure that medical personnel are aware of the material(s) involved and take precautions to protect themselves.

SODIUM NITRATE S:480

SYNONYMS: CHILE SALTPETER; CUBIC NITER; EINECS No. 231-554-3; NITRATINE; NITRATE de SODIUM (French); NITRATO SODICO (Spanish); NITRIC ACID, SODIUM SALT; SODA NITER; SODIUM NITRATE (1:1)

IDENTIFICATION

CAS:	7631-99-4
ERG ID:	UN 1498
ERG Guide:	140
ERG Guide Category:	OXIDIZERS
Molecular Formula:	$NNaO_3$
RTECS No.:	WC5600000
Properties:	Crystalline colorless to white solid. Odorless. Sinks and mixes with water.

HEALTH AND SAFETY INFORMATION

Irritates eyes, skin, and respiratory tract. Repeated or prolonged exposures may cause blood or blood vessel problems, headache, and mental impairment.

RECOMMENDATIONS FOR RESPIRATOR SELECTION

At any concentrations above the NIOSH REL, or where there is no REL, at any detectable concentration: SCBAF:PD,PP (any self-contained breathing apparatus that has a full facepiece and is operated in a pressure-demand or other positive-pressure mode); or SAF:PD,PP:ASCBA (any supplied-air respirator that has a full facepiece and is operated in a pressure-demand or other positive-pressure mode in combination with an auxiliary, self-contained breathing apparatus operated in a pressure-demand or other positive-pressure mode).

Escape: GMFOV [any air-purifying, full-facepiece respirator (gas mask) with a chin-style, front- or back-mounted organic vapor canister]; or SCBAE (any appropriate escape-type, self-contained breathing apparatus).

CHEMICAL REACTIVITY

A powerful oxidizer; violent reaction with reducing agents, powdered aluminum, gunpowder, metals, wood, and other organic materials, resulting in combustion or spontaneous combustion.

FIRE INFORMATION

Combustion products include toxic nitrogen oxides. Noncombustible solid, but can increase the combustibility of other materials and increase fire activity. If material or contaminated runoff enters waterways, notify downstream users of potentially contaminated waters. Notify local health and fire officials and pollution control agencies. Use extinguishing agent appropriate for surrounding fire. From a secure, explosion-proof location, use water spray to cool exposed containers.

FEMA Hazard Classifications:

Health Hazards (Blue) 1
Flammability (Red) 0
Reactivity (Yellow) 1
Special Situations .... OXY

FIRST AID

Move victim to fresh air. Call emergency medical care. Apply artificial respiration if victim is not breathing. Administer oxygen if breathing is difficult. Remove and isolate contaminated clothing and shoes. In case of contact with substance, immediately flush skin or eyes with running water for at least 20 minutes. Keep victim warm and quiet. Ensure that medical personnel are aware of the material(s) involved and take precautions to protect themselves.

SODIUM NITRITE S:500

SYNONYMS: ANTI-RUST; B-239; C-521; DIAZOTIZING SALTS; ECRINITRIT; EEC No. 007-010-00-4; EINECS No. 231-555-9; ENTEC 336 CORROSION INHIBITOR/DISPERSANT; ENTEC

966 Sodium Nitrite

338 CORROSION INHIBITOR; FILMERINE; NATRIUM NITRIT (German); NCI-C02084; NITRITE de SODIUM (French); NITRITO SODICO (Spanish); NITROUS ACID, SODIUM SALT; SUSITAN SODNY (Czech)

IDENTIFICATION

CAS:	7632-00-0
ERG ID:	UN 1500
ERG Guide:	140
ERG Guide Category:	OXIDIZERS
Molecular Formula:	$NNaO_2$
RTECS No.:	RA 1225000
Properties:	Noncombustible white or pale yellow crystals, granules, stick or powder. Odorless. Sinks and mixes with water.

HEALTH AND SAFETY INFORMATION

Poisonous. Overexposure can cause death. Repeated or prolonged exposure can cause the formation of methemoglobin; blood and blood vessel problems.

RECOMMENDATIONS FOR RESPIRATOR SELECTION

At any concentrations above the NIOSH REL, or where there is no REL, at any detectable concentration: SCBAF:PD,PP (any self-contained breathing apparatus that has a full facepiece and is operated in a pressure-demand or other positive-pressure mode); or SAF:PD,PP:ASCBA (any supplied-air respirator that has a full facepiece and is operated in a pressure-demand or other positive-pressure mode in combination with an auxiliary, self-contained breathing apparatus operated in a pressure-demand or other positive-pressure mode).

Escape: GMFOV [any air-purifying, full-facepiece respirator (gas mask) with a chin-style, front- or back-mounted organic vapor canister]; or SCBAE (any appropriate escape-type, self-contained breathing apparatus).

CHEMICAL REACTIVITY

Decomposes explosively at temperatures above 1000°F (538°C) A strong oxidizer; violent reaction with reducing agents and combustibles. Reacts violently (may cause fire and explosions) with acids (forms nitric oxide vapors), ammonium salts and heat, butadiene, cyanides, lithium, phthalic acid, thiosulfate, sodium amide, and many other substances. Under certain conditions the chemical can also be a strong reducing agent. Friction and material may ignite organic matter.

FIRE INFORMATION

Noncombustible but can increase the combustibility of other materials and intensify fire activity. If material or contaminated runoff enters waterways, notify downstream users of potentially contaminated waters. Notify local health and fire officials and pollution control agencies. Use extinguishing agent appropriate for surrounding fires. From a secure, explosion-proof location, use water spray to cool exposed containers.

FEMA Hazard Classifications:

Health Hazards (Blue) 2
Flammability (Red) 0
Reactivity (Yellow) 1
Special Situations . . . . OXY

FIRST AID

Move victim to fresh air. Call emergency medical care. Apply artificial respiration if victim is not breathing. Administer oxygen if breathing is difficult. Remove and isolate contaminated clothing and shoes. In case of contact with substance, immediately flush skin or eyes with running water for at least 20 minutes. Keep victim warm and quiet. Ensure that medical personnel are aware of the material(s) involved and take precautions to protect themselves.

ENVIRONMENTAL INFORMATION

Clean Water Act: Section 311 Hazardous Substances/RQ (same as CERCLA).
EPCRA Section 304: Reportable Quantity (RQ): CERCLA, 100 lb (45.4 kg).
EPCRA Section 313: Form R *de minimis* concentration reporting level: 1.0%.

SODIUM PENTACHLOROPHENTATE S:505

SYNONYMS: ALKALISALZE von PENTACHLORPHENOL (German); ALKALIZOUTEN van PENTACHLOORFENOL (Dutch); DORMANT FUNGICIDE; DOWCIDE G-ST; EINECS No. 205-025-2; EPA PESTICIDE CHEMICAL CODE 063003; MYSTOX D; NAPCHLOR-G; NATRIUMPENTACHLORPHENOLAT (German); PENTACLOROFENATO SODICO (Spanish); PHENOL, PENTACHLOROPHENOL, SODIUM SALT; PENTACHLOROPHENOXY SODIUM; PENTAPHENATE; SANTOBRITE; SODIUM PCP; SODIUM PENTALCHLOROPHENOL; SODIUM PENTALCHLOROPHENOL, SODIUM SALT; WEEDBEADS; WITOPHEN N

IDENTIFICATION

CAS:	131-52-2
ERG ID:	UN 2567
ERG Guide:	154
ERG Guide Category:	SUBSTANCES-TOXIC AND/OR CORROSIVE (NONCOMBUSTIBLE)
Molecular Formula:	C_6Cl_5NaO
RTECS No.:	SM6650000
Properties:	Noncombustible crystalline solid; beads or powder. Buff colored. Phenolic odor. Very soluble in water.

HEALTH AND SAFETY INFORMATION

Poisonous. Corrosive to eyes, skin, and respiratory tract. Inhalation may cause pulmonary edema, which can be delayed for several hours; there is a risk of death in serious cases. Can cause liver and kidney damage.

RECOMMENDATIONS FOR RESPIRATOR SELECTION

At any concentrations above the NIOSH REL, or where there is no REL, at any detectable concentration: SCBAF:PD,PP (any self-contained breathing apparatus that has a full facepiece and is operated in a pressure-demand or other positive-pressure mode); or SAF:PD,PP:ASCBA (any supplied-air respirator that has a full facepiece and is operated in a pressure-demand or other positive-pressure mode in combination with an auxiliary, self-contained breathing apparatus operated in a pressure-demand or other positive-pressure mode).

Escape: GMFOV [any air-purifying, full-facepiece respirator (gas mask) with a chin-style, front- or back-mounted organic vapor canister]; or SCBAE (any appropriate escape-type, self-contained breathing apparatus).

CHEMICAL REACTIVITY

Incompatible with strong oxidizing agents.

FIRE INFORMATION

Combustion products include highly toxic sodium oxides. If material or contaminated runoff enters waterways, notify downstream users of potentially contaminated waters. Notify local health and fire officials and pollution control agencies. Use dry chemicals, alcohol-resistant foam, carbon dioxide.

NFPA Hazard Classifications: Not listed

FIRST AID

Move victim to fresh air; call emergency medical care. Apply artificial respiration if victim is not breathing. Do not use mouth-to-mouth method if victim ingested or inhaled the substance; induce artificial respiration with the aid of a pocket mask equipped with a one-way valve or other proper respiratory medical device. Administer oxygen if breathing is difficult. Remove and isolate contaminated clothing and shoes. In case of contact with substance, immediately flush skin or eyes with running water for at least 20 minutes. For minor skin contact, avoid spreading material on unaffected skin. Keep victim warm and quiet. Effects of exposure (inhalation, ingestion, or skin contact) to substance may be delayed. Ensure that medical personnel are aware of the material(s) involved and take precautions to protect themselves.

ENVIRONMENTAL INFORMATION

EPA Hazardous Waste Number (RCRA No.): Listed as "None"
EPCRA Section 313: Form R *de minimis* concentration reporting level: 1.0%.
Marine Pollutant (49CFR, Subchapter 172.101, Appendix B)

SODIUM PHOSPHOTUNGSTATE S:523

SYNONYMS: EINECS No. 257-132-9; PHOSPHOTUNGSTIC ACID, SODIUM SALT; SODIUM TUNGSTEN HYDROXIDE OXIDE PHOSPHATE; SODIUM TUNGSTOPHOSPHATE; SODIUM-12-TUNGSTOPHOSPHATE; 12-TUNGSTOFOSFATO SODICO (Spanish); TUNGSTOPHOSPHORIC ACID, SODIUM SALT

IDENTIFICATION

CAS:	51312-42-6
ERG ID:	UN 3086 as toxic solid, oxidizing, n.o.s.
ERG Guide:	141
ERG Guide Category:	OXIDIZERS-TOXIC (SOLID)
Molecular Formula:	$Na_4O_2 \cdot O_5P_2 \cdot O_{36}W_{12} \cdot 18H_2O$
RTECS No.:	TH5775000
Properties:	Noncombustible powder. Yellowish white. Soluble in water.

HEALTH AND SAFETY INFORMATION

Tungsten compounds are moderately toxic. Irritates eyes, skin, and respiratory tract. May affect central nervous system. Long-term effect may include weight loss and behavior and blood changes.

Exposure Limits:	**TWA**	**STEL**
ACGIH:	1 mg/m^3 as tungsten	3 mg/m^3 as tungsten
NIOSH:	1 mg/m^3 as tungsten	3 mg/m^3 as tungsten
OSHA:	None	

RECOMMENDATIONS FOR RESPIRATOR SELECTION: NIOSH as soluble tungsten compounds.

10 mg/m³: HiEF (any air-purifying respirator with a high-efficiency particulate filter);or SA (any supplied-air respirator).
2.5 ppm: SA:CF (any supplied-air respirator operated in a continuous-flow mode).
50 mg/m³: HiEF (any air-purifying, full-facepiece respirator with a high-efficiency particulate filter); or SCBAF (any self-contained breathing apparatus with a full facepiece); or SAF (any supplied-air respirator with a full facepiece).

Emergency or Planned Entry into Unknown Concentrations or IDLH Conditions:
SCBAF:PD,PP (any self-contained breathing apparatus that has a full facepiece and is operated in a pressure-demand or other positive-pressure mode); or SAF:PD,PP:ASCBA (any supplied-air respirator that has a full facepiece and is operated in a pressure-demand or other positive-pressure mode in combination with an auxiliary, self-contained breathing apparatus operated in a pressure-demand or other positive-pressure mode).

Escape: HiEF (any air-purifying, full-facepiece respirator with a high-efficiency particulate filter); or SCBAE (any appropriate escape-type, self-contained breathing apparatus).

Note: Only nonoxidiziable sorbents should be used (not charcoal).

CHEMICAL REACTIVITY

A strong oxidizing agent. Decomposes to phosphorous oxides and phosphine. Avoid moisture.

FIRE INFORMATION

Combustion products include toxic sodium and phosphorus oxides. If material or contaminated runoff enters waterways, notify downstream users of potentially contaminated waters. Notify local health and fire officials and pollution control agencies. Use dry chemicals, alcohol-resistant foam, carbon dioxide.

NFPA Hazard Classifications: Not listed

FIRST AID

Move victim to fresh air. Call emergency medical care. Apply artificial respiration if victim is not breathing. Administer oxygen if breathing is difficult. Remove and isolate contaminated clothing and shoes. In case of contact with substance, immediately flush skin or eyes with running water for at least 20 minutes. Keep victim warm and quiet. Ensure that medical personnel are aware of the material(s) involved and take precautions to protect themselves.

SODIUM THIOCYANATE S:590

SYNONYMS: EEC No. 615-004-00-3; EINECS No. 208-754-4; HAIMASED; NATRIUMRHODANID (German); SCYAN; SODIUM ISOTHIOCYANATE; SODIUM RHODANATE; SODIUM RHODANIDE; SODIUM SULFOCYANATE; SODIUM SULFOCYANIDE; SODIUM THIOSULFONATE; SODIUM THIOCYANIDE; THIOCYANATE SODIUM; TIOCIANATO SODICO (Spanish); THIOCYANIC ACID, SODIUM SALT; USAF EK-T-434; ZOUTAN van THIOCYAANZUUR (Dutch)

IDENTIFICATION

CAS:	540-72-7
ERG ID:	UN 3288
ERG Guide:	151
ERG Guide Category:	SUBSTANCES-TOXIC (NONCOMBUSTIBLE)
Molecular Formula:	CNS•Na; CNNaS; CHNS•Na
RTECS No.:	XL2275000
Properties:	Noncombustible, colorless hygroscopic crystalline solid, or white powder. Soluble in water.

HEALTH AND SAFETY INFORMATION

IDLH: 25 mg/m^3 as CN
OSHA Table Z-1-A Air Contaminant as CN

Exposure Limits:	**TWA**	**CEILING**
ACGIH:[1]	4.7 ppm; 5 mg/m^3 as CN	
NIOSH:		4.7 ppm; 5 mg/m^3 [10 min] as CN
OSHA:	5 mg/m^3 as CN	

[1] Skin contact contributes significantly in overall exposure.

RECOMMENDATIONS FOR RESPIRATOR SELECTION: NIOSH/OSHA as potassium cyanide

25 mg/m^3: SA (any supplied-air respirator); or SCBAF (any self-contained breathing apparatus with a full facepiece).

Emergency or Planned Entry into Unknown Concentrations or IDLH Conditions: SCBAF:PD,PP (any self-contained breathing apparatus that has a full facepiece and is operated in a pressure-demand or other positive-pressure mode); or SAF:PD,PP:ASCBA (any supplied-air respirator that has a full facepiece and is operated in a pressure-demand or other positive-pressure mode in combination with an auxiliary, self-contained breathing apparatus operated in a pressure-demand or other positive-pressure mode).

Escape: GMFOVHiE [any air-purifying, full-facepiece respirator (gas mask) with a chin-style, front-or back-mounted canister having a high-efficiency particulate filter]; or SCBAE (any appropriate escape-type, self-contained breathing apparatus).

FIRE INFORMATION

Combustion products include toxic nitrogen, sodium and sulfur oxides. Noncombustible solid. If material or contaminated runoff enters waterways, notify downstream users of potentially contaminated waters. Notify local health and fire officials and pollution control agencies. Use dry chemicals, alcohol-resistant foam, carbon dioxide. From a secure, explosion-proof location, use water spray to cool exposed containers.

FEMA Hazard Classifications:

Health Hazards (Blue) 2
Flammability (Red) 0
Reactivity (Yellow) 0

FIRST AID

Move victim to fresh air. Call emergency medical care. Apply artificial respiration if victim is not breathing. Do not use mouth-to-mouth method if victim ingested or inhaled the substance; induce artificial respiration with the aid of a pocket mask equipped with a one–way valve or other proper respiratory medical device. Administer oxygen if breathing is difficult. Remove and isolate contaminated clothing and shoes. In case of contact with substance, immediately flush skin or eyes with running water for at least 20 minutes. For minor skin contact, avoid spreading material on unaffected skin. Keep victim warm and quiet. Effects of exposure (inhalation, ingestion, or skin contact) may be delayed. Ensure that medical personnel are aware of the material(s) involved and take precautions to protect themselves.

ENVIRONMENTAL INFORMATION as cyanide compounds

Clean Air Act: Hazardous Air Pollutants (Title I, Part A, Section 112).
Clean Water Act: Section 307 Priority Pollutants as cyanide, total.
EPA Hazardous Waste Number (RCRA No.): P030 as cyanides soluble salts and complexes, n.o.s.
RCRA Section 261 Hazardous Constituents. as cyanides, soluble salts and complexes, n.o.s.
EPCRA (Section 313): X+CN- where X = H+ or any other group where a formal dissociation may occur. For example, KCN or Ca(CN)2. Form R *de minimis* concentration reporting level: 1.0%. Form R Toxic Chemical Category Code: N016.
Marine Pollutant (49CFR, Subchapter 172.101, Appendix B) as cyanide mixtures, cyanide solutions, or cyanides, inorganic, n.o.s.

SORBIC ACID S:602

SYNONYMS: ACIDO SORBICO (Spanish); (2-BUTENYLIDENE)ACETIC ACID; CROTYLIDENE ACETIC ACID; EINECS No. 203-768-7; HEXADIENIC ACID; HEXADIENOIC ACID; 2,4-HEXADIENOIC ACID; trans-trans-2,4-HEXADIENOIC ACID; 1,3-PENTADIENE-1-CARBOXYLIC ACID; 2-PROPENYLACRYLIC ACID; SORBINSAEURE (German); SORBINSYRE (French); SORBISTAT

IDENTIFICATION

CAS:	110-44-1
Molecular Formula:	$C_6H_8O_2$
RTECS No.:	WG2100000
Properties:	Combustible needles or powdery solid. Characteristic odor. Slightly soluble in water.

HEALTH AND SAFETY INFORMATION

Toxic if swallowed. Corrosive to eyes, skin, and respiratory tract. Inhalation of dust may cause pulmonary edema, which can be delayed for several hours; there is a risk of death in serious cases.

CHEMICAL REACTIVITY

Contact with strong oxidizers may cause fire and explosions. Organic acids are incompatible with sulfuric acid, caustics, ammonia, aliphatic amines, alkanolamines, isocyanates, alkylene oxides, epichlorohydrin.

FIRE INFORMATION

Combustible solid. Combustion may produce irritants and toxic vapors. If material or contaminated runoff enters waterways, notify downstream users of potentially contaminated waters. Notify local health and fire officials and pollution control agencies. Use dry chemicals, alcohol-resistant foam, carbon dioxide. From a secure, explosion-proof location, use water spray to cool exposed containers.

NFPA Hazard Classifications: Not listed

Flash Point: 260°F (126°C) o.c.

FIRST AID

Move victim to fresh air. Call emergency medical care. Apply artificial respiration if victim is not breathing. Administer oxygen if breathing is difficult. Remove and isolate contaminated clothing and shoes. In case of contact with substance, immediately flush skin or eyes with running water for at least 20 minutes. Wash skin with soap and water. Keep victim warm and quiet. Effects of exposure (inhalation, ingestion, or skin contact) to substance may be delayed. Ensure that medical personnel are aware of the material(s) involved and take precautions to protect themselves.

SORBITOL S:604

SYNONYMS: A-625/641; CHOLAXINE; CLUCITOL; DIAKARMON; EINECS No. 200-061-5; GLUCITOL; D-GLUCITOL; GULITOL; L-GULITOL; 1,2,3,4,5,6,-HEXANNEHEXOL; HEXAHYDRIC ALCOHOL; HYDREX® 100; KARION; LIPONIC 70-NC; NIVITIN; SIONIT; SIONON; SORBELITE C; SORBIDEL; SORBITE; SORBITOL (ECIG); SORBOL; SORBO®; SORBOL; SORBOSTYL; SORVILANDE

IDENTIFICATION

CAS:	50-70-4
Molecular Formula:	$C_6H_{14}O_6$
RTECS No.:	LZ4290000
Properties:	Colorless liquid or white powder. Odorless. Sweet taste. Sinks and mixes with water.

CHEMICAL REACTIVITY

Reacts with strong oxidizing agents. Incompatible with sulfuric acid, nitric acid, caustics, aliphatic amines, isocyanates.

FIRE INFORMATION:

FEMA Hazard Classifications:

Health Hazards (Blue) 0

Flammability (Red) 1
Reactivity (Yellow) 0

Flash Point: 542°F (283°C) o.c.

FIRST AID

Move victim to fresh air. Call emergency medical care. Apply artificial respiration if victim is not breathing. Administer oxygen if breathing is difficult. Remove and isolate contaminated clothing and shoes. In case of contact with substance, immediately flush skin or eyes with running water for at least 20 minutes. Wash skin with soap and water. Keep victim warm and quiet. Effects of exposure (inhalation, ingestion, or skin contact) to substance may be delayed. Ensure that medical personnel are aware of the material(s) involved and take precautions to protect themselves.

STANNIC CHLORIDE S:605

SYNONYMS: CLORURO ESTANNICO (Spanish); EEC No. 050-001-00-5; EINECS No. 231-588-9; ETAIN (TETRACHLORURE d') (French); LIBAVIUS FUMING SPIRIT; STAGNO (TETRACLORURO di) (Italian); STANNANE, TETRACHLORO; STANNIC CHLORIDE, ANHYDROUS; STANNIC TETRACHLORIDE; TETRACHLOROSTANNANE; TETRACHLOROTIN; TIN CHLORIDE ($SnCl_4$); TIN PERCHLORIDE; TIN TETRACHLORIDE; TIN TETRACHLORIDE, ANHYDROUS; TIN(IV) CHLORIDE; TIN(IV) TETRACHLORIDE; TINTETRACHLORIDE (Dutch); ZINNTETRACHLORID (German)

IDENTIFICATION

CAS:	7646-78-8
ERG ID:	UN 1827
ERG Guide:	137
ERG Guide Category:	WATER REACTIVE-CORROSIVE
Molecular Formula:	Cl_4Sn
RTECS No.:	XP8750000
Properties:	Crystals, lumpy solid, or hygroscopic liquid. Colorless to yellow liquid; white to yellow solid; fumes in moist air. Acrid, pungent odor. Soluble in water; generates heat and forming tin oxide and hydrochloric acid.

HEALTH AND SAFETY INFORMATION

Corrosive to eyes, skin, and respiratory tract; skin or eye contact can cause burns and blindness. Inhalation may cause pulmonary edema, which can be delayed for several hours; there is a risk of death in serious cases. Long-term or prolonged exposure can cause brain, liver, and kidney damage.

IDLH: 100 mg/m^3 as tin, inorganic compounds.

OSHA Table Z-1-A Air Contaminant as tin, inorganic compounds

Exposure Limits:	**TWA**
ACGIH:	2 mg/m^3 as tin
NIOSH:	2 mg/m^3 as tin
OSHA:	2 mg/m^3 as tin

RECOMMENDATIONS FOR RESPIRATOR SELECTION: NIOSH as inorganic tin

10 mg/m^3: DM (any dust and mist respirator).
20 mg/m^3: DMXSQ (any dust and mist respirator except single-use and quarter mask respirators); or SA (any supplied-air respirator).
50 mg/m^3: SA:CF (any supplied-air respirator operated in a continuous-flow mode); or PAPRDM (any powered, air-purifying respirator with a dust and mist filter).
100 mg/m^3: HiEF (Any air-purifying, full-facepiece respirator with a high-efficiency particulate filter); or SCBAF (any self-contained breathing apparatus with a full facepiece); or SAF (any supplied-air respirator with a full facepiece).

Emergency or Planned Entry into Unknown Concentrations or IDLH Conditions: SCBAF:PD,PP (any self-contained breathing apparatus that has a full facepiece and is operated in a pressure-demand or other positive-pressure mode); or SAF:PD,PP:ASCBA (Any supplied-air respirator that has a full facepiece and is operated in a pressure-demand or other positive-pressure mode in combination with an auxiliary, self-contained breathing apparatus operated in a pressure-demand or other positive pressure mode.

Escape: HiEF (Any air-purifying, full-facepiece respirator with a high-efficiency particulate filter); or SCBAE (Any appropriate escape-type, self-contained breathing apparatus).

CHEMICAL REACTIVITY

Reacts violently with oxidizers, potassium, sodium, bromine trifluoride, chlorine, ethylene oxide, metal nitrates (explosive). Moist air forms hydrochloric acid vapors. Evolves heat, tin oxide, and hydrochloric acid on contact with moisture. Corrosive to metals in the presence of moisture.

FIRE INFORMATION

Noncombustible. Combustion products include toxic fumes of chlorides and tin. Vapors are heavier than air and will collect in low areas. If material or contaminated runoff enters waterways, notify downstream users of potentially contaminated waters. Notify local health and fire officials and pollution control agencies. Use dry chemicals, alcohol-resistant foam, carbon dioxide.

NFPA Hazard Classifications:

Health Hazards (Blue) 3
Flammability (Red) 0
Reactivity (Yellow) 1

FIRST AID

Move victim to fresh air. Call emergency medical care. Apply artificial respiration if victim is not breathing. Do not use mouth-to-mouth method if victim ingested or inhaled the substance; induce artificial respiration with the aid of a pocket mask equipped with a one-way valve or other proper respiratory medical device. Administer oxygen if breathing is difficult. Remove and isolate contaminated clothing and shoes. In case of contact with substance, immediately flush skin or eyes with running water for at least 20 minutes. For minor skin contact, avoid spreading material on unaffected skin. Removal of solidified molten material from skin requires medical assistance. Keep victim warm and quiet. Effects of exposure (inhalation, ingestion, or skin contact) to substance may be delayed. Ensure that medical personnel areaware of te material(s) involved and take precautions to protect themselves.

ENVIRONMENTAL INFORMATION as tin

RCRA List of inorganic and organic constituents: 40CFR258, Appendix 2
RCRA Ground Water Monitoring List: Suggested test methods(s) (PQL μg/L); (total) 7870(8,000)

STARCH DUST S:630

SYNONYMS: ALMIDON (Spanish); AMAIZO W 13; AMYLOMAIZE VII; AMYLUM; AQUAPEL (POLYSACCHARIDE); ARGO BRAND CORN STARCH; ARROWROOT STARCH; CLARO 5591; CLEAREL; CLEARJEL; CORN PRODUCTS; CPC 3005; CPC 6448; EINECS No. 232-679-6; FARINEX 100; GALACTASOL A; GENVIS; HRW 13; KEESTAR; MAIZENA; MARANTA; MELOGEL; MELUNA; OK PRE-GEL; PENFORD GUM 380; RICE STARCH; SORGHUM GUM; STARAMIC 747; a-STARCH; STARCH, CORN; STARCH DUST; STA-RX 1500; TAPIOCA STARCH; TAPON; TROGUM; W-GUM; W-13 STABILIZER

IDENTIFICATION

CAS:	9005-25-8
Molecular Formula:	Carbohydrate polymer composed of 25% amylose and 75% amylpectin
RTECS No.:	GM5090000
Properties:	Flammable powder or granules. White. Slight odor. Tasteless.

HEALTH AND SAFETY INFORMATION

Irritates eyes, skin, and respiratory tract. May cause allergies.

OSHA Table Z-1-A Air Contaminant

Exposure Limits:	**TWA**
ACGIH: (1)	10 mg/m^3 (inhalable particulate); 3 mg/m^3 (respirable particulate)

NIOSH: 10 mg/m³ (total dust); 5 mg/m³ (respirable fraction)
OSHA: 15 mg/m³ (total dust); 5 mg/m³ (respirable fraction)

[1] For inhalable (total) particulate matter containing no asbestos and <1% crystalline silica. Not classifiable as a human carcinogen.

RECOMMENDATIONS FOR RESPIRATOR SELECTION

At any concentrations above the NIOSH REL, or where there is no REL, at any detectable concentration: SCBAF:PD,PP (any self-contained breathing apparatus that has a full facepiece and is operated in a pressure-demand or other positive-pressure mode); or SAF:PD,PP:ASCBA (any supplied-air respirator that has a full facepiece and is operated in a pressure-demand or other positive-pressure mode in combination with an auxiliary, self-contained breathing apparatus operated in a pressure-demand or other positive-pressure mode).

Escape: GMFOVHiE [any air-purifying, full-facepiece respirator (gas mask) with a chin-style, front- or back-mounted organic vapor canister having a high-efficiency particulate filter]; or SCBAE (any appropriate escape-type, self-contained breathing apparatus).

CHEMICAL REACTIVITY

Reacts with oxidizers, acids, iodine, alkalis. Can form an explosive mixture with air; flammable and explosive when exposed to flame. Contact with chlorates may cause a violent reaction.

FIRE INFORMATION

If material or contaminated runoff enters waterways, notify downstream users of potentially contaminated waters. Notify local health and fire officials and pollution control agencies. Use any extinguishing agent suitable for surrounding fire. From a secure, explosion-proof location, use water spray to cool exposed containers.

NFPA Hazard Classifications: Not listed

FIRST AID

Move victim to fresh air. Call emergency medical care. Apply artificial respiration if victim is not breathing. Administer oxygen if breathing is difficult. Remove and isolate contaminated clothing and shoes. In case of contact with substance, immediately flush skin or eyes with running water for at least 20 minutes. Wash skin with soap and water. Keep victim warm and quiet. Effects of exposure (inhalation, ingestion, or skin contact) to substance may be delayed. Ensure that medical personnel are aware of the material(s) involved and take precautions to protect themselves.

STRONTIUM CHROMATE S:646

SYNONYMS: CROMATO de ESTRONCIO (Spanish); CHROMIC ACID, STRONTIUM SALT; C.I. PIGMENT YELLOW 32; DEEP LEMON YELLOW; EEC No. 024-009-00-4; EINECS No. 232-142-6;

978 Strontium Chromate

STRONTIUM CHROMATE; STRONTIUM CHROMATE 12170; STRONTIUM CHROMATE A; STRONTIUM CHROMATE(VI); STRONTIUM CHROMATE X-2396; STRONTIUM YELLOW; STRONZIO CROMATO (Italian)

IDENTIFICATION

CAS:	7789-06-2
ERG ID:	UN 2811
ERG Guide:	154
ERG Guide Category:	SUBSTANCES-TOXIC AND/OR CORROSIVE (NON-COMBUSTIBLE)
Molecular Formula:	CrO_4Sr
RTECS No.:	GB3240000
Properties:	Solid powder or crystals. Yellow. Odorless. Sinks in water; slowly soluble.

HEALTH AND SAFETY INFORMATION

Highly toxic if swallowed. Irritates, eyes, skin, and respiratory tract. Prolonged or repeated exposure may result in chrome sores of the skin and nasal passages. May cause allergies. A carcinogen. Skin contact contributes significantly to overall exposure.

IARC Evaluation: Group 1; sufficient animal evidence.
NTP: Carcinogen.
IDLH: 15 mg/m^3 as chromium(VI) compound
NIOSH Criteria Document: NIOSH 76-129 Chromium(VI)

Exposure Limits:	**TWA**	**CEILING**
ACGIH: (1)	0.0005 mg/m^3 as chromate	
NIOSH: (2)	0.001mg/m^3 as chromates	
OSHA:		0.1 as chromates

(1) Suspected human carcinogen
(2) Potential occupational carcinogen

RECOMMENDATIONS FOR RESPIRATOR SELECTION: NIOSH

Any detectable concentration: SCBAF:PD,PP (any self-contained breathing apparatus that has a full facepiece and is operated in a pressure-demand or other positive-pressure mode); or SAF:PD,PP:ASCBA (any supplied-air respirator that has a full facepiece and is operated in a pressure-demand or other positive-pressure mode in combination with an auxiliary, self-contained breathing apparatus operated in a pressure-demand or other positive pressure mode).

ESCAPE: HiEF (any air-purifying, full-facepiece respirator with a high-efficiency particulate filter); or SCBAE (any appropriate escape-type, self-contained breathing apparatus).

CHEMICAL REACTIVITY

Reacts violently with hydrazine. Incompatible with organic or other readily oxidizable materials such as wood, paper, sulfur, aluminum, and plastics. Produces hazardous solution in water. Avoid contact with water, acids, and bases.

FIRE INFORMATION

Poisonous fumes produced when heated. Firefighting gear (including SCBA) may not provide adequate protection. If exposure occurs, remove and isolate gear immediately and thoroughly decontaminate personnel. If material or contaminated runoff enters waterways, notify downstream users of potentially contaminated waters. Notify local health and fire officials and pollution control agencies. Use dry chemicals, foam, carbon dioxide. From a secure, explosion-proof location, use water spray to cool exposed containers.

FEMA Hazard Classifications

Health Hazards (Blue) 1
Flammability (Red) 1
Reactivity (Yellow) 0
Special Notice .... Water

Flash Point: Not flammable

FIRST AID

Move victim to fresh air. Call emergency medical care. Apply artificial respiration if victim is not breathing. Administer oxygen if breathing is difficult. In case of contact with substance, immediately flush skin or eyes with running water for at least 20 minutes. Remove and isolate contaminated clothing and shoes. Ensure that medical personnel are aware of the material(s) involved and take precautions to protect themselves.

ENVIRONMENTAL INFORMATION as strontium chromate

Clean Water Act: Section 311 Hazardous Substances/RQ (same as CERCLA)
EPCRA Section 304 Reportable Quantity (RQ): CERCLA, 10 lb (4.54 kg)

ENVIRONMENTAL INFORMATION as chromium and compounds, and chromium(VI) compounds

CLEAN AIR ACT: Hazardous Air Pollutants (Title I, Part A, Section 112).
CLEAN WATER ACT: Toxic Pollutant (Section 401.15); Section 307 Toxic Pollutants as chromium and compounds.
RCRA Section 261 Hazardous Constituents: waste number not listed.
EPCRA (Section 313): Includes any unique chemical substances that contains chromium as part of that chemical's infrastructure. Form R *de minimus* concentration reporting level: Chromium (VI) compounds: 0.1%.; Chromium(III) compounds: 1.0%. Form R Toxic Chemical Category Code: N090.
CALIFORNIA'S PROPOSITION 65: Carcinogen as chromium (hexavalent compounds).

STYRENE-ACRYLONITRILE COPOLYMER S:652

SYNONYMS: ACRILAFIL; ACRYLONITRILE-STYRENE COPOLYMER; ACRYLONITRILE-STYRENE POLYMER; ACRYLONITRILE-STYRENE RESIN; ACS; AS 61CL; BAKELITE RMD 4511; CEVIAN HL; COPOLIMERO de ESTIRENO-ACRILONITRILO (Spanish); DIALUX; ESTYRENE AS; EMU; KOSTIL; LITAC; LURAN; LUSTRAN; POLYSTYRENE-ACRYLONITRILE; 2-PROPENENITRILE POLYMER with ETHENYLBENZENE; REXENE 106; SANREX; SN 20; STYREN-ACRYLONITRILEPOLYMER; STYRENE-ACRYLONITRILE COPOLYMER; TERULAN KP 2540; TYRIL

IDENTIFICATION

CAS:	9003-54-7
Molecular Formula:	$(C_8H_8{\cdot}C_3H_3N)_x$
RTECS No.:	AT6978000
Properties:	Solid pellets. Sweet aromatic odor. Insoluble in water.

HEALTH AND SAFETY INFORMATION

Poisonous if swallowed. Long-term exposure may cause enlarged thyroid; nose sores and bleeding. High doses may decrease fertility in women and men. A questionable carcinogen.

IARC Evaluation: Group 3; no adequate human data; insufficient animal data.
IDLH: 25 mg/m^3 as CN
OSHA Table Z-1-A Air Contaminant as CN

Exposure Limits:	**TWA**	**CEILING**
ACGIH:[1]		4.7 ppm; 5 mg/m^3
NIOSH:		4.7 ppm; 5 mg/m^3 [10 min]
OSHA:	5 mg/m^3	

[1] Skin contact contributes significantly in overall exposure.

Exposure Limits:	**TWA**	**CEILING**
ACGIH: [1]	5 mg/m^3 as cyanide	
NIOSH: [1]		4.7 ppm; 5 mg/m^3 per 10 minutes as cyanides
OSHA: [1]	5 mg/m^3 as cyanides	

[1] Skin contact contributes significantly in overall exposure

RECOMMENDATIONS FOR RESPIRATOR SELECTION: NIOSH/OSHA as Cyanide

25 mg/m^3: SA (any supplied-air respirator); or SCBAF (any self-contained breathing apparatus with a full facepiece).

Emergency or Planned Entry into Unknown Concentrations or IDLH Conditions: SCBAF:PD,PP (any self-contained breathing apparatus that has a full facepiece and is operated in a pressure-demand or other positive-pressure mode); or SAF:PD,PP:ASCBA (any supplied-air respirator that has a full facepiece and is operated in a pressure-demand or other positive-pressure mode in combination with an auxiliary, self-contained breathing apparatus operated in a pressure-demand or other positive-pressure mode).

Escape: GMFOVHiE [any air-purifying, full-facepiece respirator (gas mask) with a chin-style, front- or back-mounted canister having a high-efficiency particulate filter]; or SCBAE (any appropriate escape-type, self-contained breathing apparatus).

CHEMICAL REACTIVITY

Reportedly very stable and nonreactive. May be incompatible with chlorosulfuric acid, powerful organic peroxides, fuming sulfuric acid.

FIRE INFORMATION

Combustion products include toxic cyanides and nitrogen oxides. If material or contaminated runoff enters waterways, notify downstream users of potentially contaminated waters. Notify local health and fire officials and pollution control agencies. Use dry chemicals, alcohol-resistant foam, carbon dioxide. From a secure, explosion-proof location, use water spray to cool exposed containers.

NFPA Hazard Classifications: Not listed

Flash Point: 716°F (380°C)

FIRST AID

Move victim to fresh air. Call emergency medical care. Apply artificial respiration if victim is not breathing. Do not use mouth-to-mouth method if victim ingested or inhaled the substance; induce artificial respiration with the aid of a pocket mask equipped with a one-way valve or other proper respiratory medical device. Administer oxygen if breathing is difficult. Remove and isolate contaminated clothing and shoes. In case of contact with substance, immediately flush skin or eyes with running water for at least 20 minutes. For minor skin contact, avoid spreading material on unaffected skin. Keep victim warm and quiet. Effects of exposure (inhalation, ingestion, or skin contact) may be delayed. Ensure that medical personnel are aware of the material(s) involved and take precautions to protect themselves.

ENVIRONMENTAL INFORMATION as cyanide compounds

CLEAN AIR ACT: Hazardous Air Pollutants (Title I, Part A, Section 112)
CLEAN WATER ACT: Section 307 Priority Pollutants as cyanide, total.
EPA HAZARDOUS WASTE NUMBER (RCRA No.): P030 as cyanides soluble salts and complexes, n.o.s.
RCRA Section 261 Hazardous Constituents. as cyanides, soluble salts and complexes, n.o.s.

EPCRA (Section 313): X+CN- where X = H+ or any other group where a formal dissociation may occur. For example, KCN or $Ca(CN)_2$. Form R *de minimus* concentration reporting level: 1.0%. Form R Toxic Chemical Category Code: N016
MARINE POLLUTANT (49CFR, Subchapter 172.101, Appendix B) as cyanide mixtures

STYRENE MONOMER S:660

SYNONYMS: ANNAMENE; BENZENE, VINYL-; CASTOLITE-AC; CEILCRETE 6650; CINNAMENE; CINNAMENOL; CINNAMOL; DIAREX HF 77; EEC No. 601-026-00-0; EINECS No. 202-851-5; ETHENYL BENZENE; ETHYLENE, PHENYL-; FAST CURE 45 EPOXY; FLAKELINE 22-HT; MONOMERO de ESTIRENO (Spanish); NCI-C02200; PHENYLETHENE; PHENYLETHYLENE; STIROLO (Italian); STYREEN (Dutch); STYREN (Czech); STYRENE (French); STYRENE MONOMER, INHIBITED; STYROL (German); STYROLE; STYROLENE; STYRON; STYROPOL; STYROPOR; VINYLBENZEN (Czech); VINYL BENZENE; VINYLBENZOL

IDENTIFICATION

CAS:	100-42-5
ERG ID:	UN 2055
ERG Guide:	128P (inhibited)
ERG Guide Category:	FLAMMABLE LIQUIDS (NONPOLAR/WATER-IMMISCIBLE)
Molecular Formula:	C_8H_8
RTECS No.:	WL3675000
Properties:	Highly flammable, colorless to light yellow, oily liquid. Sharp, disagreeable; sweet floral odor at lower concentrations. Floats on water; insoluble. Freezes at -23°F (-131°C).

HEALTH AND SAFETY INFORMATION

Very irritating to skin, eyes, and respiratory tract; which may be accompanied by a persistent metallic taste. Can cause headache, fatigue, drowsiness, decreased physical coordination, depression, lung inflammation. May cause cancer; damage to the fetus, kidney, and liver; decrease fertility in women; nervous system diseases; muscular disorders with tingling in the fingers and toes; headache and upset stomach. Repeated low levels of exposure may affect balance, reflexes, memory, concentration, and ability to learn. Styrene has been found to produce lung tumors in mice and cause changes in the genetic material of laboratory organisms.

IARC Evaluation: Group 2B, possible human carcinogen; human evidence, inadequate; animal evidence, limited.
IDLH: 700 ppm; 10,000 ppm may be fatal in 30–60 minutes.
Odor Threshold: 3.44 ppm
OSHA Table Z-1-A Air Contaminant
OSHA Table Z-2 Air Contaminant

Exposure Limits:	TWA	CEILING	STEL
ACGIH:[1]	20 ppm; 85 mg/m³		40 ppm;170 mg/m³)
NIOSH:	50 ppm; 215 mg/m³	100 ppm;425 mg/m³	
OSHA:	100 ppm	200 ppm, 600 ppm [5 min max peak in any 3 hours]	

[1] Skin contact contributes significantly to overall exposure. Not classifiable as a human carcinogen.

RECOMMENDATIONS FOR RESPIRATOR SELECTION: NIOSH

500 ppm: CCROV * [any chemical cartridge respirator with organic vapor cartridge(s)]; or SA * (any supplied-air respirator).
700 ppm: SA:CF * (any supplied-air respirator operated in a continuous-flow mode); or CCRFOV [any chemical cartridge respirator with a full facepiece and organic vapor cartridge(s)]; or GMFOV [any air-purifying, full-facepiece respirator (gas mask) with a chin-style, front- or back-mounted organic vapor canister]; or PAPROV * [any powered, air-purifying respirator with organic vapor cartridge(s)]; or SCBAF (any self-contained breathing apparatus with a full facepiece); or SAF (any supplied-air respirator with a full facepiece).

Emergency or Planned Entry into Unknown Concentrations or IDLH Conditions: SCBAF:PD,PP (any self-contained breathing apparatus that has a full facepiece and is operated in a pressure-demand or other positive-pressure mode); or SAF:PD,PP:ASCBA (any supplied-air respirator that has a full facepiece and is operated in a pressure-demand or other positive-pressure mode in combination with an auxiliary, self-contained breathing apparatus operated in a pressure-demand or other positive-pressure mode).

Escape: GMFOV [any air-purifying, full-facepiece respirator (gas mask) with a chin-style, front- or back-mounted organic vapor canister]; or SCBAE (any appropriate escape-type, self-contained breathing apparatus).

* Substance reported to cause eye irritation or damage; may require eye protection.

CHEMICAL REACTIVITY

forms explosive mixture with air. If inhibitor (often tert-butylcatechol) is not present in adequate concentrations, polymerization may occur, blocking relief valves and lead to container explosions. Temperatures above 150°F (66°C) speed up polymerization. Incompatible with oxidizers, acids, rust, catalysts for vinyl polymerization, such as peroxides, strong acids, and aluminum chloride. Corrodes copper, copper alloys, and dissolves rubber. Attacks some plastics, rubber, and coatings. May accumulate static electrical charges, and may cause ignition of its vapors. May accumulate static electricity, and cause ignition of its vapors.

FIRE INFORMATION

May react with itself blocking relief valves leading to tank explosions. Vapors are heavier than air and will collect in low areas. Vapors in confined areas may explode when exposed to fire. Vapors may travel long distances to ignition sources and flash back. Storage containers and parts of containers may rocket great distances, in many directions. If material or contaminated runoff enters waterways, notify downstream users of potentially contaminated waters. Notify local health and fire officials and pollution control agencies. For small fire use dry chemicals, AFFF, foam, carbon dioxide.Specially trained personnel operating from a safe distance can fight fire using foam or dry chemicals if available in sufficient amounts. Under favorable conditions experienced crews can use coordinated fog streams to sweep the flames off the surface of the burning liquid. Keep exposures cool to protect against re-ignition. Do not direct straight streams into the liquid. From a secure, explosion-proof location, use water spray in large quantities to cool exposed containers.If cooling streams are ineffective (venting sound increases in volume and pitch, vent sound suddenly increases and/or unexpectedly stops, tank discolors, or shows signs of deforming), withdraw immediately to a secure location.

NFPA Hazard Classifications: (Styrene monomer, inhibited)

Health Hazards (Blue) 2
Flammability (Red) 3
Reactivity (Yellow) 2

Flash Point: 93°F (34°C) o.c.; 88°F (31°C) c.c.
Flammable Limits in Air: LEL 0.9%; UEL 6.8%
Autoignition Temperature: 914°F (490°C)

FIRST AID

Move victim to fresh air. Call emergency medical care. Apply artificial respiration if victim is not breathing. Administer oxygen if breathing is difficult. Remove and isolate contaminated clothing and shoes. In case of contact with substance, immediately flush skin or eyes with running water for at least 20 minutes. Wash skin with soap and water. Keep victim warm and quiet. Ensure that medical personnel are aware of the material(s) involved and take precautions to protect themselves.

ENVIRONMENTAL INFORMATION

Clean Air Act: Hazardous Air Pollutants (Title I, Part A, Section 112).
Clean Water Act: Section 311 Hazardous Substances/RQ (same as CERCLA); Section 313 Priority Chemicals.
RCRA Ground Water Monitoring List: Suggested test method(s) (PQL ug/L): 8020(1); 8240(5).
Safe Drinking Water Act: MCL, 0.1 mg/L; MCLG, 0.1 mg/L.; 40CFR141.32 public notification;
EPCRA Section 304: Reportable Quantity (RQ): CERCLA, 1,000 lb (454 kg).
EPCRA Section 313: Form R *de minimis* concentration reporting level: 0.1%.
Marine Pollutant (49CFR, Subchapter 172.101, Appendix B).

STYRENE OXIDE S:680

SYNONYMS: EPOXYETHYLBENZENE (8CI); EPOXYSTYRENE; α,β-EPOXYSTYRENE;

EPOXYSTYRENE; NCI-C54977; PHENETHYLENE OXIDE; 1-PHENYL-1,2-EPOXYETHANE; PHENYLETHYLENE OXIDE; PHENYLOXIRANE; 1-PHENYLOXIRANE; 2-PHENYLOXIRANE; STYRENE EPOXIDE; STYRENE-7,8-OXIDE; STYRYL OXIDE

IDENTIFICATION

CAS:	96-09-3
ERG ID:	UN 1993 as combustible liquid, n.o.s.
ERG Guide:	128
ERG Guide Category:	FLAMMABLE LIQUIDS (NON POLAR/WATER IMMISCIBLE)
Molecular Formula:	C_8H_8O
RTECS No.:	CZ96250000
Properties:	Combustible liquid. Colorless to yellow. Sweet odor. Sinks in water; practically insoluble.

HEALTH AND SAFETY INFORMATION

Corrosive to eyes. Irritates skin and respiratory tract. Prolonged or repeated exposures may cause liver and kidney damage; muscle and nerve disorders; tingling in fingers and toes. Toxic by ingestion, inhalation and skin absorption. Skin contact contributes significantly to overall exposure. Prolonged or repeated exposures can cause kidney damage. May cause cancer.

IARC Evaluation: Group 2A; animal evidence sufficient; insufficient human evidence.

RECOMMENDATIONS FOR RESPIRATOR SELECTION

At any concentrations above the NIOSH REL, or where there is no REL, at any detectable concentration: SCBAF:PD,PP (any self-contained breathing apparatus that has a full facepiece and is operated in a pressure-demand or other positive-pressure mode); or SAF:PD,PP:ASCBA (any supplied-air respirator that has a full facepiece and is operated in a pressure-demand or other positive-pressure mode in combination with an auxiliary, self-contained breathing apparatus operated in a pressure-demand or other positive-pressure mode).

Escape: GMFOV [any air-purifying, full-facepiece respirator (gas mask) with a chin-style, front- or back-mounted organic vapor canister]; or SCBAE (any appropriate escape-type, self-contained breathing apparatus).

CHEMICAL REACTIVITY

Forms explosive mixture with air. Contact with oxidizers may cause fire and explosions. Contact with acids, caustics, or some salts may cause violent polymerization.

FIRE INFORMATION

Combustion products include toxic cyanides and nitrogen oxides. If material or contaminated runoff enters waterways, notify downstream users of potentially contaminated waters. Notify local health and fire officials and pollution control agencies. Use dry chemicals, alcohol-resistant foam, carbon dioxide. From a secure, explosion-proof location, use water spray to cool exposed containers. Vapors are heavier than air and will collect in low areas. Vapors in confined areas may explode when exposed to fire. Vapors may travel long distances to ignition sources and flash back. Storage containers and parts of containers may rocket great distances, in many directions. If material or contaminated runoff enters waterways, notify downstream users of potentially contaminated waters. Notify local health and fire officials and pollution control agencies. Specially trained personnel operating from a safe distance can fight fire using foam or dry chemicals if available in sufficient amounts. On small fires use dry chemicals, foam, AFFF, carbon dioxide. From a secure, explosion-proof location, use water spray to cool exposed containers. If cooling streams are ineffective (venting sound increases in volume and pitch, vent sound suddenly increases and/or unexpectedly stops, tank discolors, or shows signs of deforming), withdraw immediately to a secure location.

NFPA Hazard Classifications

Health Hazards (Blue) 2
Flammability (Red) 2
Reactivity (Yellow) 0

Flash Point: 165°F (74°C) o.c.

FIRST AID

Move victim to fresh air. Call emergency medical care. Apply artificial respiration if victim is not breathing. Do not use mouth-to-mouth method if victim ingested or inhaled the substance; induce artificial respiration with the aid of a pocket mask equipped with a one-way valve or other proper respiratory medical device. Administer oxygen if breathing is difficult. Remove and isolate contaminated clothing and shoes. In case of contact with substance, immediately flush skin or eyes with running water for at least 20 minutes. For minor skin contact, avoid spreading material on unaffected skin. Keep victim warm and quiet. Effects of exposure (inhalation, ingestion, or skin contact) may be delayed. Ensure that medical personnel are aware of the material(s) involved and take precautions to protect themselves.

ENVIRONMENTAL INFORMATION

CLEAN AIR ACT: Hazardous Air Pollutants (Title I, Part A, Section 112)
EPCRA Section 304 Reportable Quantity (RQ): CERCLA, 1 lb (0.454 kg).
EPCRA Section 313 Form R deminimus concentration reporting level: 0.1%.
CALIFORNIA'S PROPOSITION 65: Carcinogen.

SUCROSE OCTAACETATE S:690

SYNONYMS: α-D-GLUCOPYRANOSIDE, 1,3,4,6-TETRA-O-ACETYL-b-Δ-FRUCTOFURANOSYL-, TETRAACETATE; EINECS No. 204-772-1; OCTAACETYLSUCROSE; OCTACETATO de SACAROSE (Spanish); OTTAACETATO di SACCAROSIO (Italian)

IDENTIFICATION

CAS: 126-14-7
Molecular Formula: $C_{28}H_{38}O_{19}$
RTECS No.: WN66200000
Properties: Combustible hygroscopic crystalline solid. White. Bitter taste. Slightly soluble in water.

HEALTH AND SAFETY INFORMATION

Irritates skin. Dust may cause eye irritation. Due to the availability of insufficient data on human effects, caution should be exercised.

CHEMICAL REACTIVITY

Contact with strong oxidizers may cause fire and explosions.

FIRE INFORMATION

Combustible solid. Combustion products include irritants and toxic vapors. If material or contaminated runoff enters waterways, notify downstream users of potentially contaminated waters. Notify local health and fire officials and pollution control agencies. Use dry chemicals, alcohol-resistant foam, carbon dioxide, water. Use water spray to cool exposed containers.

NFPA Hazard Classifications: Not listed

FIRST AID

Move victim to fresh air. Call emergency medical care. Apply artificial respiration if victim is not breathing. Administer oxygen if breathing is difficult. Remove and isolate contaminated clothing and shoes. In case of contact with substance, immediately flush skin or eyes with running water for at least 20 minutes. Wash skin with soap and water. Keep victim warm and quiet. Effects of exposure (inhalation, ingestion, or skin contact) to substance may be delayed. Ensure that medical personnel are aware of the material(s) involved and take precautions to protect themselves.

SULFAMIC ACID S:700

SYNONYMS: ACIDE SULFAMIDIQUE (French); ACIDO SULFAMICO (Spanish); AMIDOSULFONIC ACID; AMIDOSULFONZUUR (Dutch); AMIDOSULFURIC ACID; AMINOSULFONIC ACID; EEC No. 016-026-00-0; EINECS No. 226-218-8; KYSELINA AMIDOSULFONOVA (Czech); KYSELINA SULFAMINOVA (Czech); SULFAMIDIC ACID; SULPHAMIC ACID

IDENTIFICATION

CAS: 5329-14-6

988 Sulfamic Acid

ERG ID: UN 2967
ERG Guide: 154
ERG Guide Category: TOXIC AND/OR CORROSIVE (NONCOMBUSTIBLE)
Molecular Formula: H_3NO_3S
RTECS No.: WO5950000
Properties: Noncombustible crystals or powder. Colorless crystals; white powder. Soluble in water; slowly hydrolyzes to ammonium bisulfate.

HEALTH AND SAFETY INFORMATION

Corrosive to eyes, skin, and respiratory tract. Inhalation may cause pulmonary edema, which can be delayed for several hours; there is a risk of death in serious cases. Similar and related chemicals may cause skin allergy and lung problems. It is unknown at this time if this chemical causes these problems.

RECOMMENDATIONS FOR RESPIRATOR SELECTION

At any concentrations above the NIOSH REL, or where there is no REL, at any detectable concentration: SCBAF:PD,PP (any self-contained breathing apparatus that has a full facepiece and is operated in a pressure-demand or other positive-pressure mode); or SAF:PD,PP:ASCBA (any supplied-air respirator that has a full facepiece and is operated in a pressure-demand or other positive-pressure mode in combination with an auxiliary, self-contained breathing apparatus operated in a pressure-demand or other positive-pressure mode).

Escape: GMFOV [any air-purifying, full-facepiece respirator (gas mask) with a chin-style, front- or back-mounted organic vapor canister]; or SCBAE (any appropriate escape-type, self-contained breathing apparatus).

CHEMICAL REACTIVITY

Slowly hydrolyzes in water to form ammonium bisulfate and heat. Violent reaction with chlorine, nitric acid, bases. Incompatible with ammonia, amines, isocyanates, alkylene oxides, epichlorohydrin, oxidizers, and water. May attack metals and coatings.

FIRE INFORMATION

Noncombustible solid. Combustion products include toxic nitrogen and corrosive sulfur oxides. If material or contaminated runoff enters waterways, notify downstream users of potentially contaminated waters. Notify local health and fire officials and pollution control agencies. Use dry chemicals, dry sand, carbon dioxide. From a secure, explosion-proof location, use water spray to cool exposed containers.

FIRST AID

Move victim to fresh air. Call emergency medical care. Apply artificial respiration if victim is not breathing. Do not use mouth-to-mouth method if victim ingested or inhaled the substance; induce artificial respiration with the aid of a pocket mask equipped with a one–way valve or other proper respiratory medical device. Administer oxygen if breathing is difficult. Remove and isolate contaminated clothing and shoes. In case of contact with substance, immediately flush skin or eyes with running water for at least 20 minutes. For minor skin contact, avoid spreading material on unaffected skin. Keep victim warm and quiet. Effects of exposure (inhalation, ingestion, or skin contact) may be delayed. Ensure that medical personnel are aware of the material(s) involved and take precautions to protect themselves.

SULFOLANE S:705

SYNONYMS: CYCLOTETRAMETHYLENE SULFONE; DIHYDROBUTADIENE SULPHONE; 1,1-DIOSSIDO di TETRAIDROTIOFENE (Italian); EEC No. 016-031-00-8; EINECS No. 204-783-1; SULFOLANE-W; SULFOLANO (Spanish); TETRAHYDROTHIOPHENE-1,1-DIOXIDE; TETRAMETHYLENE SULFONE; THIOPHENE, TETRAHYDRO-,1,1-DIOXIDE; THIOCYCLOPENTANE-1,1-DIOXIDE; THIOLANSULFON

IDENTIFICATION

CAS:	126-33-0
Molecular Formula:	$C_4H_8O_2S$
RTECS No.:	XN0700000
Properties:	Combustible, oily liquid. White or cream colored. Weak oily odor. Solidifies and sinks and mixes with water. Solid at 81°F (27°C).

HEALTH AND SAFETY INFORMATION

Mildly irritating to eyes. Toxic if swallowed. Due to the availability of insufficient data on human effects, caution should be exercised.

CHEMICAL REACTIVITY

Contact with strong oxidizers can cause fire and explosions.

FIRE INFORMATION

Combustion products include toxic sulfur oxides. Vapors are heavier than air and will collect in low areas. Vapors in confined areas may explode when exposed to fire. If material or contaminated runoff enters waterways, notify downstream users of potentially contaminated waters. Notify local health and fire officials and pollution control agencies. Use dry chemicals, alcohol-resistant foam, carbon dioxide. From a secure, explosion-proof location, use water spray to cool exposed containers.

NFPA Hazard Classifications:

Health Hazards (Blue) 2
Flammability (Red) 1
Reactivity (Yellow) 0

Flash Point: 350°F (177°C) o.c.; 330°F (166°C) c.c.

FIRST AID

Move victim to fresh air. Call emergency medical care. Apply artificial respiration if victim is not breathing. Administer oxygen if breathing is difficult. Remove and isolate contaminated clothing and shoes. In case of contact with substance, immediately flush skin or eyes with running water for at least 20 minutes. Wash skin with soap and water. Keep victim warm and quiet. Effects of exposure (inhalation, ingestion, or skin contact) to substance may be delayed. Ensure that medical personnel are aware of the material(s) involved and take precautions to protect themselves.

ENVIRONMENTAL INFORMATION

TSCA: 40CFR712.30(*e*)14 as sulphones; 40CFR716.120(*d*)14 as sulphones.

SULFUR S:710

SYNONYMS: AZUFRE (Spanish); BENSULFOID; BRIMSTONE; COLLOIDAL-S; COLLOIDAL SULFUR; COLLOKIT; COLSUL; COROSUL D; COROSUL S; COSAN; COSAN 80; CRYSTEX; EINECS NO. 231-722-6; ELEMENTAL SULFUR; ENXOFRE (Portugxese); FLOUR SULPHUR; FLOWERS OF SULPHUR; GROUND VOCLE SULPHUR; HEXASUL; KOLOFOG; KOLOSPRAY; KUMULUS; MAGNETIC 70; MAGNETIC 90; MAGNETIC 95; MICROFLOTOX; PRECIPITATED SULFUR; SOFRIL; SOLFO (Italian); SPERLOX-S; SPERSUL; SPERSUL THIOVIT; SUBLIMED SULFUR; SULFIDAL; SULFORON; SULFUR, ELEMENTAL; SULFUR COLLOID; SULKOL; SULPHUR; SUPER COSAN; SULSOL; SCHWEFEL (German); SOUFRE (French); TECHNETIUM TC 99M; TESULOID; THIOLUX; THIOVIT; ZOLFO (Italian); ZWAVEL (Dutch)

IDENTIFICATION

CAS:	7704-34-9
ERG ID:	UN 1350; UN 2448 (molten)
ERG Guide:	133
ERG Guide Category:	FLAMMABLE SOLIDS
Molecular Formula:	S_8
RTECS No.:	WS4250000
Properties:	Flammable, yellow solid in various forms; or combustible molten liquid. Molten sulfur [temperature approx 280°F (138°C) to 300°F (149°C)] is a yellow to reddish brown liquid. Characteristic faint, rotten-egg odor. Sinks in water; insoluble.

HEALTH AND SAFETY INFORMATION

Irritating to skin, eyes, and respiratory tract; molten material will cause burns. Inhalation of large quantities of fumes can cause paralysis of respiratory organs.

IDLH: 100 as sulfur dioxide (molten sulfur).
OSHA Table Z-1-A Air Contaminant (molten) as sulfur dioxide.

Exposure Limits:	**TWA**	**STEL**
ACGIH:	2 ppm; 5.2 mg/m^3	5 ppm; 13 mg/m^3
NIOSH:	2 ppm; 5 mg/m^3	5 ppm; 13 mg/m^3
OSHA:	5 ppm; 13 mg/m^3	

RECOMMENDATIONS FOR RESPIRATOR SELECTION: NIOSH

At any concentrations above the NIOSH REL, or where there is no REL, at any detectable concentration: SCBAF:PD,PP (any self-contained breathing apparatus that has a full facepiece and is operated in a pressure-demand or other positive-pressure mode); or SAF:PD,PP:ASCBA (any supplied-air respirator that has a full facepiece and is operated in a pressure-demand or other positive-pressure mode in combination with an auxiliary, self-contained breathing apparatus operated in a pressure-demand or other positive-pressure mode).

Escape: GMFOV [any air-purifying, full-facepiece respirator (gas mask) with a chin-style, front- or back-mounted organic vapor canister]; or SCBAE (any appropriate escape-type, self-contained breathing apparatus).

CHEMICAL REACTIVITY

Molten liquid forms sulfur dioxide with air. Reacts violently with strong oxidizers. Forms explosive, shock-sensitive, or pyrophoric mixtures with ammonia, ammonium nitrate, bromates, calcium carbide, charcoal, chlorates, hydrocarbons, iodates, and iron. Reacts violently with halogen compounds, sodium, tin, uranium, and other compounds. Attacks steel when moist. May accumulate static electrical charges in dry or liquid forms; can cause ignition of volatile vapors.

FIRE INFORMATION

Combustible solid. Fires may be difficult to extinguish. Hot molten sulfur will form solid crust as it cools, liquid below crust will remain hot for and extended period of time. Flames of burning sulfur may be difficult to see in daylight. Tanks of molten sulfur may contain toxic and flammable hydrogen sulfide under pressure. Combustion produces large quantities of highly toxic sulfur dioxide gas. Contaminants may cause buildup of hydrogen sulfide gas. If material or contaminated runoff enters waterways, notify downstream users of potentially contaminated waters. Notify local health and fire officials and pollution control agencies. Use foam or water spray. From a secure, explosion-proof location, use water spray to cool exposed containers. If cooling streams are ineffective (venting sound increases in volume and pitch, vent sound suddenly increases and/or unexpectedly stops, tank discolors, or shows signs of deforming), withdraw immediately to a secure location.

NFPA Hazard Classifications: (molten)

Health Hazards (Blue) 2
Flammability (Red) 1
Reactivity (Yellow) 0

Flash Point: 405°F (207°C)

Flammable Limits in Air: LEL 3.3%; UEL 46.0% as hydrogen sulfide.
Autoignition Temperature: 450°F (232°C).

FIRST AID

Move victim to fresh air. Call emergency medical care. Apply artificial respiration if victim is not breathing. Administer oxygen if breathing is difficult. Remove and isolate contaminated clothing and shoes. In case of contact with substance, immediately flush skin or eyes with running water for at least 20 minutes. Removal of solidified molten material from skin requires medical assistance. Keep victim warm and quiet. Ensure that medical personnel are aware of the material(s) involved and take precautions to protect themselves.

SULFUR DIOXIDE S:715

SYNONYMS: BISULFITE; DIOXIDO de AZUFE (Spanish); EINECS No. 231-195-2; FERMENICIDE; SCHWEFELDDIOXYD (German); SIARKI DWUTLENEK (Polish); STCC 4904290; SULFUROUS ACID ANHYDRIDE; SULFUROUS ANIIYDRIDE; SULFUR OXIDE; SULFUROUS OXIDE; SULPHUR DIOXIDE; SULPHUROUS ACID ANHYDRIDE; SULPHUROUS ANHYDRIDE; SULPHUROUS OXIDE

IDENTIFICATION

CAS:	7446-09-5
ERG ID:	UN 1079
ERG Guide:	125
ERG Guide Category:	GASES–CORROSIVE
Molecular Formula:	O_2S
RTECS No.:	WS4550000
Properties:	Nonflammable, colorless, poisonous gas; or, compressed, liquefied gas [A liquid below 14°F (-10°C)]. Irritating, suffocating, pungent odor. Partly soluble in water; forms corrosive and toxic sulfurous acid.

HEALTH AND SAFETY INFORMATION

Poisonous. Inhaling the gas can cause death. Skin and eye contact causes severe burns and blindness. Contact with liquid may cause frostbite.

IDLH: 100 ppm. Odor is not a reliable indicator of the presence of toxic amounts of gas.
Odor Threshold: 0.708 ppm
OSHA Table Z-1-A Air Contaminant
OSHA Process Safety Management of Highly Hazardous Chemicals (29CFR 1910.119, Appendix A), Threshold Quantity: 1000 pounds

Exposure Limits:	**TWA**	**STEL**
ACGIH:	2 ppm; 5.2 mg/m^3	5 ppm; 13 mg/m^3
NIOSH:	2 ppm; 5 mg/m^3	5 ppm; 13 mg/m^3
OSHA	5 ppm; 13 mg/m^3	

RECOMMENDATIONS FOR RESPIRATOR SELECTION: NIOSH

20 ppm: CCRS *[any chemical cartridge respirator with cartridge(s) providing protection against the compound of concern]; or SA * (any supplied-air respirator); or SCBAF (any self-contained breathing apparatus with a full facepiece).
50 ppm: SA:CF *(any supplied-air respirator operated in a continuous-flow mode); or PAPRS * [any powered, air-purifying respirator with cartridge(s) providing protection against the compound of concern].
100 ppm: CCRFS (any chemical cartridge respirator with a full facepiece and cartridge(s) providing protection against the compound of concern); or GMFS [any air-purifying, full-facepiece respirator (gas mask) with a chin-style, front- or back-mounted canister providing protection against the compound of concern]; or PAPRTS * (any powered, air-purifying respirator with a tight-fitting facepiece and cartridge(s) providing protection against the compound of concern); or SAT:CF * (Any supplied-air respirator that has a tight-fitting facepiece and is operated in a continuous-flow mode); or SCBAF (any self-contained breathing apparatus with a full facepiece); or SAF (any supplied-air respirator with a full facepiece).

EMERGENCY OR PLANNED ENTRY INTO UNKNOWN CONCENTRATIONS OR IDLH CONDITIONS: SCBAF:PD,PP (any self-contained breathing apparatus that has a full facepiece and is operated in a pressure-demand or other positive-pressure mode); or SAF:PD,PP:ASCBA (Any supplied-air respirator that has a full facepiece and is operated in a pressure-demand or other positive-pressure mode in combination with an auxiliary, self-contained breathing apparatus operated in a pressure-demand or other positive pressure mode.

Escape: GMFSHiE [any air-purifying, full-facepiece respirator (gas mask) with a chin-style, front- or back-mounted canister providing protection against the compound of concern and having a high-efficiency particulate filter]; or SCBAE (any appropriate escape-type, self-contained breathing apparatus).

* Substance reported to cause eye irritation or damage; may require eye protection.

CHEMICAL REACTIVITY

Contact with water forms sulfurous acid. Violent reaction with caustics, anhydrous ammonia, and alcohols. Incompatible with aliphatic amines, alknolamines, aromatic amines, amides, organic anhydrides, isocyanates, vinyl acetate, alkylene oxides, powdered alkali metals, and epichlorohydrin. Contact with copper, bronze, or alkali metals may cause fire and explosions. Decomposes above 140°F (60°C), forming toxic and corrosive oxides of sulfur. Aqueous solution attacks some plastics, rubber, and coatings. Reacts with aluminum.

FIRE INFORMATION

Firefighting gear (including SCBA) does not provide adequate protection. If exposure occurs, remove and isolate gear immediately and thoroughly decontaminate personnel. Gases are heavier than air and will collect in low areas. Gases in confined areas may explode when exposed to fire. Gases may travel long distances to ignition sources and flash back. Storage containers and parts of containers may rocket great distances, in many directions. If material or contaminated runoff enters waterways, notify downstream users of potentially contaminated waters. Notify local health and fire officials and pollution control agencies. Do not extinguish the fire unless the flow of the gas can be stopped and any remaining gas is out of the line. Specially trained personnel may use fog lines to cool exposures and let the fire burn itself out. Use dry chemicals, foam, carbon dioxide. From a secure, explosion-proof location, use water spray to cool exposed containers. If cooling streams are ineffective (venting sound increases in volume and pitch, tank discolors, or shows any signs of deforming), withdraw immediately to a secure position.

NFPA Hazard Classifications:

Health Hazards (Blue) 3
Flammability (Red) 0
Reactivity (Yellow) 0

FIRST AID

Move victim to fresh air. Call emergency medical care. Apply artificial respiration if victim is not breathing. Do not use mouth-to-mouth method if victim ingested or inhaled the substance; induce artificial respiration with the aid of a pocket mask equipped with a one–way valve or other proper respiratory medical device. Administer oxygen if breathing is difficult. Remove and isolate contaminated clothing and shoes. In case of contact with liquefied gas, thaw frosted parts with lukewarm water. Keep victim warm and quiet. Keep victim under observation. Effects of contact or inhalation may be delayed. Ensure that medical personnel are aware of the material(s) involved and take precautions to protect themselves.

ENVIRONMENTAL INFORMATION

Clean Air Act: Accidental Release Prevention/Flammable substances, (Section 112[r], Table 3), TQ = 5,000 lb (2,270 kg).
EPCRA Section 302 Extremely Hazardous Substances: TPQ = 500 lb (227 kg).
EPCRA Section 304: Reportable Quantity (RQ): EHS, 1 lb (0.454 kg).

SULFURIC ACID S:740

SYNONYMS: ACIDE SULFURIQUE (French); ACIDO SOLFORICO (Italian); ACIDO SULFURICO (Spanish); BATTERY ACID; BETZ ENTEC; BOV; CARRO'S ACID; CHAMBER ACID; CHROMIC 5; COPPERLITE RD-25; CUBATH MD ADDITION; CUBATH M-HY; CUBATH M MAKE-UP; DIPPING ACID; EEC No. 016-020-00-8; EINECS No. 231-639-5; ELECTROLYTE ACID; ENPOLATE CONDITIONER 470; ENTEC 763 FUEL OIL B-20 FIXER; FERTILIZER ACID; HYDROOT; HYDROGEN SULFATE; KODAK RAPID FIXER PART B; KOVAR BRIGHT DIP (RDX-555); KTI PHOTORESIST STRIPPER C.S.; LAUDER TRAY CLEANER CONCENTRATE; MATTING ACID; NANOSTRIP; NORDHAUSEN ACID; OIL OF VITRIOL; 1030 ACID CLEANER; PIRANHA ETCH; PREPOSIT ETCH 746; PREPOSIT ETCH 748; RT-2 STRIPPING SOLUTION; SCHWEFELSAEURELOESUNGEN (German); SEL-REX CU BATH M LO; SPIRIT OF SULFUR;

STRIPPER 3; STRIPPER 5; SULFURIC ACID, AQUEOUS; SULPHURIC ACID; VITRIOL BROWN OIL; ZWAVELZUUROPLOSSINGEN (Dutch)

IDENTIFICATION

CAS:	7664-93-9
ERG ID:	UN 1830
ERG Guide:	137
ERG Guide Category:	WATER REACTIVE-CORROSIVE
Molecular Formula:	H_2O_4S
RTECS No.:	WS5600000
Properties:	Nonflammable, thick, oily liquid. Colorless to dark brown. Odorless, unless heated (choking odor). Sinks and mixes violently with water. Freezes at 50°F (10°C).

HEALTH AND SAFETY INFORMATION

Poisonous. Inhaling vapors or swallowing the material can cause death. Corrosive to eyes, skin, and respiratory tract; skin and eye contact causes severe burns and blindness. Inhalation may cause pulmonary edema, which can be delayed for several hours; there is a risk of death in serious cases. After long exposure workers lose sensitivity to irritating action.

IDLH: 15 mg/m^3
Odor Threshold: 0.15 ppm; 1 mg/m^3
OSHA Table Z-1-A Air Contaminant
NIOSH Criteria Document: NIOSH 74-128

Exposure Limits:	**TWA**	**STEL**
ACGIH:[(1)]	1 mg/m^3	3 mg/m^3
NIOSH:	1 mg/m^3	
OSHA:	1 mg/m^3	

[(1)] A suspected human carcinogen when sulfuric acid is contained in strong, inorganic acid mists.

RECOMMENDATIONS FOR RESPIRATOR SELECTION: NIOSH/OSHA as sulfuric acid

25 mg/m^3: SA:CF (any supplied-air respirator operated in a continuous-flow mode); or PAPRAGHiE (any powered, air-purifying respirator with acid gas cartridge(s) in combination with a high-efficiency particulate filter).
50 mg/m^3: CCRFAGHiE (any chemical cartridge respirator with a full facepiece and acid gas cartridge(s) in combination with a high-efficiency particulate filter); or GMFAGHiE [any air-purifying, full-facepiece respirator (gas mask) with a chin-style, front- or back-mounted acid gas canister having a high-efficiency particulate filter]; or SCBAF (any self-contained breathing apparatus with a full facepiece); or SAF (any supplied-air respirator with a full facepiece).
80 mg/m^3: SAF:PD,PP (any supplied-air respirator that has a full facepiece and is operated in a pressure-demand or other positive-pressure mode).

Emergency or Planned Entry into Unknown Concentrations or IDLH Conditions: SCBAF:PD,PP (any self-contained breathing apparatus that has a full facepiece and is operated in a pressure-demand or other positive-pressure mode); or SAF:PD,PP:ASCBA (any supplied-air respirator that has a full facepiece and is operated in a pressure-demand or other positive-pressure mode in combination with an auxiliary, self-contained breathing apparatus operated in a pressure-demand or other positive-pressure mode).

Escape: GMFAGHiE [any air-purifying, full-facepiece respirator (gas mask) with a chin-style, front- or back-mounted acid gas canister having a high-efficiency particulate filter]; or SCBAE (any appropriate escape-type, self-contained breathing apparatus).

Note: Substance causes eye irritation or damage; eye protection needed.

CHEMICAL REACTIVITY

A strong oxidizer that can react violently with risk of fire and explosion with many substances including reducing agents, organic and combustible substances, and bases. Incompatible with nonoxidizing mineral acids, organic acids, bases, acrylates, aldehydes, alcohols, alkylene oxides, ammonia, aliphatic amines, alkanolamines, aromatic amines, amides, chlorates, epichlorohydrin, fulminates, glycols, isocyanates, ketones, metals (powdered), organic anhydrides, perchlorates, picrates, substituted allyls, phenols and cresols, and water. Also incompatible with acetic anhydride, acetone cyanhydrin, acetonitrile, acrolein, acrylonitrile, allyl alcohol, allyl chloride, 2-aminoethanol, ammonium hydroxide, aniline, bromine pentafluoride, n-butyraldehyde, caprolactum solution, carbides, cesium acetylene carbide, chlorine trifluoride, chlorosulfonic acid, cuprous nitride, diisobutylene, ethylene cyanohydrin, ethylene diamine, ethylene glycol, ethyleneimine, hydrochloric acid, iodine heptafluoride, iron, isoprene, lithium silicide, mercuric nitride, mesityl oxide, nitric acid, p-nitrotoluene, perchloric acid, phosphorus, potassium tert-butoxide, potassium chlorate, potassium permanganate, propiolactone (β-), propylene oxide, pyridine, rubidium acetylene, carbide, silver permanganate, sodium, sodium carbonate, sodium chlorate, sodium hydroxide, styrene monomer, vinyl acetate, Attacks most metals and some plastics, rubber, and coatings.

FIRE INFORMATION

Firefighting gear, including SCBA, does not provide adequate protection. If exposure occurs, remove and isolate gear immediately and thoroughly decontaminate personnel. Combustion products include toxic sulfur oxides. Noncombustible liquid, but capable of igniting finely divided combustible materials. Vapors are heavier than air and will collect in low areas. Storage containers and parts of containers may rocket great distances, in many directions. If material or contaminated runoff enters waterways, notify downstream users of potentially contaminated waters. Notify local health and fire officials and pollution control agencies. Use dry chemicals, dry sand. Do not use water-based extinguishers. From a secure, explosion-proof location, use water spray to cool exposed containers.

NFPA Hazard Classifications:

Health Hazards (Blue) 3
Flammability (Red) 0
Reactivity (Yellow) 2
Special Situations . WATER REACTIVE

FIRST AID

Move victim to fresh air. Call emergency medical care. Apply artificial respiration if victim is not breathing. Do not use mouth-to-mouth method if victim ingested or inhaled the substance; induce artificial respiration with the aid of a pocket mask equipped with a one–way valve or other proper respiratory medical device. Administer oxygen if breathing is difficult. Remove and isolate contaminated clothing and shoes. In case of contact with substance, immediately flush skin or eyes with running water for at least 20 minutes. For minor skin contact, avoid spreading material on unaffected skin. Removal of solidified molten material from skin requires medical assistance. Keep victim warm and quiet. Effects of exposure (inhalation, ingestion, or skin contact) to substance may be delayed. Ensure that medical personnel are aware of the material(s) involved and take precautions to protect themselves.

ENVIRONMENTAL INFORMATION

Clean Water Act: Section 311 Hazardous Substances/RQ (same as CERCLA); Section 313 Priority Chemicals.
EPCRA Section 302 Extremely Hazardous Substances: TPQ = 1,000 lb (454 kg).
EPCRA Section 304: Reportable Quantity (RQ): EHS, CERCLA, 1,000 lb (454 kg).
EPCRA Section 313: (acid aerosols including mists, vapors, gas, fog, and other airborne species of any particle size); Form R *de minimis* concentration reporting level: 0.1%.

SULFURYL CHLORIDE S:800

SYNONYMS: CHLOROSULFURIC ACID; CHLORURE de SULFURYLE (French); DICLORURO de SULFURILO (Spanish); EEC No. 016-016-00-6; EINECS No. 232-245-6; SULFONYL CHLORIDE; SULFURIC CHLORIDE; SULFURIC OXYCHLORIDE

IDENTIFICATION

CAS:	7791-25-5
ERG ID:	UN 1834
ERG Guide:	137
ERG Guide Category:	WATER REACTIVE-CORROSIVE
Molecular Formula:	Cl_2 SO_2
RTECS No.:	WT4870000
Properties:	Watery liquid. Colorless to light yellow. Acrid, pungent odor. Mixes and reacts violently with water forming toxic hydrochloric and sulfuric acids.

HEALTH AND SAFETY INFORMATION

Poison. Breathing the vapors can cause death. Corrosive to eyes, skin, and respiratory tract; can cause severe burns and blindness. Inhalation may cause pulmonary edema, which can be delayed for several hours; there is a risk of death in serious cases.

RECOMMENDATIONS FOR RESPIRATOR SELECTION

At any concentrations above the NIOSH REL, or where there is no REL, at any detectable concentration: SCBAF:PD,PP (any self-contained breathing apparatus that has a full facepiece and is operated in a pressure-demand or other positive-pressure mode); or SAF:PD,PP:ASCBA (any supplied-air respirator that has a full facepiece and is operated in a pressure-demand or other positive-pressure mode in combination with an auxiliary, self-contained breathing apparatus operated in a pressure-demand or other positive-pressure mode).

Escape: GMFOV [any air-purifying, full-facepiece respirator (gas mask) with a chin-style, front- or back-mounted organic vapor canister]; or SCBAE (any appropriate escape-type, self-contained breathing apparatus).

CHEMICAL REACTIVITY

Water contact produces sulfuric acid and hydrogen chloride vapors. Reacts violently with air, caustics, alkali metals, ethyl ether, lead dioxide, organic compounds, and phosphorus. Incompatible with red phosphorus. Forms an explosive compound with dimethyl sulfoxide. Acids formed by reaction with moisture attack metals and liberate flammable hydrogen gas.

FIRE INFORMATION

Firefighting gear (including SCBA) does not provide adequate protection. If exposure occurs, remove and isolate gear immediately and thoroughly decontaminate personnel. Vapors are heavier than air and will collect in low areas. Combustion products include highly toxic fumes of hydrogen chloride and sulfur oxides. If material or contaminated runoff enters waterways, notify downstream users of potentially contaminated waters. Notify local health and fire officials and pollution control agencies. Use dry chemicals, alcohol-resistant foam, carbon dioxide. If possible, do not allow water to come in contact with the material.

NFPA Hazard Classifications:

Health Hazards (Blue) 3
Flammability (Red) 0
Reactivity (Yellow) 2
Special Situations .. WATER REACTIVE

FIRST AID

Move victim to fresh air. Call emergency medical care. Apply artificial respiration if victim is not breathing. Do not use mouth-to-mouth method if victim ingested or inhaled the substance; induce artificial respiration with the aid of a pocket mask equipped with a one-way valve or other proper respiratory medical device. Administer oxygen if breathing is difficult. Remove and isolate contaminated clothing and shoes. In case of contact with substance, immediately flush skin or eyes with running water for at least 20 minutes. For minor skin contact, avoid spreading material on unaffected skin. Removal of solidified molten material from skin requires medical assistance. Keep victim warm and quiet. Effects of exposure (inhalation, ingestion, or skin contact) to substance may be delayed. Ensure that medical personnel are aware of the material(s) involved and take precautions to protect themselves.

TEREPHTHALIC ACID T:025

SYNONYMS: ACIDE TEREPHTHALIQUE (French); ACIDO TEREFTALICO (Spanish); AMOCO TA-22; BENZENE-p-DICARBOXYLIC ACID: p-BENZENEDICARBOXYLIC ACID; 1,4-BENZENEDICARBOXYLIC ACID; EINECS No. 202-830-0; KYSELINA TERFTALOVA (Czech); p-PHTHALIC ACID; TA 12; TA-33MP; TEREPHTHALSAEURE (German); TPA

IDENTIFICATION

CAS:	100-21-0
ERG ID:	UN 2811
ERG Guide:	154
ERG Guide Category:	SUBSTANCES-TOXIC AND/OR CORROSIVE
Molecular Formula:	$C_8H_6O_4$
RTECS No.:	WZ0875000
Properties:	Combustible, crystals or powder. White. Insoluble in water.

HEALTH AND SAFETY INFORMATION

Irritates eyes. Toxic by ingestion. Due to the availability of insufficient data on human effects, caution should be exercised.

Exposure Limits: TWA

ACGIH: 10 mg/m^3

RECOMMENDATIONS FOR RESPIRATOR SELECTION

At any concentrations above the NIOSH REL, or where there is no REL, at any detectable concentration: SCBAF:PD,PP (any self-contained breathing apparatus that has a full facepiece and is operated in a pressure-demand or other positive-pressure mode); or SAF:PD,PP:ASCBA (any supplied-air respirator that has a full facepiece and is operated in a pressure-demand or other positive-pressure mode in combination with an auxiliary, self-contained breathing apparatus operated in a pressure-demand or other positive-pressure mode).

Escape: GMFOVHiE [any air-purifying, full-facepiece respirator (gas mask) with a chin-style, front- or back-mounted organic vapor canister having a high-efficiency particulate filter]; or SCBAE (any appropriate escape-type, self-contained breathing apparatus).

CHEMICAL REACTIVITY

Reacts violently with strong oxidizers. Dust explosion possible if in powder or granular form, mixed with air. Organic acids are incompatible with sulfuric acid, caustics, ammonia, aliphatic amines, alkanolamines, isocyanates, alkylene oxides, epichlorohydrin. High-voltage static electricity buildup and discharge must be avoided when significant quantities of dust are present.

FIRE INFORMATION

Combustion may produce irritants and toxic vapors. Vapors are heavier than air and will collect in low areas. Vapors in confined areas may explode when exposed to fire. If material or contaminated runoff enters waterways, notify downstream users of potentially contaminated waters. Notify local health and fire officials and pollution control agencies. Use dry chemicals, alcohol-resistant foam, carbon dioxide. From a secure, explosion-proof location, use water spray to cool exposed containers.

NFPA Hazard Classifications:

Health Hazards (Blue) 0
Flammability (Red) 1
Reactivity (Yellow) 0

Flash Point: 500°F (260°C) o.c.
Autoignition Temperature: 925°F (496°C)

FIRST AID

Move victim to fresh air. Call emergency medical care. Apply artificial respiration if victim is not breathing. Administer oxygen if breathing is difficult. Remove and isolate contaminated clothing and shoes. In case of contact with substance, immediately flush skin or eyes with running water for at least 20 minutes. Wash skin with soap and water. Keep victim warm and quiet. Effects of exposure (inhalation, ingestion, or skin contact) to substance may be delayed. Ensure that medical personnel are aware of the material(s) involved and take precautions to protect themselves.

ENVIRONMENTAL INFORMATION

RCRA Land Ban Waste
RCRA Universal Treatment Standards: Wastewater (mg/L), 0.055; Nonwastewater (mg/kg), 28.

TEREPHTHALOYL DICHLORIDE T:030

SYNONYMS: 1,4-BENZENEDICARBONYL CHLORIDE; 1,4-BENZENEDICARBONYL DICHLORIDE; DICLORURO de TEREFTALOILO (Spanish); EINECS No. 202-829-5; p-PHENYLENEDICARBONYL DICHLORIDE; p-PHTHALOYL CHLORIDE; p-PHTHALOYL DICHLORIDE; TEREPHTHALIC ACID CHLORIDE; TEREPHTHALIC ACID DICHLORIDE; TEREPHTHALIC DICHLORIDE

IDENTIFICATION

CAS:	100-20-9
ERG ID:	UN 2811
ERG Guide:	154
ERG Guide Category:	SUBSTANCES-TOXIC AND/OR CORROSIVE
Molecular Formula:	$C_8H_4Cl_2O_2$
RTECS No.:	WZ1850000
Properties:	Combustible, crystalline solid. White. Musty odor. Soluble in water.

HEALTH AND SAFETY INFORMATION

Dust is corrosive to eyes, skin, and respiratory tract. Inhalation may cause pulmonary edema, which can be delayed for several hours; there is a risk of death in serious cases.

RECOMMENDATIONS FOR RESPIRATOR SELECTION

At any concentrations above the NIOSH REL, or where there is no REL, at any detectable concentration: SCBAF:PD,PP (any self-contained breathing apparatus that has a full facepiece and is operated in a pressure-demand or other positive-pressure mode); or SAF:PD,PP:ASCBA (any supplied-air respirator that has a full facepiece and is operated in a pressure-demand or other positive-pressure mode in combination with an auxiliary, self-contained breathing apparatus operated in a pressure-demand or other positive-pressure mode).

Escape: GMFOVHiE [any air-purifying, full-facepiece respirator (gas mask) with a chin-style, front- or back-mounted organic vapor canister having a high-efficiency particulate filter]; or SCBAE (any appropriate escape-type, self-contained breathing apparatus).

CHEMICAL REACTIVITY

Contact with strong oxidizers may cause fire and explosions.

FIRE INFORMATION

Combustion products include toxic fumes of chlorides. If material or contaminated runoff enters waterways, notify downstream users of potentially contaminated waters. Notify local health and fire officials and pollution control agencies. Use dry chemicals, alcohol-resistant foam, carbon dioxide. From a secure, explosion-proof location, use water spray to cool exposed containers.

NFPA Hazard Classifications

Health Hazards (Blue) 3

1002 Terpinoline

Flammability (Red) 1
Reactivity (Yellow) 0

Flash Point: 356°F (180°C) o.c.

FIRST AID

Move victim to fresh air. Call emergency medical care. Apply artificial respiration if victim is not breathing. Do not use mouth-to-mouth method if victim swallowed or inhaled the substance; induce artificial respiration with the aid of a pocket mask equipped with a one-way valve or other proper respiratory medical device. Administer oxygen if breathing is difficult. Remove and isolate contaminated clothing and shoes. In case of contact with substance, immediately flush skin or eyes with running water for at least 20 minutes. For minor skin contact, avoid spreading material on unaffected skin. Keep victim warm and quiet. Effects of exposure (inhalation, ingestion, or skin contact) to substances may be delayed. Ensure that medical personnel are aware of the material(s) involved and take precautions to protect themselves.

TERPINOLINE T:035

SYNONYMS: CYCLOHEXENE, 1-METHYL-4-(1-METHYLETHYLIDENE)-; EINECS No. 209-578-0; P-MENTHA-1,4(8)-DIENE; 1-METHYL-4-(1-METHYLETHYLIDENE)CYCLOHEXENE; 1,4(8)-TERPADIENE

IDENTIFICATION

CAS:	586-62-9
ERG ID:	UN 2541
ERG Guide:	128
ERG Guide Category:	FLAMMABLE LIQUIDS (NONPOLAR/WATER-IMMISCIBLE)
Molecular Formula:	$C_{10}H_{16}$
RTECS No.:	WZ6870000
Properties:	Highly flammable liquid. Water-white to pale amber. Insoluble in water.

HEALTH AND SAFETY INFORMATION

May irritate eyes, skin, and respiratory tract. Toxic if swallowed.

RECOMMENDATIONS FOR RESPIRATOR SELECTION

At any concentrations above the NIOSH REL, or where there is no REL, at any detectable concentration: SCBAF:PD,PP (any self-contained breathing apparatus that has a full facepiece and is

operated in a pressure-demand or other positive-pressure mode); or SAF:PD,PP:ASCBA (any supplied-air respirator that has a full facepiece and is operated in a pressure-demand or other positive-pressure mode in combination with an auxiliary, self-contained breathing apparatus operated in a pressure-demand or other positive-pressure mode).

Escape: GMFOV [any air-purifying, full-facepiece respirator (gas mask) with a chin-style, front- or back-mounted organic vapor canister]; or SCBAE (any appropriate escape-type, self-contained breathing apparatus).

CHEMICAL REACTIVITY

Forms explosive mixture with air. Contact with oxidizers can cause fire and explosions. Dienes form an explosive mixture with nitric oxide.

FIRE INFORMATION

Combustion may produce irritants and toxic vapors. Vapors are heavier than air and will collect in low areas. Vapors in confined areas may explode when exposed to fire. Vapors may travel long distances to ignition sources and flash back. Storage containers and parts of containers may rocket great distances, in many directions. If material or contaminated runoff enters waterways, notify downstream users of potentially contaminated waters. Notify local health and fire officials and pollution control agencies. Use dry chemicals, foam, carbon dioxide. From a secure, explosion-proof location, use water spray to cool exposed containers.

FEMA Hazard Classifications:

Health Hazards (Blue) 1
Flammability (Red) 2
Reactivity (Yellow) 0

Flash Point: 100°F (38°C) c.c.

FIRST AID

Move victim to fresh air. Call emergency medical care. Apply artificial respiration if victim is not breathing. If breathing is difficult, give oxygen. Remove and isolate contaminated clothing and shoes. In case of contact with substance, immediately flush skin or eyes with running water for at least 20 minutes. Wash skin with soap and water. Keep victim warm and quiet. Ensure that medical personnel are aware of the material(s) involved and take precautions to protect themselves.

p–TERPHENYL

T:040

SYNONYMS: p-DIPHENYLBENZENE; 1,4-DIPHENYLBENZENE; EINECS No. 202-205-2; 4-PHENYLBIPHENYL; 4-PHENYLDIPHENYL; SANTOWAX; p-TERFENILO (Spanish); p-TRIPHENYL

IDENTIFICATION

CAS:	92-94-4; 26140-60-3 (all isomers)
ERG ID:	UN 2811
ERG Guide:	154
ERG Guide Category:	SUBSTANCES-TOXIC AND/OR CORROSIVE
Molecular Formula:	$C_{18}H_{14}$
RTECS No.:	WZ6475000
Properties:	Combustible leaflets or needles. White or light yellow. Insoluble in water.

HEALTH AND SAFETY INFORMATION

IDLH: 500 mg/m^3
OSHA Table Z-1-A Air Contaminant as terphenyls

Exposure Limits: CEILING

ACGIH:	0.53 ppm; 5 mg/m^3
NIOSH:	0.5 ppm; 5 mg/m^3
OSHA:	1 ppm; 9 mg/m^3

RECOMMENDATIONS FOR RESPIRATOR SELECTION

At any concentrations above the NIOSH REL, or where there is no REL, at any detectable concentration: SCBAF:PD,PP (any self-contained breathing apparatus that has a full facepiece and is operated in a pressure-demand or other positive-pressure mode); or SAF:PD,PP:ASCBA (any supplied-air respirator that has a full facepiece and is operated in a pressure-demand or other positive-pressure mode in combination with an auxiliary, self-contained breathing apparatus operated in a pressure-demand or other positive-pressure mode).

Escape: GMFOVHiE [any air-purifying, full-facepiece respirator (gas mask) with a chin-style, front- or back-mounted organic vapor canister having a high-efficiency particulate filter]; or SCBAE (any appropriate escape-type, self-contained breathing apparatus).

CHEMICAL REACTIVITY

Contact with strong oxidizing agents may cause fire and explosions.

FIRE INFORMATION

Combustion may produce irritants and toxic vapors. Vapors are heavier than air and will collect in low areas. Vapors in confined areas may explode when exposed to fire. If material or contaminated runoff enters waterways, notify downstream users of potentially contaminated waters. Notify local health and fire officials and pollution control agencies. Use dry chemicals, alcohol-resistant foam, carbon dioxide. From a secure, explosion-proof location, use water spray to cool exposed containers.

NFPA Hazard Classifications:

Health Hazards (Blue) 0
Flammability (Red) 1
Reactivity (Yellow) 0

Flash Point: 405°F (207°C) o.c.

FIRST AID

Move victim to fresh air. Call emergency medical care. Apply artificial respiration if victim is not breathing. Administer oxygen if breathing is difficult. Remove and isolate contaminated clothing and shoes. In case of contact with substance, immediately flush skin or eyes with running water for at least 20 minutes. Wash skin with soap and water. Keep victim warm and quiet. Effects of exposure (inhalation, ingestion, or skin contact) to substance may be delayed. Ensure that medical personnel are aware of the material(s) involved and take precautions to protect themselves.

TETRABUTYL TITANATE T:045

SYNONYMS: 1-BUTANOL, TITANIUM(4+) SALT; BUTYL TITANATE; BUTYL TITANATE MONOMER; EINECS No. 227-006-8; TETRABUTYL ORTHOTITANATE; TITANIUM BUTOXIDE; TITANIUM TETRABUTANOLATE; TITANIUM TETRABUTOXIDE; TITANATO de TERTRABUTILO (Spanish); TITANIUM BUTOXIDE; ORTHOTITANIC ACID, TETRABUTYL ESTER

IDENTIFICATION

CAS:	5593-70-4
ERG ID:	NA 1933
ERG Guide:	128
ERG Guide Category:	FLAMMABLE LIQUIDS (NONPOLAR/WATER-IMMISCIBLE)
Molecular Formula:	$C_{16}H_{36}O_4Ti$
RTECS No.:	XR1585000

1006 Tetrabutyl Tinanate

Properties: Combustible liquid. Colorless to light yellow. Weak, alcohollike (butanol) odor. May float or sink in water; reacts forming heat and flammable butanol fumes and titanium dioxide.

HEALTH AND SAFETY INFORMATION

Irritates skin and respiratory tract. May affect central nervous system.; headache, sleepiness. Long-term or prolonged exposure may cause skin problems.

At any concentrations above the NIOSH REL, or where there is no REL, at any detectable concentration: SCBAF:PD,PP (any self-contained breathing apparatus that has a full facepiece and is operated in a pressure-demand or other positive-pressure mode); or SAF:PD,PP:ASCBA (any supplied-air respirator that has a full facepiece and is operated in a pressure-demand or other positive-pressure mode in combination with an auxiliary, self-contained breathing apparatus operated in a pressure-demand or other positive-pressure mode).

Escape: GMFOV [any air-purifying, full-facepiece respirator (gas mask) with a chin-style, front- or back-mounted organic vapor canister]; or SCBAE (any appropriate escape-type, self-contained breathing apparatus).

CHEMICAL REACTIVITY

Forms explosive mixture with air. Reacts with water to form butanol and titanium dioxide. Contact with strong oxidizers may cause fire and explosions..

FIRE INFORMATION

Vapors are heavier than air and will collect in low areas. Vapors in confined areas may explode when exposed to fire. If material or contaminated runoff enters waterways, notify downstream users of potentially contaminated waters. Notify local health and fire officials and pollution control agencies. Do not use water-based extinguishers. Use dry chemicals, dry sand, carbon dioxide. From a secure, explosion-proof location, use water spray to cool exposed containers, but do not spray material with water. If cooling streams are ineffective (venting sound increases in volume and pitch, tank discolors, or shows any signs of deforming), withdraw immediately to a secure position.

FEMA Hazard Classifications

Health Hazards (Blue) 0
Flammability (Red) 2
Reactivity (Yellow) 0

Flash Point: 170°F (77°C) c.c
Flammable Limits in Air: LEL 2%; UEL 12%

FIRST AID

Move victim to fresh air. Call emergency medical care. Apply artificial respiration if victim is not breathing. If breathing is difficult, give oxygen. Remove and isolate contaminated clothing and shoes. In case of contact with substance, immediately flush skin or eyes with running water for at least 20 minutes. Wash skin with soap and water. Keep victim warm and quiet. Ensure that medical personnel are aware of the material(s) involved and take precautions to protect themselves.

TETRACHLOROETHYLENE T:060

SYNONYMS: CARBON BICHLORIDE; DOW-PER; EEC No. 602-028-00-4; EINECS No. 204-825-9; ETHENE, TETRACHLORO-; ENT 1,860; ETHYLENE TETRACHLORIDE; J-100; NEMA; PERCHLOR; PERCHLOROETHYLENE; PERCLENE; PERK; PERKLONE; PERSEC; SURDIP STRIPER RDX-522; TETRACAP; 1,1,2,2-TETRACHLOROETHENE; 1,1,2,2,-TETRACHLOROETHYLENE TETRALENO; TETRACLOROETILENO (Spanish); TETRALEX; TETROPIL

IDENTIFICATION

CAS:	127-18-4
ERG ID:	UN 1897
ERG Guide:	160
ERG Guide Category:	HALOGENATED SOLVENTS
Molecular Formula:	C_2Cl_4
RTECS No.:	KX38500000
Properties:	Nonflammable, colorless, watery liquid. Solid at −8°F (−22°C). Sweet, chloroformlike odor. Sinks in water; insoluble. Decomposition products include toxic hydrogen chloride.

HEALTH AND SAFETY INFORMATION

Irritating to eyes, skin, nose, and lungs. May cause cancer, central nervous system depression, impaired memory, numbness of arms and legs, impaired vision, kidney and liver damage, and dermatitis from repeated skin contact. Several studies suggest that some women who work in the dry-cleaning industry have more menstrual problems and other problems than women not exposed to this chemical.

IARC Evaluation: Group 2B, possible human carcinogen; human evidence, inadequate; animal evidence, sufficient.
NTP: Anticipated human carcinogen.
IDLH: 150 ppm ; 1030.5 mg/m^3; potential human carcinogen.
Odor Threshold: 6.17 ppm
NIOSH Criteria Document: NIOSH 76-185.

1008 Tetrachloroethylene

OSHA Table Z-1-A Air Contaminant as perchloroethylene
OSHA Table Z-2 Air Contaminant

Exposure Limits:	**TWA**	**CEILING**	**STEL**
ACGIH: (1)	5 ppm; 170 mg/m^3		100 ppm; 685 mg/m^3
NIOSH: (2)			
OSHA: (3)	100 ppm	200 ppm; 300 ppm 5-min max peak in any 3 hrs	

(1) Animal carcinogen
(2) Potential human carcinogen; limit exposure to lowest feasible concentration.
(3) Potential occupational carcinogen; limit exposure to lowest feasible concentration.

RECOMMENDATIONS FOR RESPIRATOR SELECTION: NIOSH

At any concentrations above the NIOSH REL, or where there is no REL, at any detectable concentration: SCBAF:PD,PP (any self-contained breathing apparatus that has a full facepiece and is operated in a pressure-demand or other positive-pressure mode); or SAF:PD,PP:ASCBA (any supplied-air respirator that has a full facepiece and is operated in a pressure-demand or other positive-pressure mode in combination with an auxiliary, self-contained breathing apparatus operated in a pressure-demand or other positive-pressure mode).

Escape: GMFOV [any air-purifying, full-facepiece respirator (gas mask) with a chin-style, front- or back-mounted organic vapor canister]; or SCBAE (any appropriate escape-type, self-contained breathing apparatus).

CHEMICAL REACTIVITY

Incompatible with strong oxidizers, finely divided metals, caustics. Stable up to 258°F (126°C); at this temperature, product decomposes forming toxic hydrogen chloride fumes.

FIRE INFORMATION

Not flammable. Toxic and corrosive gases, including hydrogen chloride, are produced when heated. Vapors are heavier than air and will collect in low areas. Vapors in confined areas may explode when exposed to fire. Vapors may travel long distances to ignition sources and flash back. Storage containers and parts of containers may rocket great distances, in many directions. If material or contaminated runoff enters waterways, notify downstream users of potentially contaminated waters. Notify local health and fire officials and pollution control agencies. Use any appropriate extinguishing agent suitable for surrounding fires. From a secure, explosion-proof location, use water spray to cool exposed containers.

NFPA Hazard Classifications:

Health Hazards (Blue) 2
Flammability (Red) 0
Reactivity (Yellow) 0

FIRST AID

Move victim to fresh air. Call emergency medical care. Apply artificial respiration if victim is not breathing. Administer oxygen if breathing is difficult. Remove and isolate contaminated clothing and shoes. In case of contact with substance, immediately flush skin or eyes with running water for at least 20 minutes. For minor skin contact, avoid spreading material on unaffected skin. Wash skin with soap and water. Keep victim warm and quiet. Ensure that medical personnel are aware of the material(s) involved and take precautions to protect themselves.

ENVIRONMENTAL INFORMATION

Clean Air Act: Hazardous Air Pollutants (Title I, Part A, Section 112); Section 307 Toxic Pollutants.
Clean Water Act: Section 307 Priority Pollutants.
EPA Hazardous Waste Number (RCRA No.): U210.
RCRA Section 261 Hazardous Constituents
RCRA Land Ban Waste
RCRA Universal Treatment Standards: Wastewater (mg/L), 0.056; Nonwastewater (mg/kg), 6.0.
RCRA Ground Water Monitoring List: Suggested test method(s) (PQL ug/L): 8010(0.5); 8240(5).
Safe Drinking Water Act: MCL, 0.005 mg/L; MCLG, zero; Regulated chemical (47 FR 9352).
EPCRA Section 304: Reportable Quantity (RQ): CERCLA, 100 lb (45.4 kg).
EPCRA Section 313: Form R *de minimis* concentration reporting level: 0.1%.
Marine Pollutant (49CFR, Subchapter 172.101, Appendix B).
California's Proposition 65: Carcinogen.

TETRACYANOQUINODIMETHANE T:063

SYNONYMS: EINECS No. 216-174-8; PROPANEDINITRILE, 2,2'-(2,5-CYCLOHEXADIENE-1,4-DIYLIDENE)BIS-(9CI); QUINODIMETHAN, TETRACYANO-; TCNQ; TETRACYANOQUINODIMETHAN(e); TETRACYANO-p-QUINODIMETHANE; 7,7,8,8-TETRACYANOQUINODIMETHANE; 7,7',8,8'-TETRACYANOQUINODIMETHANE

IDENTIFICATION

CAS:	1518-16-7
ERG ID:	UN 2811 as toxic solid, organic, n.o.s.

1010 Tetracyanoquinodimethane

ERG Guide: 154
ERG Guide Category: SUBSTANCES-TOXIC AND/OR CORROSIVE (NONCOMBUSTIBLE)
Molecular Formula: $C_{12}H_4N_4$
RTECS No.: XB6825000
Properties: Crystalline solid. White. Soluble in water.

HEALTH AND SAFETY INFORMATION

Nitriles are organic cyanides; may be poisonous if swallowed.

IDLH: 25 mg/m^3 as CN
OSHA Table Z-1-A Air Contaminant as CN
NIOSH Criteria Document: NIOSH 77-108 CYANIDE

Exposure Limits:	TWA	CEILING
ACGIH: [1]		4.7 ppm; 5 mg/m^3 as CN
NIOSH:		4.7 ppm; 5 mg/m^3 [10 min] as CN
OSHA:	5 mg/m^3 as CN	

[1] Skin contact contributes significantly to overall exposure.

RECOMMENDATIONS FOR RESPIRATOR SELECTION: NIOSH/OSHA as cyanides

25 mg/m^3: SA (any supplied-air respirator); or SCBA (any self-contained breathing apparatus).

Emergency or Planned Entry into Unknown Concentrations or IDLH Conditions:
SCBAF:PD,PP (any self-contained breathing apparatus that has a full facepiece and is operated in a pressure-demand or other positive-pressure mode); or SAF:PD,PP:ASCBA (any supplied-air respirator that has a full facepiece and is operated in a pressure-demand or other positive-pressure mode in combination with an auxiliary, self-contained breathing apparatus operated in a pressure-demand or other positive-pressure mode)

Escape: GMFSHiE [any air-purifying, full-facepiece respirator (gas mask) with a chin-style, front- or back-mounted canister providing protection against the compound of concern and having a high-efficiency particulate filter]; or SCBAE (any appropriate escape-type, self-contained breathing apparatus).

FIRE INFORMATION

Combustion products include toxic nitrogen oxides and cyanide vapors. If material or contaminated runoff enters waterways, notify downstream users of potentially contaminated waters.Notify local health and fire officials and pollution control agencies. Use dry chemicals, alcohol-resistant foam, carbon dioxide. From a secure, explosion-proof location, use water spray to cool exposed containers. If cooling streams are ineffective (venting sound increases in volume and pitch, tank discolors, or shows any signs of deforming), withdraw immediately to a secure position.

NFPA Hazard Classifications: Not listed

FIRST AID

Move victim to fresh air; call emergency medical care. Apply artificial respiration if victim is not breathing. Do not use mouth-to-mouth method if victim ingested or inhaled the substance; induce artificial respiration with the aid of a pocket mask equipped with a one-way valve or other proper respiratory medical device. Administer oxygen if breathing is difficult. Remove and isolate contaminated clothing and shoes. In case of contact with substance, immediately flush skin or eyes with running water for at least 20 minutes. For minor skin contact, avoid spreading material on unaffected skin. Keep victim warm and quiet. Effects of exposure (inhalation, ingestion, or skin contact) to substance may be delayed. Ensure that medical personnel are aware of the material(s) involved and take precautions to protect themselves.

ENVIRONMENTAL INFORMATION as cyanide compounds

Clean Air Act: Hazardous Air Pollutants (Title I, Part A, Section 112).
Clean Water Act: Section 307 Priority Pollutants as cyanide, total.
EPA Hazardous Waste Number (RCRA No.): P030 as cyanides soluble salts and complexes, n.o.s.
RCRA Section 261 Hazardous Constituents as cyanides, soluble salts, and complexes, n.o.s.
EPCRA (Section 313): X+CN-where X = H+ or any other group where a formal dissociation may occur. For example, KCN or Ca(CN)2. Form R *de minimis* concentration reporting level: 1.0%. Form R Toxic Chemical Category Code: N016.
Marine Pollutant (49CFR, Subchapter 172.101, Appendix B) as cyanide mixtures, cyanide solutions, or cyanides, inorganic, n.o.s.

TETRAETHYLENE PENTAMINE T:067

SYNONYMS: AMINO ETHYL-1,2-ETHANEDIAMINE, 1,4,7,10,13,-PENTAAZATRIDECANE; D.E.H. 26; 1,11-DIAMINO-3,6,9 TRIAZAUNDECANE; EEC No. 612-060-00-0; EINECS No. 203-986-2; 1,2-ETHANEDIAMINE, n-(2-AMINOETHYL)-n'-[2-(2-AMINOETHYL)ETHYL]-; 1,2-ETHANEDIAMINE, n-(2-AMINOETHYL)-n'-(2-AMINOETHYL) AMINOETHYL-; 1,4,7,10,13-PENTAAZATRIDECANE;

IDENTIFICATION

CAS:	112-57-2
ERG ID:	UN 2320

1012 Tetraethylene Pentamine

ERG Guide:	153
ERG Guide Category:	SUBSTANCES-TOXIC AND/OR CORROSIVE (COMBUSTIBLE)
Molecular Formula:	$C_8H_{23}N_5$
RTECS No.:	KH8585000
Properties:	Combustible, viscous, hygroscopic liquid. Yellow. Disagreeable, ammonia-like odor. May float or sink in water; soluble.

HEALTH AND SAFETY INFORMATION

Corrosive to eyes, skin, and respiratory tract. Inhalation may cause pulmonary edema, which can be delayed for several hours; there is a risk of death in serious cases. May sensitize the skin.

RECOMMENDATIONS FOR RESPIRATOR SELECTION

At any concentrations above the NIOSH REL, or where there is no REL, at any detectable concentration: SCBAF:PD,PP (any self-contained breathing apparatus that has a full facepiece and is operated in a pressure-demand or other positive-pressure mode); or SAF:PD,PP:ASCBA (any supplied-air respirator that has a full facepiece and is operated in a pressure-demand or other positive-pressure mode in combination with an auxiliary, self-contained breathing apparatus operated in a pressure-demand or other positive-pressure mode).

Escape: GMFOV [any air-purifying, full-facepiece respirator (gas mask) with a chin-style, front- or back-mounted organic vapor canister]; or SCBAE (any appropriate escape-type, self-contained breathing apparatus).

CHEMICAL REACTIVITY

Incompatible with acids, organic anhydrides, halogenated hydrocarbons, isocyanates, nitroparaffins, vinyl acetate, acrylates, substituted allyls, alkylene oxides, epichlorohydrin, ketones, aldehydes, alcohols, glycols, mercury, phenols, cresols, caprolactam solution, and strong oxidizers. Attacks aluminum, copper, lead, tin, zinc, and alloys. May attack some forms of plastics.

FIRE INFORMATION

Combustion products include toxic nitrogen oxides. Vapors are heavier than air and will collect in low areas. Vapors in confined areas may explode when exposed to fire. If material or contaminated runoff enters waterways, notify downstream users of potentially contaminated waters. Notify local health and fire officials and pollution control agencies. Use dry chemicals, alcohol-resistant foam, carbon dioxide. From a secure, explosion-proof location, use water spray to cool exposed containers. If cooling streams are ineffective (venting sound increases in volume and pitch, tank discolors, or shows any signs of deforming), withdraw immediately to a secure position.

NFPA Hazard Classifications:

Health Hazards (Blue) 2
Flammability (Red) 1
Reactivity (Yellow) 0

Flash Point: 325°F (163°C) o.c.
Flammable Limits in Air: (estimated) LEL 0.8%; UEL 4.6%
Autoignition Temperature: 610°F (321°C)

FIRST AID

Move victim to fresh air and call emergency medical care. Apply artificial respiration if victim is not breathing. Do not use mouth-to-mouth method if victim swallowed or inhaled the substance; induce artificial respiration with the aid of a pocket mask equipped with a one-way valve or other proper respiratory medical device. If breathing is difficult, give oxygen. In case of contact with substance, immediately flush skin or eyes with running water for at least 20 minutes. Remove and isolate contaminated clothing and shoes. Keep victim warm and quiet. Effects of exposure (inhalation, ingestion, or skin contact) to substances may be delayed. Ensure that medical personnel are aware of the material(s) involved and take precautions to protect themselves.

TETRAFLUOROETHYLENE T:070

SYNONYMS: EINECS No. 204-126-9; ETHENE, TETRAFLUORO-; ETHYLENE, TETRAFLUORO-; FLUROPLAST 4; MS-122; PERFLUOROETHENE; PERFLUOROETHYLENE; STCC 4905783; TETRAFLUOROETHYLENE TELOMER; TETRAFLUOROETHENE; TETRAFLUOROETILINO (Spanish); TFE; TEFLON MONOMER

IDENTIFICATION

CAS:	116-14-3
ERG ID:	UN 1081
ERG Guide:	116P (inhibited)
ERG Guide Category:	GASES (FLAMMABLE-UNSTABLE)
Molecular Formula:	C_2F_4
RTECS No.:	KX4010000
Properties:	Extremely flammable, colorless, compressed gas. Odorless. Insoluble in water.

1014 Tetrafluoroethylene

HEALTH AND SAFETY INFORMATION

Irritating to eyes, skin, and respiratory system. Prolonged or repeated exposure can cause liver and kidney damage, and lung irritation. Contact with the liquid can cause frostbite.

IDLH: 250 mg/m^3 as F
IARC Evaluation: Group 3, unclassifiable substance; no human data; no animal data.
OSHA Process Safety Management of Highly Hazardous Chemicals (29CFR 1910.119, Appendix A), Threshold Quantity: 5,000 pounds.

Exposure Limits:	TWA
ACGIH:	2.5 mg/m^3 as F
NIOSH:	2.5 mg/m^3 as F
OSHA:	2.5 mg/m^3 as F

RECOMMENDATIONS FOR RESPIRATOR SELECTION: NIOSH/OSHA as fluorine

1 ppm: SA* (any supplied-air respirator).
2.5 ppm: SA:CF* (any supplied-air respirator operated in a continuous-flow mode).
5 ppm: SCBAF (any self-contained breathing apparatus with a full facepiece); or SAF (any supplied-air respirator with a full facepiece).
25 ppm: SAF:PD,PP (any supplied-air respirator that has a full facepiece and is operated in a pressure-demand or other positive-pressure mode).

Emergency or Planned Entry into Unknown Concentrations or IDLH Conditions: SCBAF:PD,PP (any self-contained breathing apparatus that has a full facepiece and is operated in a pressure-demand or other positive-pressure mode); or SAF:PD,PP:ASCBA (any supplied-air respirator that has a full facepiece and is operated in a pressure-demand or other positive-pressure mode in combination with an auxiliary, self-contained breathing apparatus operated in a pressure-demand or other positive-pressure mode).

Escape: GMFS [end of service life indicator (ESLI) required] (any air-purifying, full-facepiece respirator (gas mask) with a chin-style, front- or back-mounted canister providing protection against the compound of concern]; or SCBAE (any appropriate escape-type, self-contained breathing apparatus).

*Substance reported to cause eye irritation or damage; may require eye protection.

RECOMMENDATIONS FOR RESPIRATOR SELECTION

At any concentrations above the NIOSH REL, or where there is no REL, at any detectable concentration: SCBAF:PD,PP (any self-contained breathing apparatus that has a full facepiece and is operated in a pressure-demand or other positive-pressure mode); or SAF:PD,PP:ASCBA (any supplied-air

respirator that has a full facepiece and is operated in a pressure-demand or other positive-pressure mode in combination with an auxiliary, self-contained breathing apparatus operated in a pressure-demand or other positive pressure mode).

Escape: GMFOV [any air-purifying, full-facepiece respirator (gas mask) with a chin-style, front- or back-mounted organic vapor canister]; or SCBAE (any appropriate escape-type, self-contained breathing apparatus).

CHEMICAL REACTIVITY

A highly reactive, flammable gas. Able to form unstable peroxides; if terpene inhibitor is not present in adequate concentrations, explosive polymerization may occur. Reacts violently with air, oxygen, oxidizers, and sulfur trioxide.

FIRE INFORMATION

Combustion products include toxic hydrogen fluoride. Gases are heavier than air and will collect in low areas. Gases in confined areas may explode when exposed to fire. Gases may travel long distances to ignition sources and flash back. Storage containers and parts of containers may rocket great distances, in many directions. If material or contaminated runoff enters waterways, notify downstream users of potentially contaminated waters. Notify local health and fire officials and pollution control agencies. Do not extinguish the fire unless the flow of the gas can be stopped and any remaining gas is out of the line. Specially trained personnel may use fog lines to cool exposures and let the fire burn itself out. Use dry chemicals, foam, carbon dioxide. From a secure, explosion-proof location, use water spray to cool exposed containers. If cooling streams are ineffective (venting sound increases in volume and pitch, tank discolors, or shows any signs of deforming), withdraw immediately to a secure position.

NFPA Hazard Classifications:

Health Hazards (Blue) 2
Flammability (Red) 4
Reactivity (Yellow) 3

Flash Point: Flammable Gas
Flammable Limits in Air: LEL 10%; UEL 50%
Autoignition Temperature: 383°F (195°C)

FIRST AID

Move victim to fresh air. Call emergency medical care. Apply artificial respiration if victim is not breathing. Administer oxygen if breathing is difficult. Remove and isolate contaminated clothing and shoes. In case of contact with liquefied gas, thaw frosted parts with lukewarm water. Keep victim warm and quiet. Ensure that medical personnel are aware of the material(s) involved and take precautions to protect themselves.

ENVIRONMENTAL INFORMATION

Clean Air Act: Accidental Release Prevention/Flammable substances, (Section 112[r], Table 3), TQ = 10,000 lb (4,540 kg).
TSCA: 40CFR799.18

TETRAHYDROFURAN T:080

SYNONYMS: BUTANE, 1,4-EPOXY-; BUTYLENE OXIDE; CYCLOTETRAMETHYLENE OXIDE; DIETHYLENE OXIDE; EEC No. 603-025-00-0; EINECS No. 203-726-8; 1,4-EPOXYBUTANE; FURANIDINE; FURAN, TETRAHYDRO-; HYDROFURAN; NCI-C60560; OXACYCLOPENTANE; OXOLAAN (Dutch); OXOLANE; TETRAHIDROFURANO (Spanish); TETRAHYDROFURAAN (Dutch); TETRAHYDROFURANNE (French); TETRAIDROFURANO (Italian); TETRAMETHYLENE OXIDE; THF; WELD-ON 711; WELD-ON P-70 PRIMER; WELD-ON P-72 PRIMER

IDENTIFICATION

CAS:	109-99-9
ERG ID:	UN 2056
ERG Guide:	127
ERG Guide Category:	FLAMMABLE LIQUIDS (POLAR/WATER MISCIBLE)
Molecular Formula:	C_4H_8O
RTECS No.:	LU5950000
Properties:	Highly flammable, colorless liquid. Faint fruity odor, acetonelike. Floats and mixes with water.

HEALTH AND SAFETY INFORMATION

Irritating to eyes, skin, nose, and lungs; prolonged contact will cause burns. Long-term exposures may include liver and kidney disease; skin drying and cracking; and possible dermatitis and skin rash.

IDLH: 2,000 ppm (5,900 mg/m^3) [10% LEL]
Odor Threshold: 3.8 ppm
OSHA Table Z-1-A Air Contaminant

Exposure Limits:	TWA	STEL
ACGIH:	200 ppm; 590 mg/m^3	250 ppm; 735 mg/m^3
NIOSH:	200 ppm; 590 mg/m^3	250 ppm; 735 mg/m^3
OSHA:	200 ppm; 590 mg/m^3	

RECOMMENDATIONS FOR RESPIRATOR SELECTION: NIOSH/OSHA

2,000 ppm: SA:CF (any supplied-air respirator operated in a continuous-flow mode); CCRFOV [any chemical cartridge respirator with a full facepiece and organic vapor cartridge(s)]; or GMFOV [any air-purifying, full-facepiece respirator (gas mask) with a chin-style, front- or back-mounted acid gas canister); or PAPROV [any powered, air-purifying respirator with organic vapor cartridge(s)]; or SCBAF (any self-contained breathing apparatus with a full facepiece); or SAF (any supplied-air respirator with a full facepiece).

Emergency or Planned Entry into Unknown Concentrations or IDLH Conditions: SCBAF:PD,PP (any self-contained breathing apparatus that has a full facepiece and is operated in a pressure-demand or other positive-pressure mode); or SAF:PD,PP:ASCBA (any supplied-air respirator that has a full facepiece and is operated in a pressure-demand or other positive-pressure mode in combination with an auxiliary, self-contained breathing apparatus operated in a pressure-demand or other positive-pressure mode).

Escape: GMFOV [any air-purifying, full-facepiece respirator (gas mask) with a chin-style, front- or back-mounted organic vapor canister]; or SCBAE (any appropriate escape-type, self-contained breathing apparatus).

Note: Substance causes eye irritation or damage; eye protection needed.

CHEMICAL REACTIVITY

Forms explosive mixture with air. Unless inhibited, prolonged storage causes the formation of unstable and explosive peroxides. Incompatible with strong acids and strong oxidizers. Attacks some plastics. May accumulate static electric charges that can result in ignition of its vapors.

FIRE INFORMATION

Vapors are heavier than air and will collect in low areas. Vapors in confined areas may explode when exposed to fire. Vapors may travel long distances to ignition sources and flash back. Storage containers and parts of containers may rocket great distances, in many directions. If material or contaminated runoff enters waterways, notify downstream users of potentially contaminated waters. Notify local health and fire officials and pollution control agencies. Use dry chemicals, foam, carbon dioxide. From a secure, explosion-proof location, use water spray to cool exposed containers.

NFPA Hazard Classifications:

Health Hazards (Blue) 2
Flammability (Red) 3
Reactivity (Yellow) 1

1018 Tetrahydrofurfuryl Alcohol

Flash Point: –4°F (120°C) o.c.; 6°F (–14°C) c.c.
Flammable Limits in Air: LEL 2.0%; UEL 11.8%
Autoignition Temperature: 610°F (321°C)

FIRST AID

Move victim to fresh air. Call emergency medical care. Apply artificial respiration if victim is not breathing. Administer oxygen if breathing is difficult. Remove and isolate contaminated clothing and shoes. In case of contact with substance, immediately flush skin or eyes with running water for at least 20 minutes. Wash skin with soap and water. Keep victim warm and quiet. Ensure that medical personnel are aware of the material(s) involved and take precautions to protect themselves.

TETRAHYDROFURFURYL ALCOHOL T:085

SYNONYMS: ALCOHOL TETRAHIDROFURFURILICO (Spanish); ALCOOL TETRAHYDROFURFURYLIQUE (French); EEC No. 603-061-00-7; EINECS No. 202-625-6; 2-HYDROXYMETHYL OXOLANE; QO THFA; TETRAHYDRO-2-FURANCARBINOL; TETRAHYDRO-2-FURANMETHANOL; TETRAHYDROFURYLALKOHOL (Czech); TETRAHYDROFURYL CARBINOL; THFA

IDENTIFICATION

CAS:	97-99-4
ERG ID:	UN 2810 as toxic liquid, organic, n.o.s.
ERG Guide:	153
ERG Guide Category:	SUBSTANCES-TOXIC AND/0R CORROSIVE (COMBUSTIBLE)
Molecular Formula:	$C_5H_{10}O_2$
RTECS No.:	LU2450000
Properties:	Combustible, hygroscopic liquid. Colorless. Mild odor. Soluble in water.

HEALTH AND SAFETY INFORMATION

Irritates eyes, skin, and respiratory tract. Toxic if swallowed. Prolonged or repeated exposure may cause kidney and liver problems.

RECOMMENDATIONS FOR RESPIRATOR SELECTION

At any concentrations above the NIOSH REL, or where there is no REL, at any detectable concentration: SCBAF:PD,PP (any self-contained breathing apparatus that has a full facepiece and is

operated in a pressure-demand or other positive-pressure mode); or SAF:PD,PP:ASCBA (any supplied-air respirator that has a full facepiece and is operated in a pressure-demand or other positive-pressure mode in combination with an auxiliary, self-contained breathing apparatus operated in a pressure-demand or other positive-pressure mode).

Escape: GMFOV [any air-purifying, full-facepiece respirator (gas mask) with a chin-style, front- or back-mounted organic vapor canister]; or SCBAE (any appropriate escape-type, self-contained breathing apparatus).

CHEMICAL REACTIVITY

Reacts with oxidizing agents. Forms explosive mixture with air. May be able to form unstable peroxides. Alcohols are incompatible with sulfuric acid, nitric acid, caustics, aliphatic amines, isocyanates.

FIRE INFORMATION

Vapors are heavier than air and will collect in low areas. Vapors in confined areas may explode when exposed to fire. Vapors may travel long distances to ignition sources and flash back. Storage containers and parts of containers may rocket great distances, in many directions. If material or contaminated runoff enters waterways, notify downstream users of potentially contaminated waters. Notify local health and fire officials and pollution control agencies. Use dry chemicals, foam, carbon dioxide, water spray. From a secure, explosion-proof location, use water spray to cool exposed containers.

NFPA Hazard Classifications

Health Hazards (Blue) 2
Flammability (Red) 2
Reactivity (Yellow) 0

Flash Point: 167°F (75°C) o.c.
Flammable Limits in Air: LEL 1.5%; UEL 9.7% @ 72-122°F (22-50°C)
Autoignition Temperature: 540°F (282°C)

FIRST AID

Move victim to fresh air. Call emergency medical care. Apply artificial respiration if victim is not breathing. If breathing is difficult, give oxygen. Remove and isolate contaminated clothing and shoes. In case of contact with substance, immediately flush skin or eyes with running water for at least 20 minutes. Wash skin with soap and water. Keep victim warm and quiet. Ensure that medical personnel are aware of the material(s) involved and take precautions to protect themselves.

TETRAHYDROPHTHALIC ANHYDRIDE T:090

SYNONYMS: ANHYDRID KYSELINY TETRAHYDROFTALOVE (Czech); EEC No. 607-099-00-5; EINECS No. 201-605-4; MALEIC ANHYDRIDE adduct of BUTADIENE; 3a,4,7,7a-TETRAHYDRO-1,3-ISOBENZOFURANDIONE; TETRAHYDROPHTHALIC ACID ANHYDRIDE; Δ4-TETRAHYDROPHTHALIC ANHYDRIDE; 1,2,3,6-TETRAHYDRO PHTHALIC ANHYDRIDE; THPA

IDENTIFICATION

CAS:	85-43-8
ERG ID:	UN 2698
ERG Guide:	156
ERG Guide Category:	SUBSTANCES-TOXIC AND/OR CORROSIVE (COMBUSTIBLE/WATER-SENSITIVE)
Molecular Formula:	$C_8H_8O_3$
RTECS No.:	GW5775000
Properties:	Combustible crystalline powder. White.

HEALTH AND SAFETY INFORMATION

Corrosive to eyes, skin, and respiratory tract. Inhalation of dust may cause pulmonary edema, which can be delayed for several hours; there is a risk of death in serious cases.

RECOMMENDATIONS FOR RESPIRATOR SELECTION

At any concentrations above the NIOSH REL, or where there is no REL, at any detectable concentration: SCBAF:PD,PP (any self-contained breathing apparatus that has a full facepiece and is operated in a pressure-demand or other positive-pressure mode); or SAF:PD,PP:ASCBA (any supplied-air respirator that has a full facepiece and is operated in a pressure-demand or other positive-pressure mode in combination with an auxiliary, self-contained breathing apparatus operated in a pressure-demand or other positive-pressure mode).

Escape: GMFOVHiE [any air-purifying, full-facepiece respirator (gas mask) with a chin-style, front- or back-mounted organic vapor canister having a high-efficiency particulate filter]; or SCBAE (any appropriate escape-type, self-contained breathing apparatus).

CHEMICAL REACTIVITY

Contact with strong oxidizers can cause fire and explosions. Anhydrides are acidic and incompatible with sulfuric acid, caustics, ammonia, aliphatic amines, alkanolamines, isocyanates, alkylene oxides,

FIRE INFORMATION

Combustion may produce irritants and toxic vapors. If material or contaminated runoff enters waterways, notify downstream users of potentially contaminated waters. Notify local health and fire officials and pollution control agencies. Use dry chemicals, alcohol-resistant foam, carbon dioxide. From a secure, explosion-proof location, use water spray to cool exposed containers.

NFPA Hazard Classifications: Not listed

Flash Point: 315°F (157°C) o.c.

FIRST AID

Move victim to fresh air. Call emergency medical care. Apply artificial respiration if victim is not breathing. Do not use mouth-to-mouth method if victim ingested or inhaled the substance; induce artificial respiration with the aid of a pocket mask equipped with a one-way valve or other proper respiratory medical device. Administer oxygen if breathing is difficult. Remove and isolate contaminated clothing and shoes. In case of contact with substance, immediately flush skin or eyes with running water for at least 20 minutes. For minor skin contact, avoid spreading material on unaffected skin. Keep victim warm and quiet. Effects of exposure (inhalation, ingestion, or skin contact) may be delayed. Ensure that medical personnel are aware of the material(s) involved and take precautions to protect themselves.

TETRAISOPROPYL TITANATE T:095

SYNONYMS: EINECS No. 208-909-6; ISOPROPYL ALCOHOL, TITANIUM(4+) SALT; ISOPROPYL ORTHOTITANATE; ISOPROPYL TITANATE(IV); 2-PROPANOL, TITANIUM(4+) SALT; TETRAISOPROPOXIDE TITANIUM; TETRAISOPROPOXYTITANIUM; TETRAISOPROPYL ORTHOTITANATE; TETRAISOPROPYL ORTHOTITANATE; TETRAKIS(ISOPROPOXY)-TITANIUM; TITANIUM(4+) ISOPROPOXIDE; TITANIUM ISOPROPYLATE; TITANIUM TETRAISOPROPANOLATE; TITANIUM TETRAISOPROPOXIDE; TITANIUM TETRAISOPROPYLATE; TITANIUM TETRA-n-PROPOXIDE; TYZOR TPt

IDENTIFICATION

CAS:	546-68-9
ERG ID:	UN 1992 as flammable liquid, toxic, n.o.s.
ERG Guide:	131
ERG Guide Category:	FLAMMABLE LIQUIDS-TOXIC
Molecular Formula:	$C_{12}H_{28}O_4Ti$
RTECS No.:	NT8060000

1022 Tetramethyl Succinonitrile

Properties: Highly flammable liquid. Fumes in moist air. Colorless to light yellow. Reacts with water or steam forming flammable and toxic vapors.

HEALTH AND SAFETY INFORMATION

Irritates eyes and skin. Toxic if swallowed. Due to the availability of insufficient data on human effects, caution should be exercised.

CHEMICAL REACTIVITY

Forms an explosive mixture with air; moist air causes material to fume. Contact with strong oxidizers and strong acids may cause fire and explosions. Decomposes rapidly in water, water vapor or steam.

FIRE INFORMATION

Vapors are heavier than air and will collect in low areas. Vapors in confined areas may explode when exposed to fire. Vapors may travel long distances to ignition sources and flash back. Storage containers and parts of containers may rocket great distances, in many directions. If material or contaminated runoff enters waterways, notify downstream users of potentially contaminated waters. Notify local health and fire officials and pollution control agencies. Use dry chemicals, foam, carbon dioxide. From a secure, explosion-proof location, use water spray to cool exposed containers.

NFPA Hazard Classifications: Not listed

Flash Point: 72°F (22°C)

FIRST AID

Move victim to fresh air. Call emergency medical care. Apply artificial respiration if victim is not breathing. Administer oxygen if breathing is difficult. Remove and isolate contaminated clothing and shoes. In case of contact with substance, immediately flush skin or eyes with running water for at least 20 minutes. Wash skin with soap and water. Keep victim warm and quiet. Effects of exposure (inhalation, ingestion, or skin contact) to substance may be delayed. Ensure that medical personnel are aware of the material(s) involved and take precautions to protect themselves.

TETRAMETHYL SUCCINONITRILE T:110

SYNONYMS: TETRAMETHYLBUTANE DINITRILE; TETRAMETHYLSUCCINIC ACID DINITRILE; TETRAMETHYL SUCCINODINITRILE; TMSN

IDENTIFICATION

CAS:	3333-52-6
ERG ID:	UN 3275 as nitriles, poisonous, flammable, n.o.s.
ERG Guide:	131
ERG Guide Category:	FLAMMABLE LIQUIDS-TOXIC
Molecular Formula:	$C_8H_{12}N_2$
RTECS No.:	WN4025000
Properties:	Combustible solid; crystallizes in plates. Colorless. Odorless. Insoluble in water.

HEALTH AND SAFETY INFORMATION

Irritates skin and eyes. Absorbed by the skin. Symptoms include headaches, nausea, convulsions, coma. Poisonous if swallowed. Forms cyanide in the body. Affects central nervous system, liver, gastrointestinal tract.

OSHA Table Z-1-A Air Contaminant

Exposure Limits:	**TWA**
ACGIH: (1)	0.5 ppm; 2.8 mg/m^3
NIOSH: (1)	0.5 ppm; 3 mg/m^3
OSHA: (1)	0.5 ppm; 3 mg/m^3

(1) Skin contact contributes significantly to overall exposure.

RECOMMENDATIONS FOR RESPIRATOR SELECTION: NIOSH/OSHA

28 mg/m^3: SA (any supplied-air respirator); or SCBA (any self-contained breathing apparatus).

Emergency or Planned Entry into Unknown Concentrations or IDLH Conditions: SCBAF:PD,PP (any self-contained breathing apparatus that has a full facepiece and is operated in a pressure-demand or other positive-pressure mode); or SAF:PD,PP:ASCBA (any supplied-air respirator that has a full facepiece and is operated in a pressure-demand or other positive-pressure mode in combination with an auxiliary, self-contained breathing apparatus operated in a pressure-demand or other positive-pressure mode).

Escape: GMFOVHiE [any air-purifying, full-facepiece respirator (gas mask) with a chin-style, front- or back-mounted organic vapor canister having a high-efficiency particulate filter]; or SCBAE (any appropriate escape-type, self-contained breathing apparatus).

CHEMICAL REACTIVITY

Contact with strong oxidizers may cause fire and explosions.

FIRE INFORMATION

Combustion products include toxic nitrogen oxides and cyanide vapors. If material or contaminated runoff enters waterways, notify downstream users of potentially contaminated waters. Notify local health and fire officials and pollution control agencies. Use dry chemicals, alcohol-resistant foam, carbon dioxide. From a secure, explosion-proof location, use water spray to cool exposed containers.

NFPA Hazard Classifications: Not listed

FIRST AID

Move victim to fresh air. Call emergency medical care. Apply artificial respiration if victim is not breathing. Do not use mouth-to-mouth method if victim ingested or inhaled the substance; induce artificial respiration with the aid of a pocket mask equipped with a one-way valve or other proper respiratory medical device. Administer oxygen if breathing is difficult. Remove and isolate contaminated clothing and shoes. In case of contact with substance, immediately flush skin or eyes with running water for at least 20 minutes. Wash skin with soap and water. Keep victim warm and quiet. Effects of exposure (inhalation, ingestion, or skin contact) to substance may be delayed. Ensure that medical personnel are aware of the material(s) involved and take precautions to protect themselves.

ENVIRONMENTAL INFORMATION as cyanide compounds

Clean Air Act: Hazardous Air Pollutants (Title I, Part A, Section 112).
Clean Water Act: Section 307 Priority Pollutants as cyanide, total.
EPA Hazardous Waste Number (RCRA No.): P030 as cyanides soluble salts and complexes, n.o.s.
RCRA Section 261 Hazardous Constituents as cyanides, soluble salts, and complexes, n.o.s.
EPCRA (Section 313): X+CN-where X = H+ or any other group where a formal dissociation may occur. For example, KCN or Ca(CN)2. Form R *de minimis* concentration reporting level: 1.0%. Form R Toxic Chemical Category Code: N016.
Marine Pollutant (49CFR, Subchapter 172.101, Appendix B) as cyanide mixtures, cyanide solutions, or cyanides, inorganic, n.o.s.

THIODIGLYCOL T:125

SYNONYMS: BIS(β-HYDROXYETHYL)SULFIDE; BIS(2-HYDROXYETHYL)SULFIDE; β,β'-DIHYDROXYDIETHYL SULFIDE; β,β'-DIHYDROXYETHYL SULFIDE; EEC No. 603-081-00-6; EINECS No. 203-874-3; ETHANOL, 2,2'-THIOBIS-; GLYECINE A; β-HYDROXYETHYL SULFIDE; KROMFAX SOLVENT; 2,2'-THIODIETHANOL; THIODIETHYLENE GLYCOL; β-THIODIGLYCOL

IDENTIFICATION

CAS:	111-48-8
ERG ID:	UN 2929 as toxic liquid, organic, n.o.s.
ERG Guide:	153
ERG Guide Category:	SUBSTANCES-TOXIC AND/OR CORROSIVE (COMBUSTIBLE)
Molecular Formula:	$C_4H_{10}O_2S$
RTECS No.:	KM2975000
Properties:	Combustible, thick liquid. Colorless. Characteristic odor. Completely soluble in water.

HEALTH AND SAFETY INFORMATION

Irritates eyes, skin, and respiratory tract. Toxic if swallowed; abdominal pain, nausea, dizziness, sleepiness, unconsciousness.

CHEMICAL REACTIVITY

Contact with strong oxidizers may cause fire and explosions. Violent reaction with strong acids, especially hydrochloric acid and many other substances; may form poisonous and toxic vapors.

FIRE INFORMATION

Combustion products include toxic sulfur dioxide. If material or contaminated runoff enters waterways, notify downstream users of potentially contaminated waters. Notify local health and fire officials and pollution control agencies. Use dry chemicals, alcohol-resistant foam, carbon dioxide. From a secure, explosion-proof location, use water spray to cool exposed containers.

NFPA Hazard Classifications

Health Hazards (Blue) 2
Flammability (Red) 1
Reactivity (Yellow) 0

Flash Point: 320°F (160°C) o.c.
Autoignition Temperature: 568°F (298°C)

FIRST AID

Move victim to fresh air. Call emergency medical care. Apply artificial respiration if victim is not breathing. Do not use mouth-to-mouth method if victim ingested or inhaled the substance; induce artificial respiration with the aid of a pocket mask equipped with a one-way valve or other proper respiratory medical device.

Administer oxygen if breathing is difficult. Remove and isolate contaminated clothing and shoes. In case of contact with substance, immediately flush skin or eyes with running water for at least 20 minutes. Wash skin with soap and water. Keep victim warm and quiet. Effects of exposure (inhalation, ingestion, or skin contact) to substance may be delayed. Ensure that medical personnel are aware of the material(s) involved and take precautions to protect themselves.

THIOGLYCOLIC ACID T:130

SYNONYMS: ACIDE THIOGLYCOLIQUE (French); ACETIC ACID, MERCAPTO-; ACETYL MERCAPTAN; ACIDE THIOGLYCOLIQUE (French); ACIDO TIOGLICOLICO (Spanish); EEC No. 607-090-00-6; EINECS No. 200-677-4; GLYCOLIC ACID, THIO-; GLYCOLIC ACID, 2-THIO-; KYSELINA MERKAPTOOCTOVA (Polish); KYSELINA THIOGLYKOLOVA (Polish); MERCAPTOACETATE; MERCAPTOACETIC ACID; a-MERCAPTOACETIC ACID; 2-MERCAPTOACETIC ACID; MERCAPTOESSIGSAEURE; (German); 2-THIOGLYCOLIC ACID; THIOGLYCOLLIC ACID; THIOVANIC ACID; USAF CB-35

IDENTIFICATION

CAS:	68-11-1
ERG ID:	UN 3265 as corrosive liquid, acidic, organic, n.o.s.
ERG Guide:	153
ERG Guide Category:	SUBSTANCES-TOXIC AND/OR CORROSIVE (COMBUSTIBLE)
Molecular Formula:	$C_2H_4O_2S$
RTECS No.:	AI5950000
Properties:	Combustible liquid. Colorless. Strong, disagreeable odor characteristic of mercaptans. Soluble in water.

HEALTH AND SAFETY INFORMATION

Extremely corrosive to eyes, skin, and respiratory tract. Inhalation may cause pulmonary edema, which can be delayed for several hours; there is a risk of death in serious cases.

OSHA Table Z-1-A Air Contaminant

Exposure Limits:	**TWA**
ACGIH: (1)	1 ppm; 3.8 mg/m^3
NIOSH: (1)	1 ppm; 4 mg/m^3
OSHA:	None

(1)Skin contact contributes significantly to overall exposure.

RECOMMENDATIONS FOR RESPIRATOR SELECTION

At any concentrations above the NIOSH REL, or where there is no REL, at any detectable concentration: SCBAF:PD,PP (any self-contained breathing apparatus that has a full facepiece and is operated in a pressure-demand or other positive-pressure mode); or SAF:PD,PP:ASCBA (any supplied-air respirator that has a full facepiece and is operated in a pressure-demand or other positive-pressure mode in combination with an auxiliary, self-contained breathing apparatus operated in a pressure-demand or other positive-pressure mode).

Escape: GMFOV [any air-purifying, full-facepiece respirator (gas mask) with a chin-style, front- or back-mounted organic vapor canister]; or SCBAE (any appropriate escape-type, self-contained breathing apparatus).

CHEMICAL REACTIVITY

Reacts with air, strong oxidizers, acids, bases, active metals (e.g., sodium, magnesium, and calcium). It is readily oxidized by air; decomposition can lead to the release of hydrogen sulfide. Organic acids are incompatible with sulfuric acid, caustics, ammonia, aliphatic amines, alkanolamines, isocyanates, alkylene oxides, epichlorohydrin..

FIRE INFORMATION

Combustion products include toxic sulfur oxides. If material or contaminated runoff enters waterways, notify downstream users of potentially contaminated waters. Notify local health and fire officials and pollution control agencies. Use dry chemicals, alcohol-resistant foam, carbon dioxide. From a secure, explosion-proof location, use water spray to cool exposed containers.

NFPA Hazard Classifications: Not listed

FIRST AID

Move victim to fresh air and call emergency medical care. Apply artificial respiration if victim is not breathing. Do not use mouth-to-mouth method if victim swallowed or inhaled the substance; induce artificial respiration with the aid of a pocket mask equipped with a one-way valve or other proper respiratory medical device. If breathing is difficult, give oxygen. In case of contact with substance, immediately flush skin or eyes with running water for at least 20 minutes. Remove and isolate contaminated clothing and shoes. Keep victim warm and quiet. Effects of exposure (inhalation, ingestion, or skin contact) to substances may be delayed. Ensure that medical personnel are aware of the material(s) involved and take precautions to protect themselves.

THIOGLYCEROL T:135

SYNONYMS: EINECS No.202-495-0; 1-MERCAPTOGLYCEROL; 1-MERCAPTO-2,3-PROPANEDIOL; 3-MERCAPTO-1,2-PROPANEDIOL; MONOTHIOGLYCEROL; α-MONOTHIOGLYCEROL; 1,2-PROPANEDOIL,3-MERCAPTO-; THIOGLYCERIN; α-THIOGLYCEROL; 1-THIOGLYCEROL; THIOVANOL; USAF B-40; USAF CB-37

IDENTIFICATION

CAS:	96-27-5
ERG ID:	UN 3071 as mercaptans, liquid, n.o.s.
ERG Guide:	131
ERG Guide Category:	FLAMMABLE LIQUIDS-TOXIC
Molecular Formula:	$C_3H_8O_2S$
RTECS No.:	TY8140000
Properties:	Combustible liquid. Colorless. Stench odor. Soluble in water.

HEALTH AND SAFETY INFORMATION

Poisonous. Odor may cause headache and nausea. Due to the availability of insufficient data on human effects, caution should be exercised.

RECOMMENDATIONS FOR RESPIRATOR SELECTION

At any concentrations above the NIOSH REL, or where there is no REL, at any detectable concentration: SCBAF:PD,PP (any self-contained breathing apparatus that has a full facepiece and is operated in a pressure-demand or other positive-pressure mode); or SAF:PD,PP:ASCBA (any supplied-air respirator that has a full facepiece and is operated in a pressure-demand or other positive-pressure mode in combination with an auxiliary, self-contained breathing apparatus operated in a pressure-demand or other positive-pressure mode).

Escape: GMFOV [any air-purifying, full-facepiece respirator (gas mask) with a chin-style, front- or back-mounted organic vapor canister]; or SCBAE (any appropriate escape-type, self-contained breathing apparatus).

CHEMICAL REACTIVITY

Violent reaction with strong oxidizers such as calcium hypochlorite. Reacts with ethylene oxide, reducing agents, alkali metals, water, steam, or acids.

FIRE INFORMATION

Combustion products include toxic sulfur oxides. Vapors are heavier than air and will collect in low areas.

Vapors in confined areas may explode when exposed to fire. If material or contaminated runoff enters waterways, notify downstream users of potentially contaminated waters. Notify local health and fire officials and pollution control agencies. Use dry chemicals, alcohol-resistant foam, carbon dioxide. From a secure, explosion-proof location, use water spray to cool exposed containers.

NFPA Hazard Classifications: Not listed

Flash Point: 234°F (112°C)

FIRST AID

Move victim to fresh air. Call emergency medical care. Apply artificial respiration if victim is not breathing. Do not use mouth-to-mouth method if victim ingested or inhaled the substance; induce artificial respiration with the aid of a pocket mask equipped with a one-way valve or other proper respiratory medical device. Administer oxygen if breathing is difficult. Remove and isolate contaminated clothing and shoes. In case of contact with substance, immediately flush skin or eyes with running water for at least 20 minutes. Wash skin with soap and water. Keep victim warm and quiet. Effects of exposure (inhalation, ingestion, or skin contact) to substance may be delayed. Ensure that medical personnel are aware of the material(s) involved and take precautions to protect themselves.

THIOPHENE T:150

SYNONYMS: CP34; DIVINYLENE SULFIDE; EINECS No. 203-729-4; HUILE H50; THIACYCLOPENTADIENE; THIAPHENE; THIOFURAM; THIOFURAN; THIOFURFURAN; THIOLE; THIOPHEN; THIOTETROLE; TIOFENO (Spanish); USAF EK-1860

IDENTIFICATION

CAS:	110-02-1
ERG ID:	UN 2414
ERG Guide:	130
ERG Guide Category:	FLAMMABLE LIQUIDS-NONPOLAR/WATER IMMISCIBLE/NOXIOUS)
Molecular Formula:	C_4H_4S
RTECS No.:	XM7350000
Properties:	Extremely flammable liquid. Clear, colorless. Benzenelike odor. Insoluble in water.

HEALTH AND SAFETY INFORMATION

Irritates eyes and skin.

RECOMMENDATIONS FOR RESPIRATOR SELECTION: NIOSH

At any concentrations above the NIOSH REL, or where there is no REL, at any detectable concentration: SCBAF:PD,PP (any self-contained breathing apparatus that has a full facepiece and is operated in a pressure-demand or other positive-pressure mode); or SAF:PD,PP:ASCBA (any supplied-air respirator that has a full facepiece and is operated in a pressure-demand or other positive-pressure mode in combination with an auxiliary, self-contained breathing apparatus operated in a pressure-demand or other positive-pressure mode).

Escape: GMFOV [any air-purifying, full-facepiece respirator (gas mask) with a chin-style, front- or back-mounted organic vapor canister]; or SCBAE (any appropriate escape-type, self-contained breathing apparatus).

CHEMICAL REACTIVITY

Forms explosive mixture with air. Reacts violently with strong oxidizers, calcium hypochloride, n-nitrosoacetanilide (explosion at room temperature), nitric acid. May accumulate static electrical charges, and may cause ignition of its vapors.

FIRE INFORMATION

Combustion products include toxic sulfur dioxide. Vapors are heavier than air and will collect in low areas. Vapors in confined areas may explode when exposed to fire. Vapors may travel long distances to ignition sources and flash back. Storage containers and parts of containers may rocket great distances, in many directions. If material or contaminated runoff enters waterways, notify downstream users of potentially contaminated waters. Notify local health and fire officials and pollution control agencies. Use dry chemicals, foam, carbon dioxide. From a secure, explosion-proof location, use water spray to cool exposed containers.

NFPA Hazard Classifications:

Health Hazards (Blue) 2
Flammability (Red) 3
Reactivity (Yellow) 0

Flash Point: 30°F (-1°C)

FIRST AID

Move victim to fresh air. Call emergency medical care. Apply artificial respiration if victim is not breathing. Administer oxygen if breathing is difficult. Remove and isolate contaminated clothing and shoes. In case of contact with substance, immediately flush skin or eyes with running water for at least 20 minutes. Wash skin with soap and water. Keep victim warm and quiet. Effects of exposure (inhalation, ingestion, or skin contact) to substance may be delayed. Ensure that medical personnel are aware of the material(s) involved and take precautions to protect themselves.

TIN CHLORIDE T:200

SYNONYMS: C.I. 77864; CLORURO ESTANNOSO (Spanish); EINECS No. 231-868-0; ETAIN (TETRACHLORURE d') (French); FASCAL; FASCAL 2004; NCI-C02722; STANNOUS CHLORIDE, DIHYDRIDE; TIN(2+) CHLORIDE, ANHYDROUS; TIN(II) CHLORIDE, ANHYDROUS; TIN CRYSTALS; TIN DICHLORIDE; TIN PROTOCHLORIDE; TIN SALTS

IDENTIFICATION

CAS:	7772-99-8
ERG ID:	NA 1759
ERG Guide:	154
ERG Guide Category:	TOXIC AND/OR CORROSIVE (NON-COMBUSTIBLE)
Molecular Formula:	Cl_2Sn
RTECS No.:	XP870000
Properties:	Crystals. Colorless. Poisonous fumes produced when heated.

HEALTH AND SAFETY INFORMATION

Irritates eyes, skin, and respiratory tract. Inhalation can cause lung disorders. Prolonged or repeated contact can cause liver and kidney damage.

IDLH: 100 mg/m^3 as inorganic tin
OSHA Table Z-1-A Air Contaminant as tin, inorganic compounds

Exposure Limits:	**TWA**
ACGIH: (1)	2 mg/m^3 as tin
NIOSH: (1)	2 mg/m^3 as tin
OSHA: (1)	2 mg/m^3 as tin

(1) Skin contact contributes significantly to overall exposure.

RECOMMENDATIONS FOR RESPIRATOR SELECTION: NIOSH/OSHA as inorganic tin

10 mg/m^3: DM (any dust and mist respirator).
20 mg/m^3: DMXSQ (any dust and mist respirator except single-use and quarter mask respirators); or SA (any supplied-air respirator).

50 mg/m^3: SA:CF (any supplied-air respirator operated in a continuous-flow mode); or PAPRDM (any powered, air-purifying respirator with a dust and mist filter).
100 mg/m^3: HiEF (any air-purifying, full-facepiece respirator with a high-efficiency particulate filter); or SCBAF (any self-contained breathing apparatus with a full facepiece); or SAF (any supplied-air respirator with a full facepiece).

Emergency or Planned Entry into Unknown Concentrations or Idlh Conditions: SCBAF:PD,PP (any self-contained breathing apparatus that has a full facepiece and is operated in a pressure-demand or other positive-pressure mode); or SAF:PD,PP:ASCBA (any supplied-air respirator that has a full facepiece and is operated in a pressure-demand or other positive-pressure mode in combination with an auxiliary, self-contained breathing apparatus operated in a pressure-demand or other positive pressure mode.

Escape: HiEF (any air-purifying, full-facepiece respirator with a high-efficiency particulate filter); or SCBAE (any appropriate escape-type, self-contained breathing apparatus).

CHEMICAL REACTIVITY

A strong reducing agent; violent reaction with oxidizers. Reacts with metal nitrates to produce an explosive reaction. Reacts violently with nitrates, potassium, sodium, bromine trifluoride. Reacts with the oxygen in air, forming insoluble vanadium oxydichloride.

FIRE INFORMATION

Combustion products include toxic hydrogen chloride. If material or contaminated runoff enters waterways, notify downstream users of potentially contaminated waters. Notify local health and fire officials and pollution control agencies. Use any extinguishing agent suitable for surrounding fires. From a secure, explosion-proof location, use water spray to cool exposed containers.

NFPA Hazard Classifications: Not listed

FIRST AID

Move victim to fresh air. Call emergency medical care. Apply artificial respiration if victim is not breathing. Do not use mouth-to-mouth method if victim ingested or inhaled the substance; induce artificial respiration with the aid of a pocket mask equipped with a one-way valve or other proper respiratory medical devise. Administer oxygen if breathing is difficult. Remove and isolate contaminated clothing and shoes. In case of contact with substance, immediately flush skin or eyes with running water for at least 20 minutes. For minor skin contact, avoid spreading material on unaffected skin. Keep victim warm and quiet. Effects of exposure (inhalation, ingestion, or skin contact) may be delayed. Ensure that medical personnel are aware of the material(s) involved and take precautions to protect themselves.

TITANIUM DIOXIDE T:225

SYNONYMS: A-FIL CREAM; ANATASE; ATLAS WHITE TITANIUM DIOXIDE; AUSTIOX; BAYERITIAN; BAYERTITAN; BAYTITAN; BROOKITE; CALCOTONE-WHITE-T; C.I. 77891; C.I. PIGMENT WHITE 6; COSMETIC WHITE C47-5175; C-WEISS 7 (German); DIOSSIDO di TITANIO (Italian); DIOXIDO de TITANIO (Spanish); EINECS No. 236-675-5; FLAMENCO; HOMBITAN; HORSE HEAD A-410; KH 360 KRONOS-CL-220; KRONOS TITANIUM DIOXIDE; LEVANOX WHITE RKB; NCI-C04240; P-25; R-680; RAYOX; RO-2; RUNA-RH20; RUTILE; RUTIOX-CR; TIOFINE; TIONA-TD; TIOXIDE; TIPAQUE; TI-PURE R 900; TITAFRANCE; TITANDIOXID (Swedish); TITANIA; TITANIC ACID ANHYDRIDE; TITANIC ANHYDRIDE; TITANIC OXIDE; TITANIUM OXIDE; TITANIUM PEROXIDE; TITANIUM WHITE; TITANOX-2100; TITANOX-RANC; TRIOXIDE(S); TRONOX; UNITANE; UNITANE 0-110; UNITANE 0-220; UNITANE OR-150; UNITANE RUTILE PIGMENT; UNIWHITE; 1700 WHITE; ZOPAQUE

IDENTIFICATION

CAS:	13463-67-7
Molecular Formula:	TiO_2
RTECS No.:	XR2275000
Properties:	Amorphous powder, crystals, or solid. White. Odorless. Tasteless. Floats on water; insoluble.

HEALTH AND SAFETY INFORMATION

Irritates the skin and respiratory tract.

IARC Evaluation: Group 3; limited animal evidence. Insufficient human evidence.
IDLH: 5,000 mg/m^3; potential human carcinogen
OSHA Table Z-1-A Air Contaminant

Exposure Limits:	**TWA**
ACGIH: [(1)]	10 mg/m^3 (total dust)
NIOSH: [(2)]	
OSHA:	15 mg/m^3 (total dust)

[(1)] Not classifiable as a human carcinogen.
[(2)] Potential human carcinogen. Limit exposure to lowest feasible concentration.

RECOMMENDATIONS FOR RESPIRATOR SELECTION: NIOSH

At any concentrations above the NIOSH REL, or where there is no REL, at any detectable concentration: SCBAF:PD,PP (any self-contained breathing apparatus that has a full facepiece and is

operated in a pressure-demand or other positive-pressure mode); or SAF:PD,PP:ASCBA (any supplied-air respirator that has a full facepiece and is operated in a pressure-demand or other positive-pressure mode in combination with an auxiliary, self-contained breathing apparatus operated in a pressure-demand or other positive-pressure mode).

Escape: HiEF (any air-purifying, full-facepiece respirator with a high-efficiency particulate filter); or SCBAE (any appropriate escape-type, self-contained breathing apparatus).

CHEMICAL REACTIVITY

Contact at high temperatures with chemically active metals may cause a violent reaction.

FIRE INFORMATION

Noncombustible. Combustion may produce irritants and toxic vapors. If material or contaminated runoff enters waterways, notify downstream users of potentially contaminated waters. Notify local health and fire officials and pollution control agencies. Use any extinguishing agent suitable for surrounding fire. Use water spray to cool exposed containers.

NFPA Hazard Classifications: Not listed

FIRST AID

Move victim to fresh air. Call emergency medical care. Apply artificial respiration if victim is not breathing. Administer oxygen if breathing is difficult. Remove and isolate contaminated clothing and shoes. In case of contact with substance, immediately flush skin or eyes with running water for at least 20 minutes. Wash skin with soap and water. Keep victim warm and quiet. Effects of exposure (inhalation, ingestion, or skin contact) to substance may be delayed. Ensure that medical personnel are aware of the material(s) involved and take precautions to protect themselves.

TITANIUM TRICHLORIDE T:233

SYNONYMS: TRICLORURO de TITANIO (Spanish); EINECS No. 231-728-9; TAC 121; TAC 131; TITANIUM(III) CHLORIDE; TITANIUM TRICHLORIDE MIXTURES (UN 2869); TITANIUM TRICHLORIDE MIXTURES, pyrophoric (UN 2441); TITANIUM TRICHLORIDE, pyrophoric (UN 2441); TITANOUS CHLORIDE; TRICHLOROTITANIUM

IDENTIFICATION

CAS: 7705-07-9
ERG ID: UN 2441/UN 2869

ERG Guide: 135 as titanium trichloride, pyrophoric; 157 as titanium trichloride, mixture

ERG Guide Category: (135) SUBSTANCES-SPONTANEOUSLY COMBUSTIBLE
(157) SUBSTANCES-TOXIC AND/OR CORROSIVE (NONCOMBUSTIBLE/WATER-SENSITIVE)

Molecular Formula: Cl_3Ti

RTECS No.: XR1924000

Properties: Deliquescent crystalline or powdery solid. Also available as a fuming liquid. Dark-violet crystals or purple powder. Colorless to light yellow liquid. Reacts with water and steam, forming hydrochloric acid fumes.

HEALTH AND SAFETY INFORMATION

Corrosive to eyes, skin, and respiratory tract. Inhalation may cause pulmonary edema, which can be delayed for several hours; there is a risk of death in serious cases.

RECOMMENDATIONS FOR RESPIRATOR SELECTION

At any concentrations above the NIOSH REL, or where there is no REL, at any detectable concentration: SCBAF:PD,PP (any self-contained breathing apparatus that has a full facepiece and is operated in a pressure-demand or other positive-pressure mode); or SAF:PD,PP:ASCBA (any supplied-air respirator that has a full facepiece and is operated in a pressure-demand or other positive-pressure mode in combination with an auxiliary, self-contained breathing apparatus operated in a pressure-demand or other positive-pressure mode).

Escape: GMFOVHiE [any air-purifying, full-facepiece respirator (gas mask) with a chin-style, front- or back-mounted organic vapor canister having a high-efficiency particulate filter]; or SCBAE (any appropriate escape-type, self-contained breathing apparatus).

CHEMICAL REACTIVITY

Must be kept in airtight packaging, or under inert gas. May ignite spontaneously in air. A strong reducing agent; violent reaction with oxidizers, chlorates, nitrates, permanganates, peroxides, aluminum chlorate, aluminum bichlorate, benzoyl peroxide, bromine dioxide, cadmium chlorate, hydrogen fluoride, red lead oxide, magnesium peroxide, mercuric oxide, nitrogen trifluoride, oxygen difluoride, ozone, potassium chlorate, potassium nitrate, tetrafluoro hydrazine, zinc permanganate. Reacts with water, water mist, or steam, forming heat and corrosive and toxic hydrochloric acid. Attacks metals in the presence of moisture.

FIRE INFORMATION

Decomposes above the melting point 824°F (440°C), forming fumes of hydrochloric acid and fumes of titanium and titanium oxides. Firefighting gear (including SCBA) may not provide adequate protection. If exposure occurs, remove and isolate gear immediately and thoroughly decontaminate personnel. Vapors are heavier than air and will collect in low areas. If material or contaminated runoff enters waterways, notify downstream users of potentially contaminated waters. Notify local health and fire officials and pollution control agencies. Use dry chemicals, dry sand, carbon dioxide. Do not use water. From a secure, explosion-proof location, use water spray to cool exposed containers.

NFPA Hazard Classifications: Not listed

FIRST AID

(135) SUBSTANCES-SPONTANEOUSLY COMBUSTIBLE

Move victim to fresh air. Call emergency medical care. Apply artificial respiration if victim is not breathing. Administer oxygen if breathing is difficult. Remove and isolate contaminated clothing and shoes. In case of contact with substance, immediately remove material from skin before washing with water; flush skin or eyes with running water for at least 20 minutes. Keep victim warm and quiet. Ensure that medical personnel are aware of the material(s) involved and take precautions to protect themselves.

(157) SUBSTANCES-TOXIC AND/OR CORROSIVE (NONCOMBUSTIBLE/WATER-SENSITIVE)

Move victim to fresh air. Call emergency medical care. Apply artificial respiration if victim is not breathing. Do not use mouth-to-mouth method if victim ingested or inhaled the substance; induce artificial respiration with the aid of a pocket mask equipped with a one-way valve or other proper respiratory medical device. Administer oxygen if breathing is difficult. Remove and isolate contaminated clothing and shoes. In case of contact with substance, immediately flush skin or eyes with running water for at least 20 minutes. For minor skin contact, avoid spreading material on unaffected skin (remove material from skin before washing). Keep victim warm and quiet. Effects of exposure (inhalation, ingestion, or skin contact) may be delayed. Ensure that medical personnel are aware of the material(s) involved and take precautions to protect themselves.

o–TOLIDINE T:237

SYNONYMS: BENZIDINE, 3,3′-DIMETHYL-; BIANISIDINE; (1,1′-BIPHENYL)-4,4′-DIAMINE, 3,3′-DIMETHYL; 3-3'-DIMETHYLBENZIDINE; 4,4′-BI-o-TOLUIDINE; CI-37230; CI AZOIC DIAZO COMPONENT 113; 4,4′-DIAMINO-3,3′-DIMETHYLDIPHENYL; DIAMINODITOLYL; DIAMINOTOLYL; DMB; 3,3′-DIMETHYLBENZIDIN; 3,3′-DIMETHYL-4,4′-BIPHENYLDIAMINE; 3,3′-DIMETHYL-4,4′-DIAMINOBIPHENYL; 3,3′-DIMETHYL-(1,1′-BIPHENYL)-4,4′-DIAMINE; 3,3′-DIMETHYLBIPHENYL-4,4′-DIAMINE; 3,3′-DIMETHYL-4,4′-DIPHENYLDIAMINE; 3,3′-DIMETHYLDIPHENYL-4,4′-DIAMINE; 4,4′-DIAMINO-3,3′-DIMETHYLBIPHENYL; 4,4′-DI-o-TOLUIDINE; EEC No. 612-041-00-7; EINECS No. 204-358-0; FAST DARK-BLUE BASE-R; ORTHOTOLIDINE; 2-TOLIDIN (German); 2-TOLIDINA (Italian); 2-TOLIDINE; o,o′-TOLIDINE; 3,3′-TOLIDINE; o-TOLIDIN

IDENTIFICATION

CAS:	119-93-7
ERG ID:	UN 2811
ERG Guide:	154
ERG Guide Category:	SUBSTANCES-TOXIC AND/OR CORROSIVE
Molecular Formula:	$C_{14}H_{16}N_2$
RTECS No.:	DD1225000
Properties:	Combustible solid. White to reddish crystals or crystalline powder. Darkens on exposure to air. Often used is paste or wet cake form. Slightly soluble in water, soluble in hot water.

HEALTH AND SAFETY INFORMATION

Toxic if inhaled or absorbed through the skin. Irritates eyes and skin. Prolonged or repeated exposures can cause kidney and bladder disease.

IARC Evaluation: Group 2B; animal positive; possibly carcinogenic to humans.
NTP: Anticipated carcinogen
IDLH: 50 ppm; potential human carcinogen.
Odor Threshold: 0.025-6.6 ppm

Exposure Limits:	**CEILING**
ACGIH: (1)	
NIOSH: (2) (3)	0.02 mg/m³ [60 min]
OSHA: (3)	

(1) Animal carcinogen. Skin contact contributes significantly to overall exposure.
(2) Potential human carcinogen. Limit exposure to lowest feasible concentration.
(3) May present a cancer risk to workers and should be handled with caution and exposure minimized.

RECOMMENDATIONS FOR RESPIRATOR SELECTION: NIOSH

At any concentrations above the NIOSH REL, or where there is no REL, at any detectable concentration: SCBAF:PD,PP (any self-contained breathing apparatus that has a full facepiece and is operated in a pressure-demand or other positive-pressure mode); or SAF:PD,PP:ASCBA (any supplied-air respirator that has a full facepiece and is operated in a pressure-demand or other positive-pressure mode in combination with an auxiliary, self-contained breathing apparatus operated in a pressure-demand or other positive-pressure mode).

Escape: GMFOVHiE [any air-purifying, full-facepiece respirator (gas mask) with a chin-style, front- or back-mounted organic vapor canister having a high-efficiency particulate filter]; or SCBAE (any appropriate escape-type, self-contained breathing apparatus).

CHEMICAL REACTIVITY

Light can affect stability of material. Contact with string oxidizers may cause fire and explosions. May ignite on contact with cellulose nitrate. Amines are incompatible with acrolein, acrylonitrile, tert-butyl nitroacetylene, ethylene oxide, isopropyl chlorocarbonate, maleic anhydride, triisobutylaluminum.

FIRE INFORMATION

Combustible solid. Combustion products include toxic nitrogen oxides. If material or contaminated runoff enters waterways, notify downstream users of potentially contaminated waters. Notify local health and fire officials and pollution control agencies. Use dry chemicals, alcohol-resistant foam, carbon dioxide. From a secure, explosion-proof location, use water spray to cool exposed containers.

NFPA Hazard Classifications: Not listed

FIRST AID

Move victim to fresh air; call emergency medical care. Apply artificial respiration if victim is not breathing. Do not use mouth-to-mouth method if victim ingested or inhaled the substance; induce artificial respiration with the aid of a pocket mask equipped with a one-way valve or other proper respiratory medical device. Administer oxygen if breathing is difficult. Remove and isolate contaminated clothing and shoes. In case of contact with substance, immediately flush skin or eyes with running water for at least 20 minutes. For minor skin contact, avoid spreading material on unaffected skin. Keep victim warm and quiet. Effects of exposure (inhalation, ingestion, or skin contact) to substance may be delayed. Ensure that medical personnel are aware of the material(s) involved and take precautions to protect themselves.

ENVIRONMENTAL INFORMATION

Clean Air Act: Hazardous Air Pollutants (Title I, Part A, Section 112)
EPA Hazardous Waste Number (RCRA No.): U328
RCRA Section 261 Hazardous Constituents
RCRA Ground Water Monitoring List: Suggested test methods(s) (PQL μg/L); 8270 (10)
EPCRA Section 304 Reportable Quantity (RQ): CERCLA, 100 lb (45.4 kg)
EPCRA Section 313: Form R *de minimis* concentration reporting level: 0.1%.
California's Proposition 65: Carcinogen.

TOLUENE T:240

SYNONYMS: ANTISAL I; ANTISAL LA; BENZENE, METHYL-; CEMSA 15; EEC No. 601-021-00-3; EINECS No. 203-625-9; EPOXY CHROMATE METAL PRIMER; EPOXY CURE AGENT; EPOXY SOLVENT CURE AGENT; GLID-GUARD EPOXY-SAFETY BLUE; GLID-GUARD HIGH SOLIDS EPOXY No. 5433; HUMISEAL 503; HUMISEAL 1B31; 3M SCOTCHKOTE ELECTRICAL COATING;

METHACIDE; METHANE, PHENYL-; METHYLBENZENE; METHYLBENZOL; NCI-C07272; PHENYLMETHANE; PYRALIN PI 2570; REDUCER DTR 602; TOLUEEN (Dutch); TOLUEN (Czech); TOLUENO (Spanish); TOLUOL; TOLUOLO; TOLU-SOL; TPD THINNER; V 183; VANDALEX 20; VANDALEX 24

IDENTIFICATION

CAS:	108-88-3
ERG ID:	UN 1294
ERG Guide:	130
ERG Guide Category:	FLAMMABLE LIQUIDS (NONPOLAR/WATER-IMMISCIBLE/NOXIOUS)
Molecular Formula:	C_7H_8
RTECS No.:	XS5250000
Properties:	Highly flammable, clear, colorless, watery, volatile liquid. Sweet, strong, pleasant, benzene-, or model gluelike odor. Only a small amount will dissolve; the rest floats on water.

HEALTH AND SAFETY INFORMATION

Irritating to eyes, skin, nose, and lungs. Chronic toluene exposures at less than 200 ppm have been associated with headache, fatigue, and nausea. Workers repeatedly exposed to 200 to 500 ppm have reported loss of coordination, memory loss, and loss of appetite. Some workers have developed reversible disorders of the optic nerves after chronic exposure sin the workplace. Blood effects and anemia have been reported; may be due to contamination by benzene. Chronic exposure due to solvent abuse can result in permanent neuropsychiatric effects. Disorders of the muscles, cardiovascular effects, renal tubular damage, and sudden death have occurred in chronic abusers of toluene (e.g., glue sniffers). Toluene may have reproductive and developmental consequences.

IARC Evaluation: Group 3, unclassifiable; inadequate human evidence; inadequate animal evidence.
IDLH: 500 ppm (1,885 mg/m^3)
Odor Threshold: 0.16 ppm
NIOSH Criteria Document: NIOSH 73-11023
OSHA Table Z-1-A Air Contaminant
OSHA Table Z-2 Air Contaminant

Exposure Limits:	**TWA**	**CEILING**	**STEL**
ACGIH: [(1)]	50 ppm; 188 mg/m^3		
NIOSH:	100 ppm; 377 mg/m^3		150 ppm; 560 mg/m^3
OSHA:	200 ppm; 754 mg/m^3	300 ppm; 500 ppm 10-min maximum peak	

[(1)] Skin contact contributes significantly in overall exposure.

1040 Toluene

RECOMMENDATIONS FOR RESPIRATOR SELECTION: NIOSH/OSHA

500 ppm: CCRFOV[any air-purifying, full-facepiece respirator (gas mask) with a chin-style, front- or back-mounted acid gas canister]; or PAPROV [any powered, air-purifying respirator with organic vapor cartridge(s)]; or GMFOV [any air-purifying, full-facepiece respirator (gas mask) with a chin-style, front- or back-mounted organic vapor canister]; or SA (any supplied-air respirator); or SCBAF (any self-contained breathing apparatus with a full facepiece).

Emergency or Planned Entry into Unknown Concentrations or IDLH Conditions: SCBAF:PD,PP (any self-contained breathing apparatus that has a full facepiece and is operated in a pressure-demand or other positive-pressure mode); or SAF:PD,PP:ASCBA (any supplied-air respirator that has a full facepiece and is operated in a pressure-demand or other positive-pressure mode in combination with an auxiliary, self-contained breathing apparatus operated in a pressure-demand or other positive-pressure mode).

Escape: GMFOV [any air-purifying, full-facepiece respirator (gas mask) with a chin-style, front- or back-mounted organic vapor canister]; or SCBAE (any appropriate escape-type, self-contained breathing apparatus).

CHEMICAL REACTIVITY

Forms explosive mixture with air. Strong oxidizers such as permanganates, dichromate, or chlorine may cause fire and explosions. Attacks some plastics, rubber, and coatings. May accumulate static electrical charges, and may cause ignition of its vapors.

FIRE INFORMATION

A significant fire hazard at room temperature. Vapors are heavier than air and will collect in low areas. Vapors in confined areas may explode when exposed to fire. Vapors may travel long distances to ignition sources and flash back. Storage containers and parts of containers may rocket great distances, in many directions. If material or contaminated runoff enters waterways, notify downstream users of potentially contaminated waters. Notify local health and fire officials and pollution control agencies. Use dry chemicals, foam, carbon dioxide. From a secure, explosion-proof location, use water spray to cool exposed containers. If cooling streams are ineffective (venting sound increases in volume and pitch, tank discolors, or shows any signs of deforming), withdraw immediately to a secure position.

NFPA Hazard Classifications:

Health Hazards (Blue) 2
Flammability (Red) 3
Reactivity (Yellow) 0

Flash Point: 55°F (13°C) o.c.; 40°F (4.4°C) c.c.

Flammable Limits in Air: LEL 1.1%; UEL 7.1%
Autoignition Temperature: 905°F (485°C)

FIRST AID

Move victim to fresh air. Call emergency medical care. Apply artificial respiration if victim is not breathing. Administer oxygen if breathing is difficult. Remove and isolate contaminated clothing and shoes. In case of contact with substance, immediately flush skin or eyes with running water for at least 20 minutes. Wash skin with soap and water. Keep victim warm and quiet. Effects of exposure (inhalation, ingestion, or skin contact) to substance may be delayed. Ensure that medical personnel are aware of the material(s) involved and take precautions to protect themselves.

ENVIRONMENTAL INFORMATION

Clean Air Act: Hazardous Air Pollutants (Title I, Part A, Section 112).
Clean Water Act: Section 307 Toxic Pollutants; Section 307 Priority Pollutants; Section 311 Hazardous Substances/RQ (same as CERCLA).
EPA Hazardous Waste Number (RCRA No.): U220
RCRA Section 261 Hazardous Constituents
RCRA Land Ban Waste
RCRA Universal Treatment Standards: Wastewater (mg/L), 0.080; Nonwastewater (mg/kg), 10.
RCRA Ground Water Monitoring List: Suggested test method(s) (PQL ug/L): 8020(2); 8240(5).
Safe Drinking Water Act: MCL, 1.0 mg/L; MCLG, 1.0 mg/L; Regulated chemical (47 FR 9352).
EPCRA Section 304: Reportable Quantity (RQ): CERCLA, 1,000 lb (454 kg).
EPCRA Section 313: Form R *de minimis* concentration reporting level: 1.0%.
California's Proposition 65: Reproductive toxin.
TSCA: 40CFR716.120(*a*)

m-TOLUENEDIAMINE T:250

SYNONYMS: AZOGEN DEVELOPER-H; BENZENEDIAMINE,AR-METHYL-; C.I. 76035; C.I. OXIDATION BASE; 2,4-DIAMINOTOLUENE; EINECS No. 202-453-1; 4-METHYL-M-PHENYLENEDIAMINE; 2,4-TOLAMINE; m-TOLILENDIAMINA (Spanish); 4-m-TOLYLENEDIAMINE; 2,4-TOLUENEDIAMINE; 2,4-TOLUENE-2,4-DIAMINE; meta-TOLUYLENEDIAMINE; m-TOLUYLENEDIAMINE; MTD; NCI-C02302; PELAGOL GREY J; TDA; ZOBA GKE; ZOGEN DEVELOPER-H

IDENTIFICATION

CAS:	95-80-7
ERG ID:	UN 1709
ERG Guide:	151

1042 m-Toluenediamine

ERG Guide Category: SUBSTANCES-TOXIC (NONCOMBUSTIBLE)
Molecular Formula: $C_7H_{10}N_2$
RTECS No.: XS9625000
Properties: Combustible, crystalline solid. Liquid at 208°F (98°C). Colorless. Odorless. Floats on water; slightly soluble.

HEALTH AND SAFETY INFORMATION

Irritates eyes, skin, and respiratory tract. Can cause central nervous system depression; may interfere with the body's ability to use oxygen; blood effects (methemaglobinemia, anemia), liver effects. Carcinogen.

IARC Evaluation: Group 2B, animal sufficient evidence; possibly carcinogenic to humans.
NTP: Listed carcinogen

Exposure Limits:

NIOSH: REL potential human carcinogen; reduce exposure to lowest feasible levels.

RECOMMENDATIONS FOR RESPIRATOR SELECTION: NIOSH

At any concentrations above the NIOSH REL, or where there is no REL, at any detectable concentration: SCBAF:PD,PP (any self-contained breathing apparatus that has a full facepiece and is operated in a pressure-demand or other positive-pressure mode); or SAF:PD,PP:ASCBA (any supplied-air respirator that has a full facepiece and is operated in a pressure-demand or other positive-pressure mode in combination with an auxiliary, self-contained breathing apparatus operated in a pressure-demand or other positive pressure mode)

Escape: GMFOV [any air-purifying, full-facepiece respirator (gas mask) with a chin-style, front- or back-mounted organic vapor canister]; or SCBAE (any appropriate escape-type, self-contained breathing apparatus).

CHEMICAL REACTIVITY

Reacts violently with strong oxidizers. Ignites on contact with red fuming nitric acid. Incompatible with aluminum, copper, zinc, brass, and bronze.

FIRE INFORMATION

Combustion products include toxic nitrogen oxides. Vapors are heavier than air and will collect in low areas. Vapors in confined areas may explode when exposed to fire. If material or contaminated runoff enters waterways, notify downstream users of potentially contaminated waters. Notify local health and fire officials and pollution control agencies. Material does not easily burn; fight surrounding fire with an agent appropriate for surrounding fires. On small fires, use dry chemicals, alcohol-resistant foam, carbon dioxide.

From a secure, explosion-proof location, use water spray to cool exposed containers. If cooling streams are ineffective (venting sound increases in volume and pitch, tank discolors, or shows any signs of deforming), withdraw immediately to a secure position.

FEMA Hazard Classifications:

Health Hazards (Blue) 2
Flammability (Red) 1
Reactivity (Yellow) 0

Flash Point: 300°F (149°C) c.c.; 410°F (210°C) o.c.
Autoignition Temperature: More than 887°F (487°C)

FIRST AID

Move victim to fresh air. Call emergency medical care. Apply artificial respiration if victim is not breathing. Do not use mouth-to-mouth method if victim ingested or inhaled the substance; induce artificial respiration with the aid of a pocket mask equipped with a one-way valve or other proper respiratory medical device. Administer oxygen if breathing is difficult. Remove and isolate contaminated clothing and shoes. In case of contact with substance, immediately flush skin or eyes with running water for at least 20 minutes. For minor skin contact, avoid spreading material on unaffected skin. Keep victim warm and quiet. Effects of exposure (inhalation, ingestion, or skin contact) may be delayed. Ensure that medical personnel are aware of the material(s) involved and take precautions to protect themselves.

ENVIRONMENTAL INFORMATION

EPA Hazardous Waste Number (RCRA No.): U221
RCRA Section 261 Hazardous Constituents
EPCRA Section 304 Reportable Quantity (RQ): CERCLA, 10 lb (4.54 kg)
EPCRA Section 313: Form R *de minimis* concentration reporting level: 1.0%.

TOLUENE-2,4-DIISOCYANATE T:260

SYNONYMS: BENZENE,2,4-DIISOCYANATOMETHYL-; BENZENE,2,4-DIISOCYANATO-1-METHYL-; CRESORCINOL DIISOCYANATE; DESMODUR T80; DI-ISOCYANATE de TOLUYLENE (French); DI-ISO-CYANATOLUENE; 2,4-DIISOCYANATO-1-METHYLBENZENE (9CI); 2,4-DIISOCYANATOTOLUENE; DIISOCYANAT-TOLUOL (German); EEC No. 615-006-00-4; EINECS No. 209-544-5; HYLENE T; HYLENE TCPA; HYLENE TLC; HYLENE TM; HYLENE TM-65; HYLENE TRF; ISOCYANIC ACID, METHYLPHENYLENE ESTER; ISOCYANIC ACID, 4-METHYL-M-PHENYLENE ESTER; 4-METHYL-PHENYLENE DIISOCYANATE; 4-METHYL-PHENYLENE ISOCYANATE; MONDUR TDS; MS-470 URETHANE COATING; NACCONATE 1OO; NCI-C50533; NIAX TDI; NIAX TDI-P;

1044 Toluene-2,4-diisocyanate

SCURANATE; TDI; 2,4-TDI; TDI-80; TOLUEEN-DIISOCYANAAT (Dutch); TOLUEN-2,4-DIISOCIANATO (Spanish); TOLUEN-DISOCIANATO (Italian); TOLUENE DIISOCYANATE; 2,4-TOLUENEDIISOCYANATE; TOLUILENODWUIZOCYJANIAN (Polish); TOLUYLENE-2,4-DIISOCYANATE; TOLYENE 2,4-DIISOCYANATE; TOLYLENE-2,4-DIISOCYANATE 2,4-TOLYLENEDIISOCYANAT E; 2,4-TOLYLENE DIISOCYANATE; TOLUENE DI-ISOCYANATE; TULUYLENE-2,4-DIISOCYANATE; TULUYLEN DIISOCYANAT (German); VORANATE

IDENTIFICATION

CAS:	584-84-9; 26471-62-5 (mixture 80:20 2,4-TDI; 2,6-TDI)
ERG ID:	UN 2078
ERG Guide:	156
ERG Guide Category:	TOXIC AND/OR CORROSIVE (COMBUSTIBLE/ WATER-SENSITIVE)
Molecular Formula:	$C_9H_6N_2O_2$
RTECS No.:	CZ6300000
Properties:	Flammable, colorless to pale yellow liquid. In solid form, found in white lumps or crystals; melts at room temperature 68–72°F (20–22°C). Sharp, sweet, fruity, pungent odor. Insoluble in water; sinks and reacts; forming carbon dioxide. TDI is found most commonly as an 80:20 mixture of 2,4-TDI:2,6-TDI.

HEALTH AND SAFETY INFORMATION

Poison. Carcinogen. Corrosive to eyes, skin, and respiratory tract. Inhalation may cause pulmonary edema, which can be delayed for several hours; there is a risk of death in serious cases. Also, absorption through the skin, or swallowing material, may cause death. Long-term exposure may cause chronic lung disease, psychological effects, central nervous system effects, chest tightness (sometimes very severe), sneezing, blue coloration and collapse, chronic obstructive bronchitis, emphysema, chemical bronchitis, asthamic syndrome, and blood changes. Sensitization may occur after exposure to spills, or other unusual concentrations. Decreased lung function has been reported from estimated exposure of 0.02 ppm for 2 years. It has also been reported that excessive loss of pulmonary function occurs at 0.0035 ppm, possibly to 0.002 ppm. Sensitization has been reported on the first exposure at concentrations below 0.05 ppm and as late as 14 years (0.06 ppm) after first exposure. Workers who chronically inhale low levels of TDI may have minimal or no respiratory symptoms, then suddenly develop asthma. Chronic workplace exposure is associated with an increased prevalence of sensitization; the reported sensitization rate has varied between 2% and 20% of workers and is dependent on the level of exposure. Sensitized persons are at risk of developing chronic asthma that may be precipitated by exposure to other chemicals.

IARC Evaluation: Group 2B, possible human carcinogen; human evidence, inadequate; animal evidence, sufficient.
NTP: Anticipated human carcinogen.

IDLH: 2.5 ppm (17.8 mg/m^3) potential human carcinogen.
Odor Threshold: 2.14 ppm ; inadequate warning for acute or chronic exposures.
OSHA Table Z-1-A Air Contaminant
NIOSH Criteria Document: NIOSH CRITERIA DOCUMENT 73-11022

Exposure Limits:	**TWA**	**CEILING**	**STEL**
ACGIH:	0.005 ppm; 0.14 mg/m^3		0.02ppm; 0.036 mg/m^3
NIOSH: [(1)]			
OSHA:		0.02 ppm; 0.14 mg/m^3	

[(1)] Potential human carcinogen

RECOMMENDATIONS FOR RESPIRATOR SELECTION: NIOSH

At any concentrations above the NIOSH REL, or where there is no REL, at any detectable concentration: SCBAF:PD,PP (any self-contained breathing apparatus that has a full facepiece and is operated in a pressure-demand or other positive-pressure mode); or SAF:PD,PP:ASCBA (any supplied-air respirator that has a full facepiece and is operated in a pressure-demand or other positive-pressure mode in combination with an auxiliary, self-contained breathing apparatus operated in a pressure-demand or other positive-pressure mode).

Escape: GMFOV [any air-purifying, full-facepiece respirator (gas mask) with a chin-style, front- or back-mounted organic vapor canister]; or SCBAE (any appropriate escape-type, self-contained breathing apparatus).

CHEMICAL REACTIVITY

May react with itself without warning, blocking relief valves; causing violent explosions. Violent reaction with primary and secondary amines, alcohols, acids, and alkalies, and may cause fire and explosions. Incompatible with ammonia, glycols, and caprolactum solution. Water contact causes violent foaming and spattering; forms carbon dioxide (which can rupture a container) and an organic base. Attacks copper and its alloys and some plastics including polyethylene, and rubber.

FIRE INFORMATION

Combustion products include toxic nitrogen oxides and cyanide vapors. Firefighting gear (including SCBA) does not provide adequate protection. If exposure occurs, remove and isolate gear immediately and thoroughly decontaminate personnel.Vapors are heavier than air and will collect in low areas. Vapors in confined areas may explode when exposed to fire. Vapors may travel long distances to ignition sources and flash back. Storage containers and parts of containers may rocket great distances, in many directions. If material or contaminated runoff enters waterways, notify downstream users of potentially contaminated waters. Notify local health and fire officials and pollution control agencies. Use dry chemicals, alcohol-resistant foam, carbon dioxide, water spray. From a secure, explosion-proof location, use water spray to cool exposed containers.

NFPA Hazard Classifications:

Health Hazards (Blue) 3
Flammability (Red) 1
Reactivity (Yellow) 3
Special Situations . . . . WATER

Flash Point: 260°F (127°C) c.c.
Flammable Limits in Air: LEL 0.9%; UEL 9.5%
Autoignition Temperature: 1150°F (621°C)

FIRST AID

Move victim to fresh air. Call emergency medical care. Apply artificial respiration if victim is not breathing. Do not use mouth-to-mouth method if victim ingested or inhaled the substance; induce artificial respiration with the aid of a pocket mask equipped with a one–way valve or other proper respiratory medical device. Administer oxygen if breathing is difficult. Remove and isolate contaminated clothing and shoes. In case of contact with substance, immediately flush skin or eyes with running water for at least 20 minutes. For minor skin contact, avoid spreading material on unaffected skin. Keep victim warm and quiet. Effects of exposure (inhalation, ingestion, or skin contact) may be delayed. Ensure that medical personnel are aware of the material(s) involved and take precautions to protect themselves.

ENVIRONMENTAL INFORMATION

Clean Air Act: Hazardous Air Pollutants (Title I, Part A, Section 112); Accidental Release Prevention/Flammable substances (Section 112[r], Table 3), TQ = 10,000 lb (4,540 kg).
EPCRA Section 302 Extremely Hazardous Substances: TPQ = 500 lb (127.5 kg).
EPCRA Section 304: Reportable Quantity (RQ): EHS, CERCLA, 100 lb (45.4 kg).
EPCRA Section 313: Form R *de minimis* concentration reporting level: 0.1%.

TRIBUTYLETHYL PHOSPHATE T:265

SYNONYMS: 2-BUTOXYETHANOL PHOSPHATE; ETHANOL, 2-BUTOXY-, PHOSPHATE (3:1); KP 140; KRONITEX KP-140; PHOSFLEX T-BEP; TBEP; TRI(2-BUTOXYETHANOL PHOSPHATE); TRIBUTOXYETHYL PHOSPHATE; TRI(2-BUTOXYETHYL) PHOSPHATE; TRIBUTYL CELLOSOLVE PHOSPHATE; TRIS(2-BUTOXYETHYL) ESTER PHOSPHORIC ACID; TRIS(2-BUTOXYETHYL) PHOSPHATE

IDENTIFICATION

CAS:	78-51-3
Molecular Formula:	$C_{18}H_{39}O_7P$
RTECS No.:	KJ9800000
Properties:	Combustible, oily liquid. Light colored. Butyllike odor.

HEALTH AND SAFETY INFORMATION

Irritates eyes and skin.

CHEMICAL REACTIVITY

Contact with strong oxidizers may cause a violent reaction; danger of fire and explosions. Esters are incompatible with nitrates, nitric acid, sulfuric acid.

FIRE INFORMATION

Combustion products include toxic phosphorus oxides. Vapors are heavier than air and will collect in low areas. Vapors in confined arcas may explode when exposed to fire. If material or contaminated runoff enters waterways, notify downstream users of potentially contaminated waters. Notify local health and fire officials and pollution control agencies. Use dry chemicals, alcohol-resistant foam, carbon dioxide. From a secure, explosion-proof location, use water spray to cool exposed containers.

NFPA Hazard Classifications: Not listed

Flash Point: 435°F (222°C)

FIRST AID

Move victim to fresh air. Call emergency medical care. Apply artificial respiration if victim is not breathing. Administer oxygen if breathing is difficult. Remove and isolate contaminated clothing and shoes. In case of contact with substance, immediately flush skin or eyes with running water for at least 20 minutes. Wash skin with soap and water. Keep victim warm and quiet. Effects of exposure (inhalation, ingestion, or skin contact) to substance may be delayed. Ensure that medical personnel are aware of the material(s) involved and take precautions to protect themselves.

TRIBUTYL PHOSPHATE T:270

SYNONYMS: BUTYL PHOSPHATE, TRI-; CELLUPHOS 4; EEC No. 015-014-00-2; EINECS No. 204-800-2; FOSFATO de TRIBUTILO (Spanish); PHOSPHORIC ACID TRIBUTYL ESTER; TBP;

1048 Tributyl Phosphate

TRIBUTILFOSFATO (Italian); TRIBUTYLE (PHOSPHATE de) (French); TRIBUTYLFOSFAAT (Dutch); TRIBUTYLPHOSPHAT (German); TRI-N-BUTYL PHOSPHATE

IDENTIFICATION

CAS:	126-73-8
ERG ID:	UN 3287
ERG Guide:	151
ERG Guide Category:	SUBSTANCES-TOXIC (NONCOMBUSTIBLE)
Molecular Formula:	$C_{12}H_{27}O_4P$
RTECS No.:	TC7700000
Properties:	Combustible, viscous liquid. Colorless to pale yellow. Almost odorless. Slightly soluble in water.

HEALTH AND SAFETY INFORMATION

Irritates eyes, skin, and respiratory tract. May be corrosive to eyes.

IDLH: 30 ppm
OSHA Table Z-1-A Air Contaminant

Exposure Limits:	**TWA**
ACGIH:	0.2 ppm; 2.2 mg/m^3
NIOSH:	0.2 ppm; 2.5 mg/m^3
OSHA:	5 mg/m^3

RECOMMENDATIONS FOR RESPIRATOR SELECTION: NIOSH

2 ppm: SA (any supplied-air respirator).
5 ppm: SA:CF (any supplied-air respirator operated in a continuous-flow mode).
10 ppm: SCBAF (any self-contained breathing apparatus with a full facepiece); or SAF (any supplied-air respirator with a full facepiece).
50 ppm: SAF:PD,PP (any supplied-air respirator that has a full facepiece and is operated in a pressure-demand or other positive-pressure mode).

Emergency or planned entry into unknown concentrations or idlh conditions: SCBAF:PD,PP (any self-contained breathing apparatus that has a full facepiece and is operated in a pressure-demand or other positive-pressure mode); or SAF:PD,PP:ASCBA (any supplied-air respirator that has a full facepiece and is operated in a pressure-demand or other positive-pressure mode in combination with an auxiliary, self-contained breathing apparatus operated in a pressure-demand or other positive pressure mode.

Escape: GMFOVHiE [any air-purifying, full-facepiece respirator (gas mask) with a chin-style, front- or back-mounted organic vapor canister having a high-efficiency particulate filter]; or SCBAE (any appropriate escape-type, self-contained breathing apparatus).

CHEMICAL REACTIVITY

Reacts with water, decomposes slowly forming butanol and phosphoric acid. Reacts violently with strong oxidizers. Incompatible with bases. May soften some plastics and elastomers.

FIRE INFORMATION

Vapors are heavier than air and will collect in low areas. Vapors in confined areas may explode when exposed to fire. If material or contaminated runoff enters waterways, notify downstream users of potentially contaminated waters. Notify local health and fire officials and pollution control agencies. Use dry chemicals, alcohol-resistant foam, carbon dioxide, fog or mist. From a secure, explosion-proof location, use water spray to cool exposed containers. If cooling streams are ineffective (venting sound increases in volume and pitch, tank discolors, or shows any signs of deforming), withdraw immediately to a secure position.

FEMA Hazard Classifications:

Health Hazards (Blue) 2
Flammability (Red) 1
Reactivity (Yellow) 0

Flash Point: 295°F (146°C) o.c.; 380°F (193°C) c.c.
Autoignition Temperature: 770°F (288°C)

FIRST AID

Move victim to fresh air. Call emergency medical care. Apply artificial respiration if victim is not breathing. Do not use mouth-to-mouth method if victim ingested or inhaled the substance; induce artificial respiration with the aid of a pocket mask equipped with a one-way valve or other proper respiratory medical device. Administer oxygen if breathing is difficult. Remove and isolate contaminated clothing and shoes. In case of contact with substance, immediately flush skin or eyes with running water for at least 20 minutes. For minor skin contact, avoid spreading material on unaffected skin. Keep victim warm and quiet. Effects of exposure (inhalation, ingestion, or skin contact) may be delayed. Ensure that medical personnel are aware of the material(s) involved and take precautions to protect themselves.

ENVIRONMENTAL INFORMATION

TSCA: 40CFR799.4360

TRIBUTYL PHOSPHITE T:275

SYNONYMS: EINECS No. 203-061-3; FOSFITO de TRIBUTILO (Spanish); FOSFITO di TRIBUTILE (Italian); PHOSPHOROUS ACID, TRIBUTYL ESTER

IDENTIFICATION

CAS:	102-85-2
ERG ID:	UN 3094 as corrosive liquid, water reactive, n.o.s.
ERG Guide:	138
ERG Guide Category:	SUBSTANCES-WATER REACTIVE
Molecular Formula:	$C_{12}H_{27}O_3P$
RTECS No.:	THO875000
Properties:	Combustible liquid. Clear. Sickly odor. Decomposes in water, forming phosphoric acid.

HEALTH AND SAFETY INFORMATION

Corrosive to eyes, skin, and respiratory tract. Inhalation may cause pulmonary edema, which can be delayed for several hours; there is a risk of death in serious cases.

RECOMMENDATIONS FOR RESPIRATOR SELECTION

At any concentrations above the NIOSH REL, or where there is no REL, at any detectable concentration: SCBAF:PD,PP (any self-contained breathing apparatus that has a full facepiece and is operated in a pressure-demand or other positive-pressure mode); or SAF:PD,PP:ASCBA (any supplied-air respirator that has a full facepiece and is operated in a pressure-demand or other positive-pressure mode in combination with an auxiliary, self-contained breathing apparatus operated in a pressure-demand or other positive-pressure mode).

Escape: GMFOV [any air-purifying, full-facepiece respirator (gas mask) with a chin-style, front- or back-mounted organic vapor canister]; or SCBAE (any appropriate escape-type, self-contained breathing apparatus).

CHEMICAL REACTIVITY

Decomposes in water, forming phosphoric acid. Reacts with strong bases and strong oxidizing agents. Attacks metals, glass, and concrete in the presence of moisture.

FIRE INFORMATION

Combustion products include toxic phosphorus oxides. Vapors are heavier than air and will collect in low areas. Vapors in confined areas may explode when exposed to fire. If material or contaminated runoff enters

waterways, notify downstream users of potentially contaminated waters. Notify local health and fire officials and pollution control agencies. Use dry chemicals, alcohol-resistant foam, carbon dioxide. From a secure, explosion-proof location, use water spray to cool exposed containers.

NFPA Hazard Classifications:

Health Hazards (Blue) 2
Flammability (Red) 1
Reactivity (Yellow) 1

Flash Point: 248°F (120°C) o.c.

FIRST AID

Move victim to fresh air. Call emergency medical care. Apply artificial respiration if victim is not breathing. Administer oxygen if breathing is difficult. Remove and isolate contaminated clothing and shoes. In case of contact with substance, immediately flush skin or eyes with running water for at least 20 minutes. Wash skin with soap and water. Keep victim warm and quiet. Effects of exposure (inhalation, ingestion, or skin contact) to substance may be delayed. Ensure that medical personnel are aware of the material(s) involved and take precautions to protect themselves.

1,1,1-TRICHLOROETHANE T:320

SYNONYMS: AEROTHENE; AEROTHENE TT; α-T; BALTHANE; BALTHANE CF; BALTHANE D; CHLOROETHANE; CHLORTEN; CHLOROTHENE; CHLOROTHENE NU; CHLOROTHENE SM SOLVENT; CHLOROTHENE VG SOLVENT; DABCO CS90; DELF; DIALEC 32; DILEX DS1; EEC No. 602-013-00-2; EINECS No. 201-166-9; ETHANA; INHIBISOL; LOCQUIC PRIMER N; MC CLEANER ZC-675; METILCLOROFORMO (Spanish); METHYLCHLOROFORM; METHYLTRICHLOROMETHANE; PC-96 SOLVENT SOLUBLE RESIST; 6200 DRUM CLEANING SOLVENT; 6500 DRUM CLEANING SOLVENT; SOLVENT 111; SOLVETHANE; α-TCE; 1,1,1-TCE; TCE; TEFLON DRY LUBE 708; TPD THINNER; TRICHLOROETHANE; α-TRICHLOROETHANE; TRICLOROETANO (Spanish); 1,1,1-TRICLOROETANO (Spanish); TRICHLOROETHANE, STABILIZED; TRI-ETHANE; TWINKLE STAINLESS STEEL CLEANER; VANDALEX 20; VANDALEX 24; ZC-211 LUBRICANT; ZEP AEROSOLVE

IDENTIFICATION

CAS:	71-55-6
ERG ID:	UN 2831
ERG Guide:	160
ERG Guide Category:	HALOGENATED SOLVENTS

1052 1,1,1-Trichloroethane

Molecular Formula:	$C_2H_3Cl_3$
RTECS No.:	KJ2975000
Properties:	Combustible, clear, colorless watery liquid. Freezes at −22°F (−30°C). Sweet, chloroformlike odor. Slightly soluble in water; sinks. Toxic hydrogen chloride gas produced in fire. Vapors are heavier than air and will collect in low areas. Combustion products include toxic hydrogen chloride.

HEALTH AND SAFETY INFORMATION

Irritating to eyes, skin, and respiratory tract. Levels above 900 ppm can cause dizziness, mental confusion, drowsiness, loss of coordination and response time; unconsciousness. Death may result. Prolonged exposure may result in irritation and dry, scaly skin, and liver and kidney damage.

IARC Evaluation: Group 3, unclassifiable; human evidence, inadequate; animal evidence, inadequate.
IDLH: 700 ppm (3822 mg/m^3) as methyl chloroform
Odor Threshold: 22.4 ppm
OSHA Table Z-1-A Air Contaminant as methyl chloroform
NIOSH Criteria Document: NIOSH 76-184

Exposure Limits:	**TWA**	**CEILING**	**STEL:**
ACGIH:	350 ppm; 1,910 mg/m^3		450 ppm; 2,460 mg/m^3
NIOSH:		350 ppm; 1,900 mg/m^3 [15 min]	
OSHA:	350 ppm; 1,900 mg/m^3		

RECOMMENDATIONS FOR RESPIRATOR SELECTION: NIOSH/OSHA

700 ppm: SA (any supplied-air respirator); or SCBAF (any self-contained breathing apparatus with a full facepiece).

Emergency or Planned Entry into Unknown Concentrations or IDLH Conditions:
SCBAF:PD,PP (any self-contained breathing apparatus that has a full facepiece and is operated in a pressure-demand or other positive-pressure mode); or SAF:PD,PP:ASCBA (any supplied-air respirator that has a full facepiece and is operated in a pressure-demand or other positive-pressure mode in combination with an auxiliary, self-contained breathing apparatus operated in a pressure-demand or other positive-pressure mode).

Escape: GMFOV [any air-purifying, full-facepiece respirator (gas mask) with a chin-style, front- or back-mounted acid gas canister]; or SCBAE (any appropriate escape-type, self-contained breathing apparatus).

Note: Substance reported to cause eye irritation or damage; may require eye protection.

CHEMICAL REACTIVITY

Reacts slowly with water forming hydrochloric acid. Violent reaction with nitrites, oxygen, sodium, and sodium hydroxide. Reacts with acetone, strong caustics, alkaline-earth, alkali metals, dinitrogen tetroxide, sodium hydroxide, amides, strong oxidizers, and potassium. Violent reaction with copper, bronze, and other metal powders. Corrosive to aluminum. Attacks some plastics, rubber, and coatings. Hot metals or ultraviolet radiation will cause decomposition to form irritating and poisonous gases.

FIRE INFORMATION

Combustion products include toxic hydrogen chloride Combustible liquid, but burns with difficulty. Vapors are heavier than air and will collect in low areas. Storage containers and parts of containers may rocket great distances, in many directions. If material or contaminated runoff enters waterways, notify downstream users of potentially contaminated waters. Notify local health and fire officials and pollution control agencies. Use dry chemicals, alcohol-resistant foam, carbon dioxide. From a secure, explosion-proof location, use water spray to cool exposed containers.

NFPA Hazard Classifications:

Health Hazards (Blue) 2
Flammability (Red) 1
Reactivity (Yellow) 0

Flammable Limits in Air: LEL 7%; UEL 12.5%
Autoignition Temperature: 932°F (500°C)

FIRST AID

Move victim to fresh air. Call emergency medical care. Apply artificial respiration if victim is not breathing. Administer oxygen if breathing is difficult. Remove and isolate contaminated clothing and shoes. In case of contact with substance, immediately flush skin or eyes with running water for at least 20 minutes. For minor skin contact, avoid spreading material on unaffected skin. Wash skin with soap and water. Keep victim warm and quiet. Ensure that medical personnel are aware of the material(s) involved and take precautions to protect themselves.

ENVIRONMENTAL INFORMATION

Clean Air Act: Hazardous Air Pollutants (Title I, Part A, Section 112); Stratospheric ozone protection (Title VI, Subpart A, Appendix A), Class I, Ozone Depletion Potential = 0.1, all isomers except 1,1,2-trichlorethane.
Clean Water Act: Toxic Pollutant (Section 401.15) as chlorinated ethanes.
EPA Hazardous Waste Number (RCRA No.): U226.
RCRA Section 261 Hazardous Constituents

RCRA Land Ban Waste
RCRA Universal Treatment Standards: Wastewater (mg/L), 0.054; Nonwastewater (mg/kg), 6.0.
RCRA Ground Water Monitoring List: Suggested test method(s) (PQL ug/L): 8240(5).
Safe Drinking Water Act: MCL, 0.2 mg/L; MCLG, 0.20 mg/L; Regulated chemical (47 FR 9352).
EPCRA Section 304: Reportable Quantity (RQ): CERCLA, 1,000 lb (454 kg).
EPCRA Section 313: Form R *de minimis* concentration reporting level: 1.0%.
TSCA: 40CFR716.120(*a*)

1,1,2–TRICHLOROETHANE T:330

SYNONYMS: β-TRICHLOROETHANE; EEC No. 602-014-00-8; EINECS No. 201-166-9; ETHANE TRICHLORIDE; ETHANE,1,1,2-TRICHLORO-; NCI-C04579; β-T; 1,1,2-TRICHLORETHANE; 1,2,2-TRICHLOROETHANE; β-TRICHLOROETHANE; β-TRICHLOROETHANE; 1,1,2-TRICLOROETANO (Spanish); TROJCHLOROETAN(1,1,2) (Polish); VINYL TRICHLORIDE

IDENTIFICATION

CAS:	79-00-5; 25323-89-1
Molecular Formula:	$C_2H_3Cl_3$
RTECS No.:	KJ3150000
Properties:	Noncombustible liquid. Colorless. Sweet, chloroformlike odor. Sinks in water; practically insoluble.

HEALTH AND SAFETY INFORMATION

Irritates eyes, skin, and respiratory tract. Inhalation may have a narcotic effect; sleepiness, unconsciousness. Liquid can remove the skin's natural oils and cause serious inflammation. Long-term or prolonged exposure can cause central nervous system, liver, and kidney damage.

IARC Evaluation: Group 3; limited animal data
IDLH: 100 ppm; potential human carcinogen
Odor Threshold: 0.5-165 ppm
OSHA Table Z-1-A Air Contaminant

Exposure Limits:	**TWA**
ACGIH:	10 ppm; 55 mg/m^3
NIOSH: (1)(2)	10 ppm; 45 mg/m^3
OSHA: (2)	10 ppm; 45 mg/m^3

(1) Potential human carcinogen.
(2) Skin contact contributes significantly in overall exposure.

RECOMMENDATIONS FOR RESPIRATOR SELECTION: NIOSH

At any concentrations above the NIOSH REL, or where there is no REL, at any detectable concentration: SCBAF:PD,PP (any self-contained breathing apparatus that has a full facepiece and is operated in a pressure-demand or other positive-pressure mode); or SAF:PD,PP:ASCBA (any supplied-air respirator that has a full facepiece and is operated in a pressure-demand or other positive-pressure mode in combination with an auxiliary, self-contained breathing apparatus operated in a pressure-demand or other positive pressure mode.

Escape: GMFOV [any air-purifying, full-facepiece respirator (gas mask) with a chin-style, front- or back-mounted acid gas canister]; or SCBAE (any appropriate escape-type, self-contained breathing apparatus).

CHEMICAL REACTIVITY

Incompatible with oxidizing materials, strong bases, liquid oxygen, chemically active metals such as sodium, potassium, magnesium, and powdered aluminum.. Attacks aluminum, some forms of plastics, rubber and coatings, steel and zinc. Heat can cause decomposition and the formation of corrosive hydrochloric acid vapors. Do not package in aluminum containers.

FIRE INFORMATION

Combustion products include corrosive hydrochloric acid vapors and phosgene. Vapors are heavier than air and will collect in low areas. If material or contaminated runoff enters waterways, notify downstream users of potentially contaminated waters. Notify local health and fire officials and pollution control agencies. Use dry chemicals, alcohol-resistant foam, carbon dioxide. From a secure, explosion-proof location, use water spray to cool exposed containers.

NFPA Hazard Classifications:

Health Hazards (Blue) 2
Flammability (Red) 1
Reactivity (Yellow) 0

Flash Point: Nonflammable, but under certain conditions where the oxygen content is increased or small quantities of combustible materials is added, this substance can form explosive mixture with air.

FIRST AID

Move victim to fresh air. Call emergency medical care. Apply artificial respiration if victim is not breathing. Administer oxygen if breathing is difficult. In case of contact with substance, immediately flush skin or eyes with running water for at least 20 minutes. Remove and isolate contaminated clothing and shoes. Ensure that medical personnel are aware of the material(s) involved and take precautions to protect themselves.

ENVIRONMENTAL INFORMATION

Clean Air Act: Hazardous Air Pollutants (Title I, Part A, Section 112);
EPA Hazardous Waste Number (RCRA No.): U227
RCRA Section 261 Hazardous Constituents
RCRA Land Ban Waste
RCRA Universal Treatment Standards: Wastewater (mg/L), 0.054; Nonwastewater (mg/kg), 6.0.
RCRA Ground Water Monitoring List: Suggested test methods(s) (PQL μg/L); 8010(0.2); 8240 (5)
Safe Drinking Water Act: MCL, 0.005 mg/L; MCLG, 0.003 mg/L; Regulated chemical (47 FR 9352).
EPCRA Section 304 Reportable Quantity (RQ): CERCLA, 100 lb (45.4 kg)
EPCRA Section 313: Form R *de minimis* concentration reporting level: 1.0%.
California's Proposition 65: Carcinogen.

TRICHLOROETHYLENE T:340

SYNONYMS: ACETYLENE TRICHLORIDE; ALGYLEN; ALTHENE; ALTHENE DG; ANAMENTH; BENZINOL; BLACOSOLV; CECOLENE; CHLORYLEA; CHORYLEN; CIRCOSOLV; CRAWHASPOL; DENSINFLUAT; DISPARIT B; DISTILLEX; DOW-TRI; DUKERON; EEC No. 602-027-00-9; EINECS No. 201-167-4; ETHINYL TRICHLORIDE; ETHYLENE TRICHLORIDE; FLECK-FLIP; FLUATE; GERMALGENE; LANADIN; LETHURIN; NARCOGEN; NARKOSOID; NCI-C04546; NIALK; PERM-A-CHLOR; PETZINOL; TCE; THRETHYLENE; TRI; TRIAD; TRIASOL; TRICHLOR; TRICHLOORETHEEN (Dutch); TRICHLORAETHEN (German); TRICHLORORAN; TRICHLORETHENE (French); TRICHLOROETHYLENE TRI (French); TRICHLOROETHENE; 1,1,2-TRICHLOROETHYLENE; TRI-CLENE; TRICLORETENE (Italian); TRICLOROETILENE (Italian); TRICLOROETILENO (Spanish); TRILENE; TRIKLONE; TRILENE; TRIMAR; TRI-PLUS; VESTROL; VITRAN; WELD-ON 3 FOR ACRYLIC; WESTROSOL

IDENTIFICATION

CAS:	79-01-6
ERG ID:	UN 1710
ERG Guide:	160
ERG Guide Category:	HALOGENATED SOLVENTS
Molecular Formula:	C_2HCl_3
RTECS No.:	KX4550000
Properties:	Combustible, clear, colorless, watery liquid; may be dyed blue. Strong, sweet, fruity, chloroformlike odor. Sinks in water; slightly soluble. Industrial grade TCE contains small amounts of stabilizers (<0.1% by weight) such as epichlorohydrin, which may increase irritant effect.

HEALTH AND SAFETY INFORMATION

Irritating to skin, eyes, nose, and lungs. Prolonged contact causes burns. Chronic exposure may cause brain damage with progressive decline in mental abilities. Prolonged or repeated application of TCE to skin causes defatting and inflammation of the skin (i.e., contact dermatitis and exfoliative dermatitis). Contact with vapor levels near 100 ppm can cause giddiness, nervous exhaustion, increased sensitivity to alcohol, including redness in the face (trichlorethylene blush), the ability to become addicted to the vapor, as well as effects of acute exposure (headache, sleepiness, nausea, vomiting, dizziness, and coughing). Higher levels can alter one's heart rate. Repeated contact with hands can cause excessive dryness, cracking, burning, loss of sense of touch, or temporary paralysis of fingers. Most of these effects may cease after exposure has stopped. TCE readily crosses the placenta and is found in fetal blood levels comparable to those of the mother; the embryotoxic, fetotoxic, or teratogenic effects are unknown.

IARC Evaluation: Group 3, unclassifiable; human evidence, inadequate; animal evidence, limited.
IDLH: 1,000 ppm (5370 mg/m^3) potential human carcinogen.
Odor Threshold: 1.36 ppm
NIOSH Criteria Document: NIOSH 73-11025
OSHA Table Z-1-A Air Contaminant
OSHA Table Z-2 Air Contaminant

Exposure Limits:	**TWA**	**CEILING**	**STEL**
ACGIH:	50 ppm; 269 mg/m^3		100 ppm; 537 mg/m^3
NIOSH: [(1)]			
OSHA:	100 ppm	200 ppm; 300 ppm [5-min max peak in any 2 hours.]	

[(1)] Potential human carcinogen. Reduce exposure to lowest feasible level.

RECOMMENDATIONS FOR RESPIRATOR SELECTION: NIOSH

At any concentrations above the NIOSH REL, or where there is no REL, at any detectable concentration: SCBAF:PD,PP (any self-contained breathing apparatus that has a full facepiece and is operated in a pressure-demand or other positive-pressure mode); or SAF:PD,PP:ASCBA (any supplied-air respirator that has a full facepiece and is operated in a pressure-demand or other positive-pressure mode in combination with an auxiliary, self-contained breathing apparatus operated in a pressure-demand or other positive-pressure mode).

Escape: GMFOV [any air-purifying, full-facepiece respirator (gas mask) with a chin-style, front- or back-mounted organic vapor canister]; or SCBAE (any appropriate escape-type, self-contained breathing apparatus).

CHEMICAL REACTIVITY

Contact with caustics form a toxic and flammable gas. Violent reaction with chemically active metals such as barium, lithium, sodium, magnesium, titanium, and beryllium. Contact with aluminum may produce a violent, self-accelerating polymerization reaction. Incompatible with acids, organic anhydrides, isocyanates, alkylene oxides, aldehydes, alcohols, glycols, phenols, cresols; caprolactum solution, epichlorohydrin, nitrogen tetroxide, metal powders, and oxygen. May accumulate static electrical charges, and may cause ignition of its vapors. At temperatures >600°F (316°C), forms serious pulmonary irritants, hydrogen chloride and phosgene; or dichloroacetylene, a neurotoxin.

FIRE INFORMATION

Combustible liquid that does not burn easily. Combustion products include toxic hydrogen chloride. Vapors are heavier than air and will collect in low areas. Storage containers and parts of containers may rocket great distances, in many directions. If material or contaminated runoff enters waterways, notify downstream users of potentially contaminated waters. Notify local health and fire officials and pollution control agencies. Use dry chemicals, foam, carbon dioxide. From a secure, explosion-proof location, use water spray to cool exposed containers.

NFPA Hazard Classifications:

Health Hazards (Blue) 2
Flammability (Red) 1
Reactivity (Yellow) 0

Flash Point: 90°F (32°C) c.c.
Flammable Limits in Air: LEL 8.0%; UEL 10.5% both at 77°F (25°C)
Autoignition Temperature: 770°F (410°C)

FIRST AID

Move victim to fresh air. Call emergency medical care. Apply artificial respiration if victim is not breathing. Administer oxygen if breathing is difficult. Remove and isolate contaminated clothing and shoes. In case of contact with substance, immediately flush skin or eyes with running water for at least 20 minutes. For minor skin contact, avoid spreading material on unaffected skin. Wash skin with soap and water. Keep victim warm and quiet. Ensure that medical personnel are aware of the material(s) involved, and take precautions to protect themselves.

ENVIRONMENTAL INFORMATION

Clean Air Act: Hazardous Air Pollutants (Title I, Part A, Section 112).
Clean Water Act: Section 307 Toxic Pollutants; Section 307 Priority Pollutants; Section 311 Hazardous Substances/RQ (same as CERCLA).

EPA Hazardous Waste Number (RCRA No.): U228; D040.
RCRA Toxicity Characteristic (Section 261.24): Maximum Concentration of Contaminants, regulatory level, 0.5 mg/L.
RCRA Section 261 Hazardous Constituents.
RCRA Land Ban Waste
RCRA Universal Treatment Standards: Wastewater (mg/L), 0.054; Nonwastewater (mg/kg), 6.0.
RCRA Ground Water Monitoring List: Suggested test method(s) (PQL ug/L): 8010(1); 8240(5).
Safe Drinking Water Act: MCL, 0.005 mg/L; MCLG, zero; Regulated chemical (47 FR 9352).
EPCRA Section 304: Reportable Quantity (RQ): CERCLA, 100 lb (45.4 kg).
EPCRA Section 313: Form R *de minimis* concentration reporting level: 1.0%.
California's Proposition 65: Carcinogen.

TRICHLOROSILANE T:360

SYNONYMS: DYNASYLAN; EEC No. 014-001-00-9; EINECS No. 233-042-5; SILANE, TRICHLORO-; SILICOCHLOROFORM; TCS; TRICHLOROMONOSILANE; TRICLOROSILANO (Spanish)

IDENTIFICATION

CAS:	10025-78-2
ERG ID:	UN 1295
ERG Guide:	139
ERG Guide Category:	WATER-REACTIVE (EMITTING FLAMMABLE AND TOXIC GASES)
Molecular Formula:	HCl_3Si
RTECS No.:	VV5950000
Properties:	Extremely flammable liquid. Colorless; fumes on contact with air. Sharp, choking odor. Sinks in water; reacts violently, forming toxic and corrosive hydrogen chloride and explosive hydrogen gas. Irritating gas is produced on contact with water. Boiling point is 90°F (32°C).

HEALTH AND SAFETY INFORMATION

Poisonous. Breathing the vapors can cause death. Skin and eye contact causes severe burns and blindness. Corrosive to eyes, skin, and respiratory tract. Inhalation may cause pulmonary edema, which can be delayed for several hours; there is a risk of death in serious cases. Prolonged or repeated contact can cause skin disease.

OSHA Process Safety Management of Highly Hazardous Chemicals (29CFR, Part 1910.119, Appendix A): Threshold Quantity: 5,000 pounds.

RECOMMENDATIONS FOR RESPIRATOR SELECTION:

Where there is no REL, at any detectable concentration: SCBAF:PD,PP (any self-contained breathing apparatus that has a full facepiece and is operated in a pressure-demand or other positive-pressure mode); or SAF:PD,PP:ASCBA (any supplied-air respirator that has a full facepiece and is operated in a pressure-demand or other positive-pressure mode in combination with an auxiliary, self-contained breathing apparatus operated in a pressure-demand or other positive pressure mode)

Escape: GMFOV [any air-purifying, full-facepiece respirator (gas mask) with a chin-style, front- or back-mounted organic vapor canister]; or SCBAE (any appropriate escape-type, self-contained breathing apparatus).

CHEMICAL REACTIVITY

Forms explosive mixture in air. Water contact produces hydrogen chloride fumes, and flammable hydrogen gas. Contact with oxidizers may cause fire and explosions. Incompatible with acids. Corrodes metals in the presence of moisture.

FIRE INFORMATION

Extremely flammable; spontaneously combusts in air. Combustion products include toxic hydrogen chloride gas. Vapors are heavier than air and will collect in low areas. Vapors in confined areas may explode when exposed to fire. Vapors may travel long distances to ignition sources and flash back. Storage containers and parts of containers may rocket great distances, in many directions. If material or contaminated runoff enters waterways, notify downstream users of potentially contaminated waters. Notify local health and fire officials and pollution control agencies. Use dry chemicals, dry sand, medium expansion AFFF, alcohol-resistant foam. Do not use water or water-based extinguishers; violent reaction forming toxic hydrogen chloride and extremely flammable and explosive hydrogen gas. From a secure, explosion-proof location, if material is not leaking, use water spray to cool closed, exposed containers.

NFPA Hazard Classifications:

Health Hazards (Blue) 3
Flammability (Red) 4
Reactivity (Yellow) 2
Special Notice .. WATER REACTIVE

Flash Point: 7°F (-14°C), o.c.
Flammable Limits in Air: LEL 1.2%; UEL 90.5%
Autoignition Temperature: 220°F (104°C)

FIRST AID

Move victim to fresh air. Call emergency medical care. Apply artificial respiration if victim is not breathing. Do not use mouth-to-mouth method if victim ingested or inhaled the substance; induce artificial respiration with the aid of a pocket mask equipped with a one-way valve or other proper respiratory medical devise. Administer oxygen if breathing is difficult. Remove and isolate contaminated clothing and shoes. In case of contact with substance, immediately flush skin or eyes with running water for at least 20 minutes. Keep victim warm and quiet. Ensure that medical personnel are aware of the material(s) involved and take precautions to protect themselves.

ENVIRONMENTAL INFORMATION

CLEAN AIR ACT: Accidental Release Prevention/Flammable substances, (Section 112[r], Table 3), TQ = 10,000 lb (4540 kg).

TRICHLORO–s–TRIAZINETRIONE T:365

SYNONYMS: ACIDO TRICLOROISOCIANURICO (Spanish); ACL 85; CBD 90; CHLOREAL; EEC No. 613-031-00-5; EINECS No. 201-782-8; FICHLOR 91; FI CLOR 91; ISOCYANURIC CHLORIDE; KYSELINA TRICHLOISOKYANUROVA (Czech); NEOCHLOR-90; NSC-405124; SYMCLOSEN; SYMCLOSENE; TRICHLORINATED ISOCYANURIC ACID; TRICHLOROCYANURIC ACID; TRICHLOROISOCYANIC ACID; TRICHLOROISOCYANURIC ACID; n,n',n"-TRICHLORO-ISOCYANURIC ACID; 1,3,5-TRICHLOROISOCYANURIC ACID; NSC-405124; 1,3,5-TRIAZINE-2,4,6(1H,3H,5H)-TRIONE, 1,3,5-TRICHLORO-; 1,3,5-TRICHLORO-S-TRIAZINE-2,4,6(1H,3H,5H)-TRIONE; 1,3,5-TRICHLORO-2,4,6-TRIOXOHEXAHYDRO-S-TRIAZINE; 1,3,5,TRICHLORO-1,2,5-TRIAZINE-2,4,6(1H,3H,5H)-TRIONE; TRICHLORO-S-TRIAZINE-2,4,6(1H,3H,5H)-TRIONE; TRICHLORO-S-TRIAZINETRIONE, DRY; TRICHLOROIMINOISOCYANURIC ACID

IDENTIFICATION

CAS:	87-90-1
ERG ID:	UN 2468 (dry)
ERG Guide:	141
ERG Guide Category:	OXIDIZERS-TOXIC (SOLID)
Molecular Formula:	$C_3Cl_3N_3O_3$
RTECS No.:	XZ1925000
Properties:	Nonflammable needles, crystals, or powder solid. White. Chlorine odor. Sinks and mixes slowly with water.

HEALTH AND SAFETY INFORMATION

Irritates eyes, skin, and respiratory tract. Eye and skin irritation can be severe. Ingestion can cause gastrointestinal tract damage; possible liver and kidney damage.

RECOMMENDATIONS FOR RESPIRATOR SELECTION

At any concentrations above the NIOSH REL, or where there is no REL, at any detectable concentration: SCBAF:PD,PP (any self-contained breathing apparatus that has a full facepiece and is operated in a pressure-demand or other positive-pressure mode); or SAF:PD,PP:ASCBA (any supplied-air respirator that has a full facepiece and is operated in a pressure-demand or other positive-pressure mode in combination with an auxiliary, self-contained breathing apparatus operated in a pressure-demand or other positive-pressure mode).

Escape: HiEF (any air-purifying, full-facepiece respirator with a high-efficiency particulate filter); or SCBAE (any appropriate escape-type, self-contained breathing apparatus).

Note: Only nonoxidiziable sorbents should be used (not charcoal).

CHEMICAL REACTIVITY

Reacts with water to form a bleach solution. A powerful oxidizer. May be thermally unstable (NFPA). Forms explosive material with nitrogen compounds, ammonia, amines, urea, etc. Contact with most foreign material, organic matter, or easily chlorinated or oxidized materials may result in fire. Avoid contact with oil, grease, sawdust, floor sweepings, other easily oxidized organic compounds.

FIRE INFORMATION

Toxic chlorine gas, nitrogen chloride, and nitrogen trichloride may be formed in fires. Nonflammable but may increase the intensity of an existing fire. Material on clothing is a fire hazard; flush immediately with water. Vapors are heavier than air and will collect in low areas. Storage containers and parts of containers may rocket great distances, in many directions. If material or contaminated runoff enters waterways, notify downstream users of potentially contaminated waters. Notify local health and fire officials and pollution control agencies. Use dry chemicals, alcohol-resistant foam, carbon dioxide. From a secure, explosion-proof location, use water spray to cool exposed containers.

NFPA Hazard Classifications:

Health Hazards (Blue) 2
Flammability (Red) 0
Reactivity (Yellow) 2
Special Notice .... OXY

Flash Point: Not flammable but may cause fire on contact with ordinary combustibles

FIRST AID

Move victim to fresh air. Call emergency medical care. Apply artificial respiration if victim is not breathing. Do not use mouth-to-mouth method if victim ingested or inhaled the substance; induce artificial respiration with the aid of a pocket mask equipped with a one-way valve or other proper respiratory medical device. Administer oxygen if breathing is difficult. Remove and isolate contaminated clothing and shoes. In case of contact with substance, immediately flush skin or eyes with running water for at least 20 minutes. For minor skin contact, avoid spreading material on unaffected skin. Keep victim warm and quiet. Effects of exposure (inhalation, ingestion, or skin contact) may be delayed. Ensure that medical personnel are aware of the material(s) involved and take precautions to protect themselves.

TRIETHANOLAMINE T:380

SYNONYMS: DALTOGEN; EINECS No. 203-049-8; NITRILO-2,2',2"-TRIETHANOL; 2,2',2"-NITRILO-TRIETHANOL; 2,2',2"-NITRILOTRIS-(ETHANOL); STEROLAMIDE; STING-KILL; TEA; THIOFACO T-35; TREA; TRIETHYLOLAMINE; TROLAMINE; TRIETANOLAMINA (Spanish); TRI(HYDROXY-TRIETHYL)AMINE; TRI(2-HYDROXYETHYL)AMINE; tris(HYDROXYETHYL)AMINE; tris(2-HYDROXY-ETHYL)AMINE; TROLAMINE

IDENTIFICATION

CAS:	102-71-6
Molecular Formula:	$C_6H_{15}NO_3$
RTECS No.:	KL9275000
Properties:	Combustible to pale yellow, oily liquid; turns brown on exposure to air and light. Mild, ammonialike odor. Sinks and mixes with water. Readily absorbs water and carbon dioxide if exposed to the atmosphere. Combustion products include toxic carbon monoxide and nitrogen oxides.

HEALTH AND SAFETY INFORMATION

Irritates eyes, skin, and respiratory tract. Inhalation may cause pulmonary edema, which can be delayed for several hours; there is a risk of death in serious cases. Long-term exposure may cause skin irritation and disease. May be an animal carcinogen and may cause occupational asthma.

Odor Threshold: 2-3 ppm

1064 Triethanolamine

Exposure Limits: **TWA**

ACGIH: 5 mg/m^3

RECOMMENDATIONS FOR RESPIRATOR SELECTION

At any concentrations above the NIOSH REL, or where there is no REL, at any detectable concentration: any self-contained breathing apparatus that has a full facepiece and is operated in a pressure-demand or other positive-pressure mode); or any supplied-air respirator that has a full facepiece and is operated in a pressure-demand or other positive-pressure mode in combination with an auxiliary, self-contained breathing apparatus operated in a pressure-demand or other positive-pressure mode).

Escape: [any air-purifying, full-facepiece respirator (gas mask) with a chin-style, front- or back-mounted canister providing protection against the compound of concern]; or any appropriate escape-type, self-contained breathing apparatus.

CHEMICAL REACTIVITY

Aqueous solution is caustic. Vigorous reaction with oxidizers such as permanganates and strong acids. Incompatible with organic anhydrides, isocyanates, vinyl acetate, acrylates, substituted allyls, alkylene oxides, epichlorohydrin, and aldehydes. Corrodes copper and its alloys. Aluminum containers should not be used for storage of water solution.

FIRE INFORMATION

Vapors are heavier than air and will collect in low areas. Vapors in confined areas may explode when exposed to fire. If material or contaminated runoff enters waterways, notify downstream users of potentially contaminated waters. Notify local health and fire officials and pollution control agencies. Use dry chemicals, alcohol-resistant foam, carbon dioxide. From a secure, explosion-proof location, use water spray to cool exposed containers.

NFPA Hazard Classifications:

Health Hazards (Blue) 2
Flammability (Red) 1
Reactivity (Yellow) 1

Flash Point: 375°F (191°C) o.c; 355°F (180°C) c.c.
Flammable Limits in Air: LEL 1.2%; UEL N/A
Autoignition Temperature: 644°F (340°C)

FIRST AID

EYE: If this chemical contacts the eyes, immediately wash them with large amounts of water, occasionally lifting the lower and upper lids. Get medical attention immediately. Contact lenses should not be worn when working with this chemical.
SKIN: If this chemical contacts the skin, immediately wash the contaminated skin with soap and water. If this chemical penetrates the clothing, immediately remove clothing, wash the skin with soap and water, and get medical attention promptly.
BREATH: If a person breathes large amounts of this chemical, move exposed person to fresh air at once. If breathing has stopped, perform mouth-to-mouth resuscitation. Keep the affected person warm and at rest. Get medical attention as soon as possible.
SWALLOW: If this chemical has been swallowed, get medical attention immediately.

ENVIRONMENTAL INFORMATION

TSCA: 40CFR716.120(*a*)

TRIETHYLAMINE T:390

SYNONYMS: (DIETHYLAMINO)ETHANE; N,N-DIETHYLETHANEAMINE; EEC No. 612-004-00-5; EINECS No. 204-469-4; ETHANAMINE,N,N-DIETHYL-; TEA; TEN; TRIAETHYLAMIN (German); TRIETILAMINA (Italian, Spanish); STCC 4907877

IDENTIFICATION

CAS:	121-44-8
ERG ID:	UN 1296
ERG Guide:	132
ERG Guide Category:	FLAMMABLE LIQUIDS-CORROSIVE
Molecular Formula:	$C_6H_{15}N$
RTECS No.:	YE0175000
Properties:	Highly flammable, colorless liquid. Fishy odor at low concentrations; ammonialike odor at high concentrations. Floats on the surface of water; moderately soluble in water.

HEALTH AND SAFETY INFORMATION

Corrosive to eyes, skin, and respiratory tract. Inhalation may cause pulmonary edema, which can be delayed for several hours; there is a risk of death in serious cases. Eye contact causes severe burns and possible blindness. Prolonged or repeated exposure can affect central nervous system.

1066 Triethylamine

Odor Threshold: 0.309 ppm
IDLH: 200 ppm (828 mg/m^3)
Odor Threshold: 0.309 ppm
OSHA Table Z-1-A Air Contaminant

Exposure Limits:	TWA	STEL
ACGIH: [(1)]	1 ppm; 4.1 mg/m^3	3 ppm; 12 mg/m^3
NIOSH:	No established REL	
OSHA:	25 ppm; 104 mg/m^3	

[(1)] Skin contact contributes significantly to overall exposure.

RECOMMENDATIONS FOR RESPIRATOR SELECTION: NIOSH

200 ppm: SA:CF (any supplied-air respirator operated in a continuous-flow mode); or SCBAF (any self-contained breathing apparatus with a full facepiece); or SAF (any supplied-air respirator with a full facepiece).

Emergency or Planned Entry into Unknown Concentrations or IDLH Conditions: SCBAF:PD,PP (any self-contained breathing apparatus that has a full facepiece and is operated in a pressure-demand or other positive-pressure mode); or SAF:PD,PP:ASCBA (any supplied-air respirator that has a full facepiece and is operated in a pressure-demand or other positive-pressure mode in combination with an auxiliary, self-contained breathing apparatus operated in a pressure-demand or other positive-pressure mode).

Escape: GMFS [any air-purifying, full-facepiece respirator (gas mask) with a chin-style, front- or back-mounted canister providing protection against the compound of concern]; or SCBAE (any appropriate escape-type, self-contained breathing apparatus).
Note: Substance reported to cause eye irritation or damage; may require eye protection.

CHEMICAL REACTIVITY

Forms explosive mixture with air. Contact with strong acids, mineral acids, and organic acids may cause violent spattering. Also incompatible with organic anhydrides, halogenated hydrocarbons, isocyanates, nitroparaffins, nitrogen tetroxide, vinyl acetate, acrylates, substituted allyls, alkylene oxides, epichlorohydrin, ketones, aldehydes, alcohols, glycols, phenols, cresols, caprolactum solution, and strong oxidizers. Attacks aluminum, copper, lead, tin, zinc, and their alloys. Liquid attacks some plastics, rubber, and coatings.

FIRE INFORMATION

Combustion products include toxic nitrogen oxides and carbon monoxide. Firefighting gear (including SCBA) does not provide adequate protection. If exposure occurs, remove and isolate gear immediately and thoroughly decontaminate personnel.Vapors are heavier than air and will collect in low areas. Vapors in confined areas may explode when exposed to fire. Vapors may travel long distances to ignition sources and flash back. Storage containers and parts of containers may rocket great distances, in many directions. If material or contaminated runoff enters waterways, notify downstream users of potentially contaminated waters. Notify local health and fire officials and pollution control agencies. Use dry chemicals, alcohol-resistant foam, carbon dioxide. From a secure, explosion-proof location, use water spray to cool exposed containers.

NFPA Hazard Classifications:

Health Hazards (Blue) 3
Flammability (Red) 3
Reactivity (Yellow) 0

Flash Point: 20°F (-7°C) o.c.; 16°F (-9°C) c.c.
Flammable Limits in Air: LEL 1.2%; UEL 8.0%
Autoignition Temperature: 482°F (250°C)

FIRST AID

Move victim to fresh air. Call emergency medical care. Apply artificial respiration if victim is not breathing. Do not use mouth-to-mouth method if victim swallowed or inhaled the substance; induce artificial respiration with the aid of a pocket mask equipped with a one–way valve or other proper respiratory medical device. If breathing is difficult, give oxygen. Remove and isolate contaminated clothing and shoes. In case of contact with substance, immediately flush skin or eyes with running water for at least 20 minutes. Keep victim warm and quiet. Effects of exposure (inhalation, ingestion, or skin contact) to substances may be delayed. Ensure that medical personnel are aware of the material(s) involved and take precautions to protect themselves.

ENVIRONMENTAL INFORMATION

Clean Air Act: Hazardous Air Pollutants (Title I, Part A, Section 112).
Clean Water Act: Section 311 Hazardous Substances/RQ (same as CERCLA).
EPA Hazardous Waste Number (RCRA No.): U404.
RCRA Section 261 Hazardous Constituents
EPCRA Section 304: Reportable Quantity (RQ): CERCLA, 5,000 lb (2,270 kg).
EPCRA Section 313: Form R *de minimis* concentration reporting level: 1.0%.

TRIETHYLBORANE T:393

SYNONYMS: BORANE, TRIETHYL-; EINECS No. 202-620-9; TRIETILBORANO (Spanish); TRIETHYLBORINE; TRIETHYLBORON

1068 Triethylborane

IDENTIFICATION

CAS:	97-94-9
ERG ID:	UN 3185 as self-heating liquid, corrosive, organic, n.o.s.
ERG Guide:	136
ERG Guide Category:	SUBSTANCES-SPONTANEOUSLY COMBUSTIBLE-TOXIC (AIR REACTIVE)
Molecular Formula:	$C_6H_{15}B$
RTECS No.:	ED2100000
Properties:	Extremely flammable, fuming liquid. Colorless. Ignites spontaneously in air. Insoluble in water; reacts explosively. Poisonous and flammable vapors produced on contact with water or steam, upon contact with air, or when heated to decomposition.

HEALTH AND SAFETY INFORMATION

Corrosive to eyes, skin, and respiratory tract. Inhalation may cause pulmonary edema, which can be delayed for several hours; there is a risk of death in serious cases.

CHEMICAL REACTIVITY

Spontaneously flammable in air, emitting toxic smoke and fumes. Reacts explosively with water or steam to produce toxic vapors and flammable hydrogen gas. Explosive reaction with halogenated hydrocarbons, oxygen. A strong reducing agent; reacts spontaneously with oxidizers; bromine, chlorine and other halogens, with danger of fires. Incompatible with organic acids, alcohols, carbon tetrachloride, and many other substances.

FIRE INFORMATION

Vapors are heavier than air and will collect in low areas, with risk of spontaneous combustion. Vapors in confined areas may explode when exposed to fire. Vapors may travel long distances to ignition sources and flash back. Storage containers and parts of containers may rocket great distances, in many directions. If material or contaminated runoff enters waterways, notify downstream users of potentially contaminated waters. Notify local health and fire officials and pollution control agencies. Do not use halogenated extinguishing agents. Use dry chemicals, foam, carbon dioxide. From a secure, explosion-proof location, use water spray to cool exposed containers. May reignite after fire is extinguished.

NFPA Hazard Classifications:

Health Hazards (Blue) 1
Flammability (Red) 3
Reactivity (Yellow) 3
Special Situations . WATER REACTIVE

Flash Point: –135°F (–93°C); Ignites spontaneously in air. (NFPA)

FIRST AID

Move victim to fresh air. Call emergency medical care. Apply artificial respiration if victim is not breathing. Do not use mouth-to-mouth method if victim ingested or inhaled the substance; induce artificial respiration with the aid of a pocket mask equipped with a one-way valve or other proper respiratory medical device. Administer oxygen if breathing is difficult. Remove and isolate contaminated clothing and shoes. In case of contact with substance, immediately flush skin or eyes with running water for at least 20 minutes. Wash skin with soap and water. Keep victim warm and quiet. Effects of exposure (inhalation, ingestion, or skin contact) to substance may be delayed. Ensure that medical personnel are aware of the material(s) involved and take precautions to protect themselves.

TRIETHYLENE DIAMINE T:397

SYNONYMS: BICYCLO(2,2,2)-1,4-DIAZAOCTANE; CHINUCLIDINE; DABCO; 1,4-DIAZABICYCLO(2,2,2)OCTANE; DABCO CRYSTAL; DABCO EG; DABCO 33LV; DABCO R-8020; DABCO S-25; D 33LV; EINECS No. 205-999-9; 1,4-ETHYLENEPIPERAZINE; QUINUCLIDINE

IDENTIFICATION

CAS:	280-57-9
Molecular Formula:	$C_6H_{12}N_2$
RTECS No.:	HM0354200
Properties:	Hygroscopic crystals. Colorless. Poisonous gases emitted when heated.

HEALTH AND SAFETY INFORMATION

Irritates eyes, skin, and respiratory tract. Skin and respiratory tract allergen and sensitizer; asthma may develop.

RECOMMENDATIONS FOR RESPIRATOR SELECTION

At any concentrations above the NIOSH REL, or where there is no REL, at any detectable concentration: SCBAF:PD,PP (any self-contained breathing apparatus that has a full facepiece and is operated in a pressure-demand or other positive-pressure mode); or SAF:PD,PP:ASCBA (any supplied-air respirator that has a full facepiece and is operated in a pressure-demand or other positive-pressure mode in combination with an auxiliary, self-contained breathing apparatus operated in a pressure-demand or other positive-pressure mode).

1070 Triethylene Glycol

Escape: GMFOVHiE [any air-purifying, full-facepiece respirator (gas mask) with a chin-style, front- or back-mounted organic vapor canister having a high-efficiency particulate filter]; or SCBAE (any appropriate escape-type, self-contained breathing apparatus).

CHEMICAL REACTIVITY

A powerful base; reacts on contact with acids. Strong bases are incompatible with amides, isocyanates, alkylene oxides, epichlorohydrin, aldehydes, alcohols, glycols, phenols, cresols, caprolactum solution. Will explode on contact with hydrogen peroxide. Will auto-ignite with carbon at 446°F (230°C). Amines reacts with cellulose nitrate. Incompatible with acrolein, acrylonitrile, tert-butyl nitroacetylene, ethylene oxide, isopropyl chlorocarbonate, maleic anhydride, triisobutylaluminum.

FIRE INFORMATION

NFPA Hazard Classifications: Not listed

FIRST AID

Move victim to fresh air. Call emergency medical care. Apply artificial respiration if victim is not breathing. Do not use mouth-to-mouth method if victim swallowed or inhaled the substance; induce artificial respiration with the aid of a pocket mask equipped with a one-way valve or other proper respiratory medical device. Administer oxygen if breathing is difficult. Remove and isolate contaminated clothing and shoes. In case of contact with substance, immediately flush skin or eyes with running water for at least 20 minutes. For minor skin contact, avoid spreading material on unaffected skin. Keep victim warm and quiet. Effects of exposure (inhalation, ingestion, or skin contact) to substances may be delayed. Ensure that medical personnel are aware of the material(s) involved and take precautions to protect themselves.

TRIETHYLENE GLYCOL T:400

SYNONYMS: DICAPROATE; DI-β-HYDROXYETHOXYETHANE; 3,6-DIOXAOCTANE-1,8-DIOL; 2,2'-[1,2-ETHANEDIYLBIS(OXY)]BISETHANOL; ETHANOL, 2,2'-(ETHYLENEDIOXY)DI-; 2,2'-ETHYLENEDIOXYDIETHANOL; 2,2'-ETHYLENEDIOXYETHANOL; ETHYLENE GLYCOL-BIS-(2-HYDROXYETHYL ETHER); ETHYLENE GLYCOL DIHYDROXYDIETHYL ETHER; GLYCOL BIS(HYDROXYETHYL) ETHER; 1,2-BIS(2-HYDROXYETHOXY)ETHANE; TEG; TRIETILENGLICOL (Spanish); TRIGEN; TRIGLYCOL

IDENTIFICATION

CAS:	112-27-6
Molecular Formula:	$C_6 H_{14}O_8$
RTECS No.:	YE4550000

Properties: Combustible liquid. Colorless. Mild odor. Sinks and mixes with water.

HEALTH AND SAFETY INFORMATION

Prolonged or repeated contact may cause irritation of the skin, and nervous system changes.

RECOMMENDATIONS FOR RESPIRATOR SELECTION

Where there is no REL, at any detectable concentration: SCBAF:PD,PP (any self-contained breathing apparatus that has a full facepiece and is operated in a pressure-demand or other positive-pressure mode); or SAF:PD,PP:ASCBA (any supplied-air respirator that has a full facepiece and is operated in a pressure-demand or other positive-pressure mode in combination with an auxiliary, self-contained breathing apparatus operated in a pressure-demand or other positive pressure mode)

Escape: GMFOV [any air-purifying, full-facepiece respirator (gas mask) with a chin-style, front- or back-mounted organic vapor canister]; or SCBAE (any appropriate escape-type, self-contained breathing apparatus).

CHEMICAL REACTIVITY

Incompatible with sulfuric acid, isocyanates, perchloric acid, and strong oxidizers.

FIRE INFORMATION

NFPA Hazard Classifications:

Health Hazards (Blue) 1
Flammability (Red) 1
Reactivity (Yellow) 0

Flash Point: 330°F (166°C) o.c.; 350°F (177°C) c.c.
Flammable Limits in Air: LEL 0.9%; UEL 9.2%
Autoignition Temperature: 700°F (371°C)

FIRST AID

EYE: If this chemical contacts the eyes, immediately wash the eyes with large amounts of water, occasionally lifting the lower and upper lids. Get medical attention immediately. Contact lenses should not be worn when working with this chemical.
SKIN: If this chemical contacts the skin, immediately wash the contaminated skin with soap and water. If this chemical penetrates the clothing, immediately remove clothing, wash the skin with soap and water, and get medical attention promptly.

BREATH: If a person breathes large amounts of this chemical, move exposed person to fresh air at once. If breathing has stopped, perform mouth-to-mouth resuscitation. Keep the affected person warm and at rest. Get medical attention as soon as possible.
SWALLOW: If this chemical has been swallowed, get medical attention immediately.

ENVIRONMENTAL INFORMATION (as glycol ethers)

CLEAN AIR ACT: Hazardous Air Pollutants (Title I, Part A, Section 112) includes mono- and di-ethers of ethylene glycol, diethyl glycol, and triethylene glycol R-(OCH2CH2)n-OR' where n = 1,2, or 3; R = alkyl or aryl groups; R' = R, H, or groups which when removed, yield glycol ethers with the structure: R-(OCH2CH)n-OH. Polymers are excluded from the glycol category.
EPCRA Section 313: Form R *de minimus* concentration reporting level: 1.0%; Form R Toxic Chemical Category Code: N230

TRIETHYLENE GLYCOL ACETATE T:402

SYNONYMS: ACETIC ACID, TRIETHYLENE GLYCOL DIESTER; EINECS No. 203-846-0; 2,2'-(ETHYLENEDIOXY)DI(ETHYL ACETATE); 2,2'-ETHYLENEDIOXYDIETHANOL DIACETATE; TRIGLYCOL, DIACETATE

IDENTIFICATION

CAS:	111-21-7
Molecular Formula:	$C_{10}H_{18}O_6$
RTECS No.:	KK9970000; YE4950000
Properties:	Combustible liquid. Colorless.

HEALTH AND SAFETY INFORMATION

Irritates eyes, skin.

CHEMICAL REACTIVITY

Contact with strong oxidizers may cause fire and explosions.

FIRE INFORMATION

NFPA Hazard Classifications: Not listed

Combustion may produce irritants and toxic vapors. Vapors are heavier than air and will collect in low areas. Vapors in confined areas may explode when exposed to fire. If material or contaminated runoff enters

waterways, notify downstream users of potentially contaminated waters. Notify local health and fire officials and pollution control agencies. Use dry chemicals, alcohol-resistant foam, carbon dioxide. From a secure, explosion-proof location, use water spray to cool exposed containers.

Flash Point: 345°F (174°C)

FIRST AID

Move victim to fresh air. Call emergency medical care. Apply artificial respiration if victim is not breathing. If breathing is difficult, give oxygen. Remove and isolate contaminated clothing and shoes. In case of contact with substance, immediately flush skin or eyes with running water for at least 20 minutes. Wash skin with soap and water. Keep victim warm and quiet. Ensure that medical personnel are aware of the material(s) involved and take precautions to protect themselves.

TRIETHYLENE TETRAMINE T:404

SYNONYMS: ARALDITE HARDENER HY 951; ARALDITE HY 951; N,N'-BIS(2-AMINOETHYL)-1,2-DIAMINOETHANE; N,N'-BIS(2-AMINOETHYL)ETHYLENEDIAMINE; N,N'-BIS(2-AMINOETHYL)-1,2-ETHYLENEDIAMINE; DEH 24; 3,6-DIAZAOCTANE-1,8-DIAMINE; EEC No. 612-059-00-5; HY 951; TECZA; TETA; 1,4,7,10-TETRAAZADECANE; TRIEN; TRIENTINE

IDENTIFICATION

CAS:	112-24-3
ERG ID:	UN 2259
ERG Guide:	153
ERG Guide Category:	SUBSTANCES-TOXIC AND/OR CORROSIVE (COMBUSTIBLE)
Molecular Formula:	$C_6H_{16}N_4$
RTECS No.:	YE6650000
Properties:	Combustible, viscous liquid. Straw colored. Ammonialike odor. Highly soluble in water.

HEALTH AND SAFETY INFORMATION

Corrosive to eyes, skin, and respiratory tract. Avoid skin contact. Prolonged skin contact can cause burns; leads to skin sensitization and allergy. Inhalation may cause pulmonary edema, which can be delayed for several hours; there is a risk of death in serious cases. Skin contact can add to overall exposure. Prolonged or repeated contact can cause liver damage, asthma.

RECOMMENDATIONS FOR RESPIRATOR SELECTION

At any concentrations above the NIOSH REL, or where there is no REL, at any detectable concentration: SCBAF:PD,PP (any self-contained breathing apparatus that has a full facepiece and is operated in a pressure-demand or other positive-pressure mode); or SAF:PD,PP:ASCBA (any supplied-air respirator that has a full facepiece and is operated in a pressure-demand or other positive-pressure mode in combination with an auxiliary, self-contained breathing apparatus operated in a pressure-demand or other positive-pressure mode).

Escape: GMFOV [any air-purifying, full-facepiece respirator (gas mask) with a chin-style, front- or back-mounted organic vapor canister]; or SCBAE (any appropriate escape-type, self-contained breathing apparatus).

CHEMICAL REACTIVITY

React with strong oxidizers; danger of fires and explosions. Reacts with nitrogen compounds, chlorinated hydrocarbons. A strong base; reacts on contact with acids. Strong bases are incompatible with amides, isocyanates, alkylene oxides, epichlorohydrin, aldehydes, alcohols, glycols, phenols, cresols, caprolactum solution. Amines reacts with cellulose nitrate. Incompatible with acrolein, acrylonitrile, tert-butyl nitroacetylene, ethylene oxide, isopropyl chlorocarbonate, maleic anhydride, triisobutyl aluminum. Attacks copper, copper alloys, cobalt, and nickel.

FIRE INFORMATION

Combustion products include toxic nitrogen oxides. Vapors are heavier than air and will collect in low areas. Vapors in confined areas may explode when exposed to fire. If material or contaminated runoff enters waterways, notify downstream users of potentially contaminated waters. Notify local health and fire officials and pollution control agencies. Use dry chemicals, alcohol-resistant foam, carbon dioxide. From a secure, explosion-proof location, use water spray to cool exposed containers.

NFPA Hazard Classifications

Health Hazards (Blue) 3
Flammability (Red) 1
Reactivity (Yellow) 0

Flash Point: 275°F (135°C)
Autoignition Temperature: 640°F (338°C)

FIRST AID

Move victim to fresh air. Call emergency medical care. Apply artificial respiration if victim is not breathing. Do not use mouth-to-mouth method if victim swallowed or inhaled the substance; induce artificial respiration with the aid of a pocket mask equipped with a one-way valve or other proper respiratory medical device. Administer oxygen if breathing is difficult. Remove and isolate contaminated clothing and shoes. In case of contact with substance, immediately flush skin or eyes with running water for at least 20 minutes. For minor skin contact, avoid spreading material on unaffected skin. Keep victim warm and quiet. Effects of exposure (inhalation, ingestion, or skin contact) to substances may be delayed. Ensure that medical personnel are aware of the material(s) involved and take precautions to protect themselves.

TRIETHYL PHOSPHATE T:406

SYNONYMS: EEC No. 015-013-00-7; EINECS No. 201-114-5; ETHYL PHOSPHATE; FOSFATO di TRIETILE (Italian); PHOSPHATE d'ETHYLE (French); PHOSPHORIC ACID, TRIETHYL ESTER; TEP

IDENTIFICATION

CAS:	78-40-0
ERG ID:	UN 3272 as esters, n.o.s.
ERG Guide:	127
ERG Guide Category:	FLAMMABLE LIQUIDS (POLAR/WATER- MISCIBLE)
Molecular Formula:	$C_6H_{15}O_4P$
RTECS No.:	TC7900000
Properties:	Combustible liquid. Colorless. Mild odor. Soluble in water.

HEALTH AND SAFETY INFORMATION

Irritates eyes, skin, and respiratory tract. May cause nerve damage.

CHEMICAL REACTIVITY

Contact with strong oxidizers may cause fire and explosions. Incompatible with strong acids, nitrates. Slight decomposition in water.

FIRE INFORMATION

Combustion products include toxic phosphorus oxides. Vapors are heavier than air and will collect in low areas. Vapors in confined areas may explode when exposed to fire. If material or contaminated runoff enters waterways, notify downstream users of potentially contaminated waters. Notify local health and fire officials and pollution control agencies. Use dry chemicals, alcohol-resistant foam, carbon dioxide. From a secure, explosion-proof location, use water spray to cool exposed containers.

NFPA Hazard Classifications:

Health Hazards (Blue) 0

Flammability (Red) 1
Reactivity (Yellow) 1

Flash Point: 240°F (116°C) o.c.; 210°F (99°C) c.c.
Flammable Limits in Air: LEL 1.7%; UEL 10%
Autoignition Temperature: 845°F (452°C)

FIRST AID

Move victim to fresh air. Call emergency medical care. Apply artificial respiration if victim is not breathing. Administer oxygen if breathing is difficult. Remove and isolate contaminated clothing and shoes. In case of contact with substance, immediately flush skin or eyes with running water for at least 20 minutes. Wash skin with soap and water. Keep victim warm and quiet. Ensure that medical personnel are aware of the material(s) involved and take precautions to protect themselves.

TRIFLUOROCHLOROETHYLENE T:408

SYNONYMS: CHLOORTRIFLUORETHYLEEN (Dutch); CHLOROTRIFLUOROETHYLENE; CTFE; DAIFLON; EINECS No. 201-201-8; ETHENE, CHLOROTRIFLUORO-; FLUOROPLAST-3; GENETRON 1113; KEL F MONOMER; R-1113; 1,1,2-TRIFLUORO-2-CHLOROETHYLENE; TRIFLUOROMONOCHLOROETHYLENE; TRIFLUOROVINYL CHLORIDE; TRITHENE

IDENTIFICATION

CAS:	79-38-9
ERG ID:	UN 1082
ERG Guide:	119P
ERG Guide Category:	GASES-TOXIC-FLAMMABLE
Molecular Formula:	C_2ClF_3
RTECS No.:	KV0525000
Properties:	Highly flammable gas. Colorless. Odorless or faint odor. Sinks and "boils" in water. Flammable, visible vapor cloud is produced.

HEALTH AND SAFETY INFORMATION

Moderately toxic by ingestion. Inhalation can cause nausea, dizziness, vomiting. Contact with liquid will cause frostbite.

OSHA Process Safety Management of Highly Hazardous Chemicals (29CFR, Part 1910.119, Appendix A): Threshold Quantity: 10,000 pounds.

CHEMICAL REACTIVITY

Forms explosive mixture with air. Contact with oxidizers will cause fires and explosion. Contact with ethylene causes dangerous polymerization. Incompatible with vinylidene chloride.

FIRE INFORMATION

Combustion products include toxic and corrosive fumes of hydrogen chloride and hydrogen fluoride. Gases in confined areas may explode when exposed to fire. Gases may travel long distances to ignition sources and flash back. Storage containers and parts of containers may rocket great distances, in many directions. If material or contaminated runoff enters waterways, notify downstream users of potentially contaminated waters. Notify local health and fire officials and pollution control agencies. Do not extinguish the fire unless the flow of the gas can be stopped and any remaining gas is out of the line. Specially trained personnel may use fog lines to cool exposures and let the fire burn itself out. Use dry chemicals, alcohol-resistant foam, carbon dioxide. From a secure, explosion-proof location, use water spray to cool exposed containers. If cooling streams are ineffective (venting sound increases in volume and pitch, tank discolors, or shows any signs of deforming), withdraw immediately to a secure position.

NFPA Hazard Classifications:

Health Hazards (Blue) . . . . --
Flammability (Red) 4
Reactivity (Yellow) 0

Flash Point: Gas
Flammable Limits in Air: LEL 8.4%; UEL 16%

FIRST AID

Move victim to fresh air. Call emergency medical care. Apply artificial respiration if victim is not breathing. Do not use mouth-to-mouth method if victim ingested or inhaled the substance; induce artificial respiration with the aid of a pocket mask equipped with a one-way valve or other proper respiratory medical device. Administer oxygen if breathing is difficult. Remove and isolate contaminated clothing and shoes. In case of contact with substance, immediately flush skin or eyes with running water for at least 20 minutes. In case of contact with liquefied gas, thaw frosted parts with lukewarm water. Keep victim warm and quiet. Keep victim under observation. Effects of contact or inhalation may be delayed. Ensure that medical personnel are aware of the material(s) involved and take precautions to protect themselves. Urinary arsenic may be a useful indicator of exposure.

ENVIRONMENTAL INFORMATION

Clean Air Act: Section 112[r], Table 3, Accidental Release Prevention/Flammable Substances (Section 68.130), TQ = 10,000 lb (4,450 kg)

TRIFLUOROMETHANE T:410

SYNONYMS: ARCTON; CARBON TRIFLUORIDE; EINECS No. 200-872-4; FLUROFORM; FLUORYL; FREON 23; FREON-23; FREON F-23; FRIGEN 23; GENETRON-23; HALOCARBON 23; METHYL TRIFLUORIDE; PROPELLENT 23; R 23; TRIFLUOMETANO (Spanish)

IDENTIFICATION

CAS:	75-46-7
ERG ID:	UN 1984 (gas); UN 3136 (refrigerated liquid)
ERG Guide:	126 (gas); 120 (refrigerated liquid)
ERG Guide Category:	GASES-COMPRESSED OR LIQUID (INCLUDING REFRIGERANT GASES); GASES-INERT (INCLUDING REFRIGERATED LIQUIDS)
Molecular Formula:	CHF_3
RTECS No.:	PB6900000
Properties:	Noncombustible, colorless, compressed liquified gas. Odorless. Insoluble in water.

HEALTH AND SAFETY INFORMATION

Vapor is heavier than air; can accumulate in low or enclosed areas; suffocation danger. Contact with the liquid can cause frostbite.

RECOMMENDATIONS FOR RESPIRATOR SELECTION

At any concentrations above the NIOSH REL, or where there is no REL, at any detectable concentration: SCBAF:PD,PP (any self-contained breathing apparatus that has a full facepiece and is operated in a pressure-demand or other positive-pressure mode); or SAF:PD,PP:ASCBA (any supplied-air respirator that has a full facepiece and is operated in a pressure-demand or other positive-pressure mode in combination with an auxiliary, self-contained breathing apparatus operated in a pressure-demand or other positive pressure mode).

Escape: GMFOV [any air-purifying, full-facepiece respirator (gas mask) with a chin-style, front- or back-mounted organic vapor canister]; or SCBAE (any appropriate escape-type, self-contained breathing apparatus).

CHEMICAL REACTIVITY

Violent reaction with barium, calcium, strontium, finely divided magnesium, and oxides of these metals. Attacks magnesium and its alloys. Decomposes when heated or on contact with alkaline earth metals (e.g., barium, strontium, calcium, and magnesium) and alkali metals (e.g., lithium, sodium, potassium, rubidium, cesium, and francium).

FIRE INFORMATION

Noncombustible gas. Notify local health and fire officials and pollution control agencies. Use extinguishing agents suitable for surrounding fires. From a secure, explosion-proof location, use water spray to cool exposed containers. If cooling streams are ineffective (venting sound increases in volume and pitch, tank discolors, or shows any signs of deforming), withdraw immediately to a secure position.

FIRST AID

(GAS) Move victim to fresh air. Call emergency medical care. Apply artificial respiration if victim is not breathing. Administer oxygen if breathing is difficult. Remove and isolate contaminated clothing and shoes. In case of contact with liquefied gas, thaw frosted parts with lukewarm water. Keep victim warm and quiet. Keep victim under observation. Ensure that medical personnel are aware of the material(s) involved and take precautions to protect themselves.

(REFRIGERATED LIQUID) Move victims to fresh air. Call emergency medical care. Apply artificial respiration if victim is not breathing. Administer oxygen if breathing is difficult. Clothing frozen to the skin should be thawed before being removed. In case of contact with liquefied gas, thaw frosted parts with lukewarm water. Keep victim warm and quiet. Ensure that medical personnel are aware of the material(s) involved and take precautions to protect themselves.

TRIISOOCTYL PHOSPHITE T:412

SYNONYMS: EINECS No. 246-614-4; PHOSPHOROUS ACID, TRIISOOCTYL ESTER

IDENTIFICATION

CAS:	25103-12-2
Molecular Formula:	$C_{24}H_{54}O_3P$
RTECS No.:	TH1150000
Properties:	Liquid. Colorless. Insoluble in water.

HEALTH AND SAFETY INFORMATION

Irritates eyes and skin.

CHEMICAL REACTIVITY

Contact with strong oxidizers may cause fire and explosions. Phosphates are incompatible with antimony petachloride, magnesium, silver nitrate, zinc acetate.

FIRE INFORMATION

Combustion products include toxic phosphorus oxides. Vapors are heavier than air and will collect in low areas. Vapors in confined areas may explode when exposed to fire. If material or contaminated runoff enters waterways, notify downstream users of potentially contaminated waters. Notify local health and fire officials and pollution control agencies. Use dry chemicals, alcohol-resistant foam, carbon dioxide. From a secure, explosion-proof location, use water spray to cool exposed containers.

NFPA Hazard Classifications: Not listed

Flash Point: 385°F (196°C) o.c.

FIRST AID

Move victim to fresh air. Call emergency medical care. Apply artificial respiration if victim is not breathing. Administer oxygen if breathing is difficult. Remove and isolate contaminated clothing and shoes. In case of contact with substance, immediately flush skin or eyes with running water for at least 20 minutes. Wash skin with soap and water. Keep victim warm and quiet. Effects of exposure (inhalation, ingestion, or skin contact) to substance may be delayed. Ensure that medical personnel are aware of the material(s) involved and take precautions to protect themselves.

TRIMELLITIC ANHYDRIDE T:414

SYNONYMS: AMOCO TMA; ANHYDROTRIMELLIC ACID; 1,2,4-BENZENETRICARBOXYLIC ACID, CYCLIC 1,2-ANHYDRIDE; 1,2,4-BENZENETRICARBOXYLIC ANHYDRIDE; 1,2,3-BENZENETRICARBOXYLIC ACID-1,2-ANHYDRIDE; 4-CARBOXYPHTHALIC ANHYDRIDE; 1,3-DIHYDRO-1,3-DIOXO-5-ISOBENZOFURANCARBOXYLIC ACID; 1,3-DIOXO-5-PHTHALAN-CARBOXYLIC ACID; DIPHENYLMETHANE-4,4'-DIISOCYANATE-TRIMELLIC ANHYDRIDE-ETHOMID HT POLYMER; EEC No. 607-097-00-4; EINECS No. 209-008-0; NCI-C56633; TMA; TMAN; TRIMELLIC ACID ANHYDRIDE; TRIMELLIC ACID-1,2-ANHYDRIDE; TRIMELLITIC ACID CYCLIC-1,2-ANHYDRIDE

IDENTIFICATION

CAS:	552-30-7
ERG ID:	UN 2928 as toxic solid, corrosive, organic, n.o.s.
ERG Guide:	154
ERG Guide Category:	SUBSTANCES-TOXIC AND/OR CORROSIVE (NON-COMBUSTIBLE)
Molecular Formula:	$C_9H_4O_5$
RTECS No.:	DC2050000

Properties: Combustible crystals or powder. Colorless. Musty odor. Practically insoluble in water.

HEALTH AND SAFETY INFORMATION

Corrosive to eyes, skin, and respiratory tract; skin contact causes severe burns. Inhalation may cause pulmonary edema, which can be delayed for several hours; there is a risk of death in serious cases. May cause asthmalike allergy; lung disease, low blood count.

Exposure Limits:	**TWA**
ACGIH:	0.04 mg/m^3
NIOSH: [(1)]	0.005 ppm; 0.04 mg/m^3
OSHA:	None

[(1)] Should be handled in the workplace as an extremely toxic substance.

RECOMMENDATIONS FOR RESPIRATOR SELECTION: NIOSH

At any concentrations above the NIOSH REL, or where there is no REL, at any detectable concentration: SCBAF:PD,PP (any self-contained breathing apparatus that has a full facepiece and is operated in a pressure-demand or other positive-pressure mode); or SAF:PD,PP:ASCBA (any supplied-air respirator that has a full facepiece and is operated in a pressure-demand or other positive-pressure mode in combination with an auxiliary, self-contained breathing apparatus operated in a pressure-demand or other positive-pressure mode).

Escape: GMFOVHiE [any air-purifying, full-facepiece respirator (gas mask) with a chin-style, front- or back-mounted organic vapor canister having a high-efficiency particulate filter]; or SCBAE (any appropriate escape-type, self-contained breathing apparatus).

CHEMICAL REACTIVITY

Reacts violently with oxidizers. Reacts slowly with water to form trimellitic acid. Dust explosion possible if finely dispersed powder mixed with air comes in contact with source of ignition.

FIRE INFORMATION

Combustion may produce irritants and toxic vapors. If material or contaminated runoff enters waterways, notify downstream users of potentially contaminated waters. Notify local health and fire officials and pollution control agencies. Use dry chemicals, alcohol-resistant foam, carbon dioxide. From a secure, explosion-proof location, use water spray to cool exposed containers.

Hazard Classifications (Based on NFPA, Mobil, Inc.)

Health Hazards (Blue) 3
Flammability (Red) 1
Reactivity (Yellow) 1

Flash Point: 440°F (227°C)

FIRST AID

Move victim to fresh air. Call emergency medical care. Apply artificial respiration if victim is not breathing. Administer oxygen if breathing is difficult. Remove and isolate contaminated clothing and shoes. In case of contact with substance, immediately flush skin or eyes with running water for at least 20 minutes. Wash skin with soap and water. Keep victim warm and quiet. Effects of exposure (inhalation, ingestion, or skin contact) to substance may be delayed. Ensure that medical personnel are aware of the material(s) involved and take precautions to protect themselves.

TRIMETHYLACETIC ACID T:416

SYNONYMS: ACIDE PIVALIQUE (French); ACIDO PIVALICO (Spanish); ACIDO TRIMETILACETICO (Spanish); 2,2-DIMETHYLPROPIONIC ACID; α,α-DIMETHYLPROPIONIC; EINECS No. 200-922-5; NEOPENTANOIC ACID; PIVALIC ACID; PIVALINEZUUR (Dutch); PIVALINSAEURE (German); PIVALINSYRE (French); tert-PROPANOIC ACID; PROPANOIC ACID; PROPANOIC ACID, 2,2-DI-METHYL-; VERSATIC-5

IDENTIFICATION

CAS:	75-98-9
ERG ID:	UN 1325
ERG Guide:	133
ERG Guide Category:	FLAMMABLE SOLIDS
Molecular Formula:	$C_5H_{10}O_2$
RTECS No.:	TO7700000
Properties:	Combustible, hygroscopic, crystalline solid. Colorless or white. Pungent odor. Floats on water; soluble.

HEALTH AND SAFETY INFORMATION

Corrosive to eyes. Slightly corrosive to skin and respiratory tract. Mildly toxic if swallowed. Due to the availability of insufficient data on human effects, caution should be exercised.

CHEMICAL REACTIVITY

Vigorous reaction with strong oxidizers. Water solution is a weak acid. Organic acids are incompatible with sulfuric acid, caustics, ammonia, aliphatic amines, alkanolamines, isocyanates, alkylene oxides, epichlorohydrin.. In the presence of moisture will slowly corrode various metals. Fine particle or dust will form an explosive mixture with air.

FIRE INFORMATION

Combustion may produce irritants and toxic and flammable vapors. Vapors are heavier than air and will collect in low areas. Vapors in confined areas may explode when exposed to fire. If material or contaminated runoff enters waterways, notify downstream users of potentially contaminated waters. Notify local health and fire officials and pollution control agencies. Use dry chemicals, foam, AFFF, carbon dioxide. From a secure, explosion-proof location, use water spray to cool exposed containers.

NFPA Hazard Classifications: Not listed

Flash Point: 147°F (64°C) unspecified method

FIRST AID

Move victim to fresh air. Call emergency medical care. Apply artificial respiration if victim is not breathing. Administer oxygen if breathing is difficult. Remove and isolate contaminated clothing and shoes. In case of contact with substance, immediately flush skin or eyes with running water for at least 20 minutes. Removal of solidified molten material from skin requires medical assistance. Keep victim warm and quiet. Ensure that medical personnel are aware of the material(s) involved and take precautions to protect themselves.

TRIMETHYLAMINE T:418

SYNONYMS: EEC No. 612-001-00-9; EINECS No. 200-875-0; n,n-DIMETHYLMETHANEAMINE; METHANAMINE, n,n-DIMETHYL-; TMA; TRIMETILAMINA (Spanish)

IDENTIFICATION

CAS:	75-50-3
ERG ID:	UN 1083 (anhydrous/gas); 1297 (aqueous solution)
ERG Guide:	118 (anhydrous/gas); 132 (aqueous solution)
ERG Guide Category:	(118) GASES-FLAMMABLE-CORROSIVE (132) FLAMMABLE LIQUIDS-CORROSIVE
Molecular Formula:	C_3H_9N
RTECS No.:	PA0350000

1084 Trimethylamine

Properties: Extremely flammable, liquefied, compressed gas. May be shipped as an aqueous solution, which is highly flammable and will generate large amounts of flammable gas. Colorless. Fish or ammonia odor. Floats and mixes and boils on water; completely soluble. Poisonous, flammable, visible vapor cloud is produced.

HEALTH AND SAFETY INFORMATION

Severely irritating to skin, eyes, and respiratory tract. Skin and eye contact causes severe burns and blindness. Inhalation may cause pulmonary edema, which can be delayed for several hours; there is a risk of death in serious cases. Contact with liquid can cause frostbite.

Odor Threshold: 0.001
OSHA Table Z-1-A Air Contaminant

Exposure Limits:	TWA	STEL
ACGIH:	5 ppm; 12 mg/m^3	15 ppm; 36 mg/m^3
NIOSH:	10 ppm; 24 mg/m^3	15 ppm; 36 mg/m^3
OSHA:	None	

RECOMMENDATIONS FOR RESPIRATOR SELECTION

At any concentrations above the NIOSH REL, or where there is no REL, at any detectable concentration: SCBAF:PD,PP (any self-contained breathing apparatus that has a full facepiece and is operated in a pressure-demand or other positive-pressure mode); or SAF:PD,PP:ASCBA (any supplied-air respirator that has a full facepiece and is operated in a pressure-demand or other positive-pressure mode in combination with an auxiliary, self-contained breathing apparatus operated in a pressure-demand or other positive-pressure mode).

Escape: GMFOV [any air-purifying, full-facepiece respirator (gas mask) with a chin-style, front- or back-mounted organic vapor canister]; or SCBAE (any appropriate escape-type, self-contained breathing apparatus).

CHEMICAL REACTIVITY

Forms explosive mixture with air. Contact with oxidizers may cause fire and explosions. A strong base; incompatible with acids, amides, isocyanates, alkylene oxides, epichlorohydrin, aldehydes, alcohols, glycols, phenols, cresols, caprolactum solution. Attacks chemically active metals: aluminum, copper, zinc, and their alloys. Reacts with mercury, forming a shock-sensitive compound.

FIRE INFORMATION

Combustion products include toxic nitrogen oxides. Firefighting gear (including SCBA) does not provide adequate protection. If exposure occurs, remove and isolate gear immediately and thoroughly decontaminate personnel. Vapors are heavier than air and will collect in low areas. Vapors in confined areas may explode when exposed to fire. Vapors may travel long distances to ignition sources and flash back. Storage containers and parts of containers may rocket great distances, in many directions. If material or contaminated runoff enters waterways, notify downstream users of potentially contaminated waters. Notify local health and fire officials and pollution control agencies. Do not extinguish the fire unless the flow of the gas can be stopped and any remaining gas is out of the line. Specially trained personnel may use fog lines to cool exposures and let the fire burn itself out. Use dry chemicals, foam, carbon dioxide. From a secure, explosion-proof location, use water spray to cool exposed containers. If cooling streams are ineffective (venting sound increases in volume and pitch, tank discolors, or shows any signs of deforming), withdraw immediately to a secure position.

NFPA Hazard Classifications:

Health Hazards (Blue) 3
Flammability (Red) 4
Reactivity (Yellow) 0

Flash Point: Flammable gas: NA; Liquid: 20°F (-7°C) o.c.; 10°F (-27°C) c.c.
Flammable Limits in Air: LEL 2.0%; UEL 11.6%
Autoignition Temperature: 374°F (190°C)

FIRST AID

GASES-FLAMMABLE-CORROSIVE–Move victim to fresh air. Call emergency medical care. Apply artificial respiration if victim is not breathing. Do not use mouth-to-mouth method if victim ingested or inhaled the substance; induce artificial respiration with the aid of a pocket mask equipped with a one-way valve or other proper respiratory medical device. Administer oxygen if breathing is difficult. Remove and isolate contaminated clothing and shoes. In case of contact with liquefied gas, thaw frosted parts with lukewarm water. Keep victim warm and quiet. Keep victim under observation. Effects of contact or inhalation may be delayed. Ensure that medical personnel are aware of the material(s) involved and take precautions to protect themselves.

FLAMMABLE LIQUIDS-CORROSIVE–Move victim to fresh air; call emergency medical care. Apply artificial respiration if victim is not breathing. Do not use mouth-to-mouth method if victim ingested or inhaled the substance; induce artificial respiration with the aid of a pocket mask equipped with a one-way valve or other proper respiratory medical device. Administer oxygen if breathing is difficult. Remove and isolate contaminated clothing and shoes. In case of contact with substance, immediately flush skin or eyes with running water for at least 20 minutes. For minor skin contact, avoid spreading material on unaffected skin. Keep victim warm and quiet. Effects of exposure (inhalation, ingestion, or skin contact) to substance may be delayed. Ensure that medical personnel are aware of the material(s) involved and take precautions to protect themselves.

ENVIRONMENTAL INFORMATION

Clean Air Act: Accidental Release Prevention/Flammable Substances (Section 112[r], Table 3), TQ = 10,000 lb (4,450 kg)
Clean Water Act: Section 311 Hazardous Substances/RQ (same as CERCLA)
EPCRA Section 304 Reportable Quantity (RQ): CERCLA, 100 lb (45.4 kg)
EPCRA Section 313: Form R *de minimis* concentration reporting level: 1.0%.
California's Proposition 65: Carcinogen.
Marine Pollutant (49CFR, Subchapter 172.101, Appendix B)

1,2,4–TRIMETHYLBENZENE T:430

SYNONYMS: ASYMMETRICAL TRIMETHYL BENZENE; BENZENE,1,2,4-TRIMETHYL-; EINECS No. 202-436-9; PSI-CUMENE; PSEUDOCUMENE; PSEUDOCUMOL; 1,2,5-TRIMETHYL BENZENE; 1,3,4-TRIMETHYLBENZENE; 1,2,4-TRIMETILBENCENO (Spanish)

IDENTIFICATION

CAS:	95-63-6
ERG ID:	UN 1993
ERG Guide:	128
ERG Guide Category:	FLAMMABLE LIQUIDS (NONPOLAR/WATER-IMMISCIBLE)
Molecular Formula:	C_9H_{12}
RTECS No.:	DC3325000
Properties:	Flammable liquid. Colorless. Distinctive aromatic odor.

HEALTH AND SAFETY INFORMATION

Irritates eyes, skin, and respiratory tract. Inhalation can cause bronchitis. Repeated or prolonged exposure may cause blood effects; may interfere with blood's clotting ability. Skin contact may contribute significantly to overall exposure.

Exposure Limits:	**TWA**
ACGIH:	25 ppm; 123 mg/m^3
NIOSH:	25 ppm; 125 mg/m^3
OSHA:	None

RECOMMENDATIONS FOR RESPIRATOR SELECTION

At any concentrations above the NIOSH REL, or where there is no REL, at any detectable concentration: SCBAF:PD,PP (any self-contained breathing apparatus that has a full facepiece and is operated in a pressure-demand or other positive-pressure mode); or SAF:PD,PP:ASCBA (any supplied-air respirator that has a full facepiece and is operated in a pressure-demand or other positive-pressure mode in combination with an auxiliary, self-contained breathing apparatus operated in a pressure-demand or other positive-pressure mode).

Escape: GMFOV [any air-purifying, full-facepiece respirator (gas mask) with a chin-style, front- or back-mounted organic vapor canister]; or SCBAE (any appropriate escape-type, self-contained breathing apparatus).

CHEMICAL REACTIVITY

Forms explosive mixture with air. Contact with oxidizers may cause fire and explosions. Incompatible with nitric acid. May accumulate static electrical charges, and may cause ignition of its vapors.

FIRE INFORMATION

Vapors are heavier than air and will collect in low areas. Vapors in confined areas may explode when exposed to fire. If material or contaminated runoff enters waterways, notify downstream users of potentially contaminated waters. Notify local health and fire officials and pollution control agencies. Use dry chemicals, alcohol-resistant foam, carbon dioxide. From a secure, explosion-proof location, use water spray to cool exposed containers.

NFPA Hazard Classifications:

Health Hazards (Blue) 0
Flammability (Red) 2
Reactivity (Yellow) 0

Flash Point: 112°F (44°C) c.c.
Flammable Limits in Air: LEL 0.9%; UEL 6.4%
Autoignition Temperature: 959°F (515°C)

FIRST AID

Move victim to fresh air. Call emergency medical care. Apply artificial respiration if victim is not breathing. If breathing is difficult, give oxygen. Remove and isolate contaminated clothing and shoes. In case of contact with substance, immediately flush skin or eyes with running water for at least 20 minutes. Wash skin with soap and water. Keep victim warm and quiet. Ensure that medical personnel are aware of the material(s) involved and take precautions to protect themselves.

ENVIRONMENTAL INFORMATION

Safe Drinking Water Act: 40CFR141.40 (*J*) monitoring at the discretion of the state.
EPCRA Section 313: Form R *de minimis* concentration reporting level: 1.0%.
TSCA: 40CFR716.120(*a*); 40CFR799.2175

TRIMETHYLENE GLYCOL T:445

SYNONYMS: 2-DEOXYGLYCEROL; 1,3-DIHYDROXYPROPANE: EINECS No. 207-997-3; 2-(HYDROXYMETHYL)ETHANOL; NSC-65426; PG; 1,3-PROPANEDIOL; PROPANE-1,3-DIOL; b-PROPYLENE GLYCOL; 1,3-PROPYLENE GLYCOL; TRIMETILENGLICOL (Spanish)

IDENTIFICATION

CAS:	504-63-2
Molecular Formula:	$C_3H_8O_2$
RTECS No.:	TY2010000
Properties:	Combustible liquid. Colorless to pale yellow. Odorless. Sweet tasting. Soluble in water.

HEALTH AND SAFETY INFORMATION

Irritates eyes, skin, and respiratory tract. Liquid removes the skin's natural oils; causes dryness, redness, itching.

CHEMICAL REACTIVITY

Forms explosive mixture with air. Glycols are incompatible with sulfuric acid, nitric acid, caustics, aliphatic amines, isocyanates, triethylborane. Reacts with acid chlorides, acid anhydrides, oxidizers, chloroformates, and reducing agents.

FIRE INFORMATION

Vapors are heavier than air and will collect in low areas. Vapors in confined areas may explode when exposed to fire. If material or contaminated runoff enters waterways, notify downstream users of potentially contaminated waters. Notify local health and fire officials and pollution control agencies. Use dry chemicals, alcohol-resistant foam, carbon dioxide. From a secure, explosion-proof location, use water spray to cool exposed containers.

NFPA Hazard Classifications

Health Hazards (Blue) 1

Flammability (Red) .... -
Reactivity (Yellow) 0

Flash Point: 165-175°F (74-79°C)
Autoignition Temperature: 752°F (400°C)

FIRST AID

Move victim to fresh air. Call emergency medical care. Apply artificial respiration if victim is not breathing. If breathing is difficult, give oxygen. Remove and isolate contaminated clothing and shoes. In case of contact with substance, immediately flush skin or eyes with running water for at least 20 minutes. Wash skin with soap and water. Keep victim warm and quiet. Ensure that medical personnel are aware of the material(s) involved and take precautions to protect themselves.

TRIMETHYLHEXAMETHYLENE DIISOCYANATE T:450

SYNONYMS: EINECS No. 241-001-8; HEXANE,1,6-DIISOCYANATO-2,2,4-TRIMETHYL-; 2,2,4-TRIMETHYLHEXA-1,6-DIYL DIISOCYANATE

IDENTIFICATION

CAS:	16938-22-0 (2,2,4-isomer)
ERG ID:	UN 2328
ERG Guide:	156
ERG Guide Category:	SUBSTANCES-TOXIC AND/OR CORROSIVE (COMBUSTIBLE/WATER-SENSITIVE)
Molecular Formula:	$C_{11}H_{18}N_2O_2$
Properties:	Combustible liquid. Colorless or yellowish. Reacts with water to produce carbon dioxide and toxic diamine.

HEALTH AND SAFETY INFORMATION

May cause irritation of eyes, skin and repiratory tract. May cause allergic reactions.

RECOMMENDATIONS FOR RESPIRATOR SELECTION

At any concentrations above the NIOSH REL, or where there is no REL, at any detectable concentration: SCBAF:PD,PP (any self-contained breathing apparatus that has a full facepiece and is operated in a pressure-demand or other positive-pressure mode); or SAF:PD,PP:ASCBA (any supplied-air respirator that has a full facepiece and is operated in a pressure-demand or other positive-pressure mode in combination with an auxiliary, self-contained breathing apparatus operated in a pressure-demand or other positive-pressure mode).

1090 Trimethylhexamethylene Diisocyanate

Escape: GMFOV [any air-purifying, full-facepiece respirator (gas mask) with a chin-style, front- or back-mounted organic vapor canister]; or SCBAE (any appropriate escape-type, self-contained breathing apparatus).

CHEMICAL REACTIVITY

Reacts with water to produce carbon dioxide and toxic diamine. May react violently with alcohols in the presence of a base. Isocyanates react with acids, acetaldehyde, caustics, ammonia, aliphatic amines, alkanolamines, amides, alcohols, glycols, caprolactum solution.

FIRE INFORMATION

Combustion products include toxic nitrogen oxides. Firefighting gear (including SCBA) may not provide adequate protection. If exposure occurs, remove and isolate gear immediately and thoroughly decontaminate personnel. Vapors are heavier than air and will collect in low areas. Vapors in confined areas may explode when exposed to fire. Storage containers and parts of containers may rocket great distances, in many directions. If material or contaminated runoff enters waterways, notify downstream users of potentially contaminated waters. Notify local health and fire officials and pollution control agencies. Use dry chemicals, foam, carbon dioxide. From a secure, explosion-proof location, use water spray to cool exposed containers.

FEMA Hazard Classifications:

Health Hazards (Blue) 2
Flammability (Red) 2
Reactivity (Yellow) 1

FIRST AID

Move victim to fresh air. Call emergency medical care. Apply artificial respiration if victim is not breathing. Do not use mouth-to-mouth method if victim ingested or inhaled the substance; induce artificial respiration with the aid of a pocket mask equipped with a one-way valve or other proper respiratory medical device. Administer oxygen if breathing is difficult. Remove and isolate contaminated clothing and shoes. In case of contact with substance, immediately flush skin or eyes with running water for at least 20 minutes. For minor skin contact, avoid spreading material on unaffected skin. Keep victim warm and quiet. Effects of exposure (inhalation, ingestion, or skin contact) may be delayed. Ensure that medical personnel are aware of the material(s) involved and take precautions to protect themselves.

ENVIRONMENTAL INFORMATION

EPCRA Section 313: Form R *de minimis* concentration reporting level: 1.0%
TSCA: 40CFR712.30(*e*)8 as isocyanates

TRIMETHYLOLPROPANE T:455

SYNONYMS: EINECS No. 201-074-9; ETHRIOL; 2-ETHYL-2-(HYDROXYMETHYL)-; ETHYLTRIMETHYLOLMETHANE; ETRIOL; ETTRIOL; HEXAGLYCERINE; 1,3-PROPANEDIOL, PROPYLIDYNETRIMETHANOL; TMP (ALCOHOL); 1,1,1-TRI(HYDROXYMETHYL)PROPANE; 1,1,1-TRIMETHYLOLPROPANE; TRIS(HYDROXYMETHYL)PROPANE; 1,1,1-TRIS(HYDROXYMETHYL)PROPANE

IDENTIFICATION

CAS:	77-99-6
Molecular Formula:	$C_6H_{14}O_3$
RTECS No.:	TY6470000
Properties:	Combustible, hygroscopic crystals. White. Soluble in water.

HEALTH AND SAFETY INFORMATION

May cause irritation of the eyes and skin. Mildly toxic if swallowed.

CHEMICAL REACTIVITY

Reacts with nitric acid, hydrogen peroxide, and other strong oxidizers. Alcohols are incompatible with sulfuric acid, nitric acid, caustics, aliphatic amines, isocyanates, triethylborane.

FIRE INFORMATION

If material or contaminated runoff enters waterways, notify downstream users of potentially contaminated waters. Notify local health and fire officials and pollution control agencies. Use dry chemicals, alcohol-resistant foam, carbon dioxide. From a secure, explosion-proof location, use water spray to cool exposed containers.

NFPA Hazard Classifications: Not listed

Flash Point: 302°F (150°C) o.c.

FIRST AID

Move victim to fresh air. Call emergency medical care. Apply artificial respiration if victim is not breathing. Administer oxygen if breathing is difficult. Remove and isolate contaminated clothing and shoes. In case of contact with substance, immediately flush skin or eyes with running water for at least 20 minutes. Wash skin with soap and water. Keep victim warm and quiet. Effects of exposure (inhalation, ingestion, or skin contact) to substance may be delayed. Ensure that medical personnel are aware of the material(s) involved and take precautions to protect themselves.

TRIOCTYL PHOSPHATE T:465

SYNONYMS: DISFLAMOLL TOF; EINECS No. 201-116-6; 2-ETHYL-1-HEXANOL PHOSPHATE; FLEXOL TOF; FOSFATO de TRIS(2-ETILHEXILO) (Spanish); FOSFATO di TRIS(2-ETILESILE) (Italian); 1-HEXANOL,2-ETHYL-, PHOSPHATE; KRONITEX TOF; NCI-C54751; PHOSPHORIC ACID, TRIS(2-ETHYLHEXYL) ESTER; TOF; TRIETHYLHEXYL PHOSPHATE; TRI(2-ETHYLHEXYL) PHOSPHATE

IDENTIFICATION

CAS: 78-42-2
Molecular Formula: $C_{24}H_{51}O_4P$
RTECS No.: MP0770000
Properties: Combustible liquid. Colorless. Slight, sharp odor. Practically insoluble in water.

HEALTH AND SAFETY INFORMATION

Irritates eyes and skin. Toxic if swallowed or by skin contact. Questionable carcinogen.

NTP: Some animal evidence of carcinogenicity; insufficient human evidence

RECOMMENDATIONS FOR RESPIRATOR SELECTION

At any concentrations above the NIOSH REL, or where there is no REL, at any detectable concentration: SCBAF:PD,PP (any self-contained breathing apparatus that has a full facepiece and is operated in a pressure-demand or other positive-pressure mode); or SAF:PD,PP:ASCBA (any supplied-air respirator that has a full facepiece and is operated in a pressure-demand or other positive-pressure mode in combination with an auxiliary, self-contained breathing apparatus operated in a pressure-demand or other positive-pressure mode).

Escape: GMFOV [any air-purifying, full-facepiece respirator (gas mask) with a chin-style, front- or back-mounted organic vapor canister]; or SCBAE (any appropriate escape-type, self-contained breathing apparatus).

CHEMICAL REACTIVITY

Contact with strong oxidizers may cause fire and explosions. Phosphates are incompatible with antimony petachloride, magnesium, silver nitrate, zinc acetate.

FIRE INFORMATION

Combustion products include toxic fumes of phosphorus. Vapors are heavier than air and will collect in low areas. Vapors in confined areas may explode when exposed to fire. If material or contaminated runoff enters waterways, notify downstream users of potentially contaminated waters.Notify local health and fire officials and pollution control agencies. Use dry chemicals, alcohol-resistant foam, carbon dioxide. From a secure, explosion-proof location, use water spray to cool exposed containers.

NFPA Hazard Classifications: Not listed.
Flash Point: 260°F (126°C) o.c.
Autoignition Temperature: 698°F (370°C)

FIRST AID

Move victim to fresh air. Call emergency medical care. Apply artificial respiration if victim is not breathing. Administer oxygen if breathing is difficult. Remove and isolate contaminated clothing and shoes. In case of contact with substance, immediately flush skin or eyes with running water for at least 20 minutes. Wash skin with soap and water. Keep victim warm and quiet. Effects of exposure (inhalation, ingestion, or skin contact) to substance may be delayed. Ensure that medical personnel are aware of the material(s) involved and take precautions to protect themselves.

ENVIRONMENTAL INFORMATION

TSCA: 40CFR716.120(*d*)3; 40CFR712.30(*e*)3 as alkyl phosphates

TRIORTHOCRESYL PHOSPHATE T:470

SYNONYMS: o-CRESYL PHOSPHATE; EEC No. 015-015-00-8; EINECS NO. 215-548-8; FOSFITO de TRICRESILO (Spanish); PHOSFLEX 179-C; PHOSPHORIC ACID, TRI-o-CRESYL ESTER; PHOSPHORIC ACID, TRIS(METHYL PHENYL) ESTER; TCP; TOCP; TOFK; o-TOLYL PHOSPHATE; TPTP; ortho-TRICRESYL PHOSPHATE; TRICRESYL PHOSPHATE, ortho-; TRI-o-CRESYL ESTER of PHOSPHORIC ACID; TRI-o-CRESYL PHOSPHATE; o-TRIKRESYLPHOSPHAT (German); TRI-2-METHYLPHENYL PHOSPHATE; TRI-2-TOLYL PHOSPHATE; TRIS(o-CRESYL)-PHOSPHATE; TRIS(o-METHYLPHENYL)PHOSPHATE; TRI-o-TOLYL PHOSPHATE; TROJKREZYLUFOSFORAN (Polish)

IDENTIFICATION

CAS:	78-30-8
ERG ID:	UN 2574
ERG Guide:	151
ERG Guide Category:	SUBSTANCES-TOXIC (NONCOMBUSTIBLE)
Molecular Formula:	$C_{21}H_{21}O_4P$
RTECS No.:	TD0350000

1094 Triorthocresyl Phosphate

Properties: Combustible, colorless to pale yellow liquid or solid [below 52°F (11°C)]. Practically odorless. Sinks in water; slightly soluble.

HEALTH AND SAFETY INFORMATION

Poisonous by skin contact, inhalation and if swallowed.

IDLH: 40 mg/m^3
OSHA Table Z-1-A Air Contaminant

Exposure Limits:	**TWA**
ACGIH: [1]	0.1 mg/m^3
NIOSH: [1]	0.1 mg/m^3
OSHA: [1]	0.1 mg/m^3

[1] Skin contact contributes significantly to overall exposure.

RECOMMENDATIONS FOR RESPIRATOR SELECTION: NIOSH

0.5 mg/m^3: DM (any dust and mist respirator).
1 mg/m^3: DMXSQ (any dust and mist respirator except single-use and quarter mask respirators); or SA (any supplied-air respirator).
2.5 mg/m^3: SA:CF (any supplied-air respirator operated in a continuous-flow mode.); PAPRDM (any powered, air-purifying respirator with a dust and mist filter).
5 mg/m^3: HiEF (any air-purifying, full-facepiece respirator with a high-efficiency particulate filter); or SAT:CF (any supplied-air respirator that has a tight-fitting facepiece and is operated in a continuous-flow mode); or PAPRTHiE (any powered, air-purifying respirator with a tight-fitting facepiece and a high-efficiency particulate filter); or SCBAF (any self-contained breathing apparatus with a full facepiece); or SAF (any supplied-air respirator with a full facepiece).
40 mg/m^3: SA:PD,PP (any supplied-air respirator operated in a pressure-demand or other positive-pressure mode).

Emergency or Planned Entry into Unknown Concentrations or IDLH Conditions:
SCBAF:PD,PP (any self-contained breathing apparatus that has a full faceplate and is operated in a pressure-demand or other positive-pressure mode); or SAF:PD,PP:ASCBA (any supplied-air respirator that has a full facepiece and is operated in a pressure-demand or other positive-pressure mode in combination with an auxiliary, self-contained breathing apparatus operated in a pressure-demand or other positive-pressure mode).

Escape: HiEF (any air-purifying, full-facepiece respirator with a high-efficiency particulate filter); or SCBAE (any appropriate escape-type, self-contained breathing apparatus).

CHEMICAL REACTIVITY

Contact with magnesium or oxidizers may cause fire and explosions. Attacks some plastics, rubbers, and coatings. Phosphates are incompatible with antimony petachloride, magnesium, silver nitrate, zinc acetate.

FIRE INFORMATION

NFPA Hazard Classifications:

Health Hazards (Blue) 2
Flammability (Red) 1
Reactivity (Yellow) 0

Flash Point: 437°F (225°C) c.c.
Autoignition Temperature: 725°F (385°C)

FIRST AID

Move victim to fresh air. Call emergency medical care. Apply artificial respiration if victim is not breathing. Do not use mouth-to-mouth method if victim ingested or inhaled the substance; induce artificial respiration with the aid of a pocket mask equipped with a one–way valve or other proper respiratory medical device. Administer oxygen if breathing is difficult. Remove and isolate contaminated clothing and shoes. In case of contact with substance, immediately flush skin or eyes with running water for at least 20 minutes. For minor skin contact, avoid spreading material on unaffected skin. Keep victim warm and quiet. Effects of exposure (inhalation, ingestion, or skin contact) may be delayed. Ensure that medical personnel are aware of the material(s) involved and take precautions to protect themselves.

TRIPHENYL PHOSPHATE T:472

SYNONYMS: CELLUFLEX TPP; DISFLAMOLL-TP; EINECS No. 204-112-2; FOSFATO de TRIFENILO (Spanish); PHENYL PHOSPHATE; PHIOSFLEX-TPP; PHOSPHORIC ACID, TRIPHENYL ESTER; TP; TPP; TRIPHENOXYPHOSPHINE OXIDE; TRIPHENYL ESTER OF PHOSPHORIC ACID

IDENTIFICATION

CAS:	115-86-6
ERG ID:	UN 2811
ERG Guide:	154
ERG Guide Category:	SUBSTANCES-TOXIC AND/OR CORROSIVE
Molecular Formula:	$C_{18}H_{15}O_4P$
RTECS No.:	TC8400000

1096 Triphenyl Phosphate

Properties: Combustible, crystalline powder. Colorless. Phenollike odor. Insoluble in water. Melting point 122°F (50°C)

HEALTH AND SAFETY INFORMATION

Irritates eyes and respiratory tract. Affects central nervous system; prolonged exposure may cause muscle weakness and paralysis. Blood effects. Toxic if swallowed or by skin contact.

IDLH: 1,000 mg/m^3
OSHA Table Z-1-A Air Contaminant

Exposure Limits:	**TWA**
ACGIH: (1)	3 mg/m^3
NIOSH:	3 mg/m^3
OSHA:	3 mg/m^3

(1) Not classified as a human carcinogen.

RECOMMENDATIONS FOR RESPIRATOR SELECTION

At any concentrations above the NIOSH REL, or where there is no REL, at any detectable concentration: SCBAF:PD,PP (any self-contained breathing apparatus that has a full facepiece and is operated in a pressure-demand or other positive-pressure mode); or SAF:PD,PP:ASCBA (any supplied-air respirator that has a full facepiece and is operated in a pressure-demand or other positive-pressure mode in combination with an auxiliary, self-contained breathing apparatus operated in a pressure-demand or other positive-pressure mode).

Escape: GMFOVHiE [any air-purifying, full-facepiece respirator (gas mask) with a chin-style, front- or back-mounted organic vapor canister having a high-efficiency particulate filter]; or SCBAE (any appropriate escape-type, self-contained breathing apparatus).

CHEMICAL REACTIVITY

Contact with strong oxidizers may cause fire and explosions. Phosphates are incompatible with antimony petachloride, magnesium, silver nitrate, zinc acetate.

FIRE INFORMATION

Combustion products include toxic fumes of phosphoric acid, phosphorus pentoxide, and carbon monoxide. Vapors are heavier than air and will collect in low areas. Vapors in confined areas may explode when exposed to fire. If material or contaminated runoff enters waterways, notify downstream users of potentially contaminated waters. Notify local health and fire officials and pollution control agencies. Use dry chemicals, alcohol-resistant foam, carbon dioxide. From a secure, explosion-proof location, use water spray to cool exposed containers.

NFPA Hazard Classifications:

Health Hazards (Blue) 2
Flammability (Red) 1
Reactivity (Yellow) 0

Flash Point: 428°F (220°C) c.c.

FIRST AID

Move victim to fresh air. Call emergency medical care. Apply artificial respiration if victim is not breathing. Administer oxygen if breathing is difficult. Remove and isolate contaminated clothing and shoes. In case of contact with substance, immediately flush skin or eyes with running water for at least 20 minutes. Wash skin with soap and water. Keep victim warm and quiet. Effects of exposure (inhalation, ingestion, or skin contact) to substance may be delayed. Ensure that medical personnel are aware of the material(s) involved and take precautions to protect themselves.

TRIPHENYL PHOSPHITE T:474

SYNONYMS: EEC No. 015-105-00-7; EINECS No. 202-908-4; EFED; FOSFITO de TRIFENILO (Spanish); PHOSPHITE de TRIPHENYLE (French); PHOSPHOROUS ACID, TRIPHENYL ESTER; TRIFENOXYFOSFIN (Czech); TRIFENYLFOSFIET (German)

IDENTIFICATION

CAS:	101-02-0
ERG ID:	UN 2811
ERG Guide:	154
ERG Guide Category:	SUBSTANCES-TOXIC AND/OR CORROSIVE
Molecular Formula:	$C_{18}H_{15}O_3P$
RTECS No.:	TH1575000
Properties:	Combustible solid or oily liquid. White to pale yellow. Pleasant odor. Insoluble in water.

HEALTH AND SAFETY INFORMATION

Irritates eyes, skin, and respiratory tract. Affects central nervous system. Prolonged or repeated exposure may cause muscle weakness, paralysis; brain damage; skin disease. Toxic if swallowed.

RECOMMENDATIONS FOR RESPIRATOR SELECTION

At any concentrations above the NIOSH REL, or where there is no REL, at any detectable concentration: SCBAF:PD,PP (any self-contained breathing apparatus that has a full facepiece and is operated in a pressure-demand or other positive-pressure mode); or SAF:PD,PP:ASCBA (any supplied-air respirator that has a full facepiece and is operated in a pressure-demand or other positive-pressure mode in combination with an auxiliary, self-contained breathing apparatus operated in a pressure-demand or other positive-pressure mode).

Escape: GMFOVHiE [any air-purifying, full-facepiece respirator (gas mask) with a chin-style, front- or back-mounted organic vapor canister having a high-efficiency particulate filter]; or SCBAE (any appropriate escape-type, self-contained breathing apparatus).

CHEMICAL REACTIVITY

Reacts with strong oxidizers; may cause fire and explosions..

FIRE INFORMATION

Combustion products include toxic fumes of phosphoric acid, phosphorus pentoxide, and carbon monoxide. Vapors are heavier than air and will collect in low areas. Vapors in confined areas may explode when exposed to fire. If material or contaminated runoff enters waterways, notify downstream users of potentially contaminated waters. Notify local health and fire officials and pollution control agencies. Use dry chemicals, alcohol-resistant foam, carbon dioxide. From a secure, explosion-proof location, use water spray to cool exposed containers.

NFPA Hazard Classifications

Health Hazards (Blue) 0
Flammability (Red) 1
Reactivity (Yellow) 0

Flash Point: 425°F (218°C) o.c.

FIRST AID

Move victim to fresh air; call emergency medical care. Apply artificial respiration if victim is not breathing. Do not use mouth-to-mouth method if victim ingested or inhaled the substance; induce artificial respiration with the aid of a pocket mask equipped with a one-way valve or other proper respiratory medical device. Administer oxygen if breathing is difficult. Remove and isolate contaminated clothing and shoes. In case of contact with substance, immediately flush skin or eyes with running water for at least 20 minutes. For minor skin contact, avoid spreading material on unaffected skin. Keep victim warm and quiet. Effects of exposure (inhalation, ingestion, or skin contact) to substance may be delayed. Ensure that medical personnel are aware of the material(s) involved and take precautions to protect themselves.

TRIPOPYLENE GLYCOL T:476

SYNONYMS: 4,8-DIOXY-UNDECANE DIOL,-1,11; EINECS No. 246-466-0; 2-[2-(2-HYDROXYPROPOXY)PROPOXY]-1-PROPANOL; PROPANOL, [(1METHYL-1,2-ETHANEDIYL)BIS(OXY)BIS-

IDENTIFICATION

CAS:	24800-44-0
ERG ID:	UN 3082
ERG Guide:	171
ERG Guide Category:	SUBSTANCES (LOW TO MODERATE HAZARD)
Molecular Formula:	$C_9H_{20}O_4$
RTECS No.:	YK6825000
Properties:	Combustible liquid. Colorless. Characteristic odor. May float or sink on water; completely soluble.

HEALTH AND SAFETY INFORMATION

May cause irritation to eyes and skin. May affect the central nervous system.

CHEMICAL REACTIVITY

Contact with strong oxidizers can cause fire and explosions. Incompatible with sulfuric acid, isocyanates, perchloric acid. Attacks some plastics, rubber, and coatings.

FIRE INFORMATION

Combustion may produce irritants and toxic vapors. Vapors are heavier than air and will collect in low areas. Vapors in confined areas may explode when exposed to fire. If material or contaminated runoff enters waterways, notify downstream users of potentially contaminated waters. Notify local health and fire officials and pollution control agencies. Use dry chemicals, alcohol-resistant foam, carbon dioxide. From a secure, explosion-proof location, use water spray to cool exposed containers.

FEMA Hazard Classifications:

Health Hazards (Blue) 0
Flammability (Red) 1
Reactivity (Yellow) 0

Flash Point: 285°F (141°C) o.c.
Flammable Limits in Air: (estimated) LEL 0.8%; UEL 5.0% (CHRIS)

FIRST AID

Move victim to fresh air. Call emergency medical care. Apply artificial respiration if victim is not breathing. Administer oxygen if breathing is difficult. In case of contact with substance, immediately flush skin or eyes with running water for at least 20 minutes. Remove and isolate contaminated clothing and shoes. Ensure that medical personnel are aware of the material(s) involved and take precautions to protect themselves.

TRIS(2–CHLOROETHYL)PHOSPHITE T:478

SYNONYMS: 2-CHLOROETHANOL PHOSPHITE (3:1); EINECS No. 205-397-6; ETHANOL, 2-CHLORO-, PHOSPHITE; FOSFITO de TRIS(2-CLOROETILO) (Spanish); PHOSPHOROUS ACID, TRIS(2-CHLOROETHYL)ESTER; TRIS(2-CHLOROETHYL)ESTER of PHOSPHORUS ACID

IDENTIFICATION

CAS:	140-08-9
ERG ID:	UN 2810
ERG Guide:	153
ERG Guide Category:	SUBSTANCES-TOXIC AND/OR CORROSIVE (COMBUSTIBLE)
Molecular Formula:	$C_6H_{12}Cl_3O_3P$
RTECS No.:	TH1925000
Properties:	Combustible liquid. Colorless. Characteristic odor. Insoluble in water. Poisonous gases emitted when heated.

HEALTH AND SAFETY INFORMATION

Poisonous if swallowed. Skin contact may add to overall exposure.

RECOMMENDATIONS FOR RESPIRATOR SELECTION

At any concentrations above the NIOSH REL, or where there is no REL, at any detectable concentration: SCBAF:PD,PP (any self-contained breathing apparatus that has a full facepiece and is operated in a pressure-demand or other positive-pressure mode); or SAF:PD,PP:ASCBA (any supplied-air respirator that has a full facepiece and is operated in a pressure-demand or other positive-pressure mode in combination with an auxiliary, self-contained breathing apparatus operated in a pressure-demand or other positive-pressure mode).

Escape: GMFOV [any air-purifying, full-facepiece respirator (gas mask) with a chin-style, front- or back-mounted organic vapor canister]; or SCBAE (any appropriate escape-type, self-contained breathing apparatus).

CHEMICAL REACTIVITY

Contact with strong oxidizers may cause fire and explosions.

FIRE INFORMATION

Combustion products include toxic fumes of chlorides and phosphorus. Vapors are heavier than air and will collect in low areas. Vapors in confined areas may explode when exposed to fire. If material or contaminated runoff enters waterways, notify downstream users of potentially contaminated waters. Notify local health and fire officials and pollution control agencies. Use dry chemicals, alcohol-resistant foam, carbon dioxide. From a secure, explosion-proof location, use water spray to cool exposed containers.

NFPA Hazard Classifications: Not listed
Flash Point: 280°F (138°C) o.c.

FIRST AID

Move victim to fresh air. Call emergency medical care. Apply artificial respiration if victim is not breathing. Do not use mouth-to-mouth method if victim ingested or inhaled the substance; induce artificial respiration with the aid of a pocket mask equipped with a one-way valve or other proper respiratory medical device. Administer oxygen if breathing is difficult. Remove and isolate contaminated clothing and shoes. In case of contact with substance, immediately flush skin or eyes with running water for at least 20 minutes. Wash skin with soap and water. Keep victim warm and quiet. Effects of exposure (inhalation, ingestion, or skin contact) to substance may be delayed. Ensure that medical personnel are aware of the material(s) involved and take precautions to protect themselves.

ENVIRONMENTAL INFORMATION

TSCA: 40CFR716.120(*a*) listed substance

TUNGSTIC ACID T:560

SYNONYMS: KYSELINA WOLFRAMOVA (Czech); ORTHOTUNGSTIC ACID; WOLFRAMIC ACID

IDENTIFICATION

CAS: 11105-11-6; 7783-03-1 (soluble tungsten)
Molecular Formula: H_2O_4W

1102 Tungstic Acid

RTECS No.: YO7840000

Properties: Noncombustible powder. Yellow. Insoluble in water.

HEALTH AND SAFETY INFORMATION

Irritates eyes, skin, and respiratory system. Repeated or prolonged inhalation of dust may cause pulmonary edema, which can be delayed for several hours; there is a risk of death in serious cases.

Exposure Limits:	**TWA**	**STEL** as tungsten soluble compounds
ACGIH:	1 mg/m^3	3 mg/m^3
NIOSH:	1 mg/m^3	3 mg/m^3
OSHA:	None	

RECOMMENDATIONS FOR RESPIRATOR SELECTION as tungsten soluble compounds

10 mg/m^3: HiE (any air-purifying, respirator with a high-efficiency particulate filter); or SA (any supplied-air respirator).
25 mg/m^3: SA:CF (any supplied-air respirator operated in a continuous-flow mode)
50 mg/m^3: HiEF (any air-purifying, full-facepiece respirator with a high-efficiency particulate filter); or SCBAF (any self-contained breathing apparatus with a full facepiece); or SAF (any supplied-air respirator with a full facepiece).

Emergency or Planned Entry into Unknown Concentrations or IDLH Conditions: SCBAF:PD,PP (any self-contained breathing apparatus that has a full facepiece and is operated in a pressure-demand or other positive-pressure mode); or SAF:PD,PP:ASCBA (any supplied-air respirator that has a full facepiece and is operated in a pressure-demand or other positive-pressure mode in combination with an auxiliary, self-contained breathing apparatus operated in a pressure-demand or other positive-pressure mode).

Escape: HiEF (any air-purifying, full-facepiece respirator with a high-efficiency particulate filter); or SCBAE (any appropriate escape-type, self-contained breathing apparatus).

CHEMICAL REACTIVITY

Incompatible with strong oxidizers, bromine pentafluoride, iodine, pentafluoride, fluorine and chlorine compounds. Mineral acids are incompatible with sulfuric acid, caustics, ammonia, aliphatic amines, alkanolamines, aromatic amines, amides, organic anhydrides, isocyanates, vinyl acetate, alkylene oxides, epichlorohydrin.

FIRE INFORMATION

Combustion products include toxic fumes of tungsten. If material or contaminated runoff enters waterways, notify downstream users of potentially contaminated waters. Notify local health and fire officials and pollution control agencies. Use extinguishers suitable for surrounding fire. Use water spray to cool exposed containers.

NFPA Hazard Classifications: Not listed

FIRST AID

Move victim to fresh air. Call emergency medical care. Apply artificial respiration if victim is not breathing. Administer oxygen if breathing is difficult. Remove and isolate contaminated clothing and shoes. In case of contact with substance, immediately flush skin or eyes with running water for at least 20 minutes. Wash skin with soap and water. Keep victim warm and quiet. Effects of exposure (inhalation, ingestion, or skin contact) to substance may be delayed. Ensure that medical personnel are aware of the material(s) involved and take precautions to protect themselves.

10–UNDECENOIC ACID U:100

SYNONYMS: ACIDO UNDEC-10-ENOICO (Spanish); DECLID; EINECS No. 203-965-8; 10-HENEDECENOIC ACID; KYSELINA UNDECYLENOVA (Czech); RENSELIN; SEVINON; UNDECYLENIC ACID; UNDECYL-10-ENIC ACID; 9-UNDECYLENIC ACID; UNDEC-10-EENZUUR (Dutch); UNDEC-10-ENOIC ACID; UNDEC-10-ENSAEURE (German); 10-UNDECYLENIC ACID

IDENTIFICATION

CAS:	112-38-9
Molecular Formula:	$C_{11}H_{20}O_2$
RTECS No.:	YQ2975000
Properties:	Combustible, crystalline solid or liquid. Light colored. Fruity odor. Insoluble in water.

HEALTH AND SAFETY INFORMATION

Pure substance irritates skin. Ingestion may cause nausea

CHEMICAL REACTIVITY

Reacts with strong bases, strong oxidizing agents, and strong reducing agents.

FIRE INFORMATION

Combustion products include irritants and toxic vapors. If material or contaminated runoff enters waterways, notify downstream users of potentially contaminated waters. Notify local health and fire officials and pollution control agencies. Use dry chemicals, alcohol-resistant foam, carbon dioxide. Use water spray to cool exposed containers.

FEMA Hazard Classifications:

Health Hazards (Blue) 1
Flammability (Red) 1
Reactivity (Yellow) 0

Flash Point: 295°F (146°C) o.c.

FIRST AID

Move victim to fresh air. Call emergency medical care. Apply artificial respiration if victim is not breathing. Administer oxygen if breathing is difficult. Remove and isolate contaminated clothing and shoes. In case of contact with substance, immediately flush skin or eyes with running water for at least 20 minutes. Wash skin with soap and water. Keep victim warm and quiet. Effects of exposure (inhalation, ingestion, or skin contact) to substance may be delayed. Ensure that medical personnel are aware of the material(s) involved and take precautions to protect themselves.

UREA U:300

SYNONYMS: AI3-01202; ALPHADRATE; AQUA CARE; AQUA CARE HP; B-I-K; CARBADERM; CARBAMIDE; CARBAMIDE ACID; CARBAMIDE RESIN; CARBONYL DIAMINE; CARBAMIMIDIC ACID; CARBONYL DIAMIDE; CARBONYLDIAMIDE; CASWELL NO. 902; EINECS No. 200-315-5; HARNSTOFF (German); ISOUREA; MOCOVINA (Czech); NCI-C02119; PRESPERSION, 75 UREA; PSEUDOUREA; SUPERCEL 3000; ULTRA MIDE; UR; UREA AMMONIUM NITRATE SOLUTION; UREA-13C; UREACIN-20; UREACIN-40; UREACIN-10 LOTION; UREAPHIL; UREOPHIL; UREPEARL; UREVERT; VARIOFORM II

IDENTIFICATION

CAS:	57-13-6
ERG ID:	UN 3259
ERG Guide:	154
ERG Guide Category:	SUBSTANCES-TOXIC AND/OR CORROSIVE (NONCOMBUSTIBLE)
Molecular Formula:	CH_4N_2O
RTECS No.:	YR6250000
Properties:	Combustible solid crystals or pellets. White. Odorless or slight ammonia odor. Saline taste. Sinks and mixes with water.

HEALTH AND SAFETY INFORMATION

Irritates eyes, skin, and respiratory tract.

Exposure Limits: **TWA**

AIHA WEEL: 10 mg/m^3

RECOMMENDATIONS FOR RESPIRATOR SELECTION

At any concentrations above the NIOSH REL, or where there is no REL, at any detectable concentration: SCBAF:PD,PP (any self-contained breathing apparatus that has a full facepiece and is operated in a pressure-demand or other positive-pressure mode); or SAF:PD,PP:ASCBA (any supplied-air respirator that has a full facepiece and is operated in a pressure-demand or other positive-pressure mode in combination with an auxiliary, self-contained breathing apparatus operated in a pressure-demand or other positive-pressure mode).

Escape: GMFOVHiE [any air-purifying, full-facepiece respirator (gas mask) with a chin-style, front- or back-mounted organic vapor canister having a high-efficiency particulate filter]; or SCBAE (any appropriate escape-type, self-contained breathing apparatus).

CHEMICAL REACTIVITY

Heat causes formation of suffocating anhydrous ammonia fumes. Combustion products include toxic nitrogen oxides. Contact with oxidizers or nitrates can cause fire and explosions. Contact with chlorinating agents including, hypochlorite bleaches, may produce explosive nitrogen trichloride. Violent reaction with ammonium nitrate, chromyl chloride, sodium perchlorate, phosphorus pentachloride. Contact with nitrosyl perchlorate can cause fire and explosions.

FIRE INFORMATION

Not combustible, but highly reactive and many chemical reactions can cause fire and explosions. If material or contaminated runoff enters waterways, notify downstream users of potentially contaminated waters. Notify local health and fire officials and pollution control agencies. Use any extinguishing agent suitable for surrounding fire. Use water spray to cool exposed containers.

FEMA Hazard Classifications:

Health Hazards (Blue) 0
Flammability (Red) 0
Reactivity (Yellow) 0

NFPA Hazard Classifications: Not listed.

FIRST AID

Move victim to fresh air; call emergency medical care. Apply artificial respiration if victim is not breathing. Do not use mouth-to-mouth method if victim ingested or inhaled the substance; induce artificial respiration with the aid of a pocket mask equipped with a one-way valve or other proper respiratory medical device. Administer oxygen if breathing is difficult. Remove and isolate contaminated clothing and shoes. In case of contact with substance, immediately flush skin or eyes with running water for at least 20 minutes. For minor skin contact, avoid spreading material on unaffected skin. Keep victim warm and quiet. Effects of exposure (inhalation, ingestion, or skin contact) to substance may be delayed. Ensure that medical personnel are aware of the material(s) involved and take precautions to protect themselves.

UREA–FORMALDEHYDE RESIN U:500

SYNONYMS: ACRISIN FS 017; AEROLITE 300; AEROLITE A 300; AEROLITE FFD; AGROFORM; AMIKOL 65; ANAFLEX; BC 20; BC 20 (POLYMER); BC 40; BC 77; BECKAMINE 21-511; BECKAMINE NF 5; BECKAMINE P 136; BECKAMINE P 138; BECKAMINE P 138-60; BECKAMINE P 196M; BEETLE 55; BEETLE 60; BEETLE 65; BEETLE 80; BEETLE 212-9; BEETLE BE 685; BEETLE BU 700; BEETLE XB 1050; BU 700; CARBAMOL; CASCAMITE; CASCO 5H; CASCO PR 335; CASCO RESIN; CASCO UL 30; CASCO WS 114-79; CASCO WS 138-43; CASCO WS 138-44; CYREZ 933; DEPREMOL M; DIAFORM UR; DIAKOL DM; DIAKOL F; DIAKOL M; DYNOMIN UI 16; DYNOMIN UM 15; EPOK U 9048; FIBRASET TC; FORMALDEHYDE COPOLYMER with UREA; FORMALDEHYDE-UREA CONDENSATE; FORMALDEHYDE-UREA COPOLYMER; FORMALDEHYDE-UREA POLYMER; FORMALDEHYDE-UREA PRECONDENSATE; FORMALDEHYDE-UREA PREPOLYMER; FORMALDEHYDE-UREA RESIN; FORMALIN-UREA COPOLYMER; GABRITE; HYGROMULL; K 0; K 17; K 17 (POLYMER); K385; K 8870; K 411-02; KARBAMOL; KARBAMOL B/M; KAURESIN K244; KAURIT 285 FL; KAURIT 420; KFS; KM 2; KM 2 (POLYMER); KNITTEX TC; KNITTEX TS; KOPREZ 87-110; KS 11; KS 35; KS 68M; KS-M 0.3P; L 195; LAREX; M 2; M 2 (POLYMER); M 60 (FORMALDEHYDE POLYMER); M 70; M 70 (POLYMER); MCH 52; MELAN 11; METHYLOLUREA RESIN; MF; MF 17; MF 27; MFPS 1; MF RESIN; MIRBANE SU 118K; MKH 52; MOULDRITE A256; MPF 2; N 50; NOXYLIN; PARAFORMALDEHYDE-UREA POLYMER; PARAFORMALDEHYDE-UREA RESIN; PIANIZOL; PIATHERM; PIATHERM D; PLASTOPAL BT; PLYAMINE HD 1129A; PLYAMINE P 364BL; POLY(METHIBIS(HYDROXYMETHYL)UREYLENE)AMER; POLYNOXYLIN; PONOXYLAN; PR 703-78; RESAMIN 155F; RESAMIN HW 505; RESIMENE X 970; RESIMENE X 975; RESIMENE X 980; RESIMINE 975; RESINA de UREA-FORMALDEHIDO (Spanish); SFK 70; SK 75; SK 75V; S-RESIN AER 20; SUMIREZ 614; SUMITEX 260; SUMITEX 810; SUMITEX NF 113; SUMITEX RESIN 810; T 101; U 963; UF 33; UF 240; UFORMITE 700; UFORMITE F 240N; UKS 72; UKS 73; ULOID 22; ULOID 100; ULOID 301; UL 52R; UMALUR; UM-G; URALITE; URALITE (POLYMER); URAMINE T101; URAMINE T105; URAMINE TSL 58; URAMITE; UREA-FORMALDEHYDE ADDUCT; UREA-FORMALDEHYDE CONDENSATE; UREA-FORMALDEHYDE COPOLYMER; UREA-FORMALDEHYDE OLIGOMER; UREA-FORMALDEHYDE POLYMER;

UREA-FORMALDEHYDE PRECONDENSATE; UREA-FORMALDEHYDE PREPOLYMER; UREAPAP W; URECOLI S; URECOLL K; URECOLL KL; URELIT C; URELIT HM; URELIT R; UREPRET UREPRET; UROFIX; UST; W 70; YUBAN 10HV; YUBAN 10S

IDENTIFICATION

CAS:	9011-05-6
Molecular Formula:	$(C_2H_4N_6O_2)_x$
RTECS No.:	YU6100000
Properties:	Granular solid. Odorless. Some urea-formaldehyde substances are soluble in water.

HEALTH AND SAFETY INFORMATION

Dust is an allergen. Inhalation may cause lung allergy.

RECOMMENDATIONS FOR RESPIRATOR SELECTION

At any concentrations above the NIOSH REL, or where there is no REL, at any detectable concentration: SCBAF:PD,PP (any self-contained breathing apparatus that has a full facepiece and is operated in a pressure-demand or other positive-pressure mode); or SAF:PD,PP:ASCBA (any supplied-air respirator that has a full facepiece and is operated in a pressure-demand or other positive-pressure mode in combination with an auxiliary, self-contained breathing apparatus operated in a pressure-demand or other positive-pressure mode).

Escape: GMFOVHiE [any air-purifying, full-facepiece respirator (gas mask) with a chin-style, front- or back-mounted organic vapor canister having a high-efficiency particulate filter]; or SCBAE (any appropriate escape-type, self-contained breathing apparatus).

CHEMICAL REACTIVITY

Reacts with oxidizing agents and caustics. Powder forms explosive mixture with air. Some urea-formaldehyde substances are capable of building up unstable peroxides and are capable of polymerization. Sealed containers can build up pressure if exposed to heat. Amines (urea) may ignite on contact with cellulose nitrate and are incompatible with acrolein, acrylonitrile, tert-butyl nitroacetylene, ethylene oxide, isopropyl chlorocarbonate, maleic anhydride, triisobutylaluminum. Powder forms explosive mixture w/air.

FIRE INFORMATION

Combustion products include toxic oxides of nitrogen. Vapors are heavier than air and will collect in low areas. Vapors in confined areas may explode when exposed to fire. Vapors may travel long distances to ignition sources and flash back. Storage containers and parts of containers may rocket great distances, in many directions. If material or contaminated runoff enters waterways, notify downstream users of potentially contaminated waters. Notify local health and fire officials and pollution control agencies. Use dry chemicals, foam, carbon dioxide. From a secure, explosion-proof location, use water spray to cool exposed containers.

NFPA Hazard Classifications: Not listed.

FIRST AID

Move victim to fresh air. Call emergency medical care. Apply artificial respiration if victim is not breathing. Administer oxygen if breathing is difficult. Remove and isolate contaminated clothing and shoes. In case of contact with substance, immediately flush skin or eyes with running water for at least 20 minutes. Wash skin with soap and water. Keep victim warm and quiet. Effects of exposure (inhalation, ingestion, or skin contact) to substance may be delayed. Ensure that medical personnel are aware of the material(s) involved and take precautions to protect themselves.

ENVIRONMENTAL INFORMATION

TSCA: 40CFR716.120(*a*)

UREA NITRATE U:600

SYNONYMS: ACIDOGEN NITRATE; EINECS No. 204-703-5; HARNSTOFFNITRAT (German); NITRATE d'UREE (French); UREA, MONONITRATE (8CI,9CI); UREUMNITRAAT (Dutch)

IDENTIFICATION

CAS:	124-47-0
ERG ID:	UN 0220 (dry or wetted with <20% water, by weight) UN 1357 (wetted with not <20% water, by weight)
ERG Guide:	112 (UN 0220); 113 (UN 1357)
ERG Guide Category:	(112) EXPLOSIVES-DIVISION 1.1, 1.2, 1.3, 1.5, 1.6; CLASS A or B (113) FLAMMABLE SOLIDS-TOXIC (WET/DESENSITIZED EXPLOSIVE)
Molecular Formula:	$CH_5N_3O_4$
RTECS No.:	YT9675000
Properties:	Flammable solid minerals or prisms. Colorless. Slightly soluble in water.

HEALTH AND SAFETY INFORMATION

A "nitro" compound. Primarily a dangerous fire and explosion risk.

CHEMICAL REACTIVITY

A powerful oxidizer. Violent, possibly explosive, reaction with strong acids, reducing agents, cyanide, esters, aluminum, powdered heavy metals, organic materials, thiocyanates, and many other compounds. Decomposes at 305°F (152°C).

FIRE INFORMATION

Combustion products include toxic nitrogen oxides. Vapors are heavier than air and will collect in low areas. Vapors in confined areas may explode when exposed to fire. Vapors may travel long distances to ignition sources and flash back. Storage containers and parts of containers may rocket great distances, in many directions. If material or contaminated runoff enters waterways, notify downstream users of potentially contaminated waters. Notify local health and fire officials and pollution control agencies. Use flooding quantities of water, from maximum possible distance. From a secure, explosion-proof location, as far away from the material as possible, use water spray to cool exposed containers. If cooling streams are ineffective (venting sound increases in volume and pitch, tank discolors, or shows any signs of deforming), withdraw immediately to a secure position.

NFPA Hazard Classifications: Not listed

FIRST AID

(112) EXPLOSIVES-DIVISION 1.1, 1.2, 1.3, 1.5, 1.6; CLASS A or B–Move victims to fresh air. Call emergency medical care. Apply artificial respiration if victim is not breathing. Administer oxygen if breathing is difficult. Remove and isolate contaminated clothing and shoes. In case of contact with substance, immediately flush skin or eyes with running water for at least 20 minutes. Ensure that medical personnel are aware of the material(s) involved and take precautions to protect themselves.

(113) FLAMMABLE SOLIDS-TOXIC (WET/DESENSITIZED EXPLOSIVE)–Move victims to fresh air. Call emergency medical care. Apply artificial respiration if victim is not breathing. Administer oxygen if breathing is difficult. Remove and isolate contaminated clothing and shoes. In case of contact with substance, immediately flush skin or eyes with running water for at least 20 minutes. Ensure that medical personnel are aware of the material(s) involved and take precautions to protect themselves.

URETHANE U:800

SYNONYMS: A 11032; AI3-00553; AETHYLCARBAMAT (German); AETHYLURETHAN (German);

1110 Urethane

CARBAMIC ACID, ETHYL ESTER; CARBAMIDSAEURE-AETHYLESTER (German); CARBAMMATO di ETILE (Italian); EEC No. 607-149-00-6; EINECS No. 200-123-1; ESTANE 5703; ETHYL CARBAMATE; ETHYLESTER-KYSELINY-KARBAMINOVE (Czech); ETHYLURETHAN; ETHYL URETHANE; o-ETHYLURETHANE; LEUCETHANE; LEUCOTHANE; NSC 746; PRACARBAMIN; PRACARBAMINE; U-COMPOUND; URETAN-ETYLOWY (Polish); URETANO (Spanish); URETHAN; X-41

IDENTIFICATION

CAS:	51-79-6
ERG ID:	UN 2811 as toxic solid, organic, n.o.s.
ERG Guide:	154
ERG Guide Category:	SUBSTANCES-TOXIC AND/OR CORROSIVE
Molecular Formula:	$C_3H_7NO_2$
RTECS No.:	FA8400000
Properties:	Combustible, crystalline solid or powder. Melting point 118°F (48°C). Nearly odorless. Salty taste. Very soluble in water.

HEALTH AND SAFETY INFORMATION

A carcinogen and experimental teratogen. Skin contact contributes significantly to overall exposure; capable of passing through the skin. High exposure can cause headaches, nausea, unconsciousness. Prolonged or repeated exposure can affect the central nervous system and may cause brain, liver, blood, and other organ damage.

IARC Evaluation: Group 2B, animal positive; possible human carcinogen.
NTP: Confirmed carcinogen

RECOMMENDATIONS FOR RESPIRATOR SELECTION

At any concentrations above the NIOSH REL, or where there is no REL, at any detectable concentration: SCBAF:PD,PP (any self-contained breathing apparatus that has a full facepiece and is operated in a pressure-demand or other positive-pressure mode); or SAF:PD,PP:ASCBA (any supplied-air respirator that has a full facepiece and is operated in a pressure-demand or other positive-pressure mode in combination with an auxiliary, self-contained breathing apparatus operated in a pressure-demand or other positive-pressure mode).

Escape: GMFOVHiE [any air-purifying, full-facepiece respirator (gas mask) with a chin-style, front- or back-mounted organic vapor canister having a high-efficiency particulate filter]; or SCBAE (any appropriate escape-type, self-contained breathing apparatus).

CHEMICAL REACTIVITY

Incompatible with *β*-napthol, camphor, menthol, strong acids, strong caustics, strong oxidizers, phenyl salicylate, resorcinol, *m*-thymol. Contact with phosphorus pentachloride forms an explosive product (SAX).

FIRE INFORMATION

Combustion products include toxic nitrogen oxides. Vapors in confined areas may explode when exposed to fire. If material or contaminated runoff enters waterways, notify downstream users of potentially contaminated waters. Notify local health and fire officials and pollution control agencies. Use dry chemicals, alcohol-resistant foam, carbon dioxide. From a secure, explosion-proof location, use water spray to cool exposed containers. If cooling streams are ineffective (venting sound increases in volume and pitch, tank discolors, or shows any signs of deforming), withdraw immediately to a secure position.

FEMA Hazard Classifications:

Health Hazards (Blue) 1
Flammability (Red) 1
Reactivity (Yellow) 0

Flash Point: 198°F (92°C) c.c.

FIRST AID

Move victim to fresh air. Call emergency medical care. Apply artificial respiration if victim is not breathing. Administer oxygen if breathing is difficult. Remove and isolate contaminated clothing and shoes. In case of contact with substance, immediately flush skin or eyes with running water for at least 20 minutes. Wash skin with soap and water. Keep victim warm and quiet. Effects of exposure (inhalation, ingestion, or skin contact) to substance may be delayed. Ensure that medical personnel are aware of the material(s) involved and take precautions to protect themselves.

ENVIRONMENTAL INFORMATION

Clean Air Act: Hazardous Air Pollutants (Title I, Part A, Section 112)
EPA Hazardous Waste Number (RCRA No.): U238
RCRA Section 261 Hazardous Constituents
EPCRA Section 304 Reportable Quantity (RQ): CERCLA, 100 lb (45.4 kg)
EPCRA Section 313: Form R *de minimis* concentration reporting level: 0.1%.
California's Proposition 65: Carcinogen; reproductive toxin.
TSCA: 40CFR799.2140 chemicals with significant new uses (SNUR).

VALERIC ACID V:002

SYNONYMS: ACIDE VALERIQUE (French); ACIDO VALERICO (Spanish); 1-BUTANECARBOXYLIC ACID; EEC No. 607-143-00-3; EINECS No. 203-677-2; FEMA NO. 3101; KYSELINA VALEROVA (Polish); PENTANOIC ACID; n-PENTANOIC ACID; PENTANSYRE (French); PROPYLACETIC ACID; VALERIANIC ACID; n-VALERIC ACID; VALERIANSAEURE (German)

IDENTIFICATION

CAS:	109-52-4
ERG ID	UN 1760
ERG Guide:	154
ERG Guide Category:	SUBSTANCES-TOXIC AND/OR CORROSIVE (NONCOMBUSTIBLE)
Molecular Formula:	$C_5H_{10}O_2$
RTECS No.:	YV6100000
Properties:	Combustible liquid. Colorless. Unpleasant odor. Very soluble in water.

HEALTH AND SAFETY INFORMATION

Corrosive to eyes, skin, and respiratory tract. Inhalation may cause pulmonary edema, which can be delayed for several hours; there is a risk of death in serious cases.

RECOMMENDATIONS FOR RESPIRATOR SELECTION

At any concentrations above the NIOSH REL, or where there is no REL, at any detectable concentration: SCBAF:PD,PP (any self-contained breathing apparatus that has a full facepiece and is operated in a pressure-demand or other positive-pressure mode); or SAF:PD,PP:ASCBA (any supplied-air respirator that has a full facepiece and is operated in a pressure-demand or other positive-pressure mode in combination with an auxiliary, self-contained breathing apparatus operated in a pressure-demand or other positive-pressure mode).

Escape: GMFOV [any air-purifying, full-facepiece respirator (gas mask) with a chin-style, front- or back-mounted organic vapor canister]; or SCBAE (any appropriate escape-type, self-contained breathing apparatus).

CHEMICAL REACTIVITY

Forms explosive mixture with air. May generate heat in water. Incompatible with sulfuric acid, caustics, ammonia, amines, isocyanates, alkylene oxides, epichlorohydrin, and strong oxidizers.

FIRE INFORMATION

Vapors are heavier than air and will collect in low areas. Vapors in confined areas may explode when exposed to fire. Vapors may travel long distances to ignition sources and flash back. Storage containers and parts of containers may rocket great distances, in many directions. If material or contaminated runoff enters waterways, notify downstream users of potentially contaminated waters. Notify local health and fire officials and pollution control agencies. Use dry chemicals, foam, carbon dioxide. From a secure, explosion-proof location, use water spray to cool exposed containers.

NFPA Hazard Classifications:

Health Hazards (Blue) 2
Flammability (Red) 1
Reactivity (Yellow) 0

Flash Point: 205°F (96°C) c.c.
Flammable Limits in Air: LEL 1.6%; UEL 7.6%
Autoignition Temperature: 707°F (375°C)

FIRST AID

Move victim to fresh air; call emergency medical care. Apply artificial respiration if victim is not breathing. Do not use mouth-to-mouth method if victim ingested or inhaled the substance; induce artificial respiration with the aid of a pocket mask equipped with a one-way valve or other proper respiratory medical device. Administer oxygen if breathing is difficult. Remove and isolate contaminated clothing and shoes. In case of contact with substance, immediately flush skin or eyes with running water for at least 20 minutes. For minor skin contact, avoid spreading material on unaffected skin. Keep victim warm and quiet. Effects of exposure (inhalation, ingestion, or skin contact) to substance may be delayed. Ensure that medical personnel are aware of the material(s) involved and take precautions to protect themselves.

VANADIUM V:005

SYNONYMS: EINECS No. 231-171-1; VENADIO (Spanish)

IDENTIFICATION

CAS:	7440-62-2; 1314-62-1 (dust, fume)
ERG ID:	UN 3285 as vanadium compound, n.o.s.
ERG Guide:	151
ERG Guide Category:	SUBSTANCES-TOXIC (NONCOMBUSTIBLE)
Molecular Formula:	V
RTECS No.:	YW1355000

1114 Vanadium

Properties: Lustrous powder, or fused, hard, lumpy solid. Bright white to light gray. Pure vanadium is a bright metal. Dust can be yellow-orange powder or dark gray flakes. Slightly radioactive. Practically insoluble in water.

HEALTH AND SAFETY INFORMATION

Dust and fume irritate eyes, skin, and respiratory tract. Can cause tongue to turn greenish black; metallic taste; cough, bronchial and lung problems. Can cause eczema. Experimental teratogen.

IDLH: 500 mg/m^3 as ferrovanadium dust (NIOSH reccomendation).
OSHA Table Z-1-A Air Contaminant as V_sO_5 fume and dust.

Exposure Limits:	**TWA**	**CEILING**	**STEL**
ACGIH:	3 mg/m^3 resp. dust		
NIOSH:	1 mg/m^3 as ferrovanadium dust	0.05 mg/m^3 as fume	3 mg/m^3 as ferrovanadium dust
OSHA:	0.5 mg/m^3 as V_sO_5 resp. dust	0.1 mg/m^3 V_sO_5 as fume	

RECOMMENDATIONS FOR RESPIRATOR SELECTION: NIOSH (as V)

0.5 mg/m^3: HiE *(any air-purifying, respirator with a high-efficiency particulate filter); or SA *(any supplied-air respirator).
1.25 mg/m^3: SA:CF (any supplied-air respirator operated in a continuous-flow mode); or PAPRHiE * (any powered, air-purifying respirator with a high-efficiency particulate filter).
2.5 mg/m^3: HiEF (any air-purifying, full-facepiece respirator with a high-efficiency particulate filter); or PAPRTHiE *(any powered, air-purifying respirator with a tight-fitting facepiece and a high-efficiency particulate filter); or SCBAF (any self-contained breathing apparatus with a full facepiece); or SAF (any supplied-air respirator with a full facepiece).
35 mg/m^3: SAF:PD,PP (any supplied-air respirator that has a full facepiece and is operated in a pressure-demand or other positive-pressure mode).

Emergency or Planned Entry into Unknown Concentrations or IDLH Conditions:
SCBAF:PD,PP (any self-contained breathing apparatus that has a full facepiece and is operated in a pressure-demand or other positive-pressure mode); or SAF:PD,PP:ASCBA (any supplied-air respirator that has a full facepiece and is operated in a pressure-demand or other positive-pressure mode in combination with an auxiliary, self-contained breathing apparatus operated in a pressure-demand or other positive-pressure mode).

Escape: HiEF (any air-purifying, full-facepiece respirator with a high-efficiency particulate filter); or SCBAE (any appropriate escape-type, self-contained breathing apparatus).

•Substance reported to cause eye irritation or damage; may require eye protection.

CHEMICAL REACTIVITY

Flammable in dust form from heat, flames, or sparks. Reacts violently with barium trifluoride, chlorine, lilthium, nitryl fluoride, and oxidizers. Dust is incompatible with peroxyformic acid (methaneperoxic acid), sulfur, chlorine trifluoride, calcium, water; reactions may be violent with danger of fire and explosions.

FIRE INFORMATION

Combustion products include toxic vanadium oxides. Noncombustible solid, but dust may increase intensity of fire when in contact with combustible materials. If material or contaminated runoff enters waterways, notify downstream users of potentially contaminated waters. Notify local health and fire officials and pollution control agencies. Use dry chemicals, alcohol-resistant foam, carbon dioxide. Use water spray to cool exposed containers.

NFPA Hazard Classifications:

Health Hazards (Blue) 3
Flammability (Red) 2
Reactivity (Yellow) 0

Flash Point: Noncombustible

FIRST AID

Move victim to fresh air. Call emergency medical care. Apply artificial respiration if victim is not breathing. Do not use mouth-to-mouth method if victim ingested or inhaled the substance; induce artificial respiration with the aid of a pocket mask equipped with a one-way valve or other proper respiratory medical device. Administer oxygen if breathing is difficult. Remove and isolate contaminated clothing and shoes. In case of contact with substance, immediately flush skin or eyes with running water for at least 20 minutes. For minor skin contact, avoid spreading material on unaffected skin. Keep victim warm and quiet. Effects of exposure (inhalation, ingestion, or skin contact) may be delayed. Ensure that medical personnel are aware of the material(s) involved and take precautions to protect themselves.

ENVIRONMENTAL INFORMATION as vanadium fume or dust.

Clean Water Act: Section 313 Priority Chemicals
RCRA Land Ban Waste
RCRA Universal Treatment Standards: Wastewater (mg/L), 4.3; Nonwastewater (mg/kg), 0.23.

RCRA Ground Water Monitoring List: Suggested test methods(s) (PQL μg/L); (total) 6010(80); 7910(2,000); 7911(40)
Safe Drinking Water Act: Priority List (55 FR 1470) as vanadium.
EPCRA Section 313: Form R *de minimis* concentration reporting level: 1.0%.

VANADIUM OXYTRICHLORIDE — V:007

SYNONYMS: EINECS No. 231-780-2; OSSOTRICLORURO DI VANADIO (Italian); OXITROCLORURO de VANADIO (Spanish); TRICHLOROOXOVANADIUM; VANADIUM OXIDE TRICHLORIDE; VANADIUM OXYCHLORIDE; VANADIUM TRICHLORIDE OXIDE; VANADIUM, TRICHLOROOXO-; VANADYL TRICHLORIDE; VANADYL CHLORIDE

IDENTIFICATION

CAS:	7727-18-6
ERG ID:	UN 2443
ERG Guide:	137
ERG Guide Category:	WATER REACTIVE-CORROSIVE
Molecular Formula:	Cl_3VO
RTECS No.:	YW2975000
Properties:	Noncombustible liquid. Lemon yellow. Sharp unpleasant odor. Sinks and mixes violently with water.

HEALTH AND SAFETY INFORMATION

Extremely irritating to eyes, skin, and respiratory tract. Inhalation can cause sensitization, asthma, and nerve damage.

IDLH: 35 mg/m^3 as vanadium
OSHA Table Z-1-A Air Contaminant as vanadium fume

Exposure Limits:	**TWA**	**CEILING**
ACGIH:	0.05 mg/m^3	
NIOSH:		0.05 mg/m^3 [15 min] as vanadium fume
OSHA:		0.1 mg/m^3 as V_2O_5

RECOMMENDATIONS FOR RESPIRATOR SELECTION: NIOSH/OSHA as vanadium

0.5 mg/m^3: HiE (any air-purifying, respirator with a high-efficiency particulate filter); or SA (any supplied-air respirator).

1.25 mg/m³: SA:CF (any supplied-air respirator operated in a continuous-flow mode); or PAPRHiE (any powered, air-purifying respirator with a high-efficiency particulate filter)
2.5 mg/m³: HiEF (any air-purifying, full-facepiece respirator with a high-efficiency particulate filter); or PAPRHiE (any powered, air-purifying respirator with a high-efficiency particulate filter); or SCBAF (any self-contained breathing apparatus with a full facepiece); or SAF (any supplied-air respirator with a full facepiece).

Emergency or planned entry into unknown concentration or IDLH conditions: SCBAF:PD,PP (any self-contained breathing apparatus that has a full facepiece and is operated in a pressure-demand or other positive-pressure mode); or SAF:PD,PP:ASCBA (any supplied-air respirator that has a full facepiece and is operated in a pressure-demand or other positive-pressure mode in combination with an auxiliary, self-contained breathing apparatus operated in a pressure-demand or other positive pressure mode).

Escape: HiEF (any air-purifying, full-facepiece respirator with a high-efficiency particulate filter); or SCBAE (any appropriate escape-type, self-contained breathing apparatus).
Note: Substance reported to cause eye irritation or damage; may require eye protection.

CHEMICAL REACTIVITY

Incompatible with water; forms hydrochloric acid solution. Air contact causes formation of corrosive hydrogen chloride fumes. In presence of moisture corrodes most metals. Reacts violently with bases, lithium, chlorine trifluoride, sodium, and potassium.

FIRE INFORMATION

Noncombustible. Combustion products include toxic vanadium oxides and chlorine. If material or contaminated runoff enters waterways, notify downstream users of potentially contaminated waters. Notify local health and fire officials and pollution control agencies. Use dry chemicals, alcohol-resistant foam, carbon dioxide. From a secure, explosion-proof location, use water spray to cool exposed containers.

NFPA Hazard Classifications: Not listed.

FIRST AID

Move victim to fresh air. Call emergency medical care. Apply artificial respiration if victim is not breathing. Do not use mouth-to-mouth method if victim ingested or inhaled the substance; induce artificial respiration with the aid of a pocket mask equipped with a one-way valve or other proper respiratory medical device. Administer oxygen if breathing is difficult. Remove and isolate contaminated clothing and shoes. In case of contact with substance, immediately flush skin or eyes with running water for at least 20 minutes. For minor skin contact, avoid spreading material on unaffected skin. Removal of solidified molten material from skin requires medical assistance. Keep victim warm and quiet. Effects of exposure (inhalation, ingestion, or skin contact) to substance may be delayed. Ensure that medical personnel are aware of the material(s) involved and take precautions to protect themselves.

ENVIRONMENTAL INFORMATION as vanadium

CLEAN WATER ACT: Section 313 Priority Chemicals
RCRA Land Ban Waste.
RCRA Universal Treatment Standards: Wastewater (mg/L), 4.3; Nonwastewater (mg/L), 0.23
RCRA Ground Water Monitoring List. Suggested test method(s) (PQL ug/L): (total) 6010(80); 7910(2,000); 7911(40)
SAFE DRINKING WATER ACT: Priority List (55 FR 1470) as vanadium
EPCRA Section 313 Form R *de minimus* concentration reporting level: 1.0%.

VINYL ACETATE V:010

SYNONYMS: ACETATE de VINYLE (French); ACETATO de VINILO (Spanish); ACETIC ACID, ETHENYL ESTER; ACETIC ACID, VINYL ESTER; ACETO de VINILO (Spanish); 1-ACETOXYETHYLENE; EEC No. 607-023-00-0; EINECS No. 203-545-4; ETHENYL ACETATE; ETHENYL ETHANOATE; ETHENYLETHANOATE; ETHONIC ACID, ETHENYL ESTER; ETHENYL ACETATE; ETHENYL ETHANOATE; EVERFLEX 81L; OCTAN WINYLU (Polish); PLYAMUL 40305-00; STCC 4907270; UNOCAL 76 RES 6206; UNOCAL 76 RES S-55; VAC; VAM; VINILE (ACETATO di) (Italian); VINNAPAS A 50; VINYLACETAAT (Dutch); VINYLACETAT (German); VINYL ACETATE H.Q.; VINYL ACETATE MONOMER; VINYL A MONOMER; VINYLE (ACETATE de) (French); VINYL ETHANOATE; VYAC; ZESET T

IDENTIFICATION

CAS:	108-05-4
ERG ID:	UN 1301
ERG Guide:	129P
ERG Guide Category:	FLAMMABLE LIQUIDS–(POLAR/WATER-MISCIBLE/ NOXIOUS)
Molecular Formula:	$C_4H_6O_2$
RTECS No.:	AK0875000
Properties:	Extremely flammable liquid. Clear, colorless. Pleasant, fruity odor, becoming sharp and irritating. Soluble in water to 2%; the rest floats.

HEALTH AND SAFETY INFORMATION

Irritates eyes, skin, and respiratory system. Removes the skin's natural oils; causes inflammation. Toxic if swallowed. High or prolonged exposure can cause skin irritation; blisters, burns and dermatitis; eye irritation; lung damage; central nervous system changes; heart and liver problems. Human mutation data reported.

IARC Evaluation: Group 3, unclassifiable substance; no human data; inadequate animal evidence.

Odor Threshold: 0.603 ppm . The ability to recognize the characteristic odor may be lost after about 2 hours of exposure.

OSHA Table Z-1-A Air Contaminant

NIOSH Criteria Document: NIOSH 78-205

Exposure Limits:	**TWA**	**CEILING**
ACGIH: [(1)]	10 ppm; 35 mg/m^3	15 ppm; 53 mg/m^3
NIOSH:		4 ppm; 15 mg/m^3 [15-min]
OSHA:	None	

[(1)] Animal carcinogen

RECOMMENDATIONS FOR RESPIRATOR SELECTION: NIOSH

40 ppm: CCRFOV [any chemical cartridge respirator with a full facepiece and organic vapor cartridge(s)]; or SA (any supplied-air respirator).
100 ppm: SA:CF (any supplied-air respirator operated in a continuous-flow mode); or PAPROV [any powered, air-purifying respirator with organic vapor cartridge(s)].
200 ppm: CCRFOV [any chemical cartridge respirator with a full facepiece and organic vapor cartridge(s)]; or GMFOV [any air-purifying, full-facepiece respirator (gas mask) with a chin-style, front- or back-mounted acid gas canister]; or PAPRTOV [any powered, air-purifying respirator with a tight-fitting facepiece and organic vapor cartridge(s)]; or SCBAF (any self-contained breathing apparatus with a full facepiece); or SAF (any supplied-air respirator with a full facepiece).
400 ppm: SA:PD,PP (any supplied-air respirator operated in a pressure-demand or other positive-pressure mode).

Emergency or Planned Entry into Unknown Concentrations or IDLH Conditions:
SCBAF:PD,PP (any self-contained breathing apparatus that has a full facepiece and is operated in a pressure-demand or other positive-pressure mode); or SAF:PD,PP:ASCBA (any supplied-air respirator that has a full facepiece and is operated in a pressure-demand or other positive-pressure mode in combination with an auxiliary, self-contained breathing apparatus operated in a pressure-demand or other positive-pressure mode).

Escape: GMFOV [any air-purifying, full-facepiece respirator (gas mask) with a chin-style, front- or back-mounted organic vapor canister]; or SCBAE (any appropriate escape-type, self-contained breathing apparatus).

Note: Substance reported to cause eye irritation or damage; may require eye protection.

CHEMICAL REACTIVITY

Forms explosive mixture with air. Polymerizes readily if not inhibited [usually hydroquinone (less than 60 days) or diphenylamine (60 days or more)]; heat can initiate reaction. Reacts violently with oxidizers, strong acids, bases, and peroxides. Incompatible with acids, ammonia, aliphatic amines, and alkanolamines. Also reacts with 2-aminoethanol, chlorosulfonic acid, ethylene diamine, ethyeneimine, ozone, and oleum. May accumulate static electrical charges, and may cause ignition of its vapors.

FIRE INFORMATION

Vapors are heavier than air and will collect in low areas. Vapors in confined areas may explode when exposed to fire. Vapors may travel long distances to ignition sources and flash back. Storage containers and parts of containers may rocket great distances, in many directions. If material or contaminated runoff enters waterways, notify downstream users of potentially contaminated waters. Notify local health and fire officials and pollution control agencies. Use dry chemicals, foam, carbon dioxide. From a secure, explosion-proof location, use water spray to cool exposed containers. If cooling streams are ineffective (venting sound increases in volume and pitch, tank discolors, or shows any signs of deforming), withdraw immediately to a secure position.

NFPA Hazard Classifications:

Health Hazards (Blue) 2
Flammability (Red) 3
Reactivity (Yellow) 2

Flash Point: 23°F (-5°C) o.c.; 18°F (-8°C) c.c.
Flammable Limits in Air: LEL 2.6%; UEL 13.4%
Autoignition Temperature: 756°F (402°C)

FIRST AID

Move victim to fresh air. Call emergency medical care. Apply artificial respiration if victim is not breathing. Administer oxygen if breathing is difficult. Remove and isolate contaminated clothing and shoes. In case of contact with substance, immediately flush skin or eyes with running water for at least 20 minutes. Wash skin with soap and water. Keep victim warm and quiet. Effects of exposure (inhalation, ingestion, or skin contact) to substance may be delayed. Ensure that medical personnel are aware of the material(s) involved and take precautions to protect themselves.

ENVIRONMENTAL INFORMATION

Clean Air Act: Hazardous Air Pollutants (Title I, Part A, Section 112); Accidental Release Prevention/Flammable substances, (Section 112[r], Table 3), TQ = 15,000 lb (6,825 kg).
Clean Water Act: Section 311 Hazardous Substances/RQ (same as CERCLA); Priority Pollutants (40CFR PART 423); Section 313 Priority Chemicals.
RCRA Ground Water Monitoring List: Suggested test method(s) (PQL ug/L): 8240(5).
CERCLA Section 302, Extremely Hazardous Substances: TPQ = 1,000 lb (455 kg).

EPCRA Section 304: Reportable Quantity (RQ): CERCLA, 5,000 lb (2,270 kg).
EPCRA Section 313: Form R *de minimis* concentration reporting level: 1.0%.
TSCA: 40CFR716.120(*a*)

VINYL BROMIDE V:012

SYNONYMS: BROMOETHENE; BROMOETHYLENE; BROMURE de VINYLE (French); BROMURO de VINILO (Spanish); EINECS No. 209-800-6; ETHENE, BROMO-; ETHYLENE, BROMO-; NCI-C50373; VINILE (BROMURO di) (Italian); VINYLBROMID (German); VINYL BROMIDE, inhibited; VINYLE (BROMURE de) (French)

IDENTIFICATION

CAS:	593-60-2
ERG ID:	UN 1085
ERG Guide:	116P
ERG Guide Category:	GASES-FLAMMABLE (UNSTABLE)
Molecular Formula:	C_2H_3Br
RTECS No.:	KU8400000
Properties:	Extremely flammable gas or liquid below 60°F (16°C). Colorless. Pleasant odor. Liquid sinks in water; practically insoluble.

HEALTH AND SAFETY INFORMATION

Irritates eyes, skin, and respiratory tract. Affects central nervous system; anesthetic; risk of unconsciousness. May cause liver and kidney problems. Contact with liquid may cause frostbite. May be a human carcinogen.

IARC Evaluation: Group 2A; sufficient animal evidence; probable carcinogen in humans.

Exposure Limits:	**TWA**
ACGIH: (1)	5 ppm; 22 mg/m^3
NIOSH: (2)	
OSHA:	None

(1) Suspected human carcinogen
(2) Potential human carcinogen. Reduce exposure to lowest feasible levels.

RECOMMENDATIONS FOR RESPIRATOR SELECTION: NIOSH

At any concentrations above the NIOSH REL, or where there is no REL, at any detectable concentration: SCBAF:PD,PP (any self-contained breathing apparatus that has a full facepiece and is operated in a pressure-demand or other positive-pressure mode); or SAF:PD,PP:ASCBA (any supplied-air

respirator that has a full facepiece and is operated in a pressure-demand or other positive-pressure mode in combination with an auxiliary, self-contained breathing apparatus operated in a pressure-demand or other positive-pressure mode).

Escape: GMFOV [any air-purifying, full-facepiece respirator (gas mask) with a chin-style, front- or back-mounted organic vapor canister]; or SCBAE (any appropriate escape-type, self-contained breathing apparatus).

CHEMICAL REACTIVITY

Forms explosive mixture with air. A strong reducing agent; violent reaction with acetylene, strong oxidizers, chlorine, fluorine. May polymerize in sunlight. Shipped as a liquefied compressed gas with 0.1% phenol added to prevent polymerization. May react with itself without warning, blocking relief valves leading to container explosion.

FIRE INFORMATION

Combustion products include toxic fumes of bromine, flammable hydrogen, and hydrogen bromide. Gases are heavier than air and will collect in low areas. Gases in confined areas may explode when exposed to fire. Gases may travel long distances to ignition sources and flash back. Storage containers and parts of containers may rocket great distances, in many directions. If material or contaminated runoff enters waterways, notify downstream users of potentially contaminated waters. Notify local health and fire officials and pollution control agencies. Do not extinguish the fire unless the flow of the gas can be stopped and any remaining gas is out of the line. Specially trained personnel may use fog lines to cool exposures and let the fire burn itself out. Use dry chemicals, foam, carbon dioxide. From a secure, explosion-proof location, use large quantities of water to cool exposed containers. If cooling streams are ineffective (venting sound increases in volume and pitch, tank discolors, or shows any signs of deforming), withdraw immediately to a secure position.

NFPA Hazard Classifications: (rerated for the year 2000)

Health Hazards (Blue) 2
Flammability (Red) 4
Reactivity (Yellow) 1

Flash Point: None (flammable gas)
Flammable Limits in Air: LEL 9%; UEL 15%
Autoignition Temperature: 986°F (530°C)

FIRST AID

Move victim to fresh air. Call emergency medical care. Apply artificial respiration if victim is not breathing. If breathing is difficult, give oxygen. Remove and isolate contaminated clothing and shoes. In case of contact with liquefied gas, thaw frosted parts with lukewarm water. Keep victim warm and quiet. Ensure that medical personnel are aware of the material(s) involved and take precautions to protect themselves.

ENVIRONMENTAL INFORMATION

Clean Air Act: Hazardous Air Pollutants (Title I, Part A, Section 112)
EPCRA Section 304 Reportable Quantity (RQ): CERCLA, 100 lb (45/4 kg)
EPCRA Section 313: Form R *de minimis* concentration reporting level: 0.1%.
California's Proposition 65: Carcinogen.

VINYL BUTYL ETHER V:013

SYNONYMS: BUTYL VINYL ETHER; BUTYL VINYL ETHER (inhibited); BVE; EINECS No. 203-860-7; ETHER, BUTYL VINYL; ETER BUTILO VINILICO (Spanish); 1-(ETHENYLOXY) BUTANE; OXYDE de BUTYLE et de VINYLE (French); VINYL-n-BUTYL ETHER

IDENTIFICATION

CAS:	111-34-2
ERG ID:	UN 2352
ERG Guide:	127P
ERG Guide Category:	FLAMMABLE LIQUIDS (POLAR/WATER- MISCIBLE)
Molecular Formula:	$C_6H_{12}O$
RTECS No.:	KN5950000
Properties:	Extremely flammable liquid. Colorless. Ether-like odor. Slightly soluble in water.

HEALTH AND SAFETY INFORMATION

Irritates eyes, skin, and respiratory tract. Skin contact contributes significantly to overall exposure.

RECOMMENDATIONS FOR RESPIRATOR SELECTION: NIOSH

At any concentrations above the NIOSH REL, or where there is no REL, at any detectable concentration: SCBAF:PD,PP (any self-contained breathing apparatus that has a full facepiece and is operated in a pressure-demand or other positive-pressure mode); or SAF:PD,PP:ASCBA (any supplied-air respirator that has a full facepiece and is operated in a pressure-demand or other positive-pressure mode in combination with an auxiliary, self-contained breathing apparatus operated in a pressure-demand or other positive-pressure mode).

Escape: GMFOV [any air-purifying, full-facepiece respirator (gas mask) with a chin-style, front- or back-mounted organic vapor canister]; or SCBAE (any appropriate escape-type, self-contained breathing apparatus).

CHEMICAL REACTIVITY

Moderately explosive by spontaneous chemical reaction. Contact with oxidizers and strong acids may cause fire and explosions. Able to form unstable peroxides, which can cause polymerization.

FIRE INFORMATION

Combustion may produce irritants and toxic vapors. Vapors are heavier than air and will collect in low areas. Vapors in confined areas may explode when exposed to fire. If material or contaminated runoff enters waterways, notify downstream users of potentially contaminated waters. Notify local health and fire officials and pollution control agencies. Use dry chemicals, alcohol-resistant foam, carbon dioxide. From a secure, explosion-proof location, use water spray to cool exposed containers.

NFPA Hazard Classifications:

Health Hazards (Blue) 2
Flammability (Red) 3
Reactivity (Yellow) 2

Flash Point: 15°F (−9°C) o.c.
Autoignition Temperature: 437°F (255°C)

FIRST AID

Move victim to fresh air. Call emergency medical care. Apply artificial respiration if victim is not breathing. Administer oxygen if breathing is difficult. Remove and isolate contaminated clothing and shoes. In case of contact with substance, immediately flush skin or eyes with running water for at least 20 minutes. Wash skin with soap and water. Keep victim warm and quiet. Ensure that medical personnel are aware of the material(s) involved and take precautions to protect themselves.

VINYL BUTYRATE, INHIBITED V:014

SYNONYMS: BUTANOIC ACID, ETHENYL ESTER; BUTIRATO de VINILO (Spanish); BUTYRATE de VINYLE (French); BUTYRIC ACID, VINYL ESTER; EINECS No. 204-609-4; VINYL BUTANOATE; VINYLESTER KYSELINY MASELNE (Polish)

IDENTIFICATION

CAS:	123-20-6
ERG ID:	UN 2838
ERG Guide:	129P
ERG Guide Category:	FLAMMABLE LIQUIDS (POLAR/WATER-MISCIBLE/NOXIOUS)
Molecular Formula:	$C_6H_{10}O_2$
RTECS No.:	ET7000000
Properties:	Flammable liquid. Slightly soluble in water.

HEALTH AND SAFETY INFORMATION

Irritates eyes, skin, and respiratory tract.

RECOMMENDATIONS FOR RESPIRATOR SELECTION

At any concentrations above the NIOSH REL, or where there is no REL, at any detectable concentration: SCBAF:PD,PP (any self-contained breathing apparatus that has a full facepiece and is operated in a pressure-demand or other positive-pressure mode); or SAF:PD,PP:ASCBA (any supplied-air respirator that has a full facepiece and is operated in a pressure-demand or other positive-pressure mode in combination with an auxiliary, self-contained breathing apparatus operated in a pressure-demand or other positive-pressure mode).

Escape: GMFOV [any air-purifying, full-facepiece respirator (gas mask) with a chin-style, front- or back-mounted organic vapor canister]; or SCBAE (any appropriate escape-type, self-contained breathing apparatus).

CHEMICAL REACTIVITY

Contact with oxidizers may cause fire and explosions. Esters are incompatible with nitrates, nitric acid, sulfuric acid.

FIRE INFORMATION

Vapors are heavier than air and will collect in low areas. Vapors in confined areas may explode when exposed to fire. If material or contaminated runoff enters waterways, notify downstream users of potentially contaminated waters. Notify local health and fire officials and pollution control agencies. Use dry chemicals, alcohol-resistant foam, carbon dioxide. From a secure, explosion-proof location, use water spray to cool exposed containers. If cooling streams are ineffective (venting sound increases in volume and pitch, tank discolors, or shows any signs of deforming), withdraw immediately to a secure position.

NFPA Hazard Classifications:

Health Hazards (Blue) 2
Flammability (Red) 3
Reactivity (Yellow) 2

Flash Point: 68°F (20°C) o.c.
Flammable Limits in Air: LEL 1.4%; UEL 8.8%

FIRST AID

Move victim to fresh air. Call emergency medical care. Apply artificial respiration if victim is not breathing. Administer oxygen if breathing is difficult. Remove and isolate contaminated clothing and shoes. In case of contact with substance, immediately flush skin or eyes with running water for at least 20 minutes. Wash skin with soap and water. Keep victim warm and quiet. Effects of exposure (inhalation, ingestion, or skin contact) to substance may be delayed. Ensure that medical personnel are aware of the material(s) involved and take precautions to protect themselves.

VINYL CHLORIDE V:015

SYNONYMS: CHLOORETHYLEEN (Dutch); CHLORETHYLEN (French); CHLOROETHENE; CHLOROETHYLENE; CHLORURE de VINYLE (French); CHLORO di VINYLE (Italian); CLORURO de VINILO (Spanish); EEC No. 602-023-00-7; EINECS No. 200-831-0; ETHENE, CHLORO-; ETHYLENE, CHLORO-; ETHYLENE MONOCHLORIDE; MONOCHLOROETHENE; MONOCHLOROETHYLENE; STCC 4905792; TROVIDUER; TROVIDUR; VC; VCL; VCM; VINILE (CLORURO di) (Italian); VINYLCHLORID (German); VINYL CHLORIDE MONOMER; VINYL C MONOMER; CHLORURE de VINYLE (French); WINYLU CHLORED (Polish)

IDENTIFICATION

CAS:	75-01-4
ERG ID:	UN 1086
ERG Guide:	116P
ERG Guide Category:	GASES-FLAMMABLE (UNSTABLE)
Molecular Formula:	C_2H_3Cl
RTECS No.:	KU9625000
Properties:	Extremely flammable, colorless gas; becomes liquid below 7°F; (−14°C). Shipped as a liquefied compressed gas. Faint, sweet, pleasant ether-like odor in high concentrations. Liquid floats and boils on water; slightly soluble; gas is insoluble in water.

HEALTH AND SAFETY INFORMATION

Irritating to eyes, skin and lungs. Contact with liquid can cause frostbite. Contact with vapor can cause irritation and rash. Absorption is possible through the skin. Long-term exposure may cause clublike swelling and shortening of finger tips. Skin may become thickened and stiff with coarse, whitish patches. Bone and joints of arms and legs may suffer damage. Liver and spleen damage occur. Not all symptoms disappear after exposure stops. Known to cause cancer in humans, following long term exposure; contact should be avoided.

IARC Evaluation: Group 1, IARC carcinogen; sufficient human evidence; sufficient animal evidence.
Odor Threshold: 0.253 ppm
OSHA Table Z-1-A Air Contaminant
OSHA Specifically Regulated Substances (29 CFR 1910.1001-1048).

Exposure Limits:	**TWA**	**CEILING**
ACGIH: (1)	5 ppm; 13 mg/m^3	
NIOSH: (2)		
OSHA:	1 ppm; 2.56 mg/m^3	5 ppm [15 min]

(1) Confirmed human carcinogen.
(2) Potential carcinogen. Limit exposure to lowest feasible concentration.

RECOMMENDATIONS FOR RESPIRATOR SELECTION: NIOSH

At any concentrations above the NIOSH REL, or where there is no REL, at any detectable concentration: SCBAF:PD,PP (any self-contained breathing apparatus that has a full facepiece and is operated in a pressure-demand or other positive-pressure mode); or SAF:PD,PP:ASCBA (any supplied-air respirator that has a full facepiece and is operated in a pressure-demand or other positive-pressure mode in combination with an auxiliary, self-contained breathing apparatus operated in a pressure-demand or other positive-pressure mode).

Escape: GMFS [any air-purifying, full-facepiece respirator (gas mask) with a chin-style, front- or back-mounted canister providing protection against the compound of concern]; or SCBAE (any appropriate escape-type, self-contained breathing apparatus).

CHEMICAL REACTIVITY

Forms explosive mixture with air. Atmospheric oxygen and various contaminants or strong oxidizers cause formation of unstable peroxides, which can initiate violent polymerization. Also able to polymerize when heated or exposed, long-term, to light. Reacts violently with strong oxidizers or oxides of nitrogen. Contact with copper or other acetylide-forming metals can form explosive compounds. Attacks iron and steel in the presence of moisture.

FIRE INFORMATION

Combustion products include toxic hydrogen chloride. Vapors are heavier than air and will collect in low areas. Vapors in confined areas may explode when exposed to fire. Vapors may travel long distances to ignition sources and flash back. Storage containers and parts of containers may rocket great distances, in many directions. If material or contaminated runoff enters waterways, notify downstream users of potentially contaminated waters. Notify local health and fire officials and pollution control agencies. Use dry chemicals, foam, carbon dioxide. From a secure, explosion-proof location, use water spray to cool exposed containers. If cooling streams are ineffective (venting sound increases in volume and pitch, tank discolors, or shows any signs of deforming), withdraw immediately to a secure position.

NFPA Hazard Classifications:

Health Hazards (Blue) 2
Flammability (Red) 4
Reactivity (Yellow) 2

Flash Point: Flammable gas (−108°F;−78°C) o.c.
Flammable Limits in Air: LEL 3.6%; UEL 33%
Autoignition Temperature: 882°F (472°C)

FIRST AID

Move victim to fresh air. Call emergency medical care. Apply artificial respiration if victim is not breathing. Administer oxygen if breathing is difficult. Remove and isolate contaminated clothing and shoes. In case of contact with liquefied gas, thaw frosted parts with lukewarm water. Keep victim warm and quiet. Ensure that medical personnel are aware of the material(s) involved and take precautions to protect themselves.

ENVIRONMENTAL INFORMATION

Clean Air Act: Hazardous Air Pollutants (Title I, Part A, Section 112); Accidental Release Prevention/Flammable substances, (Section 112[r], Table 3), TQ = 10,000 lb (4,540 kg).
Clean Water Act: Section 307 Toxic Pollutants; Section 313 Priority Chemicals.
EPA Hazardous Waste Number (RCRA No.): U043; D043.
RCRA Toxicity Characteristic (Section 261.24): Maximum Concentration of Contaminants, regulatory level, 0.2 mg/L.
RCRA Section 261 Hazardous Constituents
RCRA Land Ban Waste
RCRA Universal Treatment Standards: Wastewater (mg/L), 0.27; Nonwastewater (mg/kg), 6.0.
RCRA Ground Water Monitoring List: Suggested test method(s) (PQL ug/L): 8010(2); 8240(10).
Safe Drinking Water Act: MCL, 0.002 mg/L; MCLG, zero; Regulated Chemical (47 FR 9352).
EPCRA Section 304: Reportable Quantity (RQ): CERCLA, 1 lb (0.454 kg).

EPCRA Section 313: Form R *de minimis* concentration reporting level: 0.1%.
California's Proposition 65: Carcinogen.

VINYL–2–CHLOROETHYL ETHER V:017

SYNONYMS: 2-CHLOROETHYL VINYL ETHER; (2-CHLOROETHOXY)ETHENE; EINECS No. 203-799-6; ETHENE, (2-CHLOROETHOXY)-; ETHER, 2-CHLOROETHYL VINYL; ETER 2-CLOROETILO VINILICO (Spanish); OSSIDO di 2-CLOROETILE e VINILE (Italian); OXYDE de 2-CHLOROETHYLE et de VINYLE (French); VINYL-b-CHLOROETHYL ETHER

IDENTIFICATION

CAS:	110-75-8
ERG ID:	UN 3271 as ethers, n.o.s.
ERG Guide:	127
ERG Guide Category:	FLAMMABLE LIQUIDS (POLAR/WATER- MISCIBLE)
Molecular Formula:	C_4H_7ClO
RTECS No.:	KN6300000
Properties:	Flammable liquid. Colorless. Slightly soluble in water.

HEALTH AND SAFETY INFORMATION

Irritates eyes, skin, and respiratory tract; irritation can be severe. Skin contact contributes significantly to overall exposure.

RECOMMENDATIONS FOR RESPIRATOR SELECTION

At any concentrations above the NIOSH REL, or where there is no REL, at any detectable concentration: SCBAF:PD,PP (any self-contained breathing apparatus that has a full facepiece and is operated in a pressure-demand or other positive-pressure mode); or SAF:PD,PP:ASCBA (any supplied-air respirator that has a full facepiece and is operated in a pressure-demand or other positive-pressure mode in combination with an auxiliary, self-contained breathing apparatus operated in a pressure-demand or other positive-pressure mode).

Escape: GMFOV [any air-purifying, full-facepiece respirator (gas mask) with a chin-style, front- or back-mounted organic vapor canister]; or SCBAE (any appropriate escape-type, self-contained breathing apparatus).

CHEMICAL REACTIVITY

Contact with oxidizers and strong acids can cause fire and explosives.

FIRE INFORMATION

Combustion products in clude toxic nitrogen oxides. Forms explosive mixture with air. Vapors are heavier than air and will collect in low areas. Vapors in confined areas may explode when exposed to fire. If material or contaminated runoff enters waterways, notify downstream users of potentially contaminated waters. Notify local health and fire officials and pollution control agencies. Use dry chemicals, alcohol-resistant foam, carbon dioxide. From a secure, explosion-proof location, use water spray to cool exposed containers.

NFPA Hazard Classifications:

Health Hazards (Blue) 2
Flammability (Red) 3
Reactivity (Yellow) 2

Flash Point: 81°F (27°C) o.c.

FIRST AID

Move victim to fresh air. Call emergency medical care. Apply artificial respiration if victim is not breathing. Administer oxygen if breathing is difficult. Remove and isolate contaminated clothing and shoes. In case of contact with substance, immediately flush skin or eyes with running water for at least 20 minutes. Wash skin with soap and water. Keep victim warm and quiet. Ensure that medical personnel are aware of the material(s) involved and take precautions to protect themselves.

ENVIRONMENTAL INFORMATION

Clean Water Act: Section 307 Priority Pollutants as 2-chloroethyl vinyl ether (mixed)
EPA Hazardous Waste Number (RCRA No.): U042
RCRA Section 261 Hazardous Constituents
RCRA Land Ban Waste
RCRA Universal Treatment Standards: Wastewater (mg/L), 0.062; Nonwastewater (mg/kg), NA
EPCRA Section 304 Reportable Quantity (RQ): CERCLA, 1,000 lb (454 kg)

VINYLCYCLOHEXENE V:018

SYNONYMS: BUTADIENE DIMER; CYCLOHEXENYLETHYLENE; CYCLOHEXENE, 4-ETHENYL-; EINECS No. 202-848-9; NCI-C54999; 1,2,3,4-TETRAHYDROSTYRENE; VINYL-1-CYCLOHEXENE; 1-VINYLCYCLOHEXENE-3; 4-VINYL-1-CYCLOHEXENE; 4-VINYL CYCLOHEXENE

IDENTIFICATION

CAS:	100-40-3
ERG ID:	UN 1993
ERG Guide:	128
ERG Guide Category:	FLAMMABLE LIQUIDS (NONPOLAR/WATER-IMMISCIBLE)
Molecular Formula:	C_8H_{12}
RTECS No.:	GW6650000
Properties:	Flammable liquid. Colorless. Floats on water.

HEALTH AND SAFETY INFORMATION

Irritates eyes, skin, and respiratory tract. Prolonged or repeated exposure may cause liver, kidney, skin and metabolic damage.

IARC Evaluation: Group 3; animal suspected
NTP: animal carcinogen

Exposure Limits: **TWA**

ACGIH: [(1)] 0.1 ppm; 0.44 mg/m^3

[(1)] Suspected human carcinogen.

RECOMMENDATIONS FOR RESPIRATOR SELECTION: NIOSH as vinylcyclohexene dioxide.

At any detectable concentration: SCBAF:PD,PP (any self-contained breathing apparatus that has a full facepiece and is operated in a pressure-demand or other positive-pressure mode); or SAF:PD,PP:ASCBA (any supplied-air respirator that has a full facepiece and is operated in a pressure-demand or other positive-pressure mode in combination with an auxiliary, self-contained breathing apparatus operated in a pressure-demand or other positive pressure mode).

Escape: GMFOV [any air-purifying, full-facepiece respirator (gas mask) with a chin-style, front- or back-mounted organic vapor canister]; or SCBAE (any appropriate escape-type, self-contained breathing apparatus).

CHEMICAL REACTIVITY

Forms explosive mixture with air. Contact with oxidizers may cause fire and explosions. Hydrolyzes in water. Incompatible with alcohols, amines, and oxidizers. Narcotic in high concentration. Protect from contact with light, air and moisture in an explosion-proof refrigerator.

FIRE INFORMATION

Vapors are heavier than air and will collect in low areas. Vapors in confined areas may explode when exposed to fire. Vapors may travel long distances to ignition sources and flash back. Storage containers and parts of containers may rocket great distances, in many directions. If material or contaminated runoff enters waterways, notify downstream users of potentially contaminated waters. Notify local health and fire officials and pollution control agencies. Use dry chemicals, foam, carbon dioxide. From a secure, explosion-proof location, use water spray to cool exposed containers.

NFPA Hazard Classifications:

Health Hazards (Blue) 0
Flammability (Red) 3
Reactivity (Yellow) 2

Flash Point: 61°F (16°C) c.c.
Autoignition Temperature: 517°C (269°C)

FIRST AID

Move victim to fresh air. Call emergency medical care. Apply artificial respiration if victim is not breathing. If breathing is difficult, give oxygen. Remove and isolate contaminated clothing and shoes. In case of contact with substance, immediately flush skin or eyes with running water for at least 20 minutes. Wash skin with soap and water. Keep victim warm and quiet. Ensure that medical personnel are aware of the material(s) involved and take precautions to protect themselves.

ENVIRONMENTAL INFORMATION

TSCA: 40CFR716.120(*a*); 40CFR712.30(*d*)

VINYL CYCLOHEXENE DIOXIDE V:020

SYNONYMS: CHISSONOX 206; DIOXIDO de VINILCICLOHEXENO (Spanish); EP-206; 1,2-EPOXY-4-(EPOXYETHYL)CYCLOHEXANE; 1-EPOXYETHYL-3-4-EPOXYCYCLOHEXANE; 3-(EPOXYETHYL)-7-OXABICYCLO(4.1.0)HEPTANE; 3-(1,2-EPOXYETHYL)-7-OXABICYCLO-(4.1.0)HEPTANE; 4-(1,2-EPOXYETHYL)-7-OXABICYCLO(4.1.0)HEPTANE; 4-(EPOXYETHYL)-7-OXABICYCLO(4.1.0)HEPTANE; 1-ETHYLENEOXY-3,4-EPOXY- CYCLOHEXANE; HYSOL K-0109; NCI-C60139; 7-OXYBICYCLO(4.1.0)HEPTANE; 3-OXIRANYL-; ERLA-2270; ERLA-2271; 3-OXIRANYL-7-OXABICYCLO(4.1.0)HEPTANE; UCET TEXTILE FINISH 11-74 (OBS.); UNOX EPOXIDE 206; 4-VINYLCYCLOHEXENE DIEPOXIDE; 4-VINYL-1,2-CYCLOHEXENE DIEPOXIDE; 1-VINYL-3-CYCLOHEXENE DIOXIDE; 4-VINYLCYCLOHEXENE DIOXIDE; 4-VINYL-1-CYCLOHEXENE DIOXIDE

IDENTIFICATION

CAS: 106-87-6
Molecular Formula: $C_8H_{12}O_2$
RTECS No.: RN8640000
Properties: Combustible liquid. Colorless. Poisonous gas produced in fire.

HEALTH AND SAFETY INFORMATION

Irritates eyes, skin, and respiratory system. Skin sensitizer. Prolonged or repeated exposure may cause cancer, mutations; male reproductive system damage; skin allergy; change in the production of white blood cells; damage to the thymus.

IARC Evaluation: Group 3, unclassifiable; animal evidence, sufficient
NTP: Resonably anticipated to be a carcinogen (7th Annual Report)

Exposure Limits:	TWA
ACGIH:(1)(2)	10 ppm; 57 mg/m^3
NIOSH::(1)(2)	10 ppm; 60 mg/m^3
OSHA:	None

(1) Skin contact contributes significantly in overall exposure.
(2) Suspected or potential human carcinogen.

RECOMMENDATIONS FOR RESPIRATOR SELECTION: NIOSH

At Concentrations above the Niosh Rel, or Any Detectable Concentration: SCBAF:PD,PP (any self-contained breathing apparatus that has a full facepiece and is operated in a pressure-demand or other positive-pressure mode); or SAF:PD,PP:ASCBA (any supplied-air respirator that has a full facepiece and is operated in a pressure-demand or other positive-pressure mode in combination with an auxiliary, self-contained breathing apparatus operated in a pressure-demand or other positive pressure mode.

Escape: GMFOV [any air-purifying, full-facepiece respirator (gas mask) with a chin-style, front- or back-mounted organic vapor canister]; or SCBAE (any appropriate escape-type, self-contained breathing apparatus).

CHEMICAL REACTIVITY

Reacts with alcohols, amines, and water.

FIRE INFORMATION

If material or contaminated runoff enters waterways, notify downstream users of potentially contaminated waters. Notify local health and fire officials and pollution control agencies. Use dry chemicals, alcohol-resistant foam, carbon dioxide. From a secure, explosion-proof location, use water spray to cool exposed containers. If cooling streams are ineffective (venting sound increases in volume and pitch, tank discolors, or shows any signs of deforming), withdraw immediately to a secure position.

Flash Point: 230°F (110°C) o.c.

FIRST AID

EYE: If this chemical contacts the eyes, immediately wash the eyes with large amounts of water, occasionally lifting the lower and upper lids. Get medical attention immediately. Contact lenses should not be worn when working with this chemical.
SKIN: If this chemical contacts the skin, immediately wash the contaminated skin with soap and water. If this chemical penetrates the clothing, immediately remove clothing, wash the skin with soap and water, and get medical attention promptly.
BREATH: If a person breathes large amounts of this chemical, move exposed person to fresh air at once. If breathing has stopped, perform mouth-to-mouth resuscitation. Keep the affected person warm and at rest. Get medical attention as soon as possible.
SWALLOW: If this chemical has been swallowed, get medical attention immediately.

VINYL ETHER V:200

SYNONYMS: DIVINIL ETER (Spanish); DIVINYL ETHER; DIVINYL ETHER, inhibited; DIVYNYL OXIDE; EINECS No. 203-720-5; ETHENYLOXYETHENE; OSSIDO di DIVINILE (Italian); 1,1'-OXYBISETHENE; OXYDE de DIVINYLE (French); VINESTHENE; VINESTHESIN; VINETHEN; VINETHENE; VINETHER; VINIDYL; VINYDAN

IDENTIFICATION

CAS:	109-93-3
ERG ID:	UN 1167
ERG Guide:	131P
ERG Guide Category:	FLAMMABLE LIQUIDS-TOXIC
Molecular Formula:	C_4H_6O
RTECS No.:	YZ6700000
Properties:	Extremely flammable liquid. Colorless. Characteristic odor. Practically insoluble in water.

HEALTH AND SAFETY INFORMATION

Vapors act as an anesthetic; possible unconsciousness. Possible mutagen.

RECOMMENDATIONS FOR RESPIRATOR SELECTION

At any concentrations above the NIOSH REL, or where there is no REL, at any detectable concentration: SCBAF:PD,PP (any self-contained breathing apparatus that has a full facepiece and is operated in a pressure-demand or other positive-pressure mode); or SAF:PD,PP:ASCBA (any supplied-air respirator that has a full facepiece and is operated in a pressure-demand or other positive-pressure mode in combination with an auxiliary, self-contained breathing apparatus operated in a pressure-demand or other positive-pressure mode).

Escape: GMFOV [any air-purifying, full-facepiece respirator (gas mask) with a chin-style, front- or back-mounted organic vapor canister]; or SCBAE (any appropriate escape-type, self-contained breathing apparatus.

CHEMICAL REACTIVITY

Forms explosive mixture with air. Reacts vigorously with oxidizing agents. Highly explosive when exposed to heat or flame. Forms peroxides when exposed to air or oxygen. Contact with air or oxygen forms unstable peroxides which can cause polymerization with formation of highly flammable acetylene gas. Uninhibited monomer may form polymer in vents and other confined spaces. May react with itself with explosive violence, blocking relief valves leading to tank explosions.

FIRE INFORMATION

Vapors are heavier than air and will collect in low areas. Vapors in confined areas may explode when exposed to fire. If material or contaminated runoff enters waterways, notify downstream users of potentially contaminated waters. Notify local health and fire officials and pollution control agencies. Use dry chemicals, alcohol-resistant foam, carbon dioxide. From a secure, explosion-proof location, use water spray to cool exposed containers. If cooling streams are ineffective (venting sound increases in volume and pitch, tank discolors, or shows any signs of deforming), withdraw immediately to a secure position.

NFPA Hazard Classifications:

Health Hazards (Blue) 2
Flammability (Red) 4
Reactivity (Yellow) 2

Flash Point: < –22°F (<–30°C) c.c.
Flammable Limits in Air: LEL 1.7%; UEL 27%
Autoignition Temperature: 680°F (360°C)

FIRST AID

Move victim to fresh air. Call emergency medical care. Apply artificial respiration if victim is not breathing.

Do not use mouth-to-mouth method if victim ingested or inhaled the substance; induce artificial respiration with the aid of a pocket mask equipped with a one-way valve or other proper respiratory medical device. Administer oxygen if breathing is difficult. Remove and isolate contaminated clothing and shoes. In case of contact with substance, immediately flush skin or eyes with running water for at least 20 minutes. Wash skin with soap and water. Keep victim warm and quiet. Effects of exposure (inhalation, ingestion, or skin contact) to substance may be delayed. Ensure that medical personnel are aware of the material(s) involved and take precautions to protect themselves.

VINYL ETHYL ETHER V:250

SYNONYMS: AGRISYNTH EVE; EINECS No. 302-718-4; ETHENE, ETHOXY; ETHER, VINYL ETHYL; ETOXYETHENE; ETHYL VINYL ETHER; EVE; ETHYL VINYL ETHER, INHIBITED; OXYDE D'ETHYLE ET de VINYLE (French); OSSIDO DI ETILE E VINILE (Italian); VINAMAR; VINIL ETIL ETER (Spanish)

IDENTIFICATION

CAS:	109-92-2
ERG ID:	UN 1302
ERG Guide:	127P
ERG Guide Category:	FLAMMABLE LIQUIDS (POLAR/WATER- MISCIBLE)
Molecular Formula:	C_4H_8O
RTECS No.:	KO0710000
Properties:	Extremely flammable, volatile liquid. Boils at 96°F (36°C). Colorless. Characteristic, etherlike odor. Floats on water; insoluble

HEALTH AND SAFETY INFORMATION

Irritates eyes, skin, and respiratory tract. Toxic if swallowed. May affect central nervous system, liver, and kidneys.

CHEMICAL REACTIVITY

Forms explosive mixture with air. Contact with oxidizers can cause fire and explosions. Incompatible with strong acids (nonoxidizing, sulfuric, nitric), ammonia, aliphatic amines, and alkanolamines. Extremely reactive. Able to form unstable peroxides and polymerize; may react with itself, blocking relief valves and leading to tank explosions.

FIRE INFORMATION

Vapors are heavier than air and will collect in low areas. Vapors in confined areas may explode when exposed to fire. Vapors may travel long distances to ignition sources and flash back. Storage containers and parts of containers may rocket great distances, in many directions. If material or contaminated runoff enters waterways, notify downstream users of potentially contaminated waters. Notify local health and fire officials and pollution control agencies. Specially trained personnel operating form a safe distance can fight fire using dry chemicals, foam. From a secure, explosion-proof location, use water spray to cool exposed containers; do not direct straight streams into the liquid. If cooling streams are ineffective (venting sound increases in volume and pitch, tank discolors, or shows any signs of deforming), withdraw immediately to a secure position.

NFPA Hazard Classifications:

Health Hazards (Blue) 2
Flammability (Red) 4
Reactivity (Yellow) 2

Flash Point: < –50°F (–46°C) c.c.
Flammable Limits in Air: LEL 1.7%; UEL 28%
Autoignition Temperature: 395°F (202°C)

FIRST AID

Move victim to fresh air. Call emergency medical care. Apply artificial respiration if victim is not breathing. Administer oxygen if breathing is difficult. Remove and isolate contaminated clothing and shoes. In case of contact with substance, immediately flush skin or eyes with running water for at least 20 minutes. Wash skin with soap and water. Keep victim warm and quiet. Ensure that medical personnel are aware of the material(s) involved and take precautions to protect themselves.

ENVIRONMENTAL INFORMATION

Clean Air Act: Accidental Release Prevention/Flammable Substances (Section 112[r], Table 3), TQ = 10,000 lb (4,540 kg)

VINYL–2–ETHYLHEXOATE V:300

SYNONYMS: EINECS No. 202-297-4; 2-ETHYLHEXANOIC ACID, VINYL ESTER; 2-ETHYLHEXOIC ACID, VINYL ESTER; 2-ETILESANOATO di VINILE (Italian); 2-ETILHEXANOATO de VINILO (Spanish); HEXANOIC ACID, 2-ETHYL-, ETHENYL ESTER; HEXANOIC ACID, 2-ETHYL-, VINYL ESTER; VINYL 2-ETHYLHEXANOATE; VINYL-2-ETHYLHEXOATE; VYNATE 2EH

IDENTIFICATION

CAS: 94-04-2

1138 Vinyl-2-Ethylhexoate

Molecular Formula: $C_{10}H_{18}O_2$
RTECS No.: MO7875000
Properties: Flammable liquid. Insoluble in water.

HEALTH AND SAFETY INFORMATION

Irritates eyes, skin, and respiratory tract.

RECOMMENDATIONS FOR RESPIRATOR SELECTION

At any concentrations above the NIOSH REL, or where there is no REL, at any detectable concentration: SCBAF:PD,PP (any self-contained breathing apparatus that has a full facepiece and is operated in a pressure-demand or other positive-pressure mode); or SAF:PD,PP:ASCBA (any supplied-air respirator that has a full facepiece and is operated in a pressure-demand or other positive-pressure mode in combination with an auxiliary, self-contained breathing apparatus operated in a pressure-demand or other positive-pressure mode).

Escape: GMFOV [any air-purifying, full-facepiece respirator (gas mask) with a chin-style, front- or back-mounted organic vapor canister]; or SCBAE (any appropriate escape-type, self-contained breathing apparatus).

CHEMICAL REACTIVITY

Contact with oxidizers, strong acids, heat, or flame can cause fire and explosions.

FIRE INFORMATION

Vapors are heavier than air and will collect in low areas. Vapors in confined areas may explode when exposed to fire. If material or contaminated runoff enters waterways, notify downstream users of potentially contaminated waters. Notify local health and fire officials and pollution control agencies. Use dry chemicals, alcohol-resistant foam, carbon dioxide. From a secure, explosion-proof location, use water spray to cool exposed containers. If cooling streams are ineffective (venting sound increases in volume and pitch, tank discolors, or shows any signs of deforming), withdraw immediately to a secure position.

NFPA Hazard Classifications:

Health Hazards (Blue) 2
Flammability (Red) 2
Reactivity (Yellow) 2

Flash Point: 165°F (74°C) o.c.

FIRST AID

Move victim to fresh air. Call emergency medical care. Apply artificial respiration if victim is not breathing. Administer oxygen if breathing is difficult. Remove and isolate contaminated clothing and shoes. In case of contact with substance, immediately flush skin or eyes with running water for at least 20 minutes. Wash skin with soap and water. Keep victim warm and quiet. Ensure that medical personnel are aware of the material(s) involved and take precautions to protect themselves.

VINYL FLUORIDE V:350

SYNONYMS: EINECS No. 200-832-6; ETHENE, FLUORO-; ETHYLENE, FLUORO-; FLUOROETHENE; FLUORETHYLENE; FLUORURO de VINILO (Spanish); MONOFLUOROETHENE; MONOFLUROETHYLENE; VINYL FLUORIDE MONOMER;

IDENTIFICATION

CAS:	75-02-5
ERG ID:	UN 1860
ERG Guide:	116P
ERG Guide Category:	GASES-FLAMMABLE (UNSTABLE)
Molecular Formula:	C_2H_3F
RTECS No.:	YZ7351000
Properties:	Extremely flammable, compressed gas. Colorless. Sweet etherlike odor. Floats and boils on water; insoluble.

HEALTH AND SAFETY INFORMATION

Irritates eyes, skin, and respiratory tract. High exposures can cause unconsciousness. Affects central nervous system. Contact with liquid may cause frostbite.

IARC: Group 2; a probably human carcinogen.

Exposure Limits:	**TWA**	**CEILING**
NIOSH:	1 ppm	5 ppm

RECOMMENDATIONS FOR RESPIRATOR SELECTION: NIOSH

10 ppm: CCROV [any chemical cartridge respirator with organic vapor cartridge(s)]; or SA (any supplied-air respirator).
25 ppm: SA:CF (any supplied-air respirator operated in a continuous-flow mode); or PAPROV [any powered, air-purifying respirator with organic vapor cartridge(s)].

50 ppm: CCRFOV [any chemical cartridge respirator with a full facepiece and organic vapor cartridge(s)]; or GMFOV [any air-purifying, full-facepiece respirator (gas mask) with a chin-style, front- or back-mounted acid gas canister]; or PAPRTOV [any powered, air-purifying respirator with a tight-fitting facepiece and organic vapor cartridge(s)]; or SCBAF (any self-contained breathing apparatus with a full facepiece); or SAF (any supplied-air respirator with a full facepiece).
200 ppm: SAF:PD,PP (any supplied-air respirator that has a full facepiece and is operated in a pressure-demand or other positive-pressure mode).

Emergency or planned entry into unknown concentration or IDLH conditions: SCBAF:PD,PP (any self-contained breathing apparatus that has a full facepiece and is operated in a pressure-demand or other positive-pressure mode); or SAF:PD,PP:ASCBA (any supplied-air respirator that has a full facepiece and is operated in a pressure-demand or other positive-pressure mode in combination with an auxiliary, self-contained breathing apparatus operated in a pressure-demand or other positive pressure mode).

Escape: GMFOV [any air-purifying, full-facepiece respirator (gas mask) with a chin-style, front- or back-mounted organic vapor canister]; or SCBAE (any appropriate escape-type, self-contained breathing apparatus).

CHEMICAL REACTIVITY

Flammable gas. May be able to form unstable peroxides and polymerize producing heat; may react with itself, blocking relief valves and leading to tank explosion. Reacts violently with oxidizers, with danger of fire and explosions. May accumulate static electrical charges, and may cause ignition of its vapors.

FIRE INFORMATION

Combustion products include toxic hydrogen fluoride gas. Gases are heavier than air and will collect in low areas. Gases in confined areas may explode when exposed to fire. Gases may travel long distances to ignition sources and flash back. Storage containers and parts of containers may rocket great distances, in many directions.If material or contaminated runoff enters waterways, notify downstream users of potentially contaminated waters. Notify local health and fire officials and pollution control agencies. Do not extinguish the fire unless the flow of the gas can be stopped and any remaining gas is out of the line. Specially trained personnel may use fog lines to cool exposures and let the fire burn itself out. Use dry chemicals, foam, carbon dioxide. From a secure, explosion-proof location, use water spray to cool exposed containers. If cylinders are exposed to excessive heat from fire or flame contact, withdraw immediately to a secure position.

NFPA Hazard Classifications:

Health Hazards (Blue) 1
Flammability (Red) 4
Reactivity (Yellow) 2

Flash Point: Flammable, compressed liquefied gas
Flammable Limits in Air: LEL 2.6%; UEL 21.7%
Autoignition Temperature: 725°F (385°C)

FIRST AID

Move victim to fresh air. Call emergency medical care. Apply artificial respiration if victim is not breathing. If breathing is difficult, give oxygen. Remove and isolate contaminated clothing and shoes. In case of contact with liquefied gas, thaw frosted parts with lukewarm water. Keep victim warm and quiet. Ensure that medical personnel are aware of the material(s) involved and take precautions to protect themselves.

ENVIRONMENTAL INFORMATION

Clean Air Act: Accidental Release Prevention/Flammable Substances (Section 112[r], Table 3), TQ = 10,000 lb (4,540 kg)
TSCA: 40CFR716.120(*a*); TSCA 8(*d*) unpublished health and safety studies 40CFR716, 47FR38800, 48FR13178, 47FR387; TSCA-CHIP available.

VINYLIDENE CHLORIDE V:400

SYNONYMS: CHLORURE de VINYLIDENE (French); CLORURO de VINILIDENO (Spanish); 1,1-DCE; 1,1-DICHLOROETHENE; 1,1-DICHLOROETHYLENE; EEC No. 602-025-00-8; EINECS No. 200-864-0; ETHENE, 1,1-DICHLORO-; ETHYLENE, 1,1-DICHLORO-; NCI-C54262; SCONATEX; VC; VDC; VINYLIDENE CHLORIDE(II); VINYLIDENE DICHLORIDE; VINYLIDENE CHLORIDE, inhibited; VINYLIDINE CHLORIDE MONOMER; UNSYM-DICHLOROETHYLENE

IDENTIFICATION

CAS:	75-35-4
ERG ID:	UN 1303
ERG Guide:	129P
ERG Guide Category:	FLAMMABLE LIQUIDS (POLAR/WATER-MISCIBLE/NOXIOUS)
Molecular Formula:	$C_2H_2Cl_2$
RTECS No.:	KV9275000
Properties:	Extremely flammable, watery liquid. Clear, colorless. Sweet chloroform-like odor. Sinks in water; slightly soluble. Flammable, poisonous vapor is produced when heated. Boiling point is 89°F (32°C).

HEALTH AND SAFETY INFORMATION

Irritates eyes, skin, and respiratory tract. Inhalation may cause pulmonary edema, which can be delayed for several hours; there is a risk of death in serious cases. Affects the central nervous system. Prolonged or repeated exposure can cause lung, liver, and kidney damage; reproductive damage in males; fetus damage.

IARC Evaluation: Group 3; animal positive
NTP: Potential human carcinogen
Odor Threshold: 190 ppm

Exposure Limits:	TWA	STEL
ACGIH: (1)	5 ppm; 20 mg/m^3	20 ppm; 79 mg/m^3
NIOSH: (2)		
OSHA:	None	

(1) Animal carcinogen
(2) Potential human carcinogen. Reduce exposure to lowest feasible level.

RECOMMENDATIONS FOR RESPIRATOR SELECTION: NIOSH

Any detectable concentration: SCBAF:PD,PP (any self-contained breathing apparatus that has a full facepiece and is operated in a pressure-demand or other positive-pressure mode); or SAF:PD,PP:ASCBA (any supplied-air respirator that has a full facepiece and is operated in a pressure-demand or other positive-pressure mode in combination with an auxiliary, self-contained breathing apparatus operated in a pressure-demand or other positive pressure mode).

Escape: GMFOV [Any air-purifying, full-facepiece respirator (gas mask) with a chin-style, front- or back-mounted organic vapor canister]; or SCBAE (any appropriate escape-type, self-contained breathing apparatus).

CHEMICAL REACTIVITY

Forms explosive mixture with air. Prolonged contact with air or contaminants can cause formation of unstable peroxides; may polymerize. Uninhibited monomer vapor may form polymer in confined spaces. May react with itself with out warning blocking release valves and causing violent explosion. Reacts violently with strong oxidizers, alkali metals, sunlight, air, copper, and heat. Copper, copper alloys, and aluminum can cause polymerization. Incompatible with nitric acid. Store in a marked, regulated area. Forms white solid deposits on storage which explode with shock or heat. Do not store in aluminum; may attack aluminum and its alloys.

FIRE INFORMATION

Combustion products include toxic phosgene and hydrogen chloride which may be more toxic than the material itself. Vapors are heavier than air and will collect in low areas. Vapors in confined areas may explode when exposed to fire. Vapors may travel long distances to ignition sources and flash back. Storage containers and parts of containers may rocket great distances, in many directions. If material or contaminated runoff enters waterways, notify downstream users of potentially contaminated waters. Notify local health and fire officials and pollution control agencies. Use dry chemicals, foam, carbon dioxide. From a secure, explosion-proof location, use water spray to cool exposed containers. If cooling streams are ineffective (venting sound increases in volume and pitch, tank discolors, or shows any signs of deforming), withdraw immediately to a secure position.

NFPA Hazard Classifications:

Health Hazards (Blue) 2
Flammability (Red) 4
Reactivity (Yellow) 2

Flash Point: –2°F (–19°C) o.c.; –19°F (–28°C) c.c..
Flammable Limits in Air: LEL 6.5%; UEL 15.5%
Autoignition Temperature: 1058°F (570°C)

FIRST AID

Move victim to fresh air. Call emergency medical care. Apply artificial respiration if victim is not breathing. Administer oxygen if breathing is difficult. Remove and isolate contaminated clothing and shoes. In case of contact with substance, immediately flush skin or eyes with running water for at least 20 minutes. Wash skin with soap and water. Keep victim warm and quiet. Effects of exposure (inhalation, ingestion, or skin contact) to substance may be delayed. Ensure that medical personnel are aware of the material(s) involved and take precautions to protect themselves.

ENVIRONMENTAL INFORMATION

Clean Air Act: Hazardous Air Pollutants (Title I, Part A, Section 112); Accidental Release Prevention/Flammable Substances (Section 112[r], Table 3), TQ = 10,000 lb (4,540 kg)
Clean Water Act: Section 307 Priority Pollutants; Section 311 Hazardous Substances/RQ (same as CERCLA); Section 313 Priority Chemicals; Toxic Pollutant (Section 401.15)
EPA Hazardous Waste Number (RCRA No.): U078; D029
RCRA Section 261 Hazardous Constituents
RCRA Toxicity Characteristic (Section 261.24), Maximum Concentration of Contaminants, regulatory level, 0.7 mg/L
RCRA Land Ban Waste
RCRA Universal Treatment Standards: Wastewater (mg/L), 0.025; Nonwastewater (mg/kg), 6.0.
RCRA Ground Water Monitoring List: Suggested test methods(s) (PQL μg/L); 8010(1); 8240(5).
Safe Drinking Water Act: MCL, 0.007 mg/L; MCLG, 0.007 mg/L; Regulated Chemical (47 FR 9352).

EPCRA Section 304 Reportable Quantity (RQ): CERCLA, 100 lb (45.4 kg)
EPCRA Section 313: Form R *de minimis* concentration reporting level: 1.0%.
Marine Pollutant (49CFR, Subchapter 172.101, Appendix B).

VINYLIDENE FLUORIDE V:450

SYNONYMS: 1,1-DIFLUOROETHENE; 1,1-DIFLUOROETHYLENE; EINECS No. 200-867-7; ETHENE, 1,1-DIFLUORO-; ETHYLENE, 1,1-DEIFLUORO-; HALOCARBON 1132A; NCI-C60208; R1132a; VDF; VINYLIDENE DIFLUORIDE; VINYLIDENE FLUORIDE

IDENTIFICATION

CAS:	75-38-7
ERG ID:	UN 1959
ERG Guide:	116P
ERG Guide Category:	GASES-FLAMMABLE (UNSTABLE)
Molecular Formula:	$C_2H_2F_2$
RTECS No.:	KW0560000
Properties:	Flammable, compressed, liquefied gas. Colorless. Faint odor. Slightly soluble in water.

HEALTH AND SAFETY INFORMATION

Irritates eyes, skin, and respiratory tract. Contact with liquid can cause frostbite. Suspected human carcinogen.

IARC Evaluation: Group 3; inadequate animal evidence
IDLH: 250 mg/m^3 as fluorine

Exposure Limits:	**TWA**
ACGIH:	2.5 mg/m^3 as fluorides
NIOSH:	2.5 mg/m^3 as fluorides
OSHA:	2.5 mg/m^3 as fluorides

RECOMMENDATIONS FOR RESPIRATOR SELECTION: NIOSH/OSHA as fluorides
12.5 mg/m^3: DM (any dust and mist respirator).
25 mg/m^3: DMXSQ $^•$ (any dust and mist respirator except single-use and quarter mask respirators); or SA $^•$ (any supplied-air respirator).
62.5 mg/m^3: SA:CF $^•$ [any supplied-air respirator operated in a continuous-flow mode)]; PAPRDM $^{•+}$ [if not present as a fume] (any powered, air-purifying respirator with a dust and mist filter).

125 mg/m³: HiEF [+] (any air-purifying, full-facepiece respirator with a high-efficiency particulate filter); or SCBAF (any self-contained breathing apparatus with a full facepiece); or SAF (any supplied-air respirator with a full facepiece).
250 mg/m³: SA:PD,PP (any supplied-air respirator operated in a pressure-demand or other positive-pressure mode).

Emergency or Planned Entry into Unknown Concentrations or IDLH Conditions: SCBAF:PD,PP (any self-contained breathing apparatus that has a full faceplate and is operated in a pressure-demand or other positive-pressure mode); or SAF:PD,PP:ASCBA (any supplied-air respirator that has a full facepiece and is operated in a pressure-demand or other positive-pressure mode in combination with an auxiliary, self-contained breathing apparatus operated in a pressure-demand or other positive-pressure mode).

Escape: HiEF [+] (any air-purifying, full-facepiece respirator with a high-efficiency particulate filter); or SCBAE (any appropriate escape-type, self-contained breathing apparatus).

[•] Substance reported to cause eye irritation or damage; may require eye protection.
[+] *Note*: May need acid gas sorbent.

CHEMICAL REACTIVITY

Contact with oxidizers can cause fire and explosions. Reacts with aluminum chloride. Capable of forming unstable peroxides, which can cause explosive polymerization. May accumulate static electricity, and cause ignition of its vapors.

FIRE INFORMATION

Combustion products include toxic hydrogen fluoride. Gases are heavier than air and will collect in low areas. Gases in confined areas may explode when exposed to fire. Gases may travel long distances to ignition sources and flash back. Storage containers and parts of containers may rocket great distances, in many directions. If material or contaminated runoff enters waterways, notify downstream users of potentially contaminated waters. Notify local health and fire officials and pollution control agencies. Do not extinguish the fire unless the flow of the gas can be stopped and any remaining gas is out of the line. Specially trained personnel may use fog lines to cool exposures and let the fire burn itself out. Use dry chemicals, foam, carbon dioxide. From a secure, explosion-proof location, use water spray to cool exposed containers. If cooling streams are ineffective (venting sound increases in volume and pitch, tank discolors, or shows any signs of deforming), withdraw immediately to a secure position.

NFPA Hazard Classifications:

Health Hazards (Blue) 1
Flammability (Red) 4
Reactivity (Yellow) 2

Flash Point: NA (Gas)
Flammable Limits in Air: LEL 5.5%; UEL 21.3%

FIRST AID

Move victim to fresh air. Call emergency medical care. Apply artificial respiration if victim is not breathing. If breathing is difficult, give oxygen. Remove and isolate contaminated clothing and shoes. In case of contact with liquefied gas, thaw frosted parts with lukewarm water. Keep victim warm and quiet. Ensure that medical personnel are aware of the material(s) involved and take precautions to protect themselves.

ENVIRONMENTAL INFORMATION

Clean Air Act: Accidental Release Prevention/Flammable Substances (Section 112[r], Table 3), TQ = 10,000 lb (4,540 kg).
TSCA: 40CFR716.120(*a*).

VINYL ISOBUTYL ETHER V:500

SYNONYMS: EINECS No. 203-678-8; ETHER, ISOBUTYL VINYL; ISOBUTOXYETHANE; ISOBUTYL VINYL ETHER; IVE; PROPANE, 1-(ETHENYLOXY)-2-METH); OSSIDO di ISOBUTILE e VINILE (Italian); OXYDE d'ISOBUTYLE et de VINYLE (French); VINOFLEX MO 400*; VINYL ISOBUTYL ETHER, inhibited

IDENTIFICATION

CAS:	109-53-5
ERG ID:	UN 1304
ERG Guide:	127P
ERG Guide Category:	FLAMMABLE LIQUIDS (POLAR/WATER- MISCIBLE)
Molecular Formula:	$C_6H_{12}O$
RTECS No.:	KO1300000
Properties:	Flammable liquid. Colorless. Slightly soluble in water.

HEALTH AND SAFETY INFORMATION

Irritates eyes, skin, and respiratory tract.

RECOMMENDATIONS FOR RESPIRATOR SELECTION

At any concentrations above the NIOSH REL, or where there is no REL, at any detectable concentration: SCBAF:PD,PP (any self-contained breathing apparatus that has a full facepiece and is operated in a pressure-demand or other positive-pressure mode); or SAF:PD,PP:ASCBA (any supplied-air

respirator that has a full facepiece and is operated in a pressure-demand or other positive-pressure mode in combination with an auxiliary, self-contained breathing apparatus operated in a pressure-demand or other positive-pressure mode).

Escape: GMFOV [any air-purifying, full-facepiece respirator (gas mask) with a chin-style, front- or back-mounted organic vapor canister]; or SCBAE (any appropriate escape-type, self-contained breathing apparatus).

CHEMICAL REACTIVITY

Forms explosive mixture with air. Reacts vigorously with oxidizing agents. A fire hazard when exposed to oxidizers, sulfuric acid, nitric acid, heat, or flame.

FIRE INFORMATION

Vapors are heavier than air and will collect in low areas. Vapors in confined areas may explode when exposed to fire. If material or contaminated runoff enters waterways, notify downstream users of potentially contaminated waters. Notify local health and fire officials and pollution control agencies. Use dry chemicals, alcohol-resistant foam, carbon dioxide. From a secure, explosion-proof location, use water spray to cool exposed containers.

NFPA Hazard Classifications:

Health Hazards (Blue) 2
Flammability (Red) 3
Reactivity (Yellow) 2

Flash Point: 15°F (–9°C) o.c.

FIRST AID

Move victim to fresh air. Call emergency medical care. Apply artificial respiration if victim is not breathing. Administer oxygen if breathing is difficult. Remove and isolate contaminated clothing and shoes. In case of contact with substance, immediately flush skin or eyes with running water for at least 20 minutes. Wash skin with soap and water. Keep victim warm and quiet. Ensure that medical personnel are aware of the material(s) involved and take precautions to protect themselves.

VINYL METHYL ETHER V:550

SYNONYMS: EEC No. 603-021-00-9; EINECS No. 203-475-4; METHYL VINYL ETHER; METHOXYETHYLENE; MVE; VINIL METIL ETER (Spanish); VINYL METHYL ETHER, INHIBITED

1148 Vinyl Methyl Ether

IDENTIFICATION

CAS:	107-25-5
ERG ID:	UN 1087
ERG Guide:	116P
ERG Guide Category:	GASES-FLAMMABLE (UNSTABLE)
Molecular Formula:	C_3H_6O
RTECS No.:	KO2300000
Properties:	Extremely flammable, compressed gas. Colorless. Sweet pleasant odor. Floats and may boil on water; slightly soluble; may decompose. A liquid below 42°F (6°C)

HEALTH AND SAFETY INFORMATION

Irritating to eyes, skin, and respiratory tract; prolonged contact with skin can cause burns. Contact with liquid can cause frostbite.

RECOMMENDATIONS FOR RESPIRATOR SELECTION

At any concentrations above the NIOSH REL, or where there is no REL, at any detectable concentration: SCBAF:PD,PP (any self-contained breathing apparatus that has a full facepiece and is operated in a pressure-demand or other positive-pressure mode); or SAF:PD,PP:ASCBA (any supplied-air respirator that has a full facepiece and is operated in a pressure-demand or other positive-pressure mode in combination with an auxiliary, self-contained breathing apparatus operated in a pressure-demand or other positive-pressure mode).

Escape: GMFOV [any air-purifying, full-facepiece respirator (gas mask) with a chin-style, front- or back-mounted organic vapor canister]; or SCBAE (any appropriate escape-type, self-contained breathing apparatus).

CHEMICAL REACTIVITY

Forms explosive mixture with air. Water contact slowly forms acetaldehyde and methyl alcohol. Forms unstable peroxides; able to polymerize; may react with itself, blocking relief valves and leading to tank explosion. Contact with acids and oxidizers may cause fire and explosions. May accumulate static electrical charges, and may cause ignition of its vapors.

FIRE INFORMATION

Gases are heavier than air and will collect in low areas. Gases in confined areas may explode when exposed to fire. Gases may travel long distances to ignition sources and flash back. Storage containers and parts of containers may rocket great distances, in many directions. If material or contaminated runoff enters waterways, notify downstream users of potentially contaminated waters. Notify local health and fire officials and pollution control agencies. Do not extinguish the fire unless the flow of the gas can be stopped and any remaining gas is out of the line. Specially trained personnel may use fog lines to cool exposures and let the fire burn itself out; otherwise, use dry chemicals, carbon dioxide. From a secure, explosion-proof location, use water spray to cool exposed containers. If cooling streams are ineffective (venting sound increases in volume and pitch, tank discolors, or shows any signs of deforming), withdraw immediately to a secure position.

NFPA Hazard Classifications:

Health Hazards (Blue) 2
Flammability (Red) 4
Reactivity (Yellow) 2

Flash Point: –69°F (–56°C) o.c.
Flammable Limits in Air: LEL 2.6%; UEL 39%
Autoignition Temperature: 545°F (285°C)

FIRST AID

Move victim to fresh air. Call emergency medical care. Apply artificial respiration if victim is not breathing. If breathing is difficult, give oxygen. Remove and isolate contaminated clothing and shoes. In case of contact with liquefied gas, thaw frosted parts with lukewarm water. Keep victim warm and quiet. Ensure that medical personnel are aware of the material(s) involved and take precautions to protect themselves.

ENVIRONMENTAL INFORMATION

Clean Air Act: Accidental Release Prevention/Flammable Substances (Section 112[r], Table 3), TQ = 10,000 lb (4,540 kg)

VINYL PROPIONATE V:600

SYNONYMS: EINECS No. 203-293-5; PROPIONATO de VINILO (Spanish); PROPANOIC ACID, ETHENYL ESTER

IDENTIFICATION

CAS:	105-38-4
ERG ID:	UN 3272 as esters, n.o.s.
ERG Guide:	127
ERG Guide Category:	FLAMMABLE LIQUIDS (POLAR/WATER- MISCIBLE)
Molecular Formula:	$C_5H_8O_2$

1150 Vinyl Propionate

RTECS No.: UF8575000
Properties: Liquid. Colorless. Floats on water; slightly soluble.

HEALTH AND SAFETY INFORMATION

Irritates eyes, skin, and respiratory tract. Mild toxicant if swallowed or inhaled.

RECOMMENDATIONS FOR RESPIRATOR SELECTION

At any concentrations above the NIOSH REL, or where there is no REL, at any detectable concentration: SCBAF:PD,PP (any self-contained breathing apparatus that has a full facepiece and is operated in a pressure-demand or other positive-pressure mode); or SAF:PD,PP:ASCBA (any supplied-air respirator that has a full facepiece and is operated in a pressure-demand or other positive-pressure mode in combination with an auxiliary, self-contained breathing apparatus operated in a pressure-demand or other positive-pressure mode).

Escape: GMFOV [any air-purifying, full-facepiece respirator (gas mask) with a chin-style, front- or back-mounted organic vapor canister]; or SCBAE (any appropriate escape-type, self-contained breathing apparatus).

CHEMICAL REACTIVITY

A very dangerous fire hazard when exposed to oxidizers, strong acids, heat, or flame.

FIRE INFORMATION

Vapors are heavier than air and will collect in low areas. Vapors in confined areas may explode when exposed to fire. If material or contaminated runoff enters waterways, notify downstream users of potentially contaminated waters. Notify local health and fire officials and pollution control agencies. Use dry chemicals, alcohol-resistant foam, carbon dioxide. From a secure, explosion-proof location, use water spray to cool exposed containers.

NFPA Hazard Classifications:

Health Hazards (Blue) 2
Flammability (Red) 3
Reactivity (Yellow) 2

Flash Point: 34°F (1°C) o.c.

FIRST AID

Move victim to fresh air. Call emergency medical care. Apply artificial respiration if victim is not breathing.

Administer oxygen if breathing is difficult. Remove and isolate contaminated clothing and shoes. In case of contact with substance, immediately flush skin or eyes with running water for at least 20 minutes. Wash skin with soap and water. Keep victim warm and quiet. Ensure that medical personnel are aware of the material(s) involved and take precautions to protect themselves.

2-VINYL PYRIDINE V:650

SYNONYMS: 2-ETHENYLPYRIDINE; PYRIDINE, 2-ETHENYL-; PYRIDINE, 2-VINYL-; α-VINYLPYRIDINE; VINYLPYRIDINE

IDENTIFICATION

CAS:	100-69-6
Molecular Formula:	C_7H_7N
RTECS No.:	UU1040000
Properties:	Combustible liquid. Colorless. Pyridinelike odor. Soluble in water.

HEALTH AND SAFETY INFORMATION

Poisonous if swallowed. Corrosive to eyes, skin, and respiratory tract. Inhalation may cause pulmonary edema, which can be delayed for several hours; there is a risk of death in serious cases.

RECOMMENDATIONS FOR RESPIRATOR SELECTION: NIOSH

At any concentrations above the NIOSH REL, or where there is no REL, at any detectable concentration: SCBAF:PD,PP (any self-contained breathing apparatus that has a full facepiece and is operated in a pressure-demand or other positive-pressure mode); or SAF:PD,PP:ASCBA (any supplied-air respirator that has a full facepiece and is operated in a pressure-demand or other positive-pressure mode in combination with an auxiliary, self-contained breathing apparatus operated in a pressure-demand or other positive-pressure mode).

Escape: GMFOV [any air-purifying, full-facepiece respirator (gas mask) with a chin-style, front- or back-mounted organic vapor canister]; or SCBAE (any appropriate escape-type, self-contained breathing apparatus).

CHEMICAL REACTIVITY

Forms explosive mixture with air. Contact with oxidizers can cause fire and explosions. Able to form unstable peroxides, which can cause explosive polymerization.

FIRE INFORMATION

Combustion products include toxic fumes of chlorides. Vapors are heavier than air and will collect in low areas. Vapors in confined areas may explode when exposed to fire. If material or contaminated runoff enters waterways, notify downstream users of potentially contaminated waters. Notify local health and fire officials and pollution control agencies. Use dry chemicals, alcohol-resistant foam, carbon dioxide. From a secure, explosion-proof location, use water spray to cool exposed containers.

NFPA Hazard Classifications: Not listed.

FIRST AID

Move victim to fresh air. Call emergency medical care. Apply artificial respiration if victim is not breathing. Administer oxygen if breathing is difficult. Remove and isolate contaminated clothing and shoes. In case of contact with substance, immediately flush skin or eyes with running water for at least 20 minutes. Wash skin with soap and water. Keep victim warm and quiet. Effects of exposure (inhalation, ingestion, or skin contact) to substance may be delayed. Ensure that medical personnel are aware of the material(s) involved and take precautions to protect themselves.

4-VINYL PYRIDINE V:700

SYNONYMS: 4-ETHENYLPYRIDINE; PYRIDINE, 4-ETHENYL-; PYRIDINE, 4-VINYL-; VINYLPYRIDINE; VINILPIRIDINA (Spanish)

IDENTIFICATION

CAS:	100-43-6
Molecular Formula:	C_7H_7N
RTECS No.:	UU1045000
Properties:	Combustible liquid. Colorless. Pyridinelike odor. Soluble in water.

HEALTH AND SAFETY INFORMATION

Poisonous if swallowed. Corrosive to eyes, skin, and respiratory tract. Inhalation may cause pulmonary edema, which can be delayed for several hours; there is a risk of death in serious cases.

RECOMMENDATIONS FOR RESPIRATOR SELECTION: NIOSH

At any concentrations above the NIOSH REL, or where there is no REL, at any detectable concentration: SCBAF:PD,PP (any self-contained breathing apparatus that has a full facepiece and is operated in a pressure-demand or other positive-pressure mode); or SAF:PD,PP:ASCBA (any supplied-air

respirator that has a full facepiece and is operated in a pressure-demand or other positive-pressure mode in combination with an auxiliary, self-contained breathing apparatus operated in a pressure-demand or other positive-pressure mode).

Escape: GMFOV [any air-purifying, full-facepiece respirator (gas mask) with a chin-style, front- or back-mounted organic vapor canister]; or SCBAE (any appropriate escape-type, self-contained breathing apparatus).

CHEMICAL REACTIVITY

Forms explosive mixture with air. Contact with oxidizers can cause fire and explosions. Able to form unstable peroxides which can cause explosive polymerization.

FIRE INFORMATION

Combustion products include toxic fumes of chlorides. Vapors are heavier than air and will collect in low areas. Vapors in confined areas may explode when exposed to fire. If material or contaminated runoff enters waterways, notify downstream users of potentially contaminated waters. Notify local health and fire officials and pollution control agencies. Use dry chemicals, alcohol-resistant foam, carbon dioxide. From a secure, explosion-proof location, use water spray to cool exposed containers.

NFPA Hazard Classifications: Not listed.

FIRST AID

Move victim to fresh air. Call emergency medical care. Apply artificial respiration if victim is not breathing. Administer oxygen if breathing is difficult. Remove and isolate contaminated clothing and shoes. In case of contact with substance, immediately flush skin or eyes with running water for at least 20 minutes. Wash skin with soap and water. Keep victim warm and quiet. Effects of exposure (inhalation, ingestion, or skin contact) to substance may be delayed. Ensure that medical personnel are aware of the material(s) involved and take precautions to protect themselves.

VINYL TOLUENE V:750

SYNONYMS: METHYL STYRENE; 3-METHYL STYRENE; 4-METHYL STYRENE; m- METHYL STYRENE; p-METHYL STYRENE; para-METHYLSTYRENE; NCI-C56406; VINILTOLUENO (Spanish); VINYL TOLUENE, INHIBITED; VINYL TOLUENE, MIXED ISOMERS; 4-VINYL TOLUENE; m-VINYL TOLUENE; meta-VINYL TOLUENE; p-VINYL TOLUENE; para-VINYL TOLUENE

1154 Vinyl Toluene

IDENTIFICATION

CAS:	25013-15-4; 622-97-9 (p-)
ERG ID:	UN 2618
ERG Guide:	130P
ERG Guide Category:	FLAMMABLE LIQUIDS (NONPOLAR/WATER-IMMISCIBLE/NOXIOUS)
Molecular Formula:	C_9H_{10}
RTECS No.:	WL5075000
Properties:	Very flammable, watery liquid. Colorless. Strong, unpleasant odor. Floats on water; insoluble.

HEALTH AND SAFETY INFORMATION

Irritates eyes, skin, and respiratory tract. Prolonged or repeated exposure can affect central nervous system, kidneys, liver.

IDLH: 400 ppm
Odor Threshold: 400 ppm
Odor Threshold: 10 ppm
OSHA Table Z-1-A Air Contaminant

Exposure Limits:	**TWA**	**STEL**
ACGIH:	50 ppm; 242 mg/m^3	100 ppm; 483 mg/m^3
NIOSH:	100 ppm; 480 mg/m^3	
OSHA:	100 ppm; 480 mg/m^3	

RECOMMENDATIONS FOR RESPIRATOR SELECTION: NIOSH/OSHA

400 ppm: CCROV [any chemical cartridge respirator with organic vapor cartridge(s)]; or PAPROV [any powered, air-purifying respirator with organic vapor cartridge(s)]; or GMFOV [any air-purifying, full-facepiece respirator (gas mask) with a chin-style, front- or back-mounted organic vapor canister]; or SA (any supplied-air respirator); or SCBAF (any self-contained breathing apparatus with a full facepiece).

Emergency or planned entry into unknown concentrations or idlh conditions: SCBAF:PD,PP (any self-contained breathing apparatus that has a full facepiece and is operated in a pressure-demand or other positive-pressure mode); or SAF:PD,PP:ASCBA (any supplied-air respirator that has a full facepiece and is operated in a pressure-demand or other positive-pressure mode in combination with an auxiliary, self-contained breathing apparatus operated in a pressure-demand or other positive pressure mode.

Escape: GMFOV [any air-purifying, full-facepiece respirator (gas mask) with a chin-style, front- or back-mounted organic vapor canister]; or SCBAE (any appropriate escape-type, self-contained breathing apparatus).

Note: Substance reported to cause eye irritation or damage; may require eye protection.

CHEMICAL REACTIVITY

Forms explosive mixture with air. A strong reducing agent; reacts violently with oxidizers, peroxides, strong acids, iron or aluminum salts. Able to form unstable peroxides and can polymerize; metallic salts can accelerate reaction.

FIRE INFORMATION

Vapors are heavier than air and will collect in low areas. Vapors in confined areas may explode when exposed to fire. Vapors may travel long distances to ignition sources and flash back. Storage containers and parts of containers may rocket great distances, in many directions. If material or contaminated runoff enters waterways, notify downstream users of potentially contaminated waters. Notify local health and fire officials and pollution control agencies. Use dry chemicals, foam, carbon dioxide. From a secure, explosion-proof location, use water spray to cool exposed containers. If cooling streams are ineffective (venting sound increases in volume and pitch, tank discolors, or shows any signs of deforming), withdraw immediately to a secure position.

NFPA Hazard Classifications:

Health Hazards (Blue) 2
Flammability (Red) 2
Reactivity (Yellow) 2

Flash Point: 137°F (58°C) o.c.; 127°F (53°C) c.c.
Flammable Limits in Air: LEL 0.8%; UEL 11%
Autoignition Temperature: 995°F (535°C)

FIRST AID

Move victim to fresh air. Call emergency medical care. Apply artificial respiration if victim is not breathing. Administer oxygen if breathing is difficult. Remove and isolate contaminated clothing and shoes. In case of contact with substance, immediately flush skin or eyes with running water for at least 20 minutes. Wash skin with soap and water. Keep victim warm and quiet. Effects of exposure (inhalation, ingestion, or skin contact) to substance may be delayed. Ensure that medical personnel are aware of the material(s) involved and take precautions to protect themselves.

ENVIRONMENTAL INFORMATION

TSCA: 40CFR716.120(*d*)10; 40CFR712.30(*e*)10, re absorption testing.

VINYL TRICHLOROSILANE V:800

SYNONYMS: A-150; EINECS No. 200-917-8; SILANE, TRICHLOROETHENYL; SILANE, TRICHLOROVINYL; SILANE, VINYL TRICHLORO 1-150; TRICHLORO(VINYL)SILANE; TRICHLOROVINYL SILICANE; UNION CARBIDE A-150; VINILTROCLOROSILANO (Spanish); VINYLSILICON TRICHLORIDE; VINYLTRICHLOROSILANE, INHIBITED; VTCS

IDENTIFICATION

CAS:	75-94-5
ERG ID:	UN 1302
ERG Guide:	155
ERG Guide Category:	SUBSTANCES-TOXIC AND/OR CORROSIVE (FLAMMABLE/WATER-SENSITIVE)
Molecular Formula:	$C_2H_3Cl_3Si$
RTECS No.:	VV6125000
Properties:	Highly flammable liquid. Colorless to pale yellow. Sharp, choking, hydrochloric acidlike odor. Reacts violently with water, forming toxic chloride and corrosive hydrochloric acid.

HEALTH AND SAFETY INFORMATION

Poisonous. Breathing the vapor can cause death. Eye and skin contact can cause burns and blindness.

RECOMMENDATIONS FOR RESPIRATOR SELECTION

At any concentrations above the NIOSH REL, or where there is no REL, at any detectable concentration: SCBAF:PD,PP (any self-contained breathing apparatus that has a full facepiece and is operated in a pressure-demand or other positive-pressure mode); or SAF:PD,PP:ASCBA (any supplied-air respirator that has a full facepiece and is operated in a pressure-demand or other positive-pressure mode in combination with an auxiliary, self-contained breathing apparatus operated in a pressure-demand or other positive-pressure mode).

Escape: GMFOV [any air-purifying, full-facepiece respirator (gas mask) with a chin-style, front- or back-mounted organic vapor canister]; or SCBAE (any appropriate escape-type, self-contained breathing apparatus).

CHEMICAL REACTIVITY

Incompatible with water; reacts violently, evolving flammable hydrogen chloride and hydrochloric acid. Reacts with surface moisture to produce hydrogen chloride, which will corrode common metals and form flammable hydrogen gas.

FIRE INFORMATION

Combustion products include toxic hydrogen chloride. Firefighting gear (including SCBA) does not provide adequate protection. If exposure occurs, remove and isolate gear immediately and thoroughly decontaminate personnel. Vapors are heavier than air and will collect in low areas. Vapors in confined areas may explode when exposed to fire. Vapors may travel long distances to ignition sources and flash back. Storage containers and parts of containers may rocket great distances, in many directions. If material or contaminated runoff enters waterways, notify downstream users of potentially contaminated waters. Notify local health and fire officials and pollution control agencies. If material is on fire and conditions permit, DO NOT EXTINGUISH. Material reacts with water but can be extinguished with low or medium expansion AFFF foam or dry chemical if available in sufficient amounts. From a secure, explosion-proof location, use water spray to cool exposed containers. If cooling streams are ineffective (venting sound increases in volume and pitch, tank discolors, or shows any signs of deforming), withdraw immediately to a secure position.

NFPA Hazard Classifications:

Health Hazards (Blue) 3
Flammability (Red) 3
Reactivity (Yellow) 2
Special Situation .... WATER

Flash Point: 70°F (21°C) o.c.; 52°F (11°C) c.c.
Flammable Limits in Air: LEL 3%; UEL NA
Autoignition Temperature: 505°F (263°C)

FIRST AID

Do not put yourself in danger by entering a contaminated area to rescue a victim. Move victim to fresh air. Call emergency medical care. Provide basic life support/CPR as needed. Decontaminate the victim as follows: Inhalation–remove victim to fresh air and give oxygen if available; Skin–remove and isolate contaminated clothing (including shoes) and wash skin with soap and large volumes of water for at least 20 minutes. Eye–rinse eyes with large volumes of water or saline for at least 20 minutes. Swallowed–do not make victim vomit. Victims should be examined by a physician as soon as possible. Ensure that medical personnel are aware of the material(s) involved and take precautions to protect themselves. Toxic effects may be delayed. For skin burns decontaminate with water and apply a clean dressing.

XYLENE X:040

SYNONYMS: ARISOLINE(+); AZ THINNER; BENZENE, DIMETHYL; DIMETHYLBENZENE; EEC No. 601-039-00-1 (m- isomer); EEC No. 601-040-00-7 (p-isomer); GLID-GUARD HIGH SOLIDS EPOXY NO. 5434; KESTER 5751 PROTECTO; KTI 45(-); KODAK PHOTORESIST DEVELOPER; 732R; KSYLEN (Polish); METHYL TOLUENE; MS-470 URETHANE COATING; MIXED XYLENES; NCI-C55232; OMR-83F (-); PROTECTO 5612; PYRALIN PI 2463; RED GLPT VARNISH; SELECTILUX; SECTIPLAST N2 NEGATIVE DEVELOPER; ULTRAMAC SOLVENT EPA; WAYCOAT; XILENO (Spanish); XILOLI (Italian); XYLENEN (Dutch); XYLOL; XYLOLE (German)

IDENTIFICATION

CAS:	1330-20-7; 95-47-6 (o-isomer); 108-38-3 (m-isomer); 106-42-3 (p-isomer)
ERG ID:	UN 1307
ERG Guide:	130
ERG Guide Category:	FLAMMABLE LIQUIDS (NONPOLAR/WATER-IMMISCIBLE/NOXIOUS)
Molecular Formula:	C_8H_{10}
RTECS No.:	ZE2100000; ZE2450000 (o-isomer); ZE2275000 (m-isomer); ZE2625000 (p-isomer)
Properties:	m-xylene is the major constituent of commercial grade xylene. Highly flammable liquid. Clear, colorless, or light colored. p-xylene is a solid (crystals) below 57°F (14°C). Sweet, aromatic, gasoline-like odor. Floats on water; insoluble.

HEALTH AND SAFETY INFORMATION

The health effects for all xylenes are similar. Irritates eyes, skin, and respiratory system. Prolonged contact with skin can lead to irritation. Repeated exposure to xylene due to solvent abuse can result in progressive and permanent neuropsychiatric manifestations. Long-term exposure has been associated with heart, liver, and kidney damage, and may cause anemia, intestinal tract disturbances, and central nervous system depression.

IARC Evaluation: Group 3, unclassifiable; human evidence, inadequate; animal evidence, inadequate.
IDLH: 900 ppm (3960 mg/m^3) all isomers.
Odor Threshold: 0.851 ppm (o- isomer); 0.324 (m- isomer); 0.49 (p- isomer)
OSHA Table Z-1-A Air Contaminant
NIOSH Criteria Document: NIOSH C-75-168

Exposure Limits:	**TWA**	**STEL**
ACGIH:	100 ppm; 434 mg/m^3	150 ppm; 651 mg/m^3

NIOSH:	100 ppm; 435 mg/m^3	150 ppm; 655 mg/m^3
OSHA:	100 ppm; 435 mg/m^3	

RECOMMENDATIONS FOR RESPIRATOR SELECTION: NIOSH/OSHA

900 ppm: CCROV [any chemical cartridge respirator with organic vapor cartridge(s)]; or PAPROV [any powered, air-purifying respirator with organic vapor cartridge(s)]; or SA (any supplied-air respirator); or SCBAF (any self-contained breathing apparatus with a full facepiece).

Emergency or Planned Entry into Unknown Concentrations or IDLH Conditions: SCBAF:PD,PP (any self-contained breathing apparatus that has a full facepiece and is operated in a pressure-demand or other positive-pressure mode); or SAF:PD,PP:ASCBA (any supplied-air respirator that has a full facepiece and is operated in a pressure-demand or other positive-pressure mode in combination with an auxiliary, self-contained breathing apparatus operated in a pressure-demand or other positive-pressure mode).

Escape: GMFOV [any air-purifying, full-facepiece respirator (gas mask) with a chin-style, front- or back-mounted organic vapor canister]; or SCBAE (any appropriate escape-type, self-contained breathing apparatus).
Note: Substance reported to cause eye irritation or damage; may require eye protection.

CHEMICAL REACTIVITY

Forms explosive mixture with air (flash point approximately 84°F/29°C). Strong oxidizers such as permanganates, dichromates, or chlorine may cause fire and explosions. Attacks many plastics, rubber, and coatings. May accumulate static electrical charges, and may cause ignition of its vapors.

FIRE INFORMATION

Vapors are heavier than air and will collect in low areas. Vapors in confined areas may explode when exposed to fire. Vapors may travel long distances to ignition sources and flash back. Storage containers and parts of containers may rocket great distances, in many directions. If material or contaminated runoff enters waterways, notify downstream users of potentially contaminated waters. Notify local health and fire officials and pollution control agencies. Use dry chemicals, foam, carbon dioxide. From a secure, explosion-proof location, use water spray to cool exposed containers. If cooling streams are ineffective (venting sound increases in volume and pitch, tank discolors, or shows any signs of deforming), withdraw immediately to a secure position.

NFPA Hazard Classifications:

Health Hazards (Blue) 2
Flammability (Red) 3
Reactivity (Yellow) 0

Flash Point: 63°F (17°C) o-isomer; 84°F (29°C) m-isomer; 81°F (27°C) p-isomer.
Flammable Limits in Air: LEL 0.9–1.1%; UEL 6.7–7.0%.
Autoignition Temperature: 986°F (530°C) m- and p- isomers; 869°F (465°C) o-isomer.

FIRST AID

Move victim to fresh air. Call emergency medical care. Apply artificial respiration if victim is not breathing. Administer oxygen if breathing is difficult. Remove and isolate contaminated clothing and shoes. In case of contact with substance, immediately flush skin or eyes with running water for at least 20 minutes. Wash skin with soap and water. Keep victim warm and quiet. Effects of exposure (inhalation, ingestion, or skin contact) to substance may be delayed. Ensure that medical personnel are aware of the material(s) involved and take precautions to protect themselves.

ENVIRONMENTAL INFORMATION

Clean Air Act: Hazardous Air Pollutants (Title I, Part A, Section 112) as xylenes (isomers and mixtures)
Clean Water Act: Section 311 Hazardous Substances/RQ 100 lb (45.4 kg); Section 313 Priority Chemicals
EPA Hazardous Waste Number (RCRA No.): U239.
RCRA Section 261 Hazardous Constituents
RCRA Land Ban Waste
RCRA Universal Treatment Standards: Wastewater (mg/L), 0.32; Nonwastewater (mg/kg), 30.
RCRA Ground Water Monitoring List: Suggested methods (PQL ug/L): (total) 8020(5); 8240(5).
SAFE DRINKING WATER ACT: xylenes (total), MCL, 10 mg/L; MCLG, 10 mg/L; Regulated chemical (47 FR 9352) as xylene.
EPCRA Section 304: Reportable Quantity (RQ): CERCLA, 100 lb (45.4 kg).
EPCRA Section 313: Form R *de minimis* concentration reporting level: 1.0%.
TSCA: 40CFR716.120(*a*); 40CFR716.120(*d*)10; 40CFR712.30(e)10

XYLENOLS X:200

SYNONYMS: DIMETHYLPHENOL; EEC No. 604-006-00-X (pure and isomers); EINECS No. 215-089-3 (pure); EINECS No. 208-395-3 (2,3-xylenol); EINECS No. 203-321-6 (2,4-xylenol); EINECS No. 202-461-5 (2,5-xylenol); EINECS No. 203-606-5 (3,5-xylenol); PHENOL, DIMETHYL-; XILENOLI (Italian); XYLENOLEN (Dutch); 2,6-XYLENOL; CRESYLIC ACID; 2-HYDROXY-M-XYLENE; XILENOL (Spanish)

IDENTIFICATION

CAS: 1300-71-6 (mixed isomers); 526-75-0 (2,3-); 576-26-1 (2,6-); 105-67-9 (2,4-); 95-65-8 (3,4-); 95-87-4 (2,5-); 108-68-9 (3,5-)

ERG ID: UN 2261
ERG Guide: 153
ERG Guide Category: SUBSTANCES-TOXIC AND/OR CORROSIVE (COMBUSTIBLE)
Molecular Formula: $C_8H_{10}O$
RTECS No.: ZE5425000; ZE5600000 (2,4-xylenol); ZE5775000 (2,5-xylenol); ZE6475000;
Properties: Combustible solid (needles, crystals), or liquid. White or light yellowish brown. Sweet tarlike odor. May float or sink in water; slightly soluble.

HEALTH AND SAFETY INFORMATION

Dust irritates eyes, skin, and respiratory tract. Extremely irritating to eyes; can cause burns and blindness. Can be absorbed through the skin, lungs. Skin contact has an anesthetic effect and cause serious burns. Prolonged or repeated exposure (based on similar compounds) may cause liver and kidney damage. Inhalation of (3,5-) isomer may cause pulmonary edema, which can be delayed for several hours; there is a risk of death in serious cases.

IDLH: 0.001 mg/m^3 (2,4-).;0.003 mg/m^3 (3,4-); 0.00004.1 mg/m^3 (3,5-) recognition in air

RECOMMENDATIONS FOR RESPIRATOR SELECTION

At any concentrations above the NIOSH REL, or where there is no REL, at any detectable concentration: SCBAF:PD,PP (any self-contained breathing apparatus that has a full facepiece and is operated in a pressure-demand or other positive-pressure mode); or SAF:PD,PP:ASCBA (any supplied-air respirator that has a full facepiece and is operated in a pressure-demand or other positive-pressure mode in combination with an auxiliary, self-contained breathing apparatus operated in a pressure-demand or other positive-pressure mode).

Escape: GMFOVHiE [any air-purifying, full-facepiece respirator (gas mask) with a chin-style, front- or back-mounted organic vapor canister having a high-efficiency particulate filter]; or SCBAE (any appropriate escape-type, self-contained breathing apparatus).

CHEMICAL REACTIVITY

Forms explosive mixture with air. Contact with strong oxidizers may cause fire and explosions.

FIRE INFORMATION

Combustible solid. Vapors are heavier than air and will collect in low areas. Vapors in confined areas may explode when exposed to fire. Storage containers and parts of containers may rocket great distances, in many directions. If material or contaminated runoff enters waterways, notify downstream users of potentially contaminated waters. Notify local health and fire officials and pollution control agencies. Use dry chemicals, foam, carbon dioxide. From a secure, explosion-proof location, use water spray to cool exposed containers. If cooling streams are ineffective (venting sound increases in volume and pitch, tank discolors, or shows any signs of deforming), withdraw immediately to a secure position.

FEMA Hazard Classifications

Health Hazards (Blue) 2
Flammability (Red) 2
Reactivity (Yellow) 0

Flash Point: 186°F (86°C) c.c.
Flammable Limits in Air: LEL 1.4%; UEL NA.
Autoignition Temperature: 1,110°F (599°C)

FIRST AID

Move victim to fresh air and call emergency medical care. Apply artificial respiration if victim is not breathing. Do not use mouth-to-mouth method if victim swallowed or inhaled the substance; induce artificial respiration with the aid of a pocket mask equipped with a one-way valve or other proper respiratory medical device. If breathing is difficult, give oxygen. In case of contact with substance, immediately flush skin or eyes with running water for at least 20 minutes. Remove and isolate contaminated clothing and shoes. Keep victim warm and quiet. Effects of exposure (inhalation, ingestion, or skin contact) to substances may be delayed. Ensure that medical personnel are aware of the material(s) involved and take precautions to protect themselves.

ENVIRONMENTAL INFORMATION

Clean Water Act: Section 311 Hazardous Substances/RQ (same as CERCLA)
EPCRA Section 304 Reportable Quantity (RQ): CERCLA, 1,000 lb (454 kg)
Marine Pollutant (49CFR, Subchapter 172.101, Appendix B) as xlyenols.

XYLIDINES X:500

SYNONYMS: ACID LEATHER BROWN 2G; ACID ORANGE 24; AMINODIMETHYLBENZENE; AMINOXYLENE; BENZENAMINE, AR,AR-DIMETHYL-; DIMETHYLANILINE; DIMETHYLAMINOBENZENE; DIMETHYLANILINE; DIMETHYLPHENYLAMINE; EEC No. 612-027-00-0; EINECS No. 215-091-4; 11460 BROWN; RESORCINE BROWN J; RESORCINE BROWN R; XYLIDIN (Dutch, German); XILIDINA (Spanish); XYLIDINE ISOMERS; XYLIDINEN (Dutch)

IDENTIFICATION

CAS:	1300-73-8 (mixed isomers); 87-62-7 (1,6-); 95-68-1 (2,4-); 95-78-3 (2,5-); 87-62-7 (2,6)
ERG ID:	UN 1711
ERG Guide:	153
ERG Guide Category:	SUBSTANCES-TOXIC AND/OR CORROSIVE (COMBUSTIBLE)
Molecular Formula:	$C_8H_{11}N$
RTECS No.:	ZE8575000
Properties:	Combustible liquid or solid.. Exists in various isometric forms varying from pale yellow to brown. Weak, ammonialike odor. Practically insoluble in water.

HEALTH AND SAFETY INFORMATION

Poisonous. Irritates eyes, skin, and respiratory tract. Possibly fatal if inhaled or absorbed through the skin; get medical attention immediately; qualified medical personnel can administer methylene blue as an antidote. Affects the central nervous system. Can form methemoglobin; affects the blood; interferes with the blood's capacity to carry oxygen. High or repeated exposure can affect the lungs, liver, and kidneys.

IDLH: 50 ppm
Odor Threshold: 0.005-0.06 ppm
OSHA Table Z-1-A Air Contaminant

Exposure Limits:	**TWA**
ACGIH: (1)	0.5 ppm; 2.5 mg/m^3
NIOSH: (2)	2 ppm; 10 mg/m^3
OSHA: (2)	5 ppm; 25 mg/m^3

(1) Animal carcinogen. Skin contact contributes to overall exposure.
(2) Skin contact contributes to overall exposure.

RECOMMENDATIONS FOR RESPIRATOR SELECTION: NIOSH

20 ppm: CCROV [any chemical cartridge respirator with organic vapor cartridge(s)]; SA (any supplied-air respirator).
50 ppm: SA:CF (any supplied-air respirator operated in a continuous-flow mode); or CCRFOV [any chemical cartridge respirator with a full facepiece and organic vapor cartridge(s)]; or GMFOV [any air-purifying, full-facepiece respirator (gas mask) with a chin-style, front- or back-mounted organic vapor canister]; or PAPROV [any powered, air-purifying respirator with an organic vapor cartridge(s)]; or SCBAF (any self-contained breathing apparatus with a full facepiece); or SAF (any supplied-air respirator with a full facepiece).

Emergency or Planned Entry into Unknown Concentrations or IDLH Conditions: SCBAF:PD,PP (any self-contained breathing apparatus that has a full facepiece and is operated in a pressure-demand or other positive-pressure mode); or SAF:PD,PP:ASCBA (any supplied-air respirator that has a full facepiece and is operated in a pressure-demand or other positive-pressure mode in combination with an auxiliary, self-contained breathing apparatus operated in a pressure-demand or other positive-pressure mode).

Escape: GMFOV [any air-purifying, full-facepiece respirator (gas mask) with a chin-style, front- or back-mounted organic vapor canister]; or SCBAE (any appropriate escape-type, self-contained breathing apparatus).

CHEMICAL REACTIVITY

Forms explosive mixture with air. Reacts vigorously with oxidizers, strong acids. Contact with hypochlorite salts forms explosive chloroamines. Attacks some forms of plastics, rubber, and coatings.

FIRE INFORMATION

Combustion products include toxic oxides of nitrogen and irritants and toxic gases. Firefighting gear (including SCBA) may not provide adequate protection. If exposure occurs, remove and isolate gear immediately and thoroughly decontaminate personnel. Vapors are heavier than air and will collect in low areas. Vapors in confined areas may explode when exposed to fire. If material or contaminated runoff enters waterways, notify downstream users of potentially contaminated waters. Notify local health and fire officials and pollution control agencies. Use dry chemicals, alcohol-resistant foam, carbon dioxide. From a secure, explosion-proof location, use water spray to cool exposed containers. If cooling streams are ineffective (venting sound increases in volume and pitch, tank discolors, or shows any signs of deforming), withdraw immediately to a secure position.

NFPA Hazard Classifications: for the o-isomer

Health Hazards (Blue) 3
Flammability (Red) 1
Reactivity (Yellow) 0

Flash Point: 205°F (96°C) c.c.
Flammable Limits in Air: LEL 1.0% for the 2,3- isomer; UEL NA

FIRST AID

Move victim to fresh air and call emergency medical care. Apply artificial respiration if victim is not breathing. Do not use mouth-to-mouth method if victim swallowed or inhaled the substance; induce artificial respiration with the aid of a pocket mask equipped with a one-way valve or other proper respiratory medical device. If breathing is difficult, give oxygen. *Speed in removing material from the skin is of extreme importance.* In case of contact with substance, immediately flush skin or eyes with running water for at least 20 minutes. Remove and isolate contaminated clothing and shoes. Keep victim warm and quiet. Effects of exposure (inhalation, ingestion, or skin contact) to substances may be delayed. Ensure that medical personnel are aware of the material(s) involved and take precautions to protect themselves.

ENVIRONMENTAL INFORMATION as 2,3,-Xylidine

EPCRA Section 313: Form R *de minimis* concentration reporting level: 1.0%.
California's Proposition 65: Carcinogen.
TSCA:40CFR716.120(*d*)10; 40CFR712.30(*e*)10 re dermal absorption testing.

ZINC Z:003

SYNONYMS: ASAREO L15; BLUE POWDER; CINC (Portugese); C.I. 77945; C.I. PIGMENT BLACK 16; C.I. PIGMENT METAL 6; EEC No. 030-001-00-1; EEC No. 030-002-00-7; EINECS No. 231-175-3; EMANAY ZINC DUST; JASAD; MERRILLITE; PASCO; TOXIC CHEMICAL CATEGORY CODE, N982; ZINC DUST; ZINCO (Italian); ZINC POWDER; ZINK (Dutch, German)

IDENTIFICATION

CAS:	7440-66-6
ERG ID:	UN 1435; UN 1436 (powder or dust)
ERG Guide:	138
ERG Guide Category:	SUBSTANCES-WATER-REACTIVE(EMITTING FLAMMABLE GASES)
Molecular Formula:	Zn
RTECS No.:	ZG8600000
Properties:	Flammable solid. Light gray powder or bluish white metallic solid. Powder can ignite spontaneously in air. Insoluble in water; the material will sink.

HEALTH AND SAFETY INFORMATION

Inhalation of dust and fume can cause metal fume fever with chills, nausea, cough, weakness.

OSHA Table Z-1-A Air Contaminant (zinc and compounds)

RECOMMENDATIONS FOR RESPIRATOR SELECTION: NIOSH

At any concentrations above the NIOSH REL, or where there is no REL, at any detectable concentration: SCBAF:PD,PP (any self-contained breathing apparatus that has a full facepiece and is operated in a pressure-demand or other positive-pressure mode); or SAF:PD,PP:ASCBA (any supplied-air

respirator that has a full facepiece and is operated in a pressure-demand or other positive-pressure mode in combination with an auxiliary, self-contained breathing apparatus operated in a pressure-demand or other positive-pressure mode).

Escape: GMFOV [any air-purifying, full-facepiece respirator (gas mask) with a chin-style, front- or back-mounted organic vapor canister]; or SCBAE (any appropriate escape-type, self-contained breathing apparatus).

CHEMICAL REACTIVITY

A strong reducing agent; reacts violently with oxidizers. Reacts with acids, caustics, fluorine, cadmium, carbon disulfide, and water forming explosive hydrogen. Contact with sulfur may cause fire and explosions. Zinc dust forms an explosive mixture with air. May accumulate static electrical charges and may cause ignition of airborne powder. Spontaneous combustion may occur if stored in a damp place.

FIRE INFORMATION

Zinc is a combustible solid. If material or contaminated runoff enters waterways, notify downstream users of potentially contaminated waters. Notify local health and fire officials and pollution control agencies. Use dry chemicals, alcohol-resistant foam, carbon dioxide. From a secure, explosion-proof location, use water spray to cool exposed containers.

FEMA Hazard Classifications:

Health Hazards (Blue) 0
Flammability (Red) 2
Reactivity (Yellow) 2

FIRST AID

Move victim to fresh air. Call emergency medical care. Apply artificial respiration if victim is not breathing. Administer oxygen if breathing is difficult. Remove and isolate contaminated clothing and shoes. In case of contact with substance, immediately flush skin or eyes with running water for at least 20 minutes. Keep victim warm and quiet. Ensure that medical personnel are aware of the material(s) involved and take precautions to protect themselves.

ENVIRONMENTAL INFORMATION

Clean Water Act: Section 307 Toxic Pollutants; Section 307 Priority Pollutants.
RCRA Universal Treatment Standards: Wastewater (mg/L), 2.61; Nonwastewater (mg/L), 5.3 TCLP.

Note: these constituents are not "underlying hazardous constituents" in characteristic wastes, according to the definition at Section 268.2(i).

RCRA Ground Water Monitoring List: Suggested test method(s) (PQL ug/L): (total) 6010(20); 7950(50).
Safe Drinking Water Act: SMCL, 5 mg/L; Priority List (55 FR 1470).
EPCRA Section 304: Reportable Quantity (RQ): CERCLA, 1,000 lb (454 kg).
EPCRA Section 313: Form R *de minimis* concentration reporting level: 1.0%; Form R Toxic Chemical Category Code: N982.

As fume or dust

Clean Water Act: Section 307 Toxic Pollutants; Section 313 Priority Chemicals.
EPCRA Section 304: Reportable Quantity (RQ): CERCLA, 1,000 lb (454 kg).
EPCRA Section 313: Form R *de minimis* concentration reporting level: 1.0%. Form R Toxic Chemical Category Code: N982.

ZINC ACETATE Z:004

SYNONYMS: ACETIC ACID, ZINC SALT; ACETIC ACID, ZINC SALT(II); AI3-04465; DIACETATO de ZINCO (Spanish); DICARBOMETHOXY ZINC; EINECS No. 209-170-2; SILTEX CL 4; ZINC(II) ACETATE; ZINC ACETATE DIHYDRATE; ZINC DIACETATE

IDENTIFICATION

CAS:	557-34-6
ERG ID:	UN 9153
ERG Guide:	171
ERG Guide Category:	SUBSTANCES (LOW TO MODERATE HAZARD)
Molecular Formula:	$C_4H_6O_4Zn$
RTECS No.:	AK1500000
Properties:	Granular solid. White. Faint vinegar odor. Astringent taste. Sinks and mixes with water.

HEALTH AND SAFETY INFORMATION

Moderately toxic if swallowed.

OSHA Table Z-1-A Air Contaminant as zinc and compounds

CHEMICAL REACTIVITY

Incompatible with strong acids, strong bases, nitrates, oxalates, phosphates, sulfides, zinc salts (i.e., zinc carbonate).

FIRE INFORMATION

Combustion products include toxic zinc oxide. If material or contaminated runoff enters waterways, notify downstream users of potentially contaminated waters. Notify local health and fire officials and pollution control agencies. Use any extinguishing agents suitable for surrounding fires. Use water spray to cool exposed containers.

FEMA Hazard Classifications:

Health Hazards (Blue) 2
Flammability (Red) 0
Reactivity (Yellow) 0

Flash Point: Not flammable

FIRST AID

Move victim to fresh air. Call emergency medical care. Apply artificial respiration if victim is not breathing. Administer oxygen if breathing is difficult. In case of contact with substance, immediately flush skin or eyes with running water for at least 20 minutes. Remove and isolate contaminated clothing and shoes. Ensure that medical personnel are aware of the material(s) involved and take precautions to protect themselves.

ENVIRONMENTAL INFORMATION

Clean Water Act: Section 307 Toxic Pollutants; Section 311 Hazardous Substances/RQ (same as CERCLA); Section 313 Priority Chemicals
EPCRA Section 304 Reportable Quantity (RQ): CERCLA, 1,000 lb (454 kg)
EPCRA Section 313: Form R *de minimis* concentration reporting level: 1.0%. Form R Toxic Chemical Category Code: N982

ZINC CARBONATE Z:005

SYNONYMS: CALAMINE; CARBINATO de ZINC (Spanish); CARBONIC ACID, ZINC SALT (1:1); C.I. 77950; EINECS No. 222-477-6; SMITHSONITE; ZINC CARBONATE(1:1); ZINC MONOCARBONATE; ZINCSPAR

IDENTIFICATION

CAS:	3486-35-9
ERG ID:	UN 9157
ERG Guide:	171
ERG Guide Category:	SUBSTANCES (LOW TO MODERATE HAZARD)

Molecular Formula: CO_3Zn
RTECS No.: FG3375000
Properties: Solid, crystalline powder. White. Odorless. Sinks in water.

HEALTH AND SAFETY INFORMATION

Moderate health hazard. Due to the availability of insufficient data on human effects, caution should be exercised.

OSHA Table Z-1-A Air Contaminant as zinc and compounds

CHEMICAL REACTIVITY

Carbonates are incompatible with calcium, lead diacetate, germanium, magnesium, mercurous chloride, silicon, silver nitrate, titanium, zinc acetate.

FIRE INFORMATION

Combustion products include toxic zinc oxide and carbon monoxide. If material or contaminated runoff enters waterways, notify downstream users of potentially contaminated waters. Notify local health and fire officials and pollution control agencies. Use extinguishing agents suitable for surrounding fires. Use water spray to cool exposed containers.

FEMA Hazard Classifications:

Health Hazards (Blue) 1
Flammability (Red) 0
Reactivity (Yellow) 0

Flash Point: Not flammable

FIRST AID

Move victim to fresh air. Call emergency medical care. Apply artificial respiration if victim is not breathing. Administer oxygen if breathing is difficult. In case of contact with substance, immediately flush skin or eyes with running water for at least 20 minutes. Remove and isolate contaminated clothing and shoes. Ensure that medical personnel are aware of the material(s) involved and take precautions to protect themselves.

ENVIRONMENTAL INFORMATION

Clean Water Act: Section 307 Toxic Pollutants; Section 311 Hazardous Substances/RQ (same as CERCLA); Section 313 Priority Chemicals

EPCRA Section 304 Reportable Quantity (RQ): CERCLA, 1,000 lb (454 kg)
EPCRA Section 313: Form R *de minimis* concentration reporting level: 1.0%. Form R Toxic Chemical Category Code: N982

ZINC CHLORIDE Z:006

SYNONYMS: BUTTER OF ZINC; CHLORURE de ZINC (French); CLORURO de ZINC (Spanish); EINECS No. 231-592-0; STCC 4966790; TINNING GLUX; ZINC BUTTER; ZINC CHLORIDE, anhydrous; ZINC CHLORIDE FUME; ZINC (CHLORURE de) (French); ZINC MURIATE; ZINCO (CLORURO di) (Italian); ZINC DICHLORIDE; ZINKCHLORID (German); ZINKCHLORIDE (Dutch)

IDENTIFICATION

CAS:	7646-85-7
ERG ID:	UN 1840 (solution); UN 2331 (anhydrous)
ERG Guide:	154
ERG Guide Category:	SUBSTANCES–TOXIC AND/OR CORROSIVE(NON-COMBUSTIBLE)
Molecular Formula:	Cl_2Zn
RTECS No.:	ZH1400000
Properties:	Noncombustible white powder or colorless crystals. Fume is a white particulate dispersed in air. Very soluble in water; forms strong corrosive acid; excess material will sink.

HEALTH AND SAFETY INFORMATION

Corrosive to eyes, skin, and respiratory tract. Inhalation may cause pulmonary edema, which can be delayed for several hours; there is a risk of death in serious cases. Prolonged or repeated exposure can cause conjunctivitis, dermatitis. Experimental teratogen and mutagen

IDLH: 50 mg/m^3 as fume
OSHA Table Z-1-A Air Contaminant

Exposure Limits:	**TWA**	**STEL**
ACGIH:	1 mg/m^3	2 mg/m^3
NIOSH:	1 mg/m^3	2 mg/m^3
OSHA:	1 mg/m^3	

RECOMMENDATIONS FOR RESPIRATOR SELECTION: NIOSH

10 mg/m^3: DMFu (any dust, mist and fume respirator); or SA (any supplied-air respirator).

25 mg/m³: SA:CF (any supplied-air respirator operated in a continuous-flow mode); or PAPRDMFu (any powered, air-purifying respirator with a dust, mist, and fume filter).
50 mg/m³: HiEF (any air-purifying, full-facepiece respirator with a high-efficiency particulate filter); or PAPRTHiE* (any powered, air-purifying respirator with a tight-fitting facepiece and a high-efficiency particulate filter); or SCBAF (any self-contained breathing apparatus with a full facepiece); or SAF (any supplied-air respirator with a full facepiece).

Emergency or Planned Entry into Unknown Concentrations or IDLH Conditions: SCBAF:PD,PP (any self-contained breathing apparatus that has a full facepiece and is operated in a pressure-demand or other positive-pressure mode); or SAF:PD,PP:ASCBA (any supplied-air respirator that has a full facepiece and is operated in a pressure-demand or other positive-pressure mode in combination with an auxiliary, self-contained breathing apparatus operated in a pressure-demand or other positive-pressure mode).

Escape: HiEF (any air-purifying, full-facepiece respirator with a high-efficiency particulate filter); or SCBAE (any appropriate escape-type, self-contained breathing apparatus).

Note: Substance reported to cause eye irritation or damage; may require eye protection.

CHEMICAL REACTIVITY

Aqueous solution is a strong acid. Solid material will absorb moisture from air, forming a concentrated, corrosive solution. Reacts with strong bases. Violent reaction with potassium, with possible explosion on impact. Attacks metals in the presence of moisture.

FIRE INFORMATION

Noncombustible solid. Combustion products include toxic chlorine and zinc oxide fumes. If material or contaminated runoff enters waterways, notify downstream users of potentially contaminated waters. Notify local health and fire officials and pollution control agencies. Use any extinguishing agent suitable for surrounding fire. Use water spray to cool exposed containers.

FEMA Hazard Classifications

Health Hazards (Blue) 1
Flammability (Red) 0
Reactivity (Yellow) 0

FIRST AID

Move victim to fresh air. Call emergency medical care. Apply artificial respiration if victim is not breathing. Do not use mouth-to-mouth method if victim ingested or inhaled the substance; induce artificial respiration with the aid of a pocket mask equipped with a one–way valve or other proper respiratory medical device.

Administer oxygen if breathing is difficult. Remove and isolate contaminated clothing and shoes. In case of contact with substance, immediately flush skin or eyes with running water for at least 20 minutes. For minor skin contact, avoid spreading material on unaffected skin. Keep victim warm and quiet. Effects of exposure (inhalation, ingestion, or skin contact) may be delayed. Ensure that medical personnel are aware of the material(s) involved and take precautions to protect themselves.

ENVIRONMENTAL INFORMATION

Clean Water Act: Section 307 Toxic Pollutants; Section 311 Hazardous Substances/RQ (same as CERCLA); Section 313 Priority Chemicals.
EPCRA Section 304: Reportable Quantity (RQ): CERCLA, 1,000 lb (454 kg).
EPCRA Section 313: Form R *de minimis* concentration reporting level: 1.0%. Form R Toxic Chemical Category Code: N982.

ZINC CYANIDE Z:012

SYNONYMS: CIANURO de ZINC (Spanish); CYANURE de ZINC (French); EINECS No. 209-162-9; STCC 4923495; ZINC DICYANIDE

IDENTIFICATION

CAS:	557-21-1
ERG ID:	UN 1713
ERG Guide:	151
ERG Guide Category:	SUBSTANCES-TOXIC (NONCOMBUSTIBLE)
Molecular Formula:	C_2N_2Zn
RTECS No.:	ZH1575000
Properties:	Noncombustible crystalline solid or powder. Colorless. Practically insoluble in water; reacts to form heat and toxic hydrogen cyanide gas.

HEALTH AND SAFETY INFORMATION

Poisonous. Irritates eyes, skin, and respiratory system. Can be absorbed through the skin; contact can contribute significantly to overall exposure.

IDLH: 25 mg/m^3 as cyanide
OSHA Table Z-1-A Air Contaminant as CN; as zinc compounds.
NIOSH Criteria Document: NIOSH 77-108 cyanide

Exposure Limits:	**TWA**	**CEILING**
ACGIH: [(1)]	5 mg/m^3 as cyanide	
NIOSH: [(1)]		4.7 ppm; 5 mg/m^3 per 10 minutes as cyanides
OSHA: [(1)]	5 mg/m^3 as cyanides	

[(1)] Skin contact contributes significantly in overall exposure

RECOMMENDATIONS FOR RESPIRATOR SELECTION: NIOSH/OSHA

25 mg/m^3: SA (any supplied-air respirator); or SCBA (any self-contained breathing apparatus).

Emergency or Planned Entry into Unknown Concentrations or IDLH Conditions: SCBAF:PD,PP (any self-contained breathing apparatus that has a full facepiece and is operated in a pressure-demand or other positive-pressure mode); or SAF:PD,PP:ASCBA (any supplied-air respirator that has a full facepiece and is operated in a pressure-demand or other positive-pressure mode in combination with an auxiliary, self-contained breathing apparatus operated in a pressure-demand or other positive-pressure mode)

Escape: GMFSHiE [any air-purifying, full-facepiece respirator (gas mask) with a chin-style, front- or back-mounted canister providing protection against the compound of concern and having a high-efficiency particulate filter]; or SCBAE (any appropriate escape-type, self-contained breathing apparatus).

CHEMICAL REACTIVITY

Contact with water, acids, or acid salts forms toxic hydrogen cyanide gas. Violent reaction with magnesium.

FIRE INFORMATION

Combustion products include toxic zinc oxide, nitrogen oxides and fumes of cyanide. If material or contaminated runoff enters waterways, notify downstream users of potentially contaminated waters. Notify local health and fire officials and pollution control agencies. Use extinguishing agent suitable for surrounding fire. Use water spray to cool exposed containers.

NFPA Hazard Classifications:

Health Hazards (Blue) 3
Flammability (Red) 0
Reactivity (Yellow) 0

FIRST AID

Move victim to fresh air. Call emergency medical care. Apply artificial respiration if victim is not breathing. Do not use mouth-to-mouth method if victim ingested or inhaled the substance; induce artificial respiration with the aid of a pocket mask equipped with a one–way valve or other proper respiratory medical device. Administer oxygen if breathing is difficult. Remove and isolate contaminated clothing and shoes. In case of contact with substance, immediately flush skin or eyes with running water for at least 20 minutes. For minor skin contact, avoid spreading material on unaffected skin. Keep victim warm and quiet. Effects of exposure (inhalation, ingestion, or skin contact) may be delayed. Ensure that medical personnel are aware of the material(s) involved and take precautions to protect themselves.

ENVIRONMENTAL INFORMATION

Clean Water Act: Section 311 Hazardous Substances/RQ (same as CERCLA); Section 307 Toxic Pollutants; Section 313 Priority Chemicals.
EPA Hazardous Waste Number (RCRA No.): P121.
RCRA Section 261 Hazardous Constituents
EPCRA Section 304: Reportable Quantity (RQ): CERCLA, 10 lb (4.54 kg).
EPCRA Section 313: (as zinc or cyanide compound) Form R *de minimis* concentration reporting level: 1.0%; Form R Toxic Chemical Category Code: N982 (zinc); Form R Toxic Chemical Category Code: N016 (cyanide).
Marine Pollutant (49CFR, Subchapter 172.101, Appendix B).

ZINC NITRATE Z:017

SYNONYMS: CELLOXAN; EINECS No. 231-943-8; NITRATE de ZINC (French); NITRIC ACID, ZINC SALT; X-4; ZINC DINITRATE; ZINC NITRATE HEXAHYDRATE

IDENTIFICATION

CAS:	7779-88-6
ERG ID:	UN 1514
ERG Guide:	140
ERG Guide Category:	OXIDIZERS
Molecular Formula:	$N_2O_6Zn \cdot 6H_2O$
RTECS No.:	ZH4772000
Properties:	Crystalline solid. White. Odorless. Astringent taste. Sinks and mixes with water. A dangerous fire and explosion hazard (ILO).

HEALTH AND SAFETY INFORMATION

Prolonged or repeated exposure to nitrates can cause dizziness, abdominal cramps, weakness, headaches, and collapse.

OSHA Table Z-1-A Air Contaminant as zinc and compounds

RECOMMENDATIONS FOR RESPIRATOR SELECTION

At any concentrations above the NIOSH REL, or where there is no REL, at any detectable concentration: SCBAF:PD,PP (any self-contained breathing apparatus that has a full facepiece and is operated in a pressure-demand or other positive-pressure mode); or SAF:PD,PP:ASCBA (any supplied-air respirator that has a full facepiece and is operated in a pressure-demand or other positive-pressure mode in combination with an auxiliary, self-contained breathing apparatus operated in a pressure-demand or other positive-pressure mode).

Escape: GMFOVHiE [any air-purifying, full-facepiece respirator (gas mask) with a chin-style, front- or back-mounted organic vapor canister having a high-efficiency particulate filter]; or SCBAE (any appropriate escape-type, self-contained breathing apparatus).

CHEMICAL REACTIVITY

A strong oxidizer, reacts with reducing agents, combustibles, phosphorus, sulfur, carbon, organic matter, powdered copper, and sulfides of metal.

FIRE INFORMATION

Combustion products include toxic oxides of nitrogen. Not combustible but will increase the intensity of surrounding fires. If material or contaminated runoff enters waterways, notify downstream users of potentially contaminated waters. Notify local health and fire officials and pollution control agencies. Use extinguishing agents suitable for surrounding fires. From a secure, explosion-proof location, use water spray to cool exposed containers. If cooling streams are ineffective (venting sound increases in volume and pitch, tank discolors, or shows any signs of deforming), withdraw immediately to a secure position. If fire becomes uncontrollable, evacuate to a radius of one-half mile (Bureau of Explosives).

FEMA Hazard Classifications:

Health Hazards (Blue) 1
Flammability (Red) 0
Reactivity (Yellow) 0
Special Situations .... OXY

FIRST AID

Move victim to fresh air. Call emergency medical care. Apply artificial respiration if victim is not breathing. Administer oxygen if breathing is difficult. Remove and isolate contaminated clothing and shoes. In case of contact with substance, immediately flush skin or eyes with running water for at least 20 minutes. Keep victim warm and quiet. Ensure that medical personnel are aware of the material(s) involved and take precautions to protect themselves.

ENVIRONMENTAL INFORMATION

Clean Water Act: Section 307 Toxic Pollutants; Section 311 Hazardous Substances/RQ (same as CERCLA); Section 313 Priority Chemicals
EPCRA Section 304 Reportable Quantity (RQ): CERCLA, 1,000 lb (454 kg)
EPCRA Section 313: Form R *de minimis* concentration reporting level: 1.0%. Form R Toxic Chemical Category Code: N511 (nitrate compounds, water dissociable; reportable only in aqueous solution).

ZINC STEARATE Z:030

SYNONYMS: COAD; DERMARONE; DIBASIC ZINC STEARATE; DIESTEARATO de ZINCO, PURO (Spanish); EINECS No. 209-151-9; HYDENSE; HYTECH; MATHE; METALLAC; METASAP 576; OCTADECANOIC ACID, ZINC SALT;; STAVINOR ZN-E; STEARIC ACID, ZINC SALT; SYNPRO STEARATE; ZINC DISTEARATE; ZINC OCTADECANOATE

IDENTIFICATION

CAS:	557-05-1
Molecular Formula:	$C_{36}H_{70}O_4Zn$
RTECS No.:	ZH5200000
Properties:	Combustible powder. White. Faint odor. Repels water; insoluble.

HEALTH AND SAFETY INFORMATION

Irritates eyes, skin, and respiratory system. High or repeated exposure causes metal fume fever, upper respiratory, and lung problems (pulmonary fibrosis).

OSHA Table Z-1-A Air Contaminant

Exposure Limits:	**TWA**
NIOSH:	10 mg/m^3 (total); 5 mg/m^3 (respirable fraction)
OSHA:	15 mg/m^3 (total); 5 mg/m^3 (respirable fraction)

RECOMMENDATIONS FOR RESPIRATOR SELECTION

At any concentrations above the NIOSH REL, or where there is no REL, at any detectable concentration: SCBAF:PD,PP (any self-contained breathing apparatus that has a full facepiece and is operated in a pressure-demand or other positive-pressure mode); or SAF:PD,PP:ASCBA (any supplied-air respirator that has a full facepiece and is operated in a pressure-demand or other positive-pressure mode in combination with an auxiliary, self-contained breathing apparatus operated in a pressure-demand or other positive-pressure mode).

Escape: GMFOV [any air-purifying, full-facepiece respirator (gas mask) with a chin-style, front- or back-mounted organic vapor canister]; or SCBAE (any appropriate escape-type, self-contained breathing apparatus).

CHEMICAL REACTIVITY

Contact with strong oxidizers may cause fire and explosions. Incompatible with (dilute) acids. Hydrophobic; repels water.

FIRE INFORMATION

Combustible solid. Combustion products include toxic zinc oxides. If material or contaminated runoff enters waterways, notify downstream users of potentially contaminated waters. Notify local health and fire officials and pollution control agencies. Use dry chemicals, water spray, alcohol-resistant foam, carbon dioxide. From a secure, explosion-proof location, use water spray to cool exposed containers. If cooling streams are ineffective (venting sound increases in volume and pitch, tank discolors, or shows any signs of deforming), withdraw immediately to a secure position.

NFPA Hazard Classifications:

Health Hazards (Blue) 0
Flammability (Red) 1
Reactivity (Yellow) 0

Flash Point: 530°F (277°C) o.c.
Autoignition Temperature: 788°F (420°C)

FIRST AID

Move victim to fresh air. Call emergency medical care. Apply artificial respiration if victim is not breathing. Administer oxygen if breathing is difficult. Remove and isolate contaminated clothing and shoes. In case of contact with substance, immediately flush skin or eyes with running water for at least 20 minutes. Removal of solidified molten material from skin requires medical assistance. Keep victim warm and quiet. Ensure that medical personnel are aware of the material(s) involved and take precautions to protect themselves.

ENVIRONMENTAL INFORMATION as zinc compounds

Clean Water Act: Section 307 Toxic Pollutants as zinc and zinc compounds
EPCRA Section 313: Includes any unique chemical substance that contains zinc as part of that chemical's infrastructure. Form R *de minimis* concentration reporting level: 1.0%. Form R Toxic Chemical Category Code: N982

ZINC SULFIDE Z:046

SYNONYMS: ALBALITH; CLEARTRAN; C.I. PIGMENT WHITE 7; EINECS No. 215-251-3; IRTRAN 2; PIGMENT WHITE 7; SACHTOLITH; SACHTOLITH HD-S; ZINC MONOSULFIDE; ZINC SULPHIDE

IDENTIFICATION

CAS:	1314-98-3
Molecular Formula:	SZn
RTECS No.:	ZH5400000
Properties:	Noncombustible powder or solid. White. Poisonous vapors produced when heated.

HEALTH AND SAFETY INFORMATION

Dust may cause irritation of the eyes, skin , mouth, and respiratory system. Zinc oxide fumes are emmited when heated; can cause metal fume fever, with chills, aching muscles, nausea, fever, dry throat, cough, and weakness.

RECOMMENDATIONS FOR RESPIRATOR SELECTION: NIOSH

At any concentrations above the NIOSH REL, or where there is no REL, at any detectable concentration: SCBAF:PD,PP (any self-contained breathing apparatus that has a full facepiece and is operated in a pressure-demand or other positive-pressure mode); or SAF:PD,PP:ASCBA (any supplied-air respirator that has a full facepiece and is operated in a pressure-demand or other positive-pressure mode in combination with an auxiliary, self-contained breathing apparatus operated in a pressure-demand or other positive-pressure mode).

Escape: GMFOVHiE [any air-purifying, full-facepiece respirator (gas mask) with a chin-style, front- or back-mounted organic vapor canister having a high-efficiency particulate filter]; or SCBAE (any appropriate escape-type, self-contained breathing apparatus).

CHEMICAL REACTIVITY

Dust may be able to form an explosive mixture with air. Incompatible with strong aluminum powder, oxidizers, barium bromate, barium chlorate, barium iodate, bromates, calcium chlorate, cerium compounds, chloric acid, chlorine, fluorine, hydrazinium nitrate, iodine, iodine monochloride, lead dioxide, magnesium chlorate, mercurous chloride, potassium bromate, potassium chlorate, potassium nitrate, silver azide, sodium bromate, sodium chlorate, sodium iodate, zinc acetate, zinc chlorate, zinc nitrate. If material or contaminated runoff enters waterways, notify downstream users of potentially contaminated waters. Notify local health and fire officials and pollution control agencies. Use dry chemicals, alcohol-resistant foam, carbon dioxide. From a secure, explosion-proof location, use water spray to cool exposed containers.

Contact with acids or water liberates toxic and flammable hydrogen sulfide. Some sulfides are spontaneously combustible. Contact with cadmium chlorate may be explosive.

FIRE INFORMATION

Combustion products include toxic zinc and sulfur oxides. If material or contaminated runoff enters waterways, notify downstream users of potentially contaminated waters. Notify local health and fire officials and pollution control agencies. Use any extinguishing agent suitable for surrounding fires. From a secure, explosion-proof location, use water spray to cool exposed containers.

NFPA Hazard Classifications: Not listed

FIRST AID

Move victim to fresh air. Call emergency medical care. Apply artificial respiration if victim is not breathing. Administer oxygen if breathing is difficult. Remove and isolate contaminated clothing and shoes. In case of contact with substance, immediately flush skin or eyes with running water for at least 20 minutes. Wash skin with soap and water. Keep victim warm and quiet. Effects of exposure (inhalation, ingestion, or skin contact) to substance may be delayed. Ensure that medical personnel are aware of the material(s) involved and take precautions to protect themselves.

ENVIRONMENTAL INFORMATION as zinc compounds

Clean Water Act: Section 307 Toxic Pollutants as zinc and zinc compounds
EPCRA Section 313: Includes any unique chemical substance that contains zinc as part of that chemical's infrastructure. Form R *de minimis* concentration reporting level: 1.0%. Form R Toxic Chemical Category Code: N982

ZIRCONOCENE DICHLORIDE Z:052

SYNONYMS: BIS(3-CYCLOPENTADIENYL)DICHLOROZIRCONIUM; DICLORURO de ZIRCONIO (Spanish); EINECS No. 215-066-8; ZIRCONIUM, DICHLORO-DI-pi-CYCLOPENTADIENYL-; ZIRCONOCENE, DICHLORIDE

IDENTIFICATION

CAS:	1291-32-3
Molecular Formula:	$C_{10}H_{10}Cl_2Zr$
RTECS No.:	ZH7525000
Properties:	Crystalline solid. Colorless. Poisonous gases emitted when heated.

HEALTH AND SAFETY INFORMATION

Dust irritates skin and respiratory system. Prolonged or repeated exposure may cause skin and lung problems. X-ray evidence of retention of zirconium in the lungs.

IDLH: 50 mg/m^3 as zirconium
OSHA Table Z-1-A Air Contaminant as zirconium compounds

Exposure Limits:	**TWA**	**STEL**
ACGIH: (1)	5 mg/m^3 as zirconium	10 mg/m^3 as zirconium
NIOSH:	5 mg/m^3 as zirconium	10 mg/m^3 as zirconium
OSHA:	5 mg/m^3 as zirconium	

(1) Not classifiable as a human carcinogen.

RECOMMENDATIONS FOR RESPIRATOR SELECTION: NIOSH/OSHA

25 mg/m^3: DM (any dust and mist respirator)
50 mg/m^3: DMXSQ (any dust and mist respirator except single-use and quarter mask respirators); or PAPRDM (any powered, air-purifying respirator with a dust and mist filter); HiEF (any air-purifying, full-facepiece respirator with a high-efficiency particulate filter); or SA (any supplied-air respirator); or SCBAF (any self-contained breathing apparatus with a full facepiece)

Emergency or Planned Entry into Unknown Concentrations or Idlh Conditions:
SCBAF:PD,PP (any self-contained breathing apparatus that has a full facepiece and is operated in a pressure-demand or other positive-pressure mode); or SAF:PD,PP:ASCBA (any supplied-air respirator that has a full facepiece and is operated in a pressure-demand or other positive-pressure mode in combination with an auxiliary, self-contained breathing apparatus operated in a pressure-demand or other positive pressure mode.

Escape: HiEF (any air-purifying, full-facepiece respirator with a high-efficiency particulate filter); or SCBAE (any appropriate escape-type, self-contained breathing apparatus).

CHEMICAL REACTIVITY

Incompatible with potassium nitrate, oxidizers. Zirconium powders may ignite spontaneously and can continue to burn under water.

FIRE INFORMATION

Combustion products include toxic zirconium and chlorine. If material or contaminated runoff enters waterways, notify downstream users of potentially contaminated waters. Notify local health and fire officials and pollution control agencies. Use dry chemicals, alcohol-resistant foam, carbon dioxide. Water may be ineffective. From a secure, explosion-proof location, use water spray to cool exposed containers.

NFPA Hazard Classifications: Not listed

FIRST AID

Move victim to fresh air. Call emergency medical care. Apply artificial respiration if victim is not breathing. Do not use mouth-to-mouth method if victim ingested or inhaled the substance; induce artificial respiration with the aid of a pocket mask equipped with a one-way valve or other proper respiratory medical device. Administer oxygen if breathing is difficult. Clothing frozen to the skin should be thawed before being removed. Remove and isolate contaminated clothing and shoes. In case of contact with substance, immediately flush skin or eyes with running water for at least 20 minutes. Keep victim warm and quiet. Keep victim under observation. Effects of contact or inhalation may be delayed. Ensure that medical personnel are aware of the material(s) involved and take precautions to protect themselves.

ZIRCONIUM SILICATE Z:200

SYNONYMS: EINECS No. 239-019-6; HYACINTH; SILICATO de ZIRCONIO (Spanish); SILICIC ACID, ZIRCONIUM(4+) SALT (1:1); ZIRCON; ZIRCONIUM(IV) SILICATE (1:1)

IDENTIFICATION

CAS:	14940-68-2
Molecular Formula:	O_4SiZr
RTECS No.:	ZH9000000
Properties:	Crystalline solid. Shades of brown, green, yellow, gray, and red. Insoluble in water.

HEALTH AND SAFETY INFORMATION

Industrial poison. Dust irritates eyes, skin, and respiratory system. High or prolonged exposure may cause lung and skin problems. Possible human carcinogen.

IDLH: 50 mg/m^3 as zirconium
OSHA Table Z-1-A Air Contaminant as zirconium compounds

Exposure Limits:	**TWA**	**STEL**
ACGIH: (1)	5 mg/m^3 as zirconium	10 mg/m^3 as zirconium

1182 Zirconium Silicate

NIOSH:	5 mg/m^3 as zirconium	10 mg/m^3 as zirconium
OSHA:	5 mg/m^3 as zirconium	

(1) Not classifiable as a human carcinogen.

RECOMMENDATIONS FOR RESPIRATOR SELECTION: NIOSH/OSHA

25 mg/m^3: DM (any dust and mist respirator)
50 mg/m^3: DMXSQ (any dust and mist respirator except single-use and quarter mask respirators); or PAPRDM (any powered, air-purifying respirator with a dust and mist filter); HiEF (any air-purifying, full-facepiece respirator with a high-efficiency particulate filter); or SA (any supplied-air respirator); or SCBAF (any self-contained breathing apparatus with a full facepiece)

Emergency or Planned Entry into Unknown Concentrations or Idlh Conditions: SCBAF:PD,PP (any self-contained breathing apparatus that has a full facepiece and is operated in a pressure-demand or other positive-pressure mode); or SAF:PD,PP:ASCBA (any supplied-air respirator that has a full facepiece and is operated in a pressure-demand or other positive-pressure mode in combination with an auxiliary, self-contained breathing apparatus operated in a pressure-demand or other positive pressure mode.

Escape: HiEF (any air-purifying, full-facepiece respirator with a high-efficiency particulate filter); or SCBAE (any appropriate escape-type, self-contained breathing apparatus).

CHEMICAL REACTIVITY

May react violently with lithium. Incompatible with acids.

FIRE INFORMATION

If material or contaminated runoff enters waterways, notify downstream users of potentially contaminated waters. Notify local health and fire officials and pollution control agencies. Use dry chemicals, alcohol-resistant foam, carbon dioxide. Use water spray to cool exposed containers.

NFPA Hazard Classifications: Not listed

FIRST AID

Move victim to fresh air. Call emergency medical care. Apply artificial respiration if victim is not breathing. Administer oxygen if breathing is difficult. Remove and isolate contaminated clothing and shoes. In case of contact with substance, immediately flush skin or eyes with running water for at least 20 minutes. Wash skin with soap and water. Keep victim warm and quiet. Effects of exposure (inhalation, ingestion, or skin contact) to substance may be delayed. Ensure that medical personnel are aware of the material(s) involved and take precautions to protect themselves.

Appendix A

SOURCES OF INFORMATION

HOTLINES

AIRS Helpline

Aerometric Information Retrieval System
Hotline: (800) 333-7909 (voice); Modem: (919) 541-5742
FAX: (919) 541-0028
WEB site: www.epa.gov/airs/airs.html
Provides a national repository for airborne pollution data in the United States and various World Health Organization (WHO) member countries. Call for other network addresses and available linkage types. Hours: Voice, 8:30 a.m. to 5:00 p.m. EST, Monday through Friday. Electronic: 24 hours, every day.

Asbestos and Small Business Ombudsman

Asbestos Information System
Office of Pesticides and Toxic Substances
U.E. Environmental Protection Agency
Phone: (703) 305-5938
Hotline: (800) 368-5888
FAX: (703) 305-5938
WEB site: www.epa.gov/earth 100/browse/C.html
Provides information on handling, abatement, and management of asbestos in schools, the workplace, and the home. Hours: 8:00 a.m. to 4:30 p.m. EST, Monday through Friday.

California Proposition 65

Phone: (916) 445-6900
WEB site: www.prop65news.com

Chemical Collection System/Request Tracking (CCS/RTS)

Phone: (202)260-3619
Provides information on properties of a number of chemicals, such as environmental and health effects and test and analysis methods. Available through the EPA.

CHEMTREC (24-hour emergency number for chemical accidents and spills)

Chemical Transportation Emergency Center
Chemical Manufacturers Association (CMA)
1300 Wilson Blvd.
Arlington, VA 22209
Hotline: (800) 424-9300; Nonemergency, (800) 262-8200
FAX: (703) 741-6037
E-mail: chemtrec@cmahq.com
WEB site: www.cmahq.com

Provides MSDS chemical profiles and emergency response information to the chemical and transportation industries and emergency services who may be called upon as first responders to chemical emergencies.

Clean Air Act

Hotline: (919) 541-0800 (English); (919) 541-1800 (Spanish)
FAX: (919) 541-0242
WEB site: catcmail@epa.gov
EPA's Clean Air Technology Center (CATC) provides general assistance and information on CAA standards. The EPA's Indoor Air Quality Information Clearinghouse provides information about indoor air problems and references to other government resources at (800) 438-4318. The Stratospheric Ozone Information Hotline, at (800) 296-1996, provides general information about regulations promulgated under Title VI of the CAA, and EPA's EPCRA Hotline, at (800) 535-0202, answers questions about accidental release prevention under CAA §112(r). In addition, the Technology Transfer Network Bulletin Board System [modem access (919) 541-5742)] includes recent CAA rules, EPA guidance documents, and updates of EPA activities.

Clean Water Act

Hotline: (202) 260-5700
EPA's Office of Water will direct callers with questions about the CWA to the appropriate EPA office. EPA also maintains a bibliographic database of Office of Water publications which can be accessed through the Ground Water and Drinking Water resource center, at (202) 260-7786.
The Storm Water Hotline number is (703) 821-4600.

Emergency Planning and Community Right-to-Know Act (EPCRA)

Phone: (703) 412-9810; TDD (800) 553-7672
Hotline: (800) 424-9346; (800) 535-0202
E-mail: tri.us@epamail.gov
Answers questions and distributes guidance regarding the emergency planning and community right-to-know regulations. Programs include the Resource Conservation and Recovery Act (RCRA), Underground Storage Tank (UST) program, the Risk Management Program (RMP), the Comprehensive Environmental Response, Compensation and Liability Act (CERCLA or Superfund), and the EPA's Oil Program. The EPCRA Hotline operates weekdays from 9:00 a.m. to 6:00 p.m., EST, excluding federal holidays. Services are also available in Spanish.

Federal Register/Code of Federal Regulations

New Orders
Superintendant of Documents
P.O. Box 371954
Pittsburgh, PA 15250-7954

National Response Center (NRC)

U.S. Coast Guard Headquarters
2100 2nd Street, S.W., Room 2611
Washington, DC 20593

Hotline: (800) 424-8802
The federal government's national communication center that receives all reports of releases involving hazardous substances and oil. Hours: 24 hours, every day.

National Technical Information Service Information Center (NTIS)

U.S. Department of Commerce
5285 Port Royal Road
Springfield, VA 22161
Phone: (703) 605-6000
Fax: (703) 605-6900
WEB site: www.ntis.gov
The federal government's central source for the sale of scientific, technical, engineering, and related business information produced by or for the U.S. government.

Pollution Prevention Information Clearinghouse (PPIC)

Hotline: (202) 260-1023
FAX: (202) 260-4659
E-mail: ppic@epa.gov
A free, nonregulatory service of the EPA dedicated to reducing or eliminating industrial pollutants through technology transfer, education, and public awareness.

Safe Drinking Water Act Hotline

Hotline: (800) 426-4791
FAX: (703) 285-1101
WEB site: hotline-sdwa@epamail.epa.gov
Answers questions and distributes guidance pertaining to drinking water regulations and related topics. Hours: 9:00 a.m. through 5:30 p.m. EST, weekdays except federal holidays.

DATABASES

ACQUIRE - Aquatic Toxicity Information Retrieval Database

Mid-Continent Ecology Division Laboratory
National Health and Environmental Effects Research Laboratory
6201 Congdon Blvd.
Duluth, MN 55804
Phone: (218) 720-5602
FAX: (218) 720-5539
Contains information on the toxic effects of 5,600 chemicals on more than 2,800 aquatic species of animals and plants, excluding birds, aquatic mammals, and bacteria. Hours: 8:00 a.m. to 4:30 p.m. CST, Monday through Friday.

ATTIC - Alternative Treatment Technology Information Center Database

Office of Research and Development
U. S. Environmental Protection Agency

Phone: (301) 816-9153
WEB site: www.epa.gov/attic
Contains several databases that are accessed through a free, public-access bulletin board. They provide information on alternative technologies for treatment of hazardous waste, including thermal, biological, chemical, and physical systems; underground storage tanks; and treatment of oil and chemical spills. It contains Treatment Technology DB, Oil/Chemical Spill DB, Treatability Study DB, Underground Storage Tanks DB, and many other documents. Hours: 24 hours, every day.

CCC - Comprehensive Chemical Contaminant Series

Electronic version of Lewis' *Sax's Dangerous Properties of Industrial Materials, 9th ed.*, Lewis' *Hawley's Condensed Chemical Dictionary*; Prager's *Environmental Contaminant Reference Databook, Vols, I, II & III*; Verschueren's *Handbook of Environmental Data on Organic Chemicals, 3rd ed.*, and Pohanish and Greene's *Hazardous Materials Handbook*, on CD-ROM, John Wiley and Sons, New York, NY, 1999.

CCINFO MSDS Series - Canadian Centre for Occcupational Health and Safety

Canadian Centre for Occupational Health and Safety (CCOHS)
250 Main St. East
Hamilton, ON Canada L8N 1H6
Phone: (800) 668-4284
FAX: (905) 572-2206
A bundle of subscription services of more than 50 databases on occupational health and safety information are fully operable in both English and French. Contains over 70,000 MSDSs as well as CHEMINFO database of hazardous information on over 1,044 chemicals. Available online through CCINFOline; on CD-ROM through CCINFOdisc.

CCRIS - Chemical Carcinogenicity Research Information System

National Library of Medicine
8600 Rockville Pike
Bethesda, MD 20894
Phone: (800) 272-4787
FAX: (301) 496-0822
Contains individual assay results and test conditions for over 5,000 chemicals in the areas of carcinogenicity, cocarcinogenicity, mutagenicity, and tumor promotion. Built and maintained by the National Cancer Institute.

CERCLIS - Comprehensive Environmental Response, Compensation, and Liability Informatiion System

Office of Solid Waste and Emergency Response (OSW)
U.S. Environmental Protection Agency
Washington, DC
Phone: (202) 260-8321; (800) 775-5037
FAX: (703) 603-9133
The Superfund database containing information on all aspects of hazardous waste sites from initial discovery to listing on the National Priorities List. Magnetic tapes are available quarterly from NTIS. Summary data under the Freedom of Information Act is available free by calling the Superfund Automated Phone System, (800) 775-5037.

CESARS - Chemical Evaluation Search & Retrieval System

Michigan State Department of Natural Resources and the Ontario Ministry of the Environment
Great Lakes and Environmental Assessment Section
Knapp's Office Centre
P.O. Box 30273
Lansing, MI 48909
Phone: (517) 373-2190; (517) 335-3308
FAX: (517) 373-9958
Provides detailed, evaluated, fully referenced profiles of more than 800 chemicals of environmental importance to the Great Lakes Basin. Available online through CCINFOline from the Canadian Centre for Occupational Health and Safety (CCOHS) and Chemical Information System (CIS); on CD-ROM through CCINFOdisc.

CHEMHAZIS - Chemical Hazards Information System

Chemical Information System (CIS)
810 Gleneagles Court, Ste. 300
Towson, MD 21286
Phone: (800) CIS-USER
FAX: (410) 296-0712
Contains information drawn from data compiled by the National Toxicology Program (NTP), which coordinates and provides information about potentially toxic chemicals with potential for human exposure to regulatory and research agencies. Contains information on 2,280 chemicals drawn from literature. The records closely resemble an MSDS format. Available online through Chemical Information System (CIS).

CHRIS - Chemical Hazard Response Information System

Office of Marine Safety, Security and Environmental Protection
U.S. Coast Guard
2100 Second St., S.W., G-MER-2
Washington, DC 20590
Phone: (800) 424-8802
FAX: (202) 426-7881
Provides emergency response and chemical handling information for more than 1,200 chemical substances that would be involved in chemical transport accidents, particularly by water. Contains information on labeling, physical and chemical properties, health and fire hazards and hazard classifications, chemical reactivity, and water pollution. Includes safety procedures for preventing emergency situations. Available online through Chemical Information System (CIS) and Canadian Centre for Occupational Health and Safety (COOHS); on CD-ROM through CCINFOdisc, SilverPlatter CHEM-BANK, as part of TOMES Plus System; on magnetic tape through TOMES Plus System.

Clu in - Clean Up Information

WEB site: www.epa.clu-in.com
Provides information on innovative treatment technologies for hazardous waste cleanup professionals. It contains site remediation technologies, the citizen's guide and newsletter on innovative treatment technologies, and regulatory information.

Earth Law

IHS Regulatory Products
15 Inverness Way East
Englewood, CO 80112
Phone: (800) 525-7052; (800) 320-5660
FAX: (303) 486-1710
A commercial subscription service offering existing and proposed EPA and OSHA environmental, health and safety legislation, rules, regulations, and notices at the state, federal, and international levels.

Envirofacts

WEB site: www.epa.gov/enviro/html/ef_home.html
Contains data from five major EPA databases.

IRIS - Integrated Risk Information System Database

U.S. Environmental Protection Agency
Office of Research and Development
26 W. Martin Luther King Dr.
Cincinnati, OH 45268
Phone: (703) 487-4650
Hotline: (513) 569-7254
Web site: www.epa.gov/iris
Provides information on how levels of exposure of hazardous chemicals affect human health. Covers levels of exposure to hazardous chemicals below which no adverse health effects are expected to occur in various segments of the human population. The reference dose and carcinogenicity assessments on IRIS can serve as guides in evaluating potential health hazards and selecting response to alleviate a potential risk to human health. Hours: 8:00 a.m. to 4:40 p.m. EST, Monday through Friday.

MEDLINE

National Library of Medicine
8600 Rockville Pike
Bethesda, MD 20894
Phone: (800) 272-4787
FAX: (301) 496-0822
A comprehensive index to world medical and health literature. Especially useful for researching toxicology of a particular chemical; for biological monitoring of a particular chemical; for linking a disease to chemical exposure, and similar medical and health matters.

NIOSHTIC - NIOSH Technical Information Center Database

Standards Development and Technology Transfer Division
Technical Information Branch
U.S. National Institute for Occupational Safety and Health 4676 Columbia Pkwy.
Cincinnati, OH 45226
Phone: (800) 356-4674
FAX: (513) 533-8575
WEB site: www.cdc.gov/niosh

Bibliographic database of more than 2,000 journals and 70,000 monographs on all aspects of occupational safety and health. Available through DIALOG Information Services, MEDLARS as part of TOXLINE, Silver Platter, and other commercial systems.

OLS - Online Library System

Public Information Center (PIC)
U.S. Environmental Protection Agency
406 M. St., SW
Washington, DC 20460
Phone: (800) 334-2405 or (919) 541-7862 for technical assistance.
FAX: (202) 260-6257
E-mail: public-access@epamail.epa.gov
A computerized list of the EPA national publications catalog and a list of departments and services. OLS includes several related databases that can be used to locate books, reports, articles, and information on a variety of topics.

OSWRCRA

WEB site: www.epa.gov/oswrcra/hazwaste
Contains information on hazardous waste identification and recycling, used oil, land disposal restrictions, permits and permitting, waste minimization and pollution prevention and radioactive mixed wastes. Waste data includes: RCRIS Treatment, storage, and disposal facilities, RCRIS Handlers with Corrective Action Activity, RCRIS RCRA Notifiers Lists, and RCRIS Other reports.

Polymers DotCom

WebLabs Publications
P.O. Box 1035
Leominster, MA 01453
Phone: (508) 345-6200
FAX: (508) 342-2020
WEB site: www.polymers.com
An Internet magazine featuring an extensive index of plastics industry Web sites; technology and business content; and online resources for plastics professionals.

Polysort

California Film Extruders and Converters Association (CFECA)
2402 Vista Nobleza
Newport Beach, CA 92660
Phone: (949) 644-7659
FAX: (949) 640-9911
E-mail: nsfox@earthlink.net
WEB site: www.polysort.com
A Web site for the polyethylene film industry. Provides links to industry associations, government site, company members, and suppliers.

Reg-Trieve Library of Federal Regulations

American Labelmark Company
Seattle WA, 1999
Phone: (800) 578-4955
FAX: (206) 443-0529
WEB site: www.ttrieve.com.

RTECS - Registry of Toxic Effects of Chemical Substances

National Library of Medicine
Specialized Information Services
8600 Rockville Pike
Bethesda, MD 20894
Phone: (800) 272-4787
FAX: (301) 480-3536
Provides toxicology data for more than 122,700 chemicals in four categories: substance identification; toxicity/biomedical effects; toxicology and carcinogenicity review; and exposure standards and regulations. Built and maintained by the National Institute for Occupational Safety and Health (NIOSH) as a segment of the Toxic Release Inventory database.

TOMES PLUS Information System

Micromedex Inc
6200 S. Syracuse Way, Ste. 300
Englewood, CO 80111-4740
Phone: (303) 486-6400
FAX: (303) 486-6464
E-mail: info@mdx.com
WEB site: www.micromedex.com
Contains a collection of proprietary and government databases on toxicology, environment, and industrial medicine and the hazardous effects related to chemical exposure in manufacturing and transportation, and gives the proper response in chemical emergency situations. Contains several proprietary files as well as files of major international journals, standard reference sources, professional specialists, chemical manufacturers, government agencies, and poison centers. Databases include Hazardtext Hazards Management, SARAtext System, HSDB (Hazardous Substance Data Bank) from NLM, CHRIS from the U.S. Coast Guard, IRIS, and New Jersey Fact Sheets.

TOXLINE - Toxicology Literature Online

National Library of Medicine
Specialized Information Services
8600 Rockville Pike
Bethesda, MD 20894
Phone: (800) 272-4787
FAX: (301) 480-3537
A centralized database to published sources of information on the pharmacological, biochemical, physiological, and toxicological effects of drugs and other chemicals.

Vermont SIRI - Safety Information Resources, Inc.

University of Vermont
E-mail: dan@siri.org
WEB site: www.haz1.siri.org
An exhaustive source of information: MSDSs; links to other MSDS sites; links to safety sites on the internet; links to occupational safety and health and environmental organizations and research programs; OSHA and EPA regulations; NIOSH databases; discussion boards; and much more.

GOVERNMENT AGENCIES

Agency for Toxic Substances and Disease Registry (ATSDR)

U.S. Public Health Service
Centers for Disease Control and Prevention (CDC)
1600 Clifton Rd., N.E., MS E-29
Atlanta, GA 30333
Phone: (404) 452-4100; (404) 639-0727; (404) 639-0700
Hotline: (888) 422-8737, 24-hour emergency number.
E-mail: atsdric@cdc.gov
WEB site: www.atsdr.atsdr.cdc.gov:/atsdrhome.html

Air Risk Information Support Center (Air RISC)

Office of Air Quality Planning and Standards (MD-13)
Air RISC
U.S. EPA
Research Triangle Park, NC 27711
Hotline: (919) 541-0888, Thursday 8:00 a.m. to 5:00 p.m., EST; Friday 8:00 a.m. to 4 p.m., EST.

Asbestos Action Program

Office of Pesticides and Toxic Substances
Federal Asbestos Task Force
U.S. Environmental Protection Agency (EPA)
401 M St., S.W.
Washington, DC 0460
Phone: (202) 382-3862
Hotline: (202) 554-1404
Coordinates public programs administered by the federal government that control potential health hazards from exposure to asbestos and other fibrous materials.

Canadian Pollution Prevention Information Clearinghouse (CPPIC)

The National Office of Pollution Prevention
Phone: (800) 667-9790
FAX: (519) 337-3486
WEB site: www.ec.gc.ca/cppic

Cancer Information Service

Office of Cancer Communication
National Cancer Institute
NCI Building 31, Room 10A1B
Bethesda, MD 20892
Phone: (800) 442-6237

Center for Environmental Information and Statistics (CEIS)

401 M St., S.W.
Washington, DC 20460
Phone: (202) 260-2090 - EPA Locator; (202) 260-2080 - Public Information Center
E-mail: ceis@epa.gov
WEB site: www.epa.gov/ceis
A one-stop source of data and information on environmental quality, status and trends, including environmental profiles, the Digital Library of Environmental Quality, and the environmental Atlas.

Center for Environmental Research Information (CERI)

U.S. Environmental Protection Agency (EPA)
Office of Research and Development (ORD)
26 W. Martin Luther King Dr.
P.O. Box 12505
Cincinnati, OH 45268
Phone: (513) 569-7562
FAX: (513) 569-7566
WEB site: www.epa.gov/ord/
CERI is the focal point for the exchange of scientific and technical environmental information produced by the Office of Research and Development (ORD), its 12 research laboratories, and associated programs nationwide.

Environment Canada

351 St. Joseph Blvd.
Hull, Quebec K1A 0H3
Phone: (819) 997-2800; (800) 668-6767
FAX: (819) 953-2225
E-mail: enviroinfo@ec.gc.ca
WEB site: www.ec.gc.ca

National Air Toxics Information Clearinghouse (NATICH)

U.S. Environmental Protection Agency (EPA)
Office of Air Quality Planning and Standards, MD-13
Research Triangle Park, NC 27711
Phone: (919) 541-0850
FAX: (919) 541-4028
Maintains a database which collects and classifies all of the information about air toxics control programs submitted by federal, state, and local agencies. The information is organized by agency,

pollutant, and emission source. Serves state and local agencies and the public. Hours: Monday through Thursday, 8:00 a.m.- 5:00 p.m., EST, Friday, 8:00 a.m.- 4:00 p.m. EST.

National Center for Environmental Publications and Information (NCEPI)

U.S. Environmental Protection Agency (EPA)
P.O. Box 42419
Cincinnati, OH 45242-2419
Phone: (513) 489-8190
FAX: (513) 891-6685
WEB site: www.epa.gov/ncepihom/index.html
The central clearinghouse for the exchange of scientific, technical, and public-oriented environmental information published by the Environmental Protection Agency. Provides copies of EPA documents and multimedia products upon request. It operates Monday through Friday, 8:00 a.m.- 5:30 p.m. EST.

National Health Information Center (NHIC)

U.S. Public Health Service
Office of Disease Prevention and Health Promotion (ODPHP)
P.O. Box 1133
Washington, DC 20013-1133
Phone: (800) 336-4797
WEB site: www.nhic-net.health.org
NHIC puts health professionals and consumers in touch with those organizations that are best able to provide answers.

National Institute of Environmental Health Sciences (NIEHS)

National Institutes of Health (NIH)
P.O. Box 12233
Research Triangle Park, NC 27709
Phone: (800) 643-4794; (919) 541-3201; Communications Office, (919) 541-2605
FAX: (919) 541-2260
E-mail: webcenter@niehs.nih.gov
WEB site: www.niehs.nih.gov
Supports and conducts basic research focusing on the interaction between humans and potentially toxic or harmful agents in the environment. Its research is the basis for preventive programs for environment-related disease and for action by regulatory agencies. Has a basic program to research problems of hazardous wastes. Has one of the world's largest research libraries. Operates the Environmental Health Clearinghouse and the National Toxicology Program. Provides research grants to Environmental Health Sciences Centers at universities and medical schools.

National Institute for Occupational Safety and Health (NIOSH)

1600 Clifton Road
Atlanta, GA 30333
Phone: (800) 356-4674
E-mail: pubstaft@cdc.gov
WEB site: www.niosh.cdc.gov/niosh

NIOSH is part of the Centers for Disease Control and Prevention (CDC) and is the only federal institute responsible for conducting research and making recommendations for the prevention of work-related illnesses and injuries.

Office of Air Quality Planning and Standards (OAQPS)

Office of Air and Radiation
U.S. Environmental Protection Agency (EPA)
523 Mutual Plaza (MD-10)
Research Triangle Park, NC 27711
Phone: (919) 541-5618
FAX: (919) 541-5663
WEB site: www.epa.gov/oar

Office of Emergency and Remedial Response (Superfund)

Emergency Response Division
U.S. Environmental Protection Agency (EPA)
Crystal Gateway One, 13th Fl.
1235 Jefferson Davis Hwy.
Arlington, VA 22202
Phone: (703) 603-8760
E-mail: superfund.info@epa.gov
WEB site: www.epa.gov/superfund

Office of Pollution Prevention and Toxics (OPPT)

U.S. Environmental Protection Agency (EPA)
401 M St., SW
Washington, DC 20460
Phone: (202) 260-3810
WEB site: www.epa.gov/opptintr/

Office of Radiation and Indoor Air (ORIA)

Office of Air and Radiation (OAR)
U.S. Environmental Protection Agency (EPA)
Waterside Mall, Rm. 201
401 M St., S.W.
Washington, DC 20460
Phone: (202) 564-9320
FAX: (202) 564-9651
WEB site: www.epa.gov/oar/

Office of Toxic Substances (OTS)

Office of Prevention, Pesticides and Toxic Substances (OPPTS)
Office of Pollution Prevention and Toxics (OPPT)
U.S. Environmental Protection Agency (EPA)

401 M St., S.W., Rm. 539
Washington, DC 20460
Phone: (202) 260-3810; (202)554-1404 TSCA Assistance Information Service
E-mail: ow-general@epamail.epa.gov
WEB site: www.epa.gov/ow

Office of Water (OW)

Office of Groundwater and Drinking Water
U.S. Environmental Protection Agency (EPA)
401 M St., S.W., Rm. 1011
Washington, DC 20460
Phone: (202) 260-5543
FAX: (202) 260-4383

E-mail: ow-general@epamail.epa.gov
WEB site: www.epa.gov/ow

Public Information Center (PIC)

U.S. Environmental Protection Agency (EPA)
401 M St. S.W., PM-3404
Washington, DC 20460
Phone: (202)260-3059
FAX: (202)260-6257
E-mail: public-access@epamail.epa.gov
Distributes general, nontechnical information about environmental issues. Refers technical information requests to the appropriate EPA office or other government agencies. Information is available on hazardous wastes, drinking water, air and water pollution, pesticides, radon, indoor air quality, Superfund, wetlands, and many other environmental topics. Hours: 9:00 a.m. to 4:30 p.m. EST, Monday through Friday.

Pollution Prevention Information Clearinghouse (PPIC)

U.S. Environmental Protection Agency (EPA)
Office of Prevention, Pesticides, and Toxic Substances (OPPTS)
401 M Street, SW (7407)
Washington, DC 20460
Phone: (202) 260-1023
FAX: (202) 260-4659
E-mail: ppic@epamail.epa.gov
WEB site: www.epa.gov/earth100
Hours: 8:30 a.m. to 4:00 p.m. EST, 24-hour voice mail.

U.S. Department of Energy

Office of Civilian Radioactive Waste Management (OCRWM)
Forrestal Bldg.
1000 Independence Ave., SW

Washington, DC 20585
Phone: (202) 586-4940
WEB site: www.rw.doe.gov

U.S. Department of Transportation (DOT)

400 7th St., SW
Washington , DC 20590
Phone: (202) 366-5580
WEB site: www.dot.gov
Responsible for setting safety standards for rail, highway, air, and water transportation and providing law enforcement and traffic management for airspace and waterways. Also regulates manufacturers of containers and transporters of hazardous materials.

U.S. Environmental Protection Agency (EPA)

401 M St., S.W.
Washington, DC 20460
Phone: (202) 260-2090 - EPA Locator; (202) 260-2080 - Public Information Center; (919) 541-2777 - Library services, North Carolina; (202) 260-5921 - EPA Headquarters Library
FAX: (202) 260-7883; (202) 260-8909-Exposure Assessment Group
WEB site: www.epa.gov
The agency's mission is to control and abate pollution in the areas of air, water, solid waste, pesticides, radiation, and toxic substances. The EPA coordinates and supports research and antipollution activities by state and local governments, private and public groups, individuals, and educational institutions.

U.S. Occupational Safety and Health Administration (OSHA)

U.S. Department of Labor
200 Constitution Ave. NW
Washington, DC 20210
Phone: (202) 693-1999
FAX: (202) 219-5986
WEB site: www.osha.gov

PROFESSIONAL AND TRADE ASSOCIATIONS

Acrylic Council, Inc.

1285 Avenue of the Americas, 35th Floor
New York, NY 10019
Phone: (212) 554-4040
FAX: (212) 554-4042
Web site: www.fabriclink.com

Alliance for the Polyurethanes Industry (API)

1300 Wilson Blvd., Ste. 800
Arlington, VA 22209
Phone: (703) 253-0656

FAX: (703) 253-0658
E-mail: polyu@socplas.org
WEB site: www.polyurethane.org

Alliance of Foam Packaging Recyclers

2124 Priest Bridge Dr., Ste.19
Crofton, MD 21114
Phone: (410) 451-8340
FAX: (410) 451-8343
WEB site: www.presstar.com

All India Plastic Manufacturers' Association (ALPMA)

A-52, Street No. 1, M..I.D.C.
Marol, Andheri (E), Mumbai-400093
Phone: 91-22-821 7324; 91-22- 821 7325
FAX: 91-22-821 6390
E-mail: info@aipma.org
www.aipma.org

American Chemical Society (ACS)

Division of Polymer Chemistry, Inc.
1155 16th Street, NW
Washington, D.C. 20036
Phone: (202) 872-4600; (800) 227-5558
FAX: (202) 776-8258
E-mail: nbyerly@vt.edu
WEB site: www.acs.org

American Conference of Governmental Industrial Hygienists (ACGIH)

1330 Kemper Meadows Dr., Ste. 600
Cincinnati, OH 45240
Phone: (513) 742-2020
FAX: (513) 742-3355
E-mail: mail@acgih.org
WEB site: www.acgih.org

American Electroplaters and Surface Finishers Society (AESF)

12644 Research Parkway
Orlando, FL 32826-3298
Phone: (407) 281-6441
FAX: (407) 281-6446
E-mail: aesf@aesf.org
WEB site: www.aesf.org

American Fiber Manufacturers Association, Inc. (AFMA)

1150 17th Street, NW, Suite 310
Washington, DC 22036
Phone: (202) 296-6508
FAX: (202) 296-3052
E-mail: afma@aol.com
WEB site: www.fibersource.com

American Industrial Hygiene Association (AIHA)

2700 Prosperity Ave., Ste. 250
Fairfax, VA 22031
Phone: (703) 849-8888
FAX: (703 207-3561
WEB site: www.aiha.org

American Plastics Council (APC)

1300 Wilson Blvd., Ste. 800
Arlington, VA 22209
Phone: (703) 253-0700
FAX: (703) 253-0701
WEB site: www.ameriplas.org
The American Plastics Council is a national trade association representing 26 of the nation's largest resin producers, including monomer and polymer production and distribution.

American Institute of Chemical Engineers (AICHE)

3 Park Ave.
New York, NY 10016-5991
Phone: (212) 591-7338; (800) 242-4363
FAX: (212) 591-8897
E-mail: xpress@aiche.org
WEB site: www.aiche.org

American Polyolefin Association, Inc. (APA)

2312 E. Mall
Ardentown, DE 19810
Phone: (302) 475-1450
FAX: (302) 475-2357
E-mail: apa@itqc.com
WEB site: www.apa-polyolefin.com

Asian Plastics Research Association (APRA)

GPO Box 2609
Canberra ACT 2601, Australia
Phone: 61-6-296 1933

FAX: 61-6-296 1937
E-mail: sira@netinfo.com.au
WEB site: www.users.netinfo.com.au/sira/aprahome

Asociacion de Investigacion de Materiales Plasticos (AIMPLAS)

Valencia Parc Technologic
Calle Gustave Eiffel, 4
Apartado de Corres 51
46980 Paterna Valencia, Spain
Phone: +34 96 136 60 40
FAX: +34 96 136 60 41
E-mail: info@aimplas.es
WEB site: www.aimplas.es

Asociacion Nacional de Industrias del Plastico (ANIPAC)

Av. Parque Chapultepec No. 66-301 3er Piso, Col. El parque
53390 Naucalpan, Edo. de Mexico
Phone: (52) 576 5547
FAX: (52) 576 5548
E-mail: anipac@inetcorp.net.mx
WEB site: www.inetcorp.net.mx/anipac
WEB site: www.aimplas.es

Association of Industrial Metalizers, Coaters and Laminators (AIMCAL)

5005 Rockside Rd., Ste. 600
Cleveland, OH 44131
Phone: (216) 573-3773
FAX: (216) 573-3783
E-mail: aimcal@aimcal.org
WEB site: www.aimcal.org

Association of Plastics Manufacturers in Europe (APME)

Avenue E van Nieuwenhuyse 4
Box 3
B-1160 Brussels, Belgium
Phone: +32 26 72 82 59
FAX: +32 26 75 39 35
E-mail: info.apme@apme.org
WEB site: www.apme.org

Association of Rotational Molders (ARM)

2000 Spring Rd., Ste.511
Oak Brook, IL 60523
Phone: (630) 571-0611

FAX: (630) 571-0616
E-mail: rotomolding@compuserve.com
WEB site: www.rotomolding.org

British Plastics Federation (BPF)

6 Bath Place, Rivington St.
London, England EC2A 3JE
Phone: 0207 457 5000
FAX: 0207 457 5045
WEB site: www.bpf.co.uk
Represents more than 400 manufacturers of polymers, additives, and machinery as well as all types of plastics processors and fabricators. Membership is divided into 15 Business Groups sharing four Market Sector Groups.

Canadian Centre for Occupational Health and Safety (CCOHS)

250 Main St. East
Hamilton, ON, Canada L8N 1H6
Phone: (800) 668-4284 (toll-free in Canada and USA)
FAX: (905) 572-2206
E-mail: custserv@ccohs.ca
WEB site: www.gala.ccohs.ca

Canadian Plastics Industry Association (CPIA)

5925 Airport Rd., ste. 500
Mississauga, ON L4V 1W1
Phone: (905) 678-7748
FAX: (905) 678-0774
E-mail: national@cpia.ca
WEB site: www.plastics.ca

Center for Excellence for Composites Manufacturing Technology (CECMT)

8401 Lake View Pkwy., Ste. 200
Kenosha, WI 53142-7403
Phone: (414) 947-8900
FAX: (414) 947-8919
A department in the Office of Naval Research.

Chemical Industry Institute of Technology (CIIT)

P.O. Box 12137
Research Triangle Park, NC 27709
Phone: (919) 558-1200
FAX: (919) 558-1300
E-mail: ciitinfo@ciit.org
WEB site: www.ciit.org

Chemical Manufacturers Association (CMA)

1300 Wilson Blvd.
Arlington, VA 22209
Hotline: (800) 424-9300; Nonemergency, (800) 262-8200
FAX: (703) 741-6037
WEB site: www.cmahq.com

Chemical Specialties Manufacturers Association (CSMA)

1913 Eye St., NW
Washington, DC 20006
Phone: (202) 872-8110
E-mail: info@asma.org

The Chlorine Institute, Inc.

2001 L Street N.W., Suite 506
Washington, DC 20036
Phone: (202) 775-2790
FAX: (202) 223-7225
WEB site: www.cl2.com

Color Pigments Manufacturer's Association, Inc.

300 N. Washington St., Ste. 102
Alexandria, VA 22314
Phone: (703) 684-4044
FAX: (703) 684-1795
WEB site: www.chemicalprocessing.com

Ethylene Oxide Industry Council

2501 M St. NW, Ste. 330
Washington, DC 20037
Phone: (202) 887-1198

European Council for Plasticisers and Intermediates (ECPI)

Avenue E. Van Nieuwenhuyse 4, Box 1
B-1160 Brussels, Belgium
Phone: 32 26 76 7260
FAX: 32 26 76 7216
E-mail: ccr@cefic.be
WEB site: www.ecpi.org

Fire Retardant Chemical Association

851 New Holland Ave., Box 3535

Lancaster, PA 17604
Phone: (717) 291-5616
FAX: (717) 295-4538

Halogenated Solvents Industry Alliance (HSIA)

2001 L St., NW
Washington, D.C. 20036
Phone: (202) 775-0232
FAX: (202) 833-0381
E-mail: info@hsia.org
WEB site: www.hsia.org

Japan Plastics Industry Federation

c/o Kaseihin Kaikan
5-18-17 Roppongi
Minato-ku, Tokyo 106-0032
Phone: +81 3 3586 9761
FAX: +81 3 3586 9760
WEB site: www.jinjapan.org/jd/org

National Association of Chemical Recyclers

1875 Connecticut Ave., NW
Suite 1200
Washington, DC 20009
Phone: (202) 986-8150
FAX: (202) 986-2021

National Association for PET Container Resources (NAPCOR)

2105 Water Ridge Parkway, Ste. 570
Charlotte, NC 28217
Phone: (704) 423-9400
FAX: (704) 423-9500
E-mail: jwindischmann@napcor.com
WEB site: www.napcor.com

National Association of Plastic Distributors (NAPD)

6333 Long Street, Suite 340
Shawnee, KS 66216
Phone: (913) 268-6273
FAX: (913) 268-6388

National Fire Protection Association (NFPA)

1 Battrymarch Park
P.O.Box 9101

Quincy, MA 02269-9101
Phone: (800) 770-3000
FAX: (617) 770-0700
E-mail: webmaster@nfpa.org
WEB site: www.nfpa.org

National Paint and Coatings Association (NPCA)

1500 Rhode Island Ave., NW
Washington, DC 20005
Phone: (202) 462-6272
FAX: (202) 462-8549
E-mail: npca@paint.org
WEB site: www.paint.org

Plastics Institute of America, Inc.

333 Aiken St.
Lowell, MA 01854
Phone: (508) 934-3130
FAX: (508) 459-9420

Plastics Institute of New Zealand (PINZ)

Level 4, Leyton House
Manukau City Shopping Centre
P.O. Box 76-378
Manukau City, NZ
Phone: 0-9-262 3773
FAX: 0-9-262 3850
E-mail: pinz@plastics.org.nz
WEB site: www.plastics.org.nz

Polyurethane Foam Association (PFA)

P.O. Box 1459
Wayne, NJ 07474-1459
Phone: (973) 633-9044
FAX: (973) 628-8986
E-mail: rtuedeka@pfa.org
WEB site: www.pfa.org

Polyurethane Manufacturers Association (PMA)

800 Roosevelt Road, Bldg.C, Ste. 20
Glen Ellyn, IL 60137-5833
Phone: (630) 858-2670
FAX: (630) 790-3095

E-mail: infor@pmahome.org
WEB site: www.pmahome.org

Society of Plastics Engineers (SPE)

14 Fairfield Drive
Brookfield, CT 06804-0403
Phone: (203) 775-0471
FAX: (203) 775-8490
E-mail: info@4spe.org
WEB site: www.4spe.org

Society of the Plastics Industry, Inc. (SPI)

1801 K Street, NW, Suite 600K
Washington, DC 20006-1301
Phone: (202) 974-5200
FAX: (202) 296-7005
WEB site: www.socplas.org

In addition to its general services, SPI has 28 business units as well as numerous key services offering programs specifically geared to the interests of particular industry segments. These special-purpose groups include the Degradable Polymers Council, which acts as a clearinghouse for research in the degradable plastics industry, and the Polymeric Materials Producers Division, which includes manufacturers of basic polymers or prepolymers for the plastics industry. Other industry segment groups which focus on particular plastic resins include the Fluoropolymers Division, Naphthalate Polymers Council, the Phenolic Division, the Polyurethane Division, the Styrene Information and Research Center, and the Vinyl Institute. SPI also has an affiliation with the American Plastics Council, which includes U.S. resin and monomer producers, plastics processors, and suppliers. Contact information for these groups is listed below.

American Plastics Council (APC)
1300 Wilson Blvd., Ste. 800
Arlington, VA 22209
Phone: (703) 253-0700
FAX: (703) 253-0701
WEB site: www.ameriplas.org

Composites Institute
600 Mamaroneck Ave.
Harrison, NY 10528
Phone: (914) 381-3572
WEB site: www.recycle.net

Degradable Polymers Council
Phone: (202) 974-5245
E-mail: dpc@socplas.org

Film and Bag Federation
P.O. Box 2811
Pittsburgh, PA 15230
Phone: (800) 438-5856
E-Mail: pbainfo@aol.com
WEB site: www.plasticbag.com

Flexible Packaging Association
1090 Vermont St., NW
Washington, DC 20005
Phone: (202) 842-3880
FAX: (202) 842-3841
WEB site: www.flexpack.org

Fluoropolymers Division (FPD)
Phone: (202) 974-5233
FAX: (202) 293-0005
WEB site: www.fluoropolymers.org

Naphthalate Polymers Council (NPC)
Phone: (202) 974-5245
WEB site: www.napcor.com

Phenolic Division
Phone: (202) 974-5233
WEB site: www.phenolics.org

Polymeric Materials Producers Division
Phone: (202) 974-5319
E-mail: pmd@socplas.org

Polystyrene Packaging Council (PSPC)
Phone: (202) 974-5321
FAX: (202) 296-7354
WEB site: www.polystyrene.org

Polyurethane Division
Phone: (212) 351-5242
E-Mail: polyu@socplas.org

Styrene Information and Research Center (SIRC)
Phone: (202) 974-5319
E-mail: sirc@socplas.org

Syndicat des Producteurs de Matieres Plastiques (SPMP)

14, rue de la Republique
92800 Puteaux, France
Phone: 01 46 53 10 53

FAX: 01 46 53 10 73
WEB site: www.spmp.proplast.org

Synthetic Organic Chemicals Manufacturer's Association (SOCMA)

900 2nd St. NE, Ste. 109
Washington, DC 20002
Phone: (202) 721-4162
FAX: (202) 296-8120
WEB site: www.socma.com
It's Association Management Center also manages several industry-specific groups in the organic chemicals field, including the Acrylonitrile Group; Alkyl Amines Group; Biphenyl Work Group; Institute for Polyacrylate Absorbents; Phosphonates Task Force; and Chlorobenzene Producers Association.

Tube and Pipe Association, International (TPFA)

833 Featherstone Rd.
Rockford, IL 61107
Phone: (815) 399-8700
FAX: (815) 339-7279
WEB site: www.tpatube.org

Plastics Industry-Related Environmental Programs and Activities

The Global Environmental Management Initiative (GEMI) is made up of a group of leading companies dedicated to fostering environmental excellence by business. GEMI promotes a worldwide business ethic for environmental management and sustainable development, to improve the environmental performance of business through example and leadership.

Center for Waste Reduction Technologies, under the aegis of the American Institute of Chemical Engineers, sponsors research on innovative technologies to reduce waste in the chemical processing industries. The primary mechanism is through funding of academic research.

The American Plastics Council is working on a life-cycle study to examine the emissions released from plastics and resins manufacturing facilities. The study will compare emissions from plastics and resins manufacturing with manufacturing of other materials, such as wood products.

The National Science Foundation and the **Environmental Protection Agency's Office of Pollution Prevention and Toxics** signed an agreement to coordinate the two agencies' programs of basic research related to pollution prevention. The collaboration will stress research in the use of less toxic chemical and synthetic feedstocks, use of photochemical processes instead of traditional ones that employ toxic reagents, use of recyclable catalysts to reduce metal contamination, and use of natural feedstocks when synthesizing chemicals in large quantities.

The Chemical Manufacturer's Association funds research on issues of interest to their members, particularly in support of their positions on proposed or possible legislation. They recently funded a study to characterize the environmental fate of organochlorine compounds.

The Responsible Care® Initiative of the Chemical Manufacturer's Association requires all members and partners to continuously improve their health, safety, and environmental performance in a manner that is responsive to the public. Launched in 1988, the Responsible Care® concepts are now being applied in 36 countries around the world. Responsible Care® is a comprehensive, performance-oriented initiative composed of ten progressive Guiding Principles and six board Codes of Management Practices. These Management Practices cover all aspects of the chemical industry's operations, from research to manufacturing, distribution, transportation, sales and marketing, and to downstream users of chemical products. Through Responsible Care®, CMA members and partners gain insight from the public through, among other means, a national Public Advisory Panel and over 250 local Community Advisory Panels. This, coupled with the fact that participation in Responsible Care® is an obligation of membership with the Chemical Manufacturer's Association, make this performance improvement initiative unique. The Synthetic Organic Chemical Manufacturer's Association whose membership consists of smaller batch and custom chemical manufacturers with typically fewer than 50 employees and less than $50 million in annual sales, encourages its members to achieve continuous performance improvement in their health, safety, and environmental programs through implementation of the chemical industry's Responsible Care® initiative. SOCMA is a partner in Responsible Care®.

The Society of the Plastics Industry has implemented two programs aimed at reducing plastic pellet loss. In 1991, SPI's Polymeric Materials Producers Division developed and endorsed a "Pellet Retention Environmental Code." Companies that sign the code commit themselves to the total containment of plastic pellets throughout the pellets' lifespan and to operating in full compliance with environmental laws and regulations pertaining to pellet containment (SPI, 1994). In 1992, SPI expanded the program to include a processor's pledge to uphold six principles to prevent the loss of resin pellets into the environment.

ISO 9000 is a series of international total quality management guidelines. After a successful independent audit of their management plans, firms are qualified to be ISO 9000 registered. In June of 1993, the International Standards Organization created a technical committee to work on new standards for environmental management systems.

REFERENCES RESOURCES

Alternative Formulations and Packaging to Reduce Use of Chlorofluorocarbons, Noyes Data Corporation, Park Ridge, NJ,1990.

Aqueous Cleaning as an Alternative to CFC and Chlorinated Solvent-Based Cleaning, Noyes Data Corporation, Park Ridge, NJ, 1991.

Assessment of Industrial Hazardous Waste Practices - Rubber and Plastic Industry, Foster D. Snell for the United States Environmental Protection Agency, Cleveland, OH, 1975.

Background Document on Clean Products Research and Implementation, United States Environmental Protection Agency, Washington, DC.

Best Management Practices for Pollution Prevention in the Textile Industry, United States Environmental Protection Agency, Office of Research and Development, September, 1995.

Book of Lists for Regulated Hazardous Substances, Government Institutes, Rockville, MD, 1996.

Case Studies from the Pollution Prevention Information Clearinghouse: Solvent Recovery, United States Environmental Protection Agency, ISM-4 (PPIC), Washington, DC, 1989.

Case Studies from the Pollution Prevention Information Center (PPIC): Electroplating, United States Environmental Protection Agency, Office of Environmental Engineering and Technology Demonstration and the Office of Pollution Prevention, Washington, DC, November 1989.

Case Studies in Waste Minimization, Government Institutes, Rockville, MD, 1991.

Case Study: Winning Ideas Reduce Waste at DOW, Nelson, K. and Lidsly, J., 1991.

CERCLA, Compliance with Other Laws Manual, Draft, OSWER Directive 9234-01, United States Environmental Protection Agency, *Washington, DC, 1988.*

Chemical Data Guide for Bulk Shipment by Water, United States Department of Transportation, United States Coast Guard, Washington, DC, 1990.

Chemical Engineers' Handbook, Perry, Robert H. and Cecil H. Chilton, New York: McGraw-Hill, New York, NY.

Chemical Fact Sheets, New York State Department of Health, Bureau of Toxic Substance Assessment, Albany, NY, various issues and dates.

Cleaner Production: An Industrial Example, Journal of Cleaner Production, volume 1, no. 1, Clements, J.W. and Thompson, J.P., 1993.

*CMA Waste Minimization Resource Manua*l, Chemical Manufacturers Association, Washington, DC, 1989.

Consolidated List of Chemicals Subject to the Emergency Planning and Community Right-to-Know Act (EPCRA) and Section 112 (r) of The Clean Air Act, As Amended, United States Environmental Protection Agency, Washington, DC, December 1996.

CRC Handbook of Lubrication; Theory and Practice Tribiology, Booser, E. R. ed., Vols. 1-2, CRC Press, Inc., Boca Raton, FL, 1985.

DCS's On-line Information Improves Resin Process Consistency, Instrumentation and Control Systems, volume 64, no. 11, Elley, D., 1991.

Degreaser System Pollution Prevention Evaluation, EPA/600/S2-90/052, United States Environmental Protection Agency, Washington, DC, 1990.

Designing Pollution Prevention into the Process - Research, Development and Engineering, Chemical Manufacturer's Association, Undated.

Designing Pollution Prevention in to the Process: Research Development and Engineering, Chemical Manufacturers Association, Washington, DC, 1993.

Development Document of Effluent Limitations Guidelines for the Organic Chemicals, Plastics and Synthetic Fibers Point Source Category, EPA 440/1-87/009, United States Environmental Protection Agency, 1987.

Development Document for Effluent Limitations Guidelines and Standards for the Organic Chemicals, Plastics, and Synthetic Fibers, Point Source Category, Volumes 1 and 2, United States Environmental Protection Agency, Office of Water Regulations and Standards, October, 1987.

Draft Polymer Exemption Guidance Manual, EPA, Office of Pollution Prevention and Toxics, United States Environmental Protection Agency, Washington, DC, March, 1995.

Electroplating Packet, United States Environmental Protection Agency, Pollution Prevention Information Center, Washington, DC.

Encyclopedia of Textiles, Fibers, and Nonwoven Fabrics, Grayson, M. (ed.), John Wiley and Sons, New York, 1984.

Encyclopedia of Chemical Processing and Design, Volume 39, McKetta, J.J. (ed.), Marcel Dekker, Inc., New York, 1992.

Encyclopedia of Associations, 33rd ed., Fisher, Carolyn A. and Carol A. Schwartz, eds., Gale Research Inc., Detroit, MI, 1998.

Encyclopedia of Polymer Science and Engineering, Volume 12, Kroschwitz, J.I. (ed.), John Wiley and Sons, New York, 1986.

Environmental Dividends: Cutting More Chemical Wastes, Dorfman, M.H. et al., INFORM Inc., New York, NY, 1995.

Environmental Fact Sheet - Plastic: The Facts About Production, Use, and Disposal, United States Environmental Protection Agency, Office of Solid Waste and Emergency Response, Washington, DC, February, 1990.

Environmental Fact Sheet - Plastic: The Facts on Source Reduction, United States Environmental Protection Agency, Office of Solid Waste and Emergency Response, Washington, DC, February, 1990.

Environmental Fact Sheet - The Facts on Degradable Plastic, United States Environmental Protection Agency, Office of Solid Waste and Emergency Response, Washington, DC, February, 1990.

Environmental Fact Sheet - The Facts on Recycling Plastic, United States Environmental Protection Agency, Office of Solid Waste and Emergency Response, Washington, DC, February, 1990.

Evaluation of the Hazard of Bulk Water Transportation of Industrial Chemicals, National Academy of Science, Washington, DC, 1973.

Facility Pollution Prevention Guide, United States Environmental Protection Agency, Office of Research and Development, Washington, DC, (EPA/600/R-92/088), May, 1992.

Fire Protection Guide to Hazardous Materials, 11th ed., National Fire Protection Association, NFPA 704 Hazard Ratings (Classifications) reprinted with permission, Quincy, MA, 1994.

Glossary of Chemical Terms, 2nd ed., Hampel, Clifford A. and Hawley, Gessner G., Van Nostrand Reinhold, New York, NY, 1982.

Guidelines for Waste Reduction and Recycling: Solvents, Oregon Department of Environmental Quality, ISM-13 (PPIC), 1989.

Guides to Pollution Prevention: The Fiberglass-Reinforced and Composite Plastics Industry, ISM-19 (PPIC), United States Environmental Protection Agency, Washington, DC, 1990.

Guides to Pollution Prevention: Research and Educational Institutions, United States Environmental Protection Agency, Washington, DC, 1990, ISM-19 (PPIC).

Handbook of Petrochemicals and Processes, Wells, G. Margaret, Aldershot, England: Gower Publishing Company, Wells, 1991.

Manufacture of Plastics: Volume 1, W.M. Smith (ed.), Reinhold Publishing Corporation, New York, 1964.

Handling Hazardous Materials, 1997 ed., Bomgardner, Paul M., ed., American Trucking Association, Alexandria, VA, 1997.

Hawley's Condensed Chemical Dictionary, 13th ed., Lewis, Richard J., Sr., Van Nostrand Reinhold, New York NY, 1998.

Hazardous Substance Resource Guide, 2nd ed., Pohanish, Richard P. and Stanley A. Greene, Gale Research, Inc., Detroit, MI, 1997.

Hazardous Waste Minimization Handbook, Lewis Publishers, Chelsea, MI, 1989.

Hazardous Substance Fact Sheets, New Jersey Department of Health, Right-to-Know Project, Trenton, NJ, various dates from 1985-1998.

Hazardous Chemical Desk Reference, 4th ed., Lewis, Richard J., Sr., Van Nostrand Reinhold, New York NY, 1997.

Hazardous Materials Handbook, Pohanish, Richard P. and Stanley A. Greene, Van Nostrand Reinhold, New York, NY, 1996.

Hazardous and Industrial Solid Waste Minimization Practices, ASTM, Philadelphia, PA,1989.

Hazardous Waste Minimization, McGraw-Hill, Inc., New York, NY,1990.

Industrial Aromatic Chemistry, Franck, H.G. and J.W. Stadelhofer, Springer-Verlag, Berlin, Germany, 1997.

Industrial Pollution Prevention for the 1990s, United States Environmental Protection Agency, Washington, DC EPA/600/S8-91/052., 1991.

Industrial Environmental Chemistry: Waste Minimization in Industrial Processes and Remediation of Hazardous Waste, Sawyer, Donald T., and Arthur E. Martell, Plenum Press, New York, NY, 1992.

Industry, Technology, and the Environment: Competitive Challenges and Business Opportunities, United States Congress, Office of Technology Assessment, OTA-ITE-586, United States Government Printing Office, Washington, DC, January 1994.

Industry and Trade Summary: Manmade Fibers, United States International Trade Commission, Washington, DC., 1995.

Introduction to Environmental Engineering and Science, Masters, G.M., Prentice-Hall, Inc., New York, 1991.

Kirk-Othmer Encyclopedia of Chemical Technology (appropriate volumes), John Wiley and Sons, New York.

Manufacture of Plastics: Volume 1, Smith, W.M. (ed.), Reinhold Publishing Corporation, New York, 1964.

McGraw-Hill Encyclopedia of Science & Technology, 6th Edition, McGraw-Hill Book Company, New York, NY, 1992.

Modern Plastics Encyclopedia, McGraw-Hill, Inc., New York, annually.

Natural and Manmade Textile Fibers: Raw Material to Finished Fabric, Linton, G. E., Duell, Sloan and Pearce, New York, 1966.

National Pretreatment Program Report to Congress, United States Environmental Protection Agency, *21W-4004, July, 1991.*

NIOSH/OSHA Occupational Health Guidelines for Chemical Hazards, R. Stricoff, Scott and Lawrence J. Partridge, Jr., eds, A.D. Little, Inc., United States Department of Health and Human Services, United States Department of Labor, Occupational Safety and Health Administration, Cincinnati, OH, 1997.

NIOSH Pocket Guide to Chemical Hazards, U. S. Department of Health and Human Services, Public Health Service, Centers for Disease Control and Prevention, National Institute for Occupational Safety and Health, Cincinnati, OH, June, 1997.

North American Emergency Response Guidebook, developed jointly by U. S. Department of Transportation Research and Special Programs Administration, Transport Canada and the Secretariat of Communications and Transportation of Mexico, U. S. Government Printing Office, Washington, DC, 1996.

Operation Clean Sweep: A Manual on Preventing Pellet Loss, Society of the Plastics Industry, Washington, DC, 1994.

Organic Building Blocks of the Chemical Industry, Szmant, H. Harry, John Wiley and Sons, New York, NY, 1989.

Plant Design and Economics for Chemical Engineers, Peters, Max S. and Klaus D. Timmerhaus , McGraw-Hill Book Company, New York.

Plastic Molding and Forming Point Source Category Effluent Limitations Guidelines; Pretreatment Standards and New Source Performance Standards, Federal Register, Final Regulation. 49 FR 49026, December 17, 1984.

Plastic Pellets in the Aquatic Environment: Sources and Recommendations, United States Environmental Protection Agency, Office of Water, Washington, DC., December, 1992.

Plastics Materials, 6th ed., Brydson, J.A., Butterworth-Heinemann Ltd., Oxford, England, 1995.

Pollution Prevention, Theodore, Louis, and Young C. McGuinn, Van Nostrand Reinhold, New York, 1992.

Pollution Prevention 1991 Progress on Reducing Industrial Pollutants, United States Environmental Protection Agency, Office of Pollution Prevention (EPA 21 P-3003), Washington, DC, October 1991.

Pollution Prevention Opportunities Checklist: Chemical Manufacturing, City of Los Angeles, FCLA-1-1 (PPIC), Los Angeles, CA.

Pollution Prevention Resource Manual, Chemical Manufacturers Association, Washington, DC, 1991.

Pollution Prevention in Industrial Processes: The Role of Process Analytical Chemistry, Breen, Joseph J., and Michael J. Dellarco, American Chemical Society, Washington, DC, 1992.

Pollution Prevention Benefits Manual: Volume 1 (Draft), United States Environmental Protection Agency, Washington, DC, 1989, WAM-1 (PPIC).

Pollution Prevention Strategies for Minimizing of Industrial Wastes in the Vinyl Chloride Monomer-Polyvinyl Chloride Industry, Randall, P.M., "Environmental Progress," volume 13, no. 4, November, 1994.

Pollution Prevention Fact Sheets: Chemical Production, FREG-1 (PPIC), United States Environmental Protection Agency, Washington, DC.

Polychlorinated Biphenyl (PWB) Information Package, TSCA Information Service, Washington, DC, April, 1993.

Polyester Film Division's Waste Minimization/Detoxification Activities, Smith, G.M., IV, Chemical Manufacturers Association Waste Minimization Workshop Proceedings, Washington, DC, 1987.

Preventing Pollution in the Chemical Industry: Five Years of Progress, Chemical Manufacturers Association, Washington, DC, 1992.

Prevention Reference Manual: User's Guide, Overview for Controlling Accidental Releases of Air Toxics, United States Environmental Protection Agency EPA/600/S8-87/028, Washington, DC, 1987.

Prevention Reference Manual: Overviews on Preventing and Controlling Accidental Releases of Selected Toxic Chemicals, United States Environmental Protection Agency EPA/600/S8-88/074, Washington, DC, 1988.

Prevention Reference Manual: Control Technologies. Volume 1: Prevention and Protection Technologies for Controlling Accidental Releases of Air Toxics, United States Environmental Protection Agency EPA/600/S8-87/039a, Washington, DC, 1987.

Prevention Reference Manual: Chemical Specific, Volume 1: Control of Accidental Releases of Hydrogen Fluoride (SCAQMD), United States Environmental Protection Agency EPA/600/S8-87/034a, Washington, DC, 1987.

Prevention Reference Manual: Chemical Specific, Volume 2: Control of Accidental Releases of Chlorine (SCAQMD), United States Environmental Protection Agency EPA/600/S8-87/034b, Washington, DC, 1987.

Prevention Reference Manual: Control Technologies. Volume 2: Post-Release Mitigation Measures for Controlling Accidental Releases of Air Toxics, United States Environmental Protection Agency EPA/600/S8-87/039b, Washington, DC, 987.

Prevention Reference Manual: Chemical Specific, Volume 3: Control of Accidental Releases of Hydrogen Cyanide (SCAQMD), United States Environmental Protection Agency EPA/600/S8-87/034c, Washington, DC, 1987.

Prevention Reference Manual: Chemical Specific, Volume 4: Control of Accidental Releases of Ammonia Cyanide (SCAQMD), United States Environmental Protection Agency EPA/600/S8-87/034d, Washington, DC, 1987.

Prevention Reference Manual: Chemical Specific, Volume 6: Control of Accidental Releases of Carbon Tetrachloride (SCAQMD), United States Environmental Protection Agency EPA/600/S8-87/034f, Washington, DC, 1987.

Prevention Reference Manual: Chemical Specific, Volume 7: Control of Accidental Releases of Chloropicrin Cyanide (SCAQMD), United States Environmental Protection Agency EPA/600/S8-87/034g, Washington, DC, 1987.

Prevention Reference Manual: Chemical Specific, Volume 8: Control of Accidental Releases of Hydrogen Fluoride, United States Environmental Protection Agency EPA/600/S8-87/034h, Washington, DC, 1987.

Prevention Reference Manual: Chemical Specific, Volume 9: Control of Accidental Releases of Chlorine, 1987, United States Environmental Protection Agency EPA/600/S8-87/034i, Washington, DC, 1987.

Prevention Reference Manual: Chemical Specific, Volume 10: Control of Accidental Releases of Hydrogen Cyanide, United States Environmental Protection Agency EPA/600-S8-87/034j, Washington, DC, 1987.

Prevention Reference Manual: Chemical Specific, Volume 11: Control of Accidental Releases of Ammonia, Pollution Prevention Resource Manual, United States Environmental Protection Agency EPA/600-S8-87/034k, Washington, DC, 1987.

Prevention Reference Manual: Chemical Specific, Volume 12: Control of Accidental Releases of Sulfur Dioxide, United States Environmental Protective Agency EPA/600/S8-87/034l, Washington, DC, 1987.

Prevention Reference Manual: Chemical Specific, Volume 13: Control of Accidental Releases of Methyl Isocyanate, United States Environmental Protection Agency EPA/600/S8-87/034m, Washington, DC, 1987.

Prevention Reference Manual: Chemical Specific, Volume 14: Control of Accidental Releases of Phosgene, United States Environmental Protection Agency EPA/600/S8-87/034n, Washington, DC, 1987.

Prevention Reference Manual: Chemical Specific, Volume 15: Control of Accidental Releases of Sulfur Trioxide, United States Environmental Protection Agency EPA/600/S8-87/034o, Washington, DC, 1987.

Principles of Polymer Systems, 4th ed., Rodriguez, F., Taylor and Francis, Washington, DC., 1996.

Proceedings of the International Workshop on Research in Pesticide Treatment/Disposal/ Waste Minimization, United States Environmental Protection Agency EPA/600-S9-91/047, Washington, DC, 1991 .

Profiting from Waste Reduction in Your Small Business, QAM-2 (PPIC), Alaska Health Project, 1988.

Rapid Guide to Chemical Incompatibilities, Pohanish, Richard P. and Stanley A. Greene, Van Nostrand Reinhold, New York, 1997.

Reducing Hazardous Waste Generation: An Evaluation and a Call for Action, National Academy Press, Washington, DC, 1985.

Registry of Toxic Effects of Chemical Substances, Sweet, Doris V. , ed., United States Department of Health and Human Services, Centers for Disease Control and Prevention, National Institute for Occupational Safety and Health, Cincinnati, OH, various dates.

Reigel's Handbook of Industrial Chemistry, 9th ed., Kent, J.(ed.), Van Nostrand Reinhold, New York, NY, 1992.

Retrospective Analysis of Compliance Strategies and Pollution Prevention in the Organic Chemicals, Plastics and Synthetic Fibers Industry, EPA, Office of the Administrator, (EPA Contract No. 68-C3-0302). United States Environmental Protection Agency, Washington, DC, 1993.

Sax's Dangerous Properties of Industrial Materials, 9th ed., Lewis, Richard J., Sr., Van Nostrand Reinhold, New York, NY, 1996.

SOCMA Pollution Prevention Study, Synthetic Organic Chemical Manufacturers Association, Washington, DC, January 1993.

Solvent Waste Reduction Alternatives, United States Environmental Protection Agency, EPA/625/4-89/021, Washington, DC, 1989.

Solvent Waste Reduction, Noyes, Data Corporation, Park Ridge, NJ, 1990.

Source Characterization and Control Technology Assessment of Methylene Choride Emissions from Eastman Kodak Company, Rochester, NY, United States Environmental Protection Agency EPA/600-S2-043., Washington, DC, 1989.

Standard Industrial Classification Manual, United States Office of Management and Budget, United States OMB, Washington, DC, annually.

A Study of Hazardous Waste Source Reduction and Recycling in Four Industry Groups in New Jersey, Commissioned by New Jersey Hazardous Waste Facilities Sitings Commission, New Jersey Hazardous Waste Facilities Sitings Commission, Trenton, NJ, April, 1987.

Sustainable Environmental Law, Environmental Law Institute, West Publishing Co., St. Paul, Minn., 1993.

Sustainable Industry: Promoting Strategic Environmental Protection in the Industrial Sector, Phase 1 Report, United States Environmental Protection Agency, Office of Emergency and Remedial Response, Washington, DC, June, 1994.

Techniques for Industrial Pollution Prevention: A Compendium for Hazardous and Non-Hazardous Waste Minimization, Overcash, Michael R., Lewis Publishers, Inc., Chelsea, MI, 1986.

The Chemical Process Industries, R. Norris Shreve, McGraw-Hill Book Company, Inc., New York, NY, 1945.

The Greening of American Business: Making Bottom-Line Sense of Environmental Responsibility, Government Institutes, Rockville, MD, 1992.

The Recycler's Manual for Business, Government, and the Environmental Community, Van Nostrand Reinhold, New York, NY, 1992.

TLVs® AND BEIs® Threshold Limit Values for Chemical Substances and Physical Agents, American Conference of Governmental Industrial Hygienists, Inc. (ACGIH), TLVs(R), reprinted by Permission, Cincinnati, OH, 1998.

Toxic Release Inventory (TRI) Public Data Release,United States Environmental Protection Agency, Office of Pollution Prevention and Toxics (EPA/745-R97-001), April 1997.

Tracking Toxic Substances at Industrial Facilities: Engineering Mass Balance Versus Materials Accounting, National Academy Press, Washington, DC, 1990.

Training Manual for NPDES Permit Writers, United States Environmental Protection Agency, Washington, DC, 1993.

United States Environmental Protection Agency, Development Document for Effluent Limitations Guidelines and Standards for the Organic Chemicals, Plastics, and Synthetic Fibers, Point Source Category, Volumes 1 and 2, EPA, Office of Water Regulations and Standards, October 1987.

Unit Operations of Chemical Engineering, 3rd ed., McCabe, W.L., and Smith, J. C., McGraw-Hill, Inc., New York, 1976.

Ward's Business Directory of U.S. Private and Public Companies, Gale Research, Inc., Detroit, MI, 1996.

Waste Minimization and Clean Technology: Waste Management Strategies for the Future, Forester, William S., and John H. Skinner, Academic Press, Inc., San Diego, CA.

Waste Oil: Reclaiming Technology, Utilization, and Disposal, Noyes Data Corporation, Park Ridge, NJ, 1989.

Waste Minimization: Implementing an Effective Program, Lewis Publishers, Chelsea, MI, 1992.

Waste Minimization Opportunity Assessment Manual, EPA/625/7-88/003, United States Environmental Protection Agency, Washington, DC, 1988.

Waste Minimization Workshop Handbook, Chemical Manufacturers Association, Washington, DC, 1987.

Waste Minimization Manual, Government Institutes, Rockville, MD, 1987.

Appendix B

OXIDIZING MATERIALS

By NAME/CAS

aluminum chlorate
aluminum nitrate 13473-90-0
amino guanidinium nitrate 103-82-4
ammonium chlorate 10192-29-7
ammonium chloride 7446-70-0
ammonium chromate 7788-98-9
ammonium dichromate 7789-09-5
ammonium iodate 12027-06-4
ammonium nitrate 6484-52-2
ammonium nitrate-phosphate mixture 57608-40-9
ammonium nitrate-sulfate mixture 6484-52-2
ammonium perchlorate 7790-98-9
ammonium permanganate 13446-10-1
ammonium peroxychromate
ammonium persulfate 7727-54-0
ammonium picrate (wet) 131-74-8
ammonium sulfamate 7773-06-0
amyl nitrate 463-04-7
ascaridole 512-85-6
barium bromate 13967-90-3
barium chlorate 13477-00-4
barium dichromate
barium hypochlorite 13477-10-6
barium iodate
barium nitrate 10022-31-8
barium perchlorate 13465-95-7
barium perchlorate trihydrate 10294-39-0
barium permanganate 7787-36-2
barium peroxide 1304-29-6
benzoyl peroxide 94-36-0
beryllium nitrate 13597-99-4
bis(p-chlorobenzoyl)peroxide 94-17-7
bromine 7726-95-6
bromine chloride 13863-41-7
bromine pentafluoride 7789-30-2
bromine trifluoride 7787-71-5
bromine(I) trifluoromethane sulfonate 70142-16-4
sec-butyl nitrate 924-43-6
t-butyl peroxybenzoate 614-45-9
cadmium nitrate 10325-94-7
cadmium nitrate tetrahydrate 10022-68-1
calcium chlorate 10137-74-3
calcium chlorite 14674-72-7
calcium chromate 13765-19-0

calcium chromate(VI), dihydrate 10060-08-9
calcium hypochlorite 7778-54-3
calcium nitrate 10124-37-5
calcium perchlorate 13477-36-6
calcium permanganate 10118-76-0
calcium peroxide 1305-79-9
cesium nitrate 7789-18-6
chloric acid solution 7790-93-4
chlorine 7787-50-5
chlorine dioxide 10049-04-4
chlorine pentafluoride 13637-63-3
chlorine trifluoride 7790-91-2
chromic acid 7738-94-5
chromic anhydride 11115-74-5
chromic chromate 24613-89-6
chromic perchlorate 13537-21-8
chromium nitrate 7789-02-8
chromium(III) oxide 1308-38-9
chromium trioxide, anhydrous 1333-82-0
chromium(III) nitrate 13548-38-4
chromium oxychloride 14977-61-8
cobalt nitrate 10141-05-6
cobalt(II) perchlorate, hexahydrate 13478-33-6
cobalt trifluoride 10026-18-3
copper chlorate 26506-47-8
copper(II) perchlorate, dihydrate 17031-32-2
cumene hydroperoxide 80-15-9
cupric nitrate (copper nitrate) 3251-23-8
cyclohexanone peroxide 12262-58-7
diacetone alcohol peroxide
N,N'-diacetyl-N,N'-dinitro-1,2-diaminoethane 922-89-4
diallyl peroxydicarbonate 34037-79-1
diaminoguanidinium nitrate 10308-83-5
diisopropylphenylhydroperoxide 26762-93-6
dimethylhexane dihydroperoxide 3025-88-5
di-t-butyl peroxide 110-05-4
dichloroisocyanuric acid 2782-57-2
ethyl nitrate 109-95-5
ferric nitrate 10421-48-4
fluorine 7782-41-4
fluorine perchlorate 10049-03-3
gallium(III) nitrate 13494-90-1
guanidine nitrate 506-93-4
hydrogen peroxide (35% to 52% by weight) 7722-84-1
indium nitrate 13770-61-1
iodic acid 7782-68-5
iodine pentafluoride 7783-66-6
iron(II) nitrate hexahydrate (1:2:6) 13520-68-8
1-isopropyl-4-methylcyclohaneperoxide 52061-60-6
isopropyl nitrate 1712-64-7

isopropyl peroxydicarbonate 105-64-6
lanthanum nitrate 10099-59-9
lead(II) cyanide 592-05-2
lead dioxide 1309-60-0
lead molybdenum chromate 12709-98-7
lead nitrate 10099-74-8
lead perchlorate 13637-76-8
lead(II) perchlorate, hexahydrate 63916-96-1
lithium bichromate
lithium chromate 14307-35-8
lithium hypochlorite 13840-33-0
lithium nitrate 7790-69-4
lithium perchlorate, trihydrate 13453-78-6
lithium peroxide 12031-80-0
magnesium bromate 7789-36-8
magnesium chlorate 7791-19-7
magnesium nitrate 10377-60-3
magnesium perchlorate 10034-81-8
magnesium peroxide 14452-57-4
manganese dioxide 1313-13-9
manganese nitrate 10377-66-9
manganese perchlorate, hexahydrate 15364-94-0
mercuric nitrate 10045-94-0
mercuric oxide 21908-33-2
mercurous nitrate 10415-75-5
mercury oxide 15829-53-5
p-methane-8-hydroperoxide 80-47-7
methyl hydroperoxide 3031-73-0
nickel nitrate 14216-75-2
nickel nitrite 17861-62-0
nickel perchlorate 13637-71-3
nickel(II) nitrate 13138-45-9
nickel(II) perchlorate, hexahydrate 13520-61-1
nitric acid 7697-37-2
nitric oxide 10102-43-9
nitro carbonitrate
nitrogen dioxide 10102-44-0
nitrogen monoxide mixed with nitrogen tetroxide 63907-41-5
nitrogen tetroxide 10544-72-6
nitrogen trifluoride 7783-54-2
nitrosyl perchlorate 15605-28-4
nitrosylsulfuric acid 7782-78-7
nitrous acid 7782-77-6
nitrosyl chloride 13444-90-1
octyl peroxide 7530-07-6
oxalic acid 6153-56-6
oxygen 7782-44-7
oxygen difluoride 7783-41-7
ozone 10028-15-6
palladium dinitrate 10102-05-3

peracetic acid 79-21-0
perchloric acid 7601-90-3
perchloryl fluoride 7616-94-6
perchloryl piperdine 768-34-3
peroxydisulfuryl difluoride 13709-32-5
peroxyformic acid 107-32-4
peroxyhexanoic acid 5106-46-7
peroxymonophosphoric acid
peroxytrifluroacetic acid 359-48-8
phthaloyl peroxide
potassium chromate(VI) 7789-00-6
potassium dichromate 7778-50-9
potassium dioxide 12030-88-5
potassium bromate 7758-01-2
potassium chlorate 1811-04-9
potassium dichloroisocyanurate 2244-21-5
potassium nitrate 7757-79-1
potassium nitrate mixed with sodium nitrite
potassium nitrite 7758-09-0
potassium pentaperoxydichromate
potassium perchlorate 7778-74-7
potassium periodate 7790-21-8
potassium permanganate 7722-64-7
potassium peroxide 17014-71-0
potassium perrhenate 10466-65-6
potassium persulfate 7727-21-1
potassium tetraperoxychromate 12331-76-9
n-propyl nitrate 627-13-4
pyrosulfuryl chloride 7791-27-7
rubidium dichromate 13446-73-6
silver nitrate 7761-88-8
sodium bromate 7789-38-0
sodium chlorate 7775-09-9
sodium chlorate borate 52623-84-4
sodium chlorite 7758-19-2
sodium chromate 7775-11-3
sodium dichlorocyanurate 2893-78-9
sodium dichromate 7789-12-0
sodium nitrate 7631-99-4
sodium nitrite 7632-00-0
sodium peracetate 64057-57-4
sodium percarbonate 15630-89-4
sodium perchlorate 7601-89-0
sodium periodate 7790-28-5
sodium permanganate 10101-50-5
sodium peroxide 1313-60-6
sodium peroxyborate
sodium persulfate 7775-27-1
sodium superoxide 12034-12-7
strontium chlorate 7791-10-8

strontium nitrate 10042-76-9
strontium perchlorate 13450-97-0
strontium peroxide 1314-18-7
sulfuric acid 7664-93-9
tetrafluorodiaziridine 17224-09-8
2-tetrahydrofuryl hydroperoxide 4676-82-8
tetranitromethane 509-14-8
thallium nitrate 10102-45-1
thorium(IV) nitrate 13823-29-5
triaminediperoxochromium(IV) 17168-85-3
trichloroisocyanuric acid 87-90-1
triphenyltin hydroperoxide 4150-34-9
2,4,6-tris(bromoamino)-1,3,5-triazine 22755-34-6
uranyl nitrate 10102-06-4
urea peroxide 124-43-6
vanadium pentoxide 1314-62-1
ytterbium nitrate 13768-67-7
yttrium(III) nitrate, hexahydrate 13494-98-9
zinc ammonium nitrite 63885-01-8
zinc bichromate
zinc bromate 13517-27-6
zinc chromate 13530-65-9
zinc chlorate 10361-95-2
zinc nitrate 7779-88-6
zinc perchlorate, hexahydrate
zinc permanganate 23414-72-4
zinc peroxide 1314-22-3
zirconium nitrate 13746-89-9

By CAS/NAME

N/A aluminum chlorate
N/A ammonium peroxychromate
N/A barium iodate
N/A barium dichromate
N/A diacetone alcohol peroxide
N/A lithium bichromate
N/A nitro carbonitrate
N/A peroxymonophosphoric acid
N/A phthaloyl peroxide
N/A potassium pentaperoxydichromate
N/A potassium nitrate mixed with sodium nitrite
N/A potassium pentaperoxydichromate
N/A sodium peroxyborate
N/A zinc bichromate
N/A zinc perchlorate, hexahydrate
79-21-0 peracetic acid
80-15-9 cumene hydroperoxide
80-47-7 p-methane-8-hydroperoxide
87-90-1 trichloroisocyanuric acid

94-17-7 bis(p-chlorobenzoyl)peroxide
94-36-0 benzoyl peroxide
103-82-4 amino guanidinium nitrate
105-64-6 isopropyl peroxydicarbonate
107-32-4 peroxyformic acid
109-95-5 ethyl nitrate
110-05-4 di-t-butyl peroxide
124-43-6 urea peroxide
131-74-8 ammonium picrate (wet)
359-48-8 peroxytrifluroacetic acid
463-04-7 amyl nitrate
506-93-4 guanidine nitrate
509-14-8 tetranitromethane
592-05-2 lead(II) cyanide
614-45-9 t-butyl peroxybenzoate
627-13-4 n-propyl nitrate
768-34-3 perchloryl piperdine
922-89-4 N,N'-diacetyl-N,N'-dinitro-1,2-diaminoethane
1304-29-6 barium peroxide
1305-79-9 calcium peroxide
1308-38-9 chromium(III) oxide
1309-60-0 lead dioxide
1313-13-9 manganese dioxide
1313-60-6 sodium peroxide
1314-18-7 strontium peroxide
1314-62-1 vanadium pentoxide
1314-22-3 zinc peroxide
1333-82-0 chromium trioxide, anhydrous
1712-64-7 isopropyl nitrate
1811-04-9 potassium chlorate
2244-21-5 potassium dichloroisocyanurate
2782-57-2 dichloroisocyanuric acid
2893-78-9 sodium dichlorocyanurate
3025-88-5 dimethylhexane dihydroperoxide
3031-73-0 methyl hydroperoxide
3251-23-8 cupric nitrate (copper nitrate)
3597-99-4 beryllium nitrate
4150-34-9 triphenyltin hydroperoxide
4676-82-8 2-tetrahydrofuryl hydroperoxide
5106-46-7 peroxyhexanoic acid
6153-56-6 oxalic acid
6484-52-2 ammonium nitrate
6484-52-2 ammonium nitrate-sulfate mixture
7446-70-0 ammonium chloride
7530-07-6 octyl peroxide
7601-89-0 sodium perchlorate
7601-90-3 perchloric acid
7616-94-6 perchloryl fluoride
7631-99-4 sodium nitrate
7632-00-0 sodium nitrite

7664-93-9 sulfuric acid
7697-37-2 nitric acid
7722-64-7 potassium permanganate
7722-84-1 hydrogen peroxide (35% to 52% by weight)
7726-95-6 bromine
7727-54-0 ammonium persulfate
7727-21-1 potassium persulfate
7738-94-5 chromic acid
7757-79-1 potassium nitrate
7758-01-2 potassium bromate
7758-09-0 potassium nitrite
7758-19-2 sodium chlorite
7761-88-8 silver nitrate
7773-06-0 ammonium sulfamate
7775-09-9 sodium chlorate
7775-11-3 sodium chromate
7775-27-1 sodium persulfate
7778-50-9 potassium dichromate
7778-54-3 calcium hypochlorite
7778-74-7 potassium perchlorate
7779-88-6 zinc nitrate
7782-41-4 fluorine
7782-44-7 oxygen
7782-68-5 iodic acid
7782-77-6 nitrous acid
7782-78-7 nitrosylsulfuric acid
7783-41-7 oxygen difluoride
7783-54-2 nitrogen trifluoride
7783-66-6 iodine pentafluoride
7787-36-2 barium permanganate
7787-50-5 chlorine
7787-71-5 bromine trifluoride
7788-98-9 ammonium chromate
7789-00-6 potassium chromate(VI)
7789-02-8 chromium nitrate
7789-09-5 ammonium dichromate
7789-12-0 sodium dichromate
7789-18-6 cesium nitrate
7789-30-2 bromine pentafluoride
7789-36-8 magnesium bromate
7789-38-0 sodium bromate
7790-21-8 potassium periodate
7790-28-5 sodium periodate
7790-69-4 lithium nitrate
7790-91-2 chlorine trifluoride
7790-93-4 chloric acid solution
7790-98-9 ammonium perchlorate
7791-10-8 strontium chlorate
7791-19-7 magnesium chlorate
7791-27-7 pyrosulfuryl chloride

10022-31-8 barium nitrate
10022-68-1 cadmium nitrate tetrahydrate
10026-18-3 cobalt trifluoride
10028-15-6 ozone
10034-81-8 magnesium perchlorate
10042-76-9 strontium nitrate
10045-94-0 mercuric nitrate
10049-03-3 fluorine perchlorate
10049-04-4 chlorine dioxide
10060-08-9 calcium chromate(VI), dihydrate
10099-59-9 lanthanum nitrate
10099-74-8 lead nitrate
10101-50-5 sodium permanganate
10102-05-3 palladium dinitrate
10102-06-4 uranyl nitrate
10102-43-9 nitric oxide
10102-44-0 nitrogen dioxide
10102-45-1 thallium nitrate
10118-76-0 calcium permanganate
10124-37-5 calcium nitrate
10137-74-3 calcium chlorate
10141-05-6 cobalt nitrate
10192-29-7 ammonium chlorate
10294-39-0 barium perchlorate trihydrate
10308-83-5 diaminoguanidinium nitrate
10325-94-7 cadmium nitrate
10361-95-2 zinc chlorate
10377-60-3 magnesium nitrate
10377-66-9 manganese nitrate
10415-75-5 mercurous nitrate
10421-48-4 ferric nitrate
10466-65-6 potassium perrhenate
10544-72-6 nitrogen tetroxide
11115-74-5 chromic anhydride
12027-06-4 ammonium iodate
12030-88-5 potassium dioxide
12031-80-0 lithium peroxide
12034-12-7 sodium superoxide
12262-58-7 cyclohexanone peroxide
12331-76-9 potassium tetraperoxychromate
12709-98-7 lead molybdenum chromate
13138-45-9 nickel(II) nitrate
13444-90-1 nitrosyl chloride
13446-10-1 ammonium permanganate
13446-73-6 rubidium dichromate
13450-97-0 strontium perchlorate
13453-78-6 lithium perchlorate, trihydrate
13465-95-7 barium perchlorate
13473-90-0 aluminum nitrate
13477-00-4 barium chlorate

13477-10-6 barium hypochlorite
13477-36-6 calcium perchlorate
13478-33-6 cobalt(II) perchlorate, hexahydrate
13494-90-1 gallium(III) nitrate
13494-98-9 yttrium(III) nitrate hexahydrate
13517-27-6 zinc bromate
13520-61-1 nickel(II) perchlorate, hexahydrate
13520-68-8 iron(II) nitrate hexahydrate (1:2:6)
13530-65-9 zinc chromate
13537-21-8 chromic perchlorate
13548-38-4 chromium(III) nitrate
13637-63-3 chlorine pentafluoride
13637-71-3 nickel perchlorate
13637-76-8 lead perchlorate
13709-32-5 peroxydisulfuryl difluoride
13746-89-9 zirconium nitrate
13765-19-0 calcium chromate
13768-67-7 ytterbium nitrate
13770-61-1 indium nitrate
13840-33-0 lithium hypochlorite
13863-41-7 bromine chloride
13967-90-3 barium bromate
14216-75-2 nickel nitrate
14307-35-8 lithium chromate
14452-57-4 magnesium peroxide
14674-72-7 calcium chlorite
14977-61-8 chromium oxychloride
15364-94-0 manganese perchlorate, hexahydrate
15605-28-4 nitrosyl perchlorate
15630-89-4 sodium percarbonate
15829-53-5 mercury oxide
17014-71-0 potassium peroxide
17168-85-3 triaminediperoxochromium(IV)
17224-09-8 tetrafluorodiaziridine
17861-62-0 nickel nitrite
21908-33-2 mercuric oxide
22755-34-6 2,4,6-tris(bromoamino)-1,3,5-triazine
23414-72-4 zinc permanganate
24613-89-6 chromic chromate
26506-47-8 copper chlorate
26762-93-6 diisopropylphenylhydroperoxide
34037-79-1 diallyl peroxydicarbonate
52061-60-6 1-isopropyl-4-methylcyclohaneperoxide
52623-84-4 sodium chlorate borate
57608-40-9 ammonium nitrate-phosphate mixture
63885-01-8 zinc ammonium nitrite
63907-41-5 nitrogen monoxide mixed with nitrogen tetroxide
63916-96-1 lead(II) perchlorate, hexahydrate
64057-57-4 sodium peracetate
70142-16-4 bromine(I) trifluoromethane sulfonate

Appendix C

POISON CONTROL CENTERS

Advice for emergency treatment for exposure to hazardous materials can be obtained from a regional Poison Control Center. Certified regional centers are certified by the American Association of Poison Control Centers (AAPCC) and are located throughout the country. In many locations, information can be obtained from hospitals that maintain centers that may not be certified by the AAPCC. Poison Control Centers will also provide referrals for specialized care when necessary.

ALABAMA

Alabama Poison Control Center, Tuscaloosa

2503 Phoenix Drive
Tuscaloosa, AL 35405
Emergency Phone: (800) 462-0800 (Alabama only); (205) 345-0600
Certified by the American Association of Poison Control Centers.
Area served: Alabama

Children's Hospital of Alabama Regional Poison Control Center

1600 7th Ave. South
Birmingham, AL 35233-1711
Emergency Phone: (800) 292-6678 (Alabama only); (205) 933-4050
Certified by the American Association of Poison Control Centers.
Area served: Alabama

ALASKA

Anchorage Poison Center

3200 Providence Dr.
P.O. Box 196604
Anchorage, AK 99519-6604
Emergency Phone: (800) 478-3193; (907) 261-3193
Area served: Alaska

ARIZONA

Arizona Poison and Drug Information Center

University of Arizona
Arizona Health Science Center, Rm 1156
1501 N. Campbell Ave.
Tucson, AZ 85724
Emergency Phone: (800) 362-0101 (Arizona only); (602) 626-6016
Certified by the American Association of Poison Control Centers.
Area served: Arizona, except Phoenix

Samaritan Regional Poison Center

Good Samaritan Regional Medical Center
1111 E. McDowell Rd., Ancillary-1

Phoenix, AZ 85006
Emergency Phone: (800) 362-0101 (Arizona only); (602) 252-3334
Certified by the American Association of Poison Control Centers.
Area served: Maricopa county (Phoenix)

ARKANSAS

Arkansas Poison and Drug Information Center

College of Pharmacy
University of Arkansas for Medical Sciences, Mail Slot 522
4301 W. Markham St.
Little Rock, AR 72205
Emergency Phone: (800) 376-4766; (800) 641-3805 (IDD/TTY)
Area served: Arkansas

CALIFORNIA

California Poison Control System, Fresno/Madera Division

Valley children's Hospital
9300 Valley Children's Place
Madera, CA 93638-8762
Emergency Phone: (800) 876-4766 (California only)
Certified by the American Association of Poison Control Centers.
Area served: Central California

California Poison Control System, Sacramento Division

UC Davis Medical Center
2315 Stockton Blvd.
Sacramento, CA 95817
Emergency Phone: (800) 876-4766 (California only)
Certified by the American Association of Poison Control Centers.
Area Served: Northern California

California Poison Control System, San Francisco Division

San Francisco General Hospital
1001 Potrero Ave., Room 1E86
San Francisco, CA 94110
Emergency Phone: (800) 876-4766 (California only)
Area Served: San Francisco Bay Area

California Poison Control System, San Diego Division

UC - San Diego Medical Center
200 West Arbor Dr.
San Diego, CA 92103-8925
Emergency Phone: (800) 876-4766 (California only)
Certified by the American Association of Poison Control Centers.
Area Served: San Diego and Imperial counties

COLORADO

Rocky Mountain Poison and Drug Center

8802 E. 9th Ave.
Denver, CO 80220-6800
Emergency Phone: (800) 332-3073 (Colorado only outside metro area); (303) 729-1123 (Denver metro only)
Certified by the American Association of Poison Control Centers.
Area served: Colorado

CONNECTICUT

Connecticut Poison Control Center

University of Connecticut Health Center
263 Farmington Ave.
Farmington, CT 06030-5365
Emergency Phone: (800) 343-2722 Connecticut only); (860) 679-3456
Certified by the American Association of Poison Control Centers.
Area served: Connecticut

DELAWARE

The Poison Control Center

3535 Market St., Ste. 985
Philadelphia, PA 19104-3309
Emergency Phone: (800) 722-7112; (215) 386-2100
Certified by the American Association of Poison Control Centers.
Area served: Delaware

DISTRICT OF COLOMBIA

National Capital Poison Center

Georgetown University Hospital
3201 New Mexico Ave., N.W., Ste. 310
Washington, DC 20016
Emergency Phone: (202) 625-3333; (202) 362-8563 (TTY)
Certified by the American Association of Poison Control Centers.
Area served: Washington, DC and metro area

FLORIDA

Florida Poison Information Center - Tampa

Tampa General Hospital
P.O. Box 1289
Tampa, FL 33601
Phone: (813) 253-4444
Emergency Phone: (800) 282-3171 (Florida only)
Certified by the American Association of Poison Control Centers.
Area served: Tampa and surrounding metro counties.

Florida Poison Information Center - Jacksonville
University Medical Center
University of Florida Health Science Center-Jacksonville
655 W. 8th St.
Jacksonville, FL 32209
Emergency Phone: (800) 282-3171 (Florida only); (904) 549-4480
Certified by the American Association of Poison Control Centers.
Area served: Northern and eastern coastal areas of Florida

Florida Poison Information Center- Miami
University of Miami School of Medicine
Department of Pediatrics
P.O. Box 016960 (R-131)
Miami, FL 33101
Emergency Phone: (800) 282-3171 (Florida only); (305) 585-5253
Certified by the American Association of Poison Control Centers.
Area served: Miami and surrounding metro counties

GEORGIA

Georgia Poison Center
Hughes Spalding Children's Hospital
80 Butler St., S.E.
P.O. Box 26066
Atlanta, GA 30335-3801
Emergency Phone: (800) 282-5846; (404) 616-9000; (404) 616-9287 (TDD)
Certified by the American Association of Poison Control Centers.
Area served: Georgia

HAWAII

Hawaii Poison Center
1319 Punahou St.
Honolulu, HI 96826
Emergency Phone: (808) 941-4411
Area served: Hawaii

IDAHO

Rocky Mountain Poison and Drug Center
8802 E. Ninth Ave.
Denver, CO 80220-6800
Emergency Phone: (800) 860-0620 (Idaho only)
Certified by the American Association of Poison Control Centers.
Area served: Idaho

ILLINOIS

Illinois Poison Center

222 S. Riverside Plaza, Ste. 1900
Chicago, IL 60606
Emergency Phone: (800) 942-5969 (Illinois only)
Certified by the American Association of Poison Control Centers.
Area served: Illinois

INDIANA

Indiana Poison Center
Methodist Hospital of Indiana
Clarian Health Partners
I-65 at 21st St.
Indianapolis, IN 46206-1367
Emergency Phone: (800) 382-9097 (Indiana only); (317) 929-2323; (317) 929-2336 (TTY)
Certified by the American Association of Poison Control Centers.
Area served: Indiana

IOWA

Iowa Poison Center
St. Luke's Regional Medical Center
2720 Stone Park Blvd.
Sioux City, IA 51104
Emergency Phone: (800) 353-2222; (712) 277-2222
Area served: Iowa

Poison Control Center
Department of Pharmaceutical Care, CC101 GH
The University of Iowa Hospitals and Clinics
200 Hawkins Dr.
Iowa City, IA 52242
Emergency Phone: (800) 272-6477 (Iowa only)
Area Served: Iowa

KANSAS

Mid-America Poison Center
University of Kansas Medical Center
3901 Rainbow Blvd., Rm. B-400
Kansas City, KS 66160-7231
Emergency Phone: (800) 332-6633 (Kansas only); (913) 588-6633; (913) 588-6639 (TDD)
Area served: Kansas

KENTUCKY

Kentucky Regional Poison Center
Medical Towers South, Ste. 572
234 E. Gray St.
Louisville, KY 40202

Emergency Phone: (502)589-8222
Certified by the American Association of Poison Control Centers.
Area served: Kentucky

LOUISIANA

Louisiana Drug and Poison Information Center

Northeast Louisiana University
School of Pharmacy, Sugar Hall
Monroe, LA 71209-6430
Emergency Phone: (800) 256-9822 (Louisiana only)
Certified by the American Association of Poison Control Centers.
Area served: Louisiana

MAINE

Maine Poison Control Center

Maine Medical Center
Department of Emergency Medicine
22 Bramhall St.
Portland, ME 04102
Emergency Phone: (800) 442-6305 (Maine only); (207) 871-2950
Area served: Maine

MARYLAND

Maryland Poison Center

University of Maryland at Baltimore
School of Pharmacy
20 N. Pine St., PH 230
Baltimore, MD 21201
Emergency Phone: (800) 492-2414 (Maryland only); (410) 528-7701; (401) 706-1858 (TDD)
Certified by the American Association of Poison Control Centers.
Area served: Maryland

National Capital Poison Center

3201 New Mexico Ave., NW, Ste. 310
Washington, DC 20016
Emergency Phone: (202) 625-3333; (202) 362-8563 (TTY)
Certified by the American Association of Poison Control Centers.
Area served: Washington, DC and metro area

MASSACHUSETTS

Massachusetts Poison Control System

300 Longwood Ave.
Boston, MA 02115
Emergency Phone: (800) 682-9211 (Massachusetts only); (617) 232-2120

Certified by the American Association of Poison Control Centers.
Area served: Massachusetts

MICHIGAN

Blodgett Regional Poison Center
1840 Wealthy S.E.
Grand Rapids, MI 49506
Hotline: (800) POISON-1; (800) 764-7661; (800) 356-3232 (TTY)
Area served: Eastern Michigan and peninsula

Regional Poison Control Center
Children's Hospital of Michigan
4160 John R. Harper Professional Office Bldg., Ste. 616
Detroit, MI 48201
Emergency Phone: (800) 462-6642 (Michigan only); (313) 745-5711; (800) 356-3232 (TDD)
Certified by the American Association of Poison Control Centers.
Area served: Southeast and thumb area of Michigan

Spectrum Health Regional Poison Center
1840 Wealth SE
Grand Rapids, MI 49506-2968
Emergency Phone: (800) 764-7661 (Michigan only); (800) 356-3232 (TTY)
Certified by the American Association of Poison Control Centers.
Area served: Eastern Michigan and Peninsula

MINNESOTA

Hennepin Regional Poison Center
Hennepin County Medical Center
701 Park Ave.
Minneapolis, MN 55415
Emergency Phone: (800) POISON7 (Minnesota and South Dakota only); (612) 347-3141; (612) 904-4691 (TDY)
Certified by the American Association of Poison Control Centers.
Area served: Minneapolis and surrounding counties

PROSAR International Poison Control Center
1295 Bandana Blvd, Ste. 335
Emergency Phone: (888) 779-7921
Area served: Does not take calls from the public.

MISSISSIPPI

Mississippi Regional Poison Control Center
University of Mississippi Medical Center
2500 N. State St.
Jackson, MS 39216
Emergency Phone: (601) 354-7660
Area served: Mississippi

MISSOURI

Regional Poison Center
Cardinal Glennon Children's Hospital
1465 S. Grand Blvd.
St. Louis, MO 63104
Emergency Phone: (800) 366-8888; (314) 772-5200
Certified by the American Association of Poison Control Centers.
Area served: Missouri

Children's Mercy Hospital Poison Control Center
2401 Gilham Rd.
Kansas City, MO 64108
Emergency Phone: (816) 234-3430
Area served: Western Missouri

MONTANA

Rocky Mountain Poison and Drug Center
8801 E. 9th Ave.
Denver, CO 80220
Emergency Phone: (800) 525-5042 (Montana only)
Certified by the American Association of Poison Control Centers.
Area served: Montana

NEBRASKA

The Poison Center
Children's Hospital
8301 Dodge St.
Omaha, NE 68114
Emergency Phone: (800) 955-9119 (Nebraska and Wyoming only); (402) 390-5555
Certified by the American Association of Poison Control Centers.
Area served: Nebraska

NEVADA

Oregon Poison Center
Oregon Health Sciences University
3181 SW Sam Jackson Park Rd.
Portland, OR 97201
Emergency Phone: (503) 494-8968
Certified by the American Association of Poison Control Centers.
Area Served: Northern Nevada

Rocky Mountain Poison and Drug Center
8801 E. 9th Avenue
Denver, CO 80220
Emergency Phone: (800) 446-6179 (Nevada only)

Certified by the American Association of Poison Control Centers.
Area served: Clark and Nye counties

NEW HAMPSHIRE

New Hampshire Poison Information Center

Dartmouth-Hitchcock Medical Center
One Medical Center Dr.
Lebanon, NH 03756
Emergency Phone: (800) 562-8236 (New Hampshire only); (603) 650-8000
Area served: New Hampshire

NEW JERSEY

New Jersey Poison Information and Education System

201 Lyons Ave.
Newark, NJ 07112
Emergency Phone: (800) POISON-1 (New Jersey only)
Certified by the American Association of Poison Control Centers.
Area served: New Jersey

NEW MEXICO

New Mexico Poison and Drug Information Center

University of New Mexico
Health Science Library, Rm. 125
Albuquerque, NM 87131-0176
Emergency Phone: (800) 432-6866 (New Mexico only); (505) 272-2222
Certified by the American Association of Poison Control Centers.
Area served: New Mexico

NEW YORK

Central New York Poison Control Center

SUNY Health Science Center
750 E. Adams St.
Syracuse, NY 13210
Emergency Phone: (800) 252-5655 (New York only); (315) 476-4766
Certified by the American Association of Poison Control Centers.
Area served: Central New York

Finger Lakes Regional Poison and Drug Information Center

University of Rochester Medical Center
601 Elmwood Ave., Box 321
Rochester, NY 14642
Emergency Phone: (800) 333-0542 (Finger Lakes Region only); (716) 275-3232; (716) 272-3854 (TTY)
Certified by the American Association of Poison Control Centers.
Area served: Finger Lakes region of New York

Hudson Valley Poison Center
Phelps Memorial Hospital Center
701 N. Broadway
Sleepy Hollow, NY 10591
Emergency Phone: (800) 336-6997 (New York only); (914) 366-3030
Certified by the American Association of Poison Control Centers.
Area served: Eastern part of New York state from New York City to Canada

Long Island Regional Poison Control Center
Winthrop University Hospital
259 First St.
Mineola, NY 11501
Emergency Phone: (516) 542-2323; (516) 663-2650; (516) 747-3323 (TDD Nassau); (516) 924-8811 (TDD Suffolk)
Certified by the American Association of Poison Control Centers.
Area served: Long Island

New York City Poison Control Center
New York City Department of Health
455 First Ave., Rm. 123
New York, NY 10016
Emergency Phone: (800) 210-3985; (212) 340-4494; (202) POISONS; (212) VENENOS; (212) 689-9014 (TDD)
Certified by the American Association of Poison Control Centers.
Area served: New York City

Western New York Regional Poison Control Center
Children's Hospital of Buffalo
219 Bryant St.
Buffalo, NY 14222
Emergency Phone: (800) 888-7655 (New York western regions only); (716) 878-7654
Certified by the American Association of Poison Control Centers.
Area served: Western New York

NORTH CAROLINA

Carolinas Poison Center
Carolinas Medical Center
5000 Airport Center Parkway, Ste. B
Charlotte, NC 28208
Emergency Phone: (800) 848-6946 (North Carolina only); (704) 355-4000
Certified by the American Association of Poison Control Centers.
Area served: North Carolina

NORTH DAKOTA

North Dakota Poison Information Center
MeritCare Medical Center
720 4th St., North
Fargo, ND 58122

Emergency Phone: (800) 732-2200 (North Dakota, Minnesota and South Dakota only); (701) 234-5575
Area served: North Dakota

OHIO

Central Ohio Poison Center

Children's Hospital
700 Children's Dr., Rm. L032
Columbus, OH 43205
Emergency Phone: (800) 682-7625 (Ohio only); (800) 762-0727 (Dayton, Ohio only); (614) 228-1323; (614) 228-2272 (TTY)
Certified by the American Association of Poison Control Centers.
Area served: Central Ohio

Cincinnati Drug & Poison Information Center

Regional Poison Control System
2368 Victory Parkway, Ste. 300
Cincinnati, OH 45206
Emergency Phone: (800) 872-5111 (Ohio only); (513) 558-5111
Certified by the American Association of Poison Control Centers.
Area served: Southwest Ohio

Greater Cleveland Poison Control Center

11100 Euclid Ave.
Cleveland, OH 44106-6010
Emergency Phone: (888) 231-4455 (Ohio only); (216) 231-4455
Area served: Cleveland and surrounding metro area

OKLAHOMA

Oklahoma Poison Control Center

Children's Hospital of Oklahoma
940 N.E. 13th St., Rm. 3512
Oklahoma City, OK 73104
Emergency Phone: (800)764-7661 (Oklahoma only); (405) 271-5454
Area served: Oklahoma

OREGON

Oregon Poison Center

Oregon Health Science University
3181 S.W. Sam Jackson Park Rd., CB550
Portland, OR 97201
Emergency Phone: (800) 452-7165 (Oregon only); (503) 494-8968
Certified by the American Association of Poison Control Centers.
Area served: Oregon

PENNSYLVANIA

Central Pennsylvania Poison Center
Penn State Geisinger Health System
The Milton S. Hershey Medical Center
MC H043 O.O. Box 850
500 University Dr.
Hershey, PA 17033-0850
Emergency Phone: (800) 521-6110; (717) 531-6111; (717) 531-8335 (TTY)
Certified by the American Association of Poison Control Centers.
Area served: Central Pennsylvania

Pittsburgh Poison Center
Children's Hospital of Pittsburgh
3705 Fifth Ave.
Pittsburgh, PA 15213
Emergency Phone: (412) 681-6669
Certified by the American Association of Poison Control Centers.
Area served: Western Pennsylvania

The Poison Control Center
3535 Market St., Ste. 985
Philadelphia, PA 19104-3309
Emergency Phone: (800) 722-7112; (215) 386-2100
Certified by the American Association of Poison Control Centers.
Area served: Southeast Pennsylvania and the Lehigh Valley

PUERTO RICO

San Jorge Children's Hospital Poison Center
258 San Jorge St.
Santurce, Puerto Rico 00912
Emergency Phone: (787) 726-5674

RHODE ISLAND

Rhode Island Poison Center
Lifespan Poison Center
Rhode Island Hospital
593 Eddy St.
Providence, RI 02903
Emergency Phone: (401) 444-5727
Certified by the American Association of Poison Control Centers.
Area served: Rhode Island

SOUTH CAROLINA

Palmetto Poison Center
University of South Carolina
College of Pharmacy

Columbia, SC 29208
Emergency Phone: (800) 922-1117 (South Carolina only); (803) 777-1117
Area served: South Carolina

SOUTH DAKOTA

Hennepin Regional Poison Center
Hennepin County Medical Center
701 Park Ave.
Minneapolis, MN 55415
Emergency Phone: (800) POISON7 (Minnesota and South Dakota only); (612) 347-3141; (612) 904-4691 (TTY)
Certified by the American Association of Poison Control Centers.
Area Served: South Dakota

TENNESSEE

Middle Tennessee Poison Center
Vanderbilt University Medical Center
The Center for Clinical Toxicology
1161 21st Ave. S.
501 Oxford House
Nashville, TN 37232-4632
Emergency Phone: (800) 288-9999 (Tennessee only); (615) 936-2034 (Greater Nashville); (615) 936-2047 (TDD)
Certified by the American Association of Poison Control Centers.
Area served: Middle Tennessee

Southern Poison Center
847 Monroe Ave., Ste. 104
Memphis, TN 38163
Emergency Phone: (800) 288-9999 (Tennessee only); (901) 528-6048
Area served: Western and Eastern Tennessee

TEXAS

Central Texas Poison Center
Scott and White Memorial Hospital
2401 South 31st St.
Temple, TX 76508
Emergency Phone: (800) POISON-1 (Texas only); (254) 724-7401
Certified by the American Association of Poison Control Centers.
Area served: Central Texas

North Texas Poison Center
Texas Poison Center Network
Parkland Health & Hospital System
5201 Harry Hines Blvd.
P.O. Box 35926
Dallas, TX 75235

Emergency Phone: (800) 764-7661 (Texas only)
Certified by the American Association of Poison Control Centers.
Area served: North Texas

Southeast Texas Poison Center
The University of Texas Medical Branch
3.112 Trauma Bldg.
Galveston, TX 77555-1175
Emergency Phone: (800) 764-7661 (Texas only)
Certified by the American Association of Poison Control Centers.
Area served: Southeast Texas

South Texas Poison Center
University of Texas Health Science Center - San Antonio
Forensic Science Building, R. 146
Department of Surgery
7703 Floyd Curl Dr.
San Antonio, TX 78284-7849
Emergency Phone: (800) 764-7661 (Texas only)
Certified by the American Association of Poison Control Centers.
Area served: South Texas

Texas Panhandle Poison Center
1501 S. Coulter
Amarillo, TX 79175
Emergency Phone: (800) 764-7661 (Texas only)
Area served: Amarillo and surrounding area

West Texas Regional Poison Center
Thomason Hospital
4815 Alameda Ave.
El Paso, TX 79998
Emergency Phone: (800) 764-7661 (Texas only)
Certified by the American Association of Poison Control Centers.
Area served: West Texas

UTAH

Utah Poison Control Center
410 Chipeta Way, Ste. 230
Salt Lake City, UT 84108
Emergency Phone: (800) 456-7707 (Utah only); (801) 581-2151
Certified by the American Association of Poison Control Centers.
Area served: Utah

VERMONT

Vermont Poison Center
Fletcher Allen Health Care
111 Colchester Ave.

Burlington, VT 05401
Emergency Phone: (877) 658-3456; (802) 658-3456
Area served: Vermont

VIRGINIA

Blue Ridge Poison Center

University of Virginia Health System
Box 437
Charlottesville, VA 22908
Emergency Phone: (800) 451-1428 (Virginia only); (804) 924-5543
Certified by the American Association of Poison Control Centers.
Area served: Western and Central Virginia

National Capital Poison Center

3201 New Mexico Ave., NW, Ste. 310
Washington, DC 20016
Emergency Phone: (202) 625-3333; (202) 362-8563 (TTY)
Certified by the American Association of Poison Control Centers.
Area served: Northern Virginia

Virginia Poison Center

Medical College of Virginia Hospitals
Virginia Commonwealth University
P.O. Box 980522
Richmond, VA 23298-0522
Emergency Phone: (800) 522-6337; (804) 828-9123
Certified by the American Association of Poison Control Centers.
Area served: Eastern and Central Virginia

WASHINGTON

Washington Poison Center

155 NE 100th St., Ste. 400
Seattle, WA 98125-8012
Emergency Phone: (800) 732-6985 (Washington only); (206) 526-2121; (800) 572-0638 (TDD); (206) 517-2394 (TDD)
Certified by the American Association of Poison Control Centers.
Area served: Washington

WEST VIRGINIA

West Virginia Poison Center

3110 McCorkle Ave., S.E.
Charleston, WV 25304
Emergency Phone: (800) 642-3625 (West Virginia only)
Certified by the American Association of Poison Control Centers.
Area served: West Virginia

WISCONSIN

Children's Hospital of Wisconsin Poison Center
P.O. Box 1997
Milwaukee, WI 53201
Emergency Phone: (800) 815-8855 (Wisconsin only); (414) 266-2222
Area served: Eastern Wisconsin

University of Wisconsin Hospital and Clinics
Poison Control Center
600 Highland Ave., Rm.F6/133
Madison, WI 53792
Emergency Phone: (800) 815-8855 (Wisconsin only); (608) 262-3702
Area served: Western Wisconsin

WYOMING

The Poison Center
Children's Hospital
8301 Dodge St.
Omaha, NE 68114
Emergency Phone: (800) 955-9119 (Nebraska and Wyoming only); (402) 354-5555
Certified by the American Association of Poison Control Centers
Area served: Wyoming

Appendix D

GLOSSARY

A

absorbent material Commercially packaged clay, kitty litter, or other material used to soak up liquid hazardous materials.

absorption Penetration of a substance across a biologic barrier (such as the skin) and into either the lymphatic system or bloodstream.

acaricide A chemical substance used to kill ticks and mites.

accident An unplanned energy transfer causing property damage and/or human injury. *See also* **incident.**

accumulative effect The effect of a chemical substance on a biologic system when the substance is being absorbed at a rate that exceeds the body's ability to eliminate it from the system. Excessive accumulation of the substance in the system can lead to toxicity.

acid Any compound containing hydrogen replaceable by metals, and having a pH of 0 to 6. Strong acids in the pH range of 0 to 2 are corrosive and will cause chemical burns to the skin, eyes, and mucous membranes. Acids turn litmus red. *See also* **pH** and **strong acids**.

acid gas A gas that forms an acid when dissolved in water.

acidosis A pathologic condition resulting from accumulation of acid in, or loss of base from, the body.

acrid Having a biting taste; sharp; pungent.

acrylics any of several commercial plastics derived from acrylic acid, methacrylic acid, or derivatives (esters) of these acids, or acrylonitrile.

active ingredient The component that actually performs the primary function of a product. Products generally contain both active and inert ingredients, and both may be harmful. For example, insecticides in spray cans contain both chemicals having pesticidal action (active) and propellent gas (inert). Active ingredients are listed on product labels as percentage by weight or as pounds per gallon of concentrate.

acute Refers to severe symptoms and a rapid onset and progressive change to an organism leading to a crisis in a relatively short period of time, measured in seconds, minutes, hours, or days, following exposure to a health hazard.

acute effect Refers to an adverse health effect that usually occurs rapidly, sometimes immediately, as a result of a single, short, significant exposure to a health hazard, without implying a degree of severity. Acute effects may include irritation, corrosivity, narcosis, and death. *See also* **chronic.**

acute exposure Refers to a single exposure to a toxic substance that results in death or severe biological harm. Acute exposures are characterized as lasting no longer than 1 day.

aerosol Fine liquid or solid particles suspended in air (dispersed in a gas), the particle size often being in the 0.01 to 100 microns range. Natural aerosols include smoke (solid particles) and fog (liquid particles). Man-made aerosols are manufactured by filling a valved container, usually a can, with a suspension (e.g., paint, insecticides, cosmetics) in a gas under pressure.

aliphatic Pertaining to an open-chain hydrocarbon compound. Substances such as methane and ethane are typical aliphatic hydrocarbons.

alkali Any acid-destroying compound having a pH of 8-14. Strong alkalis (or bases) in the pH range of 12 to 14 are considered corrosive and will cause chemical burns to the skin, eyes, and mucous membranes. Alkalis turn litmus blue. Widely used industrial alkali substances include sodium hydroxide, sodium carbonate, potassium hydroxide, and potassium carbonate. Common household products include DRANO™ and lye. *See also* **acid**, **base**, and **pH**.

alkaloid An organic nitrogen base, of vegetable origin, usually toxic.

alkyd A type of thermosetting polyester resin used as coatings and made by reacting ethylene glycol or glycerol and phthalic anhydride, with a modifier, usually a drying oil or fatty acid such as linseed oil. The name comes from the combination of the words alcohol and acid.

allergen A substance that causes the body to produce an antibody and which results in an **allergy** in hypersensitive people.

allergy A hypersensitive reaction of body tissues to specific substances. In similar concentrations and circumstances, these same substances do not affect other persons. Allergic reactions in the workplace tend to affect the skin (*see* **dermatitis**) and lungs (*see* **asthma**).

ambient An environment's surrounding conditions.

amines Refers to ammonia compounds.

amnesia Total or partial memory loss.

analgesia Insensibility to pain without loss of consciousness.

anemia A deficiency of the blood caused by reduced red blood cell count or a reduction in the amount of hemoglobin per unit volume of blood.

anesthesia Total or partial loss of sensation with or without loss of consciousness.

angina pectoris Pain in chest caused by inadequate supply of blood to the heart.

anhydrous Containing no water.

anorexia Loss or reduction of appetite for food.

anosmia Total or partial loss of the sense of smell.

anoxia A reduction in the quantity of oxygen supplied by blood to cells or tissues.

antidote A remedy to relieve, prevent, or counteract the effects of a poison; that which counteracts anything noxious.

anuria Absence of urine in the bladder caused by the failure of the kidneys to produce urine. This is a possible symptom of chlorate or inorganic mercury poisoning.

apathy Reduced emotions with lack of interest in outside stimuli.

apnea Temporary cessation of breathing. A possible symptom of poisoning.

aquatic toxicology A branch of toxicology that deals with water pollution and its ecological effects.

aqueous Watery or waterlike.

aromatic compound Pertaining to a molecular ring structure hydrocarbon compound, characterized by their unsaturated ring structure, the presence of the benzene nucleus.

arrhythmia Disturbed heartbeat.

ataxia Unsteady walk or shaky movements due to neurological problems. May be a symptom of poisoning.

asphyxia Difficulty in breathing or respiratory arrest; suffocation.

asphyxiant Refers to a substance, usually a vapor or gas, that can cause suffocation, unconsciousness, or death by preventing the blood from carrying oxygen. Most simple asphyxiants (which have no inherent toxicity) are harmful to the body only when they become so concentrated that oxygen in the air is reduced (normally about 21 percent) to dangerous levels (18% or lower).

asthenia Reduced physical and psychological strength.

asthma Respiratory problem characterized by attacks of wheezing, shortness of breath, and/or coughing and resulting in difficult breathing due to contraction of air passages.

ataxia Loss or failure of muscular coordination, voluntary movement, or muscle control.

atrophy A loss of weight, volume, and activity of an organ, tissue, or cell; shrinkage.

autoignition temperature The minimum temperature at which a substance will ignite spontaneously, or cause self-sustained combustion in the absence of any heated element, spark, or flame. The closer the auto-ignition temperature is to room temperature, the greater the risk of fire.

awareness level (trained) First responders at the awareness level are those persons who, in the course of their normal duties, may be the first on the scene of an emergency involving hazardous materials. First responders at the awareness level are expected to recognize hazardous materials presence, protect themselves, call for trained personnel, and secure the area (ANSI/NFPA 472)

B

bactericidal Destructive to bacteria. A pesticide used to control bacteria is called a bactericide.

Bakelite a proprietary name of Union Carbide for phenolic resins made by condensation of phenol or cresol with formaldehyde. Named after Leo H. Baekland (1863-1944), who studied and developed phenol-formaldehyde polymers in 1907.

base A substance that reacts with acids to form salts and water. All bases create solutions having a pH of more than 7.0, the neutral point, and may be corrosive to skin and other human tissue. The terms **alkali** and **caustic** are closely related in meaning. *See also* **acid, alkali**, and **pH**.

BEI Biological Exposure Index. The maximum recommended value of a substance in blood, urine, or exhaled air, recommended by the ACGIH. *See* "Threshold Limit Values and Biological Exposure Indices," published by the ACGIH, for an explanation.

benign Not harmful.

bile A yellow-green, bitter fluid secreted by the liver. Also called gall.

bld Blood effects. A toxicology term describing the effect on all blood elements, including oxygen carrying or releasing capacity, pH, protein, and electrolytes.

blepharospasm Abnormal contraction of eyelid muscles.

BOD Biological oxygen demand. A test that measures the dissolved oxygen consumed by microbial life while assimilating and oxidizing the organic matter present in organic waste discharges. This test permits calculation of the effect of the discharges on the oxygen resources of the receiving water.

boiling point The temperature at which a product changes from a liquid to a vapor at normal atmospheric pressure (760 mm of Hg). Mixtures may not have a specific boiling point. As a general rule, material safety data sheets (MSDS) carry the initial boiling point or a boiling range for a mixture.

bradycardia A decrease in the heartbeat rate to less than 60 beats per minute.

breakthrough time The time from initial chemical contact to detection.

breathing zone sample An air sample collected from the area around the nose of a worker to assess exposure to airborne contaminants.

bronchoconstriction Contraction with narrowing of bronchia.

bronchospasm Spasmodic contraction of the muscles surrounding the bronchia.

bulk density Mass of powdered or granulated solid material per unit of volume.

by-product Any material, other than the principal product, generated as a consequence of an industrial process.

C

C Symbol for Celsius or Centigrade, a unit of temperature in which the interval between the freezing point of water and the boiling point is divided into 100 units, or degrees, with 0° representing the freezing point and 100° the boiling point.

canister A personal air cleaning device usually worn by the user. The canister contains sorbents, catalysts, or other filter materials designed to remove gases, vapors, and liquid and solid particles from air drawn through it.

carcinogen Any substance causing the promotion or initiation of malignant or benign neoplasia (cancer) in humans or animals. A material is considered carcinogenic if, (a) it has been found to be a carcinogen or potential carcinogen by the International Agency for Research on Cancer (IARC); or, (b) if it is listed in the latest edition of the "Annual Report on Carcinogens," published by the National Toxicology Program (NTP); or, (c) it is regulated by OSHA as a carcinogen.

carcinoma A malignant tumor; a type of cancer.

cardio A medical prefix that refers to the heart.

cardiovascular A medical term that refers to the heart and blood vessel system.

catalyst A substance that changes the rate of a chemical reaction, but that is neither changed nor consumed by the reaction.

cataract A disease of the eye in which the lens becomes gray-white and loses its clearness.

cathartic Substance that aids bowel movement and stimulates evacuation of the intestine.

caustic Any strongly alkaline substance that has a corrosive effect on living tissue. *See* **alkali**.

central nervous system (CNS) Refers to the brain and spinal cord, the main network of coordination and control for the entire body. Chemicals acting on the brain may cause CNS depression with symptoms of dizziness, headache, and drowsiness; higher exposure may cause unconsciousness, coma, and death.

characteristic hazardous waste An RCRA regulated waste classified as "hazardous" because of its ignitability, corrosivity, reactivity, or toxicity as determined by the Toxicity Characteristic Leachate Procedure (TCLP). It has an EPA Waste Code ranging from "D001" to "D043."

chemical Any element, chemical compound, or mixture of elements and/or compounds.

chemical burn Similar to a thermal burn form heat or fire, but caused by contact with a chemical substance.

chemical family A group of single elements or groups of compounds having a common chemical structure and name. Also known as chemical class. As used on a material safety data sheet (MSDS), more than one chemical family may be used if applicable.

chemical formula The chemical makeup of a substance using accepted written symbols. Although several kinds of formulas are used to indicate chemical constitution and physical structure, the **molecular formula** showing the actual kinds and numbers of atoms that comprise a molecule of chemical substance is most commonly used in material safety data sheets (MSDS).

chemical hygiene plan A written action plan required by OSHA's regulation 29 CFR 1910, *Occupational Exposures to Hazardous Chemicals in Laboratories.*

chemically active metals Usually refers (but not restricted) to chemicals such as sodium, potassium, beryllium, calcium, powdered aluminum, zinc, and magnesium. These metals can cause violent reactions with certain other substances and materials.

chemical name The scientific designation of a chemical as outlined by the Chemical Abstract Service (CAS) rules of nomenclature and the nomenclature system developed by the International Union of Pure and Applied Chemistry. Also defined by the OSHA Hazard Communication Standard (HCS) as a name which will clearly identify the chemical for the purpose of conducting a hazard evaluation. By this definition, more than one valid name for many chemicals is permitted. However, when chemicals with more than one valid name appears on the OSHA (29 CFR 1910 Table Z-1) or CERCLA (40 CFR 302.4) lists, this name is generally used as the chemical name on material safety data sheets (MSDS). If a chemical does not appear on these lists, then the most common chemical name should be used. Trade names that adequately identify a chemical may be used.

chemical manufacturer Defined in the OSHA Hazard Communication Standard (HCS) as an employer with a workplace where chemical(s) are produced for use or distribution.

chemical reaction Any chemical change, regardless of rate, or whether it occurs naturally or induced by man. There are many types of chemical reactions, including decomposition, explosion, combustion, condensation, polymerization, neutralization.

chemical resistance The ability of a material to resist chemical reaction.

chlorofluorocarbons (CFCs) A group of chemicals that depletes the earth's protective ozone layer. Chemical substance often used as refrigerants, solvents, and propellants in spray cans. CFCs are not destroyed in the lower atmosphere; they drift into the upper atmosphere, where their chlorine components are released and destroy the ozone layer.

chloracne A severe, acnelike affliction of the skin resulting from excessive exposure to certain chlorinated or halogenated chemical compounds, such as carbon tetrachloride, chloroform, trichloroethylene, biphenyls, dioxins, napthalenes, and DDT.

chronic Refers to a change to a organism over a long period of time, measured in weeks, months, or years following repeated exposure to a health hazard.

chronic effect of overexposure Refers to an adverse health effect that develops slowly over a long period of time or from prolonged exposure to a health hazard without implying a degree of severity.

chronic toxicity Refers to permanent and irreversible health effects resulting from prolonged exposure to a toxic substance.

closed cup Test for the flash point of a substance.

cirrhosis Chronic, progressive illness affecting the structure and function of the liver.

co-carcinogen Any substance, not itself carcinogenic, capable of enhancing the carcinogenic effect of another substance.

Code of Federal Regulations (CFR) A publication of the regulations promulgated under United States federal law. The CFR is divided into titles as follows:

Title 29: OSHA regulations, including the Hazard Communication Standard (HCS).
Title 40: EPA regulations, including TSCA.
Title 49: DOT regulations.

Changes to the regulations are published in the "Federal Register" (FR).

colic A sharp, crampy, and possibly painful disorder of the abdomen resulting from blockage, twisting, or muscle spasm.

collapsus Rapid decrease in strength or collapse of an organ.

colorants a general term for additive dyes or pigments that give permanent color to plastics.

coma A state of deep unconsciousness from which a victim cannot be wakened by external stimulants.

combustible liquid A material having a flash point at or above 100°F (37.8°C), but below 200°F (93.3°C), except that this term does not include any liquid mixture that has one or more components with a flash point above 200°F (93.3°C), which make up 99% or more of the total volume of the mixture as determined by tests listed in 49 CFR, Sec. 173.115(d). Exceptions to this are found in 49 CFR, Sec. 173.115(b).

common name Any designation or identification such as a code name, code number, trade name, or generic name used to identify a chemical other than by its chemical name. (OSHA)

component An ingredient or constituent part.

composite a combination or blend of two or more solid materials, generally a thermosetting plastic or other polymer matrix with property-improving high-strength reinforcing fibers such as glass fibers, graphite, silicon carbide.

compound A substance consisting of two or more elements that have united chemically.

compressed gas Any material or mixture having in the container a pressure exceeding 40 psi at 70°F (21.1°C), or a pressure exceeding 104 psi at 130°F (54.4°C), regardless of the pressure at 70°F (21.1°C); or any liquid flammable material having a vapor pressure exceeding 40 psi absolute pressure at 100°F (37.8°C) as determined by the American National Standard Method of Tests for Vapor Pressure of Petroleum Products (Reid Method) Z11.44-1973 (ASTM-American Society for Testing Materials D 323-72).

congenital A condition that begins to develop in the uterus and is existing at birth.

congestion Abnormal accumulation of blood in the vessels of tissue, an organ, or other part of the body.

conjunctivitis Irritation and inflammation of the conjunctiva, a part of the inner lining of the eyelids.

container Defined by OSHA as any bag, barrel, bottle, box, can, cylinder, drum, reaction vessel, storage tank, or the like that contains a hazardous chemical. Pipes and piping systems and engines, fuel tanks, or other operating systems in a vehicle, are not considered containers. (29 CFR)

contraindication Any condition that renders some particular treatment of disease improper or undesirable.

convulsions Violent, involuntary spasms or muscle contractions.

copolymer A high-polymeric substance or system, usually an elastomer, that contains two or more different kinds of monomeric units used for producing plastics of predetermined and improved properties that are different from those of the respective homopolymers.

corrosive material Any liquid or solid with pH ranges of 2–6 or 12–14, that cause visible destruction or irreversible alteration of living tissue, or a liquid that has a severe corrosion rate on steel. To determine whether a material is to be destructive or to cause irreversible alteration of human skin, refer to the method described in Appendix A of 49 CFR Part 173. A liquid is considered to have a severe corrosion rate if its corrosion rate exceeds 0.250 inch per year (IPY) on steel (Society of Automotive Engineers (SAE) 1020) at a test temperature of 130°F (54.4°C). Also see test described in NACE Standard TM-01-69.

cryogenic liquid Defined by DOT as a refrigerated, liquefied gas having a boiling point colder than -130°F (-90°C) at one atmosphere, absolute.

curing A process such as condensation, polymerization, or vulcanization involving changes to the physical properties of a material by the application of other chemicals and/or heat

cutaneous Pertaining to the skin. *See* **dermal**.

cyanosis A bluish discoloration of the skin, lips, and mucous membrane, resulting from lack of oxygen in the blood hemoglobin, or excessive concentration of reduced hemoglobin in the blood.

D

decomposition Breakdown of a material or substance into parts, elements, or simpler compounds that may be caused by heat, chemical reaction, electrolysis, decay, biodegradation, or other process.

decontamination The removal of hazardous substances from employees and their equipment to the extent necessary to preclude the occurrence of foreseeable adverse health effects.

degradation The destructive effect a chemical may have on chemical-protective clothing, reducing its strength and flexibility and permitting a direct route to skin contact.

delayed hazard The potential to cause an adverse effect which may not appear until a long period of time. Carcinogenicity, teratogenicity , and certain **target organ**/system effects are examples of delayed hazards. (ANSI)

deliquescent Substance which absorbs moisture from the air to the point of becoming liquid.

delirium A mental state of great excitement or confusion marked by confusion, speech disorders, anxiety, and often hallucinations.

density The mass (weight) per unit volume of a substance. For example, lead has much greater density than aluminum.

dermal Pertaining to the skin. See **cutaneous.**

dermatitis Inflammation of the skin from any cause.

dermatosis Generic name for all skin disorders.

desquamation Abnormal elimination of surficial layers of skin in small flakes.

diarrhea Abnormally frequent discharge of loose, watery feces from the large intestine (colon). May be a symptom of poisoning.

diuretic That which increases volume of urine.

dose The amount of a chemical substance or drug to which a person has been exposed or has been absorbed into the body.

dysplasia Malformation; abnormal development.

dyspnea Difficulty or labored breathing. Shortness of breath. A possible symptom of poisoning.

dysuria Painful or difficult urination.

E

easily oxidized materials A broad range of materials that includes combustible materials, organic materials, paper, wood, sulfur, aluminum, acetic acid, alcohols, fuels, oils, plastics, hydrazine, acetic anhydride, sulfuric acid, and many other chemicals. Easily oxidized materials can cause violent reactions with other substances and materials. *See also* **oxidizer, strong oxidizers.**

eczema Not a distinct disease but a general medical description for swelling of the skin, or rash of unknown cause.

edema Swelling caused by infiltration of fluid into the tissues or intercellular spaces.

EIS Environmental impact statement. The results of a study to determine the probable effects of a proposed activity on the surrounding environment.

elastomer A term used to describe any synthetic material that has the essential properties of natural rubber, namely flexibility, strength, and extendability; and is capable of being converted from a thermoplastic to a thermosetting state by cross-linking.

element The simplest form of a pure substance that cannot be broken down into simpler substances by chemical means.

embryotoxic Toxic effect on the embryo.

emphysema A condition of the lungs characterized by dilation or destruction of the pulmonary areola. An illness characterized by plugged passageways and a difficulty in exhaling.

employer Defined by OSHA as a person (including contractor or subcontractor) engaged in business, where chemicals are either used, distributed, or produced for use or distribution.

encephalopathy Generic name given to illnesses affecting the brain in general.

environment Includes water, air, and land and the interrelationship which exists among and between water, air, and land and all living things. (EPA)

EP Extraction procedure toxicity characteristics. Toxicity test performed on RCRA wastes.

epidemiology Science concerned with the study of disease in a general population. Determination of the incidence (rate of occurrence) and distribution of a particular disease (as by age, sex, or occupation), which may provide information about the cause of the disease.

epoxy resin A synthetic thermoset resin made from epichlorohydrin and bisphenol A, or glycerol.

ergonomics The activity dealing with interactions between workers and their total working environment and their stresses relating to the elements of the environment, their tools, and equipment.

erythema Abnormal flushing or redness of the skin due to increase in blood flow.

etiologic agents Airborne microorganisms capable of causing disease. These agents are the only nonchemical materials regulated by the U.S. Department of Transportation (DOT).

etiology The science and study of all the factors that contribute to the development of a disease.

euphoria Intense feeling of well-being or elation.

evaporation rate A measure of the time required for a given amount of a substance to evaporate, compared with the time required for an equal amount of butyl acetate or ether to evaporate.

exothermic Heat producing.

expectoration Expulsion from the mouth of secretions from the respiratory tract.

explosive Any chemical compound, mixture, or device that produces a sudden, almost instantaneous release of pressure, gas, and heat when subjected to sudden shock, ignition source, pressure, or high temperature.

explosive limits Defined by the NFPA as the boundary-line mixture of vapor or gas with air, which, if ignited will just propagate the flame. They are known as the "lower and upper explosive limits," and are usually expressed in terms of percentage by volume of gas or vapor in air. Same as **flammable limits**. *See also* **LEL** and **UEL.**

exposure People, property, or the environment that are subjected to the harmful effects of a hazardous material. Defined by OSHA as meaning an employee subjected to a hazardous chemical in the course of employment through a ROUTE OF ENTRY (e.g., ingestion , inhalation, skin contact, or absorption) and includes potential (e.g., accidental or possible) exposure.

exposure limits Concentrations of substances and conditions under which it is believed that nearly all workers may be repeatedly exposed, day after day, without adverse effects. Standard limits are established by ACGIH and OSHA.

F

F Degree Fahrenheit, a unit for measuring temperature. On the Fahrenheit scale, water boils at 212°F or greater.

feedstocks Raw materials used to make polymers, such as gaseous or liquid petroleum-derived hydrocarbons: benzene, crude oil, ethylene, acetylene, naphthalene, natural gas, toluene, xylene.

Federal Register (FR) A daily publication of all U.S. government documents required by law. It is the daily supplement to the Code of Federal Regulations (CFR).

fibrillation Rapid and chaotic contractions of many individual muscle fibers of the heart in the area of the ventricles, capable of causing cardiac arrest.

fibrosis Chronic lung affliction or scarring of the lung caused by an unusual increase of fibrous tissue, causing progressive respiratory problems, and often occurring after exposure to certain chemical substances.

fire diamond A visual hazard rating system of the National Fire Protection Association (NFPA). Provides general information about inherent hazards of materials: Health, Flammability, Reactivity, and Special.

flammable Catches on fire and burns rapidly. The National Fire Protection Association and the U.S. Department of Transportation define a flammable liquid as having a flash point below 100°F (37.8°C).

flammable aerosol An aerosol that, when tested by the method described in 16 CFR 1500.45, yields a flame projection exceeding 18 inches at full valve opening, or a flashback (a flame extending back to the valve) at any degree of valve opening. (29 CFR)

flammable limits Range of gas or vapor concentrations (percent by volume) in air which will burn or explode if an ignition source is present. (also see **LEL** and **UEL**)

flash back A phenomenon characterized by vapor ignition and flame travelling back to the source of the vapor.

flash point The minimum temperature at which a substance gives off flammable vapors that in contact with spark or flame will easily ignite and burn rapidly. The flash point is established by one of the two following methods: in a **closed cup** (also see **Pensky-Martens**), or inside its container; or in an **open cup**, or near the surface of the liquid. The lower the flash point of a liquid, the higher the risk of fire.

flatulence Accumulation of gas in the digestive tract.

fluorosis Characteristic chronic intoxication caused by fluorine and its derivatives.

foreseeable emergency Any potential occurrence such as, but not limited to, equipment failure, rupture of containers, or failure of control equipment, that could result in an uncontrolled release of a hazardous chemical into the workplace. (OSHA)

Form R Forms that must be completed annually for the EPA by reporting industrial facilities for chemicals used above the threshold amount.

fume A suspension of very fine solid particle in air, or vapors from a volatile liquid.

G

gas An airlike, formless fluid having the property of uniformly distributing itself throughout a space in air. A state of matter.

gastric lavage Washing out the stomach.

gastrointestinal (GI) Refers to the organs of stomach, intestines, and/or other organs from mouth to anus.

gene Heredity-carrying material that is part of the chromosome.

gestation The development of the fetus in the uterus from conception to birth; during pregnancy.

glycosuria Abnormal presence of glucose in the urine.

granulomatosis Pulmonary lesion characterized by the formation of small nodules.

granulometry Indicates the size of powder, and usually expressed in microns (1 micron is 0.000001 meters, or about 1,000 times smaller than a grain of sand). Particles of less than 10 microns are capable of deep penetration and becoming deposited in the respiratory tract. Larger particles can deposit in the upper respiratory area such as the bronchia, and although they may be expulsed, they can be dissolved by the organism. Particle size makes it possible to deduce the method to be adopted for corrective action (e.g, ventilation, respirator).

H

halogens Refers to inorganic compounds containing astatine, bromine, chlorine, fluorine, and iodine.

hazardous classes A collection of terms established the United Nations Committee of Experts to categorize hazardous materials. The specific categories are: flammable liquids, explosives, gases, oxidizers, radioactive materials, corrosives, flammable solids, poisonous and infectious substances, and dangerous substances.

hazard warning The OSHA definition means any words, pictures, symbols, or combination thereof appearing on a label or other appropriate form of warning that conveys the hazards of the chemical(s) in the container(s).

hazardous materials Refers generally to hazardous substances, petroleum, natural gas, synthetic gas, acutely toxic chemicals, and other toxic chemicals. Substances or materials which have been determined to be capable of posing an unreasonable risk to health, safety, and property.

hazardous waste Defined by RCRA as any solid or combination of solid wastes, that because of its physical, chemical, or infectious properties may pose a health hazard when improperly managed. It must possess at least one of four characteristics–ignitability, corrosivity, reactivity, or toxicity–or appear on special EPA lists.

hematemesis Vomiting of blood.

hematoma Localized bleeding into tissue.

hematuria Presence of blood in the urine.

hemolysis Destruction of red blood cells, releasing hemoglobin.

hematopoietic system System responsible for formation of blood cells (includes bone marrow and lymphatic organs).

hemorrhage Loss of a significant amount blood. Can be external or internal bleeding.

hepatic Referring to the liver.

hepatoma Tumor of the liver.

hepatomegalia Increase in liver volume.

hepatotoxic Refers to a substance that is toxic to the liver.

hot zone The area immediately surrounding the incident site. Appropriate protective clothing and equipment must be worn by all personnel in the hot zone. AWARENESS LEVEL and OPERATIONAL LEVEL trained personel are not permitted in the hot zone.

hydrate Chemical substance combined with water in a specific proportion.

hydrocarbons Chemical compounds that consist entirely of carbon and hydrogen.

hydrochlorofluorocarbons (HCFCs) A group of chemicals having a weaker effect on the ozone layer and developed as temporary substitues for ozone-depleting chemicals, including **CFCs**.

hydrolysis Chemical change to a substance in an aqueous environment leading to the formation of new products.

hygroscopic Substances with a tendency to absorb moisture from the air.

hyperkeratosis Increased thickness of the cornified layer of the epidermis, such as a corn.

hyperpigmentation Excessive pigmentation of the skin.

hyperplasia Abnormal growth of normal tissue.

hyperreflexia Excessive reflexes.

hypertension Increased blood pressure.

hypotension Decreased blood pressure.

hypothermia Lowering of body temperature to below normal.

I

icterus Yellow coloration of the skin and mucosa.

ignition temperature The minimum temperature required to initiate or cause self-sustained combustion independent of a heat source.

incident An unplanned event that could have resulted in an accident or which diminishes efficiency or production. *See also* **accident.**

inhibitor A chemical that is added to another substance to prevent unwanted chemical change from occurring.

immediate hazard A hazard with immediate effects. *See* **accute effects**.

immunosuppression Decrease in the immune response.

incompatibility Indicates whether a material can be placed in contact with certain other products or materials. The direct contact of incompatible materials can cause dangerous reactions and give off heat and toxic vapors.

inflammation The response of the tissues of the body to injury, infection, or irritation. Its chief symptoms are redness, heat, swelling, and pain.

ingestion Taking in by the mouth; swallowing.

inhalation Breathing into the lungs of a (contaminated) substance in the form of a gas, vapor, fume, mist, or dust.

inorganic chemical Those chemical substances not containing carbon.

irritant A substance or material capable of causing irritation to organs and body parts.

insoluble Products that cannot be dissolved in each other.

interaction Modification of the toxic effects of one substance by another. Depending on the substances involved, the effects of interaction can be amplified or mitigated.

irritability Tendency to anger.

irritating material Any substance that upon contact with fire or air produces irritating fumes.

J

jaundice Yellowing of the skin or eyes caused by too much bilirubin in the blood, an indication of liver diseases, biliary obstructions, and hemolysis.

K

keratitis Inflammation of the cornea of the eye.

keratoconjunctivitis Inflammation of the cornea and the conjunctiva of the eye.

L

laryngitis Inflammation of the larynx.

lacrimation Production or discharge of excess tears from the eyes.

lacrimator A chemical substance that causes the secretion of excess tears from the eyes.

LEL Lower Explosive Limit of a vapor or gas; the lowest concentration (lowest percentage of the substance in air) that will produce a flash of fire when an ignition source (heat, arc, or flame) is present. At lower concentrations, the mixture is too "lean" to burn. Also see UEL.

lethargy Deep and prolonged sleep, or extreme indifference.

M

malaise A feeling of general discomfort, distress, or uneasiness, an out-of-sorts feeling.

manganism Chronic intoxication caused by manganese and its derivatives.

maximum contaminant levels (MCLs) The maximum level of a contaminant permissible in a public water system.

methemoglobinemia Presence of abnormal concentrations of methomoglobin in the blood, resulting in **cyanosis**.

miscible Products capable of mixing completely and staying mixed with each other under normal conditions.

mist Liquid droplets suspended in air.

molecular weight Weight (mass) of a molecule based on the sum of the atomic weights of the atoms that make up the molecule.

monomer A natural or synthetic substance composed of simple molecular repeating units that are capable of chemically reacting (polymerizing) with like or unlike molecules to form a polymer.

mucous membranes Membranes lining body cavities and covered by a viscous substance called mucous.

myosis Reduction in diameter of pupils of the eye.

melting point The temperature at which a product changes from the solid to liquid state at normal atmospheric pressure (760 mm Hg).

Montreal Protocol A 1987 agreement made by 25 countries to reduce the production and consumption of ozone-depleting chemicals. Since 1987, additional countries have joined this agreement.

mutagen A chemical substance or physical effect capable of inducing transmissible changes in the genetic material of a living cell that results in physical and functional changes in the descendants. Depending on the type of cells affected, ova, or spermazoids, both male and female can be affected. Mutations can lead to birth defects, miscarriage, or cancer.

N

narcosis A stupor, drowsiness, arrested activity, or unconsciousness produced by the influence of narcotics or other chemical substances. Artificially induced sleep.

narcotic Any substance that induces **narcosis**.

National Fire Protection Association (NFPA) This international membership organization promotes/improves fire protection and prevention and establishes safeguards against loss of life and property by fire. Best known in industry for the National Fire Codes. Among these codes is the NFPA 704M, the code for classifying substances according to their fire and explosion hazard (as they might be encountered under fire or related emergency conditions) using the familiar diamond-shaped label or placard with appropriate color, numbers, or symbols.

nausea The urge to vomit; a feeling of sickness in the stomach.

necrosis Cellular or tissue death.

neoplasm Presence of a new growth of abnormal cells. A tumor.

nephritis Inflammation of the kidney.

nephropathy Any affliction of the kidneys.

nephrotoxic Toxic to the kidneys.

neuropathy Any affliction of the nervous system.

neurotoxic Toxic to nerve cells and the nervous system. The effect may produce emotional or behavioral abnormalities.

neuritis Inflammation of a nerve.

neural Refers to a nerve or the nervous system.

neutralize To eliminate potential hazards by inactivating strong acids, caustics, and oxidizers. For example, acid spills can be neutralized by adding an appropriate amount of **caustic** or **alkali** substance.

NIOSH National Institute for Occupational Safety and Health. Tests equipment, evaluates and approves respirators, conducts studies of workplace hazards, and proposes standards to OSHA.

nitrogen oxides (NO_x) Gases associated with the breakdown of the earth's protective ozone layer that are released primarily from the burning of fossil fuels.

nonliquefied compressed gas DOT describes as a gas, other than gas in solution, which under the charged pressure is entirely gaseous at a temperature of 70°F. (Code of Federal Regulations, Title 49, Department of Transportation).

O

operations level (trained) First responders at the operational level are those persons who respond to releases or potential releases of hazardous materials as part of the initial response to the incident for the purpose of protecting nearby persons, the environment, or property from the effects of the release. They shall be trained to respond in a defensive fashion to control the release from a safe distance and keep it from spreading (ANSI/NFPA 472)

optic nerve Nerve running from the eye to the centers of vision in the brain.

oral Used or taken into the body through the mouth.

organic compound A class of chemical compounds containing mainly carbon atoms.

organic peroxide An organic compound containing the bivalent -O-O- structure and which may be considered a structural derivative of hydrogen peroxide where one or more of the hydrogen atoms have been replaced by organic radicals. (49 CFR 173.151)

oxidizing agent or oxidizer A substance other than an explosive such as chlorate, permanganate, inorganic peroxide, or a nitrate, that yields oxygen or other gases readily, thereby causing fire of other, usually organic, materials. Perchlorates, peroxides, permanganates, chlorates, and nitrates are examples of oxidizers.

ozone layer The protective layer of molecules surrounding the earth. The ozone layer reduces the amount of high-erergy ultraviolet radiation that reaches the earth's surface.

P

partition coefficient The ratio of a substance's distribution between oil and water when they are in contact. A value of less than 1 indicates better solubility of the substance in oils and greases. Such a product is therefore likely to be absorbed by the skin. A value greater than 1 indicates a better solubility in water and, therefore will be absorbed by the mucosa. This information can be useful in assessing first aid requirements and can help in the selection of proper protective equipment. Also known as the coefficient of water/oil distribution.

palpitation Perception of one's own heartbeat. A racing, irregular beat, or pounding of the heart.

paralysis The loss of function and/or feeling.

Pensky-Martens A **closed cup** method for determining **flash point**.

peritonitis Inflammation of the peritoneum or the membrane lining the abdominal cavity and the organs contained within it.

peroxide Compounds containing two oxygen atoms bound together. Often explosive.

permissible exposure limit (PEL) A legal occupational limit of exposure established by OSHA's regulatory authority. The limit of allowable exposure to a chemical contaminant expressed as a time-weighted average (TWA) concentration during a workday of 8 hours within, or as a maximum concentration never to be exceeded either instantaneously (**ceiling**) or a short-term exposure limit (**STEL**) during any maximum period of 15 minutes. The exposure, inhalation, or dermal permissible exposure limit specified in 29 CFR Part 1910, Subparts G and Z. These concentrations are expressed in parts per million (ppm) and or milligrams of the product per cubic meter of air (mg/m3) at 25°C. Exposure limits published in 29 CFR 1910.1000.

personal protection equipment (PPE) Safety equipment designed to protect parts or all of the body from workplace hazards. Such protective equipment includes chemical resistant clothing, gloves, respirators, and eye protection.

pH A symbol representing the concentration in hydrogen ions (H+) in aqueous solution. This logarithmic scale is expressed as a numerical value usually between 0 and 14. A pH of 7 indicates a neutral or noncorrosive substance. A pH between 0 and 7 indicates greater acidity. A pH between 7 and 14 indicates greater alkalinity. A pH of 0 (very acid), or 14 (very basic) are highly corrosive. The symbol is useful in the identification of the appropriate type of protective equipment necessary for handling a chemical material.

pharyngitis Inflammation of the throat or pharynx.

phlegm Thick mucous produced in the breathing passages.

photophobia Unpleasant to painful feeling in the eyes, caused by light.

photosensitive Substances that change in the presence of light.

photosensitization Abnormal reaction of the skin to sunlight.

physical hazard A chemical for which there is scientifically valid evidence that it is a combustible liquid, a compressed gas, explosive, flammable, an organic peroxide, an oxidizer, pyrophoric, unstable (reactive) or water-reactive.

plasticizer an organic liquid that is added to plastic mixtures to act as an internal lubricant to facilitate the manufacturing process and make the final product permanently softer and more flexible.

pneumoconiosis Chronic affliction of the lungs due to the inhaling of certain types of dust.

pneumonia Acute infection of the lung characterized by inflammation.

pneumonitis (chemical) Inflammation of the lung resulting from chemical irritation.

pneumopathy Any pulmonary affliction.

point source contamination Contamination to the environment from a specific source such as a smokestack or sewer pipe.

Poison Control Centers In the United States, a network of centers staffed 24 hours a day and equipped to provide expert advice on the proper management of human poisoning.

polychlorinated biphenyl (PCBs) A pathogenic and teratogenic industrial compound used as a heat-transfer agent, primarily in transformers. PCBs may accumulate in human or animal tissue. PCBs are banned from production in the United States.

polymerization A chemical reaction in which one or more small molecules combine to form larger molecules. A hazardous polymerization is such a reaction that takes place at a rate that releases large amounts of energy. If hazardous polymerization can occur with a given material, the MSDS usually will list conditions that could start the reaction and–since the material usually will list conditions that could start the reaction, and since the material usually contains a polymerization inhibitor–the length of time during which the inhibitor will be effective. The heat given off or the expansion in volume, or both, caused by the polymerization reaction could cause the container holding the product to break and the unpolymerized residual product to be spilled.

polyneuritis Inflammation of several nerves.

polyuria Elimination of an abnormally large amount of urine over a given period.

procarcinogen Product that must be changed by the organism in order to become a carcinogen.

promoter (of carcinogenesis) Substance capable of enhancing the carcinogenic effect of another substance.

prostration Extreme exhaustion, physical or mental.

pulmonary edema A buildup of fluid in the lungs caused by congestive heart failure, lung damage, side effects of drugs, infections, or kidney failure.

pyroforic Any chemical substance or mixture that ignites spontaneously in dry or moist air at or below 130°F (54.4°C).

psychosis A group of mental illnesses characterized by a change in personality and loss of contact with reality.

pyrophoric A chemical that will ignite spontaneously in air at a temperature of 130°F (54.4°C) or below.

R

radioactive material Any material, or combination of materials, that spontaneously emits ionizing, and having a specific activity greater than 0.002 microcuries per gram.

radionuclides Radioactive-decay particles emitted from natural and manufactured sources, including cosmic rays, X-rays, radon, and coal-fired utilities.

radon A colorless, naturally occurring, radioactive, inert gaseous element formed by radioactive decay of radium atoms in soil or rocks.

reactive Unstable.

reactivity Chemical reaction with the release of energy. Undesirable effects such as pressure buildup, temperature increase, formation of noxious, toxic or corrosive by-product, may occur because of the reactivity of a substance to heating, burning, direct contact with other materials, or other conditions in use or in storage.

reducing agent In a reducing reaction (which occurs simultaneously with an oxidation reaction) the reducing agent is the chemical or substance which (1) combines with oxygen, or (2) loses electrons to the reaction.

Registry of Toxic Effects of Chemical Substances (RTECS) Published by NIOSH, RTECS is a compendium of the known toxic and biological effects of more than 100,000 chemical substances.

regulated material A substance or material that is subject to regulations promulgated by any government agency.

remedial action The actual construction or implementation phase of a Superfund site cleanup that follows remedial design.

reportable quantity ((RQ) An amount of a hazardous substance or "extremely hazardous substance "that, if released, requires notification to the National Response Center (800/424-8802) under the emergency release reporting requirements of the Emergency Planning and Community Right-to-Know Act (EPCRA) or under those of CERCLA (Superfund). For the purpose of the Department of Transportation, RQ means the quantity specified for the substance in the Appendix to the Hazardous Materials Table. For the purposes of SARA Title III, RQ means, for any CERCLA hazardous substance the reportable quantity established therein for such a substance; for any other substance the RQ is one pound.

respirator A device worn by a person to filter dust particles or gas out of surrounding air before inhalation air.

rhinitis Inflammation of mucosa of the nasal passages.

risk assessment A detailed examination of any activity or functioning system in which potential adverse effects and their probabilities are calculated, and the various risks are quantified or measured. (F.M. Renshaw, Rohm and Haas Company, Bristol, PA.).

risk management The process, derived through system safety principles, whereby management decisions are made concerning control and minimization of hazards and acceptance of residual risks.

rodenticide Chemical substances used to kill mice, rats, and other rodents.

route of entry The means or natural route by which hazardous chemicals or other contaminants can penetrate the body. For example, the skin (by cutaneous absorption), the digestive system (by ingestion), or the respiratory system (by inhalation).

routes of exposure Also known as routes of entry, any one of the ways by which substances enter the body, such as through the skin, ingestion, eye contact, or by breathing.

S

salivation Excessive secretion of saliva.

sensitization Defense reaction by the organism following exposure to a contaminant, resulting in an allergy.

sesitizer A chemical that causes a substantial proportion of exposed people or animals to develop an allergic reaction in normal tissue after repeated exposure to the chemical.

shock An abnormal condition resulting from not enough blood flowing through the body, causing reduced blood pressure and interference with bodily functions.

short term exposure limit (STEL) The upper limit of exposure to a substance to which a worker can be exposed for a continuous period of up to 15 minutes without adverse effects to health, performance, and work efficiency.

silicosis Chronic disease of the lungs (fibrosis) provoked by inhaling dust of crystalline silica. See **pneumoconiosis**.

solid waste Any garbage, refuse, or sludge, including solid, liquid, semisolid, or contained gaseous material resulting from industrial, commercial, agricultural, and mining operations, and community activities (excluding material in domestic sewage); discharges subject to regulation as point sources under the Federal Water Pollution Control Act, or any nuclear material or byproduct regulated under the Atomic Energy Act of 1954.

solubility The property of a substance describing the degree to which one material may be completely mixed or dissolved in another material. The degree of solubility of most substances increases with the rise in temperature; however, in the case of organic salts of calcium the substance may be more soluble in cold than in hot solvents.

solution Any homogeneous liquid mixture of two or more chemical compounds or elements that will not undergo any segregation normal to transportation. [CFR, Title 49, DOT]

solvent A chemical liquid, capable of dissolving another substance. A term generally used to describe organic solvents.

spasm Involuntary muscle contraction.

specific gravity The weight of a material compared to the weight of an equal volume of water is an expression of the density (or heaviness) of a material. Insoluble materials with specific gravity of less than 1.0 will float in (or on) water. Insoluble materials with specific gravity greater than 1.0 will sink in water. Most (but not all) flammable liquids have specific gravity less than 1.0 and, if not soluble, will float on water–an important consideration for fire suppression.

spill Another name for a leak. The methods, equipment, and precautions that should be used to control or clean up a leak or spill.

stability Relates to a material's ability to resist change in form or chemical nature (e.g., xylene decomposes when strongly heated and gives off toxic fumes). For MSDS purposes, a material is stable if it remains in the same form under expected and reasonable conditions of storage or use. Conditions that may cause instability (dangerous change) are stated; for example, temperature above 150°F (66°C); shock from dropping.

stabilizers A general term for any substance or mixture additive used to preserve chemical equilibrium and hold physical properties, and to slow or retard degradation. Stabilizers include antioxidants, inhibitors, emulsifiers, etc.

Standard Transportation Commodity Code (STCC) A seven- digit identification number commonly used for materials shipped by rail. Numbers beginning with "49-" are hazardous materials.

stool Discharge from the bowels.

stratosphere That part of the earth's atmosphere, located above the troposphere, containing the ozone layer.

strong acids Refers, but not restricted to, chemicals such as hydrochloric, sulfuric, and nitric acids. Strong acids can cause violent reactions with certain other substances and materials.

strong bases Refers, but not restricted, to chemicals such as sodium hydroxide and potassium hydroxide. Strong bases can cause violent reactions with certain other substances and materials.

strong oxidizers Refers, but not restricted, to chemicals such as chlorine, bromine, and fluorine and many of their compounds. Strong oxidizers can cause violent reactions with certain other substances and materials. *See* **oxidizers, oxidizing agent, easily oxidized materials**.

sublimate To go directly from the solid to the gaseous state without passing through the liquid state.

Superfund Amendments and Reauthorization Act of 1986 (SARA) Title III of SARA establishes the first national program of emergency planning for dealing with hazardous chemicals. Includes detailed provisions for community planning and annual submission of information about hazardous chemicals to the EPA, states, and local communities.

suspected carcinogen A substance which is known to cause cancer in test animals but is only suspected of causing cancer in humans. Also referred to as an experimental or potential carcinogen.

systemic toxicity Adverse effects caused by a substance that affects the body as a whole rather than local or individual parts or organs.

T

tachycardia Increased speed of heart beat.

tachypnea Abnormally rapid breathing.

teratogen A substance that has been demonstrated to cause birth defects by causing malformations in the fetus. Teratogenic contaminants can be qualified as being "proven" when an effect has been shown in humans, "possible" when an effect has been shown in animals or suspected in humans, and "suspected" when an effect is suspected in animals.

tetanic Refers to persistence in a muscle contraction.

thermoplastic Describes the property of repeatedly softening or fusing when heated and of hardening again when cooled. The softening or fusing temperature varies with polymer type and grade.

thermosetting Describes the property of becoming permanently rigid when heated or cured with catalysts, ultraviolet light, or other factors.

thio Containing a sulfur atom.

Threshold Planning Quantity (TPQ) The quantity of a SARA extremely hazardous substance present at a facility above which the facility's owner/operator must give emergency planning notification to the SERC and LEPC.

Tier I or Tier II Describes hazardous substances inventory forms required under SARA TITLE III. These forms report quantities and locations of hazardous substances to various state agencies and planning committees.

Time-Weighted Average (TWA). The average concentration of a substance in air over the total time of exposure, usually expressed as an 8-hour day.

tinnitus Ringing in one or both ears. May be a sign of hearing injury.

Title III The third part of SARA, also known as EPCRA, the Emergency Planning and Community Right-to-Know Act of 1986. Specifies requirements for organizing the planning process at the state and local levels for extremely hazardous substances; minimum plan content; requirements for fixed facility owners and operators to inform officials about extremely hazardous substances present at the facilities; and mechanisms for making information about extremely hazardous substances available to citizens.

toxic Any substance capable of causing human injury or damage to living body tissue, impairment to the central nervous system, severe illness, and, in severe cases, death. A poison.

toxicology The branch of chemistry that deals with poisons.

toxicologist One who studies the nature, effects, and detection of poisons and the treatment of poisoning.

toxic pollutant Pollutants that, after discharge and upon exposure, cause adverse health effects.

Toxic Release Inventory (TRI) An EPA database of release quantities by facilities of a growing list of chemicals and chemical categories into the nation's air, water, and land, and transfers to off-site locations for treatment, disposal, or recycling and energy recovery.

trade name The manufacturer's commercial name for a chemical substance or product.

tumorigenic That which causes tumors.

U

UEL Upper Explosive Limit of a vapor or gas; the highest concentration (highest percentage of the substance in air) that will produce a flash of fire when an ignition source (heat, arc, or flame) is present. At higher concentrations, the mixture is too "rich" to burn. *See also* **LEL**.

UST Underground Storage Tank regulated under RCRA. A tank used to store CERCLA-regulated chemicals or petroleum products, with 10% or more of its volume underground, having connected piping.

ulceration Creation of an ulcer.

V

vapor density Indicator of the number of times that the vapors of a substance are heavier or lighter than air. Vapor density measurement is taken at the boiling point. If the vapor density is greater than 1, the vapor will tend to collect at floor level. If the vapor density is less than 1, the vapor will rise in air.

vapor pressure When a substance evaporates, its vapors create a pressure in the surrounding atmosphere; therefore, vapor pressure is a measurement of how readily a liquid or a solid mixes with air at its surface. This measurement is expressed in millimeters of mercury (mm of Hg), at 68°F (20°C) and normal atmospheric pressure (760 mm Hg). A vapor pressure above 760 mm indicates a substance in the gaseous state. The higher a product's vapor pressure , the more it tends to evaporate, resulting in a higher concentration of the substance in air and therefore increasing the likelihood of breathing in it.

vascular constriction Constriction with narrowing of blood vessels.

vascular dilation Dilation of the blood vessels.

vertigo Dizziness; giddiness.

volatiles A substance, usually a liquid, that easily vaporizes or evaporates to form a gas or vapor.

W

waste code Identifyer assigned by the EPA consisting of a single letter (D, F, P, U, K) and three numbers in the format Dxxx. This code identifies the type of hazardous waste stream being reported.

waste management Describes activities undertaken by facilities to treat, recycle, or otherwise manage generated waste, including disposal and energy recovery.

waste stream Under RCRA, solid or liquid materials containing hazardous materials and generated as waste.

Appendix E

SYNONYM AND TRADE NAME INDEX

Appendix F

CAS NUMBER-CROSS INDEX

CAS Number	Entry
N/A	A:085
N/A	C:008
N/A	C:011
N/A	C:475
50-00-0	F:120
50-70-4	S:604
51-79-6	U:800
56-23-5	C:200
56-81-5	G:160
56-93-9	B:095
57-13-6	U:300
57-55-6	P:680
60-24-2	M:154
60-29-7	E:260
60-35-5	A:015
62-38-4	P:143
62-53-3	A:440
62-73-7	D:106
64-17-5	E:160
64-18-6	F:140
64-19-7	A:020
64-67-5	D:262
65-85-0	C:222
66-25-1	H:098
67-56-1	M:300
67-63-0	I:160
67-64-1	A:040
67-66-3	C:320
67-68-5	D:320
68-11-1	T:130
68-12-2	D:300
71-23-8	P:665
71-36-3	B:360
71-43-2	B:078
71-55-6	T:320
74-82-8	M:220
74-85-1	E:190
74-86-2	A:080
74-87-3	M:320
74-90-8	H:240
74-95-3	D:050
74-98-6	P:640
75-01-4	V:015
75-02-5	V:350
75-04-7	E:170
75-05-8	A:060
75-07-0	A:010
75-09-2	M:360
75-15-0	C:160
75-20-7	C:018
75-21-8	E:250
75-25-2	B:246
75-28-5	I:087
75-31-0	I:200
75-34-3	D:094
75-35-4	V:400
75-38-7	V:450
75-44-5	P:165
75-46-7	T:410
75-50-3	T:418
75-52-5	N:220
75-56-9	P:730
75-65-0	B:364
75-69-4	F:160
75-71-8	F:180
75-75-2	M:227
75-78-5	D:289
75-79-6	M:490
75-94-5	V:800
75-98-9	T:416
76-13-1	F:300
76-16-4	H:076
77-58-7	D:075
77-71-4	D:305
77-73-6	D:118
77-78-1	D:315
77-90-7	A:077
77-99-6	T:455
78-30-8	T:470
78-40-0	T:406
78-42-2	T:465
78-51-3	T:265
78-59-1	I:114
78-63-7	D:287
78-67-1	A:700
78-78-4	I:112
78-79-5	I:120
78-83-1	I:096
78-93-3	M:380
78-94-4	M:494
78-95-5	C:235
79-00-5	T:330

79-01-6	T:340
79-06-1	A:087
79-08-3	B:244
79-09-4	P:660
79-10-7	A:089
79-21-0	P:170
79-38-9	T:408
79-41-4	M:215
79-46-9	N:237
80-05-7	B:135
80-10-4	D:420
80-15-9	C:709
80-43-3	I:250
80-47-7	M:224
80-48-8	M:487
80-52-4	M:147
80-62-6	M:450
83-32-9	A:003
84-66-2	D:258
84-69-5	D:276
84-74-2	D:072
84-76-4	D:335
84-77-5	D:124
85-43-8	T:090
85-44-9	P:287
85-68-7	B:375
87-62-7	X:500
87-90-1	T:365
88-60-8	B:440
88-69-7	I:400
89-72-5	B:675
89-84-9	A:083
91-22-5	Q:500
91-66-7	D:140
91-93-0	D:022
91-94-1	D:090
91-97-4	D:283
92-52-4	D:390
92-54-6	P:161
92-93-3	N:149
92-94-4	T:040
93-58-3	M:313
93-89-0	E:182
93-96-9	M:316
94-04-2	V:300
94-36-0	D:030
94-51-9	D:540
95-47-6	X:040
95-48-7	C:680
95-49-8	C:340
95-63-6	T:430
95-65-8	X:200
95-68-1	X:500
95-78-3	X:500
95-80-7	T:250
95-87-4	X:200
96-05-9	A:099
96-08-2	D:630
96-09-3	S:680
96-24-2	C:323
96-27-5	T:135
96-33-3	M:290
96-48-0	B:860
96-49-1	E:194
97-61-0	M:492
97-63-2	E:294
97-86-9	B:550
97-88-1	B:525
97-94-9	T:393
97-96-1	E:186
97-99-4	T:085
98-00-0	F:360
98-01-1	F:354
98-12-4	C:880
98-29-3	B:420
98-54-4	B:700
98-82-8	C:705
98-86-2	A:065
98-94-2	D:285
98-95-3	N:145
99-06-9	H:195
99-63-8	I:118
99-87-6	C:980
100-20-9	T:030
100-21-0	T:025
100-40-3	V:018
100-41-4	E:180
100-42-5	S:660
100-43-6	V:700
100-46-9	B:082
100-47-0	P:133
100-51-6	B:080
100-69-6	V:650
100-74-3	E:298
100-97-0	H:095
101-02-0	T:474
101-14-4	M:330
101-25-7	D:328
101-68-8	M:340
101-77-9	D:028

About the Authors

Richard P. Pohanish has been active in the environmental field since 1980, president and publisher of Chem-Data Systems, co-founder of Chemtox, Inc., and co-author of *Hazardous Chemical Safety Guide for the Machining and Metal Working Industries*. He has published numerous articles and professional reference works, including, with Greene, four other books and two CD-ROMs for the environmental safety and safety field.

Stanley A. Greene is the co-author of *Hazardous Chemical Safety Guide for the Machining and Metal Working Industries* and has been a publishing professional for more than 25 years. He is the former owner of Warman Publishing Company, and worked for Lawyer's Cooperative Publishing Company, The Legal Intelligencer, and the Auerbach Corporation. Greene also worked for a major chemical company and has extensive experience in the research of scientific and regulatory aspects of chemicals, especially those of major health and environmental concern.